An Introduction to
Relativistic Quantum Field Theory

Silvan S. Schweber
Brandeis University

Foreword by

Hans A. Bethe
Cornell University

DOVER PUBLICATIONS, INC.
Garden City, New York

Bibliographical Note

This Dover edition, first published in 2005, is an unabridged republication of the corrected second (1962) printing of the work originally published by Harper & Row, Publishers, New York, in 1961.

Library of Congress Cataloging-in-Publication Data

Schweber, S. S. (Silvan S.)
An introduction to relativistic quantum field theory / Silvan S. Schweber ; foreword by Hans A. Bethe.
p. cm.
Originally published: New York, N.Y. : Harper & Row, c1961.
ISBN-13: 978-0-486-44228-0
ISBN-10: 0-486-44228-4 (pbk)
1. Quantum field theory. I. Title.

QC174.45.S33 2005
530.14'3—dc22

2004065500

Manufactured in the United States by LSC Communications
4500056560
www.doverpublications.com

To Hans Bethe
With admiration and affection

Table of Contents

Part One: The One-Particle Equations

Part Two: Second Quantization

Part Three: The Theory of Interacting Fields

Part Four: Formal Developments

Foreword

It is always astonishing to see one's children grow up, and to find that they can do things which their parents can no longer fully understand. This book is a good example. It was first conceived by Dr. Frederic de Hoffmann and myself as merely a short introduction to the rather simple-minded calculations on π mesons in Volume II of the old book *Mesons and Fields*, published in 1955. In Dr. Schweber's hands Volume I, even then, had developed into a thorough textbook on renormalization in field theory. It has now become a comprehensive treatise on field theory in general.

In the six years since the publication of the two-volume *Mesons and Fields* field theory has made spectacular progress. Some of this progress was stimulated by experiment, e.g., by the discovery that parity is not conserved in weak interactions. Much of it, however, consisted in a deeper search into the foundations of field theory, trying to answer the central question of relativistic quantum theory which Schweber poses himself in Chapter 18 of this book: Do solutions of the renormalized equations of quantum electrodynamics or any meson theories exist? This search has led tò the axiomatic approach to quantum field theory which is probably the most promising and solid approach now known, and which is described in Chapter 18.

About half of the present book is devoted to the interaction between fields. This new book contains a thorough discussion of renormalization theory, starting from the general principles and leading to quantitative results in the case of electrodynamics. I do not know of any other treatment of this subject which is equally complete and rigorous. The physicist who is interested in applications of field theory will be happy about the good discussion of the theory of Chew and Low of π-meson scattering, which theory has been so successful in explaining the π-meson phenomena at low energy and which has superseded the methods presented in Volume II *Mesons* of the older book.

The book emphasizes general principles, such as symmetry, invariance, isotopic spin, etc., and develops the theory from these principles. It is never satisfied with superficial explanations. The student who really wants to know and understand field theory, and is willing to work for it, will find great satisfaction in this book.

H. A. Bethe

Ithaca, N. Y.
March 1961

Preface

The present book is an outgrowth of an attempted revision of Volume I of *Mesons and Fields* which Professors Bethe, de Hoffmann and the author had written in 1955. The intent at the outset was to revise some of the contents of that book and to incorporate into the new edition some of the changes which have occurred in the field since 1955. Unfortunately, due to the pressure of other duties, Drs. Bethe and de Hoffmann could not assist in the revision. By the time the present author completed his revision, what emerged was essentially a new text. With the gracious consent of Drs. Bethe and de Hoffmann, it is being published under a single authorship.

The motivation of the present book, however, is still the same as for the volume *Fields* on which it is based, in part: to present in a simple and self-contained fashion the modern developments of the quantum theory of fields. It is intended primarily as a textbook for a graduate course. Its aim is to bring the student to the point where he can go to the literature to study the most recent advances and start doing research in quantum field theory. Needless to say, it is also hoped that it will be of interest to other physicists, particularly solid state and nuclear physicists wishing to learn field theoretic techniques.

The desire to make the book reasonably self-contained has resulted in a lengthier manuscript than was originally anticipated. Because it was my intention to present most of the concepts underlying modern field theory, it was, nonetheless, decided to include most of the material in book form. In order to keep the book to manageable length, I have not included the Schwinger formulation of field theory based on the action principle. Similarly, only certain aspects of the rapidly growing field of the theory of dispersion relations are covered. It is with a mention of the Mandelstam representation for the two-particle scattering amplitude that the book concludes. However, some of the topics not covered in the chapters proper are alluded to in the problem section.

Notation

For the reader already accustomed to a variety of different notations, an indication of our own notation might be helpful. We have denoted by an overscore the operation of complex conjugation so that $\bar{a}$ denotes the complex conjugate of a. Hermitian conjugation is denoted by an asterisk: $(a^*)_{ij} = \overline{a_{ji}}$. Our space-time metric $g_{\mu\nu}$ is such that $g_{00} = -g_{11} = -g_{22} = -g_{33} = 1$, and we have differentiated between covariant and contravariant tensors. Our Dirac matrices satisfy the commutation rules $\gamma^\mu\gamma^\nu + \gamma^\nu\gamma^\mu = 2g^{\mu\nu}$. The adjoint of a Dirac spinor u is denoted by $\tilde{u}$, with $\tilde{u} = u^*\gamma^0$.

Acknowledgments

It is my pleasant duty to here record my gratitude to Drs. George Sudarshan, Oscar W. Greenberg and A. Grossman who read some of the early chapters and gave me the benefit of their criticism, and to Professor S. Golden and my other academic colleagues for their encouragement. I am particularly grateful to Professor Kenneth Ford, who read most of the manuscript and made many valuable suggestions for improving it. I am indebted to Drs. Bethe and de Hoffmann for their consent to use some of the material of Volume I of *Mesons and Fields*, to the Office of Naval Research for allowing me to undertake this project in the midst of prior commitments and for providing the encouragement and partial support without which this book could not have been written.

I am also grateful to Mrs. Barbara MacDonald for her excellent typing of the manuscript; to Mr. Paul Hazelrigg for his artful execution of the engravings; and to The Colonial Press Inc. for the masterly setting and printing of a difficult manuscript. I would like to thank particularly the editorial staff of the publisher for efficient and accurate editorial help and for cheerful assistance which made the task of seeing the manuscript through the press a more pleasant one.

Above all, I am deeply grateful to my wife, who offered constant warm encouragement, unbounded patience, kind consideration and understanding during the trying years while this book was being written.

For the second printing of this first edition, I have had an invaluable list of corrections from Professor Eugene P. Wigner, of Princeton University, and from others, for which I am sincerely thankful. Most of these have been incorporated in this edition.

SILVAN S. SCHWEBER

Lincoln, Mass.
August, 1962

Part One

THE ONE-PARTICLE EQUATIONS

1

Quantum Mechanics and Symmetry Principles

1a. Quantum Mechanical Formalism

Quantum Mechanics, as usually formulated, is based on the postulate that all the physically relevant information about a physical system at a given instant of time is derivable from the knowledge of the state function of the system. This state function is represented by a ray in a complex Hilbert space, a ray being a direction in Hilbert space: If $|\Psi\rangle$ is a vector which corresponds to a physically realizable state, then $|\Psi\rangle$ and a constant multiple of $|\Psi\rangle$ both represent this state. It is therefore customary to choose an arbitrary representative vector of the ray which is normalized to one to describe the state. If $|\Psi\rangle$ is this representative, the normalization condition is expressed as $\langle\Psi \mid \Psi\rangle = 1$, where $\langle\chi \mid \Psi\rangle = \overline{\langle\Psi \mid \chi\rangle}$ denotes the scalar product of the vectors $|\chi\rangle$ and $|\Psi\rangle$.[1] If the states are normalized, only a constant factor of modulus one is left undetermined and two vectors which differ by such a phase factor represent the same state. The system of states is assumed to form a linear manifold and this linear character of the state vectors is called the superposition principle. This is perhaps the fundamental principle of quantum mechanics.

A second postulate of quantum mechanics is that to every measurable (i.e., observable) property, α, of a system corresponds a self-adjoint operator $a = a^*$ with a complete set of orthonormal eigenfunctions $|a'\rangle$ and real eigenvalues a', i.e.,

$$a \mid a'\rangle = a' \mid a'\rangle \tag{1}$$

$$\langle a' \mid a''\rangle = \delta_{a'a''} \tag{2}$$

$$\sum_{a'} |a'\rangle \langle a'| = 1 \tag{3}$$

The symbol $\delta_{a'a''}$ is to be understood as the Kronecker symbol if a' and a'' lie in the discrete spectrum and as the Dirac δ function, $\delta(a' - a'')$, if either or both lie in the continuous spectrum. Similarly, the summation

[1] We shall also use the notation (f, g) to denote the scalar product; $\bar{\lambda}$ denotes the complex conjugate of λ.

sign in the completeness relation Eq. (3) is to be regarded as an integration over the continuous spectrum.

It is further postulated that if a measurement is performed on the system to determine the value of the observable α, the probability of finding the system, described by the state vector $|\Psi\rangle$, to have α with the value a' is given by $|\langle a' \mid \Psi\rangle|^2$. In other words $\langle a' \mid \Psi\rangle$ is the probability amplitude of observing the value a'. A measurement on a system will, in general, perturb the system and, thus, alter the state vector of the system. If as a result of a measurement on a system we find that the observable α has the value a' the (unnormalized) vector describing the system after the measurement is $|a'\rangle \langle a' \mid \Psi\rangle$. An immediate repetition of the measurement will thus again yield the value a' for the observable α. These statements are, strictly speaking, only correct for the case of an observable with a nondegenerate discrete eigenvalue spectrum. These rules, however, can easily be extended to more complex situations.

A measurement of the property α thus channels the system into a state which is an eigenfunction of the operator a. However, only the probability of finding the system in a particular eigenstate is theoretically predictable given the state vector $|\Psi\rangle$ of the system. If this state vector is known, measurements then allow the verification of the predicted probabilities. A measurement of the first kind (i.e., measurements which if repeated immediately give identical results) can also (and perhaps more appropriately) be regarded as the way to prepare a system in a given state.

It is usually the case that several independent measurements must be made on the system to determine its state. It is therefore assumed in quantum mechanics that it is always possible to perform a complete set of compatible independent measurements, i.e., measurements which do not perturb the values of the other observables previously determined. The results of all possible compatible measurements can be used to characterize the state of the system, as they provide the maximum possible information about the system. Necessary and sufficient conditions for two measurements to be compatible or simultaneously performable is that the operators corresponding to the properties being measured commute. A maximal set of observables which all commute with one another defines a "complete set of commuting operators" [Dirac (1958)]. There is only one simultaneous eigenstate belonging to any set of eigenvalues of a complete set of commuting observables.

The act of measurement is thus fundamental to the formulation and interpretation of the quantum mechanical formalism. An analysis of various kinds of physical measurements at the microscopic level reveals that almost every such physical measurement can be described as a collision process. One need only recall that such quantities as the energy of stationary states or the lifetime of excited states can be obtained from scattering cross sections. The realization of the central role of collision processes in quantum mechanics was of the utmost importance in the recent development of field theory. It also accounts, in part, for the intensive study of the quantum theory of scattering in the past decade.

A collision process consists of a projectile particle impinging upon a target particle, interacting with it, and thereby being scattered. Now initially the projectile particle is far removed from the target. If the force between the particles is of finite range, as is almost always the case, the projectile particle will travel initially as a free particle. Similarly, after it has interacted with the target the scattered particle is once again outside the range of the force field and thus travels as a free particle to the detector. A scattering experiment measures the angular distribution, energy, and other compatible observables of the scattered particles far away from the target, for projectile particles prepared in known states. Thus in making theoretical predictions, the statistical interpretation has only to be invoked for initial and final states of *freely* moving particles or groups of particles in stationary states. Therein lies the importance of collision phenomena from a theoretical standpoint: It is never necessary to give an interpretation of the wave function when the particles are close together and interacting strongly. These remarks also indicate the reason for studying the wave mechanical equations describing freely moving particles which take up Part One of this book.

The postulates introduced thus far allow us to deduce the fact that to every realizable state there corresponds a unique ray in Hilbert space. For if there were several distinct rays which correspond to a single distinct state, then if $|\Psi_1\rangle$, $|\Psi_2\rangle$, etc. are normalized representatives of these rays, by Schwartz's inequality $|(\Psi_1, \Psi_2)|^2 < 1$, i.e., the transition probability from $|\Psi_1\rangle$ to $|\Psi_2\rangle$ is less than one, which cannot be if they represent the same state. Therefore $|\Psi_1\rangle$, $|\Psi_2\rangle$, etc. must be constant multiples of each other. It may, however, be the case that there exist rays in Hilbert space which do not correspond to any physically realizable state. This situation occurs in relativistic field theories or in the second quantized formulation of quantum mechanics. In each of these cases the Hilbert space of rays can be decomposed into orthogonal subspaces $\mathcal{H}_A$, $\mathcal{H}_B$, $\mathcal{H}_C \cdots$ such that the relative phase of the component of a vector in each of the subspaces is arbitrary and not measurable. In other words, if we denote by $|A, l\rangle$ the basis vectors which span the Hilbert space $\mathcal{H}_A$, and by $|B, j\rangle$ the basis vectors which span $\mathcal{H}_B$, etc., then no physical measurement can differentiate between the vector

$$\sum_l a_l \,|\, A, l\rangle \oplus \sum_j b_j \,|\, B, j\rangle \oplus \cdots$$

and the vector

$$\sum_l a_l e^{i\alpha} \,|\, A, l\rangle \oplus \sum_j b_j e^{i\beta} \,|\, B, j\rangle \oplus \cdots$$

where α, β, $\cdots$ are arbitrary phase factors. The phenomenon responsible for the breakup of the Hilbert space into several incoherent orthogonal subspaces is called a superselection rule [Wick (1952), Wigner (1952a), Bargmann (1953)]. A superselection rule corresponds to the existence of an operator which is not a multiple of the identity and which commutes with all observables. If the Hilbert space of states, $\mathcal{H}$, decomposes for example into two orthogonal subspaces, $\mathcal{H}_A$ and $\mathcal{H}_B$, such that the relative phases of the components of the state vector in the two subspaces is completely arbitrary, then the expectation value of a Hermitian operator that has matrix elements between these two subspaces is likewise arbitrary when taken for a state with nonvanishing components in $\mathcal{H}_A$ and $\mathcal{H}_B$. Now for a quantity to be measurable it must surely have a well-defined expectation value in any state. Therefore, a Hermitian operator which connects two such orthogonal subspaces cannot be measurable. An example of this phenomenon is the Hilbert space which consists of the states of $1, 2, 3, \cdots, n, \cdots$ particles each carrying electric charge e. The orthogonal subsets then consist of the subspaces with definite total charge and a Hermitian operator connecting subspaces with different total charge cannot be observable. The superselection rule operating in this case is the charge conservation law, or its equivalent statement: gauge invariance of the first kind (Sec. 7*g*).

An equivalent formulation of the above consists in the statement that all rays within a single subspace are realizable but a ray which has components in two or more subspaces is not. If not all rays are realizable, then clearly no measurement can give rise to these nonrealizable states. They cannot therefore be eigenfunctions of any Hermitian operator which corresponds to an observable property of the system. To be observable a Hermitian operator must therefore satisfy certain conditions (superselection rules). Ordinary elementary quantum mechanics operates in a single coherent subspace, so that it is possible to distinguish between any two rays and all self-adjoint operators are then observable.

Quantum mechanics next postulates that the position and momentum operators of a particle obey the following commutation rules:

$$[q_l, p_j] = i\hbar\delta_{lj} \quad (l, j = 1, 2, 3) \tag{4}$$

For a particle with no internal degrees of freedom, it is a mathematical theorem [Von Neumann (1931)] that these operators are irreducible, meaning that there exists no subspace of the entire Hilbert space which is left invariant under these operators. This property is equivalent to the statements that any operator which commutes with both $\mathbf{p}$ and $\mathbf{q}$ is a multiple of the identity and that every operator is a function of $\mathbf{p}$ and $\mathbf{q}$. The description of the system in terms of the observables $\mathbf{p}$ and $\mathbf{q}$ is complete.

Finally, quantum mechanics postulates that the dynamical behavior of the system is described by the Schrödinger equation

$$i\hbar\partial_t \mid ; t\rangle = H \mid ; t\rangle \tag{5}$$

where $\partial_t = \partial/\partial t$ and H, the Hamiltonian operator of the system, corresponds to the translation operator for infinitesimal time translations. By this is meant the following: Assume that the time evolution of the state vector can be obtained by the action of an operator $U(t, t_0)$ on the initial state $\mid ; t_0\rangle$ such that

$$\mid t\rangle = U(t, t_0) \mid t_0\rangle \tag{6a}$$

$$U(t_0, t_0) = 1 \tag{6b}$$

Conservation of probability requires that the norm of the vector $\mid t\rangle$ be constant in time:

$$\begin{aligned}\langle t \mid t\rangle &= \langle t_0 \mid t_0\rangle \\ &= \langle t_0 \mid U^*(t, t_0)\, U(t, t_0) \mid t_0\rangle\end{aligned} \tag{7}$$

and therefore that

$$U^*(t, t_0)\, U(t, t_0) = 1 \tag{8a}$$

This does not yet guarantee that U is unitary. For this to be the case, the following equation must also hold:

$$U(t, t_0)\, U^*(t, t_0) = 1 \tag{8b}$$

This condition will hold if U satisfies the group property:

$$U(t, t_1)\, U(t_1, t_0) = U(t, t_0) \tag{9}$$

If, in Eq. (9), we set $t = t_0$, and assume its validity for $t_0 < t_1$, we then obtain

$$U(t_0, t_1)\, U(t_1, t_0) = 1 \tag{10a}$$

whence

$$U(t_0, t_1) = U^{-1}(t_1, t_0) \tag{10b}$$

and multiplying (10a) on the left by $U^*(t_0, t_1)$ using (8) we obtain

$$U(t_1, t_0) = U^*(t_0, t_1) = U^{-1}(t_1, t_0) \tag{10c}$$

so that U is unitary.

If we let t be infinitesimally close to t_0, with $t - t_0 = \delta t$ then to first order in δt we may write

$$U(t_0 + \delta t, t_0) = 1 - \frac{i}{\hbar} H\delta t \tag{11}$$

In order that U be unitary, H must be Hermitian. The dimension of H is that of an energy. Equation (6a) for the infinitesimal case thus reads

$$|t_0 + \delta t\rangle - |t_0\rangle = -\frac{i}{\hbar} H\delta t \mid t_0\rangle \tag{12a}$$

which in the limit as $\delta t \to 0$ becomes Eq. (5) since, by definition,

$$\lim_{\delta t \to 0} (\delta t)^{-1}(|t + \delta t\rangle - |t\rangle) = \partial_t \mid t\rangle \tag{12b}$$

1b. Schrödinger and Heisenberg Pictures

In the previous remarks about quantum mechanics, we have defined the state of the system at a given time t by the results of all possible experiments on the system at that time. This information is contained in the state vector $\mid t\rangle_S = |\Psi_S(t)\rangle$. The evolution of the system in time is then described by the time dependence of the state vector which is governed by the Schrödinger equation

$$H_S \mid \Psi_S(t)\rangle = i\hbar\partial_t \mid \Psi_S(t)\rangle \tag{13}$$

The operators corresponding to physical observables, F_S, are time-independent; they are the same for all time with $\partial_t F_S = 0$. This defines the Schrödinger picture and the subscript S identifies the picture [Dirac (1958)].

Although the operators are time-independent, their expectation value in any given state will in general be time-dependent. Call

$$\langle F_S\rangle = \langle\Psi_S(t) \mid F_S \mid \Psi_S(t)\rangle \tag{14}$$

then

$$i\hbar \frac{d}{dt}\langle F_S\rangle = \langle\Psi_S(t) \mid [F_S, H_S] \mid \Psi_S(t)\rangle \tag{15}$$

In the Schrödinger picture we call, by definition, $\dot{F}_S$ that operator for which

$$\langle\dot{F}_S\rangle = \frac{d}{dt}\langle F_S\rangle \tag{16}$$

Let us next perform a time-dependent unitary transformation $V(t)$ on $|\Psi_S(t)\rangle$ which transforms it into the state vector

$$|\Phi(t)\rangle = V(t) \mid \Psi_S(t)\rangle \tag{17a}$$

$$V(t)\, V^*(t) = V^*(t)\, V(t) = 1 \tag{17b}$$

$$V^*(t) = V^{-1}(t) \tag{17c}$$

Using Eqs. (13) and (17a) we find that $|\Phi(t)\rangle$ obeys the following equation:

$$i\hbar\partial_t \mid \Phi(t)\rangle = [i\hbar\partial_t V(t) \cdot V^{-1}(t) + V(t)\, H_S V^{-1}(t)] \mid \Phi(t)\rangle \tag{18}$$

If we choose the time-dependent unitary operator, V, to satisfy

$$-i\hbar\partial_t V(t) = V(t)\, H_S V^{-1}(t) \cdot V(t) \tag{19}$$

the transformed state, $|\Phi_H\rangle$, will then be time-independent, i.e., $\partial_t \mid \Phi_H\rangle = 0$. The operator $V(t)$ being unitary, the expectation value of the operator F_S in terms of $|\Phi_H\rangle$ is given by

$$\langle F\rangle = (\Psi_S(t), F_S\Psi_S(t)) = (V(t)\,\Psi_S(t), V(t)\,F_S\Psi_S(t))$$
$$= (\Phi_H, V(t)\,F_S V^{-1}(t)\,\Phi_H) \tag{20}$$

We define $F_H(t)$ by

$$F_H(t) = V(t)\,F_S V^{-1}(t) \tag{21}$$

so that $F_H(t)$ has the same expectation value in terms of $|\Phi_H\rangle$ as F_S had in terms of $|\Psi_S\rangle$. Differentiating Eq. (17b) with respect to time, we obtain

$$(\partial_t V)\,V^* + V\partial_t V^* = 0 \tag{22}$$

which, together with Eq. (19) and its Hermitian adjoint, implies that the time dependence of $F_H(t)$ is given by

$$\partial_t F_H(t) = (\partial_t V)\cdot V^* F_H(t) + F_H(t)\,V\partial_t V^*$$
$$= \frac{i}{\hbar}[VH_S V^*, F_H(t)]$$
$$= \frac{i}{\hbar}[H_H, F_H(t)] \tag{23}$$

This last equation together with the time independence of the state vector $\partial_t\,|\,\Phi_H\rangle = 0$ defines the Heisenberg picture. The state vector in the Heisenberg picture is the same for all time; the operators on the other hand are time-dependent. The state vector $|\Phi_H\rangle$ describes the entire history of the system, i.e., the results of all possible experiments on the system throughout its history. However, if an actual experiment is performed on the system, the state vector will be changed. Although the Heisenberg state vector $|\Phi_H\rangle$ does not depend on the time, it may be specified by the results it predicts for some experiment at a given time. Thus, we could specify $|\Phi_H\rangle$ as that state vector which corresponds to the Schrödinger state vector at time $t = 0$, i.e., $|\Phi_H\rangle = |\Psi_S(0)\rangle$.

For a closed system, for which H_S is time-independent, the unitary operator which effects the transition from the Schrödinger to the Heisenberg picture is explicitly given by

$$V(t) = \exp\left(\frac{i}{\hbar}Ht\right) \tag{24}$$

where we have assumed the pictures to coincide at time $t = 0$. Note that for a closed system $H_S = H_H = H$.

1c. Nonrelativistic Free-Particle Equation

It is often convenient for the description of a physical system to introduce a particular co-ordinate system in Hilbert space, i.e., to choose a representation. Since every observable is assumed to have a complete set of eigenfunctions which spans a subspace of the Hilbert space of state vectors, these eigenfunctions can be used as a basis in that subspace.

The representation in which $\mathbf{q}$, the position operator, is diagonal is called the position or $\mathbf{q}$ representation; that in which the momentum operator $\mathbf{p}$ is diagonal, the $\mathbf{p}$ representation. In the $\mathbf{q}$ representation the state vector $|\Psi\rangle$ is specified by its components along the basis vectors $|\mathbf{q}\rangle$, the eigenfunctions of the position operator. The components $\langle \mathbf{q} \mid \Psi\rangle$ have a direct physical meaning: $|\langle \mathbf{q} \mid \Psi\rangle|^2 \, d\mathbf{q}$ is the probability that if a position measurement is carried out, the particle will be found between $\mathbf{q}$ and $\mathbf{q} + d\mathbf{q}$. The eigenfunctions satisfy the equation

$$\mathbf{q} \mid \mathbf{q}'\rangle = \mathbf{q}' \mid \mathbf{q}'\rangle \tag{25}$$

and the spectrum of the operator $\mathbf{q}$ consists of the points in Euclidean three-space. The eigenfunctions $|\mathbf{q}'\rangle$ are not normalizable as they correspond to eigenvalues in the continuous spectrum, but are normalized to a δ function

$$\langle \mathbf{q}' \mid \mathbf{q}''\rangle = \delta^{(3)}(\mathbf{q}' - \mathbf{q}'') \tag{26}$$

A physical state is represented by a normalizable state vector and corresponds to a wave packet. Thus a particle localized around $\mathbf{q}_0'$ can be represented by a vector

$$|\,\rangle = \int d\mathbf{q}' \, f(\mathbf{q}' - \mathbf{q}_0') \mid \mathbf{q}'\rangle \tag{27}$$

where f is peaked around $\mathbf{q}_0'$. The normalization condition for this vector $|\,\rangle$ is

$$\langle\,|\,\rangle = \int d\mathbf{q}' \, \bar{f}(\mathbf{q}') \, f(\mathbf{q}') \tag{28}$$

so that $|\,\rangle$ will be normalizable if f is square integrable. The completeness relation can be written as follows:

$$\int |\mathbf{q}'\rangle \, d\mathbf{q}' \, \langle \mathbf{q}'| = 1 \tag{29}$$

The representation of the operator $\mathbf{p}$ in this representation is obtained by using the commutation rules $[q_l, p_j] = i\hbar\delta_{lj}$. If we take the $\mathbf{q}'$, $\mathbf{q}''$ matrix element of the commutator and use (29) we obtain

$$\begin{aligned}\langle \mathbf{q}'' \mid [q_r, p_s] \mid \mathbf{q}'\rangle &= i\hbar\delta_{rs} \, \delta^{(3)}(\mathbf{q}' - \mathbf{q}'') \\ &= (q''_r - q'_r) \, \langle \mathbf{q}'' \mid p_s \mid \mathbf{q}'\rangle \end{aligned} \tag{30}$$

whence, by recalling that $x\delta'(x) = -\delta(x)$ we find that

$$\langle \mathbf{q}'' \mid p_s \mid \mathbf{q}'\rangle = -i\hbar \frac{\partial}{\partial q''_s} \langle \mathbf{q}'' \mid \mathbf{q}'\rangle \tag{31}$$

Similarly, the momentum representation is characterized by basis vectors which have the following properties:

$$\mathbf{p} \mid \mathbf{p}'\rangle = \mathbf{p}' \mid \mathbf{p}'\rangle \tag{32}$$

$$\langle \mathbf{p}' \mid \mathbf{p}''\rangle = \delta^{(3)}(\mathbf{p}' - \mathbf{p}'') \tag{33}$$

$$\int |\mathbf{p}'\rangle\, d\mathbf{p}'\, \langle \mathbf{p}'| = 1 \tag{34}$$

The representation of the operator $\mathbf{q}$ is now given by

$$\langle \mathbf{p}'' \mid q_r \mid \mathbf{p}' \rangle = i\hbar \frac{\partial}{\partial p''_r} \langle \mathbf{p}'' \mid \mathbf{p}' \rangle \tag{35}$$

The unitary transformation function $\langle \mathbf{q}' \mid \mathbf{p}' \rangle$ which permits us to transform from the $\mathbf{q}$ representation to the $\mathbf{p}$ representation is obtained by taking the scalar product of (32) with the bra $\langle \mathbf{q}'|$

$$\begin{aligned} \mathbf{p}' \langle \mathbf{q}' \mid \mathbf{p}' \rangle = \langle \mathbf{q}' \mid \mathbf{p} \mid \mathbf{p}' \rangle &= \int \langle \mathbf{q}' \mid \mathbf{p} \mid \mathbf{q}'' \rangle\, d\mathbf{q}''\, \langle \mathbf{q}'' \mid \mathbf{p}' \rangle \\ &= -i\hbar \nabla_{q'} \langle \mathbf{q}' \mid \mathbf{p}' \rangle \end{aligned} \tag{36}$$

Solving this differential equation, we obtain

$$\langle \mathbf{q}' \mid \mathbf{p}' \rangle = \lambda e^{\frac{i}{\hbar}\mathbf{p}' \cdot \mathbf{q}'} \tag{37}$$

where the constant λ is determined up to a constant factor of modulus one to be $(2\pi\hbar)^{-3/2}$ from the requirement that

$$\int \langle \mathbf{q}' \mid \mathbf{p}' \rangle\, d\mathbf{p}'\, \langle \mathbf{p}' \mid \mathbf{q}'' \rangle = \langle \mathbf{q}' \mid \mathbf{q}'' \rangle = \delta^{(3)}(\mathbf{q}' - \mathbf{q}'') \tag{38}$$

The wave function $\Psi(\mathbf{q}') = \langle \mathbf{q}' \mid \Psi \rangle$ in configuration space is thus related to the momentum space wave function $\Phi(\mathbf{p}') = \langle \mathbf{p}' \mid \Psi \rangle$ by the familiar Fourier transformation

$$\begin{aligned} \Psi(\mathbf{q}') = \langle \mathbf{q}' \mid \Psi \rangle &= \int d\mathbf{p}'\, \langle \mathbf{q}' \mid \mathbf{p}' \rangle \langle \mathbf{p}' \mid \Psi \rangle \\ &= \frac{1}{(2\pi\hbar)^{3/2}} \int d\mathbf{p}'\, e^{\frac{i}{\hbar}\mathbf{q}' \cdot \mathbf{p}'}\, \Phi(\mathbf{p}') \end{aligned} \tag{39}$$

For a nonrelativistic free particle, the Hamiltonian operator H is essentially determined by the requirements that it be translationally and rotationally invariant and that it transform like an energy under Galilean transformations. Translation invariance implies that H does not depend on the position $\mathbf{q}$ of the particle. It is therefore only a function of $\mathbf{p}$, and in order to be rotationally invariant it can only be a function $\mathbf{p}^2$. Galilean relativity then requires that

$$H = \frac{\mathbf{p}^2}{2m} \tag{40}$$

where m is the mass of the particle. The eigenfunctions $|E\rangle$ of H are determined by the equation

$$H \mid E' \rangle = E' \mid E' \rangle \tag{41}$$

The completeness and orthonormality relations now read

$$\int dE' \mid E' \rangle \langle E' | = 1 \tag{42a}$$

$$\langle E' \mid E'' \rangle = \delta(E' - E'') \tag{42b}$$

The explicit form of the energy eigenfunctions can easily be derived in the **q** representation by solving the equation

$$\begin{aligned} \langle \mathbf{q}' \mid \frac{\mathbf{p}^2}{2m} \mid E' \rangle &= -\frac{\hbar^2}{2m} \nabla_{\mathbf{q}}{}^2 \langle \mathbf{q}' \mid E' \rangle \\ &= E' \langle \mathbf{q}' \mid E' \rangle \end{aligned} \tag{43}$$

Since H and $\mathbf{p}$ commute, simultaneous eigenfunctions of these two operators can be obtained. One verifies that for the case of a particle moving in one dimension

$$\langle q \mid E, p \rangle = C(E)\, e^{\frac{i}{\hbar}\sqrt{2mE}\, q} \tag{44}$$

is such an eigenfunction with eigenvalues E and $p = \sqrt{2mE}$. The normalization constant $C(E)$, determined so that Eq. (42b) holds, is found to be

$$|C(E)|^2 = \sqrt{\frac{m}{8\pi^2\hbar^2 E}} \tag{45}$$

Thus

$$\langle q \mid E, p \rangle = (2\pi v \hbar)^{-1/2}\, e^{\frac{i}{\hbar}\sqrt{2mE}\, q} \tag{46}$$

where a constant phase factor of modulus one has been omitted and where v is the velocity of the particle: $v = \sqrt{2E/m}$. Note that the probability of finding the particle described by $|E\rangle$, to have its position co-ordinate between q and $q + dq$ is given by $|\langle q \mid E \rangle|^2\, dq$ which is proportional to dq/v, that is to the time spent in the interval dq.

Finally in the Schrödinger picture the time evolution of the particle is governed by the equation

$$i\hbar\partial_t \mid ; t\rangle = \frac{1}{2m} \mathbf{p}^2 \mid ; t\rangle \tag{47}$$

which in the **q** representation, with $\langle \mathbf{q} \mid ; t\rangle = \Psi(\mathbf{q}; t)$, reads

$$i\hbar\partial_t \Psi(\mathbf{q}; t) = -\frac{\hbar^2}{2m} \nabla_{\mathbf{q}}{}^2\, \Psi(\mathbf{q}; t) \tag{48}$$

The steps leading to this last equation can be summarized by saying that in the energy-momentum relation for a nonrelativistic free particle

$$E = \frac{1}{2m} \mathbf{p}^2 \tag{49}$$

E is replaced by the operator $i\hbar\partial_t$ and $\mathbf{p}$ by $-i\hbar$ times the gradient operator, i.e.,

$$E \to i\hbar\partial_t \tag{50}$$

$$\mathbf{p} \to -i\hbar\nabla \tag{51}$$

and the resulting expression is to operate on $\Psi(\mathbf{q}; t)$ the wave function describing the particle.

The solution of (47) is given by

$$| \, ; t\rangle = e^{-\frac{i}{\hbar}\frac{\mathbf{p}^2}{2m}(t-t_0)} | \, ; t_0\rangle \tag{52}$$

Thus the time displacement operator $U(t, t_0)$ is here given by exp $\left[-\frac{i}{\hbar}\frac{\mathbf{p}^2}{2m}(t - t_0)\right]$. In the $\mathbf{q}$ representation, we may write

$$\langle \mathbf{q} \, | \, ; t\rangle = \int d\mathbf{q}_0 \, \langle \mathbf{q} \, | \, U(t, t_0) \, | \, \mathbf{q}_0\rangle \langle \mathbf{q}_0 \, | \, ; t_0\rangle \tag{53}$$

or equivalently

$$\Psi(\mathbf{q}; t) = \int d\mathbf{q}_0 \, K(\mathbf{q}t; \mathbf{q}_0t_0) \, \Psi(\mathbf{q}_0; t_0) \tag{54}$$

where

$$\begin{aligned} K(\mathbf{q}t; \mathbf{q}_0t_0) &= \langle \mathbf{q} \, | \, U(t, t_0) \, | \, \mathbf{q}_0\rangle \\ &= \frac{1}{(2\pi\hbar)^3} \int d\mathbf{p} \, e^{\frac{i}{\hbar}\mathbf{p}\cdot(\mathbf{q}-\mathbf{q}_0)} \, e^{-\frac{i}{\hbar}\frac{\mathbf{p}^2}{2m}(t-t_0)} \end{aligned} \tag{55}$$

From the fact that $U(t, t) = 1$ it follows that $K(\mathbf{q}t; \mathbf{q}_0t) = \delta^{(3)}(\mathbf{q} - \mathbf{q}_0)$, which is clearly satisfied by (55), as required in order that (54) be an identity for $t = t_0$. Now Eq. (52) is defined only for $t \geqslant t_0$, so that K is similarly only defined for $t \geqslant t_0$. It is convenient to require that $K = 0$ for $t < t_0$. We can incorporate this boundary condition by writing

$$K(\mathbf{q}t; \mathbf{q}_0t_0) = \theta(t - t_0) \, \langle \mathbf{q} \, | \, U(t, t_0) \, | \, \mathbf{q}_0\rangle \tag{56}$$

where $\theta(t)$ is a step function defined as follows:

$$\begin{aligned} \theta(t) &= 1 \quad \text{if } t > 0 \\ &= 0 \quad \text{if } t < 0 \end{aligned} \tag{57}$$

so that

$$\frac{d\theta(t)}{dt} = \delta(t) \tag{58}$$

The differential equation obeyed by K is now easily derived from Eq. (56)

$$\begin{aligned} i\hbar\partial_t K(\mathbf{q}t; \mathbf{q}_0t_0) &= i\hbar\delta(t - t_0) \, \langle \mathbf{q} \, | \, U(t, t_0) \, | \, \mathbf{q}_0\rangle + i\hbar\theta(t - t_0) \, \langle \mathbf{q} \, | \, \partial_t U(t, t_0) \, | \, \mathbf{q}_0\rangle \\ &= i\hbar\delta(t - t_0) \, \delta^{(3)}(\mathbf{q} - \mathbf{q}_0) - \frac{\hbar^2}{2m} \nabla_{\mathbf{q}}^2 \, K(\mathbf{q}t; \mathbf{q}_0t_0) \end{aligned} \tag{59}$$

since $U(t, t) = 1$. K is the Green's function which solves the Cauchy problem for the nonrelativistic free-particle Schrödinger equation.

1d. Symmetry and Quantum Mechanics

In the above "derivation" of the Schrödinger equation for a free particle, the requirement that the Hamiltonian be invariant under certain transformations played an important role. We shall here analyze, at somewhat greater length, the role played by invariance principles in the formulation of quantum mechanics [Bargmann (1953), Wick (1959); see especially Wigner (1949), (1955), (1956), (1957); also Hagedorn (1959), and Wightman (1959b)].

The possibility of abstracting laws of motion from the chaotic set of events which surround us stems from the following circumstances:

(*a*) given a physical system it is possible to isolate a manageable set of relevant initial conditions, and more importantly,

(*b*) given the same set of initial conditions the resulting motion of the system will be the same no matter where and when these conditions are realized (at least in our neighborhood of the universe).

In the language of symmetry principles (*b*) is the statement that the laws of nature are independent of the position of the observer or, equivalently, that the laws of motion are covariant with respect to displacements in space and time, i.e., with respect to the transformations

$$\mathbf{x} \to \mathbf{x} + \mathbf{a} \tag{60a}$$

$$t \to t + \tau \tag{60b}$$

Experiments have also yielded the fact that space is isotropic so that the orientation in space of an event is an irrelevant initial condition and this principle can be translated into the statement that the laws of motion are invariant under spatial rotations. Newton's law of motion further indicated that the state of motion, as long as it is uniform with constant velocity, is likewise an irrelevant initial condition. This is the principle of Galilean invariance which asserts that the laws of nature are independent of the velocity of the observer, and more precisely, that the laws of motion of classical mechanics are invariant with respect to Galilean transformations. These symmetry principles are usually stated in terms of two observers, O and O', who are in a definite relation to each other. For example, observer O may be moving with constant velocity relative to O' in such a way that the relation of the labels of the points of space and the reading of the clocks in their respective co-ordinate systems is given by the following equations:

$$\mathbf{x}' = \mathbf{x} - \mathbf{v}t \tag{61a}$$

$$t' = t \tag{61b}$$

The principle of Galilean invariance then asserts that the "laws of nature" are the same for the two observers, i.e., that the form of the equations of motion is the same for both observers. The equations of motion must therefore be covariant with respect to the transformations (61a) and (61b). Two observers using inertial co-ordinate systems (i.e., one in which the laws of motion are the same) are said to be equivalent.

The aforementioned invariance principles were experimentally established and may have limited applicability. Thus, Lorentz invariance has replaced the principle of Galilean invariance and the discovery of the nonconservation of parity in weak interactions has re-emphasized that an invariance principle and its consequences must be experimentally verified.

At the macroscopic level, the notion of an invariance principle can be made precise and explicit with the help of the concept of the complete description of a physical system. By the latter is meant a specification of the trajectories of all particles together with a full description of all fields at all points of space for all time. The equations of motion then allow one to determine whether the system could in fact have evolved in the way specified by the complete description. As stated by Haag [unpublished, but quoted in Wigner (1956)], an invariance principle then requires that the following three postulates be satisfied:

1. It should be possible to translate a complete description of a physical system from one co-ordinate system into every equivalent co-ordinate system.
2. The translation of a dynamically possible description should again be dynamically possible.
3. The criteria for the dynamic possibility of a complete description should be identical for equivalent observers.

Postulate 2 is equivalent to the statement that a possible motion to one observer must also appear possible to any other observer, and postulate 3 to the statement of the form invariance of the equation of motion.

In a quantum mechanical framework, postulate 1 remains as stated. It implies that there exists a well-defined connection and correspondence between the labels attributed to the space-time points by each observer, between the vectors each observer attributes to a given physical system, and between observables of the system. Postulate 2 is usually formulated in terms of transition probabilities, and states that the transition probability is independent of the frame of reference. In other words, different equivalent observers make the same prediction as to the outcome of an experiment carried out on a system. Note that this system will be in a different relation to each of the observers. Observer O will attribute the vector $|\Psi_O\rangle$ to the state of the system, whereas observer O' will describe the state of this same system by a vector $|\Psi_{O'}\rangle$. We shall, however, assume that given two systems S_O and $S_{O'}$ which are in the same relation to each of the two observers (i.e., the values of the observable of system S_O as measured by observer O are the same as the values of the observables of $S_{O'}$ as measured by observer O'), the observers will describe the state of their respective systems by the same vector. We shall call the vector $|\Psi_{O'}\rangle$ the translation of the vector $|\Psi_O\rangle$. Stated mathematically, postulate 2 asserts that if $|\Psi_O\rangle$ and $|\Phi_O\rangle$ are two states and $|\Psi_{O'}\rangle$ and $|\Phi_{O'}\rangle$ their translations, then

$$|(\Psi_O, \Phi_O)|^2 = |(\Psi_{O'}, \Phi_{O'})|^2 \tag{62}$$

If all rays in Hilbert space are distinguishable, it then follows from Eq. (62) as a mathematical theorem [Wigner (1959)] that the correspondence $|\Psi_O\rangle \to |\Psi_{O'}\rangle$ is effected by a unitary or an antiunitary[2] operator, $U(O', O)$, i.e.,

$$|\Psi_{O'}\rangle = U(O', O) \mid \Psi_O\rangle \tag{63}$$

where U depends on the co-ordinate systems between which it affects the correspondence and $U(O', O) = I$ if $O' = O$. Postulate 3 now asserts that U can only depend on the relation of the two co-ordinate systems and not on the intrinsic properties of either one. For example, for Lorentz transformations, $U(O', O)$ must be identical with $U(O''', O'')$ if observer O''' is in the same relation to O'' as observer O' is to O, i.e., if O''' arises from O'' by the same Lorentz transformation, L, by which O' arises from O. If this were not so there would be an intrinsic difference between the frames O', O and O''', O''. The operator U is completely determined up to a factor of modulus unity by the transformation, L, which carries O into O'. We write

$$|\Psi_{O'}\rangle = U(L) \mid \Psi_O\rangle \tag{64}$$

with $U(L) = \mathrm{I}$ if L is the identity transformation, i.e., if O and O' are the same co-ordinate systems. If we consider three equivalent frames, then we must obtain the same state by going from the first frame O to the second $O' = L_1O$ and then to the third $O'' = L_2O'$, and by going directly from the first to the third frame $O'' = L_3O$,

$$L_3 = L_2L_1 \tag{65}$$

Hence

$$\begin{aligned} |\Psi_{O''}\rangle &= U(L_2)\, U(L_1) \mid \Psi_O\rangle \\ &= U(L_3) \mid \Psi_O\rangle \end{aligned} \tag{66}$$

from which it follows

$$U(L_3) = \omega(L_2, L_1)\, U(L_2)\, U(L_1) \tag{67a}$$

$$U(1) = \mathrm{I} \tag{67b}$$

where ω is a number of modulus one which can depend on L_1 and L_2 and arises because of the indeterminate factor of modulus one in the state vectors. A set of Us which satisfy (67a) and (67b) are said to form a

[2] Recall that an operator U is said to be unitary if for every pair of vectors Ψ, χ in the Hilbert space, $\mathcal{H}$, $(U\Psi, U\chi) = (\Psi, \chi)$ and every vector Φ in $\mathcal{H}$ can be written in the form $U\varphi$, φ in $\mathcal{H}$. It follows from this definition and the properties of the scalar product that U is linear and has an inverse U^{-1} which is equal to its Hermitian adjoint $U^{-1} = U^*$. An operator U is said to be antiunitary if for every Ψ and χ in $\mathcal{H}$

$$(U\Psi, U\chi) = \overline{(\Psi, \chi)} = (\chi, \Psi)$$

An antiunitary operator is therefore antilinear: $U[\alpha|\Psi\rangle + \beta|\chi\rangle] = \bar{\alpha}U|\Psi\rangle + \bar{\beta}U|\chi\rangle$. Note that the product of two antiunitary operators is unitary.

(unitary or antiunitary) "representation up to a factor" of the group of transformations under which the observers are equivalent. For special relativity, for example, this group is the group of inhomogeneous Lorentz transformations. One is thus led to the mathematical problem of determining all the representations up to a factor of the group of interest.

It now follows from postulate 2 and from the fact that all the frames which can be reached by the symmetry transformation are equivalent for the description of the system, that together with $|\Psi_O\rangle$, $U(L) \mid \Psi_O\rangle$ must also be a possible state of the system as described by observer O. Thus, a relativity invariance requires the vector space describing the possible states of a quantum mechanical system to be invariant under all relativity transformations, i.e., it must contain together with every $|\Psi\rangle$ all transforms $U(L) \mid \Psi\rangle$ where L is any relativity transformation. This is the active view of formulating relativistic invariance [Bargmann (1953)] and it deals only with the transformed states of a single observer. Note that for the symmetry transformations which can be obtained continuously from the identity (i.e., no inversions), the transformed states can always be obtained from the original state by an actual physical operation on the system. Consider for example a Lorentz transformation along the x-axis with velocity v. The transformed state, which arises from the state $|\Psi_O\rangle$ is given by $U(v) \mid \Psi_O\rangle$. This is the state of the system as seen by observer O'. It is, however, also a possible state of the system as seen by O and which can be realized by giving the system a velocity $-v$ along the x-axis. If one deals with an inversion, e.g., time inversion, no such operation is in general possible. The invariance of the theory under this symmetry operation then essentially postulates the existence of this transformed state without necessarily giving a procedure for its realization.

For quantum mechanical applications, the importance of determining all unitary representations of a relativity group comes from the fact that the knowledge of such a unitary representation can in effect replace the wave equation for the system. For if in the above discussion in the frame O, we used a description of our system in the Heisenberg picture in which the Heisenberg state $|\Psi\rangle_H$ coincides with $|\Psi(0)\rangle_S$, the Schrödinger state at time $t = 0$, then the Schrödinger state vector at time t_0 can be obtained by transforming to a frame O' for which $t' = t - t_0$ while all other coordinates remain unchanged. If L is this transformation then

$$|\Psi(t_0)\rangle_S = U(L) \mid \Psi(0)\rangle_S \tag{68}$$

Thus a determination of all unitary representations of the inhomogeneous Lorentz group [Wigner (1939), Bargmann (1948), Shirokov (1958a, b)] is equivalent to a determination of all possible relativistic wave equations.

To clarify these concepts further, we consider in the next section the representations of the three- and four-dimensional rotation group.

1e. Rotations and Intrinsic Degrees of Freedom

The relation between the labels of the points of three-dimensional space for two observers whose co-ordinate systems are rotated with respect to one another about a common origin, is given by

$$\mathbf{x}' = R\mathbf{x} \tag{69a}$$

or

$$x'_i = \sum_{k=1}^{3} r_{ik}x_k = r_{ik}x_k \tag{69b}$$

(We use the summation convention over repeated indices.) We call R a rotation. The length of a vector and the angle between vectors are preserved under rotations, i.e.,

$$x'_iy'_i = x_iy_i \tag{70}$$

therefore

$$RR^T = R^TR = I \tag{71}$$

and rotations are represented by orthogonal matrices. It follows from (71) that

$$\det RR^T = \det R^T \det R = (\det R)^2 = 1$$

so that $\det R = \pm 1$. A rotation for which $\det R = +1$ is called a proper rotation, one for which $\det R = -1$ an improper one. An example of the latter is an inversion of the co-ordinate system about the origin represented by

$$R_- = \begin{pmatrix} -1 & 0 & 0 \\ 0 & -1 & 0 \\ 0 & 0 & -1 \end{pmatrix} \tag{72}$$

with $(R_-)^2 = +1$. R_- corresponds to a transition from a right-handed to a left-handed co-ordinate system. Every improper rotation R' with $\det R' = -1$ can be written in the form $(R'R_-)\,R_-$, i.e., as the inversion R_- followed by a proper rotation, since $\det R'R_- = \det R' \det R_- = (-1)^2 = +1$. The set of all proper rotations in Euclidean three-space forms a group: the rotation group. The group of all rotations together with reflections is called the orthogonal group. Since each element of the group can be specified by three continuously varying parameters (e.g., the direction cosines of the axis about which the rotation takes place and the angle of rotation), the rotation group is a continuous three-parameter group. The number of parameters of a group is called the dimension of the group. We wish to determine all the representations of the rotation group.

In general, a representation of a group G is a mapping (correspondence) which associates to every element g of G a linear operator T_g in a certain vector space V, such that group multiplication is preserved and the identity e of G is mapped into the identity I in V.[3] That is, if e, g_1, g_2, g_3, etc., are the elements of G and if to these elements are associated the linear operators T_e, T_{g_1}, T_{g_2}, $\cdots$ etc. in V, these operators are said to form a representation of the group G if

$$T_e = I \tag{73a}$$

and

$$T_{g_1}T_{g_2} = T_{g_1g_2} \tag{73b}$$

If T_g is represented by a matrix one speaks of a matrix representation. In quantum mechanics one is actually interested in a ray correspondence in which case T_g and $\exp(i\alpha_g) \cdot T_g$, with α_g an arbitrary real constant, represent the same correspondence. In this case Eq. (73b) is replaced by $T_{g_1}T_{g_2} = \omega(g_1, g_2)\, T_{g_1g_2}$. It has, however, been shown by Wigner (1959) that one can determine from a ray correspondence an essentially unique vector correspondence by a suitable normalization. Bargmann (1954) has furthermore shown that for the groups of interest for physical applications (rotation, Galilean, Lorentz group) with a suitable choice of T_g (recall that T_g and $T_g \exp i\alpha$ represent the same correspondence), ω is either equal to ± 1 (restricted Lorentz group, rotation group) or it can be expressed by a fairly simple expression (Galilean group).

A subspace V_1 of V is said to be invariant under the representation T_g if all vectors, v, in V_1 are transformed by T_g into vectors, v', again in V_1, and this for all T_g. If the only subspaces of V which are invariant under the representation $g \to T_g$ consist of the entire space and the subspace consisting of the null vector alone, we say that the representation is irreducible.

It is a theorem, which we state without proof [see Gel'fand (1956) or Wigner (1959)], that it is always possible to define a scalar product in V such that the representations of the rotation group in V are unitary,[4] i.e., such that the operators T_g are all unitary: $T_g^* = T_g^{-1} = T_{g^{-1}}$. Furthermore, the study of such unitary representations for compact groups can be reduced to the study of irreducible representations. For if there exists a subspace V_1 of V invariant under T_g, then the orthogonal complement of V_1, $V_1^\perp$, i.e., the set of all vectors orthogonal to V_1, is also invariant under T_g. *Proof:* If v_1 is an element of V_1 and w an element of $V_1^\perp$, since T_g is unitary we have $0 = (v_1, w) = (T_gv_1, T_gw)$. Now by assumption, T_gv_1 is again an element v_1' of V_1, therefore $(v_1', T_gw) = 0$ for arbitrary T_g,

[3] We shall always only consider continuous representations, i.e., representations T_g such that (v, T_gw) is a continuous function of g for every pair of vectors v, w in V, now assumed to have a scalar product defined in it.

[4] The possibility of introducing this scalar product depends in an essential manner on the finiteness of the group volume, i.e., on the compactness of the group, for it involves an integration over the group manifold. Loosely speaking, a matrix group is said to be compact if the matrix elements of every group element (i.e., of the matrix representing this group element) are bounded. This is clearly the case for the rotation group but is not the case for the Lorentz group. [Compare Eq. (1.78) and Eq. (2.10).]

and therefore for all T_g. Hence the set of vectors $T_g w$ for all w in $V_1^\perp$ are elements of $V_1^\perp$, and $V_1^\perp$ is therefore invariant under T_g. Thus V has been split into two invariant subspaces. In many cases, this process can be continued until one deals with only irreducible representations. For compact groups (and therefore for the rotation group in particular) it is known [see, e.g., Pontrjagin (1946)] that this inductive process of decomposing invariant subspaces into invariant subspaces terminates: The irreducible representations are all finite dimensional and every representation is a direct sum[5] of irreducible finite dimensional representations. Finally it should be noted that one is only interested in inequivalent irreducible representations. Two representations T and T' are said to be equivalent if there exists a one-to-one correspondence, $v \leftrightarrow v'$, between the vectors of the representation spaces such that if v corresponds to v' the vector $T_g v$ corresponds to $T_g' v'$ for all g and all pairs of vectors v, v'. This one-to-one correspondence can be represented by a (unitary) operator M, i.e., $v' = Mv$ and $v = M^{-1}v'$. For equivalent representations $MT_g v = {}$ $= T_g' v' = MT_g M^{-1} v'$ for all v. Two representations are thus equivalent if there exists an M such that $T_g' = MT_g M^{-1}$. Two equivalent representations can be considered as the realizations of the same representation in terms of two different bases in the vector space.

Now every rotation is a rotation about some axis so that a rotation can be specified by giving the axis of rotation about which the rotation is made and the magnitude of the angle of rotation. A rotation can thus be represented by a vector $\boldsymbol{\lambda}$, where the direction of the vector specifies the direction of the axis of rotation and the length of the vector the magnitude of the angle of rotation. A rotation about the *1*-axis is thus represented by a vector $(\lambda, 0, 0)$, a rotation about the *2*-axis by $(0, \lambda, 0)$, etc. It is evident that if $\boldsymbol{\lambda} = (\lambda_1, \lambda_2, \lambda_3)$ is a rotation vector then $|\boldsymbol{\lambda}| \leqslant \pi$ and that the set of all rotations fill a sphere of radius π. Distinct points in the interior of this sphere correspond to distinct rotations, whereas points diametrically opposed on the surface of the sphere correspond to the same rotation and must be identified. A group element can thus be considered a function of $\boldsymbol{\lambda}$, $g = g(\boldsymbol{\lambda})$ and similarly for a representation, $T_g = T(\boldsymbol{\lambda})$. Now $\boldsymbol{\lambda} = \mathbf{0}$ corresponds to the identity operation so that

$$T(\mathbf{0}) = I \tag{74}$$

Infinitesimal rotations about an axis will play a fundamental role in the following. Their importance derives from the fact they generate one-parameter subgroups and that any finite rotation can be constructed out

[5] If D, D_1, D_2 are three square matrices, D is said to be the direct sum of D_1 and D_2 if

$$D = \begin{pmatrix} D_1 & 0 \\ 0 & D_2 \end{pmatrix}$$

and one writes $D = D_1 \oplus D_2$.

of a succession of infinitesimal ones. It is to be noted that infinitesimal rotations commute with one another whereas finite rotations in general do not. Let $R^{(3)}(\theta)$ be a rotation through the angle θ about the *3*-axis, and let us define

$$A_3 = \frac{d}{d\theta} R^{(3)}(\theta)\bigg|_{\theta=0} \tag{75}$$

One calls A_3 the generator for an infinitesimal rotation about the *3*-axis. Note that for ϵ infinitesimal we may write

$$R^{(3)}(\epsilon) = 1 + A_3\epsilon + \text{terms of order } \epsilon^2 \tag{76}$$

Now a rotation through the angle θ about the *3*-axis, $R^{(3)}(\theta)$, can be considered to occur in n steps, each step consisting of a rotation through an angle θ/n. We may therefore write

$$\begin{aligned} R^{(3)}(\theta) &= \lim_{n\to\infty} \left(1 + \frac{\theta}{n} A_3\right)^n \\ &= e^{\theta A_3} \end{aligned} \tag{77}$$

We can define the generators for infinitesimal rotations about the *1*- and *2*-axes in a similar fashion. Explicitly, since

$$R^{(3)}(\theta) = \begin{pmatrix} \cos\theta & \sin\theta & 0 \\ -\sin\theta & \cos\theta & 0 \\ 0 & 0 & 1 \end{pmatrix} \tag{78}$$

$$A_3 = \frac{d}{d\theta} R^{(3)}(\theta)\bigg|_{\theta=0} = \begin{pmatrix} 0 & 1 & 0 \\ -1 & 0 & 0 \\ 0 & 0 & 0 \end{pmatrix} \tag{79a}$$

and similarly,

$$A_1 = \begin{pmatrix} 0 & 0 & 0 \\ 0 & 0 & 1 \\ 0 & -1 & 0 \end{pmatrix} \qquad A_2 = \begin{pmatrix} 0 & 0 & -1 \\ 0 & 0 & 0 \\ 1 & 0 & 0 \end{pmatrix} \tag{79b}$$

One verifies that the generators A_i ($i = 1, 2, 3$) satisfy the following commutation rules among themselves:

$$[A_l, A_j] = -\epsilon_{ljk} A_k \tag{80}$$

where ϵ_{ljk} is the totally antisymmetric tensor of rank three which is equal to $+1$ if ljk is an even permutation of 123, -1 if ljk is an odd permutation of 123, and zero otherwise. It should be noted that the reflection operator R_-, Eq. (72), commutes with all rotations

$$[R_-, A_i] = 0 \quad \text{for } i = 1, 2, 3 \tag{81}$$

Rotations about an axis form a commutative one-parameter subgroup of the group of rotations. In general, a one-parameter subgroup, $a(t)$, of a group G, is a "curve" in the group (i.e., a continuous function from the real line into G) such that

$$a(t)\, a(s) = a(t + s) \tag{82}$$

Clearly $a(0) = e$, the identity, $a(-t) = [a(t)]^{-1}$ and $a(s)\, a(t) = a(t)\, a(s)$. For the groups we shall be considering (rotation, Lorentz) the neighborhood of the identity (infinitesimal transformations) can be "filled up" with segments of one-parameter subgroups such that two such segments only have the identity in common. By a segment we mean a set $a(t)$ for $|t|$ less than some constant. Consider next the tangent to the curve at the identity, i.e., the element

$$\alpha = \left(\frac{da(t)}{dt}\right)_{t=0} \tag{83}$$

which is the analogue of Eq. (75) above. If the curves $a(t)$ and $b(t)$ have tangents α and β respectively, then the curve $c(t) = a(t)\, b(t)$ has a tangent $\alpha + \beta$. The set of tangents is a vector space under addition and multiplication by scalars and it is closed under a bracket operation denoted by $[\alpha\beta]$ and defined by

$$[\alpha\beta] = \frac{d^2}{ds\, dt} (a(s)\, b(t)\, a^{-1}(s)\, b^{-1}(t)) \Big|_{\substack{s=0 \\ t=0}} \tag{84}$$

$[\alpha\beta]$ has the property that it is antisymmetric in α and β, linear in each factor, and satisfies Jacobi's identity:

$$[\alpha\beta] = -[\beta\alpha] \tag{85a}$$

$$[\alpha(\beta + \gamma)] = [\alpha\beta] + [\alpha\gamma] \tag{85b}$$

$$[[\alpha\beta]\gamma] + [[\gamma\alpha]\beta] + [[\beta\gamma]\alpha] = 0 \tag{85c}$$

This vector space with the product operation just introduced is called the Lie algebra of the group. The dimension of the Lie algebra is equal to the dimension of the group. To every element α of the Lie algebra there corresponds a unique one-parameter group $a(t) = \exp \alpha t$ [compare Eq. (77) above]. For matrix groups the bracket operation, $[\alpha\beta]$, corresponds to taking the commutator of α and β, i.e., $[\alpha\beta] = [\alpha, \beta] = \alpha\beta - \beta\alpha$. In this case if we denote a linearly independent set of elements of the Lie algebra of dimension n by α_i, $(i = 1, 2, \cdots n)$, the closure property is expressed by the relation

$$[\alpha_i, \alpha_j] = \sum_{k=1}^{n} c_{ij}{}^k \alpha_k \tag{86}$$

where the $c_{ij}{}^k$ are constants, the so-called structure constants, which are characteristics of the group. A heuristic proof of Eq. (86) can be constructed as follows: We have to show that the left-hand side of (86) belongs to our Lie algebra, and consequently can be expressed as a linear combination of the α_is. Consider the element

$$c(s, t) = e^{s\alpha_i}\, e^{t\alpha_j}\, e^{-s\alpha_i}\, e^{-t\alpha_j} \quad \text{(no summation over } i, j) \tag{87a}$$

which for s, t infinitesimal becomes

$$c(s, t) = 1 + st\,[\alpha_i, \alpha_j] + \cdots \tag{87b}$$

Note that $c(s, t)$ is uniquely determined by the parameters s, t. For s, t infinitesimal, it must have the representation

$$\begin{aligned} c(s, t) &= 1 + s\left(\sum_{k=1}^{n} d_1{}^k\alpha_k\right) + t\left(\sum_{k=1}^{n} d_2{}^k\alpha_k\right) \\ &\quad + st\left(\sum_{k=1}^{n} d_3{}^k\alpha_k\right) + \cdots \end{aligned} \tag{88}$$

Since $c(0, t) = c(s, 0) = 1$, $d_1{}^k = d_2{}^k = 0$ for all k, so that comparing both expansions we have

$$[\alpha_i, \alpha_j] = \sum_{k=1}^{n} d_3{}^k\alpha_k \tag{89}$$

which proves (86), since the $d_3{}^k$ clearly depend on i and j and can be written as $c_{ij}{}^k$.

A representation of the Lie algebra is a correspondence, $\alpha \to A(\alpha)$, which associates to each element α of the algebra a linear operator $A(\alpha)$ in a vector space V, such that

$$A(\alpha + \beta) = A(\alpha) + A(\beta) \tag{90a}$$

$$A(c\alpha) = cA(\alpha) \tag{90b}$$

$$A([\alpha\beta]) = [A(\alpha), A(\beta)] = A(\alpha)\,A(\beta) - A(\beta)\,A(\alpha) \tag{90c}$$

i.e., the bracket operation is mapped into commutator which automatically satisfies Eq. (85c). A representation of the Lie algebra of a group will uniquely determine a representation of the group. Let us illustrate these remarks with the rotation group.

The Lie algebra of the rotation group is generated by the three linearly independent operators A_1, A_2, A_3 satisfying Eq. (80) and these operators generate the one-parameter subgroups of rotations about the three spatial axes. An infinitesimal rotation about $\boldsymbol{\epsilon}$ through the angle $|\boldsymbol{\epsilon}|$ can be represented by

$$\begin{aligned} R(\boldsymbol{\epsilon}) &= 1 + \epsilon_1 A_1 + \epsilon_2 A_2 + \epsilon_3 A_3 + O(\epsilon^2) \\ &= 1 + \epsilon_i A_i + \cdots \end{aligned} \tag{91}$$

For a representation we shall write

$$T(\boldsymbol{\epsilon}) = T(\epsilon_1, \epsilon_2, \epsilon_3) = I + \epsilon_1 M_1 + \epsilon_2 M_2 + \epsilon_3 M_3 + O(\epsilon^2) \tag{92}$$

where the M_is constitute a representation of the generators of the Lie algebra and satisfy the commutation rules

$$[M_i, M_j] = -\epsilon_{ijk} M_k \tag{93}$$

Let us next show that $T(\boldsymbol{\lambda})$ for arbitrary $\boldsymbol{\lambda}$ is completely determined by the generators M_1, M_2, M_3 and by $\boldsymbol{\lambda}$, and is given in terms of these quantities by

$$T(\lambda_1, \lambda_2, \lambda_3) = e^{\lambda_1 M_1 + \lambda_2 M_2 + \lambda_3 M_3} \tag{94}$$

Proof: Since two rotations about the same axis commute

$$R(s\boldsymbol{\lambda})\, R(t\boldsymbol{\lambda}) = R((s + t)\, \boldsymbol{\lambda}) \tag{95}$$

and therefore, similarly

$$T(s\boldsymbol{\lambda})\, T(t\boldsymbol{\lambda}) = T((s + t)\, \boldsymbol{\lambda}) \tag{96}$$

Upon differentiating both sides of this last equation with respect to s, replacing on the right side the differentiation with respect to s by one with respect to t, and setting $s = 0$ thereafter we obtain using Eq. (92)

$$\begin{aligned} \frac{d}{dt} T(t\boldsymbol{\lambda}) &= \frac{d}{ds} T(s\boldsymbol{\lambda}) \cdot T(t\boldsymbol{\lambda}) \Big|_{s=0} \\ &= (M_1\lambda_1 + M_2\lambda_2 + M_3\lambda_3)\, T(t\boldsymbol{\lambda}) \end{aligned} \tag{97}$$

Equation (97) is a differential equation determining $T(t\boldsymbol{\lambda})$. The solution of this differential equation which satisfies the boundary condition $T(\mathbf{0}) = I$, Eq. (74), is precisely given by Eq. (94).

For unitary representations, the requirement that the Ts be unitary implies that the M_j are skew-Hermitian, that is

$$M_j^* = -M_j \tag{98}$$

The operators $J_l = -iM_l$ are thus Hermitian and satisfy the familiar commutation rules of angular momenta:

$$[J_l, J_m] = i\epsilon_{lmn} J_n \tag{99}$$

Now the problem of finding all irreducible representations of the rotation group is equivalent to finding all the possible sets of matrices J_1, J_2, J_3 which satisfy the commutation rules (99). Clearly every irreducible representation of a continuous group will also be a representation in the neighborhood of the identity (infinitesimal transformations) although the converse is not necessarily true. In general, if we find all the irreducible representations of the group G in the neighborhood of the identity, i.e., find all the representations of the infinitesimal generators, then we can obtain all the irreducible representations of the entire group by exponentiation, Eq. (94). However, it is possible that some of the irreducible representations of G obtained in this manner are not continuous over the whole group but are continuous only in the neighborhood of the identity. These discontinuous representations must then be discarded.

In the theory of group representations by complex matrices, Schur's lemma [see Wigner (1959)] is of fundamental importance. It asserts that the necessary and sufficient condition for a representation to be irreducible is that the only operators which commute with all the matrices of the representation be multiples of the identity operator. Suppose that the Lie algebra of a group G contains an element A which commutes with all other elements of the Lie algebra. Let $g \to T(g)$ be a representation of G in a vector space V. The operators $(dT(g(s))/ds)_{s=0} = \alpha$ form a representation of the Lie algebra of G. The operator which corresponds to A in this representation commutes with all other operators α, and consequently commutes with all operators $T(g)$ (such commuting elements will be called invariants of the group). Because of Schur's lemma, then, a representation is irreducible if and only if the vector space on which the representation is defined is spanned by a manifold of eigenfunctions belonging to a *single* eigenvalue of this commuting operator. Conversely, if we find all the independent invariants of the group and construct a representation whose representation space is spanned by eigenfunctions belonging to the same eigenvalue of each of the invariants, then this representation will be irreducible, since each of the invariants is a multiple of the identity in this representation and by definition there are no other operators which commute with all the elements of the group. To each set of eigenvalues of all the invariants there thus corresponds one and only one irreducible representation. The problem of classifying the irreducible representations of the group is therefore reduced to finding the eigenvalue spectra of the invariants of the group.

For the proper rotation group, $J^2 = J_1^2 + J_2^2 + J_3^2$ commutes with each of the generators and it therefore is an invariant of the group. Its eigenvalues, as is well known from the theory of angular momenta, are $j(j+1)$ where $j = 0, \frac{1}{2}, 1, \frac{3}{2}, 2, \cdots$. Every irreducible representation is thus characterized by a positive integer or half-integer value including 0, the dimension of the representation being $2j + 1$ and for each j, integer or half-integer, there is an irreducible representation. In order to classify the irreducible representations of the orthogonal group we note that T_-, the linear operator corresponding to the inversion operation R_-, commutes with all rotations. By Schur's lemma, in every irreducible representation it must be a constant multiple of the identity. An irreducible representation of the orthogonal group is thus classified by a pair of indices (j, t) where the second index is the eigenvalue of T_- in that representation. For integer j, one has $t = \pm 1$ (since $T_-^2 = I$) and there exist two different irreducible representations of the orthogonal group for each integer j. For one of these $T_- = +I$ and for the other $T_- = -I$.

For $j = 0$ the representation is one dimensional, every group element is mapped into the identity and the infinitesimal generators are identically zero. We call the representation for which $T_- = +I$ the scalar representation, that for which $T_- = -I$ the pseudoscalar.

For $j = \frac{1}{2}$ the representation of the rotation group is two dimensional and the infinitesimal generators $M_j^{(1/2)}$ can be represented as $\frac{1}{2}i$ times the (Hermitian) Pauli matrices, σ_i

$$\sigma_1 = \begin{pmatrix} 0 & 1 \\ 1 & 0 \end{pmatrix} \qquad \sigma_2 = \begin{pmatrix} 0 & -i \\ i & 0 \end{pmatrix} \qquad \sigma_3 = \begin{pmatrix} 1 & 0 \\ 0 & -1 \end{pmatrix} \tag{100}$$

which satisfy

$$\sigma_l \sigma_j = \delta_{lj} + i\epsilon_{ljk}\sigma_k \tag{101}$$

The $j = \frac{1}{2}$ representation for a rotation through an angle θ about the 3-axis is thus given by

$$\begin{aligned} T_3^{(1/2)}(\theta) = T^{(1/2)}(0, 0, \theta) = e^{\frac{1}{2}i\theta\sigma_3} &= \sum_{n=0}^{\infty} \frac{1}{n!} \left(\tfrac{1}{2}i\theta\right)^n \sigma_3^n \\ &= \left[1 - \frac{1}{2!}\left(\frac{\theta}{2}\right)^2 + \frac{1}{4!}\left(\frac{\theta}{2}\right)^4 \pm \cdots\right] \\ &\quad + i\sigma_3\left[\frac{\theta}{2} - \frac{1}{3!}\left(\frac{\theta}{2}\right)^3 + \frac{1}{5!}\left(\frac{\theta}{2}\right)^5 \pm \cdots\right] \\ &= \cos\frac{\theta}{2} + i\sigma_3 \sin\frac{\theta}{2} \\ &= \begin{pmatrix} e^{i\frac{\theta}{2}} & 0 \\ 0 & e^{-i\frac{\theta}{2}} \end{pmatrix} \end{aligned} \tag{102a}$$

Similarly the $j = \frac{1}{2}$ representation for a rotation through an angle θ about the x and y axes are given by

$$\begin{aligned} T_1^{(1/2)}(\theta) = T^{(1/2)}(\theta, 0, 0) = e^{\frac{1}{2}i\theta\sigma_1} &= \cos\frac{\theta}{2} + i\sigma_1 \sin\frac{\theta}{2} \\ &= \begin{pmatrix} \cos\frac{\theta}{2} & i\sin\frac{\theta}{2} \\ i\sin\frac{\theta}{2} & \cos\frac{\theta}{2} \end{pmatrix} \end{aligned} \tag{102b}$$

$$\begin{aligned} T_2^{(1/2)}(\theta) = T^{(1/2)}(0, \theta, 0) = e^{\frac{1}{2}i\theta\sigma_2} &= \cos\frac{\theta}{2} + i\sigma_2 \sin\frac{\theta}{2} \\ &= \begin{pmatrix} \cos\frac{\theta}{2} & \sin\frac{\theta}{2} \\ -\sin\frac{\theta}{2} & \cos\frac{\theta}{2} \end{pmatrix} \end{aligned} \tag{102c}$$

Note that the $T_i^{(1/2)}(\theta)$, $(i = 1, 2, 3)$, are unitary matrices of determinant one. We also note that a rotation through the angle 2π about any axis yields

$$T_i^{(1/2)}(\theta + 2\pi) = -T_i^{(1/2)}(\theta) \tag{103}$$

The representation is therefore two-valued, and the correspondence from elements of the group to T is given by $R(\boldsymbol{\lambda}) \to \pm T(\boldsymbol{\lambda})$. Since for quantum mechanical applications we are interested only in representations up to a factor, these two-valued representations are permissible.

For $j = 1$ the representation is three dimensional and the previously determined matrices A_i, Eqs. (79a) and (79b), can be taken as the matrix representation for the infinitesimal generators $M^{(1)}{}_i$. The usual quantum mechanical representation of the J_i for $j = 1$

$$J_1 = \frac{1}{\sqrt{2}}\begin{pmatrix} 0 & 1 & 0 \\ 1 & 0 & 1 \\ 0 & 1 & 0 \end{pmatrix} \qquad J_2 = \frac{1}{\sqrt{2}}\begin{pmatrix} 0 & -i & 0 \\ i & 0 & -i \\ 0 & i & 0 \end{pmatrix}$$

$$J_3 = \begin{pmatrix} 1 & 0 & 0 \\ 0 & 0 & 0 \\ 0 & 0 & -1 \end{pmatrix} \tag{104}$$

is unitarily equivalent to the representation given by the $-iA_j$. The J_i correspond to a basis $-\frac{1}{\sqrt{2}}(x - iy)$, z, $\frac{1}{\sqrt{2}}(x + iy)$ instead of the usual Cartesian basis: (x, y, z).

A quantity, ξ, which under a rotation of the co-ordinate system

$$\mathbf{x}' = R\mathbf{x} \tag{105}$$

transforms according to

$$\xi' = T^{(j)}(R)\,\xi \tag{106}$$

is said to be a scalar for $j = 0$, a spinor of rank 1 for $j = \frac{1}{2}$, a vector for $j = 1$, etc. For an infinitesimal rotation through an angle ϵ about the lth axis, the transformation rule (106) becomes

$$\xi' = (1 + \epsilon M_l{}^{(j)})\,\xi \tag{107}$$

A scalar is thus a one-component object which under rotations $\mathbf{x} \to R\mathbf{x}$ transforms according to $\xi \to \xi' = \xi$. Similarly, a spinor of rank 1 is a two-component object

$$\xi = \begin{pmatrix} \xi_1 \\ \xi_2 \end{pmatrix} \tag{108}$$

which under an infinitesimal rotation ϵ about the lth axis

$$x_j \to x'_j = x_j + \epsilon\epsilon_{ljk}x_k \tag{109}$$

transforms according to

$$\xi \to \xi' = (1 + \tfrac{1}{2}i\epsilon\sigma_l)\,\xi \tag{110}$$

For a rotation through an arbitrary finite angle, as previously noted, a rank 1 spinor is transformed by a 2×2 unitary matrix of determinant 1. Finally a vector is a three-component object

$$\xi = \begin{pmatrix} \xi_1 \\ \xi_2 \\ \xi_3 \end{pmatrix} \tag{111}$$

the components ξ_i, $i = 1, 2, 3$, of which, under the rotation (105) transforms as the co-ordinates themselves.

The classification of tensors and spinors under inversion is as follows: For j integral we have two kinds of objects, those transforming under an inversion according to $T_- = I$ and those transforming according to $T_- = (-)I$. We call an object a pseudo quantity if it transforms according to $T_-^{(j)} = (-1)^{j+1}$. Thus a pseudoscalar is a quantity which under inversion transforms according to $\xi \to \xi' = -\xi$. Similarly a pseudovector (or axial vector) is a quantity which under the inversion (72) transforms according to $\boldsymbol{\xi} \to \boldsymbol{\xi}' = \boldsymbol{\xi}$. For spinors the situation is somewhat more involved and will be taken up after we have introduced the notion of adjoints.

The adjoint of a spinor ξ is constructed in the usual manner by taking the transposed complex conjugate. Thus for $j = \frac{1}{2}$ the adjoint spinor of ξ is given by ξ^*

$$\xi^* = (\bar{\xi}_1 \quad \bar{\xi}_2) \tag{112}$$

and under an infinitesimal rotation about the lth axis it transforms according to

$$\xi^* \to \xi^{*\prime} = \xi^*(1 - \tfrac{1}{2}i\epsilon\sigma_l) \tag{113}$$

We next define a scalar product for spinors which will allow us to combine spinors. We define the scalar product of two spinors χ and ξ as

$$\chi^*\xi = \sum_{i=1}^{2} \bar{\chi}_i\xi_i \tag{114}$$

By combining spinors we can obtain new quantities which have definite transformation properties under rotations. Thus the quantity $\chi^*\xi$ under the infinitesimal rotation (109) transforms according to

$$\begin{aligned} \chi^*\xi \to \chi'^*\xi' &= \chi^*(1 - \tfrac{1}{2}i\epsilon\sigma_l)\,(1 + \tfrac{1}{2}i\epsilon\sigma_l)\,\xi \quad \text{(not summed over } l) \\ &= \chi^*\xi + \text{terms of } O(\epsilon^2) \end{aligned} \tag{115}$$

that is, as a scalar. The proof for a finite rotation is just as simple, since $T^{(1/2)}(\boldsymbol{\lambda})$ for arbitrary $\boldsymbol{\lambda}$ is represented by a unitary 2×2 matrix, so that $\chi^*\xi \to \chi^{*\prime}\xi' = \chi^*T^{(1/2)*}\,T^{(1/2)}\xi = \chi^*\xi$. Similarly one verifies that the quantity $\chi^*\sigma_j\xi$ transforms under (109) like a vector:

$$\begin{aligned} \chi^*\sigma_j\xi \to \chi^{*\prime}\sigma_j\xi' &= \chi^*(1 - \tfrac{1}{2}i\epsilon\sigma_l)\,\sigma_j(1 + \tfrac{1}{2}i\epsilon\sigma_l)\,\xi \\ &= \chi^*\sigma_j\xi - \tfrac{1}{2}i\epsilon\chi^*[\sigma_l, \sigma_j]\,\xi \\ &= \chi^*\sigma_j\xi + \epsilon\epsilon_{ljk}\,\chi^*\sigma_k\xi \end{aligned} \tag{116}$$

which is the requisite transformation law for a vector. Note that even though the spinor representations are two-valued, both $\chi^*\xi$ and $\chi^*\sigma_i\xi$ return to their original values for a rotation of 2π about an axis. An observable quantity, although not representable by a spinor (since the latter changed sign under a rotation by 2π), can be represented by a bilinear expression in spinor quantities, since the latter has unique transformation properties under rotations.

Let us now turn briefly to the inversion properties of spinors [Cartan (1938)]. For this purpose it is convenient to first consider reflections about a plane and in particular reflections about the co-ordinate planes. Consider the reflection

$$\begin{aligned} x_1 &\to x'_1 = -x_1 \\ x_2 &\to x'_2 = x_2 \\ x_3 &\to x'_3 = x_3 \end{aligned} \tag{117a}$$

or

$$\mathbf{x}' = R_{1-}\mathbf{x} \qquad R_{1-} = \begin{pmatrix} -1 & 0 & 0 \\ 0 & 1 & 0 \\ 0 & 0 & 1 \end{pmatrix} \tag{117b}$$

with

$$R_{1-}^{\ 2} = 1 \tag{118}$$

One readily verifies that R_{1-} has the following commutation rules with the infinitesimal generators for rotations, A_i:

$$\begin{aligned} [R_{1-}, A_3]_+ = [R_{1-}, A_2]_+ &= 0 \\ [R_{1-}, A_1] &= 0 \end{aligned} \tag{119}$$

where $[C, D]_+$ denotes the anticommutator of C and D, i.e.,

$$[C, D]_+ = CD + DC \tag{120}$$

The operator corresponding to R_{1-} in any representation must satisfy the same commutation rules with the infinitesimal generators for that representation, i.e.,

$$\begin{aligned} [T_{1-}, M_2]_+ = [T_{1-}, M_3]_+ &= 0 \\ [T_{1-}, M_1] &= 0 \end{aligned} \tag{121}$$

For the $j = \frac{1}{2}$ representation, these commutation rules together with $T_{1-}^{\ 2} = 1$ imply that under this reflection the spinor ξ is transformed according to

$$\xi \to \xi' = +\sigma_1\xi \tag{122}$$

where we have arbitrarily chosen the + sign in front of σ_1, i.e., $T_{1-}^{(1/2)} = +\sigma_1$. Actually, since a spinor transforms according to a two-valued representation of the rotation group, we can have $T_{1-}^{\ 2} = \pm 1$ (since we may consider two inversions as a rotation through the angle 2π), so that not only $\pm\sigma_1$ but also $\pm i\sigma_1$ can be chosen as the representation for the inversion operator. We first consider the case when the factor multiplying σ_1 is equal to ± 1 [see in this connection Yang (1950b)]. One verifies that under the inversion $\mathbf{x} \to -\mathbf{x}$ the spinor ξ is transformed according to

$$\xi \to \xi' = \pm i\xi \tag{123}$$

(since $R_- = R_{3-} \cdot R_3(\pi)$, which for the $j = \frac{1}{2}$ representation yields $(\pm\sigma_3)(i\sigma_3) = \pm i$).

More generally let $\mathbf{n}$ be a unit vector and call P the plane perpendicular to $\mathbf{n}$ passing through the origin. If we decompose the vector $\mathbf{x}$ into a component parallel and a component perpendicular to $\mathbf{n}$

$$\mathbf{x} = (\mathbf{x} \cdot \mathbf{n})\,\mathbf{n} - \mathbf{n} \times (\mathbf{n} \times \mathbf{x}) \tag{124}$$

then a reflection about the plane P corresponds to the transformation

$$\begin{aligned} \mathbf{x} \to \mathbf{x}' &= -\mathbf{n}(\mathbf{x} \cdot \mathbf{n}) - \mathbf{n} \times (\mathbf{n} \times \mathbf{x}) \\ &= \mathbf{x} - 2\mathbf{n} \cdot (\mathbf{x} \cdot \mathbf{n}) \end{aligned} \tag{125}$$

$\mathbf{x}'$ is the mirror reflection of $\mathbf{x}$ with respect to the plane P. Under this reflection a spinor ξ is transformed according to

$$\xi \to \xi' = \boldsymbol{\sigma} \cdot \mathbf{n}\xi = N\xi \tag{126a}$$

$$N = \boldsymbol{\sigma} \cdot \mathbf{n}; \qquad N^2 = 1 \qquad \overline{N} = N^T \tag{126b}$$

where we have arbitrarily chosen the sign of N as $+\boldsymbol{\sigma} \cdot \mathbf{n}$. We shall call the transformed spinor, $N\xi$, ξ_N, i.e.,

$$\xi_N = N\xi \tag{127}$$

We next define the matrix C

$$C = i\sigma_2 = \begin{pmatrix} 0 & 1 \\ -1 & 0 \end{pmatrix} \tag{128}$$

with the properties

$$C^2 = -1 \tag{129a}$$

$$C\bar{\sigma}_i C = \sigma_i \tag{129b}$$

and the spinor η

$$\eta = iC\bar{\xi} \tag{130}$$

where by $\bar{\xi}$ we mean the column spinor with the components $\bar{\xi}_1$, $\bar{\xi}_2$. Under a reflection about P, η will transform according to

$$\eta \to \eta_N = iC\bar{\xi}_N \tag{131}$$

Since $\bar{\xi}_N = \overline{N}\bar{\xi}$, using (129a), (129b), and (130) we have

$$\begin{aligned} \eta_N &= iC\overline{N}\bar{\xi} = -iC\overline{N}CC\bar{\xi} = -iNC\bar{\xi} \\ &= -N\eta \end{aligned} \tag{132}$$

Thus an η type spinor transforms under reflections differently from a ξ type one. These two kinds of spinors cannot be reduced to each other, as there exists no linear transformation which can transform a type ξ spinor into a type η spinor. For if there existed such a transformation D for which $\xi = D\eta$ and $\xi_N = D\eta_N$ then D would have to anticommute with N, from which it follows by successive appropriate choices of the plane P that D must anticommute with each σ_i. But this is only possible for $D = 0$. We shall call a ξ type spinor a spinor of the first kind and an η type spinor, one of the second kind. If ξ, ξ' and η, η' are spinors of the first and second type respectively, one readily verifies that $\xi'^*\xi$ is a scalar under inversion; similarly for $\eta'^T C\xi$ and $\eta'^*\eta$. (On the other hand $\xi'^T\xi$,

$\eta^* C\xi \cdots$ etc. are not scalars, since their values are changed by proper rotations.) Quantities like $\eta'^*\xi$, $\eta'^T C\eta$, $\xi'^T C\xi$ transform like pseudoscalars under inversions. The quantities $\xi'^*\boldsymbol{\sigma}\xi$, $\eta'^*\boldsymbol{\sigma}\eta$ and $\eta'^T C\boldsymbol{\sigma}\xi$ are pseudovectors, whereas $\eta'^*\boldsymbol{\sigma}\xi$, $\eta'^T C\sigma\eta$ and $\xi'^T C\boldsymbol{\sigma}\xi$ transform like vectors under inversions.

These notions are readily generalized to scalar, spinor and vector fields. Thus we call a (three-dimensional) scalar field, $\xi(\mathbf{x})$, a function which under the rotation $\mathbf{x} \to \mathbf{x}' = R\mathbf{x}$ transforms according to

$$\xi(\mathbf{x}) \to \xi'(\mathbf{x}') = T^{(0)}(R)\,\xi(\mathbf{x}) = \xi(\mathbf{x}) \tag{133a}$$

or equivalently

$$\xi'(\mathbf{x}) = T^{(0)}(R)\,\xi(R^{-1}\mathbf{x}) = \xi(R^{-1}\mathbf{x}) \tag{133b}$$

i.e., the transformed quantity has the same (numerical) value at physically the same point as the original quantity. Similarly a vector valued function $\xi_i(\mathbf{x})$ transforms under a rotation $\mathbf{x} \to \mathbf{x}' = R\mathbf{x}$ according to the rule

$$\xi_i(\mathbf{x}) \to \xi'_i(\mathbf{x}') = r_{ik}\xi_k(\mathbf{x}) \tag{134a}$$

or equivalently

$$\xi'(\mathbf{x}) = T^{(1)}(R)\,\xi(R^{-1}x) = R\xi(R^{-1}\mathbf{x}) \tag{134b}$$

$$\xi'_i(\mathbf{x}) = r_{ik}\xi_k(R^{-1}\mathbf{x}) \tag{134c}$$

The significance of these considerations for quantum mechanics lies in the fact that the description of the spin (intrinsic angular momentum) of a particle can be incorporated into nonrelativistic quantum mechanics by the requirement that the wave function describing such a particle be a multicomponent object which under rotations transforms according to an irreducible representation of the three-dimensional rotation group. A massive spinless particle is represented by a wave function which under rotations transforms like a scalar. A nonrelativistic spin $\frac{1}{2}$ particle with its two degrees of freedom of spin-up and spin-down is described by a spinor wave function, and in general a particle with spin s by a $2s + 1$ component wave function. Let us consider the motivation for this in greater detail for the particular case of a spin $\frac{1}{2}$ particle in which case the wave function $\psi(\mathbf{x})$ is a two-component spinor

$$\psi(\mathbf{x}) = \begin{pmatrix} \psi_1(\mathbf{x}) \\ \psi_2(\mathbf{x}) \end{pmatrix} \tag{135}$$

Within the vector space of the possible states of this system we can introduce the following scalar product:

$$\begin{aligned}(\psi, \chi) &= \sum_{i=1}^{2} \int d^3x\, \langle \psi \mid \mathbf{x}i \rangle \langle \mathbf{x}i \mid \chi \rangle \\ &= \sum_{i=1}^{2} \int d^3x\, \bar{\psi}_i(\mathbf{x})\, \chi_i(\mathbf{x})\end{aligned} \tag{136}$$

Under a rotation $\mathbf{x}' = R\mathbf{x}$, the wave function $\langle \mathbf{x}i \mid \psi\rangle = \psi_i(\mathbf{x})$ transforms according to

$$\psi'(\mathbf{x}') = T^{(1/2)}(R)\,\psi(\mathbf{x}) \tag{137a}$$

or

$$\psi'(\mathbf{x}) = T^{(1/2)}(R)\,\psi(R^{-1}\mathbf{x}) \tag{137b}$$

Let us define the linear operator $U(R)$ in Hilbert space by

$$\langle \mathbf{x}i \mid U(R) \mid \psi\rangle = \sum_{j=1}^{2} T^{(1/2)}{}_{ij}(R)\,\psi_j(R^{-1}\mathbf{x}) \tag{138a}$$

or equivalently

$$|\psi'\rangle = U(R) \mid \psi\rangle \tag{138b}$$

This can be regarded as the relation between the states ascribed to the system by two observers with frames $\mathbf{x}$ and $\mathbf{x}'$ respectively, where $\mathbf{x}' = R\mathbf{x}$. If the observers are equivalent and the theory is to be rotationally invariant, by our previous remarks, the $U(R)$s are then unitary operators. They constitute, in fact, a unitary up-to-a-factor representation of the rotation group in the Hilbert space of states with scalar product defined by (136). For an infinitesimal rotation ϵ about the lth axis

$$(R(\epsilon)\mathbf{x})_j = x_j + \epsilon\epsilon_{ljk}x_k \tag{139a}$$

$$(R^{-1}(\epsilon)\mathbf{x})_j = x_j - \epsilon\epsilon_{ljk}x_k \tag{139b}$$

U can be written as follows:

$$U = \mathrm{I} + \frac{i}{\hbar}\,\epsilon D_l \tag{140}$$

where D_l is a self-adjoint operator on the Hilbert space, and therefore corresponds to an observable property of the system. Let us determine the infinitesimal generators D_l explicitly in configuration space. Using Eqs. (137b), (138a), and (139a, b) we obtain

$$\psi'(x_j) = \left(\mathrm{I} + \frac{i}{\hbar}\,\epsilon\mathfrak{D}_l(\mathbf{x})\right)\psi(x_j) \tag{141a}$$

$$= (1 + \tfrac{1}{2}i\epsilon\sigma_l)\,\psi(x_j - \epsilon\epsilon_{ljk}x_k) \tag{141b}$$

where in this last equation $\mathfrak{D}_l(\mathbf{x})$ represents the matrix whose matrix elements $\mathfrak{D}_{l,ij}(\mathbf{x})$ are defined by the right-hand side of the equation

$$\langle \mathbf{x}i \mid D_l \mid \mathbf{x}'j\rangle = \delta^{(3)}(\mathbf{x} - \mathbf{x}')\,\mathfrak{D}_{l,ij}(\mathbf{x}) \tag{142}$$

Expanding the right-hand side of (141b) in a Taylor series in ϵ about $\epsilon = 0$, we find, keeping only terms of order ϵ,

$$\psi'(x_j) = \left[1 + i\epsilon\left(i\epsilon_{ljk}x_k\frac{\partial}{\partial x_j} + \frac{1}{2}\sigma_l\right) + O(\epsilon^2)\right]\psi(x_j) \tag{143}$$

The generator is thus given by

$$\mathfrak{D}_l(\mathbf{x}) = +i\hbar\epsilon_{ljk}x_k\frac{\partial}{\partial x_j} + \frac{\hbar}{2}\sigma_l \tag{144}$$

and defines the lth component of the total angular momentum of the particle. The particle thus has an (intrinsic) angular momentum $\frac{\hbar}{2}\sigma$ over and above its orbital angular momentum $(\mathbf{r} \times \mathbf{p})_l = -i\hbar\epsilon_{lkj}x_k\partial_j$.

Some of the methods developed for the rotation group can easily be applied to a consideration of the group of spatial translation:

$$x'_i = x_i + a_i \tag{145}$$

Because this group is commutative, all of its irreducible unitary representations are one dimensional, and if we write

$$|\psi'\rangle = U(\mathbf{a}) \mid \psi\rangle \tag{146}$$

then

$$U(\mathbf{a}) = e^{-i\mathbf{a}\cdot\mathbf{p}} \tag{147}$$

where $\mathbf{p}$ is the momentum operator. Explicitly

$$\begin{aligned}\psi'(\mathbf{x}) = \langle \mathbf{x} \mid \psi' \rangle &= \int d^3y \, \langle \mathbf{x} \mid U(\mathbf{a}) \mid \mathbf{y}\rangle \langle \mathbf{y} \mid \psi\rangle \\ &= e^{-\mathbf{a}\cdot\nabla}\,\psi(x) = \psi(\mathbf{x} - \mathbf{a})\end{aligned} \tag{148}$$

as expected. The momentum operator p_l is thus the generator for infinitesimal translations in the lth direction. We have already established that the Hamiltonian operator is the generator for infinitesimal time translations.

For a classification of the representations of the inhomogeneous rotation group, i.e., the group which leaves invariant the quadratic form $(\mathbf{x} - \mathbf{y})^2$ (a typical transformation is $\mathbf{x} \to \mathbf{x}' = R\mathbf{x} + \mathbf{a}$ which includes a translation $\mathbf{a}$ as well as a rotation), the reader is referred to the lectures of Pauli (1956).

1f. The Four-Dimensional Rotation Group

It has been suggested recently that some of the symmetries exhibited by the strange particles could be understood in terms of a classification whereby to these particles are attributed certain internal degrees of freedom for which the "internal space" is a four-dimensional Euclidean space. It is therefore important to know the representations of the four-dimensional rotation group, i.e., the group of real linear transformations which leave invariant the quadratic form $x_1^2 + x_2^2 + x_3^2 + x_4^2$ [see e.g., Pauli (1956), Kleppner (1958), Racah (1959), Roman (1960)]. There are now six infinitesimal generators corresponding to rotations in each of the six co-ordinate planes. We shall call $M_{\nu\mu} = -M_{\mu\nu}$ the (Hermitian) infinitesimal generators for a rotation in the $\mu\nu$ plane ($\mu, \nu = 1, 2, 3, 4$) and we introduce the following notation:

$$M_{23} = M_1 \qquad M_{31} = M_2 \qquad M_{12} = M_3 \tag{149a}$$

$$M_{41} = N_1 \qquad M_{42} = N_2 \qquad M_{43} = N_3 \tag{149b}$$

By obtaining explicit matrix representations for the infinitesimal generators for co-ordinate transformations, one verifies that they satisfy the following commutation rules:

$$[M_l, M_j] = i\epsilon_{ljk}M_k \tag{150}$$

$$[M_l, N_j] = [N_l, M_j] = i\epsilon_{ljk}N_k \tag{151}$$

$$[N_l, N_j] = i\epsilon_{ljk}M_k \tag{152}$$

so that if we define the operators

$$K_i = \tfrac{1}{2}(M_i + N_i) \tag{153}$$

$$L_i = \tfrac{1}{2}(M_i - N_i) \tag{154}$$

these operators have the property that **K** and **L** commute with one another

$$[K_i, L_j] = 0 \tag{155}$$

and that the components of **K** and **L** satisfy angular momentum commutation rules:

$$[K_l, K_m] = i\epsilon_{lmn}K_n \tag{156}$$

$$[L_j, L_k] = i\epsilon_{jkm}L_m \tag{157}$$

There are now two invariants

$$F = \mathbf{K}^2 + \mathbf{L}^2 \tag{158a}$$

$$G = \mathbf{K}^2 - \mathbf{L}^2 \tag{158b}$$

which correspond to $\frac{1}{2}(\mathbf{M}^2 + \mathbf{N}^2)$ and $\mathbf{M} \cdot \mathbf{N}$ respectively, and one readily verifies that they commute with all the infinitesimal generators.[6] We can therefore label each irreducible representation by a pair of indices (k, l) where $\mathbf{K}^2 = k(k + 1)I$ and $\mathbf{L}^2 = l(l + 1)I$ for the irreducible representation labeled by k and l. The indices k and l can take the values $0, \frac{1}{2}, 1, \frac{3}{2}, \cdots$ etc. A unitary representation (k, l) has dimension $(2k + 1) \cdot (2l + 1)$. Therefore, the representation $(0, 0)$ corresponds to the scalar representation, $(0, \frac{1}{2})$ and $(\frac{1}{2}, 0)$ to two-component spinors, $(\frac{1}{2}, \frac{1}{2})$ to the four-component vector representation, etc. The operators corresponding to the generators for the spinor representations are as follows

$$M_l^{(1/2,0)} = \tfrac{1}{2}\sigma_l \qquad M_l^{(0,1/2)} = +\tfrac{1}{2}\sigma_l \tag{159}$$

$$N_l^{(1/2,0)} = \tfrac{1}{2}\sigma_l \qquad N_l^{(0,1/2)} = -\tfrac{1}{2}\sigma_l \tag{160}$$

If ξ and η are spinors which transform according to the representations $(\frac{1}{2}, 0)$ and $(0, \frac{1}{2})$ respectively, then $\xi^*\xi$ and $\eta^*\eta$ are scalars, and the quantity $(i\xi^*\eta, \xi^*\sigma_i\eta)$ is a four-vector.

[6] Note that in terms of **K** and **L** the generators break up into two independent sets and thus, in the neighborhood of the identity, the four-dimensional rotation group can be considered as the direct product of two three-dimensional rotation groups. By the direct product, $G \otimes K$, of two groups with elements $g_1, g_2, \cdots$ and $k_1, k_2 \cdots$ is meant the set of all ordered pairs (g_i, k_l) with the multiplication of pairs defined by (g_r, k_s), (g_t, k_u), (g_rg_t, k_sk_u).

When reflections are included the operators **K** and **L** are no longer independent. Thus if we choose as the basic reflection the operation under which $x_4 \to x_4$ and $x_i \to -x_i$, $i = 1, 2, 3$, and if R_- is the operator which corresponds to this reflection, one verifies that

$$R_-\mathbf{N} + \mathbf{N}R_- = 0 \tag{161a}$$

$$R_-\mathbf{M} - \mathbf{M}R_- = 0 \tag{161b}$$

or, equivalently

$$R_-\mathbf{K} = \mathbf{L}R_- \tag{162a}$$

$$R_-\mathbf{L} = \mathbf{K}R_- \tag{162b}$$

To determine the irreducible representations when reflections are included, we note that if we denote by $|k, m_k; l, m_l\rangle$ the $(2k + 1)\,(2l + 1)$ basis functions which span an irreducible representation space and for which

$$\begin{aligned}(K_1 \pm iK_2)\,|\,k, m_k; l, m_l\rangle &= \sqrt{(k \mp m_k)\,(k \pm m_k + 1)}\,|\,k, m_k \pm 1; l, m_l\rangle \\ K_3\,|\,k, m_k; l, m_l\rangle &= m_k\,|\,k, m_k; l, m_l\rangle\end{aligned} \tag{163}$$

and

$$\begin{aligned}(L_1 \pm iL_2)\,|\,k, m_k; l, m_l\rangle &= \sqrt{(l \mp m_l)\,(l \pm m_l + 1)}\,|\,k, m_k; l, m_l \pm 1\rangle \\ L_3\,|\,k, m_k; l, m_l\rangle &= m_l\,|\,k, m_k; l, m_l\rangle\end{aligned} \tag{164}$$

then Eqs. (162a) and (162b) assert that $R_-\,|\,k, m_k; l, m_l\rangle$ is an eigenfunction of L_3 with eigenvalue m_k and an eigenfunction of K_3 with eigenvalue m_l. From similar considerations of the operation of K_1, K_2 and L_1 on L_2 on $R_-\,|\,k, m_k; l, m_l\rangle$ one verifies that

$$R_-\,|\,k, m_k; l, m_l\rangle = \lambda\,|\,l, m_l; k, m_k\rangle \tag{165}$$

where λ is a constant which may depend on l, m_l, k, m_k. It therefore follows that in the case $k \neq l$ the $2(2l + 1)\,(2k + 1)$ basis vectors $|k, m_k; l, m_l\rangle$ and $|l, m_l; k, m_k\rangle$ together span a manifold which is invariant under R_-. The representation $(k, l) \oplus (l, k)$ for $k \neq l$ is therefore irreducible for the group including reflections. For the spinor representations, $(0, \frac{1}{2}) \oplus (\frac{1}{2}, 0)$ is now irreducible and it is to be noted that for the enlarged group a spinor is a four-component object. (An analogous doubling occurs for the homogeneous Lorentz group.) The irreducible representations (k, k) of the proper group are also irreducible representations of the extended group.

For an interesting application of the representations of the four-dimensional rotation group to the quantum mechanical description of the motion of a nonrelativistic electron in a pure Coulomb field and to the question of the origin of the degeneracy of the eigenvalue spectrum in this case, the reader is referred to the articles by Fock (1936) and Bargmann (1936).

2

The Lorentz Group

2a. Relativistic Notation

Before embarking upon a description of the relativistic one-particle equations, let us briefly introduce the relativistic notation that will be used throughout the book.

We shall denote the space-time co-ordinates by x^μ (which as a four-vector is denoted by a light-faced x) with the understanding that $x^0 = ct$, $x^1 = x$, $x^2 = y$ and $x^3 = z$; $x = \{x^0, \mathbf{x}\}$. We shall use a metric tensor $g_{\mu\nu}$ with components

$$g_{00} = -g_{11} = -g_{22} = -g_{33} = +1$$
$$g_{\mu\nu} = 0 \quad \text{for } \mu \neq \nu \tag{1}$$

We must therefore distinguish between covariant and contravariant vectors. A contravariant vector (one which transforms like the co-ordinate vector x^μ) is denoted by v^μ, a covariant one (which transforms like the gradient) by v_μ, and similarly for tensors. In general, Greek indices will be used to denote the components (0, 1, 2, 3) of a space-time tensor whereas Latin indices will be used to denote spatial components only (1, 2, 3). The raising and lowering of indices is defined in terms of the metric tensor, with

$$v_\mu = g_{\mu\nu}v^\nu \tag{2a}$$

$$v^\mu = g^{\mu\nu}v_\nu \tag{2b}$$

where a repeated Greek index implies a summation over that index from 0 to 3, i.e.,

$$g_{\mu\nu}v^\nu = \sum_{\nu=0}^{3} g_{\mu\nu}v^\nu \tag{3}$$

The tensor $g^{\mu\nu}$ is defined by the equation

$$g^{\mu\nu}g_{\mu\sigma} = \delta^\nu{}_\sigma \tag{4}$$

where $\delta^\nu{}_\sigma$ is the Kronecker symbol: $\delta^\mu{}_\nu = 1$ if $\mu = \nu$ and $\delta^\mu{}_\nu = 0$ otherwise.

Actually $g^{\mu\nu} = g_{\mu\nu}$. Note that the raising or lowering of the indices of a four-vector changes the sign of the space components of that vector but leaves the time component unchanged. The Lorentz invariant scalar product of two vectors p_μ, x_μ is defined by

$$\begin{aligned} g^{\mu\nu}p_\mu x_\nu &= p_\mu x^\mu = p \cdot x \\ &= x_0 p_0 - \mathbf{x} \cdot \mathbf{p} = x^0 p^0 - \mathbf{x} \cdot \mathbf{p} \end{aligned} \tag{5}$$

A vector v is called time-like if $v \cdot v = v^2 > 0$, space-like if $v^2 < 0$ and a null vector if $v^2 = 0$.[1] The relativistic energy-momentum relation for a free particle, $E^2 = \mathbf{p}^2c^2 + m^2c^4$ when written in terms of the (time-like) energy-momentum four-vector of the particle, $p = \{E/c, \mathbf{p}\}$, reads $p^2 = m^2c^2$. The operator relation $E \to i\hbar\partial_t$, $\mathbf{p} \to -i\hbar\nabla$ can now be written as

$$p_\mu \to i\hbar \frac{\partial}{\partial x^\mu} = i\hbar\partial_\mu \tag{6}$$

In this connection it should be noted that since x^μ is a contravariant vector $\partial/\partial x^\mu = \partial_\mu$ is a covariant vector with

$$\partial_\mu = \frac{\partial}{\partial x^\mu} = \left\{\frac{\partial}{\partial ct}, \nabla\right\} \tag{7a}$$

and

$$\partial^\mu = \frac{\partial}{\partial x_\mu} = \left\{\frac{\partial}{\partial ct}, -\nabla\right\} \tag{7b}$$

The wave operator, or the D'Alembertian operator as it is often called,

$$\Box = \frac{1}{c^2}\frac{\partial^2}{\partial t^2} - \nabla^2 \tag{8}$$

can thus be written as $\Box = g_{\mu\nu}\partial^\mu\partial^\nu = \partial^\mu\partial_\mu$.

We shall have frequent occasion to make use of the step functions $\theta(a)$ and $\epsilon(a)$ which are defined as follows:

$$\begin{aligned} \epsilon(a) &= +1 \quad \text{if } a > 0 \\ &= -1 \quad \text{if } a < 0 \end{aligned} \tag{9a}$$

$$\begin{aligned} \theta(a) = \tfrac{1}{2}(1 + \epsilon(a)) &= 1 \quad \text{if } a > 0 \\ &= 0 \quad \text{if } a < 0 \end{aligned} \tag{9b}$$

Finally we shall almost always use natural units in which c, the velocity of light, and $\hbar$, Planck's constant divided by 2π, are set equal to one: $\hbar = c = 1$. In this system of units, energy, mass, inverse length, and inverse time all have the same dimension.

[1] Note that by v^2 we mean $v_0^2 - \mathbf{v}^2$.

2b. The Homogeneous Lorentz Group

We shall briefly recall in this section a few facts about homogeneous Lorentz transformations. For two frames in relative motion along the x^1 axis, the Lorentz transformation relating them is

$$\begin{aligned} x'^0 &= \gamma(x^0 - \beta x^1) = x^0 \cosh u - x^1 \sinh u \\ x'^1 &= \gamma(x^1 - \beta x^0) = x^1 \cosh u - x^0 \sinh u \\ x'^2 &= x^2 \\ x'^3 &= x^3 \end{aligned} \tag{10}$$

where

$$\gamma = (1 - \beta^2)^{-1/2} \qquad \beta = v/c \tag{11a}$$

$$\tanh u = \beta \tag{11b}$$

and v is the relative velocity of the two frames. This Lorentz transformation leaves the quadratic form x^2 invariant. Note also that at time $x_0 = 0$ the origins of the two co-ordinate systems coincide. The most general homogeneous Lorentz transformation between two co-ordinate systems is a linear transformation

$$x'^\mu = \Lambda^\mu{}_\nu x^\nu \tag{12}$$

or in matrix form $x' = \Lambda x$, which leaves invariant the quadratic form $x_\mu x^\mu$, i.e., for which $x^2 = x'^2$. The transformation coefficients $\Lambda^\mu{}_\nu$ are all real. The condition that the quadratic form x^2 be invariant requires that

$$\Lambda_\mu{}^\nu \Lambda^\mu{}_\lambda = \Lambda^{\nu\mu}\Lambda_{\lambda\mu} = \delta^\nu{}_\lambda \tag{13a}$$

We can write this last equation in the form

$$\Lambda^\mu{}_\nu g_{\mu\rho} \Lambda^\rho{}_\sigma = g_{\nu\sigma} \tag{13b}$$

or in matrix form, with $\Lambda^\mu{}_\nu = (\Lambda^T)^\nu{}_\mu$

$$\Lambda^T g \Lambda = g \tag{13c}$$

where the superscript T denotes the transposed matrix. It follows from this equation that $\det \Lambda = \pm 1$ and therefore that for every homogeneous Lorentz transformation there exists an inverse transformation. Since the product of two Lorentz transformations is again a Lorentz transformation, the set of all homogeneous Lorentz transformations form a group.

The group of Lorentz transformations contains a subgroup which is isomorphic to the three-dimensional rotation group. This subgroup consists of all the $\Lambda^\nu{}_\mu$ of the form

$$\Lambda(R) = \begin{pmatrix} 1 & 0 \\ 0 & R \end{pmatrix} \tag{14}$$

where R is a 3×3 matrix with $RR^T = R^T R = 1$. We call such a Λ a spatial rotation. Every homogeneous Lorentz transformation can be decomposed as follows:

$$\Lambda = \Lambda(R_2)\,\Lambda(l_1)\,\Lambda(R_1) \tag{15}$$

where $\Lambda(R_1)$ and $\Lambda(R_2)$ are spatial rotations and $\Lambda(l_1)$ a Lorentz transformation in the x^1 direction.

If we set $\sigma = \nu = 0$ in Eq. (13b), we then obtain

$$(\Lambda^0{}_0)^2 = 1 + \sum_{i=1}^{3} (\Lambda^i{}_0)^2 \geqslant 1 \tag{16}$$

so that $\Lambda^0{}_0 \geqslant 1$ or $\Lambda^0{}_0 \leqslant -1$. A Lorentz transformation for which $\Lambda^0{}_0 \geqslant 1$ is called an orthochronous Lorentz transformation. A Lorentz transformation is orthochronous if and only if it transforms every positive time-like vector into a positive time-like vector. The set of all orthochronous Lorentz transformations forms a group: the orthochronous Lorentz group. The set of all Λ can be divided into four subsets according to whether $\det \Lambda$ equals plus or minus one and $\Lambda^0{}_0$ is greater than one or less than minus one. The subset with $\det \Lambda = +1$ and $\Lambda^0{}_0 \geqslant 1$ is called the group of restricted homogeneous Lorentz transformations. The restricted homogeneous Lorentz group is a six-parameter continuous group. The other subsets can be obtained by adjoining to the restricted Lorentz group the following three transformations:

1. *Space inversion:* $x_0 \to x_0,\ \mathbf{x} \to -\mathbf{x}$

$$\Lambda(i_s) = \begin{pmatrix} 1 & 0 & 0 & 0 \\ 0 & -1 & 0 & 0 \\ 0 & 0 & -1 & 0 \\ 0 & 0 & 0 & -1 \end{pmatrix} \tag{17}$$

2. *Time inversion:* $x_0 \to -x_0,\ \mathbf{x} \to \mathbf{x}$.

$$\Lambda(i_t) = \begin{pmatrix} -1 & & & \\ & 1 & & \\ & & 1 & \\ & & & 1 \end{pmatrix} \tag{18}$$

3. *Space-time inversion:* $x \to -x$

$$\Lambda(i_{st}) = \Lambda(i_s)\,\Lambda(i_s)$$

$$= \begin{pmatrix} -1 & & & \\ & -1 & & \\ & & -1 & \\ & & & -1 \end{pmatrix} \tag{19}$$

These subsets are disjoint and are not continuously connected.

As in the case of the rotation group, we can easily determine the form of the generators of an infinitesimal Lorentz transformation. For an infinitesimal Lorentz transformation

$$\Lambda^\mu{}_\nu = \delta^\mu{}_\nu + \epsilon\lambda^\mu{}_\nu \tag{20}$$

in order that Eqs. (13a, b, c) be satisfied, we must require that

$$\lambda^{\mu\nu} = -\lambda^{\nu\mu} \tag{21}$$

which is a necessary as well as sufficient condition for $\lambda^{\mu\nu}$ to correspond to an infinitesimal Lorentz transformation. The infinitesimal transformation which is the inverse of $\Lambda^{\mu\nu}$ is thus $\Lambda^{\nu\mu}$.

The explicit matrix representation of a restricted homogeneous Lorentz transformation in the x^1 direction (rotation in x^0x^1 plane) is given by

$$\Lambda(10, u) = \begin{pmatrix} \cosh u & -\sinh u & 0 & 0 \\ -\sinh u & \cosh u & 0 & 0 \\ 0 & 0 & 1 & 0 \\ 0 & 0 & 0 & 1 \end{pmatrix} \tag{22}$$

The infinitesimal generator $\mathfrak{M}^{10}$ for this rotation is defined as

$$\mathfrak{M}^{10} = \frac{d}{du} \Lambda(10, u) \bigg|_{u=0} \tag{23a}$$

and is exhibited by

$$\mathfrak{M}^{10} = \begin{pmatrix} 0 & -1 & 0 & 0 \\ -1 & 0 & 0 & 0 \\ 0 & 0 & 0 & 0 \\ 0 & 0 & 0 & 0 \end{pmatrix} \tag{23b}$$

Similarly the infinitesimal generators $\mathfrak{M}^{20}$ and $\mathfrak{M}^{30}$ for rotations in the "20" and "30" planes respectively, are exhibited by

$$\mathfrak{M}^{20} = \begin{pmatrix} 0 & 0 & -1 & 0 \\ 0 & 0 & 0 & 0 \\ -1 & 0 & 0 & 0 \\ 0 & 0 & 0 & 0 \end{pmatrix} \qquad \mathfrak{M}^{30} = \begin{pmatrix} 0 & 0 & 0 & -1 \\ 0 & 0 & 0 & 0 \\ 0 & 0 & 0 & 0 \\ -1 & 0 & 0 & 0 \end{pmatrix} \tag{24}$$

The infinitesimal generators for rotations in the x^i-x^j plane, i.e., spatial rotations, are

$$\mathfrak{M}^{12} = \begin{pmatrix} 0 & 0 & 0 & 0 \\ 0 & 0 & 1 & 0 \\ 0 & -1 & 0 & 0 \\ 0 & 0 & 0 & 0 \end{pmatrix} \qquad \mathfrak{M}^{23} = \begin{pmatrix} 0 & 0 & 0 & 0 \\ 0 & 0 & 0 & 0 \\ 0 & 0 & 0 & 1 \\ 0 & 0 & -1 & 0 \end{pmatrix}$$

$$\mathfrak{M}^{31} = \begin{pmatrix} 0 & 0 & 0 & 0 \\ 0 & 0 & 0 & -1 \\ 0 & 0 & 0 & 0 \\ 0 & 1 & 0 & 0 \end{pmatrix} \tag{25}$$

We define $\mathfrak{M}^{\mu\nu} = -\mathfrak{M}^{\nu\mu}$. An arbitrary infinitesimal Lorentz transformation can be written as

$$\Lambda(\omega) = I + \tfrac{1}{2}\omega^{\mu\nu}\mathfrak{M}_{\mu\nu} \tag{26}$$

where $\omega^{\mu\nu} = -\omega^{\mu\nu}$. A finite rotation in the $\mu\nu$ plane (in the sense μ to ν), is again obtained by exponentiation:

$$\Lambda(\mu\nu; u) = e^{u\mathfrak{M}^{\mu\nu}} \tag{27}$$

One verifies that the infinitesimal generators, $\mathfrak{M}_{\mu\nu}$, satisfy the following commutation rules:

$$[\mathfrak{M}_{\mu\nu}, \mathfrak{M}_{\rho\sigma}] = g_{\mu\rho}\mathfrak{M}_{\nu\sigma} + g_{\nu\sigma}\mathfrak{M}_{\mu\rho} - g_{\mu\sigma}\mathfrak{M}_{\nu\rho} - g_{\nu\rho}\mathfrak{M}_{\mu\sigma} \tag{28}$$

Thus if the four indices $\mu\nu\rho\sigma$ are all the same or all different the matrices commute. On the other hand if one index is common to both matrices, say $\sigma = \mu$, the right-hand side is proportional to $\mathfrak{M}_{\rho\nu}$.

If $D(\Lambda)$ is any representation of the restricted Lorentz group, we shall call the infinitesimal generators for that representation $M_{\mu\nu}$. Therefore if Λ is of the form given by Eq. (26)

$$D(\omega) = \mathrm{I} + \tfrac{1}{2}\omega^{\mu\nu}M_{\mu\nu} \tag{29}$$

Since the $M_{\mu\nu}$ are representations of the generators of Lie algebra, they satisfy the same commutation rules as the $\mathfrak{M}_{\mu\nu}$, viz.,

$$[M_{\mu\nu}, M_{\rho\sigma}] = g_{\mu\rho}M_{\nu\sigma} + g_{\nu\sigma}M_{\mu\rho} - g_{\nu\rho}M_{\mu\sigma} - g_{\mu\sigma}M_{\nu\rho} \tag{30}$$

The problem of finding the representations of the restricted Lorentz group is equivalent to finding all the representations of the commutation rules (30). The following important fact about representations of the restricted homogeneous Lorentz group will be used in the sequel [see Van der Waerden (1932), Bargmann (1947), Naĭmark (1957)]. The group has both finite and infinite dimensional irreducible representations. However, the only finite *unitary* representation is the one-dimensional trivial representation $\Lambda \rightarrow 1$. The finite dimensional irreducible representation of the restricted group can be labeled by two discrete indices which can take on as values the positive integers, the positive half-odd integers, and zero. That this is so can be seen as follows. Let us define the operators

$$\mathbf{M} = (M_{32}, M_{13}, M_{21}) \tag{31}$$

$$\mathbf{N} = (M_{01}, M_{02}, M_{03}) \tag{32}$$

Their commutation rules are

$$[M_i, M_j] = \epsilon_{ijk}M_k \tag{33a}$$

$$[N_i, N_j] = -\epsilon_{ijk}M_k \tag{33b}$$

$$[M_i, N_j] = \epsilon_{ijk}N_k \tag{33c}$$

From these operators we can construct the operators $\mathbf{M}^2 - \mathbf{N}^2 = \frac{1}{2}M_{\mu\nu}M^{\mu\nu}$ and $\frac{1}{8}\epsilon^{\mu\nu\rho\sigma}M_{\mu\nu}M_{\rho\sigma} = -\mathbf{M}\cdot\mathbf{N}$,[2] which commute with all the M_i and N_i. They are therefore the invariants of the group and they are multiples of the identity in any irreducible representation. The representations can thus be labeled by the values of these operators in the given representation.

[2] $\epsilon^{\mu\nu\rho\sigma}$ is the antisymmetric tensor of rank four, which equals $+1$ if $\mu\nu\rho\sigma$ is an even permutation of 0123, equals -1 if $\mu\nu\rho\sigma$ is an odd permutation of 0123, and equals zero if the indices $\mu\nu\rho\sigma$ are not distinct.

To make the range of values of the label more transparent let us introduce the following operators:

$$J_l = \tfrac{1}{2}i(M_l + iN_l) \tag{34}$$

and

$$K_l = \tfrac{1}{2}i(M_l - iN_l) \tag{35}$$

which satisfy the following commutation rules:

$$[J_k, J_l] = i\epsilon_{klm}J_m \tag{36}$$

$$[K_l, K_m] = i\epsilon_{lmn}K_n \tag{37}$$

$$[J_l, K_m] = 0 \tag{38}$$

It follows from these commutation rules that a finite dimensional irreducible representation space, $V^{jj'}$ can be spanned by a set of $(2j + 1)\,(2j' + 1)$ basis vectors $|jm; j'm'\rangle$ where j, m, j', m' are integers or half-odd integers, $-j \leqslant m \leqslant j$, $-j' \leqslant m' \leqslant j'$ and in terms of which the **J** and **K** operators have the following representation:

$$\begin{aligned} J_\pm \,|\, j, m; j', m'\rangle &= (J_1 \pm iJ_2)\,|\, j, m; j', m'\rangle \\ &= \sqrt{(j \mp m)\,(j \pm m + 1)}\,|\, j, m \pm 1; j', m'\rangle \\ J_3 \,|\, j, m; j', m'\rangle &= m \,|\, j, m; j', m'\rangle \end{aligned} \tag{39}$$

and

$$\begin{aligned} K_\pm \,|\, j, m; j', m'\rangle &= (K_1 \pm iK_2)\,|\, j, m; j', m'\rangle \\ &= \sqrt{(j' \mp m')\,(j' \pm m' + 1)}\,|\, j, m; j', m' \pm 1\rangle \\ K_3 \,|\, j, m; j', m'\rangle &= m' \,|\, j, m; j', m'\rangle \end{aligned} \tag{40}$$

The matrix representing any particular Lorentz transformation $D^{(jj')}(\Lambda)$ is now easily obtained from this representation and it will be noticed that it is not, in general, unitary. The representations we have obtained are all finite dimensional. More generally, the irreducible representations of the restricted Lorentz group can be specified by a pair of indices (j_0, ν) where j_0 is integer or half-integer and positive and ν is a complex number. If $\nu^2 = (j_0 + n)^2$ for some integer n, then the representation is finite dimensional. If, however, $\nu^2 \neq (j_0 + n)^2$ for any integer n, then the representation is infinite dimensional. We shall actually only be concerned with the finite dimensional representations since for physical applications we are interested in the classification of quantities which have a finite number of components and which transform according to a finite dimensional representation of the Lorentz group. The unitary infinite dimensional irreducible representations are, however, of relevance in the classification of the irreducible unitary representation of the inhomogeneous Lorentz group (see Sec. 2c). The reader interested in a unified derivation of all the irreducible representations of the restricted Lorentz group is referred to the very readable review article of Naĭmark (1957).

To summarize, there are a denumerable infinity of nonequivalent finite dimensional (in general nonunitary) irreducible representations. These can be labeled by two non-negative indices (j, j') where $j, j' = 0, \frac{1}{2}, 1, \frac{3}{2}, \cdots$. The dimension of the representation is $(2j + 1)(2j' + 1)$ and $D^{(j,j')}$ is single-valued if $j + j'$ is integer and double-valued otherwise. In an irreducible representation the value of the invariant $\frac{1}{2}(\mathbf{M}^2 - \mathbf{N}^2)$ is $-\{j(j + 1) + j'(j' + 1)\}$ times the unit matrix of dimension $(2j + 1)(2j' + 1)$ and the value of the second invariant is $i\{j(j + 1) - - j'(j' + 1)\}$ times the unit matrix. The basis vectors spanning the representation space $D^{\left(\frac{n}{2},\frac{m}{2}\right)}$ and $D^{\left(\frac{m}{2},\frac{n}{2}\right)}$ can be so chosen that the representation matrices for $D^{\left(\frac{n}{2},\frac{m}{2}\right)}$ are just the complex conjugate of those for $D^{\left(\frac{m}{2},\frac{n}{2}\right)}$.

A quantity which transforms under $D^{(0,0)}$ is called a scalar, one which transforms under $D^{(1/2,1/2)}$ a four-component vector, one which transforms under the $(\frac{1}{2}, 0)$ representation a two-component spinor. A quantity which transforms under $(0, \frac{1}{2})$ is called a conjugate spinor. For the $D^{(0,1/2)}$ and the $D^{(1/2,0)}$ representations an explicit matrix representation of the infinitesimal generators can be given in terms of the Pauli matrices with

$$M_j^{(1/2,0)} = -\tfrac{1}{2}i\sigma_j \qquad M_j^{(0,1/2)} = -\tfrac{1}{2}i\sigma_j \tag{41a}$$

$$N_j^{(1/2,0)} = -\tfrac{1}{2}\sigma_j \qquad N_j^{(0,1/2)} = +\tfrac{1}{2}\sigma_j \tag{41b}$$

These are clearly inequivalent representations, since there exists no 2×2 matrix which anticommutes with all the σ_i. A two-component spinor, ξ, transforms under a spatial rotation as in the three-dimensional situation. For example, under an infinitesimal rotation about the lth axis,

$$\xi = \xi' = (1 + \tfrac{1}{2}i\epsilon\sigma_l)\xi \tag{42}$$

Under an infinitesimal Lorentz transformation in the x^i direction, this spinor transforms according to

$$\xi \to \xi' = (1 + \tfrac{1}{2}\epsilon\sigma_i)\xi \tag{43}$$

Note, however, that the quantity $\xi^*\xi$ is not a scalar. This is a consequence of the fact that the representation $D^{(1/2,0)}$ is not unitary. On the other hand if ξ' is a spinor which transforms according to $D^{(0,1/2)}$, one verifies that $\xi^*\xi'$ is a scalar. Quite generally if the representations U and U^{*-1} are equivalent, this implies the existence of a nondegenerate matrix, B, such that $U = B^{-1}U^{*-1}B$. This in turn implies that if $|\psi\rangle$ is a vector which transforms according to U, namely, $|\psi\rangle \to |\psi'\rangle = U\,|\,\psi\rangle$, then $\langle\psi\,|\,B^*\,|\,\psi\rangle$ is a scalar invariant. The proof is as follows:

$$\langle\psi'\,|\,B^*\,|\,\psi'\rangle = \langle\psi\,|\,U^*B^*U\,|\,\psi\rangle = \langle\psi\,|\,B^*U^{-1}U\,|\,\psi\rangle = \langle\psi\,|\,B^*\,|\,\psi\rangle \tag{44}$$

The converse is also true. For a unitary representation clearly $B = 1$.

We shall not consider the representations of the full Lorentz group including the inversion operations. A complete and simple discussion of the finite dimensional irreducible representations of the full Lorentz group may be found in Heine (1957) [see also Watanabe (1951, 1955) and Shirokov (1960a, b)]. We shall, however, at the appropriate places discuss the inversion properties of the relativistic wave functions and operators describing free particles. We here note that the commutation rules of the operators for the inversions I_s, I_t, I_{st}, with the generators are:

$$[I_{st}, M_i] = [I_{ts}, N_i] = 0 \tag{45a}$$

$$[I_t, N_i]_+ = [I_t, M_i] = 0 \tag{45b}$$

$$[I_s, N_i]_+ = [I_s, M_i] = 0 \tag{45c}$$

To conclude this section, we briefly consider the vectors which span an irreducible representation of the orthochronous, improper (i.e., det $\Lambda = -1$) Lorentz group. This is the group obtained by adjoining the space inversion operation to the elements of the restricted group. It follows from the commutation rules (45c) that

$$I_s\mathbf{K} = \mathbf{J}I_s \tag{46a}$$

$$I_s\mathbf{J} = \mathbf{K}I_s \tag{46b}$$

so that the basis vectors $|jm; j'm'\rangle$ for $j \neq j'$ which transform under a restricted Lorentz transformation according to $D^{(j,j')}$ do not transform into one another under I_s. In fact, in view of (46)

$$J_3(I_s \mid jm; j'm'\rangle) = m'(I_s \mid jm; j'm'\rangle) \tag{47}$$

so that $I_s \mid jm; j'm'\rangle$ behaves like a base vector $\lambda \mid j'm'; jm\rangle$ where λ is a constant which depends on j, j', m, m'. The vectors $|j, m; j', m'\rangle$ and $I_s \mid j, m; j', m'\rangle$ thus transform under restricted Lorentz transformations under different irreducible representations and are therefore orthogonal to one another. To obtain a vector space invariant under the improper orthochronous Lorentz group it is therefore necessary to take the $2(2j + 1)(2j' + 1)$ linearly independent vectors $|jm; j'm'\rangle$ and $|j'm'; jm\rangle$ together. We thus expect the vector space $V^{jj'} \oplus V^{j'j}$ to be an irreducible vector space for the representations of the improper orthochronous group. This is indeed the case for $j \neq j'$.

2c. The Inhomogeneous Lorentz Group

An inhomogeneous Lorentz transformation, $L = \{a, \Lambda\}$, is defined by

$$x'_\mu = (Lx)_\mu = \Lambda_\mu{}^\nu x_\nu + a_\mu \tag{48}$$

i.e., as the product operation of a translation by a real vector a_μ and a homogeneous Lorentz transformation, Λ, the translation being performed after the homogeneous Lorentz transformation. It can conveniently be represented by the following matrix equation:

$$\begin{pmatrix} \Lambda^0{}_0 & \Lambda^0{}_1 & \Lambda^0{}_2 & \Lambda^0{}_3 & a^0 \\ \Lambda^1{}_0 & \Lambda^1{}_1 & \Lambda^1{}_2 & \Lambda^1{}_3 & a^1 \\ \Lambda^2{}_0 & \Lambda^2{}_1 & \Lambda^2{}_2 & \Lambda^2{}_3 & a^2 \\ \Lambda^3{}_0 & \Lambda^3{}_1 & \Lambda^3{}_2 & \Lambda^3{}_3 & a^3 \\ 0 & 0 & 0 & 0 & 1 \end{pmatrix} \begin{pmatrix} x^0 \\ x^1 \\ x^2 \\ x^3 \\ 1 \end{pmatrix} = \begin{pmatrix} x'^0 \\ x'^1 \\ x'^2 \\ x'^3 \\ 1 \end{pmatrix} \tag{49}$$

where the last co-ordinate, 1, has no physical significance and is left invariant by the transformation. The product of two inhomogeneous Lorentz transformations $\{a_1, \Lambda_1\}$ and $\{a_2, \Lambda_2\}$ is given by

$$\{a_1, \Lambda_1\}\,\{a_2, \Lambda_2\} = \{a_1 + \Lambda_1 a_2, \Lambda_1\Lambda_2\} \tag{50}$$

The inhomogeneous Lorentz transformations form a ten-parameter continuous group. The generators for infinitesimal translations are the Hermitian operators p_μ, and their commutation relations with the *Hermitian* generators[3] for "rotations" in the x^μ-x^ν plane, $M_{\mu\nu} = -M_{\nu\mu}$ are

$$[M_{\mu\nu}, p_\sigma] = i(g_{\nu\sigma}p_\mu - g_{\mu\sigma}p_\nu) \tag{51}$$

The commutation rules of these generators with themselves are

$$[p_\mu, p_\nu] = 0 \tag{52}$$

$$[M_{\mu\nu}, M_{\rho\sigma}] = -i(g_{\mu\rho}M_{\nu\sigma} - g_{\nu\rho}M_{\mu\sigma} + g_{\mu\sigma}M_{\rho\nu} - g_{\nu\sigma}M_{\rho\mu}) \tag{53}$$

The problem of classifying all the irreducible unitary representations of the inhomogeneous Lorentz group [Wigner (1939), Bargmann (1948), Shirokov (1958a, b)] can again be formulated in terms of finding all the representations of the commutation rules (51), (52), (53) by self-adjoint operators. The first task is to find all the invariants of the group. Clearly only scalar operators can be invariants of the group and we are thus confronted with the problem of constructing the scalar quantities which commute with p_μ and $M_{\mu\nu}$. Let us define the following quantities

$$\begin{aligned} v_{\mu\nu\rho} &= p_\mu M_{\nu\rho} + p_\nu M_{\rho\mu} + p_\rho M_{\mu\nu} \\ &= M_{\nu\rho}p_\mu + M_{\rho\mu}p_\nu + M_{\mu\nu}p_\rho \end{aligned} \tag{54}$$

and the pseudovector

$$w_\sigma = \tfrac{1}{2}\epsilon_{\sigma\mu\nu\lambda}M^{\mu\nu}p^\lambda \tag{55}$$

so that

$$(w^0, w^1, w^2, w^3) = (v^{321}, v^{320}, v^{130}, v^{210}) \tag{56}$$

or in vector notation

$$w^0 = \mathbf{p} \cdot \mathbf{M} \tag{57a}$$

$$\mathbf{w} = p_0\mathbf{M} - \mathbf{p} \times \mathbf{N} \tag{57b}$$

[3] We have appended a factor $-i$ to our previous definition of the infinitesimal generators which makes the M_l's Hermitian and the N_l's anti-Hermitian.

We note that

$$w_\sigma p^\sigma = 0 \tag{58}$$

The commutation rules of w_μ are:

$$[M_{\mu\nu}, w_\rho] = i(g_{\nu\rho}w_\mu - g_{\mu\rho}w_\nu) \tag{59}$$

$$[w_\mu, p_\nu] = 0 \tag{60}$$

$$[w_\mu, w_\nu] = i\epsilon_{\mu\nu\rho\sigma}w^\rho p^\sigma \tag{61}$$

One then verifies that the following scalar operators

$$P = p^\mu p_\mu \tag{62}$$

and

$$\begin{aligned} W &= \tfrac{1}{6}v^{\mu\nu\rho}v_{\mu\nu\rho} = -w^\mu w_\mu \\ &= \tfrac{1}{2}M_{\mu\nu}M^{\mu\nu}p_\sigma p^\sigma - M_{\mu\sigma}M^{\nu\sigma}p^\mu p_\nu \end{aligned} \tag{63}$$

commute with all the infinitesimal generators, $M_{\mu\nu}$ and p_μ. They are therefore multiples of the identity for every irreducible representation of the inhomogeneous Lorentz group and their eigenvalues can be used to classify the irreducible representations.

It is convenient for the classification of the unitary representations of the inhomogeneous Lorentz group to choose a definite basis in the vector space on which the representations are defined. To define a basis, we select from among the infinitesimal operators of the group a complete set of commuting operators. Many different sets of commuting operators can, of course, be constructed. These different sets will then give rise to equivalent representations. We could, for example, choose as a complete set the operators $M_{\mu\nu}M^{\mu\nu}$, $\epsilon^{\mu\nu\rho\sigma}M_{\mu\nu}M_{\rho\sigma}$, $\mathbf{M}^2$ and M_3 but such a choice would not be translationally invariant. A complete commuting set that is translationally invariant consists of the operators p_μ and of one of the components of w_μ, say w_3. We adopt this set for our subsequent discussion. The eigenvalue spectrum of these operators then specifies the range of the variables labeling the basis vectors. Furthermore we note that for an irreducible vector space, only three of the four momenta are independent since p^2 is an invariant of the group and has a constant value in an irreducible representation. The basis functions for an irreducible representation can thus be written as $|p'_0, p'_1, p'_2, p'_3; \zeta\rangle$ where p'^2 is equal to some constant and ζ is the variable corresponding to the w_3 eigenvalue. It is important to note that although we have chosen a complete set of commuting operators from among the operators of the group, this set will not in general be a complete set of commuting *observables* for a physical system. There will be in general other invariant operators (such as the total charge and nucleonic charge, for example) which commute with the group operators and whose eigenvalues together with p_μ' and ζ characterize *states* of the system. Therefore, the basis vectors of an irreducible representation more generally can be written as $|p'; \zeta; \alpha\rangle$ where α denotes certain invariant parameters which in physical applications are the eigenvalues of those operators which must be added to the set (p_μ, w_3) to make it a complete set of observables. In what follows we often suppress the dependence on α of the basis vectors $|p; \zeta; \alpha\rangle$. Note that for such a basis the operation of translation is very simple. The set of all four-dimensional translations is a commutative subgroup of the inhomogeneous Lorentz group. Since it is commutative, the irreducible unitary representations of this subgroup are all one dimensional and are obtained by exponentiation. The operator corresponding to the translation by the four-vector a_μ is given by

$$U(a) = \exp(-ia_\mu p^\mu) \tag{64}$$

In an irreducible representation, the operation of translation by a thus corresponds to multiplying each basis vector $|p'; \zeta\rangle$ by $\exp(-ia_\mu p'^\mu)$.

The irreducible representations of the inhomogeneous Lorentz group can now be classified according to whether p_μ is a space-like, time-like, or null vector, or p_μ is equal to zero. For this last case, $p_\mu = 0$, the complete system of unitary representations coincides with the complete system of (infinite dimensional) unitary representations of the homogeneous group [Bargmann (1947), Naĭmark (1957)]. They will not be considered further as they do not seem to have any correspondence with physical systems except for the important case of the (trivial) identity representation which is one dimensional.

The representations of principal interest for physical applications are those for which $p^2 = m^2 =$ positive constant, and those for which $p^2 = 0$. Let us first discuss the case $p^2 = m^2$. In that case, $p_0/|p_0|$, the sign of the energy, commutes with all the infinitesimal generators and is therefore an invariant of the group. There are thus two irreducible representations for each value of P and W, one for each sign of $p_0/|p_0|$. An irreducible vector space, for $p_0 > 0$, is spanned by basis vectors all belonging to the same eigenvalue m^2 of p^2 and having $p_0 = +\sqrt{\mathbf{p}^2 + m^2}$. We can therefore write $|p, \zeta\rangle$ as $|\mathbf{p}, \zeta\rangle$.

In order to obtain the spectrum of w_3 in an irreducible representation we consider Eq. (61) within the manifold obtained from linear combinations of vectors $|\mathbf{p}', \zeta\rangle$ with fixed $\mathbf{p}'$. No difficulty or ambiguity arises since p^μ and w^σ commute. It is convenient, furthermore, to make a Lorentz transformation to the "rest frame" in which $\mathbf{p}' = 0$, $p'_0 = m$. In the rest frame

$$\begin{aligned} w^\mu &= m(0, M_{23}, M_{31}, M_{12}) \\ &= m(0, S_1, S_2, S_3) \end{aligned} \tag{65}$$

with

$$[S_k, S_l] = i\epsilon_{klm} S_m \tag{66}$$

The S_l obey angular momentum commutation rules. The eigenvalues of $\mathbf{S}^2$ are therefore $s(s + 1)$ where $s = 0, \frac{1}{2}, 1, \frac{3}{2}, 2, \cdots$, and the S_i are the generators of an irreducible $2s + 1$ dimensional representation of the three-dimensional rotation group. In the rest frame, $\mathbf{w}$ is thus equal to m times the total angular momentum. This is only true for the case $m \neq 0$, for only in that case are we able to make a Lorentz transformation to the rest frame. For an irreducible representation, a basis vector therefore has $2s + 1$ components, i.e., ζ takes the values $\zeta = 1, \cdots, 2s + 1$, or stated equivalently there are $2s + 1$ independent states for a given momentum vector p_μ, with $p^2 = m^2$ and $p_0 = +\sqrt{\mathbf{p}^2 + m^2} > 0$.

We can next define a Lorentz invariant scalar product within the vector space by integrating over the set of p consistent with $p^2 = m^2$, $p_0 = +\sqrt{\mathbf{p}^2 + m^2}$, and summing over the index ζ

$$(\chi, \psi) = \sum_{\zeta=1}^{2s+1} \int \frac{d^3p}{p_0} \langle \chi \mid \mathbf{p}, \zeta \rangle \langle \mathbf{p}, \zeta \mid \psi \rangle$$

$$= \sum_{\zeta=1}^{2s+1} \int \frac{d^3p}{p_0} \bar{\chi}(\mathbf{p}, \zeta)\, \psi(\mathbf{p}, \zeta) \tag{67}$$

where d^3p/p_0 is the invariant measure on the hyperboloid $p^2 = m^2$ and $\sum_{\zeta}^{2s+1} \bar{\chi}(\mathbf{p}, \zeta)\, \psi(\mathbf{p}, \zeta)$ must be a scalar. The vector space thus equipped with a scalar product is a Hilbert space.

Before continuing the mathematical analysis of the unitary representations of the inhomogeneous Lorentz group we shall pause to inquire into their relevance for physical applications. [Haag (1955), Wigner (1956); see also Newton (1949)]. For this purpose consider the description of an elementary particle. What is meant by an elementary particle is certainly not clear and the elucidation of this concept is one of the foremost problems of theoretical physics today. Intuitively, one calls a particle of mass m and spin s an elementary particle, if for time durations large compared with its natural unit of time, $\hbar/mc^2$, it can be considered as an irreducible entity and not the union of the other particles. For such a system it is natural to require that it should not be possible to decompose its states into linear subsets which are each invariant under Lorentz transformations: all the states of the system must be obtainable from linear combinations of the Lorentz transform of any *one* state. For if there were linear subsets, each of which is invariant under Lorentz transformations, then this would imply that there is a relativistically invariant distinction between these sets of states of the system and one would logically call each subset of relativistically invariant states a different "elementary system." Quite generally, a system is called an "elementary system" if its manifold of states forms a set which is as small as possible consistent with the superposition principle and which is invariant under Lorentz transformations. The manifold of states of an elementary system therefore constitutes a representation space for an irreducible representation of the inhomogeneous

Lorentz group. Note that this definition implies that composite systems such as a helium atom in its ground state or an α particle, for example, are also elementary systems. However, a helium atom in two or more states of excitation would not be an elementary system in the above sense, since in this situation one can select a smaller set of states for which the superposition principle applies and which can be characterized in a relativistically invariant manner. It is clearly of interest to inquire what are the position operators and other global observables of an elementary system and furthermore what their meaning is. The answer [Newton (1949)] is that the operators corresponding to these observables can be found on the basis of quite general, invariant theoretic principles. For example, the position observables obtained on this basis correspond to the position co-ordinates of the center-of-mass and the momentum observables to the total momentum of the system. The momentum operators are furthermore real multiples of the infinitesimal translation operators.

An elementary system is therefore one which has definite transformation properties (its states transform under an irreducible representation of the inhomogeneous Lorentz group), and more specifically its transformation properties are those usually ascribed to a particle. Whether one calls such a system an elementary particle or not depends on whether it is useful (or possible) to ascribe to it the property of being structureless and not composed of other particles. This last property clearly depends on how small a distance can be probed by experimental means, e.g., by means of high energy scattering. Whether a particle is called elementary or composite is therefore a function of how tightly bound the constituents are. The present experimental findings, in particular the Stanford electron-nucleon scattering experiments [Hofstadter (1957)], indicate that even the stable fundamental particles (electron, proton) are not elementary in the aforementioned sense of being structureless. They are, however, elementary systems in that the states of such an isolated particle form an invariant manifold and all the states can be obtained as linear combinations of the Lorentz transforms of any *one* state. The manifold of states for the particle can be characterized by a parameter m, the mass of the particle, a parameter s the spin of the particle, and certain other invariant parameters such as the electric and nucleonic charge. Furthermore, the dependence on the kinematic variables of the wave function describing the particle is determined (apart from the equivalences) by the irreducible representations labeled (m, s). Note however that nothing is said in this description of the configuration of whatever entities may constitute our "elementary particle." What is determined by the invariant theoretical methods is the kinematic description of the free isolated particle given its mass and spin. But this, however, is what is needed in the description of the initial and final states of the particles in a scattering experiment when they are far apart and do not interact with one another. The particles are prepared in states of definite mass, momentum, spin and charge, and the detector again only records their mass, momentum, charge and spin states. The wave functions describing these situations are therefore precisely the ones which are obtained by the invariant theoretic methods.

To summarize the preceding discussion, it has been shown that an irreducible representation of the type $p^2 > 0$, $p_0 > 0$ is labeled by two indices (m, s), where m is a positive number and s is integer or half-integer. The index m characterizes the mass of the elementary system, the index s the angular momentum in its rest frame, i.e., the spin of the elementary system. The fact that the irreducible representation is infinite dimensional is just the expression of the fact that each elementary system is capable of assuming infinitely many linearly independent states. For each (m, s), and a given sign of the energy there is one and only one irreducible representation of the inhomogeneous Lorentz group to within unitary equivalence. For s half-integral the representation is double-valued. The cases $s = 0, \frac{1}{2}, 1$ will be of foremost interest to us. For $s = 0$, the representation space is spanned by the positive energy solutions of the relativistically covariant equation for a spin 0 particle: the Klein-Gordon equation; for $s = \frac{1}{2}$ by the positive energy solutions of the Dirac equation; and for $s = 1$ by the positive energy solutions of the Proca equation.

All these equations can be cast into a certain canonical form [see in this connection Foldy (1956)] by the following considerations. We have previously remarked that in the Schrödinger picture the knowledge of the translation operators is equivalent to the knowledge of the equation of motion of the system. Now we have in fact determined the representations of this operator. For a space-time translation $x_\mu \rightarrow x_\mu + a_\mu$ it is given by Eq. (64). Therefore for an elementary system of mass m and spin s, whose manifold of states spans the vector space of the irreducible representation labeled by (m, s), a time translation $a_\mu = (\tau, 0, 0, 0)$ of the state $|\psi\rangle$ gives rise to the state $U(\tau) \mid \psi\rangle$, where

$$\begin{aligned} \langle \mathbf{p}s \mid U(\tau) \mid \psi\rangle &= e^{-ip_0\tau}\, \psi(\mathbf{p}, s) \\ &= \psi(\mathbf{p}, s; \tau) \end{aligned} \tag{68}$$

Here the right-hand side corresponds to the Schrödinger state at time τ if $\psi(\mathbf{p}, s)$ corresponded to the Schrödinger state at $\tau = 0$ (Heisenberg state). The time evolution of the elementary system is thus governed by the differential equation

$$\begin{aligned} i\partial_\tau \psi(\mathbf{p}, s; \tau) &= p_0 \psi(\mathbf{p}, s; \tau) \\ &= \sqrt{\mathbf{p}^2 + m^2}\, \psi(\mathbf{p}, s; \tau) \end{aligned} \tag{69}$$

We next turn our attention to the second class of representations which are of physical interest, namely the mass zero case. If the invariant P is equal to zero, $P = 0$ and $p_\mu \neq 0$, i.e., the case of zero rest mass particles, there arise two different types of representations.

The first corresponds to the case when $W = -w_\mu w^\mu = 0$, i.e., $P = 0$ and $W = 0$. In that case, these two quantum numbers do not suffice to characterize the representation. However, since both p_μ and w_μ are now null vectors and since $w^\mu p_\mu = 0$, Eq. (58), we must have $w_\mu = \lambda p_\mu$. The eigenvalues of the operator λ, which operator turns out to be essentially the spin of the particle, can then be used to label the representation. To establish the interpretation of λ as the spin of the particle we note that if $w_\mu = \lambda p_\mu$, it then follows from this equation, together with Eqs. (57a) and (57b), that

$$\lambda = \mathbf{M} \cdot \mathbf{p}/p_0 \tag{70}$$

Since $p_0{}^2 = \mathbf{p}^2$, λ is the component of angular momentum along the direction of motion of the particle, i.e., its helicity. For a given momentum vector p_μ there exist now two independent states if $\lambda \neq 0$, which correspond to two different states of polarization (helicity). If $\lambda = 0$, there exists only one state.

The second type of representation arises when $W \neq 0$ but equal to α^2 where α is a real number. For a given momentum vector there then exist infinitely many different states of polarization which can be described by a continuous variable. We shall treat both cases simultaneously.

We first of all note that for a massless particle there does not exist any co-ordinate system in which all except one component of p_μ vanish. There is a frame, however, in which p_μ takes the form $p_\mu = (p, 0, 0, p)$. In this frame, calling

$$w_1 + iw_2 = \lambda_+ \tag{71a}$$

$$w_1 - iw_2 = \lambda_- \tag{71b}$$

and

$$w_0 = p\lambda \tag{71c}$$

where λ by Eq. (55) is equal to $M_{12} = M_3$. We note that we can write W as

$$\begin{aligned} W &= -w_\mu w^\mu = -(w_0{}^2 - w_1{}^2 - w_2{}^2 - w_3{}^2) \\ &= (w_1 + iw_2)\,(w_1 - iw_2) \\ &= \lambda_+\lambda_- \end{aligned} \tag{72}$$

since in this co-ordinate system $w_0 = w_3$, and $[w_1, w_2] = 0$, i.e., w_1 and w_2 commute. Using Eq. (61) and the fact that $p_\mu = (p, 0, 0, p)$, the commutation rules of the λs are found to be the following:

$$\left.\begin{aligned} [\lambda_+, \lambda] &= -\lambda_+ \\ [\lambda_-, \lambda] &= +\lambda_- \end{aligned}\right\} \tag{73a}$$

$$[\lambda_+, \lambda_-] = 0 \tag{73b}$$

Let us denote the eigenfunctions of λ and W by $|\alpha; \beta\rangle$,

$$W \mid \alpha, \beta\rangle = \alpha \mid \alpha, \beta\rangle \tag{74a}$$

$$\lambda \mid \alpha, \beta\rangle = \beta \mid \alpha, \beta\rangle \tag{74b}$$

To determine the eigenvalue spectrum we note, using Eq. (73a), that

$$[\lambda_+, \lambda] \mid \alpha, \beta\rangle = (\beta\lambda_+ - \lambda\lambda_+) \mid \alpha, \beta\rangle$$
$$= -\lambda_+ \mid \alpha, \beta\rangle \tag{75}$$

so that

$$\lambda(\lambda_+ \mid \alpha, \beta\rangle) = (\beta + 1)\,(\lambda_+ \mid \alpha, \beta\rangle) \tag{76}$$

and similarly

$$\lambda\{\lambda_- \mid \alpha, \beta\rangle\} = (\beta - 1)\,\{\lambda_- \mid \alpha, \beta\rangle\} \tag{77}$$

so that $\lambda_\pm \mid \alpha, \beta\rangle$ belongs to the eigenvalue $\beta \pm 1$ of λ. The eigenvalue spectrum of λ is therefore of the form

$$\beta = n_0 + n \tag{78}$$

where $n = 0, \pm 1, \pm 2$ with $1 > n_0 \geqslant 0$. Since $\lambda = M_3$, λ is actually the generator for a spatial three-dimensional rotation in the hyperplane perpendicular to p_μ. Since a rotation through 2π leaves the basis functions of a single-valued representation unchanged, or for the case of a two-valued representation multiplies the basis functions by -1, n_0 must be equal to zero for single-valued representations and equal to $\frac{1}{2}$ for double-valued representations. Now within an irreducible representation only those $|\alpha, \beta\rangle$ can occur which belong to the same eigenvalue of W, i.e., only $|\alpha, \beta\rangle$s with the same α. If we now relabel the β variable by n, then for an irreducible representation

$$\langle \alpha, n \mid \lambda \mid \alpha, m\rangle = (n_0 + n)\,\delta_{nm} \tag{79}$$

and similarly

$$\langle \alpha, n \mid \lambda_+ \mid \alpha, m\rangle = a_n\delta_{n,m+1} \tag{80a}$$

$$\langle \alpha, n \mid \lambda_- \mid \alpha, m\rangle = b_n\delta_{n,m-1} \tag{80b}$$

It follows that

$$\alpha = \langle \alpha, n \mid W \mid \alpha, n\rangle = \langle \alpha, n \mid \lambda_+\lambda_- \mid \alpha, n\rangle$$
$$= \langle \alpha, n \mid \lambda_+ \mid \alpha, n-1\rangle\,\langle \alpha, n-1 \mid \lambda_- \mid \alpha, n\rangle$$
$$= a_n b_n \tag{81}$$

For a unitary representation w_μ is Hermitian and therefore $(\lambda_+)^* = \lambda_-$, so that $a_n = \bar{b}_n$ and $\alpha = |a_n|^2 \geqslant 0$. If now $a_n = b_n = 0$, then $\alpha = 0$ for all n; therefore $\lambda_+ = \lambda_- = 0$ and, consequently, $w_1 = w_2 = 0$ and $w_\mu = \lambda p_\mu$. Note that when $\alpha = 0$, λ commutes with all the generators and it is thus an additional invariant of the group. This has the consequence that as far as the spin variable is concerned the representations are all one dimensional. Thus, for each integral or half-integral value of λ, there exist two irreducible representations (for a given sign of p_0): in one $w_\mu = \lambda p_\mu$, in the other $w_\mu = -\lambda p_\mu$. In the case $\lambda = 0$ there exists only one state. Therefore particles with a nonzero spin and zero rest mass have only two directions of polarization no matter how large their spin is in contrast to $2s + 1$ states for a particle with nonzero rest mass and spin s.[4] The photon is a

[4] This fact is discussed at greater length in Section 5*b*.

representative example of this phenomenon. Its spin is 1 yet it has only two directions of polarization.

The representations with $W = \alpha^2 > 0$ are infinite-dimensional in the spin variable and would correspond to particles with a continuous spin. We shall not consider them further since they do not seem to be realized in Nature. Their properties have, however, been investigated by Wigner (1947) and by Bargmann and Wigner [Bargmann (1948)]. The representations for $P = 0$, $W = 0$, $\lambda = 0, \frac{1}{2}, 1, \cdots$ etc. are realized in Nature for the case $\lambda = \frac{1}{2}$ (the neutrino) and $\lambda = 1$ which corresponds to the photon. An explicit determination of the representations for $P = 0$, $W = 0$, and λ arbitrary, have been obtained by Fronsdal (1959).

Except for the case $p^2 < 0$, we have enumerated all the irreducible representations of the Lorentz group. We shall not consider the representations for $p^2 < 0$ for the following reason: For these representations the representation space is spanned by basis vectors $|p', \zeta\rangle$ with $p'^2 < 0$. The energy p_0 of a particle corresponding to such a representation would therefore have the unphysical property that it could become arbitrarily large and negative by suitable Lorentz transformation. This clearly cannot be the case for a physical particle. In this connection it should be noted that for the representations of relevance for the description of physical particles the energy spectrum is positive and bounded below by zero, $p_0 \geqslant 0$, as should be the case for an actual particle. Furthermore, these representations have a well-defined and reasonable nonrelativistic limit. This last property is incidentally not shared by the representations with $p^2 < 0$.

3

The Klein-Gordon Equation

3a. Historical Background

When Schrödinger wrote down the nonrelativistic equation now bearing his name, he also formulated the corresponding relativistic equation. Subsequently, the identical equation was proposed independently by Gordon (1926a, b), Fock (1926a, b), Klein (1926), Kudar (1926), and de Donder and Van Dungen [de Donder (1926)]. The equation is derived by inserting the operator substitutions $E \to i\hbar\partial_t$, $\mathbf{p} \to -i\hbar\nabla$ into the relativistic relation between the energy and momentum for a free particle

$$E^2 = c^2\mathbf{p}^2 + \mu^2c^4 \tag{1}$$

where μ is the mass of the particle. This procedure yields

$$-\hbar^2 \frac{\partial^2\phi(\mathbf{x}, t)}{\partial t^2} = (-\hbar^2c^2\nabla^2 + \mu^2c^4)\, \phi(\mathbf{x}, t) \tag{2a}$$

or using natural units ($\hbar = c = 1$) and the Dirac notation $\phi(x) = \langle x \mid \phi\rangle$

$$(\Box + \mu^2)\, \langle x \mid \phi\rangle = 0 \tag{2b}$$

Equation (2a) has become known as the Klein-Gordon equation. The amplitude $\phi(x)$ is a one-component scalar quantity which under an inhomogeneous Lorentz transformation, $x' = \Lambda x + a$, transforms according to

$$\phi'(x') = \phi(x) \tag{3a}$$

or equivalently

$$\phi'(x) = \phi(\Lambda^{-1}(x - a)) \tag{3b}$$

We shall say that ϕ describes a scalar particle if under a spatial inversion $x_0 \to x_0$, $\mathbf{x} \to -\mathbf{x}$, $\phi \to \phi$, and that ϕ describes a pseudoscalar particle if under this spatial inversion $\phi \to -\phi$.

In order to give a physical interpretation to the Klein-Gordon equation, by analogy with the nonrelativistic equation, one might try to define a probability density, ρ, and a probability current, $\mathbf{j}$, in such a way that a continuity equation holds between them. One is then led to the following expressions for ρ and $\mathbf{j}$:

$$\rho = \frac{i\hbar}{2\mu c^2}(\bar{\phi}\partial_t\phi - \partial_t\bar{\phi}\cdot\phi) = \frac{i\hbar}{2\mu c}(\bar{\phi}\partial_0\phi - \partial_0\bar{\phi}\cdot\phi) \tag{4}$$

$$j_l = \frac{\hbar}{2\mu i}(\bar{\phi}\partial_l\phi - \partial_l\bar{\phi}\cdot\phi) \tag{5}$$

which, by virtue of Eq. (2a), satisfy

$$\nabla\cdot\mathbf{j} + \partial_t\rho = 0 \tag{6}$$

The constants appearing in the density and current have been so determined that these expressions reduce to the usual expressions for the Schrödinger theory in the nonrelativistic limit. If in the expression for ρ we substitute for $i\hbar\partial_t\phi$, $E\phi$, we obtain

$$\rho = \frac{E}{\mu c^2}\bar{\phi}\phi \tag{7}$$

which for $E \approx \mu c^2$ indeed reduces to the expression for the probability density in nonrelativistic quantum mechanics. It is, however, to be noted that in general ρ may assume negative as well as positive values, because Eq. (2a) is of second order in the time variable and therefore ϕ and $\partial_t\phi$ can be prescribed arbitrarily at some time t_0. Also, since ϕ and $\partial_t\phi$ are functions of the space co-ordinates $\mathbf{x}$, ρ can be positive in some regions and negative in others. It is thus difficult to think of ρ as a conventional probability density. Because of this possibility of negative ρ values, the Klein-Gordon equation fell into disrepute for about seven years after it was first proposed. It was only in 1934 that Pauli and Weisskopf re-established the validity of the equation by reinterpreting it as a field equation in the same sense as Maxwell's equations for the electromagnetic field and quantizing it.

3b. Properties of Solutions of K-G Equation

We next show that there exist relativistic situations for which the probability interpretation is still applicable. From (7) it is to be expected that this will be the case when the particle is free or moving in an extremely weak external field. To investigate this situation, let us obtain the solutions of the Klein-Gordon equation. It admits of plane wave solutions

$$\phi(x) = e^{-\frac{i}{\hbar}(p^0x^0 - \mathbf{p}\cdot\mathbf{x})} \tag{8}$$

if

$$cp_0 = E = \pm\sqrt{c^2\mathbf{p}^2 + \mu^2c^4} \tag{9}$$

Either sign of the square root leads to a solution. This is a consequence of the fact that the equation is covariant under all Lorentz transformations which leave invariant the quadratic form $p_\mu p^\mu = \mu^2c^2$. The transformation $p_0 \to -p_0$ is clearly one such transformation. The occurrence of negative energy solutions does not present any difficulty for a free particle. The particle is originally in a positive energy state with $E = c\sqrt{\mathbf{p}^2 + \mu^2c^2}$ and in the absence of any interaction it will always remain in a positive energy state. Furthermore, from (7), we note that for a free particle with positive energy $\rho > 0$ and remains positive-definite for all times by virtue of the equations of motion. We conclude that a consistent theory can be developed for a free particle if we adopt the manifold of positive energy solutions as the set of states which are physically realizable by a free particle. The equation of motion for a positive energy amplitude $\phi(x)$ can be taken to be

$$i\hbar\partial_t\phi(x) = \sqrt{\mu^2c^4 - \hbar^2c^2\nabla^2}\,\phi(x) \tag{10}$$

If we define the three-dimensional Fourier transform of $\phi(x)$ as

$$\phi(x) = \int d^3k e^{i\mathbf{k}\cdot\mathbf{x}}\,\chi(\mathbf{k}, x_0) \tag{11}$$

the square root operator in (10) is then to be interpreted as follows:

$$\sqrt{\mu^2c^2 - \hbar^2\nabla^2}\,\phi(x) = \int d^3k e^{i\mathbf{k}\cdot\mathbf{x}}\sqrt{\mu^2c^2 + \hbar^2\mathbf{k}^2}\,\chi(\mathbf{k}, x_0) \tag{12}$$

Note that $\chi(\mathbf{k}, x_0)$ satisfies the equation

$$i\hbar\partial_0\chi(\mathbf{k}, x_0) = \hbar\omega(\mathbf{k})\,\chi(\mathbf{k}, x_0)$$

$$\omega(\mathbf{k}) = c\sqrt{\frac{\mu^2c^2}{\hbar^2} + \mathbf{k}^2} \tag{13}$$

A concise covariant description of the manifold of positive energy solutions is the statement that it consist of all $\phi(x)$ of the form

$$\phi(x) = \frac{\sqrt{2}}{(2\pi)^{3/2}}\int d^4k e^{-ik\cdot x}\,\delta(k^2 - \mu^2)\,\theta(k_0)\,\Phi(k) \tag{14}$$

where, for convenience, we have introduced certain numerical factors and we have set $\hbar = c = 1$. The delta function $\delta(k^2 - \mu^2)$ guarantees that the Klein-Gordon equation is satisfied by $\phi(x)$ and the step function $\theta(k_0)$, which requires that $k_0 > 0$, guarantees that the energy of the particle is positive. Equation (14) can be somewhat simplified by carrying out the integration over k_0 using

$$\delta(k^2 - \mu^2) = \frac{1}{2\omega(\mathbf{k})}\{\delta(k_0 - \omega(\mathbf{k})) + \delta(k_0 + \omega(\mathbf{k}))\} \tag{15}$$

where $\omega(\mathbf{k}) = \sqrt{\mathbf{k}^2 + \mu^2}$. The $\theta(k_0)$ in (14) restricts the contribution to that of the first term only and we thus obtain

$$\phi(x) = \frac{1}{(2\pi)^{3/2}}\int_+ \frac{d^3k}{\sqrt{2}\,k_0}e^{-ik\cdot x}\,\Phi(k) \tag{16}$$

where on the right-hand side $k_0 = \omega(\mathbf{k}) = \sqrt{\mathbf{k}^2 + \mu^2}$, so that $\Phi(k)$ is really only a function of k^1, k^2, k^3. We shall therefore write $\Phi = \Phi(\mathbf{k})$. The set of positive energy solutions forms a linear vector space which can be made into a Hilbert space by defining a suitable scalar product. We define the scalar product of two positive energy Klein-Gordon amplitudes as

$$(\phi, \psi)_t = i \int_t d^3x(\bar{\phi}(x)\, \partial_0\psi(x) - \partial_0\bar{\phi}(x) \cdot \psi(x)) \tag{17a}$$

$$= i \int_t d^3x \bar{\phi}(x) \overleftrightarrow{\partial_0}\psi(x) \tag{17b}$$

Note that even though (17a) contains time derivatives, if ψ and ϕ obey Eq. (10), then all quantities appearing in the scalar product can be made to refer to quantities defined at a single time t, as in ordinary nonrelativistic quantum mechanics, and the scalar product written as

$$(\phi, \psi)_t = \int d^3x\{\bar{\phi}(x)\, \sqrt{\mu^2 - \nabla^2}\, \psi(x) + \sqrt{\mu^2 - \nabla^2}\, \bar{\phi}(x) \cdot \psi(x)\} \tag{17c}$$

This scalar product is conserved in time if ϕ and ψ obey the Klein-Gordon equation. Furthermore, it possesses all the properties usually required of a scalar product, namely:

$$(\phi, \psi) = \overline{(\psi, \phi)} \tag{18a}$$

$$(\phi_1 + \phi_2, \psi) = (\phi_1, \psi) + (\phi_2, \psi) \tag{18b}$$

$$(\phi, \phi) \geqslant 0 \tag{18c}$$

the equality sign in (18c) holding if and only if $\phi \equiv 0$. The positive-definiteness of (ϕ, ϕ) if $\phi \not\equiv 0$ is made evident by going to momentum space, where the scalar product takes the very simple form

$$(\psi, \phi) = \int_+ \frac{d^3k}{k_0} \bar{\Psi}(\mathbf{k})\, \Phi(\mathbf{k}) \tag{19}$$

where again $k_0 = +\omega(\mathbf{k})$. The relativistic invariance of the scalar product is also made explicit by (19) since Ψ and Φ as defined by Eq. (14) are scalars (since $\delta(k^2 - \mu^2)$, $\theta(k_0)$, $k \cdot x$ and d^4k are invariants) and d^3k/k_0 is the invariant measure element over the hyperboloid $k^2 = \mu^2$.

In the definition (17) of the scalar product the integral is taken over all space at the time $ct = x_0$. It can be generalized to an integral over a space-like surface which then exhibits the relativistic invariance of the scalar product directly in configuration space. A space-like surface, σ, is defined by the condition that no two points on it can be connected by a light signal: for any two points x, y on σ, $(x - y)^2$ is always space-like, i.e., $(x - y)^2 < 0$. If we denote by $n^\mu(x)$ the unit normal to σ at the point x, then for a space-like surface $n_\mu(x)\, n^\mu(x) = +1$ for all x on σ. A plane t = constant is a special case for which for all x, $n^\mu(x) = (1, 0, 0, 0)$. The (pseudo) vector surface element $d\sigma^\mu(x) = n^\mu(x)\, d\sigma$ for an arbitrary three-dimensional surface S in space-time has the components $d\sigma^\mu = \{dx^1dx^2dx^3, dx^0dx^2dx^3, dx^0dx^1dx^3, dx^0dx^1dx^2\}$.

Gauss's theorem in four-space can be written as

$$\int_V d^4x \partial^\mu F_\mu(x) = \int_S d\sigma^\mu(x)\, F_\mu(x) \tag{20}$$

where S is the surface bounding the volume V. If $G = G(\sigma)$ is a function of the space-like surface σ, we define the invariant operation $\delta/\delta\sigma(x)$, which depends on σ and on x, as

$$\frac{\delta}{\delta\sigma(x)}\, G(\sigma) = \lim_{\Omega\to 0} \frac{G(\sigma') - G(\sigma)}{\Omega(x)} \tag{21}$$

where σ' is a space-like surface which differs from σ by an infinitesimal deformation in the neighborhood of the space-time point x, and $\Omega(x)$ is the four-dimensional volume enclosed between σ and σ' (see Fig. 3.1) which

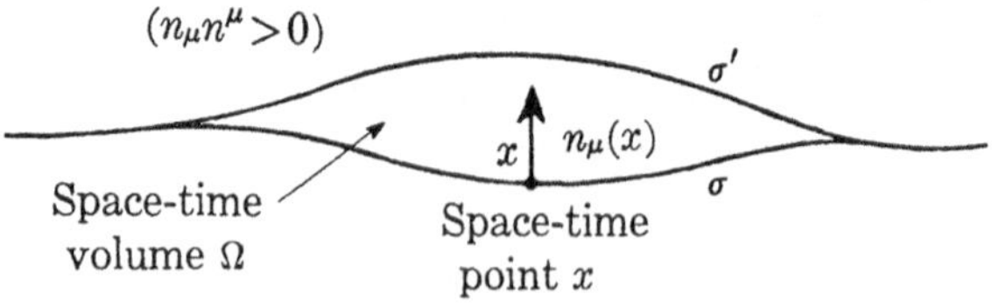

Fig. 3.1

in the limit goes over into the point x. For the particular case that $G(\sigma) = \int_\sigma F_\mu(x')\, d\sigma^\mu(x')$

$$\frac{\delta}{\delta\sigma(x)}\, G(\sigma) = \lim_{\Omega\to 0} \frac{\int_{\sigma'} F_\mu d\sigma^\mu - \int_\sigma F_\mu d\sigma^\mu}{\Omega(x)} \tag{22}$$

Gauss's theorem, Eq. (20), can now be applied to the numerator which is equal to the surface integral over the surface bounding $\Omega(x)$, so that in the limit as Ω shrinks to the point x

$$\frac{\delta}{\delta\sigma(x)}\, G(\sigma) = \frac{\partial F_\mu(x)}{\partial x_\mu} = \partial^\mu F_\mu(x) \tag{23}$$

We now rewrite the scalar product (17) in invariant form as follows:

$$\begin{aligned}(\phi, \chi)_\sigma &= i \int_\sigma d\sigma^\mu(x)\, \{\bar{\phi}(x)\partial_\mu\chi(x) - \partial_\mu\bar{\phi}(x) \cdot \chi(x)\} \\ &= i \int_\sigma d\sigma^\mu(x)\, \bar{\phi}(x)\overleftrightarrow{\partial_\mu}\chi(x)\end{aligned} \tag{24}$$

This expression clearly reduces to (17) for the case that σ is a plane surface t = constant. The fact that the scalar product is conserved in time can now be established by noting that the scalar product (24) is in fact independent of σ if ψ and χ obey the Klein-Gordon equation. *Proof:*

$$\frac{\delta}{\delta\sigma(x)}(\phi, \chi)_\sigma = i\partial^\mu\{\bar{\phi}(x)\,\partial_\mu\chi(x) - \partial_\mu\bar{\phi}(x)\cdot\chi(x)\}$$
$$= i\{\bar{\phi}(x)\,(\Box + \mu^2)\,\chi(x) - (\Box + \mu^2)\,\bar{\phi}(x)\cdot\chi(x)\}$$
$$= 0 \tag{25}$$

The scalar product thus does not depend on the particular space-like surface σ chosen for its evaluation, if ψ and χ obey the Klein-Gordon equation. In particular, we can choose the surface t = constant, in which case (24) reduces to (17). It likewise follows from Eq. (25) that $(\psi, \chi)_t = (\psi, \chi)_{t_0}$. The fact that the Klein-Gordon amplitudes transform like scalars under proper Lorentz transformation implies that $\mathbf{j}$ and ρ given by (4)–(5) are the components of a four-vector $j^\mu = \{\rho, \mathbf{j}\}$. The norm of the wave function is the square root of $\int d\sigma^\mu j_\mu$ and is therefore an invariant.

It should be noted that the scalar product (17) or (24) is only defined for wave-packet solutions of the Klein-Gordon equations, i.e., for ϕs such that $(\phi, \phi) < \infty$ and only such vectors make up the Hilbert space. A plane wave solution is a limiting case of a wave packet and is not an element of the Hilbert space. We can, however, adopt the following invariant continuum normalization for these solutions:

$$(\phi_{\mathbf{p}}, \phi_{\mathbf{p}'}) = p_0\delta^{(3)}(\mathbf{p} - \mathbf{p}') \tag{26}$$

With this convention a positive energy plane wave solution of momentum $\mathbf{p}$ has the form

$$\phi_{\mathbf{p}}(x) = \frac{1}{\sqrt{2(2\pi)^3}}\,e^{-ip\cdot x} \tag{27a}$$

with

$$p_0 = \sqrt{\mathbf{p}^2 + \mu^2} \tag{27b}$$

The completeness relation for these solutions reads

$$\sum_{\mathbf{p}} \phi_{\mathbf{p}}(x)\,\bar{\phi}_{\mathbf{p}}(x') = \frac{1}{2(2\pi)^3}\int \frac{d^3p}{p_0}\cdot e^{-ip\cdot(x-x')} \tag{28}$$

where it is to be noted that even for $x_0 = x_0'$, the right-hand side of (28) does not reduce to a δ function.

The Hilbert space, $\mathcal{H}^{KG}$, of positive energy solutions of the Klein-Gordon equation forms a representation space for an irreducible representation of the inhomogeneous Lorentz group. This representation is fixed by the following identification of the infinitesimal generators:

$$P_\mu = p_\mu \tag{29a}$$

$$M_{\mu\nu} = i\left(p_\mu\frac{\partial}{\partial p^\nu} - p_\nu\frac{\partial}{\partial p^\mu}\right) \tag{29b}$$

One verifies that the right-hand side of Eqs. (29a) and (29b) indeed satisfy the commutation rules (2.51)–(2.53). Under a proper inhomogeneous

Lorentz transformation, $x' = \Lambda x + a$, a state $|\phi\rangle$ of the system transforms according to

$$U^{(\mu,0)}(\Lambda, a) \mid \phi\rangle = |\phi'\rangle \tag{30}$$

where the $U^{(\mu,0)}$s form an irreducible unitary representation of the inhomogeneous Lorentz group corresponding to the mass μ and spin 0, on the Hilbert space $\mathcal{H}^{KG}$. More explicitly

$$\begin{aligned}\langle x \mid U^{(\mu,0)}(\Lambda, a) \mid \phi\rangle &= \langle x \mid \phi'\rangle \\ &= \langle \Lambda^{-1}(x - a) \mid \phi\rangle \end{aligned} \tag{31a}$$

which in terms of the amplitudes becomes

$$\phi'(x) = \phi(\Lambda^{-1}(x - a)) \tag{31b}$$

and in momentum space

$$\phi'(k) = e^{ik\cdot a}\, \phi(\Lambda^{-1}k) \tag{32}$$

The unitary property of the $U^{(\mu,0)}(\Lambda, a)$ is exhibited by the formulae

$$(U\phi, U\chi) = \int \frac{d^3k}{k_0}\, \overline{e^{ika}\, \phi(\Lambda^{-1}k)}\, e^{ika}\, \chi(\Lambda^{-1}k) \tag{33a}$$

$$= \int \frac{d^3k'}{k_0'}\, \bar{\phi}(k')\, \chi(k') \tag{33b}$$

$$= (\phi, \chi) \tag{33c}$$

where in going from Eq. (33a) to (33b) we have explicitly made use of the invariance of the measure element $d\Omega(k) = d^3k/k_0 = d\Omega(\Lambda^{-1}k)$.

3c. The Position Operator

The fact that the manifold of realizable states contains only positive energy solutions, with a scalar product defined by (17) or (19), has several important consequences concerning the operators representing physical observables. Thus within the scalar product (17) the operator $\mathbf{x} = i\nabla_{\mathbf{p}}$ is no longer Hermitian since

$$(\psi, \mathbf{x}\phi) = i \int \frac{d^3p}{p_0}\, \bar{\Psi}(\mathbf{p})\, \nabla_{\mathbf{p}}\Phi(\mathbf{p}) \tag{34a}$$

$$= \int \frac{d^3p}{p_0} \left[\left(-i\nabla_{\mathbf{p}} + \frac{i\mathbf{p}}{\mathbf{p}^2 + \mu^2}\right) \bar{\Psi}(\mathbf{p})\right] \Phi(\mathbf{p}) \tag{34b}$$

$$\neq (\mathbf{x}\psi, \phi) \tag{34c}$$

where in going from (34a) to (34b) a surface integral has been discarded. The operator $\mathbf{x}$ cannot therefore be interpreted as the position operator, since it is not self-adjoint and hence does not correspond to a measurable property of the system. It follows that the Klein-Gordon wave function

$\phi(x)$ cannot be called a probability amplitude for finding the particle at $\mathbf{x}$ at time x_0. In order to answer the question, "What is the probability of finding a Klein-Gordon particle at some point $\mathbf{y}$ at time y_0?" we must first find a Hermitian operator which can properly be called a position operator, and secondly find its eigenfunctions, $\Psi_{\mathbf{y},y_0}(x)$. The probability amplitude for finding at time $y^0 = x^0$, at the point $\mathbf{y}$, a particle with wave function $\phi(x)$ is then given according to the general principle of quantum mechanics by the matrix element (Ψ_y, ϕ).

The simplest way to obtain a Hermitian position operator is to define the Hermitian part of $i\nabla_{\mathbf{p}}$ as the position operator, i.e.,

$$\mathbf{x}_{op} = i\nabla_{\mathbf{p}} - \frac{1}{2}\frac{i\mathbf{p}}{\mathbf{p}^2 + \mu^2} \tag{35}$$

It turns out that this $\mathbf{x}_{op}$ is an acceptable position operator. It agrees with the definition of the center-of-mass in relativistic mechanics [Papapetrou (1939), Pryce (1948), Møller (1949a, b), (1952)]. It is also the position operator obtained by Newton and Wigner [Newton (1949)] in a derivation based on the imposition of certain natural physical requirements on localized states. Newton and Wigner have shown that in the relativistic situation the position operator and its eigenfunctions, the "localized wave functions," are determined by the following requirements:

1. That the set of all states localized at time 0 at $\mathbf{y} = 0$ form a linear manifold invariant under spatial rotations about the origin, spatial inversions and time inversions.
2. That if a state $\Psi_{\mathbf{y}}$ is localized at some point $\mathbf{y}$, then a spatial displacement shall make it orthogonal to the set of states localized at the point $\mathbf{y}$.
3. That the infinitesimal operators of the Lorentz group be applicable on the localized states. Condition (3) is a regularity condition.

For the Klein-Gordon case, let $\Psi_0(\mathbf{k})$ be the state localized at the origin at time $x_0 = 0$. Now in momentum space, the space displacement operator is simply multiplication by $\exp(-i\mathbf{k}\cdot\mathbf{a})$, so that the displaced state localized at $\mathbf{y}$ at time $y^0 = 0$ is $\exp(-i\mathbf{k}\cdot\mathbf{y})\cdot\Psi_0(\mathbf{k})$. This displaced state, by condition (2) above, must be orthogonal to $\Psi_0(\mathbf{k})$, i.e.,

$$(\Psi_{\mathbf{y}}, \Psi_0) = \delta^{(3)}(\mathbf{y}) = \int \frac{d^3k}{k_0}\,|\Psi_0(\mathbf{k})|^2\, e^{-i\mathbf{k}\cdot\mathbf{y}}$$

$$= \frac{1}{(2\pi)^3}\int \frac{d^3k}{k_0}\, k_0 e^{-i\mathbf{k}\cdot\mathbf{y}} \tag{36}$$

Hence $|\Psi_0(\mathbf{k})|^2 = (2\pi)^{-3}\, k_0$. If we allow only functions satisfying the regularity condition (3), the localized wave function about the origin at time 0 is

$$\Psi_0(\mathbf{k}) = (2\pi)^{-3/2}\, k_0^{1/2} \tag{37}$$

and the state localized about $\mathbf{y}$ at time $y_0 = 0$ is

$$\Psi_{\mathbf{y},0}(\mathbf{k}) = (2\pi)^{-3/2} e^{-i\mathbf{k}\cdot\mathbf{y}} k_0^{1/2} \tag{38}$$

The configuration space localized wave function $\psi_{\mathbf{y},0}(x)$ is obtained from $\Psi_{\mathbf{y},0}(\mathbf{k})$ by substituting the latter into (16). Thus

$$\begin{aligned} \psi_{\mathbf{y},0}(\mathbf{x}, 0) &= \frac{1}{\sqrt{2}\,(2\pi)^{3/2}} \int \frac{d^3k}{k_0} e^{+i\mathbf{k}\cdot(\mathbf{x}-\mathbf{y})} k_0^{1/2} \\ &= \text{constant} \cdot \left(\frac{\mu}{r}\right)^{5/4} H^{(1)}{}_{5/4}(i\mu r) \; ; r = |\mathbf{x} - \mathbf{y}| \end{aligned} \tag{39}$$

where $H^{(1)}{}_{5/4}$ denotes the Hankel function of the first kind of order $\frac{5}{4}$. The first thing to note about this localized eigenfunction is that it is not a δ function as in the nonrelativistic case since it is different from zero for $\mathbf{x} \neq \mathbf{y}$. The extension in space of $\psi_{\mathbf{y},0}(\mathbf{x})$ is of the order of $1/\mu$ (i.e., $\hbar/\mu c$); for large values of r, $\psi_{\mathbf{y},0}$ drops off exponentially. The explanation for this is that the Hilbert space $\mathcal{H}^{KG}$ contains only positive energy solutions and δ functions cannot be built out of these. A second point to be emphasized is that the localized states are not Lorentz covariant. They only possess the maximum symmetry properties corresponding to a plane t = constant in space-time.

One now verifies that the localized wave function $[\exp(-i\mathbf{k}\cdot\mathbf{y})] \cdot k_0^{1/2}$ is an eigenfunction of the operator (35) with eigenvalue $\mathbf{y}$:

$$\begin{aligned} \mathbf{x}_{\text{op}}\{e^{-i\mathbf{k}\cdot\mathbf{y}} k_0^{1/2}\} &= i\left\{\nabla_{\mathbf{k}} - \frac{1}{2}\frac{\mathbf{k}}{k_0^2}\right\}\{e^{-i\mathbf{k}\cdot\mathbf{y}} k_0^{1/2}\} \\ &= \mathbf{y}\{e^{-i\mathbf{k}\cdot\mathbf{y}} k_0^{1/2}\} \end{aligned} \tag{40}$$

which justifies calling $\mathbf{x}_{\text{op}}$ the position operator. The components of position operator $\mathbf{x}_{\text{op}} = \mathbf{q}$ commute with one another

$$[q_i, q_j] = 0 \tag{41}$$

and their commutation rules with the momentum operators are as expected

$$[q_l, p_j] = i\delta_{lj} \tag{42}$$

Under spatial rotations $\mathbf{q}$ transforms like a vector and under a spatial translation by the amount $\mathbf{a}$ it transforms into $\mathbf{q} + \mathbf{a}$. The time derivative of the position operator is

$$\frac{d}{dt}\mathbf{q} = i[H, \mathbf{q}] = i[p_0, \mathbf{q}] = \frac{\mathbf{p}}{p_0} \tag{43}$$

where the right-hand side will be recognized as the operator for the velocity of the particle. Finally, if we have a particle in some state $\Phi(\mathbf{k})$ at time $t = 0$, the probability amplitude that a position measurement at time $t = 0$ will find the particle at $\mathbf{y}$ is given by

$$(\psi_{\mathbf{y}}, \phi) = \frac{1}{(2\pi)^{3/2}} \int \frac{d^3k}{k_0} e^{-i\mathbf{k}\cdot\mathbf{y}} k_0^{1/2}\, \Phi(\mathbf{k}) \tag{44}$$

3d. Charged Particles

The formalism discussed thus far describes a spin zero neutral particle. A free electrically-charged spin 0 particle is described by essentially the identical formalism except that the amplitude describing such a charged particle is labeled by an extra dynamical variable: the charge e of the particle.

In the presence of an electromagnetic field the Klein-Gordon equation for a negatively charged particle is modified by making the usual gauge invariant replacements

$$\mathbf{p} \to \mathbf{p} - \frac{e}{c}\,\mathbf{A} \tag{45a}$$

$$p_0 \to p_0 - \frac{e}{c}\,A_0 \tag{45b}$$

where $\mathbf{A}$ and A_0 are the vector and scalar potentials of the electromagnetic field. The Klein-Gordon equation then becomes

$$(i\hbar\partial_t - eA_0(x))^2\,\phi(x) = (-i\hbar c\boldsymbol{\nabla} - e\mathbf{A}(x))^2\,\phi(x) + \mu^2c^4\phi(x) \tag{46}$$

The probability density associated with this equation is given by

$$\rho = \frac{i\hbar}{2\mu c^2}\,(\bar{\phi}\partial_t\phi - \partial_t\bar{\phi}\cdot\phi) - \frac{e}{\mu c^2}\,A_0\bar{\phi}\phi \tag{47}$$

The interpretation of the Klein-Gordon equation in the presence of an external field[1] is no longer as simple as in the case of the free particle. Consider, for example, a charged spin zero particle being scattered by a potential which vanishes except for a finite time interval T. The wave function for the incident particle is a superposition of positive energy solutions of the free Klein-Gordon equation, since it is to represent a real positive energy particle. However, as a result of the action of the potential, it is possible for the wave function after the time T has elapsed to have negative energy components, implying a nonvanishing probability for finding the particle to be in a negative energy state, to which, a priori, it is difficult to give a physical interpretation.

One might think that the situation when the external field is time-independent does not present such difficulties since under these circumstances Eq. (46) is separable with respect to $\mathbf{x}$ and t and the solutions are then of the form

$$\phi(\mathbf{x}, t) = u(\mathbf{x})e^{-iEt/\hbar} \tag{48}$$

with

$$\rho(\mathbf{x}) = \frac{E - eA_0}{\mu c^2}\,\bar{u}u \tag{49}$$

[1] The subsequent remarks are also valid for a neutral particle in the presence of a force field.

so that stationary states are possible. In particular, for the Coulomb field $eA_0 = -Ze^2/r$, the solutions can readily be obtained. [See, for example, Schiff (1949).] There are, as might be expected, solutions with $E > 0$ and corresponding solutions with $E < 0$. A particle initially in a state with $E > 0$ would then always remain in this state unless externally perturbed. However, even in this case the physical interpretation runs into difficulties since the expression (49) for the probability density becomes negative for sufficiently small r where the motion is essentially relativistic. In that region the one-particle interpretation breaks down: A wholly consistent relativistic one-particle theory can be put forth only for free particles. However, even though it is not possible to give a completely satisfactory physical interpretation for the Klein-Gordon equation (46) in the presence of external field, nonetheless the solutions of the equation (46) will be of physical relevance in the field theoretical reinterpretation of the equation. Let us therefore briefly consider some of the properties of these solutions for the case a time-independent magnetic field: $A_0 = 0$, $\mathbf{A} = \mathbf{A}(\mathbf{x})$. The positive energy solutions for a particle of charge $-e$ can then be characterized as solutions of

$$i\hbar\partial_t\phi(-e, +) = \sqrt{\mu^2c^4 + \hbar^2(-ic\boldsymbol{\nabla} - e\mathbf{A})^2}\,\phi(-e, +) \qquad (50)$$

and similarly the negative energy solutions satisfy the equation

$$i\hbar\partial_t\phi(-e, -) = -\sqrt{\mu^2c^4 + \hbar^2(-ic\boldsymbol{\nabla} - e\mathbf{A})^2}\,\phi(-e, -) \quad (51)$$

Under the operation of complex conjugation Eq. (51) is transformed into

$$i\hbar\partial_t\overline{\phi(-e, -)} = \sqrt{\mu^2c^4 + \hbar^2(+ic\boldsymbol{\nabla} - e\mathbf{A})^2}\,\overline{\phi(-e, -)} \qquad (52)$$

which is the equation obeyed by a positive energy amplitude for a particle of charge $+e$, i.e., $\bar{\phi}(-e, -) = \text{constant} \cdot \phi(+e, +)$. The amplitude $\bar{\phi}(-e, -)$ thus describes a positive energy particle of charge $+e$, called the "antiparticle." One calls $\bar{\phi}(-e, -)$ the charge conjugate solution. For neutral particles a similar situation obtains except that it is now possible for neutral particles to be their own antiparticles. One can therefore differentiate between two types of neutral particles depending on whether the particles are their own antiparticles or not. This distinction will be considered in greater detail when we consider the quantized version of the theory.

We conclude this section by mentioning that there exists another approach to the interpretation of the single-particle Klein-Gordon equation which employs a two-component wave function in a two-dimensional charge space equipped with an indefinite metric [Sakata (1940), Heitler (1943), Case (1954)]. The norm of the state vector is $+1$ for a positively charged particle and -1 for a negatively charged one. For a review of this work the reader is referred to the article of Feshbach and Villars [Feshbach (1958)].

4

The Dirac Equation

4a. Historical Background

In 1928 Dirac discovered the relativistic equation which now bears his name while trying to overcome the difficulties of negative probability densities of the Klein-Gordon equation. For a long time after its discovery, it was believed that the Dirac equation was the only valid relativistic wave equation for particles with mass. It was only after Pauli and Weisskopf reinterpreted the Klein-Gordon equation as a field theory in 1934 that this widely held belief was shaken. Even now the Dirac equation has special importance because it describes particles of spin $\frac{1}{2}$, and both electrons and protons have spin $\frac{1}{2}$. Many others of the "elementary particles," including the neutron, the μ mesons, and probably all the presently known hyperons (the Λ, Σ, and Ξ particles) have spin $\frac{1}{2}$. In fact, it is a theoretical conjecture that all the "elementary particles" found in Nature obeying Fermi statistics have spin $\frac{1}{2}$. The π mesons, discovered in 1947, were the first nonzero mass particles having a different spin, namely zero.

The reasoning which led Dirac to the Dirac equation [Dirac (1928)] was as follows: If we wish to prevent the occurrence of negative probability densities, we must then avoid time derivatives in the expression for ρ. The wave equation must therefore not contain time derivatives higher than first order. Relativistic covariance, furthermore, requires that there be essentially complete symmetry in the treatment of the spatial and time components. We must therefore also require that only first-order spatial derivatives appear in the wave equation. Thus the Dirac wave function must satisfy a first-order linear differential equation in all four co-ordinates. The linearity is required in order that the superposition principle of quantum mechanics hold. Finally we must also require that ψ obey the equation

$$\left(\Box + \frac{m^2c^2}{\hbar^2}\right)\psi(x) = 0 \tag{1}$$

if it is to describe a free particle of mass m, since this equation implies that the energy momentum relation for a free particle $p^2 = m^2c^2$ is satisfied, and that in the correspondence limit classical relativity is valid.

A similar situation obtains in electrodynamics, where Maxwell's equations are of first order and connect the components of the field quantities. The electrodynamic wave equation is of second order, with no mass term appearing, and implies that photons have zero mass. The wave equation is furthermore satisfied by every component of the electric and magnetic field intensity.

Let us therefore assume that ψ consists of N components ψ_l, $l = 1, \cdots N$, where the number N is as yet unspecified; it will turn out to be four. The most general first-order linear equation is then one which expresses the time derivative of one component as a linear combination of all the components as well as their spatial derivatives. Inserting the appropriate dimensional factors, the most general equation possible is

$$\frac{1}{c}\frac{\partial \psi_l}{\partial t} + \sum_{k=1}^{3}\sum_{n=1}^{N} \alpha^k{}_{ln}\frac{\partial \psi_n}{\partial x^k} + \frac{imc}{\hbar}\sum_{n=1}^{N} \beta_{ln}\psi_n = 0 \quad l = 1, 2, \cdots N \tag{2}$$

Assuming the homogeneity of space-time, the $\alpha^k{}_{ln}$ and β_{ln} are dimensionless constants, independent of the space-time co-ordinates x^0, x^1, x^2, x^3. A natural way to simplify these equations is to use matrix notation which reduces them to the following equation:

$$\frac{1}{c}\frac{\partial \psi}{\partial t} + \sum_{k=1}^{3} \alpha^k \frac{\partial \psi}{\partial x^k} + \frac{imc}{\hbar}\beta\psi = 0 \tag{3}$$

In this equation ψ is a column matrix of N rows, and α^1, α^2, α^3, and β are matrices of N rows and columns. Equation (3) is known as the Dirac equation.

We next seek the expressions for the density and current which go with Eq. (3). Since we wish to retain the conventional definition for the density ρ, we set

$$\rho = \sum_{n=1}^{N} \bar{\psi}_n\psi_n = \sum_{n=1}^{N} |\psi_n|^2 \tag{4a}$$

or in matrix notation

$$\rho = \psi^*\psi \tag{4b}$$

where ψ^* denotes the Hermitian adjoint of ψ, and hence is a row matrix, consisting of one row and N columns. The expression (4) for the density is clearly positive-definite, thus satisfying the main requirements of Dirac. We further require the density ρ to satisfy a continuity equation

$$\partial_t\rho + \nabla \cdot \mathbf{j} = 0 \tag{5}$$

(where $\mathbf{j}$ is yet to be determined) so that the usual probability interpretation, it is hoped, will be applicable. The quantity ψ^* satisfies the equation

$$\frac{1}{c}\frac{\partial \psi^*}{\partial t} + \sum_{k=1}^{3} \frac{\partial \psi^*}{\partial x^k} (\alpha^k)^* - \frac{imc}{\hbar} \psi^* \beta^* = 0 \tag{6}$$

obtained by taking the Hermitian adjoint of Eq. (3). As above, the superscript * denotes the Hermitian adjoint which for the matrices α and β means transposed conjugate, e.g.,

$$(\beta^*)_{ln} = \overline{\beta_{nl}} \tag{7}$$

The interchange in (6) of ψ and β is necessary, since ψ^* is a row matrix so that α^* and β^* must follow it (instead of preceding it).

Now a continuity equation, similar in structure to Eq. (5), can be derived from Eqs. (3) and (6) by multiplying the former by ψ^* on the left, multiplying the latter by ψ on the right, and adding the two. This results in the equation

$$0 = \frac{1}{c}\frac{\partial}{\partial t}(\psi^*\psi) + \sum_{k=1}^{3} \left(\frac{\partial \psi^*}{\partial x^k} (\alpha^k)^* \psi + \psi^* \alpha^k \frac{\partial \psi}{\partial x^k}\right) + \frac{imc}{\hbar}(\psi^*\beta\psi - \psi^*\beta^*\psi) \tag{8}$$

If we wish to identify (5) and (8) we must make the last terms of (8) vanish, since they contain no derivatives. This can be done if we require

$$\beta^* = \beta \tag{9}$$

i.e., that β be a Hermitian matrix. To identify the second set of terms of (8) with a divergence, we further require that

$$\alpha^{k*} = \alpha^k \tag{10}$$

In other words, α and β must both be Hermitian matrices. Another way of arriving at this result is to rewrite Eq. (3) in Hamiltonian form

$$\begin{aligned} i\hbar\partial_t\psi &= H\psi \\ &= (-ic\boldsymbol{\alpha} \cdot \boldsymbol{\nabla} + \beta mc^2)\psi \end{aligned} \tag{11}$$

It is then clear that the αs and β must be Hermitian if H is to be Hermitian. The comparison of Eqs. (5) and (8) then shows that

$$j^k = c\psi^*\alpha^k\psi \tag{12}$$

In order to derive further properties of the α and β matrices, we must next see what conditions are imposed by the requirement that Eq. (1) be satisfied. For this purpose we multiply Eq. (3) by the operator,

$$\frac{1}{c}\frac{\partial}{\partial t} - \sum_{k=1}^{3} \alpha^k \frac{\partial}{\partial x^k} - \frac{imc}{\hbar}\beta$$

which has the effect of introducing second derivatives. The terms with ∂_t or mixed derivatives between space and time cancel and we obtain

$$\frac{1}{c^2}\frac{\partial^2\psi}{\partial t^2} = \sum_{k=1}^{3}\sum_{l=1}^{3}\frac{1}{2}(\alpha^k\alpha^l + \alpha^l\alpha^k)\frac{\partial^2\psi}{\partial x^k\partial x^l} - \frac{m^2c^2}{\hbar^2}\beta^2\psi + \frac{imc}{\hbar}\sum_{k=1}^{3}(\alpha^k\beta + \beta\alpha^k)\frac{\partial\psi}{\partial x^k} \quad (13)$$

We have symmetrized the $\alpha^k\alpha^l$ term, which is permissible since $\partial/\partial x^k$ and $\partial/\partial x^l$ commute. To agree with the Klein-Gordon equation, the right-hand side of (13) must reduce to

$$\nabla^2\psi - \frac{m^2c^2}{\hbar^2}\psi$$

This imposes the following conditions:

$$\tfrac{1}{2}(\alpha^k\alpha^l + \alpha^l\alpha^k) = \delta^{kl} \quad (14)$$

$$\alpha^k\beta + \beta\alpha^k = 0 \quad (15)$$

$$(\alpha^k)^2 = \beta^2 = 1 \quad (16)$$

$$k = 1, 2, 3$$

i.e., that the αs as well as any α^k and β anticommute, and that the square of all four matrices is unity. For practical applications, it is not necessary to represent the αs and β explicitly; it is sufficient to know that they are Hermitian and that their properties are described by (14) through (16). In fact, it is usually best not to express the matrices explicitly when working problems. An explicit representation can, however, be easily obtained. We first note that the dimension N must be even. *Proof:* Rewrite Eq. (15) as follows:

$$\beta\alpha^k = -\alpha^k\beta = (-I)\,\alpha^k\beta \quad (17)$$

where I is the unit matrix. Then take the determinant of both sides of Eq. (17) to obtain, since $\det(-I) = (-1)^N$,

$$(\det\beta)(\det\alpha^k) = (-1)^N \det\alpha^k \cdot \det\beta \quad (18)$$

Hence since $\det\alpha$, $\det\beta \neq 0$ [Eq. (16)], $(-1)^N = 1$ and N must be even. We next give a slightly more complicated proof which, in addition, exhibits an important property of the α and β matrices, namely that their trace vanishes. Since the α and β matrices are Hermitian, they can be diagonalized. Note, however, that not all the αs and β can simultaneously be diagonalized since they anticommute with one another. Let us choose a representation in which β is diagonal, so that

$$\beta = \begin{pmatrix} b_1 & & & & \\ & \cdot & & & \\ & & \cdot & & \\ & & & \cdot & \\ & & & & b_N \end{pmatrix} \quad (19)$$

Since $\beta^2 = I$, $b_i^2 = 1$ and $b_i = \pm 1$ $(i = 1, 2, \cdots N)$. Furthermore, since $\beta^2 = (\alpha^k)^2 = 1$, each of these matrices has an inverse, so that Eq. (17) may be rewritten as

$$(\alpha^k)^{-1}\beta\alpha^k = -\beta \tag{20}$$

Taking the trace of this last equation and using the property of the trace that Tr (AB) = Tr (BA) we obtain

$$\text{Tr}\,((\alpha^k)^{-1}\,\beta\alpha^k) = \text{Tr}\,(\beta\alpha^k(\alpha^k)^{-1}) = \text{Tr}\,\beta = -\text{Tr}\,\beta \tag{21}$$

hence

$$\text{Tr}\,\beta = 0 \tag{22}$$

Similarly

$$\text{Tr}\,\alpha^k = 0 \tag{23}$$

If now in Eq. (19) there are m of the b_i equal to $+1$ and n of the b_i equal to -1, then, since β is N dimensional: $m + n = N$. On the other hand, the requirement that Tr $\beta = 0$ implies that $m - n = 0$, that is $m = n$. Therefore, $N = 2m$ and the αs and β must be even-dimensional matrices. We will show in the next section that the number of dimensions is necessarily a multiple of four. If I denotes the unit 2×2 matrix, and σ^k are the Pauli matrices [see Chap. 1, Eq. (100)] then the 4×4 matrices

$$\alpha^k = \begin{pmatrix} 0 & \sigma^k \\ \sigma^k & 0 \end{pmatrix} \qquad \beta = \begin{pmatrix} \text{I} & 0 \\ 0 & -\text{I} \end{pmatrix} \tag{24}$$

satisfy all our conditions: they are Hermitian and can be seen to anticommute by using the anticommutative properties of the σs. This particular representation is convenient for the discussion of the nonrelativistic limit of the Dirac equation (Sec. 4*d*).

Let us finally put the Dirac equation in covariant form. When the Dirac equation is written as in (3), the spatial derivatives are multiplied by a matrix whereas the time derivative is not. To eliminate this distinction, let us multiply Eq. (3) by β on the left to obtain

$$-i\hbar\beta\partial_0\psi - i\hbar\sum_{k=1}^{3}\beta\alpha^k\partial_k\psi + mc\psi = 0 \tag{25}$$

We can make this equation look even more symmetrical by introducing the matrices γ^μ, with

$$\gamma^0 = \beta \tag{26}$$

$$\gamma^k = \beta\alpha^k \quad (k = 1, 2, 3) \tag{27}$$

Note that with these definitions γ^0 is Hermitian, with $(\gamma^0)^2 = +1$, and the γ^ks are anti-Hermitian, i.e., $(\gamma^k)^* = -\gamma^k$, with $(\gamma^k)^2 = -1$ so that the γ matrices satisfy the following commutation rules:

$$\gamma^\mu\gamma^\nu + \gamma^\nu\gamma^\mu = 2g^{\mu\nu}I \tag{28}$$

In terms of the γ matrices, Eq. (25) now reads

$$\left(-i\gamma^\mu\partial_\mu + \frac{mc}{\hbar}\right)\psi = \left(-i\gamma\cdot\partial + \frac{mc}{\hbar}\right)\psi = 0 \qquad (29)$$

where our summation convention has been reintroduced. With (29) we have written the Dirac equation in a covariant form where space and time derivatives are treated alike. Feynman (1949a) has introduced the so-called "dagger" notation[1] to simplify the equation still further. He denotes by $\not{p}$ the quantity

$$\not{p} = \gamma\cdot p = \gamma^\mu p_\mu = \gamma_\mu p^\mu = \gamma^0 p^0 - \boldsymbol{\gamma}\cdot\mathbf{p} \qquad (30)$$

where γ_μ is defined by

$$\gamma_\mu = g_{\mu\nu}\gamma^\nu \qquad (31)$$

With this notation and using natural units, the Dirac equation then reads

$$(-i\not{\partial} + m)\psi = 0 \qquad (32)$$

where

$$\not{\partial} = \gamma^\mu\partial_\mu = \gamma^0\partial^0 + \boldsymbol{\gamma}\cdot\boldsymbol{\nabla} \qquad (33)$$

The current and density can be expressed in terms of the γ matrices as follows: If we multiply (27) by β on the left, we find that $\beta\gamma^k = \alpha^k$ so that (12) becomes

$$j^k = c\psi^*\beta\gamma^k\psi \qquad (34)$$

In terms of the "adjoint" wave function $\tilde{\psi}$ defined by

$$\tilde{\psi} = \psi^*\beta = \psi^*\gamma^0 \qquad (35)$$

the expression for the current becomes

$$j^k = c\tilde{\psi}\gamma^k\psi \qquad (36)$$

The expression for the density may be rewritten analogously in terms of the γ matrices

$$j^0 = c\rho = c\psi^*\gamma^0\gamma^0\psi = c\tilde{\psi}\gamma^0\psi \qquad (37)$$

The equation satisfied by the adjoint $\tilde{\psi} = \psi^*\gamma^0$ is obtained from Eq. (6) by inserting a factor $\gamma^0\gamma^0 = 1$ to the right of ψ^* in each term, and using Eqs. (9), (10), and (27). In natural units, it is given by

$$i\partial_\mu\tilde{\psi}\gamma^\mu + m\tilde{\psi} = 0 \qquad (38)$$

4b. Properties of the Dirac Matrices

The γ matrices form a set of hypercomplex numbers which satisfy the commutation rules $\gamma^\mu\gamma^\nu + \gamma^\nu\gamma^\mu = 2g^{\mu\nu}$. To study their properties [Pauli

[1] In the literature the dagger symbol is often indicated by printing the letter in boldface italics.

(1935, 1936), Good (1955)], it is not necessary to assume any hermiticity property for them and in this section we shall, in fact, not assume any such property.

Consider the sixteen elements

$$\begin{array}{c} I \\ i\gamma^1 \quad i\gamma^2 \quad i\gamma^3 \quad \gamma^0 \\ i\gamma^2\gamma^3 \quad i\gamma^3\gamma^1 \quad i\gamma^1\gamma^2 \quad \gamma^0\gamma^1 \quad \gamma^0\gamma^2 \quad \gamma^0\gamma^3 \\ i\gamma^0\gamma^2\gamma^3 \quad i\gamma^0\gamma^1\gamma^3 \quad i\gamma^0\gamma^1\gamma^2 \quad \gamma^1\gamma^2\gamma^3 \\ i\gamma^0\gamma^1\gamma^2\gamma^3 \end{array}$$

All other products of γ matrices can, by using the commutation rules, be reduced to one of these sixteen elements. The factor i has been so inserted that the square of each element is $+1$. We shall denote the elements of the above array by Γ_l, $l = 1, 2, \cdots 16$. We note that the product of any two elements is always a third, apart from a factor ± 1 or $\pm i$. For each Γ_l, except for $\Gamma_1 = I$ we can always find a Γ_j such that $\Gamma_j\Gamma_l\Gamma_j = -\Gamma_l$. The proof consists in exhibiting the element Γ_j for each Γ_l. Thus for $l = 2, \cdots 5$, i.e., for the elements of the second line of the above array $\Gamma_j = i\gamma^0\gamma^1\gamma^2\gamma^3$; for the third line, one of the second line Γs, e.g., for the element $i\gamma^2\gamma^3$, $\Gamma_j = i\gamma^2$ since $(i\gamma^2)\,(i\gamma^2\gamma^3)\,(i\gamma^2) = -(i\gamma^2\gamma^3)$; for the fourth line $i\gamma^0\gamma^1\gamma^2\gamma^3$; and for the fifth γ^0, for example. It follows that for $i \neq 1$, the trace of any Γ_i is zero since $-\text{Tr}\,\Gamma_i = \text{Tr}\,\Gamma_j\Gamma_i\Gamma_j = \text{Tr}\,\Gamma_i\Gamma_j{}^2 = \text{Tr}\,\Gamma_i$. The sixteen elements Γ_i, $i = 1, 2, \cdots 16$ are therefore linearly independent, in other words the equation $\sum_{i=1}^{16} a_i\Gamma_i = 0$ holds only if $a_i = 0$, $i = 1, 2, \cdots 16$. *Proof:* Take the trace of $\sum_{i=1}^{16} a_i\Gamma_i = 0$ to obtain $a_1 = 0$. Similarly, by multiplying the equation by each Γ_i successively, and taking the trace we obtain $a_i = 0$. Q.E.D. From this, we conclude that it is not possible to represent the γs by matrices whose dimension is less than 4×4, since it is impossible to construct 16 linearly independent matrices from such matrices. Conversely, since there exist precisely 16 linearly independent 4×4 matrices (since the number of elements of a 4×4 matrix is 16), we can represent the γ by 4×4 matrices. This representation (apart from equivalences) is in fact irreducible.[2] Any other representation can always be brought into the form

[2] A set of hypercomplex numbers satisfying $\gamma^\mu\gamma^\nu + \gamma^\nu\gamma^\mu = 2g^{\mu\nu}$ is called a Clifford algebra. A Clifford algebra exists for any space endowed with a metric $g^{\mu\nu}$. If the dimension of the space is n, the dimension of the algebra (i.e., the number of Γ_is) is 2^n. When n is even, $n = 2m$, only the identity commutes with all the elements of the algebra, so that apart from equivalences there exists one and only one irreducible representation of the algebra and that by $r \times r$ matrices with $r = 2^m$. Furthermore, the representation can always be chosen real. For the Dirac case we have $n = 4$, $m = 2$ and $r = 2^2 = 4$.

$$\Gamma^i = \begin{pmatrix} \gamma^i & & & 0 \\ & \cdot & & \\ & & \cdot & \\ 0 & & & \gamma^i \end{pmatrix}$$

with the γ^i's 4×4 matrices. In the following we shall always assume the γs to be 4×4 (irreducible) matrices.

From the linear independence of the Γ_is it follows that every 4×4 matrix, X, can be written in the form

$$X = \sum_{i=1}^{16} x_i \Gamma_i \tag{39}$$

with

$$x_i = \tfrac{1}{4}\mathrm{Tr}\,(X\Gamma_i) \tag{40}$$

Since the γ matrices are irreducible, Schur's lemma asserts that any 4×4 matrix which commutes with each of the γ^μ is a constant multiple of the identity. An algebraic proof of this assertion is as follows: Let X be the matrix which commutes with each of the γ^μs and therefore also with all the Γ_i. If we represent X in the form

$$X = x_j \Gamma_j + \sum_{l \neq j} x_l \Gamma_l \quad (\Gamma_j \neq I) \tag{41}$$

let Γ_i be an element for which $\Gamma_i \Gamma_j \Gamma_i = -\Gamma_j$. By assumption $\Gamma_i X \Gamma_i = X$, so that upon multiplying (41) on the right and left by Γ_i we obtain

$$\begin{aligned} X &= -x_j \Gamma_j + \sum_{l \neq j} x_l \Gamma_i \Gamma_l \Gamma_i \\ &= -x_j \Gamma_j + \sum_{l \neq j} x_l (\pm 1)\, \Gamma_l \end{aligned} \tag{42}$$

where the factor ± 1 is a consequence of the fact that Γ_i and Γ_l either commute or anticommute with one another. Multiplying Eqs. (41) and (42) by Γ_j and taking the trace, it follows that $x_j = -x_j = 0$. Since Γ_j is arbitrary, except for not being the identity, the only nonvanishing coefficient in the expansion (41) is x_1, which was to be proved.

The fundamental theorem concerning the γ matrices, which will be used repeatedly in the subsequent work is: Given two sets of 4×4 matrices γ^μ and γ'^μ which satisfy the commutation rules

$$\gamma^\mu \gamma^\nu + \gamma^\nu \gamma^\mu = 2g^{\mu\nu} \tag{43a}$$

$$\gamma'^\mu \gamma'^\nu + \gamma'^\nu \gamma'^\mu = 2g^{\mu\nu} \tag{43b}$$

there then exists a nonsingular matrix S, such that

$$\gamma'^\mu = S \gamma^\mu S^{-1} \tag{44}$$

Explicitly S is given by

$$S = \sum_{i=1}^{16} \Gamma_i' F \Gamma_i \tag{45}$$

where F is an arbitrary 4×4 matrix which can be so chosen that S is nonsingular. The Γ_i' are the set of 16 linearly independent matrices constructed in the same manner from the γ'^μ as the Γ_i were constructed from the γ^μ. To prove the theorem, we note that if $\Gamma_i\Gamma_j = \epsilon_{ij}\Gamma_k$, where $\epsilon_{ij} = \pm 1, \pm i$, then $\Gamma_i\Gamma_j\Gamma_i\Gamma_j = \epsilon_{ij}^2\Gamma_k^2 = \epsilon_{ij}^2$ so that $\Gamma_j\Gamma_i = \epsilon_{ij}^2\Gamma_i\Gamma_j = \epsilon_{ij}^3\Gamma_k$. Note that the same number ϵ_{ij} arise in the primed system, $\Gamma_i'\Gamma_j' = \epsilon_{ij}\Gamma_k'$, since their value depends only on the commutation rules, which are the same for both sets of matrices. Now since ϵ_{ij} is either ± 1 or $\pm i$, $\epsilon_{ij}^4 = +1$; if S is given by Eq. (45)

$$\Gamma_j'S\Gamma_j = \sum_i \Gamma_j'\Gamma_i'F\Gamma_i\Gamma_j = \sum_i \epsilon_{ij}^4\Gamma_k'F\Gamma_k$$

$$= \sum_k \Gamma_k'F\Gamma_k \tag{46}$$

since the sum over i of $\Gamma_i\Gamma_j$ for fixed j just ranges through the 16 elements of the algebra, so that the sum over i can be replaced by a sum over k, whence

$$\Gamma_j'S\Gamma_j = S \tag{47}$$

Since the γs are irreducible, by Schur's lemma S is nonsingular. Furthermore S is uniquely determined up to a factor. For suppose there were two such Ss, say S_1 and S_2 such that $\gamma'^\mu = S_1\gamma^\mu S_1^{-1}$ and $\gamma'^\mu = S_2\gamma^\mu S_2^{-1}$, then by equating γ'^μ in these equations we find that $\gamma^\mu = S_2^{-1}S_1\gamma^\mu(S_2^{-1}S_1)^{-1}$ so that $S_2^{-1}S_1$ commutes with all the γ^μ and therefore is a constant multiple of the identity, so that $S_2 = cS_1$. It is often convenient to impose the normalization condition $\det S = 1$, which then determines S except for an arbitrary factor of $\sqrt[4]{1}$, i.e., ± 1 or $\pm i$.

An interesting special case of Eq. (45) obtains when $\gamma'^\mu = \gamma^\mu$, so that $S\gamma^\mu = \gamma^\mu S$ and S is a constant multiple of the identity: $S = cI$. The $\rho\sigma$ matrix element of Eq. (45) then reads

$$c\delta_{\rho\sigma} = \sum_{i=1}^{16} \sum_{\rho',\sigma'=1}^{4} (\Gamma_i)_{\rho\rho'} F_{\rho'\sigma'}(\Gamma_i)_{\sigma'\sigma} \tag{48}$$

Since this equation is true for arbitrary F, we have

$$\sum_{i=1}^{16} (\Gamma_i)_{\rho\rho'} (\Gamma_i)_{\sigma'\sigma} = b_{\rho'\sigma'}\delta_{\rho\sigma} \tag{49}$$

where $b_{\rho'\sigma'}$ is some constant. To determine this constant, let $\rho = \sigma$ and sum over ρ:

$$\sum_{i=1}^{16} \sum_{\rho=1}^{4} (\Gamma_i)_{\sigma'\rho}(\Gamma_i)_{\rho\rho'} = \sum_{i=1}^{16} (\Gamma_i^2)_{\sigma'\rho'} = 16\delta_{\sigma'\rho'} = 4b_{\rho'\sigma'} \tag{50}$$

whence $b_{\rho'\sigma'} = 4\delta_{\sigma'\rho'}$ and

$$\sum_{i=1}^{16} (\Gamma_i)_{\sigma\rho} (\Gamma_i)_{\rho'\sigma'} = 4\delta_{\sigma\sigma'}\delta_{\rho\rho'} \tag{51}$$

We conclude this section on the general properties of the γ matrices by noting that many of their properties could have been derived using the fact that Γ_is form a group of 32 elements. Such a finite group can always be represented by unitary matrices. For such a unitary representation the Γ matrices are in addition Hermitian since $\Gamma_i^2 = \Gamma_i^*\Gamma_i = I$, hence $\Gamma_i = \Gamma_i^*$.

4c. Relativistic Invariance

We are now in a position to investigate the Lorentz invariance of the Dirac equation and to establish its connection with the representation of the inhomogeneous Lorentz group discussed in Chapter 2.

The physics expressed by any relativistic equation, and by the Dirac equation in particular, must be independent of the Lorentz frame that is used. Hence to be a true description of the physics, the equation itself must display this same invariance with respect to the choice of co-ordinates. We will now show that under an inhomogeneous Lorentz transformation $x' = \Lambda x + a$, $\Lambda^T g \Lambda = g$, the Dirac equation will be form invariant if we define

$$\begin{aligned} \psi'(x') &= S(\Lambda)\,\psi(x) \\ &= S(\Lambda)\,\psi(\Lambda^{-1}(x' - a)) \end{aligned} \tag{52}$$

and if $S(\Lambda)$ satisfies certain conditions [Eq. (59) below]; $S(\Lambda)$ is a 4×4 matrix which operates on the components of ψ, i.e., Eq. (52) written out explicitly reads

$$\psi'_\alpha(x) = \sum_{\beta=1}^{4} S_{\alpha\beta}(\Lambda)\,\psi_\beta(\Lambda^{-1}(x - a)) \tag{53}$$

The form invariance of the Dirac equation is the statement that in the new frame ψ' obeys the equation

$$(-i\gamma^\mu \partial'_\mu + m)\,\psi'(x') = 0 \tag{54}$$

where $\partial'_\mu = \partial/\partial x'^\mu$. Note that the γ matrices remain unaltered under Lorentz transformation. Now the Dirac equation

$$(-i\gamma^\mu \partial_\mu + m)\,\psi(x) = 0 \tag{55}$$

can, by means of Eq. (52) and the fact that

$$\frac{\partial}{\partial x^\mu} = \frac{\partial x'^\nu}{\partial x^\mu}\frac{\partial}{\partial x'^\nu} = \Lambda^\nu{}_\mu \partial'_\nu \tag{56}$$

be re-expressed in the form

$$-i\Lambda^\nu{}_\mu \gamma^\mu \frac{\partial S^{-1}\psi'(x')}{\partial x'^\nu} + mS^{-1}\psi'(x') = 0 \tag{57}$$

Hence, multiplying Eq. (57) by S on the left yields

$$-iS(\Lambda^{\nu}{}_{\mu}\gamma^{\mu})\, S^{-1}\partial'_{\nu}\psi'(x') + m\psi'(x') = 0 \tag{58}$$

so that the Dirac equation will be form invariant under Lorentz transformations provided that

$$S(\Lambda)^{-1}\,\gamma^{\lambda}S(\Lambda) = \Lambda^{\lambda}{}_{\mu}\gamma^{\mu} \tag{59}$$

This is the aforementioned condition on $S(\Lambda)$. One now easily verifies using the fact that $\Lambda^{T}g\Lambda = g$ that $\gamma'^{\lambda} = \Lambda^{\lambda}{}_{\mu}\gamma^{\mu}$ obeys the same commutation rules as γ^{μ}, viz., $\gamma'^{\mu}\gamma'^{\nu} + \gamma'^{\nu}\gamma'^{\mu} = 2g^{\mu\nu}$, so that there exists a nonsingular S such that Eq. (59) is satisfied; in fact, we have shown that S is uniquely determined up to a factor by Eq. (59).

A further condition on $S(\Lambda)$ can be determined as follows: Since the matrix elements $\Lambda^{\mu}{}_{\nu}$ are real, taking the Hermitian adjoint of Eq. (59) we obtain

$$\begin{aligned}(\Lambda^{\mu}{}_{\nu}\gamma^{\nu})^{*} &= (\Lambda^{\mu 0}\gamma^{0} - \sum_{k=1}^{3}\Lambda^{\mu k}\gamma^{k})^{*} \\ &= \Lambda^{\mu 0}\gamma^{0} + \sum_{k=1}^{3}\Lambda^{\mu k}\gamma^{k} \\ &= (S^{-1}\gamma^{\mu}S)^{*} \end{aligned} \tag{60}$$

since γ^{0} is Hermitian and γ^{k} anti-Hermitian. These hermiticity properties can be expressed concisely by the relation

$$(\gamma^{\mu})^{*} = \gamma^{0}\gamma^{\mu}\gamma^{0} \tag{61}$$

or equivalently $\gamma^{\mu} = \gamma^{0}\gamma^{\mu *}\gamma^{0}$. Hence, multiplying Eq. (60) by γ^{0} on the left and right we obtain:

$$\gamma^{0}\left(\Lambda^{\mu 0}\gamma^{0} + \sum_{k=1}^{3}\Lambda^{\mu k}\gamma^{k}\right)\gamma^{0} = \Lambda^{\mu}{}_{\nu}\gamma^{\nu} = \gamma^{0}(S^{-1}\gamma^{\mu}S)^{*}\,\gamma^{0} \tag{62a}$$

$$= (\gamma^{0}S^{*}\gamma^{0})\,\gamma^{\mu}(\gamma^{0}S^{*}\gamma^{0})^{-1} \tag{62b}$$

since $(\gamma^{0})^{-1} = \gamma^{0}$. Now the middle member of Eq. (62a) is, by virtue of Eq. (59), also equal to $S^{-1}\gamma^{\mu}S$, so that equating (62b) to $S^{-1}\gamma^{\mu}S$ we find that

$$(S\gamma^{0}S^{*}\gamma^{0})\,\gamma^{\mu}(S\gamma^{0}S^{*}\gamma^{0})^{-1} = \gamma^{\mu} \tag{63}$$

so that $S\gamma^{0}S^{*}\gamma^{0}$ commutes with all the γ^{μ}s; it is therefore a constant multiple of the identity:

$$S\gamma^{0}S^{*}\gamma^{0} = bI \tag{64a}$$

or equivalently

$$S\gamma^{0}S^{*} = b\gamma^{0} \tag{64b}$$

$$S^{*}\gamma^{0} = b\gamma^{0}S^{-1} \tag{64c}$$

where b is a constant. Since γ^{0} is Hermitian and the left side of (64b) is likewise Hermitian, it follows that b must be real: $b = \bar{b}$. If we prescribe a normalization for S such that det $S = 1$, then $b^4 = 1$ and hence $b = \pm 1$. We next inquire when is $b = +1$ and when is it -1. To this end we consider the following identity:

$$\begin{aligned} S^*S &= S^*\gamma^0\gamma^0 S = b\gamma^0 S^{-1}\gamma^0 S \\ &= b\gamma^0\Lambda^0{}_\nu\gamma^\nu = b\left(\Lambda^0{}_0 I - \sum_{k=1}^{3}\Lambda^{0k}\gamma^0\gamma^k\right) \\ &= b\left(\Lambda^{00}I - \sum_{k=1}^{3}\Lambda^{0k}\alpha^k\right) \end{aligned} \tag{65}$$

in which Eqs. (64c) and (59) have been used. Now S^*S is a product of a nonsingular matrix times its Hermitian adjoint; hence S^*S is positive-definite since $S \neq 0$. Its eigenvalues are real (since S^*S is Hermitian) and likewise positive-definite, so that Tr $(S^*S) > 0$. Since Tr $\alpha^k = 0$, we obtain

$$\text{Tr } S^*S = 4b\Lambda^{00} > 0 \tag{66}$$

Thus if $\Lambda^{00} \leqslant -1$, $b = -1$ and if $\Lambda^{00} \geqslant 1$, $b = +1$.

Consider next the transformation properties of the adjoint spinor, $\tilde{\psi} = \psi^*\gamma^0$. Since $\psi' = S\psi$, $(\psi')^* = \psi^*S^*$, therefore $\tilde{\psi}'$ which is defined by

$$\tilde{\psi}' = (\psi')^*\gamma^0 \tag{67}$$

(the γs are unchanged in the transformation) is, by virtue of Eq. (64), equal to

$$\begin{aligned} \tilde{\psi}' &= \psi^*S^*\gamma^0 = b\psi^*\gamma^0 S^{-1} \\ &= b\tilde{\psi}S^{-1} \end{aligned} \tag{68}$$

Hence $\tilde{\psi}' = \tilde{\psi}S^{-1}$ for those transformations which do not reverse the sign of the time ($\Lambda^0{}_0 \geqslant 1$), whereas $\tilde{\psi}' = -\tilde{\psi}S^{-1}$ for those transformations which do include a time reversal ($\Lambda^0{}_0 \leqslant -1$).

The meaning of Eq. (68) is somewhat clarified if we consider the transformation properties of the current

$$j^\mu = \tilde{\psi}\gamma^\mu\psi \tag{69}$$

In order that the physical consequences of the Dirac equation be the same in every Lorentz frame, we must require that j^μ transform like a four vector, i.e., transform as

$$\begin{aligned} j'^\mu &= \Lambda^\mu{}_\nu j^\nu = \Lambda^\mu{}_\nu\tilde{\psi}\gamma^\nu\psi \\ &= \psi^*\gamma^0 S^{-1}\gamma^\mu S\psi \end{aligned} \tag{70}$$

Now we should obtain the same value for j'^μ by calculating it using the wave functions in the primed system, but with the same γ matrices, so that

$$j'^\mu = \psi'^*\gamma^0\gamma^\mu\psi' \tag{71}$$

Since $\psi'^* = \psi^*S^*$ we may rewrite Eq. (71) as follows:

$$j'^\mu = \psi^*S^*\gamma^0\gamma^\mu S\psi \tag{72}$$

Comparing Eqs. (70) and (72) we see that the relation $S^*\gamma^0 = \gamma^0 S^{-1}$ must be satisfied. This is identical to the relation (64) for a Lorentz transformation without time reversal. However, Eqs. (64) and (66) further indicate that j^μ is really a pseudovector under time reversal, since it changes sign under such an inversion.

Let us next exhibit the actual form of S for a given Lorentz transformation. We consider first the case of proper orthochronous homogeneous Lorentz transformations for which $\det \Lambda = +1$ and $\Lambda^0{}_0 \geqslant 1$. It is sufficient to investigate the infinitesimal transformations, since a finite transformation can be obtained by exponentiation. For the infinitesimal Lorentz transformation, $\Lambda = I + \epsilon\lambda$,

$$x'^\mu = x^\mu + \epsilon\lambda^\mu{}_\nu x^\nu \tag{73a}$$

$$\lambda^{\mu\nu} = -\lambda^{\nu\mu} \tag{73b}$$

we write $S(\Lambda)$ to first order in ϵ as follows:

$$S(I + \epsilon\lambda) = I + \epsilon T \tag{74}$$

and

$$[S(I + \epsilon\lambda)]^{-1} = S^{-1}(I + \epsilon\lambda) = I - \epsilon T \tag{75}$$

For the infinitesimal situation, Eq. (59) can be rewritten as

$$\begin{aligned} S^{-1}\gamma^\mu S &= (I - \epsilon T)\,\gamma^\mu(I + \epsilon T) = \gamma^\mu + \epsilon(\gamma^\mu T - T\gamma^\mu) \\ &= \Lambda^\mu{}_\nu\gamma^\nu = \gamma^\mu + \epsilon\lambda^\mu{}_\nu\gamma^\nu \end{aligned} \tag{76}$$

whence T must be such that

$$\gamma^\mu T - T\gamma^\mu = \lambda^\mu{}_\nu\gamma^\nu \tag{77}$$

T is uniquely defined up to the addition of constant multiple of the identity as a consequence of Eq. (77). For if there were two such Ts, from (77) their difference commutes with all the γ^μs, hence their difference must be a constant multiple of the identity. The normalization $\det S = 1$ requires that $\det(I + \epsilon T) = 1 + \epsilon \operatorname{Tr} T = 1$, i.e., $\operatorname{Tr} T = 0$. By the addition to T of a suitable multiple of the identity, we can always make $\operatorname{Tr} T = 0$, so that this choice determines T uniquely and makes $\det S = 1$. One readily establishes that

$$T = \tfrac{1}{8}\lambda^{\mu\nu}(\gamma_\mu\gamma_\nu - \gamma_\nu\gamma_\mu) \tag{78}$$

satisfies Eq. (77) and the requirement that its trace vanish. This is therefore the required T.

For an infinitesimal rotation through an angle ϵ about the x^1 axis, $\lambda^{23} = -\lambda^{32} = +1$ with all other $\lambda^{\mu\nu} = 0$. The generator for such a transformation by Eq. (78), is

$$T(R_1) = \tfrac{1}{2}\gamma_2\gamma_3 \tag{79}$$

With the representation (24) for the γ matrices

$$T(R_1) = -\frac{1}{2}\alpha_2\alpha_3 = +\frac{i}{2}\begin{pmatrix}\sigma_1 & 0\\ 0 & \sigma_1\end{pmatrix} = +\frac{i}{2}\Sigma_1 \tag{80}$$

The $S(\theta)$ corresponding to a rotation through the angle θ about the x^1 axis is therefore given by

$$\begin{aligned} S(R_1) &= e^{\theta T(R_1)} = e^{+i\frac{\theta}{2}\Sigma_1} \\ &= \cos\frac{\theta}{2} + i\Sigma_1 \sin\frac{\theta}{2} \end{aligned} \tag{81}$$

There are two Ss which differ in sign for every θ, since $S(\theta + 2\pi) = -S(\theta)$. Note however that $S^*S = +1$ for a rotation.

For an infinitesimal Lorentz transformation in the x^1 direction, $\lambda^{10} = = -\lambda^{01} = -1$ and

$$T(L_1) = \tfrac{1}{2}\gamma^0\gamma^1 = \tfrac{1}{2}\alpha^1 \tag{82}$$

The generator for a finite Lorentz transformation along the x^1 axis, with $\tanh \omega = v/c$, is

$$\begin{aligned} S(L_1) &= e^{\frac{\omega}{2}\alpha_1} \\ &= \cosh\frac{\omega}{2} + \alpha_1 \sinh\frac{\omega}{2} \end{aligned} \tag{83}$$

There is no ambiguity in the sign of S in this case due to half-angles. For a space inversion $x'^k = -x^k$, $x'^0 = x^0$, $S(i_s)$ must satisfy the equations

$$S^{-1}(i_s)\, \gamma^0 S(i_s) = \gamma^0 \tag{84a}$$

$$S^{-1}(i_s)\, \gamma^k S(i_s) = -\gamma^k \tag{84b}$$

A possible choice for $S(i_s)$ is therefore either $\pm\gamma^0$ or $\pm i\gamma^0$. For a time inversion ($x'^k = x^k$, $x'^0 = -x^0$), $S(i_t)$ must satisfy

$$S^{-1}(i_t)\, \gamma^0 S(i_t) = -\gamma^0 \tag{85a}$$

$$S^{-1}(i_t)\, \gamma^k S(i_t) = \gamma^k \tag{85b}$$

so that $\pm\gamma_5\gamma^0$ or $\pm i\gamma_5\gamma^0$, where $\gamma_5 = \gamma^0\gamma^1\gamma^2\gamma^3$, are possible choices for $S(i_t)$. We defer answering the question as to which of these $S(i_t)$ and $S(i_s)$ may be selected until after having discussed the notion of charge conjugation.

Let us next briefly consider the relation of a Dirac wave function $\psi(x)$ to quantities transforming irreducibly under the proper homogeneous Lorentz group. In our discussion in Chapter 2, we called a spinor a two-component object which transforms according to the irreducible representation $\{\frac{1}{2}, 0\}$ of the homogeneous Lorentz group, and a conjugate spinor one transforming according to the $\{0, \frac{1}{2}\}$ representation. More specifically, the infinitesimal generator for a rotation about the ith axis, M_i, was $\frac{1}{2}\sigma_i$, and the generator N_l for a Lorentz transformation about the lth axis was $\frac{1}{2}i\sigma_l$ for the $\{\frac{1}{2}, 0\}$ representation and $-\frac{1}{2}i\sigma_l$ for the $\{0, \frac{1}{2}\}$ representation. If we only consider proper homogeneous Lorentz transformations, without space inversions, two-component spinors would then be sufficient to describe particles of spin $\frac{1}{2}$. However, under a space inversion $\mathbf{M} \to \mathbf{M}$, $\mathbf{N} \to -\mathbf{N}$ so that a two-component spinor and its transform under a space inversion do not transform the same way under Lorentz transformations. In order to have a theory which is invariant under spatial inversions, four-component spinors are necessary. These four-component spinors are direct sums of $\{\frac{1}{2}, 0\}$ and $\{0, \frac{1}{2}\}$ spinors, since an inversion carries a spinor belonging to the $\{\frac{1}{2}, 0\}$ representation into one transforming according to $\{0, \frac{1}{2}\}$. A representation of the γ matrices, which makes these remarks explicit, is given by

$$\gamma^0 = \begin{pmatrix} 0 & I \\ I & 0 \end{pmatrix} \qquad \boldsymbol{\gamma} = \begin{pmatrix} 0 & \boldsymbol{\sigma} \\ -\boldsymbol{\sigma} & 0 \end{pmatrix} \tag{86a}$$

$$\gamma_5 = \gamma^0\gamma^1\gamma^2\gamma^3 = i\begin{pmatrix} I & 0 \\ 0 & -I \end{pmatrix} \tag{86b}$$

With this specialization of the γ matrices, the $S(\Lambda)$ for restricted Lorentz transformations can be written in the form

$$S(\Lambda) = \begin{pmatrix} S(\Lambda) & 0 \\ 0 & S'(\Lambda) \end{pmatrix} \tag{87}$$

where S and S' are 2×2 matrices. The four-component Dirac spinor ψ splits into two pairs of two-component spinors (ψ_1, ψ_2) and (ψ_3, ψ_4) which transform separately under Lorentz transformation. Under inversions the pairs (ψ_1, ψ_2) and (ψ_3, ψ_4) are simply interchanged. The systematic treatment of the Dirac equation in terms of such two-component spinors is due to Van der Waerden (1929). For a detailed exposition of this treatment the reader is referred to the articles of Bade and Jehle [Bade (1953)], and Cap (1955).

The results obtained above, concerning the transformation properties of Dirac spinors under Lorentz transformations, allow us to ascribe definite transformation properties to bilinear quantities constructed from such spinors. Thus $\tilde{\psi}(x)\,\psi(x)$ is a scalar covariant since

$$\tilde{\psi}\psi = \tilde{\psi}S^{-1}S\psi = \tilde{\psi}'\psi' \tag{88}$$

It is, however, a pseudoscalar under time inversion. We have already established the vector character of $\tilde{\psi}\gamma_\mu\psi$. If we define the matrix

$$\sigma^{\mu\nu} = \frac{1}{2i}(\gamma^\mu\gamma^\nu - \gamma^\nu\gamma^\mu) = -\sigma^{\nu\mu} \tag{89}$$

then $\tilde{\psi}\sigma^{\mu\nu}\psi$ transforms like an antisymmetric tensor of second rank, since for transformations with $\Lambda^0{}_0 \geqslant 1$

$$\begin{aligned} \tilde{\psi}'\sigma^{\mu\nu}\psi' &= \tilde{\psi}S^{-1}\sigma^{\mu\nu}S\psi \\ &= \Lambda^\mu{}_\rho\Lambda^\nu{}_\delta\tilde{\psi}\sigma^{\rho\delta}\psi \end{aligned} \tag{90}$$

The quantity $\tilde{\psi}\gamma^\lambda\gamma^\mu\gamma^\nu\psi$ for $\lambda < \mu < \nu$ transforms like a tensor of the third rank, antisymmetric in all three indices, since

$$\tilde{\psi}'\gamma^\lambda\gamma^\mu\gamma^\nu\psi' = \sum_{\rho<\sigma<\tau} \begin{vmatrix} \Lambda^\lambda{}_\rho & \Lambda^\lambda{}_\sigma & \Lambda^\lambda{}_\tau \\ \Lambda^\mu{}_\rho & \Lambda^\mu{}_\sigma & \Lambda^\mu{}_\tau \\ \Lambda^\nu{}_\rho & \Lambda^\nu{}_\sigma & \Lambda^\nu{}_\tau \end{vmatrix} \tilde{\psi}\gamma^\rho\gamma^\sigma\gamma^\tau\psi \tag{91}$$

There are four independent nonvanishing components of this tensor, conveniently denoted by the one missing γ. These nonvanishing components may also be written as $\tilde{\psi}\gamma^5\gamma^\mu\psi$, where

$$-\gamma^5 = \gamma_5 = \frac{1}{4!}\epsilon^{\mu\nu\rho\sigma}\gamma_\mu\gamma_\nu\gamma_\rho\gamma_\sigma = \frac{1}{4!}\epsilon_{\mu\nu\rho\sigma}\gamma^\mu\gamma^\nu\gamma^\rho\gamma^\sigma$$
$$= \gamma^0\gamma^1\gamma^2\gamma^3 \tag{92}$$

is an anti-Hermitian matrix which anticommutes with all the γ^μ and whose square is minus one

$$\gamma^\mu\gamma^5 + \gamma^5\gamma^\mu = 0 \tag{93a}$$
$$(\gamma^5)^2 = -1 \tag{93b}$$

With the help of this matrix we can write

$$\gamma^5\gamma^\lambda = -\frac{1}{3!}\epsilon^{\lambda\mu\nu\rho}\gamma_\mu\gamma_\nu\gamma_\rho \tag{94}$$

The four products $\tilde{\psi}\gamma_5\gamma^\mu\psi$, $\mu = 0, 1, 2, 3$ behave like the components of a four-vector in all respects, except that multiplication by a true four-vector (e.g., a co-ordinate vector) yields a pseudoscalar rather than a scalar. One, therefore, refers to $\tilde{\psi}\gamma_5\gamma^\mu\psi$ as a pseudovector (sometimes also called the dual vector).

Finally the quantity $\tilde{\psi}\gamma_5\psi$ transforms as follows:

$$\tilde{\psi}'\gamma_5\psi' = b\tilde{\psi}S^{-1}\gamma_5 S\psi$$
$$= \frac{1}{4!}b\Lambda^\mu{}_\alpha\Lambda^\nu{}_\beta\Lambda^\rho{}_\epsilon\Lambda^\sigma{}_\delta\epsilon_{\mu\nu\rho\sigma}\tilde{\psi}\gamma^\alpha\gamma^\beta\gamma^\epsilon\gamma^\delta\psi$$
$$= b(\det\Lambda)\,\tilde{\psi}\gamma_5\psi \tag{95}$$

since

$$(\det\Lambda)\,\epsilon_{\alpha\beta\gamma\delta} = \epsilon_{\mu\nu\rho\sigma}\Lambda^\mu{}_\alpha\Lambda^\nu{}_\beta\Lambda^\rho{}_\gamma\Lambda^\sigma{}_\delta \tag{96}$$

Hence $\tilde{\psi}\gamma_5\psi$ transforms like a scalar under proper Lorentz transformations, but like a pseudoscalar under space inversion.

The knowledge of the transformation properties of the bilinear forms $\tilde{\psi}\Gamma_i\psi$ will enable us later to construct interaction functions which are invariant in form. As an example, the scalar covariant $\tilde{\psi}\psi(x)$ would lend itself to multiplication by a scalar function $\chi(x)$, the form $\tilde{\psi}\gamma_5\psi(x)$ to multiplication by the pseudoscalar function $\phi(x)$, and the form $\tilde{\psi}\gamma_\mu\psi(x)$ to multiplication by the vector $A^\mu(x)$ which may, for example, represent the electromagnetic field. Interaction with the latter may also take the form $\tilde{\psi}\sigma^{\mu\nu}\psi F_{\mu\nu}$, known as the Pauli moment interaction where the $F_{\mu\nu}(x)$s are the components of the electromagnetic field tensor.

We next define the invariant scalar product of two Dirac spinors $\psi(x)$ and $\phi(x)$ at time $ct = x_0$ by

$$(\psi, \phi)_t = \int_t d^3x \tilde{\psi}(x)\, \gamma^0 \phi(x)$$

$$= \int_t d^3x \sum_{\alpha=1}^{4} \tilde{\psi}_\alpha(x)\, \phi_\alpha(x) \tag{97}$$

This scalar product can be generalized to a space-like surface by the expression

$$(\psi, \phi)_\sigma = \int_\sigma d\sigma^\mu(x)\, \tilde{\psi}(x)\, \gamma_\mu \phi(x) \tag{98}$$

This scalar product is independent of σ if ψ and ϕ obey the Dirac equation, hence we may choose for σ a plane t = *constant.* The proof of the independence of $(\psi, \phi)_\sigma$ of σ consists in showing that

$$\frac{\delta}{\delta\sigma(x)} (\psi, \phi)_\sigma = \partial^\mu(\tilde{\psi}(x)\, \gamma_\mu \phi(x))$$

$$= (\partial^\mu \tilde{\psi} \gamma_\mu)\, \phi + \tilde{\psi}(\gamma^\mu \partial_\mu \phi)$$

$$= +im(\tilde{\psi}\phi - \tilde{\psi}\phi) = 0 \tag{99}$$

It also follows from this that the scalar product is conserved in time. Within this scalar product the mapping $|\,\psi\rangle \rightarrow |\,\psi'\rangle$ for an arbitrary Lorentz transformation which does not reverse the sign of the time, is induced by a unitary operator $U(a, \Lambda)$

$$U(a, \Lambda)\,|\,\psi\rangle = |\,\psi'\rangle \tag{100}$$

with

$$\langle x\,|\,U(a, \Lambda)\,|\,\psi\rangle = \langle x\,|\,\psi'\rangle = \mathfrak{U}_{(a,\Lambda)}\, \psi(x)$$

$$= S(\Lambda)\, \psi(\Lambda^{-1}(x - a)) \tag{101}$$

The unitarity is proved from the fact that

$$(U(a, \Lambda)\, \psi, U(a, \Lambda)\, \phi)$$

$$= \int_\sigma d\sigma^\mu(x)\, \tilde{\psi}(\Lambda^{-1}(x - a))\, S(\Lambda)^{-1}\, \gamma_\mu S(\Lambda)\, \phi(\Lambda^{-1}(x - a))$$

$$= \int_\sigma d\sigma^\mu(x)\, \tilde{\psi}(\Lambda^{-1}(x - a))\, \Lambda_\mu{}^\nu \gamma_\nu \phi(\Lambda^{-1}(x - a))$$

$$= \int_\sigma d\sigma^\nu(\Lambda^{-1}(x - a))\, \tilde{\psi}(\Lambda^{-1}(x - a))\, \gamma_\nu \phi(\Lambda^{-1}(x - a))$$

$$= (\psi, \phi) \tag{102}$$

For an infinitesimal transformation we write

$$\mathfrak{U}_{(a,\Lambda)} = 1 + \frac{i}{\hbar}\, \epsilon \mathfrak{D} \tag{103}$$

where $\mathfrak{D}$, the infinitesimal generator, is determined using Eqs. (73) and (74) by Eq. (101):

$$\left(1 + \frac{i}{\hbar}\epsilon\mathfrak{D}\right)\psi(x) = (I + \epsilon T)\,\psi(x - \epsilon\lambda x)$$

$$= (I + \epsilon T)\left(\psi(x) - \epsilon(\lambda x)^\mu \frac{\partial\psi(x)}{\partial x^\mu} + \cdots\right)$$

$$= \psi(x) + \epsilon(T - \lambda^\mu{}_\nu x^\nu \partial_\mu)\,\psi(x) + \cdots \tag{104}$$

so that

$$\mathfrak{D} = -i\hbar(T - \lambda^\rho{}_\sigma x^\sigma \partial_\rho) \tag{105}$$

For an infinitesimal rotation about the x^3 axis, $T = \frac{1}{2}i\Sigma_3$ and only $\lambda^{21} = -\lambda^{12} = +1$ are different from zero, so that

$$\mathfrak{D}_3 = \frac{\hbar}{2}\Sigma_3 + (\mathbf{r} \times \mathbf{p})_3 \tag{106}$$

which we can define as the *3*-component of the total angular momentum of the particle. Thus, a Dirac particle has in addition to its orbital momentum an intrinsic angular momentum $\mathbf{\Sigma}$ of magnitude $\hbar/2$. It is to be noted that the spin operator $\frac{1}{2}\mathbf{\Sigma}$ is not a constant of the motion, since $[H, \mathbf{\Sigma}] \neq 0$. The same is true for the orbital angular momentum; however, the total angular momentum $\mathbf{J} = \frac{\hbar}{2}\mathbf{\Sigma} + (\mathbf{r} \times \mathbf{p})$ is a constant of the motion. Actually, this definition of the angular momentum is based on the assumption that the operator $\mathbf{r}$ is the position operator for a Dirac particle. We shall see below that this is not so and that a different angular momentum operator can be defined in terms of the position operator which has the property that the orbital and spin parts are separately constants of the motion.

4d. Solutions of the Dirac Equation

The Dirac equation admits of plane wave solutions of the form

$$\psi(x) = e^{-ip\cdot x/\hbar}\,u(p) \tag{107}$$

where $u(p)$ is a four-component spinor which satisfies the equation

$$(\not{p} - mc)\,u(p) = 0 \tag{108}$$

The scalar product of two spinors u and u' is denoted by

$$u^*u' = \sum_{\alpha=1}^{4} \bar{u}_\alpha u'_\alpha = \tilde{u}\gamma^0 u' \tag{109}$$

The Hamiltonian $H = c\boldsymbol{\alpha}\cdot\mathbf{p} + \beta mc^2$ is Hermitian within this scalar product:

$$u'^*Hu = \sum_{\alpha,\beta=1}^{4} \bar{u}'_\alpha H_{\alpha\beta} u_\beta = (Hu')^*u \tag{110}$$

since $\alpha = \alpha^*$ and $\beta = \beta^*$; its eigenvalues are therefore real. Equation (108) is a system of four-linear homogeneous equations for the components u_α, $\alpha = 1, 2, 3, 4$, for which nontrivial solutions exist only if $\det(\not{p} - mc) = (p^2 - m^2c^2)^2 = 0$. Solutions therefore only exist if $p^2 = m^2c^2$, i.e., only if $p_0 = \pm\sqrt{\mathbf{p}^2 + m^2c^2}$. Let $u_+(\mathbf{p})$ be a solution for $cp_0 = E(\mathbf{p}) = +c\sqrt{\mathbf{p}^2 + m^2c^2}$ so that $u_+(\mathbf{p})$ satisfies the equation

$$(c\boldsymbol{\alpha} \cdot \mathbf{p} + \beta mc^2)\, u_+(\mathbf{p}) = E(\mathbf{p})\, u_+(\mathbf{p}) \tag{111}$$

If we write $u_+ = \begin{pmatrix} u_1 \\ u_2 \end{pmatrix}$ where u_1 and u_2 each have two components, and adopt the representation (24) for the $\boldsymbol{\alpha}$ and β matrices, we find that u_1 and u_2 obey the following equations:

$$c\boldsymbol{\sigma} \cdot \mathbf{p} u_2 + mc^2 u_1 = E(\mathbf{p})\, u_1 \tag{112a}$$

$$c\boldsymbol{\sigma} \cdot \mathbf{p} u_1 - mc^2 u_2 = E(\mathbf{p})\, u_2 \tag{112b}$$

Since $E(\mathbf{p}) + mc^2 \neq 0$

$$u_2 = c\,\frac{\boldsymbol{\sigma} \cdot \mathbf{p}}{E(\mathbf{p}) + mc^2}\, u_1 \tag{113}$$

and substituting this value of u_2 back into (112a), we find that

$$\left(\frac{c^2(\boldsymbol{\sigma} \cdot \mathbf{p})^2}{E(\mathbf{p}) + mc^2} + mc^2\right) u_1 = E(\mathbf{p})\, u_1 \tag{114}$$

Since $(\boldsymbol{\sigma} \cdot \mathbf{p})^2 = \mathbf{p}^2$ and

$$\frac{c^2\mathbf{p}^2}{E(\mathbf{p}) + mc^2} = \frac{E^2(\mathbf{p}) - m^2c^4}{E(\mathbf{p}) + mc^2} = E(\mathbf{p}) - mc^2 \tag{115}$$

Equation (112b) is identically satisfied. There are therefore two linearly independent positive energy solutions for each momentum $\mathbf{p}$, which correspond, for example, to choosing u_1 equal to $\begin{pmatrix} 1 \\ 0 \end{pmatrix}$ or $\begin{pmatrix} 0 \\ 1 \end{pmatrix}$. This can also be seen somewhat differently. The Hamiltonian operator $H = c\boldsymbol{\alpha} \cdot \mathbf{p} + \beta mc^2$ commutes with the Hermitian operator

$$s(\mathbf{p}) = \frac{\boldsymbol{\Sigma} \cdot \mathbf{p}}{|\mathbf{p}|} \tag{116}$$

where

$$\boldsymbol{\Sigma} = \begin{pmatrix} \boldsymbol{\sigma} & 0 \\ 0 & \boldsymbol{\sigma} \end{pmatrix} \tag{117}$$

$s(\mathbf{p})$ is called the helicity operator or simply the helicity of the particle, and physically corresponds to the spin of the particle parallel to the direction of motion. The solutions can therefore be chosen to be simultaneous eigenfunctions of H and $s(\mathbf{p})$. Since $s^2(\mathbf{p}) = 1$, the eigenvalues of $s(\mathbf{p})$ are ± 1. For a given momentum and sign of the energy, the solutions can therefore be classified according to the eigenvalues $+1$ or -1 of

$s(\mathbf{p})$. A similar classification can be made for the negative energy solutions for which $p_0 = -\sqrt{\mathbf{p}^2 + m^2c^2}$ and where for a given momentum there are again two linearly independent solutions which correspond to the eigenvalue $+1$ and -1 of $s(\mathbf{p})$. Summarizing, for a given four-momentum p there are four linearly independent solutions of the Dirac equation, characterized by $p_0 = \pm cE(\mathbf{p})$, $s(\mathbf{p}) = \pm 1$.

An explicit form for two linearly independent solutions for positive energy and momentum $\mathbf{p}$ is given by

$$u_+^{(1)}(\mathbf{p}) = \sqrt{\frac{E(\mathbf{p}) + mc^2}{2E(\mathbf{p})}} \begin{bmatrix} 1 \\ 0 \\ \dfrac{c\boldsymbol{\sigma} \cdot \mathbf{p}}{E(\mathbf{p}) + mc^2} \begin{pmatrix} 1 \\ 0 \end{pmatrix} \end{bmatrix} \tag{118a}$$

$$u_+^{(2)}(\mathbf{p}) = \sqrt{\frac{E(\mathbf{p}) + mc^2}{2E(\mathbf{p})}} \begin{bmatrix} 0 \\ 1 \\ \dfrac{c\boldsymbol{\sigma} \cdot \mathbf{p}}{E(\mathbf{p}) + mc^2} \begin{pmatrix} 0 \\ 1 \end{pmatrix} \end{bmatrix} \tag{118b}$$

where the normalization constant is determined by the requirement that $u^*u = 1$. Note that these two solutions are orthogonal to each other, i.e.,

$$u_+^{(r)*}(\mathbf{p})\, u_+^{(s)}(\mathbf{p}) = \delta_{rs} \quad r, s = 1, 2 \tag{119}$$

The above solutions are not eigenfunctions of $s(\mathbf{p})$. Positive energy solutions corresponding to a definite helicity are obtained by noting that the equation $s(\mathbf{p})\, u_+^{(\pm)}(\mathbf{p}) = \pm u_+^{(\pm)}(\mathbf{p})$ implies that

$$\boldsymbol{\sigma} \cdot \mathbf{n} u_1^{(\pm)} = \pm u_1^{(\pm)} \tag{120a}$$

$$\boldsymbol{\sigma} \cdot \mathbf{n} u_2^{(\pm)} = \pm u_2^{(\pm)} \tag{120b}$$

where $u_1^{(\pm)}$ and $u_2^{(\pm)}$ are the upper and lower components respectively of $u_+^{(\pm)}$, and $\mathbf{n}$ is the unit vector in the direction of $\mathbf{p}$, $\mathbf{n} = \mathbf{p}/|\mathbf{p}|$. Hence, the normalized $u_1^{(\pm)}$ are given by

$$u_1^{(+)} = \frac{1}{\sqrt{2(n_3 + 1)}} \begin{pmatrix} n_3 + 1 \\ n_1 + in_2 \end{pmatrix} \tag{121a}$$

and

$$u_1^{(-)} = \frac{1}{\sqrt{2(n_3 + 1)}} \begin{pmatrix} -n_1 + in_2 \\ n_3 + 1 \end{pmatrix} \tag{121b}$$

A normalized positive energy eigenfunction with helicity $+1$ is thus given by

$$u_+^{(+)}(\mathbf{p}) = \frac{1}{\sqrt{2(n_3 + 1)}} \sqrt{\frac{E(\mathbf{p}) + mc^2}{2E(\mathbf{p})}} \begin{bmatrix} n_3 + 1 \\ n_1 + in_2 \\ \dfrac{c|\mathbf{p}|}{E(\mathbf{p}) + mc^2} \begin{pmatrix} n_3 + 1 \\ n_1 + in_2 \end{pmatrix} \end{bmatrix} \tag{122}$$

Incidentally, we note that in the nonrelativistic limit the components u_2 of a positive energy solution are of order v/c times u_1, and therefore small.

Proof: In the nonrelativistic limit the norm of u_2 is of order $(v/c)^2$ that of u_1 since

$$u_2{}^*u_2 = \frac{c^2\mathbf{p}^2}{(E(\mathbf{p}) + mc^2)^2}\, u_1{}^*u_1$$
$$\approx \left(\frac{mv}{2mc}\right)^2 u_1{}^*u_1 \approx \frac{1}{4}\left(\frac{v}{c}\right)^2 u_1{}^*u_1 \tag{123}$$

In the limiting case of a particle at rest, $\mathbf{p} = 0$, the four linearly independent spinors, which in this case can also be taken as eigenfunctions of Σ_3, are

$$\begin{array}{cccc} \multicolumn{2}{c}{p_0 = mc^2} & \multicolumn{2}{c}{p_0 = -mc^2} \\ \Sigma_3 = +1 & \Sigma_3 = -1 & \Sigma_3 = +1 & \Sigma_3 = -1 \\ \begin{pmatrix}1\\0\\0\\0\end{pmatrix} & \begin{pmatrix}0\\1\\0\\0\end{pmatrix} & \begin{pmatrix}0\\0\\1\\0\end{pmatrix} & \begin{pmatrix}0\\0\\0\\1\end{pmatrix} \end{array} \tag{124}$$

4e. Normalization and Orthogonality Relations: Traces

It is convenient to normalize the Dirac spinors so that $u_+{}^*(\mathbf{p})\, u_+(\mathbf{p}) = E(\mathbf{p})/mc^2$ rather than $u_+{}^*(\mathbf{p})\, u_+(\mathbf{p}) = 1$ since such a normalization is invariant, both sides transforming like the fourth component of a vector. This convention makes the normalization in terms of the adjoint spinor very simple. Since

$$p_0 u = (c\boldsymbol{\alpha} \cdot \mathbf{p} + \beta mc^2)\, u \tag{125a}$$
$$p_0 u^* = u^*(c\boldsymbol{\alpha} \cdot \mathbf{p} + \beta mc^2) \tag{125b}$$

upon multiplying (125a) by $u^*\beta$ on the left and (125b) by βu on the right and adding, we obtain

$$2p_0 u^*\beta u = 2mc^2 u^* u \tag{126}$$

since β and $\boldsymbol{\alpha}$ anticommute. With the normalization $u^*u = |p_0|/mc^2$, recalling that $\tilde{u} = u^*\beta$, we obtain

$$\tilde{u}u = \frac{p_0}{|p_0|} = \epsilon(p_0) \tag{127}$$

$\tilde{u}u$ is thus equal to plus or minus one depending on the sign of the energy. It is usual to write $\tilde{u}u = \epsilon$ with $\epsilon = +1$ for positive energy spinors, $\epsilon = -1$ for negative energy spinors. Incidentally, the equation satisfied by the adjoint spinor is obtained by multiplying (125b) by β from the right; in Feynman notation

$$\tilde{u}(\mathbf{p})\cdot(\not{p} - mc) = 0 \tag{128}$$

One can derive the magnitude of the operators Γ_i in the state $u(\mathbf{p})$ by techniques similar to those used for the derivation of Eq. (126). Thus to compute $\tilde{u}(\mathbf{p})\,\gamma^\mu u(\mathbf{p})$, for example, multiply the equation $(\not{p} - mc)\,u = 0$ by $\tilde{u}\gamma^\mu$ from the left and Eq. (128) by $\gamma^\mu u$ from the right, and add to obtain

$$\begin{aligned} 2mc\tilde{u}(\mathbf{p})\,\gamma^\mu u(\mathbf{p}) &= \tilde{u}(\mathbf{p})\,(\not{p}\gamma^\mu + \gamma^\mu\not{p})\,u(\mathbf{p}) \\ &= p_\nu\tilde{u}(\mathbf{p})\,(\gamma^\nu\gamma^\mu + \gamma^\mu\gamma^\nu)\,u(\mathbf{p}) \\ &= 2p^\mu\tilde{u}(\mathbf{p})\,u(\mathbf{p}) \qquad (129) \end{aligned}$$

whence $\tilde{u}(\mathbf{p})\,\gamma^\mu u(\mathbf{p}) = \epsilon(p)\,p^\mu/mc$, if the us are normalized so that $\tilde{u}(\mathbf{p})\,u(\mathbf{p}) = \epsilon(p)$. Similarly one readily verifies that $\tilde{u}(\mathbf{p})\,\gamma_5 u(\mathbf{p}) = 0$. These methods can be extended to compute the matrix elements of Γ_i between different initial and final states. Consider for example the matrix element $\tilde{u}(\mathbf{p}_2)\gamma_5 u(\mathbf{p}_1)$. Proceeding as before, except that we now consider the equations obeyed by $u(\mathbf{p}_1)$ and $\tilde{u}(\mathbf{p}_2)$, we derive

$$2mc\tilde{u}(\mathbf{p}_2)\,\gamma_5 u(\mathbf{p}_1) = \tilde{u}(\mathbf{p}_2)\,(\gamma_5\not{p}_1 + \not{p}_2\gamma_5)\,u(\mathbf{p}_1) \qquad (130)$$

Since γ_5 anticommutes with γ^μ, we may rewrite the right-hand side as

$$\tilde{u}(\mathbf{p}_2)\,\gamma_5 u(\mathbf{p}_1) = \frac{1}{2mc}\,(p_1{}^\mu - p_2{}^\mu)\,\tilde{u}(\mathbf{p}_2)\,\gamma_5\gamma_\mu u(\mathbf{p}_1) \qquad (131)$$

For $p_1 = p_2$ we recover the result that the matrix element vanishes. If p_2 is negative, i.e., if $u(\mathbf{p}_2)$ is a negative energy solution, it will be observed that the resulting matrix element is particularly large.

Consider next the orthogonality relation between the spinors. If we denote by $u_+{}^r(\mathbf{p})$, $r = 1, 2$, the two positive energy solutions corresponding to the momentum $\mathbf{p}$ and helicity $+1$ or -1, then $u_+{}^1$ and $u_+{}^2$ are orthogonal since they belong to different eigenvalues of the helicity operator:

$$\tilde{u}_+{}^r(\mathbf{p})\,u_+{}^s(\mathbf{p}) = \delta_{rs} \quad r, s = 1, 2 \qquad (132)$$

The solutions, $u_-(-\mathbf{p})$, of momentum $-\mathbf{p}$ and negative energy are likewise orthogonal to $u_+{}^r(\mathbf{p})$. They satisfy the equations

$$(\not{p} + mc)\,u_-(-\mathbf{p}) = 0 \qquad (133a)$$

$$\tilde{u}_-(-\mathbf{p})\,(\not{p} + mc) = 0 \qquad (133b)$$

Upon multiplying the equation $(\not{p} - mc)\,u_+(\mathbf{p}) = 0$ by $\tilde{u}_-(-\mathbf{p})$ on the left, we obtain

$$\tilde{u}_-(-\mathbf{p})\,\not{p}u_+(\mathbf{p}) = mc\tilde{u}_-(-\mathbf{p})\,u_+(\mathbf{p}) \qquad (134)$$

whereas, multiplying (133b) by $u_+(\mathbf{p})$, we obtain

$$\tilde{u}_-(-\mathbf{p})\,\not{p}u_+(\mathbf{p}) = -mc\tilde{u}_-(-\mathbf{p})\,u_+(\mathbf{p}) \qquad (135)$$

whence $\tilde{u}_-(-\mathbf{p})\,u_+(\mathbf{p}) = 0$. If we denote the two linearly independent and orthogonal negative energy solutions for momentum $-\mathbf{p}$ by $v^s(\mathbf{p})$, $s = 1, 2$:

$$u_-{}^s(-\mathbf{p}) = v^s(\mathbf{p}) \qquad (136)$$

then the orthonormality relations for these reads

$$\tilde{v}^r(\mathbf{p})\, v^s(\mathbf{p}) = -\delta_{rs} \tag{137}$$

Collecting all orthonormality relations we have

$$\tilde{u}_+{}^r(\mathbf{p})\, u_+{}^s(\mathbf{p}) = -\tilde{v}^r(\mathbf{p})\, v^s(\mathbf{p}) = \delta_{rs} \tag{138a}$$

$$\tilde{v}^r(\mathbf{p})\, u_+{}^s(\mathbf{p}) = \tilde{u}_+{}^r(\mathbf{p})\, v^s(\mathbf{p}) = 0 \tag{138b}$$

Because of the orthogonality and normalization relations, these solutions satisfy the relation

$$\sum_{r=1}^{2} \{u_{+\alpha}{}^r(\mathbf{p})\, \tilde{u}_{+\beta}{}^r(\mathbf{p}) - v_\alpha{}^r(\mathbf{p})\, \tilde{v}_\beta{}^r(\mathbf{p})\} = \delta_{\alpha\beta} \quad \alpha, \beta = 1, 2, 3, 4 \tag{139}$$

The order of the factors in this equation should be noted; it corresponds to an outer product, $\otimes$, of u by $\tilde{u}$ which is a 4×4 matrix. Similarly our normalization equations (138a) and (138b) imply that

$$\sum_{r=1}^{2} \tilde{u}_+{}^r(\mathbf{p})\, u_+{}^r(\mathbf{p}) - \tilde{v}^r(\mathbf{p})\, v^r(\mathbf{p}) = 4 \tag{140}$$

For notational convenience we introduce the spinors w defined by

$$w^r(\mathbf{p}) = u_+{}^r(\mathbf{p}) \qquad r = 1, 2 \tag{141a}$$

$$w^{r+2}(\mathbf{p}) = v_+{}^r(\mathbf{p}) = u_-{}^r(-\mathbf{p}) \quad r = 1, 2 \tag{141b}$$

in which case the orthogonality relations (138) read

$$\tilde{w}^m(\mathbf{p})\, w^n(\mathbf{p}) = \epsilon^m \delta_{mn} \quad (m, n = 1, 2, 3, 4) \tag{142}$$

with $\epsilon^m = +1$ for $m = 1, 2$ and $\epsilon^m = -1$ for $m = 3, 4$. Equations (139) and (140) can now be recast into the form

$$\sum_{m=1}^{4} \epsilon^m w^m(\mathbf{p}) \otimes \tilde{w}^m(\mathbf{p}) = I \tag{143a}$$

$$\sum_{m=1}^{4} \epsilon^m \tilde{w}^m(\mathbf{p})\, w^m(\mathbf{p}) = 4 \tag{143b}$$

In the computation of scattering cross sections involving spin $\frac{1}{2}$ particles, it is often necessary to sum over intermediate spin states, and in particular over only those intermediate states associated with positive energy, or those associated with negative energy. Let the sum of interest be of the form

$$\Omega = \sum_{r=1}^{2} (\tilde{f}Qw^r)\, (\tilde{w}^r Pg)$$

$$= \sum_{r=1}^{2} \sum_{\alpha,\beta=1}^{4} (\tilde{f}_\alpha Q_{\alpha\beta} w_\beta{}^r) \left(\sum_{\rho,\sigma=1}^{4} \tilde{w}^r{}_\rho P_{\rho\sigma} g_\sigma \right) \tag{144}$$

where Q and P are operators (products of γ matrices), f and g are spinors, and the index r under the summation sign indicates that we are to sum only over the two states w^r of positive energy. The case where the intermediate states have negative energy can be worked out in a similar way.

We would therefore like to find covariant projection operators which we could insert in the right-hand side of Eq. (144) so that the summation, which presently runs only over the two states $w^r(\mathbf{p})$ $(r = 1, 2)$, could be extended to include all four states in order that we may use Eq. (143a) to simplify the expression. We want the projection operator in question to have the property that it multiplies the spinor w by $+1$ for positive energy states and by zero for negative states. We may construct such an operator by recalling that in terms of our spinors, w, the Dirac equation reads (with $\hbar = c = 1$)

$$(\not{p} - m)\, w^r(\mathbf{p}) = 0 \quad r = 1, 2 \tag{145a}$$

$$(\not{p} + m)\, w^r(\mathbf{p}) = 0 \quad r = 3, 4 \tag{145b}$$

These equations suggest that we define the following projection operator for positive energy states of momentum $\mathbf{p}$:

$$\Lambda_+(\mathbf{p}) = \frac{\not{p} + m}{2m} \tag{146}$$

which has the following properties:

$$\Lambda_+(\mathbf{p})\, w^r(\mathbf{p}) = w^r(\mathbf{p}) \quad r = 1, 2 \tag{147a}$$

$$\Lambda_+(\mathbf{p})\, w^r(\mathbf{p}) = 0 \qquad r = 3, 4 \tag{147b}$$

$$[\Lambda_+(\mathbf{p})]^* = \gamma^0 \Lambda_+(\mathbf{p})\, \gamma^0 \tag{147c}$$

and

$$[\Lambda_+(\mathbf{p})]^2 = \left(\frac{\not{p} + m}{2m}\right)^2 = \frac{p^2 + 2m\not{p} + m^2}{4m^2} = \frac{\not{p} + m}{2m} = \Lambda_+(\mathbf{p}) \tag{148}$$

since for a free particle

$$\not{p}^2 = \gamma^\mu p_\mu \gamma^\nu p_\nu = \tfrac{1}{2}(\gamma^\mu \gamma^\nu + \gamma^\mu \gamma^\nu)\, p_\mu p_\nu = p^2 = m^2 \tag{149}$$

By virtue of Eqs. (143) and (147) one readily finds that

$$\Lambda_+(\mathbf{p}) = \frac{\not{p} + m}{2m} = \sum_{r=1}^{2} w^r(\mathbf{p}) \otimes \tilde{w}^r(\mathbf{p}) \tag{150}$$

Note further that $\Lambda_+ \epsilon^r$ has the same properties as Λ_+ alone, since for negative states, when $\epsilon^r = -1$, Λ_+ operating on these gives 0. This enables us to rewrite Eq. (144) as follows:

$$\Omega = \sum_{m=1}^{4} (\tilde{f} Q \Lambda_+ \epsilon^r w^r)\, (\tilde{w}^r P g) \tag{151}$$

Using (143b), Eq. (151) reduces to

$$\Omega = (\tilde{f} Q \Lambda_+ P g) \tag{152}$$

and we have accomplished our objective of evaluating the sum over intermediate states.

If we are interested in negative energy states only, we can define an analogous projection operator

$$\Lambda_-(\mathbf{p}) = \frac{m - \not{p}}{2m} \tag{153}$$

whose properties are

$$\Lambda_-(\mathbf{p})\, w^r(\mathbf{p}) = w^r(\mathbf{p}) \quad \text{for } r = 3, 4 \tag{154a}$$

$$\Lambda_-(\mathbf{p})\, w^r(\mathbf{p}) = 0 \quad \text{for } r = 1, 2 \tag{154b}$$

$$(\Lambda_-(\mathbf{p}))^2 = \Lambda_-(\mathbf{p}) = -\sum_{r=3}^{4} w^r(\mathbf{p}) \otimes \tilde{w}^r(\mathbf{p}) \tag{154c}$$

It should be noted that the sum of Λ_+ and Λ_- is the identity operator

$$\Lambda_+(\mathbf{p}) + \Lambda_-(\mathbf{p}) = I \tag{155}$$

and also that

$$\Lambda_+(\mathbf{p})\, \Lambda_-(\mathbf{p}) = \Lambda_-(\mathbf{p})\, \Lambda_+(\mathbf{p}) = 0 \tag{156}$$

The probability of an event whose amplitude is $M = \tilde{w}_f Q w_i$ is proportional to the absolute value squared of M, i.e., to

$$\begin{aligned} |M|^2 = MM^* &= (\tilde{w}_f Q w_i)\,(w_i^* Q^* \tilde{w}_f^*) \\ &= (\tilde{w}_f Q w_i)\,(\tilde{w}_i \gamma^0 Q^* \gamma^0 w_f) \end{aligned} \tag{157}$$

Here Q^* is the Hermitian adjoint (complex conjugate transpose) of the operator Q. Note that by virtue of Eq. (61), that part of Q which is composed simply of products of γ matrices—call it Q_M—satisfies $\gamma^0 Q_M^* \gamma^0 = Q_M$. The nonmatrix part of Q must have i changed into $-i$ wherever i appears. Let us write

$$Q' = \gamma^0 Q^* \gamma^0 \tag{158}$$

so that

$$|M|^2 = (\tilde{w}_f Q w_i)\,(\tilde{w}_i Q' w_f) \tag{159}$$

In many problems one is not interested in the spin in the final state. One is then to sum over the two final spin states. By the methods of the last section, we can accomplish this summation by introducing the appropriate projection operators. If the initial and final states correspond to positive energy spinors, $w_i(\mathbf{p})$ and $w_f(\mathbf{p}')$ of momentum $\mathbf{p}$ and $\mathbf{p}'$, respectively, we obtain

$$\begin{aligned} \left.\begin{matrix}\text{sum of } |M|^2 \text{ over} \\ \text{final spin states}\end{matrix}\right\} &= \sum_{r=1,2} (\tilde{w}_i Q' w_f^r)\,(\tilde{w}_f^r Q w_i) \\ &= \sum_{r=1}^{4} (\tilde{w}_i Q' \Lambda_+(\mathbf{p}')\, \epsilon^r w_f^r \tilde{w}_f^r Q w_i) = \tilde{w}_i Q' \Lambda_+(\mathbf{p}')\, Q w_i \end{aligned} \tag{160}$$

If the incident state is unpolarized, we then average over the initial spin states, obtaining

$$\left.\begin{array}{l}\text{average of } |M|^2 \text{ over} \\ \text{initial and sum} \\ \text{over final spin} \\ \text{states}\end{array}\right\} = \tfrac{1}{2} \sum_{r=1}^{2} \tilde{w}_i^r Q' \Lambda_+(\mathbf{p}')\, Q w_i^r$$

$$= \tfrac{1}{2} \sum_{r=1}^{4} \sum_{\alpha\beta=1}^{4} (\tilde{w}_i^r)_\alpha\, (Q'\Lambda_+(\mathbf{p}')\, Q\Lambda_+(\mathbf{p}))_{\alpha\beta}\, (w_i^r)_\beta\, \epsilon^r$$

$$= \tfrac{1}{2} \sum_{\alpha,\,\beta=1}^{4} (Q'\Lambda_+(\mathbf{p}')\, Q\Lambda_+(\mathbf{p}))_{\alpha\beta}\, \delta_{\alpha\beta}$$

$$= \tfrac{1}{2}\mathrm{Tr}\, (Q'\Lambda_+(\mathbf{p}')\, Q\Lambda_+(\mathbf{p})) \tag{161}$$

If we call

$$L = Q'\Lambda_+(\mathbf{p}')\, Q\Lambda_+(\mathbf{p}) \tag{162}$$

then L can be represented in terms of a linear combination of the 16 linearly independent matrices, Γ_i, as discussed previously. Thus

$$L = \sum_{l=1}^{16} a_l \Gamma_l = a_1 I + a_2 \gamma_1 + \cdots \tag{163}$$

If we take the trace of both sides of (163), only the term with the unit matrix can contribute, so that

$$\mathrm{Tr}\, L = 4a_1 \tag{164}$$

In practice, the following two properties of the trace of a product of matrices are useful:

1. The trace of an odd number of γ matrices vanishes. To prove this, recall the elementary property of the trace,

$$\mathrm{Tr}\, (ABC) = \mathrm{Tr}\, (CAB) \tag{165}$$

(or any cyclic permutation). Now we have previously indicated the existence of the matrix γ_5 with the property that

$$\gamma_5 \gamma_\mu + \gamma_\mu \gamma_5 = 0; \; (\gamma_5)^2 = -I \tag{166a}$$

or, alternatively,

$$\gamma_5 \gamma_\mu (\gamma_5)^{-1} = -\gamma_\mu \tag{166b}$$

whence

$$\gamma_5 \gamma_{\mu_1} \gamma_{\mu_2} \cdots \gamma_{\mu_n} (\gamma_5)^{-1} = (-1)^n\, \gamma_{\mu_1} \gamma_{\mu_2} \cdots \gamma_{\mu_n} \tag{167}$$

Taking the trace of both sides of Eq. (167) and using (165) immediately yields

$$(-1)^n\, \mathrm{Tr}\, (\gamma_{\mu_1} \gamma_{\mu_2} \cdots \gamma_{\mu_n}) = \mathrm{Tr}\, (\gamma_{\mu_1} \cdots \gamma_{\mu_n}) \tag{168}$$

so that the trace of an odd number of γ matrices vanishes. In particular, we note that

$$\text{Tr}\,\gamma_\mu = 0 \tag{169}$$

Similarly, we have

$$\text{Tr}\,\gamma_5 = -\text{Tr}\,[\gamma_\mu\gamma_5(\gamma_\mu)^{-1}] = -\text{Tr}\,\gamma_5 = 0 \tag{170}$$

2. If n is even, the commutation rules can always be used to reduce the expression to one involving only $n - 2$ factors. Take, for example, the case of $n = 2$. By virtue of the commutation rules

$$\gamma_\mu\gamma_\nu + \gamma_\nu\gamma_\mu = 2g_{\mu\nu} \tag{171}$$

we have

$$\begin{aligned}\text{Tr}\,(\gamma_\mu\gamma_\nu) &= \text{Tr}\,(\gamma_\nu\gamma_\mu) = \tfrac{1}{2}\text{Tr}\,(\gamma_\nu\gamma_\mu + \gamma_\nu\gamma_\mu) \\ &= g_{\mu\nu}\,\text{Tr}\,I = 4g_{\mu\nu}\end{aligned} \tag{172}$$

since $\text{Tr}\,I = 4$. Similarly, one readily establishes that

$$\text{Tr}\,(\gamma_\mu\gamma_\nu\gamma_\rho\gamma_\sigma) = 4g_{\mu\sigma}g_{\nu\rho} - 4g_{\sigma\nu}g_{\rho\mu} + 4g_{\sigma\rho}g_{\mu\nu} \tag{173}$$

These results then imply that if A and B commute

$$\text{Tr}\,(\not{A}\not{B}) = 4A \cdot B = 4A_\mu B^\mu \tag{174a}$$

$$\text{Tr}\,\not{A} = 0 \tag{174b}$$

etc. For further methods of evaluating traces, see Duffin (1950) and Caianiello (1952).

4f. Foldy-Wouthuysen Representation

Although we have derived many properties of the Dirac equation, we have not as yet given a physical interpretation to the operators appearing in the theory. The fact is that the Dirac equation in the form described above does not lend itself easily to a simple interpretation. Consider for example the operator, $\dot{\mathbf{x}}$,

$$\dot{\mathbf{x}} = \frac{i}{\hbar}\,[H, \mathbf{x}] = c\boldsymbol{\alpha} \tag{175}$$

which one might want to call the velocity operator. Since $\alpha_i^2 = 1$ the absolute magnitude of the "velocity" in any given direction is always c, which is not physically reasonable. Furthermore, since $[\alpha_1, \alpha_2] \neq 0$, it would seem that, when the velocity in any one direction is defined, the velocity in the other two directions cannot be simultaneously defined. But this would deny the existence of velocity measurements. One must conclude that there must exist another representation of the Dirac equation in which the physical interpretation is more transparent. That this must be the case can also be inferred from the following argument. For a positive energy Dirac particle there are two independent states associated with each value of the momentum. These correspond to the two possible directions of the spin. According to quantum mechanics, each such pair of physical states is to be represented by exactly two vectors in Hilbert space. There exists therefore a redundancy in the representation of these vectors in the usual formulation of the Dirac theory where the corresponding wave functions have four components. It must therefore be possible to find a transformation such that the wave functions of a free Dirac particle of definite momentum have just two components, as in the nonrelativistic Pauli theory [Pauli (1927)]. This problem was solved by Becker (1945) to order v^2/c^2, and exactly by Foldy and Wouthuysen [Foldy (1950); see also Tani (1951)] who noticed that the essential reason why four-component spinors are necessary to describe a solution in the Dirac representation is that the Hamiltonian contains operators, specifically the operators α^i, which in the representation (24) have matrix elements which connect the upper and lower components of the wave function. An operator which connects upper and lower components will be called "odd." If it were possible to perform a canonical transformation on the Hamiltonian $H = c\boldsymbol{\alpha} \cdot \mathbf{p} + \beta mc^2$ which brings it into a form which is free of odd operators, it would then be possible to represent the solutions by two-component spinors.

Let us, with Foldy and Wouthuysen, make a canonical transformation e^{iS}, where S is Hermitian, such that

$$\psi \rightarrow e^{iS}\psi = \psi' \tag{176a}$$

$$H \rightarrow e^{iS} H e^{-iS} = H' \tag{176b}$$

We choose S of the form

$$S = -\left(\frac{i}{2mc}\right)\beta\boldsymbol{\alpha} \cdot \mathbf{p}w\left(\frac{|\mathbf{p}|}{mc}\right) \tag{177}$$

where w is a real function to be determined so that H' is free of odd operators. Now

$$\begin{aligned} H' &= e^{iS}(c\boldsymbol{\alpha} \cdot \mathbf{p} + \beta mc^2)\, e^{-iS} = e^{iS}\beta(c\beta\boldsymbol{\alpha} \cdot \mathbf{p} + mc^2)\, e^{-iS} \\ &= e^{iS}\beta e^{-iS}\beta(c\boldsymbol{\alpha} \cdot \mathbf{p} + \beta mc^2) \end{aligned} \tag{178}$$

since S commutes with $\beta\boldsymbol{\alpha} \cdot \mathbf{p}$. Furthermore, since

$$\beta(\beta\boldsymbol{\alpha} \cdot \mathbf{p})^n = (-1)^n(\beta\boldsymbol{\alpha} \cdot \mathbf{p})^n\beta \tag{179}$$

we deduce that

$$\begin{aligned} \beta e^{-iS} &= \beta \sum_{n=0}^{\infty}\left(\frac{-1}{2mc}\right)^n(\beta\boldsymbol{\alpha} \cdot \mathbf{p})^n w^n \\ &= \sum_{n=0}^{\infty}\left(\frac{+1}{2mc}\right)^n(\beta\boldsymbol{\alpha} \cdot \mathbf{p})^n w^n\beta = e^{iS}\beta \end{aligned} \tag{180}$$

so that

$$\begin{aligned} H' &= e^{2iS}(c\boldsymbol{\alpha}\cdot\mathbf{p}+\beta mc^2) \\ &= \left[\cos\left(\frac{|\mathbf{p}|}{mc}w\right)+\frac{\beta\boldsymbol{\alpha}\cdot\mathbf{p}}{|\mathbf{p}|}\sin\left(\frac{|\mathbf{p}|}{mc}w\right)\right](c\boldsymbol{\alpha}\cdot\mathbf{p}+\beta mc^2) \\ &= \beta\left[mc^2\cos\left(\frac{|\mathbf{p}|}{mc}w\right)+c|\mathbf{p}|\sin\left(\frac{|\mathbf{p}|}{mc}w\right)\right] \\ &\quad+\frac{\boldsymbol{\alpha}\cdot\mathbf{p}}{|\mathbf{p}|}\left[|\mathbf{p}|c\cos\left(\frac{|\mathbf{p}|}{mc}w\right)-mc^2\sin\left(\frac{|\mathbf{p}|}{mc}w\right)\right] \end{aligned} \tag{181}$$

If we now choose w such that the coefficient of the $\boldsymbol{\alpha}\cdot\mathbf{p}$ term vanishes, i.e., set

$$w=\frac{mc}{|\mathbf{p}|}\tan^{-1}\left(\frac{|\mathbf{p}|}{mc}\right) \tag{182}$$

H' will then be free of odd operators, and have the form

$$\begin{aligned} H' &= \beta\left[mc^2\frac{mc}{\sqrt{\mathbf{p}^2+m^2c^2}}+|\mathbf{p}|c\frac{|\mathbf{p}|}{\sqrt{\mathbf{p}^2+m^2c^2}}\right] \\ &= \beta c\sqrt{\mathbf{p}^2+m^2c^2}=\beta E(\mathbf{p}) \end{aligned} \tag{183}$$

In the representation (24) in which β is diagonal, the equation $H'\psi'=E'\psi'$ now has solutions where the upper components represent positive energies and the lower components negative energies, since if we write

$$\psi'=\psi'_+ + \psi'_- \tag{184}$$

where

$$\psi'_+ = \tfrac{1}{2}(1+\beta)\,\psi' \tag{185a}$$

$$\psi'_- = \tfrac{1}{2}(1-\beta)\,\psi' \tag{185b}$$

then

$$H'\psi'_+ = E(\mathbf{p})\,\psi'_+ \tag{186a}$$

$$H'\psi'_- = -E(\mathbf{p})\,\psi_-' \tag{186b}$$

Note that $\psi'_+(\psi'_-)$ is now essentially a two-component wave function, since its lower (upper) components are identically zero. This can also be seen by noting that with the choice (182) for w the transformation function $\exp iS$ can be written as

$$e^{iS}=\left[\frac{2E(\mathbf{p})}{E(\mathbf{p})+mc^2}\right]^{1/2}\frac{1}{2}\left(\frac{c\gamma\cdot\mathbf{p}+mc^2}{E(\mathbf{p})}+1\right) \tag{187}$$

Operating on the positive energy spinor $u_+(\mathbf{p})$, since $(c\gamma\cdot\mathbf{p}+mc^2)\,u_+(\mathbf{p}) = \beta E(\mathbf{p})\,u_+(\mathbf{p})$, it yields

$$e^{iS}\,u_+(\mathbf{p})=\left[\frac{2E(\mathbf{p})}{E(\mathbf{p})+mc^2}\right]^{1/2}\frac{1}{2}(1+\beta)\,u_+(\mathbf{p}) \tag{188}$$

The operator

$$\tfrac{1}{2}(1+\beta) = \begin{pmatrix} 1 & 0 & 0 & 0 \\ 0 & 1 & 0 & 0 \\ 0 & 0 & 0 & 0 \\ 0 & 0 & 0 & 0 \end{pmatrix} \tag{189}$$

annihilates the lower components, so that exp iS operating on the positive energy spinors given by Eqs. (118a) and (118b) results in the spinors

$$\begin{pmatrix} 1 \\ 0 \\ 0 \\ 0 \end{pmatrix} \quad \text{and} \quad \begin{pmatrix} 0 \\ 1 \\ 0 \\ 0 \end{pmatrix} \tag{190}$$

Besides the Hamiltonian, there are two other transformed operators of particular interest. If we inquire as to the operator $\mathbf{X}$ in the Dirac representation, whose transform in the Foldy-Wouthuysen representation is $\mathbf{x}$, the answer is

$$\mathbf{X} = e^{-iS}\,\mathbf{x}e^{iS} = \mathbf{x} + i\hbar c\,\frac{\beta\boldsymbol{\alpha}}{2E(\mathbf{p})} - i\hbar c^2\,\frac{\beta(\boldsymbol{\alpha}\cdot\mathbf{p})\,\mathbf{p} - i[\boldsymbol{\Sigma}\times\mathbf{p}]\,|\mathbf{p}|}{2E(\mathbf{p})\,(E(\mathbf{p})+mc^2)\,|\mathbf{p}|} \tag{191}$$

This operator has the property that

$$[X_i, X_j] = 0 \tag{192}$$

$$[X_l, p_j] = i\hbar\delta_{lj} \tag{193}$$

and furthermore[3]

$$\frac{i}{\hbar}\,[H, \mathbf{X}] = \frac{d}{dt}\mathbf{X} = \frac{\mathbf{p}}{E(\mathbf{p})}\,\frac{\beta mc^2 + \boldsymbol{\alpha}\cdot\mathbf{p}c}{E(\mathbf{p})} \tag{194}$$

so that within the manifold of positive energy solutions the time derivative of $\mathbf{X}$ is the operator $\mathbf{p}/E(\mathbf{p})$, which can be identified with the velocity of the particle. This operator, $\mathbf{X}$, is identical with the "position operator" for a spin $\frac{1}{2}$ system derived by Newton and Wigner [Newton (1949)]. It transforms like a vector under rotations and its eigenfunctions are "localized wave functions" satisfying all the requirements enumerated in Section 3*c*. Furthermore, $\mathbf{X}$ transforms a positive energy wave function into a positive energy wave function. If we denote by $\psi_{\mathbf{y},\,y^0}(x)$ the wave function localized at time y^0 [for an explicit form see Newton (1949)], the probability amplitude for finding at time $x^0 = y^0$ a particle described by the wave function $f(x)$ at the position $\mathbf{y}$ is then

$$(\psi_{\mathbf{y}y^0}, f) = \int_{x^0} d\sigma^\mu(x)\,\bar{\psi}_{y^0\mathbf{y}}(x)\,\gamma_\mu f(x) \tag{195}$$

We have seen that in the Dirac representation, the orbital angular momentum $\mathbf{x}\times\mathbf{p}$ and the spin angular momentum $\frac{1}{2}\boldsymbol{\Sigma}$ are not separately

[3] It is easiest to establish this relation in the F-W representation and then to transform back to the Dirac representation.

constants of the motion although their sum is. One, however, verifies that the operator $\mathbf{X} \times \mathbf{p}$ and the operator

$$\mathbf{\Sigma}_M = \mathbf{\Sigma} - \frac{i\beta c(\boldsymbol{\alpha} \times \mathbf{p})}{E(\mathbf{p})} - \frac{c^2\mathbf{p} \times (\mathbf{\Sigma} \times \mathbf{p})}{E(\mathbf{p})\,(E(\mathbf{p}) + mc^2)} \tag{196}$$

where

$$e^{iS}\,\mathbf{\Sigma}_M e^{-iS} = \mathbf{\Sigma} = \frac{1}{2i}\,(\boldsymbol{\alpha} \times \boldsymbol{\alpha}) = \begin{pmatrix} \boldsymbol{\sigma} & 0 \\ 0 & \boldsymbol{\sigma} \end{pmatrix} \tag{197}$$

are separately constants of the motion. This is easily verified in the Foldy-Wouthuysen representation where the transforms of these operators are $\mathbf{x} \times \mathbf{p}$ and $\mathbf{\Sigma}$ respectively. These clearly commute with $H' = \beta E(\mathbf{p})$ since $E(\mathbf{p})$ depends only on $\mathbf{p}^2$ and is therefore a scalar under three-dimensional rotations so that $\mathbf{x} \times \mathbf{p}$ commutes with it, and $\mathbf{\Sigma}$ and β commute. Foldy and Wouthuysen have called $\mathbf{\Sigma}_M$ the "mean spin operator."

The Foldy-Wouthuysen representation is particularly useful for the discussion of the nonrelativistic limit of the Dirac equation, since the operators representing physical quantities are in a one-to-one correspondence with the operators of the Pauli theory. There exists another limit which is of considerable interest, namely the ultrarelativistic limit where the mass of the particle can be neglected in comparison to its kinetic energy, or equivalently where mc can be neglected in comparison with $|\mathbf{p}|$, the momentum of the particle. A form of the Dirac equation which has this property is obtained by choosing w such that the coefficient of the term in β in Eq. (181) vanishes and only the term in $\boldsymbol{\alpha} \cdot \mathbf{p}$ remains. The form of w which accomplishes this is

$$w' = -\frac{mc}{|\mathbf{p}|} \tan^{-1} \frac{mc}{|\mathbf{p}|} \tag{198a}$$

$$e^{\pm iS'} = \frac{E(\mathbf{p}) + |\mathbf{p}|c \mp \beta \dfrac{\boldsymbol{\alpha} \cdot \mathbf{p}}{|\mathbf{p}|} mc^2}{2E(\mathbf{p})} \tag{198b}$$

and the Hamiltonian in the new representation is

$$H'' = \frac{\boldsymbol{\alpha} \cdot \mathbf{p}}{|\mathbf{p}|} E(\mathbf{p}) \tag{199}$$

This transformation leads to a form of the Hamiltonian for which states of positive and negative helicity are separately described by two-component equations [Cini (1958), Bose (1959)].

4g. Negative Energy States

We have noted that the Dirac equation admits of negative energy solutions. Their physical interpretation presented a great deal of difficulty for some time; for example, a negative energy particle, if it existed, would be accelerated in the direction opposite to that of the external force. One would not have to be concerned with what the negative energy solutions represent if there was no probability for transitions from positive to negative energy states, for then a particle in a positive energy state would always remain in a positive state. But the fact is that the Dirac equation does predict a finite probability for a transition from a positive to a negative energy state under the influence of an external field. Such transitions, with the release of the energy necessary to guarantee energy conservation are, however, never observed. In 1930, Dirac (1930) resolved this difficulty by suggesting the so-called "hole theory" which he formulated as follows: "Assume that nearly all the negative energy states are occupied, with one electron in each state in accordance with the exclusion principle of Pauli." If this is so, the exclusion principle also makes it impossible for positive energy electrons to make transitions to negative energy states (since they are all occupied) unless they are first emptied by some means. Such "an unoccupied negative energy state will now appear as something with positive energy, since to make it disappear, i.e., to fill it up, we should have to add to it an electron with negative energy." Similarly, the "hole" would have a charge opposite of that of the positive energy particle.

Thus, once again, one concludes that a one-particle theory makes sense only in the absence of interactions, i.e., for an isolated free particle. To interpret the theory in the presence of interactions one is forced to a many-particle formulation in which the number of particles is not conserved, i.e., to a field theory.

When Dirac first proposed his theory, it was believed that the world was made up of protons and negatons and so it was natural to hope the "something" would be identifiable as a proton (the larger inertia, it was argued, being due to the presence of the electrons in negative energy states). This possibility was soon disproved [Oppenheimer (1930b); Weyl (1931), p. 234] and it was shown that the "something" had to have a mass equal to that of the particle described by the positive energy solutions. Since by 1931 no such "positon" (i.e., a particle with the same mass as the negaton but with a charge $+|e|$) had been found experimentally, Pauli in his article in the *Handbuch der Physik* considered this prediction of the theory as a shortcoming of the Dirac formulation. However, by the time his article appeared in print in 1933, the theory was vindicated, because Anderson (1932) had found experimentally that positons existed.

In the original formulation of the hole theory, Dirac envisaged an electron distribution of infinite density everywhere in the world, with a perfect vacuum having all states of negative energy occupied, and all states of positive energy unoccupied. One difficulty encountered with this assumption involves the enormous charge density contributed by the negative energy states. In the modern version of the theory as developed by Heisenberg (1934a, b), Kramers (1937), and others, this difficulty has been dealt with and we are no longer forced to accept this asymmetry in charge. In fact, in these newer formulations there is a complete symmetry between positive and negative charges and the states of one positon or one negaton are both positive energy states.

Dirac's hole theory predicts the possibility of pair creation, if enough energy is supplied by a γ ray (or otherwise) to lift an electron from a negative energy state to a positive energy state, i.e., if $h\nu > 2mc^2$. In fact, the theory gives rise to an electromagnetic effect even if $h\nu < 2mc^2$ because the electromagnetic field will still cause a redistribution of charge and thus give rise to a "polarization of the vacuum." These polarization effects have been established experimentally, particularly by the fine structure of the positronium ground state, through a contribution to the Lamb shift, and by the level shifts of mesic atoms.

The term *positronium* is applied to a positon and a negaton bound together by their Coulomb forces. In the ground state, these two particles have no angular momentum, except for their intrinsic spin, so that we may have a triplet ($S = 1$) or singlet ($S = 0$) ground state. According to theory [Yang (1950a), Wolfenstein (1952)], the two particles can annihilate each other and emit two γ rays when they are in the singlet state, but not when in the triplet state. In the triplet state, annihilation is only possible by emission of three γ rays. The lifetime of this state is about a thousand times longer than that of the singlet state. These facts have been confirmed experimentally [Deutsch (1951a), Pond (1952)]. The two states are separated in energy [Deutsch (1951b)] by about 9×10^{-4} ev, and about half of this separation is due to the virtual annihilation and re-creation of the electron-positron pair. This process can only occur in the singlet state and is the inverse of the vacuum polarization effect, which consists first of virtual creation and then annihilation of electron-positron pairs.

In the Lamb shift, there is also a term which occurs due to the fact (as discussed above) that the vacuum slightly modifies the electric field around the proton. This term, first calculated by Uehling (1935), amounts to 27 megacycles for the ground state of the hydrogen atom. Since theory and experiment check to within 0.1 megacycles, this again indicates that the polarization of the vacuum is a real effect.

This fact has been further verified by the level shifts in mesic atoms, i.e., a π^- or μ^- meson moving in a Bohr orbit around a proton or a light nucleus. Consider the π^--p system. Just as for the hydrogen atom, there will be in the present case deviations from the energy levels predicted by the Klein-Gordon equation, the appropriate equation for the π meson, for a particle in a Coulomb field due to electromagnetic radiative corrections (and, to a much smaller extent, deviations due to the finite size of the nucleus, specifically meson-nuclear interactions, etc.). The electromagnetic radiative corrections comprise two effects, one arising from the difference between the self-energy of the bound and free π meson, and the other from the polarization of the vacuum by the field of the nucleus. The polarization of the vacuum in the present situation, as is the case in general, is actually due to the possibility of creating pairs of oppositely charged particles of any kind, that is, it includes effects due to π^+-π^-, μ^+-μ^- pairs, as well as due to negaton-positon pairs. However, in first approximation, each type of particle ($\pi^\pm$, $\mu^\pm$, $e^\pm$) contributes to this effect an amount which is inversely proportional to the square of the mass of the particle so that, in fact, it is the redistribution of the electronic and positonic charges by the potential which gives rise to the observable effects of the vacuum polarization.

For mesic atoms, the contribution to the level shift due to the vacuum polarization by the field of the nucleus is actually much larger (in the ratio of the square of the meson mass to the electron mass, i.e., three orders of magnitudes) than that due to the meson self-energy. Furthermore, since in a mesic atom the meson and the nucleus are much closer together on the average than in the corresponding electronic atom, the magnitude of the level shift due to vacuum polarization is also much larger in this case. Thus, for the $1S$ level of the π^--p system the level shift is about -10 ev (as compared with about -10^{-2} ev for the hydrogen atom), and in the π^--Mg atom the level is lowered by about 1.4×10^3 ev (the shift being proportional to Z^2). That the levels are lowered by the vacuum polarization can be understood qualitatively on the basis of the following picture: the "bare" nucleus, interacting with the "sea" surrounds itself by a neutral cloud of negatons and positons. Some of the positons get displaced to infinity leaving an excess of negative charge in the neighborhood of the nucleus (within a sphere of radius $\sim\hbar/m_ec$). At large distances, i.e., distances greater than $\hbar/m_ec$, the meson sees the (renormalized) charge Ze; however, at distances less than $\hbar/m_ec$ the effective charge is greater than Ze. For the low-lying levels with small orbits, the energy is therefore lowered. The finite size of the nucleus, on the other hand, raises the $1S$ level (but leaves all other levels relatively unchanged), and this raising of the level is roughly of the same order of magnitude for light nuclei. That level shifts of this order of magnitude do occur in mesic atoms has been demonstrated by the experiments of Stearns *et al.* and others [see the review articles of de Benedetti (1956), Stearns (1957), and West (1958)], again demonstrating the reality of the vacuum polarization effect. In addition, these experimental results clearly point out the interrelation of the interaction of the electromagnetic field with all charged particles found in Nature.

The vacuum polarization effect in the electrostatic interaction of charged particles (such as two protons, for example) also implies that deviations from pure Coulomb scattering should exist. In spite of the smallness of these effects and relatively large errors in the available low-energy p-p scattering data, Foldy and Eriksen [Foldy (1955)] have concluded that the data substantiates the existence of a vacuum polarization effect, on the assumption of a Yukawa shape for the nuclear potential.

The above examples illustrate some of the phenomena predicted by the hole theory. It should again be emphasized that the hole theory is a many-particle theory (in fact, an infinitely many-particle theory) and that the one-particle Dirac theory outlined thus far cannot account for the aforementioned phenomena. A practical difficulty encountered in calculating with the hole theory is that the wave function which describes even a single negaton must take into account all the filled negative energy states. Such a wave function in configuration space must therefore be an infinite dimensional determinant. We shall see later in Chapter 8 that it is possible to formulate the hole theory in a concise and simple fashion using field theoretic methods.

We conclude this section by obtaining the relation between the description in terms of positive energy positonic states and that in terms of emptied negative energy states. For this purpose consider the process of pair creation by two photons. In the negaton-positon description we have initially a state consisting of two photons of energy-momentum $k_{1\mu}$ and $k_{2\mu}$, respectively. In the final state we have a positon of energy-momentum $p_{+\mu}$, and a negaton of energy-momentum $p_{-\mu}$. Energy-momentum conservation for the process requires that

$$p_{+\mu} + p_{-\mu} = k_{1\mu} + k_{2\mu} \tag{200}$$

In the hole theoretic description, on the other hand, in the initial state we have the two photons present and the negative-energy state q_μ occupied, whereas in the final state the state q_μ is empty and there is a negatively charged electron in the state $p_{-\mu}$. Energy-momentum conservation now asserts that

$$k_{1\mu} + k_{2\mu} + q_\mu = p_{-\mu} \tag{201}$$

On comparing (200) and (201) we obtain $p_{+\mu} = -q_\mu$. Thus the positon has a momentum which is opposite to that of the emptied "hole" and an energy equal to $-q_0$, where q_0 is the energy of the hole. Since q_0 is negative, p_{+0} is positive. Similarly, the angular momentum of the hole is opposite to that of the positon.

4h. Dirac Equation in External Field—Charge Conjugation

Although we have noted that a consistent one-particle interpretation for the Dirac equation can be given only in the absence of interactions, nonetheless we shall find that the solutions of the Dirac equation in external fields play an important role in the mathematical formulation of the field theory. Furthermore, we shall establish that the quantized field theory indicates that in the nonrelativistic limit the amplitude — from which the observables of a one-particle system in the presence of a weak, slowly varying field can be calculated — obeys to first approximation the Dirac equation for a particle in this external field.

If a spin $\frac{1}{2}$ particle has a charge $-e$ and interacts with an external electromagnetic field specified by a four-vector potential $A_\mu(x)$, which in the Lorentz gauge satisfies the gauge equation $\partial^\mu A_\mu(x) = 0$, then the Dirac equation describing its motion (neglecting radiative corrections) is obtained by the gauge invariant replacement

$$p_\mu \to p_\mu - \frac{e}{c} A_\mu$$

in (29). In the presence of an external classically prescribed field the Dirac equation is thus

$$\gamma^\mu \left(i\hbar\partial_\mu - \frac{e}{c} A_\mu(x) \right) \psi(x) = mc\psi(x) \tag{202a}$$

or

$$\left(\not{p} - \frac{e}{c} \not{A} - mc \right) \psi = 0 \tag{202b}$$

This equation can be transformed into a second-order equation which is similar in form to the Klein-Gordon equation in the presence of an external electromagnetic field. Define the spinor χ by the equation

$$\psi = \frac{1}{mc} \left(\not{p} - \frac{e}{c} \not{A} + mc \right) \chi \tag{203}$$

In term of χ, Eq. (202) becomes

$$\left(\not{p} - \frac{e}{c} \not{A} \right) \left(\not{p} - \frac{e}{c} \not{A} \right) \chi = m^2c^2\chi \tag{204}$$

the left-hand side of which can be rewritten as follows, recalling that $\sigma^{\mu\nu}$ is defined as $\sigma^{\mu\nu} = -\sigma^{\nu\mu} = \frac{1}{2i} (\gamma^\mu\gamma^\nu - \gamma^\nu\gamma^\mu)$:

$$\gamma^\mu\gamma^\nu \left(p_\mu - \frac{e}{c} A_\mu \right) \left(p_\nu - \frac{e}{c} A_\nu \right) = (g^{\mu\nu} - i\sigma^{\mu\nu}) \left(p_\mu - \frac{e}{c} A_\mu \right) \left(p_\nu - \frac{e}{c} A_\nu \right)$$

$$= g^{\mu\nu} \left(p_\mu - \frac{e}{c} A_\mu \right) \left(p_\nu - \frac{e}{c} A_\nu \right) - \frac{i}{2} (\sigma^{\mu\nu} - \sigma^{\nu\mu}) \left(p_\mu - \frac{e}{c} A_\mu \right) \left(p_\nu - \frac{e}{c} A_\nu \right)$$

$$= \left(p - \frac{e}{c} A \right)^2 - \frac{i}{2} \sigma^{\mu\nu} \left[p_\mu - \frac{e}{c} A_\mu, p_\nu - \frac{e}{c} A_\nu \right] \tag{205}$$

Since

$$[p_\mu, A_\nu] = i\hbar \frac{\partial A_\nu}{\partial x^\mu} \tag{206}$$

and the electromagnetic field tensor $F_{\mu\nu}$ is given by

$$F_{\mu\nu} = \partial_\nu A_\mu - \partial_\mu A_\nu \tag{207}$$

upon evaluating the commutator in Eq. (205), we finally obtain the following expression for Eq. (204)

$$\left\{\left(p_\mu - \frac{e}{c} A_\mu\right)\left(p^\mu - \frac{e}{c} A^\mu\right) + \frac{1}{2}\frac{e\hbar}{c}\,\sigma^{\mu\nu}F_{\mu\nu}\right\}\chi = m^2c^2\chi \tag{208}$$

The additional term

$$\tfrac{1}{2}\sigma^{\mu\nu}F_{\mu\nu} = \boldsymbol{\Sigma}\cdot\boldsymbol{\mathcal{H}} - i\boldsymbol{\alpha}\cdot\boldsymbol{\mathcal{E}} \tag{209}$$

corresponds to the interaction of a spin moment $\frac{1}{2}\hbar\boldsymbol{\Sigma}$ with the magnetic field $\boldsymbol{\mathcal{H}}$, and of an electric moment with the electric field $\boldsymbol{\mathcal{E}}$. The Dirac equation thus predicts that a charged spin $\frac{1}{2}$ particle has a magnetic moment of one Bohr magneton. We shall see later that radiative corrections (the effects of the interaction of the particle with the vacuum fluctuations of other fields) alters this prediction and gives rise to corrections to the value of the magnetic moment.

Returning to Eq. (208), it will be noted that it is a second-order equation in the four-component spinor χ. It thus has twice as many solutions as we want. However, Eq. (208) contains only the operator $\sigma^{\mu\nu}$ with which the operator γ_5 commutes. There exist, therefore, solutions of (208) for which $i\gamma_5\chi = \lambda\chi$. Furthermore, since $(i\gamma_5)^2 = I$, $\lambda^2 = 1$ and $\lambda = \pm 1$. In the representation (24) for the γ matrices, the eigenfunctions of $i\gamma_5$ with eigenvalue $+1$ are of the form

$$\chi_+ = \begin{pmatrix} v \\ -v \end{pmatrix} \tag{210a}$$

while those belonging to (-1) are of the form

$$\chi_- = \begin{pmatrix} u \\ u \end{pmatrix} \tag{210b}$$

If we restrict ourselves to solutions of (208) which belong either to $\lambda = +1$ or to $\lambda = -1$ we can then make a one-to-one between these solutions and the solutions of the first-order Dirac equation (202). Let us select the set for which $i\gamma_5\chi_+ = \chi_+$. The one-to-one correspondence is then exhibited by demonstrating that for each ψ there is an unique χ_+. By multiplying Eq. (203) by $\frac{1}{2}(1 + i\gamma_5)$, using the fact that $i\gamma_5\chi_+ = \chi_+$, we find that

$$\chi_+ = \tfrac{1}{2}(1 + i\gamma_5)\,\psi \tag{211}$$

which is the desired relation. We also note from (210a) that χ_+ can be specified by the two-component spinor v. By noticing that we can write σ_{30} in the form

$$\begin{aligned}\sigma_{30} &= \frac{1}{i}\,\gamma_{30} \\ &= i\gamma_1\gamma_2(\gamma_0\gamma_1\gamma_2\gamma_3) = -i\sigma_{12}\gamma_5\end{aligned} \tag{212}$$

we may therefore replace σ_{30} when operating on χ_+ by σ_{12}; similarly, σ_{10} and σ_{20} can be replaced by σ_{23} and σ_{13}, respectively, when operating on χ_+. In the representation (24) for the γ matrices the equation satisfied by the two-component spinor v is therefore

$$\left[\left(i\hbar\partial - \frac{e}{c}A\right)^2 + \boldsymbol{\sigma} \cdot (\mathcal{H} + i\mathcal{E})\right]v = m^2c^2v \tag{213}$$

Feynman and Gell-Mann [Feynman (1958)] have made the proposal that electrons be described by the two-component spinor v satisfying the second-order equation (213). Positons would again be made to correspond to the negative energy solutions of (213). In all theories where negatons and positons are created or destroyed in pairs, it can be shown [Brown (1958)] that a formulation in terms of the spinors v is equivalent to one in terms of ψ. It does not, however, seem possible to incorporate such a description into a quantized field theoretic formalism based on a Lagrangian formalism [Kibble and Polkinghorne (1958)], although a Feynman path integral description [Feynman (1948)] can be based on it. We will not pursue this description of electrons further.

Actually, the electromagnetic properties of a Dirac particle are best exhibited by a transformation to a Foldy-Wouthuysen (F-W) representation. In the presence of interaction, the generator for the transformation can only be obtained [Foldy (1950)] as a power series expansion in powers of the Compton wave length of the particle. The transformed Hamiltonian in the F-W representation is therefore likewise obtainable only as a power series in the same parameter. We here quote the result to order $(\hbar/mc)^2$

$$\begin{aligned}\Big\{\beta mc^2 &+ \frac{\beta}{2m}\left(\mathbf{p} - \frac{e}{c}\mathbf{A}\right)^2 - eA^0 - \frac{e\hbar}{2mc}\beta\boldsymbol{\sigma} \cdot \mathcal{H} \\ &+ \frac{e\hbar}{8mc^2}\left[\boldsymbol{\sigma} \cdot \left(\mathbf{p} - \frac{e}{c}\mathbf{A}\right) \times \mathcal{E} - \boldsymbol{\sigma} \cdot \mathcal{E} \times \left(\mathbf{p} - \frac{e}{c}\mathbf{A}\right)\right] \\ &- \frac{e\hbar^2}{8m^2c^2}\operatorname{div}\mathcal{E} + \cdots\Big\}\psi' = i\hbar\partial_t\psi'\end{aligned} \tag{214}$$

The term $\beta\left(\mathbf{p} - \frac{e}{c}\mathbf{A}\right)^2 - eA^0$ represents the interaction of a point charge with the electromagnetic field, the term $-(e\hbar/2mc)\,\beta\boldsymbol{\sigma} \cdot \mathcal{H}$ that of a magnetic moment of one Bohr magneton, $e\hbar/2mc$, with the magnetic field. The term in the braces represents a spin-orbit coupling term. It arises from the fact that the motion of the magnetic moment gives rise to an electric moment for the particle which then interacts with the electric field. The Darwin term $\frac{e^2\hbar^2}{8m^2c^2}\operatorname{div}\mathcal{E}$ corresponds to a correction to the direct point charge interaction due to the fact that in the Foldy-Wouthuysen representation the particle is not concentrated at a point but is spread out over a volume with a radius whose magnitude is roughly $\hbar/mc$. For a positive energy state the wave function in the Foldy-Wouthuysen representation is then identical with the nonrelativistic Pauli wave function for a spin $\frac{1}{2}$ particle [see e.g., Bethe and Salpeter (Bethe 1957)].

It has been shown by Pauli (1941) that the Dirac equation (202) in the presence of an external field can be further modified so as to represent a particle having an arbitrary magnetic moment, by adding to it the term $\frac{1}{2}\frac{\mu}{\hbar c}\sigma^{\rho\nu}F_{\rho\nu}$

$$\left\{\gamma\cdot\left(i\partial - \frac{e}{\hbar c}A\right) - \frac{1}{2}\frac{\mu}{\hbar c}\sigma^{\rho\nu}F_{\rho\nu}\right\}\psi = mc\psi \tag{215a}$$

whereupon the particle behaves as if it had an "anomalous" moment $\mu\, e\hbar/2mc$ addition to its "normal" moment $e\hbar/2mc$. Foldy (1952) has investigated the degree to which one can add further interaction terms to Eq. (202) without destroying its relativistic covariance and gauge invariance, assuming that the interaction is (*a*) linear in the electromagnetic fields (the assumption of weak fields), (*b*) does not vanish in the limit of vanishing momentum of the Dirac particle and hence does not involve derivatives of the wave function (the assumption of quasi-static fields), and (*c*) depends on the potentials and their derivatives only at x. He finds that the most general equation is

$$\left\{-i\gamma^\nu\partial_\nu - \frac{mc}{\hbar}\right\}\psi(x) - \frac{1}{\hbar c}\sum_{n=0}^{\infty}\left[\epsilon_n\Box^n\gamma^\nu A_\nu + \tfrac{1}{2}\mu_n\sigma^{\rho\nu}\Box^n F_{\rho\nu}\right]\psi(x) = 0 \tag{215b}$$

where $\Box$ is the D'Alembertian operator. The coefficients ϵ_n and μ_n are constants characterizing the interaction, with ϵ_0 the charge and μ_0 the anomalous magnetic moment of the particle. The higher terms represent direct interactions between the Dirac particle and the external charge and current distribution j_μ since $\Box A_\mu = -j_\mu$. An equation of the form (215b) containing terms up to $n = 1$ will be derived from field theory for the description of a charged spin $\frac{1}{2}$ particle interacting with a weak, slowly-varying electromagnetic field. The constants ϵ_1 and μ_1 are introduced by the electromagnetic effects of the "cloud" of virtual quanta which surround the particles.

There are several problems for which the Dirac equation (202a) can be solved exactly. The most important of these are:

1. The Coulomb potential [Dirac (1928), Darwin (1928), Gordon (1928), Mott (1929), Hylleraas (1955)];

2. The case of a homogeneous magnetic field extending over all space [Rabi (1928), Huff (1931), Sauter (1931), Johnson (1949)];

3. The field of an electromagnetic plane wave [Volkow (1935)].

These are treated in many references and we refer the reader to the *Handbuch* article of Bethe and Salpeter [Bethe (1957)] or to Akhiezer and Berezetski [Akhiezer (1953)] for a detailed discussion.

The solution in the Coulomb field has important applications, particularly to the energy levels of the hydrogen atom and to the calculation of x-ray spectra due to the K and L electrons of the heavy elements.

In the case of the hydrogen atom, the energy levels E_{nj} are described by the principal quantum number n and the total angular momentum quantum number j with

$$E_{n,j} = mc^2 \left\{1 + \frac{\alpha^2 Z^2}{(n' + \sqrt{(j + \frac{1}{2})^2 - \alpha^2 Z^2})^2}\right\}^{1/2}$$

$$n' = 0, 1, 2, \cdots ; \quad j = \frac{1}{2}, \frac{3}{2}, \cdots ; \quad \alpha = \frac{e^2}{4\pi\hbar c} \cong \frac{1}{137}$$

$$E_{n,j} \approx mc^2 - \frac{1}{2}\alpha^2 mc^2 \frac{Z^2}{n^2}\left[1 + \frac{\alpha^2 Z^2}{n}\left(\frac{1}{j + \frac{1}{2}} - \frac{3}{4n}\right) + \cdots\right]$$

$$n = n' + j + \tfrac{1}{2} = 1, 2, \cdots \qquad (216)$$

If the Dirac matrices are defined as in Section 4*a*, the "large" components of the wave function, ψ_1 and ψ_2, will have an orbital momentum $l_1 = j + \frac{1}{2}$ (or $l_1 = j - \frac{1}{2}$), and the small components, ψ_3 and ψ_3, will then have orbital momentum $l_2 = j - \frac{1}{2}$ (or $l_2 = j_2 + \frac{1}{2}$). Thus large and small components have opposite parity, and that of the large components, $(-1)^{l_1}$, is designated as the parity of the state.

The observed fine structure of the levels of hydrogen and hydrogen-like atoms, particularly He^+, is in good agreement with Dirac's theory. This agreement includes the degeneracy of levels with the same j and different l except for one case, namely, the $S_{1/2}$ and $P_{1/2}$ levels. In this case, Lamb (1947) found the $2S_{1/2}$ state of hydrogen to be about 1058 megacycles higher than the $2P_{1/2}$ state, a separation which has been explained by radiative corrections to the simple Dirac theory to an accuracy of 0.1 megacycles. [For a detailed description of the hydrogen spectrum and its interpretation, see Series (1957).]

The theory of the x-ray levels of heavy elements has not as yet been worked out to such perfection. The contributions other than the Lamb shift and vacuum polarization to the K-absorption edge have been noted by Brenner and Brown [Brenner (1953)], (e.g., the energy from the interaction between the two K-electrons, the energy from the interaction of the K-electron with all the outer electrons). These contributions have been carefully evaluated by Cohen (1954, 1960) who obtains wave functions by carrying out a relativistic self-consistent field calculation. For mercury the best experimental value for the K-energy is 6107.7 ± 0.6 Ry. The discrepancy between the experimental value and the theoretical one including all effects of order αmc^2 except for the Lamb shift is approximately 30 Ry. The Lamb shift of a K-electron in mercury, exclusive of the vacuum polarization, has been calculated by Brown, Langer, and Schaffer [Brown (1959)], who find a value of 41 Ry for it. The vacuum polarization effect gives rise to a contribution of −3 Ry [Wichmann (1956)], so that the Lamb shift is 38 Ry. Inclusion of the Lamb shift therefore significantly improves agreement between experiment and theory. The theory of the L shell energy has as yet not been worked out to the same accuracy because the calculation of the Lamb shift for 2s electrons in mercury is very complicated. Preliminary results for the 2s-1s energy difference [Cohen (1960)] give a value of 5035.2 Ry as compared with an experimental value of 5018 Ry.

The theory of electron scattering by a Coulomb field was originally worked out by Mott in 1929 [Mott (1929); see also Mott (1949)]. In recent years the experimental data on the scattering of high energy electrons by nuclei has been the source of the most accurate and detailed information about nuclear charge sizes and radial shapes. This is due to the fact that the nature of the electron-nucleon and electron-nuclear interaction is reasonably well understood theoretically (it is primarily electromagnetic) and that experimentally mono-energetic beams of high intensity electrons are readily obtainable. Ford and Hill [Ford (1955)] have reviewed all the methods of examining nuclear charge distributions and the results of electron-nuclear scattering experiments prior to 1955. The excellent review article of Hofstadter (1957) contains a complete account of the theory of electron scattering as well as a presentation of the experimental data on high energy electron-nucleon and electron-nuclear scattering and its theoretical inference. [See also Ravenhall (1958).] We refer the reader to these articles for a detailed exposition.

We conclude this section with a derivation of certain relations between the solutions for a charge $+e$ and those for a charge $-e$. To this end, consider the set of matrices $\gamma_\mu{}^T$, $\mu = 0, 1, 2, 3$. By taking the transpose of the commutation rules satisfied by the γ matrices, $[\gamma^\mu, \gamma^\nu]_+ = 2g^{\mu\nu}$ we see that $\gamma_\mu{}^T$ satisfies the same commutation rules, namely:

$$\gamma_\mu{}^T\gamma_\nu{}^T + \gamma_\nu{}^T\gamma_\mu{}^T = 2g_{\mu\nu} \tag{217}$$

Since the 4×4 representation of the γ matrices is irreducible, by the fundamental lemma of Section 4b there exists a nonsingular matrix B such that

$$\gamma^{\mu T} = B^{-1}\gamma^\mu B \tag{218}$$

Furthermore B can be chosen to be unitary. By taking the transpose of Eq. (218) one verifies that B^TB^{-1} commutes with all the γ^μs and hence $B^TB^{-1} = aI$ where a is a constant. By taking the inverse and complex conjugate of the equation $B^TB^{-1} = aI$, recalling that B is assumed unitary, one finds that $a = \pm 1$, so that $B = \pm B^T$. To decide whether the + or − sign is correct, we note that with the choice $B = -B^T$

$$\gamma^\mu B = BB^{-1}\gamma^\mu B = -B^T\gamma^{\mu T} = -(\gamma^\mu B)^T \tag{219a}$$

and similarly

$$\gamma^5 B = BB^{-1}\gamma^5 B = -B^T(\gamma^3\gamma^2\gamma^1\gamma^0)^T = -(\gamma^5 B)^T \tag{219b}$$

so that B, $B\gamma^\mu$ and $B\gamma^5$ are antisymmetric. These constitute a set of six antisymmetric matrices. In a similar fashion, one verifies that the ten matrices $B\gamma^5\gamma^\mu$, $B\sigma^{\mu\nu}$ are symmetric. Conversely, with the choice $B = +B^T$ the ten matrices $B\gamma^5\gamma^\mu$, $B\sigma^{\mu\nu}$ are antisymmetric and the matrices B, $B\gamma^\mu$ and $B\gamma^5$ are symmetric. The latter is, however, not possible since the ten matrices $B\gamma^5\gamma^\mu$, $B\sigma^{\mu\nu}$ are linearly independent and there are only six linearly independent 4×4 antisymmetric matrices. We therefore conclude [Pauli (1935)] that $B = -B^T$. If we introduce the matrix

$$C = -\gamma_5 B \tag{220}$$

it has the property that it is unitary

$$C^{-1} = -B^{-1}\gamma_5^{-1} = B^{-1}\gamma_5 = -B^*\gamma_5^* = (-\gamma_5 B)^* = C^* \tag{221}$$

and from (219b) that it is antisymmetric

$$C^T = -C \tag{222}$$

Furthermore, under the similarity transformation induced by C, the γ^μs are transformed into minus their transpose:

$$\begin{aligned} C^{-1}\gamma^\mu C &= -B^{-1}\gamma_5\gamma^\mu\gamma_5 B = -B^{-1}\gamma^\mu B \\ &= -\gamma^{\mu T} \end{aligned} \tag{223}$$

Consider now the Dirac equation in the presence of an electromagnetic field:

$$(i\gamma^\mu\partial_\mu - m)\,\psi = e\gamma^\mu A_\mu\psi \tag{224a}$$

$$(i\partial_\mu\tilde{\psi}\gamma^\mu + m\tilde{\psi}) = -e\tilde{\psi}\gamma^\mu A_\mu \tag{224b}$$

We note that if we take the transpose of Eq. (224b) and replace $\gamma^{\mu T}$ by $-C^{-1}\gamma^\mu C$, we obtain the following equation:

$$(i\gamma^\mu\partial_\mu - m)\,C\tilde{\psi}^T = -e\gamma^\mu A_\mu C\tilde{\psi}^T \tag{225}$$

Thus if ψ describes the motion of a particle of charge e, magnetic moment μ, then $C\tilde{\psi}^T$ describes that of a particle of charge $-e$, magnetic moment $-\mu$. One calls $C\tilde{\psi}^T$ the charge conjugate spinor and denotes it by

$$\psi^c = C\tilde{\psi}^T \tag{226}$$

It should be noted that in the one-particle theory under consideration the operation of charge conjugation, U_c, under which

$$U_c\psi = \psi^c = C\tilde{\psi}^T \tag{227}$$

is antiunitary. If the operation of charge conjugation is to be relativistically invariant then ψ^c and ψ must have the same transformation properties under Lorentz transformations, i.e.,

$$\psi'(x') = S(\Lambda)\,\psi(x) \tag{228a}$$

$$\psi'^c(x') = S(\Lambda)\,\psi^c(x) \tag{228b}$$

Inserting the expression (226) for ψ^c into (228b) and recalling that $\tilde{\psi}' = \pm\tilde{\psi}S(\Lambda)^{-1}$ (+ for $L\uparrow$ and − for $L\downarrow$), we infer that $\pm(S^{-1})^T\,\tilde{\psi}^T = SC\tilde{\psi}^T$ or that

$$S(\Lambda)^T = C^{-1}S^{-1}(\Lambda)\,C \qquad \text{for } \Lambda\epsilon L\uparrow \text{ (i.e., when } \Lambda^{00} \geqslant 1) \tag{229a}$$

$$S(\Lambda)^T = -C^{-1}S^{-1}(\Lambda)\,C \quad \text{for } \Lambda\epsilon L\downarrow \text{ (i.e., when } \Lambda^{00} \leqslant -1) \tag{229b}$$

Equations (229a) and (229b) can be used to narrow the choices of possible $S(i_s)$s and $S(i_t)$s. For example, if we choose for $S(i_s)$ the matrix γ^0, then it will be noted that since $\gamma^{0T} = \gamma^0 = (\gamma^0)^{-1}$, (229a) cannot be satisfied. Hence, we must choose $\pm i\gamma^0$ for $S(i_s)$ [Racah (1937)]. For time inversions, the choice $S(i_t) = \pm i\gamma_5 C = \pm iB$ guarantees that (229b) is satisfied.

Consider finally the properties of the charge conjugate spinors in the case of free particles. We know from our previous remarks that a negaton is described by a wave function $u(p)$ which in momentum space obeys

$$(\not p - m)\,u(p) = 0 \tag{230}$$

Similarly, we have inferred that the amplitude for finding a positon of energy-momentum $+p$ is expressed in terms of a negative energy spinor of momentum $-\mathbf{p}$, i.e., by the wave function of the electron to whose absence (in the sea of negative energy states) it corresponds. It satisfies the equation

$$(-\not p - m)\,v(p) = 0 \tag{231a}$$

and

$$\tilde{v}(p)\,(-\not p - m) = 0 \tag{231b}$$

Again taking the transpose of Eq. (231b), and using (223), we find that the spinor

$$C\tilde{v}^T(p) = u^c(p) \tag{232}$$

satisfies the equation $(\not p - m)\,u^c(p) = 0$. The charge conjugate spinor therefore refers to a particle of energy-momentum $+p$. Using the property of the matrix C, we may invert this last equation and write

$$v(p) = C[\tilde{u}^c(p)]^T \tag{233}$$

5

The Zero Mass Equations

In this chapter we consider the two mass zero equations of greatest interest for physical applications at the microscopic level: the neutrino equation and the photon (Maxwell's) equation. We shall not consider the equation for a graviton.

5a. The Two-Component Theory of the Neutrino

This theory was first proposed by Weyl in 1929 [Weyl (1929)] to describe a particle of mass zero and spin $\frac{1}{2}$. It was discussed in Pauli's *Handbuch* article but was rejected by him because of its noninvariance under space reflections. When recently experiments revealed that parity is not conserved in β decay, Lee and Yang [Lee (1957a)], who were responsible for the original suggestion that parity might not be conserved and indicated ways of testing this hypothesis, Landau (1957), and Salam (1957) proposed that neutrinos obey the Weyl equation to account for the observed nonconservation of parity.

The wave equation obeyed by the two-component amplitude, $\phi(x)$, describing neutrinos is

$$i\hbar\partial_t\phi(x) = -ci\hbar\boldsymbol{\sigma}\cdot\boldsymbol{\nabla}\phi(x) \tag{1}$$

or, using the notation $\mathbf{p} = -i\hbar\boldsymbol{\nabla}$,

$$i\hbar\partial_t\phi = c\boldsymbol{\sigma}\cdot\mathbf{p}\phi \tag{2}$$

where σ_i are the 2×2 Pauli matrices. Using arguments similar to those used in showing that the Dirac equation is form invariant under restricted inhomogeneous Lorentz transformations, one can prove the form invariance of the Weyl equation. For notational convenience we introduce the matrix $\sigma_0 = I$ and write Eq. (2) in the form

$$\sigma^\mu p_\mu\phi = 0 \tag{3}$$

This equation will be form invariant if $\phi(x)$ transforms under restricted inhomogeneous Lorentz transformations $x' = \Lambda x + a$, $p' = \Lambda^{-1}p$ according to

$$\phi'(x') = S(\Lambda)\,\phi(x) \tag{4}$$

where $S(\Lambda)$ is a two-by-two matrix which satisfies the equation

$$S(\Lambda)^*\,\sigma^\mu S(\Lambda) = \Lambda^\mu{}_\nu \sigma^\nu \tag{5}$$

The Weyl equation in the primed system will then be

$$\sigma^\mu p'_\mu \phi'(x') = 0 \tag{6}$$

The form of $S(\Lambda)$ for particular Lorentz transformations can be ascertained as in the Dirac case. For a Lorentz transformation along the ith axis one finds

$$S(i, \omega) = e^{-\frac{1}{2}\omega\sigma_i}; \qquad \tanh\omega = \frac{v}{c} \tag{7}$$

and for a rotation through an angle θ about the jth axis

$$S(j; \theta) = e^{\frac{1}{2}i\theta\sigma_j} \tag{8}$$

The spinor ϕ thus transforms under homogeneous Lorentz transformations according to the $D^{(1/2,0)}$ irreducible representation of the homogeneous Lorentz group. There exists another wave equation describing a spin $\frac{1}{2}$ mass zero particle, the wave function ψ of which transforms according to the $D^{(0,1/2)}$ representation of the homogeneous Lorentz group. It is

$$i\hbar\partial_t\psi(x) = +i\hbar c\boldsymbol{\sigma}\cdot\nabla\psi(x) \tag{9}$$

Under a homogeneous Lorentz transformation along the ith axis ψ transforms according to

$$\psi' = e^{+\frac{1}{2}\omega\sigma_i}\psi \tag{10}$$

As noted previously, the representations $D^{(1/2,0)}$ and $D^{(0,1/2)}$ are no longer irreducible if we include space inversions. Under a space inversion $D^{(1/2,0)} \rightleftarrows D^{(0,1/2)}$ and only the direct sum of these representations $D^{(1/2,0)} \oplus D^{(0,1/2)}$ is irreducible. Hence (1) and (9) are not covariant under space inversions.

The plane wave solutions of (1) are of the form

$$\phi(x) = e^{-ip\cdot x}\,u(p) \tag{11}$$

with $u(p)$ a two-component spinor which satisfies the equation

$$p_0 u = \boldsymbol{\sigma}\cdot\mathbf{p}u \tag{12}$$

Multiplying both sides by $\boldsymbol{\sigma}\cdot\mathbf{p}$, we obtain

$$(p_0^2 - \mathbf{p}^2)\,u = 0 \tag{13}$$

so that nonvanishing solutions, u, exist only when $p_0 = \pm|\mathbf{p}|$: neutrinos travel with the speed of light. We further note that Eq. (12) implies that the solutions for a given sign of the energy correspond to a definite orientation of the spin with respect to the direction of motion. The experimental evidence from β decay indicates that the neutrino always has its spin antiparallel to its direction of motion and obeys the wave equation

(9), the plane wave solutions of which are given by $\psi = \exp(-ipx)\, v(p)$ with

$$p_0 v(p) = -\boldsymbol{\sigma} \cdot \mathbf{p} v(p) \tag{14}$$

We call the solution with $p_0 = +|\mathbf{p}|$ of this equation a neutrino state. In this state the spin is antiparallel to $\mathbf{p}$ and can be represented by a left-hand screw as in Figure 5.1. The solution with $p_0 = -|\mathbf{p}|$ has the spin

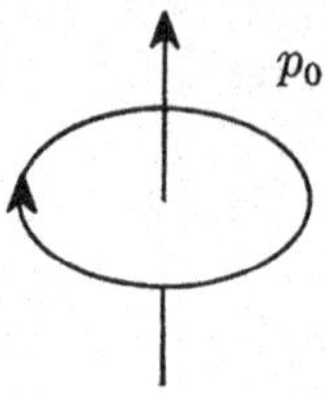

Fig. 5.1

parallel to $\mathbf{p}$ and can be represented by a right-hand screw (Fig. 5.2). In a hole theoretic interpretation of the negative energy states the antineutrino will then have a momentum opposite to that of the negative energy state which is vacated. Hence the relation of the spin direction to the momentum of the antineutrino is represented by a right-handed screw.

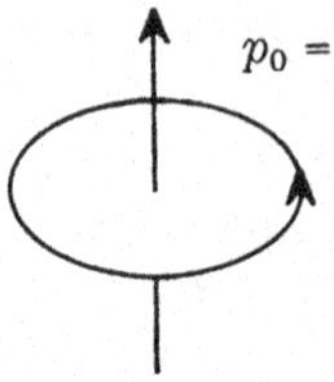

Fig. 5.2

The noninvariance of this two-component theory can now be graphically represented. Under the operation of space inversion ($\mathbf{p} \to -\mathbf{p}$, $\mathbf{x} \to -\mathbf{x}$, $\boldsymbol{\sigma} \to \boldsymbol{\sigma}$) the state of energy $p_0 = +|\mathbf{p}|$, momentum $\mathbf{p}$, and helicity $\boldsymbol{\sigma} \cdot \mathbf{p}/|\mathbf{p}| = +1$ is carried into a state with $p_0 = +|\mathbf{p}|$, momentum $-\mathbf{p}$, and helicity -1, but no such state exists for the two-component neutrino (see Fig. 5.3). This is another way of stating the noninvariance under

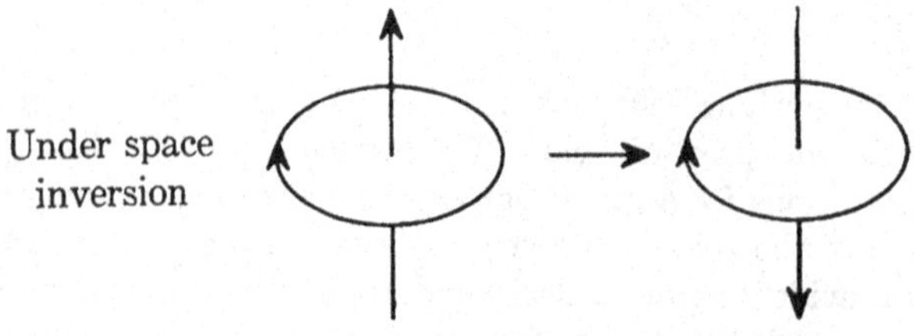

Fig. 5.3

space inversion of the two-component theory. Alternatively the lack of invariance of (1) and (9) can be inferred from the fact $\mathbf{p}$ is a polar vector, whereas $\boldsymbol{\sigma}$ is an axial vector, so that $\boldsymbol{\sigma} \cdot \mathbf{p}$ is a pseudoscalar under inversions.

The Weyl two-component theory is equivalent to a Dirac (four-component) description in which some constraints have been imposed on the amplitudes. To determine these constraints let us consider the solutions of the Dirac equation for a massless particle:

$$i\gamma_\mu \partial^\mu \psi(x) = 0 \tag{15}$$

which in Hamiltonian form reads

$$i\partial_t \psi(x) = -i\gamma^0 \boldsymbol{\gamma} \cdot \nabla \psi(x) \tag{16}$$

With the representation

$$\gamma^0 = \begin{pmatrix} I & 0 \\ 0 & -I \end{pmatrix} \qquad \gamma_5 = -i \begin{pmatrix} 0 & I \\ I & 0 \end{pmatrix} \tag{17}$$

$$\boldsymbol{\gamma} = \begin{pmatrix} 0 & \boldsymbol{\sigma} \\ -\boldsymbol{\sigma} & 0 \end{pmatrix} \tag{18}$$

one notes that

$$\boldsymbol{\Sigma} = \begin{pmatrix} \boldsymbol{\sigma} & 0 \\ 0 & \boldsymbol{\sigma} \end{pmatrix} = i\gamma_5 \gamma^0 \boldsymbol{\gamma} \tag{19}$$

so that the Hamiltonian can be written in the form

$$H = i\gamma_5 \boldsymbol{\Sigma} \cdot \mathbf{p} = i\gamma_5 |\mathbf{p}| s(\mathbf{p}) \tag{20}$$

The eigenfunctions of H and $s(\mathbf{p})$ are therefore also eigenfunctions of $i\gamma_5$. Now the four linearly independent solutions of $Hu = p_0 u$ in the case in which one takes the z axis as the direction of $\mathbf{p}$ are given by:

$$\begin{array}{lcccc} \text{Helicity:} & +1 & -1 & +1 & -1 \\ & \begin{pmatrix} 1 \\ 0 \\ 1 \\ 0 \end{pmatrix} & \begin{pmatrix} 0 \\ 1 \\ 0 \\ -1 \end{pmatrix} & \begin{pmatrix} -1 \\ 0 \\ 1 \\ 0 \end{pmatrix} & \begin{pmatrix} 0 \\ 1 \\ 0 \\ 1 \end{pmatrix} \\ & \multicolumn{2}{c}{\textit{positive energy}} & \multicolumn{2}{c}{\textit{negative energy}} \end{array} \tag{21}$$

The eigenvalue of $i\gamma_5$ for these solutions are summarized in Table 5.1.

TABLE 5.1

p_0	*Helicity*	*Eigenvalue of* $i\gamma_5$
+	+1	+1
+	−1	−1
−	+1	−1
−	−1	+1

Thus, the eigenvalue of $i\gamma_5$ for a positive energy solution is the same as that of the helicity operator, whereas for a negative energy solution the eigenvalue of $i\gamma_5$ is -1 times that of the helicity operator. We have seen that a neutrino is described by a two-component equation, the plane wave solutions of which have the property that for $p_0 = +|\mathbf{p}|$ the helicity is -1, and for $p_0 = -|\mathbf{p}|$ the helicity is $+1$. But this is also the property of the plane wave solutions of (16) upon which the constraint:

$$\psi = -i\gamma_5\psi \tag{22}$$

has been placed. This constraint is invariant for restricted Lorentz transformations. Alternatively, if ψ is a four-component spinor satisfying Eq. (16), the spinor ψ_n defined by

$$\psi_n = \tfrac{1}{2}(1 - i\gamma_5)\,\psi \tag{23}$$

automatically satisfies the conditions that

$$i\gamma_5\psi_n = -\psi_n \tag{24}$$

In a representation where

$$\gamma^0 = \begin{pmatrix} 0 & \mathrm{I} \\ \mathrm{I} & 0 \end{pmatrix} \qquad \boldsymbol{\gamma} = \begin{pmatrix} 0 & \boldsymbol{\sigma} \\ -\boldsymbol{\sigma} & 0 \end{pmatrix} \tag{25a}$$

$$\gamma_5 = i\begin{pmatrix} \mathrm{I} & 0 \\ 0 & -\mathrm{I} \end{pmatrix} \tag{25b}$$

the spinor ψ_n is essentially a two-component quantity

$$\psi_n = \begin{pmatrix} \psi_n' \\ 0 \end{pmatrix} \tag{26}$$

since the projection operator $\frac{1}{2}(1 - i\gamma_5)$ annihilates the two lower components. The two-component spinor ψ_n' satisfies the equation

$$-\boldsymbol{\sigma}\cdot\mathbf{p}\psi_n' = p_0\psi_n' \tag{27}$$

obtained by multiplying Eq. (16) by $\frac{1}{2}(1 - i\gamma_5)$.

One verifies that for the four-component mass zero theory there exists a Foldy-Wouthuysen transformation which brings the Hamiltonian in a form free of odd operators:

$$H' = \beta c|\mathbf{p}| \tag{28}$$

The canonical transformation which accomplishes this is

$$\begin{aligned} e^{iS} &= e^{\frac{\pi}{4}\beta\frac{\boldsymbol{\alpha}\cdot\mathbf{p}}{|\mathbf{p}|}} \\ &= \frac{1}{\sqrt{2}}\left(1 + \frac{\beta\boldsymbol{\alpha}\cdot\mathbf{p}}{|\mathbf{p}|}\right) \end{aligned} \tag{29}$$

Again a position operator can be defined in the Dirac representation by the equation

$$\mathbf{X} = e^{-iS}\,\mathbf{x}e^{iS} \tag{30}$$

It has the desired properties that (*a*) it transforms like a vector under rotations; (*b*) $[X_i, X_j] = 0$; (*c*) $[X_l, p_j] = i\delta_{lj}$; and (*d*) the time derivative of X_i is $p_i/c|\mathbf{p}|$ on the positive energy states. On the other hand in the two-component theory it can be shown [see, e.g., Fronsdal (1959)] that there does not exist a position operator which transforms like a vector under rotation.

5*b*. The Polarization States of Mass Zero Particles

The intrinsic angular momentum of a particle with zero rest mass is parallel to its direction of motion. If we associate an internal motion with the spin, then this motion is perpendicular to the velocity ("transverse polarization"). Furthermore, for the mass zero case the statement that the helicity, $s(\mathbf{p})$, is ± 1 is relativistically invariant in contrast to the case of a particle with nonvanishing rest mass. The angular momentum of a particle with a finite rest mass can also be parallel to the velocity and there exist states of transverse polarization, but these are not Lorentz invariant statements: if the velocity and spin are parallel in one frame, they are not parallel in another co-ordinate system. In particular, in the frame in which the particle is at rest the angular momentum is clearly not parallel to the velocity, since the latter is zero. Now every particle with a finite rest mass can be viewed from a co-ordinate system in which it is at rest. Therefore for such a particle the statement that the spin is parallel to the direction of motion cannot be valid for all observers and the particles must have other states of polarization. However, a mass zero particle traveling with the velocity of light cannot be viewed from a co-ordinate system in which it is at rest. This accounts for the difference between particles with zero rest mass which have only two directions of polarization no matter what their spin is, and particles with finite rest mass which have $2s + 1$ states of polarization if their spin is s. Wigner has analyzed this difference somewhat more fully [Wigner (1956, 1957)]. His analysis follows.

Consider a particle at rest with a given direction of polarization, say the z axis. One notes that one obtains the same state whether one first imparts a velocity v to the particle in the z direction and then rotates the state or whether one first rotates the state and then imparts to it a velocity in the direction of the spin. This is illustrated in Figure 5.4. The simplest way to verify this assertion is to view the particle from different Lorentz frames. Thus in the Lorentz frame moving with a velocity $-v$ with respect to the one in which the particle is at rest, the particle appears moving with a velocity $+v$ in the z direction. Similarly if a particle which has its spin pointing in the z direction in its rest frame is viewed from a rotated co-ordinate system, for example, one obtained by making a rota-

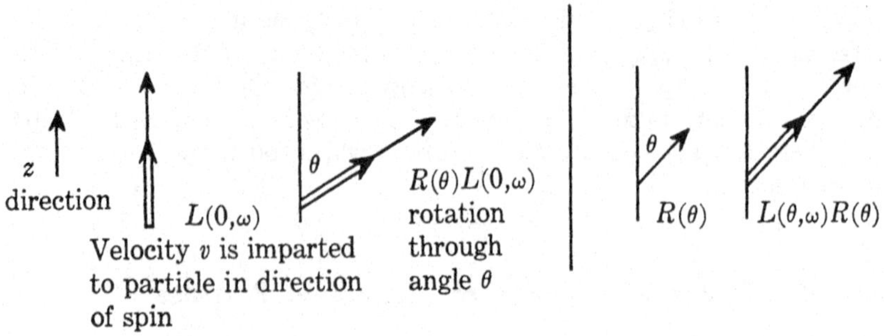

The short arrow denotes the spin, the double arrow the velocity of the particle

Fig. 5.4

tion through the angle θ about the x axis, the particle will appear to have its spin pointing in the $z'y'$ plane making an angle θ with the z' axis. These, of course, are merely illustrations of the fact that all the states of the system can be obtained by looking at a "standard state" from various co-ordinate systems. To define a standard state one chooses an arbitrary, but fixed, Lorentz frame and stipulates that in this Lorentz frame the particles have definite properties when in the standard state. In the above (and in the following) the standard state is that state in which the particle is at rest and has its spin pointing in the z direction. (Such a standard state therefore exists only for particles with nonvanishing rest mass.) Every Lorentz frame then defines a state of the system, namely that state in which the particle appears when viewed from this Lorentz frame. Two states of the system are then identical if and only if the Lorentz frames which define them are identical. Two states will be approximately the same if the Lorentz transformation which defines them differs by an infinitesimal amount, i.e., by a Lorentz transformation close to the identity. Note that the comparison of the states is independent of the properties of the particles, such as mass and spin, and depends only on the Lorentz transformation involved. The matrix[1]

$$L(0, \omega) = \begin{pmatrix} \cosh \omega & 0 & \sinh \omega \\ 0 & 1 & 0 \\ \sinh \omega & 0 & \cosh \omega \end{pmatrix} \tag{31}$$

characterizes the transformation to a Lorentz frame in which the particle moves with velocity v in the z direction and has its spin pointing in the z direction. The transformed frame moves with velocity $-v$ along the z direction with respect to the original Lorentz frame in which the standard

[1] Since the x' axis plays no role in the following consideration, it is suppressed and the three rows and columns of L refer to the t, y, z axes.

state is defined, so that $c \tanh \omega = v$; recall Eq. (2.10). It will also be recalled that the matrix corresponding to a rotation through the angle θ in the yz plane is

$$R(\theta) = \begin{pmatrix} 1 & 0 & 0 \\ 0 & \cos\theta & \sin\theta \\ 0 & -\sin\theta & \cos\theta \end{pmatrix} \tag{32}$$

Let $\mathbf{n}$ be a unit vector lying in the zy plane such that the angle between $\mathbf{n}$ and the z axis is θ, $\mathbf{e}_3 \cdot \mathbf{n} = \cos\theta$. The co-ordinate system which moves with velocity $-v$ in the direction $\mathbf{n}$ is then obtained by the transformation

$$L(\theta, \omega) = R(\theta)\, L(0, \omega)\, R(-\theta) \tag{33}$$

Proof: The transformation $R(-\theta)$ rotates the co-ordinate system so that the z' axis lies along $\mathbf{n}$, $L(0, \omega)$ then gives it a velocity $-v$ along $\mathbf{n}$ and $R(\theta)$ makes the z and z' axis parallel again. Now to obtain the state of a particle which moves in the direction $\mathbf{n}$ and is polarized in this direction, we first rotate the co-ordinate system through an angle $+\theta$ (so that the particle has its spin direction making an angle θ with the z' axis) and then impart a velocity $-v$ in the direction $\mathbf{n}$, that is, we perform the transformation

$$\begin{aligned} T(\theta, \omega) &= L(\theta, \omega)\, R(\theta) \\ &= R(\theta)\, L(0, \omega)\, R(-\theta)\, R(\theta) = R(\theta)\, L(0, \omega) \\ &= \begin{pmatrix} \cosh\omega & 0 & \sinh\omega \\ \sinh\omega \sin\theta & \cos\theta & \sin\theta \cosh\omega \\ \cos\theta \sinh\omega & -\sin\theta & \cos\theta \cosh\omega \end{pmatrix} \end{aligned} \tag{34}$$

which characterizes the above state of the particle. From the equality

$$R(\theta)\, T(0, \omega) = R(\theta)\, L(0, \omega) \tag{35}$$

it follows that the same state can be obtained by first imparting the velocity $-v$ to the co-ordinate system and then performing the rotation. The resulting state appears the same because the two co-ordinate systems are identical and they view the same particle. The result (34) implies that the statement "the particle has its spin parallel or antiparallel to its velocity" is invariant with respect to rotation. It is clearly also invariant with respect to further increase of the velocity in the z direction. Consider next the motion of a particle moving in the z direction with a velocity v and spin parallel to the z direction. When viewed from a co-ordinate system, l', moving in the $-y$ direction with velocity $v' = c \tanh \omega'$, the particle will now appear to have velocity pointing between the y and z axis and its spin will no longer be parallel to its direction of motion, unless the velocity v is close to the velocity of light. Stated more precisely, the states generated by $L(0, \omega)$ for large ω (i.e., when $v \sim c$) are such that when viewed from a different Lorentz frame which is not moving too rapidly in the direction of motion of the particle they have the property that spin and velocity are still nearly parallel. In the limiting case of particles moving with the velocity of light, i.e., particles with zero rest mass, the statement that spin and velocity are parallel is invariant under all Lorentz transformations.

The state seen by an observer moving in the Lorentz frame l' is generated by a transformation corresponding to

$$L(\tfrac{1}{2}\pi, \omega')\, L(0, \omega) = \begin{pmatrix} \cosh\omega' & \sinh\omega' & 0 \\ \sinh\omega' & \cosh\omega' & 0 \\ 0 & 0 & 1 \end{pmatrix} \begin{pmatrix} \cosh\omega & 0 & \sinh\omega \\ 0 & 1 & 0 \\ \sinh\omega & 0 & \cosh\omega \end{pmatrix}$$

$$= \begin{pmatrix} \cosh\omega'\cosh\omega & \sinh\omega' & \cosh\omega'\sinh\omega' \\ \sinh\omega'\cosh\omega & \cosh\omega' & \sinh\omega'\sinh\omega \\ \sinh\omega & 0 & \cosh\omega \end{pmatrix} \tag{36}$$

To bring this transformation into the form of Eq. (34), corresponding to a particle moving with its spin parallel to the velocity, we must multiply it on the right by $R(\epsilon)$ (i.e., the spin must be rotated before any motion is imparted) with the angle ϵ given by

$$\tan\epsilon = \frac{\tanh\omega'}{\tanh\omega} = \frac{v'}{v}\left(1 - \frac{v^2}{c^2}\right)^{1/2} \tag{37}$$

The angle ϵ is called the angle between spin and velocity. Note that for large v, the velocity v carries the spin with itself, and the angle between the direction of motion and spin direction is very small in the moving coordinate system. The above statements also prove that for a mass zero particle the statement that spin and velocity are parallel is invariant under all proper Lorentz transformations. The fact that there are two states of opposite helicity (rather than one) for such mass zero particles is then required by invariance under spatial reflections. In fact, as we have seen if the theory is not invariant under reflections so that the mirror image state does not exist, there is only one state of polarization for a mass zero particle. On the other hand, recall that for particles with mass the fact that there are two states of polarization follows from the invariance of the theory under *proper* Lorentz transformations (Chap. 2).

5c. The Photon Equation

We may adopt as the "wave function" for a single photon the vector valued function $A_i(x)$, $i = 1, 2, 3$, which satisfies the equation

$$i\hbar c\partial_0 \mathbf{A}(x) = \hbar\sqrt{-\nabla^2}\,\mathbf{A}(x) \tag{38a}$$

and the subsidiary (gauge) condition,

$$\nabla \cdot \mathbf{A}(x) = 0 \tag{38b}$$

Its physical interpretation is obtained from the Fourier transform $\chi(\mathbf{k})$ of $\mathbf{A}(x)$ defined by

$$\mathbf{A}(x) = \int_{k_0>0} \frac{d^3k}{k_0} e^{-ik\cdot x} \chi(\mathbf{k}) \tag{39}$$

In Eq. (39) $k_0 = |\mathbf{k}|$ so that Eq. (38) is satisfied; and $|\chi(\mathbf{k})|^2 d^3k$ is then proportional to the probability that the photon have a momentum between $\mathbf{k}$ and $\mathbf{k} + d\mathbf{k}$. If we introduce the three linearly independent vectors $\mathbf{e}_1(\mathbf{k})$, $\mathbf{e}_2(\mathbf{k})$ and $\mathbf{k}/|\mathbf{k}|$ with

$$\mathbf{e}_i(\mathbf{k}) \cdot \mathbf{e}_j(\mathbf{k}) = \delta_{ik} \quad i, j = 1, 2 \tag{40a}$$

$$\mathbf{e}_i(\mathbf{k}) \cdot \mathbf{k} = 0 \quad i, j = 1, 2 \tag{40b}$$

then the transversality condition (38b) implies that

$$\mathbf{A}(x) = \sum_{i=1,2} \int_+ \frac{d^3k}{k_0} \chi_i(\mathbf{k})\, \mathbf{e}_i(\mathbf{k})\, e^{-ik\cdot x} \tag{41}$$

where $\chi_i(\mathbf{k})$ is the amplitude for the photon to have momentum $\mathbf{k}$ and polarization i. Within the scalar product

$$(\chi, \phi) = \sum_{i=1}^{2} \int \frac{d^3k}{k_0} \bar{\chi}_i(\mathbf{k})\, \phi_i(\mathbf{k}) \tag{42}$$

by invoking arguments similar to those for the spin 0 case, one verifies that it is not possible to obtain any localized state for a photon [Newton (1949)]. The reason for this, roughly speaking, is that only the momenta in the plane normal to the "direction" of the assumed localized (three-vector) wave function are available, so that it is not possible to localize the photon in the direction of its polarization. Strictly speaking, therefore, only a momentum space interpretation exists for the photon wave function. We defer until Chapter 9 the presentation of a more covariant formulation of the one-photon amplitude.

Part Two

SECOND QUANTIZATION

6

Second Quantization: Nonrelativistic Theory

In this section we establish the equivalence between the ordinary wave mechanical description of a system of n identical particles and a certain operator formalism which has become known as "second quantization." The importance of this formalism derives from the fact that it permits performing calculations which automatically take into account the combinatorial aspects arising from the particular statistics (Bose-Einstein or Fermi-Dirac) obeyed by the particles. Secondly, it allows an extension of ordinary nonrelativistic quantum mechanics to systems for which the number of particles is no longer a constant of the motion. Such an extension is in fact necessary to describe the physical phenomena encountered in the relativistic domain.

The equivalence of the operator (second-quantized) description and the ordinary Schrödinger theory description of a system of n particles was established by Jordan and Klein [Jordan (1927)] for particles obeying Bose statistics and by Jordan and Wigner [Jordan (1928b)] for particles obeying Fermi statistics. The operator formalism was subsequently reformulated in "Fock space" by Fock (1932) who gave the generalization of the ordinary Schrödinger wave mechanics to systems for which the number of particles is not a constant of the motion. Fock's work was based on some previous investigations along similar lines by Landau and Peierls [Landau (1930)] who gave a configuration space treatment of the quantized electromagnetic field interacting with matter. Landau and Peierls' treatment was in turn suggested earlier by Oppenheimer, Heisenberg, and Pauli [Heisenberg (1930)] who sketched a configuration space treatment of quantized field theories. Our own presentation is based, in part, on some unpublished lectures by Bargmann (1951) and the papers of Fock (1932) and of Pirenne (1949).

6a. Permutations and Transpositions

We here briefly recall certain simple properties of permutations. A permutation is an operation which correlates a set of n ordered objects, e.g., the set $x_1, x_2, \cdots x_n$, with the same set of objects in a different order. We represent the operation which maps x_1 into x_{α_1}, x_2 into x_{α_2}, etc. by

$$P = \begin{pmatrix} 1 & 2 & \cdots & n \\ \alpha_1 & \alpha_2 & \cdots & \alpha_n \end{pmatrix} \tag{1}$$

with $Px_j = x_{\alpha_j}$, where the set $\{\alpha_1, \alpha_2 \cdots \alpha_n\}$ is the same as $\{1, \cdots n\}$ except for order. In this notation for P, the image of j, α_j, lies below j. This property is independent of the order in which the columns are written, so that they may be rearranged at will. There are $n!$ possible permutations of n distinguishable objects. If

$$Q = \begin{pmatrix} \alpha_1 & \alpha_2 & \cdots & \alpha_n \\ \beta_1 & \beta_2 & \cdots & \beta_n \end{pmatrix} \tag{2}$$

then the product QP is defined as

$$QP = \begin{pmatrix} 1 & 2 & 3 & \cdots & n \\ \beta_1 & \beta_2 & \beta_3 & \cdots & \beta_n \end{pmatrix} \tag{3}$$

with $QPx_j = x_{\beta_j}$. The identity permutation, which leaves the order of the set unchanged, is

$$E = \begin{pmatrix} 1 & 2 & 3 & \cdots & n \\ 1 & 2 & 3 & \cdots & n \end{pmatrix} \tag{4}$$

Every permutation, P, has an inverse P^{-1}, such that $PP^{-1} = P^{-1}P = E$. Clearly the inverse of P, Eq. (1), is

$$P^{-1} = \begin{pmatrix} \alpha_1 & \alpha_2 & \cdots & \alpha_n \\ 1 & 2 & \cdots & n \end{pmatrix} \tag{5}$$

The set of all permutations on n objects forms a group usually called the symmetric group, the subgroups of which are called permutation groups. A transposition, T, is a permutation in which only two elements are interchanged,

$$T = \begin{pmatrix} 1 & 2 & \cdots & k & \cdots & l & \cdots & n \\ 1 & 2 & \cdots & l & \cdots & k & \cdots & n \end{pmatrix} \tag{6}$$

Every permutation can be decomposed into a product of transpositions. This decomposition is not unique. However, the even or oddness of the number of transpositions into which a permutation can be decomposed is unique. Note that P^{-1} is even or odd according to whether P is even or odd. A permutation may thus be further characterized by its parity, δ_P, which is $+1$ for an even permutation and -1 for an odd permutation. If P is written in the form $\prod_i^n T_i$, where T_i are transpositions, and Q as $\prod_j^m T_j'$, then clearly $PQ = \prod_{j,i}^{nm} T_iT_j'$ and

$$\delta_P\delta_Q = \delta_Q\delta_P = \delta_{QP} = \delta_{PQ} \tag{7}$$

so that the product of two even permutations is even, the product of two odd permutations is even, while the product of an even and an odd permutation is odd. Also since the identity is even, $\delta_P \delta_{P^{-1}} = \delta_{P^{-1}} \delta_P = \delta_E = 1$ and

$$\delta_{P^{-1}} = \delta_P \tag{8}$$

6b. Symmetric and Antisymmetric Wave Functions

We shall here present a concise quantum mechanical description of a system of identical, indistinguishable particles. In a quantum mechanical formulation, we call particles indistinguishable if the expectation value of any physical observable is unaltered by the exchange of the particles, i.e., by a permutation of their labels. It is well known that such a system of identical particles is described by a wave function which is either symmetrical (Bose particles) or antisymmetrical (Fermi particles) in the exchange of two particles.

Let $|\Psi\rangle$, $|\Phi\rangle$ be state vectors which describe a system of n particles. Their scalar product in configuration space is defined as

$$\begin{aligned}(\Psi, \Phi) &= \sum_{s_1 s_2 \cdots s_n} \int d\mathbf{x}_1 \int d\mathbf{x}_2 \cdots \int d\mathbf{x}_n \langle \Psi \mid \mathbf{x}_1 s_1, \cdots \mathbf{x}_n s_n \rangle \langle \mathbf{x}_1 s_1, \cdots, \mathbf{x}_n s_n \mid \Phi \rangle \\ &= \sum_{s_1 \cdots s_n} \int d\mathbf{x}_1 \int d\mathbf{x}_2 \cdots \int d\mathbf{x}_n \bar{\Psi}(\mathbf{x}_1 s_1, \cdots \mathbf{x}_n s_n) \Phi(\mathbf{x}_1 s_1, \cdots \mathbf{x}_n s_n) \end{aligned} \tag{9}$$

where s_i is the spin variable of the ith particle. In the following we suppress the dependence of the wave function on the spin variables and assume that the variable $\mathbf{x}_i$ includes the description of the spin if the particles have spin. Similarly we assume the integration symbol to include, if required, a summation over spin indices. Under a permutation, P: $P\mathbf{x}_i = \mathbf{x}_{\alpha_i}$, the wave function $\Psi(\mathbf{x}) = \langle \mathbf{x} \mid \Psi \rangle = \langle \mathbf{x}_1, \mathbf{x}_2, \cdots \mathbf{x}_n \mid \Psi \rangle$ is transformed into a new wave function $\Psi'(\mathbf{x}) = \Psi(P\mathbf{x})$. This transformed state is induced by a linear operation U_P

$$U_P \mid \Psi \rangle = \mid \Psi' \rangle \tag{10a}$$

such that

$$\langle \mathbf{x} \mid U_P \mid \Psi \rangle = \langle \mathbf{x} \mid \Psi' \rangle = \langle P\mathbf{x} \mid \Psi \rangle \tag{10b}$$

so that

$$\langle \mathbf{x} \mid U_P \mid \mathbf{x}' \rangle = \langle P\mathbf{x} \mid \mathbf{x}' \rangle = \delta(P\mathbf{x}_1 - \mathbf{x}_1) \cdots \delta(P\mathbf{x}_n - \mathbf{x}_n') \tag{11}$$

We shall sometimes denote by $\mathcal{P}$ the operator U_P in configuration space and write

$$\langle P\mathbf{x} \mid \Psi \rangle = \mathcal{P} \langle \mathbf{x} \mid \Psi \rangle = \mathcal{P}\Psi(\mathbf{x}) = \Psi(P\mathbf{x}) \tag{12}$$

Since the Jacobian of the transformation $\mathbf{x} \rightarrow P\mathbf{x}$ is one, it follows that $\langle \Phi' \mid \Psi' \rangle = \langle \Phi \mid \Psi \rangle$ so that U_P is a unitary operator. The U_Ps, in fact, form a representation of the symmetric group by unitary operators. If the particles are indistinguishable, then the aforementioned criterion for indistinguishability requires that, if O is any observable of our n particle system, then for any permutation P

$$\begin{aligned}(\Psi, O\Psi) &= (\Psi', O\Psi') = (U_P\Psi, OU_P\Psi) \\ &= (\Psi, U_P^{-1}OU_P\Psi) \qquad (13a)\end{aligned}$$

i.e., that

$$O = U_P^{-1}OU_P \qquad (13b)$$

or equivalently that $[O, U_P] = 0$. This last equation expresses the invariance of O under permutations of the labels of the particles, and O must therefore be a symmetric function of the particle observables.

We can now give a precise characterization of a symmetric or antisymmetric wave function. If T is any transposition, we shall say a wave function is symmetric if

$$\langle \mathbf{x}_1, \mathbf{x}_2 \cdots \mathbf{x}_n \mid U_T \mid \Psi_s \rangle = \langle \mathbf{x}_1, \mathbf{x}_2, \cdots \mathbf{x}_n \mid \Psi_s \rangle \qquad (14a)$$

or equivalently if

$$\mathfrak{I}\Psi_s(\mathbf{x}_1, \cdots \mathbf{x}_n) = \Psi_s(\mathbf{x}_1, \cdots \mathbf{x}_n) \qquad (14b)$$

and antisymmetric if

$$\langle \mathbf{x}_1, \mathbf{x}_2, \cdots \mathbf{x}_n \mid U_T \mid \Psi_a \rangle = -\langle \mathbf{x}_1, \mathbf{x}_2, \cdots \mathbf{x}_n \mid \Psi_a \rangle \qquad (15a)$$

$$\mathfrak{I}\Psi_a(\mathbf{x}_1, \mathbf{x}_2, \cdots \mathbf{x}_n) = -\Psi_a(\mathbf{x}_1, \mathbf{x}_2, \cdots \mathbf{x}_n) \qquad (15b)$$

More generally, for any permutation P, a symmetric wave function has the property that

$$\langle \mathbf{x}_1, \mathbf{x}_2, \cdots \mathbf{x}_n \mid U_P \mid \Psi_s \rangle = \langle \mathbf{x}_1, \cdots \mathbf{x}_n \mid \Psi_s \rangle \qquad (16)$$

whereas for an antisymmetric state

$$\langle \mathbf{x}_1, \mathbf{x}_2, \cdots \mathbf{x}_n \mid U_P \mid \Psi_a \rangle = \delta_P \langle \mathbf{x}_1, \mathbf{x}_2 \cdots \mathbf{x}_n \mid \Psi_a \rangle \qquad (17)$$

where δ_P is the parity of the permutation P. It is a consequence of the unitary character of U_P that the manifold of symmetric states is orthogonal to the manifold of antisymmetric states, since $(\Psi_s, \Psi_a) = (U_T\Psi_s, U_T\Psi_a) = -(\Psi_s, \Psi_a) = 0$. Furthermore, no perturbation can take a symmetric state into an antisymmetric state, since for any observable, O, on a system of n identical particle $U_POU_P^{-1} = O$, so that $(\Psi_s, O\Psi_a) = (U_T\Psi_s, U_TOU_T^{-1}U_T\Psi_a) = -(\Psi_s, O\Psi_a) = 0$. Finally since the Hamiltonian is invariant under U_P, i.e., $H = U_PHU_P^{-1}$, the symmetry character of a state is preserved in time.

In order to construct symmetric and antisymmetric wave functions from arbitrary functions of n variables $\mathbf{x}_1, \mathbf{x}_2 \cdots \mathbf{x}_n$, we define the following operators

$$S = \frac{1}{n!} \sum_P U_P \qquad (18)$$

$$A = \frac{1}{n!} \sum_P \delta_P U_P \tag{19}$$

where the summation is over the $n!$ elements of the symmetric group of order n. S is called the *symmetrizer* and A the *antisymmetrizer*, for reasons which will become clear shortly. We can unify the discussion of their properties by writing

$$\Lambda = \frac{1}{n!} \sum_P \lambda_P U_P \tag{20}$$

where when $\lambda_P = 1$, $\Lambda = S$, and when $\lambda_P = \delta_P$, $\Lambda = A$. The properties of Λ are as follows:

(*a*) Λ is Hermitian
(*b*) $\Lambda U_P = U_P \Lambda$ for any permutation
(*c*) $\Lambda^2 = \Lambda$, i.e., Λ is a projection operator

The proof of these assertions is quite simple. To prove *a* we note that since the U_Ps form a unitary representation of the symmetric group, $U_P U_{P^{-1}} = U_E = I$ so that $U_{P^{-1}} = [U_P]^{-1}$, and therefore (since the Us are unitary) $U_P^* = U_{P^{-1}}$. Therefore, using the property $\delta_P = \delta_{P^{-1}}$, we have

$$\Lambda^* = \frac{1}{n!} \sum_P \lambda_{P^{-1}} U_{P^{-1}} \tag{21}$$

But since the set of permutations P form a group, we may replace the summation over P by a summation over P^{-1}, hence $\Lambda^* = \Lambda$. Similarly, to prove *b* we note that $\lambda_P^2 = 1$, so that using Eq. (7)

$$\Lambda U_P = \frac{1}{n!} \sum_Q \lambda_Q U_Q U_P = \frac{1}{n!} \lambda_P^2 \sum_Q \lambda_Q U_{QP}$$

$$= \frac{1}{n!} \lambda_P \sum_Q \lambda_{QP} U_{QP} \tag{22}$$

Again because of the group property, the elements QP, for fixed P, when Q ranges through the group exhaust the group, so that $\sum_Q = \sum_{QP}$ and $\Lambda U_P = \lambda_P \Lambda$. One verifies that similarly $U_P \Lambda = \lambda_P \Lambda$, so that *b* is proved. Finally, to prove *c*, we use property *b* to write

$$\Lambda^2 = \Lambda \frac{1}{n!} \sum_Q \lambda_Q U_Q$$

$$= \frac{1}{n!} \sum_Q \lambda_Q^2 \Lambda = \Lambda \tag{23}$$

since $\sum_Q 1 = n!$. The symmetrizer or antisymmetrizer is thus a projection operator. Given any vector $|\Psi\rangle$ in the Hilbert space, S (A) projects it into the subspace of symmetric (antisymmetric) states. *Proof:*

$U_T S \mid \Psi\rangle = S \mid \Psi\rangle$ by property (2) of S. Similarly $U_T A \mid \Psi\rangle = -A \mid \Psi\rangle$ since $\delta_T = -1$ for any transposition T.

These properties of Λ can often be used to simplify the computation of normalization integrals and expectation values of observables for symmetric or antisymmetric states. Thus if $|\Psi\rangle = \Lambda \mid \Phi\rangle$, where $|\Phi\rangle$ is a vector without symmetry properties, then the expectation value of an observable O in the state $|\Psi\rangle$ is given by $(\Psi, O\Psi) = (\Lambda\Phi, O\Lambda\Phi) = (\Phi, O\Lambda^2\Phi) = (\Phi, O\Lambda\Phi)$, since O and U_P commute. It is therefore necessary only to use one properly symmetrized wave function, $\Lambda \mid \Phi\rangle$, in computing the expectation value.

6c. Occupation Number Space

For many applications, a very useful characterization of an n particle system is in terms of one-particle observables. Such a description consists in stating the probability amplitudes for finding n_1 particles in an eigenstate $|\lambda'\rangle$, n_2 in an eigenstate $|\lambda''\rangle$, etc., where $|\lambda\rangle$ is an eigenfunction of a complete set of observables for a one-particle system with eigenvalue λ, and where $n_1 + n_2 + n_3 + \cdots = n$. To effect such a description, we first construct properly symmetrized basis functions. Let $|\lambda'\rangle, |\lambda''\rangle, \cdots$, etc. be a complete orthonormal set of one-particle states, so that if $\langle \mathbf{x} \mid \lambda\rangle = g_\lambda(\mathbf{x})$

$$\begin{aligned}\langle \lambda \mid \lambda'\rangle &= \delta_{\lambda\lambda'} \\ &= \int \langle \lambda \mid \mathbf{x}\rangle\, d\mathbf{x}\, \langle \mathbf{x} \mid \lambda'\rangle = \int \bar{g}_\lambda(\mathbf{x})\, g_{\lambda'}(\mathbf{x})\, d\mathbf{x} \end{aligned} \tag{24}$$

We next define the functions

$$G_{\lambda_1\lambda_2 \cdots \lambda_n}(\mathbf{x}_1, \mathbf{x}_2, \cdots \mathbf{x}_n) = g_{\lambda_1}(\mathbf{x}_1)\, g_{\lambda_2}(\mathbf{x}_2) \cdots g_{\lambda_n}(\mathbf{x}_n) \tag{25}$$

Their orthogonality properties within the scalar product (9) are expressed by the following relation

$$(G_{\lambda_1\lambda_2 \cdots \lambda_n}, G_{\lambda'_1\lambda'_2, \cdots \lambda'_n}) = \delta_{\lambda_1\lambda'_1}\delta_{\lambda_2\lambda'_2} \cdots \delta_{\lambda_n\lambda'_n} \tag{26}$$

The set of G_λs forms a complete orthonormal set in n variables if the set $\{g_\lambda\}$ is complete.

Now instead of specifying $G_{\lambda_1\lambda_2 \cdots \lambda_n}$ by the set $\{\lambda_1, \cdots \lambda_n\}$, we may specify G by stating how many of the eigenvalues $\lambda', \lambda'', \cdots$ repeat, which ones and how often. Such a specification of G is in terms of an infinite sequence of integers $n_1, n_2 \cdots$ which are called occupation numbers, such that n_1 is the number of times $g_{\lambda'}$ (irrespective of what its argument is) occurs in $G_{\lambda_1\lambda_2 \cdots \lambda_n}$; n_2 the number of times $g_{\lambda''}$ (irrespective of its argument) occurs in G; n_i the number of times $g_{\lambda^{(i)}}$ occurs. Clearly

$$\sum_{i=1}^{\infty} n_i = n \tag{27}$$

This is not, however, an unique specification since $U_P G$ where U_P is an arbitrary permutation will have the same occupation numbers. This does not give rise to any difficulties, since for physical applications we are only interested in expansions in terms of properly symmetrized or antisymmetrized functions, i.e., in terms of functions $\mathcal{G}$ which satisfy $\Lambda\mathcal{G} = \mathcal{G}$. We therefore have to form linear combinations of Gs which have the correct symmetry properties or, alternatively, we may apply Λ on each of the Gs, and group together all the resulting functions which are the same. Let us call $\mathcal{G}_{\{n_1, n_2, n_3, \cdots\}}$ such a properly symmetrized function. Clearly the $G_{\lambda_1\lambda_2\cdots\lambda_n}$s which will occur in a given $\mathcal{G}$ are exactly those which can be obtained from one another by a permutation U_P. Stated differently, a properly symmetrized function $\mathcal{G}_{\{n\}}$ can be obtained from any one of the functions $G_{\lambda_1\lambda_2\cdots\lambda_n}$ corresponding to a specified sequence $\{n\} = \{n_1, n_2 \cdots\}$, by applying Λ on it, i.e.,

$$\mathcal{G}_{\{n_1, n_2 \cdots\}}(\mathbf{x}_1, \mathbf{x}_2 \cdots \mathbf{x}_n) = \langle \mathbf{x}_1, \mathbf{x}_2, \cdots \mathbf{x}_n \mid \Lambda \mid G_{\lambda_1\lambda_2\cdots\lambda_n}\rangle \tag{28}$$

A sequence $\{n_1, n_2, n_3 \cdots\}$ with $n_1 + n_2 + \cdots = n$ uniquely specifies a properly symmetrized or antisymmetrized $\mathcal{G}_{\{n_1, n_2 \cdots\}}$. Such a $\mathcal{G}_{\{n_1, n_2 \cdots\}}$ is, however, not normalized although clearly $(\mathcal{G}_{\{n\}}, \mathcal{G}_{\{n'\}}) = 0$ unless $\{n\} = \{n'\}$, i.e., unless $n_1 = n_1'$, $n_2 = n_2'$, $\cdots$, etc. We shall denote by $\Phi^{(n)}_{\{n_1, n_2, \cdots\}}$ a properly normalized $\mathcal{G}$, i.e.,

$$\begin{aligned}\Phi^{(n)}_{\{n_1, n_2, \cdots\}}(\mathbf{x}_1, \cdots \mathbf{x}_n) &= N_{\{n\}}\langle \mathbf{x}_1, \mathbf{x}_2 \cdots \mathbf{x}_n \mid \Lambda \mid G_{\lambda_1\lambda_2\cdots\lambda_n}\rangle \\ &= \langle \mathbf{x}_1, \mathbf{x}_2, \mathbf{x}_3, \cdots \mathbf{x}_n \mid n_1, n_2, \cdots n_i, \cdots\rangle \end{aligned}\tag{29}$$

where $N_{\{n\}}$ is a normalization constant to be determined from the requirement that

$$\begin{aligned}(\Phi^{(n)}_{\{n_1, n_2, \cdots\}}, \Phi^{(n)}_{\{n_1, n_2, \cdots\}}) &= 1 \\ &= N^2_{\{n\}}(G_{\lambda_1\lambda_2\cdots\lambda_n}, \sum_P \frac{1}{n!}\lambda_P U_P G_{\lambda_1\lambda_2\cdots\lambda_n})\end{aligned} \tag{30}$$

This formula can be further simplified by noting that since the $G_{\lambda_1\lambda_2\cdots\lambda_n}(\mathbf{x}_1 \cdots \mathbf{x}_n)$ are product wave functions

$$\begin{aligned}(U_P G)_{\lambda_1\lambda_2\cdots\lambda_n}(\mathbf{x}_1 \cdots \mathbf{x}_n) &= G_{\lambda_1\lambda_2\cdots\lambda_n}(\mathbf{x}_{\alpha_1}, \cdots \mathbf{x}_{\alpha_n}) \\ &= \langle \mathbf{x}_{\alpha_1} \mid \lambda_1\rangle \cdots \langle \mathbf{x}_{\alpha_n} \mid \lambda_n\rangle = \langle \mathbf{x}_1 \mid \lambda_{\beta_1}\rangle \cdots \langle \mathbf{x}_n \mid \lambda_{\beta_n}\rangle \\ &= G_{\lambda_{\beta_1}\lambda_{\beta_2}\cdots\lambda_{\beta_n}}(\mathbf{x}_1, \cdots \mathbf{x}_n) = G_{P^{-1}\{\lambda_1, \lambda_2 \cdots \lambda_n\}}(\mathbf{x}_1, \cdots \mathbf{x}_n)\end{aligned} \tag{31}$$

where the set $\{\lambda_{\beta_1}, \cdots \lambda_{\beta_n}\}$ is obtained from the set $\{\lambda_1, \cdots \lambda_n\}$ by the permutation P^{-1}. We may therefore rewrite Eq. (30) as follows:

$$N^{-2}_{\{n\}} = \frac{1}{n!}\sum_P \lambda_P(G_{\lambda_1, \lambda_2, \cdots \lambda_n}, G_{\lambda_{\beta_1}\lambda_{\beta_2}\cdots\lambda_{\beta_n}}) \tag{32}$$

where the summation now runs over all the permutations of the set $\{\beta_1, \beta_2, \cdots \beta_n\}$.

6d. The Symmetric Case

Up to this point our discussion was quite general and applied equally well to the symmetric as to the antisymmetric situation. However, from now until Section 6g we shall only consider the case of symmetric wave functions. To compute the normalization factor for the symmetric case, we note that if $G_{\lambda_1\lambda_2 \cdots \lambda_n}$ is such that all the λs are different, i.e., $n_i \leqslant 1$, then only the identity permutation will contribute in the sum on the right-hand side of Eq. (32) and $N_{\{n\}}$ will be equal to $\sqrt{n!}$. More generally, if in $G_{\lambda_1 \cdots \lambda_n}$, $g_{\lambda^{(i)}}$ occurs n_i times, then in the summation in Eq. (32) only those permutations which exchange particles in the *same* state contribute. The number of such permutations is $n_1!n_2! \cdots$, therefore

$$N_{\{n_1, n_2, \cdots\}} = \sqrt{\frac{n!}{n_1!\, n_2!\, n_3! \cdots}} \tag{33}$$

so that

$$\begin{aligned}\Phi^{(n)}_{\{n_1, n_2, \cdots n_i, \cdots\}}(\mathbf{x}_1, \mathbf{x}_2, \cdots \mathbf{x}_n) &= \langle \mathbf{x}_1, \mathbf{x}_2, \mathbf{x}_3, \cdots \mid n_1, n_2, \cdots n_i \cdots\rangle \\ &= \sqrt{\frac{n!}{n_1!\, n_2! \cdots}}\, \frac{1}{n!} \sum_P \mathcal{P}(g_{\lambda_1}(\mathbf{x}_1) \cdots g_{\lambda_n}(\mathbf{x}_n))\end{aligned} \tag{34}$$

where $\mathcal{P}$ is the permutation operator which interchanges the label of the xs and the summation runs over all $n!$ permutation of the n labels. The vectors $|n_1, n_2 \cdots\rangle$ for all possible sequences $\{n\}$ with $\sum_i n_i = n$ form a basis which span the Hilbert space of the realizable states of a system of n identical particles obeying Bose statistics. The completeness of the set is expressed by the relation

$$\sum_{\{n\}} |n_1, n_2, \cdots, n_i, \cdots\rangle \langle n_1, n_2, \cdots, n_i, \cdots| = 1 \tag{35}$$

where the summation is over all sequences $\{n_1, n_2 \cdots\}$ with $\sum_{i=1}^{\infty} n_i = n$. The orthogonality relation is expressed by Eq. (26) and in the present notation reads

$$\langle n_1, n_2, \cdots n_i, \cdots \mid n_1', n_2', \cdots n_i', \cdots\rangle = \delta_{n_1n_1'}\delta_{n_2n_2'} \cdots \delta_{n_in_i'} \cdots \tag{36}$$

An arbitrary state vector, $|\Psi(t)\rangle$, describing an n-particle system obeying Bose statistics can be expanded in terms of these basis vectors

$$|\Psi\rangle = \sum_{\{n\}} |n_1, n_2, \cdots n_i, \cdots\rangle \langle n_1, n_2, \cdots n_i, \cdots \mid \Psi(t)\rangle \tag{37}$$

The expansion coefficients, $\langle n_1, n_2, \cdots n_i, \cdots \mid \Psi(t)\rangle$, are then the probability amplitude for finding at time t the system with n_1 particles in the state $|\lambda'\rangle$, n_2 in the state $|\lambda''\rangle$, n_i in the state $|\lambda^{(i)}\rangle$, etc.

The transformation function $\langle \mathbf{x}_1, \mathbf{x}_2, \mathbf{x}_3, \cdots \mathbf{x}_n; n \mid n_1, n_2, \cdots\rangle$ is a "permanent," i.e., a determinant where all the permutations are added together with a plus sign only. It can be expanded along a row or a column in a manner quite similar to determinants. These expansion formulae, which will prove very useful later, are

$$\langle \mathbf{x}_1, \mathbf{x}_2, \cdots \mathbf{x}_n; n \mid n_1, n_2 \cdots n_i, \cdots \rangle = \sum_{i=1}^{\infty} \sqrt{\frac{n_i}{n}} \langle \mathbf{x}_1 \mid \lambda^{(i)} \rangle \cdot \langle \mathbf{x}_2, \mathbf{x}_3, \cdots \mathbf{x}_n; n-1 \mid n_1, n_2, \cdots n_i - 1, \cdots \rangle \quad (38)$$

and

$$\langle \mathbf{x}_1, \mathbf{x}_2, \cdots \mathbf{x}_n; n \mid n_1, n_2, \cdots n_i, \cdots \rangle = \sum_{r=1}^{n} \frac{1}{\sqrt{n_i n}} \langle \mathbf{x}_r \mid \lambda^{(i)} \rangle \langle \mathbf{x}_1, \mathbf{x}_2, \cdots \mathbf{x}_{r-1}, \mathbf{x}_{r+1}, \cdots \mathbf{x}_n; n-1 \mid n_1, n_2, \cdots n_i - 1, \cdots \rangle \quad (39)$$

The proof of (38) is based on the fact that every permutation P on n things allows a decomposition of the form $P = QT_m$ where Q is a permutation of the set $(2, 3, \cdots n)$ and T_m is a transposition of 1 and m. The symmetrizer $\mathcal{S}$ can therefore be written in the form

$$\mathcal{S} = \frac{1}{n!} \sum_P \mathcal{P} = \frac{1}{n} \frac{1}{(n-1)!} \sum_{m,\,Q} \mathfrak{Q}\mathfrak{T}_m = \mathcal{S}' \frac{1}{n} \sum_{m=1}^{n} \mathfrak{T}_m \quad (40)$$

where $\mathcal{S}'$ is the symmetrizer of the set $(2, \cdots, n)$. Therefore, since G is a product wave function

$$\begin{aligned} \mathfrak{T}_m G_{\lambda_1\lambda_2 \cdots \lambda_n}(\mathbf{x}_1, \cdots \mathbf{x}_n) &= G_{\lambda_1\lambda_2\lambda_3 \cdots \lambda_n}(\mathbf{x}_m, \mathbf{x}_2, \cdots \mathbf{x}_{m-1}, \mathbf{x}_1, \mathbf{x}_{m+1}, \cdots \mathbf{x}_n) \\ &= G_{\lambda_m\lambda_2 \cdots \lambda_{m-1}\lambda_1\lambda_{m+1} \cdots \lambda_n}(\mathbf{x}_1, \mathbf{x}_2, \cdots \mathbf{x}_n) \\ &= \langle \lambda_m \mid \mathbf{x}_1 \rangle G'_{\lambda_2 \cdots \lambda_n}(\mathbf{x}_2 \cdots \mathbf{x}_n) \end{aligned} \quad (41)$$

where G' is a function of $n - 1$ variables, whose occupation numbers are $n_1, n_2, \cdots n_i - 1, \cdots$, if those of $G_{\lambda_1\lambda_2 \cdots \lambda_n}$ were $n_1, n_2, \cdots n_i, \cdots$ and $\lambda_m = \lambda^{(i)}$. We may therefore write

$$\begin{aligned} \Phi^{(n)}_{\{n_1, n_2, \cdots\}}(\mathbf{x}_1, \mathbf{x}_2, \cdots \mathbf{x}_n) &= N_{\{n\}} \mathcal{S} G_{\lambda_1\lambda_2 \cdots \lambda_n}(\mathbf{x}_1, \cdots \mathbf{x}_n) \\ &= N_{\{n\}} \frac{1}{n} \sum_{m=1}^{n} \mathcal{S}'(\langle \lambda_m \mid \mathbf{x}_1 \rangle G'_{\lambda_2 \cdots \lambda_{m-1}\lambda_1\lambda_{m+1} \cdots \lambda_n}(\mathbf{x}_2 \cdots \mathbf{x}_n)) \\ &= \sum_{i=1}^{\infty} \sqrt{\frac{n_i}{n}} \langle \lambda^{(i)} \mid \mathbf{x}_1 \rangle \Phi^{(n-1)}_{\{n_1, n_2 \cdots n_i - 1, \cdots\}}(\mathbf{x}_2, \cdots \mathbf{x}_n) \end{aligned} \quad (42)$$

where the factor $\sqrt{n_i/n}$ arises from the difference between the normalization factors for $\Phi^{(n)}$ and $\Phi^{(n-1)}$

$$N^{(n)}_{\{n_1, n_2, \cdots n_i \cdots\}} = \sqrt{\frac{n}{n_i}} N^{(n-1)}_{\{n_1, n_2, \cdots n_i - 1, \cdots\}} \quad (43)$$

and the fact that the sum over m, $\sum_m$ has been replaced by a summation over occupied states, $\sum_i n_i$, to which it is clearly equivalent. If we rewrite Eq. (42) in the Dirac notation given by Eq. (34), Eq. (38) then follows. The proof of Eq. (39) is quite similar. Since $S = QS$ for any permutation Q, it follows that $\mathcal{S} = \mathcal{S}\mathcal{S}'$ where $\mathcal{S}'$ is the symmetrizer of the set $(2, \cdots n)$. We may therefore write, assuming that $|\lambda^{(i)}\rangle$ occurs in $G_{\lambda_1\lambda_2\cdots\lambda_n}$, i.e., assuming $n_i \neq 0$

$$\begin{aligned}\Phi^{(n)}_{\{n_1, n_2, \cdots n_i, \cdots\}}(\mathbf{x}_1, \cdots \mathbf{x}_n) &= N_{\{n\}}\mathcal{S}G_{\lambda_1\lambda_2\cdots\lambda_n}(\mathbf{x}_1, \mathbf{x}_2, \cdots \mathbf{x}_n) \\ &= N_{\{n\}}\mathcal{S}(\langle\lambda^{(i)} \mid \mathbf{x}_1\rangle\, G'(\mathbf{x}_2 \cdots \mathbf{x}_n)) \qquad (44)\end{aligned}$$

where G' has occupation numbers $n_1, n_2, \cdots n_i - 1, \cdots$. Therefore, inserting $\mathcal{S}\mathcal{S}'$ for $\mathcal{S}$ we obtain

$$\begin{aligned}&\Phi^{(n)}_{\{n_1, n_2, \cdots n_i \cdots\}}(\mathbf{x}_1, \mathbf{x}_2 \cdots \mathbf{x}_n) \\ &\qquad = \sqrt{\frac{n}{n_i}}\,\mathcal{S}(\langle\lambda^{(i)} \mid \mathbf{x}_1\rangle\, \Phi_{\{n_1n_2 \cdots n_i-1, \cdots\}}(\mathbf{x}_2, \mathbf{x}_3, \cdots \mathbf{x}_n)) \qquad (45)\end{aligned}$$

Using Eq. (40) we then finally obtain

$$\begin{aligned}\Phi^{(n)}_{\{n_1, n_2, \cdots n_i, \cdots\}}(\mathbf{x}_1, \cdots \mathbf{x}_n) &= \sum_{l=1}^{n} \frac{1}{\sqrt{n_i n}} \langle\lambda^{(i)} \mid \mathbf{x}_l\rangle \\ &\quad \cdot \Phi^{(n-1)}_{\{n_1, n_2 \cdots n_i - 1 \cdots\}}(\mathbf{x}_1, \mathbf{x}_2, \cdots \mathbf{x}_{l-1}, \mathbf{x}_{l+1}, \cdots \mathbf{x}_n) \qquad (46)\end{aligned}$$

which is Eq. (39).

6e. Creation and Annihilation Operators

We now introduce the operator a_i defined so that

$$a_i \mid n_1, n_2, \cdots n_i \cdots\rangle = \sqrt{n_i} \mid n_1, n_2, \cdots, n_i - 1, \cdots\rangle \qquad (47)$$

The operator a_i is called an annihilation operator, since it destroys a particle in the state $|\lambda^{(i)}\rangle$. It has matrix elements

$$\begin{aligned}&\langle n_1', n_2', \cdots, n_i', \cdots; n' \mid a_i \mid n_1, n_2, \cdots n_i, \cdots; n\rangle \\ &\qquad = \sqrt{n_i}\, \delta_{n_1n_1'}\delta_{n_2n_2'} \cdots \delta_{n_i-1, n_i'} \cdots \delta_{n', n-1} \qquad (48)\end{aligned}$$

which vanish unless $n' = n - 1$. Note that a_i is defined in the Hilbert space $\mathcal{H}^{(0)} \oplus \mathcal{H}^{(1)} \oplus \mathcal{H}^{(2)} \oplus \cdots$, where $\mathcal{H}^{(n)}$ is the Hilbert space for an n-particle system. It follows from their definition that the operators a_i and a_j commute

$$[a_i, a_j] = 0 \qquad (49)$$

We next introduce the adjoint operator a_i^*, which is defined by the equation

$$(\Psi, a_i\Phi) = (a_i^*\Psi, \Phi) \qquad (50)$$

so that

$$a_i^* \mid n_1, n_2, \cdots n_i, \cdots\rangle = \sqrt{n_i + 1} \mid n_1, n_2, \cdots n_i + 1, \cdots\rangle \tag{51}$$

The operator a_i^* is called a creation operator, since operating on an n-particle state it transforms it into an $n + 1$ particle state by increasing the number of particles in the state $|\lambda^{(i)}\rangle$ by one. It has matrix elements

$$\langle n_1', n_2', \cdots n_i' \cdots ; n' \mid a_i^* \mid n_1, n_2, \cdots n_i, \cdots ; n\rangle$$
$$= \sqrt{n_i + 1}\, \delta_{n', n+1} \delta_{n_1 n_1'} \cdots \delta_{n_i+1, n_i'} \cdots \tag{52}$$

which vanish unless $n' = n + 1$. We next note that

$$[a_i^*, a_j^*] = 0 \tag{53}$$

whereas the commutation rules between a_i and a_j^* are

$$[a_i, a_j^*] = \delta_{ij} \tag{54}$$

which is verified by explicitly computing the left-hand side when operating on an arbitrary state $|n_1, n_2, \cdots n_i, \cdots n_j \cdots\rangle$. Finally, we observe that the Hermitian operator $N_i = a_i^* a_i$, when operating on $|n_1, n_2, \cdots n_i, \cdots\rangle$, yields $n_i \mid n_1, n_2, \cdots n_i, \cdots\rangle$, that is, it tells us how many particles are in the state $|\lambda^{(i)}\rangle$. We call

$$N_i = a_i^* a_i \tag{55}$$

the number operator for the state i, and the operator

$$N = \sum_{i=1}^{\infty} N_i \tag{56}$$

the total number operator, since operating on a state $|n_1, n_2, \cdots\rangle$ it yields $\sum_{i=1}^{\infty} n_i \mid n_1 n_2 \cdots\rangle = n \mid n_1, n_2, \cdots\rangle$.

We can characterize our presentation up to this point somewhat more abstractly in terms of the creation and annihilation operators a_i^* and a_i. By virtue of the commutation rules of a_i^* and a_j, the different number operators commute: $[N_i, N_j] = 0$. We can therefore define a representation in which all the N_is are diagonal. We denote the basis vectors for this representation by $|n_1, n_2, \cdots n_i, \cdots\rangle$. We note that the commutation rules between N_i and the creation and annihilation operators are:

$$[N_i, a_i^*] = a_i^* \tag{57}$$

$$[N_i, a_i] = -a_i \tag{58}$$

so that

$$\langle \cdots n_i' \cdots \mid [N_i, a_i^*] \mid \cdots n_i \cdots\rangle = \langle \cdots n_i' \cdots \mid a_i^* \mid \cdots n_i \cdots\rangle \tag{59a}$$

and therefore

$$(n_i' - n_i - 1) \langle \cdots n_i' \cdots \mid a_i^* \mid \cdots n_i \cdots\rangle = 0 \tag{59b}$$

which asserts that the matrix element of a_i^* vanishes unless $n_i' = n_i + 1$ and $n_j = n_j'$, $j \neq i$. We next inquire as to the eigenvalues of N_i. By virtue of the commutation rules (57), it follows that if $|\nu_i\rangle$ is an eigenfunction of N_i with eigenvalue ν_i, then $a_i^* \mid \nu_i\rangle$ is an eigenfunction of N_i with eigenvalue $\nu_i + 1$. *Proof*:

$$\begin{aligned} N_i a_i^* \mid \nu_i\rangle &= (a_i^* N_i + [N_i, a_i^*]) \mid \nu_i\rangle \\ &= (\nu_i + 1)\, a_i^* \mid \nu_i\rangle \end{aligned} \tag{60}$$

Using (58) one verifies that $a_i \mid \nu_i\rangle$ is an eigenfunction of N_i with eigenvalue $\nu_i - 1$. Similarly $(a_i)^m \mid \nu_i\rangle$ is an eigenfunction of N_i with eigenvalue $\nu_i - m$, where m is a positive integer. Thus either N_i has negative eigenvalues (since $\nu_i - m$ can be made negative for sufficiently large m) or for some m, $(a_i)^m \mid \nu_i\rangle$ must be the null vector. However, N_i is positive-semidefinite ($\geqslant 0$) since it is the product of an operator times its Hermitian adjoint, so that its eigenvalues can never be negative. In fact for an arbitrary non-null vector $|\Phi\rangle$, for which $(\Phi, \Phi) > 0$, we have

$$\begin{aligned} (\Phi, N_i\Phi) &= (\Phi, a_i^* a_i \Phi) = (a_i\Phi, a_i\Phi) \\ &= ||a_i\Phi||^2 \geqslant 0 \end{aligned} \tag{61}$$

from the definition of the scalar product in Hilbert space. The equality sign in Eq. (61) holds only for that state $|\Phi_0\rangle$ for which $a_i \mid \Phi_0\rangle = 0$. Clearly the eigenvalue of N_i for that state is 0. Therefore the eigenvalues of N_i are the positive integers and zero. The state, $|\Phi_0\rangle = |0\rangle$, for which $N_i \mid 0\rangle = 0$ for all i is called the *no-particle state*. It is defined by the equation

$$a_i \mid 0\rangle = 0 \quad \text{for all } i \tag{62}$$

We next introduce the operators

$$\psi(\mathbf{x}) = \sum_{i=1}^{\infty} \langle \mathbf{x} \mid \lambda^{(i)}\rangle\, a_i \tag{63}$$

and

$$\psi^*(\mathbf{x}) = \sum_{i=1}^{\infty} \langle \lambda^{(i)} \mid \mathbf{x}\rangle\, a_i^* \tag{64}$$

They have the property that they no longer depend on the particular basis, $\langle \mathbf{x} \mid \lambda^{(i)}\rangle$, that was chosen for the one-particle states. Their commutation rules with one another are

$$\begin{aligned} [\psi(\mathbf{x}), \psi^*(\mathbf{y})] &= \sum_{i,j}^{\infty} \langle \mathbf{x} \mid \lambda^{(i)}\rangle \langle \lambda^{(j)} \mid \mathbf{y}\rangle [a_i, a_j^*] \\ &= \sum_{i}^{\infty} \langle \mathbf{x} \mid \lambda^{(i)}\rangle \langle \lambda^{(i)} \mid \mathbf{y}\rangle = \langle \mathbf{x} \mid \mathbf{y}\rangle \\ &= \delta(\mathbf{x} - \mathbf{y}) \end{aligned} \tag{65}$$

and

$$[\psi(\mathbf{x}), \psi(\mathbf{y})] = [\psi^*(\mathbf{x}), \psi^*(\mathbf{y})] = 0 \tag{66}$$

In terms of these operators, the total number operator may be written as

$$N = \sum_i a_i^* a_i = \sum_{ij} \int \langle \lambda^{(i)} \mid \mathbf{x} \rangle \, d\mathbf{x} \, \langle \mathbf{x} \mid \lambda^{(j)} \rangle \, a_j^* a_i$$
$$= \int d\mathbf{x} \, \psi^*(\mathbf{x}) \, \psi(\mathbf{x}) \tag{67}$$

The operator $\psi^*(\mathbf{x}) \, \psi(\mathbf{x})$ may be interpreted as the particle density operator. We next note that we may rewrite Eq. (38) as follows

$$\langle \mathbf{x}_1, \mathbf{x}_2, \mathbf{x}_3, \cdots \mathbf{x}_n; n \mid n_1, n_2, \cdots n_i \cdots \rangle = \sum_{i=1}^{\infty} \frac{1}{\sqrt{n}} \langle \mathbf{x}_1 \mid \lambda^{(i)} \rangle$$
$$\cdot \langle \mathbf{x}_2, \mathbf{x}_3, \cdots \mathbf{x}_n; n - 1 \mid a_i \mid n_1, n_2, \cdots n_i \cdots \rangle$$
$$= \frac{1}{\sqrt{n}} \langle \mathbf{x}_2 \mathbf{x}_3 \cdots \mathbf{x}_n; n - 1 \mid \psi(\mathbf{x}_1) \mid n_1, n_2, \cdots n_i \cdots \rangle \tag{68}$$

and by induction

$$\langle \mathbf{x}_1, \mathbf{x}_2, \cdots \mathbf{x}_n; n \mid n_1, n_2, \cdots n_i \cdots \rangle$$
$$= \frac{1}{\sqrt{n!}} \langle 0 \mid \psi(\mathbf{x}_n) \cdots \psi(\mathbf{x}_2) \, \psi(\mathbf{x}_1) \mid n_1, n_2, \cdots n_i \cdots \rangle \tag{69}$$

It now follows from Eq. (69) that

$$S \mid \mathbf{x}_1, \mathbf{x}_2, \cdots \mathbf{x}_n \rangle = \frac{1}{\sqrt{n!}} \psi^*(\mathbf{x}_1) \, \psi^*(\mathbf{x}_2) \cdots \psi^*(\mathbf{x}_n) \mid 0 \rangle \tag{70}$$

where it is to be noted that since the creation operators $\psi^*(\mathbf{x}_i)$ commute with one another, the state obtained from the right-hand side of (70) is symmetric.

Let us briefly consider the physical interpretation of the operators $\psi(\mathbf{x})$ and $\psi^*(\mathbf{x})$. It is clear that the state $\psi^*(\mathbf{x}) \mid 0 \rangle$ is a one-particle state since $\psi^*(\mathbf{x}) \mid 0 \rangle$ is an eigenfunction of N with eigenvalue 1. *Proof*:

$$[N, \psi^*(\mathbf{x})] = \int d\mathbf{x}' \, \{\psi^*(\mathbf{x}') \, \psi(\mathbf{x}') \, \psi^*(\mathbf{x}) - \psi^*(\mathbf{x}) \, \psi^*(\mathbf{x}') \, \psi(\mathbf{x}')\}$$
$$= \int d\mathbf{x}' \, \psi^*(\mathbf{x}') \, \delta(\mathbf{x}' - \mathbf{x})$$
$$= \psi^*(\mathbf{x}) \tag{71}$$

so that

$$N\psi^*(\mathbf{x}) \mid 0 \rangle = [N, \psi^*(\mathbf{x})] \mid 0 \rangle = 1 \psi^*(\mathbf{x}) \mid 0 \rangle \tag{72}$$

since $N \mid 0 \rangle = 0$. Similarly, one verifies that $\psi^*(\mathbf{x}_1) \, \psi^*(\mathbf{x}_2) \mid 0 \rangle$ is a two-particle state, etc. Now the probability amplitude for finding the particle in the state $\psi^*(\mathbf{x}) \mid 0 \rangle$ at the position $\mathbf{x}_1$ is given by

$$\langle \mathbf{x}_1 \mid \psi^*(\mathbf{x}) \mid 0 \rangle = \frac{1}{\sqrt{1!}} \langle 0 \mid \psi(\mathbf{x}_1) \, \psi^*(\mathbf{x}) \mid 0 \rangle$$
$$= \delta(\mathbf{x} - \mathbf{x}_1) \, \langle 0 \mid 0 \rangle = \delta(\mathbf{x} - \mathbf{x}_1) \tag{73}$$

where, in obtaining the second line of this last equation, we have used Eqs. (70) and (65) and have assumed the no-particle state to be normalized, $\langle 0 \mid 0 \rangle = 1$. The operator $\psi^*(\mathbf{x})$ thus creates a particle localized at $\mathbf{x}$. Similarly, $\psi(\mathbf{x})$ destroys a particle localized at the point $\mathbf{x}$. Let us compute the norm of the vector $\psi^*(\mathbf{x}) \mid 0 \rangle$. Its square is equal to

$$\langle 0 \mid \psi(\mathbf{x})\, \psi^*(\mathbf{x}) \mid 0 \rangle = \delta(\mathbf{0}) \tag{74}$$

so that the state $\psi^*(\mathbf{x}) \mid 0 \rangle$ is not normalizable and it is therefore not a vector in the Hilbert space. But this was to be expected since $\psi^*(\mathbf{x}) \mid 0 \rangle$ is an eigenfunction of the position operator and thus an eigenfunction of the continuous spectrum, hence not normalizable. A physical one-particle state will correspond to a wave packet $|\Psi\rangle = \int d\mathbf{x}\, f(\mathbf{x})\, \psi^*(\mathbf{x}) \mid 0 \rangle$, with $f(\mathbf{x})$ square integrable. One verifies that such a state is indeed normalizable with the square of its norm equal to

$$\begin{aligned} ||\Psi||^2 = (\Psi, \Psi) &= \int d\mathbf{x} \int d\mathbf{x}'\, \bar{f}(\mathbf{x})\, f(\mathbf{x}')\, \langle 0 \mid \psi(\mathbf{x})\, \psi^*(\mathbf{x}') \mid 0 \rangle \\ &= \int d\mathbf{x}\, |\, f(\mathbf{x})\, |^2 < \infty \end{aligned} \tag{75}$$

6f. Fock Space

Let us next obtain the representations of the operators $\psi(\mathbf{x})$ and $\psi^*(\mathbf{x})$ in configuration space. Consider first the destruction operator $\psi(\mathbf{x})$. We wish to obtain an explicit formula for the matrix element $\langle \mathbf{x}_1, \mathbf{x}_2, \cdots \mathbf{x}_n \mid \psi(\mathbf{x}) \mid \Psi \rangle$ where the state $|\Psi\rangle$ is an arbitrary vector in $\mathcal{H}^{(0)} \oplus \mathcal{H}^{(1)} \oplus \mathcal{H}^{(2)} \cdots$. To do this we write

$$\begin{aligned} &\langle \mathbf{x}_1, \mathbf{x}_2, \cdots \mathbf{x}_{n-1}; n-1 \mid \psi(\mathbf{x}) \mid \Psi \rangle \\ &\quad = \sum_{i=0}^{\infty} \sum_{\{n'\}} \langle \mathbf{x} \mid \lambda^{(i)} \rangle \langle \mathbf{x}_1 \cdots \mathbf{x}_{n-1}; n-1 \mid a_i \mid n_1', n_2' \cdots n_i' \rangle \\ &\qquad \cdot \langle n_1', n_2', \cdots n_i' \cdots \mid \Psi \rangle \\ &\quad = \sum_{i=0}^{\infty} \sum_{\{n'\}} \langle \mathbf{x} \mid \lambda^{(i)} \rangle \sqrt{n_i'}\, \langle \mathbf{x}_1, \mathbf{x}_2, \cdots \mathbf{x}_{n-1}; n-1 \mid n_1', n_2', \cdots n_i' - 1 \rangle \\ &\qquad \langle n_1', n_2', \cdots n_i' \cdots \mid \Psi \rangle \end{aligned} \tag{76}$$

In obtaining the right-hand side of this last equation, we have used Eq. (47) and the completeness of the basis vectors $|n_1', n_2', \cdots n_i' \cdots \rangle$. Although the completeness relation implies a summation over all sequences $\{n_1', n_2', \cdots\}$ with $\sum_{i=0}^{\infty} n_i' = n' = 0, 1, 2, \cdots$ the matrix element $\langle \mathbf{x}_1, \mathbf{x}_2, \cdots \mathbf{x}_{n-1} \mid a_i \mid n_1', n_2' \cdots n_i' \cdots \rangle$ vanishes unless $\sum_{i=1}^{\infty} n_i' = n$ so that the summation over $\{n'\}$ is only over all sequences $\{n_1', n_2', \cdots n_i' \cdots\}$

such that $\sum_{i=1}^{\infty} n_i' = n$. We next use the completeness of the one-particle basis functions to rewrite Eq. (38) as follows:

$$\int d\mathbf{x}\, \langle \lambda^{(i)} \mid \mathbf{x}\rangle \langle \mathbf{x}, \mathbf{x}_2, \mathbf{x}_3 \cdots, \mathbf{x}_n; n \mid n_1, n_2, \cdots n_i, \cdots\rangle$$
$$= \sum_{j=1}^{\infty} \sqrt{\frac{n_j}{n}} \int d\mathbf{x}\, \langle \lambda^{(i)} \mid \mathbf{x}\rangle \langle \mathbf{x} \mid \lambda^{(j)}\rangle$$
$$\langle \mathbf{x}_2, \mathbf{x}_3, \cdots \mathbf{x}_n; n-1 \mid n_1, n_2, \cdots n_i - 1 \cdots\rangle$$
$$= \sqrt{\frac{n_i}{n}} \langle \mathbf{x}_2, \mathbf{x}_3, \cdots \mathbf{x}_n; n-1 \mid n_1, n_2, \cdots n_i - 1 \cdots\rangle \tag{77a}$$

or equivalently

$$\sqrt{n_i} \langle \mathbf{x}_1, \mathbf{x}_2, \cdots \mathbf{x}_{n-1}; n-1 \mid n_1, n_2 \cdots n_i - 1 \cdots\rangle$$
$$= \sqrt{n} \int d\mathbf{x}\, \langle \lambda^{(i)} \mid \mathbf{x}\rangle \langle \mathbf{x}, \mathbf{x}_1, \cdots \mathbf{x}_{n-1}; n \mid n_1, n_2, \cdots n_i \cdots\rangle \tag{77b}$$

Substituting this last expression into the right-hand side of Eq. (76), we obtain

$$\langle \mathbf{x}_1, \mathbf{x}_2, \mathbf{x}_3, \cdots \mathbf{x}_{n-1}; n-1 \mid \psi(\mathbf{x}) \mid \Psi\rangle = \sqrt{n} \sum_i \sum_{\{n\}} \int d\mathbf{y}\, \langle \mathbf{x} \mid \lambda^{(i)}\rangle \langle \lambda^{(i)} \mid \mathbf{y}\rangle$$
$$\cdot \langle \mathbf{y}, \mathbf{x}_1, \mathbf{x}_2, \cdots \mathbf{x}_{n-1}; n-1 \mid n_1, n_2 \cdots n_i\rangle \langle n_1, n_2, \cdots n_i \cdots \mid \Psi\rangle$$
$$= \sqrt{n} \sum_i \int d\mathbf{y}\, \langle \mathbf{x} \mid \lambda^{(i)}\rangle \langle \lambda^{(i)} \mid \mathbf{y}\rangle \langle \mathbf{y}, \mathbf{x}_1, \mathbf{x}_2, \cdots \mathbf{x}_{n-1}; n \mid \Psi\rangle$$
$$= \sqrt{n} \int d\mathbf{y}\, \langle \mathbf{x} \mid \mathbf{y}\rangle \langle \mathbf{y}, \mathbf{x}_1, \mathbf{x}_2, \cdots \mathbf{x}_{n-1}; n \mid \Psi\rangle$$
$$= \sqrt{n} \langle \mathbf{x}, \mathbf{x}_1, \mathbf{x}_2, \cdots \mathbf{x}_{n-1}; n \mid \Psi\rangle \tag{78}$$

which is the desired formula. Similarly, using the expansion formula (39), we derive that

$$\langle 0 \mid \psi^*(\mathbf{x}) \mid \Psi\rangle = 0 \tag{79a}$$

$$\langle \mathbf{x}_1, \mathbf{x}_2, \cdots \mathbf{x}_{n+1}; n+1 \mid \psi^*(\mathbf{x}) \mid \Psi\rangle$$
$$= \sum_{i=1}^{\infty} \sum_{\{n\}} \sqrt{n_i + 1}\, \langle \lambda^{(i)} \mid \mathbf{x}\rangle \langle \mathbf{x}_1, \mathbf{x}_2, \cdots \mathbf{x}_{n+1}; n+1 \mid n_1, n_2, \cdots n_i + 1 \cdots\rangle$$
$$\langle n_1, n_2, \cdots n_i \cdots \mid \Psi\rangle$$
$$= \sqrt{n} \sum_i \sum_{l=1}^{n} \sum_{\{n\}} \frac{1}{\sqrt{n+1}} \langle \mathbf{x}_l \mid \lambda^{(i)}\rangle \langle \lambda^{(i)} \mid \mathbf{x}\rangle \langle \mathbf{x}_1, \mathbf{x}_2, \cdots \mathbf{x}_{l-1}, \mathbf{x}_{l+1}, \cdots \mathbf{x}_{n+1}; n|$$
$$\cdot | n_1, n_2, \cdots n_i \cdots\rangle \langle n_1, n_2, \cdots n_i, \cdots \mid \Psi\rangle$$
$$= \frac{1}{\sqrt{n+1}} \sum_{l=1}^{n} \langle \mathbf{x}_l \mid \mathbf{x}\rangle \langle \mathbf{x}_1, \mathbf{x}_2, \cdots \mathbf{x}_{l-1}, \mathbf{x}_{l+1}, \cdots \mathbf{x}_{n+1}; n \mid \Psi\rangle$$
$$= \frac{1}{\sqrt{n+1}} \sum_{l=1}^{n} \delta(\mathbf{x}_l - \mathbf{x}) \langle \mathbf{x}_1, \mathbf{x}_2, \cdots \mathbf{x}_{l-1}, \mathbf{x}_{l+1}, \cdots \mathbf{x}_{n+1}; n \mid \Psi\rangle \tag{79b}$$

Equivalently, since $|\Psi\rangle$ is an arbitrary state we infer from Eqs. (78) and (79b) that

$$\psi^*(\mathbf{x}) \mid \mathbf{x}_1, \cdots \mathbf{x}_{n-1}; n-1\rangle = \sqrt{n} \mid \mathbf{x}, \mathbf{x}_1, \cdots \mathbf{x}_{n-1}; n\rangle \tag{78a}$$

$$\psi(\mathbf{x}) \mid \mathbf{x}_1, \cdots \mathbf{x}_{n+1}; n+1\rangle = \frac{1}{\sqrt{n+1}} \sum_{l=1}^{n} \delta(\mathbf{x}_l - \mathbf{x}) \mid \mathbf{x}_1, \cdots \mathbf{x}_{l-1}, \mathbf{x}_{l+1}, \cdots \mathbf{x}_{n+1}; n\rangle \tag{79c}$$

The physical meaning of these equations becomes somewhat more transparent if we adopt the following notational convention. Let $|\Psi\rangle$ be a vector in $\mathcal{H}^{(0)} \oplus \mathcal{H}^{(1)} \oplus \mathcal{H}^{(2)} \oplus \cdots$, i.e., it may represent a system whose number of particles is not a constant of the motion. The physical interpretation of this vector identifies the component $\langle \mathbf{x}_1, \mathbf{x}_2, \cdots \mathbf{x}_n; n \mid \Psi\rangle$ in $\mathcal{H}^{(n)}$ as the probability amplitude for finding our system to consist of n particles whose positions are $\mathbf{x}_1, \mathbf{x}_2, \cdots \mathbf{x}_n$. We can specify the vector $|\Psi\rangle$ by giving its component in the various subspaces and represent it as

$$|\Psi\rangle \to \begin{bmatrix} \langle 0 \mid \Psi\rangle \\ \langle \mathbf{x}_1 \mid \Psi\rangle \\ \langle \mathbf{x}_1, \mathbf{x}_2 \mid \Psi\rangle \\ \vdots \\ \langle \mathbf{x}_1, \mathbf{x}_2, \cdots \mathbf{x}_n; n \mid \Psi\rangle \\ \vdots \end{bmatrix} = \begin{bmatrix} \Psi^{(0)} \\ \Psi^{(1)}(\mathbf{x}_1) \\ \Psi^{(2)}(\mathbf{x}_1, \mathbf{x}_2) \\ \vdots \\ \Psi^{(n)}(\mathbf{x}_1, \mathbf{x}_2, \cdots \mathbf{x}_n) \\ \vdots \end{bmatrix} \tag{80}$$

The scalar product of two vectors $|\Psi\rangle$ and $|\Phi\rangle$ is defined as

$$\begin{aligned} \langle \Phi \mid \Psi\rangle &= \sum_{n=0}^{\infty} \int d\mathbf{x}_1 \int d\mathbf{x}_2 \cdots \int d\mathbf{x}_n \, \langle \Phi \mid \mathbf{x}_1, \mathbf{x}_2 \cdots \mathbf{x}_n; n\rangle \langle \mathbf{x}_1, \mathbf{x}_2, \cdots \mathbf{x}_n; n \mid \Psi\rangle \\ &= \overline{\Phi^{(0)}} \Psi^{(0)} + \sum_{n=1}^{\infty} \int d\mathbf{x}_1 \cdots \int d\mathbf{x}_n \, \overline{\Phi^{(n)}(\mathbf{x}_1, \mathbf{x}_2, \cdots \mathbf{x}_n)} \, \Psi^{(n)}(\mathbf{x}_1, \cdots \mathbf{x}_n) \end{aligned} \tag{81}$$

Equation (78) now reads

$$\psi(\mathbf{x}) \begin{bmatrix} \Psi^{(0)} \\ \Psi^{(1)}(\mathbf{x}_1) \\ \Psi^{(2)}(\mathbf{x}_1, \mathbf{x}_2) \\ \vdots \\ \Psi^{(n)}(\mathbf{x}_1, \mathbf{x}_2, \cdots \mathbf{x}_n) \\ \vdots \end{bmatrix} = \begin{bmatrix} \sqrt{1}\, \Psi^{(1)}(\mathbf{x}) \\ \sqrt{2}\, \Psi^{(2)}(\mathbf{x}, \mathbf{x}_1) \\ \sqrt{3}\, \Psi^{(3)}(\mathbf{x}, \mathbf{x}_1, \mathbf{x}_2) \\ \vdots \\ \sqrt{n+1}\, \Psi^{(n+1)}(\mathbf{x}, \mathbf{x}_1, \mathbf{x}_2, \cdots \mathbf{x}_n) \\ \vdots \end{bmatrix} \tag{82}$$

and similarly Eq. (79) reads

$$\psi^*(\mathbf{x})\begin{bmatrix}\Psi^{(0)}\\ \Psi^{(1)}(\mathbf{x}_1)\\ \Psi^{(2)}(\mathbf{x}_1, \mathbf{x}_2)\\ \cdot\\ \cdot\\ \cdot\\ \Psi^{(n)}(\mathbf{x}_1, \mathbf{x}_2, \cdots \mathbf{x}_n)\\ \cdot\\ \cdot\\ \cdot\end{bmatrix} =$$

$$\begin{bmatrix} 0 \\ (1)^{-1/2}\,\delta(\mathbf{x}-\mathbf{x}_1)\,\Psi^{(0)} \\ (2)^{-1/2}\,\{\delta(\mathbf{x}-\mathbf{x}_1)\,\Psi^{(1)}(\mathbf{x}_2)+\delta(\mathbf{x}-\mathbf{x}_2)\,\Psi^{(1)}(\mathbf{x}_1)\} \\ (n)^{-1/2}\sum_{l=1}^{n}\delta(\mathbf{x}_l-\mathbf{x})\,\Psi^{(n-1)}(\mathbf{x}_1, \cdots \mathbf{x}_{l-1}, \mathbf{x}_{l+1}, \cdots \mathbf{x}_n) \\ \cdot \\ \cdot \\ \cdot \end{bmatrix} \tag{83}$$

That in Eq. (83) the $n = 0$ component of the vector $\psi^*(\mathbf{x}) \mid \Psi\rangle$ is zero follows from the fact that it is defined as $\langle 0 \mid \psi^*(\mathbf{x}) \mid \Psi\rangle$, and this matrix element vanishes since the no-particle state is defined by $0 = \psi(\mathbf{x}) \mid 0\rangle$, or, taking the adjoint of this equation, by $0 = \langle 0 \mid \psi^*(\mathbf{x})$.

The above representation which is characterized by the number operator N being diagonal, is known as a Fock space or configuration space description. The representations (82) and (83) for the operators $\psi(\mathbf{x})$ and $\psi^*(\mathbf{x})$ clearly demonstrate their property of being creation and annihilation operators. For example, $\psi(\mathbf{x})$ transforms a vector, $|\,\rangle$, whose only nonvanishing component is $\langle \mathbf{x}_1, \cdots \mathbf{x}_n; n|\,\rangle$ into a vector $\psi(\mathbf{x}) \mid\,\rangle$ whose only nonvanishing component is $\langle \mathbf{x}_1, \mathbf{x}_2, \cdots \mathbf{x}_{n-1}; n-1 \mid \psi(\mathbf{x}) \mid\,\rangle$, i.e., it transforms an n-particle state into an $n - 1$ particle state. Note also that the representation (82), (83) does not alter the symmetry property of the states: it carries symmetric states into symmetric states. Furthermore one verifies that with these representations of the operators $\psi(\mathbf{x})$ and $\psi^*(\mathbf{x})$, the commutation rules, Eqs. (65) and (66), are identically satisfied.

6g. The Antisymmetric Case

The description of an assembly of fermions is quite similar to that presented in the preceding sections. The situation is in fact simpler, since the properly antisymmetrized basis vectors $\mathcal{G}_{\{n_1, n_2, \cdots\}}(\mathbf{x}_1, \cdots \mathbf{x}_n)$ are such that $n_i \leqslant 1$, i.e., n_i can only be equal to 0 or 1. To prove this assertion, assume that the $G_{\lambda_1\lambda_2\cdots\lambda_n}$[1] from which $\mathcal{G}_{\{n_1, n_2, \cdots\}}$ is constructed by application of the antisymmetrizer A, is such that $n_i \geqslant 2$. This means that $G_{\lambda_1\lambda_2\cdots\lambda_n}(\mathbf{x}_1, \cdots \mathbf{x}_n)$ is of the form $\langle \mathbf{x}_1 \mid \lambda_1\rangle \cdots \langle \mathbf{x}_r \mid \lambda^{(i)}\rangle \cdots \langle \mathbf{x}_s \mid \lambda^{(i)}\rangle \cdots$, i.e., at least two of the eigenvalues, say the sth and rth, are the same. Let now $\mathfrak{I}_{rs}$ be the transposition which exchanges the rth and sth co-ordinate, then $\mathfrak{I}_{rs}G_{\lambda_1\lambda_2\cdots\lambda_n}(\mathbf{x}_1, \cdots \mathbf{x}_n) = G_{\lambda_1\lambda_2\cdots\lambda_n}(\mathbf{x}_1, \cdots \mathbf{x}_n)$, whence

$$\begin{aligned}\mathcal{G}_{\{n_1, n_2, \cdots\}} &= \mathfrak{A}G_{\lambda_1\lambda_2\cdots\lambda_n} = \mathfrak{A}\mathfrak{I}_{rs}G_{\lambda_1\lambda_2\cdots\lambda_n} \\ &= -\mathfrak{A}G_{\lambda_1\lambda_2\cdots\lambda_n} = 0 \end{aligned} \tag{84}$$

where we have used property b (see Sec. 6b) of the antisymmetrizer, namely that $\mathfrak{I}\mathfrak{A} = -\mathfrak{A}$ for any transposition $\mathfrak{I}$. Therefore $\mathcal{G}$ vanishes unless all the λ_is are different, i.e., unless $n_i = 0, 1$. The normalization constant is therefore $N^{(n)}_{\{n_1, n_2, \cdots n_i \cdots\}} = \sqrt{n!}$ and the properly antisymmetrized basis vectors are

$$\begin{aligned}\langle \mathbf{x}_1, \mathbf{x}_2, \cdots \mathbf{x}_n; n \mid n_1, n_2, \cdots n_i\rangle &= \frac{1}{\sqrt{n!}}\sum_P \delta_P \mathcal{P}(g_{\lambda_1}(\mathbf{x}_1) \cdots g_{\lambda_n}(\mathbf{x}_n)) \\ &= \frac{1}{\sqrt{n!}}\begin{vmatrix} \langle \mathbf{x}_1 \mid \lambda_1\rangle & \langle \mathbf{x}_1 \mid \lambda_2\rangle \cdots & \langle \mathbf{x}_1 \mid \lambda_n\rangle \\ \langle \mathbf{x}_2 \mid \lambda_1\rangle & \langle \mathbf{x}_2 \mid \lambda_2\rangle \cdots & \langle \mathbf{x}_2 \mid \lambda_n\rangle \\ \cdot & \cdot & \cdot \\ \cdot & \cdot & \cdot \\ \cdot & \cdot & \cdot \\ \langle \mathbf{x}_n \mid \lambda_1\rangle & \langle \mathbf{x}_n \mid \lambda_2\rangle \cdots & \langle \mathbf{x}_n \mid \lambda_n\rangle \end{vmatrix}\end{aligned} \tag{85}$$

In determinantal form, it is obvious that no two g_λs can be the same, since if this were so two columns would be identical and the determinant would vanish.

The expansion formulae, which are the analogue of Eqs. (38) and (39) for the symmetric case, are now just the ordinary expansions of a determinant about a row or column. They are, if $\lambda_m = \lambda^{(i)}$

$$\begin{aligned}\langle \mathbf{x}_1, \mathbf{x}_2, \cdots \mathbf{x}_n; n \mid n_1, n_2, n_3 \cdots n_i \cdots\rangle &= \sum_{m=1}^{n} \frac{1}{\sqrt{n}} (-1)^{m-1} \langle \mathbf{x}_1 \mid \lambda_m\rangle \\ &\quad \langle \mathbf{x}_2, \mathbf{x}_3, \cdots \mathbf{x}_n; n-1 \mid n_1, n_2, \cdots n_i - 1 \cdots\rangle \\ = \frac{1}{\sqrt{n}}\sum_{i=1}^{\infty} \langle \mathbf{x}_1 \mid \lambda^{(i)}\rangle\, n_i(-1)^{s_i} &\langle \mathbf{x}_2, \mathbf{x}_3, \cdots \mathbf{x}_n; n-1 \mid n_1, \cdots n_i - 1, \cdots\rangle \end{aligned} \tag{86}$$

where

$$s_j = \sum_{k=1}^{j-1} n_k \tag{87}$$

is equal to the number of occupied states up to the jth. The second expansion formula is

[1] For convenience we shall assume $\lambda_1 < \lambda_2 < \lambda_3 \cdots < \lambda_n$.

$$\langle \mathbf{x}_1, \mathbf{x}_2, \cdots \mathbf{x}_n; n \mid n_1, n_2, \cdots n_i \cdots \rangle = \frac{1}{\sqrt{n}} \sum_{l=1}^{n} (-1)^{l-1} \langle \mathbf{x}_l \mid \lambda^{(i)} \rangle$$

$$\cdot \langle \mathbf{x}_1, \mathbf{x}_2, \cdots \mathbf{x}_{l-1}, \mathbf{x}_{l+1} \cdots \mathbf{x}_n; n-1 \mid n_1, \cdots n_i - 1, \cdots \rangle \quad (88)$$

We now define the destruction operator a_i by the equation

$$a_i \mid n_1, n_2, \cdots, n_i, \cdots \rangle = (-1)^{s_i} n_i \mid n_1, n_2, \cdots, n_i - 1, \cdots \rangle \quad (89)$$

and the adjoint operator is given by

$$a_i^* \mid n_1, n_2, \cdots, n_i, \cdots \rangle = (-1)^{s_i} (1 - n_i) \mid n_1, n_2, \cdots, n_i + 1, \cdots \rangle \quad (90)$$

The factors $(1 - n_i)$ in the definition of the adjoint guarantee that if the state $|\lambda^{(i)}\rangle$ is already occupied a second particle cannot be put into that state.

The commutation rules are now verified to be

$$[a_i, a_j]_+ = [a_i^*, a_j^*]_+ = 0 \quad (91a)$$

$$[a_i, a_j^*]_+ = \delta_{ij} \quad (91b)$$

($[A, B]_+ = AB + BA$ is the anticommutator of A and B), since

$$a_i a_i^* \mid n_1, n_2, \cdots n_i, \cdots \rangle = (-1)^{2s_i} (1 - n_i)(n_i + 1) \mid n_1, \cdots n_i, \cdots \rangle \quad (92)$$

$$a_i^* a_i \mid n_1, n_2, \cdots, n_i \cdots \rangle = (-1)^{2s_i} (2 - n_i) n_i \mid n_1, \cdots n_i \cdots \rangle \quad (93)$$

and $n_i^2 = n_i$, since $n_i = 0, 1$. Note that $a_i^2 = a_i^{*2} = 0$ which is the operator expression of the fact that no state can have an occupation number greater than one. The number operator for the ith state is again given by $N_i = a_i^* a_i$. It now has the additional property of being a projection operator since

$$N_i^2 = a_i^* a_i a_i^* a_i = a_i^*(1 - a_i^* a_i) a_i$$

$$= a_i^* a_i = N_i \quad (94)$$

so that its eigenvalues are 0 and 1 as expected. If we define the operators

$$\psi(\mathbf{x}) = \sum_i \langle \mathbf{x} \mid \lambda^{(i)} \rangle a_i \quad (95a)$$

$$\psi^*(\mathbf{x}) = \sum_i \overline{\langle \mathbf{x} \mid \lambda^{(i)} \rangle} a_i^* \quad (95b)$$

these will satisfy the following commutation rules:

$$[\psi(\mathbf{x}), \psi^*(\mathbf{x}')]_+ = \delta(\mathbf{x} - \mathbf{x}') \quad (96a)$$

$$[\psi(\mathbf{x}), \psi(\mathbf{y})]_+ = [\psi^*(\mathbf{x}), \psi^*(\mathbf{y})]_+ = 0 \quad (96b)$$

Their Fock space representations are

$$\langle \mathbf{x}_1, \mathbf{x}_2, \cdots \mathbf{x}_n; n \mid \psi(\mathbf{x}) \mid \Psi \rangle = (\psi(\mathbf{x}) \Psi)^{(n)}(\mathbf{x}_1, \mathbf{x}_2, \cdots \mathbf{x}_n)$$

$$= \sqrt{n+1} \langle \mathbf{x}, \mathbf{x}_1, \cdots \mathbf{x}_n; n+1 \mid \Psi \rangle$$

$$= \sqrt{n+1}\, \Psi^{(n+1)}(\mathbf{x}, \mathbf{x}_1, \mathbf{x}_2, \cdots \mathbf{x}_n) \quad (97)$$

and

$$\langle \mathbf{x}_1, \mathbf{x}_2, \cdots \mathbf{x}_n; n \mid \psi^*(\mathbf{x}) \mid \Psi\rangle = (\psi^*(\mathbf{x})\,\Psi)^{(n)}(\mathbf{x}_1, \mathbf{x}_2, \cdots \mathbf{x}_n)$$

$$= \frac{1}{\sqrt{n}} \sum_{l=1}^{n} (-1)^{l-1}\, \delta(\mathbf{x} - \mathbf{x}_l)\, \Psi^{(n-1)}(\mathbf{x}_1, \mathbf{x}_2, \cdots \mathbf{x}_{l-1}, \mathbf{x}_{l+1}, \cdots \mathbf{x}_n) \quad (98)$$

Finally, as in the symmetric case, we can rewrite Eq. (85) as follows:

$$\langle \mathbf{x}_1, \mathbf{x}_2, \cdots \mathbf{x}_n; n \mid n_1, n_2, \cdots n_j, \cdots\rangle$$

$$= \frac{1}{\sqrt{n!}} \langle 0 \mid \psi(\mathbf{x}_n) \cdots \psi(\mathbf{x}_1) \mid n_1, n_2, \cdots n_j \cdots\rangle \quad (99a)$$

whence

$$A \mid \mathbf{x}_1, \mathbf{x}_2, \cdots \mathbf{x}_n; n\rangle = \frac{1}{\sqrt{n!}} \psi^*(\mathbf{x}_n) \cdots \psi^*(\mathbf{x}_1) \mid 0\rangle \quad (99b)$$

6h. Representation of Operators

We next obtain the representation in occupation number space of an operator, F, which does not change the number of particles and whose configuration space representation is known:

$$\langle \mathbf{x}_1', \mathbf{x}_2', \cdots \mathbf{x}_n' \mid F \mid \mathbf{x}_1, \cdots \mathbf{x}_n\rangle$$

$$= \delta_{nn'} F(n; \mathbf{x}_1, \mathbf{x}_2, \cdots \mathbf{x}_n)\, \delta(\mathbf{x}_1 - \mathbf{x}_1') \cdots \delta(\mathbf{x}_n - \mathbf{x}_n') \quad (100)$$

Using the completeness relation for the basis vectors $|\mathbf{x}_1, \mathbf{x}_2, \cdots \mathbf{x}_n\rangle$ we obtain the following representation for F in occupation number space

$$\langle n_1', n_2' \cdots \mid F \mid n_1, n_2 \cdots\rangle$$

$$= \int d\mathbf{x}_1 \cdots \int d\mathbf{x}_n \int d\mathbf{x}_1' \cdots \int d\mathbf{x}_n' \langle n_1', n_2' \cdots \mid \mathbf{x}_1' \cdots \mathbf{x}_n'\rangle$$

$$\cdot \langle \mathbf{x}_1', \mathbf{x}_2', \cdots \mathbf{x}_n' \mid F \mid \mathbf{x}_1, \mathbf{x}_2, \cdots \mathbf{x}_n\rangle \langle \mathbf{x}_1, \mathbf{x}_2, \cdots \mathbf{x}_n \mid n_1, n_2 \cdots n_i \cdots\rangle$$

$$= \int d\mathbf{x}_1 \cdots \int d\mathbf{x}_n \langle n_1', \cdots n_i' \cdots \mid \mathbf{x}_1 \cdots \mathbf{x}_n\rangle$$

$$F(\mathbf{x}_1, \mathbf{x}_2, \cdots \mathbf{x}_n; n) \langle \mathbf{x}_1, \mathbf{x}_2, \cdots \mathbf{x}_n \mid n_1, n_2 \cdots\rangle$$

$$= \frac{1}{n!} \int d\mathbf{x}_1 \cdots \int d\mathbf{x}_n \langle n_1', \cdots n_i' \cdots \mid \psi^*(\mathbf{x}_n) \cdots \psi^*(\mathbf{x}_1) \mid 0\rangle$$

$$F(n; \mathbf{x}_1, \mathbf{x}_2 \cdots \mathbf{x}_n) \cdot \langle 0 \mid \psi(\mathbf{x}_1) \cdots \psi(\mathbf{x}_n) \mid n_1, n_2, \cdots\rangle \quad (101)$$

where $n_1' + n_2' + \cdots = n_1 + n_2 + \cdots = n$ and where in deriving the last equation we have used Eq. (69) or (99), depending on whether we are dealing with the symmetric or antisymmetric case. Since $\psi(\mathbf{x}_n) \cdots \psi(\mathbf{x}_1) \mid n\rangle$ has a nonvanishing component only on the no-particle state, we can reinstate the summation over complete sets over $\mathcal{H}^{(1)}, \mathcal{H}^{(2)}, \cdots$, etc. to rewrite Eq. (101) as follows

$$\langle n_1', n_2' \cdots \mid F \mid n_1, n_2, \cdots\rangle = \frac{1}{n!} \int d\mathbf{x}_1 \cdots \int d\mathbf{x}_n$$

$$\cdot \langle n_1', n_2', \cdots \mid \psi^*(\mathbf{x}_n) \cdots \psi^*(\mathbf{x}_1)\, F(n; \mathbf{x}_1, \cdots \mathbf{x}_n)\, \psi(\mathbf{x}_1) \cdots \psi(\mathbf{x}_n) \mid n_1, n_2 \cdots\rangle$$

$$(102)$$

Consider now the case that F is a sum of one-particle operators, i.e.,

$$F(\mathbf{x}_1, \mathbf{x}_2, \mathbf{x}_3, \cdots \mathbf{x}_n; n) = \sum_{i=1}^{n} f(\mathbf{x}_i) \tag{103}$$

Substituting this expression into the right-hand side of Eq. (102), we obtain

$$\langle n_1', n_2' \cdots | F | n_1, n_2, \cdots \rangle = \frac{1}{n!} \int d\mathbf{x}_1 \cdots \int d\mathbf{x}_n$$

$$\cdot \langle n_1', n_2' \cdots | \psi^*(\mathbf{x}_n) \cdots \psi^*(\mathbf{x}_1) \sum_{i=1}^{n} f(\mathbf{x}_i) \psi(\mathbf{x}_1) \cdots \psi(\mathbf{x}_n) | n_1, n_2 \cdots \rangle \tag{104}$$

Now the term

$$\int d\mathbf{x}_1 \cdots \int d\mathbf{x}_n \langle n_1' n_2' \cdots | \psi^*(\mathbf{x}_n) \cdots \psi^*(\mathbf{x}_1) f(\mathbf{x}_n) \psi(\mathbf{x}_1) \cdots \psi(\mathbf{x}_n) | n_1 n_2 \cdots \rangle \tag{105}$$

can be simplified since the integration over $\int d\mathbf{x}_1$ yields the number operator N, which operating on the state to the right of it,

$$\psi(\mathbf{x}_2) \cdots \psi(\mathbf{x}_n) | n_1 n_2 \cdots n_i \cdots \rangle$$

gives 1 times this state, since the latter is a one-particle state. Similarly the integration over $\int d\mathbf{x}_2$ then yields a factor 2 from the number operator operating on the two-particle state to the right of it. Therefore integrating over the variables $\mathbf{x}_1, \mathbf{x}_2, \cdots \mathbf{x}_{n-1}$ finally yields

$$\int d\mathbf{x}_n (n-1)! \langle n_1' n_2' \cdots | \psi^*(\mathbf{x}_n) f(\mathbf{x}_n) \psi(\mathbf{x}_n) | n_1 n_2 \cdots \rangle$$

Every term in the summation can be reduced to this form by using the commutation or anticommutation rules to commute the operators $\psi^*(\mathbf{x}_i)$ and $\psi(\mathbf{x}_i)$ which are integrated together with $f(\mathbf{x}_i)$ to the extreme left and extreme right, respectively. Therefore

$$\begin{aligned}\langle n_1', n_2' \cdots | F | n_1, n_2 \cdots \rangle &= \frac{1}{n} \sum_{i=1}^{n} \langle n_1' n_2' \cdots | \int d\mathbf{x}_i \psi^*(\mathbf{x}_i) f(\mathbf{x}_i) \psi(\mathbf{x}_i) | n_1 n_2 \cdots \rangle \\ &= \langle n_1' n_2' \cdots | \int d\mathbf{x} \psi^*(\mathbf{x}) f(\mathbf{x}) \psi(\mathbf{x}) | n_1 n_2 \cdots \rangle \end{aligned} \tag{106}$$

so that if F has the representation (103) in the n-particle configuration space, the operator F in Fock space has the form

$$F = \int d\mathbf{x} \psi^*(\mathbf{x}) f(\mathbf{x}) \psi(\mathbf{x}) \tag{107}$$

Similarly, if F is a two-particle operator of the form

$$F(n; \mathbf{x}_1, \mathbf{x}_2, \cdots \mathbf{x}_n) = \sum_{i<j} V(\mathbf{x}_i, \mathbf{x}_j) = \tfrac{1}{2} \sum_{\substack{i,j \\ i \neq j}} V(\mathbf{x}_i, \mathbf{x}_j) \tag{108}$$

one easily derives that

$$\langle n_1', n_2' \cdots | F | n_1, n_2 \cdots \rangle$$

$$= \frac{1}{n(n-1)} \langle n_1', n_2' \cdots | \sum_{i<j} \int d\mathbf{x}_i \int d\mathbf{x}_j$$

$$\cdot \psi^*(\mathbf{x}_i)\, \psi^*(\mathbf{x}_j)\, V(\mathbf{x}_i, \mathbf{x}_j)\, \psi(\mathbf{x}_j)\, \psi(\mathbf{x}_i) \,|\, n_1, n_2 \cdots \rangle$$

$$= \langle n_1', n_2' \cdots | \tfrac{1}{2} \int d\mathbf{x}' \int d\mathbf{x}\, \psi^*(\mathbf{x}')\, \psi^*(\mathbf{x})\, V(\mathbf{x}, \mathbf{x}')\, \psi(\mathbf{x})\, \psi(\mathbf{x}') \,|\, n_1, n_2 \cdots \rangle \tag{109a}$$

so that

$$F = \tfrac{1}{2} \int d\mathbf{x}' \int d\mathbf{x}\, \psi^*(\mathbf{x}')\, \psi^*(\mathbf{x})\, V(\mathbf{x}, \mathbf{x}')\, \psi(\mathbf{x})\, \psi(\mathbf{x}') \tag{109b}$$

Note the order of the operators in this expression: $\psi^*(\mathbf{x}')\, \psi^*(\mathbf{x})\, \psi(\mathbf{x})\, \psi(\mathbf{x}')$. This order is of importance since it implies that there is *no* self-interaction (i.e., that a term of the form $V(\mathbf{x}_i, \mathbf{x}_i)$ does not occur in the expansion (108)), and secondly it guarantees the hermiticity of the operator F in Fock space.

Thus in Fock space the Hamiltonian which corresponds in the n-particle subspace to the ordinary Schrödinger theory Hamiltonian describing a system of n particles interacting through two body forces, is given by

$$H = -\frac{\hbar^2}{2m} \int d\mathbf{x}\, \psi^*(\mathbf{x})\, \nabla^2 \psi(\mathbf{x})$$

$$+ \tfrac{1}{2} \int d\mathbf{x} \int d\mathbf{x}'\, \psi^*(\mathbf{x}')\, \psi^*(\mathbf{x})\, V(\mathbf{x}, \mathbf{x}')\, \psi(\mathbf{x})\, \psi(\mathbf{x}')$$

$$= H_0 + H_I \tag{110}$$

Since the explicit verification of this assertion gives an insight into the meaning of the Fock space formalism, we here carry out the proof for the H_0 terms, for the case of fermions, i.e., for the case that ψ, ψ^* satisfy anticommutation rules.

We wish to compute the action of H_0 on an arbitrary state $|\Psi\rangle$ which in Fock space is given by

$$|\Psi\rangle = \begin{bmatrix} \Psi^{(0)} \\ \Psi^{(1)}(\mathbf{x}_1) \\ \cdot \\ \cdot \\ \cdot \\ \Psi^{(n)}(\mathbf{x}_1, \mathbf{x}_2, \cdots \mathbf{x}_n) \\ \cdot \\ \cdot \\ \cdot \end{bmatrix} \tag{111}$$

Using Eqs. (97) and (98), we find

$$-\frac{\hbar^2}{2m}\int d\mathbf{x}\,\psi^*(\mathbf{x})\,\nabla^2\psi(\mathbf{x})\begin{bmatrix}\Psi^{(0)}\\ \Psi^{(1)}(\mathbf{x}_1)\\ \Psi^{(2)}(\mathbf{x}_1,\mathbf{x}_2)\\ \cdot\\ \cdot\\ \cdot\\ \Psi^{(n)}(\mathbf{x}_1\cdots\mathbf{x}_n)\\ \cdot\\ \cdot\\ \cdot\end{bmatrix}$$

$$= -\frac{\hbar^2}{2m}\int d\mathbf{x}\,\psi^*(\mathbf{x})\begin{bmatrix}\sqrt{1}\,\nabla_{\mathbf{x}}{}^2\Psi^{(1)}(\mathbf{x})\\ \sqrt{2}\,\nabla_{\mathbf{x}}{}^2\Psi^{(2)}(\mathbf{x},\mathbf{x}_1)\\ \sqrt{3}\,\nabla_{\mathbf{x}}{}^2\Psi^{(3)}(\mathbf{x},\mathbf{x}_1,\mathbf{x}_2)\\ \cdot\\ \cdot\\ \cdot\\ \sqrt{n}\,\nabla_{\mathbf{x}}{}^2\Psi^{(n)}(\mathbf{x},\mathbf{x}_1,\cdots\mathbf{x}_n)\\ \cdot\\ \cdot\\ \cdot\end{bmatrix} \tag{112a}$$

$$= \begin{bmatrix}0\\ -\frac{\hbar^2}{2m}\int d\mathbf{x}\,\delta(\mathbf{x}-\mathbf{x}_1)\,\nabla_{\mathbf{x}}{}^2\Psi^{(1)}(\mathbf{x})\\ -\frac{\hbar^2}{2m}\int d\mathbf{x}\,\{\delta(\mathbf{x}-\mathbf{x}_1)\,\nabla_{\mathbf{x}}{}^2\Psi^{(2)}(\mathbf{x},\mathbf{x}_2)\\ \qquad -\delta(\mathbf{x}-\mathbf{x}_2)\,\nabla_{\mathbf{x}}{}^2\Psi^{(2)}(\mathbf{x},\mathbf{x}_1)\}\\ \cdot\\ \cdot\\ \cdot\\ -\frac{\hbar^2}{2m}\int d\mathbf{x}\,\{\delta(\mathbf{x}-\mathbf{x}_1)\,\nabla_{\mathbf{x}}{}^2\Psi^{(n)}(\mathbf{x},\mathbf{x}_2,\cdots\mathbf{x}_n)\\ -\delta(\mathbf{x}-\mathbf{x}_2)\,\nabla_{\mathbf{x}}{}^2\Psi^{(n)}(\mathbf{x},\mathbf{x}_1,\mathbf{x}_3,\cdots\mathbf{x}_n)\\ \pm\cdots+(-1)^{n-1}\,\delta(\mathbf{x}-\mathbf{x}_n)\,\nabla_{\mathbf{x}}{}^2\Psi^{(n)}(\mathbf{x},\mathbf{x}_1,\cdots\mathbf{x}_{n-1})\}\\ \cdot\\ \cdot\\ \cdot\end{bmatrix} \tag{112b}$$

$$
= \begin{bmatrix} 0 \\ -\frac{\hbar^2}{2m} \nabla_{\mathbf{x}_1}{}^2 \Psi^{(1)}(\mathbf{x}_1) \\ -\frac{\hbar^2}{2m} \{\nabla_{\mathbf{x}_1}{}^2 \Psi^{(2)}(\mathbf{x}_1, \mathbf{x}_2) - \nabla_{\mathbf{x}_2}{}^2 \Psi^{(2)}(\mathbf{x}_2, \mathbf{x}_1)\} \\ \cdot \\ \cdot \\ \cdot \\ -\frac{\hbar^2}{2m} \{\nabla_{\mathbf{x}_1}{}^2 \Psi^{(n)}(\mathbf{x}_1, \mathbf{x}_2 \cdots \mathbf{x}_n) - \nabla_{\mathbf{x}_2}{}^2 \Psi^{(n)}(\mathbf{x}_2, \mathbf{x}_1, \mathbf{x}_3, \cdots \mathbf{x}_n) \\ \qquad \pm \cdots + (-1)^{n-1} \nabla_{\mathbf{x}_n}{}^2 \Psi^{(n)}(\mathbf{x}_n, \mathbf{x}_1, \cdots \mathbf{x}_{n-1})\} \\ \cdot \\ \cdot \\ \cdot \end{bmatrix} \tag{112c}
$$

$$
= \begin{bmatrix} 0 \\ -\frac{\hbar^2}{2m} \nabla_{\mathbf{x}_1}{}^2 \Psi^{(1)}(\mathbf{x}_1) \\ -\frac{\hbar^2}{2m} \sum_{i=1}^{2} \nabla_{\mathbf{x}_i}{}^2 \Psi^{(2)}(\mathbf{x}_1, \mathbf{x}_2) \\ \cdot \\ \cdot \\ \cdot \\ -\frac{\hbar^2}{2m} \sum_{i=1}^{n} \nabla_{\mathbf{x}_i}{}^2 \Psi^{(n)}(\mathbf{x}_1, \mathbf{x}_2, \cdots \mathbf{x}_n) \\ \cdot \\ \cdot \\ \cdot \end{bmatrix} \tag{112d}
$$

In going from Eq. (112c) to Eq. (112d) we have made use of the antisymmetry of the function $\Psi(\mathbf{x}_1, \mathbf{x}_2 \cdots \mathbf{x}_n)$, e.g., $\Psi^{(n)}(\mathbf{x}_n, \mathbf{x}_1, \cdots \mathbf{x}_{n-1}) = = (-1)^{n-1}\Psi(\mathbf{x}_1, \cdots \mathbf{x}_n)$. In a similar manner, one verifies that in the n-particle subspace the equation $H \mid \Psi(t)\rangle = i\hbar\partial_t \mid \Psi(t)\rangle$ takes the form

$$
\left\{-\frac{\hbar^2}{2m}\sum_{i=1}^{n} \nabla_i{}^2 + \sum_{i<j} V(\mathbf{x}_i, \mathbf{x}_j)\right\} \Psi^{(n)}(\mathbf{x}_1, \mathbf{x}_2 \cdots \mathbf{x}_n; t) = i\hbar\partial_t \Psi^{(n)}(\mathbf{x}_1, \cdots \mathbf{x}_n; t) \tag{113}
$$

Thus in each subspace the appropriate Schrödinger equation for the given number of particle is reproduced. The n-particle Schrödinger theory may thus be characterized as the solution of

$$
H \mid \Psi\rangle = i\hbar\partial_t \mid \Psi\rangle \tag{114}
$$

for which $|\Psi\rangle$ is also an eigenfunction of N, the number operator, with eigenvalue n

$$N \mid \Psi\rangle = n \mid \Psi\rangle \tag{115}$$

It is possible to find a $|\Psi\rangle$ which has this property since for the Hamiltonian Eq. (110) N and H commute or, in other words, for the theory under consideration the number of particles is a constant of the motion. The state vector describing this situation is clearly of the form

$$|\Psi(t)\rangle = \frac{1}{\sqrt{n!}} \int d\mathbf{x}_1 \cdots \int d\mathbf{x}_n \Psi^{(n)}(\mathbf{x}_1, \cdots \mathbf{x}_n; t)\, \psi^*(\mathbf{x}_n) \cdots \psi^*(\mathbf{x}_1) \mid 0\rangle \tag{116}$$

From the fact that the ψ^* operators anticommute with one another, it follows that the amplitude $\Psi^{(n)}(\mathbf{x}_1, \cdots \mathbf{x}_n; t)$ is an antisymmetric function of $\mathbf{x}_1, \mathbf{x}_2, \cdots \mathbf{x}_n$. Furthermore,

$$\Psi^{(n)}(\mathbf{x}_1, \mathbf{x}_2, \cdots \mathbf{x}_n; t) = \frac{1}{\sqrt{n!}} \langle 0 \mid \psi(\mathbf{x}_1) \cdots \psi(\mathbf{x}_n) \mid \Psi(t)\rangle \tag{117}$$

For the case $n = 2$ the explicit verification of this equation proceeds as follows: Using the anticommutation rules (96) and the fact that $\psi(\mathbf{x}) \mid 0\rangle = 0$ for all $\mathbf{x}$, we obtain

$$\begin{aligned}
\frac{1}{2!} \langle 0 \mid \psi(\mathbf{x}_1)\, \psi(\mathbf{x}_2) \mid \Psi\rangle &= \frac{1}{2!} \int d\mathbf{x}_1' \int d\mathbf{x}_2' \langle 0 \mid \psi(\mathbf{x}_1)\, \psi(\mathbf{x}_2)\, \psi^*(\mathbf{x}_2')\, \psi^*(\mathbf{x}_1') \mid 0\rangle \\
&\qquad \cdot \Psi^{(2)}(\mathbf{x}_1', \mathbf{x}_2') \\
&= \frac{1}{2!} \int d\mathbf{x}_1' \int d\mathbf{x}_2' \langle 0 \mid \psi(\mathbf{x}_1) (-\psi^*(\mathbf{x}_2')\, \psi(\mathbf{x}_2) \\
&\qquad + \delta(\mathbf{x}_2 - \mathbf{x}_2'))\, \psi^*(\mathbf{x}_1) \mid 0\rangle\, \Psi(\mathbf{x}_1', \mathbf{x}_2') \\
&= \frac{1}{2!} \int d\mathbf{x}_1' \int d\mathbf{x}_2' \{\delta(\mathbf{x}_1 - \mathbf{x}_1')\, \delta(\mathbf{x}_2 - \mathbf{x}_2') \\
&\qquad - \delta(\mathbf{x}_1 - \mathbf{x}_2')\, \delta(\mathbf{x}_2 - \mathbf{x}_1')\}\, \Psi(\mathbf{x}_1', \mathbf{x}_2') \\
&= \Psi^{(2)}(\mathbf{x}_1, \mathbf{x}_2)
\end{aligned} \tag{118}$$

where again the antisymmetry of $\Psi^{(2)}(\mathbf{x}_1, \mathbf{x}_2)$ has been used.

Similar techniques can be used to derive the equation of motion of the state vector $|\Psi\rangle$. Thus, for example,

$$\begin{aligned}
&H_0 \mid \Psi\rangle \\
&= \frac{1}{\sqrt{n!}} \int d\mathbf{x}_1 \cdots \int d\mathbf{x}_n\, \psi^*(\mathbf{x}_n) \cdots \psi^*(\mathbf{x}_1) \mid 0\rangle \sum_{i=1}^{n} -\frac{\hbar^2}{2m} \nabla_i^2 \Psi^{(n)}(\mathbf{x}_1 \cdots \mathbf{x}_n)
\end{aligned} \tag{119}$$

where the commutation rules have been used to bring the operator $\psi(\mathbf{x})$ in H_0 to stand to the right of all the creation operators $\psi^*(\mathbf{x}_1) \cdots \psi^*(\mathbf{x}_n)$.

However, since $\psi(\mathbf{x}) \mid 0\rangle = 0$, this term will not contribute and the result of commuting this factor across is precisely Eq. (119). A similar procedure can be applied for the H_I term. Taking the scalar product of $H \mid \Psi\rangle$ with $\langle \mathbf{x}_1, \mathbf{x}_2 \cdots \mathbf{x}_n|$ then yields Eq. (113).

Although we have illustrated the formalism with a Hamiltonian for which the number of particles is a constant of the motion, it should be clear that it permits an immediate generalization to Hamiltonians for which N is no longer a constant of the motion (e.g., a Hamiltonian such that H_I has different number of ψ and ψ^* operators). It therefore encompasses situations wherein particles are created or destroyed, such as for example is the case in β decay or in the production of a meson in a high energy nucleon-nucleon collision. In these situations the state vector $|\Psi\rangle$ is of the form

$$|\Psi\rangle = \Psi^{(0)} \mid 0\rangle + \frac{1}{\sqrt{1!}} \int d\mathbf{x}_1\, \psi^*(\mathbf{x}_1)\, \Psi^{(1)}(\mathbf{x}_1) \mid 0\rangle + \cdots$$
$$+ \frac{1}{\sqrt{n!}} \int d\mathbf{x}_1 \cdots \int d\mathbf{x}_n\, \psi^*(\mathbf{x}_n) \cdots \psi^*(\mathbf{x}_1)\, \Psi^{(n)}(\mathbf{x}_1 \cdots \mathbf{x}_n) \mid 0\rangle + \cdots \tag{120}$$

and the equations of motion would couple the amplitude for n particles $\Psi^{(n)}$, to that for m particles $\Psi^{(m)}$, with $m \neq n$. We shall give an explicit example of this situation in Section 12*a*.

6*i*. Heisenberg Picture

The presentation thus far was given in the Schrödinger picture: all the operators were time-independent and the state vectors time-dependent. The Heisenberg picture is defined by the following unitary transformation

$$|\Psi_H\rangle = e^{\frac{i}{\hbar}Ht} |\Psi_S(t)\rangle \tag{121}$$

$$\psi_H(\mathbf{x}, t) = e^{\frac{i}{\hbar}Ht}\, \psi(\mathbf{x})\, e^{-\frac{i}{\hbar}Ht} \tag{122}$$

The unitary transformation is such that the Schrödinger and Heisenberg pictures coincide at $t = 0$. In the subsequent equations we shall drop the subscript H on the operators, since their dependence on the time t will imply that they are Heisenberg operators. The equal-time commutation rules of the Heisenberg operators are:

$$[\psi(\mathbf{x}, t), \psi^*(\mathbf{x}', t)]_\pm = e^{\frac{i}{\hbar}Ht} [\psi(\mathbf{x}), \psi^*(\mathbf{x}')]_\pm\, e^{-\frac{i}{\hbar}Ht} = \delta^{(3)}(\mathbf{x} - \mathbf{x}') \tag{123a}$$

$$[\psi(\mathbf{x}, t), \psi(\mathbf{x}', t)]_\pm = [\psi^*(\mathbf{x}, t), \psi^*(\mathbf{x}', t)]_\pm = 0 \tag{123b}$$

The unequal-time commutators are in general no longer c-numbers and are much more difficult to evaluate. For the particular case of noninteracting particles, i.e., $H_I = 0$, an explicit representation can be obtained, and it will be presented in Section 6*j*.

The equation of motion of the Heisenberg operator is given by the equation

$$i\hbar\partial_t\psi(\mathbf{x}, t) = [\psi(\mathbf{x}, t), H] \tag{124}$$

which follows from taking the time derivative of (122). For the particular case of a Hamiltonian of the form

$$H = \int d\mathbf{x}\, \psi^*(\mathbf{x})\, \mathcal{H}(\mathbf{x})\, \psi(\mathbf{x}) + \tfrac{1}{2}\int d\mathbf{x} \int d\mathbf{x}'\, \psi^*(\mathbf{x}')\, \psi^*(\mathbf{x})\, U(\mathbf{x}, \mathbf{x}')\, \psi(\mathbf{x})\, \psi(\mathbf{x}') \tag{125a}$$

where $\mathcal{H}(\mathbf{x})$ is time-independent (e.g., $\mathcal{H}(\mathbf{x}) = -\frac{\hbar^2}{2m}\nabla^2 + V(\mathbf{x})$ where $V(\mathbf{x})$ is the potential of some external field) and $U(\mathbf{x}, \mathbf{x}') = U(\mathbf{x}', \mathbf{x})$

$$i\hbar\partial_t\psi(\mathbf{x}, t) = \mathcal{H}(\mathbf{x})\, \psi(\mathbf{x}, t) + \int d\mathbf{x}'\, \psi^*(\mathbf{x}', t)\, U(\mathbf{x}', \mathbf{x})\, \psi(\mathbf{x}', t)\, \psi(\mathbf{x}, t) \tag{126}$$

To illustrate the derivation of the right-hand side of this last equation, consider the second term arising from $[\psi(\mathbf{x}, t), H_I]$. Since the Hamiltonian is time-independent, $H = H(0) = H(t)$, we can also write

$$H = \int d\mathbf{x}\, \psi^*(\mathbf{x}, t)\, \mathcal{H}(\mathbf{x})\, \psi(\mathbf{x}, t) + \tfrac{1}{2}\int d\mathbf{x} \int d\mathbf{x}'\, \psi^*(\mathbf{x}', t)\, \psi^*(\mathbf{x}, t)\, U(\mathbf{x}, \mathbf{x}')\, \psi(\mathbf{x}, t)\, \psi(\mathbf{x}', t) \tag{125b}$$

so that, using the commutation rules (123),

$$\begin{aligned}
&[\psi(\mathbf{y}, t), \int d\mathbf{x} \int d\mathbf{x}'\, \psi^*(\mathbf{x}', t)\, \psi^*(\mathbf{x}, t)\, U(\mathbf{x}, \mathbf{x}')\, \psi(\mathbf{x}, t)\, \psi(\mathbf{x}', t)] \\
&\quad = \int d\mathbf{x} \int d\mathbf{x}'\, [\psi(\mathbf{y}, t), \psi^*(\mathbf{x}', t)\, \psi^*(\mathbf{x}, t)]\, U(\mathbf{x}, \mathbf{x}')\, \psi(\mathbf{x}, t)\, \psi(\mathbf{x}', t)
\end{aligned} \tag{127}$$

since $[\psi(\mathbf{y}, t), \psi(\mathbf{x}, t)\, \psi(\mathbf{x}', t)] = 0$ for the symmetric as well as the antisymmetric case. Now

$$\begin{aligned}
&[\psi(\mathbf{y}, t), \psi^*(\mathbf{x}', t)\, \psi^*(\mathbf{x}, t)] \\
&\quad = \psi(\mathbf{y}, t)\, \psi^*(\mathbf{x}', t)\, \psi^*(\mathbf{x}, t) - \psi^*(\mathbf{x}', t)\, \psi^*(\mathbf{x}, t)\, \psi(\mathbf{y}, t) \\
&\quad = \delta(\mathbf{y} - \mathbf{x}')\, \psi^*(\mathbf{x}, t) \mp \psi^*(\mathbf{x}', t)\, \psi(\mathbf{y}, t)\, \psi^*(\mathbf{x}, t) - \psi^*(\mathbf{x}', t)\, \psi^*(\mathbf{x}, t)\, \psi(\mathbf{y}, t) \\
&\quad = \delta(\mathbf{y} - \mathbf{x}')\, \psi^*(\mathbf{x}, t) \mp \delta(\mathbf{x} - \mathbf{y})\, \psi^*(\mathbf{x}', t)
\end{aligned} \tag{128}$$

$$\begin{pmatrix} - = \text{F.D} \\ + = \text{B.E} \end{pmatrix}$$

so that finally

$$\begin{aligned}
&[\psi(\mathbf{y}, t), H_I] \\
&\quad = \tfrac{1}{2}\int d\mathbf{x} \int d\mathbf{x}'\, \{\psi^*(\mathbf{x}, t)\, \delta(\mathbf{y} - \mathbf{x}')\, U(\mathbf{x}, \mathbf{x}')\, \psi(\mathbf{x}, t)\, \psi(\mathbf{x}', t) \\
&\qquad \mp \psi^*(\mathbf{x}', t)\, \delta(\mathbf{x} - \mathbf{y})\, U(\mathbf{x}, \mathbf{x}')\, \psi(\mathbf{x}, t)\, \psi(\mathbf{x}', t)\} \\
&\quad = \tfrac{1}{2}\int d\mathbf{x}'\, \psi^*(\mathbf{x}', t)\, (U(\mathbf{x}', \mathbf{y}) + U(\mathbf{y}, \mathbf{x}'))\, \psi(\mathbf{x}', t)\, \psi(\mathbf{y}, t)
\end{aligned} \tag{129}$$

which together with the assumed symmetry of $U(\mathbf{x}, \mathbf{x}') = U(\mathbf{x}', \mathbf{x})$ yields the second term of (126).

When there is no interaction between the particles, i.e., when $H_I = 0$, the equation of motion for the operator is formally identical to the Schrödinger equation for a one-particle system. The Schrödinger wave function has, however, become replaced by an operator satisfying certain commutation rules. Since this Schrödinger equation itself was obtained by a process of "first quantization" whereby the variables $\mathbf{p}$ and $\mathbf{q}$ in the classical Hamiltonian were replaced by operators satisfying the commutation rules $[q_j, p_l] = i\hbar\delta_{jl}$, this second replacement of a commuting function (the wave function) by an operator is then called "second quantization." This is the origin of the appellation "second quantization" for the formalism under review. However, our derivation should make it clear that this nomenclature is misleading since the operator and the wave function formulation are completely equivalent.

6j. Noninteracting Multiparticle Systems

As a first illustration of the operator methods, we here give the second-quantized description of a system of n particles which move freely without interaction with one another. The Hamiltonian describing this system is

$$H = -\frac{\hbar^2}{2m} \int \psi^*(\mathbf{x}, t)\, \nabla^2\psi(\mathbf{x}, t)\, d\mathbf{x} \tag{130}$$

the total momentum operator is given by

$$\mathbf{P} = \frac{\hbar}{2i} \int d\mathbf{x}\, \{\psi^*(\mathbf{x}, t)\, \nabla\psi(\mathbf{x}, t) - \nabla\psi^*(\mathbf{x}, t) \cdot \psi(\mathbf{x}, t)\} \tag{131}$$

and the number operator by

$$N = \int \psi^*(\mathbf{x}, t)\, \psi(\mathbf{x}, t)\, d\mathbf{x} \tag{132}$$

The operators ψ and ψ^* satisfy the commutation rules:

$$\begin{gathered} [\psi(\mathbf{x}, t), \psi^*(\mathbf{x}', t)]_\pm = \delta(\mathbf{x} - \mathbf{x}') \\ [\psi(\mathbf{x}, t), \psi(\mathbf{x}', t)]_\pm = [\psi^*(\mathbf{x}, t), \psi^*(\mathbf{x}', t)]_\pm = 0 \end{gathered} \tag{133}$$

and the following equations of motion:

$$i\hbar\partial_t\psi(\mathbf{x}, t) = [\psi(\mathbf{x}, t), H] = -\frac{\hbar^2}{2m} \nabla^2\psi(\mathbf{x}, t) \tag{134}$$

Since $[H, \mathbf{P}] = 0$, it is convenient to introduce a description of the system in terms of plane wave states, each characterized by a wave vector $\mathbf{k}$ (and possibly a polarization or spin index). This description in terms of plane waves is accomplished by introducing the decomposition

$$\psi(\mathbf{x}, t) = \frac{1}{\sqrt{V}} \sum_{\mathbf{k}} a_{\mathbf{k}} e^{i(\mathbf{k}\cdot\mathbf{x} - \omega_{\mathbf{k}} t)} \tag{135}$$

with $\omega_{\mathbf{k}} = \hbar\mathbf{k}^2/2m$, so that $\psi(\mathbf{x}, t)$ satisfies (134). We assume the system to be in a large box of volume $V = L^3$, and impose periodic boundary conditions. The allowed (discrete) values of k_i are then $k_i = n_i \cdot 2\pi/L$ where $n_i = 0, \pm 1, \pm 2$, etc. In the limit as $V \to \infty$, the sum over the discrete $\mathbf{k}$ values is replaced by an integral over $d\mathbf{k}$ with $\frac{1}{V}\sum_{\mathbf{k}} \to \frac{1}{(2\pi)^3}\int d\mathbf{k}$.

The commutation rules of the a^*, a operators are:

$$[a_{\mathbf{k}}, a_{\mathbf{k}'}^*] = \delta(\mathbf{k} - \mathbf{k}') \tag{136a}$$

$$[a_{\mathbf{k}}, a_{\mathbf{k}'}] = [a_{\mathbf{k}}^*, a_{\mathbf{k}'}^*] = 0 \tag{136b}$$

For the system under consideration, we can exhibit the commutation rules for the Heisenberg operators at different times:

$$\begin{aligned}
[\psi(\mathbf{x}, t), \psi^*(\mathbf{x}', t)]_{\pm} &= \frac{1}{V}\sum_{\mathbf{k}\mathbf{k}'} [a_{\mathbf{k}}, a_{\mathbf{k}'}^*]\, e^{i(\mathbf{k}\cdot\mathbf{x} - \omega_{\mathbf{k}} t)}\, e^{-i(\mathbf{k}'\cdot\mathbf{x}' - \omega_{\mathbf{k}'} t)} \\
&= \frac{1}{V}\sum_{\mathbf{k}} e^{i\mathbf{k}\cdot(\mathbf{x}-\mathbf{x}')}\, e^{-i\omega\mathbf{k}(t-t')} \\
&= \frac{1}{(2\pi)^3}\int d\mathbf{k}\, e^{i\mathbf{k}\cdot(\mathbf{x}-\mathbf{x}') - i\omega_{\mathbf{k}}(t-t')}
\end{aligned} \tag{137}$$

The observables (130)–(132) when expressed in terms of the a, a^* operators are:

$$H = \sum_{\mathbf{k}} \frac{\hbar\mathbf{k}^2}{2m} a_{\mathbf{k}}^* a_{\mathbf{k}} \tag{138}$$

$$\mathbf{P} = \sum_{\mathbf{k}} \hbar\mathbf{k} a_{\mathbf{k}}^* a_{\mathbf{k}} \tag{139}$$

$$N = \sum_{\mathbf{k}} a_{\mathbf{k}}^* a_{\mathbf{k}} \tag{140}$$

For a system of n noninteracting fermions with spin $\frac{1}{2}$, let the ground state be denoted by $|\Phi_0(n)\rangle$. It is that state for which each one-particle state starting from the $\mathbf{k} = 0$ state is successively filled, until all particles have been allocated. It is characterized by $n_{\mathbf{k}} = 1$ for $|\mathbf{k}| \leqslant k_F$ where k_F is the Fermi momentum, with $\frac{4\pi}{3}\hbar^3 k_F{}^3 V = nh^3$, and $n_{\mathbf{k}} = 0$ for $|\mathbf{k}| > k_F$. We assume the Fermi sphere to be filled in a spherically symmetric fashion. The resulting state then also has zero total momentum. An abstract characterization of the ground state is therefore $a_{\mathbf{k}} \mid \Phi_0(n)\rangle = 0$ for $|\mathbf{k}| > k_F$. The ground state energy is

$$(\Phi_0(n), H\Phi_0(n)) = 2 \sum_{|\mathbf{k}| \leqslant k_F} \frac{\hbar^2 \mathbf{k}^2}{2m} n_{\mathbf{k}}$$
$$= 2 \frac{V}{(2\pi)^3} \int_0^{k_F} d\mathbf{k} \frac{\hbar^2 \mathbf{k}^2}{2m} = 2 \frac{\hbar^2 V}{2m(2\pi)^3} 4\pi \int_0^{k_F} dk\, k^4$$
$$= 2 \frac{\hbar^2 k_F{}^5}{5(2\pi)^2 m} V \tag{141}$$

The factor of 2 is due to the spin. Qualitatively we note that the role of the Fermi statistics is to assert that particles with relative momentum p cannot come closer together than a distance $\hbar/p$: If the allowable volume to the system is V and it consists of n particles, the average separation between particles is approximately $a \sim (3V/4\pi n)^{1/3}$ since each "occupies" a volume $\frac{4}{3}\pi a^3 = V/n$. The average momentum per particle is therefore $\hbar/a \sim (3V/4\pi n)^{1/3}$ and the average energy per particle $p^2/2m \simeq \simeq (\hbar^2/2m)\,(4\pi N/3V)^{2/3}$ so that the total energy is approximately $np^2/2m$, which in turn is proportional to $Vk_F{}^5(\hbar^2/2m)$.

For a Bose system the ground state of the noninteracting system is the state in which all the particles are packed in the $\mathbf{k} = 0$ state. The ground state is therefore characterized by $n_0 = n$, $n_{\mathbf{k}} = 0$ if $\mathbf{k} \neq 0$, so that $a_{\mathbf{k}} \mid \Phi_0(n)\rangle = 0$ unless $\mathbf{k} = 0$. Its energy and momentum are zero.

6k. Hartree-Fock Method

As another illustration of the second-quantized formalism we shall describe the Hartree-Fock approximation as applied to a system of n electrons in an external field. The basic assumption of this approximation method is that the state vector describing this system can be approximated by a single properly antisymmetrized product wave function. In other words, the system is assumed to be describable in occupation number space by a state vector $|n_1, n_2, \cdots n_i, \cdots\rangle$ with $\sum_i^\infty n_i = n$ where for the present we do not specify the one-particle wave functions $\varphi_i(\mathbf{x}) = \langle \mathbf{x} \mid {}_i\rangle$ to which the n_i refer, beyond requiring that they form a complete set and that they be eigenfunctions of a complete set of commuting one-particle observables. The Hamiltonian describing the system is assumed to be of the form

$$H = \int d\mathbf{x}\, \psi^*(\mathbf{x})\, \mathcal{H}(\mathbf{x})\, \psi(\mathbf{x}) + \tfrac{1}{2} \int d\mathbf{x} \int d\mathbf{x}'\, \psi^*(\mathbf{x}')\, \psi^*(\mathbf{x})\, U(\mathbf{x}, \mathbf{x}')\, \psi(\mathbf{x})\, \psi(\mathbf{x}')$$
$$= H_0 + H_I \tag{142}$$

with

$$\mathcal{H}(\mathbf{x}) = -\frac{\hbar^2}{2m} \nabla_{\mathbf{x}}{}^2 + V(\mathbf{x}) \tag{143}$$

$$U(\mathbf{x}, \mathbf{x}') = \frac{e^2}{|\mathbf{x} - \mathbf{x}'|} \tag{144}$$

where $V(\mathbf{x})$ describes the external field, e.g., the Coulomb field of a nucleus. Introducing the creation and annihilation operators a_i, a_i^* through the equation

$$\psi(\mathbf{x}) = \sum_{i=1}^{\infty} \varphi_i(\mathbf{x})\, a_i \tag{145}$$

we can rewrite H in terms of these as follows:

$$H = \sum_{ij} h_{ij} a_i^* a_j + \tfrac{1}{2} \sum_{ijkl} a_i^* a_j^* a_k a_l u_{jikl} \tag{146}$$

where

$$h_{ij} = \int d\mathbf{x}\, \bar{\varphi}_i(\mathbf{x})\, \mathcal{H}(\mathbf{x})\, \varphi_j(\mathbf{x}) \tag{147}$$

$$u_{ijkl} = \int d\mathbf{x} \int d\mathbf{x}'\, \bar{\varphi}_i(\mathbf{x})\, \bar{\varphi}_j(\mathbf{x}')\, U(\mathbf{x}, \mathbf{x}')\, \varphi_k(\mathbf{x})\, \varphi_l(\mathbf{x}') \tag{148}$$

The expectation value of the Hamiltonian in the state $|n_1, n_2, \cdots n_i \cdots\rangle$ is readily evaluated. Consider first the term:

$$\langle n_1, n_2, \cdots n_i \cdots | H_0 | n_1, n_2 \cdots\rangle = \sum_{ij} \langle n_1, n_2 \cdots | a_i^* a_j | n_1, n_2 \cdots\rangle\, h_{ij} \tag{149}$$

Since a_i^* creates a particle in the state i, and a_j destroys a particle in the state j, the vector $a_i^* a_j | n_1, n_2, \cdots n_i \cdots n_j \cdots\rangle$ is apart from a phase factor equal to $n_j(1 - n_i) | n_1, n_2, \cdots n_i + 1, \cdots n_j - 1 \cdots\rangle$ and will therefore be orthogonal to the vector $|n_1, n_2 \cdots n_i \cdots n_j \cdots\rangle$ unless $i = j$. Thus, in the sum over j only the term $j = i$ contributes and

$$\langle n_1 n_2 \cdots | H_0 | n_1 n_2 \cdots\rangle = \sum_{i=1}^{\infty} n_i h_{ii} \tag{150}$$

Similarly we note that when evaluating

$$\langle n_1 n_2 \cdots | H_I | n_1 n_2 \cdots\rangle = \tfrac{1}{2} \sum_{ijkl} \langle n_1 n_2 \cdots | a_i^* a_j^* a_k a_l | n_1 n_2 \cdots\rangle\, u_{jikl} \tag{151}$$

the matrix element on the right-hand side is different from zero only when $j = l$ and $i = k$ or when $i = l$ and $j = k$, so that

$$\begin{aligned}\langle n_1 n_2 \cdots &| H_I | n_1 n_2 \cdots\rangle \\ &= \tfrac{1}{2} \sum_{ij} \langle n_1 n_2 \cdots | a_i^* a_j^* a_i a_j | n_1 n_2 \cdots\rangle\, u_{jiij} \\ &\quad + \tfrac{1}{2} \sum_{ij} \langle n_1 n_2 \cdots | a_i^* a_j^* a_j a_i | n_1 n_2 \cdots\rangle\, u_{jiji} \\ &= \tfrac{1}{2} \sum_{ij} \langle n_1 n_2 \cdots | a_i^* a_j^* a_i a_j | n_1 n_2 \cdots\rangle\, (u_{jiij} - u_{jiji})\end{aligned} \tag{152}$$

where the last step follows from the fact that $a_i a_j = -a_j a_i$.

Using the anticommutation rules, we have

$$\begin{aligned} a_i{}^*a_j{}^*a_ia_j &= -a_i{}^*a_ia_j{}^*a_j + a_i{}^*a_j\delta_{ij} \\ &= -N_iN_j + N_i\delta_{ij} \end{aligned} \tag{153}$$

so that finally

$$\langle n_1n_2 \cdots | H_I | n_1n_2 \cdots \rangle = \tfrac{1}{2}\sum_{ij} n_in_j(u_{jiji} - u_{jiij}) \tag{154}$$

and

$$\langle n_1n_2 \cdots | H | n_1n_2 \cdots \rangle = \sum_i h_{ii}n_i + \tfrac{1}{2}\sum_{ij} n_in_j(u_{jiji} - u_{jiij}) \tag{155}$$

If we denote the one-particle wave functions by $\varphi_i(\mathbf{x}, \zeta)$ (we now exhibit the spin variable explicitly), then

$$h_{jj} = \sum_{\zeta=1}^{2} \int d^3x\, \bar{\varphi}_j(\mathbf{x}, \zeta)\left[-\frac{\hbar^2}{2m}\nabla^2 + V(\mathbf{x})\right]\varphi_j(\mathbf{x}, \zeta) \tag{156}$$

$$u_{jiji} = \sum_{\zeta,\zeta'} \int d^3x \int d^3x'\, \bar{\varphi}_j(\mathbf{x}, \zeta)\, \bar{\varphi}_i(\mathbf{x}', \zeta')\, U(\mathbf{x}, \mathbf{x}')\, \varphi_j(\mathbf{x}, \zeta)\, \varphi_i(\mathbf{x}', \zeta') \tag{157}$$

$$u_{jiij} = \sum_{\zeta,\zeta'} \int d^3x \int d^3x'\, \bar{\varphi}_j(\mathbf{x}, \zeta)\, \bar{\varphi}_i(\mathbf{x}', \zeta')\, U(\mathbf{x}, \mathbf{x}')\, \varphi_i(\mathbf{x}, \zeta)\, \varphi_j(\mathbf{x}', \zeta') \tag{158}$$

Note that u_{jiij} is different from zero only if the states i and j correspond to states in which the electrons have parallel spin. Since the Coulomb interaction $U(\mathbf{x}, \mathbf{x}') = e^2/|\mathbf{x} - \mathbf{x}'|$ is spin-independent, the summation over the spin variables ζ, ζ' can be carried out; therefore, if the states $|i\rangle$ and $|j\rangle$ are states in which the electrons have antiparallel spin, the states will be orthogonal in the spin scalar product. If we introduce the density matrix

$$\begin{aligned} \rho_{\zeta\zeta'}(\mathbf{x}, \mathbf{x}') &= \langle n_1n_2 \cdots | \psi_\zeta{}^*(\mathbf{x})\, \psi_{\zeta'}(\mathbf{x}') | n_1n_2 \cdots \rangle \\ &= \sum_i n_i\bar{\varphi}_i(\mathbf{x}, \zeta)\, \varphi_i(\mathbf{x}', \zeta') = \overline{\rho_{\zeta'\zeta}(\mathbf{x}', \mathbf{x})} \end{aligned} \tag{159}$$

in terms of which the ordinary charge density can be written as

$$\begin{aligned} \rho(\mathbf{x}) &= \sum_i \sum_\zeta n_i\, |\varphi_i(\mathbf{x}, \zeta)|^2 \\ &= \sum_\zeta \rho_{\zeta\zeta}(\mathbf{x}, \mathbf{x}) \end{aligned} \tag{160}$$

the interaction term then takes the form

$$\begin{aligned} \langle n_1n_2 \cdots | H_I | n_1n_2 \cdots \rangle &= \frac{e^2}{2}\int d^3x \int d^3x'\, \frac{\rho(\mathbf{x})\,\rho(\mathbf{x}')}{|\mathbf{x} - \mathbf{x}'|} \\ &\quad - \frac{e^2}{2}\sum_{\zeta\zeta'}\int d^3x \int d^3x'\, \frac{|\rho_{\zeta\zeta'}(\mathbf{x}, \mathbf{x}')|^2}{|\mathbf{x} - \mathbf{x}'|} \end{aligned} \tag{161}$$

These terms have the following interpretation. The first clearly corresponds to the Coulomb interaction between the charge clouds. Note that it includes the Coulomb self-interaction of each electron. The second term (the exchange interaction between electrons of parallel spin) corrects for this and is, in addition, the contribution to the energy which arises due to the correlations introduced between the particles by the Pauli exclusion principle.

The Hartree-Fock method [Hartree (1928), Fock (1930), Slater (1930)] now consists in finding the one-electron wave function making up $\langle \mathbf{x}_1, \mathbf{x}_2 \cdots \mathbf{x}_n \mid n_1 n_2 \cdots \rangle$ which minimize $\langle n_1 n_2 \cdots \mid H \mid n_1 n_2 \cdots \rangle$. It is also further required that these one-particle wave functions be orthonormal. If we call

$$\nu_{ij} = \sum_{\zeta} \int d^3\mathbf{x}\, \bar{\varphi}_i(\mathbf{x}, \zeta)\, \varphi_j(\mathbf{x}, \zeta) \tag{162}$$

then the φ_is are determined so that $\langle n_1 n_2 \cdots \mid H \mid n_1 n_2 \cdots \rangle$ is stationary with respect to variations of the φ_is, subject to the constraints (162). If the λ_{ij} are the Lagrangian multipliers for the constraints, the variational equation is

$$\delta\{\langle n_1 n_2 \cdots \mid H \mid n_1 n_2 \cdots \rangle - \sum_{ij} \lambda_{ij}\nu_{ij}\} = 0 \tag{163}$$

The equations for the determination of the φs which result from the variation with respect to φ_j^* are

$$-\frac{\hbar^2}{2m} \nabla^2 \varphi_j(\mathbf{x}, \zeta) - V(\mathbf{x})\, \varphi_j(\mathbf{x}, \zeta) + e^2 \int d^3x\, \frac{\rho(\mathbf{x}')}{|\mathbf{x} - \mathbf{x}'|}\, \varphi_j(\mathbf{x}, \zeta)$$

$$- e^2 \sum_{\zeta'} \int d^3x'\, \frac{\rho_{\zeta'\zeta}(\mathbf{x}', \mathbf{x})\, \varphi_j(\mathbf{x}', \zeta')}{|\mathbf{x} - \mathbf{x}'|} = \sum_{l} \lambda_{jl}\varphi_l(\mathbf{x}, \zeta) \tag{164}$$

These are Fock's equations. By a linear unitary transformation on the φs it is always possible to choose solutions such that λ_{jl} is a diagonal matrix. We therefore need only consider the λ_{jj} which we call ϵ_j. The right-hand side of Eq. (164) then becomes equal to $\epsilon_j \varphi_j(\mathbf{x}, \zeta)$. These equations must then be solved self-consistently for the $\varphi_j(\mathbf{x}, \zeta)$s. A detailed and particularly lucid interpretation of these equations has been given by Slater (1951), to whose paper the interested reader is referred.

There exists an important system for which the Hartree-Fock equations can be solved exactly for the ground state vector. The system consists of electrons moving in a box of volume V in which there exists a uniform background of positive charge of density $+en/V$. The solutions $\varphi_i(\mathbf{x}, \zeta)$ are then plane waves. (We shall first consider the situation when there are n_+ electrons with spin-up and n_- electrons with spin-down, with $n = n_+ + n_-$. Later we shall specialize to the case that $n_+ = n_- = n/2$.) To verify this assertion, we assume that the solutions are such that the phase space of both the spin-up and spin-down particles is filled up in a spherically symmetric manner until the Fermi surface at k_{F+} (for the spin-up particles) and at k_{F-} (for the spin-down particles) is reached

$$\tfrac{4}{3}\pi k_{F\pm}{}^3 V = n_\pm (2\pi\hbar)^3 \tag{165}$$

(in the following we set $\hbar = 1$). By assumption our wave functions are of the form

$$\varphi_{\mathbf{k}}(\mathbf{x}, \zeta) = \frac{1}{\sqrt{V}}\, e^{i\mathbf{k}\cdot\mathbf{x}} \begin{Bmatrix} \alpha(\zeta) \\ \beta(\zeta) \end{Bmatrix} \tag{166}$$

where $k_i = n_i 2\pi/L$, $n_i = 0, \pm 1, \pm 2, \cdots$ and α, β are the usual spin-up, spin-down wave functions. The density matrix for the spin-up electrons is

$$\begin{aligned} \rho^+{}_{\zeta\zeta'}(\mathbf{x}, \mathbf{x}') &= \frac{1}{V} \sum_{|\mathbf{k}| < k_{F+}} e^{-i\mathbf{k}\cdot(\mathbf{x}-\mathbf{x}')}\, \bar{\alpha}(\zeta)\, \alpha(\zeta') \\ &\rightarrow \frac{1}{(2\pi)^3} \int_0^{k_{F+}} k^2 dk \int_0^{2\pi} d\phi \int_{-1}^{+1} d\mu\, e^{-i|\mathbf{k}|\,|\mathbf{x}-\mathbf{x}'|\mu}\, \bar{\alpha}(\zeta)\, \alpha(\zeta') \\ &= \frac{n_\pm}{V} \left(3\, \frac{\sin k_{F+}r - k_{F+}r \cos k_{F+}r}{(k_{F+}r)^3} \right) \bar{\alpha}(\zeta)\, \alpha(\zeta') \qquad (167a) \\ &= \frac{n_\pm}{V}\, Q(\eta_+)\, \bar{\alpha}(\zeta)\, \alpha(\zeta) \qquad (167b) \end{aligned}$$

where $r = |\mathbf{x} - \mathbf{x}'|$, $\eta_\pm = k_{F\pm}r$ and n_+ is the number of electrons with spin-up. The function Q is defined by Eqs. (167a) and (167b). Similarly, the density matrix for spin-down electrons is

$$\rho^-{}_{\zeta\zeta}(\mathbf{x}, \mathbf{x}') = \frac{n_-}{V}\, Q(\eta_-)\, \bar{\beta}(\zeta)\, \beta(\zeta') \tag{168}$$

Note that $Q(\eta) = 1$ for $\eta = 0$, i.e., as $r \rightarrow 0$, so that the electronic charge density is uniform in the volume V.

Consider now an electron in the state $\varphi_j(\mathbf{x}, \zeta)$ which corresponds to a momentum $\mathbf{k}$ and spin-up. In the Fock equations, the exchange term now becomes

$$-e^2 \frac{n_\pm}{V} \int d^3x'\, \frac{Q(k_{F+}\,|\,\mathbf{x} - \mathbf{x}'\,|)}{|\mathbf{x} - \mathbf{x}'|}\, \frac{1}{\sqrt{V}}\, e^{i\mathbf{k}\cdot\mathbf{x}'}\, \alpha(\zeta) \tag{169}$$

since only ρ^+ contributes due to the orthogonality of the spin states α and β and we have carried out the summation over ζ'. The exchange term can be rewritten, by changing integration variables from $\mathbf{x}'$ to $\mathbf{x} - \mathbf{x}'$, as follows:

$$\left(e^2 \frac{n_+}{V} \int d^3r\, \frac{Q(k_{F+}r)}{r}\, e^{-i\mathbf{k}\cdot\mathbf{r}} \right) \frac{1}{\sqrt{V}}\, e^{i\mathbf{k}\cdot\mathbf{x}}\, \alpha(\zeta) \tag{170}$$

The integral in Eq. (170) can be evaluated [Dirac (1930, 1936)] to yield

$$\frac{1}{3} \int_0^\infty d\eta\, Q(\eta) \sin \alpha_+\eta = \frac{\alpha_+}{2} \left\{ 1 + \frac{1 - \alpha_+{}^2}{2\alpha_+} \log \left| \frac{1 + \alpha_+}{1 - \alpha_+} \right| \right\} \tag{171}$$

where $\alpha_\pm = k/k_{F\pm}$. The Fock equation is therefore satisfied with a plane wave function. Since the interaction with the background of positive charge cancels the electronic Coulombic interactions, the equation that φ_k satisfies is

$$\frac{\hbar^2\mathbf{k}^2}{2m}\varphi_{\mathbf{k}}(\mathbf{x},\zeta) - \frac{e^2k_{F+}}{\pi}\left\{1 + \frac{1-\alpha_+{}^2}{2\alpha_+}\log\left|\frac{1+\alpha_+}{1-\alpha_+}\right|\right\}\varphi_{\mathbf{k}}(\mathbf{x},\zeta) = \epsilon_{\mathbf{k}}\varphi_{\mathbf{k}}(\mathbf{x},\zeta) \tag{172}$$

The parameter $\epsilon_{\mathbf{k}}$ is therefore given by the expression

$$\epsilon_{\mathbf{k}} = \frac{\hbar^2\mathbf{k}^2}{2m} - \frac{e^2k_{F+}}{2\pi}\left\{2 + \frac{1-\alpha_+{}^2}{\alpha_+}\log\left|\frac{1+\alpha_+}{1-\alpha_+}\right|\right\} \tag{173}$$

The total energy, $\langle H\rangle$, for the case $n_+ = n = \frac{n}{2}$, $k_{F\pm} = k_F$ is readily evaluated since

$$\langle H_0\rangle = 2\frac{\hbar^2}{2m}\sum_{|\mathbf{k}|<k_F}\mathbf{k}^2 = \frac{V}{(2\pi)^3}\frac{4\pi\hbar^2}{5m}k_F{}^5 \tag{174}$$

and the exchange energy of the spin-up electrons is

$$-\frac{1}{2}e^2\sum_{\zeta\zeta'}\int d^3x\int d^3x'\,\frac{|\rho_{\zeta\zeta'}(\mathbf{x},\mathbf{x}')|^2}{|x-x'|} = -2\pi e^2V\left(\frac{n_+}{V}\right)^2\frac{1}{k_F{}^2}\int_0^\infty d\eta\;\eta\;|\;Q(\eta)\;|^2$$

$$= -\frac{9}{2}\pi e^2\,V\left(\frac{n_+}{V}\right)^2\frac{1}{k_F{}^2} \tag{175}$$

The total exchange energy is twice this amount. The total energy for the ground state is therefore given by

$$\langle H\rangle = E = \frac{V}{(2\pi)^3}\left\{\frac{4\pi\hbar^2}{5m}k_F{}^5 - 2e^2k_F{}^4\right\} \tag{176}$$

If one considers the valence electrons in a metal as quasi-free and describes them as free electrons moving through a positive background interacting with one another through their Coulomb field, then the electronic specific heat as computed from (176) is in poor agreement with experiment. A somewhat better treatment of the correlation energy [Gell-Mann and Brueckner (Gell-Mann 1957a, b)] improves the agreement.

7

Relativistic Fock Space Methods

The Fock space formalism outlined in the previous section for the nonrelativistic theory can readily be extended to the relativistic domain to describe an assembly of free relativistic particles. Once again we shall find that connected with such a second-quantized description in space-time are "field" operators which obey certain commutation rules and which satisfy field equations. In Section 7*b* we shall see how the relativistic transformation of the particle amplitudes determine the transformation properties of the field operators. The chapter then takes up the equivalence of the second-quantized formalism for spin 0 particles with that obtained from the "quantization" of the Klein-Gordon field. It concludes with the extension of the second-quantized and field theoretic methods to the case of charged spin 0 particles.

7a. The Neutral Spin 0 Boson Case

Consider a system of n free neutral spin 0 particles of mass μ. In the absence of interaction the state of such a system can be specified by the momenta $\mathbf{k}_1, \cdots, \mathbf{k}_n$ and corresponding energies $\omega(\mathbf{k}_1), \cdots \omega(\mathbf{k}_n)$, $\omega(\mathbf{k}) = \sqrt{\mathbf{k}^2 + \mu^2}$, of the particles. It is well known that an assembly of identical spin 0 or integer spin particles is described by a state vector which must be symmetric under the exchange of the particles. Hence, our system can be described by a vector

$$|\mathbf{k}_1, \mathbf{k}_2, \cdots \mathbf{k}_n\rangle = U_T \mid \mathbf{k}_1, \mathbf{k}_2, \cdots \mathbf{k}_n\rangle = S\{|\mathbf{k}_1\rangle \mid \mathbf{k}_2\rangle \cdots \mid \mathbf{k}_n\rangle\} \quad (1)$$

where U_T is an arbitrary transposition of the labels of the particles and S the symmetrizer.

To obtain a second-quantized description of our system, we could proceed in complete analogy with the nonrelativistic development. Briefly, we would adopt as the one-particle set $\{g_\lambda(x)\}$, the set of positive energy solutions of the Klein-Gordon equation since the physically realizable states of a relativistic spin 0 particle correspond to precisely these states. Relativistic covariance suggests that the time co-ordinate, x_0, be included in $g_\lambda(x)$. In our subsequent discussion we shall assume the index λ to stand for a specification of the energy and linear momentum of the particle, so that

$$g_\lambda(x) = \langle x \mid \mathbf{k}, \omega(\mathbf{k})\rangle = \frac{1}{\sqrt{2(2\pi)^3}} e^{-ik\cdot x} \tag{2}$$

These are eigenfunctions of the continuous spectrum and therefore not normalizable. It is more convenient from a mathematical point of view to deal with a discrete set of normalizable functions. This is also dictated by the physical situations actually encountered. For example, in the description of the initial states of scattering experiments, one has to describe the states of the particles by wave packets since the localization of the particles is known only to within a region of space of finite extent. Let us therefore denote by $\{f_\alpha(x)\}$ a complete set of normalizable "wave-packet" positive energy solutions of the Klein-Gordon equation:

$$(\Box + \mu^2) f_\alpha(x) = 0$$

$$f_\alpha(x) = \frac{2}{\sqrt{2(2\pi)^3}} \int d^4k\, \theta(k)\, \delta(k^2 - \mu^2)\, e^{-ikx}\, \tilde{f}_\alpha(k) \tag{3}$$

with $(f_\alpha, f_\alpha) < \infty$. Let us further assume that they are pairwise orthogonal in the scalar product (3.24)

$$(f_\alpha, f_\beta) = i \int d\sigma^\mu(x)\, \bar{f}_\alpha(x)\, \overleftrightarrow{\partial_\mu} f_\beta(x) = \delta_{\alpha\beta} \tag{4}$$

which holds for all times, if it holds for any one time. In the limit of a plane wave $\alpha \to \mathbf{k}$, $\beta \to \mathbf{k}'$, $\delta_{\alpha\beta} \to k_0\delta^{(3)}(\mathbf{k} - \mathbf{k}')$. The completeness relation is:

$$\sum_\alpha f_\alpha(x)\, \bar{f}_\alpha(x') = \sum_\alpha \langle x \mid \alpha\rangle \langle \alpha \mid x'\rangle = \frac{1}{2(2\pi)^3} \int_{k_0>0} \frac{d^3k}{k_0} e^{-ik(x-x')} \tag{5}$$

$$k_0 = \omega(\mathbf{k}) = \sqrt{\mathbf{k}^2 + \mu^2}$$

An n-particle state can again be characterized by a set of occupation numbers $n_1, n_2, \cdots$ which specify the number of particles having quantum numbers $\alpha', \alpha'', \cdots$. The transformation function from occupation number space to configuration space is now given by

$$\langle x_1, x_2, \cdots x_n; n \mid n_1, n_2 \cdots\rangle = \sqrt{\frac{n!}{n_1!\, n_2! \cdots n!}} \frac{1}{n!} \sum_P \mathcal{P}(f_{\alpha_1}(x_1) \cdots f_{\alpha_n}(x_n)) \tag{6}$$

Creation and annihilation operators a_β^* and a_β, which increase and decrease by one the number of particles in the state β, can again be defined by the equations:

$$a_\beta \mid \cdots, n_\beta, \cdots\rangle = \sqrt{n_\beta} \mid \cdots, n_\beta - 1, \cdots\rangle \tag{7a}$$

$$a_\beta^* \mid \cdots, n_\beta, \cdots\rangle = \sqrt{n_\beta + 1} \mid \cdots, n_\beta + 1, \cdots\rangle \tag{7b}$$

They satisfy the following commutation rules:

$$[a_\alpha, a_\beta^*] = \delta_{\alpha\beta} \tag{8a}$$

$$[a_\alpha, a_\beta] = [a_\alpha^*, a_\beta^*] = 0 \tag{8b}$$

The total number of particles operator is $\sum_\alpha a_\alpha^* a_\alpha$; it has the positive integers and zero as eigenvalues. We can next introduce the operators

$$\phi^{(+)}(x) = \sum_\alpha f_\alpha(x)\, a_\alpha \tag{9a}$$

$$\phi^{(-)}(x) = \sum_\alpha \bar{f}_\alpha(x)\, a_\alpha^* = [\phi^{(+)}(x)]^* \tag{9b}$$

which no longer depend on the one-particle basis $\{f_\alpha\}$ originally chosen to define occupation number space. They satisfy, by virtue of the commutation rules (8) of the a operators, the following commutation rules:

$$[\phi^{(\pm)}(x), \phi^{(\pm)}(x')] = 0 \tag{10a}$$

$$[\phi^{(+)}(x), \phi^{(-)}(x')] = \sum_\alpha f_\alpha(x)\, \bar{f}_\alpha(x') = \frac{1}{2(2\pi)^3} \int_{k_0>0} \frac{d^3k}{k_0}\, e^{-ik(x-x')} \tag{10b}$$

The fact that the right-hand side of these commutation rules is no longer a δ function, even for equal times $x_0 = x'_0$, is to be noted. It is a direct consequence of the fact that the set of one-particle states $\{f_\alpha\}$ comprises only positive energy solutions of the Klein-Gordon equation. We could next obtain the representation of these operators and thus complete our Fock space representation. Let us, however, proceed somewhat differently.

Intuitively, what is meant by the no-particle state, by a one-particle state and, in general, by an n-particle state seems evident. By a single particle we mean an entity of mass μ and spin 0 which has the property that the events caused by it are localized in space. The following gedanken experiments [Haag (1955)] could therefore be used to single out the one-particle states: Take two Geiger counters which are connected in coincidence and are separated by distance d and have resolving time $\delta t \ll d/c$. Consider all such coincidence arrangements for arbitrary position, orientation, and velocity of the apparatus, and for all separations d greater than some $d_{\min}$. The one-particle states are then those which give a negative result in all these experiments. The distance $d_{\min}$ is arbitrary but will, of course, be chosen as small as possible. Similarly the no-particle state can be excluded in terms of an experiment with a single Geiger counter, and many-particle states by more elaborate coincidence arrangements. Let us introduce vectors corresponding to these states. We shall denote by $|0\rangle$ the no-particle or vacuum state which is normalized so that

$$\langle 0 \mid 0 \rangle = 1 \tag{11}$$

We denote by $|\mathbf{k}\rangle$ a one-particle state of momentum $\mathbf{k}$, energy $k_0 = \omega(\mathbf{k})$. It is normalized invariantly in such a way that

$$\langle \mathbf{k}' \mid \mathbf{k}\rangle = k_0\delta^{(3)}(\mathbf{k} - \mathbf{k}') \tag{12}$$

In general, we denote the n-particle state consisting of n particles with momenta $\mathbf{k}_1, \cdots \mathbf{k}_n$, energy $\omega(\mathbf{k}_1), \cdots \omega(\mathbf{k}_n)$, by $|\mathbf{k}_1, \mathbf{k}_2, \cdots \mathbf{k}_n\rangle$. It satisfies the following orthonormality relations

$$\langle \mathbf{k}_1', \mathbf{k}_2' \cdots \mathbf{k}_m' \mid \mathbf{k}_1, \cdots, \mathbf{k}_n\rangle = \delta_{nm}\frac{1}{n!}\sum_P k_{10}\delta(\mathbf{k}_1 - \mathbf{k}_{\alpha_1}') \cdots k_{n0}\delta(\mathbf{k}_n - \mathbf{k}_{\alpha_n}') \tag{13}$$

where the summation is over all the permutations of the set $\{\alpha_1, \alpha_2, \cdots \alpha_n\}$. The totality of these states forms a complete set, the completeness relation being

$$\begin{aligned} I = |0\rangle\langle 0| &+ \int \frac{d^3k}{k_0}|\mathbf{k}\rangle\langle\mathbf{k}| + \cdots \\ &+ \int \frac{d^3k_1}{k_{10}} \cdots \int \frac{d^3k_n}{k_{n0}} |\mathbf{k}_1, \mathbf{k}_2 \cdots \mathbf{k}_n\rangle\langle \mathbf{k}_1, \cdots \mathbf{k}_n| + \cdots \end{aligned} \tag{14}$$

The proof of this completeness relation consists in showing that the identity operator as expressed above, when operating on a state $|\mathbf{k}_1', \mathbf{k}_2', \cdots \mathbf{k}_n'\rangle$ reproduces this state. That this is so follows from the orthonormality relations (13). *Proof*: Using (13) we find

$$\begin{aligned} |\mathbf{k}_1', \cdots \mathbf{k}_m'\rangle &= \int \frac{d^3k_1}{k_{10}} \cdots \int \frac{d^3k_m}{k_{m0}} |\mathbf{k}_1, \cdots \mathbf{k}_m\rangle\langle \mathbf{k}_1, \cdots \mathbf{k}_m \mid \mathbf{k}_1', \cdots, \mathbf{k}_m'\rangle \\ &= |\mathbf{k}_1', \cdots \mathbf{k}_m'\rangle \end{aligned} \tag{15}$$

We next define the creation and annihilation operators $a_\mathbf{k}^*$ and $a_\mathbf{k}$ by the equations

$$a(\mathbf{k}) \mid \mathbf{k}'\rangle = a_\mathbf{k} \mid \mathbf{k}'\rangle = k_0\delta(\mathbf{k} - \mathbf{k}') \mid 0\rangle \tag{16a}$$

$$a^*(\mathbf{k}) \mid 0\rangle = a_\mathbf{k}^* \mid 0\rangle = \mid \mathbf{k}\rangle; \quad [a_\mathbf{k}]^* = a_\mathbf{k}^* \tag{16b}$$

and, more generally,

$$|\mathbf{k}_1, \mathbf{k}_2, \cdots \mathbf{k}_n\rangle = \frac{1}{\sqrt{n!}} a^*(\mathbf{k}_1)\, a^*(\mathbf{k}_2) \cdots a^*(\mathbf{k}_n) \mid 0\rangle \tag{17}$$

The vacuum state has the further property that

$$a_\mathbf{k} \mid 0\rangle = 0 \quad \text{for all } \mathbf{k} \tag{18}$$

The symmetric character of the state $|\mathbf{k}_1, \mathbf{k}_2 \cdots \mathbf{k}_n\rangle$ requires that $a^*(\mathbf{k}_i)$ commute with $a^*(\mathbf{k}_j)$ for all $\mathbf{k}_i$ and $\mathbf{k}_j$, whence

$$[a_\mathbf{k}^*, a_{\mathbf{k}'}^*] = 0 \tag{19}$$

and by taking the adjoint of this equation we deduce that

$$[a_\mathbf{k}, a_{\mathbf{k}'}] = 0 \tag{20}$$

which, of course, just reflects the symmetry property of the adjoint vectors

$$\langle \mathbf{k}_1, \mathbf{k}_2, \cdots \mathbf{k}_n| = \frac{1}{\sqrt{n!}} \langle 0 \mid a_{\mathbf{k}_1} \cdots a_{\mathbf{k}_n} \tag{21}$$

The normalization of the one-particle states, Eq. (12), now demands that

$$\begin{aligned} \langle \mathbf{k}' \mid \mathbf{k} \rangle &= \langle 0 \mid a_{\mathbf{k}'} a_{\mathbf{k}}^* \mid 0 \rangle \\ &= \langle 0 \mid [a_{\mathbf{k}'}, a_{\mathbf{k}}^*] \mid 0 \rangle = k_0 \delta^{(3)}(\mathbf{k} - \mathbf{k}') \end{aligned} \tag{22}$$

so that if we require the commutator of a and a^* to be a c-number, then

$$[a_{\mathbf{k}}, a_{\mathbf{k}'}^*] = k_0 \delta^{(3)}(\mathbf{k} - \mathbf{k}') \tag{23}$$

One verifies that the orthonormality conditions (13) are in fact generally satisfied with these commutation rules.

The operator for the number of particles having momenta in the range Δ between $\mathbf{l}$ and $\mathbf{l} + \Delta\mathbf{l}$ is

$$n(\Delta) = \int_{\Delta} \frac{d^3k}{k_0} a^*(\mathbf{k})\, a(\mathbf{k}) \tag{24}$$

for operating on the state $|\mathbf{k}_1, \cdots \mathbf{k}_n\rangle$ it has as its eigenvalue the number of particles with momenta in the range Δ. *Proof*: Its commutation rules with $a_{\mathbf{k}}$ and $a_{\mathbf{k}}^*$ are

$$[n(\Delta), a_{\mathbf{k}}] = -a_{\mathbf{k}} \delta(\mathbf{k} \subset \Delta) \tag{25a}$$

$$[n(\Delta), a_{\mathbf{k}}^*] = a_{\mathbf{k}}^* \delta(\mathbf{k} \subset \Delta) \tag{25b}$$

where $\delta(\mathbf{k} \subset \Delta)$ is equal to $+1$ if $\mathbf{k}$ lies in Δ, and zero otherwise. These commutation rules allow us to verify the interpretation of the operator $n(\Delta)$. As an example consider the two-particle state

$$|\mathbf{k}_1, \mathbf{k}_2\rangle = \frac{1}{\sqrt{2!}} a_{\mathbf{k}_1}^* a_{\mathbf{k}_2}^* \mid 0\rangle \tag{26}$$

using the commutation rules (23), we find

$$\begin{aligned} &n(\Delta)\, a_{\mathbf{k}_1}^* a_{\mathbf{k}_2}^* \mid 0\rangle \\ &\quad = [n(\Delta), a_{\mathbf{k}_1}^*]\, a_{\mathbf{k}_2}^* \mid 0\rangle + a_{\mathbf{k}_1}^*[n(\Delta), a_{\mathbf{k}_2}^*] \mid 0\rangle + a_{\mathbf{k}_1}^* a_{\mathbf{k}_2}^* n(\Delta) \mid 0\rangle \\ &\quad = [\delta(\mathbf{k}_1 \subset \Delta) + \delta(\mathbf{k}_2 \subset \Delta)]\, a_{\mathbf{k}_1}^* a_{\mathbf{k}_2}^* \mid 0\rangle \end{aligned} \tag{27}$$

since $n(\Delta) \mid 0\rangle = 0$. Hence the eigenvalue of $n(\Delta)$ in this two-particle state is 2, 1, or 0, depending on whether $\mathbf{k}_1$ and $\mathbf{k}_2$ are both in Δ; only $\mathbf{k}_1$ or $\mathbf{k}_2$ but not both are in Δ; or neither $\mathbf{k}_1$ nor $\mathbf{k}_2$ is in Δ. The total number of particles can be defined as

$$N = \int \frac{d^3k}{k_0} a^*(\mathbf{k})\, a(\mathbf{k}) \tag{28}$$

and its commutation rules with a and a^* are:

$$[N, a_{\mathbf{k}}] = -a_{\mathbf{k}} \tag{29a}$$

$$[N, a_{\mathbf{k}}^*] = +a_{\mathbf{k}}^* \tag{29b}$$

so that

$$N \mid 0\rangle = 0 \tag{30a}$$

$$N \mid \mathbf{k}_1, \mathbf{k}_2 \cdots \mathbf{k}_n\rangle = n \mid \mathbf{k}_1, \mathbf{k}_2, \cdots \mathbf{k}_n\rangle \tag{30b}$$

Since the particles are assumed free and not to interact with one another, the total energy of the system is the sum of the energies of the individual particles and the total momentum the sum of the momenta of the individual particles. One can therefore construct the total energy operator by multiplying the number of particles operator in the state of momentum $\mathbf{k}$, $a_{\mathbf{k}}{}^*a_{\mathbf{k}}$, by the energy, $\omega_{\mathbf{k}}$, of a particle in such a state and summing over all states $\int d\Omega_{\mathbf{k}}$. The energy operator is therefore given by

$$H = \int \frac{d^3k}{k_0}\, \omega(\mathbf{k})\, a_{\mathbf{k}}{}^*a_{\mathbf{k}} \tag{31}$$

with the property that

$$H \mid \mathbf{k}_1, \mathbf{k}_2, \cdots \mathbf{k}_n\rangle = \sum_{j=1}^{n} \omega(\mathbf{k}_j) \mid \mathbf{k}_1, \cdots \mathbf{k}_n\rangle \tag{32}$$

Similarly, the total momentum operator is given by

$$\mathbf{P} = \int \frac{d^3k}{k_0}\, \mathbf{k} a_{\mathbf{k}}{}^*a_{\mathbf{k}} \tag{33}$$

and

$$\mathbf{P} \mid \mathbf{k}_1, \mathbf{k}_2, \cdots \mathbf{k}_n\rangle = \sum_{i=1}^{n} \mathbf{k}_i \mid \mathbf{k}_1, \cdots \mathbf{k}_n\rangle \tag{34}$$

We note that the vacuum has zero as its energy eigenvalue and it is the state of lowest energy; it is also a state of zero momentum. Every other state is characterized by a total energy which is positive and greater than zero. The commutation rules of H and $\mathbf{P}$ with $a_{\mathbf{k}}$ and $a_{\mathbf{k}}{}^*$ are:

$$[H, a_{\mathbf{k}}] = -\omega_{\mathbf{k}}a_{\mathbf{k}} \qquad [H, a_{\mathbf{k}}{}^*] = \omega_{\mathbf{k}}a_{\mathbf{k}}{}^* \tag{35}$$

$$[\mathbf{P}, a_{\mathbf{k}}] = -\mathbf{k}a_{\mathbf{k}} \qquad [\mathbf{P}, a_{\mathbf{k}}{}^*] = \mathbf{k}a_{\mathbf{k}}{}^* \tag{36}$$

It follows from Eq. (35) that if $|E\rangle$ is an eigenstate of H with eigenvalue E, then $a_{\mathbf{k}} \mid E\rangle$ is eigenfunction of H with energy $E - \omega_{\mathbf{k}}$. *Proof:*

$$\begin{aligned} Ha_{\mathbf{k}} \mid E\rangle &= a_{\mathbf{k}}H \mid E\rangle + [H, a_{\mathbf{k}}] \mid E\rangle \\ &= (E - \omega_{\mathbf{k}})\, a_{\mathbf{k}} \mid E\rangle \end{aligned} \tag{37}$$

Similarly one verifies that $a_{\mathbf{k}}{}^* \mid E\rangle$ is a state with energy $E + \omega_{\mathbf{k}}$. In a similar manner one also verifies that if $|\mathbf{P}'\rangle$ is an eigenstate of the total momentum operator with eigenvalue $\mathbf{P}'$, then $a_{\mathbf{k}} \mid \mathbf{P}'\rangle$ is an eigenfunction of $\mathbf{P}$ with momentum $\mathbf{P}' - \mathbf{k}$ and $a_{\mathbf{k}}{}^* \mid \mathbf{P}'\rangle$ one with momentum $\mathbf{P}' + \mathbf{k}$. These facts further substantiate our interpretation of the operators $a_{\mathbf{k}}$

and $a_{\mathbf{k}}^*$ as annihilation and creation operators for a particle of energy $\omega_{\mathbf{k}}$ and momentum $\mathbf{k}$. The operator $a_{\mathbf{k}}^*(a_{\mathbf{k}})$ indeed adds (subtracts) an amount of momentum $\mathbf{k}(-\mathbf{k})$, and energy $\omega_{\mathbf{k}}(-\omega_{\mathbf{k}})$ to the state on which it operates.

The operators $a_{\mathbf{k}}$ and $a_{\mathbf{k}}^*$ as we have defined them are time-independent; they are therefore Schrödinger operators. The state vectors describing the possible states will therefore be time-dependent, $|\,\rangle = |\Phi(t)\rangle$. An arbitrary state, $|\Phi(t)\rangle$ can, using the completeness relation (14), be expanded in terms of the basis vectors

$$|\Phi(t)\rangle = \Phi^{(0)}(t)\,|\,0\rangle + \frac{1}{\sqrt{1!}}\int \frac{d^3k}{k_0}\,\Phi^{(1)}(\mathbf{k};t)\,a_{\mathbf{k}}^*\,|\,0\rangle + \cdots$$
$$+ \frac{1}{\sqrt{n!}}\int \frac{d^3k_1}{k_{10}}\cdots\int\frac{d^3k_n}{k_{n0}}\,\Phi^{(n)}(\mathbf{k}_1,\mathbf{k}_2,\cdots\mathbf{k}_n;t)\,a_{\mathbf{k}_1}^*\cdots a_{\mathbf{k}_n}^*\,|\,0\rangle + \cdots \tag{38}$$

where the expansion coefficient

$$\Phi^{(n)}(\mathbf{k}_1,\mathbf{k}_2,\cdots\mathbf{k}_n;t) = \langle\mathbf{k}_1,\mathbf{k}_2,\cdots\mathbf{k}_n\,|\,\Phi(t)\rangle \tag{39}$$

the Fock space components of $|\Phi(t)\rangle$ in the n-particle subspace, is a symmetric function of $\mathbf{k}_1,\cdots,\mathbf{k}_n$. It is the probability amplitude for finding n particles with momenta $\mathbf{k}_1,\cdots\mathbf{k}_n$, energy $\omega(\mathbf{k}_1),\cdots\omega(\mathbf{k}_n)$ at time t. For a system of free particles $|\Phi(t)\rangle$ will satisfy $i\partial_t\,|\,\Phi(t)\rangle = H\,|\,\Phi\rangle$, from which one easily obtains the time dependence of the amplitude $\Phi^{(n)}(\mathbf{k}_1,\cdots\mathbf{k}_n;t)$ as follows:

$$\begin{aligned} i\partial_t\langle\mathbf{k}_1,\cdots\mathbf{k}_n\,|\,\Phi(t)\rangle &= \langle\mathbf{k}_1,\mathbf{k}_2,\cdots,\mathbf{k}_n\,|\,H\,|\,\Phi(t)\rangle \\ &= \sum_{i=1}^{n}\omega(\mathbf{k}_i)\,\langle\mathbf{k}_1,\mathbf{k}_2,\cdots\mathbf{k}_n\,|\,\Phi(t)\rangle \end{aligned} \tag{40}$$

so that

$$\Phi^{(n)}(\mathbf{k}_1,\mathbf{k}_2,\cdots\mathbf{k}_n;t) = e^{-i\sum_i\omega(\mathbf{k}_i)(t-t_0)}\,\Phi^{(n)}(\mathbf{k}_1,\cdots\mathbf{k}_n;t_0) \tag{41}$$

If $|\Phi(t)\rangle$ is the state vector of a system consisting of single free particle, i.e., $N\,|\,\Phi\rangle = |\Phi\rangle$, then its only nonvanishing Fock space amplitude is $\Phi^{(1)}(\mathbf{k};t) = \langle\mathbf{k}\,|\,\Phi(t)\rangle$. This amplitude is the Klein-Gordon amplitude studied extensively in Chapter 3 where it was denoted by $\Phi(\mathbf{k},t)$.

A Fock space representation of the commutation rules (19), (20), and (23) can easily be obtained as follows: From the definition (17) of the basis vectors $|\mathbf{k}_1,\mathbf{k}_2,\cdots\mathbf{k}_n\rangle$ we obtain

$$a_{\mathbf{k}}^*\,|\,\mathbf{k}_1,\mathbf{k}_2,\cdots\mathbf{k}_n\rangle = \sqrt{n+1}\,|\,\mathbf{k},\mathbf{k}_1,\cdots\mathbf{k}_n\rangle \tag{42}$$

Similarly using the commutation rules (23) and the definition of the vacuum $a_{\mathbf{k}}\,|\,0\rangle = 0$, we find that

$$a_{\mathbf{k}} \mid \mathbf{k}_1, \cdots \mathbf{k}_n\rangle = \frac{1}{\sqrt{n!}} a_{\mathbf{k}} a_{\mathbf{k}_1}^* \cdots a_{\mathbf{k}_n}^* \mid 0\rangle$$

$$= \frac{1}{\sqrt{n!}} \sum_{i=1}^{n} k_0 \delta(\mathbf{k} - \mathbf{k}_i)\, a_{\mathbf{k}_1}^* \cdots a_{\mathbf{k}_{i-1}}^* a_{\mathbf{k}_{i+1}}^* \cdots a_{\mathbf{k}_n}^* \mid 0\rangle$$

$$+ \frac{1}{\sqrt{n!}} a_{\mathbf{k}_1}^* \cdots a_{\mathbf{k}_n}^* a_{\mathbf{k}} \mid 0\rangle$$

$$= \frac{1}{\sqrt{n}} \sum_{i=1}^{n} k_0 \delta(\mathbf{k} - \mathbf{k}_i) \mid \mathbf{k}_1, \cdots \mathbf{k}_{i-1}, \mathbf{k}_{i+1}, \cdots \mathbf{k}_n\rangle \tag{43}$$

The role of the equation $a_{\mathbf{k}} \mid 0\rangle = 0$ in arriving at Eq. (43) is to be noted. From the adjoint of Eqs. (42) and (43) we infer that

$$\langle \mathbf{k}_1, \cdots \mathbf{k}_n; n \mid a_{\mathbf{k}} \mid \Phi\rangle = \sqrt{n+1}\, \langle \mathbf{k}, \mathbf{k}_1 \cdots \mathbf{k}_n; n+1 \mid \Phi\rangle \tag{44}$$

$$\langle \mathbf{k}_1, \mathbf{k}_2, \cdots \mathbf{k}_n; n \mid a_{\mathbf{k}}^* \mid \Phi\rangle$$

$$= \frac{1}{\sqrt{n}} \sum_{i=1}^{n} k_0 \delta^{(3)}(\mathbf{k} - \mathbf{k}_i)\, \langle \mathbf{k}_1, \cdots, \mathbf{k}_{i-1}, \mathbf{k}_{i+1}, \cdots \mathbf{k}_n; n-1 \mid \Phi\rangle \tag{45}$$

These are the Fock space representations of the operator a and a^*; they clearly guarantee that the commutation rules are satisfied. Equations (44) and (45) are said to define a representation of the commutation rules (19), (20), and (23) in $\mathcal{H}^{(0)} \oplus \mathcal{H}^{(1)} \oplus \mathcal{H}^{(2)} \oplus \cdots$. This representation is irreducible since a (a^*) operating on an n-particle state creates an $n-1$ ($n+1$) particle state and there are thus no invariant subspaces. This representation is characterized by the fact that a no-particle state exists and that an operator for the number of particles can be defined for it. Besides the above representation there exist a host (at least $\aleph_0$ of them!) of other irreducible representations [Gärding (1954), Wightman (1955), Haag (1955); see also Friedrichs (1953)]. For none of these other irreducible representations, does there exist a no-particle state or a number of particles operator. The representation space in these cases is spanned by basis vectors which all have an infinite number of particles.

The existence of these other representations is a consequence of the fact that our system has an infinite number of degrees of freedom. To make this clear, consider the set of commutation rules:

$$[a_\alpha^*, a_\beta^*] = [a_\alpha, a_\beta] = 0 \tag{46a}$$

$$[a_\alpha, a_\beta^*] = \delta_{\alpha\beta} \quad \alpha = 1, 2, \cdots N \tag{46b}$$

For N finite, it is an extension of the previously mentioned theorem of Von Neumann (1931) that there exists, apart from equivalences, only one irreducible representation of the operators a_α, a_α^*. This is made more explicit by the introduction of the operators

$$q_\alpha = \frac{1}{\sqrt{2}}(a_\alpha + a_\alpha{}^*) \tag{47a}$$

$$p_\alpha = \frac{1}{i\sqrt{2}}(a_\alpha - a_\alpha{}^*) \tag{47b}$$

so that the q_α and p_α are Hermitian operators which satisfy the following commutation rules:

$$[q_\alpha, q_\beta] = [p_\alpha, p_\beta] = 0 \tag{48a}$$

$$[q_\alpha, p_\beta] = i\delta_{\alpha\beta} \tag{48b}$$

For N finite, the vector space on which these operators are defined can be spanned by configuration space vectors $|q_1, q_2, \cdots q_N\rangle$. The number of such vectors is $\aleph_0{}^n$ which is of the same power as $\aleph_0$, i.e., it is still denumerable. However, if $N \to \infty$ the space is then spanned by a noncountable number ($\aleph_0{}^{\aleph_0} = \aleph_1$) of such basis vectors.[1] There still exists a single irreducible representation of the commutation rules (23) for $N \to \infty$ on a separable Hilbert space, and this is precisely the Fock space representation we have been using. It follows from the assumption of the existence of a no-particle state. In general, however, when $N \to \infty$ the equation $a_{\mathbf{k}} \mid 0\rangle = 0$ is no longer a consequence of the commutation rules and many inequivalent representations are possible.

7b. Lorentz Invariance

To discuss the Lorentz covariance of the second-quantized theory, it is convenient to introduce a somewhat more relativistic notation. Where formerly we wrote $|\mathbf{k}_1, \mathbf{k}_2, \cdots \mathbf{k}_n\rangle$, we shall now write $|k_1, k_2, \cdots k_n\rangle$ for our basis vectors. In this notation the energy of the ith particle, k_{0i}, is always to be taken as positive and equal to $+(\mathbf{k}_i{}^2 + \mu^2)^{1/2}$; this was tacitly understood in our previous notation. Similarly we shall write $a_k = a(k)$ for $a_{\mathbf{k}}$.

We have seen that in the Schrödinger picture the state is specified in terms of the results of possible measurements on the system at time t. Such a description picks out a particular Lorentz frame and is, therefore, not covariant. In the Heisenberg picture, on the other hand, the state of the system is the same for all time. For the discussion of the relativistic invariance the Heisenberg picture has, therefore, decided advantages.

Consider an assembly of free particles described by the Heisenberg vector $|\Phi\rangle$. The amplitudes $\langle 0 \mid \Phi\rangle$, $\langle k_1 \mid \Phi\rangle$, $\cdots$, $\langle k_1, k_2, \cdots k_n \mid \Phi\rangle$, etc.

[1] A Hilbert space which is spanned by a nondenumerable basis is called a nonseparable Hilbert space, in contradistinction to a separable Hilbert space which can be spanned by a denumerable basis. The Hilbert space of ordinary quantum mechanics is of the separable variety.

are then the probability amplitudes for finding at any time no particles, one particle with four momentum $k_1, \cdots$, n particles with momenta $k_1, k_2, \cdots k_n$, respectively. Under an inhomogeneous Lorentz transformation $\{a, \Lambda\}$, $|\Phi\rangle \to U(a, \Lambda) \mid \Phi\rangle$ where U is a unitary or antiunitary operator, now to be defined on $\mathcal{H}^{(1)} \oplus \mathcal{H}^{(2)} \oplus \cdots \oplus \mathcal{H}^{(n)} \oplus \cdots$. In Chapter 3 we have seen that under a proper, orthochronous Lorentz transformation the one-particle amplitude $\Phi^{(1)}(k)$ transforms according to

$$\Phi^{(1)}(k) \to e^{ik\cdot a}\, \Phi^{(1)}(\Lambda^{-1}k) \tag{49a}$$

or equivalently

$$\langle k \mid U(a, \Lambda) \mid \Phi\rangle = e^{ik\cdot a}\, \langle \Lambda^{-1}k \mid \Phi\rangle \tag{49b}$$

Let us therefore define the transformation law for an n-particle amplitude as

$$\begin{aligned}\langle k_1, k_2, \cdots k_n \mid U(a, \Lambda) \mid \Phi\rangle &= e^{i\sum_l^n k_l\cdot a}\, \langle \Lambda^{-1}k_1, \cdots \Lambda^{-1}k_n \mid \Phi\rangle \\ &= e^{i\sum_l^n k_l\cdot a}\, \Phi^{(n)}(\Lambda^{-1}k_1, \cdots \Lambda^{-1}k_n)\end{aligned} \tag{50}$$

i.e.,

$$\langle k_1, k_2, \cdots k_n \mid U(a, \Lambda) = \langle \Lambda^{-1}k_1, \cdots \Lambda^{-1}k_n \mid e^{i\sum_{j=1}^n k_j\cdot a} \tag{51}$$

Since every element $\{a, \Lambda\}$ of the restricted Lorentz group ($\det \Lambda = 1$, $\Lambda^0{}_0 \geqslant 1$) can be written as the square of some element $\{b, \Lambda'\}$, i.e., $\{a, \Lambda\} = \{b, \Lambda'\}\,\{b, \Lambda'\}$, and the square of a unitary as well as of an antiunitary operator is unitary, $U(a, \Lambda)$ is unitary for $\{a, \Lambda\}$, an element of the restricted Lorentz group. For restricted Lorentz transformations therefore

$$U^*(a, \Lambda) = U(a, \Lambda)^{-1} = U(-\Lambda^{-1}a, \Lambda^{-1}) \tag{52}$$

and upon taking the adjoint of Eq. (51)

$$U(a, \Lambda) \mid k_1, \cdots, k_n\rangle = e^{+i\sum_j^n \Lambda k_j\cdot a} \mid \Lambda k_1, \cdots, \Lambda k_n\rangle \tag{53}$$

Thus in the active point of view, $U(a, \Lambda)$ transforms a one-particle state of momentum k into a state of momentum Λk and thereafter displaces it by the amount a. This is what is expected, since in the Lorentz transformation $x' = \Lambda x + a$, the homogeneous part is carried out first and thereafter a translation through a is performed. This interpretation of Eq. (53) follows from the fact that the generator for an infinitesimal translation is the momentum operator, so that a translation of the state $|\Lambda k\rangle$ through a space-time distance a is effected by multiplying it by $\exp i\Lambda k \cdot a$.

The transformation property of the vacuum is characterized by

$$U(a, \Lambda) \mid 0\rangle = \mid 0\rangle \tag{54}$$

since the vacuum must look identical to all observers. The vacuum is thus invariant under all Lorentz transformations. This is a consistent characterization only if the vacuum is a state of zero energy and zero momentum. The above properties of the amplitudes, which reflect the physical characteristics of the particles and of the vacuum, are now sufficient to determine the transformation properties of the operators a_k and a_k^*. Using the representations (44), (45), we note that

$$\begin{aligned}
\langle k_1, k_2, \cdots k_n \mid U(a, \Lambda)\, a_k \mid \Phi\rangle
&= \langle \Lambda^{-1}k_1, \cdots \Lambda^{-1}k_n \mid a_k \mid \Phi\rangle\, e^{i\sum_j^n k_j\cdot a} \\
&= \sqrt{n+1}\,\langle k, \Lambda^{-1}k_1, \cdots \Lambda^{-1}k_n \mid \Phi\rangle\, e^{i\sum_j^n k_j\cdot a} \\
&= \langle k_1, k_2 \cdots k_n \mid e^{-i\Lambda k\cdot a}\, a(\Lambda k)\, U(a, \Lambda) \mid \Phi\rangle
\end{aligned} \tag{55}$$

Upon inserting the factor $UU^{-1} = I$, we can rewrite Eq. (55) so as to read:

$$\begin{aligned}
\langle k_1, k_2 \cdots k_n \mid U(a, \Lambda)\, a_k\, U^{-1}(a, \Lambda)\, U(a, \Lambda) \mid \Phi\rangle \\
= \langle k_1, k_2, \cdots k_n \mid e^{-i\Lambda k\cdot a}\, a(\Lambda k)\, U(a, \Lambda) \mid \Phi\rangle
\end{aligned} \tag{56}$$

Since this last equation holds for arbitrary $|\Phi\rangle$

$$U(a, \Lambda)\, a_k\, U(a, \Lambda)^{-1} = e^{-i\Lambda k\cdot a}\, a(\Lambda k) \tag{57}$$

We have previously established that for orthochronous Lorentz transformations $U(a, \Lambda)$ is unitary, so that taking the adjoint of (57) we obtain for the transformation law of a_k^* under orthochronous Lorentz transformations:

$$U(a, \Lambda)\, a_k^*\, U(a, \Lambda)^{-1} = e^{i\Lambda k\cdot a}\, a^*(\Lambda k) \tag{58}$$

This is consistent with Eq. (53) if we recall the definition of the basis vectors, Eq. (17). In particular, as noted above, the one-particle state $|k\rangle$ is transformed under an orthochronous homogeneous Lorentz transformation into the state

$$U(\Lambda) \mid k\rangle = U(\Lambda)\, a_k^* \mid 0\rangle = U(\Lambda)\, a_k^*\, U(\Lambda)^{-1} \mid 0\rangle = |\Lambda k\rangle \tag{59}$$

and similarly under a pure translation

$$U(a) \mid k\rangle = e^{ik\cdot a} \mid k\rangle \tag{60}$$

We defer the treatment of the inversion properties of the field operators till after the configuration space operators have been introduced.

We can obtain an explicit representation of the unitary operators corresponding to translations by noting that for an arbitrary real function $f(k) = \bar{f}(k)$, of the four momentum k, $(k_0 = \omega_k)$,

$$e^{i\int d\Omega(k) f(k) a_k^* a_k}\, a_q e^{-i\int d\Omega(k) f(k) a_k^* a_k} = e^{-if(q)}\, a_q \tag{61}$$

where $d\Omega(k) = d^3k/k_0$. *Proof:* Consider the operator

$$A_q(\lambda) = e^{i\lambda\int d\Omega(k) f(k) a_k^* a_k}\, a_q e^{-i\lambda\int d\Omega(k) f(k) a_k^* a_k} \tag{62}$$

When $\lambda = 0$

$$A_q(0) = a_q \tag{63}$$

and $A_q(1)$ is to be determined. By differentiating (62) with respect to λ we find that $A_q(\lambda)$ obeys the following differential equation

$$\frac{dA_q(\lambda)}{d\lambda} = e^{i\lambda \int d\Omega(k) f(k) a_k^* a_k}\, i\lambda \int d\Omega(k)\, f(k)\, [a_k^* a_k, a_q]\, e^{-i\lambda \int d\Omega(k) f(k) a_k^* a_k}$$
$$= -i\lambda f(q)\, A_q(\lambda) \tag{64}$$

The solution of this differential equation which satisfies the initial condition $A_q(0) = a_q$ is

$$A_q(\lambda) = e^{-i\lambda f(q)}\, a_q \tag{65}$$

Setting $\lambda = 1$, we obtain (61).

We note that for f real, the operator $F = \int d\Omega(k)\, f(k)\, a_\beta^* a_\beta$ is Hermitian, so that $\exp iF$ is unitary. Hence, comparing (61) and the transformation laws for the operator a, a^* under translations, we find that

$$U(b, I) = e^{i\int \frac{d^3k}{k_0} k_\mu a_k^* a_k b^\mu} = e^{iP_\mu b^\mu} \tag{66}$$

where

$$P_\mu = \int \frac{d^3k}{k_0}\, k_\mu a_k^* a_k \tag{67}$$

is the energy-momentum four-vector of the field: $P_\mu = (H, \mathbf{P})$. That P_μ has indeed the right transformation properties under homogeneous Lorentz transformation to be called a four-vector is easily verified:

$$U(\Lambda)\, P_\mu\, U(\Lambda)^{-1} = \int d\Omega(k)\, k_\mu a_{\Lambda k}^* a_{\Lambda k}$$
$$= \int d\Omega(\Lambda k)\, k_\mu a_{\Lambda k}^* a_{\Lambda k} = \int d\Omega(k)\, (\Lambda^{-1}k)_\mu\, a_k^* a_k$$
$$= (\Lambda^{-1})_\mu{}^\nu\, P_\nu \tag{68}$$

Finally, we note that

$$[P_\mu, P_\nu] = 0 \tag{69}$$

so that a representation in which all the components of P_μ are diagonal can be defined. Summarizing, the energy-momentum operators are invariant under translations and they are the generators for infinitesimal space-time translations.

7c. Configuration Space

We next introduce the configuration space Heisenberg operator $\phi^{(+)}(x)$ defined by

$$\phi^{(+)}(x) = \int_{k_0>0} \frac{d^3k}{k_0} \frac{1}{\sqrt{2(2\pi)^3}}\, e^{-ik\cdot x}\, a_k \tag{70}$$

Note that $\phi^{(+)}(x)$ is a destruction operator for a particle. The superscript ${}^{(+)}$ indicates that $\phi^{(+)}$ contains only positive frequencies in its temporal-

Fourier analysis, i.e., only factors of the form $\exp(-ik_0x_0)$ with $k_0 > 0$. Since $k_0 = +\sqrt{\mathbf{k}^2 + \mu^2}$, $\phi^{(+)}(x)$ obeys the Klein-Gordon equation

$$(\Box + \mu^2)\,\phi^{(+)}(x) = 0 \tag{71}$$

The creation operator $\phi^{(-)}(x)$ defined by the equation

$$\phi^{(-)}(x) = \frac{1}{\sqrt{2(2\pi)^3}} \int_{k_0>0} \frac{d^3k}{k_0}\, e^{ik\cdot x}\, a_k^* \tag{72a}$$

$$= [\phi^{(+)}(x)]^* \tag{72b}$$

contains only negative frequencies [indicated by the superscript $^{(-)}$] and also obeys the Klein-Gordon equation

$$(\Box + \mu^2)\,\phi^{(-)}(x) = 0 \tag{73}$$

The commutation rules of these operators, using (19), (20), and (23) are:

$$[\phi^{(\pm)}(x), \phi^{(\pm)}(x')] = 0 \tag{74a}$$

$$[\phi^{(+)}(x), \phi^{(-)}(y)] = \frac{1}{2(2\pi)^3} \int_{k_0>0} \frac{d^3k}{k_0}\, e^{-ik\cdot(x-y)} \tag{74b}$$

$$= i\Delta^{(+)}(x - y) \tag{74c}$$

We have designated the right-hand side of the commutator by $\Delta^{(+)}(x - y)$, i.e.,

$$\Delta^{(+)}(x) = -\frac{i}{2(2\pi)^3} \int_{k_0>0} \frac{d^3k}{k_0}\, e^{-ik\cdot x} \tag{75a}$$

$$= -\frac{i}{(2\pi)^3} \int_{-\infty}^{+\infty} d^4k\, \theta(k_0)\, \delta(k^2 - \mu^2)\, e^{-ik\cdot x} \tag{75b}$$

Similarly, we find that

$$[\phi^{(-)}(x), \phi^{(+)}(y)] = -\frac{1}{2(2\pi)^3} \int_{k_0>0} \frac{d^3k}{k_0}\, e^{ik\cdot(x-y)} \tag{76a}$$

$$= i\Delta^{(-)}(x - y) \tag{76b}$$

where

$$\Delta^{(-)}(x) = \frac{i}{2(2\pi)^3} \int_{k_0>0} \frac{d^3k}{k_0}\, e^{ik\cdot x} \tag{77a}$$

$$= \frac{i}{(2\pi)^3} \int d^4k e^{-ik\cdot x}\, \theta(-k_0)\, \delta(k^2 - \mu^2) \tag{77b}$$

Comparing (75) and (77), we note that

$$-\Delta^{(-)}(-x) = \Delta^{(+)}(x) \tag{78}$$

It is clear from the representations (75b) and (77b) that $\Delta^{(\pm)}$ are invariant under proper homogeneous Lorentz transformation

$$\Delta^{(\pm)}(\Lambda x) = \Delta^{(\pm)}(x) \tag{79}$$

and that they are solutions, in fact invariant solutions, of the Klein-Gordon equation:

$$(\Box + \mu^2)\, \Delta^{(\pm)}(x) = 0 \tag{80}$$

The following property of $\Delta^{(\pm)}$ will be used repeatedly in the subsequent discussion: If $f^{(+)}(x)$ is any positive energy solution of the Klein-Gordon equation, then its value at the space-time point x is expressed in terms of its initial values $\varphi(x)$ on some earlier space-like surface σ, $f^{(+)}(x') = \varphi(x')$ when $x'\epsilon\sigma$, and the value of its normal derivative on σ, $n^\mu\partial'_\mu f^{(+)}(x') = n^\mu\partial'_\mu\varphi^{(+)}(x')$, $x'\epsilon\sigma$, by the formula

$$f^{(+)}(x) = \int_\sigma d\sigma^\mu(x') \left\{\frac{\partial\Delta^{(+)}(x - x')}{\partial x'^\mu}\, \varphi^{(+)}(x') - \Delta^{(+)}(x - x')\, \frac{\partial\varphi^{(+)}(x')}{\partial x'^\mu}\right\} \tag{81}$$

where σ is earlier than x. *Proof:* Consider the expression

$$F(x) = -\int_{\sigma'} d\sigma^\mu(x')\, \Delta^{(+)}(x - x')\, \overleftrightarrow{\partial_\mu} f^{(+)}(x') \tag{82}$$

where σ' is some arbitrary space-like surface. Since $f^{(+)}$ and $\Delta^{(+)}$ both satisfy the Klein-Gordon equation, by Gauss's theorem F does not depend on the space-like surface σ' chosen to evaluate it. Let us therefore choose σ' to be the plane $x_0 = x_0'$, i.e., a plane through the point x^0. We can then evaluate F by substituting for $f^{(+)}$ and $\Delta^{(+)}$ their Fourier representations. Thus with

$$f^{(+)}(x) = \int \frac{d^3k}{k_0}\, e^{-ik\cdot x}\, \tilde{f}^{(+)}(k) \tag{83}$$

and (75a), we find

$$F = f^{(+)}(x) \tag{84}$$

If next we choose for σ' the space-like surface σ on which the initial data is prescribed, we obtain (81). If, in particular, $f^{(+)}(x) = \Delta^{(+)}(x - x')$, Eq. (81) then reads

$$\Delta^{(+)}(x - x') = \int_\sigma d\sigma^\mu(x'') \left\{\frac{\partial\Delta^{(+)}(x - x'')}{\partial x''^\mu}\, \Delta^{(+)}(x'' - x') - \Delta^{(+)}(x - x'')\, \frac{\partial\Delta^{(+)}(x'' - x')}{\partial x''^\mu}\right\} \tag{85}$$

Analogous expressions hold for the negative energy solutions of the Klein-Gordon equation, for example,

$$f^{(-)}(x) = \int_\sigma d\sigma^\mu(x')\, \{\partial'_\mu\Delta^{(-)}(x - x') \cdot f^{(-)}(x') - \Delta^{(-)}(x - x')\, \partial'_\mu f^{(-)}(x')\} \tag{86}$$

Note incidentally that since positive and negative energy solutions of the Klein-Gordon equations are orthogonal:

$$\int_\sigma d\sigma^\mu(x')\, \Delta^{(\pm)}(x - x')\, \overleftrightarrow{\partial}_\mu f^{(\mp)}(x') = 0 \tag{87}$$

so that $\Delta^{(+)}$ acts as a projection operator onto the positive energy solutions and $\Delta^{(-)}$ as projection operator onto the manifold of negative energy solutions.

The number of particle operator at time x_0, $N(x_0)$, expressed in terms of the $\phi^{(\pm)}(x)$ operators is given by

$$N(x_0) = i \int_{x_0} d^3x \left\{\phi^{(-)}(x)\, \frac{\partial\phi^{(+)}(x)}{\partial x^0} - \frac{\partial\phi^{(-)}(x)}{\partial x^0}\, \phi^{(+)}(x)\right\} \tag{88}$$

This operator can be generalized to a space-like surface, in which case it has the form

$$N(\sigma) = i \int_\sigma d\sigma^\mu(x) \left\{\phi^{(-)}(x)\, \frac{\partial\phi^{(+)}(x)}{\partial x^\mu} - \frac{\partial\phi^{(-)}(x)}{\partial x^\mu}\, \phi^{(+)}(x)\right\} \tag{89}$$

However, since $\phi^{(\pm)}(x)$ obeys the Klein-Gordon equation, $\delta N(\sigma)/\delta\sigma(x) = 0$, so that $N(\sigma)$ is independent of the surface σ and is constant in time; we therefore write N for $N(x_0)$. One verifies by direct substitution of Eqs. (70) and (72) into (88) that the latter reduces to the expression $\int d\Omega(k)\, a_k{}^* a_k = N$. The vacuum $|0\rangle$ can now be characterized by the equation

$$\phi^{(+)}(x) \mid 0\rangle = 0 \quad \text{for all } x \tag{90}$$

The state $\phi^{(-)}(x) \mid 0\rangle$ is a one-particle state; this is clear from the representation of $\phi^{(-)}$ in terms of $a_k{}^*$. This can also be verified directly in configuration space by using the commutation rules between N and $\phi^{(-)}$:

$$\begin{aligned}
&[N, \phi^{(-)}(x)] \\
&= i \int_\sigma d\sigma_\mu(x') \left\{\phi^{(-)}(x') \left[\frac{\partial\phi^{(+)}(x')}{\partial x'_\mu}, \phi^{(-)}(x)\right] - \frac{\partial\phi^{(-)}(x')}{\partial x'_\mu}\, [\phi^{(+)}(x'), \phi^{(-)}(x)]\right\} \\
&= -\int_\sigma d\sigma^\mu(x') \left\{\phi^{(-)}(x')\, \frac{\partial\Delta^{(+)}(x' - x)}{\partial x'^\mu} - \frac{\partial\phi^{(-)}(x')}{\partial x'^\mu}\, \Delta^{(+)}(x' - x)\right\} \\
&= \int_\sigma d\sigma^\mu(x') \left\{\frac{\partial\Delta^{(-)}(x - x')}{\partial x'^\mu}\, \phi^{(-)}(x') - \Delta^{(-)}(x - x')\, \frac{\partial\phi^{(-)}(x')}{\partial x'^\mu}\right\} \\
&= \phi^{(-)}(x)
\end{aligned} \tag{91}$$

and similarly,

$$[N, \phi^{(+)}(x)] = -\phi^{(+)}(x) \tag{92}$$

Since the no-particle state is an eigenfunction of N with eigenvalue 0, we have

$$N\phi^{(-)}(x) \mid 0\rangle = [N, \phi^{(-)}(x)] \mid 0\rangle = \phi^{(-)}(x) \mid 0\rangle \tag{93}$$

More generally, the vector

$$|x_1, x_2, \cdots x_n\rangle = (n!)^{-1/2}\, \phi^{(-)}(x_1) \cdots \phi^{(-)}(x_n) \mid 0\rangle \tag{94}$$

is an eigenfunction of N with eigenvalue n, and is an n-particle basis vector. The n-particle Fock space amplitude of the vector $|\Phi\rangle$

$$\Phi^{(n)}(x_1, \cdots x_n) = (n!)^{-1/2} \langle 0 \mid \phi^{(+)}(x_1) \cdots \phi^{(+)}(x_n) \mid \Phi\rangle \tag{95}$$

satisfies the Klein-Gordon equation in each of its co-ordinates

$$\begin{aligned}(\Box_i + \mu^2)\, &\Phi^{(n)}(x_1, x_2, \cdots x_n) \\ &= (n!)^{-1/2} \langle 0 \mid \phi^{(+)}(x_1) \cdots (\Box_i + \mu^2)\, \phi^{(+)}(x_i) \cdots \phi(x_n) \mid \Phi\rangle = 0 \\ &\qquad i = 1, 2, \cdots\end{aligned} \tag{96}$$

since $\phi^{(+)}(x_i)$ obeys the Klein-Gordon equation. Furthermore, $\Phi^{(n)}(x_1, x_2 \cdots x_n)$ contains only positive frequencies in its time dependence on the co-ordinates $x_{10}, x_{20}, \cdots x_{n0}$, and is symmetric under the exchange of particle co-ordinates. It is thus a sum of symmetrized products of positive energy solutions of the Klein-Gordon equation. This is precisely what we would expect on the basis of our considerations at the beginning of this chapter. The Fock space amplitude

$$\begin{aligned}\Phi^{(n)}(x_1, \cdots x_n) &= \langle x_1, x_2, \cdots x_n \mid \Phi\rangle \\ &= \sum_{\{n_i \mid \sum_i n_i = n\}} \langle x_1, x_2, \cdots \mid n_1, n_2, \cdots n_i \cdots\rangle \langle n_1, n_2, \cdots n_i \cdots \mid \Phi\rangle\end{aligned} \tag{97}$$

is indeed seen using Eq. (6), the explicit form of the transformation function $\langle x_1, x_2, \cdots x_n \mid n_1, n_2, \cdots n_i \cdots\rangle$, to be a sum of symmetrized products of one-particle *positive* energy functions. An arbitrary Heisenberg state can be expanded in terms of the Fock space basis function as follows:

$$\begin{aligned}|\Phi\rangle = \Phi^{(0)} \mid 0\rangle &+ i \int_\sigma d\sigma^\mu(x) \left(\frac{\partial\phi^{(-)}(x)}{\partial x^\mu} - \phi^{(-)}(x)\, \frac{\partial}{\partial x^\mu}\right) \Phi^{(1)}(x) \mid 0\rangle \\ &+ \cdots + (i)^n \int_\sigma d\sigma^{\mu_1}(x_1) \cdots \int d\sigma^{\mu_n}(x_n) \\ &\prod_{i=1}^{n} \left(\frac{\partial\phi^{(-)}(x_i)}{\partial x_i^{\mu_i}} - \phi^{(-)}(x_i)\, \frac{\partial}{\partial x_i^{\mu_i}}\right) \Phi^{(n)}(x_1, \cdots x_n) \mid 0\rangle + \cdots\end{aligned} \tag{98}$$

The physical interpretation of the Fock space component $\Phi^{(n)}(x_1, \cdots x_n)$ when $x_{10} = x_{20} = \cdots = x_{n0}$ is that it is the probability amplitude for finding n particles at time $x_0 = x_{10} = \cdots = x_{n0}$. It is not the probability amplitude for finding the particles at positions $\mathbf{x}_1, \cdots \mathbf{x}_n$ at time x^0, for the reasons already discussed in the one-particle Klein-Gordon theory. To answer questions regarding probabilities of finding the particles with specified positions, our previous investigations of the position operator and localized wave functions for a free Klein-Gordon particle suggest that we introduce the following operator:

$$\phi_1(q) = \int_{k_0 > 0} \frac{d^3k}{k_0} \frac{1}{(2\pi)^{3/2}} (k_0)^{1/2}\, e^{-ik\cdot q}\, a_k \tag{99}$$

For then if $|\Phi^1\rangle$ corresponds to a state of one particle of momentum $\mathbf{k}$, i.e., $|\Phi^1\rangle = |k\rangle = a_k^* \mid 0\rangle$, the matrix element

$$\langle 0 \mid \phi_1(q) \mid \Phi^1\rangle = \frac{1}{(2\pi)^{3/2}} (k_0)^{1/2} e^{ik\cdot q} \tag{100}$$

is precisely the probability amplitude for finding our particle at the position $\mathbf{q}$ at time q^0. We therefore interpret the state $\phi_1^*(q) \mid 0\rangle$ as corresponding to a state of a single particle localized at $\mathbf{q}$ at time q^0, and the state $(n!)^{-1/2} \phi_1^*(q^0, \mathbf{q}_1) \phi_1^*(q^0, \mathbf{q}_2) \cdots \phi_1^*(q^0, \mathbf{q}_n) \mid 0\rangle$ as corresponding to a state with n spin 0 particles of mass μ localized at position $\mathbf{q}_1, \cdots \mathbf{q}_n$ at time q^0. The equal-time commutation rules of these operators are:

$$\begin{aligned}[\phi_1(q^0, \mathbf{q}), \phi_1^*(q^0, \mathbf{q}')] &= \frac{1}{(2\pi)^3} \int \frac{d^3k}{k_0} \int \frac{d^3l}{l_0} (k_0 l_0)^{1/2} e^{i(k_0 - l_0)q_0} \\ &\quad \cdot e^{-i\mathbf{k}\cdot\mathbf{q}} e^{i\mathbf{l}\cdot\mathbf{q}'} [a_k, a_l^*] \\ &= \delta^{(3)}(\mathbf{q} - \mathbf{q}')\end{aligned} \tag{101}$$

We can therefore define the operator N_V for the number of particles in the three-dimensional volume V at time q^0 as

$$N_V(q_0) = \int_V d^3q\, \phi_1^*(q_0, \mathbf{q})\, \phi_1(q_0, \mathbf{q}) \tag{102}$$

By virtue of the δ-function commutation rules,

$$[N_V(q_0), \phi_1(q^0, \mathbf{q})] = -\phi_1(q^0, \mathbf{q})\, \delta(\mathbf{q} \subset V) \tag{103a}$$

$$[N_V(q_0), \phi_1^*(q^0, \mathbf{q})] = +\phi_1^*(q^0, \mathbf{q})\, \delta(\mathbf{q} \subset V) \tag{103b}$$

so that $N_V(q_0)$ operating on $\phi_1^*(q^0, \mathbf{q}_1) \cdots \phi_1^*(q^0, \mathbf{q}_n) \mid 0\rangle$ reproduces this state with an eigenvalue equal to the number of particles with positions $\mathbf{q}_i$ in V.

The configuration space representation of the operators $\phi^{(\pm)}(x)$ can again be obtained by noting that by virtue of the commutation rules (74) and the definitions of the basis vectors, Eqs. (90) and (94),

$$\phi^{(-)}(x) \mid x_1, \cdots x_n\rangle = \sqrt{n+1} \mid x, x_1, x_2, \cdots x_n\rangle \tag{104}$$

$$\phi^{(+)}(x) \mid x_1, \cdots, x_n\rangle = \frac{1}{\sqrt{n}} \sum_{j=1}^{n} i\Delta^{(+)}(x - x_j) \mid x_1, \cdots x_{j-1}, x_{j+1}, \cdots x_n\rangle \tag{105}$$

whence, since

$$\overline{\Delta^{(+)}}(x) = \Delta^{(-)}(x) = -\Delta^{(+)}(-x) \tag{106}$$

we find, by taking the adjoint of Eqs. (104) and (105), that

$$\langle x_1, \cdots x_n \mid \phi^{(+)}(x) \mid \Phi\rangle = \sqrt{n+1}\, \langle x, x_1, \cdots x_n \mid \Phi\rangle \tag{107}$$

$$\langle x_1, \cdots x_n \mid \phi^{(-)}(x) \mid \Phi\rangle = \frac{1}{\sqrt{n}} \sum_{j=1}^{n} i\Delta^{(+)}(x_j - x) \langle x_1, \cdots x_{j-1}, x_{j+1} \cdots x_n \mid \Phi\rangle \tag{108}$$

Having obtained the Fock space representations of the operators $\phi^{(+)}(x)$ and $\phi^{(-)}(x)$, we next analyze their transformation properties under Lorentz transformations. For restricted Lorentz transformations, these are easily deduced from those of the a_k and a_k^* operators

$$U(a, \Lambda)\, \phi^{(+)}(x)\, U(a, \Lambda)^{-1}$$

$$= \frac{1}{\sqrt{2(2\pi)^3}} \int_+ d\Omega(k)\, e^{-ik\cdot x}\, U(a, \Lambda)\, a_k\, U(a, \Lambda)^{-1}$$

$$= \frac{1}{\sqrt{2(2\pi)^3}} \int_+ d\Omega(k)\, e^{-i\Lambda k\cdot \Lambda x}\, e^{-i\Lambda k\cdot a}\, a_{\Lambda k}$$

$$= \frac{1}{\sqrt{2(2\pi)^3}} \int_+ d\Omega(k')\, e^{-ik'(\Lambda x + a)}\, a_{k'} = \phi^{(+)}(\Lambda x + a) \quad (109)$$

and similarly,

$$U(a, \Lambda)\, \phi^{(-)}(x)\, U(a, \Lambda)^{-1} = \phi^{(-)}(\Lambda x + a) \quad (110)$$

Let us next consider the transformation properties under inversions. Under a space inversion $x \to x' = i_s x$ ($x_0' = x_0$, $\mathbf{x}' = -\mathbf{x}$) the one-particle amplitude $\Phi^{(1)}(x)$ transforms according to

$$\langle x \mid \Phi' \rangle = \langle x \mid U(i_s) \mid \Phi \rangle = \eta_S \langle i_s x \mid \Phi \rangle \quad (111)$$

where $\eta_S = \pm 1$. The $+$ sign corresponds to a scalar particle and the $-$ sign to a pseudoscalar particle. This operator $U(i_s)$ is unitary since

$$(\Phi', \Psi') = i \int d^3x\, \langle \Phi \mid i_s x \rangle\, \overleftrightarrow{\partial_0}\, \langle i_s x \mid \Psi \rangle$$

$$= i \int d^3x\, \langle \Phi \mid x \rangle\, \overleftrightarrow{\partial_0}\, \langle x \mid \Psi \rangle = (\Phi, \Psi) \quad (112)$$

We define the transformation rule for the n-particle amplitude as follows:

$$\langle x_1, x_2, \cdots x_n \mid U(i_s) \mid \Phi \rangle = (\eta_S)^n \langle i_s x_1, \cdots i_s x_n \mid \Phi \rangle \quad (113)$$

and that of the vacuum as

$$U(i_s) \mid 0 \rangle = \mid 0 \rangle \quad (114)$$

To derive the transformation law of the field operators $\phi^{(+)}$ and $\phi^{(-)}$ we proceed as in the derivation of Eq. (57). We apply $U(i_s)$ to $\phi^{(+)}(x)$ and get

$$\langle x_1, x_2, \cdots x_n \mid U(i_s)\, \phi^{(+)}(x) \mid \Phi \rangle = (\eta_S)^n \langle i_s x_1, \cdots i_s x_n \mid \phi^{(+)}(x) \mid \Phi \rangle$$

$$= (\eta_S)^n \sqrt{n+1}\, \langle x, i_s x_1, \cdots i_s x_n \mid \Phi \rangle$$

$$= \eta_S \sqrt{n+1}\, \langle i_s x, x_1 \cdots x_n \mid U(i_s) \mid \Phi \rangle$$

$$= \eta_S \langle x_1, \cdots, x_n \mid \phi^{(+)}(i_s x)\, U(i_s) \mid \Phi \rangle \quad (115)$$

whence

$$U(i_s)\, \phi^{(+)}(x)\, U(i_s)^{-1} = \eta_S \phi^{(+)}(i_s x) \quad (116)$$

and similarly, using the property of $\Delta^{(+)}$, that $\Delta^{(+)}(i_s x) = \Delta^{(+)}(x)$

$$U(i_s)\,\phi^{(-)}(x)\,U(i_s)^{-1} = \eta_S\phi^{(-)}(i_s x) \tag{117}$$

The transformation properties of the operators a_k and a_k^* under a spatial inversion are therefore given by

$$U(i_s)\,a_{\mathbf{k}}\,U(i_s)^{-1} = \eta_S a_{-\mathbf{k}} \tag{118a}$$

$$U(i_s)\,a_{\mathbf{k}}\,U(i_s)^{-1} = \eta_S a_{-\mathbf{k}}^* \tag{118b}$$

Under a spatial inversion, the one-particle state of momentum $\mathbf{k}$, energy $\omega(\mathbf{k})$, is transformed into a state of momentum $-\mathbf{k}$, energy $\omega(-\mathbf{k}) = \omega(\mathbf{k})$.
Proof:

$$\begin{aligned} U(i_s)\,|\,\mathbf{k}\rangle &= U(i_s)\,a_{\mathbf{k}}^*\,U(i_s)^{-1}\,U(i_s)\,|\,0\rangle = a_{-\mathbf{k}}^*\,|\,0\rangle \\ &= |\,-\mathbf{k}\rangle \end{aligned} \tag{119}$$

The Hamiltonian is invariant under the operation $U(i_s)$ since $U(i_s)\,HU(i_s)^{-1} = H$; however, the total momentum $\mathbf{P}$ transforms like a polar vector

$$U(i_s)\,\mathbf{P}\,U(i_s)^{-1} = \int_+ \frac{d^3k}{k_0}\,\mathbf{k}a_{-\mathbf{k}}^*a_{-\mathbf{k}} = -\mathbf{P} \tag{120a}$$

or equivalently

$$U(i_s)\,\mathbf{P} = -\mathbf{P}U(i_s) \tag{120b}$$

An explicit representation of the operator $U(i_s)$ is given by

$$U(i_s) = e^{i\frac{\pi}{2}\int d\Omega(k)\,(\eta_S a_{\mathbf{k}}^* a_{\mathbf{k}} - a_{\mathbf{k}}^* a_{-\mathbf{k}})} \tag{121}$$

[Federbush (1958a)]. To prove this representation, we note that $N = \int d\Omega(k)\,a_{\mathbf{k}}^*a_{\mathbf{k}}$ commutes with $\int a_{\mathbf{k}}^*a_{-\mathbf{k}}d\Omega(k)$, so that we may write

$$U(i_s) = e^{-i\frac{\pi}{2}\int d\Omega(k)a_{\mathbf{k}}^*a_{-\mathbf{k}}}\,e^{i\frac{\pi}{2}\eta_s N} \tag{122}$$

We have seen that

$$e^{i\frac{\pi}{2}\eta_s N}\,a_{\mathbf{k}}e^{-i\frac{\pi}{2}\eta_s N} = e^{i\frac{\pi}{2}\eta_s}\,a_{\mathbf{k}} = i\eta_S a_{\mathbf{k}} \tag{123}$$

(since $\eta_S = \pm 1$) and, using the formula

$$\begin{aligned} e^{i\lambda B}Ae^{-i\lambda B} = A + \frac{i\lambda}{1!}[B, A] &+ \frac{(i\lambda)^2}{2!}[B, [B, A]] \\ &+ \frac{(i\lambda)^3}{3!}[B, [B, [B, A]]] + \cdots \end{aligned} \tag{124}$$

obtained by making a Taylor expansion in λ of the left-hand side about $\lambda = 0$, we find that

$$\begin{aligned} e^{-i\frac{\pi}{2}\int d\Omega(k)a_{\mathbf{k}}^*a_{-\mathbf{k}}}\,a_k e^{i\frac{\pi}{2}\int d\Omega(k)a_{\mathbf{k}}^*a_{-\mathbf{k}}} \\ = a_{\mathbf{k}}\cos\frac{\pi}{2} - ia_{-\mathbf{k}}\sin\frac{\pi}{2} = -ia_{-\mathbf{k}} \end{aligned} \tag{125}$$

whence

$$U(i_s)\, a_{\mathbf{k}}\, U(i_s)^{-1} = \eta_S a_{-\mathbf{k}} \qquad \text{Q.E.D.} \quad (126)$$

Since two successive inversions correspond to reverting to the original co-ordinate system ($i_s^2 = 1$), $U(i_s)^2 = I$ so that $U(i_s) = U^{-1}(i_s) = U^*(i_s)$, and $U(i_s)$ is Hermitian. The fact that $U(i_s)$ is Hermitian implies that it corresponds to an observable of the system. However, since it anti-commutes with **P** it cannot be included in our complete set of observables describing the system, except for the case that the particles are at rest. For example, the state of a single particle at rest is an eigenstate of $U(i_s)$ with eigenvalue η_S:

$$\begin{aligned} U(i_s)\, a_0^* \mid 0\rangle &= U(i_s)\, a_0^*\, U(i_s)^{-1} \mid 0\rangle \\ &= \eta_S a_0^* \mid 0\rangle \end{aligned} \qquad (127)$$

One calls this factor η_S the "intrinsic" parity of the particle. Similarly the n-particle state $(a_0^*)^n \mid 0\rangle$ is an eigenfunction of $U(i_s)$ with eigenvalue $(\eta_S)^n$.

Let us next consider time inversions. If under a time inversion, $x' = i_t x$ ($x_0' = -x_0$, $\mathbf{x}' = +\mathbf{x}$), a one-particle state of positive energy is to be carried into a one-particle state of positive energy, then the operator $U(i_t)$ which induces the transformation must be antiunitary [Wigner (1932)]. For if the transformation were unitary, the one-particle amplitude would transform according to

$$\langle x \mid \Phi'\rangle = \langle x \mid U(i_t) \mid \Phi\rangle = \eta_T \langle i_t x \mid \Phi\rangle \qquad (128)$$

where η_T is ± 1 and is the time inversion parity of the particle. Consider then the transformation property of the one-particle amplitude when $|\Phi\rangle$ corresponds to a one-particle state of momentum $\mathbf{k}$, $|\Phi\rangle = |\mathbf{k}\rangle$. Since $\langle x \mid k\rangle = [2(2\pi)^3]^{-1/2} \exp(-ik \cdot x)$, the transformed wave function is

$$\langle x \mid \Phi'\rangle = \frac{1}{\sqrt{2(2\pi)^3}} e^{ik_0x_0} e^{i\mathbf{k}\cdot\mathbf{x}} \qquad (129)$$

which is a negative energy solution of the Klein-Gordon equation. Therefore, in order that the final wave function again be a positive energy solution, we require that

$$\begin{aligned} \langle x \mid \Phi'\rangle &= \langle x \mid U(i_t) \mid \Phi\rangle = \eta_T \overline{\langle i_t x \mid \Phi\rangle} \\ &= \eta_T \langle \Phi \mid i_t x\rangle \end{aligned} \qquad (130)$$

i.e., that $U(i_t)$ be antiunitary. The vacuum has again the property that it is invariant under this operation

$$\langle \Phi \mid U(i_t) \mid 0\rangle = \langle 0 \mid \Phi\rangle \quad \text{for all } |\Phi\rangle \qquad (131)$$

If we define the transformation rule for the n-particle amplitude as

$$\begin{aligned} \langle x_1, x_2, \cdots x_n \mid U(i_t) \mid \Phi\rangle &= (\eta_T)^n \overline{\langle i_t x_1, \cdots i_t x_n \mid \Phi\rangle} \\ &= (\eta_T)^n \langle \Phi \mid i_t x_1, i_t x_2, \cdots i_t x_n\rangle \end{aligned} \qquad (132)$$

one derives that

$$U(i_t)\,\phi^{(+)}(x)\,U(i_t)^{-1} = \eta_T\phi^{(+)}(i_t x) \tag{133a}$$

$$U(i_t)\,\phi^{(-)}(x)\,U(i_t)^{-1} = \eta_T\phi^{(-)}(i_t x) \tag{133b}$$

where the operator $U(i_t)$ is antiunitary, i.e.,

$$U(i_t)\,\alpha\,|\,\Psi\rangle = \bar{\alpha}\,U(i_t)\,|\,\Psi\rangle \tag{134a}$$

$$U(i_t)\,(|\Psi\rangle + |\Phi\rangle) = U(i_t)\,|\,\Psi\rangle + U(i_t)\,|\,\Phi\rangle \tag{134b}$$

$$(\Phi, \Psi) = (U(i_t)\,\Psi, U(i_t)\,\Phi) \tag{134c}$$

It is often convenient to decompose antiunitary operators into the product of a conjugation K and a unitary operator. By a conjugation K we mean an operator with the property that $K^2 = 1$ and $(K\Phi, K\Psi) = (\Psi, \Phi)$. If the operator U is antiunitary, then clearly $UK = \mathfrak{V}$ is unitary, so that an arbitrary antiunitary operator U can always be written in the form $U = \mathfrak{V}K$ where $\mathfrak{V}$ is unitary.

To clarify the meaning of the antiunitary operator $U(i_t)$, let us derive the transformation of the momentum space operators a_k, a_k^* from those of $\phi^{(+)}(x)$ and $\phi^{(-)}(x)$. Using the representation (70) and the property (134a) of an antiunitary operator, we find that

$$U(i_t)\,\phi^{(+)}(x)\,U(i_t)^{-1} = \frac{1}{\sqrt{2(2\pi)^3}}\int_{k_0>0}\frac{d^3k}{k_0}\,e^{+ik_0x_0}\,e^{-i\mathbf{k}\cdot\mathbf{x}}\,U(i_t)\,a_{\mathbf{k}}\,U(i_t)^{-1}$$

$$= \eta_T\,\frac{1}{\sqrt{2(2\pi)^3}}\int_{k_0>0}\frac{d^3k}{k_0}\,e^{ik_0x_0}\,e^{+i\mathbf{k}\cdot\mathbf{x}}\,a_{\mathbf{k}} \tag{135a}$$

whence

$$U(i_t)\,a_{\mathbf{k}}\,U(i_t)^{-1} = \eta_T a_{-\mathbf{k}} \tag{135b}$$

and similarly, using (72) and (133a), we derive that

$$U(i_t)\,a_{\mathbf{k}}^*\,U(i_t)^{-1} = \eta_T a_{-\mathbf{k}}^* \tag{135c}$$

One verifies that in the Schrödinger picture, under time reversal each state of a physical system at time t is mapped into a state at time $-t$ in which the velocities of the particles are reversed. Invariance under time inversion implies that such a state exists. This is the translation into quantum mechanics of the classical notion of invariance under time reversal which asserts that, for Lagrangians which do not depend on the time explicitly (and contain only even powers of the momenta), if $q_i(t)$, $p_i(t)$ is a possible motion for the system, then so is $q_i(-t)$, $-p_i(-t)$. Loosely speaking, time reversal invariance asserts that if a motion is possible in one spatial direction there exists another possible motion over the same spatial path, but in the reversed order.

We conclude this section by considering the Hermitian field operator

$$\phi(x) = \phi^{(+)}(x) + \phi^{(-)}(x) \tag{136a}$$

$$\phi(x) = \phi^*(x) \tag{136b}$$

This operator when expressed in terms of the operators $a_{\mathbf{k}}$ and $a_{\mathbf{k}}{}^*$ has the form

$$\phi(x) = \frac{1}{\sqrt{2(2\pi)^3}} \int_{k_0>0} \frac{d^3k}{k_0} (a_{\mathbf{k}} e^{-ik\cdot x} + a_{\mathbf{k}}{}^* e^{ik\cdot x}) \tag{137}$$

and obeys the Klein-Gordon equation:

$$(\Box + \mu^2)\, \phi(x) = 0 \tag{138}$$

The operators $\phi^{(+)}(x)$ and $\phi^{(-)}(x)$ are the positive and negative frequency parts of the operator $\phi(x)$ and can be recovered from the latter by the following Lorentz invariant process:

$$\phi^{(\pm)}(x) = \int_\sigma \left\{ \frac{\partial\Delta^{(\pm)}(x - x')}{\partial x'^\mu}\, \phi(x') - \Delta^{(\pm)}(x - x')\, \frac{\partial \phi(x')}{\partial x'^\mu} \right\} d\sigma^\mu(x') \tag{139}$$

Alternatively, if we Fourier analyze $\phi(x)$

$$\phi(x) = \int d^4k e^{-ik\cdot x}\, \delta(k^2 - \mu^2)\, \tilde{\phi}(k) \tag{140}$$

then

$$\phi^{(\pm)}(x) = \int d^4k e^{-ik\cdot x}\, \theta(\pm k_0)\, \delta(k^2 - \mu^2)\, \tilde{\phi}(k) \tag{141}$$

The transformation properties of $\phi(x)$ are trivially obtained from those of $\phi^{(+)}(x)$ and $\phi^{(-)}(x)$. Under orthochronous proper Lorentz transformations ϕ transforms as follows:

$$U(a, \Lambda)\, \phi(x)\, U(a, \Lambda)^{-1} = \phi(\Lambda x + a) \tag{142}$$

Under space and time inversions it transforms according to

$$U(i_s)\, \phi(x)\, U(i_s)^{-1} = \eta_S \phi(i_s x) \tag{143}$$

$$U(i_t)\, \phi(x)\, U(i_t)^{-1} = \eta_T \phi(i_t x) \tag{144}$$

where $U(i_t)$ is antiunitary.

The commutation rules of $\phi(x)$ are:

$$[\phi(x), \phi(x')] = [\phi^{(+)}(x) + \phi^{(-)}(x), \phi^{(+)}(x') + \phi^{(-)}(x')]$$

$$= i\Delta^{(+)}(x - x') + i\Delta^{(-)}(x - x') \tag{145a}$$

$$= i\Delta(x - x') \tag{145b}$$

where

$$\Delta(x) = -\frac{i}{2(2\pi)^3} \int_{k_0>0} \frac{d^3k}{k_0} (e^{-ikx} - e^{+ik\cdot x})$$

$$= -\frac{1}{(2\pi)^3} \int_{k_0>0} \frac{d^3k}{k_0} e^{i\mathbf{k}\cdot\mathbf{x}} \sin k_0 x_0 \tag{146a}$$

$$= -\frac{i}{(2\pi)^3} \int d^4k\, \epsilon(k_0)\, \delta(k^2 - \mu^2)\, e^{-ik\cdot x} \tag{146b}$$

The step in going from (146a) to (146b) uses the fact that

$$\epsilon(k_0)\,\delta(k^2 - \mu^2) = \frac{1}{2|k_0|}\{\delta(k_0 - \omega_{\mathbf{k}}) - \delta(k_0 + \omega_{\mathbf{k}})\} \tag{147}$$

where $\epsilon(k_0) = \theta(k_0) - \theta(-k_0) = k_0/|k_0|$ is equal to $+1$ for $k_0 > 0$ and -1 for $k_0 < 0$. The important properties of $\Delta(x)$ which can be deduced from the above representation are

$$\Delta(-x) = -\Delta(+x) \tag{148a}$$

$$\Delta(\Lambda x) = \Delta(x) \tag{148b}$$

$$(\square + \mu^2)\,\Delta(x) = 0 \tag{148c}$$

The singular function $\Delta(x)$ can be characterized as an odd, invariant solution of the Klein-Gordon equation (see below). The commutator is therefore an invariant c-number. The invariance of the commutator can also be inferred from the validity of the following equations:

$$\begin{aligned} U(a, \Lambda)\,[\phi(x), \phi(x')]\,U(a, \Lambda)^{-1} &\\ &= U(a, \Lambda)\,\Delta(x - x')\,U(a, \Lambda)^{-1} = \Delta(x - x') \\ &= [\phi(\Lambda x + a), \phi(\Lambda x' + a)] = \Delta(\Lambda x + a - \Lambda x' - a) \\ &= \Delta(\Lambda(x - x')) = \Delta(x - x') \end{aligned} \tag{149}$$

Conversely, the requirement that the commutator be an invariant c-number fixes, apart from a constant, the value of the commutator. To prove this, let us denote by $F(x, x')$ the right-hand side of the commutator

$$[\phi(x), \phi(x')] = F(x, x') \tag{150}$$

where by assumption F is an invariant c-number. Invariance under translations requires that for arbitrary a

$$F(x, x') = F(x + a, x' + a) \tag{151}$$

so that F can only be a function of $x - x'$. Invariance under proper homogeneous Lorentz transformations requires that

$$F(x) = F(\Lambda x) \tag{152}$$

Since the operator $\phi(x)$ satisfies the Klein-Gordon equation, F must likewise satisfy the Klein-Gordon equation:

$$(\square_x + \mu^2)\,[\phi(x), \phi(x')] = 0 = (\square_x + \mu^2)\,F(x - x') \tag{153}$$

Finally, since the commutator has the property that

$$[\phi(x), \phi(x')] = -[\phi(x'), \phi(x)] \tag{154}$$

F must be odd:

$$F(x - x') = -F(x' - x) \tag{155a}$$

or equivalently

$$F(x) = -F(-x) \tag{155b}$$

$F(x)$ is thus an odd invariant solution of the Klein-Gordon equation.

Now any $F(x)$ satisfying the Klein-Gordon equation may be written in the form

$$F(x) = \int d^4k\, e^{-ikx}\, \delta(k^2 - \mu^2)\, \tilde{F}(k) \tag{156}$$

Note that $\tilde{F}(k)$ is only defined on the "mass shell" $k^2 = \mu^2$, i.e., on the two hyperboloids $k_0 = \pm\sqrt{\mathbf{k}^2 + \mu^2}$. The possible values of the vector k are thus all time-like. The Lorentz invariance of $F(x)$ now asserts that

$$\begin{aligned} F(x) = F(\Lambda x) &= \int \delta(k^2 - \mu^2)\, e^{-ik\cdot\Lambda x}\, \tilde{F}(k)\, d^4k \\ &= \int \delta((\Lambda^{-1}k)^2 - \mu^2)\, e^{-i\Lambda^{-1}k\cdot x}\, \tilde{F}(k)\, d^4k \\ &= \int \delta(k^2 - \mu^2)\, e^{-ik\cdot x}\, \tilde{F}(\Lambda k)\, d^4k \end{aligned} \tag{157}$$

whence

$$\tilde{F}(k) = \tilde{F}(\Lambda k) \tag{158}$$

so that $\tilde{F}(k)$ can only be a function of k^2 and $\epsilon(k_0)$, the two invariants which can be constructed from the time-like vector k. We write

$$F(k) = F_1(k^2) + \epsilon(k_0)\, F_2(k^2) \tag{159}$$

Because of the $\delta(k^2 - \mu^2)$ factor in (156) only the value of $F_1(k^2)$ and $F_2(k^2)$ at $k^2 = \mu^2$ enter in the definition of $F(x)$, so that we may write

$$F(x) = \int d^4k\, \delta(k^2 - \mu^2)\, (F_1 + \epsilon(k_0)\, F_2)\, e^{-ik\cdot x} \tag{160}$$

where $F_{1,2} = F_{1,2}(\mu^2)$ are constants. The requirement that $F(x) = -F(-x)$ next yields the condition that

$$\begin{aligned} -F(-x) &= -\int d^4k\, \delta(k^2 - \mu^2)\, (F_1 + \epsilon(k_0)\, F_2)\, e^{ik\cdot x} \\ &= -\int d^4k\, \delta(k^2 - \mu^2)\, (F_1 - \epsilon(k_0)\, F_2)\, e^{-ik\cdot x} \\ &= F(x) = \int d^4k\, \delta(k^2 - \mu^2)\, (F_1 + \epsilon(k_0)\, F_2)\, e^{-ik\cdot x} \end{aligned} \tag{161a}$$

therefore

$$F_1 = -F_1 = 0 \tag{161b}$$

so that

$$F(x) = F_2 \int d^4k\, e^{-ik\cdot x}\, \delta(k^2 - \mu^2)\, \epsilon(k) \tag{162}$$

Finally, since the operator ϕ is Hermitian, $F(x)$ must have the property that

$$\bar{F}(x) = F(-x) = -F(x) \tag{163}$$

obtained by taking the Hermitian adjoint of (150). The constant F_2 is therefore pure imaginary. Q.E.D.

From the invariance of $\Delta(x)$ under proper homogeneous Lorentz transformations and the fact that $\Delta(x)$ is odd, we deduce that $\Delta(x)$ vanishes for x^2 space-like, i.e., for $x^2 < 0$. *Proof:* The Lorentz invariance implies that Δ is a function of x^2 only for $x^2 < 0$, and of x^2 and $\epsilon(x_0)$, the sign of the time, for $x^2 \geqslant 0$, i.e., on and within the light cone. (Recall that $\epsilon(x_0)$ has an invariant meaning only on and within the light cone.) Thus for space-like x, $\Delta(x) = f_1(x^2)$, but a function of x^2 cannot be odd; hence $f_1(x^2) = 0$, which was to be demonstrated. Within the light cone $\Delta(x)$ must have the form $\Delta(x) = \epsilon(x_0) f_2(x^2)$ to be odd and invariant. Alternatively, we note that the rotational invariance of the equal-time commutator $\Delta(0, \mathbf{x}) = \Delta(0, R\mathbf{x})$ implies that $\Delta(0, \mathbf{x}) = \Delta(0, \mathbf{x}^2)$. This together with the oddness requirement, $\Delta(-x) = -\Delta(x)$, yields

$$\Delta(0, \mathbf{x}) = 0 \tag{164}$$

i.e., that the equal-time commutator vanishes:

$$[\phi(x^0, \mathbf{x}), \phi(x^0, \mathbf{x}')] = 0 \tag{165}$$

Since by a suitable Lorentz transformation we can always transform a space-like vector x so that its image $x' = \Lambda x$ has $x'_0 = 0$, it follows from $\Delta(x) = \Delta(\Lambda x)$ that $\Delta(x) = 0$ for all space-like x.

The other important property of $\Delta(x)$ which is readily deduced from the representation (146a) is that

$$\left(\frac{\partial \Delta(x)}{\partial x_0}\right)_{x_0=0} = -\delta^{(3)}(\mathbf{x}) \tag{166}$$

Since $\Delta(x)$ satisfies a second-order hyperbolic differential equation, it is uniquely determined by the differential equation:

$$(\Box + \mu^2)\, \Delta(x) = 0 \tag{167}$$

and the two initial conditions at $x_0 = 0$:

$$(a) \qquad \Delta(\mathbf{x}, 0) = 0 \tag{168a}$$

$$(b) \qquad \left(\frac{\partial \Delta(x)}{\partial x_0}\right)_{x_0=0} = -\delta^{(3)}(\mathbf{x}) \tag{168b}$$

The Δ function enables us to solve the Cauchy, or initial value problem for the Klein-Gordon equation. Let $g(x)$ be a wave-packet solution of the Klein-Gordon equation taking on values $g_0(x)$ and $n_\mu(x)\, \partial^\mu g_0(x)$ on a space-like surface σ_0 with normal $n_\mu(x)$; then the value of $g(x)$ on any later space-like surface is given by

$$g(x) = \int_{\sigma_0} \{\Delta(x - x')\, n_\mu \partial'^\mu g_0(x') - n_\mu \partial'^\mu \Delta(x - x') \cdot g_0(x')\}\, d\sigma(x') \tag{169}$$

Proof: For a function $F_\mu(x)$ which vanishes for $|\mathbf{x}| \to \infty$, Gauss's theorem asserts that

$$\int_{\sigma} F_\mu(x')\, d\sigma^\mu(x') - \int_{\sigma_0} F_\mu(x')\, d\sigma^\mu(x') = \int_{\Omega} \partial'_\mu F^\mu(x')\, d^4x' \quad (170)$$

where Ω is the space-time volume between σ and σ_0. Let

$$F_\mu(x') = \Delta(x - x')\, \partial'_\mu g(x') - \partial'_\mu \Delta(x - x') \cdot g(x') \quad (171)$$

Since g and Δ satisfy the Klein-Gordon equation, $\partial_\mu F^\mu(x) = 0$. Take σ to be the plane $x_0 = x_0' = constant$, so that the normal to σ is $n_\mu(x) = = (1, 0, 0, 0)$. Using the properties (164) and (166) of the Δ function, we obtain

$$\int_{\sigma} F_\mu(x')\, d\sigma^\mu(x') = g(x) \quad (172)$$

whence by (170), recalling that $\partial_\mu F^\mu = 0$,

$$g(x) = \int_{\sigma_0} F_\mu(x')\, d\sigma^\mu(x') \quad \text{Q.E.D.} \quad (173)$$

Besides $\Delta(x)$, there exists another invariant function, $\Delta^{(1)}(x)$, which satisfies the Klein-Gordon equation but which is even. It has the following properties:

$$(\Box + \mu^2)\, \Delta^{(1)}(x) = 0 \quad (174a)$$

$$\Delta^{(1)}(x) = \Delta^{(1)}(-x) \quad (174b)$$

$$\left(\frac{\partial \Delta^{(1)}(x)}{\partial x_0}\right)_{x_0=0} = 0 \quad (174c)$$

Its integral representation is

$$\Delta^{(1)}(x) = \frac{1}{(2\pi)^3} \int_{k_0>0} \frac{d^3k}{k_0}\, e^{i\mathbf{k}\cdot\mathbf{x}} \cos k_0 x_0 \quad (175a)$$

$$= \frac{1}{(2\pi)^3} \int_{-\infty}^{+\infty} d^4k\, e^{-ik\cdot x}\, \delta(k^2 - \mu^2) \quad (175b)$$

The (generalized) functions $\Delta(x)$ and $\Delta^{(1)}(x)$ can be expressed in terms of elementary functions by evaluating the integral representations (146a) and (175a). Thus, with $|\mathbf{x}| = r$, we find that $\Delta(x)$ may be written in the form

$$\Delta(\mathbf{x}, x_0) = -\frac{1}{(2\pi)^3} \int_{k_0>0} \frac{d^3k}{k_0}\, e^{i\mathbf{k}\cdot\mathbf{x}} \sin k_0 x_0$$

$$= -\frac{1}{2\pi^2} \int_0^\infty \frac{k^2 dk}{\sqrt{k^2 + \mu^2}}\, \frac{\sin kr \sin k_0 x_0}{kr}$$

$$= \frac{1}{2\pi^2} \frac{1}{r} \frac{\partial}{\partial r} \int_0^\infty \frac{dk}{\sqrt{k^2 + \mu^2}} \cos kr \sin k_0 x_0$$

$$= \frac{1}{4\pi} \frac{1}{r} \frac{\partial}{\partial r} F(r, x_0) \quad (176)$$

Making the transformation $k = \mu \sinh y$, we then obtain[2]

$$F(r, x_0) = \frac{1}{\pi} \int_{-\infty}^{+\infty} \frac{dk}{\sqrt{k^2 + \mu^2}} \cos kr \sin (\sqrt{k^2 + \mu^2}\, x_0)$$

$$= \begin{cases} +J_0(\mu\sqrt{x_0^2 - r^2}) & \text{for } x_0 > r \\ 0 & \text{for } -r < x_0 < r \\ -J_0(\mu\sqrt{x_0^2 - r^2}) & \text{for } x_0 < -r \end{cases} \tag{177}$$

By carrying out the indicated differentiations in Eq. (176), $\Delta(x)$ may be written in a form which exhibits its covariance properties as well as its singularities [see also the appendix of Schwinger (1949a)]:

$$\Delta(x) = -\frac{1}{2\pi} \epsilon(x_0) \left\{ \delta(x^2) - \frac{\mu^2}{2} \theta(x^2) \frac{J_1(\mu\sqrt{x^2})}{\mu\sqrt{x^2}} \right\} \tag{178}$$

where $\theta(y) = 1$ if $y > 0$, and $\theta(y) = 0$ if $y < 0$. For small x^2, i.e., near the light cone, the following expansion holds:

$$\Delta(x) = -\frac{1}{2\pi} \epsilon(x_0) \left\{ \delta(x^2) - \frac{\mu^2}{2} \theta(x^2) + \cdots \right\} \tag{179}$$

where the terms which vanish for $x^2 \to 0$ have been omitted; $\Delta(x)$ therefore has a delta-function singularity as well as a finite discontinuity (jump) on the light cone.

Using procedures similar to the ones outlined above for $\Delta(x)$, one readily verifies that

$$\Delta^{(1)}(\mathbf{x}, x_0) = \frac{1}{4\pi} \frac{1}{r} \frac{\partial}{\partial r} F^{(1)}(r, x_0) \tag{180}$$

where

$$F^{(1)}(r, x_0) = -\frac{1}{\pi} \int_{-\infty}^{+\infty} \frac{dk}{\sqrt{k^2 + \mu^2}} \cos k_0 x_0 \cos kr$$

$$= \begin{cases} N_0(\mu\sqrt{x_0^2 - r^2}) & \text{for } |x_0| > r \\ -iH_0^{(1)}(i\mu\sqrt{r^2 - x_0^2}) & \text{for } r > |x_0| \end{cases} \tag{181}$$

Note therefore that $\Delta^{(1)}$ does not vanish outside the light cone. For large space-like distances (i.e., $r \gg x_0$) $\Delta^{(1)}$ falls off exponentially. The $\Delta^{(1)}$ function can also be written in a form similar to Eq. (178) [see the appendix of Schwinger (1949a)]. We here give only its representation near the light cone so as to exhibit its singularities

$$\Delta^{(1)}(x) = -\frac{1}{2\pi^2} \left\{ \mathrm{P} \frac{1}{x^2} - \frac{\mu^2}{2} \log \left[\frac{\gamma}{2} (\mu^2 |x^2|)^{1/2} \right] + \frac{\mu^2}{4} + \cdots \right\} \tag{182}$$

[2] Recall that $J_0(z) = \frac{1}{\pi} \int_{-\infty}^{+\infty} \sin (z \cosh \beta)\, d\beta$; $\left(\frac{z}{2}\right)^\nu \sum_{l=0}^{\infty} \left(\frac{iz}{2}\right)^{2l} [l!\, \Gamma(\nu + l + 1)]^{-1} = J_\nu(z)$ so that in the neighborhood of $z = 0$, $J_0(z) \approx 1 - \frac{z^2}{4} + 0(z^4)$; recall also that $J_1(z) = -\frac{\partial}{\partial z} J_0(z)$ and that $N_0(z) = -\frac{1}{\pi} \int_{-\infty}^{+\infty} \cos (z \cosh \beta)\, d\beta$.

where P denotes the principal value, $\gamma = 1.78107 \cdots$ is Euler's constant, and where we have again neglected the terms which vanish in the limit $x^2 \to 0$.

7d. Connection with Field Theory

All the formalism developed up to this point has been based on the properties of the particles which make up the system under consideration: their spin, their mass, the transformation properties of their amplitudes, and their Bose-Einstein character. We shall now show that this second-quantized formalism is equivalent to a certain quantization procedure on a classical c-number field amplitude which obeys the Klein-Gordon equation.

Our experience with the quantization of point mechanical systems suggests that, in order to establish the aforementioned equivalence, we should look for a description of the theory in terms of canonical variables $\pi(\mathbf{x}, t)$ and $\phi(\mathbf{x}, t)$ whose commutation rules we expect to be of the form $[\pi(\mathbf{x}, t), \phi(\mathbf{x}', t)] = -i\hbar\delta^{(3)}(\mathbf{x} - \mathbf{x}')$. The latter would be the analogue of the commutation rules $[q_r, p_s] = i\hbar\delta_{rs}$, for a system with a continuous infinity of degrees of freedom, such as a field.

We note that if we define the operator

$$\begin{aligned}\pi(x) &= \partial_0\phi(x) \\ &= -\frac{i}{\sqrt{2(2\pi)^3}} \int\limits_{k_0=\omega_{\mathbf{k}}} d^3k\, (e^{-ik\cdot x}\, a_{\mathbf{k}} - e^{+ik\cdot x}\, a_{\mathbf{k}}^*) \end{aligned} \tag{183}$$

that $\pi(x)$ has delta-function equal-time commutation rules with $\phi(x)$, i.e.,

$$\begin{aligned}[\pi(x), \phi(x')]\Big|_{x_0=x_0'} &= \frac{\partial}{\partial x^0}[\phi(x), \phi(x')]\Big|_{x_0'=x_0} \\ &= -\frac{i}{(2\pi)^3}\int_{-\infty}^{+\infty} d^3k\, e^{i\mathbf{k}\cdot\mathbf{x}-\mathbf{x}'} \cos k_0(x_0 - x'_0)\Big|_{x_0'=x_0} \\ &= -i\delta^{(3)}(\mathbf{x} - \mathbf{x}') \end{aligned} \tag{184}$$

Thus, in some sense to be made precise later, the $\pi(x)$ defined in this manner and $\phi(x)$ are conjugate quantities. Let us next try to express H in terms of π and ϕ. To that end we note that we can obtain expressions for $a_{\mathbf{k}}$ and $a_{\mathbf{k}}^*$ in terms of π and ϕ by inverting Eqs. (137) and (183). For example, Eq. (183) for $\pi(x)$ can be rewritten in the form

$$\pi(x) = \frac{-i}{\sqrt{2(2\pi)^3}} \int\limits_{k_0=\omega_{\mathbf{k}}} d^3k\, \{a_{\mathbf{k}}e^{-i\omega_{\mathbf{k}}x_0} - a_{-\mathbf{k}}^*\, e^{i\omega_{\mathbf{k}}x_0}\}\, e^{i\mathbf{k}\cdot\mathbf{x}} \tag{185}$$

whence

$$a_{\mathbf{k}}\, e^{-i\omega_{\mathbf{k}}x_0} - a_{-\mathbf{k}}^*\, e^{i\omega_{\mathbf{k}}x_0} = \frac{i\sqrt{2}}{(2\pi)^{3/2}} \int d^3x\, \pi(x)\, e^{-i\mathbf{k}\cdot\mathbf{x}} \tag{186}$$

Similarly one derives from Eq. (137) that

$$a_{\mathbf{k}}\, e^{-i\omega_{\mathbf{k}}x_0} + a_{-\mathbf{k}}{}^*\, e^{i\omega_{\mathbf{k}}x_0} = \frac{\sqrt{2}\,\omega_{\mathbf{k}}}{(2\pi)^{3/2}} \int d^3x\, \phi(x)\, e^{-i\mathbf{k}\cdot\mathbf{x}} \tag{187}$$

so that

$$a_{\mathbf{k}} = \frac{\omega_k}{\sqrt{2(2\pi)^3}} \int d^3x\, e^{+ik\cdot x}\, \phi(x) + \frac{i}{\sqrt{2(2\pi)^3}} \int d^3x\, e^{+ik\cdot x}\, \pi(x) \tag{188a}$$

$$a_{\mathbf{k}}{}^* = \frac{\omega_k}{\sqrt{2(2\pi)^3}} \int d^3x'\, e^{-ik\cdot x'}\, \phi(x') - \frac{i}{\sqrt{2(2\pi)^3}} \int d^3x'\, e^{-ik\cdot x'}\, \pi(x') \tag{188b}$$

If we substitute these values into $H = \int d\Omega(k)\; a_{\mathbf{k}}{}^* a_{\mathbf{k}} \omega_{\mathbf{k}}$ and note that in the expression (188a) and (188b) the times x_0, x'_0 are arbitrary, so that for simplicity we may choose $x_0 = x_0'$, we obtain

$$\begin{aligned} H = \frac{1}{2(2\pi)^3} \int d^3k\, \omega_{\mathbf{k}}{}^2 \int d^3x\, e^{i\mathbf{k}\cdot\mathbf{x}}\, \phi(x) \cdot \int d^3x'\, e^{-i\mathbf{k}\cdot\mathbf{x}'}\, \phi(x') \\ + \frac{1}{2(2\pi)^3} \int d^3k \int d^3x\, e^{i\mathbf{k}\cdot\mathbf{x}}\, \pi(x) \int d^3x'\, e^{i\mathbf{k}\cdot\mathbf{x}'}\, \pi(x') + H_c \end{aligned} \tag{189}$$

where H_c corresponds to the cross terms in π and ϕ [see Eq. (192) below]. Since $\omega_{\mathbf{k}}{}^2 = \mathbf{k}^2 + \mu^2$

$$(\mathbf{k}^2 + \mu^2) \int d^3x\, e^{-i\mathbf{k}\cdot\mathbf{x}}\, \phi(x) = \int d^3x\, \{(\mu^2 - \nabla^2)\, e^{-i\mathbf{k}\cdot\mathbf{x}}\}\, \phi(x) \tag{190a}$$

$$= \int d^3x\, e^{-i\mathbf{k}\cdot\mathbf{x}}\, (\mu^2 - \nabla^2)\, \phi(x) \tag{190b}$$

The justification for the integration by part is as follows: going from Eq. (190a) to (190b) is valid for the matrix elements of these operator equations between arbitrary *normalizable* vectors $|\Psi\rangle$ and $|\Phi\rangle$. Carrying out the $\mathbf{k}$ integration we obtain

$$H = \tfrac{1}{2} \int d^3x\, \{\phi(x)\, (\mu^2 - \nabla^2)\, \phi(x) + \pi^2(x)\} + H_c \tag{191}$$

where

$$\begin{aligned} H_c &= \frac{i}{2(2\pi)^3} \int d^3k\, \omega_{\mathbf{k}} \int d^3x \int d^3x'\, e^{-ik\cdot(x-x')}\, [\phi(x), \pi(x')]\Big|_{x_0 = x_0'} \\ &= -\tfrac{1}{2} \int d^3k\, \omega_{\mathbf{k}}\, V \qquad (V = \text{quantization volume}) \end{aligned} \tag{192}$$

The term H_c is thus a c-number, infinite in magnitude, behaving as $\int k^3dk$ for large k. Its presence is due to the fact that when we write the Hamiltonian in the form (191) and perform a decomposition of ϕ and π into creation and annihilations operators ($\phi^{(\pm)}$ and $\partial_0\phi^{(\pm)}$), there will occur in $H - H_c$ creation operators standing to the right of destruction operators.

Operating on the no-particle state, $H - H_c$ would then not give zero (due to the presence of creation operators standing to the right of destruction operators), but an energy of the amount $+\frac{1}{2}\int d^3k\, \omega_{\mathbf{k}}$. The cross-term H_c readjusts the energy so that the no-particle state has zero energy. Alternatively, we could write

$$H = \frac{1}{2} \int d^3x\, N\{\pi^2(x) + \phi(x)\,(\mu^2 - \nabla^2)\,\phi(x)\} \tag{193}$$

where the operator N is defined as follows: Acting on a product of creation and annihilation operators, it rewrites the product in *"normal form"* with all creation operators standing to the left of all destruction operators, with the understanding that the rearrangement is to be made as if all commutators were zero. Thus

$$N(\phi^{(-)}(y)\,\phi^{(+)}(x)) = N(\phi^{(+)}(x)\,\phi^{(-)}(y)) = \phi^{(-)}(y)\,\phi^{(+)}(x) \tag{194}$$

By definition, the distributive law is valid for the normal product operation N, i.e.,

$$N[\phi^{(+)}(x)\,\{\phi^{(+)}(y) + \phi^{(-)}(z)\}] = N(\phi^{(+)}(x)\,\phi^{(+)}(y)) + N(\phi^{(+)}(x)\,\phi^{(-)}(z)) \tag{195}$$

The normal product is also sometimes denoted by colons; for example, instead of the notation of Eq. (193) we shall also write

$$H = \frac{1}{2} \int d^3x : (\pi^2(x) + \nabla\phi(x) \cdot \nabla\phi(x) + \mu^2\phi^2(x)) : \tag{196}$$

where we have performed an integration by parts on the $\nabla^2\phi$ term in (193).

In a manner analogous to the procedure outlined for the Hamiltonian operator we can express the total momentum operator in terms of $\pi(x)$ and $\phi(x)$, in which case

$$\mathbf{P} = \frac{1}{2} \int d^3x : \pi(x)\,\nabla\phi(x) + \nabla\phi(x) \cdot \pi(x) : \tag{197}$$

Note that the difference between $\mathbf{P}$ in normal form and the form without the N operation is a c-number which is equal to $\frac{1}{2}\int d^3k\, \mathbf{k}$, and that this term vanishes by symmetry.

In the Schrödinger picture, the second-quantized description of a system of free particles can be characterized by the Hamiltonian

$$H = \frac{1}{2} \int d^3x : \pi^2(\mathbf{x}) + \nabla\phi(\mathbf{x}) \cdot \nabla\phi(\mathbf{x}) + \mu^2\phi^2(\mathbf{x}) : \tag{198}$$

where $\pi(\mathbf{x}) = \pi(\mathbf{x}, 0)$ and $\phi(\mathbf{x}) = \phi(\mathbf{x}, 0)$ are the field operators in the Schrödinger picture. (We again have taken the Schrödinger and Heisenberg pictures to coincide at $t = 0$). They satisfy the following canonical commutation rules

$$[\phi(\mathbf{x}), \pi(\mathbf{x}')] = i\delta^{(3)}(\mathbf{x} - \mathbf{x}') \tag{199}$$

The Schrödinger equation which determines the evolution of the system in time is $i\partial_t \mid \Psi(t)\rangle = H \mid \Psi(t)\rangle$, where $|\Psi(t)\rangle$ is a vector in $\mathcal{H}^{(0)} \oplus \mathcal{H}^{(1)} \oplus \oplus \mathcal{H}^{(2)} \oplus \cdots$.

We will now show that the quantized theory in the form just presented can also be obtained by considering the quantity $\phi(x)$ as a classical field amplitude obeying the field equation $(\Box + \mu^2)\,\phi(x) = 0$ and "quantizing" this classical field theory. By this is meant the following: We can regard a field as a system with an infinite number of degrees of freedom, with the value of the field amplitude at each point of space corresponding to one such degree of freedom. If it were possible to introduce a variable conjugate to the field amplitude at each point of space, then the imposition of the quantum condition would proceed as in the transition from classical to quantum mechanics for mechanical systems with a finite number of degrees of freedom; namely, by the rule that the classical Poisson brackets of the canonical variables go over into $i/\hbar$ times the commutator of these variables, which become noncommuting operators in the quantum theory.

Now the simplest way to introduce conjugate variables is in terms of a Lagrangian

$$L = \int d^3x \, \mathcal{L}(\phi, \phi_\mu) \tag{200}$$

where $\mathcal{L} = \mathcal{L}(\phi, \phi_\mu)$ is called the Lagrangian density and $\phi_\mu = \partial_\mu \phi$. We then require that the action integral

$$I = \int_{t_1}^{t_2} d^4x \, \mathcal{L}(\phi, \phi_\mu) \tag{201}$$

be stationary for arbitrary variations of the field quantities which vanish at the end points, i.e., variations such as $\delta\phi(t_2, \mathbf{x}) = \delta\phi(t_1, \mathbf{x}) = 0$ where t_1 and t_2 are the time limits in the integration of (201). We shall see below that the necessary and sufficient conditions for I to be stationary are that the Euler equation

$$\frac{\partial \mathcal{L}}{\partial \phi} - \frac{\partial}{\partial x^0}\frac{\partial \mathcal{L}}{\partial \phi_0} - \sum_{i=1}^{3} \frac{\partial}{\partial x_i}\frac{\partial \mathcal{L}}{\partial \phi_i} = 0 \tag{202}$$

be satisfied. The Lagrangian is so chosen that the Euler equations (202) coincide with the field equations. This action principle is the natural extension of Hamilton's principle in ordinary mechanics, since the aggregate of the values which ϕ and ϕ_0 take on can be thought of as corresponding to the co-ordinates and velocities occurring in particle mechanics.

Let us carry out the variation of (201) with respect to ϕ

$$\delta I = \int_\Omega d^4x \left[\frac{\partial \mathcal{L}}{\partial \phi(x)}\,\delta\phi(x) + \frac{\partial \mathcal{L}}{\partial \phi_\mu(x)}\,\delta\phi_\mu(x)\right] \tag{203}$$

where Ω denotes the four-volume of integration over which (201) is carried out. The shape of Ω is left arbitrary for the present. The second term depends on $\delta\phi(x)$ through the defining equation $\phi_\mu = \partial_\mu\phi$. Integrating this second term by parts we find

$$\delta I = \int_\Omega d^4x \left[\frac{\partial \mathcal{L}}{\partial \phi} - \frac{\partial}{\partial x^\mu} \left(\frac{\partial \mathcal{L}}{\partial \phi_\mu} \right) \right] \delta\phi + \int_\Sigma d\sigma(x)\, \delta\phi \left(\frac{\partial \mathcal{L}}{\partial \phi_\mu} \right) n_\mu(x) \tag{204}$$

where Σ denotes the surface bounding Ω and $n_\mu(x)$ is the outward normal to this surface at x. Now the action principle requires that $\delta I = 0$ for all variations $\delta\phi$, subject only to the restriction that $\delta\phi = 0$ on the boundary Σ of Ω. Since $\delta\phi$ inside Ω is arbitrary, $\delta I = 0$ requires that

$$\frac{\partial \mathcal{L}}{\partial \phi(x)} - \frac{\partial}{\partial x^\mu} \frac{\partial \mathcal{L}}{\partial \phi_\mu(x)} = 0 \tag{205}$$

i.e., that the Euler equations are satisfied. These are identified with the field equations by a suitable choice of $\mathcal{L}$. This choice, incidentally, is not unique. For example, the addition of a four-divergence to the Lagrangian density does not alter the field equations. *Proof:* Consider a new Lagrangian density, $\mathcal{L}'$, obtained from $\mathcal{L}$ by adding a four-divergence

$$\mathcal{L}' = \mathcal{L} + \frac{\partial F_\mu(\phi)}{\partial x_\mu} \tag{206}$$

where $F_\mu = F_\mu(\phi)$. The term $\int d^4x\, \partial^\mu F_\mu(\phi)$ in the action principle can by Gauss's theorem be transformed into an integral over the surface Σ so that its variation vanishes identically. If F_μ depends on ϕ_ν as well as on ϕ, this is still true provided $\mathcal{L}'$ does not depend on the second derivatives of ϕ. Although the physical content of two Lagrangians which differ from each other by a four-divergence is the same, the Hamiltonian and canonical formalism will in general be different for them.

To develop the canonical formalism further by analogy with the case of a system of particles, it is desirable to work with a denumerable infinity of degrees of freedom rather than a continuous infinity. To achieve this, one can consider the three-dimensional continuum to be divided into little cubical cells, each so small that presumably no important physical quantity varies appreciably inside a cube. These cells can then be labeled by a discrete variable. The average value of ϕ and $\partial_0\phi$ over the individual cells can then be adopted as the variables describing the system [see for example, Wentzel (1949)]. Alternatively, we can expand the real field variable $\phi(x)$ into a complete *real* orthonormal set of function $\varphi_n(\mathbf{x})$ which satisfy:

$$\int d^3x\, \varphi_m(\mathbf{x})\, \varphi_n(\mathbf{x}) = \delta_{mn} \tag{207a}$$

$$\sum_n \varphi_n(\mathbf{x})\, \varphi_n(\mathbf{x}') = \delta^{(3)}(\mathbf{x} - \mathbf{x}') \tag{207b}$$

i.e., we write

$$\phi(x) = \sum_n q_n(t)\, \varphi_n(\mathbf{x}) \qquad q_n(t) = \int d^3x\, \varphi_n(\mathbf{x})\, \phi(x) \tag{208a}$$

$$\phi_0(x) = \sum_n \dot{q}_n(t)\, \varphi_n(\mathbf{x}) \qquad \dot{q}_n(t) = \int d^3x\, \varphi_n(\mathbf{x})\, \phi_0(x) \tag{208b}$$

We next show that the Euler equations, which are valid at every x, can be replaced by a denumerably infinite set of equations in q_n and $\dot{q}_n$. We note that

$$\frac{\partial L}{\partial q_n} = \int d^3x \left(\frac{\partial \mathcal{L}}{\partial \phi} \frac{\partial \phi}{\partial q_n} + \sum_{k=1}^{3} \frac{\partial \mathcal{L}}{\partial \phi_k} \frac{\partial \phi_k}{\partial q_n} \right)$$

$$= \int d^3x \left(\frac{\partial \mathcal{L}}{\partial \phi} - \sum_{k=1}^{3} \frac{\partial}{\partial x^k} \frac{\partial \mathcal{L}}{\partial \phi_k} \right) \varphi_n(\mathbf{x}) \tag{209}$$

and similarly,

$$\frac{\partial L}{\partial \dot{q}_n} = \int d^3x \frac{\partial \mathcal{L}}{\partial \dot{\phi}} \frac{\partial \dot{\phi}}{\partial \dot{q}_n} = \int d^3x \frac{\partial \mathcal{L}}{\partial \dot{\phi}} \varphi_n(\mathbf{x}) \tag{210}$$

so that if (202) is satisfied

$$\frac{\partial L}{\partial q_n} - \frac{d}{dt} \frac{\partial L}{\partial \dot{q}_n} = 0 \quad n = 0, 1, 2, \cdots \tag{211}$$

Conversely, if the equations (211) hold, so that

$$\int d^3x \left(\frac{\partial \mathcal{L}}{\partial \phi} - \frac{\partial}{\partial x^\mu} \frac{\partial \mathcal{L}}{\partial \phi_\mu} \right) \varphi_n(\mathbf{x}) = 0 \tag{212}$$

then the completeness of the set $\{\varphi_n\}$ allows us to infer that the field equations (211) are satisfied. In complete analogy with ordinary mechanics we can now define momenta canonically conjugate to the $q_n(t)$ by

$$p_n(t) = \frac{\partial L}{\partial \dot{q}_n(t)} = \int \frac{\partial \mathcal{L}}{\partial \phi_0(x)} \varphi_n(\mathbf{x})\, d^3x$$

$$= \int \pi(x)\, \varphi_n(\mathbf{x})\, d^3x \tag{213}$$

where we have defined $\pi(x)$, the momentum canonically conjugate to $\phi(x)$ by

$$\pi(x) = \frac{\partial \mathcal{L}}{\partial \phi_0(x)} \tag{214}$$

The Hamiltonian can then be defined as

$$H = \sum_n p_n(t)\, \dot{q}_n(t) - L = \sum_n \int d^3x\, \pi(\mathbf{x}, t)\, \varphi_n(\mathbf{x}) \int d^3x'\, \phi_0(\mathbf{x}', t)\, \varphi_n(\mathbf{x}') - L$$

$$= \int d^3x \left(\frac{\partial \mathcal{L}}{\partial \phi_0} \phi_0 - \mathcal{L} \right) = \int d^3x\, \mathcal{H} \tag{215}$$

where $\mathcal{H} = \pi\phi_0 - \mathcal{L} = \mathcal{H}(\phi, \phi_i, \pi)$ is the Hamiltonian density. We can also write H in the form

$$H = \int d^3x \left(\frac{\partial \mathcal{L}}{\partial \phi_0} \phi^0 - \mathcal{L} g^{00} \right) = \int d^3x\, T^{00} \tag{216}$$

where T^{00} is the (0, 0) component of a tensor, the canonical energy-momentum tensor, defined by

$$T^{\mu\nu} = \phi^{\mu} \frac{\partial \mathcal{L}}{\partial \phi_{\nu}} - \mathcal{L} g^{\mu\nu} \tag{217}$$

If $\mathcal{L}$ does *not* explicitly depend on the space-time co-ordinates x, then by virtue of the equations of motion (205)

$$\begin{aligned} \frac{dT^{\mu\nu}}{dx^{\nu}} &= \phi^{\mu} \left(\frac{\partial}{\partial x^{\nu}} \frac{\partial \mathcal{L}}{\partial \phi_{\nu}} \right) + \frac{\partial \phi^{\mu}}{\partial x^{\nu}} \frac{\partial \mathcal{L}}{\partial \phi_{\nu}} - \frac{d\mathcal{L}}{dx_{\mu}} \\ &= \phi^{\mu} \left(\frac{\partial}{\partial x^{\nu}} \frac{\partial \mathcal{L}}{\partial \phi_{\nu}} \right) + \frac{\partial^2 \phi}{\partial x_{\mu} \partial x^{\nu}} \frac{\partial \mathcal{L}}{\partial \phi_{\nu}} - \left(\frac{\partial \mathcal{L}}{\partial \phi} \phi^{\mu} + \frac{\partial \mathcal{L}}{\partial \phi_{\lambda}} \frac{\partial^2 \phi}{\partial x_{\mu} \partial x^{\lambda}} \right) \\ &= \phi^{\mu} \left(\frac{\partial}{\partial x^{\nu}} \frac{\partial \mathcal{L}}{\partial \phi_{\nu}} - \frac{\partial \mathcal{L}}{\partial \phi} \right) = 0 \end{aligned} \tag{218}$$

[3]

If we integrate the equation $d_{\nu} T^{\mu\nu} = 0$ over all three-dimensional space, we infer that

$$\frac{d}{dx^0} \int d^3x \, T^{\mu 0} + \int d^3x \sum_{i=1}^{3} \frac{d}{dx^i} T^{\mu i} = 0 \tag{219}$$

so that if $T^{\mu k}$ vanishes at large spatial distances (which will be the case if the fields have this property) then the second term in (219) vanishes, since by Gauss's theorem it can be cast into a surface integral at infinity. We therefore obtain the four conservation laws

$$\frac{d}{dx^0} \int d^3x \, T^{\mu 0} = 0 \quad \mu = 0, 1, 2, 3 \tag{220}$$

which assert that the quantities $\int d^3x \, T^{\mu 0}$ ($\mu = 0, 1, 2, 3$) are constants of the motion. Equation (220) for $\mu = 0$ corresponds to the conservation of energy, i.e., $d_0 H = 0$, while for $\mu = 1, 2, 3$ it corresponds to the conservation of the three quantities

$$\begin{aligned} P^k &= \int d^3x \, T^{k0} \\ &= \int d^3x \frac{\partial \mathcal{L}}{\partial \phi} \phi^k = \int d^3x \, \pi(x) \, \phi^k(x) \end{aligned} \tag{221}$$

which upon symmetrization read

$$P^k = \tfrac{1}{2} \int d^3x \, \{\pi(x) \, \phi^k(x) + \phi^k(x) \, \pi(x)\} \tag{222}$$

The quantities P^k can be identified with the three components of the total momentum of the field.

In analogy to (215) we may define the momentum density of the field

[3] If $\mathcal{L}$ does depend explicitly on the co-ordinates x, Eq. (218) would read $d_{\nu} T^{\mu\nu} = -\partial^{\mu} \mathcal{L}$. An explicit dependence of $\mathcal{L}$ on x corresponds to the presence of external, prescribed fields with which ϕ interacts.

as T^{k0}. A natural definition of the angular momentum density of the field about the origin is then given by the expression

$$m_{kl0} = x_k T_{l0} - x_l T_{k0} \tag{223}$$

so that the total angular momentum is $M_{kl} = \int d^3x \, m_{kl0}$. In order that the quantity M_{kl} be the space part of a conserved angular momentum tensor $M_{\mu\nu}$

$$M_{\mu\nu} = \int d^3x \, (x_\mu T_{\nu 0} - x_\nu T_{\mu 0}) \tag{224}$$

we must have

$$\partial^\rho m_{\mu\nu\rho} = 0 \tag{225a}$$

$$m_{\mu\nu\rho} = x_\mu T_{\nu\rho} - x_\nu T_{\mu\rho} \tag{225b}$$

since an argument similar to that leading to Eq. (220) would then show that $\int d^3x \, m_{\mu\nu 0}$ is conserved. However, by virtue of (225b)

$$\begin{aligned} \partial^\rho m_{\mu\nu\rho} &= \partial^\rho (x_\mu T_{\nu\rho} - x_\nu T_{\mu\rho}) \\ &= T_{\nu\mu} - T_{\mu\nu} \end{aligned} \tag{226}$$

Thus $M_{\mu\nu}$ will be a conserved tensor if $T_{\mu\nu} = T_{\nu\mu}$, i.e., if the energy-momentum tensor is symmetric. This symmetry is, in general, not a consequence of its definition (217). It is, however, always possible to symmetrize the energy-momentum tensor [Belinfante (1939, 1940)] by making use of the fact that we can always add a four-divergence to the Lagrangian without altering the physical content of the theory. Thus although the energy density, for example, is not uniquely determined, the total energy of the field is a well-defined and unique quantity. The three space-space (k, l) components of the antisymmetric angular momentum tensor are the angular momenta of the system, whereas the three space-time $(0, k)$ components are related to the co-ordinates of the center-of-mass of the system [Møller (1949, 1952), Pryce (1948)].

For the spin 0 field under consideration, the Lagrangian density characterizing the noninteracting field can be taken to be

$$\mathfrak{L} = -\tfrac{1}{2}(\mu^2\phi^2 - g_{\rho\nu}\partial^\rho\phi \cdot \partial^\nu\phi) = -\tfrac{1}{2}(\mu^2\phi^2 - \phi_\nu\phi^\nu) \tag{227}$$

The proof that this is a possible Lagrangian consists in showing that it gives rise to the correct field equation, $(\Box + \mu^2)\,\phi = 0$. That this is so is readily verified:

$$\begin{aligned} 0 = \frac{\partial \mathfrak{L}}{\partial \phi} - \frac{\partial}{\partial x_\lambda}\frac{\partial \mathfrak{L}}{\partial \phi^\lambda} &= -\mu^2\phi - \frac{\partial}{\partial x_\lambda}\phi_\lambda = -\left(\mu^2 + \frac{\partial^2}{\partial x_\lambda \partial x^\lambda}\right)\phi \\ &= -(\Box + \mu^2)\,\phi \end{aligned} \tag{228}$$

Note that the Lagrangian density is a scalar covariant under proper Lorentz transformations (in fact under all Lorentz transformation since it is quadratic in ϕ). *Proof:* under the Lorentz transformation $x \to x' = \Lambda x + a$, $\phi \to \phi'(x') = \phi(x)$, therefore $\mathfrak{L}'(x') = \mathfrak{L}(x)$. The momentum canonically conjugate to ϕ (not to be confused with the total linear momentum $\mathbf{P}$) is

$$\pi(x) = \frac{\partial \mathcal{L}}{\partial \phi_0} = \phi_0(x) \tag{229}$$

The Hamiltonian corresponding to this Lagrangian is therefore

$$\begin{aligned} H &= \int d^3x\, \mathcal{H} \\ &= \int d^3x\, \{\pi(x)\, \phi_0(x) - \mathcal{L}\} = \tfrac{1}{2} \int \{\pi^2(x) + \nabla\phi \cdot \nabla\phi(x) + \mu^2\phi^2(x)\} \end{aligned} \tag{230}$$

This expression is positive-definite, as is to be expected of an energy density. This explains our choice of sign in the expression (227) for the Lagrangian. Note that the form of the Hamiltonian is identical in structure with that derived from our consideration of the second-quantized theory. The expression for the canonical energy-momentum tensor derived from (217) and (227) is

$$T^{\mu\nu} = \phi^\mu\phi^\nu + \tfrac{1}{2}g^{\mu\nu}(\mu^2\phi^2 - \phi_\lambda\phi^\lambda) \tag{231}$$

which, it will be noted, is symmetric as it stands. This is connected with the fact that the Klein-Gordon field has no intrinsic angular momentum, i.e., has spin 0. We shall return to this point in our later general discussion of classical field theories in Section 7*g*.

To quantize the classical field theory we impose the quantum conditions

$$\begin{aligned} [q_n(t), q_m(t)] &= [p_n(t), p_m(t)] = 0 \\ [q_n(t), p_m(t)] &= i\hbar\delta_{nm} \end{aligned} \tag{232}$$

By multiplying these expressions by $\varphi_n(\mathbf{x})$ and $\varphi_m(\mathbf{x})$, summing over n and m, and using the completeness of the set $\{\varphi_m\}$, they yield the commutation relations for the operators $\phi(x)$ and $\pi(x)$ defined by Eqs. (208) and (213), where now the q_ns and p_ns are considered to be operators:

$$[\phi(\mathbf{x}, x^0), \phi(\mathbf{x}', x^0)] = [\pi(\mathbf{x}, x^0), \pi(\mathbf{x}', x^0)] = 0 \tag{233}$$

$$\begin{aligned} [\phi(\mathbf{x}, x^0), \pi(\mathbf{x}', x^0)] &= [\phi(\mathbf{x}, x^0), \phi_0(\mathbf{x}', x^0)] \\ &= i\hbar c\delta^{(3)}(\mathbf{x} - \mathbf{x}') \end{aligned} \tag{234}$$

(Note that the time co-ordinates of the operators are the same!) Thus quantization has made operators out of the field variables and they satisfy the commutation rules (233)–(234) which are identical with (184). We have therefore shown the equivalence of the second-quantized theory for a system of free noninteracting particles and the quantization of the classical Klein-Gordon field.

Conversely, this equivalence allows us to reinterpret the expressions for the Lagrangian, the energy-momentum tensor, the angular momentum, etc. as operator expressions. In order to make the transition from the classical to the quantum level unambiguous as to the order of the noncommuting factors $\pi(x)$, $\phi(x)$ (whose order is arbitrary at the classical level), we shall adopt the rule that the normal product of these expressions is to be taken.

The commutator of H with π and ϕ is easily computed.[4] Thus, since $\phi(\mathbf{x}, x_0)$ commutes with $\phi(\mathbf{x}', x_0)$ and $\nabla\phi(\mathbf{x}', x_0)$

$$\begin{aligned}
[\phi(x), H] &= \tfrac{1}{2} \int d^3x' \, [\phi(\mathbf{x}, x^0), \pi^2(\mathbf{x}', x_0)] \\
&= \tfrac{1}{2} \int d^3x' \, \{\pi(\mathbf{x}', x_0) \, [\phi(x), \pi(\mathbf{x}', x_0)] \\
&\quad + [\phi(x), \pi(\mathbf{x}', x_0)] \, \pi(\mathbf{x}', x_0)\} \\
&= i\hbar c\pi(x) \qquad (235)
\end{aligned}$$

Similarly,

$$\begin{aligned}
[\pi(x), H] &= \tfrac{1}{2} \int d^3x' \, [\pi(x), \nabla\phi(\mathbf{x}', x_0) \cdot \nabla\phi(\mathbf{x}', x_0) + \mu^2\phi^2(\mathbf{x}', x_0)] \\
&= i\hbar c(\nabla^2\phi(x) - \mu^2\phi(x)) \qquad (236)
\end{aligned}$$

This agrees with the definition of H as the time displacement operator:

$$i\hbar c\partial_0 F(x) = [F(x), H] \qquad (237)$$

since Eqs. (235) and (236) then state that

$$\phi_0(x) = \pi(x) \qquad (238a)$$

and

$$\begin{aligned}
\partial_0\pi(x) &= \nabla^2\phi(x) - \mu^2\phi(x) \\
&= \partial_0{}^2\phi(x) \qquad (238b)
\end{aligned}$$

and these equations are equivalent to the equations of motions. Conversely, if H is the time displacement operator [i.e., Eq. (237) holds], the requirement that the equations derived by computing $[\phi(x), H]$ and $[\partial_0\phi, H]$ agree with the equations of motions obtained from the variational principle implies that the canonical commutation rules (233)–(234) must hold. Actually, one verifies that if the commutation rules (233)–(234) were replaced by anticommutation rules, the latter would also guarantee the equality between (237) and the equation of motions. The commutation rules can be regarded as a consistency requirement on the canonical formalism. However, it should be emphasized that the canonical formalism does not differentiate between quantization with commutation or anticommutation rules. Some physical principle must decide on the correct choice for the system under consideration.

[4] In computing $[H, \phi]$ and $[H, \pi]$ we neglect the fact that the normal product of the Hamiltonian is to be taken, since the difference between the normal and non-normal ordered Hamiltonian for the quadratic Hamiltonian at hand is a c-number which commutes with ϕ and π.

Consider next the commutation rules of P_k with $\pi(x)$ and $\phi(x)$:

$$[\phi(x), P_k] = \tfrac{1}{2} \int d^3\mathbf{x}' \, [\phi(x), \pi(x')\, \phi_k(x') + \phi_k(x')\, \pi(x')]_{x_0' = x_0}$$

$$= i\hbar c \frac{\partial \phi(x)}{\partial x^k} \tag{239a}$$

$$[\pi(x), P_k] = i\hbar c \frac{\partial \pi(x)}{\partial x^k} \tag{239b}$$

from which it follows that for an arbitrary operator $F = F(\phi, \pi)$

$$i\hbar c \frac{\partial F}{\partial x^k} = [F, P_k] \tag{240}$$

so that P_k is the generator for infinitesimal spatial translations. In relativistic notation (237) and (240) read

$$i\hbar c \partial_\mu F = [F, P_\mu] \tag{241}$$

7e. The Field Aspect

The fact that the second-quantized formulation can be considered as a quantized field theory suggests that besides the particle description there should also be a field description of the quantized theory. More precisely stated, our development to this point concerned itself with the particle aspect of the field system, i.e., with a description in terms of amplitudes $\langle k_1, k_2 \cdots k_n \mid \Psi\rangle$ or $\langle x_1, \cdots x_n \mid \Psi\rangle$ for finding a given number of particles with specified momenta or positions. But we could also inquire as to the probability for finding our field system at time x_0 in the field configuration $\chi(\mathbf{x})$, where $\chi(\mathbf{x})$ is some prescribed function of $\mathbf{x}$. For the Hermitian (neutral) field which we are considering, we can answer this question by obtaining the eigenfunctions $|\chi(\mathbf{x})\rangle$ of the field operator $\phi(\mathbf{x})$:

$$\phi(\mathbf{x}) \mid \chi(\mathbf{x})\rangle = \chi(\mathbf{x}) \mid \chi(\mathbf{x})\rangle \tag{242}$$

Thus $\langle\chi(\mathbf{x}) \mid \Psi(t)\rangle$ is then the amplitude for finding our field system characterized by the state vector $|\Psi(t)\rangle$ in the field configuration $\chi(\mathbf{x})$ at time t. (For the sake of simplicity we are working in the Schrödinger picture.) The amplitude $\langle\chi(\mathbf{x}) \mid \Psi\rangle = \Psi\{\chi\}$ is a functional of χ.

We next introduce the notion of the functional derivative of a functional $F\{\chi\}$. We define

$$\frac{\delta F\{\chi\}}{\delta\chi(\mathbf{x})} = \lim_{\kappa\to 0} \lim_{\epsilon\to 0} \frac{1}{\epsilon} [F\{\chi(\mathbf{y}) + \epsilon\delta_\kappa(\mathbf{y} - \mathbf{x})\} - F\{\chi(\mathbf{y})\}] \tag{243}$$

where δ_κ is any well-behaved function which in the limit $\kappa \to 0$ tends to the Dirac δ function. The limits are to be taken in the order indicated. For example, if

$$F\{\chi\} = \int d^3x\, f(\mathbf{x})\, \chi(\mathbf{x}) \tag{244}$$

then

$$\frac{\delta F\{\chi\}}{\delta\chi(\mathbf{x}')} = f(\mathbf{x}') \tag{245}$$

Alternatively, we may write

$$\delta F\{\chi\} = \int \frac{\delta F\{\chi\}}{\delta\chi(\mathbf{x})}\, \delta\chi(\mathbf{x})\, d^3x \tag{246}$$

so that, for example, since

$$\delta\chi(\mathbf{x}) = \int \delta^3(\mathbf{x} - \mathbf{x}')\, \delta\chi(\mathbf{x}')\, d^3x' \tag{247}$$

we find that

$$\frac{\delta\chi(\mathbf{x})}{\delta\chi(\mathbf{x}')} = \delta^{(3)}(\mathbf{x} - \mathbf{x}') \tag{248}$$

By considering the difference between

$$\frac{\delta}{\delta\chi(\mathbf{x})}\left[\chi(\mathbf{x}')\, F\{\chi\}\right] = \left[\delta^{(3)}(\mathbf{x} - \mathbf{x}') + \frac{\delta}{\delta\chi(\mathbf{x})}\right] F\{\chi\} \tag{249}$$

and $\chi(\mathbf{x}')\, \dfrac{\delta}{\delta\chi(\mathbf{x})}\, F\{\chi\}$, we derive that

$$\left[\chi(\mathbf{x}),\, -i\frac{\delta}{\delta\chi(\mathbf{x}')}\right] = +i\delta^{(3)}(\mathbf{x} - \mathbf{x}') \tag{250}$$

which then suggests that in the representation in which $\phi(\mathbf{x})$ is diagonal, $\pi(\mathbf{x}')$ has the representation $i\hbar\delta/\delta\chi(\mathbf{x}')$. The proof is very similar to that outlined in Chapter 1 for the operators p and q which satisfy $[q, p] = i\hbar$. We therefore write

$$\langle\chi(\mathbf{x}) \mid \pi(\mathbf{x}') \mid \Psi\rangle = -i\hbar\, \frac{\delta}{\delta\chi(\mathbf{x}')}\, \langle\chi(\mathbf{x}) \mid \Psi\rangle \tag{251}$$

Let us next obtain the explicit dependence on $\chi(\mathbf{x})$ for the no-particle state $|0\rangle$. This state is characterized, as we have seen, by the fact that $a_{\mathbf{k}} \mid 0\rangle = 0$ for all $\mathbf{k}$. From (188a) we find that the vacuum can also be characterized by the relation

$$\int d^3x\, e^{i\mathbf{k}\cdot\mathbf{x}}\, (\omega_{\mathbf{k}}\phi(\mathbf{x}) + i\pi(\mathbf{x})) \mid 0\rangle = 0 \quad \text{for all } \mathbf{k} \tag{252}$$

or equivalently

$$\omega_{\mathbf{k}}\phi(\mathbf{x}) + i\pi(\mathbf{x}) \mid 0\rangle = 0 \quad \text{for all } \mathbf{x} \tag{253}$$

where $\omega_{\mathbf{k}} = \sqrt{\mu^2 - \nabla^2}$. If we take the scalar product of this equation with the bra $\langle\chi(\mathbf{x}')|$, we then obtain

$$\left\{\omega_{\mathbf{k}}\chi(\mathbf{x}) + \hbar\, \frac{\delta}{\delta\chi(\mathbf{x})}\right\} \langle\chi(\mathbf{x}') \mid 0\rangle = 0 \tag{254}$$

which can be considered as a functional differential equation for the no-particle state amplitude $\langle\chi(\mathbf{x}') \mid 0\rangle$. The solution of (254) is

$$\Phi_0\{\chi\} = \langle \chi(\mathbf{x}) \mid 0 \rangle = \lambda e^{-\frac{1}{2\hbar}\int \chi(\mathbf{x}')\sqrt{\mu^2 - \nabla_{\mathbf{x}'}^2}\,\chi(\mathbf{x}')d^3x'} \tag{255}$$

where λ is some constant to be determined by the normalization condition $\langle 0 \mid 0 \rangle = 1$. This method of functionals can be extended to the theory of interacting fields and is closely related to the Feynman formulation of quantum field theory, wherein the amplitudes for given processes are expressed as sums over classical histories. The reader is referred to the articles of Fock (1934), Symanzik (1954), and Schwinger (1954b) as well as the review article of Novozilov and Tolub (1958) for an exposition and application of this technique to field theoretic problems.

7f. The Charged Scalar Field

We next consider the generalization of the second-quantized and field theoretic formalism to the case that the system consists of oppositely charged particles, both having spin 0 and mass μ. A state of the system is now characterized by stating the number of positively and negatively charged particles, their momenta and energies. Since oppositely charged particles are distinguishable, the basis vector $|p_1, p_2 \cdots p_m; q_1, \cdots q_n\rangle$ corresponding to the presence of m positively charged particles of charge $+e$, momenta $\mathbf{p}_1, \mathbf{p}_2, \cdots \mathbf{p}_m$, energies $\omega(\mathbf{p}_1), \omega(\mathbf{p}_2), \cdots \omega(\mathbf{p}_m)$ and n negatively charged particles of charge $-e$, momenta $\mathbf{q}_1, \mathbf{q}_2, \cdots \mathbf{q}_n$, energies $\omega(\mathbf{q}_1)$, $\omega(\mathbf{q}_2), \cdots \omega(\mathbf{q}_n)$, will be symmetric only under the interchange of the $\mathbf{p}$s and $\mathbf{q}$s separately. We shall refer to the particles of charge $+e$ as the "particles," and those of charge $-e$ as the "antiparticles."

Let us denote by $a_\mathbf{k}$, $a_\mathbf{k}^*$ the annihilation and creation operators for a positively charged particle of momentum $\mathbf{k}$, and by $b_\mathbf{k}$, $b_\mathbf{k}^*$ the corresponding operators for the negatively charged particles. They satisfy the commutation rules:

$$[a_\mathbf{k}, a_{\mathbf{k}'}^*] = k_0\delta(\mathbf{k} - \mathbf{k}')$$
$$[a_\mathbf{k}, a_{\mathbf{k}'}] = [a_\mathbf{k}^*, a_{\mathbf{k}'}^*] = 0 \tag{256}$$
$$[b_\mathbf{k}, b_{\mathbf{k}'}^*] = k_0\delta(\mathbf{k} - \mathbf{k}')$$
$$[b_\mathbf{k}, b_{\mathbf{k}'}] = [b_\mathbf{k}^*, b_{\mathbf{k}'}^*] = 0 \tag{257}$$

where $k_0 = +\omega_\mathbf{k}$. Since the a and b operators refer to different degrees of freedom of our system, we assume them to commute with one another:

$$[a_\mathbf{k}, b_{\mathbf{k}'}^*] = [a_\mathbf{k}, b_{\mathbf{k}'}] = 0 \tag{258a}$$
$$[a_\mathbf{k}^*, b_{\mathbf{k}'}] = [a_\mathbf{k}^*, b_{\mathbf{k}'}^*] = 0 \tag{258b}$$

The basis vectors expressed in terms of the creation operators are

$$|\mathbf{p}_1, \mathbf{p}_2, \cdots \mathbf{p}_m; \mathbf{q}_1, \cdots \mathbf{q}_n\rangle = \frac{1}{\sqrt{n!\,m!}}\, a_{\mathbf{p}_1}^* \cdots a_{\mathbf{p}_n}^* b_{\mathbf{q}_1}^* \cdots b_{\mathbf{q}_n}^* \mid 0\rangle \tag{259}$$

where the no-particle state $|0\rangle$ now has the property that

$$b_{\mathbf{k}} \mid 0\rangle = a_{\mathbf{k}} \mid 0\rangle = 0 \quad \text{for all } \mathbf{k} \tag{260}$$

The number operators N_+, N_- for the number of positively and negatively charged particles, respectively, are given by the following expressions:

$$N_+ = \int_+ \frac{d^3k}{k_0} a_{\mathbf{k}}^* a_{\mathbf{k}} \tag{261a}$$

$$N_- = \int_+ \frac{d^3k}{k_0} b_{\mathbf{k}}^* b_{\mathbf{k}} \tag{261b}$$

By arguments familiar by now, using the commutation rules and the property (260) of the vacuum, we derive that

$$N_+ \mid \mathbf{p}_1, \cdots \mathbf{p}_m; \mathbf{q}_1, \cdots, \mathbf{q}_n\rangle = m \mid \mathbf{p}_1, \cdots \mathbf{p}_m; \mathbf{q}_1, \cdots, \mathbf{q}_n\rangle \tag{262a}$$

$$N_- \mid \mathbf{p}_1, \cdots \mathbf{p}_m; \mathbf{q}_1, \cdots, \mathbf{q}_n\rangle = n \mid \mathbf{p}_1, \cdots \mathbf{p}_m; \mathbf{q}_1, \cdots, \mathbf{q}_n\rangle \tag{262b}$$

The completeness relation for the basis vectors is

$$\begin{aligned} I = |0\rangle\langle 0| &+ \int \frac{d^3p}{p_0} |\mathbf{p};\rangle\langle\mathbf{p};| + \int \frac{d^3q}{q_0} |;\mathbf{q}\rangle\langle;\mathbf{q}| \\ &+ \cdots + \int \frac{d^3p_1}{p_{10}} \cdots \int \frac{d^3p_m}{p_{m0}} \int \frac{d^3q_1}{q_{10}} \cdots \int \frac{d^3q_n}{q_{n0}} \\ &|\mathbf{p}_1, \cdots \mathbf{p}_m; \mathbf{q}_1, \cdots \mathbf{q}_n\rangle\langle\mathbf{p}_1, \cdots \mathbf{p}_m; \mathbf{q}_1, \cdots \mathbf{q}_n| + \cdots \end{aligned} \tag{263}$$

The expansion of an arbitrary vector $|\Phi\rangle$ in terms of the basis vectors is obtained by applying both sides of this completeness relation on the vector $|\Phi\rangle$. The Fock space amplitude $\langle\mathbf{p}_1, \cdots \mathbf{p}_m; \mathbf{q}_1, \cdots \mathbf{q}_n \mid \Phi\rangle$ now corresponds to the probability of finding m positive mesons with momenta $\mathbf{p}_1, \cdots \mathbf{p}_m$ and n negative mesons with momenta $\mathbf{q}_1, \cdots \mathbf{q}_n$. The Fock space representations of the operators a, a^*, b and b^* are trivial generalizations of the ones previously determined for the neutral field and will not be repeated here.

There exists in the theory an observable corresponding to the total charge of the system, which is represented by the Hermitian operator

$$\begin{aligned} Q &= e(N_+ - N_-) = Q^* \\ &= e \int_+ \frac{d^3k}{k_0} (a_{\mathbf{k}}^* a_{\mathbf{k}} - b_{\mathbf{k}}^* b_{\mathbf{k}}) \end{aligned} \tag{264}$$

Its commutation rules with the a and b operators are:

$$[Q, a_{\mathbf{k}}] = -ea_{\mathbf{k}} \qquad [Q, a_{\mathbf{k}}^*] = ea_{\mathbf{k}}^* \tag{265a}$$

$$[Q, b_{\mathbf{k}}] = eb_{\mathbf{k}} \qquad [Q, b_{\mathbf{k}}^*] = -eb_{\mathbf{k}}^* \tag{265b}$$

The operators corresponding to the total energy and total momenta are now given by

$$H = \int_+ \frac{d^3k}{k_0} \omega_{\mathbf{k}}(a_{\mathbf{k}}^* a_{\mathbf{k}} + b_{\mathbf{k}}^* b_{\mathbf{k}}) \tag{266}$$

$$\mathbf{P} = \int_+ \frac{d^3k}{k_0} \mathbf{k}(a_{\mathbf{k}}^* a_{\mathbf{k}} + b_{\mathbf{k}}^* b_{\mathbf{k}}) \tag{267}$$

or, more succinctly,

$$P_\mu = \int_+ \frac{d^3k}{k_0} k_\mu(a_{\mathbf{k}}^* a_{\mathbf{k}} + b_{\mathbf{k}}^* b_{\mathbf{k}}) \tag{268}$$

We next introduce the operators

$$\phi^{(+)}(x) = \frac{1}{\sqrt{2(2\pi)^3}} \int_+ \frac{d^3k}{k_0} b_{\mathbf{k}} e^{-ik\cdot x} \tag{269a}$$

$$\phi^{(-)}(x) = \frac{1}{\sqrt{2(2\pi)^3}} \int_+ \frac{d^3k}{k_0} a_{\mathbf{k}}^* e^{ik\cdot x} \tag{269b}$$

$$[\phi^{(+)}(x)]^* = \frac{1}{\sqrt{2(2\pi)^3}} \int_+ \frac{d^3k}{k_0} b_{\mathbf{k}}^* e^{ik\cdot x} \tag{269c}$$

$$[\phi^{(-)}(x)]^* = \frac{1}{\sqrt{2(2\pi)^3}} \int_+ \frac{d^3k}{k_0} a_{\mathbf{k}} e^{-ik\cdot x} \tag{269d}$$

which, since $k_0 = +\sqrt{\mathbf{k}^2 + \mu^2}$, all satisfy the Klein-Gordon equation. We next combine these operators to define the field operators:

$$\phi(x) = \phi^{(+)}(x) + \phi^{(-)}(x) = \frac{1}{\sqrt{2(2\pi)^3}} \int_+ \frac{d^3k}{k_0} (b_{\mathbf{k}} e^{-ik\cdot x} + a_{\mathbf{k}}^* e^{+ik\cdot x}) \tag{270}$$

and

$$\phi^*(x) = [\phi^{(+)}(x) + \phi^{(-)}(x)]^* = \frac{1}{\sqrt{2(2\pi)^3}} \int_+ \frac{d^3k}{k_0} (b_{\mathbf{k}}^* e^{ik\cdot x} + a_{\mathbf{k}} e^{-ik\cdot x}) \tag{271}$$

which obey the equations of motion

$$(\square + \mu^2)\, \phi(x) = 0 \tag{272a}$$

and

$$(\square + \mu^2)\, \phi^*(x) = 0 \tag{272b}$$

It should be noted that the operator $\phi(x)$ is *not* Hermitian. Using the previously established properties of $\Delta^{(\pm)}$ and the fact that ϕ, ϕ^* obey the Klein-Gordon equation, we can recover the positive and negative frequency parts of the operators ϕ, ϕ^* by the operation

$$\phi^{(\pm)}(x) = \int_\sigma d\sigma^\mu(x') \, \{\partial'_\mu \Delta^{(\pm)}(x - x') \cdot \phi(x') - \Delta^{(\pm)}(x - x')\, \partial_\mu{}' \phi(x')\} \tag{273a}$$

and

$$\phi^{*(\pm)}(x) = \int_\sigma d\sigma^\mu(x')\, \{\partial_\mu{}'\Delta^{(\pm)}(x - x') \cdot \phi^*(x') - \Delta^{(\pm)}(x - x')\, \partial_\mu{}'\phi^*(x')\} \tag{273b}$$

Comparing with Eqs. (269a) and (269b), we deduce that

$$\phi^{*(+)}(x) = [\phi^{(-)}(x)]^* \tag{274a}$$

$$\phi^{*(-)}(x) = [\phi^{(+)}(x)]^* \tag{274b}$$

The configuration space basis vectors are

$$|x_1, x_2, \cdots x_m; y_1, \cdots y_n\rangle$$
$$= \frac{1}{\sqrt{n!\, m!}} \phi^{(-)}(x_1) \cdots \phi^{(-)}(x_m)\, \phi^{*(-)}(y_1) \cdots \phi^{*(-)}(y_n) \mid 0\rangle \tag{275}$$

and $\langle x_1, \cdots x_m; y_1, \cdots y_n \mid \Phi\rangle$, if $x_{10} = \cdots = x_{m0} = y_{10} = \cdots = y_{n0} = \tau$, corresponds to the probability amplitude for finding m positive mesons and n negative ones at time τ.

The operators ϕ and ϕ^* were combined from the a and b operators so that they have simple commutation rules with the total charge operator.[5] In view of Eqs. (270) and (265), $\phi(x)$ has the following commutation rules with Q

$$[Q, \phi(x)] = +e\phi(x) \tag{276}$$

so that $\phi(x)$ creates an amount of charge e. *Proof*: If $|Q'\rangle$ is an eigenstate of the total charge operator Q, with eigenvalue Q', then $\phi(x) \mid Q'\rangle$ is an eigenfunction of Q with eigenvalue $Q' + e$:

$$Q\{\phi(x) \mid Q'\rangle\} = \phi(x)\, Q \mid Q'\rangle + [Q, \phi(x)] \mid Q'\rangle$$
$$= (Q' + e)\, \{\phi(x) \mid Q'\rangle\} \tag{277}$$

Similarly, since

$$[Q, \phi^*(x)] = -e\phi^*(x) \tag{278}$$

$\phi^*(x)$ is a destruction operator for an amount of charge e:

$$Q\phi^*(x) \mid Q'\rangle = \phi^*(x)\, Q \mid Q'\rangle + [Q, \phi^*(x)] \mid Q'\rangle$$
$$= (Q' - e)\, \phi^*(x) \mid Q'\rangle \tag{279}$$

An operator which destroys an amount of charge e must clearly be built up from the operators a and b^* as shown by Eq. (270); for we can reduce the total charge of the system either by destroying a charge $+e$, or by adding a charge $-e$ to it.

The commutation rules of the operators are easily computed using the representation (270)–(271) and the commutation rules given by Eqs. (256)–(258). One finds:

[5] The operator $\phi(x) = \alpha\phi^{(+)}(x) + \beta\phi^{(-)}(x)$ would, of course, obey the same commutation rules with Q as ϕ defined by Eq. (270). The invariance of the commutation rules for the operators $\phi(x)$ and $\phi^*(x)$ under Lorentz transformation would then further require that $|\alpha|^2 = |\beta|^2 = 1$ [e.g., see Eqs. (314) and (324). The particular choice of phases corresponding to Eq. (270) will be discussed later in this section.

$$[\phi(x), \phi^*(x')] = \frac{1}{2(2\pi)^3} \int \frac{d^3k}{k_0} (e^{-ik\cdot(x-x')} - e^{ik\cdot(x-x')}) = i\Delta(x - x') \tag{280a}$$

$$[\phi^*(x), \phi(x')] = i\Delta(x - x') \tag{280b}$$

$$[\phi(x), \phi(x')] = [\phi^*(x), \phi^*(x')] = 0 \tag{280c}$$

We can also express the observables N_+, N_- in terms of the operators $\phi^{(\pm)}$, $\phi^{*(\pm)}$ as follows:

$$N_- = i \int_\sigma d\sigma^\mu(x) \{\phi^{*(-)}(x)\, \partial_\mu\phi^{(+)}(x) - \partial_\mu\phi^{*(-)}(x) \cdot \phi^{(+)}(x)\} \tag{281a}$$

$$N_+ = i \int_\sigma d\sigma^\mu(x) \{\phi^{(-)}(x)\, \partial_\mu\phi^{*(+)}(x) - \partial_\mu\phi^{(-)}(x) \cdot \phi^{*(+)}(x)\} \tag{281b}$$

so that the total charge can be rewritten in the form

$$Q = e(N_+ - N_-) \tag{282a}$$

$$= -ie : \int_\sigma d\sigma^\mu(x)\, (\phi^*(x)\, \partial_\mu\phi(x) - \partial_\mu\phi^*(x) \cdot \phi(x)) : \tag{282b}$$

The total charge Q is conserved, since

$$\delta Q/\delta\sigma(x) = 0 \tag{283}$$

To verify that the expression (282b) indeed reduces to the form (282a) with N_+ and N_- expressed as in Eq. (281), we note that the normal product expansion of Eq. (282b), using the rules (194)–(195), is given by

$$\begin{aligned} Q = -ie \int d\sigma^\mu(x)\, \{&(\phi^{*(-)}(x)\, \partial_\mu\phi^{(+)}(x) - \partial_\mu\phi^{*(-)}(x) \cdot \phi^{(+)}(x)) \\ &- (\phi^{(-)}(x)\, \partial_\mu\phi^{*(+)}(x) - \partial_\mu\phi^{(-)}(x) \cdot \phi^{*(+)}(x)) \\ &+ (\phi^{*(-)}(x)\, \partial_\mu\phi^{(-)}(x) - \partial_\mu\phi^{*(-)}(x) \cdot \phi^{(-)}(x)) \\ &+ (\phi^{*(+)}(x)\, \partial_\mu\phi^{(+)}(x) - \partial_\mu\phi^{*(+)}(x)\, \phi^{(+)}(x))\} \end{aligned} \tag{284}$$

The first two terms correspond to $e(N_+ - N_-)$. We next show that the third and fourth terms vanish when integrated over σ, due to the orthogonality of positive and negative energy solutions of the Klein-Gordon equation. Consider the fourth term, for example. Since $\phi^{*(+)}$ and $\phi^{(+)}$ satisfy the Klein-Gordon equation, the surface σ over which the integration is carried out can be chosen the plane x^0 = constant. Substituting for $\phi^{*(+)}$ and $\phi^{(+)}$ their representation (269d) and (269a), we find

$$\begin{aligned} i \int d\sigma^\mu(x)\, &(\phi^{*(+)}(x)\, \partial_\mu\phi^{(+)}(x) - \partial_\mu\phi^{*(+)}(x) \cdot \phi^{(+)}(x)) \\ &= \int d^3x \int_+ \frac{d^3k}{k_0} \int_+ \frac{d^3k'}{k'_0}\, b_{\mathbf{k}} a_{\mathbf{k}'} e^{-i(k_0+k_0')x_0}\, e^{+i(\mathbf{k}+\mathbf{k}')\cdot\mathbf{x}} \cdot (k'_0 - k_0) \\ &= 0 \end{aligned} \tag{285}$$

since the integration over $\mathbf{x}$ yields a factor $\delta(\mathbf{k} + \mathbf{k}')$ which makes $k_0 = k_0'$; and similarly for the third term.

The expression (282b) for the total charge suggests that we can define the current operator, $j_\mu(x)$, as

$$j_\mu(x) = -ie : \phi^*(x)\, \partial_\mu\phi(x) - \partial_\mu\phi^*(x) \cdot \phi(x) : \tag{286}$$

It is conserved, i.e., $\partial_\mu j^\mu(x) = 0$, by virtue of the equations of motions (272) and the total charge

$$Q = \int_\sigma j^\mu(x)\, d\sigma_\mu(x) \tag{287}$$

is therefore time-independent. This current operator, as noted above, can be analyzed into two parts: one part, which we call the "normal" part, which is diagonal in the $N_\pm$ representation, and a second part which we call the "fluctuation" part. This fluctuation part consists of terms which are the product of two creation or annihilation operators. Although the presence of such pair creation and annihilation terms appears surprising at first, Bohr and Rosenfeld (1950) have shown that these terms are essential to make $\int_V j_0(x)\, d^3x = Q_V$, the charge in volume V, correspond to what one actually would measure in the laboratory with classical charge and current distributions.

It is instructive to compute the expectation value of the charge density, $j_0(x)$, in the state

$$\begin{aligned} |y;\rangle &= \frac{1}{(2\pi)^{3/2}} \int \frac{d^3k}{k_0}\, a_{\mathbf{k}}{}^* e^{-ik\cdot y} \sqrt{k_0}\, |\, 0\rangle \\ &= (\psi_y(x), \phi^{(-)}(x))\, |\, 0\rangle \end{aligned} \tag{288}$$

corresponding to a positive meson localized at the point $\mathbf{y}$ at time y^0. Only the normal part of $j_0(x)$ will contribute, since the fluctuation part operating on the state $|y;\rangle$ will either annihilate it (the terms in $\phi^{*(+)}\phi^{(+)}$) or transform it into a three-particle state (the terms in $\phi^{*(-)}\phi^{(-)}$) which is orthogonal to the one-meson state $|y;\rangle$. Thus

$$\begin{aligned} \langle y;\,|\, j_0(x)\,|\, y;\rangle &= ie\,\langle y;\,|\, \phi^{(-)}(x)\, \partial_0\phi^{*(+)}(x) - \partial_0\phi^{(-)}(x) \cdot \phi^{*(+)}(x)\,|\, y;\rangle \\ &= e\left\{\bar{\psi}_y(x)\, \frac{\partial}{\partial x^0}\, \psi_y(x) - \frac{\partial}{\partial x^0}\, \bar{\psi}_y(x) \cdot \psi_y(x)\right\} \end{aligned} \tag{289}$$

which does not vanish when $\mathbf{x} \neq \mathbf{y}$. Our particles are therefore not point charges, but have a charge distribution of the order of the Compton wave length, $\hbar/\mu c$.

To obtain the connection with the field quantization procedure previously outlined for the case of the neutral field, we note that the commutation rules (280) and the property (168) of the Δ function imply that

$$[\partial_0\phi(x), \phi^*(x')]_{x_0 = x_0'} = -i\delta^{(3)}(\mathbf{x} - \mathbf{x}') \tag{290a}$$

$$[\partial_0\phi^*(x), \phi(x')]_{x_0 = x_0'} = -i\delta^{(3)}(\mathbf{x} - \mathbf{x}') \tag{290b}$$

We therefore define the momentum canonically conjugate to ϕ to be

$$\pi(x) = \partial_0\phi^*(x) \tag{291}$$

and the momentum canonically conjugate to ϕ^* as

$$\pi^*(x) = \partial_0\phi(x) \tag{292}$$

Proceeding in complete analogy with the neutral case, one verifies that the Hamiltonian can be expressed in the form

$$H = \int d^3x : \pi^*(x)\,\pi(x) + \nabla\phi^*(x)\cdot\nabla\phi(x) + \mu^2\phi^*(x)\,\phi(x) : \tag{293}$$

which can be derived from a Lagrangian density

$$\mathcal{L} = :\partial_\lambda\phi^*(x)\cdot\partial^\lambda\phi(x) - \mu^2\phi^*(x)\,\phi(x): \tag{294}$$

The expression for the energy-momentum tensor is now

$$T_{\mu\nu} = :\partial_\mu\phi^*(x)\,\partial_\nu\phi(x) + \partial_\nu\phi^*(x)\,\partial_\mu\phi(x): - g_{\mu\nu}\mathcal{L} \tag{295}$$

and that for the charge density is

$$\rho(x) = j_0(x) = -ie : \phi^*(x)\,\pi^*(x) - \phi(x)\,\pi(x) : \tag{296}$$

We also note that the current operator can be written in the form

$$j^\mu(x) = +ie : \frac{\partial\mathcal{L}}{\partial\phi_\mu(x)}\,\phi(x) - \frac{\partial\mathcal{L}}{\partial\phi_\mu{}^*(x)}\,\phi^*(x) : \tag{297}$$

We shall see in Section 7g that the existence of the current j^μ is a consequence of the invariance of the Lagrangian under gauge transformations wherein

$$\phi \longrightarrow e^{i\alpha}\,\phi \tag{298a}$$

$$\phi^* \longrightarrow e^{-i\alpha}\,\phi^* \tag{298b}$$

We will also show that this four-vector indeed plays the role of the charge-current in the interaction of the charged scalar field with the electromagnetic field. The form (297) incidentally indicates that if the field is Hermitian $\phi^* = \phi$, then j^μ vanishes identically.

The classical Lagrangian for the complex field ϕ which upon quantization yields the theory we have outlined in its second-quantized form above, is given by

$$\mathcal{L} = \partial_\lambda\bar{\phi}\partial^\lambda\phi - \mu^2\bar{\phi}\phi \tag{299}$$

The field amplitude ϕ, and its complex conjugate $\bar{\phi}$ are to be considered and treated as independent classical field variables. The variation with respect to ϕ yields the equation of motion for $\bar{\phi}$

$$\frac{\delta\mathcal{L}}{\delta\phi} = \frac{\partial\mathcal{L}}{\partial\phi} - \frac{\partial}{\partial x^\mu}\frac{\partial\mathcal{L}}{\partial\phi_\mu} = (\Box + \mu^2)\,\bar{\phi}(x) = 0 \tag{300a}$$

and the variation with respect to $\bar{\phi}$ the field equation for ϕ

$$\frac{\delta\mathcal{L}}{\delta\bar{\phi}} = (\Box + \mu^2)\,\phi(x) = 0 \tag{300b}$$

The quantization procedure outlined previously, however, was formulated in terms of *real* fields which were expanded into a complete *real* orthonormal set of functions. In order to apply this same procedure in the present situation, we decompose ϕ into a real and imaginary part:

$$\phi(x) = \frac{1}{\sqrt{2}} (\phi^{(1)}(x) + i\phi^{(2)}(x)) \tag{301a}$$

$$\bar{\phi}(x) = \frac{1}{\sqrt{2}} (\phi^{(1)}(x) - i\phi^{(2)}(x)) \tag{301b}$$

where $\phi^{(1)}$ and $\phi^{(2)}$ are real, and we consider the latter as the dynamical variables. The classical Lagrangian in terms of $\phi^{(1)}$ and $\phi^{(2)}$ has the form

$$\mathfrak{L} = \tfrac{1}{2} \sum_{j=1}^{2} (\partial_\rho\phi^{(j)}\partial^\rho\phi^{(j)} - \mu^2\phi^{(j)}\phi^{(j)}) \tag{302}$$

corresponding to two noninteracting scalar fields. The momenta canonically conjugate to the $\phi^{(i)}$ are

$$\frac{\partial\mathfrak{L}}{\partial{\phi_0}^{(j)}} = \pi^{(j)}(x) = {\phi^{(j)}}_0(x) \tag{303}$$

and the Hamiltonian density is

$$\begin{aligned} \mathfrak{H} = \mathfrak{H}(\nabla\phi^{(j)}, \phi^{(j)}, \pi^{(j)}) &= \sum_{j=1}^{2} \pi^{(i)}\phi^{(i)} - \mathfrak{L} \\ &= \tfrac{1}{2} \sum_{j=1}^{2} (\pi^{(j)2}(x) + \nabla\phi^{(j)}(x) \cdot \nabla\phi^{(j)}(x) + \mu^2\phi^{(j)2}(x)) \end{aligned} \tag{304}$$

The expansion in terms of the complete orthonormal set of real functions, $\varphi_n(\mathbf{x})$ defined by (207a) and (207b), now is

$$\phi^{(i)}(x) = \sum_n {q^{(i)}}_n(t)\, \varphi_n(\mathbf{x}); \quad i = 1, 2 \tag{305}$$

and $L = \int d^3x\, \mathfrak{L}$ can be considered a function of the ${q^{(i)}}_n$ and ${\dot{q}^{(i)}}_n$. The momenta conjugate to the ${q^{(i)}}_n$ are

$$\begin{aligned} {p^{(i)}}_n(t) = \frac{\partial L}{\partial {q^{(i)}}_n(t)} &= \int d^3x\, \frac{\partial\mathfrak{L}}{\partial{\phi_0}^{(i)}(x)}\, \varphi_n(\mathbf{x})\, d^3x \\ &= \int d^3x\, \pi^{(i)}(x)\, \varphi_n(\mathbf{x}) \end{aligned} \tag{306}$$

Finally, the quantum conditions are expressed in terms of the commutation rules:

$$[{q^{(l)}}_n(t), {p^{(j)}}_m(t)] = i\delta_{lj}\delta_{nm} \tag{307a}$$

$$[{q^{(l)}}_n(t), {q^{(j)}}_m(t)] = [{p^{(l)}}_n(t), {p^{(j)}}_m(t)] = 0 \tag{307b}$$

where the ${p^{(l)}}_n$ and ${q^{(l)}}_n$ are now Hermitian operators. The equal-time commutation rules of the (Hermitian) operators $\pi^{(l)}$ and $\phi^{(l)}$ are then

$$[\pi^{(l)}(x), \phi^{(j)}(x')]_{x_0=x'_0} = -i\delta_{lj}\delta^{(3)}(\mathbf{x} - \mathbf{x}') \tag{308}$$

with all other equal-time commutators vanishing. In terms of the (non-Hermitian) operators π, π^*, ϕ and ϕ^* defined by

$$\phi(x) = \frac{1}{\sqrt{2}}\left(\phi^{(1)}(x) + i\phi^{(2)}(x)\right) \tag{309a}$$

$$\phi^*(x) = \frac{1}{\sqrt{2}}\left(\phi^{(1)}(x) - i\phi^{(2)}(x)\right) \tag{309b}$$

$$\begin{aligned}\pi(x) &= \frac{\partial\mathfrak{L}}{\partial\phi_0} = \frac{\partial\mathfrak{L}}{\partial\phi^{(1)}{}_0} - i\,\frac{\partial\mathfrak{L}}{\partial\phi^{(2)}{}_0} = \frac{1}{\sqrt{2}}\left(\phi^{(1)}{}_0(x) - i\phi^{(2)}{}_0(x)\right)\\ &= \frac{1}{\sqrt{2}}\left(\pi^{(1)}(x) - i\pi^{(2)}(x)\right)\end{aligned} \tag{309c}$$

$$\begin{aligned}\pi^*(x) &= \frac{\partial\mathfrak{L}}{\partial\phi^*{}_0} = \frac{\partial\mathfrak{L}}{\partial\phi^{(1)}{}_0} + i\,\frac{\partial\mathfrak{L}}{\partial\phi^{(2)}{}_0} = \frac{1}{\sqrt{2}}\left(\phi^{(1)}{}_0(x) + i\phi^{(2)}{}_0(x)\right)\\ &= \frac{1}{\sqrt{2}}\left(\pi^{(1)}(x) + i\pi^{(2)}(x)\right)\end{aligned} \tag{309d}$$

the commutation rules become:

$$[\phi(x), \pi(x')]_{x_0=x'_0} = i\delta^{(3)}(\mathbf{x} - \mathbf{x}') \tag{310a}$$

$$[\phi^*(x), \pi^*(x')]_{x_0=x'_0} = i\delta^{(3)}(\mathbf{x} - \mathbf{x}') \tag{310b}$$

in agreement with our previously established commutation rules.

The transformation laws of the field operators under proper Lorentz transformations are:

$$U(a, \Lambda)\,\phi(x)\,U(a, \Lambda)^{-1} = \phi(\Lambda x + a) \tag{311a}$$

$$U(a, \Lambda)\,\phi^*(x)\,U(a, \Lambda)^{-1} = \phi^*(\Lambda x + a) \tag{311b}$$

Under the operation of space inversion, we require that particles be transformed into *particles* and the sign of their momenta be changed in accordance with the classical interpretation of a spatial inversion. Therefore, under inversion $\phi(x)$ and $\phi^*(x)$ are transformed according to the rule:

$$U(i_s)\,\phi(x)\,U(i_s)^{-1} = \eta_P\phi(i_s x) = \eta_P\phi(x_0, -\mathbf{x}) \tag{312a}$$

$$U(i_s)\,\phi^*(x)\,U(i_s)^{-1} = \bar{\eta}_P\phi^*(i_s x) = \bar{\eta}_P\phi^*(x_0, -\mathbf{x}) \tag{312b}$$

where $U(i_s)$ is linear and unitary, and η_P is a phase factor. It will be noted that the operation $U(i_s)$ maps the positive (negative) frequency parts of field operators into positive (negative) frequency parts (i.e., it maps $\phi^{(\pm)}(x)$ into $\eta_P\phi^{(\pm)}(i_s x)$, as is verified by operating with the projection operator $\int d\sigma^\mu(x)\,\Delta^{(\pm)}(x' - x)\,\overleftrightarrow{\partial}_\mu$ on both sides of (312)). Now the vacuum can be characterized by the equation $\phi^{(+)}(x)\,|\,0\rangle = \phi^{*(+)}(x)\,|\,0\rangle = 0$; hence it follows from the aforementioned property of $U(i_s)$ that it is possible to postulate that the vacuum is invariant under $U(i_s)$,

$$U(i_s) \mid 0\rangle = |0\rangle \tag{313}$$

which makes the definition of $U(i_s)$ unique. Note further that, in order that the Lagrangian be invariant, i.e., that $U(i_s)\, \mathcal{L}(x)\, U(i_s)^{-1} = \mathcal{L}(i_s x)$, $|\eta_P|^2$ must be equal to $+1$. Alternatively, if the parity operation is to map the operator algebra ϕ, ϕ^* (which is defined by the field equations and the commutation rules) into itself, it has to preserve the defining relations, i.e., field equations and commutation rules. For the latter to remain invariant we must have $|\eta_P|^2 = +1$, since

$$\begin{aligned} i\Delta(x - x') &= U(i_s)\, [\phi(x), \phi^*(x')]\, U(i_s)^{-1} = |\eta_P|^2\, [\phi(i_s x), \phi^*(i_s x')] \\ &= i|\eta_P|^2\, \Delta(i_s(x - x')) = i|\eta_P|^2\, \Delta(x - x') \end{aligned} \tag{314}$$

We shall say that the theory is invariant under space inversion if the phase factor η_P can be so chosen that the operation of inversion as defined with these phases commutes with the Hamiltonian

$$U(i_s)\, \mathcal{H}(x)\, U(i_s)^{-1} = \mathcal{H}(i_s x) \tag{315}$$

and preserves the defining relations, i.e., the commutation rules. This is equivalent to the statement that there exists no prediction of the theory which enables one to distinguish between left and right. For the theory under consideration where the degeneracy of the one-particle states is completely specified by the spin (which is zero) and the particle-antiparticle character ($+$ or $-$ charge), we may choose $\eta_P = \pm 1$. The phase $\eta_P = +1$ corresponds to the case that the particles are "scalar," the phase $\eta_P = -1$ to the case that they are "pseudoscalar." We next show that a characteristic of Bose fields is that particles and antiparticles have the same parity. To demonstrate this we introduce a decomposition of the field operators in terms of a complete set of one-particle eigenstates of the energy and total angular momentum [Sachs (1952)]. That is, we expand $\phi(x)$ in terms of the spherical solution $\varphi_s(\mathbf{x})$ of $(\nabla^2 + \mathbf{k}^2)\, \varphi_s = 0$ corresponding to a definite parity and angular momentum:

$$\varphi_s(r) = j_l(|\mathbf{k}|r)\, Y_l^m(\theta, \phi) \quad (r = |\mathbf{x}|) \tag{316}$$

where the Y_l^m are spherical harmonics so chosen that $\overline{Y_l^m} = (-1)^m\, Y_l^{-m}$ and $j_l(|\mathbf{k}|r)$ are the spherical Bessel functions which are regular everywhere. The φ_s are assumed to vanish on the surface of a very large sphere of radius R (i.e., such that $|\mathbf{k}|R \gg 1$), a condition which leads to discrete values for $|\mathbf{k}|$. Furthermore, the j_l are assumed to be normalized within this volume with $\frac{2}{R}\int_0^R (|\mathbf{k}|r)^2\, j_l^2(|\mathbf{k}|r)\, dr = 1$. The decomposition of $\phi(x)$ is then given by

$$\begin{aligned} \phi(x) = \sum_{|\mathbf{k}|,\, l,\, m} &\{a(|\mathbf{k}|, l, m)\, Y_l^m(\theta, \phi)\, e^{-i\omega_{\mathbf{k}}t} \\ &+ b^*(|\mathbf{k}|, l, m)\, \overline{Y_l^m}(\theta, \phi)\, e^{i\omega_{\mathbf{k}}t}\}\, |\mathbf{k}| \sqrt{\frac{2}{R}}\, j_l(|\mathbf{k}|r) \end{aligned} \tag{317}$$

The operator $a^*(|\mathbf{k}|, l, m)$ $(b^*(|\mathbf{k}|, l, m))$ is then a creation operator for a positively (negatively) charged particle with energy $\omega_{\mathbf{k}} = \sqrt{\mathbf{k}^2 + \mu^2}$, a total (orbital) angular momentum of $l(l+1)\hbar^2$ and a 3-component of angular momentum of $m\hbar$. One verifies that the Hamiltonian operator H and the total angular momentum operator $\mathbf{M} = \int d^3x\, \mathbf{x} \times \mathbf{G}$ ($\mathbf{G}$ is the momentum density operator $G_i = T_{0i}$) can be simply expressed in terms of $a(|\mathbf{k}|, l, m)$, $b(|\mathbf{k}|, l, m)$ and their adjoints, and that in fact H, $\mathbf{M}^2$ and M_3 are diagonal in the representation in which $N^+_{|\mathbf{k}|lm} = a^*_{|\mathbf{k}|lm}a_{|\mathbf{k}|lm}$ and $N^-_{|\mathbf{k}|lm} = b^*_{|\mathbf{k}|lm}b_{|\mathbf{k}|lm}$, the number operator for particles in the state $|\mathbf{k}|lm$, are diagonal. The relation between creation operators for states of definite linear momentum and those for states of definite angular momentum is obtained by comparing (269) and (317), e.g.,

$$a^*(\mathbf{k}) = \frac{\omega_{\mathbf{k}}}{|\mathbf{k}|}\sqrt{\frac{2R}{\pi}} \sum_{l=0}^{\infty} \sum_{m=-l}^{l} i^l \bar{Y}_l^m(\mathbf{k})\, a^*(|\mathbf{k}|, l, m) \tag{318}$$

Now under inversions

$$\begin{aligned} U(i_s)\,\phi(x)\,U(i_s)^{-1} &= \sum_{|\mathbf{k}|lm} \{U(i_s)\,a(|\mathbf{k}|, l, m)\,U(i_s)^{-1}\,Y_l^m(\theta, \phi)\,e^{-i\omega_{\mathbf{k}}t} \\ &\quad + U(i_s)\,b^*(|\mathbf{k}|, l, m)\,U(i_s)^{-1}\,\bar{Y}_l^m(\theta, \phi)\,e^{i\omega_{\mathbf{k}}t}\}\,|\mathbf{k}|\sqrt{\frac{2}{R}}\,j_l(|\mathbf{k}|r) \\ &= \eta_P\phi(i_sx) = \eta_P \sum_{|\mathbf{k}|lm} \{a(|\mathbf{k}|, l, m)\,(-1)^l\,Y_l^m(\theta, \phi)\,e^{-i\omega_{\mathbf{k}}t} \\ &\quad + b^*(|\mathbf{k}|, l, m)\,(-1)^l\,\bar{Y}_l^m(\theta, \phi)\,e^{-i\omega_{\mathbf{k}}t}\}\,|\mathbf{k}|\sqrt{\frac{2}{R}}\,j_l(|\mathbf{k}|r) \end{aligned} \tag{319}$$

so that using the orthonormality of the Y_l^ms and $j_l(|\mathbf{k}|r)$s we obtain

$$U(i_s)\,a_{|\mathbf{k}|lm}\,U(i_s)^{-1} = \eta_P(-1)^l\,a_{|\mathbf{k}|lm} \tag{320a}$$

$$U(i_s)\,b_{|\mathbf{k}|lm}\,U(i_s)^{-1} = \eta_P(-1)^l\,b_{|\mathbf{k}|lm} \tag{320b}$$

Thus the reflection properties of the one-particle state $a^*(|\mathbf{k}|, l, m)\,|\,0\rangle$ associated with a pseudoscalar field ($\eta_P = -1$) are such that the inverted state contains an extra minus sign compared to that expected from its orbital motion alone, implying a negative parity. The one antiparticle state $b^*(|\mathbf{k}|lm)\,|\,0\rangle$ likewise obtains the factor $\eta_P(-1)^l$ under inversions, so that particle and antiparticle have the same "intrinsic" parity.

The operation of (weak or Wigner) time reversal [in the nomenclature of Lüders (1954, 1957) operation of motion reversal] is induced by the antiunitary operator $U(i_t)$ with

$$U(i_t)\,\phi(x)\,U(i_t)^{-1} = \eta_T\phi(i_tx) \tag{321a}$$

$$U(i_t)\,\phi^*(x)\,U(i_t)^{-1} = \bar{\eta}_T\phi(i_tx) \tag{321b}$$

$$U(i_t)^2 = 1 \tag{321c}$$

The operator $U(i_t)$ must be antiunitary in order to preserve the invariance of the commutation rules. For assume that time inversion is induced by a unitary operator $\mathcal{U}(i_t)$ with the above properties. Under this operation the commutation rules would then transform as follows:

$$\begin{aligned} i\Delta(x - x') &= \mathcal{U}(i_t)\,[\phi(x), \phi^*(x')]\,\mathcal{U}(i_t)^{-1} \\ &= |\eta_T|^2\,[\phi(i_tx), \phi^*(i_tx')] = i\Delta(i_t(x - x'))\,|\eta_T|^2 \\ &= -i\,|\eta_T|^2\,\Delta(x - x') \end{aligned} \tag{322}$$

which leads to a contradiction, since the first and third line of Eq. (322) cannot be simultaneously satisfied. On the other hand, if $U(i_t)$ is antiunitary

$$U(i_t)\,i\,U(i_t)^{-1} = -i \tag{323}$$

so that

$$\begin{aligned} U(i_t)\,[\phi(x), \phi^*(x')]\,U(i_t)^{-1} &= -i\bar{\Delta}(x - x') = -i\Delta(x - x') \\ &= -i\,|\eta_T|^2\,\Delta(x - x') \end{aligned} \tag{324}$$

and the commutation rules will be invariant if $|\eta_T|^2$ is chosen to equal $+1$. One verifies that with this choice the free-field Lagrangian is likewise invariant; similarly $U(i_t)\,H\,U(i_t)^{-1} = H$, so that the theory is invariant under time inversion. Note, however, that

$$U(i_t)\,\mathbf{P}\,U(i_t)^{-1} = -\mathbf{P} \tag{325}$$

$$U(i_t)\,\mathbf{M}\,U(i_t)^{-1} = -\mathbf{M} \tag{326}$$

We next define the linear, unitary operation of particle-antiparticle conjugation U_c by the property that

$$U_c\,|\,\mathbf{p}_1, \mathbf{p}_2, \cdots \mathbf{p}_m; \mathbf{q}_1, \cdots \mathbf{q}_n\rangle = \eta(n, m)\,|\,\mathbf{q}_1, \cdots \mathbf{q}_n; \mathbf{p}_1, \cdots \mathbf{p}_m\rangle \tag{327}$$

$$U_c\,|\,0\rangle = |0\rangle \tag{328}$$

where $\eta = \eta(n, m)$ is a phase factor. The operator U_c changes particles into antiparticles without changing their momenta, and leaves the vacuum unchanged. In particular, operating on a one-particle state

$$\begin{aligned} U_c a_{\mathbf{p}_1}^*\,|\,0\rangle &= U_c\,|\,\mathbf{p}_1;\,\rangle = U_c a_{\mathbf{p}_1}^*\,U_c^{-1}\,|\,0\rangle \\ &= \eta_C\,|\,;\mathbf{p}_1\rangle = \eta_C b_{\mathbf{p}_1}^*\,|\,0\rangle \end{aligned} \tag{329}$$

so that we may define

$$U_c a_{\mathbf{k}}^*\,U_c^{-1} = \eta_C b_{\mathbf{k}}^* \tag{330}$$

and

$$U_c a_{\mathbf{k}}\,U_c^{-1} = \bar{\eta}_C b_{\mathbf{k}} \tag{331}$$

where η_C, which satisfies $|\eta_C|^2 = +1$, is the charge conjugation parity of the particle. Alternatively, we can define the charge conjugate operator $\phi_c(x)$ by

$$\phi_c(x) = U_c\phi(x)\,U_c^{-1} = \eta_C\phi^*(x) \tag{332a}$$

and

$$\phi_c^*(x) = U_c\phi^*(x)\,U_c^{-1} = \bar{\eta}_C\phi(x) \tag{332b}$$

Equation (332) together with (328) can be considered as defining U_c. Because of our definition of $\mathcal{L}$ in terms of normal products, $\mathcal{L}$ is invariant under charge conjugation:

$$U_c \mathcal{L} U_c^{-1} = \mathcal{L} \tag{333}$$

The same holds true for the commutation rules, implying the invariance of the theory under charge conjugation. Note however that the total charge operator Q anticommutes with U_c:

$$U_c Q U_c^{-1} = -Q \tag{334}$$

In terms of the charge conjugation operation, a real field representing particles which are identical with their antiparticles, can be characterized by

$$\phi_c(x) = \phi^*(x) = \phi(x) \tag{335}$$

Such a relation between the operator ϕ and the charge conjugate operator ϕ_c guarantees the identity of the particle and antiparticle states. For such a field Eqs. (335) and (332) imply that $\bar{\eta}_C = \eta_C$ so that η_C must be real and equal to ± 1. A similar argument indicates that for self-charge conjugate fields, η_P and η_T are real and can only equal ± 1.

It is interesting to note that for a Bose field U_c and $U(i_s)$ commute since

$$\begin{aligned} U_c U(i_s)\, \phi(x)\, U(i_s)^{-1} U_c^{-1} &= U(i_s)\, U_c \phi(x)\, U_c^{-1} U(i_s)^{-1} \\ &= \eta_C \eta_P \phi^*(i_s x) \end{aligned} \tag{336}$$

This is true only for integer spin particles.

7g. Conservation Laws and Lagrangian Formalism

In our discussion of the charged scalar field we have already briefly indicated how the Lagrangian formalism is generalized to encompass fields with more than one component. Let us denote by $\phi_r(x)$ ($r = 1, 2, \cdots n$), an arbitrary, such multicomponent field. The change of the classical Lagrangian density, $\mathcal{L} = \mathcal{L}(\phi_r, \phi_{r\mu})$, due to variations with respect to ϕ_r is then given by

$$\delta\mathcal{L} = \sum_{r=1}^{n} \left\{ \frac{\partial \mathcal{L}}{\partial \phi_r} - \frac{\partial}{\partial x^\mu} \frac{\partial \mathcal{L}}{\partial \phi_{r\mu}} \right\} \delta\phi_r + \sum_{r=1}^{n} \frac{\partial}{\partial x^\mu} \left(\frac{\partial \mathcal{L}}{\partial \phi_{r,\mu}} \delta\phi_r \right) \tag{337}$$

The action principle which requires that $\delta I = \delta \int_\Omega \mathcal{L} d^4x = 0$ for arbitrary variations $\delta\phi_r$ which vanish on the boundary Σ of Ω implies that

$$\frac{\partial \mathcal{L}}{\partial \phi_r} - \frac{\partial}{\partial x^\mu} \frac{\partial \mathcal{L}}{\partial \phi_{r\mu}} = 0 \quad r = 1, 2, \cdots n \tag{338}$$

which are the field equations. For a set of fields which satisfy the equations of motion, the change in $\mathcal{L}$, $\delta\mathcal{L}$, due to variations $\delta\phi_r$ of ϕ_r is therefore given by

$$\delta\mathfrak{L} = \sum_{r=1}^{n} \frac{\partial}{\partial x^\mu}\left(\frac{\partial\mathfrak{L}}{\partial\phi_{r,\mu}}\,\delta\phi_r\right) \tag{339}$$

This equation enables us to derive conservation laws by considering particular variations of the ϕ_r [Pauli (1941), Hill (1951)].

Consider first the case that $\mathfrak{L}$ does not depend on the space-time coordinates explicitly and that the variation $\delta\phi_r$ is induced by an infinitesimal translation

$$x \to x' = x + \epsilon a \tag{340}$$

under which

$$\begin{aligned}\phi_r(x) \to \phi_r'(x') &= \phi_r(x + \epsilon a)\\ &= \phi_r(x) + \delta\phi_r(x)\end{aligned} \tag{341}$$

The variation $\delta\phi_r$ to first order in ϵ, is given by

$$\delta\phi_r(x) = \phi_r'(x') - \phi_r(x) = \epsilon\left(\frac{\partial\phi'_r}{\partial\epsilon}\right)_{\epsilon=0} + O(\epsilon^2) = \epsilon a^\mu\partial_\mu\phi_r(x) \tag{342}$$

Similarly the variation $\delta\mathfrak{L}$ of the Langrangian is given by

$$\begin{aligned}\delta\mathfrak{L} &= \mathfrak{L}(\phi'_r(x'), \phi'_{r\mu}(x')) - \mathfrak{L}(\phi_r(x), \phi_{r\mu}(x)) = \epsilon\left(\frac{\partial\mathfrak{L}(\phi'(x'))}{\partial\epsilon}\right)_{\epsilon=0}\\ &= \epsilon a^\mu\partial_\mu\mathfrak{L}(\phi_r(x), \phi_{r\mu}(x))\end{aligned} \tag{343}$$

Equation (339) therefore states that

$$\epsilon a^\mu\left\{\partial_\mu\mathfrak{L} - \sum_{r=1}^{n}\partial_\nu\left(\frac{\partial\mathfrak{L}}{\partial\phi_{r,\mu}}\,\partial_\mu\phi_r\right)\right\} = 0 \tag{344}$$

Since a^μ is arbitrary, Eq. (344) implies that

$$\partial^\nu\left(\sum_{r=1}^{n}\frac{\partial\mathfrak{L}}{\partial\phi_{r,\nu}}\frac{\partial\phi_r}{\partial x^\mu} - g_{\mu\nu}\mathfrak{L}\right) = 0 \tag{345a}$$

$$= \frac{\partial}{\partial x_\nu}\,T'_{\mu\nu} = 0 \tag{345b}$$

which is the conservation law for the canonical energy-momentum tensor. The tensor character of $T'_{\mu\nu}$ under proper Lorentz transformation is a consequence of the assumed transformation property of $\mathfrak{L}$, namely that it transforms like a scalar under proper Lorentz transformation. From Eq. (345) the constancy in time of the total energy-momentum four-vector

$$P_\mu = \int_\sigma d\sigma_\nu(x)\,T'_\mu{}^\nu \tag{346}$$

is derived as indicated previously in Eqs. (219) and (220).

Considerations similar to the above can be applied to the case where the variation of the ϕ_r is induced by an infinitesimal homogeneous Lorentz transformation

$$x_\mu \to x'_\mu = (\Lambda x)_\mu = (\delta_\mu{}^\nu + \epsilon\lambda_\mu{}^\nu)\, x_\nu$$
$$\lambda^{\mu\nu} = -\lambda^{\nu\mu} \tag{347}$$

under which the ϕ_r are assumed to transform as follows:

$$\phi_r(x) \to \phi_r'(x') = \sum_{s=1}^{r} B_r{}^s(\Lambda)\, \phi_s(x)$$
$$\approx \sum_{s=1}^{r} (\delta_r{}^s + \tfrac{1}{2}\epsilon b_r{}^{s\mu\nu}\lambda_{\mu\nu})\, \phi_s(x) \tag{348a}$$

with

$$b_r{}^{s\mu\nu} = -b_r{}^{s\nu\mu} \tag{348b}$$

The Lorentz covariance of the Lagrangian ($\mathcal{L}$ is a scalar under Lorentz transformation) asserts that

$$\mathcal{L}'(\phi'_r(x'), \phi'_{r\mu}(x')) = \mathcal{L}(\phi_r(x), \phi_{r\mu}(x))$$
$$= \mathcal{L}((B^{-1}\phi')_r(x'), (B^{-1}\phi'_\mu)_r(x')) \tag{349}$$

i.e., $\mathcal{L}$ and $\mathcal{L}'$ have the same numerical value at physically the same point. One verifies that Eq. (339) for variations induced by an infinitesimal homogeneous Lorentz transformation implies that the tensor $m'^{\mu\nu\rho}$

$$m'^{\mu\nu\rho} = \sum_{r,s=1}^{n} b_r{}^{s\nu\mu}\phi_s \frac{\partial\mathcal{L}}{\partial\phi_{r,\rho}} + x^\nu T'^{\mu\rho} - x^\mu T'^{\nu\rho} \tag{350}$$

is conserved and satisfies the equation

$$\partial_\rho m'^{\mu\nu\rho} = 0 \tag{351}$$

We can therefore define the six time-independent quantities

$$M'^{\mu\nu} = \int_\sigma m'^{\mu\nu\rho}\, d\sigma_\rho(x) \tag{352}$$

which form the components of an antisymmetric tensor whose space components correspond to the components of the angular momentum of the system and whose time components are related to the center-of-mass coordinates of the system [Pryce (1948), Møller (1949, 1952)]. The first term of Eq. (350) corresponds to the spin angular momentum density and the term $x^\nu T'^{\mu\rho} - x^\mu T'^{\nu\rho}$ to the orbital angular momentum density.

Belinfante (1939) has shown how to redefine the canonical energy-momentum tensor, $T'^{\mu\nu}$, such that the new tensor, $T^{\mu\nu}$, is always symmetric, $T^{\mu\nu} = T^{\nu\mu}$, and such that

$$\frac{\partial T_{\mu\nu}}{\partial x_\nu} = \frac{\partial T'_{\mu\nu}}{\partial x_\nu} = 0 \tag{353}$$

Furthermore, the angular momentum tensor defined from the latter by the equation

$$m_{\mu\nu\rho} = x_\mu T_{\nu\rho} - x_\nu T_{\mu\rho} \tag{354}$$

has the property that

$$\frac{\partial m_{\mu\nu\rho}}{\partial x_\rho} = \frac{\partial m'_{\mu\nu\rho}}{\partial x_\rho} = 0 \tag{355}$$

and therefore that

$$\int_\sigma m_{\mu\nu\rho}\, d\sigma^\rho(x) = \int_\sigma m'_{\mu\nu\rho}\, d\sigma^\rho(x) \tag{356}$$

It follows from Eq. (356) that the tensor $m_{\mu\nu\rho}$ has the same physical consequences as $m'_{\mu\nu\rho}$, i.e., they give rise to the same six constants of the motion. Incidentally, note that if ϕ is a scalar field, then $b_r^{s\mu\nu} = 0$ and it follows from (350) and (351) that the canonical energy-momentum tensor is already symmetric.

The ten conserved quantities $\int T^{\mu\nu}\, d\sigma_\nu = P^\mu$ and $M^{\mu\nu} = \int m^{\mu\nu\rho}\, d\sigma_\rho$ are the consequence of the invariance of $\mathfrak{L}$ under inhomogeneous Lorentz transformation. The Lagrangian may have additional invariance properties which in turn will give rise to additional conservation theorems.

Consider, for example, the case where $\mathfrak{L}$ corresponds to the description of complex fields characterized by field variables ϕ_r, ϕ_r^*. Since $\mathfrak{L}$ is real (Hermitian in the quantized theory), only the combination $\phi_r^*\phi_r$ or $\partial^\mu\phi_r^*\partial_\mu\phi_r$ occurs in the Lagrangian. Therefore, $\mathfrak{L}$ is invariant under the gauge transformation wherein

$$\phi_r(x) \to \phi_r'(x) = e^{i\alpha}\, \phi_r(x) \approx (1 + i\alpha)\, \phi_r(x) \tag{357}$$

where α is a real infinitesimal constant. Since by assumption $\mathfrak{L}$ is invariant under this transformation

$$\delta\mathfrak{L} = 0 \tag{358}$$

and

$$\delta\phi_r = \phi_r' - \phi_r = +i\alpha\phi_r \tag{359a}$$

$$\delta\phi_r^* = \phi_r'^* - \phi_r^* = -i\alpha\phi_r^* \tag{359b}$$

it follows from Eq. (339) that

$$\delta\mathfrak{L} = 0 = i\alpha \frac{\partial}{\partial x^\mu} \sum_{r=1}^{n} \left(\frac{\partial\mathfrak{L}}{\partial\phi_{r,\mu}}\, \phi_r - \frac{\partial\mathfrak{L}}{\partial\phi^*_{r,\mu}}\, \phi_r^* \right) \tag{360a}$$

$$= \partial_\mu j^\mu(x) \tag{360b}$$

where

$$j^\mu(x) = i\alpha \sum_{r=1}^{n} \left(\frac{\partial\mathfrak{L}}{\partial\phi_{r,\mu}}\, \phi_r - \frac{\partial\mathfrak{L}}{\partial\phi^*_{r,\mu}}\, \phi_r^* \right) \tag{361}$$

is the conserved electric charge-current density. The proof that the quantity so defined is indeed the charge-current density will be given later in Chapter 10 when we show that it indeed acts as the source for electromagnetic fields in accordance with Maxwell's equations.

More generally if a Lagrangian is invariant (i.e., $\delta\mathfrak{L} = 0$) under the gauge transformation

$$\phi_r \to \phi_r' = e^{i\epsilon q_r}\,\phi_r \approx (1 + i\epsilon q_r)\,\phi_r \tag{362a}$$

$$\delta\phi_r = i\epsilon q_r \phi_r \tag{362b}$$

where ϵ is an infinitesimal quantity and q_r a number characteristic of the field ϕ_r, then Eq. (339) asserts that the four-current

$$J_\mu(x) = \sum_{r=1}^{n} \frac{\partial \mathfrak{L}}{\partial \phi_{r,\mu}(x)}\, q_r \phi_r(x) \tag{363}$$

is divergence free,

$$\partial^\mu J_\mu(x) = 0 \tag{364}$$

Equation (364) then defines a constant of the motion C

$$\begin{aligned} C &= \int_\sigma J_\mu(x)\, d\sigma^\mu(x) = \int J_0(x)\, d^3x \\ &= \int \sum_{r=1}^{n} \frac{\partial \mathfrak{L}}{\partial \phi_{r,0}}\, q_r \phi_r(x)\, d^3x \\ &= \sum_{r=1}^{n} \int d^3x\, q_r \pi_r(x)\, \phi_r(x) \end{aligned} \tag{365}$$

For the gauge transformation (357), C corresponds to the total charge, Q.

7h. The Pion System

As is well known, there exist in Nature three (unstable) particles, the π^+, π^-, and π^0 mesons, whose masses are approximately $273m_e$, and whose spins are all zero. More precisely, the masses of the charged mesons are the same, $\mu(\pi^+) = \mu(\pi^-) = \mu_\pm = 273.27m_e$; whereas the mass of the π^0 is $264.37 \pm 0.6m_e$, the latter value being obtained from the data on the capture of π^- by protons. The spin of the charged pions is determined to be zero from detailed balance arguments applied to the reaction $p + p \to \to d + \pi^+$. The neutral pion has also spin zero, since it decays into two γ rays which is not possible for an odd spin particle (spins higher than 1 are excluded for theoretical reasons). From the fact the reaction $\pi^- + d \to 2n$ is observed, one infers that the π^- has negative parity (relative to the nucleon which by convention has a positive parity). On the assumption that the π^+ is the antiparticle of the π^-, it must have the same parity as the π^-, i.e., negative parity. All three particles are unstable, the charged mesons decaying with the emission of a charged μ meson and a neutrino, $\pi^\pm \to \mu^\pm + \nu$, in $2.56 \pm 0.05 \times 10^{-8}$ second. The π^0 decays into two γ rays (and into $e^+ + e^- + \gamma$, 1.45% of the time) and has a lifetime of $<10^{-17}$ second.[6] Much of the theoretical interpretation of pion inter-

[6] For the detailed description of the π mesons, together with the interpretation of the experiments leading to the above assignments of spin and parity, consult Bethe and de Hoffmann (1955).

actions is based on the conjecture that the three observed particles are three charged states of one kind of "elementary" particle. The difference in mass between charged and neutral pions is then assumed to arise from electromagnetic self-energy effects. On this assumption, neglecting all interactions (which give rise not only to the mass differences between them, but also to their instability), the Lagrangian describing this field system is

$$\mathfrak{L} = -\tfrac{1}{2} : \mu_0{}^2\phi_3{}^2 - \phi_{3\mu}\phi_3{}^\mu : - : \mu_\pm{}^2\phi^*\phi - \phi^*{}_\mu\phi^\mu : \tag{366}$$

where we have assumed the masses of the charged pions to be the same $\mu_+ = \mu_- = \mu_\pm$; ϕ_3 is the (Hermitian) field operator describing the neutral π^0 field and ϕ, ϕ^* are the (non-Hermitian) operators describing the charged field. Upon the introduction of the Hermitian field operators ϕ_1, ϕ_2 with

$$\phi = \frac{1}{\sqrt{2}}(\phi_1 + i\phi_2) \tag{367a}$$

$$\phi^* = \frac{1}{\sqrt{2}}(\phi_1 - i\phi_2) \tag{367b}$$

$\mathfrak{L}$ takes the form

$$\mathfrak{L} = -\tfrac{1}{2}\sum_{j=1}^{3}(\mu_j{}^2\phi_j\phi_j - \phi_{j\mu}\phi_j{}^\mu) \tag{368}$$

with $\mu_1 = \mu_2 = \mu_\pm$ and $\mu_3 = \mu_0$. The momenta canonically conjugate to the ϕ_i are given by

$$\pi_i(x) = \frac{\partial\mathfrak{L}}{\partial\phi_{i0}(x)} = \phi_{i0}(x) \tag{369}$$

so that the equal-time commutation rules read:

$$[\phi_j(x), \pi_l(x')]_{x_0=x_0'} = i\delta_{jl}\delta^{(3)}(\mathbf{x} - \mathbf{x}') \tag{370}$$

all other commutation relations giving zero. The current-density operator of the field is given by the expression

$$j^\mu = +ie : \left(\frac{\partial\mathfrak{L}}{\partial\phi_\mu}\phi - \frac{\partial\mathfrak{L}}{\partial\phi^*{}_\mu}\phi^*\right) : \tag{371}$$

with

$$\begin{aligned} j_0 = \rho &= -ie : \pi^*\phi^* - \pi\phi : \\ &= e : \pi_2\phi_1 - \pi_1\phi_2 : \end{aligned} \tag{372}$$

This current is conserved, $\partial_\mu j^\mu = 0$, and the existence of this conserved four-vector can be considered as a consequence of the invariance of $\mathfrak{L}$ under the transformation

$$\begin{aligned} \phi &\to e^{i\alpha}\,\phi \\ \phi^* &\to e^{-i\alpha}\,\phi \\ \phi_3 &\to \phi_3 \end{aligned} \tag{373}$$

If we consider the fields ϕ_i, $i = 1, 2, 3$ as the components of a vector $\boldsymbol{\phi}$ in some "charge space," then it will be observed that $\mathfrak{L}$ is invariant under rotations about the 3-axis in "charge space." Call T_3 the operator which generates an infinitesimal rotation about the 3-axis, i.e.,

$$e^{-i\epsilon T_3}\,\phi_1\,e^{+i\epsilon T_3} = \phi_1 + \epsilon\phi_2 \tag{374a}$$

$$e^{-i\epsilon T_3}\,\phi_2\,e^{+i\epsilon T_3} = -\epsilon\phi_1 + \phi_2 \tag{374b}$$

$$e^{-i\epsilon T_3}\,\phi_3\,e^{+i\epsilon T_3} = \phi_3 \tag{374c}$$

Since for infinitesimal ϵ

$$e^{-i\epsilon T_3}\,\phi_i(x)\,e^{+i\epsilon T_3} \cong \phi_i(x) - i\epsilon[T_3, \phi_i(x)] \tag{375}$$

by comparing Eqs. (374) and (375), we deduce that T_3 must be such that

$$\begin{aligned} i[T_3, \phi_1] &= -\phi_2 \\ i[T_3, \phi_2] &= \phi_1 \\ i[T_3, \phi_3] &= 0 \end{aligned} \tag{376}$$

A possible choice for T_3 is therefore

$$\begin{aligned} T_3 &= -\int \{\pi_1(x)\,\phi_2(x) - \pi_2(x)\,\phi_1(x)\}\,d^3x \\ &= \frac{1}{e}Q = \frac{1}{e}\int d^3x\,\rho(x) \end{aligned} \tag{377}$$

If the masses $\mu_\pm$ and μ_0 are taken to be the same, then clearly $\mathfrak{L}$ will be invariant under arbitrary rotation in the charge space. The components T_i, $i = 1, 2, 3$, of the vector $\mathbf{T}$ with

$$\mathbf{T} = -\int d^3x\,\boldsymbol{\pi}(x) \times \boldsymbol{\phi}(x) \tag{378}$$

are the generators for the rotations about the ith axis in charge space. Since $\mathbf{T}$ and H commute, $\mathbf{T}$ is a constant of the motion. The commutation rules of the operators T_i are easily deduced with the help of Eq. (370), with the result that

$$\begin{aligned} [T_l, T_j] &= \int d^3x \sum_{r,s,m,n=1}^{3} \epsilon_{lrs}\epsilon_{jmn}[\pi_r(x)\,\phi_s(x), \pi_m(x')\,\phi_n(x')]_{x_0=x_0'} \\ &= i\epsilon_{ljm}T_m \end{aligned} \tag{379}$$

The operators T_i are therefore isomorphic to the angular momentum operators, and both $\mathbf{T}^2$ and T_3 can be chosen as diagonal. One calls $\mathbf{T}$ the isotopic spin of the field and the "charge space" isotopic spin space. From the fact that $T_3 = \frac{1}{e}Q$, the one-particle eigenstate of $\mathbf{T}^2$ and T_3 can immediately be written down. Thus $\phi_3(x)\,|\,\Phi_0\rangle$, the state of one neutral π^0 meson ($|\Phi_0\rangle$ is the no-particle state for which $T_3\,|\,\Phi_0\rangle = \mathbf{T}^2\,|\,\Phi_0\rangle = 0$), is an eigenstate of T_3 with eigenvalue 0 and of $\mathbf{T}^2$ with eigenvalue 2; i.e., if the eigenvalue of $\mathbf{T}^2$ is denoted by $t(t+1)$ then $t = 1$. Similarly, $(\phi_1 + i\phi_2) \mid \Phi_0\rangle$ is an eigenfunction of T_3 with eigenvalue $+1$ and $(\phi_1 - i\phi_2) \mid \Phi_0\rangle$ is an eigenfunction of T_3 with eigenvalue -1, both states being eigenfunction of $\mathbf{T}^2$ with $t = +1$. In other words, the one $\pi^+(\pi^-)$ state is an eigenfunction of the *3*-component of isotopic spin with eigenvalue $+1(-1)$. One therefore speaks of the π-meson system as having total isotopic spin $+1$, the three different states with $t_3 = 0, \pm 1$ corresponding to the three different charge states. The above considerations will play an important role when we analyze the interactions between mesons and nucleons.

A more formal derivation of the properties of the isotopic spin of the meson field is as follows: If $\mu_i = \mu$, the Lagrangian (368) is invariant under arbitrary rotation in isotopic spin space, and in particular, it is invariant under the infinitesimal rotation by an amount λ about the ith axis for which

$$\phi_j(x) \rightarrow \phi_j'(x) = \phi_j(x) + \lambda\epsilon_{ijk}\phi_k(x) \tag{380}$$

[a summation over k from 1 to 3 is implied in (380)]. The variation of ϕ_j is given by

$$\delta\phi_j = \phi_j' - \phi_j = \lambda\epsilon_{ijk}\phi_k \tag{381}$$

and since $\mathcal{L}$ is invariant under this rotation, Eq. (339) yields the conservation law

$$\sum_{j,k=1}^{3} :\partial_\mu\left(\frac{\partial\mathcal{L}}{\partial\phi_{j,\mu}}\epsilon_{ijk}\phi_k\right): = 0 \tag{382}$$

which, using the fact that,

$$\frac{\partial\mathcal{L}}{\partial\phi_{j,\mu}} = \phi_j^\mu \tag{383a}$$

can be rewritten as follows:

$$\sum_{j,k=1}^{3} :\partial_\mu(\phi_j^\mu\epsilon_{ijk}\phi_k): = 0 \tag{383b}$$

The isotopic spin vector

$$\mathfrak{I}_i^\mu = \sum_{j,k=1}^{3} \epsilon_{ijk}\phi_j^\mu\phi_k \tag{384}$$

is therefore conserved in space-time, i.e.,

$$\partial_\mu\mathfrak{I}_i^\mu = 0; \quad i = 1, 2, 3 \tag{385}$$

so that the vector $\mathbf{T}$ with components

$$T_i = \int d\sigma_\mu\, \mathfrak{I}_i^\mu \tag{386}$$

is constant in time. We conclude this chapter with some further comments on the descriptions of the states of the meson system in isotopic spin space.

Recall that we have defined the meson field operators $\phi_j(x)$ by

$$\phi_j(x) = \frac{1}{\sqrt{2(2\pi)^3}} \int \frac{d^3k}{k_0} (a_j(k)\, e^{-ik\cdot x} + a_j^*(k)\, e^{+ik\cdot x}) \tag{387}$$

so that $\phi = \frac{1}{\sqrt{2}}(\phi_1 + i\phi_2)$ corresponds to a creation operator for a positive meson and a destruction operator for a negatively charged meson, $\phi^* = \frac{1}{\sqrt{2}}(\phi_1 - i\phi_2)$ corresponds to the creation operator for a negatively charged meson, etc., and ϕ_3 to the field operator of neutral mesons. Hence the operator $\frac{1}{\sqrt{2}}(a_1(\mathbf{k}) - ia_2(\mathbf{k}))^*$ is a creation operator for a positive meson of momentum $\mathbf{k}$. With this definition

$$\begin{aligned} T_3 &= \epsilon_{3jl} \int d^3x\, \phi_j(x)\, \pi_l(x) \\ &= \frac{1}{2} \int \frac{d^3k}{k_0} \{(a_{1\mathbf{k}} - ia_{2\mathbf{k}})^* (a_{1\mathbf{k}} - ia_{2\mathbf{k}}) \\ &\qquad - (a_{1\mathbf{k}} + ia_{2\mathbf{k}})^* (a_{1\mathbf{k}} + ia_{2\mathbf{k}})\} \qquad (388a) \\ &= N_+ - N_- = \frac{1}{e} Q \qquad (388b) \end{aligned}$$

In terms of the operators $a_{i\mathbf{k}}$, $\mathbf{T}$ is given by

$$T_j = -i \int \frac{d^3k}{k_0}\, \epsilon_{jlm} a_l^*(\mathbf{k})\, a_m(\mathbf{k}) \tag{389}$$

Since $[T_j, T_l] = i\epsilon_{jlm}T_m$, the isotopic spin operators are isomorphic to the angular momentum operators. The operators

$$\begin{aligned} T_\pm &= T_1 \pm iT_2 \\ &= -\int \frac{d^3k}{k_0} \{\pm(a_{1\mathbf{k}} \mp ia_{2\mathbf{k}})^*\, a_{3\mathbf{k}} \mp a_{3\mathbf{k}}^*(a_{1\mathbf{k}} \pm ia_{2\mathbf{k}})\} \qquad (390) \end{aligned}$$

again play the role of "step-up" and "step-down" operators. The classification of the states of n mesons in terms of eigenstates of T^2 and T_3 proceeds in complete analogy with the angular momentum case. Consider for example the two-meson situation. Clearly the state $|\pi^+(\mathbf{k}_1), \pi^+(\mathbf{k}_2)\rangle$, consisting of two positively charged mesons of momenta $\mathbf{k}_1$ and $\mathbf{k}_2$ (hence $Q = 2$), is an eigenstate of T_3 with eigenvalue 2 and of T^2 with eigenvalue $2(2 + 1)$. (We shall sometimes suppress the momentum dependence of the state and write $|2\pi^+\rangle$.)

We can therefore write

$$|T = 2, T_3 = 2; \mathbf{k}_1, \mathbf{k}_2\rangle = \frac{1}{\sqrt{2!}} \frac{1}{\sqrt{2}} (a_{1\mathbf{k}_1} - ia_{2\mathbf{k}_1})^* \frac{1}{\sqrt{2}} (a_{1\mathbf{k}_2} - ia_{2\mathbf{k}_2})^* \,|\,0\rangle \tag{391}$$

where the notation on the left is self explanatory: $|T = 2,\ T_3 = 2;\ \mathbf{k}_1, \mathbf{k}_2\rangle$ is a two-meson state with total isotopic spin $T = 2$, $T_3 = 2$. The state $|T = 2,\ T_3 = 1;\ \mathbf{k}_1, \mathbf{k}_2\rangle$ is now obtained by the process of applying T_- on $|2, 2;\ \mathbf{k}_1, \mathbf{k}_2\rangle$. With the phase convention that

$$T_\pm \mid t, t_3\rangle = \sqrt{(t \mp t_3)\,(t \pm t_3 + 1)} \mid t, t_3 \pm 1\rangle \tag{392}$$

we find

$$\begin{aligned}
&T_- \mid 2, 2; \mathbf{k}_1, \mathbf{k}_2\rangle \\
&\quad = 2 \mid 2, 1; \mathbf{k}_1, \mathbf{k}_2\rangle = \sqrt{2} \int \frac{d^3k}{k_0} \\
&\qquad \left[\frac{1}{\sqrt{2}}\,(a_{1\mathbf{k}} + ia_{2\mathbf{k}})^*\, a_{3\mathbf{k}} - a_{3\mathbf{k}}{}^* \frac{1}{\sqrt{2}}\,(a_{1\mathbf{k}} - ia_{2\mathbf{k}})\right] \mid \pi^+(\mathbf{k}_1), \pi^+(\mathbf{k}_2)\rangle \\
&\quad = -\sqrt{2}\,(|\pi^+(\mathbf{k}_2); \pi^0(\mathbf{k}_1)\rangle + |\pi^+(\mathbf{k}_1), \pi^0(\mathbf{k}_2)\rangle)
\end{aligned} \tag{393}$$

In a similar fashion,

$$\begin{aligned}
T_- \mid 2, 1; \mathbf{k}_1, \mathbf{k}_2\rangle &= \sqrt{6} \mid 2, 0; \mathbf{k}_1, \mathbf{k}_2\rangle \\
&= T_- \frac{-1}{\sqrt{2}}\,(|\pi^+(\mathbf{k}_2); \pi^0(\mathbf{k}_1)\rangle + |\pi^+(\mathbf{k}_1); \pi^0(\mathbf{k}_2)\rangle) \\
&= \{2 \mid\ ; \pi^0(\mathbf{k}_1), \pi^0(\mathbf{k}_2)\rangle - |\pi^+(\mathbf{k}_2);\ ;\ \pi^-(\mathbf{k}_1)\rangle \\
&\quad - |\pi^+(\mathbf{k}_1);\ ;\ \pi^-(\mathbf{k}_2)\rangle\}
\end{aligned} \tag{394}$$

and so forth. There will be three states with $T = 1$. The one with $T_3 = 1$ (and hence $Q = 1$) must clearly be a linear combination of the states $|\pi^+, \pi^0\rangle$, i.e.,

$$|1, 1; \mathbf{k}_1, \mathbf{k}_2\rangle = a \mid \pi^+(\mathbf{k}_1); \pi^0(\mathbf{k}_2)\rangle + b \mid \pi^+(\mathbf{k}_2); \pi^0(\mathbf{k}_1)\rangle \tag{395}$$

This state must be orthogonal to the state $|2, 1; \mathbf{k}_1, \mathbf{k}_2\rangle$ from which one infers that $a = -b$, so that

$$|1, 1; \mathbf{k}_1, \mathbf{k}_2\rangle = \frac{1}{\sqrt{2!}}\,\{|\pi^+(\mathbf{k}_1); \pi^0(\mathbf{k}_2)\rangle - |\pi^+(\mathbf{k}_2); \pi^0(\mathbf{k}_1)\rangle\} \tag{396}$$

Proceeding as before,

$$\begin{aligned}
|1, 0; \mathbf{k}_1, \mathbf{k}_2\rangle &= \frac{1}{\sqrt{2}}\, T_- \mid 1, 1; \mathbf{k}_1, \mathbf{k}_2\rangle \\
&= \frac{1}{\sqrt{2!}}\,\{|\pi^+(\mathbf{k}_1); \pi^-(\mathbf{k}_2)\rangle - |\pi^-(\mathbf{k}_1); \pi^+(\mathbf{k}_2)\rangle\}
\end{aligned} \tag{397}$$

$$\begin{aligned}
|1, -1; \mathbf{k}_1, \mathbf{k}_2\rangle &= \frac{1}{\sqrt{2}}\, T_- \mid 1, 0; \mathbf{k}_1, \mathbf{k}_2\rangle \\
&= \frac{1}{\sqrt{2!}}\,\{|\pi^0(\mathbf{k}_2), \pi^-(\mathbf{k}_1)\rangle - |\pi^-(\mathbf{k}_2), \pi^0(\mathbf{k}_1)\rangle\}
\end{aligned} \tag{398}$$

There is one state with $T = 0$. Since it must have $T_3 = 0$ and hence zero charge, it is a linear combination of the states $|\pi_1^+, \pi_2^-\rangle$, $|\pi_1^-, \pi_2^+\rangle$, and $|\pi_1^0, \pi_2^0\rangle$. It must be orthogonal to the states $|2, 0; \mathbf{k}_1, \mathbf{k}_2\rangle$ and $|1, 0; \mathbf{k}_1, \mathbf{k}_2\rangle$, which then determines the linear combination as

$$|0, 0; \mathbf{k}_1, \mathbf{k}_2\rangle = \frac{1}{\sqrt{2!\,3}} \{|\pi^+(\mathbf{k}_1); \pi^-(\mathbf{k}_2)\rangle + |\pi^+(\mathbf{k}_2); \pi^-(\mathbf{k}_1)\rangle - |\pi^0(\mathbf{k}_1); \pi^0(\mathbf{k}_2)\rangle\} \quad (399)$$

8

Quantization of the Dirac Field

8a. The Commutation Rules

The free charged spin $\frac{1}{2}$ field is described by a four-component complex spinor ψ which satisfies the Dirac equation, and which under Lorentz transformations transforms according to

$$\psi_\alpha(x) \rightarrow \psi_\alpha'(x') = \sum_\beta S_{\alpha\beta}(\Lambda)\,\psi_\beta(\Lambda^{-1}(x'-a)); \quad \alpha = 1, \cdots, 4 \tag{1}$$

For the purpose of applying the canonical formalism we shall consider the four components as independent dynamical variables. The chief difference between the Dirac Lagrangian and those discussed thus far is that the former is of first order in $\partial_\mu \psi$ in order that it lead to a first-order differential equation for ψ. This then necessitates a slight modification of the Hamiltonian formalism.

The Lagrangian density is taken to be

$$\mathfrak{L} = -\tfrac{1}{2}\bar{\psi}(-i\gamma^\mu \partial_\mu + m)\,\psi - \tfrac{1}{2}(i\partial_\mu \bar{\psi}\gamma^\mu + m\bar{\psi})\,\psi \tag{2}$$

We have here denoted the reciprocal Compton wave length of the Dirac particle by m. We shall retain this notation in the future to distinguish a spin $\frac{1}{2}$ particle from a spin 0 particle, whose mass we shall continue to denote by μ. The equations of motion are obtained by the independent variations of ψ and $\bar{\psi}$. In particular,

$$\frac{\partial \mathfrak{L}}{\partial \bar{\psi}} = -\tfrac{1}{2}(-i\gamma^\mu \partial_\mu + m)\,\psi - \tfrac{1}{2}m\psi \tag{3}$$

$$\frac{\partial}{\partial x^\mu}\frac{\partial \mathfrak{L}}{\partial(\partial_\mu \bar{\psi})} = -\tfrac{1}{2}i\gamma^\mu \partial_\mu \psi \tag{4}$$

so that the Euler variational equation reads

$$(-i\gamma^\mu \partial_\mu + m)\,\psi = 0 \tag{5}$$

the desired Dirac equation. The correct equation for the adjoint spinor follows similarly from the variation of ψ. Incidentally, $\mathfrak{L}$ vanishes if the wave equations are satisfied.

Using the general expression for the charge-current vector, Eq. (7.361), we obtain the following current vector for the Dirac field:

$$j^\mu(x) = \epsilon\bar{\psi}(x)\,\gamma^\mu\psi(x) \tag{6}$$

which is conserved by virtue of the equations of motion. In particular, it follows from the conservation law $\partial_\mu j^\mu(x) = 0$ that

$$\frac{\partial}{\partial x^0}\int \bar{\psi}\gamma^0\psi\, d^3x = 0 \tag{7}$$

corresponding to the statement that the total charge is conserved.

The fact that $\mathfrak{L}$ is of first order in $\partial_0\psi$ now implies that the canonically conjugate momenta are not independent of the field variables ψ and $\bar{\psi}$. The canonical Lagrangian formalism is therefore not immediately applicable and the passage to the Hamiltonian cannot be effected straightforwardly. We can bypass this difficulty by recalling that all quantities of physical interest can be derived from the energy-momentum tensor, Eq. (7.345), which in the present case, remembering that $\mathfrak{L} = 0$, is

$$T^{\mu\nu} = \frac{i}{2}\left(\bar{\psi}\gamma^\nu\frac{\partial\psi}{\partial x_\mu} - \frac{\partial\bar{\psi}}{\partial x_\mu}\gamma^\nu\psi\right) \tag{8}$$

This tensor is not symmetric but can readily be symmetrized [see, for example, Pauli (1933), p. 235]. It, however, yields the same energy-momentum four-vector as the symmetrized tensor, namely,

$$\begin{aligned} P^\mu &= \int T^{\mu 0}\, d^3x = \frac{i}{2}\int d^3x\left(\bar{\psi}\gamma^0\frac{\partial\psi}{\partial x_\mu} - \frac{\partial\bar{\psi}}{\partial x_\mu}\gamma^0\psi\right) \\ &= i\int d^3x\left(\bar{\psi}\gamma^0\frac{\partial}{\partial x_\mu}\psi\right) \end{aligned} \tag{9}$$

The last form of Eq. (9) is obtained by partial integration, dropping the surface terms for $\mu = 1, 2, 3$ and using (7) for $\mu = 0$. By virtue of the equations of motion, we obtain for the total energy, or Hamiltonian, the expression

$$\begin{aligned} H = P^0 &= \int d^3x\, T^{00} = \int d^3x\, \bar{\psi}(x)\,(-i\boldsymbol{\gamma}\cdot\boldsymbol{\nabla} + m)\,\psi(x) \\ &= \int d^3x\, \psi^*(x)\,(-i\boldsymbol{\alpha}\cdot\boldsymbol{\nabla} + \beta m)\,\psi(x) \end{aligned} \tag{10}$$

The quantization of the Dirac spin $\frac{1}{2}$ field can be carried out by analogy to the quantization of the scalar field. We expect that, in an expansion of the Schrödinger field operators $\psi(\mathbf{x})$, $\bar{\psi}(\mathbf{x})$ in terms of the complete set of eigenfunctions $w_n(\mathbf{x})$ of the free-particle Dirac equation, the expansion coefficients will be interpretable as creation and annihilation operators. The index n of w_n stands for a specification of a complete set of commuting one-particle observables—here the energy, momentum, and spin of the particle. These functions $w_n(\mathbf{x})$ are the spinors $w^r(\mathbf{p})$ $(r = 1, 2, 3, 4)$ previously introduced in Chapter 4 by Eq. (141), multiplied by their plane wave factor exp $(i\mathbf{p}\cdot\mathbf{x})$ thus,

$$w_r(\mathbf{x}) = w^r(\mathbf{p})\, e^{i\mathbf{p}\cdot\mathbf{x}} \quad \text{for } r = 1, 2$$
$$w_r(\mathbf{x}) = v^r(\mathbf{p})\, e^{-i\mathbf{p}\cdot\mathbf{x}} \quad \text{for } r = 3, 4 \tag{11}$$

In the Schrödinger picture the expansion of the operators $\psi(\mathbf{x})$ and $\tilde{\psi}(\mathbf{x})$ is then given by

$$\psi(\mathbf{x}) = \frac{1}{\sqrt{V}} \sum_n \sqrt{\frac{m}{|E_n|}}\, b_n w_n(\mathbf{x}) \tag{12a}$$

$$\tilde{\psi}(\mathbf{x}) = \frac{1}{\sqrt{V}} \sum_n \sqrt{\frac{m}{|E_n|}}\, b_n{}^* \tilde{w}_n(\mathbf{x}) \tag{12b}$$

the sum running over all states of positive as well as negative energy. The factor $(m/|E_n|)^{1/2}$ is included due to our normalization of the spinors: $\tilde{w}w = \epsilon$.

Consider now the Hamiltonian operator of the field which, according to Eq. (10), is given by

$$H = \int d^3x\, \tilde{\psi}(\mathbf{x})\, (-i\gamma\cdot\nabla + m)\, \psi(\mathbf{x}) \tag{13}$$

Introducing the expressions (11) and (12) into H, we obtain

$$H = \sum_m b_m{}^* b_m E_m \tag{14}$$

where the E_m are the eigenvalues of the Dirac operator $\boldsymbol{\alpha}\cdot\mathbf{p} + \beta m$. If we were to use Bose commutation rules for the quantization of the theory, we then would run into two related difficulties. The first consists of the fact that if we interpret $b_n{}^*b_n$ as a number operator, then the energy of the field can be positive as well as negative, since the eigenvalues E_n can be positive as well as negative and $b_n{}^*b_n \geqslant 0$. Indeed, if the number operators can have any positive integer eigenvalues, as they do according to Bose commutation rules [see, e.g., Chapter 6, Eqs. (60)–(62)], then states of arbitrarily high negative energy would be allowed, which does not make any physical sense. Actually, we know from experiment that spin $\frac{1}{2}$ particles obey the Pauli exclusion principle, so that not more than one particle can be in any state m (m includes the specification of energy, momentum, and spin). The quantization with commutators, leading to Bose statistics, is therefore incorrect, since it allows for any number of particles to be present in a state.

We have noted in Chapter 6 that a quantization scheme which incorporates the Pauli principle has been developed by Jordan and Wigner [Jordan (1928b)]. In this treatment the operators, instead of satisfying commutation rules, satisfy "anticommutation" rules,

$$[b_n, b_m]_+ = [b_n{}^*, b_m{}^*]_+ = 0 \tag{15}$$

$$[b_n, b_m{}^*]_+ = \delta_{nm} \tag{16}$$

The number operator for the state m is still given by

$$N_m = b_m{}^*b_m \tag{17}$$

but it now follows from the commutation rules (15)–(16) that

$$\begin{aligned} N_m{}^2 &= b_m{}^*b_mb_m{}^*b_m \\ &= b_m{}^*(1 - b_m{}^*b_m)\, b_m = b_m{}^*b_m \\ &= N_m \end{aligned} \tag{18}$$

so that the possible occupation number for the states can only be zero or one, in accordance with the Pauli principle. That no more than one particle can be fitted into any state is already indicated by the commutation rule (15) which requires that the application on any state vector of two creation operators for the same m gives the null vector.

Although we have satisfied the Pauli principle by using anticommutation rules, we are still faced with the difficulty that the Hamiltonian is not positive-definite. This is eliminated by Dirac's idea that in the vacuum state all the negative energy states are occupied. If we write out the index m explicitly as $(\mathbf{p}, r)$, where r runs over the four solutions corresponding to a given $\mathbf{p}$ ($r = 1, 2$ for positive energy; $r = 3, 4$ for negative energy), then in the limit of continuous $\mathbf{p}$ values, the expansion for $\psi(\mathbf{x})$, Eq. (12), becomes

$$\psi(\mathbf{x}) = \int \frac{d^3p}{(2\pi)^{3/2}} \left(\frac{m}{E_\mathbf{p}}\right)^{1/2} \left\{\sum_{r=1}^{2} b_r(\mathbf{p})\, w^r(\mathbf{p})\, e^{i\mathbf{p}\cdot\mathbf{x}} + \sum_{r=3}^{4} b_r(-\mathbf{p})\, w^r(\mathbf{p})\, e^{-i\mathbf{p}\cdot\mathbf{x}}\right\} \tag{19}$$

with

$$E_\mathbf{p} = +\sqrt{\mathbf{p}^2 + m^2} \tag{20}$$

If we denote the number operators by

$$N_r{}^{(+)}(\mathbf{p}) = b_r{}^*(\mathbf{p})\, b_r(\mathbf{p}) \qquad \text{for } r = 1, 2 \tag{21}$$

$$N_r{}^{(-)}(\mathbf{p}) = b_{r+2}{}^*(\mathbf{p})\, b_{r+2}(\mathbf{p}) \quad \text{for } r = 1, 2 \tag{22}$$

then our Hamiltonian can be rewritten in the following form:

$$H = \int d^3p \sum_{r=1}^{2} E_\mathbf{p}\{N_r{}^{(+)}(\mathbf{p}) - N_r{}^{(-)}(\mathbf{p})\} \tag{23}$$

Similarly, we may express the total charge[1] in terms of the number operators as follows:

$$\begin{aligned} Q &= -e \int d^3x\, \bar{\psi}(\mathbf{x})\, \gamma^0\psi(\mathbf{x}) \\ &= -e \int d^3p \sum_{r=1}^{2} \{N_r{}^{(+)}(\mathbf{p}) + N_r{}^{(-)}(\mathbf{p})\} \end{aligned} \tag{24}$$

[1] The assignment of charge in Eq. (24) is such that the particle will have charge $-e$, and the antiparticle charge $+e$. The theory in this form is therefore immediately applicable to electrons.

Now according to Dirac's hole theory, the vacuum, $|0\rangle$, is characterized by the fact that all the negative energy states are occupied (i.e., $N_r^{(-)}(\mathbf{p}) \mid 0\rangle = |0\rangle$ for all $\mathbf{p}$ and all r) and all positive energy states are empty (i.e., $N_r^{(+)}(\mathbf{p}) \mid 0\rangle = 0$ for all $\mathbf{p}$ and all r). The energy and total charge of the vacuum is therefore given by

$$E_0 = -\sum_{r=1}^{2} \int d^3p\, E_{\mathbf{p}}$$

$$Q_0 = -e \sum_{r=1}^{2} \int d^3p \tag{25}$$

These quantities are clearly infinite. However, by Dirac's hypothesis they are unobservable. Furthermore, only the difference between H and E_0 and Q and Q_0 is observable. Let us therefore define the observable energy H' and the observable total charge Q' as follows:

$$H' = H - E_0 = \int d^3p \sum_{r=1}^{2} \{N_r^{(+)}(\mathbf{p}) - N_r^{(-)}(\mathbf{p})\}\, E_{\mathbf{p}} + \int d^3p \sum_{r=1}^{2} E_{\mathbf{p}}$$

$$= \int d^3p \sum_{r=1}^{2} \{N_r^{(+)}(\mathbf{p}) + (1 - N_r^{(-)}(\mathbf{p}))\}\, E_{\mathbf{p}} \tag{26}$$

and

$$Q' = Q - Q_0 = -e \int d^3p \sum_{r=1}^{2} \{N_r^{(+)}(\mathbf{p}) - (1 - N_r^{(-)}(\mathbf{p}))\} \tag{27}$$

Equations (26) and (27) are the mathematical statements of our previous discussion of the hole theory. They indicate explicitly that contributions to the total charge and energy from the negative energy states occur only when $N_r^{(-)}(\mathbf{p})$ is zero, that is, when the state is *unoccupied.* In the case of the charge, the contribution to Q from an empty negative energy state is of the opposite sign ($+e$) from that of the positive energy states. On the other hand, the contribution to H from an unoccupied negative energy state $\mathbf{p}r$ is $+E_{\mathbf{p}}$; hence, H' is now positive-semidefinite. Thus, the absence of a particle in a negative energy state corresponds to the presence of a positive energy particle of opposite charge, called the "positon" in the case of electrons, while in general it is called an "antiparticle."

It should be emphasized that the Dirac hole theory and the subtraction procedure it entails are only possible due to the quantization with anticommutation rules, which imply that every negative energy state is occupied by only one particle. The lack of positive-definiteness of the energy in the classical theory is a feature of all half-integer spin theories. In fact, if with Pauli (1940a) we impose the following physical requirements on a quantized theory:

1. that the commutator of two physical observables pertaining to space-time points which are separated by a space-like distance commute (causality condition), and

2. that the energy of the system be positive-semidefinite,

then a connection between spin and statistics can be established. It can be shown [Pauli (1940a, b)] that the quantization of integral spin theories according to Jordan-Wigner anticommutation rules (leading to Fermi-Dirac statistics) violates the first postulate, whereas the quantization of half-integral spin according to Bose commutation quantization rules (leading to Einstein-Bose statistics) violates the second postulate.

The justification for the first postulate stems from the fact that in the quantum theory the lack of commutativity of two observable operators implies that these cannot be measured simultaneously with arbitrary accuracy. However, measurements at points which are separated by space-like distances can never perturb one another, since relativistic causality demands that signals (energy) cannot be propagated with a velocity greater than that of light. Hence, the commutator of observables with space-like connections must vanish. The second postulate is self-evident. We shall give a modern version of the proof of the connection between spin and statistics in Chapter 18.

It has become customary to redefine the operators for the negative energy states,[2] and call

$$\begin{aligned} b_{r+2}(-\mathbf{p}) &= d_r^*(\mathbf{p}) \quad (r = 1, 2) \\ b_{r+2}^*(-\mathbf{p}) &= d_r(\mathbf{p}) \quad (r = 1, 2) \end{aligned} \tag{28}$$

so that the commutation rules (15)–(16) now read:

$$\begin{aligned} [d_r^*(\mathbf{p}), d_s(\mathbf{p}')]_+ &= \delta_{rs}\delta^{(3)}(\mathbf{p} - \mathbf{p}') \\ [d_r(\mathbf{p}), b_s(\mathbf{p}')]_+ &= [d_r^*(\mathbf{p}), b_s(\mathbf{p}')]_+ = 0 \\ [d_r(\mathbf{p}), b_s^*(\mathbf{p}')]_+ &= [d_r^*(\mathbf{p}), b_s(\mathbf{p}')]_+ = 0 \end{aligned} \tag{29}$$

In terms of these operators, the number operator $1 - N_r^{(-)}(-\mathbf{p})$ can be expressed as follows:

$$\begin{aligned} N_r^{(-)\prime}(\mathbf{p}) &= 1 - N_r^{(-)}(-\mathbf{p}) = 1 - b_{r+2}^*(-\mathbf{p})\, b_{r+2}(-\mathbf{p}) \\ &= b_{r+2}(-\mathbf{p})\, b_{r+2}^*(-\mathbf{p}) = d_r^*(\mathbf{p})\, d_r(\mathbf{p}) \end{aligned} \tag{30}$$

The operators d^*, d are now creation and annihilation operators for the antiparticles, and $N_r^{(-)\prime}(\mathbf{p})$ (we shall drop the prime hereafter) is the number operator for the antiparticles. The expansion of the ψ operator is then given by

$$\psi(x) = \frac{1}{(2\pi)^{3/2}} \int d^3p \sqrt{\frac{m}{E_{\mathbf{p}}}} \sum_{r=1}^{2} \{b_r(\mathbf{p})\, w^r(\mathbf{p})\, e^{i\mathbf{p}\cdot\mathbf{x}} + d_r^*(\mathbf{p})\, v^r(\mathbf{p})\, e^{-i\mathbf{p}\cdot\mathbf{x}}\} \tag{31}$$

[2] This is possible because the operators satisfy anticommutation rules and the latter remain invariant under the interchange: $b \leftrightarrow b^*$.

where $v^r(\mathbf{p})$ is a negative energy spinor of momentum $-\mathbf{p}$. Thus $\psi(\mathbf{x})$ is a creation operator for particles as well as a destruction operator for antiparticles. In this formulation of the hole theory, the vacuum state, $|\Phi_0\rangle$, is now characterized by

$$N_r^{(+)}(\mathbf{p}) \mid \Phi_0\rangle = N_r^{(-)}(\mathbf{p}) \mid \Phi_0\rangle = 0 \quad \text{for all } r \text{ and all } \mathbf{p} \tag{32}$$

It thus has zero energy, charge, and momentum, and is the state of no particles and that of lowest energy. It is an invariant under all Lorentz transformations. Alternatively, we may characterize it by

$$b_r(\mathbf{p}) \mid \Phi_0\rangle = d_r(\mathbf{p}) \mid \Phi_0\rangle = 0 \quad \text{for all } r \text{ and all } \mathbf{p} \tag{33}$$

Note that in this formulation one no longer speaks of a "sea."

The commutation rules between the number operators $N_r^{(\pm)}(\mathbf{p})$ and the operators $b_r(\mathbf{p})$, $b_r^*(\mathbf{p})$ are

$$[N_r^{(+)}(\mathbf{p}), b_s(\mathbf{p}')] = -\delta_{rs}\delta^{(3)}(\mathbf{p} - \mathbf{p}')\, b_n(\mathbf{p}) \tag{34}$$

$$[N_r^{(+)}(\mathbf{p}), b_s^*(\mathbf{p}')] = \delta_{rs}\delta^{(3)}(\mathbf{p} - \mathbf{p}')\, b_r^*(\mathbf{p}) \tag{35}$$

$$0 = [N_r^{(-)}(\mathbf{p}), b_s(\mathbf{p}')] = [N_r^{(-)}(\mathbf{p}), b_s^*(\mathbf{p}')] \tag{36}$$

Similar expressions hold for the commutation rules of $N^{(\pm)}$ and the antiparticle operators. The state $b_s^*(\mathbf{p}) \mid \Phi_0\rangle$ is therefore a one-particle state of spin s and momentum $\mathbf{p}$; $b_{s_1}^*(\mathbf{p}_1)\, b_{s_2}^*(\mathbf{p}_2) \mid \Phi_0\rangle$ a two-particle state, etc.; $d_{t_1}^*(\mathbf{q}_1) \mid \Phi_0\rangle$ a one antiparticle state, and so forth. The basis vectors for the representation in which the number operators are diagonal are

$$|\mathbf{p}_1 s_1, \cdots \mathbf{p}_m s_m; \mathbf{q}_1 t_1, \cdots ; \mathbf{q}_n t_n\rangle$$

$$= \frac{1}{\sqrt{n!m!}} b_{s_1}^*(\mathbf{p}_1) \cdots b_{s_m}^*(\mathbf{p}_m)\, d_{t_1}^*(\mathbf{q}_1) \cdots d_{t_n}^*(\mathbf{q}_n) \mid \Phi_0\rangle \tag{37}$$

with

$$\langle \mathbf{p}'_1 s'_1, \cdots \mathbf{p}'_{m'} s'_{m'}; \mathbf{q}'_1 t'_1, \cdots \mathbf{q}'_{n'} t_{n'} \mid \mathbf{p}_1 s_1, \cdots \mathbf{p}_m s_m; \mathbf{q}_1 t_1, \cdots \mathbf{q}_n t_n\rangle$$

$$= \frac{\delta_{mm'}\delta_{nn'}}{m!n!} \det |\delta^{(3)}(\mathbf{p}_i - \mathbf{p}_j')\, \delta_{s_i s_j}| \cdot \det |\delta^{(3)}(\mathbf{q}_k - \mathbf{q}_e')\, \delta_{t_k t_e}| \tag{38}$$

8b. Configuration Space

The Heisenberg operator, $\psi(x)$, is obtained from the Schrödinger operator $\psi(\mathbf{x})$ by the time-dependent unitary transformation

$$\psi(x) = e^{iH_0x^0/\hbar}\psi(\mathbf{x})\, e^{-iH_0x^0/\hbar} \tag{39}$$

Using procedures familiar by now, we find

$$\psi(x) = \frac{1}{(2\pi)^{3/2}} \int d^3p \left(\frac{m}{E(\mathbf{p})}\right)^{1/2} \sum_{r=1}^{2} \{b_r(\mathbf{p})\, w^r(\mathbf{p})\, e^{-ip\cdot x} + d_r^*(\mathbf{p})\, v^r(\mathbf{p})\, e^{ip\cdot x}\} \tag{40}$$

where $p_0 = +E(\mathbf{p})$, so that $\psi(x)$ obeys the Dirac equation

$$(-i\gamma^\mu \partial_\mu + m)\,\psi(x) = 0 \tag{41}$$

Similarly,

$$\bar{\psi}(x) = \frac{1}{(2\pi)^{3/2}} \int d^3p \left(\frac{m}{E(\mathbf{p})}\right)^{1/2} \sum_{r=1}^{2} \{b_r^*(\mathbf{p})\, \bar{w}^r(\mathbf{p})\, e^{ip\cdot x} + d_r(\mathbf{p})\, \bar{v}^r(\mathbf{p})\, e^{-ip\cdot x}\} \tag{42}$$

$$i\partial_\mu \bar{\psi}(x)\, \gamma^\mu + m\bar{\psi}(x) = 0 \qquad (43)^3$$

The positive and negative frequency parts of these operators are given by the following expressions:

$$\psi^{(+)}(x) = \frac{1}{(2\pi)^{3/2}} \int_{p_0>0} d^3p \left(\frac{m}{E(\mathbf{p})}\right)^{1/2} \sum_{r=1}^{2} b_r(\mathbf{p})\, w^r(\mathbf{p})\, e^{-ip\cdot x} \tag{44}$$

which is a destruction operator for a fermion,

$$\psi^{(-)}(x) = \frac{1}{(2\pi)^{3/2}} \int_{p_0>0} d^3p \left(\frac{m}{E(\mathbf{p})}\right)^{1/2} \sum_{r=1}^{2} d_r^*(\mathbf{p})\, v^r(\mathbf{p})\, e^{ip\cdot x} \tag{45}$$

which is a creation operator for an antifermion,

$$\bar{\psi}^{(-)}(x) = \frac{1}{(2\pi)^{3/2}} \int_{p_0>0} d^3p \left(\frac{m}{E(\mathbf{p})}\right)^{1/2} \sum_{r=1}^{2} b_r^*(\mathbf{p})\, \bar{w}^r(\mathbf{p})\, e^{ip\cdot x} \tag{46}$$

which is a creation operator for a fermion, and

$$\bar{\psi}^{(+)}(x) = \frac{1}{(2\pi)^{3/2}} \int_{p_0>0} d^3p \left(\frac{m}{E(\mathbf{p})}\right)^{1/2} \sum_{r=1}^{2} d_r(\mathbf{p})\, \bar{v}(\mathbf{p})\, e^{-ip\cdot x} \tag{47}$$

which is a destruction operator for an antifermion. It follows from Eqs. (44)–(47) that

$$\widetilde{\psi^{(+)}}(x) = \bar{\psi}^{(-)}(x) \tag{48}$$

$$\widetilde{\psi^{(-)}}(x) = \bar{\psi}^{(+)}(x) \tag{49}$$

In terms of the configuration space operators, the vacuum $|\Phi_0\rangle$ is characterized by

$$\psi^{(+)}(x) \mid \Phi_0\rangle = \bar{\psi}^{(+)}(x) \mid \Phi_0\rangle = 0 \tag{50}$$

Their commutation rules read:

$$[\psi^{(+)}(x), \psi^{(-)}(x')]_+ = [\bar{\psi}^{(-)}(x), \bar{\psi}^{(+)}(x')]_+ = 0$$

$$[\bar{\psi}^{(+)}(x), \psi^{(+)}(x')]_+ = [\bar{\psi}^{(-)}(x), \psi^{(-)}(x')]_+ = 0 \tag{51}$$

[3] The discussion of the relativistic invariance is facilitated by the introduction of operators $b'_r(\mathbf{p}) = \sqrt{E(\mathbf{p})}\, b_r(\mathbf{p})$, $d_s(\mathbf{p}) = \sqrt{E(\mathbf{p})}\, d_s(\mathbf{p})$ so that in (40)–(42) the integration is over the hyperboloid $p_0^2 - \mathbf{p}^2 = m^2$, $p_0 > 0$, using the invariant measure d^3p_0/p_0. The (invariant) commutation rules of the primed operators are $[b_r'(\mathbf{p}), b_s'^*(\mathbf{p}')]_+ = \delta_{rs}\delta^{(3)}(\mathbf{p} - \mathbf{p}')\, p_0$ etc. with $p_0 = E(\mathbf{p})$.

whereas

$$[\psi_\alpha^{(+)}(x), \tilde{\psi}_\beta^{(-)}(x')]_+ = \frac{1}{(2\pi)^3} \int d^3p \int d^3p' \left(\frac{m^2}{E(\mathbf{p})\,E(\mathbf{p}')}\right)^{1/2} \sum_{r,s=1}^{2} [b_r(\mathbf{p}), b_s^*(\mathbf{p}')]_+ \, w_\alpha^r(\mathbf{p})\, \tilde{w}_\beta^s(\mathbf{p}')\, e^{-ip\cdot x + ip'\cdot x'}$$

$$= \frac{1}{2(2\pi)^3} \int \frac{d^3p}{E(\mathbf{p})} (\not{p} + m)_{\alpha\beta}\, e^{-ip\cdot(x-x')}$$

$$= \frac{1}{2(2\pi)^3} (i\gamma^\mu\partial_\mu + m)_{\alpha\beta} \int_{p_0>0} \frac{d^3p}{p_0}\, e^{-ip(x-x')}$$

$$= i(i\gamma^\mu\partial_\mu + m)_{\alpha\beta}\, \Delta^{(+)}(x - x'; m) \tag{52}$$

In the derivation of this last expression we have made use of Eqs. (29), (4.150), and (7.75). We shall denote the frequently occurring expression $-(i\gamma^\mu\partial_\mu + m)\,\Delta(x)$ by $S(x)$, so that with

$$-(i\gamma^\mu\partial_\mu + m)\,\Delta(x - x'; m) = S(x - x') \tag{53}$$

the commutation rules (52) become

$$[\psi_\alpha^{(+)}(x), \tilde{\psi}_\beta^{(-)}(x')]_+ = -iS_{\alpha\beta}^{(+)}(x - x') \tag{54}$$

where

$$S^{(+)}(x) = -(i\gamma^\mu\partial_\mu + m)\,\Delta^{(+)}(x) \tag{55}$$

Similarly, one derives that

$$[\psi_\alpha^{(-)}(x), \tilde{\psi}_\beta^{(+)}(x')]_+ = -iS_{\alpha\beta}^{(-)}(x - x') \tag{56}$$

where

$$S^{(-)}(x) = -(i\gamma^\mu\partial_\mu + m)\,\Delta^{(-)}(x) \tag{57}$$

and, more generally, that

$$[\psi_\alpha(x), \tilde{\psi}_\beta(x')]_+ = -iS_{\alpha\beta}(x - x') \tag{58}$$

with

$$S(x) = S^{(+)}(x) + S^{(-)}(x) \tag{59}$$

The equal-time commutation rules are given by

$$[\psi_\alpha(x), \tilde{\psi}_\beta(x')]_+ \Big|_{x_0 = x'_0} = -\gamma^0_{\alpha\beta}\partial_0\Delta(x - x')\Big|_{x_0 = x_0'} = +\gamma^0_{\alpha\beta}\delta^{(3)}(\mathbf{x} - \mathbf{x}') \tag{60}$$

recalling that $\Delta(0, \mathbf{x}) = 0$ and $\partial_0\Delta(x)\Big|_{x_0=0} = -\delta^{(3)}(\mathbf{x})$. Hence, multiplying (60) by γ^0 we obtain

$$[\psi_\alpha(x), \psi_\beta^*(x')]_+ \Big|_{x_0 = x'_0} = \delta^{(3)}(\mathbf{x} - \mathbf{x}')\,\delta_{\alpha\beta} \tag{61}$$

which are the canonical anticommutation rules for the Dirac field. The properties of $S(x)$ which will be of importance in the following are

$$\begin{aligned}(i\partial\!\!\!/ - m)\, S(x) &= -(i\partial\!\!\!/ - m)\,(i\partial\!\!\!/ + m)\,\Delta(x) \\ &= +(\Box + m^2)\,\Delta(x) \\ &= 0 \end{aligned} \tag{62}$$

[consistent with Eq. (58) and the fact that $\psi(x)$, $\tilde{\psi}(x)$ obey Dirac equations] and

$$S(x)\Big|_{x^0=0} = i\gamma^0\delta(\mathbf{x}) \tag{63}$$

The differential equation (62) and the initial condition (63) determine S uniquely. The singular function S plays the role of a Green's function for the solution of the initial value problem. *Proof:* Gauss's theorem asserts that

$$\int_\Omega \partial^\mu F_\mu(x)\, d^4x = \int_\Sigma d\sigma^\mu(x')\, F_\mu(x') \tag{64}$$

where Σ is the surface enclosing Ω; if we choose for F

$$F_\mu(x') = S(x - x')\,\gamma_\mu\psi(x') \tag{65}$$

then, since S and ψ satisfy the Dirac equation, $\partial^\mu F_\mu(x) = 0$, so that

$$\int_{\sigma_1} S(x - x')\,\gamma^\mu\psi(x')\, d\sigma_\mu(x') = \int_{\sigma_2} S(x - x')\,\gamma^\mu\psi(x')\, d\sigma_\mu(x') \tag{66}$$

where σ_1 and σ_2 are space-like surfaces. (We have assumed that the contribution from the integration over the surface joining σ_1 and σ_2 vanishes. We suppose this surface to be at a large spatial distance where the fields vanish.) If we choose for σ_2 the flat surface x^0 = *constant*, then using (63) we obtain

$$\psi(x) = -i\int_{\sigma_1} S(x - x')\,\gamma^\mu\psi(x')\, d\sigma_\mu(x') \tag{67}$$

The same value is obtained for any other space-like surface through x^0 since

$$\frac{\delta}{\delta\sigma(x)}\int_\sigma S(x - x')\,\gamma^\mu\psi(x')\, d\sigma_\mu(x') = 0 \tag{68}$$

The reader should verify that $S^{(+)}$ and $S^{(-)}$ are again projection operators for the positive and negative frequencies of $\psi(x)$.

If we define the charge conjugate operator ψ_c by

$$\psi_c(x) = C\tilde{\psi}^T(x) \tag{69}$$

where C is the unitary matrix defined previously with the property that

$$C^{-1}\gamma_\mu C = -\gamma_\mu{}^T \tag{70a}$$

$$C^T = -C \tag{70b}$$

then

$$\psi(x) = C[\tilde{\psi}_c(x)]^T \tag{71}$$

so that

$$\psi^{(-)}(x) = C[\widetilde{\psi_c{}^{(+)}}(x)]^T \tag{72}$$

$$\tilde{\psi}^{(+)}(x) = [C^{-1}\psi_c{}^{(+)}(x)]^T \tag{73}$$

The interpretation of the operators $\psi_c^{(\pm)}(x)$ is made explicit by their expansion in terms of the d operators, thus, using Eq. (4.233) we find

$$\begin{aligned}\psi_c^{(+)}(x) &= C[\bar{\psi}^{(+)}(x)]^T = \frac{1}{(2\pi)^{3/2}} \int d^3p \left(\frac{m}{E_{\mathbf{p}}}\right)^{1/2} \sum_{r=1}^{2} d_r(\mathbf{p})\, C\bar{v}_r^T(\mathbf{p})\, e^{-ip\cdot x} \\ &= \frac{1}{(2\pi)^{3/2}} \int d^3p \left(\frac{m}{E_{\mathbf{p}}}\right)^{1/2} \sum_{r=1}^{2} d_r(\mathbf{p})\, u_{cr}(\mathbf{p})\, e^{-ip\cdot x} \qquad (74)\end{aligned}$$

and similarly,

$$\widetilde{\psi_c^{(+)}}(x) = \frac{1}{(2\pi)^{3/2}} \int d^3p \left(\frac{m}{E_{\mathbf{p}}}\right)^{1/2} \sum_{r=1}^{2} d_r^*(\mathbf{p})\, \tilde{u}_{cr}(\mathbf{p})\, e^{-ip\cdot x} \qquad (75)$$

so that $\psi_c^{(+)}$ is a destruction operator and $\widetilde{\psi_c^{(+)}}$ a creation operator for an antiparticle. Their commutation rules are:

$$[\psi_{c\alpha}^{(+)}(x), \widetilde{\psi_{c\beta}^{(+)}}(x')]_+ = -iS_{\alpha\beta}^{(+)}(x - x') \qquad (76)$$

which follows from the fact that

$$[C^{-1}S^{(\pm)}(-x)\, C]^T = -S^{(\mp)}(x) \qquad (77)$$

More generally, one verifies that

$$[\psi_c(x), \bar{\psi}_c(x')]_+ = -iS(x - x') \qquad (78)$$

since

$$[C^{-1}S(-x)\, C]^T = -S(x) \qquad (79)$$

The number operators expressed in terms of the configuration space operators are then given by

$$N^{(+)} = \int d^3x\, \widetilde{\psi^{(+)}}(x)\, \gamma^0\psi^{(+)}(x) = \int_\sigma d\sigma^\mu(x)\, \widetilde{\psi^{(+)}}(x)\, \gamma_\mu\psi^{(+)}(x) \qquad (80)$$

$$N^{(-)} = \int_\sigma d\sigma^\mu(x)\, \widetilde{\psi_c^{(+)}}(x)\, \gamma_\mu\psi_c^{(+)}(x) \qquad (81)$$

The expression for the total charge

$$\begin{aligned}Q &= \int d\sigma^\mu(x)\, j_\mu(x) \\ &= -e(N^{(+)} - N^{(-)}) \qquad (82)\end{aligned}$$

together with Eq. (6), the classical expression for the current, suggest the following definition for the current operator:

$$j_\mu(x) = -e : \bar{\psi}(x)\, \gamma_\mu\psi(x) : = -eN(\bar{\psi}(x)\, \gamma_\mu\psi(x)) \qquad (83)$$

where the normal product for fermion operators is defined as follows: Acting on a product of creation and annihilation operators, it rewrites the expression so that all creation operators stand to the left of all destruction operators, the rearrangement being effected as if all *anticommutators* vanish. Thus N includes the change of sign which arises when the order of anticommuting field variables is changed. For example,

$$N(\psi^{(+)}(x)\,\tilde{\psi}^{(-)}(y)) = -N(\tilde{\psi}^{(-)}(y)\,\psi^{(+)}(x)) = -\tilde{\psi}^{(-)}(y)\,\psi^{(+)}(x) \tag{84}$$

$$N(\psi^{(+)}(x)\,\psi^{(+)}(y)) = -N(\psi^{(+)}(y)\,\psi^{(+)}(x)) = -\psi^{(+)}(y)\,\psi^{(+)}(x)$$
$$= \psi^{(+)}(x)\,\psi^{(+)}(y) \tag{85}$$

The distributive law is defined to be valid for the normal product operation.

Conversely, using the definition of normal product and the commutation rules, any operator expression can be rewritten in terms of normal products. For example, the decomposition $\tilde{\psi}_\alpha(x)\,\psi_\beta(y)$ into normal products is accomplished as follows:

$$\begin{aligned}\tilde{\psi}_\alpha(x)\,\psi_\beta(y) &= (\tilde{\psi}_\alpha{}^{(+)}(x) + \tilde{\psi}_\alpha{}^{(-)}(x))\,(\psi_\beta{}^{(+)}(y) + \psi_\beta{}^{(-)}(y)) \\ &= \tilde{\psi}_\alpha{}^{(+)}(x)\,\psi_\beta{}^{(+)}(y) - \psi_\beta{}^{(-)}(y)\,\tilde{\psi}_\alpha{}^{(+)}(x) - iS_{\beta\alpha}^{(-)}(y-x) \\ &\qquad + \tilde{\psi}_\alpha{}^{(-)}(x)\,\psi_\beta{}^{(+)}(y) + \tilde{\psi}_\alpha{}^{(-)}(x)\,\psi_\beta{}^{(-)}(y) \\ &= N(\tilde{\psi}_\alpha{}^{(+)}(x)\,\psi_\beta{}^{(+)}(y) + \psi_\alpha{}^{(+)}(x)\,\psi_\beta{}^{(-)}(y) + \tilde{\psi}_\alpha{}^{(-)}(x)\,\psi_\beta{}^{(+)}(y) \\ &\qquad + \psi_\alpha{}^{(-)}(x)\,\psi_\beta{}^{(-)}(y)) - iS_{\beta\alpha}^{(-)}(y-x) \\ &= N(\tilde{\psi}_\alpha(x)\,\psi_\beta(y)) - iS_{\beta\alpha}^{(-)}(y-x) \end{aligned} \tag{86}$$

In the last line of this derivation we have made use of the distributive law. Using similar algebraic manipulations, one easily derives that

$$\begin{aligned}\psi_\alpha(x)\,\tilde{\psi}_\beta(y) &= N(\psi_\alpha(x)\,\tilde{\psi}_\beta(y)) - iS_{\alpha\beta}^{(+)}(x-y) \\ \psi(x)\,\psi(y) &= N(\psi(x)\,\psi(y)) \\ \tilde{\psi}(x)\,\tilde{\psi}(y) &= N(\tilde{\psi}(x)\,\tilde{\psi}(y))\end{aligned} \tag{87}$$

The vacuum expectation value of any normal product vanishes, so that the vacuum expectation value of the current operator as defined by Eq. (83) vanishes. This is what we would expect of any satisfactory definition of a current operator. Note that this is not the case for the expression $e\tilde{\psi}\gamma_\mu\psi(x)$ whose vacuum expectation value, using (86), is given by

$$\begin{aligned}-e\langle\tilde{\psi}(x)\,\gamma_\mu\psi(x)\rangle_0 &= -e\lim_{x\to x'}\sum_{\alpha\beta}(\Phi_0, \tilde{\psi}_\alpha(x)\,\psi_\beta(x')\,\Phi_0)\,(\gamma_\mu)_{\alpha\beta} \\ &= -e\lim_{x\to x'}\sum_{\alpha\beta} S_{\beta\alpha}^{(-)}(x-x')\,(\gamma_\mu)_{\alpha\beta} \\ &= -e\,\mathrm{Tr}\,(\gamma_\mu S^{(-)}(0))\end{aligned} \tag{88}$$

which is infinite. This infinity for $\mu = 0$ is in fact equal to the contribution of the charge of the "sea" (i.e., of all the occupied negative energy states). The decomposition of the current operator into creation and annihilation operators

$$\begin{aligned}j_\mu(x) = -e(&\widetilde{\psi^{(+)}}(x)\,\gamma_\mu\psi^{(+)}(x) - \widetilde{\psi_c{}^{(+)}}(x)\,\gamma_\mu\psi_c{}^{(+)}(x) \\ &+ C^{-1}\psi_c{}^{(+)}(x)\,\gamma_\mu\psi^{(+)}(x) + \widetilde{\psi^{(+)}}(x)\,\gamma_\mu C\widetilde{\psi_c{}^{(+)}}(x))\end{aligned} \tag{89}$$

again indicates the presence of fluctuation terms, i.e., pair creation and annihilation terms. These terms vanish when integrated over a space-like surface due to the orthogonality of the positive and negative energy solutions of the Dirac equation, so that $\int_\sigma j_\mu(x)\, d\sigma^\mu(x)$ reduces to the expression (82) for the total charge. [See, e.g., Rosenfeld (1953, 1955).]

The basis vectors $(m!n!)^{-1/2}\, \widetilde{\psi^{(+)}}(x_1) \cdots \widetilde{\psi^{(+)}}(x_m)\, \widetilde{\psi_c^{(+)}}(y_1) \cdots \widetilde{\psi_c^{(+)}}(y_n) \mid \Phi_0\rangle$ for all x and y on σ, span the states in which there exist m particles and n antiparticles on σ. A general state of particles and antiparticles is then of the form

$$|\Psi\rangle = \Psi^{(0,0)} \mid 0\rangle + \cdots + (n!m!)^{-1} \prod_{j=1}^{m} \left(\int d\sigma^{\mu_j}(x_j)\, \widetilde{\psi^{(+)}}(x_j)\, \gamma_{\mu_j} \right)$$
$$\prod_{k=1}^{n} \left(\int d\sigma^{\nu_k}(y_k)\, (\widetilde{\psi_c^{(+)}}(y_k)\, \gamma_{\nu_k} \right) \Psi^{(m,n)}(x_1, x_2, \cdots x_m; y_1 \cdots y_n) \mid 0\rangle + \cdots \tag{90}$$

where it is understood that the m particle, n antiparticle amplitude $\Psi^{(m,n)}$ has a four-valued spin index for each of its arguments, on which the γ matrices act [see e.g., Eq. (92) below]. The amplitude $\Psi^{(m,n)}(x_1 \cdots ; \cdots y_1)$ is antisymmetric in the xs and ys separately, satisfies the Dirac equation in each of its variables and contains only positive frequencies since

$$\begin{aligned} \Psi^{(m,n)}&(x_1, \cdots x_m; y_1 \cdots y_n) \\ &= \langle 0 \mid N(\psi(x_1) \cdots \psi(x_m)\, \psi_c(y_1) \cdots \psi_c(y_m)) \mid \Psi\rangle \\ &= \langle 0 \mid \psi^{(+)}(x_1) \cdots \psi^{(+)}(x_m)\, \psi_c^{(+)}(y_1) \cdots \psi_c^{(+)}(y_m) \mid \Psi\rangle \end{aligned} \tag{91}$$

An irreducible representation of the commutation rules which is characterized by the existence of a no-particle state is given by

$$\begin{aligned} (\psi_\alpha^{(+)}(x)\, \Psi)^{(m,n)}&\, (x_1\alpha_1, \cdots x_m\alpha_m; y_1\beta_1, \cdots y_n\beta_n) \\ &= \sqrt{m+1}\, \Psi^{(m+1,n)}(x\alpha, x_1\alpha_1, \cdots x_m\alpha_m; y_1\beta_1, \cdots y_n\beta_n) \end{aligned} \tag{92}$$

$$\begin{aligned} (\widetilde{\psi_\alpha^{(+)}}(x)\, \Psi)^{(m,n)}&\, (x_1\alpha_1, \cdots x_m\alpha_m; y_1\beta_1, \cdots y_n\beta_n) \\ &= \frac{1}{\sqrt{m}} \frac{1}{i} \sum_{j=1}^{m} \sum_{\alpha=1}^{4} (-1)^{j+1}\, S^{(+)}_{\alpha_j\alpha}(x_j - x) \\ &\quad \Psi^{(m-1,n)}(x_1\alpha_1, \cdots x_{j-1}\alpha_{j-1}, x_{j+1}\alpha_{j+1} \cdots x_m\alpha_m; y_1\beta_1 \cdots y_n\beta_n) \end{aligned} \tag{93}$$

and

$$\begin{aligned} (\psi_{c\beta}^{(+)}(y)\, \Psi)^{(m,n)}&\, (x_1\alpha_1, \cdots x_m\alpha_m; y_1\beta_1 \cdots y_n\beta_n) \\ &= \sqrt{n+1}\, (-1)^m\, \Psi^{(m,n+1)}(x_1\alpha_1, \cdots x_m\alpha_m; y\beta, y_1\beta_1 \cdots y_n\beta_n) \end{aligned} \tag{94}$$

$$\begin{aligned} (\widetilde{\psi_{c\beta}^{(+)}}(y)\, \Psi)^{(m,n)}&\, (x_1\alpha_1, \cdots x_m\alpha_m; y_1\beta_1 \cdots y_n\beta_n) \\ &= \frac{1}{\sqrt{n}} \sum_{j=1}^{n} (-1)^{j+m+1}\, S^{(+)}_{\beta_j\beta}(y_j - y) \\ &\quad \Psi^{(m,n-1)}(x_1\alpha_1, \cdots x_m\alpha_m; y_1\beta_1, \cdots y_{j-1}\beta_{j-1}, y_{j+1}\beta_{j+1}, \cdots y_n\beta_n) \end{aligned} \tag{95}$$

The factors $(-1)^m$ in the representation of the charge conjugate operators guarantee that the anticommutation rules $\{\psi_c^{(+)}(x), \psi^{(+)}(x')\} = 0$, etc., are satisfied. [For other representations of the commutation rules, see Wightman (1955).]

The connection between the Dirac theory for a single isolated free particle described by a positive energy spinor and the field theory under discussion is easily established. If $|\Psi\rangle$ is the vector describing the particle, then

$$\Psi^{(1,0)}(x) = (\Phi_0, \psi(x)\, \Psi) \tag{96}$$

is the only nonvanishing amplitude. It satisfies the Dirac equation since $\psi(x)$ does; $\Psi^{(1,0)}(x)$ is thus the "one-particle" Dirac wave function which was discussed in detail in Chapter 4. Note that since $\psi^{(+)}(x) \mid \Phi_0\rangle = \tilde{\psi}^{(+)}(x) \mid \Phi_0\rangle = 0$, upon inserting Eq. (40) into (96) we obtain

$$\Psi^{(1,0)}(x) = \frac{1}{(2\pi)^{3/2}} \sum_{r=1}^{2} \int d^3p \sqrt{\frac{m}{E_{\mathbf{p}}}}\, (\Phi_0, b_r(\mathbf{p})\, \Psi)\, w^r(\mathbf{p})\, e^{-ip\cdot x} \tag{97}$$

so that $\Psi^{(1,0)}(x)$ is a superposition of only positive energy solution of the Dirac equation, in accordance with our previous discussion. It is here an automatic consequence of the fact that the vacuum is the state of lowest energy. Similarly, if $|\Psi\rangle$ is a state of one antiparticle, the amplitude $\Psi^{(0,1)}(x)$

$$\begin{aligned}\Psi^{(0,1)}(x) &= (\Phi_0, \psi_c(x)\, \Psi) \\ &= \frac{1}{(2\pi)^{3/2}} \sum_{r=1}^{2} \int d^3q \sqrt{\frac{m}{E_{\mathbf{q}}}}\, (\Phi_0, d_r(\mathbf{q})\, \Psi)\, u_c^r(\mathbf{q})\, e^{-iq\cdot r}\end{aligned} \tag{98}$$

is the probability amplitude for finding the antiparticle. Note that $\Psi^{(0,1)}(x)$ satisfies the Dirac equation and is also a superposition of positive energy solutions, since the time dependence is $\exp -iq_0x_0$ with $q_0 > 0$.

8c. Transformation Properties

In our discussion in Chapter 4 of the relativistic invariance of the Dirac equation, we saw that if under inhomogeneous Lorentz transformations, $x' = \Lambda x + a$, the one-particle wave function $\Psi^{(1,0)}(x)$ transformed according to

$$\Psi'^{(1,0)}(x') = S(\Lambda)\, \Psi^{(1,0)}(\Lambda^{-1}(x' - a)) \tag{99}$$

where $S(\Lambda)$ is a 4×4 nonsingular matrix operating on the spinor indices of $\Psi^{(1,0)}$ which satisfies

$$S^{-1}\gamma^\lambda S = \Lambda^\lambda{}_\nu \gamma^\nu \tag{100a}$$

$$\det S = 1 \tag{100b}$$

then the one-particle Dirac equation was form invariant. We shall now determine the transformation properties of the field operators assuming that the transformation law of the (m, n) amplitude is

$$\langle x_1\alpha_1, \cdots y_n\beta_n \mid U(a, \Lambda) \mid \Psi\rangle = (U(a, \Lambda)\,\Psi)^{(m,n)}(x_1\alpha_1 \cdots x_m\alpha_m; y_1\beta_1 \cdots y_n\beta_n)$$

$$= \sum_{\{\alpha'\}\{\beta'\}} \prod_{j=1}^{m} S(\Lambda)_{\alpha_j\alpha'_j} \prod_{k=1}^{n} S(\Lambda)_{\beta_k\beta'_k}$$

$$\Psi^{(m,n)}(\Lambda^{-1}(x_1 - a), \alpha_1', \cdots; \cdots \Lambda^{-1}(y_n - a), \beta'_n) \quad (101)$$

Just as in the situation discussed in Chapter 7, we shall here adopt the Schrödinger type of transformation wherein the wave function (i.e., the state vector) is transformed, but the *same* set of operators are used by the observers in both S and S'. An identical procedure as that used for the scalar field now yields the result

$$(U(a, \Lambda)\,\psi_\alpha(x)\,U(a, \Lambda)^{-1}\,U(a, \Lambda)\,\Psi)^{(m,n)}\,(x_1, \cdots x_m; y_1 \cdots y_n)$$

$$= \sum_{\alpha'=1}^{4} S(\Lambda)^{-1}_{\alpha\alpha'}\,(\psi_{\alpha'}(\Lambda x + a)\,U(a, \Lambda)\,\Psi)^{(m,n)}\,(x_1, \cdots x_n; y_1 \cdots y_n) \quad (102)$$

where the identity

$$S(\Lambda)\,S^{(+)}(\Lambda^{-1}(y - a) - x)\,C = S^{(+)}(y - (\Lambda x + a))\,CS(\Lambda)^{-1T} \quad (103)$$

which is valid for restricted Lorentz transformation, and the representations (92)–(95) for the operators $\psi^{(+)}$ and $\widetilde{\psi_c^{(+)}}$ have been used. The proof of (103) is as follows: From (55) and (52) it follows that

$$S^{(+)}(x) = -\frac{1}{(2\pi)^3} \int_{p_0>0} \frac{d^3p}{p_0}\,(\not p + m)\,e^{-ip\cdot x} \quad (104)$$

whence

$$S^{(+)}(\Lambda^{-1}(y - a) - x)\,C = -\frac{1}{(2\pi)^3} \int \frac{d^3p}{p_0}\,(\not p + m)\,Ce^{-i\Lambda p\cdot\Lambda(\Lambda^{-1}(y-a)-x)}$$

$$= -\frac{1}{(2\pi)^3} \int \frac{d^3p}{p_0}\,[\gamma\cdot(\Lambda^{-1}p) + m]\,Ce^{ip\cdot(y-\Lambda x-a)} \quad (105)$$

Using (100), we find that

$$\gamma\cdot(\Lambda^{-1}p)\,C = \Lambda_{\nu\mu}p^\nu\gamma^\mu C = S(\Lambda)^{-1}\,\gamma_\nu S(\Lambda)\,Cp^\nu$$

$$= S(\Lambda)^{-1}\,\gamma_\nu C[S(\Lambda)^T]^{-1}\,p^\nu \quad (106)$$

so that

$$S^{(+)}(\Lambda^{-1}(y - a) - x)\,C$$

$$= -\frac{1}{(2\pi)^3}\,S(\Lambda)^{-1} \int \frac{d^3p}{p_0}\,(\not p + m)\,e^{-ip\cdot(y-\Lambda x-a)}CS(\Lambda)^{-1T} \quad (107)$$

which is Eq. (103).

Thus, for restricted Lorentz transformations

$$U(a, \Lambda)\, \psi_\alpha(x)\, U(a, \Lambda)^{-1} = \sum_{\alpha=1}^{4} S(\Lambda)^{-1}_{\alpha\alpha'}\, \psi_{\alpha'}(\Lambda x + a) \tag{108}$$

Upon taking the adjoint of (108) and multiplying by γ^0, we derive that

$$U(a, \Lambda)\, \bar{\psi}(x)\, U(a, \Lambda)^{-1} = \bar{\psi}(\Lambda x + a)\, S(\Lambda) \tag{109}$$

from which it also follows that

$$U(a, \Lambda)\, \psi_c(x)\, U(a, \Lambda)^{-1} = S(\Lambda)^{-1}\, \psi_c(\Lambda x + a) \tag{110}$$

Equations (108)–(110) guarantee the covariance of the free-field Lagrangian and commutation rules under homogeneous Lorentz transformations.

For a space inversion, $x \to x' = i_s x$, we define the transformation law of the field operators as

$$U(i_s)\, \psi(x)\, U(i_s)^{-1} = \eta_P \gamma_0 \psi(i_s x) \tag{111a}$$

$$U(i_s)\, \bar{\psi}(x)\, U(i_s)^{-1} = \bar{\eta}_P \bar{\psi}(i_s x)\, \gamma_0 \tag{111b}$$

The free-field Lagrangian as well as the free-field commutation rules will be invariant under $U(i_s)$ as defined above if $|\eta_P|^2 = 1$ and $U(i_s)$ is unitary. Now two reflections in succession may be considered either as a reversion to the original co-ordinate system or as a rotation through 2π. Since under a rotation through 2π a Fermi field transforms according to $\psi \to -\psi$, the square of η_P can be taken as either $+1$ or -1, i.e., $\eta_P{}^2 = \pm 1$, so that $\eta_P = \pm 1$ or $\pm i$ [Yang and Tiomno (Yang 1950)]. The vacuum is defined to be invariant under $U(i_s)$

$$U(i_s) \mid 0\rangle = |0\rangle \tag{112}$$

With the choice $\eta_P = \pm i$ the charge conjugate operator ψ_c will have the same transformation properties as ψ [Racah (1937)], except that the phase factor η becomes replaced by $\bar{\eta}$

$$U(i_s)\, \psi_c(x)\, U(i_s)^{-1} = \bar{\eta}_P \gamma_0 \psi_c(i_s x) \tag{113}$$

We shall always impose this further condition on the theory so that for us η_P will always be either $\pm i$.

The transformation law of the operators $b_s(\mathbf{p})$ and $d_s(\mathbf{p})$, where s labels the two states of definite z component of the spin corresponding to Eqs. (4.118a) and (4.118b), is obtained by substituting the expansions (40) and (41) into (111a) and (111b); one finds

$$U(i_s)\, b_s(\mathbf{p})\, U(i_s)^{-1} = \eta_P b_s(-\mathbf{p}) \tag{114a}$$

$$U(i_s)\, d_s(\mathbf{p})\, U(i_s)^{-1} = -\bar{\eta}_P d_s(-\mathbf{p}) \tag{114b}$$

where use has been made of the following properties of the solutions of the Dirac equation:

$$\gamma^0 w_r(-\mathbf{p}) = w_r(\mathbf{p}) \tag{115}$$

$$\gamma^0 v_r(-\mathbf{p}) = -v_r(\mathbf{p}) \tag{116}$$

Under the operation of space inversion the one-particle states are therefore transformed as follows:

$$U(i_s)\, b_s^*(\mathbf{p}) \mid 0\rangle = \bar{\eta}_P b_s^*(-\mathbf{p}) \mid 0\rangle = \bar{\eta}_P|-\mathbf{p}, s;\rangle \tag{117}$$

$$U(i_s)\, d_t^*(\mathbf{q}) \mid 0\rangle = -\eta_P d_t^*(-\mathbf{q}) \mid 0\rangle = -\eta_P|\,;\, -\mathbf{q}, t\rangle \tag{118}$$

A space inversion thus transforms a particle of momentum $\mathbf{p}$ into one of momentum $-\mathbf{p}$ and leaves the spin direction unchanged. The operator $b_\uparrow^*(\mathbf{p})$ corresponding to the creation of a particle in a state of definite helicity with the spin pointing in the same direction as the momentum would therefore be transformed under $U(i_s)$ into $\bar{\eta}_P b_\downarrow^*(-\mathbf{p})$, the creation operator for a particle of momentum $-\mathbf{p}$ and spin antiparallel to the direction of motion. Note the minus sign in Eq. (114b). It has the consequence that the state containing one particle and one antiparticle in an S state of orbital momentum will, under a space inversion, acquire a factor $-|\eta_P|^2 = -1$; in other words, such a state has a negative parity relative to the vacuum. Fermions and antifermions therefore have opposite intrinsic parities in contradistinction to the Bose case, where particle and antiparticle have the same intrinsic parity.

The transformation laws (111) for the field operators $\psi(x)$ and $\bar{\psi}(x)$ imply definite transformation properties for the bilinear form $\bar{\psi}(x)\, O_i\psi(x)$ where $O_i = 1,\ \gamma_\mu,\ \sigma_{\mu\nu},\ \gamma_5\gamma_\mu,\ \gamma_5$. These are summarized in Table 8.1. In the latter ϵ_i is defined by

$$\begin{aligned} U(i_s)\, \bar{\psi}(x)\, O_i\psi(x)\, U(i_s) &= |\eta_P|^2\, \bar{\psi}(i_sx)\, \gamma_0 O_i \gamma_0 \psi(i_sx) \\ &= \epsilon_i \bar{\psi}(i_sx)\, O_i\psi(i_sx) \end{aligned} \tag{119}$$

TABLE 8.1

$i =$	*Scalar*	*Vector*	*Tensor*	*Axial Vector*	*Pseudo-scalar*
$O_i =$	1	γ_μ	$\sigma_{\mu\nu}$	$\gamma_5\gamma_\mu$	γ_5
$\epsilon_i =$	1	$+1 \quad \mu = 0$ $-1 \quad \mu = 1, 2, 3$	$+1 \quad (\mu, \nu = 1, 2, 3)$ $-1 \quad (\mu \text{ or } \nu = 0)$	$-1 \quad \mu = 0$ $+1 \quad \mu = 1, 2, 3$	-1

Under the operation of charge conjugation, whereby every state is mapped into the state where all particles are replaced by their antiparticles with the same energy, momentum and spin, the field operators are defined to transform according to

$$U_c b_s^*(\mathbf{p})\, U_c^{-1} = \bar{\eta}_C d_s^*(\mathbf{p}) \tag{120a}$$

$$U_c d_s^*(\mathbf{p})\, U_c^{-1} = \eta_C b_s^*(\mathbf{p}) \tag{120b}$$

or equivalently

$$U_c\psi(x)\,U_c^{-1} = \eta_C\psi_c(x) = \eta_C C\tilde{\psi}^T(x) \tag{120c}$$

and

$$U_c\tilde{\psi}(x)\,U_c^{-1} = \bar{\eta}_C\tilde{\psi}_c(x) = -\bar{\eta}_C\psi^T(x)\,C^* \tag{120d}$$

The invariance of the commutation rules requires that η_C have modulus one and that U_c is unitary. In checking the commutation rules use is made of the relation:

$$C^{-1}S(x - x')\,C = -S^T(x' - x) \tag{121}$$

which is proved in the same way as Eq. (103). The vacuum is postulated to be invariant under the operation U_c, i.e., $U_c\,|\,0\rangle = |0\rangle$.

The theory will therefore be invariant under charge conjugation if the phase factor η_C can be so chosen that the Hamiltonian H is invariant under U_c, i.e., if $U_cHU_c^{-1} = H$. Since under the operation of charge conjugation a product of operators is transformed into the product of the Hermitian adjoint operators, the final order of the factors is in general different from that from which one started. In particular, the Hamiltonian expressed as in Eq. (9) or (10) is not invariant. To make the theory charge conjugation invariant one must either antisymmetrize the factors in the Hamiltonian [Lüders (1957), Schwinger (1951a)] or alternatively postulate a normal ordering, i.e., one must write

$$H = \frac{i}{2}\int d^3x\,\{\tfrac{1}{2}[\tilde{\psi}\gamma^0, \partial_0\psi]_- - \tfrac{1}{2}[\partial_0\tilde{\psi}\gamma^0, \psi]_-\} \tag{122a}$$

or

$$H = \frac{i}{2}\int d^3x : \tilde{\psi}\gamma^0\partial_0\psi - \partial_0\tilde{\psi}\gamma^0 \cdot \psi : \tag{122b}$$

For the free-field theory the two methods are completely equivalent. For the case of interacting fields in the Heisenberg picture, where the notion of normal product cannot be as straightforwardly defined, the process of antisymmetrization is usually invoked to guarantee the invariance of the theory under charge conjugation. We leave it as an exercise to the reader to obtain the transformation properties of the bilinear forms $\tilde{\psi}O_i\psi$ under charge conjugation. We only note that under charge conjugation

$$U_cj_\mu(x)\,U_c^{-1} = -j_\mu(x) \tag{123}$$

We define the transformation law of the field under time reversal to be

$$U(i_t)\,\psi_\alpha(x)\,U(i_t)^{-1} = \eta_T\sum_{\alpha'=1}^{4}(C^{-1}\gamma_5)_{\alpha\alpha'}\,\psi_{\alpha'}(i_tx) \tag{124a}$$

$$U(i_t)\,\tilde{\psi}_\alpha(x)\,U(i_t)^{-1} = \bar{\eta}_T\sum_{\alpha'=1}^{4}\tilde{\psi}_{\alpha'}(i_tx)\,(\gamma_5C)_{\alpha'\alpha} \tag{124b}$$

To preserve the invariance of the commutation rules, $U(i_t)$ must be antiunitary and η_T must have modulus one. One then also verifies that H is invariant, in the sense that $U(i_t)\, \mathcal{H}(x)\, U(i_t)^{-1} = \mathcal{H}(i_t x)$ and $U(i_t)\, H(t)\, U(i_t)^{-1} = H(-t)$. The transformation properties of the bilinear covariants [Lüders (1952a)] under time inversion are as follows:

$$U(i_t)\, \bar{\psi}(x)\, \psi(x)\, U(i_t)^{-1} = \bar{\psi}(i_t x)\, \psi(i_t x) \tag{125}$$

$$U(i_t)\, \bar{\psi}(x)\, \gamma_5 \psi(x)\, U(i_t)^{-1} = -\bar{\psi}(i_t x)\, \gamma_5 \psi(i_t x) \tag{126}$$

$$U(i_t)\, \bar{\psi}(x) \begin{Bmatrix} i\gamma_5\gamma^0 \\ i\gamma_5\gamma^i \end{Bmatrix} \psi(x)\, U(i_t)^{-1} = \bar{\psi}(i_t x) \begin{Bmatrix} i\gamma_5\gamma^0 \\ -i\gamma_5\gamma^i \end{Bmatrix} \psi(i_t x) \tag{127}$$

$$U(i_t)\, \bar{\psi}(x) \begin{Bmatrix} \gamma^0 \\ \gamma^i \end{Bmatrix} \psi(x)\, U(i_t)^{-1} = \bar{\psi}(i_t x) \begin{Bmatrix} \gamma^0 \\ -\gamma^i \end{Bmatrix} \psi(i_t x) \tag{128}$$

$$U(i_t)\, \bar{\psi}(x) \begin{Bmatrix} \sigma^{0j} \\ \sigma^{jk} \end{Bmatrix} \psi(x)\, U(i_t)^{-1} = \bar{\psi}(i_t x) \begin{Bmatrix} \sigma^{0j} \\ -\sigma^{jk} \end{Bmatrix} \psi(i_t x) \tag{129}$$

In the derivation the antiunitary character of $U(i_t)$ plays an important role in that

$$\begin{aligned} U(i_t)\, \bar{\psi}(x)\, O_i \psi(x)\, U(i_t)^{-1} &= U(i_t)\, \bar{\psi}(x)\, U(i_t)^{-1}\, U(i_t)\, O_i \psi(x)\, U(i_t)^{-1} \\ &= U(i_t)\, \bar{\psi}(x)\, U(i_t)^{-1}\, \bar{O}_i U(i_t)\, \psi(x)\, U(i_t)^{-1} \\ &= \bar{\psi}(i_t x)\, \gamma_5 C \bar{O}_i C^{-1} \gamma_5 \psi(i_t x) \end{aligned} \tag{130a}$$

Since

$$\bar{O}_i = O_i^{*T} = -C^{-1} O_i^* C \tag{130b}$$

the right-hand side in Eq. (130a) is determined from a computation of $-\gamma_5 O^* \gamma_5$.

8*d*. The Field Theoretic Description of Nucleons

There is much evidence, based on the analysis of the properties of nuclei, that to the approximation that electromagnetic effects and weak interactions can be neglected, protons and neutrons have identical properties. They both have spin $\frac{1}{2}$, they have the same space parity (arbitrarily taken as $+1$) and have very nearly the same mass. [Their mass difference is conjectured to be attributable to electromagnetic effects (Feynman 1954). The difference in their magnetic moment is observable only in the presence of the electromagnetic field which breaks down their "similar" character and, of course, allows a differentiation between neutron and proton.] The nuclear forces (in the approximation that electromagnetic effects are neglected) are known to be the same for any two nucleons in the same spin and orbital angular momentum state. This suggests that neutron and proton are two states of one kind of particle, the nucleon. [For a historical review of the subject see Bethe and de Hoffmann (Bethe 1955).] Let us see how this can be formalized. Consider a system of nucleons described by the Hamiltonian

$$H = \int d^3x : \{\tilde{p}(x) (-i\gamma \cdot \partial + m_p) p(x) + \tilde{n}(x) (-i\gamma \cdot \partial + m_n) n(x)\} : \tag{131}$$

where p, $\tilde{p}$ and n, $\tilde{n}$ are the spinor operators describing the protons and neutrons, respectively. Their anticommutation rules are:

$$\{p(x), \tilde{p}(x')\}_{x_0=x'_0} = \{n(x), \tilde{n}(x')\}_{x_0=x'_0} = +\delta(\mathbf{x} - \mathbf{x}') \tag{132}$$

all other equal-time anticommutators vanishing. Such a description implies the existence of antinucleons. These, in fact, have been observed experimentally.

We next introduce the operator

$$\psi = \begin{pmatrix} p \\ n \end{pmatrix} \tag{133a}$$

which we shall assume to transform like a two-component spinor in a three-dimensional Euclidian "charge" space, with the adjoint of ψ defined as

$$\tilde{\psi} = (\tilde{p} \quad \tilde{n}) \tag{133b}$$

We shall in the subsequent exposition consider the γ matrices as outer products of the four-dimensional γ matrices and the unit matrix in charge space, i.e., consider them to form a reducible eight-dimensional representation

$$\text{"}\gamma\text{"} = \gamma \otimes \begin{pmatrix} 1 & 0 \\ 0 & 1 \end{pmatrix} \tag{134}$$

If we now assume the equality of the proton and neutron mass $m_n = m_p = m$, H can then be rewritten as

$$H = \int d^3x\, \tilde{\psi}(x) (-i\gamma \cdot \partial + m) \psi(x) \tag{135}$$

If we introduce the 2×2 Pauli matrices τ_i, with τ_3 given by

$$\tau_3 = \begin{pmatrix} 1 & 0 \\ 0 & -1 \end{pmatrix} \tag{136}$$

then $\frac{1}{2}(1 + \tau_3)$ is a projection operator for proton states since

$$\tfrac{1}{2}(1 + \tau_3) \begin{pmatrix} p \\ n \end{pmatrix} = \begin{pmatrix} 1 & 0 \\ 0 & 0 \end{pmatrix} \begin{pmatrix} p \\ n \end{pmatrix} = \begin{pmatrix} p \\ 0 \end{pmatrix} \tag{137}$$

and, similarly, $\frac{1}{2}(1 - \tau_3)$ is a projection operator for neutron states. The total charge for the system is given by

$$Q = e \int d^3x\, \tfrac{1}{2}[\tilde{p}, \gamma^0 p] \tag{138}$$

which can be rewritten in terms of ψ as

$$\begin{aligned} Q &= \tfrac{1}{2}e \int d^3x\, [\tilde{\psi}, \tfrac{1}{2}(1 + \tau_3)\, \gamma^0 \psi] \\ &= e \int : \tilde{\psi} \tfrac{1}{2}(1 + \tau_3)\, \gamma^0 \psi : d^3x \end{aligned} \tag{139}$$

Under an infinitesimal rotation, ϵ, about the lth axis, ψ transforms according to

$$\psi \to (1 + \tfrac{1}{2}i\epsilon\tau_l)\,\psi \tag{140}$$

the required transformation law for a spinor in a Euclidian three-space. It then follows that H is a scalar in charge space since it is invariant under rotations. (Recall Chapter 1 where the transformation properties of the bilinear quantities $\psi^* O_i \psi$ under rotations were discussed.) The unitary transformation, $\exp i\epsilon T_l$, which generates an infinitesimal rotation about the lth axis can be determined from the requirement that

$$\begin{aligned} e^{-i\epsilon T_l}\psi\, e^{+i\epsilon T_l} &= \psi - i\epsilon[T_l, \psi] \\ &= (1 + \tfrac{1}{2}i\epsilon\tau_l)\,\psi \end{aligned} \tag{141}$$

that is, that

$$-\,[T_l, \psi] = \tfrac{1}{2}\tau_l\psi \tag{142}$$

A possible choice for T is

$$\begin{aligned} T_j &= \tfrac{1}{2}\int d^3x : \psi^*\tau_j\psi : \\ &= \tfrac{1}{2}\int d^3x\, [\bar{\psi}\gamma^0, \tfrac{1}{2}\tau_j\psi] = \tfrac{1}{2}\int d\sigma^\mu(x)\, [\bar{\psi}\gamma_\mu, \tfrac{1}{2}\tau_j\psi] \end{aligned} \tag{143}$$

Upon comparing Eqs. (143) and (139), we note that the expression for the total charge Q may be rewritten in terms of T_3 as follows:

$$\frac{Q}{e} = \frac{1}{2}\int d\sigma^\mu : \bar{\psi}\gamma_\mu\psi : + T_3 \tag{144}$$

where the expression $\int d\sigma^\mu : \bar{\psi}\gamma_\mu\psi :$ will be recognized as the total number of nucleons minus the total number of antinucleons. One verifies that the one-particle states are eigenstates of T_3. With an obvious notation, one finds:

$$\begin{aligned} T_3\,|\,p\rangle &= \tfrac{1}{2}\,|\,p\rangle \\ T_3\,|\,\bar{p}\rangle &= -\tfrac{1}{2}\,|\,\bar{p}\rangle \\ T_3\,|\,n\rangle &= -\tfrac{1}{2}\,|\,n\rangle \\ T_3\,|\,\bar{n}\rangle &= \tfrac{1}{2}\,|\,\bar{n}\rangle \end{aligned} \tag{145}$$

The commutation rules of the operators T_i are determined from those of ψ, and one establishes that

$$[T_l, T_j] = i\epsilon_{ljk}T_k \tag{146}$$

so that once again the Ts are isomorphic to the angular momentum operators, with similar consequences [see, e.g., Malenka (1957)]. As in the mesonic case, the T_is are constants of the motion since they are the space integrals of the fourth component of a conserved current. From the invariance of the Lagrangian under rotations in "isotopic spin space" (the space in which ψs are considered to transform as spinors), under which

$$\psi \to \psi + \tfrac{1}{2} i\epsilon\tau_l\psi \tag{147}$$

or

$$\delta\psi = \tfrac{1}{2} i\epsilon\tau_l\psi \tag{148a}$$

$$\delta\tilde{\psi} = -\tfrac{1}{2} i\epsilon\tilde{\psi}\tau_l \tag{148b}$$

we infer the following conservation laws [recall Eq. (7.339)]:

$$0 = \partial^\mu \left(\frac{\partial \mathfrak{L}}{\partial(\partial^\mu \psi)} \delta\psi - \frac{\partial \mathfrak{L}}{\partial(\partial^\mu \tilde{\psi})} \delta\tilde{\psi} \right) = 0$$

$$= \partial_\mu : \tilde{\psi}\gamma^\mu\tau_l\psi : \tag{149}$$

Stated differently, Eq. (149) asserts that the isocurrent

$$\mathfrak{J}_l^\mu = \tfrac{1}{2} : \tilde{\psi}\gamma^\mu\tau_l\psi : \tag{150}$$

is conserved in space-time, $\partial_\mu \mathfrak{J}_i^\mu = 0$ for $i = 1, 2, 3$ so that

$$T_i = \int_\sigma d\sigma_\mu \mathfrak{J}_i^\mu \tag{151}$$

is a constant of the motion.

9

Quantization of the Electromagnetic Field

9a. Classical Lagrangian

A gauge invariant Lagrangian for the electromagnetic field is given by[1]

$$\mathcal{L} = -\tfrac{1}{4}F_{\mu\nu}(x)\, F^{\mu\nu}(x) \tag{1}$$

where $F^{\mu\nu}$ is the electromagnetic field tensor which is related to the potentials by

$$F^{\mu\nu}(x) = \frac{\partial A^\mu(x)}{\partial x_\nu} - \frac{\partial A^\nu(x)}{\partial x_\mu} = A^{\mu,\nu}(x) - A^{\nu,\mu}(x) \tag{2}$$

Here $A^\mu = (\phi, \mathbf{A})$, $F^{0k} = \mathcal{E}^k$ is the electric field vector, $F^{kl} = \epsilon_{klj}\mathcal{H}^j$ is the magnetic field vector, and $\mathcal{L} = -\tfrac{1}{2}(\mathcal{E}^2 - \mathcal{H}^2)$. The variation of the Lagrangian $\mathcal{L}$ with respect to the *potentials* then yields Maxwell's equations,

$$\frac{\partial F^{\mu\nu}(x)}{\partial x^\nu} = 0 \tag{3}$$

which in terms of the potentials read

$$\Box A^\mu(x) - \partial^\mu \chi(x) = 0 \tag{4a}$$

$$\chi(x) = \partial^\nu A_\nu(x) \tag{4b}$$

In classical theory we may choose the Lorentz gauge $\chi = 0$; Maxwell's equations are then equivalent to $\Box A^\mu(x) = 0$ with $\chi = 0$.

The difficulty with the above Lagrangian is that the momentum canonically conjugate to A^0 vanishes identically, so that the Hamiltonian theory must be amended. In the quantum theory, where A^0 and π^0 become operators satisfying certain commutation rules, the fact that π^0 vanishes presents further difficulties.

To bypass this difficulty, the Lagrangian that is often written for the electromagnetic field is due to Fermi (1929, 1930, 1932), and is given by

[1] Throughout this book we shall use Heaviside-Lorentz rationalized units, so that $e^2/4\pi\hbar c = \alpha = 1/(137)$. The potentials here used are related to the Gaussian unit ones, by a factor of $1/\sqrt{4\pi}$. In the Gaussian system of units, the Lagrangian (1) would read $\mathcal{L} = -(16\pi)^{-1} F_{\mu\nu}F^{\mu\nu}$.

$$\mathcal{L} = -\frac{1}{4} F_{\mu\nu}F^{\mu\nu} - \frac{1}{2}\left(\frac{\partial A_\mu}{\partial x_\mu}\right)^2 \tag{5}$$

It is not gauge invariant, due to the presence of the χ^2 term. It is, however, clearly relativistically invariant. The use of this Lagrangian then yields the following equations of motion upon variation with respect to A_μ:

$$\partial_\mu(\partial^\nu A^\mu - \partial^\mu A^\nu) - \partial^\nu(\partial_\mu A^\mu) = 0 \tag{6a}$$

or

$$\Box A^\nu = 0 \tag{6b}$$

These are not yet equivalent to Maxwell equations. In order that they be equivalent we must impose subsidiary conditions, which are that

$$\chi = 0 \quad \text{at } t = 0 \tag{7a}$$

and

$$\frac{\partial \chi}{\partial t} = 0 \quad \text{at } t = 0 \tag{7b}$$

Then, by virtue of the equation of motion $\Box\chi = 0$ which follows from Eq. (6), $\chi = 0$ for all times. We shall later discuss the effect of this subsidiary condition in the quantum theoretical case.

The Lagrangian (5) may be rewritten in the form

$$\mathcal{L} = -\frac{1}{2}\left(\frac{\partial A^\mu}{\partial x_\nu}\right)\left(\frac{\partial A_\mu}{\partial x^\nu}\right) - \frac{1}{2}\frac{\partial}{\partial x^\mu}\left(A^\nu \frac{\partial A^\mu}{\partial x^\nu}\right) - \frac{1}{2}\chi^2 + \frac{1}{2}A^\nu \frac{\partial \chi}{\partial x^\nu} \tag{8}$$

which is equal to

$$\mathcal{L} = -\tfrac{1}{2}A^{\mu,\nu}A_{\mu,\nu} \tag{9}$$

when $\chi = 0$ and when one neglects the divergence term. The momentum canonically conjugate to A_μ, using (9), is now

$$\pi^\mu = \frac{\partial \mathcal{L}}{\partial A_{\mu,0}} = -A^{\mu,0} \tag{10}$$

so that the Hamiltonian H is given by the following expression:

$$H = \int \mathcal{H}\, d^3x \tag{11a}$$

$$\mathcal{H} = -\tfrac{1}{2}\pi^\mu\pi_\mu - \tfrac{1}{2}\sum_{k=1}^{3} A_{\mu,}{}^{k}A^{\mu,k} \tag{11b}$$

The energy is thus similar in structure to the superposition of four-scalar fields. It is, however, not positive-definite since the $\mu = 0$ component contributes a negative-definite quantity to H. This is because we have as yet not made use of the subsidiary condition. The latter can in fact be used to cast the Hamiltonian into a positive-definite form, namely, the familiar $\tfrac{1}{2}(\mathcal{E}^2 + \mathcal{H}^2)$ [see Gupta (1950a) and below].

It should be noted that all the above considerations were based on the use of the potentials as the field variables. Particularly in the quantum theoretical case, this becomes a source of difficulties, since the As are not uniquely defined; only their four-dimensional curl, $A_{\mu,\nu} - A_{\nu,\mu}$, is observable, i.e., only the electric and magnetic *fields* are observable. Therefore, in a description using only the potentials A_μ, we have the freedom of making gauge transformations whereby

$$A_\mu \to A_\mu + \frac{\partial \Lambda}{\partial x^\mu} \tag{12}$$

since the field tensor $F_{\mu\nu}$ is invariant under this gauge transformation. In Eq. (12) Λ is an arbitrary function which, due to the subsidiary condition $\chi = 0$ must satisfy $\Box\Lambda = 0$. The requirement of *gauge invariance* is that all physically observable field quantities must remain invariant under the transformation (12).

9b. Quantization: The Gupta-Bleuler Formalism

The canonical commutation rules read:

$$[\pi^\mu(x), A_\nu(x')]_{x_0 = x'_0} = +i\hbar c\delta^\mu{}_\nu \delta(\mathbf{x} - \mathbf{x}') \tag{13a}$$

or equivalently

$$[A^{\mu,0}(x), A_\nu(x')]_{x_0 = x'_0} = -i\hbar c\delta^\mu{}_\nu \delta(\mathbf{x} - \mathbf{x}') \tag{13b}$$

An argument similar to that leading to the commutation rules for the scalar field now indicates that the covariant commutation rules for the electromagnetic potentials are

$$[A_\mu(x), A_\nu(x')] = -i\hbar c g_{\mu\nu} D(x - x') \tag{14}$$

where $D(x - x')$ is equal to $\Delta(x - x')$ with $\mu = 0$. The difference in sign in the right-hand side of (14) for the time component as compared to the space components should be noted. Explicitly, $D(x)$ is given by [Jordan and Pauli (Jordan 1928a)]

$$\begin{aligned} D(x) &= -\frac{1}{(2\pi)^3} \int_{k_0>0} \frac{d^3k}{k_0} e^{i\mathbf{k}\cdot\mathbf{x}} \sin k_0 x_0 = -\frac{1}{2\pi^2|\mathbf{x}|} \int_0^\infty \sin k|\mathbf{x}| \sin k x_0 \, dk \\ &= -\frac{1}{4\pi|\mathbf{x}|} \{\delta(|\mathbf{x}| - x_0) - \delta(|\mathbf{x}| + x_0)\} \\ &= -\frac{1}{2\pi} \epsilon(x_0)\, \delta(x^2) \end{aligned} \tag{15}$$

The even singular function $D^{(1)}$ is given by[2]

[2] Note $\lim\limits_{\epsilon\to 0+} i \int_0^\infty e^{-i(x - i\epsilon)k}\, dk = \lim\limits_{\epsilon \to 0+} \frac{1}{x - i\epsilon} = P\frac{1}{x} + i\pi\delta(x)$

$$D^{(1)}(x) = \frac{1}{(2\pi)^3}\int_{k_0>0}\frac{d^3k}{k_0}e^{i\mathbf{k}\cdot\mathbf{x}}\cos k_0x_0 = \frac{1}{2\pi^2|\mathbf{x}|}\int_0^\infty dk\,\sin k|\mathbf{x}|\cos kx_0$$
$$= \frac{1}{4\pi^2|\mathbf{x}|}\left\{\mathrm{P}\frac{1}{|\mathbf{x}|-x_0}+\mathrm{P}\frac{1}{|\mathbf{x}|+x_0}\right\}$$
$$= -\frac{1}{2\pi^2}\mathrm{P}\frac{1}{x^2} \tag{16}$$

where P denotes the fact that the principal value is to be taken when integrating over the singularity at $x^2 = 0$. With the help of the commutation rules (13), (14), one verifies that

$$[H, A_\mu(x)] = -\tfrac{1}{2}\int[\pi^\nu(x')\,\pi_\nu(x'), A_\mu(x)]\,d^3x'$$
$$= i\hbar c\pi_\mu(x) \tag{17}$$

Since the right-hand side of this equation is equal to $i\hbar cA_{\mu,0} = i\hbar c\partial_0A_\mu$, the commutation rules are consistent with the interpretation of the Hamiltonian (11) as the time translation operator.

Recalling that there are four linearly independent plane wave solutions of the equation $\Box A_\mu = 0$ for each three-dimensional momentum $\mathbf{k}$, the expansion of the operator $A_\mu(x)$ is given by

$$A_\mu(x) = \sqrt{\frac{\hbar c}{2(2\pi)^3}}\int\frac{d^3k}{k_0}\sum_{\lambda=0}^{3}\epsilon^{(\lambda)}{}_\mu(\mathbf{k})\,\{a^{(\lambda)}(\mathbf{k})\,e^{-ik\cdot x}+a^{(\lambda)*}(\mathbf{k})\,e^{ik\cdot x}\}$$
$$k_0 = |\mathbf{k}| \tag{18}$$

where the $\epsilon^{(\lambda)}{}_\mu(\mathbf{k})$, $\lambda = 0, 1, 2, 3$, are four (linearly independent) unit polarization vectors which can be so chosen that they form an orthonormal system with

$$\epsilon^{(\lambda)}{}_\mu\epsilon^{(\lambda')\mu} = g^{\lambda\lambda'} \tag{19}$$

where $g^{\lambda\lambda'}$ is the metric tensor. In Eq. (18) the sum over (λ) is an ordinary sum (not the Lorentz invariant scalar product) over the four solutions corresponding to the different polarizations. We shall also consider the operators

$$a_\mu(\mathbf{k}) = \sum_{\lambda=0}^{3}\epsilon^{(\lambda)}{}_\mu(\mathbf{k})\,a^{(\lambda)}(\mathbf{k}) \tag{20}$$

$$a_\mu{}^*(\mathbf{k}) = \sum_{\lambda=0}^{3}\epsilon^{(\lambda)}{}_\mu(\mathbf{k})\,a^{(\lambda)*}(\mathbf{k}) \tag{21}$$

which satisfy the commutation rules:

$$[a_\mu(\mathbf{k}), a_\nu{}^*(\mathbf{k}')] = -g_{\mu\nu}k_0\delta^{(3)}(\mathbf{k}-\mathbf{k}');\ k_0 = |\mathbf{k}| \tag{22a}$$
$$[a_\mu(\mathbf{k}), a_\nu(\mathbf{k}')] = [a_\mu{}^*(\mathbf{k}), a_\nu{}^*(\mathbf{k}')] = 0 \tag{22b}$$

as is verified from Eq. (14). The Hamiltonian when expressed in terms of these operators is given by

$$H = -\int \frac{d^3k}{k_0} \hbar k_0 a_\mu^*(\mathbf{k})\, a^\mu(\mathbf{k}) \tag{23}$$

To obtain the physical interpretation of the quantized theory, by analogy with the procedure used in the quantization of the scalar field, let us suppose the operators $a_\mu(\mathbf{k})$ for $\mu = 0, 1, 2, 3$, to be destruction operators, and the operators $a_\mu^*(\mathbf{k})$, $\mu = 0, 1, 2, 3$, to be creation operators. In addition, let us assume the existence of a vacuum state, $|0\rangle$, characterized by $a_\mu(\mathbf{k}) \mid 0\rangle = 0$ for all $\mathbf{k}$ and all μ. These assumptions, however, lead to difficulties, since under such circumstances the vacuum expectation value of the operator $a_0(\mathbf{k})\, a_0^*(\mathbf{k})$ is equal to

$$\begin{aligned}\langle 0 \mid a_0(\mathbf{k})\, a_0^*(\mathbf{k}') \mid 0\rangle &= \langle 0 \mid [a_0(\mathbf{k}), a_0^*(\mathbf{k}')] \mid 0\rangle \\ &= -g_{00}\delta(\mathbf{k} - \mathbf{k}')\, k_0 \end{aligned} \tag{24}$$

so that the state $\int f(\mathbf{k})\, a_0(\mathbf{k})\, d\mathbf{k} \mid 0\rangle$, with $\int |f(\mathbf{k})|^2\, d\mathbf{k} < \infty$, has a negative norm. The vector space spanned by the basis vectors $a_0^*(\mathbf{k}_1) \cdots a_3^*(\mathbf{k}_{3n}) \mid 0\rangle$ is therefore no longer a Hilbert space within the above scalar product, since nonvanishing vectors with zero and negative norm exist. A probabilistic interpretation is therefore not immediately applicable.

Alternatively, we could suppose that $a_l(\mathbf{k})$ for $l = 1, 2, 3$ and $a_0^*(\mathbf{k})$ are destruction operators, whereas the operator $a_l^*(\mathbf{k})$ for $l = 1, 2, 3$ and $a_0(\mathbf{k})$ are creation operators. This avoids the difficulty of negative norms but leads to another difficulty. With this interpretation the Hamiltonian H does not have a lower bound, i.e., states of arbitrarily high negative energy are possible. For example, the state $a_0(\mathbf{k}) \mid 0\rangle$ which contains one time-like photon then has an energy which is negative:

$$\begin{aligned} Ha_0(\mathbf{k}) \mid 0\rangle &= [H, a_0(\mathbf{k})] \mid 0\rangle \\ &= -k_0 a_0(\mathbf{k}) \mid 0\rangle \end{aligned} \tag{25}$$

Thus both interpretations of the operators $a_\mu(\mathbf{k})$ and $a_\mu^*(\mathbf{k})$ lead to difficulty. However, we have not as yet imposed any subsidiary condition so that in fact the theory considered so far does not correspond to the Maxwell theory. Stated differently, $\langle A_\mu(x)\rangle_\Psi = \langle \Psi \mid A_\mu(x) \mid \Psi\rangle$, the expectation value of the potential in the state $|\Psi\rangle$, does not satisfy the Lorentz condition (i.e., $\partial_\mu\langle A^\mu\rangle_\Psi$ is not necessarily zero), so that Maxwell's equations are not necessarily satisfied by $\langle F_{\mu\nu}(x)\rangle_\Psi$.

In the classical field theory, the Lorentz condition $A^\mu{}_{,\mu}(x) = 0$ insured that the field equations $\Box A_\mu = 0$ correspond to Maxwell's equations and guaranteed the positive-definite character of the total energy. In the quantized theory, however, we no longer can impose the Lorentz condition as an operator identity, since it would lead to a contradiction of the commutation rules because

$$\left[\frac{\partial A_\mu(x)}{\partial x_\mu}, A_\nu(x')\right] = -i\hbar c \frac{\partial D(x - x')}{\partial x^\nu} \neq 0$$

To circumvent this difficulty, what was usually done until about 1949 was to give up the commutation rules (14) and (22) for the 0- and 3-component of A_μ. One thus had to eliminate the longitudinal and time-like part of the vector A_μ, using the subsidiary condition $A^\mu{}_{,\mu} \mid \Psi\rangle = 0$ for the possible states, $|\Psi\rangle$, of the field. Only the remaining transverse parts of the potential were then considered as dynamical variables and only these were quantized. In the presence of charges this corresponds to the separation of the *instantaneous* Coulomb interaction from the rest of the retarded transverse electromagnetic interaction. This, however, is not a manifestly relativistically invariant procedure, since for a moving observer the Coulomb interaction (longitudinal waves) and the transverse waves become mixed again. [See in this connection Zumino (1960). A presentation of quantum electrodynamics in the Coulomb gauge is outlined there and a proof of its relativistic invariance sketched.]

Gupta (1950a) and Bleuler (1950) have developed a method which justifies using all four potentials on the same footing, exhibits the relativistic invariance explicitly, and allows the consistent use of the commutation rules (22) and (14). In their formulation, the operators $a_\mu(\mathbf{k})$ ($\mu = 0, 1, 2, 3$) are destruction operators and the operators $a_\mu{}^*(\mathbf{k})$ ($\mu = 0, 1, 2, 3$) are creation operators with the following Fock space representation:

$$(a_\mu(\mathbf{k})\,\Psi)^{(n)}_{\mu_1\mu_2\cdots\mu_n}(\mathbf{k}_1, \mathbf{k}_2, \cdots \mathbf{k}_n) = \sqrt{n+1}\,\Psi^{(n+1)}_{\mu\mu_1\cdots\mu_n}(\mathbf{k}, \mathbf{k}_1, \cdots \mathbf{k}_n) \quad (26a)$$

$$(a_\mu{}^*(\mathbf{k})\,\Psi)^{(n)}_{\mu_1\cdots\mu_n}(\mathbf{k}_1, \mathbf{k}_2, \cdots \mathbf{k}_n) = -\frac{1}{\sqrt{n}}\sum_{j=1}^{n} g_{\mu\mu_j}k_0\delta^{(3)}(\mathbf{k}-\mathbf{k}_j)$$
$$\cdot\,\Psi^{(n-1)}_{\mu_1\mu_2\cdots\mu_{j-1}\mu_{j+1}\cdots\mu_n}(\mathbf{k}_1, \mathbf{k}_2, \cdots \mathbf{k}_{j-1}, \mathbf{k}_{j+1}, \cdots \mathbf{k}_n) \quad (26b)$$

The vacuum state is characterized by

$$a_\mu(\mathbf{k}) \mid 0\rangle = 0 \quad \mu = 0, 1, 2, 3 \text{ for all } \mathbf{k} \quad (27a)$$

or equivalently by

$$A_\mu{}^{(+)}(x) \mid 0\rangle = 0 \quad \text{for all } x \quad (27b)$$

The basis vectors $\prod_j a_{\mu_j}{}^*(\mathbf{k}_j) \mid 0\rangle$ span a linear vector space which we shall denote by $\mathcal{G}$. In $\mathcal{G}$ a bilinear form, $(\Psi, \chi)_G$, is defined by

$$(\Psi, \chi)_G = \sum_{n=0}^{\infty} (-1)^n \int \frac{d^3k_1}{k_{10}} \cdots \int \frac{d^3k_n}{k_{n0}} \overline{\Psi^{(n)}_{\mu_1\mu_2\cdots\mu_n}(\mathbf{k}_1, \mathbf{k}_2, \cdots \mathbf{k}_n)}$$
$$\cdot\,\chi^{(n)\mu_1\mu_2\cdots\mu_n}(\mathbf{k}_1, \mathbf{k}_2 \cdots, \mathbf{k}_n) \quad (28a)$$

where

$$\Psi^{(n)}_{\mu_1\mu_2\cdots\mu_n}(\mathbf{k}_1, \mathbf{k}_2, \cdots \mathbf{k}_n) = \langle 0 \mid a_{\mu_1}(\mathbf{k}_1)\,a_{\mu_2}(\mathbf{k}_2) \cdots a_{\mu_n}(\mathbf{k}_n) \mid \Psi\rangle \quad (28b)$$

$$\chi^{(n)}_{\mu_1\mu_2\cdots\mu_n}(\mathbf{k}_1, \mathbf{k}_2, \cdots \mathbf{k}_n) = \langle 0 \mid a_{\mu_1}(\mathbf{k}_1)\,a_{\mu_2}(\mathbf{k}_2) \cdots a_{\mu_n}(\mathbf{k}_n) \mid \chi\rangle \quad (28c)$$

It is then postulated that all expectation values of operators must be computed using the bilinear form, $(\Psi, \chi)_G$. Within this "Gupta" scalar product, states with an odd number of time-like photons have a negative norm. The above indefinite bilinear form ("Gupta" scalar product) is to be distinguished from the scalar product (Ψ, χ) which is defined as

$$(\Psi, \chi) = \sum_{h=0}^{\infty} \int \frac{d^3k_1}{k_{10}} \cdots \int \frac{d^3k_n}{k_{n0}}$$

$$\sum_{\mu_1, \cdots \mu_n = 0}^{3} \overline{\Psi^{(n)}_{\mu_1\mu_2 \cdots \mu_n}(\mathbf{k}_1, \mathbf{k}_2, \cdots \mathbf{k}_n)}\, \chi^{(n)}_{\mu_1\mu_2 \cdots \mu_n}(\mathbf{k}_1, \mathbf{k}_2, \cdots \mathbf{k}_n) \quad (29)$$

and which makes $\mathcal{G}$ into a Hilbert space. The indefinite bilinear form $(\Psi, \chi)_G$ can be expressed in terms of the scalar product (Ψ, χ) as follows [Wightman (1958), Pandit (1959)]:

$$(\Psi, \chi)_G = (\Psi, \eta\chi) \quad (30)$$

where η is a linear operator. Comparing Eqs. (28) and (30), we deduce that

$$(\eta\Psi)^n_{\mu_1\mu_2 \cdots \mu_n}(\mathbf{k}_1, \cdots, \mathbf{k}_n) = \sum_{\nu_1, \cdots \nu_n = 0}^{3} \prod_{j=1}^{n} (-g_{\mu_j\nu_j})\, \Psi^{(n)}_{\nu_1 \cdots \nu_n}(\mathbf{k}_1, \cdots \mathbf{k}_n) \quad (31)$$

and also that

$$\eta^2 = 1 \quad (32a)$$

$$\eta = \eta^* \quad (32b)$$

so that η is Hermitian. It should be noted that within the indefinite form (28) the operator $A_\mu(x)$ is self-adjoint, i.e.,

$$(\chi, a_\mu(\mathbf{k})\, \Psi)_G = (a_\mu^*(\mathbf{k})\, \chi, \Psi)_G \quad (33)$$

as can be verified by using the representations (26a) and (26b). Furthermore, it obeys the following commutation rules with η:

$$[A_l(x), \eta] = 0; \quad l = 1, 2, 3 \quad (34a)$$

$$[A_0(x), \eta]_+ = 0 \quad (34b)$$

which are useful if one works with the scalar product $(\chi, \eta\Psi)$.

An important observation in the Gupta-Bleuler development was the recognition that the subsidiary condition $A^\mu{}_{,\mu}(x) \mid \Psi\rangle = 0$, for all x, that the physical states should satisfy, is too stringent. In fact, if one imposed this subsidiary condition as such there would be no states which would satisfy it. For not only does it require that certain kinds of photons are not present but, in addition, that these same photons cannot be emitted. Gupta and Bleuler thus proposed that the Lorentz condition be modified so that it holds only for the *destruction* part of the operator, namely, that

$$\left(\frac{\partial A_\mu{}^{(+)}(x)}{\partial x_\mu}\right) \mid \Psi\rangle = 0 \quad \text{for all } x \quad (35a)$$

or equivalently that

$$L(\mathbf{k}) \mid \Psi\rangle = k^\mu a_\mu(\mathbf{k}) \mid \Psi\rangle = 0 \quad \text{for all } \mathbf{k};\ k_0 = |\mathbf{k}| \quad (35b)$$

for every physical state $|\Psi\rangle$ of the electromagnetic field. Since the operator $\partial_\mu A^\mu(x)$ satisfies $\Box \partial_\mu A^\mu(x) = 0$, its decomposition into positive and negative frequencies, $\partial_\mu A^{\mu(\pm)}(x)$, is relativistically invariant. We shall denote by $\mathcal{P}$ the manifold of states $|\Psi\rangle$ which satisfy the subsidiary condition, i.e., the set of vectors $|\Psi\rangle$ such that $L(\mathbf{k}) = k^\mu a_\mu(\mathbf{k})$ operating on them vanishes. The subsidiary condition $L(\mathbf{k}) \mid \Psi\rangle = 0$ is then sufficient to insure that the photons are polarized transversely, and does not destroy the correspondence with classical electrodynamics. For example, if we denote by

$$\Psi^{(1)}{}_\mu(\mathbf{k}) = \langle 0 \mid a_\mu(\mathbf{k}) \mid \Psi\rangle_G \tag{36}$$

the probability amplitude for finding in the state $|\Psi\rangle$, which is in $\mathcal{P}$, one photon of momentum $k = \{k_0, \mathbf{k}\}$ with $k_\mu k^\mu = 0$ and polarization μ, then the subsidiary condition $L(\mathbf{k}) \mid \Psi\rangle = 0$ requires that

$$k^\mu \Psi^{(1)}{}_\mu(\mathbf{k}) = 0 \tag{37}$$

That this guarantees that the photon is polarized perpendicularly to its direction of motion is shown as follows: We recall that since we are describing the electromagnetic field in terms of potentials, the theory is not unique, for we may always perform gauge transformations under which

$$A_\mu(x) \to A_\mu(x) + \partial_\mu \Lambda(x) \tag{38a}$$

with

$$\Box \Lambda = 0 \tag{38b}$$

where Λ is a nonoperator function. If we denote the Fourier transform of $\Lambda(x)$ by

$$\Lambda(x) = \sqrt{\frac{\hbar c}{2(2\pi)^3}}\, i \int \frac{d^3k}{k_0} \left(\Lambda_+(\mathbf{k})\, e^{-ik\cdot x} + \Lambda_-(\mathbf{k})\, e^{+ik\cdot x}\right) \tag{39a}$$

$$\Lambda_+(\mathbf{k}) = \overline{\Lambda_-(\mathbf{k})} \tag{39b}$$

then in momentum space Eq. (38) reads

$$a_\mu(\mathbf{k}) \to a_\mu(\mathbf{k}) + k_\mu \Lambda_+(\mathbf{k}) \tag{40}$$

This freedom of making gauge transformations implies that if we add to $\Psi^{(1)}{}_\mu(\mathbf{k})$ a function of the form $k_\mu \Lambda_+(\mathbf{k})$, then $\Psi^{(1)}{}_\mu(\mathbf{k}) + k_\mu \Lambda_+(\mathbf{k})$ describes the same physical state as did $\Psi^{(1)}{}_\mu(\mathbf{k})$. In other words, the amplitudes $\Psi^{(1)}{}_\mu(\mathbf{k})$ and $\Psi^{(1)}{}_\mu(\mathbf{k}) + k_\mu \Lambda_+(\mathbf{k})$ are equivalent (and indistinguishable). But if this is so, then by a gauge transformation whereby

$$\Psi^{(1)}{}_\mu(\mathbf{k}) \to \Psi^{(1)}{}_\mu(\mathbf{k}) - \frac{k_\mu}{k_0} \Psi^{(1)}{}_0(\mathbf{k}) \tag{41}$$

(i.e., $\Lambda_+(\mathbf{k}) = -\Psi^{(1)}{}_0(\mathbf{k})/k_0$) we can always obtain a one-photon amplitude whose time component is zero, but which describes the same physical state.[3] For such an amplitude, the transversality condition now reads

[3] Classically, these statements correspond to the fact that the radiation gauge, div $\mathbf{A} = 0$, $\phi = 0$ can be obtained from the Lorentz gauge $A_{r,}{}^{\mu} = 0$ by a gauge transformation.

$$\sum_{i=1}^{3} k_i \Psi^{(1)}{}_i(\mathbf{k}) = \mathbf{k} \cdot \mathbf{\Psi}^{(1)}(\mathbf{k}) = 0 \tag{42}$$

which indeed states that the vector potential (wave function) corresponding to the situation where one photon is present is transverse, i.e., $\mathbf{k} \cdot \mathbf{\Psi}^{(1)}(\mathbf{k}) = 0$. This demonstration of the transversality can similarly be carried through for the n-photon configuration.

The freedom in the description of the one-photon amplitude implied by Eq. (40), namely, that the amplitude $\Psi^{(1)}{}_\mu(\mathbf{k}) + k_\mu \Lambda_+(\mathbf{k})$ is equivalent to $\Psi^{(1)}{}_\mu(\mathbf{k})$ also implies that any amplitude proportional to k_μ is equivalent to zero (since it can be "gauged" away). Thus, the only physically meaningful part of $\Psi^{(1)}{}_\mu(\mathbf{k})$ is that part of $\Psi^{(1)}{}_\mu(\mathbf{k})$ which is orthogonal to k_μ (due to the subsidiary condition) and not proportional to k_μ (due to the freedom of gauge transformation), i.e., the space-like part of $\Psi^{(1)}{}_\mu(\mathbf{k})$. We may therefore take (and it is simplest to do so) $\Psi'^{(1)}{}_\mu(\mathbf{k}) = \Psi^{(1)}{}_\mu(\mathbf{k}) - k_\mu \Psi^{(1)}{}_0(\mathbf{k})/k_0$, i.e., the amplitude whose time component vanishes and for which $\mathbf{k} \cdot \mathbf{\Psi}'^{(1)}(\mathbf{k}) = 0$, as the representative of the equivalence class of one-photon amplitudes. This choice guarantees that a physically realizable one-photon state (i.e., a state whose only nonvanishing amplitude is $\Psi'^{(1)}{}_\mu(\mathbf{k})$) has a positive-definite norm:

$$(\Psi, \Psi)_G = -\int \frac{d^3k}{k_0} \overline{\Psi'^{(1)}{}_\mu(\mathbf{k})}\, \Psi'^{(1)\mu}(\mathbf{k}) \geqslant 0 \tag{43}$$

with the equality sign holding only when $\Psi'^{(1)}{}_\mu(\mathbf{k}) = 0$. Similar statements can be made about the n-photon amplitudes. A vector $|\Psi\rangle$ belonging to $\mathcal{P}$ has the property that its Fock space components are "transverse," i.e., satisfy the condition

$$k^\mu \Psi^{(n)}_{\mu\mu_1 \cdots \mu_{n-1}}(\mathbf{k}, \mathbf{k}_1, \cdots \mathbf{k}_{n-1}) = 0 \tag{44a}$$

Due to their Bose character, the amplitudes $\Psi^{(n)}_{\mu_1\mu_2 \cdots \mu_n}(\mathbf{k}_1, \cdots \mathbf{k}_n)$ are symmetric under the exchange $(\mathbf{k}_j\mu_j) \leftrightarrow (\mathbf{k}_i\mu_i)$, so that

$$(k_i)^{\mu_i} \Psi^{(n)}_{\mu_1\mu_2 \cdots \mu_i \cdots \mu_n}(\mathbf{k}_1, \cdots, \mathbf{k}_i, \cdots \mathbf{k}_n) = 0 \tag{44b}$$

for all $i(i = 1, 2, \cdots n)$. The tensors $\Psi^{\mu_1 \cdots \mu_n}(\mathbf{k}_1, \cdots \mathbf{k}_n)$ are therefore either null or space-like in each index μ_i (being orthogonal to the time-like vector $(k_i)_{\mu_i}$). Within $\mathcal{P}$, the Gupta scalar product is therefore certainly positive-semidefinite, i.e., $(\Psi, \Psi)_G \geqslant 0$. The freedom of gauge transformations again implies that any amplitude $\Psi^{(n)}_{\mu_1 \cdots \mu_n}$ which is a null vector in the index μ_i, i.e., is proportional to k_{μ_i} (and by symmetry to all the k_{μ_i}) is equivalent to zero, i.e., does not represent a physical situation. We therefore conclude that realizable states have a positive-definite norm within the Gupta scalar product. This will make possible the physical interpretation of the theory.

The energy-momentum four-vector of the field expressed in terms of momentum space operators, is

$$P^\mu = -\int \frac{d^3k}{k_0} a_\nu^*(\mathbf{k})\, a^\nu(\mathbf{k})\, \hbar k_\mu \tag{45}$$

Let us show that the expression $-a_\nu^*(\mathbf{k})\, a^\nu(\mathbf{k})$ is actually positive-definite when operating on a vector in $\mathcal{P}$. To this end, we shall choose a special co-ordinate system in which the polarization vectors, $\epsilon_\mu^{(\lambda)}$ are such that

$$k^\mu \epsilon^{(1)}{}_\mu(\mathbf{k}) = k^\mu \epsilon^{(2)}{}_\mu(\mathbf{k}) = 0 \tag{46a}$$

$$-k^\mu \epsilon^{(3)}{}_\mu(\mathbf{k}) = k^\mu \epsilon^{(0)}{}_\mu(\mathbf{k}) = 1 \tag{46b}$$

A particular way of satisfying these relations is, for example, to rotate the co-ordinate system so that $k^0 = |\mathbf{k}|$, $k^1 = |\mathbf{k}|$ and $k^2 = k^3 = 0$, i.e., such that

$$k_\mu = |\mathbf{k}|\,(1, 1, 0, 0) \tag{47a}$$

and to choose

$$\begin{aligned} \epsilon^{(0)}{}_\mu(\mathbf{k}) &= (1, 0, 0, 0) \\ \epsilon^{(1)}{}_\mu(\mathbf{k}) &= (0, 0, 1, 0) \\ \epsilon^{(2)}{}_\mu(\mathbf{k}) &= (0, 0, 0, 1) \\ \epsilon^{(3)}{}_\mu(\mathbf{k}) &= (0, 1, 0, 0) \end{aligned} \tag{47b}$$

With these polarization unit-vectors, recalling that

$$a_\mu(\mathbf{k}) = \sum_{\lambda=1}^{4} \epsilon^{(\lambda)}{}_\mu(\mathbf{k})\, a^{(\lambda)}(\mathbf{k}) \tag{48}$$

the Lorentz condition (34) becomes

$$L(\mathbf{k}) \mid \Psi\rangle = k^\mu a_\mu(\mathbf{k}) \mid \Psi\rangle = 0 = |\mathbf{k}|\,[a^{(0)}(\mathbf{k}) - a^{(3)}(\mathbf{k})] \mid \Psi\rangle = 0 \tag{49a}$$

or

$$a^{(0)}(\mathbf{k}) \mid \Psi\rangle = a^{(3)}(\mathbf{k}) \mid \Psi\rangle \tag{49b}$$

for any physically realizable state $|\Psi\rangle$. Hence, using Eqs. (48) and (49) and the fact that the polarization vectors are real, we deduce the fact that for any $|\Psi\rangle$ in $\mathcal{P}$

$$\begin{aligned} (\Psi, -a_\mu^*(\mathbf{k})\, a^\mu(\mathbf{k})\, \Psi)_G &= (\Psi, -\sum_{\lambda\lambda'} a^{(\lambda)*}(\mathbf{k})\, a^{(\lambda')}(\mathbf{k})\, g_{\lambda\lambda'}\Psi)_G \\ &= +(\Psi, (a^{(1)*}(\mathbf{k})\, a^{(1)}(\mathbf{k}) + a^{(2)*}(\mathbf{k})\, a^{(2)}(\mathbf{k}))\, \Psi)_G \geqslant 0 \end{aligned} \tag{50}$$

as only the transverse photons contribute.

Thus we see that the energy-momentum four-vector is in fact positive-semidefinite in $\mathcal{P}$ by virtue of the subsidiary condition. This result becomes more transparent if we note that in the co-ordinate system specified by Eq. (42) the subsidiary condition restricts the admissible states of the system to those which contain an equal number of longitudinal and time-like photons of the same momentum, since $\langle\Psi \mid a^{(0)*}(\mathbf{k})\, a^{(0)}(\mathbf{k}) \mid \Psi\rangle = \langle\Psi \mid a^{(3)*}(\mathbf{k})\, a^{(3)}(\mathbf{k}) \mid \Psi\rangle$, by (49b). Now the time-like photons contribute a negative energy (momentum) to the total field energy (momentum) so that their contribution cancels that of the longitudinal photons.

Let us consider somewhat more closely the states in $\mathcal{P}$. We note that if $|\Psi\rangle$ is any physically realizable state, i.e., one which satisfies $L(\mathbf{k})\,|\,\Psi\,\rangle = 0$ *and* contains only transverse photons, then the vector

$$|\Psi'\rangle = \left\{1 + \int \frac{d^3k_1}{k_{10}}\, g^{(1)}(\mathbf{k}_1)\, L^*(\mathbf{k}_1) + \cdots \right.$$
$$\left. + \int \frac{d^3k_1}{k_{10}} \cdots \int \frac{d^3k_n}{k_{n0}}\, g^{(n)}(\mathbf{k}_1, \mathbf{k}_2, \cdots \mathbf{k}_n)\, L^*(\mathbf{k}_1) \cdots L^*(\mathbf{k}_n) + \cdots \right\} |\,\Psi\rangle \tag{51}$$

is again in $\mathcal{P}$, i.e., $|\Psi'\rangle$ is such that $L(\mathbf{k})\,|\,\Psi'\rangle = 0$ for all $\mathbf{k}$. This follows from the fact that $L(\mathbf{k})$ and $L^*(\mathbf{k}')$ commute

$$\begin{aligned}[L(\mathbf{k}), L^*(\mathbf{k}')] &= k_\mu k_\nu'[a^\mu(\mathbf{k}), a^{\nu *}(\mathbf{k}')] \\ &= -k_0 k^2 \delta(\mathbf{k} - \mathbf{k}') = 0\end{aligned} \tag{52}$$

since $k^2 = 0$. In fact, the states obtained from the physically realizable states (i.e., the ones containing only transverse photons) by the process indicated in Eq. (51) exhaust all the vectors in $\mathcal{P}$. Furthermore, the scalar product of two such states $|\chi'\rangle$, $|\Phi'\rangle$ where

$$|\chi'\rangle = \left\{1 + \sum_{n=0}^{\infty} \prod_{i=1}^{n} \left(\int \frac{d^3k_i}{k_{i0}}\, L^*(\mathbf{k}_i) \right) \cdot h^{(n)}(\mathbf{k}_1, \mathbf{k}_2, \cdots \mathbf{k}_n) \right\} |\,\chi\rangle$$
$$|\Phi'\rangle = \left\{1 + \sum_{h=0}^{\infty} \prod_{i=1}^{n} \left(\int \frac{d^3k_i}{k_{i0}}\, L^*(\mathbf{k}_i) \right) \cdot g^{(n)}(\mathbf{k}_1, \mathbf{k}_2, \cdots \mathbf{k}_n) \right\} |\,\Phi\rangle \tag{53}$$

where $|\Phi\rangle$ and $|\chi\rangle$ are physical states containing only transverse photons, is equal to

$$\langle \chi' \mid \Phi' \rangle_G = \langle \chi \mid \Phi \rangle_G \tag{54}$$

since $L(\mathbf{k})\,|\,\Phi\rangle = \langle \Phi\,|\,L^*(\mathbf{k}) = 0$. The scalar product of $|\chi'\rangle$ and $|\Phi'\rangle$ thus depends only on the projection of $|\chi'\rangle$ and $|\Phi'\rangle$ on the manifold, $\mathcal{H}$, of physically realizable states. As noted above, $\mathcal{H}$ has the property that the norm of all vectors in it is positive. The probabilistic interpretation is therefore applicable to the vectors in $\mathcal{H}$. The above result further suggests the following rule for the interpretation of an arbitrary state $|\chi'\rangle$ in $\mathcal{P}$: Project it into $\mathcal{H}$; its projection, $|\chi\rangle$ (which is either a state containing only transverse photons, or the vacuum or the null vector), represents a state which is physically realizable (if $|\chi\rangle$ is not the null vector) and to which $|\chi'\rangle$ is equivalent: $|\chi\rangle$ and $|\chi'\rangle$ give rise to the same observable consequences. The equivalence of the state $|\chi\rangle$ and $|\chi'\rangle$ corresponds precisely to the lack of uniqueness that exists in the definition of $\chi^{(n)}_{\mu_1\mu_2 \cdots \mu_n}(\mathbf{k}_1, \cdots \mathbf{k}_n)$ due to the freedom of making gauge transformations, but now expressed in terms of the state vectors.

To see this more explicitly, let us evaluate the expectation value of $A_\mu(x)$ in the states $|\chi\rangle$ and $|\chi'\rangle$. To this end we note that

$$\langle\chi' \mid a_\mu(\mathbf{k}) \mid \chi'\rangle_G$$
$$= \langle\chi \mid \left\{1 + \sum_{n=1}^{\infty} \prod_{i=1}^{n} \left(\int \frac{d^3k_i}{k_{i0}} L(\mathbf{k}_i)\right) \bar{h}^{(n)}(\mathbf{k}_1, \cdots \mathbf{k}_n)\right\}$$
$$a_\mu(\mathbf{k}) \left\{1 + \sum_{n=1}^{\infty} \prod_{i=1}^{n} \left(\int \frac{d^3k_i}{k_{i0}} L^*(\mathbf{k}_i)\right) h^{(n)}(\mathbf{k}_1, \cdots \mathbf{k}_n)\right\} \mid \chi\rangle_G$$
$$= \langle\chi \mid a_\mu(\mathbf{k}) \left\{1 + \sum_{n=1}^{\infty} \prod_{i=1}^{n} \left(\int \frac{d^3k_i}{k_{i0}} L^*(\mathbf{k}_i)\right) h^{(n)}(\mathbf{k}_1, \cdots \mathbf{k}_n)\right\} \mid \chi\rangle_G$$
$$= \langle\chi \mid a_\mu(\mathbf{k}) + k_\mu \Lambda_+(\mathbf{k}) \mid \chi\rangle_G \tag{55}$$

where

$$\Lambda_+(\mathbf{k}) = h^{(1)}(\mathbf{k}) \tag{56}$$

In obtaining (56) we have made use of the fact that $a_\mu(\mathbf{k})$ and $L(\mathbf{k}')$ commute, that $L(\mathbf{k}) \mid \chi\rangle = \langle\chi \mid L^*(\mathbf{k}) = 0$, and of the commutation rules

$$[a_\mu(\mathbf{k}), L^*(\mathbf{k}')] = k_\mu k_0 \delta^{(3)}(\mathbf{k} - \mathbf{k}') \tag{57}$$

Similarly,

$$\langle\chi' \mid a_\mu^*(\mathbf{k}) \mid \chi'\rangle_G = \langle\chi \mid a_\mu^*(\mathbf{k}) + k_\mu \Lambda_-(\mathbf{k}) \mid \chi\rangle_G$$

with

$$\Lambda_-(\mathbf{k}) = \bar{h}^{(1)}(\mathbf{k}) \tag{58}$$

so that

$$\langle\chi' \mid A_\mu(x) \mid \chi'\rangle_G = \langle\chi \mid A_\mu(x) + \partial_\mu \Lambda(x) \mid \chi\rangle_G \tag{59a}$$

with

$$\Box \Lambda = 0 \tag{59b}$$

Thus, the equivalence of the state vectors $|\chi\rangle$ and $|\chi'\rangle$ reflects the fact that a description in terms of potentials is always arbitrary to within a gauge transformation whereby $A_\mu \to A_\mu + \partial_\mu \Lambda$. The Gupta scalar product, however, insures that all equivalent wave functions will give rise to the same physical consequences. In particular, the expectation value of the field observables $F_{\mu\nu} = \partial_\nu A_\mu - \partial_\mu A_\nu$ is clearly the same in the state $|\chi'\rangle$ as in the state $|\chi\rangle$. We may, if we wish, choose the state $|\chi\rangle$ containing no time-like and longitudinal photons, as the representative of the equivalence class. This, however, corresponds to picking a particular gauge. Thus the vacuum state $|0\rangle$ characterized by $a_\mu(\mathbf{k}) \mid 0\rangle = 0$ ($\mu = 0, 1, 2, 3$) can be considered as the "representative" of the vacuum states possible within the theory. These other "vacua" differ from the state $|0\rangle$ by the presence of (equal numbers of) time-like and longitudinal photons. Nevertheless, the expectation value of any gauge invariant quantity will be the same for all these different vacuum states.

The Gupta formalism thus has the virtue that the states with time-like and longitudinal photons are included and well defined within the framework of the theory (as must be the case for the proper mathematical definition of the operators $A_\mu(x)$, $\mu = 0, 1, 2, 3$), but nonetheless they do not contribute to anything physically observable (in the free-field situation).

In concluding this section, we note that the gauge transformation $A_\mu \to A'_\mu(x) = A_\mu(x) + \partial_\mu \Lambda$ with $\Box \Lambda = 0$ is generated by the following unitary transformation:

$$U_\Lambda(\sigma) = \frac{1}{\hbar c} \int_\sigma d\sigma^\mu(x) \left(\frac{\partial \Lambda(x)}{\partial x^\mu} \chi(x) - \Lambda(x) \frac{\partial \chi(x)}{\partial x^\mu} \right) \tag{60}$$

with

$$\chi(x) = \partial_\mu A^\mu(x) \tag{61a}$$

and

$$A'_\mu(x) = e^{iU_\Lambda} A_\mu(x)\, e^{-iU_\Lambda} = A_\mu(x) + \frac{i}{1!} [U_\Lambda, A_\mu(x)] + \cdots \tag{61b}$$

Proof: We first observe that $U_\Lambda(\sigma)$ does not in fact depend on σ since $0 = (\delta/\delta\sigma(x))\, U_\Lambda(\sigma)$, as χ and Λ satisfy $\Box\chi = \Box\Lambda = 0$. Furthermore, since the commutator of U_Λ and $A_\mu(x)$

$$\begin{aligned} [U_\Lambda, A_\mu(x)] &= \int d\sigma^\rho(x') \left(\frac{\partial \Lambda(x')}{\partial x'^\rho} \partial'_\mu D(x - x') - \Lambda(x') \frac{\partial \partial'_\mu D(x - x')}{\partial x'^\rho} \right) \\ &= \partial_\mu \Lambda(x) \end{aligned} \tag{62}$$

is a *c*-number, only the first two terms in (61b) contribute. Q.E.D. An observable (which is necessarily gauge invariant) can therefore be characterized as one which commutes with U_Λ, or more particularly with $L(\mathbf{k})$ and $L^*(\mathbf{k})$. An observable has, therefore, the property that it takes a vector in $\mathcal{P}$ into another vector in $\mathcal{P}$: it leaves $\mathcal{P}$ invariant.

We close this section with a statement of the commutation rules between the field intensities, which are:

$$\begin{aligned} &[F_{\mu\nu}(x), F_{\lambda\rho}(x')] \\ &\qquad = i\hbar c \{g_{\mu\lambda}\partial_\nu\partial_\rho' + g_{\nu\rho}\partial_\mu\partial_\lambda' - g_{\nu\lambda}\partial_\mu\partial_\rho' - g_{\mu\rho}\partial_\nu\partial_\lambda'\}\, D(x - x') \end{aligned} \tag{63}$$

The reader is referred to the classic papers of Bohr and Rosenfeld [Bohr (1933, 1950)] as well as the review article of Corinaldesi (1953) for the implication of these commutation rules on the possibility of performing compatible measurements of the components of the electric and magnetic fields in small regions of space-time. [See also Rosenfeld (1953, 1955).]

9c. Transformation Properties

We here briefly note the transformation properties of the operator $A_\mu(x)$. Under proper Lorentz transformations it transforms as a vector with

$$U(a, \Lambda)\, A_\mu(x)\, U(a, \Lambda)^{-1} = (\Lambda^{-1})_\mu{}^\nu\, A_\nu(\Lambda x + a) \tag{64}$$

where U is a unitary operator within the Gupta scalar product, i.e., for arbitrary $|\Psi\rangle$, $|\Phi\rangle$

$$(U(a, \Lambda)\, \Psi, U(a, \Lambda)\, \Phi)_G = (\Psi, \Phi)_G \tag{65}$$

The vacuum for the electromagnetic field (in a particular gauge) is further required to be invariant, i.e., $U(a, \Lambda) \mid 0\rangle = |0\rangle$. The transformation law for the field under space inversion which guarantees the invariance of the commutation rules and Lagrangian is:

$$U(i_s)\, A_\mu(x)\, U(i_s)^{-1} = \epsilon_{(\mu)} A_\mu(i_s x) \tag{66}$$

where $\epsilon_{(\mu)}$ can equal ± 1 and may depend on μ. (The electromagnetic field being a Hermitian field, the space inversion parity must be real.) The value of $\epsilon_{(\mu)}$ is fixed by a consideration of the interaction of the electromagnetic field with charged matter. In order that Maxwell's equations $\Box A_\mu(x) = j_\mu(x)$ be invariant, it is clear that A_μ must transform like the current vector j_μ. We have already noted that the latter transforms like an ordinary vector (recall, e.g., the transformation property of $:\tilde{\psi}\gamma_\mu\psi:$), so that $\epsilon_{(0)} = +1$ and $\epsilon_{(i)} = -1$, $i = 1, 2, 3$.

Under charge conjugation $A_\mu(x)$ transforms according to

$$U_c A_\mu(x)\, U_c^{-1} = -A_\mu(x) \tag{67}$$

which rule, as we shall see, guarantees the invariance under charge conjugation of the Lagrangian of charged fields interacting with the electromagnetic field.

For the theory to be invariant under Wigner time inversion, $A_\mu(x)$ must transform according to

$$U(i_t) \begin{Bmatrix} A_0(x) \\ A_i(x) \end{Bmatrix} U(i_t)^{-1} = \begin{Bmatrix} A_0(\ x) \\ -A_i(i_t x) \end{Bmatrix} \tag{68}$$

where $U(i_t)$ is antiunitary. In Eq. (68) the time parity has been so chosen that A_μ transforms like a current operator j_μ. This choice has the consequence that $U(i_t)$ transforms a creation operator for a photon of momentum $\mathbf{k}$ and a given circular polarization into a creation operator of a photon of momentum $-\mathbf{k}$ and the same circular polarization.

Part Three

THE THEORY OF INTERACTING FIELDS

10

Interaction Between Fields

10a. Symmetries and Interactions

We have in the previous chapters considered only systems of free non-interacting particles. We have seen that such systems could be described in terms of field operators which obey definite commutation rules and whose equations of motion could be derived from a Lagrangian. The Lagrangian formalism and the canonical quantization procedure summarized the content of the second-quantized theory and were completely equivalent to it. It is only when we consider interactions between fields that the Lagrangian approach achieves a status of its own: it is in fact the only known simple method for introducing interactions between the particles and for which a quantization procedure can be formulated. The most elegant formulation of quantum field theory based on a Lagrangian is due to Schwinger (1951; 1953a, b). It makes no reference to the classical fields and deals only with the quantized operators. It is, however, not without ambiguities since it is difficult to give a precise mathematical formulation to concepts such as the variations of operators. We shall therefore adhere to our former procedure and base our consideration on Lagrangians originally defined in terms of classical *c*-number fields.

Consider first the classical theory and in particular the case of a spin $\frac{1}{2}$ field, $\psi(x)$, interacting with a spin 0 neutral scalar field, $\phi(x)$. The theory will be relativistically invariant if we require $\mathfrak{L}$, the Lagrangian density of the field system, to be invariant under restricted Lorentz transformations. More precisely, if under a Lorentz transformation $x \to x' = \Lambda x + a$,

$$\psi(x) \to S(\Lambda)\, \psi(\Lambda^{-1}(x' - a)) = \psi'(x') \tag{1a}$$

$$\phi(x) \to \phi(\Lambda^{-1}(x' - a)) = \phi'(x') \tag{1b}$$

then the invariance requirement is that $\mathfrak{L}$ is a scalar

$$\mathfrak{L}'(\psi'(x'), \phi'(x')) = \mathfrak{L}(\psi(x), \phi(x)) \quad \text{(scalarity)} \tag{2}$$

i.e., $\mathfrak{L}$ and $\mathfrak{L}'$ have the same value at physically the same point, and that $\mathfrak{L}$ preserves its functional form under Lorentz transformation:

$$\mathfrak{L}'(\psi'(x), \phi'(x)) = \mathfrak{L}(\psi'(x'), \phi'(x')) \quad \text{(form invariance)} \tag{2b}$$

Equation (2b) guarantees that the equations of motion expressed in terms of the new variables are of precisely the same functional form as in terms of the old variables. Since the Lagrangian is directly related to such observable quantities as the Hamiltonian and the action, we require it to be real at the classical level and Hermitian in the quantized theory. We further require that the Lagrangian density contain no derivatives of field quantities higher than the first, so that the field equations are at most of second order.

The interaction between the fields is introduced by adding to the Lagrangian of the free uncoupled fields, $\mathfrak{L}_0$, an interaction term, $\mathfrak{L}_I$, which must satisfy the above requirements of hermiticity, relativistic invariance, and the limitation that it contain no space-time derivatives of field quantities higher than the first. The strength of the interaction term in the Lagrangian is measured by the magnitude of a multiplicative factor, called the coupling constant.

We shall distinguish between local couplings where the interaction term is built up from field quantities referring to the same space-time point, e.g., $\mathfrak{L}_I = G\tilde{\psi}(x)\,\psi(x)\,\phi(x)$, and nonlocal couplings where this is not the case, e.g., $\mathfrak{L}_I = G \int \tilde{\psi}(x)\,\psi(x)\,F(x-x')\,\phi(x')\,d^4x'$ where $F(x-x')$ is a prescribed scalar function which characterizes the space-time "region" over which the interaction takes place. We also distinguish between direct and derivative couplings. We shall call a coupling *direct* if no derivatives of field quantities are contained in $\mathfrak{L}_I$, whereas a derivative coupling is one where derivatives of field quantities appear in $\mathfrak{L}_I$.

The simplest way to introduce a local relativistically invariant interaction between a scalar field ϕ and a spinor field ψ is to couple the invariant $\tilde{\psi}\psi$ with ϕ and write at the classical level, $\mathfrak{L}_I = G \int d^4x\,\tilde{\psi}(x)\,\psi(x)\,\phi(x)$, where G is the coupling constant. To determine the dimensionality of the coupling constant, recall that the field quantities are normalized in terms of certain free-field expressions. For example, for the scalar boson field, the quantity

$$\frac{1}{2}\int d^3x \left\{\mu^2\phi^2 + \frac{1}{c^2}\dot{\phi}^2 + (\nabla\phi)^2\right\} c^2$$

(where μ is the inverse Compton wave length of the particle) has the dimension of an energy so that ϕ^2 has the dimensions of $\hbar^2/\mu V$ where V is a volume. Similarly, for the Dirac field $\int d^3x\,\tilde{\psi}(-i\hbar c\gamma\cdot\partial + mc^2)\,\psi$ is an energy, so that $\tilde{\psi}\psi V$ is dimensionless. The quantity $G\int d^3x\,\tilde{\psi}\psi\phi$ must again have the dimension of an energy, from which one infers that $G^2/\hbar c$ is dimensionless.

In general, one can differentiate between two types of interactions, those wherein the coupling constants have (in natural units with $\hbar = c = 1$) the dimension of a length raised to the zero or a negative power and those where the coupling constants have the dimension of a length raised to a positive power. In the system of natural units ($\hbar = c = 1$), the spin 0 Bose field has the dimension of inverse length (L^{-1}) while a spin $\frac{1}{2}$ Fermi-Dirac field has the dimension $L^{-3/2}$, the Lagrangian having the dimension of an inverse four-dimensional volume. The simplest of couplings are, then, those composed of algebraic functions of the field operators wherein only dimensionless coupling constants appear. A theory of this kind contains no intrinsic standard of length: it tacitly asserts the infinite divisibility of the space-time manifold. The constants μ_0 and m_0 which appear in the (uncoupled) Lagrangian are identified as the masses of the field quanta in the absence of interaction. The latter, however, are not observable. These dimensional parameters cannot therefore be used to determine the standard of length. The actual masses of the particles that the theory is to describe, of course, do establish an absolute scale of length and mass. However, we shall see that the quantum theory of fields as presently formulated cannot account for the observed mass spectrum of fundamental particles. The actual (observed) masses of the particles are introduced into the theory by a technique known as mass renormalization (see Chapter 15). In this approach the mass constants of the particles are regarded as *phenomenological* manifestations of an unknown physical agency, the neglect of whose inclusion into the theory produces the failure of the present description. This same physical agency is assumed to establish the absolute scale of length and of mass. If one adopts this viewpoint, it is then reasonable to postulate that the coupling term should not embody a unit of length that finds its dynamical origin outside of the domain of physical experience to which the theory is applicable. For interacting spin 0 and spin $\frac{1}{2}$ fields only two types of couplings terms are then admitted by this principle,[1] namely, couplings of the form $\tilde{\psi}\Gamma\psi\phi$ and couplings of the form ϕ^4. Note that it excludes derivative couplings. Most of the considerations of this book will be primarily concerned with the above kinds of couplings.

For the case of a pseudoscalar field ϕ, the coupling which guarantees the invariance of $\mathfrak{L}_I$ under proper Lorentz transformations as well as under spatial inversions is of the form $\mathfrak{L}_I(x) = G\tilde{\psi}\gamma_5\psi\phi(x)$ resulting from combining the pseudoscalar $\tilde{\psi}\gamma_5\psi$ with ϕ, so that $\tilde{\psi}\gamma_5\psi\phi$ is invariant under inversions. The field ϕ might, for example, describe the neutral π-meson field, and ψ the nucleon field. If we wish to describe the interaction of charged as well as neutral mesons with nucleons in such a way that the prediction of the theory will be charge independent, then the interaction term must be invariant under rotations in isotopic spin space. The simplest such interaction is one which couples the nucleon isotopic vector $\tilde{\psi}\boldsymbol{\tau}\gamma_5\psi$ with the meson isotopic vector $\boldsymbol{\phi}$, where ϕ_3 describes the neutral meson field and

[1] This principle, incidentally, is related to the principle of renormalizability (see Chapters 15–17).

$\frac{1}{\sqrt{2}}(\phi_1 \pm i\phi_2)$ the charged meson field. The interaction term in the Lagrangian would then be of the form $\mathfrak{L}_I = G\tilde{\psi}\gamma_5\boldsymbol{\tau}\psi \cdot \boldsymbol{\phi} = \sum_{i=1}^{3} G\tilde{\psi}\gamma_5\tau_i\psi\phi_i$. This interaction is an example of how an invariance requirement determines the form of $\mathfrak{L}_I$. In the present chapter we shall investigate the general structure of the Lagrangian as determined from the requirement that it be invariant under certain symmetry operation. We shall primarily consider the electromagnetic interactions and the pion nucleon interaction and more briefly comment on the present-day description of the other interactions. However, before delving further into the mathematical description of interacting quantized fields, it might be well to consider briefly the basis for such a description of physical phenomena.

Our earlier work has established that to every particle (and its antiparticle) can be associated a quantized field. Each such particle, when moving freely in space, is characterized by a mass, a spin, an electric charge and possibly other quantum numbers, such as nucleonic charge and hypercharge. A classification in terms of mass and spin follows from the assumption that the particles can be described by a formalism based on the general principles of quantum mechanics and special relativity. In a Lagrangian description of the quantized fields, the other "quantum" numbers are associated with the invariance of the Lagrangian under certain gauge transformations.

It is a fundamental assumption of the quantum theory of fields, as presently formulated, that the structure and observed behavior of all forms of matter can be accounted for and described in terms of a *limited* number of fundamental particles interacting in definite ways. These "fundamental" particles seem to fall into five groups:[2]

A. *The Baryons:* These are particles whose masses lie between the nucleon mass and the deuteron. They are all fermions and obey the rigorous conservation law known as the conservation of baryons or the conservation of heavy particles. There are eight such particles known at present. Their properties are listed in Table 10.1.

The six particles listed in Table 10.1 which are heavier than the neutron or proton are called *hyperons*. All the known hyperons have spin $\frac{1}{2}$ and are baryons. The antiparticles of the nucleons and of the Λ and Σ^- particles have in fact been observed. The decay processes of the unstable hyperons, it will be noted, are always such as to include one hyperon or nucleon among the decay products.

[2] For a complete review of the experimental situation up to 1958, the reader is referred to the book of Jackson (1958) and the reviews of Gell-Mann (1957d), Franzinetti and Morpugo (1957), Dalitz (1957), Ashkin (1959). For a review of the theoretical models to account for this, the articles of Nishijima (1956), Kemmer, Polkinghorne, and Pursey [Kemmer (1959)] are recommended. See also the Proceedings of the CERN, Kiev and Rochester High Energy Conference of 1958, 1959, and 1960.

TABLE 10.1

THE BARYONS

Type	*Particle*	*Mass (in electron masses)*	*Mean Life (in seconds)*	*Charge (in multiple of e)*	*Common Decay Mode*
Nucleon (N)	p (proton)	1836.12 ± 0.04	stable $> 4 \times 10^{23}$	+1	
	n (neutron)	1838.65 ± 0.04	1050 ± 200	0	$p + e^- + \bar{\nu}$
Hyperon (Y)	Λ	2182.5 ± 0.4	$2.77 \pm 0.2 \times 10^{-10}$	0	$N + \pi$
	Σ^0	2329.9 ± 0.3	0.1×10^{-10}	0	$\Lambda + \gamma$
	Σ^+	2327.9 ± 0.6	$0.83 \pm 0.06 \times 10^{-10}$	+1	$N + \pi$
	Σ^-	2341.6 ± 0.8	$1.67 \pm 0.17 \times 10^{-10}$	−1	$n + \pi^-$
	Ξ^-	2585 ± 6	$\sim 3 \times 10^{-10}$	−1	$\Lambda + \pi^-$
	Ξ^0	$\sim 2587 \pm 10$	$\sim 10^{-10}$	0	$\Lambda + \pi^0$

B. *Mesons:* This group of particles includes the π and K mesons whose main characteristics are tabulated in Table 10.2. The mesons are known to be bosons, and very likely they all have spin 0. They interact strongly with baryons.

TABLE 10.2

THE MESONS

Type	*Particle*	*Mass (in electron masses)*	*Mean Life (in seconds)*	*Decay Modes*
Pion	π^0	264.3 ± 0.3	$.4 \times 10^{-15} > \tau > 10^{-18}$	$2\gamma, \gamma + e^+ + e^-$
	$\pi^\pm$	273.3 ± 0.1	2.56×10^{-8}	$\mu^\pm + \nu$ $e^\pm + \nu$
Kaon	$K^0, \overline{K^0}$		$(K_1^0)\ 1 \times 10^{-10}$ $(K_2^0)\ 9 \times 10^{-7}$	2π $3\pi, \mu + \nu + \pi, e + \nu + \pi$
	$K^\pm$		1×10^{-8}	$\{2\pi, 3\pi, \mu + \nu$ $\mu + \nu + \pi, e + \nu + \pi$

C. *Leptons* (light fermions): This group of particles includes the muons, the electrons and the neutrinos. They are known to be fermions and to have spin $\frac{1}{2}$. Their properties are tabulated in Table 10.3. They interact weakly with one another and with baryons and mesons.

TABLE 10.3

THE LEPTONS

Type	*Particles*	*Mass (in electron masses)*	*Mean Life*	*Decay Mode*
Neutrinos	$\nu, \bar{\nu}$	0	stable	—
Electron	e^-, e^+	1	stable	—
Muon	μ^-, μ^+	206.9 ± 0.1	2.22×10^{-6}	$e + \nu + \bar{\nu}$

D. *The Photon:* Zero rest mass particle, which is a stable, has spin 1 and is coupled to all charged particles. It is responsible for the electromagnetic interactions among such particles.

E. *The Graviton:* The particle presumably associated with gravitational phenomena.

As alluded to above, the interactions amongst these particles also seem to have a natural classification as follows:

1. The strong interactions: consisting of the interactions among the baryons, antibaryons and mesons. These interactions include the forces responsible for the production and scattering of the baryons, pions and K-mesons. They are characterized by large coupling constants of the order of magnitude 1 or larger.

2. Electromagnetic interactions: through which the photon is linked to all charged particles (real or virtual) and which are characterized by a coupling constant $(e^2/4\pi\hbar c) = (1/137)$.

3. Weak interactions: responsible for the β-decay phenomena, the slow decay of hyperons and mesons, the absorption of μ^- mesons in nuclear matter and the decay of the muon. They are characterized by coupling constants of the order 10^{-13}.

4. Gravitational interactions: which are characterized by Newton's gravitational constant G, which has the dimension of inverse square mass with $(G/\hbar c)\, m_p{}^2 \sim 10^{-39}$; m_p = proton mass.

As mentioned above, there exists very strong evidence which suggests that an absolute conservation law operates for the baryons. It states that the number of baryons minus the number of antibaryons remains constant in any interaction. This conservation law is usually called the law of heavy particle conservation. It accounts, for example, for the nonoccurrence of such energetically possible processes as $\mathrm{p} \rightarrow \mathrm{e}^+ + \nu + \bar{\nu}$, $\mathrm{n} \rightarrow \mathrm{e}^+ + \mathrm{e}^- + \nu$, and thus more generally for the stability of matter. For the leptons, a similar conservation law seems to operate and is known as the law of conservation of leptons. It states that in any reaction involving leptons the number of leptons minus the number of antileptons is conserved. The experimental evidence indicates that it is satisfied with the assignment of e^-, μ^- and ν as the "leptons" and e^+, μ^+ and $\bar{\nu}$ as the "antileptons."

Besides the aforementioned laws of heavy particle and lepton conservations, it is known that to a very high accuracy all reactions between the fundamental particles conserve

(*a*) energy and linear momentum,
(*b*) angular momentum,
(*c*) electric charge.

These conservation laws can, as noted previously, be translated into symmetry and invariance principles which any theory attempting to describe the interaction between particles must satisfy. Thus, (*a*) the conservation of energy and momentum is related to the homogeneity of space and time and the invariance of the theory under space and time translations;

(*b*) the conservation of angular momentum (isotropy of space-time) with invariance under proper Lorentz transformation;

(*c*) the conservation of charge with gauge invariance.

The Lagrangian describing the field systems associated with the fundamental particles will, therefore, be required to be invariant under inhomogeneous proper Lorentz transformations and certain gauge transformations, the latter to account for charge, baryon and lepton conservation. It is presently also conjectured, and it is consistent with experiments, that all interactions are such that they are invariant under the operation of time reversal [Landau (1957), Wigner (1957)].

The above-enumerated conservation laws are at present considered to be exact ones (at least in our neighborhood of the universe), i.e., they are obeyed in all interactions among particles. In addition to these exact conservation laws, there are a number of conservation laws which seem to be obeyed by only some of the interactions. Thus the strong interactions between baryons and mesons conserve isotopic spin. [This is deduced from the charge independence of the nucleon-nucleon and pion-nucleon interaction and seems also to be substantiated in kaon-baryon interaction, see e.g., Dalitz (1957), Lee (1955), Feldman (1956).] Very probably parity is also conserved in all strong interactions. Electromagnetic interaction, on the other hand, only conserves the *3*-component of isotopic spin, and parity. The weak interactions conserve neither parity nor isotopic spin (the latter is, in fact, not defined for leptons). This brief discussion indicates that there seems to be a connection between the strength of the interaction and the number of conservation laws obeyed or, conversely, the weakness of the interaction and the number of conservation laws it violates. This connection has not been fully elucidated up to date. It has, however, been incorporated in much of the current description of the interaction between the fundamental particles.

Much progress has been made in the past few years by adopting the viewpoint that nature is most easily described by a sequence of approximations [see Pais (1952a), Gell-Mann (1956a)]. In the first of these, the electromagnetic and weak interactions are "turned off," i.e., assumed nonexistent. The leptons and the photon are then completely noninteracting. Under these circumstances baryons and mesons undergo interactions characterized by the aforementioned conservation laws (baryon, total and *3*-component of isotopic spin, strangeness, *PC*, etc.). Decays involving leptons and photons cannot occur in this approximation. In the second approximation, the electric charge of the particles is turned on, so that both the strong and electromagnetic interactions are effective, but still not the weak interactions. The processes involving baryons and mesons are now modified by electromagnetic effects with the consequent breakdown of conservation of total isotopic spin (charge independence), and decays involving photons, such as $\pi^0 \to 2\gamma$ and $\Sigma^0 \to \Lambda + \gamma$, are now permitted. The leptons, however, remain uncoupled except for electromagnetic interactions. In the final approximation, which is as exact a description of matter as conceived at present (except for the inclusion of gravitational effects), the weak interactions are turned on and the "slow" decays of the hyperons, kaons, muons, pions and of the neutron can occur, as well as the absorption of negative muons in matter.

It should be emphasized at this point that the above description of the "fundamental particles" is clearly not the final one. It associates with each particle a field without any attempts at making a distinction between "elementary" or "composite" fundamental particles. It is, of course, possible within the framework of field theory to conjecture that, for example, pions are really composite systems composed of a nucleon and an antinucleon tightly bound together [Fermi and Yang (Fermi 1949); see in this connection also Heisenberg (1957a), Sakata (1956), Lévy (1954), Markov (1956), Okun (1958, 1959)], or that the hyperons are bound states of one or more kaons bound to nucleons [Goldhaber (1956), Christy (1957)]. In fact, a Fermi-Yang scheme would "explain" the pseudoscalar nature of the π meson since the system of a nucleon-antinucleon bound in an S state has a parity opposite to that of the vacuum. The basic difficulty with all such approaches is that there exists at present no really satisfactory way to calculate properties of strongly interacting systems. In a scheme such as Fermi's, and its extension by Sakata (1956) and Okun (1958), in which all the mesons (pions and kaons) as well as the Σ and Ξ are considered composite, these composite systems have a binding energy which is comparable to the masses of the free particles. It may be the case that in their interactions at not too high energies such composite particles can indeed be considered as "elementary," and therefore that a description which assigns a field to each fundamental particle may have a certain approximate validity. Our present theoretical understanding of meson-nucleon phenomena based on this (approximate) picture lends credence to this view.

10*b*. Restrictions Due to Space-Time Symmetries

The remarks in Section *a* of this chapter indicated that the invariance properties of physical laws may be divided into two groups. The first group contains those symmetries which appear to hold universally, and the second consists of those symmetries which are exhibited by restricted classes of phenomena. An example of a symmetry from the second group is isotopic spin invariance which is a characteristic of the strong interactions, but which is violated by the weaker interactions. The first group of symmetries can be further subdivided into two classes: those associated with the space-time symmetries consisting of

(a) space-time translations
(b) proper Lorentz transformation
(c) (Wigner) time inversion,

and those associated with

(a) electric charge conservation
(b) conservation of heavy particles.

We have seen that conservation of charge can be related to the invariance of the Lagrangian under a certain kind of gauge transformation of the charged fields. Attempts have been made to relate the conservation of heavy particles to certain gauge transformation of the heavy particle field operators. However, no really satisfactory scheme has been put forward thus far. We shall describe one such attempt in Section *e* of this chapter.

In this section we shall consider the restrictions on the form of the interaction Lagrangian for local field theories imposed by the space-time symmetries. We shall deal immediately with the quantized theory.

Let us again consider first the case of a spinor field $\psi(x)$ interacting with a neutral scalar boson field $\phi(x)$. We assume that in a particular Lorentz frame our field system can be characterized by a Lagrangian density $\mathcal{L} = \mathcal{L}(\phi, \psi)$. The statement that the field system consists of a spinor and scalar field means that ψ and ϕ transform according to

$$U(a, \Lambda)\, \phi(x)\, U(a, \Lambda)^{-1} = \phi(\Lambda x + a) \tag{3a}$$

$$U(a, \Lambda)\, \psi(x)\, U(a, \Lambda)^{-1} = S(\Lambda)^{-1}\, \psi(\Lambda x + a) \tag{3b}$$

for any inhomogeneous Lorentz transformation $\{a, \Lambda\}$, under which $x \to x' = \Lambda x + a$. Here $U(a, \Lambda)$ is a unitary representation of the inhomogeneous Lorentz group on the Hilbert space of state vector of the system. As noted previously (Chapter 2), the form of $U(a, \Lambda)$ for pure translations is

$$U(a, 1) = e^{ia_\mu P^\mu} \tag{4}$$

where $P_\mu = \int d\sigma_\nu T_\mu{}^\nu$ and $T_{\mu\nu}$ is the energy-momentum four-vector of the field system; that of $U(a, \Lambda)$ for homogeneous Lorentz transformation is

$$U(0, \Lambda) = e^{\frac{1}{2}i\Lambda^{\mu\nu}M_{\mu\nu}} \tag{5}$$

where $M_{\mu\nu}$ is the angular momentum tensor of the field. The existence of the operators P_μ and $M_{\mu\nu}$ is thus seen to be a consequence of Lorentz invariance. The operators P_μ and $M_{\mu\nu}$ are the generators of infinitesimal translations and Lorentz transformation, respectively, and therefore satisfy the following commutation rules:

$$[P_\mu, P_\nu] = 0 \tag{6}$$

$$[P_\mu, M_{\kappa\lambda}] = i(g_{\mu\kappa}P_\lambda - g_{\mu\lambda}P_\kappa) \tag{7}$$

$$[M_{\mu\nu}, M_{\kappa\lambda}] = i(g_{\kappa\mu}M_{\lambda\nu} + g_{\lambda\nu}M_{\kappa\mu} - g_{\lambda\mu}M_{\kappa\nu} - g_{\kappa\nu}M_{\lambda\mu}) \tag{8}$$

For infinitesimal transformations

$$U(a, 1) \approx 1 + ia^\mu P_\mu \tag{9a}$$

$$U(0, \Lambda) \approx 1 + \tfrac{1}{2}i\Lambda^{\mu\nu}M_{\mu\nu} \tag{9b}$$

Eq. (3) becomes

$$(1 + ia^\mu P_\mu)\, \phi(x)\, (1 - ia^\mu P_\mu) + O(a^2) \approx \phi(x) + a^\mu \partial_\mu \phi(x) + O(a^2)$$

$$\approx \phi(x) + ia^\mu[P_\mu, \phi(x)] + \cdots \tag{10}$$

whence

$$[P_\mu, \phi(x)] = -i\partial_\mu \phi(x) \tag{11}$$

Similarly,

$$[M_{\mu\nu}, \phi(x)] = i(x_\mu \partial_\nu - x_\nu \partial_\mu)\, \phi(x) \tag{12}$$

$$[P_\mu, \psi_\alpha(x)] = -i\partial_\mu \psi_\alpha(x) \tag{13}$$

$$[M_{\mu\nu}, \psi_\alpha(x)] = i(x_\mu \partial_\nu - x_\nu \partial_\mu)\, \psi_\alpha(x) + i \sum_{\beta=1}^{4} (\sigma_{\mu\nu})_{\alpha\beta}\, \psi_\beta(x) \tag{14a}$$

where

$$\sigma_{\mu\nu} = \frac{1}{2i}(\gamma_\mu\gamma_\nu - \gamma_\nu\gamma_\mu) \tag{14b}$$

It should at this point be noted that the above relativity transformations as well as those considered previously in the field free case were of the "Schrödinger type" in which the state vectors are transformed ($|\Phi\rangle \to$ $\to |\Phi'\rangle = U(a, \Lambda) \mid \Phi\rangle$), but the "same" operators (field observables) are used.

For some purposes in the following it is convenient to deal with "Heisenberg type" relativity transformations in which the operators rather than the state vectors are transformed. For the Heisenberg type transformation, an observable $O(x)$ under the relativity transformation $x \to x' =$ $= \Lambda x + a$ transforms as $O(x) \to O'(x)$. The connection between Schrödinger and Heisenberg type transformations is as follows: if $|\Phi'\rangle$ is the transform of the state vector $|\Phi\rangle$ in the Schrödinger type transformation, then

$$(\Phi, O'(x)\, \Phi) = (\Phi', O(x)\, \Phi') \tag{15}$$

Hence if U is unitary, since $|\Phi'\rangle = U(a, \Lambda) \mid \Phi\rangle$,

$$\begin{aligned} O'(x) &= U^{-1}(a, \Lambda)\, O(x)\, U(a, \Lambda) \\ &= U(-\Lambda^{-1}a, \Lambda^{-1})\, O(x)\, U(-\Lambda^{-1}a, \Lambda^{-1})^{-1} \\ &= S_O(\Lambda)\, O(\Lambda^{-1}(x - a)) \end{aligned} \tag{16a}$$

or equivalently

$$O'(x') = S_O(\Lambda)\, O(x) \tag{16b}$$

The fact that $\mathcal{L}$ is to transform as a scalar can then be expressed by the statement that the transformed Lagrangian $\mathcal{L}'$ is to have the same value at the same physical point, i.e.,

$$\begin{aligned}\mathcal{L}'[\psi'(x'), \phi'(x')] &= \mathcal{L}(\psi(x), \phi(x)) \\ &= \mathcal{L}(S(\Lambda)^{-1}\, \psi'(x'), \phi'(x')) \qquad (17a)\end{aligned}$$

or equivalently by

$$U^{-1}(a, \Lambda)\, \mathcal{L}(\psi(x'), \phi(x'))\, U(a, \Lambda) = \mathcal{L}(\psi(x), \phi(x)) \qquad (17b)$$

We have already established that the free-field part of the Lagrangian, $\mathcal{L}_0$, consisting of the sum of the free Dirac field Lagrangian [Eq. (8.2)] and that of the free spin 0 scalar field [Eq. (7.227)], transforms as a scalar under proper Lorentz transformation. The action integral $\int d^4x\, \mathcal{L}_0(x)$ for the free field is therefore an invariant. Turning to the interaction term between a neutral, hence Hermitian, scalar field and a charged spinor field, one verifies that the following local coupling terms transform like scalars under *proper* Lorentz transformations:

$G\bar{\psi}(x)\, \psi(x)\, \phi(x)$	(scalar coupling)
$G\bar{\psi}(x)\, \gamma_5\psi(x)\, \phi(x)$	(pseudoscalar coupling)
$G\bar{\psi}(x)\, \gamma_\mu\psi(x)\, \partial^\mu\phi(x)$	(vector coupling)
$iG\bar{\psi}(x)\, \gamma_5\gamma_\mu\psi(x)\, \partial^\mu\phi(x)$	(pseudovector coupling)

The factor i in the pseudovector coupling has been inserted to insure the hermiticity of the expression. In this connection recall that γ_5 is anti-Hermitian, with $\gamma_5{}^* = (\gamma_0\gamma_1\gamma_2\gamma_3)^* = -\gamma_5$ and $\gamma_\mu{}^* = \gamma^0\gamma_\mu\gamma^0$, so that

$$\begin{aligned}(\bar{\psi}\gamma_5\psi)^* &= (\psi^*\gamma_0\gamma_5\psi)^* = \psi^*\gamma_5{}^*\gamma_0{}^*\psi \\ &= \bar{\psi}\gamma_5\psi \qquad (18)\end{aligned}$$

$$\begin{aligned}(\bar{\psi}\gamma_\mu\gamma_5\psi)^* &= \psi^*\gamma_5{}^*\gamma_\mu{}^*\gamma_0{}^*\psi \\ &= \bar{\psi}\gamma_5\gamma_\mu\psi = -\bar{\psi}\gamma_\mu\gamma_5\psi \qquad (19)\end{aligned}$$

Let us next consider the restrictions imposed by the requirement that the theory be invariant under (Wigner) time inversion. An important theorem due to Pauli (1955) and Lüders (1957) [this discovery was essentially anticipated by Shell (1948) and by Schwinger (1951)], and currently known as the *TCP* theorem, asserts that within the framework of relativistically invariant *local* field theories, assuming the usual connection between spin and statistics, invariance under time reversal is equivalent to invariance under U_PU_C, i.e., the combined operation of charge conjugation (U_C) and space inversion (U_P). In a Lagrangian formulation, the *TCP* theorem is a result of the assumed invariance under proper Lorentz transformation of $\mathcal{L}$, the hermiticity of $\mathcal{L}$, the locality of the theory, and the assumption that particles of integral spin (bosons) must obey Bose-Einstein statistics and those of half-integral spin (fermions) must obey

Fermi-Dirac statistics, i.e., the particles obey the usual connection with statistics. We shall give a detailed proof of this theorem in Chapter 18. We here briefly state the content of the assumption of invariance under the operation CP in somewhat more physical terms. To this end we recall that invariance under the operation of space inversion U_P implied that $[H, U_P] = 0$, and this is equivalent to the conservation of a quantum number called parity. If the Hamiltonian, H, is invariant under a spatial inversion, the theory cannot distinguish between left and right. In the active viewpoint, the invariance of the theory is the statement that the mirror reflection of any state of the system is also a possible state of the same system. It is now firmly established that the weak interactions are not invariant under U_P. After this important insight by Lee and Yang (1957), Landau (1957) suggested that even though conservation of parity may be violated, the operation of U_PU_C may still be an exact symmetry operation under which the physical laws are invariant. This would say that the state obtained from the mirror reflection of the state of a system of particles together with the replacing of each particle by its antiparticle is always a possible state of the corresponding system of antiparticles.

The transformation properties of the spinor field $\psi(x)$ under charge conjugation and space inversion are *defined* as:

$$U_C\psi(x)\,U_C^{-1} = \eta_C C\tilde{\psi}^T(x) \tag{20a}$$

$$U_P\psi(x)\,U_P^{-1} = \eta_P\gamma_0\psi(i_sx) \tag{20b}$$

where $|\eta_P|^2 = |\eta_C|^2 = 1$ and $C^{-1}\gamma^\mu C = -\gamma^{\mu T}$. Note that we need not specify the phase factor appearing in the right side of Eqs. (20a) and (20b) beyond the requirement that $|\eta_P|^2 = |\eta_C|^2 = +1$, since we shall consider only bilinear covariants formed from ψ and $\tilde{\psi}$. The neutral field $\phi(x)$ under U_PU_C transforms according to

$$U_PU_C\phi(x)\,(U_PU_C)^{-1} = n\phi(i_sx) \tag{21}$$

In view of the neutrality of the boson field, n is real with $n^2 = +1$. The case $n = +1$ corresponds to a scalar field, the case $n = -1$ to a pseudoscalar field. The transformation properties of the derivatives of the field operators are

$$U_PU_C\partial_\mu\phi(x)\,(U_PU_C)^{-1} = n\epsilon_\mu\partial'_\mu\phi(i_sx) \tag{22}$$

where $\epsilon_\mu = g_{\mu\mu}$, i.e., $\epsilon_0 = +1$, $\epsilon_i = -1$, $i = 1, 2, 3$.

A local theory is said to be CP invariant (and by Lüders' theorem T invariant) if there exists a unique choice of phases η and n such that the Lagrangian is invariant under U_PU_C. An equivalent criterion for the CP invariance of the theory is that there exist a unique choice of phases such that the Hamiltonian H is invariant under U_PU_C. In other words, the theory will be CP invariant if $H = U_CU_PH(U_CU_P)^{-1}$ or, equivalently, if $L = U_CU_PL(U_CU_P)^{-1}$.

We next show that if CP is to be conserved in the interaction between the spinor field $\psi(x)$ and the boson field $\phi(x)$, then the interaction term must consist either of a term of the form $\tilde{\psi}\psi\phi$ or of any combination of $\tilde{\psi}\gamma_5\psi\phi$, $\tilde{\psi}\gamma_\mu\psi\partial^\mu\phi$ and $\tilde{\psi}\gamma_\mu\gamma_5\psi\partial^\mu\phi$ [Feinberg (1957)]. To this end let us consider the transformation properties of the bilinear forms $\tilde{\psi}\psi$, $\tilde{\psi}\gamma_\mu\psi$, $\tilde{\psi}\gamma_5\psi$ and $\tilde{\psi}\gamma_5\gamma_\mu\psi$ under CP. Using (20), we find that

$$U_C\tilde{\psi}_\alpha(x)\, O_{\alpha\beta}\psi_\beta(x)\, U_C^{-1} = C^{-1T}{}_{\alpha\delta}\psi_\delta(x)\, C_{\beta\rho}\tilde{\psi}_\rho(x)\, O_{\alpha\beta} \tag{23a}$$

$$= -\psi_\delta(x)\, C^{-1}{}_{\delta\alpha}O_{\alpha\beta}C_{\beta\rho}\tilde{\psi}_\rho(x) \tag{23b}$$

$$= \tilde{\psi}_\rho(x)\, [C^{-1}OC]_{\delta\rho}\, \psi_\delta \tag{23c}$$

In going from Eq. (23a) to Eq. (23b) we have made use of the fact that $C^TC^{-1} = -I$. In arriving at Eq. (23c) we have used the equal-time anticommutation rules and have neglected a term involving $\delta^{(3)}(\mathbf{0})$ which arises from these commutation rules. In order that no such ambiguity arise, we shall assume that all spinor expressions have been symmetrized in the following way:

$$\tilde{\psi}_\alpha(x)\, O_{\alpha\beta}\psi_\beta(x) \to \tfrac{1}{2}[(\tilde{\psi}(x)\, O)_\beta, \psi_\beta(x)] = \tfrac{1}{2}[\tilde{\psi}_\alpha(x), \psi_\beta(x)]\, O_{\alpha\beta} \tag{24}$$

Note that the commutator $\frac{1}{2}[\tilde{\psi}(x)\, O, \psi(x)]$, the symmetrized form of $\tilde{\psi}O\psi(x)$, is a one-component object, in contradistinction to the commutator $[\psi(x), \tilde{\psi}(x')]_+$ which is a 16-component object. The transformation properties of the symmetrized product under charge conjugation are given by

$$U_C[\tilde{\psi}_\alpha(x), \psi_\beta(x)]\, U_C^{-1}O_{\alpha\beta} = [\tilde{\psi}_\rho(x), \psi_\delta(x)]\, (C^{-1}OC)_{\delta\rho} \tag{25}$$

Table 10.4 lists the transforms of the sixteen covariants. Combining these

TABLE 10.4

O	$C^{-1}OC$	$(C^{-1}OC)^T$	$\gamma_0(C^{-1}OC)^T\gamma_0$
I	I	I	I
γ_μ	$-\gamma_\mu{}^T$	$-\gamma_\mu$	$-\epsilon_\mu\gamma_\mu$
γ_5	$+\gamma_5{}^T$	γ_5	$-\gamma_5$
$\gamma_5\gamma_\mu$	$+(\gamma_5\gamma_\mu)^T$	$\gamma_5\gamma_\mu$	$-\epsilon_\mu\gamma_5\gamma_\mu$
$\sigma_{\mu\nu}$	$-\sigma_{\mu\nu}{}^T$	$-\sigma_{\mu\nu}$	$-\epsilon_\nu\epsilon_\mu\sigma_{\mu\nu}$

results with those obtained from the application of U_P, we find

$$U_CU_P[\tilde{\psi}(x)\, O, \psi(x)]\, (U_CU_P)^{-1} = [\tilde{\psi}(i_sx)\, \gamma_0(C^{-1}OC)^T\gamma_0, \psi(i_sx)] \tag{26}$$

Hence, writing $U_{CP} = U_CU_P$,

$$U_{CP}[\tilde{\psi}(x), \psi(x)]\, U_{CP}^{-1} = [\tilde{\psi}(i_sx), \psi(i_sx)] \tag{27a}$$

$$U_{CP}[\tilde{\psi}(x)\, \gamma_5, \psi(x)]\, U_{CP}^{-1} = -[\tilde{\psi}(i_sx)\, \gamma_5, \psi(i_sx)] \tag{27b}$$

$$U_{CP}[\tilde{\psi}(x)\, \gamma_\mu, \psi(x)]\, U_{CP}^{-1} = -\epsilon_\mu[\tilde{\psi}(i_sx)\, \gamma_\mu, \psi(i_sx)] \tag{27c}$$

$$U_{CP}[\tilde{\psi}(x)\, \gamma_5\gamma_\mu, \psi(x)]\, U_{CP}^{-1} = -\epsilon_\mu[\tilde{\psi}(i_sx)\, \gamma_5\gamma_\mu, \psi(i_sx)] \tag{27d}$$

The various interaction terms thus transform according to:

$$U_{CP}[\bar{\psi}(x), \psi(x)]\, \phi(x)\, U_{CP}^{-1} = n[\bar{\psi}(i_s x), \psi(i_s x)]\, \phi(i_s x) \tag{28a}$$

$$U_{CP}[\bar{\psi}(x)\, \gamma_5, \psi(x)]\, \phi(x)\, U_{CP}^{-1} = -n[\bar{\psi}(i_s x)\, \gamma_5, \psi(i_s x)]\, \phi(i_s x) \tag{28b}$$

$$U_{CP}[\bar{\psi}(x)\, \gamma_\mu, \psi(x)]\, \partial^\mu\phi(x)\, U_{CP}^{-1} = -n[\bar{\psi}(i_s x)\, \gamma_\mu, \psi(i_s x)]\, \partial'^\mu\phi(i_s x) \tag{28c}$$

$$U_{CP}[\bar{\psi}(x)\, \gamma_5\gamma_\mu, \psi(x)]\, \partial^\mu\phi(x)\, U_{CP}^{-1} = -n[\bar{\psi}(i_s x)\, \gamma_5\gamma_\mu, \psi(i_s x)]\, \partial'^\mu\phi(i_s x) \tag{28d}$$

It will be noted that these terms split into two groups according to whether there is a factor $+n$ or $-n$ on the right-hand side. This proves our former assertion.

A further simplification occurs if one notes that the theory described by the Lagrangian $\mathfrak{L} = \mathfrak{L}_0 + \mathfrak{L}_I$, where $\mathfrak{L}_I$ is a linear combination of any of the four terms (28), is invariant under the gauge transformation $\psi \to [\exp(i\alpha)]\, \psi$, $\bar{\psi} \to [\exp(-i\alpha)]\, \bar{\psi}$ so that the quantity $[\bar{\psi}\gamma_\mu, \psi]$ is conserved, i.e., $\partial_\mu[\bar{\psi}\gamma^\mu, \psi] = 0$. Now the interaction term $[\bar{\psi}\gamma_\mu, \psi]\, \partial^\mu\phi$ can be rewritten in the form $\partial^\mu([\bar{\psi}\gamma_\mu, \psi]\, \phi) - (\partial^\mu[\bar{\psi}\gamma_\mu, \psi])\phi$, which is equal to $\partial^\mu([\bar{\psi}\gamma_\mu, \psi]\, \phi)$ by virtue of the above conservation law (or equivalently by virtue of the equations of motion). The remaining term is of the form of a four-divergence. We have indicated previously that two Lagrangians which differ by a four-divergence are equivalent insofar as the observable predictions of the theory are concerned. Hence we need not consider the vector coupling term. Stated differently, the vector coupling term can always be transformed away by a unitary transformation [Okubo (1954)]. We thus conclude that in the interaction of a boson spin 0 field ϕ with a spinor field ψ, if we restrict ourselves to local trilinear couplings, the coupling term can only be of the form

$$\mathfrak{L}_I = \tfrac{1}{2}G[\bar{\psi}(x), \psi(x)]\, \phi(x) \tag{29a}$$

or of the form

$$\mathfrak{L}_I = \tfrac{1}{2}G_1[\bar{\psi}(x)\, \gamma_5, \psi(x)]\, \phi(x) + \tfrac{1}{2}G_2[\bar{\psi}(x)\, \gamma_5\gamma_\mu, \psi(x)]\, \partial^\mu\phi(x) \tag{29b}$$

if the theory is to be Lorentz and CP invariant. It is interesting to note that the above restriction on interactions imposed by CP is precisely the same as if parity were conserved, i.e., as if we required the Lagrangian to be covariant under U_P separately. In this case the scalar interaction again does not mix with pseudoscalar or pseudovector.

The same considerations apply to the case of the interaction of a real spin 1 boson-vector field ϕ_μ with the spinor field ψ. For local interactions the $\mathfrak{L}_I$ term must be formed by combining spinor covariants and ϕ_μ or $F_{\mu\nu} = \partial_\mu\phi_\nu - \partial_\nu\phi_\mu$. The latter quantity transforms under U_{PC} according to

$$U_{CP}\phi_\mu(x)\, U_{CP}^{-1} = n'\epsilon_\mu\phi_\mu(i_s x) \tag{30a}$$

and

$$U_{CP}F_{\mu\nu}(x)\, U_{CP}^{-1} = n'\epsilon_\mu\epsilon_\nu F_{\mu\nu}(i_s x) \tag{30b}$$

The scalar interaction terms $\tilde{\psi}\gamma^\mu\psi\phi_\mu$, $i\tilde{\psi}\gamma^\mu\gamma_5\psi\phi_\mu$, $\tilde{\psi}\sigma^{\mu\nu}\psi F_{\mu\nu}$, where $\sigma^{\mu\nu} = \frac{1}{2i}[\gamma^\mu, \gamma^\nu]$, $\tilde{\psi}\sigma^{\mu\nu}\gamma_5\psi F_{\mu\nu}$ under CP then transform according to:

$$U_{CP}[\tilde{\psi}(x)\,\gamma^\mu, \psi(x)]\,\phi_\mu(x)\,U_{CP}^{-1} = -n'[\tilde{\psi}(i_s x)\,\gamma^\mu, \psi(i_s x)]\,\phi_\mu(i_s x) \tag{31a}$$

$$U_{CP}[\tilde{\psi}(x)\,\gamma^\mu\gamma_5, \psi(x)]\,\phi_\mu(x)\,U_{CP}^{-1} = -n'[\tilde{\psi}(i_s x)\,\gamma^\mu\gamma_5, \psi(i_s x)]\,\phi_\mu(i_s x) \tag{31b}$$

$$U_{CP}[\tilde{\psi}(x)\,\sigma^{\mu\nu}, \psi(x)]\,F_{\mu\nu}(x)\,U_{CP}^{-1} = -n'[\tilde{\psi}(i_s x)\,\sigma^{\mu\nu}, \psi(i_s x)]\,F_{\mu\nu}(i_s x) \tag{31c}$$

$$U_{CP}[\tilde{\psi}(x)\,\sigma^{\mu\nu}\gamma_5, \psi(x)]\,F_{\mu\nu}(x)\,U_{CP}^{-1} = n'[\tilde{\psi}(i_s x)\,\sigma^{\mu\nu}\gamma_5, \psi(i_s x)]\,F_{\mu\nu}(i_s x) \tag{31d}$$

The interaction can thus be a linear combination of the terms $\tilde{\psi}\gamma_\mu\psi\phi^\mu$, $\tilde{\psi}\sigma_{\mu\nu}\psi F^{\mu\nu}$ and $\tilde{\psi}\gamma_\mu\gamma_5\psi\phi^\mu$, or it can be $\tilde{\psi}\sigma_{\mu\nu}\gamma_5\psi F^{\mu\nu}$.

We conclude this section with a brief inquiry as to what the requirements of Lorentz and CP invariance imply about the structure of the Lagrangian of the free (noninteracting) fields [see, e.g., Soloviev (1958)]. Using the relations (27), one readily establishes that the most general local free-field Dirac Lagrangian which is Lorentz and CP invariant and bilinear in ψ and $\tilde{\psi}$ is given by

$$\begin{aligned}\mathcal{L}(x) = {} & \tfrac{1}{2}ia(\tilde{\psi}(x)\,\gamma^\mu\partial_\mu\psi(x) - \partial_\mu\tilde{\psi}(x)\,\gamma^\mu\psi(x)) \\ & + \tfrac{1}{2}b(\tilde{\psi}(x)\,\gamma_5\gamma^\mu\partial_\mu\psi(x) - \partial_\mu\tilde{\psi}(x)\,\gamma_5\gamma^\mu\psi(x)) - \tfrac{1}{4}m[\tilde{\psi}(x), \psi(x)]\end{aligned} \tag{32}$$

where a and b are real constants and the factor i is introduced to guarantee the hermiticity of $\mathcal{L}(x)$. This Lagrangian is clearly *not* invariant under space inversion since under the operation U_P the terms involving γ_5 will change sign. The equations of motion of the field operators ψ and $\tilde{\psi}$ which follow from the above Lagrangian are

$$(-i\Gamma^\mu\partial_\mu + m)\,\psi(x) = 0 \tag{33a}$$

with

$$\Gamma^\mu = (a - ib\gamma_5)\,\gamma^\mu \tag{33b}$$

The requirement that $\psi(x)$ describe particles of mass m can be translated into the requirement that ψ obey the equation

$$(\Box + m^2)\,\psi = 0 \tag{34}$$

The Γ^μ matrices must therefore satisfy the anticommutation rules:

$$\Gamma^\mu\Gamma^\nu + \Gamma^\nu\Gamma^\mu = 2g^{\mu\nu} \tag{35}$$

The substitution of (33b) into (35), together with the known commutation rules of γ_μ, γ_5, namely $[\gamma^\mu, \gamma^\nu]_+ = 2g^{\mu\nu}$ and $[\gamma^\mu, \gamma^5]_+ = 0$, indicate that the choice $a^2 - b^2 = 1$ will guarantee that Eqs. (33b) and (35) are satisfied, in which case Eq. (33a) is nothing but an "ordinary" Dirac equation with a different representation of the γ matrices. A particular choice of a and b which satisfies $a^2 - b^2 = 1$ is clearly $a = \sqrt{2}$, $b = 1$.

Finally we note that the Lagrangian for a charged boson field, in order to be Lorentz and CP invariant, must be written in the form

$$\mathcal{L} = -\tfrac{1}{2}\mu^2[\phi^*(x), \phi(x)]_+ + \tfrac{1}{2}[\partial_\mu\phi^*(x), \partial^\mu\phi(x)]_+ \tag{36}$$

i.e., a symmetrization of the bilinear boson factors has to be introduced. We shall sometimes indicate the requisite symmetrization by a dot and thus write

$$\mathcal{L} = -\mu^2\phi^* \cdot \phi + \partial_\mu\phi^* \cdot \partial^\mu\phi \tag{37}$$

where

$$\phi^* \cdot \phi = \tfrac{1}{2}[\phi^*, \phi]_+ \tag{38a}$$

To simplify the typography, we shall often indicate the necessary symmetrization in the fermion case also by a dot and write

$$\tfrac{1}{2}[\bar{\psi}O, \psi] = \bar{\psi}O \cdot \psi \tag{38b}$$

10c. Electromagnetic Interactions

In Chapter 7, Section *g*, we indicated that when the field equations are satisfied, the invariance of the Lagrangian under the transformation $\phi_r \to \phi_r' = \phi_r + \delta\phi_r$ implied[3] the existence of a conserved current J_μ

$$J^\mu = \sum_r \left(\frac{\partial\mathcal{L}}{\partial\phi_{r,\mu}}\,\delta\phi_r + \frac{\partial\mathcal{L}}{\partial\phi^*_{r,\mu}}\,\delta\phi_r^*\right) \tag{39a}$$

$$\partial_\mu J^\mu = 0 \tag{39b}$$

We further established that the invariance of the (Hermitian) Lagrangian density $\mathcal{L}$ under the gauge transformations

$$\phi_r = e^{i\alpha}\phi_r \tag{40a}$$

$$\phi_r^* = e^{-i\alpha}\phi_r^* \tag{40b}$$

gave rise, by virtue of (39), to a conserved current which could be identified as the electric charge-current density of the field system. Conversely, the imposition of the requirement that the Lagrangian describing electrically charged fields be invariant under the gauge transformation:

$$\phi_r = e^{+ie\chi}\phi_r \tag{41a}$$

$$\phi_r^* = e^{-ie\chi}\phi_r^* \tag{41b}$$

(known as the charge gauge) assigns to particle and antiparticle equal and opposite charges, the magnitude of which is not fixed. Since $\mathcal{L}$ is Hermitian, gauge invariance implies that only the combination $\phi^*_r\phi_r$ occurs in the Lagrangian.

Yang and Mills [Yang (1954); see also Utiyama (1956)] have suggested that the above gauge invariance requirements be extended to gauge transformations where the phase factor χ is an arbitrary function of space-time, $\chi = \chi(x)$, so that the relative phase of ϕ_r at two different space-time points is completely arbitrary. In order to make the Lagrangian still invariant

[3] In the subsequent until Eq. (47), the operator ϕ_r denotes an arbitrary field operator, not just the meson field operator.

under this wider transformation, it is necessary to introduce a new field, $A_\mu(x)$, which has a vector character and which under the above gauge transformation transforms according to

$$A_\mu \to A_\mu' = A_\mu + \partial_\mu \chi(x) \tag{42}$$

In addition, the A_μ field must be coupled to the ϕ_r field only through the replacement

$$\partial_\mu \phi_r(x) \to \left(\partial_\mu - i\frac{e}{c} A_\mu(x)\right) \phi_r(x) \tag{43a}$$

and

$$\partial_\mu \phi_r^*(x) \to \left(\partial_\mu + i\frac{e}{c} A_\mu(x)\right) \phi_r^*(x) \tag{43b}$$

in the free-field Lagrangian. This extended gauge invariance (also called invariance under gauge transformations of the second kind) essentially determines the interaction of charge fields with the electromagnetic field, $F_{\mu\nu}$ where $F_{\mu\nu} = \partial_\nu A_\mu - \partial_\mu A_\nu$. The way one arrives at (43) is by noting that under the gauge transformation:

$$\phi_r \to \phi_r' = e^{ie\chi(x)} \phi_r \tag{44a}$$

$$\phi_r^* \to \phi_r'^* = e^{-ie\chi(x)} \phi_r^* \tag{44b}$$

$$A_\mu \to A_\mu' = A_\mu + \partial_\mu \chi(x) \tag{44c}$$

the expression $(\partial_\mu - ieA_\mu)\,\phi_r$ has the property that

$$\begin{aligned}(\partial_\mu - ieA_\mu')\,\phi_r' &= e^{ie\chi}(\partial_\mu - ieA_\mu - ie\partial_\mu\chi + ie\partial_\mu\chi)\,\phi_r \\ &= e^{ie\chi}(\partial_\mu - ieA_\mu)\,\phi_r \end{aligned} \tag{45}$$

i.e., the phase function behaves as if it were a constant so that the expression $\phi_r^*(\partial_\mu - ieA_\mu)\,\phi_r$ is invariant under the gauge transformation (44). Therefore, if in the Lagrangian for the free-matter fields, $\mathfrak{L}_{\text{matter}}$, the replacement

$$\partial_\mu \phi_r \to (\partial_\mu - ieA_\mu)\,\phi_r \tag{46a}$$

$$\partial_\mu \phi_r^* \to (\partial_\mu + ieA_\mu)\,\phi_r^* \tag{46b}$$

is made for all the charged fields and to this Lagrangian is added the Lagrangian for the uncoupled electromagnetic field $\mathfrak{L}_{\text{em}} = -\frac{1}{4}F_{\mu\nu}F^{\mu\nu}$, then the total Lagrangian $\mathfrak{L} = \mathfrak{L}'_{\text{matter}} + \mathfrak{L}_{\text{em}}$ will be invariant under the transformation (44). Note that this extended invariance holds only if the quanta of the electromagnetic field have a mass which is identically zero. If the quanta associated with the A_μ field have a mass m, then the free-field Lagrangian of the A_μ field is $-\frac{1}{4}F_{\mu\nu}F^{\mu\nu} - m^2 A_\mu A^\mu$. The theory is then not invariant under extended gauge transformations since the $m^2 A_\nu A^\nu$ term is not.

Incidentally, in the special case where the charged field is interacting with a prescribed external electromagnetic field $A_\mu^e(x)$, the variation with respect to A_μ^e is no longer arbitrary. The correct description for this situation is in terms of the following Lagrangian:

$$\mathcal{L} = \mathcal{L}_{\text{matter}} - j^\mu(x)\, A_\mu^e(x) \tag{47}$$

where $\mathcal{L}_{\text{matter}}$ is the Lagrangian for the charged-matter field and j^μ the appropriate expression for its current in the presence of an electromagnetic field.

As an illustration of these remarks, consider the case of the charged Klein-Gordon field interacting with the Maxwell field. We have previously noted that the Lagrangian density for this field is

$$\mathcal{L}_{\text{m}} = -\mu^2\phi^* \cdot \phi + \partial_\mu\phi^* \cdot \partial^\mu\phi \tag{48}$$

According to the above, in order to obtain the correct Lagrangian which describes the interaction of the Maxwell field with this charged boson field, we must replace ∂_μ by (46a) and (46b) in the appropriate factors of $\mathcal{L}_{\text{m}}$ and then add to it the Lagrangian for the Maxwell field. We thus obtain for the total Lagrangian of the system the following expression:

$$\mathcal{L} = -\frac{1}{2}\frac{\partial A_\mu}{\partial x_\nu}\frac{\partial A^\mu}{\partial x^\nu} - \mu^2\phi^* \cdot \phi + \left(\frac{\partial \phi^*}{\partial x^\mu} + ieA_\mu\phi^*\right) \cdot \left(\frac{\partial \phi}{\partial x_\mu} - ieA^\mu\phi\right) \tag{49}$$

The variation of $\mathcal{L}$ with respect to A_μ now yields Maxwell's equations:

$$-\frac{\partial}{\partial x^\nu}\frac{\partial \mathcal{L}}{\partial(\partial_\nu A^\mu)} = \Box A_\mu = -\frac{\partial \mathcal{L}}{\partial A^\mu} \tag{50}$$

whereas variation with respect to ϕ and ϕ^* yields the Klein-Gordon equation in the presence of the electromagnetic field. Equation (50) now indicates that the correct definition of the current, as the source of the electromagnetic field, is given by

$$-\frac{\partial \mathcal{L}}{\partial A^\mu} = j_\mu \tag{51}$$

In the present example j_μ is given by

$$j_\mu = +ie\left[\left(\frac{\partial \phi^*}{\partial x^\mu} + ieA_\mu\phi^*\right) \cdot \phi - \phi^* \cdot \left(\frac{\partial \phi}{\partial x^\mu} - ieA_\mu\phi\right)\right] \tag{52}$$

This current is again conserved by virtue of the equations of motion. It is also the conserved current obtained as a consequence of the invariance of the total Lagrangian, Eq. (49), under gauge transformations of the first kind, i.e., under transformations

$$\phi \to e^{i\chi}\phi \tag{53a}$$

$$\phi^* \to e^{-i\chi}\phi^* \tag{53b}$$

where χ is a constant phase factor. From the fact that j_μ is conserved, i.e., that $\partial^\mu j_\mu = 0$, it follows that the quantity $Q = \int_\sigma d\sigma^\mu(x)\, j_\mu(x)$ is independent of σ and is a constant of the motion. Using the equal-time commutation rules:

$$[\phi(x), \pi(x')]_{x_0=x_0'} = \left[\phi(x), \frac{\partial \mathcal{L}}{\partial \phi_0(x')}\right]_{x_0=x_0'}$$
$$= [\phi(x), \partial_0'\phi^*(x') + ieA_0(x')\, \phi^*(x')]_{x_0=x_0'}$$
$$= i\delta^{(3)}(\mathbf{x} - \mathbf{x}') \tag{54a}$$

$$[\phi^*(x), \pi^*(x')]_{x_0=x_0'} = \left[\phi^*(x), \frac{\partial \mathcal{L}}{\partial \phi_0^*(x')}\right]_{x_0=x_0'}$$
$$= [\phi^*(x), \partial_0'\phi(x') - ieA_0(x')\, \phi(x')]_{x_0=x_0'}$$
$$= i\delta^{(3)}(\mathbf{x} - \mathbf{x}') \tag{54b}$$

one verifies that

$$[Q, \phi(x)] = -e\phi(x) \tag{55a}$$
$$[Q, \phi^*(x)] = +e\phi^*(x) \tag{55b}$$

Equations (55a) and (55b) are established by choosing the surface σ over which Q is defined to be the surface x_0 = constant. These again give an interpretation of ϕ and ϕ^* as charge creation and annihilation operators and of Q as the total charge operator. We shall expand on these remarks in our discussion of the charged Dirac field.

The expression for the Lagrangian density (49) is usually written in the form:

$$\mathcal{L} = \mathcal{L}_{\text{m}} + \mathcal{L}_{\text{em}} + \mathcal{L}_I \tag{56a}$$

where $\mathcal{L}_{\text{em}}$ is the Lagrangian for the free electromagnetic field and $\mathcal{L}_I$ is the interaction term given by

$$\mathcal{L}_I = -ie\left(\frac{\partial \phi^*}{\partial x^\mu} \cdot \phi - \phi^* \cdot \frac{\partial \phi}{\partial x^\mu}\right) A^\mu + e^2 A_\mu A^\mu \phi^* \cdot \phi \tag{56b}$$

This term is of the form $j_\mu(x)\, A^\mu(x)$ with j_μ given by Eq. (52). The occurrence of the term $e^2 A_\mu A^\mu \phi^* \cdot \phi$ is characteristic of the boson field in interaction with the electromagnetic field. It corresponds, in the nonrelativistic limit, to the $\mathbf{A}^2$ term in the Schrödinger equation in the presence of an electromagnetic field.

A Hamiltonian formalism can again be developed using the methods presented in Section 7*f*. As noted above, the momenta canonically conjugate to the ϕ variables now depend on the electromagnetic field.

The Lagrangian density for the Dirac field interacting with the electromagnetic field is:

$$\mathcal{L} = -\frac{1}{2}\frac{\partial A_\mu}{\partial x_\nu} \cdot \frac{\partial A^\mu}{\partial x^\nu} - \frac{1}{2i}\left[\tilde{\psi}\gamma^\mu \cdot \left(\frac{\partial}{\partial x^\mu} - ieA_\mu\right)\psi + im\tilde{\psi} \cdot \psi\right]$$
$$+ \frac{1}{2i}\left[\left(\frac{\partial \tilde{\psi}}{\partial x^\mu} + ieA_\mu\tilde{\psi}\right)\gamma^\mu \cdot \psi - im\tilde{\psi} \cdot \psi\right]$$
$$= \mathcal{L}_{\text{m}} + \mathcal{L}_{\text{em}} + e\tilde{\psi}\gamma^\mu \cdot \psi A_\mu \tag{57}$$

where $\mathcal{L}_{\text{m}}$ is the Lagrangian for the free Dirac field. The variation of $\mathcal{L}$ with respect to the electromagnetic potentials now yields

$$\Box A_\mu = -\frac{\partial \mathfrak{L}}{\partial A^\mu} = -e\tilde{\psi}\gamma_\mu \cdot \psi \tag{58}$$

Similarly, the variations with respect to ψ and $\tilde{\psi}$ result in the Dirac equation for $\tilde{\psi}$ and ψ in the presence of the electromagnetic field

$$(i\gamma_\mu\partial^\mu + e\gamma_\mu A^\mu(x))\,\psi(x) - m\psi(x) = 0 \tag{59a}$$

$$(i\partial^\mu - eA^\mu(x))\,\tilde{\psi}(x)\,\gamma_\mu + m\tilde{\psi}(x) = 0 \tag{59b}$$

Equation (58) indicates that the expression for the current in the presence of an electromagnetic field, in the quantized theory, is given by

$$j_\mu(x) = -\frac{e}{2}\,[\tilde{\psi}(x)\,\gamma_\mu, \psi(x)] \tag{60}$$

which is formally identical to the expression for the current in the non-interacting field case, Eq. (8.83). It is, however, not the same, due to the fact that in the expression (58), ψ and $\tilde{\psi}$ obey Eqs. (59a) and (59b) in which the electromagnetic potentials occur, whereas in the expression (8.83) ψ obeys the free-field equation.

Equations (58) and (59) are the equations of motion of the field operators in the Heisenberg picture; the state vector $|\Psi\rangle$ of the system is then time-independent. The canonical formalism (together with the usual connection of spin and statistics) allows one to infer that the equal-time commutation rules of operators are:[4]

$$[\psi(x), \tilde{\psi}(x')]_{+\; x_0 = x_0'} = \gamma_0\delta^{(3)}(\mathbf{x} - \mathbf{x}') \tag{61a}$$

$$[\psi(x), \psi(x')]_{+\; x_0 = x_0'} = [\tilde{\psi}(x), \tilde{\psi}(x')]_{+\; x_0 = x_0'} = 0 \tag{61b}$$

$$[A_\mu(x), A_{\nu,0}(x')]_{x_0 = x_0'} = -i\hbar c g_{\mu\nu}\delta^{(3)}(\mathbf{x} - \mathbf{x}') \tag{61c}$$

$$[\psi(x), A_\mu(x')]_{x_0 = x_0'} = [\tilde{\psi}(x), A_\mu(x')]_{x_0 = x_0'} = 0 \tag{61d}$$

By virtue of the relativistic covariance of the theory, these commutation rules can be generalized to arbitrary space-like separations and then read:

$$[\psi(x), \tilde{\psi}(x')]_+ = -iS(x - x') \qquad \text{for } (x - x')^2 < 0 \tag{62a}$$

$$[A_\mu(x), A_\nu(x')] = -i\hbar c g_{\mu\nu} D(x - x') \quad \text{for } (x - x')^2 < 0 \tag{62b}$$

with all other commutators or anticommutators vanishing.

Note that it is now impossible to write down the commutation rules valid for all times; this would require the knowledge of the solutions of the equations of motion (58), (59) for all times. (This is, in fact, the problem to be solved!) The physical interpretation of the operators ψ, $\tilde{\psi}$ and A_μ, as well as the calculation of the predictions of the quantized theory, are now much more difficult than in the free-field case. One important statement can, however, still be made about the operators ψ and $\tilde{\psi}$. From the fact that the current operator [see Eq. (60)]

[4] For explicitness we have here included the factors $\hbar$, c in the commutation rules.

$$j_\mu(x) = -\frac{e}{2}\,[\bar{\psi}(x)\,\gamma_\mu, \psi(x)]$$

is divergence-free, $\partial^\mu j_\mu(x) = 0$, it follows that the total charge operator Q

$$Q = -\frac{e}{2}\int_\sigma d\sigma^\mu(x')\,[\bar{\psi}(x')\,\gamma_\mu, \psi(x')] \tag{63}$$

is constant in time, i.e., it will have the same (operator) value independent of the particular space-like surface chosen in the right-hand side of (63). This fact allows us to compute the commutation rules of $\psi(x)$ and $\bar{\psi}(x)$ with Q by choosing for the space-like surface σ one which contains the space-time point x, so that the commutation rules (61) and (62) hold. One then verifies that:

$$[Q, \psi(x)] = +e\psi(x) \tag{64a}$$

$$[Q, \bar{\psi}(x)] = -e\bar{\psi}(x) \tag{64b}$$

A familiar argument then allows us to infer from these commutation rules that the Heisenberg operator $\psi(x)$ destroys an amount of charge $-e$ or creates an amount of charge $+e$, and that $\bar{\psi}$ destroys an amount of charge $+e$ or creates an amount of charge $-e$.

It is interesting to note that if, in our consideration of the charged spinor field, we had proceeded as in Section *b* of this chapter and considered all possible interaction terms involving only A_μ and $F_{\mu\nu}$ (but not their derivatives), then as indicated in Section *b*, the interactions would fall into two separate groups by virtue of the required invariance under CP. The interaction, if of first degree in A^μ, can be either of the form

$$\mathcal{L}_I' = e\bar{\psi}\gamma_\mu \cdot \psi A^\mu + \mu'\bar{\psi}\sigma_{\mu\nu} \cdot \psi F^{\mu\nu} + e'\bar{\psi}\gamma_\mu\gamma_5 \cdot \psi A^\mu \tag{65a}$$

(where one notes that the term in e' does not conserve parity), or it can be of the form

$$\mathcal{L}_I'' = \mu''\bar{\psi}\sigma_{\mu\nu}\gamma_5 \cdot \psi F^{\mu\nu} \tag{65b}$$

The requirement of gauge invariance of the second type implies that the interaction must be of the form $\mathcal{L}_I'$ with $e' = 0$. The Lagrangian $\mathcal{L}_{\text{Dirac}} + \mathcal{L}_{\text{em}} + \mathcal{L}_I''$ is clearly not gauge invariant since $\mathcal{L}_{\text{Dirac}}$ is not. Similarly the term $e'\bar{\psi}\gamma_\mu\gamma_5 \cdot \psi A^\mu$ is not gauge invariant. Extended gauge invariance together with CP invariance limits the form of the interaction to

$$\mathcal{L}_I = e\bar{\psi}\gamma_\mu \cdot \psi A^\mu + \mu'\bar{\psi}\sigma_{\mu\nu} \cdot \psi F^{\mu\nu} \tag{66}$$

which it will be noted is invariant under space inversion and thus implies parity conservation. The experimental evidence is that both P and C are indeed conserved in electromagnetic interactions [Lee (1957)].

The interaction $\mu_0\bar{\psi}\sigma_{\mu\nu}\psi F^{\mu\nu}$ has been used to describe an additional interaction (in addition to the interaction $j_\mu A^\mu$) of a Dirac particle (e.g., a nucleon) with the electromagnetic field. It can be considered as the interaction between the electromagnetic field and an anomalous moment μ' of the fermion with $\gamma_k\gamma_l\ (k, l = 1, 2, 3) = -\sigma_i\ (j \neq k \neq i)$ denoting the magnetic moment and $\gamma_0\gamma_k$ the electric moment. The interaction $\mu_0\bar{\psi}\sigma_{\mu\nu}\psi F^{\mu\nu}$ is known as the Pauli moment interaction (recall Section 4h). It can be used to describe in first order and in the nonrelativistic region the interaction which an electron, proton, or neutron has with an external electromagnetic field due to the fact that its magnetic moment is not exactly a Bohr magneton $e\hbar/2mc$ (where e and m are the appropriate charge and mass of the particle). This interaction, however, leads to divergences in higher orders of perturbation theory in the quantized theory, as do all derivative coupling interactions.

It turns out that in actuality only coupling terms of the form $j_\mu A^\mu$ are necessary to account for the observed interaction of photons with charged particles. That this is possible is a consequence of the fact that, in a field theoretic description, the anomalous magnetic moment of spin $\frac{1}{2}$ particles is not an intrinsic property of these particles but is due to their interactions with other fields (electromagnetic for the electron, mesic for the nucleon) [Wick (1935)]. Thus the only electromagnetic interactions which ever have to be considered are the familiar interactions with charges and currents of the form $j_\mu A^\mu$. The apparent absence of other couplings is sometimes referred to as "the principle of minimal electromagnetic interactions" [Gell-Mann (1956a)]. We shall see that a minimal coupling implies that although electromagnetic interactions no longer conserve T^2, i.e., the total isotopic spin, they still conserve T_3. The conservation of T_3 is equivalent to charge conservation when only electromagnetic interactions are considered. However, this conservation law takes on independent significance in the presence of other interactions.

The previous remarks are readily extended to the case of the pion and nucleon system. Consider first the nucleon system. Since the charge of the neutron is zero, the electromagnetic field interacts only with the proton field. This is accomplished by requiring that the electromagnetic field is coupled to the nucleon field $\psi = \begin{pmatrix}\psi_p \\ \psi_n\end{pmatrix}$ by the replacements

$$\partial_\mu\psi \to (\partial_\mu + ie\tfrac{1}{2}(1 + \tau_3)\, A_\mu)\, \psi \tag{67a}$$

$$\widetilde{\partial_\mu\psi} \to (\partial_\mu\bar{\psi} - ie\bar{\psi}\tfrac{1}{2}(1 + \tau_3)\, A_\mu) \tag{67b}$$

Since $\frac{1}{2}(1 + \tau_3)\,\psi = \begin{pmatrix}\psi_p \\ 0\end{pmatrix}$, clearly only the proton field is coupled to electromagnetic field in (67). Note also that the sign of ieA_μ which is added to ∂_μ is opposite from that in the expressions (43). This is because the protons, which are considered the "particles," have charge $+e$ and are associated with the field ψ_p. With the replacements (67) in the free-field nucleon Lagrangian ψ_p will create an amount of charge $-e$ and destroy an amount $+e$. This is readily verified from the commutation rules of ψ_p and $Q = \int_\sigma j^\mu d\sigma_\mu(x)$, where the current in the expression for Q is

$j_\mu = -\dfrac{\partial \mathfrak{L}}{\partial A^\mu} = \dfrac{e}{2}[\bar{\psi}\gamma_\mu\tfrac{1}{2}(1 + \tau_3), \psi]$, the source of the electromagnetic field.

Similarly for the pion field. Here the 1 and 2 components of $\boldsymbol{\phi}$ are to be coupled to the electromagnetic field. To obtain the form of the coupling in this case, we first of all note that the total isotopic spin for the meson field can be written in the form

$$T_i = -\int d\sigma^\mu \sum_{j,\,k=1}^{3} \partial_\mu\phi_j \cdot \epsilon_{ijk}\phi_k = -\sum_{j=1}^{3} \int d\sigma^\mu \partial_\mu\phi_j \cdot (t_i\boldsymbol{\phi})_j$$

where the matrix elements of the operator t_i are given by $(t_i)_{jk} = \epsilon_{ijk}$. The total charge can therefore be expressed as:[5]

$$Q = -e\int d\sigma^\mu \partial_\mu\boldsymbol{\phi} \cdot t_3\boldsymbol{\phi} \tag{68a}$$

$$= -e\int d\sigma^\mu(\partial_\mu\phi_1\ \partial_\mu\phi_2\ \partial_\mu\phi_3) \cdot \begin{pmatrix} 0 & 1 & 0 \\ -1 & 0 & 0 \\ 0 & 0 & 0 \end{pmatrix}\begin{pmatrix} \phi_1 \\ \phi_2 \\ \phi_3 \end{pmatrix} \tag{68b}$$

$$= -e\int d\sigma^\mu(\partial_\mu\phi_1 \cdot \phi_2 - \partial_\mu\phi_2 \cdot \phi_1) = ie\int d\sigma^\mu(\partial_\mu\phi^* \cdot \phi - \phi^* \cdot \partial_\mu\phi) \tag{68c}$$

We next show that the interaction with the electromagnetic field is correctly introduced by the gauge invariant replacement

$$\partial_\mu\boldsymbol{\phi} \longrightarrow (\partial_\mu - et_3A_\mu)\,\boldsymbol{\phi} \tag{69}$$

in the free-field Lagrangian. To verify this, we note that:

$$\begin{aligned}
\tfrac{1}{2}[\mu^2\boldsymbol{\phi} \cdot \boldsymbol{\phi} &- (\partial_\mu - et_3A_\mu)\,\boldsymbol{\phi} \cdot (\partial^\mu - et_3A^\mu)\,\boldsymbol{\phi}] \\
&= \tfrac{1}{2}[\mu^2\boldsymbol{\phi} \cdot \boldsymbol{\phi} - \partial_\mu\boldsymbol{\phi} \cdot \partial^\mu\boldsymbol{\phi}] - e(\partial_\mu\phi_1 \cdot \phi_2 - \partial_\mu\phi_2 \cdot \phi_1)\,A^\mu \\
&\quad - \tfrac{1}{2}e^2(\phi_1^2 + \phi_2^2)\,A_\mu A^\mu \\
&= [\mu^2\phi^* \cdot \phi - \partial_\mu\phi^* \cdot \partial^\mu\phi] + \tfrac{1}{2}[\mu^2\phi_3^2 - \partial_\mu\phi_3 \cdot \partial^\mu\phi_3] \\
&\quad + ie(\partial_\mu\phi^* \cdot \phi - \phi^* \cdot \partial_\mu\phi)\,A^\mu - e^2\phi^* \cdot \phi A_\mu A^\mu
\end{aligned} \tag{70}$$

which is the required form for the interaction of a charged field ϕ, ϕ^* and a neutral field ϕ_3 with the electromagnetic field. It will be observed that for the nucleon system, as well as for the pion system, the current interacting with the electromagnetic field is such that it commutes with the 3-component of the total isotopic spin. In particular, for the pion field, the current is of the form, $e\partial_\mu\boldsymbol{\phi} \cdot t_3\boldsymbol{\phi} + e^2(\boldsymbol{\phi} \cdot t_3{}^*t_3\boldsymbol{\phi})$ which is only invariant under arbitrary rotations about the 3-axis in isotopic spin space. Similarly, the nucleon current $\tfrac{1}{2}\bar{\psi}\gamma_\mu(1 + \tau_3) \cdot \psi$ separates into two parts, one of which transforms like a scalar in isotopic spin space, $\bar{\psi}\gamma_\mu \cdot \psi$, and the

[5] Note that the dot inside the product $\partial_\mu\boldsymbol{\phi} \cdot t_3\boldsymbol{\phi}$ actually means two things. Firstly, it implies that the scalar product in isotopic spin space of the vectors $\partial_\mu\boldsymbol{\phi}$ and $t_3\boldsymbol{\phi}$ is to be taken, and secondly that the resulting expression is to be symmetrized in the boson operators.

other of which transforms like the $\mathcal{3}$-component of a vector, $\bar{\psi}\gamma_\mu\tau_3 \cdot \psi$. This current is therefore again only invariant under rotations about the $\mathcal{3}$-axis in isotopic spin space.

10*d*. The Meson-Nucleon Interaction

The idea that the neutron and proton could be looked upon as corresponding to two states of the same particle was originally suggested by Heisenberg (1932). Later Cassen and Condon (1936) introduced an "isotopic" spin parameter to describe these two charge states. In this description a proton is a "nucleon" whose $\mathcal{3}$-component of isotopic spin, T_3, is equal to $+\frac{1}{2}$ and a neutron one with $T_3 = -\frac{1}{2}$. The charge of the nucleon, in units of e, is then given by $q = T_3 + \frac{1}{2}$. One thus speaks of the neutron and proton as belonging to an isotopic doublet characterized by an isotopic spin $T = \frac{1}{2}$. The equality of the neutron-proton and proton-proton interaction, provided that the nucleons are in the same spin and orbital angular momentum state (as indicated by much experimental evidence), leads immediately to the concept of a total isotopic spin which is conserved in nucleon-nucleon interactions [see, e.g., Bethe (1955)]. A direct implication of this for the current field theoretic description, in which the interaction between nucleons is ascribed to the exchange of (virtual) mesons, is that all strong interactions must also satisfy the same conservation law.

The fact that all the strongly interacting particles seem to fall into groups with relatively small mass differences between the members in each group, forms the basis for the assignment of the particles into "charge" or "isotopic spin" multiplets which are characterized by an isotopic spin quantum number T, and have a multiplicity $2T + 1$. The components of the multiplets (to which correspond the different particles in the group) are characterized by values of T_3 ranging from $-T$ to T. Thus, for example, we have seen that the three states of a pion can be made to correspond to the three components of a $T = 1$ multiplet with the following assignments:

TABLE 10.5

Particle	T	T_3
π^+	1	+1
π^0	1	0
π^-	1	−1

Similarly, the kaons can be grouped into two pairs of doublets ($T = \frac{1}{2}$), the K^+ and K^0 constituting one such doublet and the K^- and $\bar{K}^0$ the other. The grouping of hyperons into "charge" multiplets is as follows: The Λ particle can be assigned an isotopic spin $T = 0$, the Σ particles an isotopic spin $T = 1$, with $T_3(\Sigma^\pm) = \pm 1$ and $T_3(\Sigma^0) = 0$ and the Ξ (cascade) particles an isotopic spin $T = \frac{1}{2}$ with $T_3(\Xi^-) = -\frac{1}{2}$ and $T_3(\Xi^0) = +\frac{1}{2}$.

The total isotopic spin of a system of strongly interacting particles is given by the vector sum of the isotopic spin vectors of all the particles (in complete analogy with the rules for addition of angular momenta). In all reactions between strongly interacting particles observed to date, the total isotopic spin as well as the *3*-component of isotopic spin is conserved. (This is true to the extent that electromagnetic effects, and a fortiori weak interactions can be neglected.) The conservation of T^2 and T_3 for the total system is the statement of the charge independence of the strong interaction.

The conservation of isotopic spin in strong interactions is identical with the requirement of invariance of the interactions under isotopic rotations. When the weak and electromagnetic interactions are neglected, the orientation of the isotopic spin is then of no physical significance. The differentiation between protons and neutrons, for example, is then a purely arbitrary process. (It, of course, becomes unambiguous in the presence of electromagnetic interactions since the proton is charged: The electromagnetic interactions lead to a violation of charge independence). Charge independence also asserts, for example, that the force between a Λ particle and a proton is the same as between a Λ particle and a neutron (in the same orbital and spin angular momentum state).

Consider next the field theoretic description of the charge-independent interaction between the strongly interacting particles. For the moment, let us restrict ourselves to the pion-nucleon system. We have indicated in Chapter 7 how, in accordance with its assignment of isotopic spin 1, the pion system can be described by a field operator $\boldsymbol{\phi}$ which transforms like a vector in isotopic spin space (ϕ_3 describes the neutral pion field and $\phi_1 \pm i\phi_2$ the charged pion fields). In Chapter 8, we sketched the description of the nucleon field by an eight-component spinor operator

$$\psi(x) = \begin{pmatrix} \psi_{\mathrm{p}}(x) \\ \psi_{\mathrm{n}}(x) \end{pmatrix} \tag{71}$$

which transforms like a two-component spinor in isotopic spin space. Let us next analyze the form of the possible interaction terms between these fields which are invariant under rotations in isotopic spin space. Consider first the possible forms of nonderivative couplings of ψ with the charged Bose fields ϕ and ϕ^*. In order to conserve charge, the interaction must clearly be of the form

$$\mathfrak{L}_I = g_1\bar{\psi}_{\mathrm{n}}O_1 \cdot \psi_{\mathrm{p}}\phi + g_2\bar{\psi}_{\mathrm{p}}O_2 \cdot \psi_{\mathrm{n}}\phi^* \tag{72}$$

Formally the requirement of charge conservation can be stated as $[\mathfrak{L}_I, Q] = 0$. That $\mathfrak{L}_I$ as given by (72) does indeed conserve charge can be established as follows: Since the operator $\phi(x)$ destroys an amount of charge $-e$ (or creates a charge $+e$), in order that $\mathfrak{L}_I$ operating on a state

$|\Psi\rangle$ of charge q does not alter the total charge of this state, the combination of spinor factors coupled to operator ϕ must be such as to create a charge $-e$ (or destroy a charge $+e$). The term $\tilde{\psi}_n O_1 \cdot \psi_p$ clearly does this. Similarly, ϕ^* creates an amount of charge $-e$ (or destroys $+e$) and $\tilde{\psi}_p O_2 \cdot \psi_n$ destroys an amount of charge $-e$ (or creates a charge $+e$).

An interaction is said to be charge-symmetric if $\mathcal{L}_I$ is invariant under the transformation:

$$\psi_n \rightarrow \psi_p \qquad \phi \rightarrow \phi^* \tag{73a}$$

$$\psi_p \rightarrow \psi_n \qquad \phi^* \rightarrow \phi \tag{73b}$$

This transformation leaves the commutation rules invariant. Under this transformation $\mathcal{L}_I$ transforms into

$$\mathcal{L}_I \rightarrow g_1 \tilde{\psi}_p O_1 \cdot \psi_n \phi^* + g_2 \tilde{\psi}_n O_2 \cdot \psi_p \phi \tag{74}$$

Invariance under charge symmetry thus requires that

$$g_1 O_1 = g_2 O_2 \tag{75}$$

The hermiticity of $\mathcal{L}_I$, $\mathcal{L}_I = \mathcal{L}_I^*$, further implies that

$$\bar{g}_1 \gamma_0 O_1^* \gamma_0 = g_2 O_2 \tag{76a}$$

$$\bar{g}_2 \gamma_0 O_2^* \gamma_0 = g_1 O_1 \tag{76b}$$

which together with (75) implies that to be charge-symmetric $\mathcal{L}_I$ must be of the form

$$\mathcal{L}_I = g \tilde{\psi}_n O \psi_p \phi^* + \bar{g} \tilde{\psi}_p \gamma_0 O^* \gamma_0 \psi_n \phi \tag{77}$$

with

$$gO = \bar{g} \gamma_0 O^* \gamma_0 \tag{78}$$

The following interactions in any combination are therefore charge-symmetric:

$$A.\ g_S(\tilde{\psi}_p \cdot \psi_n \phi^* + \tilde{\psi}_n \cdot \psi_p \phi) \tag{79}$$

$$B.\ g_{PS}(\tilde{\psi}_p \gamma_5 \cdot \psi_n \phi^* + \tilde{\psi}_n \gamma_5 \cdot \psi_p \phi) \tag{80}$$

$$C.\ g_V(\tilde{\psi}_p \gamma_\mu \cdot \psi_n \partial^\mu \phi^* + \tilde{\psi}_n \gamma_\mu \cdot \psi_p \partial^\mu \phi) \tag{81}$$

$$D.\ i g_{PV}(\tilde{\psi}_p \gamma_5 \gamma_\mu \cdot \psi_n \partial^\mu \phi^* + \tilde{\psi}_n \gamma_5 \gamma_\mu \cdot \psi_p \partial^\mu \phi) \tag{82}$$

The requirement of CP invariance will next be investigated. Under CP, the spinor fields transform as before, whereas the boson field ϕ transforms according to:

$$U_{CP} \phi(x)\, U_{CP}^{-1} = n \phi^*(i_s x) \tag{83}$$

$$U_{CP} \phi^*(x)\, U_{CP}^{-1} = n \phi(i_s x) \tag{84}$$

where we have chosen the space parity n real. Under CP, therefore, the interaction (A) is multiplied by n, whereas interactions (B), (C), and (D) are multiplied by $-n$. Thus invariance under CP leads to the same restrictions for charge-symmetric interactions as it did for the interactions described in Section a. That is, the requirement of a unique value for n

implies that the scalar interaction does not mix with the vector, pseudoscalar or pseudovector interaction.

To obtain an interaction of the nucleon field with the pion field which will be charge independent (i.e., invariant under rotations in isotopic spin space), we must require that

$$[\mathcal{L}_I, T_i] = 0 \quad i = 1, 2, 3 \tag{85}$$

where

$$T_i = \int d\sigma^\mu \tilde{\psi} \tfrac{1}{2}\tau_i \cdot \psi + \int d\sigma^\mu \boldsymbol{\phi} \cdot \partial_\mu t_i \boldsymbol{\phi} \tag{86}$$

is the ith component of the total isotopic spin of the meson-nucleon system. (Recall T_i is the generator for an infinitesimal rotation about the ith axis in isotopic spin space.) This clearly will be the case if we combine the isotopic spin vector $\tilde{\psi}O\boldsymbol{\tau} \cdot \psi$ with the iso-vector $\boldsymbol{\phi}$ to obtain an interaction Lagrangian

$$\mathcal{L}_I = \sum_{j=1}^{3} g\tilde{\psi}O\tau_j\psi\phi_j \tag{87}$$

This can also be written in the form:

$$\begin{aligned}\mathcal{L}_I &= \sqrt{2}\, g[\tilde{\psi}\tau_- O \cdot \psi(\phi_1 + i\phi_2) + \tilde{\psi}\tau_+ O \cdot \psi(\phi_1 - i\phi_2)] + g\tilde{\psi}\tau_3 O \cdot \psi\phi_3 \\ &= \sqrt{2}\, g[\tilde{\psi}_n O \cdot \psi_p \phi + \tilde{\psi}_p O \cdot \psi_n \phi^*] + g[\tilde{\psi}_p O \cdot \psi_p - \tilde{\psi}_n O \cdot \psi_n]\, \phi_3 \end{aligned} \tag{88}$$

where $\tau_\pm = \frac{1}{2}(\tau_1 \pm i\tau_2)$. Note, therefore, that in a charge independent theory with a Yukawa coupling, i.e., of the form $\tilde{\psi}O\psi\phi$, the coupling of the nucleon field to the neutral meson field is $\sqrt{2}$ times weaker than to the charged meson field and, furthermore, the coupling constants measuring the interaction of the neutron and proton fields with the neutral pion field are equal in magnitude but opposite in sign.

If we restrict ourselves to nonderivative Yukawa type couplings, the most general coupling of the nucleon field to the pion field which satisfies the requirement of Lorentz invariance and CP invariance only is of the form:

$$\begin{aligned}\mathcal{L}_I &= g[\tilde{\psi}_p\gamma_5 \cdot \psi_n\phi^* + \tilde{\psi}_n\gamma_5 \cdot \psi_p\phi] + g_3'\tilde{\psi}_p\gamma_5 \cdot \psi_p\phi_3 + g_3''\tilde{\psi}_n\gamma_5 \cdot \psi_n\phi_3 \\ &= g[\tilde{\psi}\tau_-\gamma_5 \cdot \psi\phi + \tilde{\psi}\tau_+\gamma_5 \cdot \psi\phi^*] + g_3'\tilde{\psi}_p\gamma_5 \cdot \psi_p\phi_3 + g_3''\tilde{\psi}_n\gamma_5 \cdot \psi_n\phi_3 \end{aligned} \tag{89}$$

where g, g_3', g_3'' are real constants. If one next imposes the requirement of charge independence, it is clear that $g_3' = -g_3'' = \dfrac{1}{\sqrt{2}}g$. The interaction thus reduces to the form $\mathcal{L}_I = \frac{1}{2}g[\tilde{\psi}\gamma_5\boldsymbol{\tau}, \psi] \cdot \boldsymbol{\phi}$ which, it should be noted, is P invariant (and which therefore conserves parity). The requirement of CP invariance together with charge independence thus implies that for a Yukawa coupling the theory will also be automatically P invariant (and hence also C invariant). The interaction of the meson-nucleon system can thus be described by the following Lagrangian:

$$\mathfrak{L} = -\tfrac{1}{2}\bar{\psi} \cdot (-i\gamma^\mu \partial_\mu + M)\,\psi - \tfrac{1}{2}(i\partial_\mu \bar{\psi}\gamma^\mu + M\bar{\psi}) \cdot \psi$$
$$- \tfrac{1}{2}(\mu^2 \boldsymbol{\phi} \cdot \boldsymbol{\phi} - \partial_\mu \boldsymbol{\phi} \cdot \partial^\mu \boldsymbol{\phi}) + \tfrac{1}{2}G \sum_{j=1}^{3} [\bar{\psi}\gamma_5 \tau_j, \psi]\,\phi_j \quad (90)$$

This Lagrangian will give rise to interactions between mesons and nucleons which are:

(*a*) Lorentz invariant
(*b*) charge independent
(*c*) *CP* invariant
(*d*) *P* invariant

By virtue of *c* and *d*, the theory is also *C* invariant (i.e., invariant under particle-antiparticle conjugation), as is indeed easily verified. It will be noted that in the approximation that we are working (i.e., neglecting the weak and electromagnetic interactions), all the particles are stable. Moreover, we have assumed that the neutron and proton have the same (bare) mass: M, and similarly that all the π mesons have the same (bare) mass: μ.

Thus far we have only considered Yukawa type couplings between nucleons and mesons. There seems to be some evidence that there exists in addition to these couplings an additional intrinsic meson-meson interaction. The only meson-meson interaction which is invariant under rotations in isotopic spin space and characterized by a dimensionless coupling constant is an interaction of the form

$$\mathfrak{L}_I' = \lambda(\boldsymbol{\phi} \cdot \boldsymbol{\phi})^2 \quad (91)$$

If we next allow the system to interact with the electromagnetic field, the Lagrangian describing this field system will, according to the principle of minimal electromagnetic interaction, be given by

$$\mathfrak{L} = -\tfrac{1}{2}(\partial_\nu A_\mu) \cdot (\partial^\nu A^\mu) - \tfrac{1}{2}\bar{\psi}\left[i\gamma^\mu\left(\partial_\mu + \frac{ie}{2}(1 + \tau_3)\,A_\mu\right) + M\right] \cdot \psi$$
$$-\tfrac{1}{2}\left[i\left(\partial_\mu \bar{\psi} - \frac{ie}{2}\bar{\psi}(1 + \tau_3)\,A_\mu\right)\gamma^\mu + M\bar{\psi}\right] \cdot \psi$$
$$-\tfrac{1}{2}[\mu^2 \boldsymbol{\phi} \cdot \boldsymbol{\phi} - (\partial_\mu - et_3 A_\mu)\,\boldsymbol{\phi} \cdot (\partial^\mu - et_3 A^\mu)\,\boldsymbol{\phi}]$$
$$+\tfrac{1}{2}G[\bar{\psi}\gamma_5 \boldsymbol{\tau}, \psi] \cdot \boldsymbol{\phi} + \lambda(\boldsymbol{\phi} \cdot \boldsymbol{\phi})^2 \quad (92)$$

It will be observed that in the presence of the electromagnetic field the Lagrangian (92) is no longer invariant under arbitrary rotations in isotopic spin space but only under rotations about the three axis. Hence T^2 is no longer conserved but only T_3. This of course is nothing but the statement that electromagnetic interactions are manifestly charge dependent and remove the degeneracy of the isotopic spin multiplets: the neutron-neutron and proton-proton forces are identical as far as the strong interactions are concerned, but obviously the Coulomb force destroys the equality. Similarly, electromagnetic effects are (presumably) responsible for the mass differences between neutron and proton and the charged and neutral mesons.

In the presence of electromagnetic interaction, the π^0 is no longer stable but can disintegrate into two photons. It is to be noted that this is not a direct interaction but is mediated by the nucleon field as follows:

$$\pi^0 \xrightarrow[\text{strong}]{} \text{p} + \bar{\text{p}} \xrightarrow[\text{em}]{} \text{p} + \gamma + \bar{\text{p}} + \gamma \xrightarrow[\text{strong}]{} 2\gamma$$

The total isotopic spin changes by one in the reaction $\pi^0 \to 2\gamma$, which illustrates the statement that total isotopic spin is no longer conserved in the presence of electromagnetic interactions. Note that according to the principle of minimal electromagnetic interaction, the coupling with the electromagnetic field is proportional to the total charge and therefore linear in T_3 (e.g., the nucleon current is proportional to $\frac{1}{2}\bar{\psi}(1 + \tau_3)\, \gamma_\mu \cdot \psi)$. The coupling term $j_\mu A^\mu$ therefore transforms in isotopic spin space like a scalar plus the 3-component of a vector. Each time the electromagnetic coupling acts in the sense of perturbation theory, the total isotopic spin of the system can change by zero or one unit. For a first-order electromagnetic process, the selection rules are therefore $|\Delta T| = 0$, or 1 and $\Delta T_3 = 0$. For a $T = 0$ to $T = 0$ transition, only the isotopic spin scalar part can contribute. For a $|\Delta T| = 1$ transition only the vector part of $j_\mu A^\mu$ can contribute to the matrix element.

10e. The Strong Interactions

The fact that at sufficiently high energy the reaction $\pi^- + \text{p}$ is observed to give rise not only to collision products such as $\pi^- + \text{p}$ or $\pi^0 + \pi^- + \text{p}$ but also to $\Lambda + \text{K}^0$, $\Sigma^- + \text{K}^+$, gives a clear indication that the pion-nucleon system cannot be considered as closed, and by inference therefore implies that the Lagrangian (92) is incomplete. A description of the interaction between pions and nucleons, and more generally between any pair of strongly interacting particles, must take into account the interaction among all of them.

There are two characteristic experimental facts about processes involving strongly interacting particles. One of these is that the law of conservation of heavy particles is rigorously satisfied. It is convenient to cast the formulation of this law somewhat differently by analogy with conservation of charge. Suppose we assign to each baryon a "baryonic" charge (sometimes also called "nucleonic" charge) equal to +1, to each antibaryon a "baryonic" charge −1, and to each particle whose mass is less than that of the nucleon a baryonic charge 0. The law of heavy particle conservation can then be restated as: Baryonic charge is an additive quantum number and is always conserved in all reactions. The second fundamental fact (as indicated by much experimental evidence) is that the strong interactions are charge independent if suitable assignments are made to the strongly interacting particles, i.e., the total and 3-component of total isotopic spin is conserved in such interactions. One is thus led to extend the isotopic spin concept to the K, Λ, Σ and Ξ particles. This extension is due to Gell-Mann (1953b, 1955a, 1957d) and independently to Nishijima (1954, 1955). As noted above, they made the assignment $T = 0$ for the Λ particle, $T = 1$ for the Σ particles, and $T = \frac{1}{2}$ for the cascade particles. Incidentally, these assignments were made at a time when both the Σ^0 and Ξ^0 particles had not as yet been discovered, nor had the fact that there exist two different neutral kaons been observed. The successful prediction of the existence of these particles constitutes one of the major successes of the Gell-Mann–Nishijima scheme.

In our discussion of the pion-nucleon system, we noted that the relation between the charge Q and the 3-component of isotopic spin for a system of mesons and nucleons is given by

$$Q/e = T_3 + \tfrac{1}{2}N \tag{93}$$

where $N = \int d\sigma^\mu \bar{\psi}\gamma_\mu \cdot \psi$ is the number of nucleons minus the number of antinucleons operator. Gell-Mann and Nishijima extended the above relation (93) to apply also to systems including kaons and hyperons by writing

$$\frac{Q}{e} = T_3 + \tfrac{1}{2}B + \tfrac{1}{2}S \tag{94}$$

where B is the baryon number (number of baryons minus number of antibaryons) and where S is called the "strangeness" number of the system and specifies the displacement from the form $Q/e = T_3 + \frac{1}{2}B$ which is applicable to π mesons and nucleons. Clearly nucleons and π mesons have strangeness 0. The assignment of strangeness to the other strongly interacting particles is a straightforward application of Eq. (94). For example, since the Λ particle is an isotopic singlet with $T = 0$, $T_3 = 0$ and has zero electric charge and nucleonic charge $B = +1$, for Eq. (94) to be satisfied $S(\Lambda) = -1$. Similarly, one verifies that $S(\Sigma) = -1$, and that $S(\Xi) = -2$. The assignment for the kaons is as follows: the K^+ and K^0 particles have strangeness $+1$, the K^- and $\overline{K}^0$ strangeness -1. These assignments for the kaons are supported by the fact that the K^+ and K^- are observed to have qualitatively different interactions with nuclei. Thus at moderate energies K^+ particles are observed to undergo only elastic and charge-exchange scattering in their interactions with nucleons (e.g., $K^+ + n \to K^0 + p$). On the other hand, it is observed that K^- particles can be absorbed with transformation to a hyperon and a pion, e.g., $K^- + n \to \pi^- + \Lambda$ or $\pi^- + \Sigma^0$. Similarly, high energy π^-s incident on hydrogen are observed to give rise to the reactions:

$$\pi^- + \mathrm{p} \rightarrow \Sigma^- + \mathrm{K}^+ \tag{95a}$$

$$\rightarrow \Sigma^0 + \mathrm{K}^0 \tag{95b}$$

but never to the reaction $\Sigma^+ + \mathrm{K}^-$.

Gell-Mann and Nishijima further postulated that the quantum number S is conserved in both strong and electromagnetic interactions. This is, of course, equivalent to the conservation of T_3 for these interactions since the charge and baryon conservation laws are universally valid. The strong, electromagnetic and weak interactions can now also be classified as follows:

(*a*) Strong interactions: the interactions between baryons and mesons which conserve T and T_3 and hence also strangeness. The coupling constants characterizing these interactions are large, of the order 1–10.

(*b*) Electromagnetic interactions: conserve T_3 (and hence S) but not T^2, and are characterized by the fine structure constant $e^2/4\pi\hbar c = 1/137$.

(*c*) Weak interactions: conserve neither T nor T_3 and therefore not S.

The rule $\Delta S = 0$ in strong and electromagnetic processes gives rise to severe restrictions on particle reactions. Thus, in all reactions involving nucleons and pions, the initial strangeness is zero. If the collision gives rise to a "strange particle" (i.e., a particle whose strangeness is different from zero), then it must be accompanied by at least one other strange particle. Pais (1952) had, in fact, been led to this law of "associated production" as a result of the analysis of the strange particle data prior to the Gell-Mann–Nishijima strangeness assignments.

In the decay processes of the Λ and $\Sigma^\pm$ particles (e.g., $\Lambda \rightarrow \mathrm{p} + \pi^-$), the total S for the final particles is zero and differs from that of the initial particle whose strangeness is -1. According to Gell-Mann and Nishijima, these decays must proceed via weak interactions and should be "slow." By "slow" we mean a lifetime long compared with $\hbar/Mc^2$, $\hbar/Mc^2$ being the characteristic time associated with the strong interactions. For M of the order of a nucleon mass $\hbar/Mc^2 \sim 10^{-22}$ second. On the other hand, in the decay $\Sigma^0 \rightarrow \Lambda + \gamma$ or $\Sigma^0 + \Lambda + e^+ + e^-$, there is no change in S, $\Delta S = 0$, and the decay should therefore proceed via electromagnetic interactions and therefore with a lifetime of the order $\left(\frac{e^2}{\hbar c}\right)\left(\frac{\hbar}{Mc^2}\right)$. The lifetimes of the hyperons decaying through a process with $\Delta S \neq 0$ all are of the order 10^{-10} second, indeed slow when compared with 10^{-22} second. The lifetime of the Σ^0 has not yet been measured, but it is known to be shorter than 10^{-15} second. Note that in the presence of only the strong and electromagnetic interactions, all the strange particles except for the Σ^0 and $\bar{\Sigma}^0$ particles are stable, such reactions as $\Lambda \rightarrow \mathrm{p} + \pi^-$ or $\mathrm{K}^+ \rightarrow \pi^+ + \pi^-$ being forbidden by conservation of strangeness.

Perhaps the most brilliant prediction of the strangeness scheme was made by Gell-Mann and Pais [Gell-Mann (1955b)], and relates to the prediction that there must be two different neutral kaons. Gell-Mann and Pais observed that although in the production of neutral kaons, the conservation of strangeness discriminates sharply between K^0 and $\bar{K}^0$, in the subsequent decay strangeness is not conserved and therefore cannot play a crucial role. The quantum number CP is, however, assumed to be conserved and so must play an important role.

Let us denote by $|K^0\rangle$ the state of a K^0 meson at rest and by $|\bar{K}^0\rangle$ the state of one $\bar{K}^0$ meson at rest. We have assumed the kaons to be spinless particles. The states $|K^0\rangle$ and $|\bar{K}^0\rangle$ are eigenstates of the strangeness operator S with eigenvalue $+1$ and -1 respectively. Furthermore, under the operation CP, they transform into each other since K^0 and $\bar{K}^0$ are antiparticles of each other. By adjusting the relative phases of these states, one can require that:

$$CP\,|\,K^0\rangle = |\bar{K}^0\rangle \tag{96a}$$

$$CP\,|\,\bar{K}^0\rangle = |K^0\rangle \tag{96b}$$

Now to describe the decay one must form eigenstates of CP rather than of S. With Gell-Mann and Pais, we therefore define the states:

$$|K^0{}_1\rangle = \frac{1}{\sqrt{2}}\,|\,K^0\rangle + \frac{1}{\sqrt{2}}\,|\,\bar{K}^0\rangle \tag{97a}$$

$$|K^0{}_2\rangle = \frac{1}{\sqrt{2}}\,|\,K^0\rangle - \frac{1}{\sqrt{2}}\,|\,\bar{K}^0\rangle \tag{97b}$$

which correspond to eigenstates of CP with eigenvalues $+1$ and -1 respectively. Conversely,

$$|K^0\rangle = \frac{1}{\sqrt{2}}\,|\,K^0{}_1\rangle + \frac{1}{\sqrt{2}}\,|\,K^0{}_2\rangle \tag{98a}$$

$$|\bar{K}^0\rangle = \frac{1}{\sqrt{2}}\,|\,K^0{}_1\rangle - \frac{1}{\sqrt{2}}\,|\,K^0{}_2\rangle \tag{98b}$$

so that one can say that the production of a K^0 meson (or a $\bar{K}^0$ meson) corresponds to the production with equal probability $(= \frac{1}{2})$ and prescribed relative phase of a "$K^0{}_1$ meson" or a "$K^0{}_2$ meson." The $K^0{}_1$ meson has a CP parity equal to $+1$, whereas the $K^0{}_2$ meson has a CP parity equal to -1. Since CP is assumed to be conserved[6] in the decay, some decay modes are available to $K^0{}_1$ which are forbidden to $K^0{}_2$, and conversely. These two particles must therefore have different lifetimes. For example, a final state containing two pions with total charge $Q = 0$ (either $2\pi^0$ or $\pi^+ + \pi^-$) and with zero total momentum is even under CP, i.e., is an

[6] Some of the qualitative features of this scheme remain even if CP is not conserved.

eigenstate of CP with eigenvalue $+1$, so that decay into 2 pions is allowed for $K^0{}_1$ and forbidden for $K^0{}_2$.

Two neutral kaons having different lifetimes have indeed been observed [Landé (1956)]. One of these decays nearly always into 2πs and has a lifetime of approximately 10^{-10} second and is to be identified with the $K^0{}_1$. The other kaon, the $K^0{}_2$, has a much longer lifetime, 10^{-8} second, and does not decay into 2πs but into 3πs (as well as other modes). Besides the prediction that neutral kaons should exhibit two different lifetimes as well as different sets of decay products, another aspect of the Gell-Mann–Pais theory [Pais and Piccioni (Pais 1955)] has been also verified. Suppose one generates a beam of xK^0 mesons; for example, by using the fast reaction $\pi^- + p \to \Lambda + K^0$. In terms of $K^0{}_1$s and $K^0{}_2$s the beam will consist of $\frac{x}{2}\,K^0{}_1$s and $\frac{x}{2}\,K^0{}_2$s. After about 10^{-9} second (in the rest frames of the mesons), nearly all the $K^0{}_1$s will have decayed but very few of the $K^0{}_2$s will have decayed since their lifetime is appreciably longer. The beam will therefore contain approximately $\frac{x}{2}\,K^0{}_2$s. These, as indicated by Eq. (97b) are not in a pure state of strangeness. They have, with equal probability, $S = +1$ and $S = -1$. Therefore, if the remaining beam were to impinge on a proton target, about $\frac{x}{4}$ of the kaons are capable of reactions like $K^0 + p \to K^+ + n$ characteristic of $S = +1$, and the other half are capable of inducing reactions like $\overline{K}^0 + n \to K^- + p$, $\overline{K}^0 + n \to \Lambda + \pi^0$ or $\overline{K}^0 + n \to \Sigma^+ + \pi^-$ characteristic of $S = -1$. This is to be contrasted with the reactions which could be induced by the original beam all of whose mesons had strangeness $+1$. This effect has indeed been demonstrated experimentally by verifying that the "stale" beam can produce Λs after 10^{-10} second.

The theoretical proposals which have been made to describe and interrelate the strong interactions are for the most part entirely speculative. We shall, therefore, only briefly outline some of the proposals that have been made. [For a more complete review see d'Espagnat and Prentki [d'Espagnat (1958)], Amati (1959), and Kemmer (1959)].

Historically, the first field theoretical formulation of the Gell-Mann–Nishijima scheme was due to d'Espagnat and Prentki [d'Espagnat (1956)]. It was based on the observation that for theoretical purposes it is convenient to classify the fundamental particles in terms of the quantum numbers B, the baryon charge, and in terms of what Schwinger (1957) later called the hyperonic charge, $Y = S + B$. The hyperonic charge of the baryons and mesons are listed in Table 10.6.

It will be noticed that all the particles have a hypercharge 0 or ± 1. Furthermore, particles with integer isotopic spin all have $Y = 0$, and those with half-integer isotopic spin have $Y = \pm 1$. (It is interesting to note

TABLE 10.6

Particle	T	T_3	Q	S	B	Y
p	$\frac{1}{2}$	$\frac{1}{2}$	1	0	1	1
n	$\frac{1}{2}$	$-\frac{1}{2}$	0	0	1	1
Λ	0	0	0	−1	1	0
Σ^+	1	1	1	−1	1	0
Σ^-	1	−1	−1	−1	1	0
Σ^0	1	0	0	−1	1	0
Ξ^-	$\frac{1}{2}$	$-\frac{1}{2}$	−1	−2	1	−1
Ξ^0	$\frac{1}{2}$	$\frac{1}{2}$	0	−2	1	−1
π^+	1	1	1	0	0	0
π^-	1	−1	−1	0	0	0
π^0	1	0	0	0	0	0
K^+	$\frac{1}{2}$	$\frac{1}{2}$	1	1	0	1
K^0	$\frac{1}{2}$	$-\frac{1}{2}$	0	1	0	1
K^-	$\frac{1}{2}$	$-\frac{1}{2}$	−1	−1	0	−1
$\overline{K}^0$	$\frac{1}{2}$	$+\frac{1}{2}$	0	−1	0	−1

that there exists an analogy in ordinary space where all the particles with integer spins, i.e., the mesons, have $B = 0$, while the particles with half-integer spin, the baryons and antibaryons, have $B = \pm 1$.) In view of this analogy, d'Espagnat and Prentki call "isobosons" particles with $T = 1$, and "isofermions" particles with $T = \frac{1}{2}$. Hypercharge conservation, i.e., the conservation of total Y in any strong reaction, is clearly equivalent to conservation of strangeness. It can also be stated as the number of isofermions minus the number of anti-isofermions is a constant.

In order to formulate the theory mathematically, d'Espagnat and Prentki assume that isotopic space is three-dimensional, and they then make the following assignments for the field operators corresponding to the various particles:

Λ: is a scalar in isotopic spin space

$\boldsymbol{\Sigma}$ and $\boldsymbol{\pi}$: are pseudovectors in isotopic spin space

K and N: are spinors of the first kind in isotopic spin space

$\bar{K}$ and Ξ: are spinors of the second kind in isotopic spin space

Note that the notion of transformation properties under inversions of isotopic spin space has been introduced in making these assignments. The statement that $\boldsymbol{\Sigma}$ is a pseudovector means that under an inversion of isotopic spin space $\boldsymbol{\Sigma} \to +\boldsymbol{\Sigma}$, and the isotopic spin parity quantum number of the Σ is $\eta_T = +1$. Similarly for the pion.

It will also be recalled (see Chapter 1, Section e) that there are two kinds of spinors in (Euclidean) three-space which are irreducible to each other. The first kind can be characterized by an inversion parity $\eta_T = +1$ and the second kind by an inversion parity $\eta_T = -i$. The Ks and nucleons (N) are assumed to be spinors of the first kind; the $\overline{K}$ mesons and Ξ particles, spinors of the second kind. If one defines the quantity Y by the relation

$$\eta_T = e^{i\frac{\pi}{2}Y} \tag{99}$$

then these assignments of hypercharge to the particle are consistent with those found in Table 10.6. (Note that the assumption $|Y| = 0, 1$ implies that pseudoscalars and vectors are not admitted.) These assignments are consistent with considering the K^+, K^0 and K^-, $\overline{K}^0$ as antiparticles of each other, since under the operation of changing a particle into its charge conjugate particle (antiparticle), the hypercharge Y changes into $-Y$, so that the charge conjugate particle (i.e., the antiparticle) is described by a spinor of the second kind if the particle was described by a spinor of the first kind, and conversely. D'Espagnat and Prentki can then relate the conservation of hypercharge to the invariance of the Lagrangian under reflections in isotopic spin space, on the assumption that only Yukawa type interactions are admitted between baryons and mesons, and that under these circumstances all possible scalars in isotopic spin space are present.

With the above assignments the most general Lagrangian which is a true scalar in isotopic spin space, contains only trilinear interaction terms, and will conserve heavy particles and strangeness, is

$$\begin{aligned}\mathcal{L}_{I\,(\text{strong})} = {} & g_1\tilde{N}\gamma_5\boldsymbol{\tau}\cdot\boldsymbol{\pi}N + g_2\tilde{\Lambda}\gamma_5\boldsymbol{\pi}\cdot\boldsymbol{\Sigma} + \text{h.a.}\\ & + ig_3\tilde{\boldsymbol{\Sigma}}\gamma_5\times\boldsymbol{\Sigma}\cdot\boldsymbol{\pi} + g_4\tilde{\Xi}\gamma_5\boldsymbol{\tau}\cdot\boldsymbol{\pi}\Xi\\ & + g_5\tilde{N}\gamma_5K\Lambda + \text{h.a.} + g_6\tilde{N}\gamma_5\boldsymbol{\tau}\cdot\boldsymbol{\Sigma}K\\ & + g_7\tilde{\Xi}\gamma_5\hat{K}\Lambda + \text{h.a.} + g_8\tilde{\Xi}\gamma_5\boldsymbol{\tau}\cdot\boldsymbol{\Sigma}\hat{K} + \text{h.a.}\end{aligned} \tag{100}$$

where

$$\hat{K} = -i\tau_2K^* \tag{101}$$

In Eq. (100) we have suppressed the dependence of the *operators* Ξ, N, Σ, Λ, π, K (corresponding to the Ξ, N (nucleon), Σ, Λ, and K particles) on the space-time co-ordinates as well as their transformation properties under Lorentz transformations. Thus, on the assumption that all baryons are spin ½ particles (as seems to be indicated by the experimental data), N, Σ, Λ, Ξ are spinors in space-time and similarly, since the mesons are known to have spin 0, the operators π and K transform like scalars under proper Lorentz transformations. Qualitatively, the form of Eq. (100) for $\mathcal{L}_{I\,(\text{strong})}$ is determined by the fact that pions have strangeness zero and therefore couple particles of the same strangeness, whereas kaons carry strangeness and therefore will only couple particles whose strangeness differs by one unit. It should be pointed out that in writing down the interaction term given by Eq. (100), certain assumptions have been made regarding the relative parity of the particles so that the interaction be invariant under P.

Since states with different strangeness are separated by a superselection rule, one can adopt the convention that Λ and N have the same parity.[7] If (as seems to be the case) the kaons are pseudoscalar, as are the pions, then whether or not the $\Sigma\Lambda\pi$ interaction term has a γ_5 in it or not depends on the relative parity of Λ and Σ. In Eq. (100) we have assumed that Λ and Σ have the same parity relative to the nucleon, the latter being taken as positive, and also that N and Ξ have the same parity. Other relative parity assignments imply the omission of the γ_5 matrix at the appropriate places. We have also assumed that the K meson parity relative to that of the Λ hyperon is odd, as suggested by the experimental evidence on the absorption of negative K mesons in helium [Dalitz (1959)]. Hermiticity and invariance under CP requires that the coupling constants should be real. In this scheme the heavy particle conservation law is now a consequence of the invariance of $\mathfrak{L}$ under the "gauge transformation"

$$\begin{aligned} N &\to e^{i\alpha}N & \tilde{N} &\to e^{-i\alpha}\tilde{N} \\ \Lambda &\to e^{i\alpha}\Lambda & \tilde{\Lambda} &\to e^{-i\alpha}\tilde{\Lambda} \\ \Sigma &\to e^{i\alpha}\Sigma & \tilde{\Sigma} &\to e^{-i\alpha}\tilde{\Sigma} \\ \Xi &\to e^{i\alpha}\Xi & \tilde{\Xi} &\to e^{-i\alpha}\tilde{\Xi} \end{aligned} \tag{102}$$

From this invariance of the Lagrangian one then infers that the operator

$$B = \int d\sigma_\mu(x)\ \{\tilde{N}\gamma^\mu \cdot N + \tilde{\Lambda}\gamma^\mu \cdot \Lambda + \tilde{\boldsymbol{\Sigma}}\gamma^\mu \cdot \boldsymbol{\Sigma} + \tilde{\Xi}\gamma^\mu \cdot \Xi\} \tag{103}$$

is a constant of the motion. $\mathfrak{L}_I$ is also invariant under the transformation

$$\begin{aligned} N &\to Ne^{i\alpha} & \Lambda &\to \Lambda \\ \Xi &\to \Xi e^{i\alpha} & \Sigma &\to \Sigma \\ K &\to Ke^{i\alpha} & \pi &\to \pi \end{aligned} \tag{104}$$

which implies the existence of the following constant of the motion

$$Y = \int d\sigma_\mu(x)\ \{\tilde{N}\gamma^\mu N - \tilde{\Xi}\gamma^\mu\Xi - i(\partial^\mu K^* \cdot K - K^* \cdot \partial^\mu K)\} \tag{105}$$

which is equal to the number of nucleons[8] minus the number of Ξ particles minus the number of K particles. The Y values of the individual field quanta are then $Y = 1$ for N and K, $Y = -1$ for Ξ, $\bar{\text{K}}$ and $\bar{\text{N}}$, and $Y = 0$ for π, Σ and Λ, which agree with the assignments in Table 10.6. Charge is then *defined* by the relation

$$Q = T_3 + \tfrac{1}{2}Y \tag{106}$$

[7] For a general discussion of the question of the relative parity and in general of the phase assignments of quantized fields under improper transformations (e.g., T, P, C) in the presence of superselection rules, see Matthews (1957); also Feinberg and Weinberg [Feinberg (1959)].

[8] To what extent the Heisenberg operator $\int d\sigma^\mu(x)\, \tilde{\psi}\gamma_\mu \cdot \psi$ in the presence of interactions can be called the number of nucleons (minus the number of antinucleons) will be explored later.

where, upon writing $Y = S + B$, the conservation of Y and B implies the conservation of S.

It is interesting to note that if multiply charged particles are excluded, the rule $|Y| \leqslant 1$ allows enough room to fit all known fundamental particles into the scheme and leaves only one unfilled slot which, if it existed, would correspond to an isobaric scalar boson. Such a particle, if it existed, and were sufficiently massive, would be very difficult to observe since it would be expected to decay with a lifetime of approximately 10^{-23} second.

The d'Espagnat-Prentki formalism is formulated in terms of a three-dimensional isotopic spin space. A natural generalization is to suppose the isotopic spin space to be a four-dimensional Euclidean space and again to assume the field operators associated with each particle to transform under rotations in isotopic spin space according to an irreducible representation of the four-dimensional rotation group (see Chapter 1f). Such a possibility was considered by Salam and Polkinghorne [Salam (1955)] who proposed that Λ transform under the (0, 0) (scalar) representation, that Σ and π under the (1, 0) (self-dual antisymmetric tensor) representation, that N and Ξ together transform under the $(\frac{1}{2}, \frac{1}{2})$ (vector) representation, and that the K and $\overline{\text{K}}$ together transform under the $(\frac{1}{2}, \frac{1}{2})$ representation. [For other possible assignments see Kemmer (1959).]

Recent experiments on the photoproduction of K mesons, as well as the theoretical interpretation of hyperfragment binding energies, seem to indicate that the coupling constant measuring the strength of the interaction of K mesons with baryons is somewhat weaker (coupling constant $\sim$1) than the coupling of pions with nucleons (for which the coupling constant G is such that $G^2/\hbar c \sim 15$). Gell-Mann (1957c) and Schwinger (1957) accordingly suggested that the strong interactions can be subdivided into two classes: the medium strong, consisting of the kaon-baryon interactions, and the very strong or pion interactions. In line with the evidence displayed by the known interactions—that the stronger the interaction the greater its number of symmetry properties—Gell-Mann and Schwinger independently postulated further symmetries for the very strong interactions, which they term "global symmetry," over and above charge independence. Global symmetry assumes that in the absence of kaon interactions the baryons form a degenerate set, all the members of which have the same mass and the same interactions with pions. In the absence of kaon interaction, the pion-baryon interaction is thus assumed to be described by an interaction Lagrangian:

$$\mathcal{L}_I = g(\tilde{N}\boldsymbol{\tau}\gamma_5 \cdot N + \tilde{\Xi}\boldsymbol{\tau}\gamma_5 \cdot \Xi + \tilde{\Lambda}\gamma_5\boldsymbol{\Sigma} + \tilde{\boldsymbol{\Sigma}}\gamma_5\Lambda + \tilde{\boldsymbol{\Sigma}} \times \gamma_5\boldsymbol{\Sigma}) \cdot \boldsymbol{\pi} \quad (107)$$

The interaction with the kaons is then stated to be responsible for the observed mass splittings of the baryons.

It would take us too far afield at this point to consider further the detailed mathematical description of the strong interactions. We shall conclude this section with two remarks concerning the strong interactions. The first concerns the motivation for trying to introduce a "global" interaction, and the second the limitations of such a scheme.

The motivation derives from an observation by Wigner (1952) on the similarities between electric and baryonic charge and an analogy between the law of conservation of electric charge and the law of conservation of baryonic charge (i.e., conservation of heavy particles). Wigner suggested that the meson field plays the same role with respect to baryonic charge as does the electromagnetic field with respect to electric charge. That is, just as the electromagnetic field "distinguishes" electrically charged and uncharged particles, the meson field distinguishes baryons and "light" particles and only interacts directly with the baryons. Carrying the analogy further, just as for electromagnetic interactions for which all the particles with the same electric charge interact in the same way with the electromagnetic field, we should expect that the interaction of all the baryons with the meson field to be the same (apart from sign differences). The global symmetry scheme of Gell-Mann and of Schwinger can be considered as a realization of these remarks of Wigner. Although a formal statement of the baryon conservation law can be obtained from such a scheme from the invariance of the Lagrangian under constant phase gauge transformations, the concept of extended gauge transformations *cannot* be made the motivation for the introduction of the meson field [Yang and Mills (Yang 1955), Utiyama (1956)].

Finally, in concluding this section, it should be pointed out that Pais (1958a, b) has indicated that any model of strong interactions with too many symmetry properties is in conflict with the experimental data. In particular, he has shown that, provided that the commonly assumed baryon mass spectrum is complete, "global" symmetry (the assumption that all pion-baryon coupling constants are the same) and "cosmic" symmetry (the assumption that K-baryon coupling constants are the same) cannot hold simultaneously without contradicting experiments. For example, if global and cosmic symmetry are assumed the reaction,

$$\pi^+ + \mathrm{p} \rightarrow \mathrm{K}^+ + \Sigma^+$$

would be forbidden to order Δ^2 (where $\Delta = (M_\Sigma - M_\Lambda)/M_\Lambda$) relative to the reaction

$$\pi^- + \mathrm{p} \rightarrow \Sigma^- + \mathrm{K}^+$$

However, both reactions are observed experimentally and have cross sections of the same order of magnitudes.

10f. The Weak Interactions

The third class of interactions among the fundamental particles are the so-called weak interactions characterized by coupling constants of the order

$GM^2/\hbar c \sim 10^{-39}$ (M = proton mass). These weak interactions are responsible for the instability of the strange particles, of the pions, the neutron, the muons as well as the absorption of μ^-s by nuclei. The only exceptions are the two strangeness conserving decays: $\pi^0 \to 2\gamma$ and $\Sigma^0 \to \Lambda + \gamma$, which are fast and are the result of electromagnetic interactions. The assumption that all weak interactions are the result of Fermi type couplings between Λ, nucleons and leptons whereby four fermions are coupled together, allows one to account for a great many of the observed facts concerning both beta decay of nuclei as well as the decays of hyperons and mesons.

Thus the decay neutron into proton, electron and antineutrino

$$\mathrm{n} \to \mathrm{p} + \mathrm{e}^- + \bar{\nu} \tag{108}$$

can be accounted for and understood if an interaction of the form

$$\mathfrak{L}_I = \sum_{i=1}^{5} G_i' \tilde{p} O_i n \; \tilde{e} O_i \nu + \text{h.a.} \tag{109}$$

exists between the particles. In Eq. (109), $O_i = (1, \gamma_5, \gamma_\mu\gamma_5, \sigma_{\mu\nu}, \gamma_\mu)$ and p, n, ν and e are the (spinor) operators corresponding to proton, neutron, neutrino, and electron respectively ($\tilde{e}$ creates an electron of charge $-e$ or destroys a positon of charge e, etc.). An interaction between four fermions such as the one given by Eq. (109) is called a Fermi coupling, after Fermi (1934), who was the first to construct a theory to account for the β decay of nuclei. More specifically, Fermi originally proposed that the interaction to account for $\mathrm{n} \to \mathrm{p} + \mathrm{e}^- + \bar{\nu}$ be the form

$$\mathfrak{L}_I = G' \tilde{p} \gamma^\mu n \; \tilde{e} \gamma_\mu \nu + \text{h.a.} \tag{110}$$

where the neutrino was assumed to have zero rest mass, and to obey the equation $-i\gamma^\mu \partial_\mu \nu(x) = 0$ in the absence of interactions.

The decay of the muon

$$\mu^\pm \to \mathrm{e}^\pm + \nu + \bar{\nu} \tag{111}$$

is also believed to be the result of a weak Fermi interaction of the form

$$\mathfrak{L}_I = \sum_{i} G_i'' \tilde{\nu} O_i \mu \; \tilde{e} O_i \nu + \text{h.a.} \tag{112}$$

Similarly the absorption of muons by nuclei:

$$\mu^- + \mathrm{p} \to \mathrm{n} + \nu$$

is assumed to be mediated by an interaction of the form

$$\mathfrak{L}_I = \sum_{i=1}^{5} G_i''' \tilde{n} O_i p \tilde{\nu} O_i \mu \tag{113}$$

It is a remarkable fact that the rates for each of the above weak interactions can be accounted for by considering only the vector and axial vector interactions ($O_i = \gamma_\mu, \gamma_5\gamma_\mu$) with coupling constants of the same order of magnitude ($G_A^{(\ )} \approx G_V^{(\ )} \approx G^{(\ )}$), and more remarkably the coupling constants for all three reactions are approximately the same, i.e., $G' \approx G'' \approx G''' \approx G$ with $GM^2/\hbar c \sim 10^{-39}$ when M is the proton mass. This last observation was first made by Klein (1948) and independently by others. It was thereafter suggested by Puppi (1949) that these processes are individual realizations of a universal Fermi interaction. He proposed a triangle, indicated in Figure 10.1, to characterize this universal Fermi

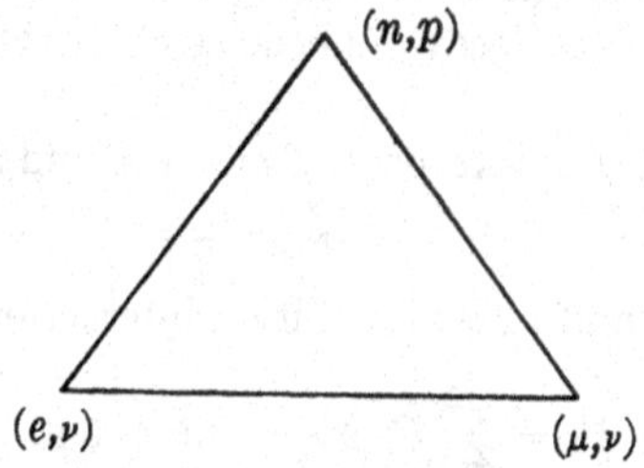

The Puppi Triangle

Fig. 10.1

interaction, and postulated that any of the pair of particles (ab) which are adjoined to the vertices of the triangle could interact with one another via a Fermi interaction of the form $(\tilde{a}Ob) \cdot (\tilde{c}Od)$ with the *same* coupling constant G.

On the basis of the picture developed thus far in this chapter, a reasonably complete field theoretic description of the interaction of the fundamental particles would be given in terms of a Lagrangian:

$$\mathfrak{L} = \mathfrak{L}_{\text{free}} + \mathfrak{L}_{\text{strong}} + \mathfrak{L}_{\text{em}} + \mathfrak{L}_{\text{weak}} \tag{114}$$

where $\mathfrak{L}_{\text{free}}$ is the Lagrangian describing the noninteracting particles (Ξ, Σ, Λ, N, K, π, μ, e, ν, γ), $\mathfrak{L}_{\text{strong}}$ represents coupling terms responsible for the strong interactions between baryons and mesons, $\mathfrak{L}_{\text{em}}$ the electromagnetic interactions between the particles, and $\mathfrak{L}_{\text{weak}}$ corresponds to the universal Fermi couplings implied by the Puppi triangle. An important by-product of the assumed existence of such a universal Fermi interaction is that the decay of the pions can be accounted for, at least in principle, as follows: The pion is coupled to the nucleons via the strong interactions; the latter can also interact via the Fermi coupling, so that the following chain of processes is possible:

$$\pi^+ \xrightarrow[\text{strong}]{} \text{p} + \bar{\text{n}} \xrightarrow[\text{weak}]{} \mu^+ + \nu \tag{115a}$$

or

$$\pi^+ \xrightarrow[\text{strong}]{} \text{p} + \bar{\text{n}} \xrightarrow[\text{weak}]{} \text{e}^+ + \nu \tag{115b}$$

which accounts qualitatively for the pion decay. Naturally, if this scheme could quantitatively account for the observed branching ratios, rates, angular distributions and correlations, such an explanation would be much more satisfying and "fundamental" than one which introduces a direct interaction of the form

$$\mathfrak{L}_I = \frac{f}{m_\pi} (\bar{\nu}\gamma^\alpha \mu \partial_\alpha \pi^* + \text{h.a.}) \tag{116}$$

for example, to account for the decay. The indications are that the aforementioned "fundamental" scheme can indeed account quantitatively for the π-μ decay [Goldberger (1958)].

The above decay modes all refer to decays wherein strangeness is conserved, $\Delta S = 0$. The weak decays of the other fundamental particles involve a change of strangeness of one unit, $\Delta S = \pm 1$, and although the mechanism has not been definitely established they are most probably due to Fermi interactions.

The experimental information on weak decays is listed in Table 10.7.

It would, of course, be possible to again devise direct interactions to account for the above decays, e.g., for the decay $\Sigma^+ \to p + \pi^0$, an interaction of the form:

TABLE 10.7

Particle	*Decay Product*	*Ratio*	*Lifetime in Sec.*
Ξ^-	$\Lambda^0 + \pi^-$	—	3×10^{-10}
Ξ^0	$\Lambda^0 + \pi^0$	—	$\sim 10^{-10}$
Σ^+	$p + \pi^0$	46 ± 6	$0.83 \pm 0.06 \times 10^{-10}$
	$n + \pi^+$	54 ± 6	
Σ^-	$n + \pi^-$	100	$1.67 \pm 0.17 \times 10^{-10}$
Λ	$p + \pi^-$	63	$2.77 \pm 0.15 \times 10^{-10}$
	$n + \pi^0$	37	
	$p + e^- + \bar{\nu}$	<1	
n	$p + e + \bar{\nu}$	100	1040
K^+	$\mu^+ + \nu$	59 ± 2	$1.22 \pm 0.01 \times 10^{-6}$
	$\pi^+ + \pi^0$	$26 + 2$	
	$\pi^+ + \pi^- + \pi^-$	5.7 ± 0.3	
	$\pi^+ + \pi^0 + \pi^0$	1.7 ± 0.3	
	$\pi^0 + e + \nu$	4.2 ± 0.4	
	$\pi^0 + \mu^+ + \nu$	4.0 ± 0.8	
K^-	$2\pi, 3\pi, \mu + \nu$		$1.22 \pm 0.01 \times 10^{-8}$
	$\pi + \mu + \nu$		
	$\pi + e + \nu$		
K^0_1	$\pi^+ + \pi^-$	78 ± 6	$0.95 \pm 0.18 \times 10^{-10}$
	$\pi^0 + \pi^0$	22 ± 6	
K^0_2	$3\pi, \mu + \nu + \pi, e + \nu + \pi$		9×10^{-7}
$\pi^\pm$	$\mu^\pm + \nu$	~ 100	2.56×10^{-8}
	$e^\pm + \nu$	$\sim 10^{-5}$	
$\mu^\pm$	$e^\pm + \bar{\nu} + \nu$	100	$2.22 \pm 0.02 \times 10^{-6}$

$$\mathfrak{L}_I = \frac{f}{M} (\bar{p}\gamma^\mu \Sigma^+ \partial_\mu \pi^0 + \text{h.a.}) \tag{117}$$

etc. Clearly, a large number of such direct interactions would have to be postulated. It is therefore natural to inquire whether it is not possible to extend the notion of a universal (primary) Fermi interaction so as to account for the decay of the strange particles. In trying to devise such a model, one notes that the decay

$$K^+ \to e^+ + \nu + \pi^0 \quad (\Delta S = 1)$$

implies ($e\nu$) are coupled to a strangeness changing pair. It need not be ($K\pi$), but could for example be ($p\Lambda$) since the latter is coupled to the $K\pi$ through the strong interactions. Since muons are also emitted in such strangeness nonconserving decays, it suggests we also need a coupling of the type $\bar{p}\Lambda\bar{\mu}\nu$. In addition, since the hyperons seem to decay primarily without lepton emission, it would seem that a weak Fermi coupling of the type $\bar{p}\Lambda\bar{p}n$ is needed, in which case a decay such as $\Lambda \to p + \pi^-$ could proceed as follows: $\Lambda \to n + p + \bar{p}$ via the weak interactions, and then $n + \bar{p} \to \pi^-$ via strong interaction, so that the end product is $\Lambda \to n + \pi^-$. Gell-Mann (1956a) [see also Costa (1955)] has suggested that the Puppi triangle be converted into the tetrahedron of Fig. 10.2 by adding a vertex

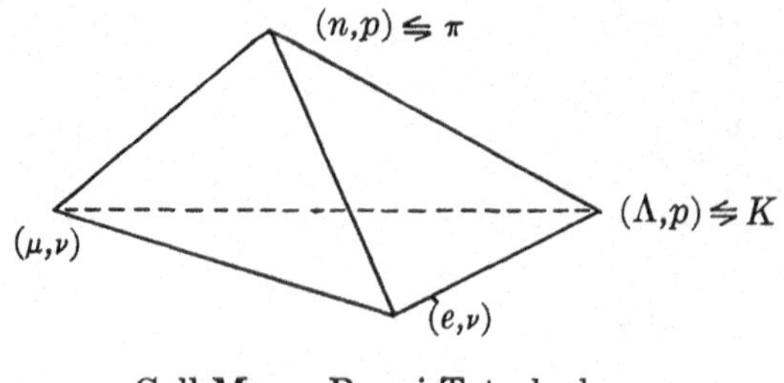

Gell-Mann–Puppi Tetrahedron

Fig. 10.2

and thus extend the idea of universal interaction. Qualitatively such a scheme is capable of accounting for all the known decays. The non-leptonic decays result from the coupling between the np vertex (which, through the strong interactions, is equivalent to a π vertex) and the Λp vertex (which, through the strong interactions, is equivalent to a K vertex).

More recently it has been noted by Feynman and Gell-Mann [Feynman (1958)] that the Fermi interactions take on a simpler form if it is assumed that they are of the nature of the interaction of a current J_i with itself, i.e., of the form $J_i{}^*J_i$ where

$$J_i = G(\bar{\nu}O_i e + \bar{\nu}O_i \mu + \bar{p}O_i n + \text{terms which change strangeness by 1}) \tag{118}$$

Let us next briefly review the Fermi description of the decays of the strangeness 0 particles and, in particular, the interaction responsible for the nuclear β decay with a view towards establishing the exact form of that interaction. The most general local four-fermion Lorentz invariant interaction is clearly of the form:

$$\mathfrak{L}_I = \sum_{i=1}^{5} G_i(\bar{p}O_i n)\,(\bar{e}O^i\nu) + \text{h.a.} \tag{119}$$

with

$$G_1 = G_S \qquad O_1 = 1 \qquad \text{(scalar)} \tag{120a}$$

$$G_2 = G_P \qquad O_2 = \gamma_5 \qquad \text{(pseudoscalar)} \tag{120b}$$

$$G_3 = G_V \qquad O_3 = \gamma_\mu \qquad \text{(vector)} \tag{120c}$$

$$G_4 = G_A \qquad O_4 = i\gamma_\mu\gamma_5 \qquad \text{(axial vector)} \tag{120d}$$

$$G_5 = \tfrac{1}{2}G_T \qquad O_5 = \sigma_{\mu\nu} \qquad \text{(tensor)} \tag{120e}$$

What linear combination is the correct one must be determined from experiments. The decipherment of this puzzle constitutes a brilliant chapter in the annals of physics. Strangely enough, even though a large amount of experimental data as well as the basic theory was known since the early part of the 1930's, it is only since the fall of 1956 that the basic pattern has emerged. The impetus and revival of interest in β decay was due to the recognition by Lee and Yang that no experiments until 1956 had demonstrated the conservation of parity in beta decay. At their suggestion, an experiment on the decay of polarized Co^{60} was carried out by Wu, Ambler, Hayward, Hoppes, and Hudson [Wu (1957)] which clearly demonstrated a correlation between the spin orientation of Co^{60} and the direction of the emitted electron. The distribution function to account for the observed angular distribution must therefore contain a term of the form $\mathbf{J}\cdot\mathbf{p}$, where $\mathbf{J}$ is the nuclear spin and $\mathbf{p}$ the emitted electron's momentum. If the β-decay interaction is invariant under space inversion, then the distribution function is a true scalar (and cannot contain a term such as $\mathbf{J}\cdot\mathbf{p}$ or $\boldsymbol{\sigma}\cdot\mathbf{p}$ where $\boldsymbol{\sigma}$ is the electron spin). Conversely, only if the β-decay interaction is not P invariant can there occur a pseudoscalar term in the distribution function. The experiment of Wu *et al.* on aligned Co^{60} nuclei thus proved conclusively the presence of terms in $\mathfrak{L}_I$ such that $P\mathfrak{L}_IP^{-1} \neq \mathfrak{L}_I$. One possible way of achieving this is to mix each parity conserving interaction:

$$G_i(\bar{p}O_i n)\,(\bar{e}O^i\nu) + \text{h.a.}$$

with a non-P-invariant one of the form:

$$G_i'(\bar{p}O_i n)\,(\bar{e}O^i\gamma_5\nu) + \text{h.a.}$$

Let us consider the vector interaction in detail. The interaction Lagrangian in this case can be written in the form:

$$\mathcal{L}_I = G_V(\bar{p}\gamma^\mu n)\left(\bar{e}\gamma_\mu \frac{1 + ib\gamma_5}{\sqrt{1 + |b|^2}}\,\nu\right) + \text{h.a.} \tag{121}$$

The requirement of *CP* invariance demands that G_V and b be real [Landau (1957)]. Experiments indicate that $b = -1$, i.e., that the neutrino operator only enters the interaction in the form $\frac{1}{2}(1 - i\gamma_5)\,\nu$. We have seen in Chapter 5 that this is equivalent to a two-component neutrino theory in which only left-handed (negative helicity) neutrinos interact with matter.

It has been shown in the past few years [see the excellent review article by Konopinski (1959)] that all the present β-decay experiments can be accounted for by an interaction containing a mixture of vector and axial vector contribution:

$$\mathcal{L}_I = G_V(\bar{p}\gamma^\mu n)\,(\bar{e}\gamma_\mu\tfrac{1}{2}(1 - i\gamma_5)\,\nu) + G_A(\bar{p}\gamma^\mu\gamma_5 n)\,(\bar{e}\gamma_\mu\tfrac{1}{2}(1 - i\gamma_5)\,\nu) \tag{122}$$

Furthermore, there exists good evidence that $|G_A| \approx |G_V|$ $\left(\text{more precisely, } \left|\frac{G_A}{G_V}\right| \simeq 1.20\right)$. We shall later on indicate how this inference is made.

We conclude this section with a brief review of the theoretical conjectures which have been made to "justify" the "*V-A*" interaction [Sudarshan and Marshak (Sudarshan 1958), Feynman and Gell-Mann (Feynman 1958), Sakurai (1958a)]. Perhaps the most satisfactory of these (in the sense that it is capable of an extension to the strong interactions) is that of Sudarshan and Marshak and of Sakurai, which requires the Lagrangian of weak interactions to be invariant under the transformation:

$$\begin{aligned}\psi &\to \gamma_5\psi \\ \bar{\psi} &\to -\bar{\psi}\gamma_5\end{aligned} \qquad m \to -m \tag{123}$$

for every fermion field *separately*.[9] This can be considered as a generalization of the requirement that the Lagrangian be invariant under the transformation:

$$\nu \to \gamma_5\nu$$

or, more generally,

$$\nu \to e^{ia\gamma_5}\nu \quad (a = \text{constant})$$

which guarantees that the mass of the neutrino is rigorously zero (and remains zero as the result of the interaction). It also extends to the full Lagrangian a symmetry of the noninteracting Lagrangian.

An interaction Lagrangian for a Fermi interaction between four fermion fields ψ_1, ψ_2, ψ_3, ψ_4, which is invariant under proper Lorentz transformation, can be written as:

$$\mathcal{L}_I = \sum_{i=1}^{5} (\bar{\psi}_1 O_i \psi_2)\,(\bar{\psi}_3(G_i O_i + G_i' O_i \gamma_5)\,\psi_4) + \text{h.a.} \tag{124}$$

[9] Invariance under mass reversal was first investigated by Tiomno (1955) who, however, only required invariance under the *simultaneous* transformation: $\psi \to \gamma_5\psi$, $m \to -m$ for all fields.

Invariance under mass reversal $\psi_i \to \gamma_5 \psi_i$, $m_i \to -m_i$ then requires that $\mathfrak{L}_I$ have the form

$$\mathfrak{L}_I = G(\bar{\psi}_1 \gamma_\mu (1 \pm i\gamma_5)\, \psi_2)\, (\bar{\psi}_3 \gamma^\mu (1 \pm i\gamma_5)\, \psi_4) \tag{125}$$

Feynman and Gell-Mann, on the other hand, postulate that in the fermion interaction (124) only the projection φ_i

$$\varphi_i = \tfrac{1}{2}(1 - i\gamma_5)\, \psi_i$$

of the field operator ψ_i appear, i.e., only those components of the fields which have the same chirality, namely those for which $\varphi_i = -i\gamma_5 \varphi_i$, interact. We shall denote the projection operator $\frac{1}{2}(1 - i\gamma_5)$ by a_-

$$a_- = \tfrac{1}{2}(1 - i\gamma_5) = a_-^2 \tag{126}$$

Then, since γ_5 is anti-Hermitian,

$$\begin{aligned} \tilde{\varphi}_i &= (\tfrac{1}{2}(1 - i\gamma_5)\, \psi_i)^* \, \gamma_0 = \tfrac{1}{2}\psi_i^*(1 - i\gamma_5)\, \gamma_0 \\ &= \bar{\psi}_i \cdot \tfrac{1}{2}(1 + i\gamma_5) \end{aligned} \tag{127}$$

We shall denote $\frac{1}{2}(1 + i\gamma_5)$ by a_+ and note that $a_+a_- = a_-a_+ = 0$. The most general local nonderivative interaction Lagrangian is then of the form:

$$\begin{aligned} \mathfrak{L}_I &= \sum_{i=1}^{5} g_i (\widetilde{a_-\psi_1} O_i a_- \psi_2)\, (\widetilde{a_-\psi_3} O_i a_- \psi_4) \\ &= \sum_{i=1}^{5} g_i (\bar{\psi}_1 a_+ O_i a_- \psi_2)\, (\bar{\psi}_3 a_+ O_i a_- \psi_4) \end{aligned} \tag{128}$$

Now $a_+ O_i a_- = O_i a_+ a_- = 0$ for $O_i = 1$, γ_5, $\sigma_{\mu\nu}$, since γ_5 and O_i commute under these circumstances. On the other hand, when $O_i = \gamma_\mu$ and $O_i = = i\gamma_\mu\gamma_5$, $a_+ O_i = O_i a_-$ so that

$$\mathfrak{L}_I = \sum_{i=V,\,A} g_i (\bar{\psi}_1 O_i a_- \psi_2)\, (\bar{\psi}_3 O_i a_- \psi_4) \tag{129}$$

but since $i\gamma_\mu\gamma_5(1 - i\gamma_5) = -\gamma_\mu(1 - i\gamma_5)$

$$\mathfrak{L}_I = G(\bar{\psi}_1 \gamma_\mu a_- \psi_2)\, (\bar{\psi}_3 \gamma^\mu a_- \psi_4) \tag{130}$$

which is the "$V - A$" interaction.

10g. The Equivalence Theorem

In the final section of this chapter we investigate the degree to which a pseudoscalar and pseudovector coupling are equivalent. Although these two types of coupling are fundamentally different, the pseudoscalar coupling being characterized by a dimensionless coupling constant, whereas the pseudovector is characterized by a coupling constant which has the dimension of a length (the pseudoscalar interaction is renormalizable whereas the pseudovector is not), nonetheless one finds that to certain orders in the coupling constant the two couplings yield similar results. [See Dyson (1948), Case (1949a, b), and Foldy (1951).]

Consider in the Schrödinger picture the Hamiltonian for the meson-nucleon field interacting through a direct pseudoscalar coupling:

$$H = \int d^3x\, \psi^*(\boldsymbol{\alpha} \cdot \mathbf{p} + \beta M)\, \psi(\mathbf{x}) + \tfrac{1}{2} \int d^3x\, \{\pi^2(\mathbf{x}) + \nabla\phi \cdot \nabla\phi(\mathbf{x}) + \mu^2\phi^2(\mathbf{x})\} + G \int d^3x\, \psi^*\beta\gamma_5\phi\psi(\mathbf{x}) \quad (131)$$

We are considering the case of a neutral meson field. The generalization to the symmetric meson theory is straightforward.

We now wish to perform a unitary transformation on the Hamiltonian

$$H' = e^{iS}He^{-iS} \quad (132)$$

where S is to be determined so that the pseudoscalar coupling term is eliminated in favor of a pseudovector coupling term. The reason for this procedure is that it makes easier the transition to a nonrelativistic Hamiltonian for the study of nonrelativistic problems. Throughout this section, unless otherwise specified, we shall be working in the Schrödinger picture.

We assume [Dyson (1948), Foldy (1951), Berger (1952)] that the Hermitian operator S can be written in the form

$$S = \int d^3x\, \psi^* s \psi(\mathbf{x}) = \int d^3x \sum_{\alpha, \beta = 1}^{4} \psi^*_\alpha s_{\alpha\beta} \psi_\beta \quad (133a)$$

with

$$s(\mathbf{x}) = i\gamma_5 w(\phi(\mathbf{x})) \quad (133b)$$

where w is a function of ϕ only, and $\gamma_5 = \gamma^0\gamma^1\gamma^2\gamma^3$ is anti-Hermitian, and its square is -1. Note that S is not a matrix, but a c-number as far as spin space is concerned, and therefore commutes with any Dirac matrix.

In order to evaluate (132), we make use of the following procedure, already used in Section 7c: Define

$$\psi'(\mathbf{x}, \lambda) = e^{iS\lambda}\psi(\mathbf{x})\, e^{-iS\lambda} \quad (134)$$

Then the operator we wish to calculate is $\psi'(\mathbf{x}, 1)$, and the untransformed operator is given by $\psi'(\mathbf{x}, 0)$. Now

$$\frac{\partial\psi'(\mathbf{x}, \lambda)}{\partial\lambda} = ie^{iS\lambda}[S, \psi(\mathbf{x})]\, e^{-iS\lambda} = i[S, \psi'(\mathbf{x}, \lambda)] \quad (135)$$

Recalling that the anticommutation rules

$$[\psi_\alpha(\mathbf{x}), \psi^*_\beta(\mathbf{x}')]_+ = \delta_{\alpha\beta}\delta^{(3)}(\mathbf{x} - \mathbf{x}')$$
$$[\psi_\alpha(\mathbf{x}), \psi_\beta(\mathbf{x}')]_+ = [\psi^*_\alpha(\mathbf{x}), \psi^*_\beta(\mathbf{x}')]_+ = 0 \quad (136)$$

are invariant under the unitary transformation (132), and that ϕ and ψ commute, we obtain, using (133a),

$$[S, \psi'_\rho(\mathbf{x}, \lambda)] = \int d^3x' \sum_{\alpha\beta} [\psi'^*_\alpha(\mathbf{x}', \lambda)\, s_{\alpha\beta}\psi'_\beta(\mathbf{x}', \lambda), \psi'_\rho(\mathbf{x}, \lambda)]$$

$$= -\sum_\beta s_{\rho\beta}\psi'_\beta(\mathbf{x}, \lambda) = -[s\psi'(\mathbf{x}, \lambda)]_\rho \tag{137}$$

Equation (135) now reads

$$\frac{\partial\psi'(\mathbf{x}, \lambda)}{\partial\lambda} = -is\psi'(\mathbf{x}, \lambda) \tag{138}$$

and upon integration yields

$$\psi'(\mathbf{x}, \lambda) = e^{-is\lambda}\psi'(\mathbf{x}, 0) \tag{139}$$

In obtaining (139), we have made use of the initial condition that $\psi'(\mathbf{x}, 0)$ be the untransformed operator. The transformed operator is therefore given by

$$e^{iS}\psi(\mathbf{x})\, e^{-iS} = e^{-is}\psi(\mathbf{x}) \tag{140}$$

Upon taking the Hermitian adjoint of this last equation, we obtain, since s as given by (133b) is Hermitian, $s = s^*$,

$$e^{iS}\psi^*(\mathbf{x})\, e^{-iS} = \psi^*(\mathbf{x})\, e^{is} \tag{141}$$

We shall find in evaluating the expression (132) that the following formula will prove very useful:

$$e^{iS}Qe^{-iS} = Q + \frac{i}{1!}[S, Q] + \frac{i^2}{2!}[S, [S, Q]] + \frac{i^3}{3!}[S, [S, [S, Q]]] + \cdots \tag{142}$$

The equality between the left and right sides of this equation is readily established by expanding $\exp(iS\lambda)\, Q \exp(-iS\lambda)$ in a Taylor series about $\lambda = 0$.

Furthermore, note that since $\phi(\mathbf{x})$ commutes with $\phi(\mathbf{x}')$ and S with any Dirac matrix

$$e^{iS}\,(\beta M + G\beta\gamma_5\phi(\mathbf{x}))_{\alpha\beta}\, e^{-iS} = (\beta M + G\beta\gamma_5\phi(\mathbf{x}))_{\alpha\beta} \tag{143}$$

where the subscripts α, β refer to the fact that we are considering the $\alpha\beta$ matrix element of the γ matrices (these matrix elements are numbers!). We therefore obtain, using (140), (141), and (143),

$$H_1 = e^{iS} \int d^3x \sum_{\alpha\beta} \psi^*_\alpha(\mathbf{x})\, (\beta M + G\beta\gamma_5\phi(\mathbf{x}))_{\alpha\beta}\, \psi_\beta(\mathbf{x})\, e^{-iS}$$

$$= \int d^3x \sum_{\alpha\beta} (\psi^*(\mathbf{x})\, e^{is})_\alpha\, (\beta M + G\beta\gamma_5\phi(\mathbf{x}))_{\alpha\beta}\, (e^{-is}\,\psi(\mathbf{x}))_\beta$$

$$= \int d^3x\, \psi^*(\mathbf{x})\, e^{is}\, (\beta M + G\beta\gamma_5\phi(\mathbf{x}))\, e^{-is}\, \psi(\mathbf{x}) \tag{144}$$

Now, recalling that β and γ_5 anticommute, we may transform the factor $\beta \exp(-is)$ into the following expression:

$$\beta\, e^{-is} = \beta\, e^{+\gamma_5 w(\phi)} = \sum_{n=0}^{\infty} \beta(\gamma_5)^n \frac{w^n}{n!}$$

$$= \sum_{n=0}^{\infty} (\gamma_5)^n (-1)^n \beta \frac{w^n}{n!} = e^{is}\,\beta \qquad (145)$$

Similarly,

$$\beta\gamma_5\, e^{-is} = e^{is}\,\beta\gamma_5 \qquad (146)$$

so that Eq. (144) can be written as

$$H_1 = \int d^3x\, \psi^*(\mathbf{x})\, e^{2is}\, (\beta M + G\beta\gamma_5\phi(\mathbf{x}))\, \psi(\mathbf{x}) \qquad (147)$$

Similarly, we obtain for the term

$$H_2 = e^{iS} \int d^3x\, \psi^*(\mathbf{x})\, \boldsymbol{\alpha} \cdot \mathbf{p}\, \psi(\mathbf{x})\, e^{-iS}$$

$$= \int d^3x\, \psi^*(\mathbf{x})\, e^{is}\, \boldsymbol{\alpha} \cdot \mathbf{p}\, e^{-is}\, \psi(\mathbf{x}) \qquad (148)$$

We now no longer can commute the factor $\exp(-is)$ through, since $\mathbf{p} = -i\boldsymbol{\nabla}$ and $\phi(\mathbf{x})$ do not commute. However, recalling that $\boldsymbol{\alpha}$ and γ_5 commute, and using (142), we may write

$$e^{is}\, \boldsymbol{\alpha} \cdot \mathbf{p}\, e^{-is} = \sum_{j=1}^{3} e^{is}\, \alpha_j\, e^{-is}\, e^{is}\, p_j\, e^{-is}$$

$$= \sum_{j=1}^{3} \alpha_j\, e^{is}\, p_j\, e^{-is}$$

$$= \boldsymbol{\alpha} \cdot \mathbf{p} - \boldsymbol{\alpha} \cdot \boldsymbol{\nabla} s \qquad (149)$$

At this point, it is useful to introduce the 4×4 matrix Σ_i, which is related to α_i and γ_5 by

$$\Sigma_i = i\gamma_5\alpha_i \qquad (150)$$

so that Eq. (148) may be rewritten as

$$H_2 = \int d^3x\, \psi^*(\mathbf{x})\, \boldsymbol{\alpha} \cdot \mathbf{p}\, \psi(\mathbf{x}) - \int d^3x\, \psi^*(\mathbf{x})\, \boldsymbol{\Sigma} \cdot \boldsymbol{\nabla} w\, \psi(\mathbf{x}) \qquad (151)$$

Finally, we have to calculate

$$H_3 = e^{iS}\, \tfrac{1}{2} \int d^3x\, \{\pi^2(\mathbf{x}) + \boldsymbol{\nabla}\phi \cdot \boldsymbol{\nabla}\phi(\mathbf{x}) + \mu^2\phi^2(\mathbf{x})\}\, e^{-iS} \qquad (152)$$

To evaluate this term we shall, with Dyson (1948), assume the unknown function $w(\phi)$ to be of the form,

$$w(\phi) = \lambda\phi(\mathbf{x}) \qquad (153)$$

where λ is a constant to be determined. Since $\phi(\mathbf{x})$ and $\phi(\mathbf{x}')$ commute, we need only compute the quantity

$$e^{iS}\, \pi(\mathbf{x})\, e^{-iS} = \pi'(\mathbf{x}) \qquad (154)$$

Using Eq. (142), we obtain

$$\pi'(\mathbf{x}) = \pi(\mathbf{x}) + i\lambda \int d^3x' \, [\psi^*(\mathbf{x}') \, i\gamma_5\psi(\mathbf{x}') \, \phi(\mathbf{x}'), \pi(\mathbf{x})] \tag{155}$$

all higher terms (repeated commutators) vanishing since the commutator of $\phi(\mathbf{x})$ and $\pi(\mathbf{x}')$,

$$[\phi(\mathbf{x}), \pi(\mathbf{x}')] = i\delta^{(3)}(\mathbf{x} - \mathbf{x}') \tag{156}$$

is a c-number. Hence,

$$\pi'(\mathbf{x}) = \pi(\mathbf{x}) - i\lambda\psi^*(\mathbf{x}) \, \gamma_5\psi(\mathbf{x}) \tag{157}$$

and Eq. (152) becomes

$$\begin{aligned} H_3 &= \tfrac{1}{2} \int d^3x \, \{\pi'^2(\mathbf{x}) + \nabla\phi \cdot \nabla\phi(\mathbf{x}) + \mu^2\phi^2(\mathbf{x})\} \\ &= \tfrac{1}{2} \int d^3x \, (\pi^2(\mathbf{x}) + \nabla\phi \cdot \nabla\phi(\mathbf{x}) + \mu^2\phi^2(\mathbf{x})) \\ &\quad - i\lambda \int d^3x \, \psi^*(\mathbf{x}) \, \gamma_5\psi(\mathbf{x}) \, \pi(\mathbf{x}) - \tfrac{1}{2}\lambda^2 \left(\int d^3x \, \psi^*(\mathbf{x}) \, \gamma_5\psi(\mathbf{x}) \right)^2 \end{aligned} \tag{158}$$

Collecting terms, we therefore obtain the following expression for the transformed Hamiltonian:

$$\begin{aligned} H' &= \int d^3x \, \psi^*(\mathbf{x}) \, (\boldsymbol{\alpha} \cdot \mathbf{p} + \beta M) \, \psi(\mathbf{x}) \\ &\quad + \tfrac{1}{2} \int d^3x \, \{\pi^2(\mathbf{x}) + \nabla\phi \cdot \nabla\phi(\mathbf{x}) + \mu^2\phi^2(\mathbf{x})\} \\ &\quad + \int d^3x \, \psi^*(\mathbf{x}) \, (e^{2is} - 1) \, \beta M\psi(\mathbf{x}) + \int d^3x \, \psi^*(\mathbf{x}) \, e^{2is} \, G\beta\gamma_5\phi(\mathbf{x}) \, \psi(\mathbf{x}) \\ &\quad - \lambda \int d^3x \, \psi^*(\mathbf{x}) \, \boldsymbol{\Sigma} \cdot \nabla\phi(\mathbf{x}) \, \psi(\mathbf{x}) - i\lambda \int d^3x \, \psi^*(\mathbf{x}) \, \gamma_5\psi(\mathbf{x}) \, \pi(\mathbf{x}) \\ &\quad - \tfrac{1}{2}\lambda^2 \left(\int d^3x \, \psi^*(\mathbf{x}) \, \gamma_5\psi(\mathbf{x}) \right)^2 \end{aligned} \tag{159}$$

Although the form (159) is exact, we see that we are not able to adjust λ so that the pseudoscalar coupling is completely eliminated. This is due to our assumption (153). The procedure adopted by Berger, Foldy, and Osborn [Berger (1952)] is to leave w arbitrary until (131) has been completely evaluated. Then w is determined in such a way that in the transformed Hamiltonian there remains no pseudoscalar coupling.

If in the expression (159) we expand the exponential factors and retain terms quadratic in ϕ, and if we choose

$$-\lambda = \frac{G}{2M} \tag{160}$$

then the pseudoscalar coupling will, in fact, be eliminated to this order, and the transformed Hamiltonian becomes

$$H' = H_N + H_M + H_I'' \tag{161a}$$

where

$$H_I'' = \frac{G}{2M} \int d^3x\, \psi^*(\mathbf{x})\, (\boldsymbol{\Sigma} \cdot \nabla\phi(\mathbf{x}) + i\gamma_5\pi(\mathbf{x}))\, \psi(\mathbf{x})$$

$$+ \frac{G^2}{2M} \int d^3x\, \psi^*(\mathbf{x})\, \beta\psi(\mathbf{x})\, \phi^2(\mathbf{x}) - \frac{G^2}{8M^2} \left(\int d^3x\, \psi^*(\mathbf{x})\, \gamma_5\psi(\mathbf{x}) \right)^2 \quad (161b)$$

We shall now briefly analyze the modified Hamiltonian (161a) and (161b). The first term in (161b) is the usual derivative coupling form of the meson-nucleon Hamiltonian. Recall that $\pi(\mathbf{x})$ is essentially $\partial_t\phi(\mathbf{x})$, so that this interaction term may be written as proportional to

$$\frac{F}{\mu}\, \tilde{\psi}(\mathbf{x})\, \gamma_5\gamma_\mu\psi(\mathbf{x})\, \partial^\mu\phi \quad (162)$$

which is the conventional way of expressing the pseudovector interaction. We have, therefore, shown the equivalence of pseudoscalar and pseudovector coupling to first order in the coupling constant. The coupling constant for the derivative interaction is related to G through

$$\frac{F}{\mu} = \frac{G}{2M} \quad (163)$$

The "two-meson" term,

$$\frac{G^2}{2M} \int d^3x\, \psi^*\beta\psi(\mathbf{x})\, \phi^2(\mathbf{x}) \quad (164)$$

is analogous to the quadratic $(e^2/2m)\,\mathbf{A}^2$ term that occurs in the nonrelativistic radiation theory. It also arises from the elimination of the virtual pair transition. It gives rise (in perturbation theory) to a large spin-independent contribution to nuclear forces [Lepore (1952)]. Its importance in any calculation of meson-nucleon scattering was first emphasized by Drell and Henley [Drell (1952)], who showed that it gave rise to strong repulsive S wave interactions of mesons with nucleons. That its effect is important can already be inferred from the fact that the pseudovector coupling term is proportional to F/μ, whereas the ϕ^2 term is proportional to $M(F/\mu)^2$.

Finally, the contact term, $(\int d^3x\, \psi^*\gamma_5\psi(\mathbf{x}))^2$, is always present in the Hamiltonian for a derivative coupling. If the Lagrangian contains the derivative coupling term (162), then the momentum conjugate to ϕ will be

$$\pi = \frac{\partial \mathcal{L}}{\partial \dot{\phi}} = \dot{\phi} - \frac{F}{\mu}\, \psi^*\gamma_5\psi \quad (165)$$

Eliminating $\dot{\phi}$ from the Hamiltonian will then give rise to the contact term, $(\int d^3x\, \psi^*\gamma_5\psi(\mathbf{x}))^2$.

The equivalence of pseudoscalar and pseudovector coupling is most clearly established by Foldy's transformation [Foldy (1951), Wentzel (1952), Berger (1952)], for which S is given by

$$S = \frac{1}{2} \int d^3x \, \psi^*(\mathbf{x}) \, \gamma_5 \psi(\mathbf{x}) \tan^{-1} \left(\frac{G\phi(\mathbf{x})}{M} \right) \tag{166}$$

With this canonical transformation, the pseudoscalar coupling is completely eliminated. The pseudovector coupling term, however, now appears with a nonlinear coefficient.

The most general equivalence theorems have been stated by Moldauer and Case [Moldauer (1953)], who have shown that to the first order in the coupling constant all possible linear interactions between scalar (or pseudoscalar) mesons and nucleons either vanish or are equivalent to the scalar (or pseudoscalar) interaction.

Equivalence theorems in the presence of an electromagnetic field have been stated by Case (1949b) and by Drell and Henley [Drell (1952)]. An extremely elegant method of deriving equivalence theorems, which makes use of the arbitrariness of the Lagrangian to within addition of four divergences, has been given by Kelly (1950).

11

The Formal Theory of Scattering

Our considerations up to this point have been formal: We have outlined a procedure for the quantization of field theories using the canonical formalism. The latter allows us to cast the quantized theory in a form analogous to ordinary quantum mechanics in which the evolution of the system is described by a Schrödinger equation:

$$H \mid \Psi\rangle = i\hbar\partial_t \mid \Psi\rangle \tag{1a}$$

$$H = H_0 + H_I \tag{1b}$$

We must next extract the physics implied by the quantized version of the theory. To this end we shall investigate soluble models (Chapter 12) as well as the application of perturbation theory to quantized field theories (Chapters 13–14).

It will be a common feature of all quantized field theories that by virtue of the interaction, H_I, each quantum (which is defined in terms of the unperturbed Hamiltonian, H_0, of the field system) acquires a persistent cloud of other quanta around it. In the limit of point (local) interaction these cloud effects give rise to divergences. The renormalization program is a procedure designed to circumvent these difficulties. It is based on the observation that by virtue of the interaction the constants appearing in the theory, such as the mass of the bare quanta and the coupling constant, are altered and cannot be identified with the corresponding measured quantities. It is the great success of the renormalization program that these redefinitions of coupling constant and mass parameters proved sufficient to circumvent all divergence difficulties in the scattering and production amplitudes, i.e., in the elements of the S matrix, in quantum electrodynamics. Present-day field theories can be characterized as being adequate to give a semiquantitative, and in the case of electrodynamics quantitative, description of *processes* involving elementary particles but shedding much less light on the problem of the *structure* of these particles. For example, we have no way of accounting for the observed mass spectrum of the elementary particles.

In this chapter we shall give a brief account of those aspects of the formal theory of scattering which will be needed in the subsequent work.

We also introduce the Dirac (or interaction) picture as well as certain expansions of the time displacement operators in that picture.

11a. Potential Scattering

We first treat the scattering of a spinless nonrelativistic particle by a potential of finite range, since it forms the simplest example of scattering theory. We are interested in describing the process in which a beam of particles collimated in direction and energy impinges upon a target and is scattered by it. (It is assumed that the intensity of the beam is such that the mutual interaction between the projectile particles is negligible.) We wish to compute the intensity of the beam after the scattering has taken place as a function of direction, assuming that the intensity of the incoming beam and the interaction potential between projectile and target are known. We shall consider the scattering of one particle at a time.

The quantum mechanical description of the process is then as follows [Eisenbud (1956)]: at time $t = 0$ there is a wave packet located around the point $\mathbf{r}_0 = (0, 0, -z_0)$ with a spatial dispersion Δx, Δy, Δz, moving to the right with an average momentum $\mathbf{p}_0 = \hbar\mathbf{k}_0$ in the z direction (with a dispersion Δk_x, Δk_y, Δk_z) with $\Delta k_i \Delta x_i \simeq 1$. In order that the particle described have a reasonably well-defined energy, we require that $\Delta k_z \ll k_0$. The collimation in space is such that at time $t = 0$, the scattering hasn't started yet, that is, there is no interaction between the projectile and target (the source of the projectile particles is far away from the target!). We therefore require that $\Delta z \ll |z_0|$ and $|z_0| \gg a$, where a is the range of the force field. In order that we may be able to differentiate between scattered and transmitted beam, the interference between scattered and incident wave should be minimized. This aspect of the preparation of the initial state is translated into the statement that Δx, $\Delta y \ll |z_0|$. It is incidentally usually the case that Δx, Δy, $\Delta z \gg a$.

At time $t = 0$ our particle is thus described by a wave function

$$\Psi(\mathbf{r}, 0) = \int e^{i\mathbf{k}\cdot(\mathbf{r}-\mathbf{r}_0)}\, \Phi(\mathbf{k} - \mathbf{k}_0)\, d^3k \tag{2}$$

where Φ is a (square integrable) function peaked around $\mathbf{k} = \mathbf{k}_0$ with a dispersion Δk_i. Consider next the time evolution of this packet. At some later time $t > 0$ we have:

$$\Psi(\mathbf{r}, t) = e^{-iHt/\hbar}\, \Psi(\mathbf{r}, 0) \tag{3a}$$

$$H = -\frac{\hbar^2}{2m}\nabla^2 + V(\mathbf{r}) \tag{3b}$$

where H is the Hamiltonian of the system. In order to evaluate this expression, we introduce auxiliary functions $\psi_{\mathbf{k}+}(\mathbf{r})$ which are defined as the solutions of:

$$\left\{-\frac{\hbar^2}{2m}\nabla^2 + V(\mathbf{r})\right\}\psi_{\mathbf{k}+}(\mathbf{r}) = E_{\mathbf{k}}\psi_{\mathbf{k}+}(\mathbf{r}) \tag{4a}$$

$$E_{\mathbf{k}} = \frac{\hbar k^2}{2m} \tag{4b}$$

having the following asymptotic behavior:

$$\lim_{|\mathbf{r}|\to\infty} \psi_{\mathbf{k}+}(\mathbf{r}) \sim e^{i\mathbf{k}\cdot\mathbf{r}} + \frac{e^{ikr}}{r} f_{\mathbf{k}+}(\theta, \phi) \tag{5}$$

That solutions with this prescribed asymptotic behavior exist is a consequence of the fact that we assume that $V(\mathbf{r})$ falls off more rapidly than $1/r$, so that for large r, $\psi_{\mathbf{k}+}$ obeys the equation $(\nabla^2 + k^2)\,\psi_{\mathbf{k}+} = 0$ which has plane wave solutions, $\exp\,(i\mathbf{k}\cdot\mathbf{r})$, as well as outgoing spherical wave solutions $(\exp ikr)/r$ to first order in $1/r$. We next show that the differential equation (4) as well as the boundary condition (5) are satisfied by the solutions of the integral equation:

$$\psi_{\mathbf{k}+}(\mathbf{r}) = e^{i\mathbf{k}\cdot\mathbf{r}} + \int G_{\mathbf{k}+}(\mathbf{r} - \mathbf{r}')\,\frac{2m}{\hbar^2}\,V(\mathbf{r}')\,\psi_{\mathbf{k}+}(\mathbf{r}')\,d^3r' \tag{6}$$

where the Green's function $G_{+\mathbf{k}}$ is given by

$$G_{\mathbf{k}+}(\mathbf{r} - \mathbf{r}') = -\frac{1}{4\pi}\frac{e^{ik|\mathbf{r}-\mathbf{r}'|}}{|\mathbf{r} - \mathbf{r}'|} = \lim_{\epsilon\to 0+} \frac{1}{(2\pi)^3}\int d^3k'\,\frac{e^{i\mathbf{k}'\cdot(\mathbf{r}-\mathbf{r}')}}{\mathbf{k}^2 - \mathbf{k}'^2 + i\epsilon} \tag{7}$$

Proof: For large values of r, since the contribution to the integral in (6) comes from r' around the origin due to the assumed properties of $V(\mathbf{r})$, we may write $|\mathbf{r} - \mathbf{r}'| \simeq r - \dfrac{\mathbf{r}\cdot\mathbf{r}'}{r} + \cdots$; hence

$$\lim_{r\to\infty} \psi_{\mathbf{k}+}(\mathbf{r}) = e^{i\mathbf{k}\cdot\mathbf{r}} - \frac{e^{ikr}}{r}\cdot\frac{1}{4\pi}\int e^{-ik\frac{\mathbf{r}\cdot\mathbf{r}'}{r}}\,\frac{2m}{\hbar^2}\,V(\mathbf{r}')\,\psi_{\mathbf{k}+}(\mathbf{r}')\,d^3r' \tag{8}$$

so that

$$f_{\mathbf{k}+}(\theta, \phi) = -\frac{1}{4\pi}\frac{2m}{\hbar^2}\int e^{-ik\frac{\mathbf{r}\cdot\mathbf{r}'}{r}}\,V(\mathbf{r}')\,\psi_{\mathbf{k}+}(\mathbf{r}')\,d^3r' \tag{9}$$

Note that if we write $f_{\mathbf{k}+} = |f_{\mathbf{k}+}|\exp i\phi_+$, then the phase function ϕ_+ is characterized by a length determined by the range of the potential.

There exists another set of solutions of Eq. (4), which we denote by $\psi_{\mathbf{k}-}(\mathbf{r})$ whose asymptotic behavior is

$$\lim_{r\to\infty} \psi_{\mathbf{k}-}(\mathbf{r}) = e^{i\mathbf{k}\cdot\mathbf{r}} + \frac{e^{-ikr}}{r} f_{\mathbf{k}-}(\theta, \phi) \tag{10}$$

corresponding to a plane wave plus incoming spherical waves. These solutions satisfy the integral equation:

$$\psi_{\mathbf{k}-}(\mathbf{r}) = e^{i\mathbf{k}\cdot\mathbf{r}} + \int G_{\mathbf{k}-}(\mathbf{r} - \mathbf{r}')\,\frac{2m}{\hbar^2}\,V(\mathbf{r}')\,\psi_{\mathbf{k}-}(\mathbf{r}')\,d^3r' \tag{11}$$

where the Green's function $G_{\mathbf{k}-}$ is given by

$$G_{\mathbf{k}-}(\mathbf{r}-\mathbf{r}') = -\frac{1}{4\pi}\frac{e^{-ik|\mathbf{r}-\mathbf{r}'|}}{|\mathbf{r}-\mathbf{r}'|} = \lim_{\epsilon\to 0+}\frac{1}{(2\pi)^3}\int d^3k\,\frac{e^{i\mathbf{k}'\cdot(\mathbf{r}-\mathbf{r}')}}{\mathbf{k}^2-\mathbf{k}'^2-i\epsilon} \tag{12}$$

In the case that H has no bound states, the set $\{\psi_{\mathbf{k}+}\}$ (or $\{\psi_{\mathbf{k}-}\}$) is complete with

$$\int \frac{d^3k}{(2\pi)^3}\,\psi_{\mathbf{k}\pm}(\mathbf{r})\,\bar{\psi}_{\mathbf{k}\pm}(\mathbf{r}') = \delta^{(3)}(\mathbf{r}-\mathbf{r}') \tag{13}$$

We may therefore expand $\Psi(\mathbf{r}, 0)$ in terms of the set $\{\psi_{\mathbf{k}+}\}$ and write

$$\Psi(\mathbf{r}, 0) = \int d^3k\, B_{\mathbf{k}}\psi_{\mathbf{k}+}(\mathbf{r}) \tag{14}$$

The motivation for the choice of the set $\{\psi_{\mathbf{k}+}\}$ will become apparent shortly. Furthermore, since at time $t = 0$, $\Psi(\mathbf{r}, 0)$ represents a packet *far away* from the scattering center, we may take for $\psi_{\mathbf{k}+}(\mathbf{r})$ its asymptotic value and write

$$\Psi(\mathbf{r}, 0) = \int d^3k\, B_{\mathbf{k}}\left(e^{i\mathbf{k}\cdot\mathbf{r}} + f_{\mathbf{k}+}(\theta,\phi)\,\frac{e^{ikr}}{r}\right) \tag{15}$$

The expansion coefficients $B_{\mathbf{k}}$ must be so chosen that Eq. (15) represents a packet localized at $\mathbf{r}_0$, with an average momentum $\mathbf{k}_0$ along the z axis. Since initially (at $t = 0$) the packet moves in a region where the potential does not act, the motion should be that of a free particle and we expect that in the expansion (15) only the plane wave part contributes. Now the position of the maximum of the amplitude $\Psi(\mathbf{r}, 0)$ due to the plane wave factor in Eq. (15) can be obtained by the method of stationary phase. If we write

$$B_{\mathbf{k}} = b(\mathbf{k})\,e^{i\beta(\mathbf{k})} \tag{16}$$

with b and β real, the maximum will occur when

$$\nabla_{\mathbf{k}}(\beta(\mathbf{k}) + \mathbf{k}\cdot\mathbf{r})\Big|_{\mathbf{k}=\mathbf{k}_0} = 0 \tag{17}$$

i.e., the position of the maximum amplitude occurs at that $\mathbf{r}$ for which Eq. (17) is satisfied. We therefore set

$$\mathbf{r}_0 = -\nabla_{\mathbf{k}}\beta(\mathbf{k})\Big|_{\mathbf{k}=\mathbf{k}_0} \tag{18}$$

For the contribution of the spherical wave we write

$$f_{\mathbf{k}+} = |f_{\mathbf{k}+}|\,e^{i\phi_+} \tag{19}$$

and the rate of change of phase for this part of the integrand is

$$\begin{aligned}\nabla_{\mathbf{k}}(\beta(\mathbf{k}) + \phi_+ + kr)\Big|_{\mathbf{k}=\mathbf{k}_0} &= \nabla_{\mathbf{k}}\phi_+\Big|_{\mathbf{k}=\mathbf{k}_0} + \boldsymbol{\epsilon}_3 r + \nabla_{\mathbf{k}}\beta(\mathbf{k})\Big|_{\mathbf{k}=\mathbf{k}_0} \\ &= \nabla_{\mathbf{k}}\phi_+\Big|_{\mathbf{k}=\mathbf{k}_0} + \boldsymbol{\epsilon}_3(r + |z_0|)\end{aligned} \tag{20}$$

where $\boldsymbol{\epsilon}_3$ is a unit vector in the z direction. We have also used the fact that $\mathbf{r}_0 = -\boldsymbol{\epsilon}_3 |z_0|$. Recalling that $\nabla_{\mathbf{k}}\phi_+ \Big|_{\mathbf{k}=\mathbf{k}_0}$ is a distance characterized by the range of the potential, hence small compared with r_0, we note that the phase of the spherical wave part is never stationary since $|\nabla_{\mathbf{k}}(\beta(\mathbf{k}) + \phi_+ + kr)| \approx (r + |z_0|) > 0$. It is, in fact, large for large $|z_0|$, so that the contribution of this part to $\Psi(\mathbf{r}, 0)$ is negligibly small due to cancellation by interference. Thus at $t = 0$:

$$\Psi(\mathbf{r}, 0) \approx \int B_{\mathbf{k}}\, e^{i\mathbf{k}\cdot\mathbf{r}}\, d^3k \tag{21}$$

and comparing Eq. (21) with (2), we find that

$$B_{\mathbf{k}} = \Phi(\mathbf{k} - \mathbf{k}_0)\, e^{-i\mathbf{k}\cdot\mathbf{r}_0} \tag{22}$$

The simplicity of this result is due to the choice of the complete set $\{\psi_{\mathbf{k}+}\}$ for the expansion of $\Psi(\mathbf{r}, 0)$: In an expansion in terms of $\{\psi_{\mathbf{k}+}\}$, $\Psi(\mathbf{r}, 0)$ has the same expansion coefficients as in an expansion in terms of plane waves $\{\exp i\mathbf{k}\cdot\mathbf{r}\}$.

Now as time proceeds the particle moves toward the scattering center and eventually comes under the influence of the force field and may be deflected. In the experimental situation the detector is located at a distance which is large compared to the range of the force field, so that in the region of observation the particle is again moving freely in its trajectory. To predict the outcome of the experiment we are therefore interested in the behavior of $\Psi(\mathbf{r}, t)$ for large t, when once again the particle is far away from the scattering center. Since the $\psi_{\mathbf{k}+}$s are eigenfunctions of H with eigenvalue $\hbar^2\mathbf{k}^2/2m$, we may write

$$\begin{aligned}\Psi(\mathbf{r}, t) &= e^{-iHt/\hbar}\,\Psi(\mathbf{r}, 0) \\ &= \int d^3k\, \Phi(\mathbf{k} - \mathbf{k}_0)\, e^{-i\mathbf{k}\cdot\mathbf{r}_0 - i\hbar\mathbf{k}^2t/2m\hbar}\, \psi_{\mathbf{k}+}(\mathbf{r})\end{aligned} \tag{23}$$

Now for sufficiently large t, we can again replace $\psi_{\mathbf{k}+}(\mathbf{r})$ by its asymptotic value, Eq. (5), so that we may write

$$\lim_{t\to\infty} \Psi(\mathbf{r}, t) = \Psi_0(\mathbf{r}, t) + \Psi_{sc}(\mathbf{r}, t) \tag{24a}$$

with

$$\Psi_0(\mathbf{r}, t) = \int d^3k\, e^{-iE_{\mathbf{k}}t/\hbar}\, B_{\mathbf{k}}\, e^{i\mathbf{k}\cdot\mathbf{r}} \tag{24b}$$

$$\Psi_{sc}(\mathbf{r}, t) = \int d^3k\, e^{-iE_{\mathbf{k}}t/\hbar}\, B_{\mathbf{k}}\, \frac{e^{ikr}}{r} f_{\mathbf{k}+}(\theta, \phi) \tag{24c}$$

We again can find the value of $\mathbf{r}$ at which $\Psi_0(\mathbf{r}, t)$ has its maximum by the method of stationary phase: $\Psi_0(\mathbf{r}, t)$ will get large contributions if

$$\nabla_{\mathbf{k}}\left(-\mathbf{k}\cdot\mathbf{r}_0 - \frac{\hbar\mathbf{k}^2}{2m}t + \mathbf{k}\cdot\mathbf{r}\right)\Bigg|_{\mathbf{k}=\mathbf{k}_0} = 0 \tag{25}$$

i.e., for those $\mathbf{r}$ for which $\mathbf{r} = (0, 0, z)$, with

$$z = -z_0 + \frac{\hbar k_0}{m} t = -z_0 + vt \tag{26}$$

$\Psi_0(\mathbf{r}, t)$ is thus the contribution to $\Psi(\mathbf{r}, t)$ corresponding to the transmitted (unscattered) packet. For Ψ_{sc} we obtain a contribution if

$$\nabla_{\mathbf{k}} \left(-\mathbf{k} \cdot \mathbf{r}_0 - \frac{\hbar \mathbf{k}^2}{2m} t + \phi_+ + kr \right) \bigg|_{\mathbf{k}=\mathbf{k}_0} = 0 \tag{27a}$$

i.e., if

$$\boldsymbol{\epsilon}_3 \left(r + |z_0| - \frac{\hbar k_0}{m} t \right) + \nabla_{\mathbf{k}} \phi_+ \bigg|_{\mathbf{k}=\mathbf{k}_0} = 0 \tag{27b}$$

Neglecting the small $\nabla_{\mathbf{k}}\phi_+$ term, the contribution to Ψ_{sc} will occur at those points r for which

$$r = -|z_0| + vt \tag{28}$$

$\Psi_{sc}(\mathbf{r}, t)$ thus represents the scattered wave. Since r is positive, Eq. (27) holds only for $t > |z_0|/v$. The scattered wave Ψ_{sc} is therefore different from zero only after the incoming wave packet has reached the scattering center, i.e., only after a time $|z_0|/v$ has elapsed. This can be taken as the nonrelativistic causality principle. Thus for $t > |z_0|/v$, in addition to the packet moving along the z axis, there is a spherical wave diverging from the origin the amplitude, $f_{\mathbf{k}+}(\theta, \phi)$, of which is a function of direction. For $t \gg |z_0|/v$, the probability for finding the particle in a direction different from the forward direction in a volume d^3r about $\mathbf{r}$ at time t is given by

$$|\Psi_{sc}(\mathbf{r}, t)|^2 \, d^3r = d^3r \left| \int \Phi(\mathbf{k} - \mathbf{k}_0) \, e^{-i\mathbf{k}\cdot\mathbf{r}_0} f_{\mathbf{k}+}(\theta, \phi) \frac{e^{ikr - iE_{\mathbf{k}}t/\hbar}}{r} \, d^3k \right|^2 \tag{29}$$

If we assume that $f_{\mathbf{k}+}$ is a not too rapidly varying function of $\mathbf{k}$ within the range of values Δk of $\mathbf{k}_0$ around which Φ is peaked, we may then write:[1]

$$|\Psi_{sc}(\mathbf{r}, t)|^2 \, d^3r = |f_{\mathbf{k}_0+}(\theta, \phi)|^2 \cdot \left| \int \Phi(\mathbf{k} - \mathbf{k}_0) \, e^{-i\mathbf{k}\cdot\mathbf{r}_0} \frac{e^{i(kr - E_{\mathbf{k}}t/\hbar)}}{r} \, d^3k \right|^2 \tag{30}$$

The probability that the particle is found within a solid angle $d\Omega$ about the direction θ, ϕ is therefore given by

$$P_\Omega \, d\Omega = d\Omega \, |f_{\mathbf{k}_0+}(\theta, \phi)|^2 \int_0^\infty dr \left| \int \Phi(\mathbf{k} - \mathbf{k}_0) \, e^{-i\mathbf{k}\cdot\mathbf{r}_0} \frac{e^{i(kr - E_{\mathbf{k}}t/\hbar)}}{r} \, d^3k \right|^2 \tag{31}$$

Finally, since the integral over $\mathbf{k}$ is different from zero only for large positive r, we can extend without any error the integration over r to range from $-\infty$ to ∞. The differential cross section for scattering, $d\sigma$, is defined

[1] For a discussion of the case when this assumption is not valid (resonance scattering), see the excellent review article by Brenig and Haag [Brenig (1959)]. Their paper is highly recommended as an introduction to the more rigorous formulation of the formal theory of scattering. It also contains an exposition of the theory of resonant scattering and a discussion of the analytic properties of the scattering amplitudes.

as the probability for scattering into unit solid angle divided by the probability per cm² that the particle is initially in a cylinder of unit cross section about the direction of the incident beam, i.e.,

$$d\sigma(\theta, \phi) = \frac{P_\Omega \, d\Omega}{P_0} \tag{32}$$

where

$$\begin{aligned} P_0 &= \frac{1}{A} \int_A dx\, dy \int_{-\infty}^{+\infty} dz \, |\Psi(\mathbf{r}, 0)|^2 \\ &= \int_{-\infty}^{+\infty} dz \left| \int B_\mathbf{k} \, e^{ikz} \, d^3k \right|^2 \end{aligned} \tag{33}$$

and where we have assumed rotational symmetry about the z axis for the initial beam. We next show that the integrals in Eqs. (31) and (33) are the same, so that

$$d\sigma(\theta, \phi) = |f_{\mathbf{k}_0+}(\theta, \phi)|^2 \tag{34}$$

Proof: We write for $E_\mathbf{k}$ in the exponent of the integral for P_Ω, the expression

$$\frac{E_\mathbf{k} t}{\hbar} = \frac{\hbar \mathbf{k}^2}{2m} t = \frac{\hbar}{2m} t[(\mathbf{k} - \mathbf{k}_0)^2 - \mathbf{k}_0^2 + 2\mathbf{k} \cdot \mathbf{k}_0] \tag{35}$$

Now $(\mathbf{k} - \mathbf{k}_0)^2 \sim (\Delta\mathbf{k})^2$ and $t \sim |z_0|/v$ so that the first term is of order $\hbar t(\Delta\mathbf{k})^2/2m \sim (\Delta\mathbf{k})^2 |z_0|/k_0$ and is much smaller than 1. That this is so, corresponds to the requirement that the packet not spread too much, for otherwise one would not be able to differentiate between the spreading of the packet and the actual scattering. The condition that the packet does not spread too rapidly is that $\delta(\Delta r)/\Delta r \ll 1$, where $\delta(\Delta r) \sim \hbar(\Delta k)\, t/m$, or that

$$\frac{\hbar(\Delta k)\, t}{m \Delta r} = \frac{(\Delta k)^2}{k_0} |z_0| \ll 1 \tag{36}$$

We therefore neglect the term $\exp\,[i\hbar(\mathbf{k} - \mathbf{k}_0)^2\, t/2m]$. The term $\exp\,(i\hbar \mathbf{k}_0^2 t/2m)$ is a phase factor which drops out upon taking the absolute value squared. Furthermore, the properties of $B_\mathbf{k}$ allow us to write within the integral

$$e^{i\mathbf{k} \cdot \mathbf{k}_0 \hbar t/m} = e^{ik_z vt} \approx e^{ikvt} \tag{37}$$

so that

$$P_\Omega \, d\Omega = d\Omega \int_{-\infty}^{+\infty} dr \left| \int B_\mathbf{k} \, e^{ik(r - vt)} \, d^3k \right|^2 |f_{\mathbf{k}_0+}(\phi, \theta)|^2 \tag{38}$$

and changing variables to $r' = r - vt$ we have proved our assertions. The replacement of k by k_z is allowed since $B_\mathbf{k}$ has its maximum value for $\mathbf{k}$ lying around $\mathbf{k}_0$, and the latter points in the z direction.

The importance of the above exposition lies in the fact it indicates a method for computing the differential cross section using time-independent methods: $f_{\mathbf{k}+}(\theta, \phi)$ is uniquely determined from the differential Eq. (4) and the boundary condition that the asymptotic form of the solution contain only outgoing spherical waves.

11b. The Lippmann-Schwinger Equations

Let us recast our previous result in a much more general form. A closer inspection of Section a of this chapter will reveal that we have assumed that the Hamiltonian can be split up in the form $H = H_0 + V$, where H_0 describes the unperturbed motion of the particles and V describes the interaction between them which is assumed to vanish if the particles are sufficiently far apart. The following is therefore not immediately applicable to field theoretic models, where V is responsible not only for the interaction between particles but also for self-interactions giving rise to cloud and self-energy effects.

In what follows we assume that H_0 has only a continuous spectrum, with

$$H_0 \mid \varphi_a\rangle = E_a \mid \varphi_a\rangle \tag{39a}$$

$$(\varphi_a, \varphi_b) = \delta(a - b) \tag{39b}$$

and that H has the same continuous spectrum as H_0, starting with the minimum value $E_a = 0$. This can always be achieved by a suitable choice of additive constants in H and H_0. We shall denote the "incoming" and "outgoing" wave eigenstates of H by $|\psi_a{}^+\rangle$ and $|\psi_a{}^-\rangle$, respectively:

$$H \mid \psi_a{}^\pm\rangle = E_a \mid \psi_a{}^\pm\rangle \tag{40}$$

They satisfy the Lippmann-Schwinger equations [Lippmann (1950); see also Gell-Mann and Goldberger (Gell-Mann 1953a)]:

$$|\psi_a{}^+\rangle = |\varphi_a\rangle + \lim_{\epsilon\to 0+} \frac{1}{E_a - H_0 + i\epsilon} V \mid \psi_a{}^+\rangle \tag{41a}$$

$$|\psi_a{}^-\rangle = |\varphi_a\rangle + \lim_{\epsilon\to 0+} \frac{1}{E_a - H_0 - i\epsilon} V \mid \psi_a{}^-\rangle \tag{41b}$$

That the solutions of these equations are eigenstates of H is trivially verified by multiplying both sides of (41) by $E_a - H_0$.[2] The eigenstate $|\psi_a{}^+\rangle$ is that solution of the scattering problem which has incoming plane waves [corresponding to the term $|\varphi_a\rangle$ in (41a)] and outgoing scattered waves [represented by the second term on the right-hand side of (41a)]. The term $+i\epsilon$ in the denominator of (41a) guarantees that only outgoing scattered waves exist [Dirac (1947)]. Similarly, $|\psi_a{}^-\rangle$ is that solution which has incoming spherical waves and outgoing plane waves and is

[2] For a discussion of the question of the normalization of vectors $|\varphi_a\rangle$ and $|\psi_a{}^\pm\rangle$ in nonrelativistic quantum mechanics and in field theory, and its relevance to the formulation of the Lippmann-Schwinger equation, see de Witt (1954) and Van Hove (1955). For nonrelativistic quantum mechanics, Eqs. (41a) and (41b) are correct as they stand, since $|\varphi_a\rangle$ and $|\psi_a{}^\pm\rangle$ are both normalized to a δ function. In field theory, due to self-interactions induced by H_I, $|\varphi_a\rangle$ and $|\psi_a{}^\pm\rangle$ differ in their normalization by a factor Z (the wave function renormalization), and the Lippmann-Schwinger equation must then be amended to take this renormalization into account. We shall discuss the necessary changes in Chapter 12.

called the outgoing solution. More precisely, in a time-dependent formulation, a wave packet constructed from states $|\psi_a{}^-\rangle$ would at time $t = -\infty$ contain incoming spherical waves plus plane waves, and at time $t = +\infty$ only plane waves.

For the case of the scattering of a particle by a local potential, $\langle \mathbf{r} \mid V \mid \mathbf{r}' \rangle = \delta^{(3)}(\mathbf{r} - \mathbf{r}')\ V(\mathbf{r})$, the equivalence of (41a) with our previous Eq. (6) is obtained by taking the scalar product of (41a) with $\langle \mathbf{r} |$:

$$\langle \mathbf{r} \mid \psi_a{}^+ \rangle = \langle \mathbf{r} \mid \varphi_a \rangle + \lim_{\epsilon \to 0+} \int d^3r' \int d^3r'' \int d^3p' \int d^3p''$$

$$\cdot \langle \mathbf{r} \mid \mathbf{p}' \rangle \left\langle \mathbf{p}' \left| \frac{1}{E_a - H_0 + i\epsilon} \right| \mathbf{p}'' \right\rangle \langle \mathbf{p}'' \mid \mathbf{r}' \rangle \langle \mathbf{r}' \mid V \mid \mathbf{r}'' \rangle \langle \mathbf{r}'' \mid \psi_a{}^+ \rangle \quad (42a)$$

or

$$\langle \mathbf{r} \mid \psi_a{}^\pm \rangle = \psi_a{}^\pm(\mathbf{r})$$

$$= \varphi_a(\mathbf{r}) + \lim_{\epsilon \to 0+} \frac{1}{(2\pi)^3} \int d^3r' \int d^3k' \frac{e^{i\mathbf{k}' \cdot (\mathbf{r} - \mathbf{r}')}}{\mathbf{k}_a{}^2 - \mathbf{k}'^2 \pm i\epsilon}$$

$$\cdot \frac{2m}{\hbar^2} V(\mathbf{r}')\, \psi_a{}^\pm(\mathbf{r}') \quad (42b)$$

which is Eq. (6), since a labels the eigenstate of H_0 and $G_{a\pm}$ is defined by precisely the integral occurring in (42b), namely Eq. (7).

The Hamiltonian H may, in addition to the above scattering solutions, have bound states (denoted by Greek subscript)

$$H \mid \psi_\beta \rangle = E_\beta \mid \psi_\beta \rangle \quad (43)$$

We shall assume that the energy of these states is less than that of any eigenstate of H_0 with the same symmetry quantum numbers. These bound state solutions, because of the hermiticity of H, are orthogonal to the continuum states

$$(\psi_\beta, \psi_a{}^\pm) = 0 \quad (44)$$

The set $\{\psi_a{}^\pm\} + \{\psi_\beta\}$ is complete in the sense that

$$\sum_a |\psi_a{}^\pm\rangle \langle \psi_a{}^\pm| + \sum_\beta |\psi_\beta\rangle \langle \psi_\beta| = 1 \quad (45)$$

For the time being, we again view these stationary state solutions as auxiliary functions with the help of which the time-dependent problem, corresponding to the evolution of the system in time, can easily be solved.

11c. The Dirac Picture

For the solution of the time-dependent problem, it is advantageous to work in the Dirac picture[3] which is related to the Schrödinger and Heisen-

[3] This picture is very closely related to the time-dependent amplitudes used in the method of variation of constants first introduced by Dirac (1926, 1927), whence the appellation.

berg pictures by unitary transformations. Let $|\Phi_S(t)\rangle$ be the time-dependent vector describing the evolution of the system in the Schrödinger picture, with

$$i\hbar\partial_t \mid \Phi_S(t)\rangle = H_S \mid \Phi_S(t)\rangle$$
$$= (H_{0S} + V_S) \mid \Phi_S(t)\rangle \tag{46}$$

We then define the Dirac picture vector (sometimes also called the interaction picture vector) $|\Psi_D(t)\rangle$ by

$$|\Psi_D(t)\rangle = e^{iH_{0S}t/\hbar} \mid \Phi_S(t)\rangle \tag{47}$$

which by virtue of Fq. (46) satisfies the following equation:

$$i\hbar\partial_t \mid \Psi_D(t)\rangle = e^{iH_{0S}t/\hbar}\, V_S\, e^{-iH_{0S}t/\hbar} \mid \Psi_D(t)\rangle$$
$$= V_D(t) \mid \Psi_D(t)\rangle \tag{48}$$

The time dependence of the Dirac picture vector $|\Psi_D(t)\rangle$ is thus determined by the interaction energy operator

$$V_D(t) = e^{iH_{0S}t/\hbar}\, V_S\, e^{-iH_{0S}t/\hbar} \tag{49}$$

If $V_D = 0$, i.e., in the absence of interaction, $|\Psi_D(t)\rangle$ is constant in time and the Dirac picture coincides with the Heisenberg picture.

The relation between a Dirac picture operator $Q_D(t)$ and the corresponding Schrödinger picture operator $Q_S(t)$ is defined so that

$$(\Psi_D(t) \mid Q_D(t) \mid \Psi_D(t)) = (\Phi_S(t) \mid Q_S(t) \mid \Phi_S(t)) \tag{50}$$

whence, using (47),

$$Q_D(t) = e^{iH_{0S}t/\hbar}\, Q_S\, e^{-iH_{0S}t/\hbar} \tag{51}$$

The time dependence of Dirac picture operators is therefore determined by the unperturbed Hamiltonian H_0, and

$$i\hbar\partial_t Q_D(t) = +[Q_D(t), H_0] \tag{52}$$

We have written $H_0 = H_{0S} = H_{0D}$. In the following we drop the subscript D on vectors and operators in the Dirac picture, since we shall deal only with quantities defined in this picture in the rest of this chapter.

We next introduce the time displacement operator $U(t, t_0)$ defined by

$$U(t, t_0) \mid \Psi(t_0)\rangle = |\Psi(t)\rangle \tag{53}$$

which satisfies the equation

$$i\hbar\partial_t U(t, t_0) = V(t)\, U(t, t_0) \tag{54}$$

and the boundary condition

$$U(t, t) = 1 \tag{55}$$

For finite t, t_0 it has the following properties:

$$U(t, t_0)\, U(t_0, t') = U(t, t') \tag{56a}$$

$$U(t, t_0) = U^{-1}(t_0, t) = U^*(t_0, t) \tag{56b}$$

which can be proved in the same way as for the corresponding Schrödinger picture time displacement operators (see Chapter 1*a*). An explicit representation of $U(t, t_0)$ is obtained by recalling that in the Schrödinger picture (with $\hbar = 1$)

$$|\Phi_S(t)\rangle = e^{-iH_S(t-t_0)} \mid \Phi_S(t_0)\rangle \tag{57}$$

so that, replacing the states $|\Phi_S(t)\rangle$ by the corresponding Dirac picture ones, we obtain

$$|\Psi(t)\rangle = e^{+iH_0t}\, e^{-iH(t-t_0)}\, e^{-iH_0t_0} \mid \Psi(t_0)\rangle \tag{58}$$

where now $H = H_0 + V(t)$, whence

$$U(t, t_0) = e^{iH_0t}\, e^{-iH(t-t_0)}\, e^{-iH_0t_0} \tag{59}$$

In the description of the scattering process in the Dirac picture the initial state $|\Phi_i\rangle$ is time-independent. It corresponds to the particles being far apart, and hence not interacting with one another. Since the state vector of the system is time-independent as long as the particles do not interact, it is convenient to consider the initial state to be specified at time $t = -\infty$, i.e., far in the past; thus $|\Phi_i\rangle = \lim_{t\to-\infty} |\Psi(t)\rangle$ where $|\Psi(t)\rangle$ is the Dirac picture state vector which describes the evolution of the system. The initial state $|\Phi_i\rangle$ is a normalizable wave packet state, and is a superposition of eigenstates of H_0. In the limit of plane waves $|\Phi_i\rangle \to |\varphi_a\rangle$. Similarly, the state vector of the particles after they have interacted and are moving freely away from each other is time-independent. It is therefore given by $\lim_{t\to+\infty} |\Psi(t)\rangle$.

The amplitude for scattering from the initial state $|\Phi_i\rangle$, to some final state $|\Phi_f\rangle$ which is again a normalizable state corresponding to the particles moving freely with specified momenta and spins, is given by

$$\lim_{t_2\to\infty} \langle\Phi_f \mid \Psi(t_2)\rangle = \lim_{t_2\to\infty}\ \lim_{t_1\to-\infty} (\Phi_f \mid U(t_2, t_1) \mid \Phi_i) \tag{60}$$

In the limit of plane wave states, i.e., when $|\Phi_i\rangle \to |\varphi_a\rangle$ and $|\Phi_f\rangle \to |\varphi_b\rangle$, we call this amplitude, S_{ba}, i.e.,

$$S_{ba} = \lim_{t_2\to\infty}\ \lim_{t_1\to-\infty} (\varphi_b \mid U(t_2, t_1) \mid \varphi_a) \tag{61}$$

It corresponds to the probability amplitude for a transition of the system from an initial state $|\varphi_a\rangle$ at time $t = -\infty$ to a state $|\varphi_b\rangle$ at time $t = +\infty$. The operator S whose matrix elements between initial and final states $|\varphi_a\rangle$ and $|\varphi_b\rangle$ correspond to the transition amplitude between these states is called the S matrix, or the scattering matrix. Such a scattering matrix was first introduced by Wheeler (1937) in connection with the problems of nuclear structure and scattering. It has been reinvestigated in great detail by Heisenberg (1943a, b) in connection with the theory of elementary particles.

The philosophy which led Heisenberg to the consideration of the S matrix can be traced to his belief [Heisenberg (1930a, 1938)] that the divergence difficulty which is inherent in all relativistic field theories might be avoided by the introduction of a new fundamental constant of the dimension of a length. Heisenberg believed that this new constant was to play the same role in limiting quantum field theory as did Planck's constant in limiting the application of classical mechanics to atomic systems.

In ordinary quantum mechanics, an atomic system is completely defined by the Hamiltonian, H, of the system. However, the assumption of a Hamiltonian implies the possibility of a continuous time displacement of the wave function which seems to contradict the existence of a fundamental length. Heisenberg therefore abandoned the notion of a Schrödinger equation and of a Hamiltonian. In attempting to discover what kind of operators and functions should replace these concepts, he was led to inquire which quantities in the present formulation of field theory will still be called observables in the future "correct" theory, i.e., which observables do not depend upon the existence of a minimal length. Heisenberg regarded the following as observable quantities, which must necessarily be described in any theory:

1. The energy and momentum of free particles;
2. The discrete energy levels of stationary closed systems;
3. The asymptotic behavior of wave functions describing collision, emission, and absorption processes (from which the cross section of such processes can be calculated).

In order to make the cross section of processes and other observables (which could easily be calculated from the Hamiltonian if it existed) the central feature of the new theory, Heisenberg introduced a certain unitary matrix S, the S matrix, from which it was hoped that the transition rates and the observable properties of the bound states could be obtained. In his original paper, this S matrix was defined as the operator which transforms the incoming state into the outgoing state. Heisenberg believed that in the future theory the S matrix would take over the role played by the Hamiltonian. Heisenberg's theory of the S matrix has been extended by Møller (1945, 1946), Stückelberg (1944a, b, 1945a, b), and particularly by Lehmann, Symanzik, and Zimmermann [Lehmann (1957)] who have given its most concise formulation within the framework of relativistic quantum field theory. We shall return to their work in Chapter 18.

The relation of the S matrix to the usual form of quantum theory and its use in calculating cross section and other observables have been discussed in many papers, particularly Lippmann and Schwinger [Lippmann (1950)]. An excellent survey of its application to scattering theory is given in the papers of Gell-Mann and Goldberger [Gell-Mann (1953a)] and of Brenig and Haag [Brenig 1959)].

In order to obtain various different expressions for the scattering amplitude S_{ba}, we first prove the following theorem:

$$\lim_{t_0 \to -\infty} U(0, t_0) \mid \varphi_a\rangle = |\psi_a^+\rangle \tag{62}$$

i.e., we shall prove that the operator $U(0, -\infty)$ transforms the state $|\varphi_a\rangle$, an eigenstate of H_0, into the state $|\psi_a^+\rangle$ which is an eigenstate of H. The proof [see, in particular, Hack (1954)] will be given in terms of a wave-packet state $|\Phi_i\rangle$:

$$|\Phi_i\rangle = \int da \mid \varphi_a\rangle \langle\varphi_a \mid \Phi_i\rangle \tag{63}$$

where the expansion coefficients cluster about some state $\hat{a}$. We shall first obtain an expression for $U(0, t_0) \mid \Phi_i\rangle$:

$$\begin{aligned} U(0, t_0) \mid \Phi_i\rangle &= e^{iHt_0} e^{-iH_0t_0} \mid \Phi_i\rangle \\ &= e^{iHt_0} \int da\, e^{-iE_at_0} \mid \varphi_a\rangle \langle\varphi_a \mid \Phi_i\rangle \end{aligned} \tag{64}$$

on which the limit $t \to -\infty$ can readily be carried out. To evaluate $\exp(iHt_0)$ operating on the state standing to the right of it in Eq. (64), we expand $|\varphi_a\rangle$ in terms of the complete set $\{|\psi_a^+\rangle\} + \{|\psi_\beta\rangle\}$:

$$|\varphi_a\rangle = \int dc \mid \psi_c^+\rangle \langle\psi_c^+ \mid \varphi_a\rangle + \sum_\beta \mid \psi_\beta\rangle \langle\psi_\beta \mid \varphi_a\rangle \tag{65}$$

[compare with the steps leading to Eq. (21)] so that

$$\begin{aligned} U(0, t_0) \mid \Phi_i\rangle = &\int dc \int da\, e^{i(E_c - E_a)t_0} \langle\varphi_a \mid \Phi_i\rangle \langle\psi_c^+ \mid \varphi_a\rangle \mid \psi_c^+\rangle \\ &+ \sum_\beta \int da\, e^{i(E_\beta - E_a)t_0} \langle\varphi_a \mid \Phi_i\rangle \langle\psi_\beta \mid \varphi_a\rangle \mid \psi_\beta\rangle \end{aligned} \tag{66}$$

If we substitute into this last expression the value of $\langle\psi_c^+ \mid \varphi_a\rangle$ obtained from Eq. (41)

$$\langle\psi_c^+ \mid \varphi_a\rangle = \delta(c - a) + \lim_{\epsilon \to 0+} \frac{\langle\psi_c^+ \mid V \mid \varphi_a\rangle}{E_c - E_a - i\epsilon} \tag{67}$$

we find

$$\begin{aligned} U(0, t_0) \mid \Phi_i\rangle = &\int da\, \langle\varphi_a \mid \Phi_i\rangle \mid \psi_a^+\rangle \\ &+ \lim_{\epsilon \to 0+} \int da\, \langle\varphi_a \mid \Phi_i\rangle \int dc_0 \int dE_c \frac{e^{i(E_c - E_a)t_0}}{E_c - E_a - i\epsilon} \langle\psi_c^+ \mid V \mid \varphi_a\rangle \mid \psi_c^+\rangle \\ &+ \sum_\beta \int da\, e^{i(E_\beta - E_a)t_0} \langle\varphi_a \mid \Phi_i\rangle \langle\psi_\beta \mid \varphi_a\rangle \mid \psi_\beta\rangle \end{aligned} \tag{68}$$

where, in the second term on the right-hand side of (68), we have assumed the energy, E_c, to be a member of the complete set of observables, c, i.e., $c = \{E_c, c_0\}$ and $\int dc = \int dE_c \int dc_0$. In the last term of (68) we can substitute for $\langle\psi_\beta \mid \varphi_a\rangle$ the expression

$$\langle\psi_\beta \mid \varphi_a\rangle = \frac{\langle\psi_\beta \mid V \mid \varphi_a\rangle}{E_\beta - E_a} \tag{69}$$

since

$$\langle \psi_\beta \mid V \mid \varphi_a \rangle = \langle \psi_\beta \mid H - H_0 \mid \varphi_a \rangle = (E_\beta - E_a) \langle \psi_\beta \mid \varphi_a \rangle \tag{70}$$

and $E_\beta - E_a \neq 0$, by assumption. We are now interested in the limit of $U(0, t_0) \mid \Phi_i \rangle$ as t_0 approaches $-\infty$. We shall show that in this limit the second and third term vanish. A heuristic proof of this assertion is as follows: We write

$$\begin{aligned} \lim_{\epsilon \to 0+} \frac{e^{i(E_c - E_a)t_0}}{E_c - E_a - i\epsilon} &= \lim_{\epsilon \to 0+} e^{i(E_c - E_a - i\epsilon)t_0}\, i \int_{-\infty}^{0} e^{i(E_c - E_a - i\epsilon)t'}\, dt' \\ &= i \int_{-\infty}^{0} e^{i(E_c - E_a - i\epsilon)(t_0 + t')}\, dt' \\ &= i \int_{-\infty}^{t_0} e^{i(E_c - E_a - i\epsilon)\tau}\, d\tau \end{aligned} \tag{71}$$

so that

$$\lim_{t_0 \to -\infty} \frac{e^{i(E_c - E_a)t_0}}{E_c - E_a - i\epsilon} = 0 \tag{72a}$$

and, therefore, the second term on the right-hand side of (68) vanishes in the limit as $t \to -\infty$. Incidentally, from Eq. (71) one also infers that

$$\lim_{t_0 \to +\infty} \frac{e^{i(E_c - E_a)t_0}}{E_c - E_a - i\epsilon} = 2\pi i \delta(E_c - E_a) \tag{72b}$$

A more rigorous way of establishing this result consists in changing variables from E_c to $x = (E_a - E_c)\, t_0$. In the limit as $t_0 \to -\infty$ the x integration ranges from $-\infty$ to $+\infty$ since E_c can be greater or less than E_a. The x integration can then be performed by closing the contour in the lower half of the complex x plane, with the result that the contribution vanishes since the only pole of the integrand lies in the upper half plane [see, e.g., Hack (1954)]. Next one notices that the denominator of the third term of Eq. (68) [after having substituted the expression (69) into it] never vanishes since E_β does not overlap the continuous spectrum. By the Riemann-Lebesgue lemma the third term of Eq. (69) therefore vanishes in the limit $t_0 \to -\infty$; so that

$$\lim_{t_0 \to -\infty} U(0, t_0) \mid \Phi_i \rangle = \int da\, \langle \varphi_a \mid \Phi_i \rangle \mid \psi_a^+ \rangle \tag{73}$$

and in the limit of a plane wave

$$\lim_{t_0 \to -\infty} U(0, t_0) \mid \varphi_a \rangle = \mid \psi_a^+ \rangle \tag{74}$$

It should be noted that the proof of Eq. (73) depended on the fact that $\langle \varphi_a \mid \Phi_i \rangle$ is a well-behaved function of a, which in turn is a consequence of the fact that we used a normalizable state $\mid \Phi_i \rangle$. This then allowed the application of the Riemann-Lebesgue lemma which guarantees the vanishing of expressions such as $\int \exp(-ixt) \cdot f(x)\, dx$ for continuous $f(x)$ in the limit as $t \to \pm\infty$.

In calculations, one often replaces the use of wave packets by the mathematical device of replacing the potential $V(t)$ by

$$V(t) \to V_\epsilon(t) = e^{-\epsilon|t|}\, V(t) \tag{75}$$

so that in the limit as $t \to -\infty$

$$\lim_{t\to-\infty} H_\epsilon(t) = \lim_{t\to-\infty} (H_0 + e^{-\epsilon|t|}\, V(t)) = H_0 \tag{76}$$

$V_\epsilon(t)$ is called the "adiabatic potential." The state $|\Psi_\epsilon(t)\rangle$ which evolves from the state $|\varphi_a\rangle$ at $t = -\infty$ under the influence of $V_\epsilon(t)$ is then obtained by solving the integral equation

$$|\Psi_\epsilon(t)\rangle = |\varphi_a\rangle - \frac{i}{\hbar}\int_{-\infty}^{t} e^{-\epsilon|t'|}\, V(t')\, |\, \Psi_\epsilon(t')\rangle\, dt' \tag{77}$$

Equation (77) is equivalent to the differential equation

$$i\hbar\partial_t\, |\, \Psi_\epsilon(t)\rangle = e^{-\epsilon|t|}\, V(t)\, |\, \Psi_\epsilon(t)\rangle \tag{78}$$

and the boundary condition that $|\Psi_\epsilon(t = -\infty)\rangle = |\varphi_a\rangle$. In particular $|\Psi_\epsilon(0)\rangle$ satisfies the equation

$$|\Psi_\epsilon(0)\rangle = |\varphi_a\rangle - \frac{i}{\hbar}\int_{-\infty}^{0} dt'\, e^{\epsilon t'}\, e^{iH_0t'}\, V\, e^{-iH_0t'}\, |\, \Psi_\epsilon(t')\rangle \tag{79}$$

An explicit representation for $|\Psi_\epsilon(0)\rangle$ can be obtained by substituting into the right-hand side of (79) the expression (77) for $|\Psi_\epsilon(t')\rangle$:

$$|\Psi_\epsilon(0)\rangle = |\varphi_a\rangle + \frac{1}{E_a - H_0 + i\epsilon}\, V\, |\, \varphi_a\rangle$$
$$+ \left(-\frac{i}{\hbar}\right)^2 \int_{-\infty}^{0} dt'\, e^{-\epsilon t'}\, e^{iH_0t'}\, V\, e^{-iH_0t'} \int_{-\infty}^{t'} dt''\, e^{\epsilon t''}\, e^{iH_0t''}\, V\, e^{-iH_0t''}\, |\, \Psi_\epsilon(t'')\rangle \tag{80}$$

Again substituting expression (77) for $|\Psi_\epsilon(t'')\rangle$ into the right-hand side of (80) and continuing this iteration procedure, we find

$$|\Psi_\epsilon(0)\rangle = |\varphi_a\rangle + \frac{1}{E_a - H_0 + i\epsilon}\, V\, |\, \varphi_a\rangle + \frac{1}{E_a - H_0 + i\epsilon}\, V\, \frac{1}{E_a - H_0 + i\epsilon}\, V\, |\, \varphi_a\rangle + \cdots$$
$$= |\varphi_a\rangle + \frac{1}{E_a - H_0 + i\epsilon}\, V\, |\, \Psi_\epsilon(0)\rangle \tag{81}$$

Hence, comparing Eqs. (81) and (41a), we find that

$$|\Psi_\epsilon(0)\rangle = |\psi_a{}^+\rangle \tag{82}$$

In other words, the solution $|\Psi_\epsilon(t)\rangle$ of the time-dependent equation (77) allows us to compute $|\psi_a{}^+\rangle$. The convergence difficulties, which are present when Eq. (77) is defined without the exp $(-\epsilon|t|)$ factor, are bypassed by using the "adiabatic" potential in all the computations and only performing the limit $\epsilon \to 0+$ at the end.

In a similar fashion, by using the complete set of states $\{|\psi_b^-\rangle\} + \{|\psi_\beta\rangle\}$, one verifies that

$$\lim_{t\to\infty} U(0, t) \mid \varphi_a\rangle = |\psi_a^-\rangle \tag{83}$$

With the help of these results, and in particular of Eqs. (83) and (74), we may rewrite the scattering amplitude S_{ba} as:

$$\begin{aligned}
S_{ba} = (\varphi_b, S\varphi_a) &= \lim_{t_2\to+\infty} \lim_{t_1\to-\infty} \lim_{|\Phi_i\rangle\to|\varphi_a\rangle} (\varphi_b \mid U(t_2, t_1) \mid \Phi_i) \\
&= \lim_{t_2\to+\infty} \lim_{t_1\to-\infty} (\varphi_b, U(t_2, 0)\, U(0, t_1)\, \varphi_a) \\
&= \lim_{t_2\to+\infty} \lim_{t_1\to-\infty} (U^*(t_2, 0)\, \varphi_b, U(0, t_1)\, \varphi_a) \\
&= \lim_{t_2\to+\infty} \lim_{t_1\to-\infty} (U(0, t_2)\, \varphi_b, U(0, t_1)\, \varphi_a) \\
&= (\psi_b^-, \psi_a^+)
\end{aligned} \tag{84}$$

Equation (84) for the elements of the S matrix will play an important role in the field theoretic applications. Its importance lies in the fact that the in and out states ψ^+, ψ^- can be defined without decomposing H into an unperturbed and perturbation part, $H_0 + V$. We also note that in the Dirac picture the S matrix can be written as:

$$S = U(\infty, -\infty) \tag{85}$$

The expression for S_{ba} can be written in still another way by noting that

$$\begin{aligned}
S_{ba} &= \lim_{t_2\to+\infty} \lim_{t_1\to-\infty} (\varphi_b, U(t_2, 0)\, U(0, t_1)\, \varphi_a) \\
&= \lim_{t_2\to+\infty} (\varphi_b, e^{iH_0t_2} e^{-iHt_2} \psi_a^+) = \lim_{t_2\to+\infty} e^{i(E_b-E_a)t_2} (\varphi_b, \psi_a^+) \\
&= \delta(a-b) + \lim_{t_2\to\infty} \frac{e^{i(E_b-E_a)t_2}}{E_b - E_a + i\epsilon} (\varphi_b, V\psi_a^+) \\
&= \delta(a-b) - 2\pi i \delta(E_a - E_b)\, R_{ba}
\end{aligned} \tag{86}$$

where

$$R_{ba} = (\varphi_b, V\psi_a^+) = (\varphi_b, R\varphi_a) \tag{87}$$

is called the ba matrix element of the reaction or R matrix. The amplitude R_{ba} satisfies the integral equation:

$$\begin{aligned}
R_{ba} = (\varphi_b, V\psi_a^+) &= (\varphi_b, V\varphi_a) + \left(\varphi_b, V \frac{1}{E_a - H_0 + i\epsilon} V\psi_a^+\right) \\
&= V_{ba} + \sum_c \left(\varphi_b, V \frac{1}{E_a - H_0 + i\epsilon} \varphi_c\right) (\varphi_c, V\psi_a^+) \\
&= V_{ba} + \sum_c \frac{V_{bc}R_{ca}}{E_a - E_c + i\epsilon}
\end{aligned} \tag{88}$$

which can be solved by iteration to yield:

$$R_{ba} = V_{ba} + \sum_c \frac{V_{bc}V_{ca}}{E_a - E_c + i\epsilon} + \cdots \tag{89}$$

The matrix R is related to an important quantity: the transition probability per unit time.

The increase per unit time of the probability that a system initially in the state a is found at time t to be in the state b, w_{ba}, is given by

$$w_{ba} = \lim_{t_0 \to -\infty} \frac{d}{dt} |(\varphi_b, U(t, t_0)\, \varphi_a)|^2 \tag{90}$$

We first note that this expression is independent of t. *Proof:*

$$\begin{aligned}
&\frac{d}{dt} |(\varphi_b, U(t, t_0)\, \varphi_a)|^2 \\
&\quad = \frac{d}{dt} \{(\varphi_b, U(t, t_0)\, \varphi_a)\, (\overline{\varphi_b, U(t, t_0)\, \varphi_a})\} \\
&\quad = \frac{2}{\hbar} \operatorname{Im} \{(\varphi_b, V(t)\, U(t, t_0)\, \varphi_a)\, (\overline{\varphi_b, U(t, t_0)\, \varphi_a})\} \\
&\quad = \frac{2}{\hbar} \operatorname{Im} \{(\varphi_b, V(t)\, U(t, 0)\, U(0, t_0)\, \varphi_a)\, (\overline{\varphi_b, U(t, 0)\, U(0, t_0)\, \varphi_a})\}
\end{aligned} \tag{91a}$$

which in the limit as $t_0 \to -\infty$ becomes equal to:

$$\begin{aligned}
w_{ba} &= \frac{2}{\hbar} \operatorname{Im} \{(\varphi_b, e^{iH_0t}\, V\, e^{-iH_0t}\, e^{iH_0t}\, e^{-iHt}\, \psi_a^+) \cdot (\overline{\varphi_b, e^{iH_0t}\, e^{-iHt}\, \psi_a^+})\} \\
&= \frac{2}{\hbar} \operatorname{Im} \{(\varphi_b, V\psi_a^+)\, (\overline{\varphi_b, \psi_a^+})\} \\
&= \frac{2}{\hbar}\, \delta(b - a) \operatorname{Im} R_{ba} + \frac{2}{\hbar} \lim_{\epsilon \to 0+} \operatorname{Im} \frac{|R_{ba}|^2}{E_a - E_b - i\epsilon} \\
&= \frac{2}{\hbar}\, \delta(b - a) \operatorname{Im} R_{ba} + \frac{2\pi}{\hbar}\, \delta(E_a - E_b)\, |R_{ba}|^2
\end{aligned} \tag{91b}$$

where we have used Eq. (41) and the following formal representation for the δ function:

$$\begin{aligned}
\delta(x) &= \lim_{\epsilon \to 0} \frac{1}{2\pi} \int_{-\infty}^{+\infty} dk\, e^{ikx - \epsilon|k|} \\
&= \lim_{\epsilon \to 0} \frac{1}{2\pi i} \left\{ \frac{1}{x - i\epsilon} - \frac{1}{x + i\epsilon} \right\} \\
&= \frac{1}{\pi} \lim_{\epsilon \to 0} \frac{\epsilon}{x^2 + \epsilon^2}
\end{aligned} \tag{92}$$

The differential cross section, σ_{ab}, for the transition $a \to b$, with $a \neq b$, is equal to the transition rate divided by the flux of incoming particles:

$$\sigma_{ab} = \frac{2\pi}{\hbar} \delta(E_a - E_b) |R_{ba}|^2/\text{Flux} \tag{93}$$

Actually one is always only interested in transitions to groups of final states, with energy between E_b and $E_b + dE_b$ and density ρ_f. The transition rate per unit time of interest is therefore given by:

$$\begin{aligned} w &= \frac{2\pi}{\hbar} \int_{\substack{\text{final states}\\ \text{of interest}}} db\, w_{ba} \\ &= \frac{2\pi}{\hbar} \int dE_b\, \delta(E_b - E_a) |R_{ba}|^2 \rho_f(E_b) \\ &= \frac{2\pi}{\hbar} |R_{ba}|^2 \rho_f(E_b) \Big|_{E_b = E_a} \end{aligned} \tag{94}$$

The cross section for the transition is again obtained by dividing by the flux.

Conservation of probability requires that the rate out of the state a must be compensated by a corresponding decrease in the amplitude of the state a or, more precisely, that:

$$\int db\, w_{ba} = 0 \tag{95}$$

where in the integration the state $b = a$ has been included. *Proof:*

$$\begin{aligned} \int db\, w_{ba} &= \lim_{t_0 \to -\infty} \frac{d}{dt} \int db\, \langle \varphi_b \mid U(t, t_0) \mid \varphi_a \rangle \langle \varphi_a \mid U(t_0, t) \mid \varphi_b \rangle \\ &= \lim_{t_0 \to -\infty} \frac{d}{dt} (\varphi_a, U(t_0, t)\, U(t, t_0)\, \varphi_a) \\ &= \lim_{t_0 \to -\infty} \frac{d}{dt} (\varphi_a, \varphi_a) = 0 \end{aligned} \tag{96}$$

Equation (95) is, of course, equivalent to the conservation of norm of the state vector $|\Psi(t)\rangle$. Substituting into Eq. (95) the expression (91b) we obtain:

$$2 \operatorname{Im} R_{aa} = -2\pi \int db\, \delta(E_a - E_b) |R_{ba}|^2 \tag{97}$$

Now in the right-hand side of Eq. (97) the contribution of a single state is negligible, so that $\operatorname{Im} R_{aa}$ is proportional to $\sum_{b \neq a} \sigma_{ab}$, i.e., to the total cross section. Equation (97) is known as the optical theorem and will be of importance in our later discussion of dispersion relations.

11*d*. Unitarity of *S* Matrix

We shall now prove that the S matrix is unitary. We have seen that:

$$U(0, \pm\infty) \mid \varphi_a \rangle = |\psi_a^{\mp}\rangle \tag{98}$$

or equivalently that:

$$U(0, \pm\infty) = \sum_a |\psi_a^{\mp}\rangle\langle\varphi_a| = \Omega^{(\mp)} \tag{99}$$

where we have designated the operator $U(0, \pm\infty)$ by $\Omega^{(\mp)}$. The latter is often called the Møller wave matrix [Møller (1945)]. It transforms the continuous spectrum solution $|\varphi_a\rangle$ of H_0, $H_0 \mid \varphi_a\rangle = E_a \mid \varphi_a\rangle$, into a continuous spectrum eigenfunction $|\psi_a^{\pm}\rangle$ of H, $H \mid \psi_a^{\pm}\rangle = E_a \mid \psi_a^{\pm}\rangle$. A concise statement of this property of $\Omega^{(\pm)}$ is obtained by noting that it is defined as:

$$\begin{aligned} \Omega^{(\pm)} &= \lim_{t\to\mp\infty} U(0, t) \\ &= \lim_{t\to\mp\infty} e^{+iHt} e^{-iH_0t} \end{aligned} \tag{100}$$

For every finite t_0, therefore,

$$e^{iHt_0}\,\Omega^{(\pm)}\,e^{-iH_0t_0} = \Omega^{(\pm)} \tag{101}$$

so that

$$H\Omega^{(\pm)} = \Omega^{(\pm)}H_0 \tag{102}$$

which is the desired formula.

The S matrix expressed in terms of $\Omega^{(+)}$ and $\Omega^{(-)}$ is given by

$$\begin{aligned} S &= U(\infty, -\infty) = U(\infty, 0)\,U(0, -\infty) \\ &= U^*(0, \infty)\,U(0, -\infty) \\ &= \Omega^{(-)*}\Omega^{(+)} \end{aligned} \tag{103}$$

It should be noted that the wave matrix, $\Omega^{(\pm)}$, is not unitary, since

$$\begin{aligned} \Omega^{(\pm)*}\Omega^{(\pm)} &= \sum_{ab} |\varphi_a\rangle\langle\psi_a^{\pm} \mid \psi_b^{\pm}\rangle\langle\varphi_b| \\ &= \sum_{ab} |\varphi_a\rangle\langle\varphi_a| = 1 \end{aligned} \tag{104}$$

but

$$\begin{aligned} \Omega^{(\pm)}\Omega^{(\pm)*} &= \sum_{ab} |\psi_a^{\pm}\rangle\langle\varphi_a \mid \varphi_b\rangle\langle\psi_b^{\pm}| \\ &= \sum_a |\psi_a^{\pm}\rangle\langle\psi_a^{\pm}| = 1 - \sum_\beta |\psi_\beta\rangle\langle\psi_\beta| \\ &= 1 - \Lambda \end{aligned} \tag{105}$$

where Λ is the projection operator onto the bound states of H. These results imply that:[4]

$$\Omega^{(\pm)*} \mid \psi_\beta\rangle = 0 \qquad \text{or} \qquad \Omega^{(\pm)*}\Lambda = 0 \tag{106a}$$

$$\Omega^{(\pm)*} \mid \psi_a^{\pm}\rangle = |\varphi_a\rangle \tag{106b}$$

[4] Note that if H_0 also has a discrete spectrum, so that the completeness relation for its eigenstates reads: $\sum_a |\varphi_a\rangle\langle\varphi_a| + \sum_\beta |\varphi_\beta\rangle\langle\varphi_\beta| = 1$, then the right-hand side of (104) would read $1 - \Lambda_0$ where Λ_0 is the projection operator onto the space spanned by the eigenfunctions of the discrete spectrum of H_0.

For the S matrix, however, we find that:

$$SS^* = [\Omega^{(-)*}\Omega^{(+)}]^*\,\Omega^{(-)*}\Omega^{(+)}$$
$$= \Omega^{(+)*}(1 - \Lambda)\,\Omega^{(+)} = \Omega^{(+)*}\Omega^{(+)}$$
$$= 1 \qquad (107)$$

and

$$SS^* = \Omega^{(-)*}\Omega^{(+)}\Omega^{(+)*}\Omega^{(-)} = 1 \qquad (108)$$

so that S is unitary; in particular, for the bd matrix element of (108) we have:

$$\sum_c S_{bc}(S^*)_{cd} = \delta(b - d) \qquad (109)$$

If we substitute into this equation the expression (86) of S in terms of R, we then find that the unitarity of the S matrix implies that R satisfies the equation:

$$R_{ba} - R^*_{ba} = 2\pi i \sum_c \delta(E_c - E_a)\,R_{bc}R^*_{cd} \quad (E_a = E_b) \qquad (110)$$

which is valid only when $E_a = E_b$. When $b = a$, Eq. (110) will be recognized to be the same as Eq. (97).

The connection between the unitarity of the S matrix and the hermiticity of H, and in particular of V, is made explicit by noting that:

$$V_{ba} = \langle\varphi_b \mid V \mid \varphi_a\rangle$$
$$= \sum_c \langle\varphi_b \mid V \mid \psi_c{}^+\rangle\,\langle\psi_c{}^+ \mid \varphi_a\rangle + \sum_\beta \langle\varphi_b \mid V \mid \psi_\beta\rangle\,\langle\psi_\beta \mid \varphi_a\rangle \qquad (111)$$

Using Eqs. (67), (69), and (87), we may rewrite Eq. (111) as follows:

$$V_{ba} = R_{ba} + \sum_c \frac{R_{bc}\overline{R_{ac}}}{E_c - E_a - i\epsilon} + \sum_\beta (E_\beta - E_b)\,\langle\varphi_b \mid \psi_\beta\rangle\,\langle\psi_\beta \mid \varphi_a\rangle \qquad (112)$$

Similarly one deduces that

$$\overline{V_{ab}} = \overline{R_{ab}} + \sum_c \frac{\overline{R_{ac}}R_{bc}}{E_c - E_b + i\epsilon} + \sum_\beta (E_\beta - E_a)\,\langle\psi_\beta \mid \varphi_a\rangle\,\langle\varphi_b \mid \psi_\beta\rangle \qquad (113)$$

If V is Hermitian, $V = V^*$, then $\overline{V_{ab}} = V_{ba}$, so that we derive:

$$R_{ba} - \overline{R_{ab}} = -\sum_c \left(\frac{R_{bc}\overline{R_{ac}}}{E_c - E_a - i\epsilon} - \frac{\overline{R_{ac}}R_{bc}}{E_c - E_b + i\epsilon}\right)$$
$$+ \sum_\beta (E_b - E_a)\,\langle\varphi_b \mid \psi_\beta\rangle\,\langle\psi_\beta \mid \varphi_a\rangle \qquad (114)$$

If $E_b = E_a$, Eq. (114) reduces to Eq. (110). Equation (112) has become known as the Low equation [Low (1955)].

We conclude this section by noting the integral equation governing the time evolution of a bound state wave function. We first of all note that $U(t, 0)$ satisfies the equation:

$$i\partial_t U(t, 0) = V(t)\, U(t, 0) \tag{115}$$

which can be integrated to read

$$U(t, 0) = \Omega^{(+)*} - i \int_{-\infty}^{t} dt'\, V(t')\, U(t', 0) \tag{116}$$

since $U(-\infty, 0) = \Omega^{(+)*}$. If $|\psi_\beta\rangle$ is a bound state vector, call

$$|\psi_\beta(t)\rangle = U(t, 0) \mid \psi_\beta\rangle \tag{117}$$

whence, recalling that $\Omega^{(+)*} \mid \psi_\beta\rangle = 0$ [Eq. (106)], we find that $|\psi_\beta(t)\rangle$ satisfies the following homogeneous integral equation:

$$|\psi_\beta(t)\rangle = -i \int_{-\infty}^{t} dt'\, V(t') \mid \psi_\beta(t')\rangle \tag{118}$$

Equation (118) is to be contrasted with the inhomogeneous integral Eq. (77) obeyed by the scattering solutions.

11e. The Reactance Matrix

In addition to ingoing and outgoing solutions $|\psi_a{}^{\pm}\rangle$, we can define standing wave solutions $|\psi_a{}^{1}\rangle$ which satisfy the equation:

$$|\psi_a{}^{1}\rangle = |\varphi_a\rangle + \mathrm{P}\frac{1}{E_a - H_0} V \mid \psi_a{}^{1}\rangle \tag{119}$$

where P denotes the principal value. We next introduce the operators $R^{(\pm)}$ and K by:

$$R^{(\pm)}{}_{ba} = (\varphi_b, R^{(\pm)}\varphi_a) = (\varphi_b, V\psi_a{}^{\pm}) \tag{120}$$

$$K_{ba} = (\varphi_b, K\, \varphi_a) = (\varphi_b, V\psi_a{}^{1}) \tag{121}$$

Note that we have previously denoted $R^{(+)}{}_{ba}$ by R_{ba}. We also define the wave operator $\Omega^{(1)}$ corresponding to standing wave solutions by

$$|\psi_a{}^{1}\rangle = \Omega^{(1)} \mid \varphi_a\rangle \tag{122}$$

Thus $\Omega^{(1)}$ has the property that operating on the inhomogeneous term in Eq. (119) it will yield the solution to this equation. Using the formal relation,

$$\frac{1}{x \pm i\epsilon} = \mathrm{P}\frac{1}{x} \mp i\pi\delta(x) \tag{123}$$

we derive from the Lippmann-Schwinger equations the fact that

$$|\psi_a{}^{\pm}\rangle = |\varphi_a\rangle + \mathrm{P}\frac{1}{E_a - H_0} V \mid \psi_a{}^{\pm}\rangle \mp i\pi\delta(E_a - H_0)\, V \mid \psi_a{}^{\pm}\rangle \tag{124}$$

Using the completeness of the states $|\varphi_a\rangle$, we note that

$$\begin{aligned} V \mid \psi_a{}^{\pm}\rangle &= \sum_b |\varphi_b\rangle \langle\varphi_b \mid V \mid \psi_a{}^{\pm}\rangle \\ &= \sum_b R^{(\pm)}{}_{ba} \mid \varphi_b\rangle \end{aligned} \tag{125}$$

so that (124) can be rewritten in the form

$$|\psi_a^{\pm}\rangle = \{|\varphi_a\rangle \mp i\pi \sum_b \delta(E_a - E_b)\, R^{(\pm)}{}_{ba} \mid \varphi_b\rangle\} + \mathrm{P}\frac{1}{E_a - H_0} V \mid \psi_a^{\pm}\rangle \tag{126}$$

Equation (126) is of the same form as (119) except that its inhomogeneous term is the bracketed quantity. Recalling that $\Omega^{(1)}$ acts as a Green's function for the integral equation (119), we write

$$|\psi_a^{\pm}\rangle = \Omega^{(1)}\ \{|\varphi_a\rangle \mp i\pi \sum_b \delta(E_a - E_b)\ R^{(\pm)}{}_{ba} \mid \varphi_b\rangle\}$$

$$= |\psi_a^{1}\rangle \mp i\pi \sum_b \delta(E_a - E_b)\, R^{(\pm)}{}_{ba} \mid \psi_b^{1}\rangle \tag{127}$$

In an analogous fashion one verifies that

$$|\psi_a^{1}\rangle = |\psi_a^{\pm}\rangle \mp i\pi \sum_b \delta(E_a - E_b)\, K_{ba} \mid \psi_a^{\pm}\rangle \tag{128}$$

Upon multiplying Eq. (127) by V and taking the scalar product with $\langle\varphi_c|$, we obtain the following relation between the R and K matrices:

$$R^{(\pm)}{}_{ba} = K_{ba} \mp i\pi \sum_c \delta(E_a - E_c)\, R^{(\pm)}{}_{ca} V_{bc} \tag{129}$$

If we define the energy shell operators $\mathbf{R}^{(\pm)}$ and $\mathbf{K}$ by

$$\mathbf{R}^{(\pm)}{}_{ba} = 2\pi\delta(E_b - E_a)\, R^{(\pm)}{}_{ba} \tag{130}$$

$$\mathbf{K}_{ba} = 2\pi\delta(E_b - E_a)\, K_{ba} \tag{131}$$

(where $\mathbf{K}$ is called the reactance matrix, and $\mathbf{R}$ the reaction matrix), Eq. (129) on the energy shell then becomes

$$\mathbf{R}^{(\pm)} = \mathbf{K} \mp \frac{i}{2}\mathbf{K}\mathbf{R}^{(\pm)} \tag{132}$$

Equation (132) is known as Heitler's equation. Note that the matrices $R^{(\pm)}$ and K have the dimension of energy (or inverse length) while $\mathbf{R}^{(\pm)}$ and $\mathbf{K}$, like S are dimensionless. The S matrix can now be defined by the relation

$$\mathbf{R}^{(+)} = S\mathbf{R}^{(-)} \tag{133}$$

or equivalently,

$$S = 1 - i\mathbf{R}^{(+)} = (1 + i\mathbf{R}^{(-)})^{-1} \tag{134a}$$

$$= \frac{1 - \frac{i}{2}\mathbf{K}}{1 + \frac{i}{2}\mathbf{K}} \tag{134b}$$

Since S is unitary it follows that $\mathbf{K}$ is Hermitian. From the orthogonality properties of the set $\{\psi_c^{\pm}\}$,

$$(\psi_c{}^{\pm}, \psi_b{}^{\pm}) = \delta(c - b) \tag{135}$$

upon taking the scalar product of Eq. (128) with $|\psi_a{}^{\pm}\rangle$, it follows that

$$(\psi_a{}^{\pm}, \psi_b{}^1) = \left(1 \pm \frac{i}{2}\mathsf{K}\right)_{ab} \tag{136}$$

so that $|\psi_a{}^{\pm}\rangle$ and $|\psi_b{}^{(1)}\rangle$ are orthogonal, unless the states a and b have the same energy. Note also that

$$(\psi_a{}^-, \psi_b{}^1) = (\psi_a{}^1, \psi_b{}^+) \tag{137}$$

The elements of the S matrix are given by

$$\begin{aligned}(\psi_a{}^-, \psi_b{}^+) &= (\psi_a{}^-, \psi_b{}^1) - i\pi \sum_c \delta(E_a - E_b)\, \mathsf{R}^{(+)}{}_{ca}(\psi_c{}^1, \psi_b{}^+) \\ &= \left(1 - \frac{i}{2}\mathsf{K}\right)_{ab} - \frac{i}{2}\left(\mathsf{R}^{(+)} - \frac{i}{2}\mathsf{K}\mathsf{R}^{(+)}\right)_{ab} \\ &= (1 - i\mathsf{R}^{(+)})_{ab} = S_{ab}\end{aligned} \tag{138}$$

verifying the result obtained by time-dependent methods. The normalization of the standing wave solutions is obtained using Eq. (136) and the completeness of the set $\{\psi_c{}^{(+)}\} + \{\psi_\beta\}$:

$$\begin{aligned}(\psi_a{}^1, \psi_b{}^1) &= \sum_c (\psi_a{}^1, \psi_c{}^+)\,(\psi_c{}^+, \psi_b{}^1) + \sum_\beta (\psi_a{}^1, \psi_\beta)\,(\psi_\beta, \psi_b{}^1) \\ &= \sum_c \left(1 - \frac{i}{2}\mathsf{K}\right)_{ac}\left(1 + \frac{i}{2}\mathsf{K}\right)_{cb} \\ &= (1 + \tfrac{1}{4}\mathsf{K}^2)_{ab}\end{aligned} \tag{139}$$

where we have used the orthogonality of the states $|\psi_a{}^1\rangle$ and $|\psi_\beta\rangle$ since they both are eigenfunctions of H corresponding to different eigenvalues $(E_a \neq E_\beta)$.

11f. The *U* Matrix

We have seen that it is advantageous to formulate scattering problems in the Dirac picture, since in that picture the scattering matrix is simply related to the time translation operator $U(t, t_0)$ by the formal expression

$$S = U(\infty, -\infty) \tag{140}$$

The U matrix obeys the integral equation:

$$U(t, t_0) = 1 - \frac{i}{\hbar}\int_{t_0}^{t} H_I(t')\, U(t', t_0)\, dt' \tag{141}$$

which is equivalent to the differential equation:

$$i\partial_t U(t, t_0) = H_I(t)\, U(t, t_0) \tag{142}$$

and the boundary condition

$$U(t, t_0) = 1 \tag{143}$$

If the Dirac, Schrödinger, and Heisenberg pictures are made to coincide at time $t = 0$, $U(t, t_0)$ is given by

$$U(t, t_0) = e^{iH_0t}\, e^{-iH(t-t_0)}\, e^{-iH_0t_0} \tag{144}$$

which expression makes evident the unitarity of $U(t, t_0)$ for finite t and t_0. More generally an operator $U(t, t_0; \tau)$ can be defined, where τ is the time at which the Dirac picture coincides with the Schrödinger one. We shall, however, have no occasion to use this more general definition.

The group property

$$U(t_1, t_2)\, U(t_2, t_0) = U(t_1, t_0) \tag{145}$$

permits us to express a finite transformation as a product of infinitesimal ones ordered along any path

$$U(t, t_0) = U(t, t_n)\, U(t_n, t_{n-1}) \cdots U(t_2, t_1)\, U(t_1, t_0) \tag{146}$$

where $U(t_{j+1}, t_j)$ is an infinitesimal transformation from t_j to t_{j+1}. Now the integral Eq. (141) for the case that the step from t_j to t_{j+1} is infinitesimal can be approximated by:

$$\begin{aligned} U(t_{j+1}, t_j) &= 1 - \frac{i}{\hbar} \int_{t_j}^{t_{j+1}} dt'\, H_I(t')\, U(t', t_j) \\ &\approx 1 - \frac{i}{\hbar} \int_{t_j}^{t_{j+1}} dt'\, H_I(t')\, U(t_j, t_j) \\ &\approx 1 - \frac{i}{\hbar} \int_{t_j}^{t_{j+1}} dt'\, H_I(t') \end{aligned} \tag{147}$$

By letting the number of steps from t_0 to t go to infinity, we obtain

$$\begin{aligned} U(t, t_0) = 1 &+ \left(-\frac{i}{\hbar}\right) \int_{t_0}^{t} dt_1\, H_I(t_1) \\ &+ \left(-\frac{i}{\hbar}\right)^2 \int_{t_0}^{t} dt_1 \int_{t_0}^{t_1} dt_2\, H_I(t_1)\, H_I(t_2) \\ &+ \left(-\frac{i}{\hbar}\right)^3 \int_{t_0}^{t} dt_1 \int_{t_0}^{t_1} dt_2 \int_{t_0}^{t_2} dt_3\, H_I(t_1)\, H_I(t_2)\, H_I(t_3) \\ &+ \cdots \end{aligned} \tag{148}$$

This formula can also be derived by successive iterations of Eq. (141); i.e., Eq. (148) is the Neumann-Liouville expansion of the integral (141). This formal expansion is known to converge for ordinary scattering problems when H_I is a bounded operator. In field theoretical applications the convergence has to be investigated for each theory separately.

Consider now the nth integral

$$I_n = \int_{t_0}^{t} dt_1 \int_{t_0}^{t_1} dt_2 \cdots \int_{t_0}^{t_{n-1}} dt_n\, H_I(t_1)\, H_I(t_2) \cdots H_I(t_n) \qquad (149)$$

Dyson (1949a, b) observed that it is essentially an integral over the whole time interval t_0 to t with the restriction that t_j be earlier than t_{j-1} $(j \leqslant n)$. In other words, we may write I_n as

$$I_n = \int_{t_0}^{t} dt_1 \int_{t_0}^{t} dt_2 \cdots \int_{t_0}^{t} dt_n\, \theta(t_1 - t_2)\, \theta(t_2 - t_3) \cdots \theta(t_{n-1} - t_n) \cdot H_I(t_1)\, H_I(t_2) \cdots H_I(t_n) \qquad (150)$$

where $\theta(t)$ is the step function with $\theta(t) = +1$ if $t > 0$ and $\theta(t) = 0$ if $t < 0$. Using the fact that if it is permissible to interchange the order of integration

$$\int_{t_0}^{t} dt_1 \cdots \int_{t_0}^{t} dt_n\, f(t_1, \cdots t_n) = \frac{1}{n!} \sum_P \int_{t_0}^{t} dt_1 \cdots \int_{t_0}^{t} dt_n\, f(t_{\alpha_1}, \cdots t_{\alpha_n}) \qquad (151)$$

where $\sum_P$ denotes the summation over all permutations of $(t_1, \cdots t_n)$, we can recast I_n in the form:

$$\begin{aligned} I_n &= \frac{1}{n!} \sum_P \int_{t_0}^{t} dt_1 \int_{t_0}^{t} dt_2 \cdots \int_{t_0}^{t} dt_n\, \theta(t_{\alpha_1} - t_{\alpha_2}) \cdots \theta(t_{\alpha_{n-1}} - t_{\alpha_n}) \cdot H_I(t_{\alpha_1}) \cdots H(t_{\alpha_n}) \\ &= \frac{1}{n!} \int_{t_0}^{t} dt_1 \int_{t_0}^{t} dt_2 \cdots \int_{t_0}^{t} dt_n\, P(H_I(t_1) \cdots H_I(t_n)) \end{aligned} \qquad (152)$$

The operator P is defined by

$$P(H_I(t_1) \cdots H_I(t_n)) = \sum_P \theta(t_{\alpha_1} - t_{\alpha_2}) \cdots \theta(t_{\alpha_{n-1}} - t_{\alpha_n})\, H_I(t_{\alpha_1}) \cdots H_I(t_{\alpha_n}) \qquad (153a)$$

and has the property that, operating on a product of time-labeled operators, it rearranges them in the same order as the time sequence of their label, the latest one in time occurring first in the product:

$$P(H_I(t_1) \cdots H_I(t_n)) = H_I(t_i) \cdots H_I(t_j) \cdots H_I(t_k) \qquad t_i > \cdots t_j \cdots > t_k \qquad (153b)$$

For equal times the P product as defined by (153a) is ambiguous since $\theta(0)$ is not defined. However, since $H_I(t)$ clearly commutes with itself, we write

$$P(H_I(t)\, H_I(t)) = H_I(t)\, H_I(t) \qquad (154)$$

It is instructive to consider the equivalence of the left- and right-hand sides of Eq. (152) in detail for the case $n = 2$:

$$\int_{t_0}^{t} dt_1 \int_{t_0}^{t} dt_2\, P(H_I(t_1)\, H_I(t_2)) = \int_{t_0}^{t} dt_1 \int_{t_0}^{t_1} dt_2\, H_I(t_1)\, H_I(t_2) + \int_{t_0}^{t} dt_1 \int_{t_1}^{t} dt_2\, H_I(t_2)\, H_I(t_1) \tag{155}$$

The region of integration for the left-hand side of Eq. (155) is the whole square in Fig. 11.1. On the other hand, the first integral on the right-hand

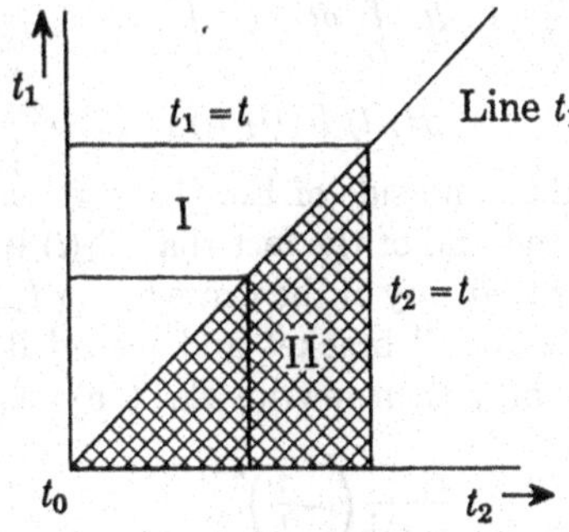

Fig. 11.1

side is to be integrated over triangle I (white area), whereas the second integral is to be integrated over the cross-hatched triangle II. Although we are not allowed to commute the operator factors $H_I(t_2)\, H_I(t_1)$, on the assumption that it is permissible to invert the order of integration, upon integrating first with respect to t_1 the limits of integration in the second integral become

$$\int_{t_0}^{t} dt_2 \int_{t_0}^{t_2} dt_1\, H_I(t_2)\, H_I(t_1) \tag{156}$$

If we now relabel the variables so that $t_1 \to t_2$ and $t_2 \to t_1$, we may write the expression (156) as

$$\int_{t_0}^{t} dt_1 \int_{t_0}^{t_1} dt_2\, H_I(t_1)\, H_I(t_2) \tag{157}$$

so that

$$\int_{t_0}^{t} dt_1 \int_{t_0}^{t} dt_2\, P(H_I(t_1)\, H_I(t_2)) = 2! \int_{t_0}^{t} dt_1 \int_{t_0}^{t_1} dt_2\, H_I(t_1)\, H_I(t_2) \tag{158}$$

The expansion (148) for $U(t, t_0)$ may therefore be rewritten as follows:

$$\begin{aligned} U(t, t_0) &= 1 + \frac{1}{1!}\left(-\frac{i}{\hbar}\right) \int_{t_0}^{t} dt_1\, P(H_I(t_1)) \\ &\quad + \frac{1}{2!}\left(-\frac{i}{\hbar}\right)^2 \int_{t_0}^{t} dt_1 \int_{t_0}^{t} dt_2\, P(H_I(t_1)\, H_I(t_2)) + \cdots \\ &= \sum_{n=0}^{\infty} \frac{1}{n!}\left(-\frac{i}{\hbar}\right)^n \int_{t_0}^{t} dt_1 \int_{t_0}^{t} dt_2 \cdots \int_{t_0}^{t} dt_n\, P(H_I(t_1)\, H_I(t_2) \cdots H_I(t_n)) \end{aligned} \tag{159}$$

or, upon formally summing the series,

$$U(t, t_0) = P\left(e^{-\frac{i}{\hbar}\int_{t_0}^{t} H_I(t')\, dt'}\right) \tag{160}$$

We can actually verify that (159) is a formal solution of Eq. (142). To this end we differentiate Eq. (159) with respect to t and obtain:

$$\frac{\partial U(t, t_0)}{\partial t} = \sum_{n=1}^{\infty} \frac{1}{n!} \left(-\frac{i}{\hbar}\right)^n \int_{t_0}^{t} dt_1 \int_{t_0}^{t} dt_2 \cdots \int_{t_0}^{t} dt_{n-1}$$
$$nH_I(t)\, P(H_I(t_1)\, H_I(t_2) \cdots H_I(t_{n-1})) \tag{161}$$

In writing down the right-hand side of Eq. (161) we have made use of the symmetry of the integrand and of the fact that $H_I(t)$ is an operator whose time label is always later than any of the times $t_1, \cdots t_{n-1}$. We have therefore taken $H_I(t)$ outside the P bracket and placed it to the left of the remaining factors. We may then rewrite Eq. (161) as follows:

$$i\hbar \frac{\partial U(t, t_0)}{\partial t} = H_I(t) \sum_{n=1}^{\infty} \frac{1}{(n-1)!} \left(-\frac{i}{\hbar}\right)^{n-1}$$
$$\cdot \int_{t_0}^{t} dt_1 \int_{t_0}^{t} dt_2 \cdots \int_{t_0}^{t} dt_{n-1}\, P(H_I(t_1)\, H_I(t_2) \cdots H_I(t_{n-1}))$$
$$= H_I(t) \sum_{n=0}^{\infty} \frac{1}{n!} \left(-\frac{i}{\hbar}\right)^n \int_{t_0}^{t} dt_1 \cdots \int_{t_0}^{t} dt_n\, P(H_I(t_1) \cdots H_I(t_n))$$
$$= H_I(t)\, U(t, t_0) \tag{162}$$

which proves that $U(t, t_0)$ is indeed a solution of (7).

In the application of the expansion (162) to scattering problems, to obtain the S matrix one must extend the initial time t_0 to $-\infty$ and the time t to $+\infty$. Care must, however, be exercised in going to the limit since for the nth order term there are $n!$ ways to go to the limit $t_0 \to -\infty$. Furthermore, unless wave packets are used for the description of the initial and final states, a prescription must be given for averaging out all terms depending periodically on t_0 and t. The simplest way to overcome these difficulties is to define an "adiabatic" U matrix by replacing $H_I(t)$ by $H_I(t) \exp(-\alpha|t|)$ and to require that the limit $\alpha \to 0$ is to be taken after the integrations have been performed. We regard this procedure as a mathematical convenience which allows us to deal with plane wave (non-normalizable) states. This convergence factor also removes the ambiguity from the orders in which the limits $t \to \pm\infty$ are performed. It is equivalent to averaging over the $n!$ orders of taking the limit.

The meaning of the introduction of the α factor is made clearer by a consideration of the second-order term in the expansion (148) for the S matrix. The second-order S-matrix element for a transition from a state $|a\rangle$ to a state $|b\rangle$ which are eigenstates of H_0, is given by

$$\langle b \mid S^{(2)} \mid a\rangle = \left(-\frac{i}{\hbar}\right)^2 \int_{-\infty}^{+\infty} dt_1 \int_{-\infty}^{t_1} dt_2 \, \langle b \mid H_I(t_1)\, H_I(t_2) \mid a\rangle \tag{163}$$

Making use of the definition

$$H_I(t) = e^{iH_0t/\hbar}\, H_{IS}\, e^{-iH_0t/\hbar} \tag{164}$$

where H_{IS} is a Schrödinger operator, upon introducing the complete set of eigenstates of H_0 between $H_I(t_1)$ and $H_I(t_2)$ we obtain:

$$\langle b \mid S^{(2)} \mid a\rangle = \left(-\frac{i}{\hbar}\right)^2 \sum_n \int_{-\infty}^{+\infty} dt_1 \int_{-\infty}^{t_1} dt_2 \; e^{i(E_b-E_n)t_1/\hbar}\, e^{i(E_n-E_a)t_2/\hbar}\, \langle b \mid H_{IS} \mid n\rangle \langle n \mid H_{IS} \mid a\rangle \tag{165}$$

The integration over t_2 is not defined at the lower limit if the state $|a\rangle$ is a plane wave state. For a wave-packet state $|\Phi_i\rangle = \int da\, c_a \mid a\rangle$, $\int da\, |c_a|^2 < \infty$, the contribution at the lower limit $t_2 = -\infty$ will vanish by virtue of the Riemann-Lebesgue lemma. Equivalently, if we replace $H_I(t)$ by $H_I(t) \exp(-\alpha|t|)$ we obtain:

$$\langle b \mid S^{(2)} \mid a\rangle = -2\pi i\delta(E_a - E_b) \sum_n \frac{\langle b \mid H_{IS} \mid n\rangle \langle n \mid H_{IS} \mid b\rangle}{E_a - E_n + i\alpha} \tag{166}$$

In general, for $b \neq a$, recalling Eq. (94):

$$\begin{aligned} \langle b \mid S \mid a\rangle = -2\pi i\delta(E_a - E_b) \Bigg\{ &\langle b \mid H_{IS} \mid a\rangle + \sum_n \frac{\langle b \mid H_{IS} \mid n\rangle \langle n \mid H_{IS} \mid a\rangle}{E_a - E_n + i\alpha} \\ &+ \sum_{n_1, n_2, \cdots n_f} \frac{\langle b \mid H_{IS} \mid n_1\rangle \langle n_1 \mid H_{IS} \mid n_2\rangle \cdots \langle n_f \mid H_{IS} \mid a\rangle}{(E_a - E_{n_1} + i\alpha)(E_a - E_{n_2} + i\alpha) \cdots (E_a - E_{n_f} + i\alpha)} \\ &+ \cdots \Bigg\} \end{aligned} \tag{167a}$$

the familiar result of "old-fashioned" perturbation theory. Equation (167a) can also be rewritten as follows:

$$\begin{aligned} \langle b \mid S \mid a\rangle &= -2\pi i\delta(E_a - E_b) \left\{ \langle b \mid H_{IS} + H_{IS} \frac{1}{E_a - H_0 + i\alpha} H_{IS} + \cdots \mid a\rangle \right\} \\ &= -2\pi i\delta(E_a - E_b)\, R_{ba} \end{aligned} \tag{167b}$$

which is Eq. (86) for the case $b \neq a$.

We again remind the reader that the formalism just outlined is not applicable to the quantized field theoretical situation since in that case the interaction H_I also causes a level shift, so that the unperturbed energy level, E_a, under the influence of H_I goes over into a level $E_a + \Delta E_a$ (with ΔE_a often infinite!). The application of (159) in that case will be discussed in detail in Chapter 12.

As an introduction to these formal methods when H_I is responsible for a level shift, we here derive a useful relation between unperturbed and perturbed eigenstates which lie in the discrete spectrum [Gell-Mann and Low

(Gell-Mann 1951) and Sucher (1957)]; for simplicity we shall assume them to be nondegenerate. Thus let $|\beta_0\rangle$ be an eigenstate of H_0 with eigenvalue ϵ

$$H_0 \mid \beta_0\rangle = \epsilon \mid \beta_0\rangle \tag{168}$$

and let $|\beta\rangle$ be an eigenstate of $H = H_0 + \lambda H_I(0)$ with eigenvalue E

$$H \mid \beta\rangle = E \mid \beta\rangle \tag{169}$$

where we have explicitly exhibited the dependence of the interaction Hamiltonian on the coupling constant, λ. Under the assumption that

$$\lim_{\lambda\to 0} \mid \beta\rangle = |\beta_0\rangle \tag{170}$$

$$\lim_{\lambda\to 0} E = \epsilon \tag{171}$$

we shall show that

$$|\beta\rangle = \lim_{\alpha\to 0} \frac{U_\alpha(0, \pm\infty) \mid \beta_0\rangle}{\langle\beta_0 \mid U_\alpha(0, \pm\infty) \mid \beta_0\rangle} \tag{172}$$

and that ΔE is given by the following expression:

$$\Delta E = E - \epsilon = \lim_{\alpha\to 0} \frac{\frac{1}{2}i\hbar\alpha\lambda \frac{\partial}{\partial\lambda} \langle\beta_0 \mid U_\alpha(\infty, -\infty) \mid \beta_0\rangle}{\langle\beta_0 \mid S_\alpha \mid \beta_0\rangle} \tag{173}$$

In (172) and (173) U_α is the "adiabatic" U matrix which satisfies the differential equation:

$$i\partial_t U_\alpha(t, t_0) = \lambda e^{-\alpha|t|} H_I(t)\, U_\alpha(t, t_0)$$

$$\alpha > 0 \tag{174}$$

Proof: Call

$$U_\alpha(0, -\infty) \mid \beta_0\rangle = |\beta_0; \alpha\rangle \tag{175}$$

Now by virtue of Eqs. (168) and (175)

$$[H_0, U_\alpha(0, -\infty)] \mid \beta_0\rangle = H_0 \mid \beta_0; \alpha\rangle - \epsilon \mid \beta_0; \alpha\rangle \tag{176}$$

On the other hand,

$$\begin{aligned}[H_0, U_\alpha(0, -\infty)] &= \Big[H_0, \sum_n \int_{-\infty}^0 dt_1 \cdots \int_{-\infty}^0 dt_n\, e^{\alpha(t_1+t_2+\cdots+t_n)} \\ &\quad \cdot \lambda^n \left(-\frac{i}{\hbar}\right)^n \frac{1}{n!} P(H_I(t_1) \cdots H_I(t_n))\Big] \\ &= -i\hbar \sum_{n=0}^{\infty} \left(-\frac{i}{\hbar}\right)^n \frac{\lambda^n}{n!} \int_{-\infty}^0 dt_1 \cdots \int_{-\infty}^0 dt_n\, e^{\alpha(t_1+t_2+\cdots+t_n)} \\ &\quad \sum_{l=1}^{n} \frac{\partial}{\partial t_l} P(H_I(t_1) \cdots H_I(t_n))\end{aligned} \tag{177}$$

the last line of Eq. (177) being a consequence of the fact that in the Dirac picture H_0 is the time displacement operator; i.e., for any operator $F(t)$ in the Dirac picture

$$\frac{i}{\hbar}[H_0, F(t)] = \frac{\partial F(t)}{\partial t} \tag{178}$$

Since the integrand in (177) is symmetric, $\sum_{i=1}^{n} \partial_{t_i} = n\partial_{t_1}$ and we may rewrite Eq. (177) as follows:

$$\begin{aligned}(H_0 - \epsilon) \mid \beta_0; \alpha\rangle &= \left\{-\sum_{n=1}^{\infty} \frac{(-i/\hbar)^{n-1}}{(n-1)!} \lambda^n \int_{-\infty}^{0} dt_1 \cdots \int_{-\infty}^{0} dt_n \right. \\ &\quad \left. e^{\alpha(t_1+t_2+\cdots+t_n)} \frac{\partial}{\partial t_1} P(H_I(t_1) \cdots H_I(t_n))\right\} \mid \beta_0\rangle \\ &= \left\{-\sum_{n=1}^{\infty} \frac{(-i/\hbar)^{n-1}}{(n-1)!} \lambda^n \int_{-\infty}^{0} dt_1 \int_{-\infty}^{0} dt_2 \cdots \int_{-\infty}^{0} dt_n \right. \\ &\quad \left[\frac{\partial}{\partial t_1} (e^{\alpha(t_1+t_2+\cdots+t_n)} P(H_I(t_1) \cdots H_I(t_n))) \right. \\ &\quad \left.\left. - \alpha e^{\alpha(t_1+t_2+\cdots t_n)} P(H_I(t_1) \cdots H_I(t_n))\right]\right\} \mid \beta_0\rangle \\ &= -\lambda H_I(0)\, U_\alpha(0, -\infty) \mid \beta_0\rangle \\ &\quad + i\alpha\hbar\lambda \frac{\partial}{\partial \lambda} U_\alpha(0, -\infty) \mid \beta_0\rangle \end{aligned} \tag{179}$$

where the second term on the right-hand side has been obtained on the assumption that the interaction term in $H(0)$ depends on λ only linearly in the form indicated by $H = H_0 + \lambda H_I(0)$. We have therefore shown that

$$(H_0 + \lambda H_I(0) - \epsilon) \mid \beta_0; \alpha\rangle = i\hbar\alpha\lambda \frac{\partial}{\partial \lambda} \mid \beta_0; \alpha\rangle \tag{180}$$

Similarly one verifies that

$$(H - \epsilon)\, U_\alpha(0, +\infty) \mid \beta_0\rangle = -i\hbar\alpha \frac{\partial}{\partial \lambda} U_\alpha(0, +\infty) \mid \beta_0\rangle \tag{181}$$

Taking the scalar product of Eq. (180) with $\langle\beta|$ and using the fact that $\langle\beta \mid H = E \langle\beta|$, we obtain the result that the level shift is given by

$$\begin{aligned}\Delta E &= E - \epsilon \\ &= i\hbar\alpha\lambda \frac{\langle\beta \mid \frac{\partial}{\partial\lambda} U_\alpha(0, -\infty) \mid \beta_0\rangle}{\langle\beta \mid U_\alpha(0, -\infty) \mid \beta_0\rangle}\end{aligned} \tag{182}$$

On the other hand, using (180), we note that

$$\begin{aligned}&\left\{H - \epsilon_0 - i\hbar\alpha\lambda \frac{\partial}{\partial\lambda}\right) \frac{U_\alpha(0, -\infty) \mid \beta_0\rangle}{\langle\beta_0 \mid U_\alpha(0, -\infty) \mid \beta_0\rangle} \\ &\quad = i\hbar\alpha\lambda \left(\frac{\partial}{\partial\lambda} \log \langle\beta_0 \mid U_\alpha(0, -\infty) \mid \beta_0\rangle\right) \frac{U_\alpha(0, -\infty) \mid \beta_0\rangle}{\langle\beta_0 \mid U_\alpha(0, -\infty) \mid \beta_0\rangle}\end{aligned} \tag{183}$$

so that if we call

$$|\beta'; \alpha\rangle = \frac{U_\alpha(0, -\infty) \mid \beta_0\rangle}{\langle\beta_0 \mid U_\alpha(0, -\infty) \mid \beta_0\rangle} \tag{184}$$

then

$$\langle\beta_0 \mid H - \epsilon_0 \mid \beta'; \alpha\rangle = i\hbar\alpha\lambda \frac{\partial}{\partial\lambda} (\log \langle\beta_0 \mid U_\alpha(0, -\infty) \mid \beta_0\rangle \tag{185}$$

On the assumption that $\lim_{\alpha\to 0} \mid \beta'; \alpha\rangle$ exists and is well defined with

$$|\beta'\rangle = \lim_{\alpha\to 0} \mid \beta'; \alpha\rangle \tag{186}$$

upon taking the limit $\alpha \to 0$ of Eq. (183) we obtain

$$\begin{aligned}(H - \epsilon_0) \mid \beta'\rangle &= \lim_{\alpha\to 0} \left(i\hbar\alpha\lambda \frac{\partial}{\partial\lambda} \log \langle\beta_0 \mid U_\alpha(0, -\infty) \mid \beta_0\rangle \right) \mid \beta'\rangle \\ &= \langle\beta_0 \mid H - \epsilon_0 \mid \beta'\rangle \mid \beta'\rangle \end{aligned} \tag{187}$$

so that $|\beta'\rangle$ is an eigenfunction of H which as $\lambda \to 0$ goes over into $|\beta_0\rangle$. Therefore $|\beta'\rangle = |\beta\rangle$. We could have carried out a similar calculation for $U(0, +\infty)$, and we would have obtained the result that in general

$$|\beta\rangle = \lim_{\alpha\to 0} \frac{U_\alpha(0, \pm\infty) \mid \beta_0\rangle}{\langle\beta_0 \mid U_\alpha(0, \pm\infty) \mid \beta_0\rangle} \tag{188}$$

It should be stressed that $U(0, +\infty) \mid \beta_0\rangle$ and $U(0, -\infty) \mid \beta_0\rangle$ converge to the same state only in the discrete spectrum. In fact, we have seen that in the continuous spectrum $U(0, \pm\infty) \mid \beta_0\rangle$ converges to $|\psi_\beta{}^{\pm}\rangle$, the in- and out-state solutions, respectively, which are different eigenstates of the total Hamiltonian.

Combining the representation (188) for $\langle\beta|$ and the expression (182) for the level shift, we obtain

$$\begin{aligned}\Delta E &= \lim_{\alpha\to 0} i\hbar\alpha\lambda \frac{\langle\beta_0 \mid U_\alpha{}^{-1}(0, +\infty) \frac{\partial}{\partial\lambda} U_\alpha(0, -\infty) \mid \beta_0\rangle}{\langle\beta_0 \mid S_\alpha \mid \beta_0\rangle} \\ &= \lim_{\alpha\to 0} i\hbar\alpha\lambda \frac{\langle\beta_0 \mid \frac{\partial}{\partial\lambda} U_\alpha(\infty, 0) \cdot U_\alpha(0, -\infty) \mid \beta_0\rangle}{\langle\beta_0 \mid S_\alpha \mid \beta_0\rangle} \\ &= \lim_{\alpha\to 0} \frac{1}{2} i\hbar\alpha\lambda \frac{\frac{\partial}{\partial\lambda} \langle\beta_0 \mid U_\alpha(\infty, -\infty) \mid \beta_0\rangle}{\langle\beta_0 \mid S_\alpha \mid \beta_0\rangle} \end{aligned} \tag{189}$$

Q.E.D.

12

Simple Field Theoretic Models

In the present chapter we shall investigate three systems of interacting fields: (1) the scalar field, interacting with a (nonrecoiling) nucleon, (2) the Lee model, and (3) the Chew model of meson-nucleon interaction. The first two examples have the property that they allow exact solutions for many of the important states of the systems. The third example, the Chew model, although not soluble, has the property that it accounts for many of the important features of low energy meson-nucleon scattering. The analysis of these simple models will give us an insight into the content as well as into the problematic aspects of quantized field theories.

12a. The Scalar Field

The simplest soluble model of a system of interacting fields consists of a neutral scalar field ("mesons") interacting with spinless fermions ("nucleons") whose energy is taken to be independent of momentum. The Hamiltonian describing this field system is:

$$H = H_0 + H_I \tag{1a}$$

$$H_0 = m_0 \int d\mathbf{p}\, \psi^*(\mathbf{p})\, \psi(\mathbf{p}) + \int d\mathbf{k}\, \omega(\mathbf{k})\, a^*(\mathbf{k})\, a(\mathbf{k})$$

$$\omega(\mathbf{k}) = \sqrt{\mathbf{k}^2 + \mu^2} \tag{1b}$$

$$H_I = \frac{\lambda}{(2\pi)^{3/2}} \int d\mathbf{p} \int \frac{d\mathbf{k}\, f(\mathbf{k}^2)}{\sqrt{2\omega(\mathbf{k})}}\, \psi^*(\mathbf{p}+\mathbf{k})\, \psi(\mathbf{p})\, (a(\mathbf{k}) + a^*(-\mathbf{k})) \tag{1c}$$

where $\psi(\mathbf{p})$, $\psi^*(\mathbf{p})$ and $a(\mathbf{k})$, $a^*(\mathbf{k})$ are the destruction and creation operators for the nucleons and mesons, respectively. They satisfy the following commutation rules:

$$[\psi(\mathbf{p}), \psi^*(\mathbf{p}')]_+ = \delta^{(3)}(\mathbf{p} - \mathbf{p}') \tag{2a}$$

$$[\psi(\mathbf{p}), \psi(\mathbf{p}')]_+ = [\psi^*(\mathbf{p}), \psi^*(\mathbf{p}')]_+ = 0 \tag{2b}$$

$$[a(\mathbf{k}), a^*(\mathbf{k}')] = \delta^{(3)}(\mathbf{k} - \mathbf{k}') \tag{3a}$$

$$[a(\mathbf{k}), a(\mathbf{k}')] = [a^*(\mathbf{k}), a^*(\mathbf{k}')] = 0 \tag{3b}$$

$$[\psi(\mathbf{p}), a(\mathbf{k})] = [\psi(\mathbf{p}), a^*(\mathbf{k})] = [\psi^*(\mathbf{p}), a(\mathbf{k})] = [\psi^*(\mathbf{p}), a^*(\mathbf{k})] = 0 \tag{4}$$

The function $f(\mathbf{k}^2)$ describes the "extension" of the nucleon and plays the role of a cutoff function. It is assumed to fall off rapidly enough for large $|\mathbf{k}|$ to make finite all the integrals encountered in the theory. Since the energy of a nucleon does not depend on its momentum, the theory does not take into account recoil effects, although, due to the translational invariance of the Hamiltonian, momentum is conserved. Stated differently, the total momentum operator

$$\mathbf{P} = \int d\mathbf{p}\, \mathbf{p}\psi^*(\mathbf{p})\, \psi(\mathbf{p}) + \int d\mathbf{k}\, \mathbf{k}a^*(\mathbf{k})\, a(\mathbf{k}) \tag{5}$$

commutes with H. The operator $\mathbf{P}$ is the generator of spatial translations with

$$e^{i\mathbf{P}\cdot\mathbf{d}}\, \psi(\mathbf{q})\, e^{-i\mathbf{P}\cdot\mathbf{d}} = \psi(\mathbf{q})\, e^{-i\mathbf{q}\cdot\mathbf{d}} \tag{6a}$$

$$e^{i\mathbf{P}\cdot\mathbf{d}}\, \psi^*(\mathbf{q})\, e^{-i\mathbf{P}\cdot\mathbf{d}} = \psi^*(\mathbf{q})\, e^{+i\mathbf{q}\cdot\mathbf{d}} \tag{6b}$$

$$e^{i\mathbf{P}\cdot\mathbf{d}}\, a(\mathbf{k})\, e^{-i\mathbf{P}\cdot\mathbf{d}} = a(\mathbf{k})\, e^{-i\mathbf{k}\cdot\mathbf{d}} \tag{6c}$$

$$e^{i\mathbf{P}\cdot\mathbf{d}}\, a^*(\mathbf{k})\, e^{-i\mathbf{P}\cdot\mathbf{d}} = a^*(\mathbf{k})\, e^{+i\mathbf{k}\cdot\mathbf{d}} \tag{6d}$$

The translated Hamiltonian H' is equal to H, i.e.,

$$\begin{aligned} H' &= e^{iP\cdot d}\, H\, e^{-i\mathbf{P}\cdot\mathbf{d}} \\ &= H \end{aligned} \tag{7}$$

This is the proof of the translational invariance of H. From (7) it follows that $[H, P] = 0$, i.e., that momentum is conserved by the Hamiltonian (1).

Due to the fact that each term in H contains at least one destruction operator which can be made to stand to the right of all creation operators, the no-particle state $|0\rangle$ defined by

$$\psi(\mathbf{p}) \mid 0\rangle = a(\mathbf{k}) \mid 0\rangle = 0 \quad \text{for all } \mathbf{p} \text{ and all } \mathbf{k} \tag{8}$$

is an eigenstate of H with eigenvalue 0. Similarly, the one-meson state $a^*(\mathbf{k}) \mid 0\rangle$ is an eigenstate of H with eigenvalue $\omega_{\mathbf{k}}$. The state $\psi^*(\mathbf{p}) \mid 0\rangle$, on the other hand, is not an eigenstate of H. Let us denote by $|\Psi^{(1)}(\mathbf{p})\rangle$ the "physical" one-nucleon state. The vector $|\Psi^{(1)}(\mathbf{p})\rangle$ is an eigenstate of H, as well as an eigenstate of $\mathbf{P}$ with eigenvalue $\mathbf{p}$,

$$\mathbf{P} \mid \Psi^{(1)}(\mathbf{p})\rangle = \mathbf{p} \mid \Psi^{(1)}(\mathbf{p})\rangle \tag{9}$$

which in the limit as λ, the coupling constant, goes to zero goes over into the state $\psi^*(\mathbf{p}) \mid 0\rangle$. To obtain an explicit representation for this state, we note that the operator

$$N_\psi = \int d\mathbf{p}\, \psi^*(\mathbf{p})\, \psi(\mathbf{p}) \tag{10}$$

commutes with H and is therefore also a constant of the motion. It corresponds to the number of heavy "quanta" present. We use the appellation "quanta" to denote the particles associated with the "bare" fields described by the Hamiltonian H_0, in contradistinction to the "physical" particles, the description of which is given by eigenstates of H. If we expand $|\Psi^{(1)}(\mathbf{p})\rangle$ in terms of "bare" states, i.e., eigenfunctions of H_0, then only states with the same eigenvalue of N_ψ can occur:

$$|\Psi^{(1)}(\mathbf{p})\rangle = \sum_{n=0}^{\infty} \int d^3q \int d^3k_1 \cdots \int d^3k_n\, c_{\mathbf{p}}^{(n)}(\mathbf{q}; \mathbf{k}_1, \mathbf{k}_2, \cdots \mathbf{k}_n) \frac{1}{\sqrt{n!}} a^*(\mathbf{k}_1) \cdots a^*(\mathbf{k}_n)\, \psi^*(\mathbf{q}) \mid 0\rangle \tag{11}$$

The amplitude $c_{\mathbf{p}}^{(n)}(\mathbf{q}; \mathbf{k}_1, \cdots \mathbf{k}_n)$ is a symmetric function of $\mathbf{k}_1, \cdots, \mathbf{k}_n$ and is the probability amplitude for finding a heavy quantum of momentum $\mathbf{q}$ together with n mesons of momenta $\mathbf{k}_1, \mathbf{k}_2, \cdots \mathbf{k}_n$:

$$c_{\mathbf{p}}^{(n)}(\mathbf{q}; \mathbf{k}_1, \mathbf{k}_2, \cdots \mathbf{k}_n) = \frac{1}{\sqrt{n!}} \langle 0 \mid a(\mathbf{k}_1) \cdots a(\mathbf{k}_n)\, \psi(\mathbf{q}) \mid \Psi^{(1)}(\mathbf{p})\rangle \tag{12}$$

Translation invariance allows us to separate off a δ-function factor from $c_{\mathbf{p}}^{(n)}(\mathbf{q}; \mathbf{k}_1, \mathbf{k}_2, \cdots \mathbf{k}_n)$ corresponding to conservation of momentum. *Proof:* The unitary operator

$$U(\mathbf{d}) = e^{i\mathbf{P}\cdot\mathbf{d}} \tag{13}$$

corresponding to a translation through $\mathbf{d}$, when inserted in (12) yields:

$$\begin{aligned} &\langle 0 \mid a(\mathbf{k}_1) \cdots a(\mathbf{k}_n)\, \psi(\mathbf{q}) \mid \Psi^{(1)}(\mathbf{p})\rangle \\ &\quad = \langle 0 \mid U^*(\mathbf{d})\, U(\mathbf{d})\, a(\mathbf{k}_1)\, U^*(\mathbf{d})\, U(\mathbf{d}) \cdots \psi(\mathbf{q})\, U^*(\mathbf{d})\, U(\mathbf{d}) \mid \Psi^{(1)}(\mathbf{p})\rangle \\ &\quad = e^{-i(\mathbf{k}_1+\mathbf{k}_2+\cdots+\mathbf{k}_n+\mathbf{q}-\mathbf{p})\cdot\mathbf{d}} \langle 0 \mid a(\mathbf{k}_1) \cdots a(\mathbf{k}_n)\, \psi(\mathbf{q}) \mid \Psi^{(1)}(\mathbf{p})\rangle \end{aligned} \tag{14}$$

Since Eq. (14) is valid for arbitrary $\mathbf{d}$, it follows that

$$\begin{aligned} &\langle 0 \mid a(\mathbf{k}_1)\, a(\mathbf{k}_2) \cdots a(\mathbf{k}_n)\, \psi(\mathbf{q}) \mid \Psi^{(1)}(\mathbf{p})\rangle \\ &\quad = \delta^{(3)}(\mathbf{p} - \mathbf{k}_1 - \mathbf{k}_2 - \cdots - \mathbf{k}_n - \mathbf{q})\, c'^{(n)}(\mathbf{q}; \mathbf{k}_1, \cdots \mathbf{k}_n) \end{aligned} \tag{15}$$

To determine the amplitudes $c^{(n)}$ or $c'^{(n)}$, we note that if $|\Psi^{(1)}(\mathbf{p})\rangle$ is to be an eigenstate of H with eigenvalue m_ψ

$$H \mid \Psi^{(1)}(\mathbf{p})\rangle = m_\psi \mid \Psi^{(1)}(\mathbf{p})\rangle \tag{16}$$

then the $c^{(n)}$s must satisfy certain equations which can be determined as follows: Consider the commutator $[a(\mathbf{k}_1)\, a(\mathbf{k}_2) \cdots a(\mathbf{k}_n)\, \psi(\mathbf{q}), H]$ evaluated between the states $|0\rangle$ and $|\Psi^{(1)}(\mathbf{p})\rangle$:

$$\langle 0 \mid [a(\mathbf{k}_1)\, a(\mathbf{k}_2) \cdots a(\mathbf{k}_n)\, \psi(\mathbf{q}), H_0 + H_I] \mid \Psi^{(1)}(\mathbf{p})\rangle$$
$$= \left(\sum_{i=1}^{n} \omega(\mathbf{k}_i) + m_0\right) \langle 0 \mid a(\mathbf{k}_1)\, a(\mathbf{k}_2) \cdots a(\mathbf{k}_n)\, \psi(\mathbf{q}) \mid \Psi^{(1)}(\mathbf{p})\rangle$$
$$+ \frac{\lambda}{(2\pi)^{3/2}} \int \frac{d^3k\, f(\mathbf{k}^2)}{\sqrt{2\omega(\mathbf{k})}} \langle 0 \mid a(\mathbf{k})\, a(\mathbf{k}_1) \cdots a(\mathbf{k}_n)\, \psi(\mathbf{q} - \mathbf{k}) \mid \Psi^{(1)}(\mathbf{p})\rangle$$
$$+ \frac{\lambda}{(2\pi)^{3/2}} \sum_{i=1}^{n} \frac{f(\mathbf{k}^2)}{\sqrt{2\omega(\mathbf{k}_i)}} \langle 0 \mid a(\mathbf{k}_1) \cdots a(\mathbf{k}_{i-1})\, a(\mathbf{k}_{i+1}) \cdots a(\mathbf{k}_n)$$
$$\cdot\, \psi(\mathbf{q} + \mathbf{k}_i) \mid \Psi^{(1)}(\mathbf{p})\rangle \qquad (17a)$$
$$= m_\psi \langle 0 \mid a(\mathbf{k}_1) \cdots a(\mathbf{k}_n)\, \psi(\mathbf{q}) \mid \Psi^{(1)}(\mathbf{p})\rangle \qquad (17b)$$

Equation (17), when expressed in terms of the amplitudes $c_{\mathbf{p}}^{(n)}(\mathbf{q}; \mathbf{k}_1, \cdots \mathbf{k}_n)$ reads

$$\left(\sum_{i=1}^{n} \omega(\mathbf{k}_i) + m_0\right) c_{\mathbf{p}}^{(n)}(\mathbf{q}; \mathbf{k}_1, \cdots \mathbf{k}_n)$$
$$+ \frac{\lambda}{(2\pi)^{3/2}} \sqrt{n+1} \int \frac{d^3k\, f(\mathbf{k}^2)}{\sqrt{2\omega(\mathbf{k})}} c_{\mathbf{p}}^{(n+1)}(\mathbf{q} - \mathbf{k}; \mathbf{k}, \mathbf{k}_1, \mathbf{k}_2, \cdots \mathbf{k}_n)$$
$$+ \frac{\lambda}{(2\pi)^{3/2}} \frac{1}{\sqrt{n}} \sum_{i=1}^{n} \frac{f(\mathbf{k}^2)}{\sqrt{2\omega(\mathbf{k}_i)}} c_{\mathbf{p}}^{(n-1)}(\mathbf{q} + \mathbf{k}_i; \mathbf{k}_1, \cdots \mathbf{k}_{i-1}, \mathbf{k}_{i+1} \cdots \mathbf{k}_n)$$
$$= m_\psi c_{\mathbf{p}}^{(n)}(\mathbf{q}; \mathbf{k}_1, \cdots \mathbf{k}_n) \qquad (18)$$

One then verifies that the ansatz

$$c_{\mathbf{p}}^{(n)}(\mathbf{q}; \mathbf{k}_1, \mathbf{k}_2, \cdots \mathbf{k}_n)$$
$$= \sqrt{Z}\, \delta^{(3)}\left(\mathbf{q} + \sum_{i=1}^{n} \mathbf{k}_i - \mathbf{p}\right) \frac{(-\lambda)^n}{\sqrt{n!}} \prod_{i=1}^{n} \frac{f(\mathbf{k}_i^2)}{\sqrt{2(2\pi)^3\, \omega^3(\mathbf{k}_i)}} \qquad (19)$$

solves this set of coupled equations if

$$m_\psi = m_0 - \lambda^2 \Delta = m_0 + \delta m$$
$$\Delta = \frac{1}{(2\pi)^3} \int d^3k \frac{|f(\mathbf{k}^2)|^2}{2\omega^2(\mathbf{k})} \qquad (20)$$

The quantity Z is a normalization constant whose value will be determined below. The one-nucleon state of momentum $\mathbf{p}$ can therefore be written as:

$$|\Psi^{(1)}(\mathbf{p})\rangle$$
$$= \sqrt{Z} \sum_{n=0}^{\infty} \frac{(-\lambda)^n}{n!} \int d^3q \int d^3k_1 \cdots \int d^3k_n\, \delta^{(3)}(\mathbf{p} - \sum_{i=1}^{n} \mathbf{k}_i - \mathbf{q})$$
$$\cdot \prod_{i=1}^{n} \frac{f(\mathbf{k}_i^2)}{\sqrt{2(2\pi)^3\, \omega^3(\mathbf{k}_i)}} a^*(\mathbf{k}_1) \cdots a^*(\mathbf{k}_n)\, \psi^*(\mathbf{q}) \mid 0\rangle \qquad (21a)$$
$$= \frac{\sqrt{Z}}{(2\pi)^3} \int d^3q \int d^3x\, e^{i(\mathbf{q}-\mathbf{p})\cdot \mathbf{x}}\, e^{\int \frac{d^3k f(\mathbf{k}^2)}{[2(2\pi)^3\omega^3(k)]^{1/2}} e^{i\mathbf{k}\cdot\mathbf{x}} a^*(\mathbf{k})} \psi^*(\mathbf{q}) \mid 0\rangle \qquad (21b)$$

Equation (21) expresses the one-particle state as a superposition of unperturbed states admixed to the "bare" one-quantum state. We will refer to these "attached" states as constituting the "cloud" which each heavy quantum acquires by virtue of the interaction H_I.

The constant Z is determined from the renormalization condition:

$$\langle \Psi^{(1)}(\mathbf{p}) \mid \Psi^{(1)}(\mathbf{p}')\rangle = \delta^{(3)}(\mathbf{p} - \mathbf{p}') \tag{22}$$

Substituting Eq. (21) into (22), we find

$$\begin{aligned} Z\Bigg\{\sum_{n=0}^{\infty} \int d^3q \int d^3k_1 \cdots \int d^3k_n\, \delta^{(3)}\Big(\mathbf{p} - \sum_{i=1}^{n} \mathbf{k}_i - \mathbf{q}\Big)\, \delta^{(3)}\Big(\mathbf{p}' - \sum_{i=1}^{n} \mathbf{k}_i - \mathbf{q}\Big) \\ \frac{(-\lambda)^{2n}}{[2(2\pi)^3]^n\, n!} \prod_{i=1}^{n} \frac{|f(\mathbf{k}_i^2)|^2}{\omega^3(\mathbf{k}_i)}\Bigg\} \\ = \delta^{(3)}(\mathbf{p} - \mathbf{p}')\, Z \sum_{n=0}^{\infty} \frac{1}{n!}\left[\frac{\lambda^2}{2(2\pi)^3} \int d^3k\, \frac{|f(\mathbf{k}^2)|^2}{\omega^3(\mathbf{k})}\right]^n \\ = \delta^{(3)}(\mathbf{p} - \mathbf{p}') \end{aligned} \tag{23}$$

whence

$$Z = \exp(-\lambda^2 L) \tag{24}$$

$$L = \frac{1}{(2\pi)^3} \int \frac{d^3k}{2\omega^3(\mathbf{k})}\, |f(\mathbf{k}^2)|^2 \tag{25}$$

For a point source, $f(\mathbf{k}^2) = 1$, and L diverges logarithmically. In this limit, the normalization constant Z which is proportional to $\langle 0 \mid \psi(\mathbf{q}) \mid \Psi^{(1)}(\mathbf{p})\rangle$, goes to zero. Since Z can be interpreted as the probability of finding a "bare" nucleon in the physical nucleon state, in the limit of a point particle, this probability vanishes. The same holds true for the other components of $|\Psi^{(1)}(\mathbf{p})\rangle$ in the Hilbert space, $\mathcal{H}_0$, spanned by the basis functions $\prod_{i=1}^{n} a^*(\mathbf{k}_i)\, \psi^*(\mathbf{q}) \mid 0\rangle$, the eigenstates of H_0. Thus $|\Psi^{(1)}(\mathbf{p})\rangle$ no longer lies in this Hilbert space, $\mathcal{H}_0$. The same situation prevails as long as $f(\mathbf{k}^2)$ does not make L converge.

To understand what is happening, consider the case that there is only one nucleon present which is localized at the origin of our co-ordinate system. The Hamiltonian corresponding to this situation can be taken to be

$$H = m_0 + \int d^3k\, \omega(\mathbf{k})\, a^*(\mathbf{k})\, a(\mathbf{k}) + \frac{\lambda}{(2\pi)^{3/2}} \int \frac{d^3k\, f(\mathbf{k}^2)}{\sqrt{2\omega(\mathbf{k})}} (a(\mathbf{k}) + a^*(-\mathbf{k})) \tag{26}$$

One again verifies that the vector

$$|\Psi\rangle = \sqrt{Z} \sum_{n=0}^{\infty} \frac{1}{n!}\left(-\int d^3k\, \frac{\lambda f(\mathbf{k}^2)}{\sqrt{2(2\pi)^3\, \omega^3(\mathbf{k})}}\, a^*(k)\right)^n \mid 0\rangle \tag{27}$$

is an eigenstate of H with eigenvalue $m_0 - \lambda^2\Delta$, and that Z is given by (24), so that the phenomenon that $Z \to 0$ if $f \to 1$ is again observed. Now the Hamiltonian (26) can be written as

$$H = \int d^3k\, \omega(\mathbf{k})\, (a(\mathbf{k}) + v(\mathbf{k}))^*\, (a(\mathbf{k}) + v(\mathbf{k})) + m_0 - \int d^3k\, |v(\mathbf{k}^2)|^2\, \omega(\mathbf{k}) \tag{28a}$$

$$= \int d^3k\, c^*(\mathbf{k})\, c(\mathbf{k})\, \omega(\mathbf{k}) + m \tag{28b}$$

where

$$v(\mathbf{k}^2) = \frac{\lambda f(\mathbf{k}^2)}{\sqrt{2(2\pi)^3\, \omega^3(\mathbf{k})}} \tag{29}$$

and

$$c(\mathbf{k}) = a(\mathbf{k}) + v(\mathbf{k}) \tag{30}$$

It will be noted that the operators $c(\mathbf{k})$, $c^*(\mathbf{k})$ satisfy the same commutation rules as $a(\mathbf{k})$. The Hamiltonian (26) when expressed in terms of operators $c^*(\mathbf{k})$, $c(\mathbf{k})$ is diagonal and has a ready physical interpretation: it describes a source (characterized by m and f) together with an assembly of free mesons which do not interact with each other or the source. We next inquire whether the operators $c(\mathbf{k})$, $c^*(\mathbf{k})$ belong to the same representation of the commutation rules as the operators $a(\mathbf{k})$, $a^*(\mathbf{k})$. The answer will be in the affirmative only if $a(\mathbf{k}) + v(\mathbf{k}) = c(\mathbf{k})$ has a no-particle state $|\Psi_0\rangle$ for which

$$c(\mathbf{k}) \mid \Psi_0\rangle = 0 \quad \text{for all } \mathbf{k} \tag{31}$$

If we adopt a Fock space representation for $|\Psi_0\rangle$ (defined in terms of the occupation number operators $N_a = \int d^3k\, a^*(\mathbf{k})\, a(\mathbf{k})$), with

$$|\Psi_0\rangle = \begin{pmatrix} \Psi_0^{(0)} \\ \Psi_0^{(1)}(\mathbf{k}_1) \\ \cdot \\ \cdot \\ \cdot \\ \Psi_0^{(n)}(\mathbf{k}_1, \mathbf{k}_2, \cdots \mathbf{k}_n) \\ \cdot \\ \cdot \\ \cdot \end{pmatrix} \tag{32a}$$

or equivalently

$$|\Psi_0\rangle = \Psi_0^{(0)} \mid 0\rangle + \sum_{n=1}^{\infty} \int d^3k_1 \cdots \int d^3k_n\, \Psi_0^{(n)}(\mathbf{k}_1, \cdots \mathbf{k}_n) \cdot \frac{1}{\sqrt{n!}}\, a^*(\mathbf{k}_1) \cdots a^*(\mathbf{k}_n) \mid 0\rangle \tag{32b}$$

the equation $c(\mathbf{k}) \mid \Psi_0\rangle = (a(\mathbf{k}) + v(\mathbf{k})) \mid \Psi_0\rangle = 0$ states that

$$\sqrt{n+1}\, \Psi_0^{(n+1)}(\mathbf{k}, \mathbf{k}_1, \cdots \mathbf{k}_n) = -v(k)\, \Psi_0^{(n)}(\mathbf{k}_1, \cdots \mathbf{k}_n) \tag{33}$$

whence

$$\Psi_0^{(n)}(\mathbf{k}_1, \mathbf{k}_2, \cdots \mathbf{k}_n) = \frac{1}{\sqrt{n!}} \prod_{i=1}^{n} (-v(\mathbf{k}_i) \cdot \Psi_0^{(0)} \tag{34}$$

The condition that $|\Psi_0\rangle$ have a finite norm is that

$$(\Psi_0, \Psi_0) = |\Psi_0^{(0)}|^2 \sum_{n=0}^{\infty} \frac{1}{n!} \left(\int d^3k \, |v(\mathbf{k})|^2 \right)^n < \infty \tag{35}$$

or that

$$\int d^3k \, |v(\mathbf{k})|^2 = L < \infty \tag{36}$$

Thus $a(\mathbf{k})$ and $a(\mathbf{k}) + v(\mathbf{k})$ will simultaneously possess a no-particle state only if $L < \infty$; otherwise $c(\mathbf{k})$ belongs to a continuous representation when $a(\mathbf{k})$ has a no-particle state. These facts are relevant in answering the question as to whether perturbation theory is applicable to the Hamiltonian (26), with H_I considered as the perturbation. It is clear that, if L diverges, perturbation theory in the ordinary sense cannot be formulated because it rests on the assumption that to the no-particle state of the uncoupled system there corresponds a no-particle state of the coupled system. [See also in this connection Van Hove (1952).]

Let us return to the theory described by the Hamiltonian (1) and investigate some of its predictions. We shall first investigate the scattering of a meson by a nucleon. To this end, we wish to find the eigenstate $|\mathbf{k}; \mathbf{p}\rangle_+$ of H with eigenvalue $m + \omega(\mathbf{k})$, which describes a meson of momentum $\mathbf{k}$ impinging on a nucleon of momentum $\mathbf{p}$ and which then scatter each other. We thus write

$$|\mathbf{k}, \mathbf{p}\rangle_+ = a_{\mathbf{k}}^* \mid \Psi^{(1)}(\mathbf{p})\rangle + |\chi\rangle_+ \tag{37}$$

where $a_{\mathbf{k}}^* \mid \Psi^{(1)}(\mathbf{p})\rangle$ describes the incoming state of a meson of momentum $\mathbf{k}$ and of the (physical) nucleon of momentum $\mathbf{p}$, and $|\chi\rangle_+$ corresponds to the scattered wave. The requirement that $|\mathbf{k}, \mathbf{p}\rangle_+$ be an eigenfunction of H

$$H \mid \mathbf{k}, \mathbf{p}\rangle_+ = (m + \omega(\mathbf{k})) \mid \mathbf{k}, \mathbf{p}\rangle_+ \tag{38}$$

implies that

$$\begin{aligned}
H(a_{\mathbf{k}}^* \mid \Psi^{(1)}(\mathbf{p})\rangle + |\chi_+\rangle) &= (a_{\mathbf{k}}^* H + [H, a_{\mathbf{k}}^*]) \mid \Psi^{(1)}(\mathbf{p})\rangle + H \mid \chi\rangle_+ \\
&= a_{\mathbf{k}}^*(m + \omega(\mathbf{k})) \mid \Psi^{(1)}(\mathbf{p})\rangle \\
&\quad + \frac{\lambda}{(2\pi)^{3/2}} \int d^3p' \, \frac{f(\mathbf{k}^2)}{\sqrt{2\omega(\mathbf{k})}} \psi^*(\mathbf{p}' + \mathbf{k}) \, \psi(\mathbf{p}') \mid \Psi^{(1)}(\mathbf{p})\rangle + H \mid \chi\rangle_+ \\
&= (m + \omega(\mathbf{k})) \, (a_{\mathbf{k}}^* \mid \Psi^{(1)}(\mathbf{p})\rangle + |\chi\rangle_+)
\end{aligned} \tag{39}$$

so that $|\chi\rangle_+$ satisfies the equation:

$$\begin{aligned}
(m + \omega(\mathbf{k}) - H) \mid \chi\rangle_+ &= \frac{\lambda}{(2\pi)^{3/2}} \int d^3p' \, \frac{f(\mathbf{k}^2)}{\sqrt{2\omega(\mathbf{k})}} \psi^*(\mathbf{p}' + \mathbf{k}) \, \psi(\mathbf{p}') \mid \Psi^{(1)}(\mathbf{p})\rangle \\
&= \frac{\lambda}{(2\pi)^{3/2}} \frac{f(\mathbf{k}^2)}{\sqrt{2\omega(\mathbf{k})}} \mid \Psi^{(1)}(\mathbf{p} + \mathbf{k})\rangle
\end{aligned} \tag{40}$$

The second line of Eq. (40) has been obtained by using the explicit representation of $|\Psi^{(1)}(\mathbf{p})\rangle$, Eq. (21a). Hence

$$|\chi\rangle_+ = \frac{1}{m + \omega_\mathbf{k} - H + i\epsilon} \frac{\lambda}{(2\pi)^{3/2}} \frac{f(\mathbf{k}^2)}{\sqrt{2\omega_\mathbf{k}}} \mid \Psi^{(1)}(\mathbf{p} + \mathbf{k})\rangle$$

$$= \frac{1}{\omega_\mathbf{k} + i\epsilon} \frac{\lambda}{(2\pi)^{3/2}} \frac{f(\mathbf{k}^2)}{\sqrt{2\omega_\mathbf{k}}} \mid \Psi^{(1)}(\mathbf{p} + \mathbf{k})\rangle \tag{41}$$

where the factor $+i\epsilon$ is to guarantee that $|\chi\rangle_+$ only contain outgoing waves. We note, however, that there is no singularity in the denominator of (41) since $\omega_\mathbf{k}$ is always greater than zero. There are, therefore, no outgoing scattered waves and there is, in fact, no scattering. This can also be seen by noting that $|\mathbf{k}, \mathbf{p}\rangle_-$, the eigenstate of H satisfying the incoming wave boundary condition, is clearly equal $|\mathbf{k}, \mathbf{p}\rangle_+$ [since the factors $\pm i\epsilon$ are irrelevant in (41)], so that

$$|\mathbf{k}, \mathbf{p}\rangle_\pm = a_\mathbf{k}{}^* \mid \Psi^{(1)}(\mathbf{p})\rangle + \frac{\lambda}{(2\pi)^{3/2}} \frac{f(\mathbf{k}^2)}{\sqrt{2\omega_\mathbf{k}{}^3}} \mid \Psi^{(1)}(\mathbf{p} + \mathbf{k})\rangle \tag{42}$$

More explicitly, the S-matrix element for the scattering from the state $\mathbf{k}, \mathbf{p}$ to the state $\mathbf{k}', \mathbf{p}'$ is given by

$$\begin{aligned} {}_-\langle \mathbf{p}', \mathbf{k}' \mid \mathbf{p}, \mathbf{k}\rangle_+ &= \langle \Psi^{(1)}(\mathbf{p}') \mid a_{\mathbf{k}'} a_\mathbf{k}{}^* \mid \Psi^{(1)}(\mathbf{p})\rangle \\ &+ \frac{\lambda}{(2\pi)^{3/2}} \frac{f(\mathbf{k}^2)}{\sqrt{2\omega_\mathbf{k}{}^3}} \langle \Psi^{(1)}(\mathbf{p}') \mid a_{\mathbf{k}'} \mid \Psi^{(1)}(\mathbf{p} + \mathbf{k})\rangle \\ &+ \frac{\lambda}{(2\pi)^{3/2}} \frac{\overline{f(\mathbf{k}'^2)}}{\sqrt{2\omega_{\mathbf{k}'}{}^3}} \langle \Psi^{(1)}(\mathbf{p}' + \mathbf{k}') \mid a_\mathbf{k}{}^* \mid \Psi^{(1)}(\mathbf{p})\rangle \\ &+ \frac{\lambda^2}{(2\pi)^3} \frac{f(\mathbf{k}^2)\overline{f(\mathbf{k}'^2)}}{\sqrt{4\omega_\mathbf{k}{}^3\omega_{\mathbf{k}'}{}^3}} \langle \Psi^{(1)}(\mathbf{p}' + \mathbf{k}') \mid \Psi^{(1)}(\mathbf{p} + \mathbf{k})\rangle \end{aligned} \tag{43}$$

The first term of (43) can be rewritten using the commutation rules for the $a_\mathbf{k}$ operators and the orthogonality relation for the dressed nucleon states, Eq. (22), to read:

$$\delta^{(3)}(\mathbf{p} - \mathbf{p}')\, \delta^{(3)}(\mathbf{k} - \mathbf{k}') + \langle \Psi^{(1)}(\mathbf{p}') \mid a_\mathbf{k}{}^* a_{\mathbf{k}'} \mid \Psi^{(1)}(\mathbf{p})\rangle$$

Also, the scalar product in the last term in Eq. (43) is equal to $\delta^{(3)}(\mathbf{p} + \mathbf{k} - \mathbf{p}' + \mathbf{k}')$. The other terms can be evaluated by noting that

$$\begin{aligned} &[H, a_{\mathbf{k}'}] \mid \Psi^{(1)}(\mathbf{p} + \mathbf{k})\rangle \\ &= \left(-\omega_{\mathbf{k}'} a_{\mathbf{k}'} - \frac{\lambda}{(2\pi)^{3/2}} \int d^3q \, \frac{f(\mathbf{k}'^2)}{\sqrt{2\omega_{\mathbf{k}'}}} \psi^*(\mathbf{q} - \mathbf{k}')\, \psi(\mathbf{q}) \right) \mid \Psi^{(1)}(\mathbf{p} + \mathbf{k})\rangle \\ &= H a_{\mathbf{k}'} \mid \Psi^{(1)}(\mathbf{p} + \mathbf{k})\rangle - m a_{\mathbf{k}'} \mid \Psi^{(1)}(\mathbf{p} + \mathbf{k})\rangle \end{aligned} \tag{44}$$

and, again using the representation (21), we can rewrite Eq. (44) as:

$$(H + \omega_{\mathbf{k}'} - m)\, a_{\mathbf{k}'} \mid \Psi^{(1)}(\mathbf{p} + \mathbf{k})\rangle = -\frac{\lambda}{(2\pi)^3} \frac{f(\mathbf{k}'^2)}{\sqrt{2\omega_{\mathbf{k}'}}} \mid \Psi^{(1)}(\mathbf{p} + \mathbf{k} - \mathbf{k}')\rangle \tag{45}$$

Since the operator $H + \omega_{\mathbf{k}'} - m$ is positive-definite in the one-nucleon subspace, it has an inverse, so that by multiplying Eq. (45) by $(H + \omega_{\mathbf{k}'} - m)^{-1}$ we obtain:

$$a_{\mathbf{k}'} \mid \Psi^{(1)}(\mathbf{p} + \mathbf{k})\rangle = -\frac{\lambda}{(2\pi)^{3/2}} \frac{f(\mathbf{k}'^2)}{\omega_{\mathbf{k}'}\sqrt{2\omega_{\mathbf{k}'}}} \mid \Psi^{(1)}(\mathbf{p} + \mathbf{k} - \mathbf{k}')\rangle \tag{46}$$

The expression for $\langle\Psi^{(1)}(\mathbf{p} + \mathbf{k}) \mid a_{\mathbf{k}}^*$ is obtained by taking the adjoint of Eq. (46). Substituting these expressions into the right-hand side of (43), we find that

$$\begin{aligned} {}_-\langle\mathbf{p}', \mathbf{k}' \mid \mathbf{p}, \mathbf{k}\rangle_+ &= {}_\pm\langle\mathbf{p}', \mathbf{k}' \mid \mathbf{p}, \mathbf{k}\rangle_\pm \\ &= \delta^{(3)}(\mathbf{p} - \mathbf{p}')\, \delta^{(3)}(\mathbf{k} - \mathbf{k}') \end{aligned} \tag{47}$$

so that the R matrix vanishes and there is no scattering: The interaction only "dresses" the nucleons but does not give rise to any further interaction between mesons and nucleons. This fact can be made explicit by performing a unitary transformation. The operator

$$U = \exp iS \tag{48a}$$

$$\begin{aligned} S &= \frac{i\lambda}{(2\pi)^{3/2}} \int d^3x\, \psi^*(\mathbf{x})\, \psi(\mathbf{x}) \int \frac{d^3k\, f(\mathbf{k}^2)}{\sqrt{2\omega^3(\mathbf{k})}} [a^*(\mathbf{k}) - a(-\mathbf{k})]\, e^{-i\mathbf{k}\cdot\mathbf{x}} \\ &= \frac{i\lambda}{(2\pi)^{3/2}} \int d^3p \int d^3k \frac{f(\mathbf{k}^2)}{\sqrt{2\omega^3(\mathbf{k})}} \psi^*(\mathbf{p})\, \psi(\mathbf{p} + \mathbf{k})\, [a^*(\mathbf{k}) - a(-\mathbf{k})] \end{aligned} \tag{48b}$$

has the property that it leaves the vacuum invariant, and operating on the bare one-nucleon state $\psi^*(\mathbf{p}) \mid 0\rangle$ will yield the properly normalized "dressed" one-nucleon state $|\Psi^{(1)}(\mathbf{p})\rangle$. This unitary operator thus attaches to the nucleon its appropriate bare meson cloud. Let us define [Greenberg (1958)] the operator:

$$\begin{aligned} \psi_d^*(\mathbf{p}) &= e^{iS}\, \psi^*(\mathbf{p})\, e^{-iS} \\ &= \psi^*(\mathbf{p}) + \frac{i}{1!} [S, \psi^*(\mathbf{p})] + \frac{i^2}{2!} [S, [S, \psi^*(\mathbf{p})]] + \cdots \\ &= \int d^3q \int \frac{d^3x}{(2\pi)^3} e^{-i(\mathbf{q}-\mathbf{p})\cdot\mathbf{x}}\, \psi^*(\mathbf{q}) \\ &\quad \cdot \exp\left[-\frac{\lambda}{(2\pi)^{3/2}} \int \frac{d^3k\, f(\mathbf{k}^2)}{\sqrt{2\omega^3(\mathbf{k})}} e^{-i\mathbf{k}\cdot\mathbf{x}} (a^*(\mathbf{k}) - a(-\mathbf{k}))\right] \end{aligned} \tag{49}$$

Operating on the vacuum state $|0\rangle$, $\psi_d^*(\mathbf{p})$ creates an eigenfunction of H with eigenvalue m which is also simultaneously an eigenfunction of $\mathbf{P}$ with eigenvalue $\mathbf{p}$. Since the transformation is unitary, the operators ψ_d and ψ_d^* again obey canonical anticommutation rules:

$$[\psi_d(\mathbf{p}), \psi_d^*(\mathbf{p}')]_+ = \delta^{(3)}(\mathbf{p} - \mathbf{p}') \tag{50}$$

Similarly, we define the "dressed" meson creation operators:

$$\begin{aligned} a_d^*(\mathbf{k}) &= e^{iS}\, a(\mathbf{k})\, e^{-iS} \\ &= a^*(\mathbf{k}) + \frac{\lambda}{(2\pi)^{3/2}} \int d^3q\, \psi^*(\mathbf{q}+\mathbf{k})\, \psi(\mathbf{q})\, \frac{f(\mathbf{k}^2)}{\sqrt{2\omega^3(k)}} \end{aligned} \quad (51)$$

The operator $a_d^*(\mathbf{k})$ has the property that operating on $|0\rangle$ it gives rise to a one-meson state of momentum $\mathbf{k}$ and energy $\omega_{\mathbf{k}}$ since

$$a_d^*(\mathbf{k}) \mid 0\rangle = a^*(\mathbf{k}) \mid 0\rangle \quad (52)$$

The Hamiltonian when expressed in terms of dressed operators is given by

$$H_d = H_{0d} + H_{Id} \quad (53)$$

with

$$H_{0d} = m \int d^3\mathbf{p}\, \psi_d^*(\mathbf{p})\, \psi_d(\mathbf{p}) + \int d^3k\, \omega_{\mathbf{k}} a_d^*(\mathbf{k})\, a_d(\mathbf{k})$$

$$H_{Id} = \frac{\lambda^2}{(2\pi)^3} \int d^3q \int d^3p \int \frac{d^3k\, |f(\mathbf{k}^2)|^2}{2\omega_{\mathbf{k}}^2} \psi_d^*(\mathbf{p}+\mathbf{k})\, \psi_d^*(\mathbf{q})\, \psi_d(\mathbf{p})\, \psi_d(\mathbf{q}+\mathbf{k}) \quad (54a)$$

Thus, in terms of dressed operators, the new interaction Hamiltonian, H_{Id}, no longer contains any self-interactions and gives rise only to an interaction between pairs of nucleons. The static potential between two nucleons is given by

$$V(\mathbf{x}-\mathbf{x}') = \frac{\lambda^2}{(2\pi)^3} \int \frac{d^3k\, |f(\mathbf{k}^2)|^2}{2\omega^2(\mathbf{k})} e^{i\mathbf{k}\cdot(\mathbf{x}-\mathbf{x}')} \quad (55)$$

since H_{Id} can also be written as

$$H_{Id} = \lambda^2 \int d^3x \int d^3x'\, \psi_d^*(\mathbf{x}')\, \psi_d^*(\mathbf{x})\, V(\mathbf{x}-\mathbf{x}')\, \psi_d(\mathbf{x})\, \psi_d(\mathbf{x}') \quad (54b)$$

where V is the potential between nucleon pairs. [Recall Eqs. (6.108–6.110).] In the limit $f(\mathbf{k}^2) \to 1$, i.e., for point nucleons, the potential becomes

$$V(\mathbf{x}-\mathbf{x}') = -\frac{\lambda^2}{8\pi} \frac{e^{-\mu|\mathbf{x}-\mathbf{x}'|}}{|\mathbf{x}-\mathbf{x}'|} \quad (56)$$

the Yukawa potential.

The simplicity of the model at hand also allows one to compute the $U(t, t_0)$ matrix in closed form and thus to analyze the workings of the adiabatic formalism. We shall conclude this section on the scalar model with a brief exposition of the time-dependent formalism for the scalar field.

We want to calculate the time displacement operator $U(t, t_0)$ in the interaction picture for the above theory, and in particular for the case that $t, t_0 \leqslant 0$. The operator $U(t, t_0)$ obeys the equation:

$$i\partial_t U(t, t_0) = H_I(t)\, U(t, t_0) \quad (57)$$

where

$$H_I(t) = e^{iH_0t}\, H_I\, e^{-iH_0t}$$

$$= \frac{\lambda}{(2\pi)^{3/2}} \int d^3p \int \frac{d^3k\, f(\mathbf{k}^2)}{\sqrt{2\omega_{\mathbf{k}}}}\, \psi^*(\mathbf{p}+\mathbf{k})\, \psi(\mathbf{p})$$

$$\cdot\, (a(\mathbf{k})\, e^{-i\omega_{\mathbf{k}}t} + a^*(-\mathbf{k})\, e^{i\omega_{\mathbf{k}}t})$$

$$- \delta m \int d^3p\, \psi^*(\mathbf{p})\, \psi(\mathbf{p}) \tag{58}$$

Note that Eq. (58) implies that the splitting of Hamiltonian into an unperturbed and a perturbed part is such that the bare quanta have mass m. This is accomplished by writing in (1) for m_0, $m_0 = m_0 + \delta m - \delta m = = m - \delta m$ and considering the $\delta m \int \psi^*(\mathbf{p})\, \psi(\mathbf{p})\, d^3p$ term part of the perturbation.

By our previous rules, the "adiabatic" U matrix is the solution of

$$i\partial_t U_\alpha(t, t_0) = H_I(\alpha; t)\, U_\alpha(t, t_0) \tag{59a}$$

$$H_I(\alpha; t) = e^{-\alpha|t|}\, H_I(t) \tag{59b}$$

To obtain a solution of (59a) we write

$$U_\alpha(t, t_0) = e^{-i\int_{t_0}^{t} H_I(\alpha;\, t')\, dt'}\, V(t, t_0) \tag{60}$$

so that by virtue of (57) the operator V satisfies the equation:

$$i\partial_t V(t, t_0) = \left\{e^{i\int_{t_0}^{t} H_I(\alpha;\, t')\, dt'}\, (H_I(\alpha; t) - i\partial_t)\, e^{-i\int_{t_0}^{t} H_I(\alpha;\, t')\, dt'}\right\} V(t, t_0) \tag{61}$$

To evaluate the right-hand side of Eq. (61), call

$$e^{i\sigma\int_{t_0}^{t} H_I(\alpha;\, t')\, dt}\, \frac{\partial e^{-i\sigma\int_{t_0}^{t} H_I(\alpha;\, t')\, dt'}}{\partial t} = D_\sigma(t, t_0) \tag{62}$$

so that

$$[D_\sigma]_{\sigma=0} = 0 \tag{63}$$

and

$$\frac{\partial D_\sigma}{\partial \sigma} = -\, ie^{i\sigma\int_{t_0}^{t} H_I(\alpha;\, t')\, dt'}\, H_I(t)\, e^{-i\sigma\int_{t_0}^{t} H_I(\alpha;\, t')\, dt'} \tag{64}$$

with

$$\left(\frac{\partial D_\sigma}{\partial \sigma}\right)_{\sigma=0} = -iH_I(t) \tag{65}$$

Similarly, we derive that

$$\frac{\partial^2 D_\sigma}{\partial \sigma^2} = e^{i\sigma\int_{t_0}^{t} H_I(\alpha;\, t')\, dt'} \int_{t_0} dt''\, [H_I(\alpha; t''), H_I(\alpha; t)]\, e^{-i\sigma\int_{t_0}^{t} H_I(\alpha;\, t')\, dt'} \tag{66}$$

The commutator in (66) is easily evaluated with the result:

$$[H_I(t''), H_I(t)] = \frac{\lambda^2}{(2\pi)^3} \int \frac{d^3k\, |f(\mathbf{k}^2)|^2}{2\omega_\mathbf{k}} (e^{i\omega_\mathbf{k}(t-t'')} - e^{-i\omega_\mathbf{k}(t-t'')})$$
$$\cdot \left\{ \int d^3p_1 \int d^3p_2\, \psi^*(\mathbf{p}_1 + \mathbf{k})\, \psi^*(\mathbf{p}_2 - \mathbf{k})\, \psi(\mathbf{p}_1)\, \psi(\mathbf{p}_2) - \int d^3p\, \psi^*(\mathbf{p})\, \psi(\mathbf{p}) \right\} \tag{67}$$

One checks that $[H_I(t''), H_I(t)]$ commutes with $H_I(t')$ so that

$$\frac{\partial^2 D_\sigma}{\partial \sigma^2} = e^{-\alpha|t|} \int_{t_0}^{t} dt''\, e^{-\alpha|t''|} \frac{\lambda^2}{(2\pi)^3} \int \frac{d^3k\, |f(\mathbf{k}^2)|^2}{2\omega_\mathbf{k}} (e^{i\omega_\mathbf{k}(t-t'')} - e^{-i\omega_\mathbf{k}(t-t'')})$$
$$\cdot \left\{ \int d^3p_1 \int d^3p_2\, \psi^*(\mathbf{p}_1 + \mathbf{k})\, \psi^*(\mathbf{p}_2 - \mathbf{k})\, \psi(\mathbf{p}_1)\, \psi(\mathbf{p}_2) - \int d^3p\, \psi^*(\mathbf{p})\, \psi(\mathbf{p}) \right\}$$
$$= R_\alpha(t) \tag{68}$$

By the same argument $\partial^3 D_\sigma/\partial\sigma^3$ and all higher derivatives vanish. Since

$$\int_0^\sigma \frac{\partial^2 D_{\sigma'}}{\partial \sigma'^2}\, d\sigma' = \frac{\partial D_\sigma}{\partial \sigma} - \left(\frac{\partial D_\sigma}{\partial \sigma}\right)_{\sigma=0} \tag{69}$$

upon substituting into this expression the values obtained for $\partial^2 D_\sigma/\partial\sigma^2$ and $(\partial D_\sigma/\partial\sigma)_{\sigma=0}$, we find that

$$\frac{\partial D_\sigma}{\partial \sigma} = -ie^{i\sigma \int_{t_0}^t H_I(\alpha;\, t')\, dt'} H_I(\alpha; t)\, e^{-i\sigma \int_{t_0}^t H_I(\alpha;\, t')\, dt'}$$
$$= \sigma R_\alpha(t) - iH_I(\alpha; t) \tag{70}$$

so that

$$D_\sigma = [D_\sigma]_{\sigma=0} + \int_0^\sigma \frac{\partial D_{\sigma'}}{\partial \sigma'}\, d\sigma'$$
$$= \frac{\sigma^2}{2} R_\alpha(t) - i\sigma H_I(\alpha; t) \tag{71}$$

Using Eqs. (61) and (71), we finally obtain that

$$i\partial_t V(t, t_0) = i\left[\frac{\partial D_\sigma}{\partial \sigma} - D_\sigma\right]_{\sigma=1} V(t, t_0)$$
$$= \tfrac{1}{2} i R_\alpha(t)\, V(t, t_0) \tag{72}$$

whence

$$V(t, t_0) = e^{\frac{1}{2}\int_{t_0}^t R_\alpha(t')\, dt'} \tag{73}$$

and

$$U_\alpha(t, t_0) = e^{-i\int_{t_0}^t H_I(\alpha;\, t')\, dt'}\, e^{\frac{1}{2}\int_{t_0}^t R_\alpha(t')\, dt'} \tag{74}$$

Finally let us evaluate the second exponential

$$\int_{t_0}^{t} R_\alpha(t')\, dt' = \tfrac{1}{2} \int_{t_0}^{t} dt'\, e^{-\alpha|t'|} \int_{t_0}^{t} dt''\, [H_I(t''), H_I(t')]\, e^{-\alpha|t''|} \tag{75}$$

for the particular case that $t_0 = -\infty$ and $t = 0$, in which case

$$\begin{aligned}
&\int_{-\infty}^{0} R_\alpha(t)\, dt' \\
&= \frac{1}{2} \frac{\lambda^2}{(2\pi)^3} \int \frac{d^3k\, |f(\mathbf{k}^2)|^2}{2\omega_\mathbf{k}} \left[\frac{1}{\alpha - i\omega_\mathbf{k}} - \frac{1}{\alpha + i\omega_\mathbf{k}} \right] \frac{1}{\alpha} \\
&\quad \cdot \left\{ \int d^3p_1 \int d^3p_2\, \psi^*(\mathbf{p}_1 + \mathbf{k})\, \psi^*(\mathbf{p}_2 - \mathbf{k})\, \psi(\mathbf{p}_1)\, \psi(\mathbf{p}_2) \right. \\
&\quad \left. - \int d^3p\, \psi^*(\mathbf{p})\, \psi(\mathbf{p}) \right\} \\
&= iT \frac{\lambda^2}{(2\pi)^3} \int \frac{d^3k\, |f(\mathbf{k}^2)|^2}{2\omega_\mathbf{k}^2} \left\{ \int d^3p_1 \int d^3p_2\, \psi^*(\mathbf{p}_1 + \mathbf{k})\, \psi^*(\mathbf{p}_2 - \mathbf{k})\, \psi(\mathbf{p}_1)\, \psi(\mathbf{p}_2) \right. \\
&\quad \left. - \int d^3p\, \psi^*(\mathbf{p})\, \psi(\mathbf{p}) \right\} + O(\alpha)
\end{aligned} \tag{76}$$

where we have denoted $1/\alpha$ by T. The latter plays the role of the large time (from $-\infty$ to 0) during which the system has evolved. Similarly,

$$\begin{aligned}
&\int_{-\infty}^{0} H_I(\alpha; t')\, dt' \\
&\qquad = \frac{i\lambda}{(2\pi)^{3/2}} \int \frac{d^3k\, f(\mathbf{k}^2)}{\sqrt{2\omega_\mathbf{k}^3}} \int d^3p\, \psi^*(\mathbf{p} + \mathbf{k})\, \psi(\mathbf{p})\, (a(\mathbf{k}) - a^*(-\mathbf{k})) \\
&\qquad - \frac{\delta m}{\alpha} \int d^3p\, \psi^*(\mathbf{p})\, \psi(\mathbf{p})
\end{aligned} \tag{77}$$

so that

$$\begin{aligned}
U_\alpha(0, -\infty) = \exp\Bigg[&\frac{\lambda i}{(2\pi)^{3/2}} \int \frac{d^3k\, f(\mathbf{k}^2)}{\sqrt{2\omega_\mathbf{k}^3}} \int d^3p \\
&\psi^*(\mathbf{p} + \mathbf{k})\, \psi(\mathbf{p})\, (a(\mathbf{k}) - a^*(-\mathbf{k})) + iT \frac{\lambda^2}{(2\pi)^3} \int \frac{d^3k\, |f(\mathbf{k}^2)|^2}{2\omega_\mathbf{k}^2} \\
&\left\{ \int d^3p_1 \int d^3p_2\, \psi^*(\mathbf{p}_1 + \mathbf{k})\, \psi^*(\mathbf{p}_2 - \mathbf{k})\, \psi(\mathbf{p}_1)\, \psi(\mathbf{p}_2) \right\} + O(\alpha) \Bigg]
\end{aligned} \tag{78}$$

Note that mass renormalization counterterms have canceled out in the above expression. This is as expected since in the unperturbed state the heavy quanta already have the right masses.

One verifies that the vector $U_\alpha(0, -\infty)\, \psi^*(\mathbf{p}) \mid 0\rangle$ is indeed in the limit as $\alpha \to 0$ an eigenstate of H. In fact, $U_\alpha(0, -\infty)$ is nothing but the unitary dressing transformation previously introduced.

12*b*. The Lee Model

The second soluble field theoretic model that we consider is the Lee model [Lee (1954), Källén and Pauli (Källén 1955b, 1957); see also Heisenberg (1957b)]. This model is very similar to the scalar field model, except that the nucleon is assumed to be able to exist in two different intrinsic states, and to be able to transform from one kind of state to the other. These intrinsic states are called, arbitrarily, the N particle and the V particle. The V particle can emit a Bose quantum called the θ particle and transform into an N particle. The N particle cannot emit a θ particle, but can absorb a θ particle and transform into a V particle. The V particle cannot absorb a θ particle. The allowed processes are summarized by the equation:

$$\mathrm{V} \leftrightarrows \mathrm{N} + \theta$$

The Lee model thus consists of three interacting fields: two fermion fields describing the V and N particles and a boson field describing the θ particles.

As in the scalar field model, the energies of the nucleons are assumed to be independent of their momenta, the particles are spinless and momentum is conserved. The θ particles are taken to be spin zero neutral bosons. The model is in some aspects simpler than the scalar field, yet it possesses interesting features which the latter does not, for example, the possibility of N-θ scattering and the occurrence of a "charge" renormalization. The Hamiltonian describing this field system in the Schrödinger picture is given by

$$H = H_0 + H_I \tag{79a}$$

where

$$H_0 = m_{\mathrm{V}0} \int d\mathbf{p}\, V^*(\mathbf{p})\, V(\mathbf{p}) + m_{\mathrm{N}0} \int d\mathbf{p}\, N^*(\mathbf{p})\, N(\mathbf{p}) + \int d\mathbf{k}\, \omega_{\mathbf{k}} a^*(\mathbf{k})\, a(\mathbf{k}) \tag{79b}$$

and

$$H_I = \frac{\lambda_0}{(2\pi)^{3/2}} \int \frac{d\mathbf{k}\, f_1(\mathbf{k}^2)}{\sqrt{2\omega_{\mathbf{k}}}} \int d\mathbf{p} \cdot \{V^*(\mathbf{p})\, N(\mathbf{p}-\mathbf{k})\, a(\mathbf{k}) + N^*(\mathbf{p}-\mathbf{k})\, a^*(\mathbf{k})\, V(\mathbf{p})\} \tag{79c}$$

The notation is self-explanatory: $N^*(\mathbf{p})$ and $N(\mathbf{p})$ are the creation and annihilation operators for an N quantum of momentum $\mathbf{p}$; $V^*(\mathbf{p})$, $V(\mathbf{p})$ and $a^*(\mathbf{k})$, $a(\mathbf{k})$ play the same role for the V quanta and θ quanta, respectively. The parameters $m_{\mathrm{V}0}$, $m_{\mathrm{N}0}$ are the masses of the quanta of the V and N field and $\omega_{\mathbf{k}} = \sqrt{\mu_0^2 + \mathbf{k}^2}$ is the energy of a θ quantum of momentum $\mathbf{k}$, mass μ_0. The parameter λ_0 characterizes the strength of the coupling between the fields and $f_1(\mathbf{k}^2) = f(\omega_{\mathbf{k}}) = \overline{f(\omega_{\mathbf{k}})}$ is a (real) cutoff function which describes the size of the region over which the interaction takes place. It is introduced to avoid the divergences connected with a point interaction of the form $H_I \sim \int V^*(\mathbf{x})\, N(\mathbf{x})\, \theta(\mathbf{x})\, d^3x$ + h.a. (Here, $\theta(\mathbf{x})$ is the Fourier transform of $a(\mathbf{k})/\sqrt{2\omega_{\mathbf{k}}}$). The neglect of recoil for the fermion quanta is characterized by the fact that the energy of the V and N quanta does not depend on their momentum. We could consider a more general theory where H_0 is of the form:

$$H_0 = \int d\mathbf{p}\, E_{V0}(\mathbf{p})\, V^*(\mathbf{p})\, V(\mathbf{p}) + \int d\mathbf{p}\, E_{N0}(\mathbf{p})\, N^*(\mathbf{p})\, N(\mathbf{p}) + \int d\mathbf{k}\, \omega(\mathbf{k})\, a^*(\mathbf{k})\, a(\mathbf{k}) \quad (80)$$

[see e.g., Van Hove (1955)]; however, this more general form of the theory, although still soluble, complicates the formal aspects without any gain in physical insight.

The field operators obey the following commutation rules:

$$[a(\mathbf{k}), a^*(\mathbf{k}')] = \delta^{(3)}(\mathbf{k} - \mathbf{k}') \quad (81a)$$

$$[a(\mathbf{k}), a(\mathbf{k}')] = [a^*(\mathbf{k}), a^*(\mathbf{k}')] = 0 \quad (81b)$$

$$[N(\mathbf{p}), N^*(\mathbf{p}')]_+ = \delta^{(3)}(\mathbf{p} - \mathbf{p}') \quad (82a)$$

$$[N(\mathbf{p}), N(\mathbf{p}')]_+ = [N^*(\mathbf{p}), N^*(\mathbf{p}')]_+ = 0 \quad (82b)$$

$$[V(\mathbf{p}), V^*(\mathbf{p}')]_+ = \delta^{(3)}(\mathbf{p} - \mathbf{p}') \quad (83a)$$

$$[V(\mathbf{p}), V(\mathbf{p}')]_+ = [V^*(\mathbf{p}), V^*(\mathbf{p}')]_+ = 0 \quad (83b)$$

$$0 = [N(\mathbf{p}), V(\mathbf{p}')]_+ = [N(\mathbf{p}), V^*(\mathbf{p}')]_+ = [N^*(\mathbf{p}), V(\mathbf{p}')]_+ = [N^*(\mathbf{p}), V^*(\mathbf{p}')]_+$$

$$[a(\mathbf{k}), N(\mathbf{p})] = [a(\mathbf{k}), N^*(\mathbf{p})] = \cdots = [a^*(\mathbf{k}), V^*(\mathbf{p})] = 0 \quad (84)$$

The interaction Hamiltonian only allows the following process

$$\mathrm{V} \leftrightharpoons \mathrm{N} + \theta$$

i.e., the destruction of a V quantum and the simultaneous creation of an N and θ quantum, or the inverse process, the absorption of a θ quantum by an N quantum to become a V quantum. The process

$$\mathrm{N} \leftrightharpoons \mathrm{V} + \theta$$

is forbidden. It will be noted that the interaction is such that momentum is conserved. A slightly different interpretation of the model consists in regarding the N particle as a neutron, the V particle as a proton and the θ particle as a π^+ meson, so that the reaction $\mathrm{p} \to \pi^+ + \mathrm{n}$ is allowed by charge conservation whereas $\mathrm{n} \to \mathrm{p} + \pi^+$ is clearly forbidden. This interpretation gives an insight on the limitations of the Lee model. In a "normal" field theory the antiparticle of the θ would appear—call it $\bar{\theta}$—which would make the reaction $\mathrm{V} + \bar{\theta} \to \mathrm{N}$ possible. Furthermore, note that the coupling term in the Lee model is of the form $N^*V\phi^{(+)}(x)$ which is not a causal coupling; it is nonlocal because the operation of taking the positive frequency part of $\phi(x)$ involves an integration over all space.

The simplicity of the Lee model is manifested by the fact that the bare and physical vacuum states and the bare and physical one N-particle and one θ-particle states coincide. That is, the "bare" vacuum $|0\rangle$ defined by the equations:

$$N(\mathbf{p}) \mid 0\rangle = V(\mathbf{p}) \mid 0\rangle = a(\mathbf{p}) \mid 0\rangle \quad \text{for all } \mathbf{p} \tag{85}$$

is also an eigenfunction of the total Hamiltonian, H, with eigenvalue 0 since $H_I \mid 0\rangle = 0$. Similarly, one checks that $N^*(\mathbf{p}) \mid 0\rangle$ is an eigenfunction of H with eigenvalue m_{N0} since $H_I N^*(\mathbf{p}) \mid 0\rangle = 0$ as each term of H_I contains a destruction operator which either commutes ($a_{\mathbf{k}}$) or anticommutes (V) with $N^*(\mathbf{p})$. Similarly for the one θ-particle state, $a^*(\mathbf{k}) \mid 0\rangle$ which is an eigenfunction of H with eigenvalue $\omega_{\mathbf{k}}$. Thus for the Lee model the θ and N quanta are the same as the "physical" particles, which we assume have ("observed") mass m_N and μ respectively. We therefore identify μ_0 and μ and m_{N0} and m_N. In what follows we shall denote the "physical" states (i.e., the eigenstates of H) either by a subscript d (for "dressed") or by boldfaced letters:

$$|0\rangle_d = |\mathbf{0}\rangle \tag{86}$$

$$|N_{\mathbf{p}}\rangle_d = |\mathbf{N}_{\mathbf{p}}\rangle = N_{\mathbf{p}}^* \mid 0\rangle \tag{87}$$

$$|\theta_{\mathbf{k}}\rangle_d = |\boldsymbol{\theta}_{\mathbf{k}}\rangle = a_{\mathbf{k}}^* \mid 0\rangle \tag{88}$$

To obtain the other eigenstates of H we note that, besides the total momentum operator $\mathbf{P}$,

$$\mathbf{P} = \int d\mathbf{p}\, \mathbf{p} N^*(\mathbf{p})\, N(\mathbf{p}) + \int d\mathbf{p}\, \mathbf{p} V^*(\mathbf{p})\, V(\mathbf{p}) + \int d\mathbf{k}\, \mathbf{k} a^*(\mathbf{k})\, a(\mathbf{k}) \tag{89}$$

the following two operators:

$$Q_1 = \int d\mathbf{p}\, (V^*(\mathbf{p})\, V(\mathbf{p}) + N^*(\mathbf{p})\, N(\mathbf{p})) \tag{90}$$

$$Q_2 = \int d\mathbf{p}\, N^*(\mathbf{p})\, N(\mathbf{p}) - \int d\mathbf{k}\, a^*(\mathbf{k})\, a(\mathbf{k}) \tag{91}$$

commute with H

$$[Q_1, H] = [Q_2, H] = 0 \tag{92}$$

and are therefore constants of the motion. The operator Q_1 corresponds to the sum of the number operators of V quanta and N quanta (i.e., to the total number of heavy quanta) and Q_2 to the difference between the number operator of N and θ quanta. The eigenstates of H can, therefore, be labeled by the eigenvalues of Q_1 and Q_2. If we expand a physical state in terms of the bare states (i.e., eigenstates of H_0):

$$|\,\rangle_d = \Phi^{(0,0,0)} \mid 0\rangle + \int d\mathbf{p}\, \Phi^{(1,0,0)}(\mathbf{p})\, V^*(\mathbf{p}) \mid 0\rangle + \cdots$$

$$+ \frac{1}{\sqrt{m!n!l!}} \int d\mathbf{p}_1 \cdots \int d\mathbf{p}_m \int d\mathbf{q}_1 \cdots \int d\mathbf{q}_n \int d\mathbf{k}_1 \cdots \int d\mathbf{k}_l$$

$$\Phi^{(m,n,l)}(\mathbf{p}_1, \cdots \mathbf{p}_m; \mathbf{q}_1, \cdots \mathbf{q}_n; \mathbf{k}_1, \cdots \mathbf{k}_l) \cdot$$

$$\cdot V^*(\mathbf{p}_1) \cdots V^*(\mathbf{p}_m) N^*(\mathbf{q}_1) \cdots N^*(\mathbf{q}_n) \cdot a^*(\mathbf{k}_1) \cdots a^*(\mathbf{k}_l) \mid 0\rangle + \cdots \tag{93}$$

then only states with the same eigenvalue of Q_1 and Q_2 can occur in the expansion.

To obtain the physical one V-particle state $|V_\mathbf{p}\rangle_d$, we note that in the limit as the coupling constant $\lambda \to 0$, $|V_\mathbf{p}\rangle_d$ must go over into $V_\mathbf{p}^* \mid 0\rangle$. The eigenvalue of Q_1 in this state is $+1$ and that of Q_2 is 0. The only other bare state which has these same eigenvalues is the state $|N_\mathbf{p}, \theta_\mathbf{k}\rangle = N_\mathbf{p}^* a_\mathbf{k}^* \mid 0\rangle$. We therefore expect that the state of one physical V particle which is an eigenfunction of H with eigenvalue m_V, the "observed" mass of the V particle, and of $\mathbf{P}$ the total momentum operator with eigenvalue $\mathbf{p}$, can be represented by the following superposition of bare states

$$|V_\mathbf{p}\rangle_d = |\mathbf{V}_\mathbf{p}\rangle = Z_V^{1/2} \left\{ V_\mathbf{p}^* \mid 0\rangle + \int d\mathbf{k}\, \Phi(\mathbf{k})\, N_{\mathbf{p}-\mathbf{k}}^* a_\mathbf{k}^* \mid 0\rangle \right\} \tag{94}$$

where $Z_V^{1/2}$ is a normalization constant. The momenta of the N and θ particles in the second term on the right-hand side are such that the state $\int d\mathbf{k}\, \Phi(\mathbf{k})\, N_{\mathbf{p}-\mathbf{k}}^* a_\mathbf{k}^* \mid 0\rangle$ is an eigenfunction of $\mathbf{P}$ with eigenvalue $\mathbf{p}$. The coefficient $\Phi(\mathbf{k})$ is determined from the requirement that $|\mathbf{V}_\mathbf{p}\rangle$ be an eigenfunction of H with eigenvalue m_V. If we take the scalar product of the equation:

$$H \mid \mathbf{V}_\mathbf{p}\rangle = m_V \mid \mathbf{V}_\mathbf{p}\rangle \tag{95}$$

with the bra $\langle V_{\mathbf{p}'}|$, recalling the orthogonality of the state of one V quantum to that having one N and one θ quantum, i.e., that $\langle V_{\mathbf{p}'} \mid N_{\mathbf{p}-\mathbf{k}}\theta_\mathbf{k}\rangle = \langle 0 \mid V_{\mathbf{p}'} N_{\mathbf{p}-\mathbf{k}}^* a_\mathbf{k}^* \mid 0\rangle = 0$, we obtain

$$\langle V_{\mathbf{p}'} \mid H_0 \mid V_\mathbf{p}\rangle + \int d\mathbf{k}\, \langle V_{\mathbf{p}'} \mid H_I \mid N_{\mathbf{p}-\mathbf{k}}\theta_\mathbf{k}\rangle\, \Phi(\mathbf{k}) = m_V \langle V_{\mathbf{p}'} \mid V_\mathbf{p}\rangle \tag{96}$$

In the evaluation of $\langle V_\mathbf{p} \mid H_I \mid N_{\mathbf{p}-\mathbf{k}}\theta_\mathbf{k}\rangle$ only the term V^*Na in H_I will contribute, so that using the commutation rules (81)–(84) we find:

$$\begin{aligned}
\langle V_{\mathbf{p}'} \mid H_I \mid N_{\mathbf{p}-\mathbf{k}}\theta_\mathbf{k}\rangle \\
&= \langle 0 \mid V_{\mathbf{p}'} H_I N_{\mathbf{p}-\mathbf{k}}^* a_\mathbf{k}^* \mid 0\rangle \\
&= \frac{\lambda_0}{(2\pi)^{3/2}} \int d\mathbf{p}'' \int \frac{d\mathbf{k}'\, f(\omega_{\mathbf{k}'})}{\sqrt{2\omega_{\mathbf{k}'}}} \langle 0 | V_{\mathbf{p}'} V_{\mathbf{p}''}^* N_{\mathbf{p}''-\mathbf{k}'} a_{\mathbf{k}'} N_{\mathbf{p}-\mathbf{k}}^* a_\mathbf{k}^* | 0\rangle \\
&= \frac{\lambda_0}{(2\pi)^{3/2}} \frac{f(\omega_\mathbf{k})}{\sqrt{2\omega_\mathbf{k}}}\, \delta^{(3)}(\mathbf{p} - \mathbf{p}')
\end{aligned} \tag{97}$$

Hence Eq. (96) becomes

$$m_{V0} + \frac{\lambda_0}{(2\pi)^{3/2}} \int \frac{d\mathbf{k}\, f(\omega_\mathbf{k})}{\sqrt{2\omega_\mathbf{k}}}\, \Phi(\mathbf{k}) = m_V \tag{98}$$

In an analogous manner, upon taking the scalar product of Eq. (95) with $\langle N_\mathbf{q}\theta_\mathbf{l}|$ one arrives at the equation:

$$(m_V - m_N - \omega_{\mathbf{k}})\,\Phi(\mathbf{k}) = \frac{\lambda_0}{(2\pi)^{3/2}} \frac{f(\omega_{\mathbf{k}})}{\sqrt{2\omega_{\mathbf{k}}}} \tag{99}$$

If $m_V - m_N - \omega_{\mathbf{k}} \neq 0$, which will be the case when $m_V < m_N + \mu$, we can then solve for $\Phi(\mathbf{k})$ in Eq. (99):

$$\Phi(\mathbf{k}) = \frac{\lambda_0}{(2\pi)^{3/2}} \frac{f(\omega_{\mathbf{k}})}{\sqrt{2\omega_{\mathbf{k}}}\,(m_V - m_N - \omega_{\mathbf{k}})} \tag{100a}$$

This corresponds to the case of a stable V particle which cannot spontaneously decay into an N and θ particle. In the case that $m_V > m_N + \mu$ the denominator can vanish. We then solve for $\Phi(\mathbf{k})$ by taking the principal value for the singularity

$$\Phi(\mathbf{k}) = \frac{\lambda_0}{(2\pi)^{3/2}}\, \mathrm{P}\, \frac{f(\omega_{\mathbf{k}})}{\sqrt{2\omega_{\mathbf{k}}}\,(m_V - m_N - \omega_{\mathbf{k}})} \tag{100b}$$

where P denotes the principal value. The reason for this prescription is that we wish to describe the state of a single particle and this state must be invariant under time inversion. Substituting this value of $\Phi(\mathbf{k})$ into (98), we obtain an equation which determines m_V

$$\begin{aligned} m_V &= m_{V0} + \frac{{\lambda_0}^2}{(2\pi)^3}\, \mathrm{P} \int d\mathbf{k}\, \frac{|f(\omega_{\mathbf{k}})|^2}{2\omega_{\mathbf{k}}(m_V - m_N - \omega_{\mathbf{k}})} \\ &= m_{V0} + \delta m_V \end{aligned} \tag{101}$$

The "wave function" normalization constant Z_V is determined from the requirement that

$${}_d\langle V_{\mathbf{p}'} \mid V_{\mathbf{p}}\rangle_d = \delta^{(3)}(\mathbf{p} - \mathbf{p}') \tag{102}$$

which, when expressed in terms of the expansion (94), yields

$$1 = Z_V \left(1 + \int d\mathbf{k}\, |\Phi(\mathbf{k})|^2\right) \tag{103}$$

If we define the function

$$\begin{aligned} F(x) &= \frac{{\lambda_0}^2}{(2\pi)^3}\, \mathrm{P} \int \frac{|f(\omega_{\mathbf{k}})|^2}{2\omega_{\mathbf{k}}(\omega_{\mathbf{k}} - x)}\, d\mathbf{k} \\ &= \frac{{\lambda_0}^2}{(2\pi)^2}\, \mathrm{P} \int_\mu^\infty \frac{|f(\omega)|^2\sqrt{\omega^2 - \mu^2}}{\omega - x}\, d\omega \end{aligned} \tag{104}$$

the eigenvalue equation (101) can then be expressed as

$$F(m_V - m_N) = m_{V0} - m_V \tag{105}$$

and the normalization constant Z_V can be written as equal

$$Z_V^{-1} = 1 + \left(\frac{dF}{dx}\right)_{x = m_V - m_N} \tag{106}$$

The function $F(x)$ has the property that

$$\frac{dF}{dx} = \frac{{\lambda_0}^2}{(2\pi)^2}\, \mathrm{P} \int_\mu^\infty d\omega\, \frac{|f(\omega)|^2\sqrt{\omega^2 - \mu^2}}{(\omega - x)^2} > 0 \tag{107}$$

so that $F(x)$ is a monotonically increasing function of x which goes to zero as $x \to -\infty$ and is continuous at $x = \mu$. The qualitative behavior of $F(x)$ is exhibited in Figures 12.1 and 12.2. The value of m_V in each case is

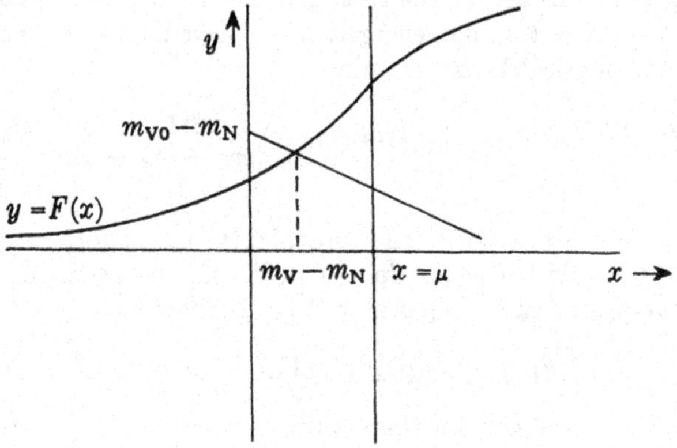

Fig. 12.1

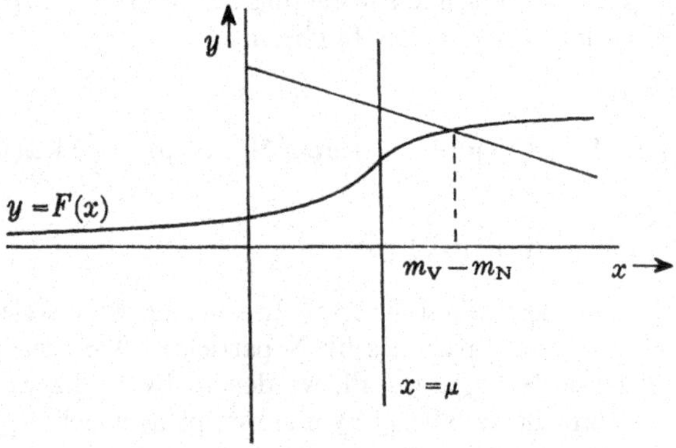

Fig. 12.2

determined by the intersection of the line $y = -x + (m_{V0} - m_N)$ and the curve $y = F(x)$. Two cases can be differentiated. If

$$-\mu + m_{V0} - m_N < F(\mu)$$

then the root of Eq. (105) occurs for $x < \mu$ and $m_V - m_N < \mu$, corresponding to a stable V particle. The above inequality can always be fulfilled by choosing m_{V0} small enough. It is interesting to note that in the stability condition $m_V - m_N < \mu$, it is the mass of the physical particle, m_V, which is the quantity deciding whether or not the reaction $V \to N + \theta$ is energetically allowed. On the other hand, if m_{V0} satisfies the inequality

$$-\mu + m_{V0} - m_N > F(\mu)$$

then there is no root of the equation $m_{V0} - m_V = F(m_V - m_N)$ below the point $x = \mu$. There is a root above this point which corresponds to an unstable V particle [Glaser and Källén (Glaser (1956–57)]. We shall only consider the Lee model for the case that the V particle is stable and the reaction $V \to N + \theta$ is not energetically allowed. In this case we can write for the physical V-particle state

$$|V_{\mathbf{p}}\rangle_d = Z_V^{1/2}\left\{|V_{\mathbf{p}}\rangle + \frac{\lambda_0}{(2\pi)^{3/2}}\int d\mathbf{k}\,\frac{f(\omega_{\mathbf{k}})}{\sqrt{2\omega_{\mathbf{k}}}\,(m_V - m_N - \omega_{\mathbf{k}})}\,|\,N_{\mathbf{p}-\mathbf{k}},\theta_{\mathbf{k}}\rangle\right\} \tag{108}$$

with Z_V given by (106). The probability of finding in the state $\int d\mathbf{p}\, g(\mathbf{p})\mid V_{\mathbf{p}}\rangle_d$, with $\int |g(\mathbf{p})|^2\, d\mathbf{p} = 1$, a "bare" quantum corresponding to the wave-packet state $\int d\mathbf{p}\, g(\mathbf{p})\, V_{\mathbf{p}}^*\mid 0\rangle$ is given by

$$\int d\mathbf{p}\int d\mathbf{p}'\,\bar{g}(\mathbf{p})\,g(\mathbf{p}')\,\langle V_{\mathbf{p}}\mid V_{\mathbf{p}'}\rangle_d = Z_V^{-1/2} \tag{109}$$

Z_V is thus the probability for this contingency.

It will be convenient for our later applications to recast our Hamiltonian in a form where H_0 describes "bare" particles with the correct physical mass, m_V. We do this by adding and subtracting the term $\delta m_V \int d\mathbf{p}\, V^*(\mathbf{p})\, V(\mathbf{p})$ to H, and write $m_V = m_{V0} + \delta m_V$ to obtain

$$H = H_0' + H_I' \tag{110a}$$

$$H_0' = m_V\int d\mathbf{p}\, V^*(\mathbf{p})\, V(\mathbf{p}) + m_N\int d\mathbf{p}\, N^*(\mathbf{p})\, N(\mathbf{p}) + \int d\mathbf{k}\,\omega(\mathbf{k})\, a^*(\mathbf{k})\, a(\mathbf{k}) \tag{110b}$$

$$H_I' = H_I - \delta m_V\int d\mathbf{p}\, V^*(\mathbf{p})\, V(\mathbf{p}) \tag{110c}$$

To obtain a further insight into the Lee model, we next compute the predicted scattering of θ particles by N particles. We are, therefore, to solve for the state $|\mathbf{N}_{\mathbf{q}}, \boldsymbol{\theta}_{\mathbf{k}}\rangle_+$ (recall, we denote by boldface symbols the physical states previously labeled by a subscript d) which is a solution of

$$H\mid \mathbf{N}_{\mathbf{q}}, \boldsymbol{\theta}_{\mathbf{k}}\rangle_+ = (\omega_{\mathbf{k}} + m_N)\mid \mathbf{N}_{\mathbf{q}}, \boldsymbol{\theta}_{\mathbf{k}}\rangle_+ \tag{111}$$

and corresponds to an incoming plane wave state, $a_{\mathbf{k}}^*\mid \mathbf{N}_{\mathbf{q}}\rangle = |N_{\mathbf{q}}, \theta_{\mathbf{k}}\rangle$, and an outgoing scattered wave, denoted by $|\chi\rangle_+$

$$\begin{aligned}|\mathbf{N}_{\mathbf{q}}, \boldsymbol{\theta}_{\mathbf{k}}\rangle_+ &= |N_{\mathbf{q}}, \theta_{\mathbf{k}}\rangle + |\chi\rangle_+\\ &= N_{\mathbf{q}}^* a_{\mathbf{k}}^*\mid 0\rangle + |\chi\rangle_+\end{aligned} \tag{112}$$

Now

$$\begin{aligned}H\mid \mathbf{N}_{\mathbf{q}}, \boldsymbol{\theta}_{\mathbf{k}}\rangle_+ &= Ha_{\mathbf{k}}^*\mid \mathbf{N}_{\mathbf{q}}\rangle + H\mid \chi\rangle_+\\ &= a_{\mathbf{k}}^* H\mid \mathbf{N}_{\mathbf{q}}\rangle + [H, a_{\mathbf{k}}^*]\mid \mathbf{N}_{\mathbf{q}}\rangle + H\mid \chi\rangle_+\end{aligned} \tag{113a}$$

Since $H\mid \mathbf{N}_{\mathbf{q}}\rangle = m_N\mid \mathbf{N}_{\mathbf{q}}\rangle$ and

$$[H, a_{\mathbf{k}}^*] = \omega_{\mathbf{k}} a_{\mathbf{k}}^* + \frac{\lambda_0}{(2\pi)^{3/2}}\int d\mathbf{p}\,\frac{f(\omega_{\mathbf{k}})}{\sqrt{2\omega_{\mathbf{k}}}}\, V^*(\mathbf{p})\, N(\mathbf{p} - \mathbf{k}) \tag{114}$$

we obtain for Eq. (113a)

$$(m_N + \omega_{\mathbf{k}} - H) \mid \chi\rangle_+ = \frac{\lambda_0}{(2\pi)^{3/2}} \frac{f(\omega_{\mathbf{k}})}{\sqrt{2\omega_{\mathbf{k}}}} V^*(\mathbf{q} + \mathbf{k}) \mid 0\rangle \quad (113b)$$

so that, if the scattered wave (at $t = +\infty$) is to have outgoing waves only,

$$|\chi\rangle_+ = \frac{\lambda_0}{(2\pi)^{3/2}} \frac{f(\omega_{\mathbf{k}})}{\sqrt{2\omega_{\mathbf{k}}}} \frac{1}{m_N + \omega_{\mathbf{k}} - H + i\epsilon} V^*(\mathbf{q} + \mathbf{k}) \mid 0\rangle \quad (115)$$

whence

$$|\mathbf{N_q}, \boldsymbol{\theta}_{\mathbf{k}}\rangle_+ = |N_{\mathbf{q}}, \theta_{\mathbf{k}}\rangle + \frac{\lambda_0}{(2\pi)^{3/2}} \frac{f(\omega_{\mathbf{k}})}{\sqrt{2\omega_{\mathbf{k}}}} \frac{1}{m_N + \omega_{\mathbf{k}} - H + i\epsilon} V^*(\mathbf{q} + \mathbf{k}) \mid 0\rangle \quad (116)$$

with

$$_+\langle \mathbf{N_{q'}}, \boldsymbol{\theta}_{\mathbf{k'}} \mid \mathbf{N_q}, \boldsymbol{\theta}_{\mathbf{k}}\rangle_+ = \delta^{(3)}(\mathbf{q}' - \mathbf{q})\, \delta^{(3)}(\mathbf{k}' - \mathbf{k}) \quad (117)$$

This normalization is the one expected from general principles (recall Section *b* of Chapter 10). It can be verified explicitly using techniques similar to those which will be needed to evaluate the *S* matrix. We therefore present these techniques at this point. Essentially the required formula is:

$$\frac{1}{x - H + i\epsilon} a_{\mathbf{k}}^* = a_{\mathbf{k}}^* \frac{1}{x - H - \omega_{\mathbf{k}} + i\epsilon} + \frac{1}{x - H + i\epsilon} [H_I, a_{\mathbf{k}}^*] \frac{1}{x - H - \omega_{\mathbf{k}} + i\epsilon} \quad (118)$$

To prove (118) we first of all note that

$$\begin{aligned} \frac{1}{x - H_0 + i\epsilon} a_{\mathbf{k}}^* &= -i \int_0^\infty d\lambda\, e^{i(x+i\epsilon - H_0)\lambda} a_{\mathbf{k}}^* \\ &= -i \int_0^\infty d\lambda\, e^{i(x+i\epsilon)} e^{-iH_0\lambda} a_{\mathbf{k}}^* e^{iH_0\lambda} e^{-iH_0\lambda} \\ &= -i \int_0^\infty d\lambda\, a_{\mathbf{k}}^* e^{-i(\omega_{\mathbf{k}} + H_0 - x - i\epsilon)\lambda} \\ &= a_{\mathbf{k}}^* \frac{1}{x - H_0 - \omega_{\mathbf{k}} + i\epsilon} \end{aligned} \quad (119)$$

Next using the algebraic identity

$$\frac{1}{a + b} = \frac{1}{a} - \frac{1}{a + b} b \frac{1}{a} \quad (120)$$

which is valid for operators as well as *c*-numbers, we write

$$\frac{1}{x - H_0 - H_I + i\epsilon} = \frac{1}{x - H_0 + i\epsilon} + \frac{1}{x - H + i\epsilon} H_I \frac{1}{x - H_0 + i\epsilon} \quad (121)$$

so that

$$\frac{1}{x - H + i\epsilon} a_{\mathbf{k}}^*$$

$$= a_{\mathbf{k}}^* \frac{1}{x - H_0 - \omega_{\mathbf{k}} + i\epsilon} + \frac{1}{x - H + i\epsilon} H_I a_{\mathbf{k}}^* \frac{1}{x - H_0 - \omega_{\mathbf{k}} + i\epsilon}$$

$$= a_{\mathbf{k}}^* \frac{1}{x - H_0 - \omega_{\mathbf{k}} + i\epsilon} + \frac{1}{x - H + i\epsilon} a_{\mathbf{k}}^* H_I \frac{1}{x - H_0 - \omega_{\mathbf{k}} + i\epsilon}$$

$$+ \frac{1}{x - H + i\epsilon} [H_I, a_{\mathbf{k}}^*] \frac{1}{x - H_0 - \omega_{\mathbf{k}} + i\epsilon} \qquad (122)$$

We then bring the second term on the right-hand side to the left side of the equation to obtain:

$$\frac{1}{x - H + i\epsilon} a_{\mathbf{k}}^*(x - H_0 - \omega_{\mathbf{k}} - H_I + i\epsilon) \frac{1}{x - H_0 - \omega_{\mathbf{k}} + i\epsilon}$$

$$= a_{\mathbf{k}}^* \frac{1}{x - H_0 - \omega_{\mathbf{k}} + i\epsilon} + \frac{1}{x - H + i\epsilon} [H_I, a_{\mathbf{k}}^*] \frac{1}{x - H_0 - \omega_{\mathbf{k}} + i\epsilon} \qquad (123)$$

Upon canceling the factor $(x - H_0 + \omega_{\mathbf{k}} + i\epsilon)^{-1}$ and thereafter multiplying by $(x + H + \omega_{\mathbf{k}} + i\epsilon)^{-1}$ from the right, we obtain Eq. (118).

Let us now make use of Eq. (118) to rewrite Eq. (116) for the in-state $|\mathbf{N}_{\mathbf{q}}, \boldsymbol{\theta}_{\mathbf{k}}\rangle_+$. We first of all re-express the bare state $V^*(\mathbf{q} + \mathbf{k}) \mid 0\rangle$ in terms of the physical V-particle state:

$$V^*(\mathbf{q} + \mathbf{k}) \mid 0\rangle = Z_V^{-1/2} \mid \mathbf{V}_{\mathbf{q}+\mathbf{k}}\rangle$$

$$- \frac{\lambda_0}{(2\pi)^{3/2}} \int \frac{d\mathbf{k}' f(\omega_{\mathbf{k}'})}{\sqrt{2\omega_{\mathbf{k}'}}\,(m_V - m_N - \omega_{\mathbf{k}'})} \mid N_{\mathbf{q}+\mathbf{k}-\mathbf{k}'}, \theta_{\mathbf{k}'}\rangle \qquad (124)$$

so that

$$\frac{1}{m_N + \omega_{\mathbf{k}} - H + i\epsilon} V^*(\mathbf{q} + \mathbf{k}) \mid 0\rangle = \frac{Z_V^{-1/2}}{m_N + \omega_{\mathbf{k}} - m_V + i\epsilon} \mid \mathbf{V}_{\mathbf{q}+\mathbf{k}}\rangle$$

$$- \frac{\lambda_0}{(2\pi)^{3/2}} \int \frac{d\mathbf{k}' f(\omega_{\mathbf{k}'})}{\sqrt{2\omega_{\mathbf{k}'}}\,(m_V - m_N - \omega_{\mathbf{k}'})} \frac{1}{m_N + \omega_{\mathbf{k}} - H + i\epsilon} a_{\mathbf{k}'}^* \mid N_{\mathbf{q}+\mathbf{k}-\mathbf{k}'}\rangle \qquad (125)$$

We now rewrite the second term with the help of Eq. (118), as follows:

$$\frac{1}{m_N + \omega_{\mathbf{k}} - H + i\epsilon} a_{\mathbf{k}'}^* \mid N_{\mathbf{q}+\mathbf{k}-\mathbf{k}'}\rangle = \frac{1}{\omega_{\mathbf{k}} - \omega_{\mathbf{k}'} + i\epsilon} \cdot$$

$$\cdot \left\{ a_{\mathbf{k}}^* \mid N_{\mathbf{q}+\mathbf{k}-\mathbf{k}'}\rangle + \frac{\lambda_0}{(2\pi)^{3/2}} \frac{f(\omega_{\mathbf{k}'})}{\sqrt{2\omega_{\mathbf{k}'}}} \frac{1}{m_N + \omega_{\mathbf{k}} - H + i\epsilon} V^*(\mathbf{q} + \mathbf{k}) \mid 0\rangle \right\} \qquad (126)$$

Combining the results of (125) and (126), we obtain:

$$\frac{1}{m_N + \omega_{\mathbf{k}} - H + i\epsilon} \mid V_{\mathbf{q}+\mathbf{k}}\rangle = \frac{1}{G_+(\omega_{\mathbf{k}})} \frac{1}{m_N + \omega_{\mathbf{k}} - m_V} \cdot$$

$$\cdot \left\{ |V_{\mathbf{q}+\mathbf{k}}\rangle + \frac{\lambda_0}{(2\pi)^{3/2}} \int \frac{d\mathbf{k}' f(\omega_{\mathbf{k}'})}{\sqrt{2\omega_{\mathbf{k}'}}\,(\omega_{\mathbf{k}} - \omega_{\mathbf{k}'} + i\epsilon)} \,|\, N_{\mathbf{q}+\mathbf{k}-\mathbf{k}'}, \theta_{\mathbf{k}'}\rangle \right\} \tag{127a}$$

where

$$G_+(\omega_{\mathbf{k}}) = \left(1 + \frac{\lambda_0^2}{(2\pi)^3} \int d\mathbf{k}' \frac{|f(\omega_{\mathbf{k}'})|^2}{(m_{\mathrm{V}} - m_{\mathrm{N}} - \omega_{\mathbf{k}'})\, 2\omega_{\mathbf{k}'}} \frac{1}{\omega_{\mathbf{k}} - \omega_{\mathbf{k}'} + i\epsilon}\right) \tag{127b}$$

The normalization integral is therefore given by

$$\begin{aligned}
{}_+\langle \mathbf{N}_{\mathbf{q}'}, \boldsymbol{\theta}_{\mathbf{k}'} \,|\, \mathbf{N}_{\mathbf{q}}, \boldsymbol{\theta}_{\mathbf{k}}\rangle_+ &= \delta^{(3)}(\mathbf{q} - \mathbf{q}')\, \delta^{(3)}(\mathbf{k} - \mathbf{k}') + \frac{\lambda_0^2}{(2\pi)^3} \frac{f(\omega) f(\omega')}{\sqrt{4\omega\omega'}} \\
&\cdot \delta^{(3)}(\mathbf{q} + \mathbf{k} - \mathbf{q}' - \mathbf{k}') \left\{ \frac{1}{G_+(\omega)} \frac{1}{m_{\mathrm{N}} + \omega - m_{\mathrm{V}}} - \frac{1}{\overline{G_+(\omega')}} \frac{1}{m_{\mathrm{N}} + \omega' - m_{\mathrm{V}}} \right\} \\
&\cdot \frac{1}{\omega - \omega' + i\epsilon} + \frac{\lambda_0^2}{(2\pi)^3} \frac{f(\omega) f(\omega')}{\sqrt{4\omega\omega'}} \frac{1}{G_+(\omega)} \frac{1}{\overline{G_+(\omega')}} \\
&\cdot \frac{1}{m_{\mathrm{N}} + \omega' - m_{\mathrm{V}}} \frac{1}{m_{\mathrm{N}} + \omega - m_{\mathrm{V}}} \cdot \delta^{(3)}(\mathbf{q} + \mathbf{k} - \mathbf{q}' - \mathbf{k}') \\
&\cdot \left\{ 1 + \frac{\lambda_0^2}{(2\pi)^3} \int d\mathbf{k}'' \frac{|f(\omega'')|^2}{2\omega''(\omega' - \omega'' - i\epsilon)\,(\omega - \omega'' + i\epsilon)} \right\}
\end{aligned} \tag{128}$$

where, for typographical convenience, we have set $\omega_{\mathbf{k}} = \omega$, $\omega_{\mathbf{k}'} = \omega'$, etc. By combining the terms within the brackets of the second term on the right-hand side, one easily verifies that it cancels the last term, so that only the first term $\delta^{(3)}(\mathbf{q} - \mathbf{q}')\, \delta^{(3)}(\mathbf{k} - \mathbf{k}')$ survives, and Eq. (117) is proved.

In a completely analogous manner one verifies that the solution of $H \,|\, \mathbf{N}_{\mathbf{q}}, \boldsymbol{\theta}_{\mathbf{k}}\rangle_- = (\omega_{\mathbf{k}} + m_{\mathrm{N}}) \,|\, \mathbf{N}_{\mathbf{q}}, \boldsymbol{\theta}_{\mathbf{k}}\rangle_-$, which satisfies the boundary condition of plane waves plus incoming waves, is given by

$$|\mathbf{N}_{\mathbf{q}}, \boldsymbol{\theta}_{\mathbf{k}}\rangle_- = |N_{\mathbf{q}}, \theta_{\mathbf{k}}\rangle + \frac{\lambda_0}{(2\pi)^{3/2}} \frac{f(\omega_{\mathbf{k}})}{\sqrt{2\omega_{\mathbf{k}}}} \frac{1}{m_{\mathrm{N}} + \omega_{\mathbf{k}} - H - i\epsilon} V^*(\mathbf{q} + \mathbf{k}) \,|\, 0\rangle \tag{129}$$

with the normalization

$${}_-\langle \mathbf{N}_{\mathbf{q}'}, \boldsymbol{\theta}_{\mathbf{k}'} \,|\, \mathbf{N}_{\mathbf{q}}, \boldsymbol{\theta}_{\mathbf{k}}\rangle_- = \delta^{(3)}(\mathbf{q}' - \mathbf{q})\, \delta^{(3)}(\mathbf{k}' - \mathbf{k}) \tag{130}$$

Comparing Eqs. (129) and (116), we note that

$$|\mathbf{N}_{\mathbf{q}}, \boldsymbol{\theta}_{\mathbf{k}}\rangle_+ = |\mathbf{N}_{\mathbf{q}}, \boldsymbol{\theta}_{\mathbf{k}}\rangle_- - 2\pi i \frac{\lambda_0}{(2\pi)^{3/2}} \frac{f(\omega_{\mathbf{k}})}{\sqrt{2\omega_{\mathbf{k}}}} \delta(m_{\mathrm{N}} + \omega_{\mathbf{k}} - H)\, V^*(\mathbf{q} + \mathbf{k}) \,|\, 0\rangle \tag{131}$$

so that the S matrix for the scattering from the state $|N_{\mathbf{q}}, \theta_{\mathbf{k}}\rangle$ to the state $|N_{\mathbf{q}'}, \theta_{\mathbf{k}'}\rangle$ is given by

$$\begin{aligned}
{}_-\langle \mathbf{N}_{\mathbf{q}'}, \boldsymbol{\theta}_{\mathbf{k}'} \,|\, \mathbf{N}_{\mathbf{q}}, \boldsymbol{\theta}_{\mathbf{k}}\rangle_+ &= \delta^{(3)}(\mathbf{q} - \mathbf{q}')\, \delta^{(3)}(\mathbf{k} - \mathbf{k}') \\
&- 2\pi i \frac{\lambda_0}{(2\pi)^{3/2}} \frac{f(\omega_{\mathbf{k}})}{\sqrt{2\omega_{\mathbf{k}}}} \delta(\omega_{\mathbf{k}} - \omega_{\mathbf{k}'})\; {}_-\langle \mathbf{N}_{\mathbf{q}'}, \boldsymbol{\theta}_{\mathbf{k}'} \,|\, V^*(\mathbf{q} + \mathbf{k}) \,|\, 0\rangle
\end{aligned} \tag{132}$$

and the R matrix by

$$\langle N_{\mathbf{q}'}, \theta_{\mathbf{k}'} \mid R \mid N_{\mathbf{q}}, \theta_{\mathbf{k}} \rangle = \frac{\lambda_0}{(2\pi)^{3/2}} \frac{f(\omega_{\mathbf{k}})}{\sqrt{2\omega_{\mathbf{k}}}} \langle \mathbf{N}_{\mathbf{q}'}, \boldsymbol{\theta}_{\mathbf{k}'} \mid V^*(\mathbf{q} + \mathbf{k}) \mid 0 \rangle$$
$$= \frac{\lambda_0{}^2}{(2\pi)^3} \frac{|f(\omega_{\mathbf{k}})|^2}{2\omega_{\mathbf{k}}} \langle 0 \mid V(\mathbf{q}' + \mathbf{k}') \frac{1}{m_{\mathrm{N}} + \omega_{\mathbf{k}} - H + i\epsilon} V^*(\mathbf{q} + \mathbf{k}) \mid 0 \rangle \tag{133}$$

where in the last expression $\omega_{\mathbf{k}} = \omega_{\mathbf{k}'}$, since we are on the energy shell. To evaluate (133) we again use Eq. (127) and find

$$\langle N_{\mathbf{q}'}, \theta_{\mathbf{k}'} \mid R \mid N_{\mathbf{q}}, \theta_{\mathbf{k}} \rangle = \frac{\lambda_0{}^2}{(2\pi)^3} \frac{|f(\omega_{\mathbf{k}})|^2}{2\omega_{\mathbf{k}}} \frac{1}{m_{\mathrm{N}} + \omega_{\mathbf{k}} - m_{\mathrm{V}}} \cdot \delta^{(3)}(\mathbf{k}' + \mathbf{q}' - \mathbf{k} - \mathbf{q})$$
$$\cdot \left[1 + \frac{\lambda_0{}^2}{(2\pi)^3} \int \frac{|f(\omega'')|^2}{2\omega''(m_{\mathrm{V}} - m_{\mathrm{N}} - \omega'')} \frac{1}{\omega_{\mathbf{k}} - \omega'' + i\epsilon} d\mathbf{k}'' \right]^{-1} \tag{134}$$

We note that in the limit of a point particle, i.e., when $f(\omega) = 1$, the integral in the denominator of Eq. (134) behaves like $\int dk/k$ for large k and hence is divergent. In this limit the R matrix therefore vanishes and there is no scattering. We may, however, rewrite the denominator using the algebraic identity

$$\frac{1}{ab} = \frac{1}{a^2} + \frac{a - b}{a^2 b} \tag{135}$$

to read

$$1 + \frac{\lambda_0{}^2}{(2\pi)^3} \int d\mathbf{k}'' \frac{|f(\omega'')|^2}{2\omega''(m_{\mathrm{V}} - m_{\mathrm{N}} - \omega'')} \frac{1}{\omega - \omega'' + i\epsilon}$$
$$= Z_{\mathrm{V}}{}^{-1} + \frac{\lambda_0{}^2}{(2\pi)^3} \int d\mathbf{k}'' \frac{|f(\omega'')|^2 (m_{\mathrm{V}} - m_{\mathrm{N}} - \omega)}{2\omega''(m_{\mathrm{V}} - m_{\mathrm{N}} - \omega'')^2 (\omega - \omega'' + i\epsilon)} \tag{136}$$

so that the R matrix can be written in the form

$$\langle N_{\mathbf{q}'}, \theta_{\mathbf{k}'} \mid R \mid N_{\mathbf{q}}, \theta_{\mathbf{k}} \rangle = \frac{Z_{\mathrm{V}} \lambda_0{}^2}{(2\pi)^3} \frac{|f(\omega_{\mathbf{k}})|^2}{2\omega_{\mathbf{k}}} \delta^{(3)}(\mathbf{k} + \mathbf{q} - \mathbf{k}' - \mathbf{q}') \frac{1}{m_{\mathrm{N}} + \omega_{\mathbf{k}} - m_{\mathrm{V}}}$$
$$\left[1 + \frac{Z_{\mathrm{V}} \lambda_0{}^2}{(2\pi)^3} \int d\mathbf{k}'' \frac{|f(\omega'')|^2 (m_{\mathrm{V}} - m_{\mathrm{N}} - \omega_{\mathbf{k}})}{2\omega''(m_{\mathrm{V}} - m_{\mathrm{N}} - \omega'')^2 (\omega_{\mathbf{k}} - \omega'' + i\epsilon)} \right]^{-1} \tag{137}$$

The R matrix will remain finite even in the limit of a point source if we define a new "renormalized" coupling constant, λ, by

$$\lambda^2 = Z_{\mathrm{V}} \lambda_0{}^2 \tag{138}$$

and choose it to be finite. The motivation for introducing a "renormalized" coupling constant in the Lee model is thus to make the scattering nonvanishing in the limit of a point source.

We could also have proceeded somewhat differently by noting that the parameter λ_0 which enters the theory must be fixed in terms of an observable quantity, for example a cross section at some fixed energy. Assume that such a "measurement" exists and we know the scattering length, a, for N-θ scattering. It will be noted that (137) predicts only S-wave scattering, so that if δ is the S-wave phase shift, then the reduced S matrix for S-wave scattering is:

$$s(k) = e^{2i\delta(k)} = 1 - 4\pi \frac{k^2}{\Delta} 2\pi i \frac{A(k)}{1 + B(k)} \tag{139a}$$

where

$$(N_{\mathbf{q}'}, \theta_{\mathbf{k}'} \mid S \mid N_{\mathbf{q}}, \theta_{\mathbf{k}}) = \delta^{(3)}(\mathbf{k}' + \mathbf{q}' - \mathbf{k} - \mathbf{q})\, \delta^{(1)}(k - k') \frac{1}{kk'} s(k) \tag{139b}$$

since

$$\delta(\omega - \omega') = \frac{1}{\Delta} \delta(k - k') \tag{140a}$$

$$\Delta = \frac{d\omega}{dk} = \frac{k}{\omega} \tag{140b}$$

and

$$A(k) = \frac{Z_V \lambda_0^2}{(2\pi)^3} \frac{|f(\omega)|^2}{2\omega} \frac{1}{m_N + \omega - m_V} \tag{141}$$

$$B(k) = \frac{\lambda_0^2}{(2\pi)^3} \int d\mathbf{k}'' \frac{|f(\omega'')|^2}{2\omega''(m_V - m_N - \omega'')(\omega - \omega'' + i\epsilon)} \tag{142}$$

From the above we derive that

$$\begin{aligned} -\frac{\tan \delta(k)}{k} &= \frac{4\pi^2 \omega A(k)}{1 + \frac{1}{2}(B(k) + \overline{B(k)})} \\ &= 4\pi^2 \frac{\lambda_0^2}{2(2\pi)^3} \frac{|f(\omega)|^2}{m_N + \omega - m_V} \\ &\quad \cdot \left[1 + \frac{\lambda_0^2}{(2\pi)^3} \mathrm{P} \int \frac{|f(\omega'')|^2}{2\omega''(m_V - m_N - \omega'')} \frac{1}{\omega - \omega''} d\mathbf{k}''\right]^{-1} \end{aligned} \tag{143}$$

The scattering length, a, which is defined as $\lim_{k \to 0} (-\tan \delta(k)/k)$, is given by

$$\begin{aligned} a &= -\frac{\tan \delta(k)}{k}\bigg|_{k=0} \\ &= \frac{\lambda_0^2}{4\pi} \frac{|f(\mu)|^2}{m_N + \mu - m_V} \cdot \left[1 + \frac{\lambda_0^2}{4\pi} \mathrm{P} \int_\mu^\infty \frac{|f(\omega'')|^2 \sqrt{\omega''^2 - \mu^2}}{(m_V - m_N - \omega'')(\mu - \omega'')} d\omega''\right]^{-1} \end{aligned} \tag{144}$$

We can now fix the quantity $\lambda_0^2/4\pi$ in terms of the "measured" quantity a, since

$$\left(\frac{\lambda_0^2}{4\pi}\right)^{-1} = \frac{1}{a} \frac{|f(\mu)|^2}{m_N + \mu - m_V} - \mathrm{P} \int_\mu^\infty d\omega'' \frac{|f(\omega'')|^2 \sqrt{\omega''^2 - \mu^2}}{(m_V - m_N - \omega'')(\mu - \omega'')} \tag{145}$$

so that, in general,

$$\frac{\tan \delta(k)}{k} = \frac{m_V - m_N - \mu}{m_V - m_N - \omega_{\mathbf{k}}} a|f(\omega_{\mathbf{k}})|^2 \cdot \Bigg[|f(\mu)|^2 + a(m_N + \mu - m_V)(\mu - \omega_{\mathbf{k}}) \, \mathrm{P} \int_\mu^\infty \frac{|f(\omega'')|^2 \sqrt{\omega''^2 - \mu^2}\, d\omega''}{(m_V - m_N - \omega'')(\omega - \omega'')(\mu - \omega'')} \Bigg] \tag{146}$$

Note that $\tan \delta(k)/k$ is now expressed in terms of "measurable" quantities only. This procedure again results in a finite, nonvanishing, scattering amplitude in the limit of a point particle, $f(\omega) \to 1$. In general, the procedure of re-expressing the scattering amplitudes in terms of observable quantities will be called "renormalization."

In general, a "renormalized" coupling constant can be defined by comparing the exact expression for the scattering amplitude with the result of lowest order perturbation theory at zero energy. It should be emphasized that this "renormalized" coupling constant is different from that defined by Eq. (138), although we shall see that it is related to it by a simple finite expression. Now the scattering amplitude for Nθ scattering to lowest order in perturbation theory is given by

$$r(k)\, \delta^{(3)}(\mathbf{q}' + \mathbf{k}' - \mathbf{q} - \mathbf{k})|_{\text{Born}} = \langle N_{\mathbf{q}'}, \theta_{\mathbf{k}'} | H_I \frac{1}{m_N + \omega_{\mathbf{k}} - H_0 + i\epsilon} H_I | N_{\mathbf{q}}, \theta_{\mathbf{k}} \rangle \tag{147}$$

so that

$$r(k)\Big|_{\text{Born}} = \frac{\lambda_0^2}{(2\pi)^3} \frac{|f(\omega)|^2}{2\omega(m_N + \omega - m_V)} \tag{148}$$

We then define the renormalized constant λ_R by the requirement that the scattering amplitude for zero momentum mesons be equal to the Born amplitude with λ_0^2 replaced by λ_R^2:

$$\begin{aligned} r(k)\Big|_{\substack{\text{exact}\\ k=0}} &= \frac{\lambda_0^2}{(2\pi)^3} \frac{|f(\mu)|^2}{2\mu(m_N + \mu - m_V)} \left[1 + \frac{\lambda_0^2}{(2\pi)^3} \mathrm{P} \int_\mu^\infty \frac{4\pi |f(\omega'')|^2 \sqrt{\omega''^2 - \mu^2}}{(m_V - m_N - \omega'')(\mu - \omega'')} \, d\omega'' \right]^{-1} \\ &= \frac{\lambda_R^2}{(2\pi)^3} \frac{|f(\mu)|^2}{2\mu(m_N + \mu - m_V)} \end{aligned} \tag{149}$$

so that

$$\lambda_R^2 = \frac{\lambda_0^2}{1 + \dfrac{\lambda_0^2}{(2\pi)^3} \mathrm{P} \displaystyle\int \frac{|f(\omega'')|^2 \, d\mathbf{k}''}{2\omega''(m_V - m_N - \omega'')(\mu - \omega'')}} \tag{150a}$$

$$\lambda_0^2 = \frac{\lambda_R^2}{1 - \dfrac{\lambda_R^2}{(2\pi)^3} \mathrm{P} \displaystyle\int \frac{|f(\omega'')|^2 \, d\mathbf{k}''}{2\omega''(m_V - m_N - \omega'')(\mu - \omega'')}} \tag{150b}$$

The coupling constant λ_R^2 is related to the coupling constant $\lambda^2 = Z_V\lambda_0^2$ by

$$\lambda^2 = \frac{\lambda_R^2}{1 - \dfrac{\lambda_R^2}{(2\pi)^3}(\mu + m_N - m_V)\displaystyle\int \frac{|f(\omega'')|^2\, d\mathbf{k}''}{2\omega''(m_V - m_N - \omega'')^2(\mu - \omega'')}} \tag{151}$$

Note that the integral in Eq. (151) is finite even in the limit $f(\omega) \to 1$, so that given λ_R^2, λ^2 can be calculated without any ambiguities. The two coupling constants are therefore on an equivalent footing as far as the subsequent work is concerned. For reasons which will become clear later, we shall work with the renormalized coupling constant defined by $\lambda^2 = Z_V\lambda_0^2$, Eq. (138),

$$\lambda^2 = \frac{\lambda_0^2}{1 + \lambda_0^2 L} \tag{152a}$$

so that

$$\lambda_0^2 = \frac{\lambda^2}{1 - \lambda^2 L} \tag{152b}$$

where

$$L = \frac{1}{(2\pi)^3}\int d\mathbf{k}\, \frac{|f(\omega)|^2}{2\omega(m_V - m_N - \omega)^2} \tag{153}$$

Note that for any value of L, i.e., for a given cutoff function, there is a maximum value of the renormalized coupling constant. This result is indicated diagrammatically in Figure 12.3.

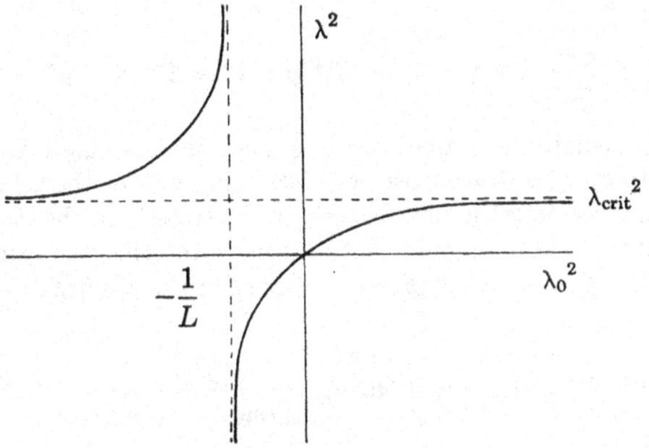

Fig. 12.3

As indicated above, the application of the formalism to a "physical" situation consists in fixing the value of λ^2 and in adjusting the cutoff function $f(\omega)$ so as to obtain agreement with "experiment." It is therefore possible that for a given λ^2 and $f(\omega)$, and equivalently a value of L, no $\lambda_0^2 > 0$ exists. It has been conjectured that such a phenomenon occurs in "realistic" field theories, e.g., quantum electrodynamics. We shall return to this point later. Figure 12.3 further indicates that for a given L and $\lambda^2 > \lambda_{crit}^2$ although a positive λ_0^2 does not exist, a negative λ_0^2 does exist. A negative λ_0^2 would, however, mean an imaginary λ_0 so that the Hamiltonian (1) is not Hermitian and therefore probabilities are neither conserved nor are they necessarily positive-definite.

The same situation occurs for fixed λ^2 in the limit as $f(\omega) \to 1$ and $L \to \infty$. In this limit λ_0^2 becomes negative and λ_0 becomes imaginary, which again implies a *non-Hermitian* Hamiltonian. This difficulty can be expressed somewhat differently by a consideration of the normalization constant Z_V, which when expressed in terms of the renormalized coupling constant is given by

$$Z_V = \frac{\lambda^2}{\lambda_0^2} = 1 - \lambda^2 L \tag{154}$$

We have seen that Z_V is the probability for finding a bare V quantum in the physical V-particle state, hence it must have a value between 0 and 1, $0 \leqslant Z_V \leqslant 1$. Equation (154), however, indicates that for $\lambda^2 L$ large enough Z_V can be negative.

A proof that Z_V must be positive can be obtained from a consideration of the vacuum expectation value of the anticommutator

$$\langle 0 \mid [V(\mathbf{p}), V^*(\mathbf{p}')]_+ \mid 0 \rangle = \delta^{(3)}(\mathbf{p} - \mathbf{p}') \tag{155}$$

Since $V(\mathbf{p}) \mid 0 \rangle = 0$, only the first term in the anticommutator contributes and

$$\sum_{|n\rangle} \langle 0 \mid V(\mathbf{p}) \mid n \rangle \langle n \mid V^*(\mathbf{p}') \mid 0 \rangle = \delta^{(3)}(\mathbf{p} - \mathbf{p}') \tag{156}$$

where $\{|n\rangle\}$ denotes a complete set of states. Let us choose for $\{|n\rangle\}$ the physical states, i.e., the eigenstates of H. Because of the selection rules (92) presumably only the physical V-particle states and the Nθ-scattering states contribute to the sum. Substituting into (156) the explicit representation for these states expressed in terms of the renormalized coupling constant:

$$|\mathbf{V}_\mathbf{p}\rangle = Z_V^{1/2} \mid V_\mathbf{p}\rangle + \frac{\lambda}{(2\pi)^{3/2}} \int d\mathbf{k} \frac{f(\omega_\mathbf{k})}{\sqrt{2\omega_\mathbf{k}}\,(m_V - m_N - \omega_\mathbf{k})} \mid N_{\mathbf{p}-\mathbf{k}}\theta_\mathbf{k}\rangle \tag{157}$$

$$\begin{aligned}|\mathbf{N}_\mathbf{q}, \boldsymbol{\theta}_\mathbf{k}\rangle_+ = {} & |N_\mathbf{q}, \theta_\mathbf{k}\rangle + Z_V^{1/2} \frac{\lambda}{(2\pi)^{3/2}} \frac{f(\omega_\mathbf{k})}{\sqrt{2\omega_\mathbf{k}}\, h(\omega_\mathbf{k} + i\epsilon)} \mid V_{\mathbf{q}+\mathbf{k}}\rangle \\ & + \frac{\lambda^2}{(2\pi)^3} \frac{f(\omega_\mathbf{k})}{\sqrt{2\omega_\mathbf{k}}\, h(\omega_\mathbf{k} + i\epsilon)} \int d\mathbf{k}' \frac{f(\omega_{\mathbf{k}'})}{\sqrt{2\omega_{\mathbf{k}'}}\,(\omega_\mathbf{k} - \omega_{\mathbf{k}'} + i\epsilon)} \mid N_{\mathbf{q}+\mathbf{k}-\mathbf{k}'}, \theta_{\mathbf{k}'}\rangle\end{aligned} \tag{158}$$

where

$$\frac{h(\omega_{\mathbf{k}} + i\epsilon)}{m_{\rm N} + \omega_{\mathbf{k}} - m_{\rm V}} = 1 + \frac{\lambda^2}{(2\pi)^3}(m_{\rm N} + \omega_{\mathbf{k}} - m_{\rm V}) \int \frac{d\mathbf{k}'\,|f(\omega_{\mathbf{k}'})|^2}{2\omega_{\mathbf{k}'}(m_{\rm V} - m_{\rm N} - \omega_{\mathbf{k}'})^2} \frac{1}{\omega_{\mathbf{k}'} - (\omega_{\mathbf{k}} + i\epsilon)} \tag{159}$$

we obtain

$$Z_{\rm V}\delta^{(3)}(\mathbf{p} - \mathbf{p}')\left(1 + \frac{\lambda^2}{(2\pi)^3}\int d\mathbf{k}\,\frac{|f(\omega_{\mathbf{k}})|^2}{2\omega_{\mathbf{k}}|h(\omega_{\mathbf{k}} + i\epsilon)|^2}\right) = \delta^{(3)}(\mathbf{p} - \mathbf{p}') \tag{160a}$$

or

$$Z_{\rm V}^{-1} = 1 + \frac{\lambda^2}{(2\pi)^3}\int d\mathbf{k}\,\frac{|f(\omega_{\mathbf{k}})|^2}{2\omega_{\mathbf{k}}|h(\omega_{\mathbf{k}} + i\epsilon)|^2} \tag{160b}$$

so that $Z_{\rm V}$ is positive and less than one, a result clearly contradicting the relation (154) for $\lambda^2 L$ large enough. It should be noted that the result $0 \leqslant Z_{\rm V} \leqslant 1$ is independent of the detailed structure of f or h since the integral on the right-hand side of (160) is positive-definite. It does, however, assume that the V-particle states and the Nθ-scattering states form a complete set in the subspace characterized by the eigenvalues $q_1 = 1$, $q_2 = 0$. Let us therefore reinvestigate the eigenvalue spectrum of H in this subspace. Before doing this, it is convenient to define a renormalized V-particle operator by the equation:

$$V_R(\mathbf{p}) = Z_{\rm V}^{-1/2}V(\mathbf{p}) \tag{161}$$

This renormalized operator has the property that it has finite matrix elements between physical states, even in the limit as $f(\omega) \to 1$. In particular,

$$\langle 0 \mid V_R(\mathbf{p}) \mid \mathbf{V}_{\mathbf{p}'}\rangle = \delta^{(3)}(\mathbf{p} - \mathbf{p}') \tag{162}$$

in contradistinction to the operator $V(\mathbf{p})$ whose matrix elements between physical states were proportional to $Z_{\rm V}^{1/2}$, where $Z_{\rm V}^{-1}$ is divergent in the limit $f \to 1$. Note that the commutation rules for the renormalized operators

$$[V_R^*(\mathbf{p}), V_R(\mathbf{p}')]_+ = Z_{\rm V}^{-1}\delta^{(3)}(\mathbf{p} - \mathbf{p}') \tag{163}$$

are more singular in the limit $f \to 1$ than the commutation rules for the unrenormalized operators. The Hamiltonian expressed in terms of renormalized operators and coupling constants is given by

$$H_0 = m_{\rm V}Z_{\rm V}\int d\mathbf{p}\,V_R^*(\mathbf{p})\,V_R(\mathbf{p}) + m_{\rm N}\int d\mathbf{p}\,N^*(\mathbf{p})\,N(\mathbf{p}) + \int d\mathbf{k}\,\omega(\mathbf{k})\,a^*(\mathbf{k})\,a(\mathbf{k}) \tag{164}$$

$$H_I = \frac{\lambda}{(2\pi)^{3/2}}\int d\mathbf{k}\,\frac{f(\omega_{\mathbf{k}})}{\sqrt{2\omega_{\mathbf{k}}}}\int d\mathbf{p}\,\{V_R^*(\mathbf{p})\,N(\mathbf{p} - \mathbf{k})\,a(\mathbf{k})$$

$$+ N^*(\mathbf{p} - \mathbf{k})\, a^*(\mathbf{k})\, V_R(\mathbf{p})\} - \delta m_V Z_V \int d\mathbf{p}\, V_R^*(\mathbf{p})\, V_R(\mathbf{p}) \quad (165)$$

In the following we drop the subscript R on the renormalized V-particle operators, since we only deal with these operators.

The eigenstates of H, corresponding to $q_1 = 1$ and $q_2 = 0$ and momentum $\mathbf{p}$, are determined by the equation:

$$H \mid \mathbf{p}\rangle = (m_N + \Omega) \mid \mathbf{p}\rangle \quad (166)$$

where Ω will be allowed to take on any value between $-\infty$ and $+\infty$, and $|\mathbf{p}\rangle$, as we have seen, is of the form

$$|\mathbf{p}\rangle = \beta V^*(\mathbf{p}) \mid 0\rangle + \int d\mathbf{k}\, \varphi(\mathbf{k})\, N^*(\mathbf{p} - \mathbf{k})\, a^*(\mathbf{k}) \mid 0\rangle \quad (167)$$

where β is some constant. Substituting this expression into (166) we obtain two conditions for β and φ

$$(m_V - m_N - \Omega - \delta m_V)\, \beta = -\frac{\lambda}{(2\pi)^{3/2}} \int d\mathbf{k}\, \frac{f(\omega_\mathbf{k})}{\sqrt{2\omega_\mathbf{k}}}\, \varphi(\mathbf{k}) \quad (168a)$$

$$(\omega_\mathbf{k} - \Omega)\, \varphi(\mathbf{k}) = -\frac{\lambda}{Z_V (2\pi)^{3/2}}\, \beta\, \frac{f(\omega_\mathbf{k})}{\sqrt{2\omega_\mathbf{k}}} \quad (168b)$$

The compatibility equation is:

$$0 = Z_V^{-1}(m_V - m_N - \Omega) \left[1 + (m_N + \Omega - m_V)\, \frac{\lambda^2}{(2\pi)^3} \int \frac{d\mathbf{k}\, |f(\omega_\mathbf{k})|^2}{2\omega_\mathbf{k}(\omega_\mathbf{k} - \Omega)\, (m_V - m_N - \omega_\mathbf{k})}\right] \quad (169)$$

or equivalently

$$h(\Omega) = 0 \quad (170)$$

where we used the value of δm_V given by Eq. (101) and the definition (159) of $h(x)$. Clearly $\Omega = m_V - m_N$ is a root corresponding to the V-particle state of energy m_V. Conversely, we could have fixed the value of δm_V in the compatibility equation (168) by requiring that there exist a root with value $\Omega = m_V - m_N$.

We also note that

$$\frac{d}{dx}\left(\frac{h(x)}{m_N + x - m_V}\right) = \frac{\lambda^2}{(2\pi)^3} \int \frac{d\mathbf{k}\, |f(\omega_\mathbf{k})|^2}{2\omega_\mathbf{k}(\omega_\mathbf{k} + m_N - m_V)}\, \frac{1}{(\omega_\mathbf{k} - x)^2} > 0 \quad (171)$$

when $m_N + \mu > m_V \geqslant 0$, and furthermore that

$$\begin{aligned} \lim_{x \to \infty} \frac{h(x)}{m_N + x - m_V} &= 1 - \frac{\lambda^2}{(2\pi)^2} \int \frac{d\mathbf{k}\, |f(\omega_\mathbf{k})|^2}{2\omega_\mathbf{k}(m_V - m_N - \omega_\mathbf{k})^2} + O\left(\frac{1}{x}\right) \\ &= 1 - \lambda^2 L \\ &= Z_V \end{aligned} \quad (172)$$

Therefore when $Z_V > 0$, the plot of $h(x)/m_N + x - m_V$ is as indicated in Figure 12.4. There is therefore only one root of $h(x) = 0$ occurring at

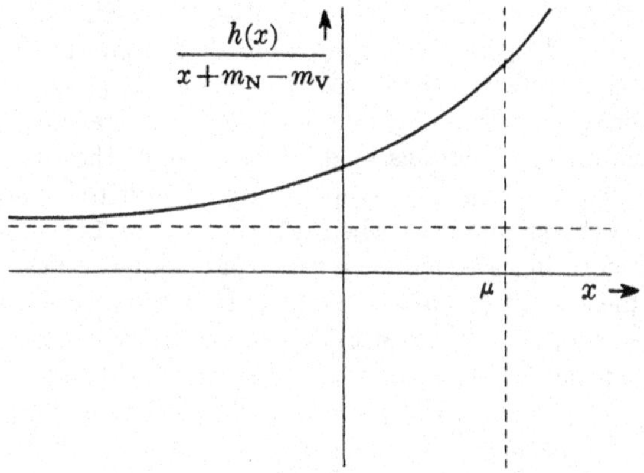

Fig. 12.4

$x = m_V - m_N$. On the other hand, when $Z_V < 0$ there is an additional root at the point $x = \Omega_0$ with $\Omega_0 < \mu$, which corresponds to a new state in the theory (see Fig. 12.5). This state, $|G\rangle$, can be constructed explicitly

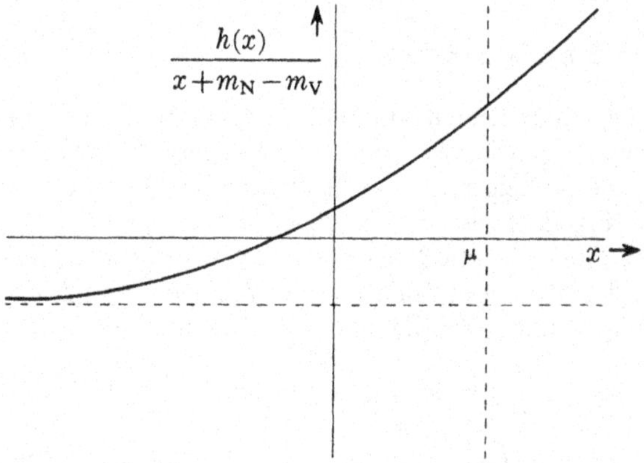

Fig. 12.5

[Källén and Pauli (Källén 1955b)] and it is found that the value of $\langle 0 \mid V_R \mid G\rangle \langle G \mid V_R^* \mid G\rangle$ is negative and given by $[h'(\Omega_0)]^{-1}$. It can further be shown that this negative term yields exactly the contribution necessary to reconcile (160) and (154). The "new" state has been dubbed the "ghost" state by Källén and Pauli. In the presence of the ghost state the only way to make the formalism consistent is to accept these states as appearing with negative norms. It turns out that such negative probabilities imply that the S matrix is not unitary, and the model when $Z_V < 0$ is not physically acceptable. Källén and Pauli [Källén (1955b)],

Heisenberg (1957b), Ferretti (1959) and others (see, for example, the discussion in the CERN High Energy Conference of 1957) have tried to give a consistent interpretation of the theory in the presence of ghost states by using a vector space with an indefinite metric for the representation of the states.[1] Such an approach has made some sense of the theory, but only at the price of giving up microscopic causality and the unitarity of the S matrix. In the subsequent, when referring to the Lee model, we shall limit ourselves to the case that Z_V is real and positive and δm_V is also real, so that no ghost problems arise. Note that this requires a cutoff, since for $f \to 1$, $Z_V < 0$, and the ghost difficulties are unavoidable.

Much attention has also been given to the Lee model for the case when $m_V > m_\theta + m_N$, i.e., when the V particle is unstable and can decay into an N and θ particle. The model can be used to test the definitions of the mass and lifetime of an unstable particle. The reader is referred to the articles of Lévy (1959a, b) for a general discussion of the description of unstable particles in quantum field theory, and in particular for the description of the unstable V particle in the Lee model [see also Glaser (1956), Araki (1957), Höhler (1958)].

12c. Other Simple Models

Besides the scalar field model and the Lee model, there exist several other simple models which are soluble in the sense that explicit representations of the one-particle states can be obtained. For several of them some of the scattering states can also be determined.

One of the most important of these is the Wentzel pair model [Wentzel (1941, 1942)]. It consists of a scalar field interacting with a source. The system is described by the following Hamiltonian:

$$H_0 = \tfrac{1}{2} \int d^3x \, \{\pi^2 + (\nabla\phi)^2 + \mu^2\phi^2\} + m_0\psi^*\psi \tag{173a}$$

$$H_I = \tfrac{1}{2}\lambda \int d^3x \int d^3x' \, U(\mathbf{x}) \, U(\mathbf{x}') : \phi(\mathbf{x}) \, \phi(\mathbf{x}') : \psi^*\psi \tag{173b}$$

with

$$[\psi, \psi^*]_+ = 1 \qquad [\psi, \psi]_+ = 0 \tag{174}$$

and

$$[\phi(\mathbf{x}), \pi(\mathbf{x}')] = i\delta^{(3)}(\mathbf{x} - \mathbf{x}') \tag{175}$$

[1] The interest of the aforementioned authors does, of course, not lie in the Lee model itself but in the question whether it is possible to give a physical interpretation to a field system for which it is necessary to represent the states in a vector space with an indefinite norm, and more particularly, whether this can be accomplished without too drastic a change of the usual rules of quantum mechanics [see, e.g., Ascoli and Minardi (1959)]. For a further appreciation of the relation of the Lee model to more realistic field theories, see the paper of Ford (1957).

and where $U(\mathbf{x})$ describes the shape of the source. It is usually assumed that $U(\mathbf{x}) = U(|\mathbf{x}|)$, in which case only S-wave mesons interact with the source. The model is completely soluble and has been discussed by many authors. In addition to Wentzel's original articles [Wentzel (1941, 1942)], the reader is referred to the paper of Klein and McCormick [Klein (1955a)] and to that of Chevalier and Rideau [Chevalier (1958)] who discuss in detail the nature of the solutions as a function of λ^2, the strength of the coupling constant. (A bound state of a real meson around the source can exist for $\lambda < 0$.) A somewhat related model has been treated by Enz (1956).

Another class of soluble models has been proposed by Ruijgrok and Van Hove [Ruijgrok (1956)]. The model is a nontrivial generalization of the Lee model which allows the heavy particles to emit and absorb an arbitrary number of quanta. The model in its simplest form describes the interaction between a heavy or V-particle field and a light or θ-particle field. The V quanta are fermions with two possible internal states, V_1 and V_2 (the N and V particles of the Lee model), and the θ quanta are spin zero bosons of mass μ. The Hamiltonian allows the following elementary transitions:

$$V_1 \leftrightarrows V_2 + \theta \quad \text{(coupling constant } \lambda_1\text{)}$$

$$V_2 \leftrightarrows V_1 + \theta \quad \text{(coupling constant } \lambda_2\text{)}$$

and is given by

$$H = H_0 + H_I$$

$$H_0 = \sum_{i=1}^{2} m_{V_i0} \int V_i^*(\mathbf{p})\, V_i(\mathbf{p})\, d\mathbf{p} + \int \omega_{\mathbf{k}} a^*(\mathbf{k})\, a(\mathbf{k})\, d\mathbf{k}$$

$$H_I = \lambda_1 \int d\mathbf{p} \int d\mathbf{k}\, \frac{f(\omega_{\mathbf{k}})}{\sqrt{2(2\pi)^3\, \omega_{\mathbf{k}}}} [V_1^*(\mathbf{p})\, V_2(\mathbf{p} - \mathbf{k})\, a(\mathbf{k}) + \text{h.a.}]$$
$$+ \lambda_2 \int d\mathbf{p} \int d\mathbf{k}\, \frac{f(\omega_{\mathbf{k}})}{\sqrt{2(2\pi)^3\, \omega_{\mathbf{k}}}} [V_2^*(\mathbf{p})\, V_1(\mathbf{p} - \mathbf{k})\, a(\mathbf{k}) + \text{h.a.}] \tag{176}$$

When $\lambda_1 = 0$ the Ruijgrok-Van Hove model becomes identical to the Lee model, whereas when $\lambda_1 = \lambda_2$ it corresponds to the scalar field model discussed in Section *a*. For arbitrary λ_1, λ_2 the one-particle states and their wave function renormalization constants can be obtained in closed form. The model has been extensively studied by Ruijgrok (1958a, b; 1959). [See also Greenberg (1958) and Lopuszanski (1959).]

The Ruijgrok-Van Hove model has certain interesting properties which have been adduced by Van Hove (1959) to raise some questions concerning the current field theoretic description of weak and strong interactions. When considered with a very large but finite cutoff, $\omega_{\max}$, the model gives rise to "observable" effects of two different types. One type, called the "weak interactions," has interaction effects of order $\mu/\omega_{\max}$ times those of the second type, which are called the strong interactions. The latter remain finite in the limit $\mu/\omega_{\max} \to 0$, i.e., in the limit the cutoff goes to infinity, whereas the weak interactions vanish in that limit. In the limit $\omega_{\max} \to \infty$ when only strong interactions are present, the system is characterized by certain symmetry properties, but these invariances are no longer exact for large but finite $\omega_{\max}$, i.e., in the presence of the weak interaction.

Such a description is in contradistinction to the customary way of describing strong and weak interactions. In the usual formulation one introduces a Lagrangian which contains as its main interaction terms, expressions corresponding to the strong interactions characterized by a "large" coupling constant and certain symmetries. The weak interactions are then included by adding to the Lagrangian "small" terms which violate the symmetries of the strong interaction terms. In the Ruijgrok-Van Hove model the Hamiltonian contains only what one would call strong interactions with the distinction between strong and weak interactions closely tied up with cutoff which operates at high energy. The distinction, incidentally, is only valid in the energy region $E \ll \omega_{\max}$, and nothing is said of the features of the model and of the nature of the interaction when $E \sim \omega_{\max}$ or $E > \omega_{\max}$.

Although the model is unrealistic, nonetheless it gives an indication of another direction in which an explanation for the interaction between elementary particles could be looked for.

12*d*. The Chew-Low Theory

The models previously considered in this chapter have the property that the one-particle states are exactly soluble. In some cases, such as the Lee model, some scattering states can also be calculated. Although these models are not directly applicable to physical phenomena, they serve important functions. For example, approximation methods can be carried out on them and compared with the exact solutions. Pathologies, such as ghost states, can be investigated completely and their origin established. Finally, they allow an insight into the structure of field theories.[2]

The next model we shall investigate, although "simple," is no longer exactly soluble. However, it is applicable to the description of low energy pion-nucleon phenomena: in fact, its importance derives from its success in accounting for low energy meson-nucleon scattering and photomeson production.

[2] An interesting paper in this connection is that of Haag and Luzzatto [Haag (1959)] who show that for the Lee model and the scalar field interacting with fixed sources, the equations of motion for the renormalized field can be formulated as differential equations involving only *finite* quantities, even in the limit of point interactions (i.e., $f(\omega) = 1$).

In some sense, which we will not attempt to justify here [see, however, Osborn (1952), Drell (1952)], the model for meson-nucleon interaction we shall be studying can be considered as the nonrelativistic limit of the γ_5 theory described by the Hamiltonian:

$$H = \int d^3x : \psi^*(\mathbf{x}) \, (\boldsymbol{\alpha} \cdot \mathbf{p} + \beta M_0) \, \psi(\mathbf{x}) : + \tfrac{1}{2} \int : \boldsymbol{\pi} \cdot \boldsymbol{\pi}(\mathbf{x}) + \nabla\boldsymbol{\phi} \cdot \nabla\boldsymbol{\phi}(\mathbf{x})$$

$$+ \mu_0{}^2 \boldsymbol{\phi} \cdot \boldsymbol{\phi}(\mathbf{x}) : + \tfrac{1}{2} G \int d^3x \, [\psi^*(\mathbf{x}) \, \boldsymbol{\tau}\gamma_5, \psi(\mathbf{x})] \cdot \boldsymbol{\phi}(\mathbf{x}) \quad (177)$$

where ψ is the (eight-component) nucleon field operator, $\boldsymbol{\phi}$ the pion field operator, and M_0 and μ_0 the bare masses of the nucleons and mesons, respectively. It can be shown that there exist successive canonical transformations which remove from this H terms which are odd in the Dirac matrices.[3] To order $1/M_0$, the transformed Hamiltonian which is free of odd Dirac matrices is given by

$$H' \approx \int d^3x : \psi^*(\mathbf{x}) \, \beta \left(M_0 + \frac{\mathbf{p}^2}{2M_0} \right) \psi(\mathbf{x}) : + H^{(0)}_{\text{meson}}$$

$$+ \frac{G}{2M_0} \frac{1}{2} \int d^3x \sum_{j=1}^{3} [\psi^*(\mathbf{x}) \, \tau_j \boldsymbol{\sigma}, \psi(\mathbf{x})] \cdot \nabla\phi_j(\mathbf{x}) + O\left(\frac{1}{M_0{}^2}\right) \quad (178)$$

We shall actually go one step further: we shall neglect recoil, i.e., we shall neglect the $\mathbf{p}^2/2M_0$ term in $H^{(0)}_{\text{nucleon}}$, and limit ourselves to the case when only one nucleon is present which is localized at the origin of our co-ordinate system. That is, we take as our Hamiltonian:

$$H = H_0 + H_I \quad (179a)$$

$$H_0 = H^{(0)}_{\text{meson}} + M_0 \psi^* \psi \quad (179b)$$

$$H_I = \frac{f_0}{\mu} \int d^3x \, \rho(\mathbf{x}) \, \psi^* (\boldsymbol{\sigma} \cdot \nabla) \, \boldsymbol{\tau} \cdot \boldsymbol{\phi}(\mathbf{x}) \, \psi$$

$$= \frac{f_0}{\mu} \int d^3x \, \rho(\mathbf{x}) \sum_{i=1}^{3} \sum_{j=1}^{3} \psi^* \sigma_i \tau_j \psi \partial_i \phi_j(\mathbf{x}) \quad (179c)$$

In (179) ψ is a four-component operator describing the nucleon. It transforms like a two-component spinor in both isotopic spin space and ordinary space; more precisely

$$\psi = \begin{pmatrix} \psi_{\text{p}} \\ \psi_{\text{n}} \end{pmatrix} \quad (180a)$$

where ψ_{p} and ψ_{n} are two-component spinors in ordinary space:

$$\psi_{\text{p}} = \begin{pmatrix} \psi_{p\uparrow} \\ \psi_{p\downarrow} \end{pmatrix} \qquad \psi_{\text{n}} = \begin{pmatrix} \psi_{n\uparrow} \\ \psi_{n\downarrow} \end{pmatrix} \quad (180b)$$

[3] This is the generalization of the Foldy-Wouthuysen transformation to the field theoretic case. Recall also Section 10g.

The latter satisfy the anticommutation rules:

$$\{\psi_p, \psi_p^*\} = \{\psi_n, \psi_n^*\} = 1 \tag{181a}$$

$$\{\psi_n, \psi_p\} = \{\psi_n^*, \psi_p^*\} = \{\psi_n, \psi_p^*\} = \cdots = \{\psi_n^*, \psi_n^*\} = 0 \tag{181b}$$

The source function $\rho(\mathbf{x})$ describes the extent of the meson-nucleon interaction region. It will be assumed spherically symmetric, i.e., $\rho(\mathbf{x}) = \rho(|\mathbf{x}|)$, and real. Furthermore, the cutoff function will be normalized so that

$$\int \rho(\mathbf{x})\, d^3x = 1 \tag{182}$$

If we introduce the Fourier transform $v(\mathbf{k}) = v(\mathbf{k}^2)$ of $\rho(\mathbf{x})$

$$v(\mathbf{k}^2) = \int e^{-i\mathbf{k}\cdot\mathbf{x}}\, \rho(\mathbf{x})\, d^3x \tag{183}$$

then, by virtue of Eq. (182), $v(0) = 1$. If R_0 is the radius of the interaction region, i.e., the region for which $\rho(\mathbf{x})$ is different from zero, then for **k**s such that $|\mathbf{k}|R_0 \gg 1$, $v(\mathbf{k}^2)$ will fall rapidly to zero. In all later applications, we shall choose

$$\begin{aligned} v(\mathbf{k}^2) &= 1 \quad \text{for } |\mathbf{k}| \le k_{max} \\ &= 0 \quad \text{for } |\mathbf{k}| \ge k_{max} \end{aligned} \tag{184}$$

where $k_{max} \sim 1/R_0$. We shall also speak of $\omega_{max} = \sqrt{\mathbf{k}^2_{max} + \mu^2}$ as the cutoff energy. The radius R_0 will always be assumed to be smaller than $1/\mu$. It will turn out to be of order $1/M$, where M is the nucleon mass.

The coupling constant $f/\mu = G/2M$ is called the pseudovector coupling constant. The theory described by the Hamiltonian (179) neglects the following effects which may have some bearing on pion-nucleon phenomena: (*a*) recoil of the nucleon, (*b*) antinucleon effects, (*c*) possible direct meson-meson interactions, (*d*) kaons and hyperon effects. In applying the model to low energy pion-nucleon interactions, it is therefore assumed that these effects are small and that in some sense they are included in the cutoff function $v(\mathbf{k}^2)$. Note that the Hamiltonian (179) is invariant under:

(*a*) rotations in space
(*b*) rotations in isotopic spin space
(*c*) time inversion
(*d*) space inversion

In fact, the form of the interaction Hamiltonian is essentially completely determined by the above invariance requirements, together with the Yukawa assumption that pions are emitted singly in the elementary interaction process: $\text{N} \leftrightarrows \text{N} + \pi$. The term $\nabla\boldsymbol{\phi}$ occurs since $\boldsymbol{\phi}$ is a pseudoscalar under spatial inversions, and the only true scalar under inversion which can be constructed from the operators $\boldsymbol{\phi}$, $\psi^*\psi$ and $\psi^*\boldsymbol{\sigma}\psi$ is $\psi^*(\boldsymbol{\sigma}\cdot\nabla)\,\boldsymbol{\tau}\cdot\boldsymbol{\phi}\psi$. The requirement that H_I be invariant under time inversion is responsible for the fact that no term in $\boldsymbol{\pi}(\mathbf{x})$, the momentum canonically conjugate to

$\boldsymbol{\phi}(\mathbf{x})$, occurs in H_I. The invariance of the theory under spatial rotations implies the constancy in time of the total angular momentum $\mathbf{J}$ of the system, where

$$J_i = \tfrac{1}{2}\psi^*\sigma_i\psi - \tfrac{1}{2}\int d^3x \sum_{jkl} [\pi_l(\mathbf{x}), \epsilon_{ijk}x_j\partial_k\phi_l(\mathbf{x})]_+ \tag{185}$$

with

$$[J_i, H] = 0 \quad i = 1, 2, 3 \tag{186}$$

Similarly, the statement that the theory is invariant under rotations of isotopic spin space can be translated into the constancy in time of the isotopic spin vector

$$T_i = \tfrac{1}{2}\psi^*\tau_i\psi + \int d^3x \sum_{jk} \epsilon_{ijk}\phi_j\pi_k \tag{187}$$

with

$$[T_i, H] = 0 \quad i = 1, 2, 3 \tag{188}$$

The theory has one further characteristic property: only P-wave mesons are coupled to the nucleon. The simplest way to see this is to expand the meson field operator, not in terms of one-particle eigenstates of the linear momentum:

$$\phi_i(x) = \frac{1}{(2\pi)^{3/2}} \int \frac{d^3k}{\sqrt{2\omega_{\mathbf{k}}}} (a_{i\mathbf{k}}\, e^{i\mathbf{k}\cdot\mathbf{x}} + a_{i\mathbf{k}}{}^*\, e^{-i\mathbf{k}\cdot\mathbf{x}}) \tag{189a}$$

$$[a_{i\mathbf{k}}, a_{j\mathbf{k}'}{}^*] = \delta_{ij}\delta^{(3)}(\mathbf{k} - \mathbf{k}') \tag{189b}$$ [4]

but in terms of a complete set of one-particle eigenstates of the total angular momentum, its $\mathcal{Z}$-component and the energy

$$\phi_i(x) = \sum_{k,l,m} \sqrt{\frac{1}{2\omega_{\mathbf{k}}}} (a_{iklm}Y_{lm}(\theta, \varphi) + (-1)^m a_{iklm}{}^*Y_{l,-m}(\theta, \varphi)) \sqrt{\frac{2k^2}{R}}\, j_l(kr); \; |\mathbf{k}| = k \tag{190a}$$ [5]

$$[a_{iklm}, a_{jk'l'm'}{}^*] = \delta_{ij}\delta_{kk'}\delta_{ll'}\delta_{mm'} \tag{190b}$$

If we substitute this expansion into H_I, only the term with $l = 1$ is large when the source size is less than the pion de Broglie wavelength since only the $l = 1$ little Bessel function $j_l(kr)$ has a finite derivative at the origin; the mesons with $l \neq 1$ do not interact with the nucleons and behave as if free.

The absence of interaction in states with $l \neq 1$ can also be understood

[4] Note we have used a different normalization of the operators a, a^* than was previously done in Chapter 7. This is the usual form for dealing with noncovariant systems.

[5] R is the radius of the quantization sphere. The $j_l(kr)$ are normalized so that

$$\frac{2k^2}{R}\int_0^R dr\, r^2 j_l^2(kr) = 1$$

in the following manner. The above model is based on the Yukawa model that the emission and absorption of mesons by the nucleon is one at a time. Since the total angular momentum is conserved in each elementary process $N \rightleftharpoons N + \pi$, the eigenvalue of the total angular momentum of the $N + \pi$ system, $(\frac{1}{2}\sigma + \mathbf{L})^2$, must be equal to the eigenvalue of the total angular momentum of nucleon, $(\frac{1}{2}\sigma)^2$, before the act of emission of the meson. Hence $l = 0$ or 1. Since parity is also conserved, the parity of the $N + \pi$ system, $(-1)^l(-1)$ [the extra factor (-1) is due to the fact that the pion has a negative intrinsic parity], must be equal to that of the nucleon, $(+1)$, so that $l = 1$. Note that these considerations are only valid for a nucleon which is maintained rigidly at rest and which does not recoil in the act of emission.

The above model for the description of low energy meson-nucleon phenomena has become known as the Chew model, after Chew who originally pointed out that the Hamiltonian (179) could, in fact, quantitatively account for the observed features of meson-nucleon scattering at low energies [Chew (1954)].

The vacuum can be defined as the eigenstate of H with eigenvalue 0, and is characterized by

$$\psi \mid 0\rangle = a_{i\mathbf{k}} \mid 0\rangle = 0 \quad \text{for all } i \text{ and all } \mathbf{k} \tag{191}$$

The one-meson state $a_{i\mathbf{k}} \mid 0\rangle$ is likewise an eigenfunction of H with eigenvalue $\omega_{\mathbf{k}}$. The simultaneous eigenfunctions of H, T^2, T_3, $J_3{}^2$ and J_3 corresponding to the presence of one meson are $\frac{1}{\sqrt{2}}(a_{1klm} \pm ia_{2klm})^* \mid 0\rangle$ and $a_{3klm}{}^* \mid 0\rangle$. The state $\frac{1}{\sqrt{2}}(a_{1klm} - ia_{2klm})^* \mid 0\rangle$ corresponds to the state of one positively-charged meson of angular momentum l, and 3-component of angular momentum m, etc. The bare and physical one-meson states are therefore identical in the present model.

The "bare" one-nucleon state $\psi^* \mid 0\rangle$ is not an eigenstate of H. We shall denote by $|s, t\rangle_+$ the eigenstates of H corresponding to the presence of a "physical" nucleon with the "observed" mass M, where s, and t denote the eigenvalue of J_3 and T_3 respectively. The state $|\frac{1}{2}; \frac{1}{2}\rangle_+$ corresponds to a proton with spin-up, $|-\frac{1}{2}, \frac{1}{2}\rangle_+$ to a proton with spin-down, $|\frac{1}{2}; -\frac{1}{2}\rangle_+$ to a neutron with spin-up, and $|-\frac{1}{2}, -\frac{1}{2}\rangle_+$ to a neutron with spin-down. These four states are degenerate with

$$H \mid s, t\rangle_+ = M \mid s, t\rangle_+ \tag{192}$$

where M is the mass of the dressed nucleon and is identified with the mass of the physical nucleon. We shall denote the bare one-nucleon states by $|s, t\rangle$.

In the following we shall be concerned only with the properties of either the one-nucleon system or the meson-nucleon system, i.e., of states which always contain one nucleon. We can therefore omit the ψ operators in H and write for the Hamiltonian of our meson-one nucleon system:

$$H = H^{(0)}_{\text{meson}} + M_0 + H_I \tag{193a}$$

$$H_I = \frac{f_0}{\mu} \int d^3x \, \rho(\mathbf{x}) \, (\boldsymbol{\sigma} \cdot \boldsymbol{\nabla}) \, \boldsymbol{\phi}(\mathbf{x}) \cdot \boldsymbol{\tau}$$
$$= \frac{f_0}{\mu} \sum_{i,j=1}^{3} d^3x \, \tau_j \sigma_i \frac{\partial \phi_j(\mathbf{x})}{\partial x_i} \rho(\mathbf{x}) \tag{193b}$$

The unperturbed state vectors $|n\rangle$ in occupation number space are now of the form $|s, t; \mathbf{k}^+_1, \cdots \mathbf{k}^+_m; \mathbf{k}^-_1, \cdots \mathbf{k}^-_n; \mathbf{k}^0_1, \cdots \mathbf{k}^0_r\rangle$ where the operator σ operates on the s (spin) variable, the τ operator on the t (isotopic) spin variable, and $\mathbf{k}^+$, $\mathbf{k}^-$, $\mathbf{k}^0$ denote the momenta of the positive, negative, and neutral mesons, respectively.

As in the models considered earlier in this chapter, the mass M in Eq. (192) is introduced into the theory "phenomenologically." One can re-express the Hamiltonian in terms of the observable mass M by writing

$$H = M_0 + H^{(0)}_{\text{meson}} + H_I \tag{194a}$$
$$= (M_0 + \Delta M) + H^{(0)}_{\text{meson}} + H_I - \Delta M \tag{194b}$$
$$= M + H^{(0)}_{\text{meson}} + (H_I - \Delta M) = M + H^{(0)}_{\text{meson}} + H_I' \tag{194c}$$

where ΔM is the mass shift introduced by the perturbation. Since M_0 is unobservable, this parameter can always be adjusted so that $M_0 + \Delta M$ is equal to the observed mass. In the following, it will be useful to consider $M + H^{(0)}_{\text{meson}} = H_0$ as the unperturbed Hamiltonian and $H_I' = H_I - \Delta M$ as the interaction Hamiltonian. With this separation of H into a perturbed and an unperturbed part, the bare one-nucleon state has the same mass, M, as the "physical" nucleon. In general, this particular separation implies that H has the same continuous spectrum as H_0. This, it will be recalled, was one of the assumptions in the Lippmann-Schwinger formalism.

An integral equation for the one-nucleon eigenstate of H, $|s, t\rangle_+$ can be obtained [Wick (1955)] by noting that if we decompose $|s, t\rangle_+$ into a vector along $|s, t\rangle$, the unperturbed state, and a vector orthogonal to $|s, t\rangle$, i.e., write[6]

$$|s, t\rangle_+ = \sqrt{Z_2} \, | \, s, t\rangle + |\chi\rangle \tag{195}$$

with

$$P_N \, | \, \chi\rangle = 0 \tag{196a}$$

$$P_N = \sum_{s,t} | \, s, t\rangle \langle s, t| \tag{196b}$$

then

$$(H_0 - M) \, | \, s, t\rangle_+ = (H_0 - M) \, | \, \chi\rangle \tag{197}$$

[6] The invariance properties of H_I guarantee that only that unperturbed one-nucleon state $|s, t\rangle$ with the same symmetry quantum numbers as $|s, t\rangle_+$ occurs in the expansion (195).

The factor $Z_2^{1/2}$ is a normalization constant. Since $(H_0 + H_I') \mid s, t\rangle_+ = M \mid s, t\rangle_+$

$$(H_0 - M) \mid \chi\rangle = -H_I' \mid s, t\rangle_+ \tag{198}$$

We also notice that, since P_N and H_0 commute, $P_N H_I' \mid s, t\rangle_+ = 0$. Recalling that the vector $|\chi\rangle$ has no component along $|s, t\rangle$, upon solving (198), we find

$$|\chi\rangle = (1 - P_N) \frac{1}{M - H_0} H_I' \mid s, t\rangle_+ \tag{199}$$

so that

$$|s, t\rangle_+ = Z_2^{1/2} \mid s, t\rangle + (1 - P_N) \frac{1}{M - H_0} H_I' \mid s, t\rangle_+ \tag{200}$$

Upon iterating this equation, we obtain the Wigner-Brillouin perturbation theoretic expansion of the physical state $|s, t\rangle_+$ in terms of bare states:

$$|s, t\rangle_+ = Z_2^{1/2}\left[1 + (1 - P_N) \frac{1}{M - H_0} H_I' + \frac{(1 - P_N)}{M - H_0} H_I' \frac{(1 - P_N)}{M - H_0} H_I' + \cdots\right] | s, t\rangle \tag{201}$$

The mass shift is now obtained from the equality

$$\begin{aligned}\langle s, t \mid H - H_0 \mid s, t\rangle_+ &= (M - M) \langle s, t \mid s, t\rangle_+ = 0 \\ &= \langle s, t \mid H_I' \mid s, t\rangle_+ = \langle s, t \mid H_I - \Delta M \mid s, t\rangle_+ \end{aligned} \tag{202}$$

whence, since $\langle s, t \mid s, t\rangle_+ = Z_2^{1/2}$

$$\Delta M = Z_2^{-1/2} \langle s, t \mid H_I \mid s, t\rangle_+ \tag{203}$$

By substituting into (203) the perturbation expansion of $|s, t\rangle_+$, Eq. (201), the contributions to ΔM in successive orders of perturbation theory are easily computed. To first order, the mass shift vanishes since H_I creates or annihilates a meson:

$$\Delta M^{(1)} = \langle s, t \mid H_I \mid s, t\rangle = 0 \tag{204}$$

More generally, $\langle s', t' \mid H_I \mid s, t\rangle = 0$. To second order the mass shift is given by

$$\Delta M^{(2)} = \langle s, t \mid H_I (1 - P_N) \frac{1}{M - H_0} H_I \mid s, t\rangle \tag{205a}$$

$$= \sum_{\substack{|n\rangle \\ \neq \text{ one nucleon} \\ \text{states}}} \frac{\langle s, t \mid H_I \mid n\rangle \langle n \mid H_I \mid s, t\rangle}{M - \epsilon_n} \tag{205b}$$

where the states $|n\rangle$ are eigenstates of H_0 with eigenvalue ϵ_n. The projection operator $1 - P_N$ in (205a) asserts that in the sum over $|n\rangle$, the one-nucleon states are to be omitted. Due to the symmetry properties of H_I only those states $|n\rangle$ with $J = \frac{1}{2}$, $J_3 = s$, and $T = \frac{1}{2}$, $T_3 = t$ contribute. The second-order mass shift is readily computed using the expression

(205a). We first note that we can write the interaction Hamiltonian in the concise form:

$$H_I = \sum_{j\mathbf{k}} (a_{j\mathbf{k}} V_{j\mathbf{k}}^{(0)} + a_{j\mathbf{k}}^* V_{j\mathbf{k}}^{(0)*}) \tag{206}$$

where

$$V_{j\mathbf{k}}^{(0)} = \frac{f_0}{\mu} \frac{iv(\mathbf{k}^2)}{\sqrt{2(2\pi)^3 \omega_\mathbf{k}}} \boldsymbol{\sigma} \cdot \mathbf{k} \tau_j \tag{207}$$

Equation (206) is obtained by substituting the expansion (189) into (193b). Now in the expression (205a) for the second-order mass shift, only the $\sum_\mathbf{k} V_{j\mathbf{k}}^{(0)*} a_{j\mathbf{k}}^*$ term of the H_I standing furthest to the right contributes, since $a_{j\mathbf{k}} \mid s, t\rangle = 0$ for all j and all $\mathbf{k}$. Similarly only the $\sum_\mathbf{k} V_{j\mathbf{k}}^{(0)} a_{j\mathbf{k}}$ term of the H_I standing furthest to the left contributes, so that

$$\Delta M^{(2)} = \langle s, t \mid \sum_{\mathbf{k}, r} \sum_{\mathbf{k}', r'} a_{r\mathbf{k}} V_{r\mathbf{k}}^{(0)} \frac{1}{M - H_0} V_{r'\mathbf{k}'}^{(0)*} a_{r'\mathbf{k}'}^* \mid s, t\rangle \tag{208a}$$

$$= \langle s, t \mid \sum_{\mathbf{k}, r} \sum_{\mathbf{k}', r'} a_{r\mathbf{k}} a_{r'\mathbf{k}'}^* V_{r\mathbf{k}}^{(0)} V_{r'\mathbf{k}'}^{(0)*} \frac{1}{M - H_0 - \omega_{\mathbf{k}'}} \mid s, t\rangle \tag{208b}$$

$$= \langle s, t \mid \sum_{\mathbf{k}, r} \frac{V_{r\mathbf{k}}^{(0)} V_{r\mathbf{k}}^{(0)*}}{-\omega_\mathbf{k}} \mid s, t\rangle \tag{208c}$$

In going from (208a) to (208b) we have used Eq. (118), the fact that $V_{r\mathbf{k}}^{(0)}$ and $a_{r\mathbf{k}}$ commute, as well as the fact that $V_{r\mathbf{k}}$ and H_0 commute. In going to (208c) we have used the commutation rules $[a_{r\mathbf{k}}, a_{r'\mathbf{k}'}^*] = \delta_{rr'} \delta^{(3)}(\mathbf{k} - \mathbf{k}')$ and the fact that $a_{r\mathbf{k}} \mid s, t\rangle = 0$. Hence to second order

$$\Delta M^{(2)} = \langle s, t \mid \frac{f_0^2}{\mu^2} \int d^3k \sum_r \frac{(\boldsymbol{\sigma} \cdot \mathbf{k})\,(\boldsymbol{\sigma} \cdot \mathbf{k})\, \tau_r \tau_r}{-\omega_\mathbf{k}} \frac{|v(\mathbf{k}^2)|^2}{2(2\pi)^3 \omega_\mathbf{k}} \mid s, t\rangle$$

$$= -\frac{3(4\pi)}{2(2\pi)^3} \frac{f_0^2}{\mu^2} \int_0^\infty k^2\, dk \frac{k^2}{k^2 + \mu^2} \mid v(\mathbf{k}^2)|^2 \tag{209}$$

The normalization constant $\sqrt{Z_2}$ is determined from the requirement that

$${}_+\langle s, t \mid s', t'\rangle_+ = \delta_{ss'} \delta_{tt'} \tag{210}$$

from which it follows that

$$Z_2 = 1 - {}_+\langle s, t \mid H_I'(1 - P_N) \frac{1}{(H_0 - M)^2} H_I' \mid s, t\rangle_+ \tag{211}$$

The normalization constant $\sqrt{Z_2}$ is also equal to

$$\sqrt{Z_2} = \langle s, t \mid s, t\rangle_+ \tag{212}$$

i.e., to the amplitude of finding a bare nucleon state in the physical nucleon state. To lowest order, Eq. (211) asserts that

$$Z_2 = 1 - Z_2 \langle s, t \mid \sum_{r, \mathbf{k}} V_{r\mathbf{k}}^{(0)} a_{r\mathbf{k}} V_{r'\mathbf{k}'}^{(0)*} a_{r'\mathbf{k}'}^* \frac{1}{\omega_{\mathbf{k}'}^2} \mid s, t \rangle$$

$$= 1 - Z_2 \left(\frac{3f_0^2}{2(2\pi)^3 \mu^2} \int d^3k \frac{v^2(\mathbf{k}^2)}{\omega_{\mathbf{k}}^3} \right) \tag{213a}$$

or

$$Z_2 = \left(1 + \frac{3f_0^2}{2(2\pi)^3 \mu^2} \int d^3k \frac{v^2(\mathbf{k}^2)}{\omega_{\mathbf{k}}^3} \right)^{-1}$$

$$\approx 1 - \frac{3f_0^2}{2(2\pi)^3 \mu^2} \int d^3k \frac{v^2(\mathbf{k}^2)}{\omega_{\mathbf{k}}^3} \pm \cdots \tag{213b}$$

for $f_0^2/\mu^2 \ll 1$. The physical one-nucleon eigenstate to first order is given by

$$|s, t\rangle_+ = |s, t\rangle + (1 - P_N) \frac{1}{M - H_0} H_I \mid s, t\rangle$$

$$= |s, t\rangle + \sum_{\mathbf{k}, r} - \frac{1}{\omega_{\mathbf{k}}} V_{r\mathbf{k}}^{(0)} a_{r\mathbf{k}}^* \mid s, t\rangle \tag{214}$$

It has a nonvanishing amplitude for the presence of a bare nucleon and a (virtual) meson. This state is properly normalized to order f_0/μ.

The higher order contributions to the mass shift, normalization constant, etc. are similarly computed. Actually, the discussion of the perturbation expansions is most easily discussed in terms of graphs [see Wick (1955) for greater detail]. Thus to the self-energy term $\Delta M^{(2)}$ is associated the diagram illustrated in Figure 12.6 corresponding to the emission and reabsorption of a meson of momentum **k**. More precisely, the nucleon is represented by a straight horizontal line, and each meson by a dashed line. A dashed line terminating on a nucleon line represents the absorption or emission of a meson. Thus in the graph corresponding to the second-order self-energy term, the vertex at a corresponds to the term $\langle n \mid H_I \mid s, t\rangle$ in (205) and represents the emission of a meson; similarly the vertex at b corresponds to the term $\langle s, t \mid H_I \mid n\rangle$ in (205). In the diagram, the initial state is exhibited at the extreme right of the diagram and the final state at the extreme left (as in the matrix element to which it corresponds). Conversely, given the diagram of Figure 12.6, one can immediately write

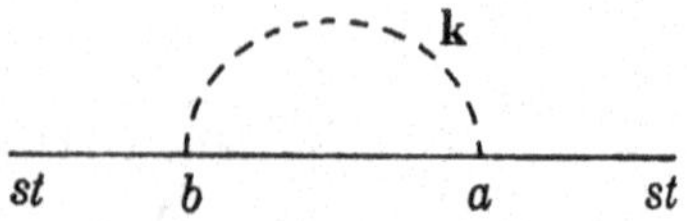

Fig. 12.6

down the corresponding matrix element. We shall not delve into the graphical methods at this point since they are lucidly set forth in the excellent review article by Wick (1955) and in Chew (1954). Graphical methods will be expounded at length for the relativistic theories (Chapter 13). Furthermore, our primary concern will be with nonperturbation theoretic methods in dealing with the Chew model.

Our next task is to compute the predictions of the Chew model concerning the scattering of mesons by nucleons. We shall first briefly present the perturbation theoretic methods and then consider at greater length the nonperturbation theoretic methods due to Low (1955), Wick (1955), and Chew and Low (1956a).

Consider first the predicted scattering in the lowest order of perturbation theory, i.e., in Born approximation. We are to calculate the amplitude for scattering of meson from an initial state $|\mathbf{k}, i; s, t\rangle = a_{i\mathbf{k}} \mid s, t\rangle$, in which the meson has a momentum $\mathbf{k}$ and isotopic spin index i, to a final state $|\mathbf{k}', j'; s', t'\rangle = a_{j'\mathbf{k}'} \mid s', t'\rangle$. Since H_0 and H_I are invariant under rotations in isotopic spin space, the transition matrix element $R_{fi} = \langle f \mid R \mid i\rangle$, where

$$R_{fi} = (H_I')_{fi} + \sum_{j(\neq i)} (H_I')_{fj} \frac{1}{E_i - E_j + i\epsilon} (H_I')_{ji} + \cdots \tag{215}$$

vanishes unless the final and initial states are states of the same total and $\mathcal{3}$-component of isotopic spin. It will therefore be useful to classify initial and final states according to the values of T^2 and T_3.

The noninteracting states which are eigenstates of T^2 and T_3 and which correspond to the presence of a nucleon and a single meson can easily be constructed. Since they are compounded from states with $T = 1$ (meson) and $T = \frac{1}{2}$ (nucleon), there will be two sets of meson-nucleon states, one set with $T = \frac{3}{2}$ and the other with $T = \frac{1}{2}$. The state with $T = \frac{3}{2}$ and $T_3 = \frac{3}{2}$ clearly corresponds to the state $|\mathrm{p}; \pi^+\rangle$, i.e., to the state composed of a proton and a positive meson. We thus write, suppressing the momenta of the particles,

$$|T = \tfrac{3}{2}, T_3 = \tfrac{3}{2}\rangle = |\tfrac{3}{2}, \tfrac{3}{2}\rangle = |\mathrm{p}; \pi^+\rangle \tag{216}$$

Since for the meson-nucleon system

$$T_- = \tau_- + \sqrt{2} \int d^3k \left[\frac{1}{\sqrt{2}} (a_{1\mathbf{k}} + ia_{2\mathbf{k}})^* a_{3\mathbf{k}} - a_{3\mathbf{k}}^* \frac{1}{\sqrt{2}} (a_{1\mathbf{k}} - ia_{2\mathbf{k}})\right] \tag{217}$$

we find

$$|\tfrac{3}{2}, \tfrac{1}{2}\rangle = \sqrt{\tfrac{2}{3}} \mid \mathrm{p}; \pi^0\rangle - \sqrt{\tfrac{1}{3}} \mid \mathrm{n}; \pi^+\rangle \tag{218}$$

$$|\tfrac{3}{2}, -\tfrac{1}{2}\rangle = \sqrt{\tfrac{1}{3}} \mid \mathrm{p}; \pi^-\rangle + \sqrt{\tfrac{2}{3}} \mid \mathrm{n}; \pi^0\rangle \tag{219}$$

$$|\tfrac{3}{2}, -\tfrac{3}{2}\rangle = |\mathrm{n}; \pi^-\rangle \tag{220}$$

where the overall phase of the states has been so chosen as to agree with the Bethe-deHoffmann choice [Bethe (1955)]. Similarly the states with

$T = \frac{1}{2}$, $T_3 = \frac{1}{2}$ must be that linear combination of $|p; \pi^0\rangle$ and $|n; \pi^+\rangle$ which is orthogonal to the $|\frac{3}{2}, \frac{1}{2}\rangle$ state, hence

$$|\tfrac{1}{2}, \tfrac{1}{2}\rangle = \sqrt{\tfrac{1}{3}}\,|\,p; \pi^0\rangle + \sqrt{\tfrac{2}{3}}\,|\,n; \pi^+\rangle \tag{221}$$

and

$$|\tfrac{1}{2}, -\tfrac{1}{2}\rangle = \sqrt{\tfrac{2}{3}}\,|\,p; \pi^-\rangle - \sqrt{\tfrac{1}{3}}\,|\,n; \pi^0\rangle \tag{222}$$

Upon inverting the above equations, we find that

$$|p; \pi^-\rangle = \sqrt{\tfrac{1}{3}}\,|\,\tfrac{3}{2}, -\tfrac{1}{2}\rangle + \sqrt{\tfrac{2}{3}}\,|\,\tfrac{1}{2}, -\tfrac{1}{2}\rangle \tag{223}$$

and

$$|p; \pi^0\rangle = \sqrt{\tfrac{2}{3}}\,|\,\tfrac{3}{2}, \tfrac{1}{2}\rangle + \sqrt{\tfrac{1}{3}}\,|\,\tfrac{1}{2}, \tfrac{1}{2}\rangle \tag{224}$$

Since the transition operator R commutes with T_i, $i = 1, 2, 3$ the scattering amplitude $R(T', T_3'; \mathbf{k}, \mathbf{k}', s', s)$ defined by

$$\langle T'', T_3''; \mathbf{k}'', s'' \,|\, R \,|\, T', T_3'; \mathbf{k}', s'\rangle = R(T', T_3'; \mathbf{k}'', \mathbf{k}', s'', s')\, \delta_{T'T''}\delta_{T_3''T_3'} \tag{225}$$

and corresponding to the scattering from an initial state with total and 3-component isotopic spin T' and T_3', meson momentum $\mathbf{k}'$ and nuclear spin s', to a final one with the same isotopic spin quantum numbers but meson momentum $\mathbf{k}''$ and nuclear spin s'', does not depend on T_3'. *Proof:*

$$\begin{aligned}\langle T', T_3' \,|\, T_-RT_+ \,|\, T', T_3'\rangle &= (T' - T_3')\,(T' + T_3' + 1)\,\langle T', T_3' + 1 \,|\, R \,|\, T', T_3' + 1\rangle \\ &= \langle T', T_3' \,|\, RT_-T_+ \,|\, T', T_3'\rangle \\ &= \langle T', T_3' \,|\, R(T^2 - T_3^2 - T_3) \,|\, T', T_3'\rangle \\ &= (T' - T_3')\,(T' + T_3' + 1)\,\langle T', T_3' \,|\, R \,|\, T', T_3'\rangle \end{aligned} \tag{226a}$$

hence

$$\langle T', T_3' + 1 \,|\, R \,|\, T', T_3' + 1\rangle = \langle T', T_3' \,|\, R \,|\, T', T_3'\rangle \tag{226b}$$

Equivalently, we can say that R is a scalar in isotopic spin space, and therefore by the Wigner-Eckhart theorem [see e.g., Rose (1957) or Edmond (1957)] the selection rules are $\Delta T = 0$, $\Delta T_3 = 0$ and the matrix element is independent of T_3'. The scattering of mesons by nucleons is thus described by amplitudes $R(T; \mathbf{k}, \mathbf{k}', s', s)$ where $T = \frac{3}{2}$ or $T = \frac{1}{2}$. The relation between the scattering amplitudes for the experimentally feasible processes

$$\begin{aligned} &\pi^+ + p \to \pi^+ + p \\ &\pi^- + p \to \pi^- + p \\ &\pi^- + p \to \pi^0 + n \qquad \text{(charge exchange)} \end{aligned}$$

and those of definite isotopic spin are

$$\begin{aligned} R(\pi^+ + p \to \pi^+ + p) &= {}_-\langle \pi^+; p \,|\, \pi^+; p\rangle_+ \\ &= {}_-\langle \tfrac{3}{2}, \tfrac{3}{2} \,|\, \tfrac{3}{2}, \tfrac{3}{2}\rangle_+ = R(3) \end{aligned} \tag{227}$$

where $R(3)$ is the abbreviation for $R(3; \mathbf{k}', \mathbf{k}, s', s)$ and we have suppressed the dependence of the scattering amplitude R on the variables other than T. Similarly,

$$\begin{aligned} R(\pi^- + \mathrm{p} \to \pi^- + \mathrm{p}) & \\ &= {}_-\langle \pi^-; \mathrm{p} \mid \pi^-; \mathrm{p}\rangle_+ \\ &= \{\sqrt{\tfrac{1}{3}}\, {}_-\langle \tfrac{3}{2}, -\tfrac{1}{2} \mid + \sqrt{\tfrac{2}{3}}\, {}_-\langle \tfrac{1}{2}, -\tfrac{1}{2} \mid\} \{\sqrt{\tfrac{1}{3}} \mid \tfrac{3}{2}, -\tfrac{1}{2}\rangle_+ + \sqrt{\tfrac{2}{3}} \mid \tfrac{1}{2}, -\tfrac{1}{2}\rangle_+\} \\ &= \tfrac{1}{3}\, {}_-\langle \tfrac{3}{2}, -\tfrac{1}{2} \mid \tfrac{3}{2}, -\tfrac{1}{2}\rangle_+ + \tfrac{2}{3}\, {}_-\langle \tfrac{1}{2}, -\tfrac{1}{2} \mid \tfrac{1}{2}, -\tfrac{1}{2}\rangle_+ \\ &= \tfrac{1}{3}R(3) + \tfrac{2}{3}R(1) \end{aligned} \tag{228}$$

and

$$\begin{aligned} R(\pi^- + \mathrm{p} \to \pi^0 + \mathrm{n}) & \\ &= {}_-\langle \pi^-; \mathrm{p} \mid \pi^0; \mathrm{n}\rangle_+ \\ &= \{\sqrt{\tfrac{1}{3}}\, {}_-\langle \tfrac{3}{2}, -\tfrac{1}{2} \mid + \sqrt{\tfrac{2}{3}}\, {}_-\langle \tfrac{1}{2}, -\tfrac{1}{2} \mid\} \{\sqrt{\tfrac{2}{3}} \mid \tfrac{3}{2}, -\tfrac{1}{2}\rangle_+ - \sqrt{\tfrac{1}{3}} \mid \tfrac{1}{2}, -\tfrac{1}{2}\rangle_+\} \\ &= \sqrt{\tfrac{2}{9}}\,(R(3) - R(1)) \end{aligned} \tag{229}$$

The differential cross sections for the processes (227)–(229) are given by

$$\frac{d\sigma_{ab}}{d\Omega} = \frac{\omega^2 |R_{ab}|^2}{(2\pi)^2} \tag{230}$$

where R_{ab} is the ab matrix element of R corresponding to the process $a \to b$.

The derivation of (230) proceeds in the following familiar fashion: It will be recalled that the transition probability per unit time from a state $|i\rangle$ to a state $|f\rangle$ is given by

$$w_{fi} = 2\pi\delta(E_f - E_i)\, |R_{fi}|^2 \tag{231}$$

We are interested only in the transition probability to a group of final states with energy around E_f and momenta between $\mathbf{k}_f$ and $\mathbf{k}_f + d\mathbf{k}_f$. The latter is given by

$$\begin{aligned} w &= \int dn_f\, 2\pi\delta(E_f - E_i)\, |R_{fi}|^2 \\ &= 2\pi \left(\frac{dn_f}{dE}\right)_{E=E_f} |R_{fi}|^2 \end{aligned} \tag{232}$$

The cross section for scattering is related to w by

$$d\sigma = w/v_i \tag{233}$$

where $v_i = \omega_i/k_i$ is the velocity of the incident meson. Now the number of final states with momenta between $\mathbf{k}$ and $\mathbf{k} + d\mathbf{k}$ is

$$dn = \frac{d\mathbf{k}}{(2\pi)^3} \tag{234}$$

so that[7]

$$\rho(E_f) = \left(\frac{dn_f}{dE}\right)_{E=E_f} = \frac{k_f^2\, dk_f\, d\Omega}{dE_f\, (2\pi)^3}$$

$$= \frac{k_f \omega_f\, d\Omega}{(2\pi)^3} \qquad (235)$$

since $E = \omega = \sqrt{\mathbf{k}^2 + \mu^2}$.

The calculation of the differential cross section thus reduces to the computation of $|R_{fi}|^2$. For practical purposes, it is somewhat easier to compute the amplitude $\langle \mathbf{k}', i'; s', t' \mid R \mid \mathbf{k}, i; s, t\rangle$ than the amplitude

$$\langle T, T_3; \mathbf{k}', s' \mid R \mid T, T_3; \mathbf{k}, s\rangle = R(T, T_3; \mathbf{k}', \mathbf{k}; s', s)$$
$$= R_{s's}(T; \mathbf{k}', \mathbf{k}) \qquad (236)$$

We shall therefore wish to relate $R_{s's}(T, T_3; \mathbf{k}', \mathbf{k})$ to $\langle \mathbf{k}', i'; s', t' \mid R \mid \mathbf{k}, i; s, t\rangle$. Of course, the standard Clebsch-Gordan coefficients of angular momentum theory could be used to transform the results computed in the i, t representation to the ones diagonal in T and T_3. The material to be presented in the next few pages is merely included for the sake of completeness.

By the standard methods of transformation theory, we note that $\langle \mathbf{k}', r'; s', t' \mid R \mid \mathbf{k}, r; s, t\rangle$

$$= \sum_{T,\,T_3} \langle r', t' \mid T, T_3\rangle \langle T, T_3; \mathbf{k}', s' \mid R \mid T, T_3; \mathbf{k}, s\rangle \langle T, T_3 \mid r, t\rangle$$

$$= \sum_{T,\,T_3} \langle r', t' \mid T, T_3\rangle R_{s's}(T; \mathbf{k}', \mathbf{k}) \langle T, T_3 \mid r, t\rangle$$

$$= \sum_{T} \langle r', t' \mid P(T) \mid r, t\rangle R_{s's}(T; \mathbf{k}', \mathbf{k}) \qquad (237)$$

where $\langle r', t' \mid T, T_3\rangle$ are Clebsch-Gordan coefficients and

$$P(T) = \sum_{T_3} |T, T_3\rangle \langle T, T_3| \qquad (238a)$$

is the projection operator onto the states with total isotopic spin T, so that

$$\langle r', t' \mid P(T) \mid r, t\rangle = \sum_{T_3} \langle r', t' \mid T, T_3\rangle \langle T, T_3 \mid r, t\rangle \qquad (238b)$$

In the following it will be convenient to make use of the isomorphism between $\mathbf{T}$ and $\mathbf{t} + \frac{1}{2}\boldsymbol{\tau}$ in the subspace consisting of one meson and one nucleon. The isotopic spin space corresponding to the meson-nucleon system is spanned by vectors $|r, t\rangle$ which can be represented by outer products of (meson) three-vectors, $|r\rangle$ and (nucleon) two-component spinors $|t\rangle$, on which the isotopic spin operators $\mathbf{t}$ (of the meson) and $\boldsymbol{\tau}$ (of the nucleon) act.

[7] For convenience, we have taken unit normalization volume $V = 1$. Otherwise the cross section which is related to w by $d\sigma = (w/\text{incident flux})$ would be given by Vw/v_i, and $dn = V d\mathbf{k}/(2\pi)^3$, but $|R_{fi}|^2 \propto 1/V^2$ so that V again cancels from the final expression for $d\sigma$.

The projection operators on the $T = \frac{3}{2}$ and $T = \frac{1}{2}$ states are then given by

$$P_T(\tfrac{3}{2}) = \frac{2 + \mathbf{t} \cdot \boldsymbol{\tau}}{3} \tag{239}$$

$$P_T(\tfrac{1}{2}) = \frac{1 - \mathbf{t} \cdot \boldsymbol{\tau}}{3} \tag{240}$$

with

$$P_T(\tfrac{3}{2}) + P_T(\tfrac{1}{2}) = 1 \tag{241}$$

Proof: The operator $\mathbf{T}^2 = (\mathbf{t} + \frac{1}{2}\boldsymbol{\tau})^2 = \mathbf{t}^2 + \mathbf{t} \cdot \boldsymbol{\tau} + \frac{1}{4}\boldsymbol{\tau}^2 = 2 + \mathbf{t} \cdot \boldsymbol{\tau} + \frac{3}{4}$ has the eigenvalue $\frac{15}{4}$ on a $T = \frac{3}{2}$ state, and the eigenvalue $\frac{3}{4}$ on a $T = \frac{1}{2}$ state, so that

$$\begin{aligned} \mathbf{t} \cdot \boldsymbol{\tau} &= 1 \qquad \text{for } T = \tfrac{3}{2} \\ &= -2 \quad \text{for } T = \tfrac{1}{2} \end{aligned}$$

By requiring that $[P_T(T)]^2 = P_T(T)$ and that (241) be satisfied, one easily constructs the explicit form (239) and (240).

The matrix element $\langle r', t' \mid P_T(T) \mid r, t\rangle$ can be further reduced by recalling that

$$\langle r' \mid t_j \mid r\rangle = -i\epsilon_{jr'r} \tag{242}$$

whence, for example,

$$\begin{aligned} \tfrac{1}{3}\langle r', t' \mid 1 - \mathbf{t} \cdot \boldsymbol{\tau} \mid r, t\rangle &= \tfrac{1}{3}\langle t' \mid \delta_{rr'} + i\epsilon_{jr'r}\tau_j \mid t\rangle \\ &= \tfrac{1}{3}\langle t' \mid \delta_{rr'} - \tfrac{1}{2}[\tau_r, \tau_{r'}] \mid t\rangle \\ &= \langle t' \mid \tfrac{1}{3}\tau_r\tau_{r'} \mid t\rangle \end{aligned} \tag{243}$$

where we have used the commutation rules $[\tau_{r'}, \tau_r] = 2i \sum_j \epsilon_{jr'r}\tau_j$ and $[\tau_{r'}, \tau_r]_+ = 2\delta_{rr'}$. Similarly,

$$\tfrac{1}{3}\langle r', t' \mid 2 + \mathbf{t} \cdot \boldsymbol{\tau} \mid r, t\rangle = \tfrac{1}{2}\langle t' \mid \tau_r\tau_{r'} + \tfrac{1}{3}\tau_{r'}\tau_r \mid t\rangle \tag{244}$$

whence

$$\langle t' \mid \tau_r\tau_{r'} \mid t\rangle = \langle r', t' \mid 2P_T(\tfrac{3}{2}) - P_T(\tfrac{1}{2}) \mid r, t\rangle \tag{245a}$$

$$\langle t' \mid \tau_{r'}\tau_r \mid t\rangle = \langle r', t' \mid 3P_T(\tfrac{1}{2}) \mid r, t\rangle \tag{245b}$$

We shall denote by $\mathfrak{J}_{1/2}$ and $\mathfrak{J}_{3/2}$ the following quantities:

$$\mathfrak{J}_{1/2}(r', r) = \tfrac{1}{3}\delta_{r'r} + \tfrac{1}{3}\,\tfrac{1}{2}[\tau_{r'}, \tau_r] \tag{246a}$$

$$\mathfrak{J}_{3/2}(r', r) = \tfrac{2}{3}\delta_{r'r} - \tfrac{1}{3}\,\tfrac{1}{2}[\tau_{r'}, \tau_r] \tag{246b}$$

They are projection operators [in the sense that $\sum_{r''} \mathfrak{J}_T(r', r'')\,\mathfrak{J}_{T'}(r'', r) = \delta_{TT'}\,\mathfrak{J}_T(r', r)$] on the $\frac{1}{2}$ and $\frac{3}{2}$ isotopic spin states in the nucleon isotopic spin space.

Completely analogous procedures can be worked out for the angular momentum states. Since $\mathbf{J}^2$ and J_3 as well as the parity operation commute with H it is convenient to classify initial and final states according to total and 3-component of the total angular momentum and to the parity of the system. Again, since R and J_i commute, the scattering amplitude

$$\begin{aligned} R(T, T_3; J, J_3; \omega_{\mathbf{k}}) &= \langle T, T_3; J, J_3, \omega(\mathbf{k}_f) \mid R \mid T, T_3; J, J_3; \omega(\mathbf{k}_i)\rangle \\ &= R(T, J; \omega(\mathbf{k})) = R_{TJ}(\omega_{\mathbf{k}}) \end{aligned} \tag{247}$$

does not depend on J_3; [$\omega = \omega(\mathbf{k}_f) = \omega(\mathbf{k}_i) = \omega_{\mathbf{k}}$ is the energy of the meson]. Therefore

$$\begin{aligned} &\langle T, T_3; \mathbf{k}', s' \mid R \mid T, T_3; \mathbf{k}, s\rangle \\ &= \sum_{JJ_3} \langle \mathbf{k}', s' \mid J, J_3\rangle \cdot \langle T, T_3; J, J_3, \omega(\mathbf{k}_f) \mid R \mid T, T_3; J, J_3; \omega(\mathbf{k}_i)\rangle \langle J, J_3 \mid \mathbf{k}, s\rangle \\ &= \sum_{JJ_3} \langle \mathbf{k}', s' \mid J, J_3\rangle \langle J, J_3 \mid \mathbf{k}, s\rangle R(T, J; \omega_{\mathbf{k}}) \\ &= \sum_{J} \langle \mathbf{k}', s' \mid P_J(J) \mid \mathbf{k}, s\rangle R_{TJ}(\omega_{\mathbf{k}}) \end{aligned} \tag{248}$$

where

$$P_J(J) = \sum_{J_3} |J, J_3\rangle \langle J, J_3| \tag{249}$$

is the projection operator on the states with total angular momentum J.

We have already noted that only $l = 1$ mesons interact with the nucleons; we shall therefore only be interested in the $P_J(\frac{3}{2})$ and $P_J(\frac{1}{2})$ projection operators since only in the states $J = \frac{3}{2}$ and $J = \frac{1}{2}$ is there any scattering. If we denote by $\mathbf{l}$ the orbital momentum of the meson and by $\frac{1}{2}\boldsymbol{\sigma}$ the spin angular momentum of the nucleon, then in the meson-nucleon subspace the field operator $\mathbf{J}$ is isomorphic to $\mathbf{l} + \frac{1}{2}\boldsymbol{\sigma}$. Whence, in complete analogy with the isotopic spin case

$$P_J(\tfrac{3}{2}) = \frac{2 + \mathbf{l} \cdot \boldsymbol{\sigma}}{3} \tag{250}$$

$$P_J(\tfrac{1}{2}) = \frac{1 - \mathbf{l} \cdot \boldsymbol{\sigma}}{3} \tag{251}$$

The states $|\mathbf{k}, s\rangle$ are now outer product of meson states $|\mathbf{k}\rangle = |\omega(\mathbf{k}), \theta, \varphi\rangle$ and nucleon spinors $|s\rangle$, where we have denoted, by θ, φ, the direction of the vector $\mathbf{k}$. The matrix element $\langle \omega(\mathbf{k}'), \theta', \varphi', s' \mid P_J(\frac{3}{2}) \mid \omega(\mathbf{k}), \theta, \varphi, s\rangle$ can be evaluated in a straightforward manner. With the previous normalization such that

$$\langle \theta\varphi \mid lm\rangle = \frac{1}{\sqrt{\omega k}} Y_{lm}(\theta, \varphi) \tag{252}$$

($k = |\mathbf{k}|$), one deduces that when $\omega(\mathbf{k}') = \omega(\mathbf{k})$, $k = k'$

$$\langle \mathbf{k}', s' \mid P_J(\tfrac{3}{2}) \mid \mathbf{k}, s\rangle = \frac{1}{4\pi k^2 \sqrt{k'\omega' k\omega}} \langle s' \mid 2\mathbf{k}' \cdot \mathbf{k} - i\mathbf{k}' \times \mathbf{k} \cdot \boldsymbol{\sigma} \mid s\rangle \tag{253}$$

and

$$\langle \mathbf{k}', s' \mid P_J(\tfrac{1}{2}) \mid \mathbf{k}, s\rangle = \frac{1}{4\pi k^2\sqrt{k'\omega' k\omega}} \langle s' \mid 2\mathbf{k}' \cdot \mathbf{k} + i\mathbf{k}' \times \mathbf{k} \cdot \boldsymbol{\sigma} \mid s\rangle \quad (254)$$

Using the fact that

$$(\boldsymbol{\sigma} \cdot \mathbf{k})\, (\boldsymbol{\sigma} \cdot \mathbf{k}') = \mathbf{k} \cdot \mathbf{k}' + i\mathbf{k} \times \mathbf{k}' \cdot \boldsymbol{\sigma} \quad (255)$$

we deduce that

$$\langle s' \mid (\boldsymbol{\sigma} \cdot \mathbf{k})\, (\boldsymbol{\sigma} \cdot \mathbf{k}') \mid s\rangle = \frac{4\pi k^2}{3} \sqrt{k'\omega' k\omega}\, \langle \theta', \varphi', s' \mid 2P_J(\tfrac{3}{2}) - P_J(\tfrac{1}{2}) \mid \theta, \varphi, s\rangle \quad (256a)$$

and

$$\langle s' \mid (\boldsymbol{\sigma} \cdot \mathbf{k}')\, (\boldsymbol{\sigma} \cdot \mathbf{k}) \mid s\rangle = 4\pi k^2\sqrt{k'\omega' k\omega}\, \langle \theta', \varphi', s' \mid P_J(\tfrac{1}{2}) \mid \theta, \varphi, s\rangle \quad (256b)$$

We also introduce the operators

$$\mathcal{J}_{1/2}(\mathbf{k}_2, \mathbf{k}_1) = \mathbf{k}_2 \cdot \mathbf{k}_1 + i\boldsymbol{\sigma} \cdot (\mathbf{k}_2 \times \mathbf{k}_1) \quad (257a)$$

$$\mathcal{J}_{3/2}(\mathbf{k}_2, \mathbf{k}_1) = 2\mathbf{k}_2 \cdot \mathbf{k}_1 - i\boldsymbol{\sigma} \cdot (\mathbf{k}_2 \times \mathbf{k}_1) \quad (257b)$$

which are projection operators on the $J = \frac{1}{2}$ and $J = \frac{3}{2}$ states in the nucleon spin space, in the sense that

$$\int d\Omega_{\mathbf{k}}\, \mathcal{J}_J(\mathbf{k}_2, \mathbf{k})\, \mathcal{J}_{J'}(\mathbf{k}, \mathbf{k}_1) = 4\pi \mathbf{k}^2\, \delta_{JJ'} \mathcal{J}_J(\mathbf{k}_2, \mathbf{k}_1) \quad (257c)$$

We are now ready to compute the transition matrix. To second order R_{fi} is given by

$$\begin{aligned} R_{fi} &= \langle \mathbf{k}', r'; s', t' \mid \Delta M^{(2)} \mid \mathbf{k}, r; s, t\rangle \\ &\quad + \langle \mathbf{k}', r'; s', t' \mid H_I \frac{1}{E_i - H_0 + i\epsilon} H_I \mid \mathbf{k}, r; s, t\rangle \end{aligned} \quad (258)$$

where $\Delta M^{(2)}$ is given by Eq. (208c) and H_I by

$$H_I = \sum_{j\mathbf{k}} (a_{j\mathbf{k}} V_{j\mathbf{k}}^{(0)} + a_{j\mathbf{k}}^* V_{j\mathbf{k}}^{(0)*}) \quad (259a)$$

$$V_{j\mathbf{k}}^{(0)} = i \frac{f_0}{\mu} \frac{v(\mathbf{k}^2)}{\sqrt{2\omega_{\mathbf{k}}(2\pi)^3}}\, \boldsymbol{\sigma} \cdot \mathbf{k}\tau_j \quad (259b)$$

Note that H_I creates or annihilates a meson; also that the operator $H_I G_{0+}(E_i)\, H_I$, where

$$G_{0+}(E) = \frac{1}{E - H_0 + i\epsilon} \quad (260)$$

must connect a one-meson state to a one-meson state. Therefore, if the H_I standing furthest to the right creates a meson, the second H_I must destroy a meson, since $G_{0+}(E_i)$ is diagonal. Similarly, if the first H_I destroys the meson in the initial state, the second H_I must create the meson in the final state. More specifically, since

$$\frac{1}{E_i - H_0 + i\epsilon} a_{i\mathbf{k}} = a_{i\mathbf{k}} \frac{1}{E_i - H_0 + \omega_{\mathbf{k}} + i\epsilon} \tag{261a}$$

$$\frac{1}{E_i - H_0 + i\epsilon} a_{i\mathbf{k}}{}^* = a_{i\mathbf{k}}{}^* \frac{1}{E_i - H_0 - \omega_{\mathbf{k}} + i\epsilon} \tag{261b}$$

and the state $|\mathbf{k}, r; s, t\rangle$ is an eigenstate of H_0 with eigenvalue $M + \omega_{\mathbf{k}}$, we have

$$R_{fi} = \langle \mathbf{k}', r'; s', t' \mid \Delta M^{(2)} + \sum_{\mathbf{q}, \mathbf{q}', j, j'} \left\{ \frac{a_{j\mathbf{q}}{}^* V_{j\mathbf{q}}^{(0)*} a_{j\mathbf{q}} V_{j\mathbf{q}}^{(0)}}{E_i - \omega_{\mathbf{k}} + \omega_{\mathbf{q}} - M + i\epsilon} + \frac{a_{j'\mathbf{q}'} V_{j'\mathbf{q}'}^{(0)} a_{j\mathbf{q}}{}^* V_{j\mathbf{q}}^{(0)*}}{E_i - \omega_{\mathbf{k}} - \omega_{\mathbf{q}} - M + i\epsilon} \right\} \mid \mathbf{k}, r; s, t\rangle \tag{262}$$

Since $E_i = \omega_{\mathbf{k}} + M$, upon using the commutation rules $[a_{j\mathbf{k}}, a_{j'\mathbf{k}'}{}^*] = \delta_{jj'}\delta^{(3)}(\mathbf{k} - \mathbf{k}')$ and the fact that $a_{j\mathbf{q}} \mid \mathbf{k}, r; s, t\rangle = a_{j\mathbf{q}} a_{r\mathbf{k}}{}^* \mid s, t\rangle = \delta_{jr}\delta^{(3)}(\mathbf{q} - \mathbf{k}) \mid s, t\rangle$, we finally obtain

$$R_{fi} = \langle s', t' \mid \Delta M^{(2)} \delta_{rr'} \delta^{(3)}(\mathbf{k} - \mathbf{k}') + \frac{V_{r'\mathbf{k}'}^{(0)*} V_{r\mathbf{k}}^{(0)}}{\omega_{\mathbf{k}}} - \frac{V_{r\mathbf{k}}^{(0)} V_{r'\mathbf{k}'}^{(0)*}}{\omega_{\mathbf{k}'}} + \sum_{\mathbf{q} j} \frac{V_{j\mathbf{q}}^{(0)} V_{j\mathbf{q}}^{(0)*}}{-\omega_{\mathbf{q}}} \delta^{(3)}(\mathbf{k} - \mathbf{k}') \delta_{rr'} \mid s, t\rangle \tag{263}$$

These four terms can be represented by the diagrams shown in Figure 12.7. The diagram 12.7*a* corresponds to the $\Delta M^{(2)}$ term acting singly. Diagram 12.7*b* corresponds to the V^*V term and it also represents the process

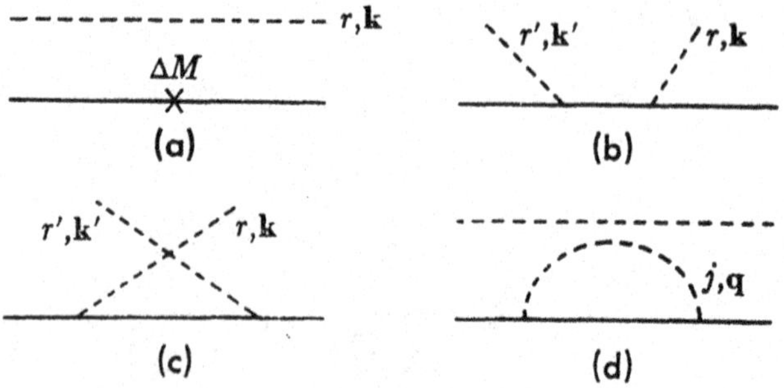

Fig. 12.7

wherein the incident meson is first destroyed and the final meson subsequently created. Diagram 12.7*c* corresponds to the process where the final meson is first created and the incident meson subsequently absorbed (VV^* term) and, finally, diagram 12.7*d* corresponds to the process where the incident meson does not interact with the nucleon, the latter emitting and reabsorbing a (virtual) meson. It will next be noticed that the $\Delta M^{(2)}$ term exactly cancels the second-order self-energy term, i.e., the term corresponding to diagram 12.7*d*, so that

$$R_{fi} = \langle s', t' \mid [V_{rk'}^{(0)*}, V_{rk}^{(0)}] \mid s, t\rangle \frac{1}{\omega_{\mathbf{k}'}}$$

$$= \frac{f_0^2}{\mu^2} \frac{v(\mathbf{k}'^2)\, v(\mathbf{k}^2)}{\sqrt{(2\pi)^6\, 4\omega_{\mathbf{k}}\omega_{\mathbf{k}'}}} \cdot \frac{1}{\omega_{\mathbf{k}'}} \tag{264a}$$

$$\cdot \langle s', t' \mid (\boldsymbol{\sigma} \cdot \mathbf{k}')\, (\boldsymbol{\sigma} \cdot \mathbf{k})\, \tau_{r'}\tau_r - (\boldsymbol{\sigma} \cdot \mathbf{k})\, (\boldsymbol{\sigma} \cdot \mathbf{k}')\, \tau_r\tau_{r'} \mid s, t\rangle \tag{264b}$$

The form of the scattering matrix in Born approximation should be noted, and, in particular, its dependence on the meson energy $\omega_{\mathbf{k}'}$. It will be noticed that R in Born approximation has a pole at $\omega_{\mathbf{k}'} = 0$ and that the residue is proportional to f_0^2. Making use of our previous results, Eqs. (247), (248) and (258), (259), we can now read off from (264b) the scattering amplitudes corresponding to scattering in states of definite total angular momentum and isotopic spin. We find, with the notation $R_{2J\,2T}(\omega)$:

$$R_{11}(\omega) = \frac{8}{3} \frac{f_0^2}{\mu^2} \frac{4\pi k^3 |v(\mathbf{k}^2)|^2}{2(2\pi)^3\, \omega_{\mathbf{k}}} \tag{265}$$

$$R_{13}(\omega) = \frac{2}{3} \frac{f_0^2}{\mu^2} \frac{4\pi k^3 |v(\mathbf{k}^2)|^2}{2(2\pi)^3\, \omega_{\mathbf{k}}} \tag{266}$$

$$R_{13}(\omega) = R_{31}(\omega) \tag{267}$$

$$R_{33}(\omega) = -\frac{4}{3} \frac{f_0^2}{\mu^2} \frac{4\pi k^3 |v(\mathbf{k}^2)|^2}{2(2\pi)^3\, \omega_{\mathbf{k}}} \tag{268}$$

Since J^2, J_3, T^2, T_3 are conserved in the scattering, the S matrix is diagonal in a representation labeled by J, J_3, T, T_3. Furthermore, since S is unitary, its eigenvalues are phase factors of modulus unity and one usually writes:

$$\langle J, T, \omega' \mid S \mid J, T, \omega\rangle = e^{2i\delta_{2J2T}(\omega)}\, \delta(\omega' - \omega) \tag{269}$$

where $\delta_{2J\,2T}$ are the phase shifts. Now, by definition $S_{fi} = (1_{fi} - 2\pi i R_{fi}) \cdot \delta(E_f - E_i)$, whence

$$\langle J', T', \omega \mid R \mid J', T', \omega\rangle = -\frac{1}{\pi} e^{i\delta_{2J2T}} \sin \delta_{2J\,2T} \tag{270}$$

In the present case, therefore,

$$e^{i\delta_{11}} \sin \delta_{11} = -\frac{8}{3} \left(\frac{f_0^2}{4\pi} \frac{|v(\mathbf{k}^2)|^2\, k^3}{\mu^2 \omega_{\mathbf{k}}}\right) \tag{271}$$

$$e^{i\delta_{13}} \sin \delta_{13} = -\frac{2}{3} \left(\frac{f_0^2}{4\pi} \frac{|v(\mathbf{k}^2)|^2\, k^3}{\mu^2 \omega_{\mathbf{k}}}\right) \tag{272}$$

$$\delta_{13} = \delta_{31} \tag{273}$$

$$e^{i\delta_{33}} \sin \delta_{33} = \frac{4}{3} \left(\frac{f_0^2}{4\pi} \frac{|v(\mathbf{k}^2)|^2\, k^3}{\mu^2 \omega_{\mathbf{k}}}\right) \tag{274}$$

Thus in Born approximation all the phase shifts are of the same order of magnitude. This is not what is observed experimentally. The low energy meson-nucleon scattering data yields phase shifts δ_{11}, δ_{13} which are small compared to δ_{33}. In fact, experimentally, the dominant feature of meson-nucleon scattering is that the scattering in the 33 state is much larger than that in 31, 13, and 11 states. Figure 12.8 shows the

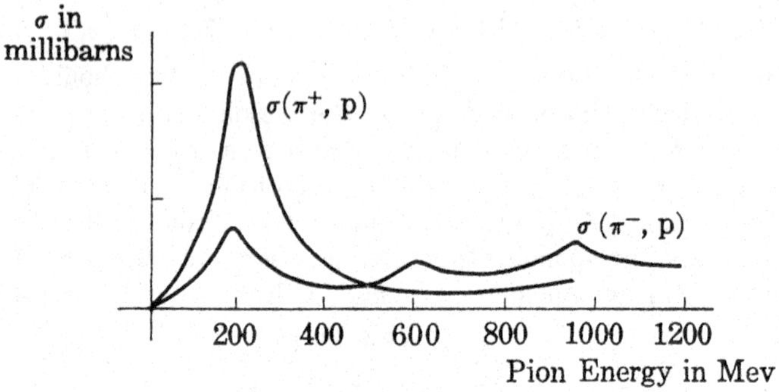

Fig. 12.8

total cross sections for π^+ and π^- on protons initially at rest between zero and 1.2 Bev. The fact that the cross sections at zero kinetic energy are very small is already an indication that the S-wave scattering is anomalously small at low energies and that the scattering is predominantly P-wave scattering. The large maximum in the π^+p scattering at approximately 190 Mev is to be noted. This is the famous 33 resonance. Its reflection in the π^-p scattering at that energy is also to be noticed.

It will be noted that the 33 phase shift as given by (274) is positive, whereas δ_{11} and δ_{13} are negative. These features are experimentally verified. The potential between meson and nucleon is therefore attractive in the $(\frac{3}{2}, \frac{3}{2})$ states and repulsive in the (3, 1), (1, 3), and (1, 1) states. It will be recalled that in nonrelativistic quantum mechanics an attractive potential gives rise to a positive phase shift which is enhanced in higher order Born approximations, whereas a repulsive potential gives rise to a negative phase shift which is diminished in higher order Born approximations.[8] This is only true in the absence of bound states. Nonetheless, these remarks suggest that closer agreement with experiment could perhaps be obtained in fourth and higher orders. Indeed fourth-order calculations have been performed by Chew (1954) which substantiate, in part, these conjectures. However, the question of whether the Chew model is in agreement with low energy meson-nucleon scattering experiments cannot be settled by perturbation theoretic calculations in view of the slowness of the convergence of the series. Thus the ratio of the fourth-order (charge-

[8] Qualitatively this is because an attractive potential will further "suck in" the wave function in higher orders, whereas a repulsive potential will further "push out" the wave function out of the potential region.

renormalized) phase shifts to the second-order (charge-renormalized) phase shifts is of order

$$\frac{\delta^{(4)}}{\delta^{(2)}} \sim \frac{f^2}{\mu^2}\frac{\omega_{\max}}{\mu}$$

For a cutoff $\omega_{\max} \sim M$ (M = the nucleon mass) and a (renormalized) coupling constant $f^2/\mu^2 = 0.08$, this ratio becomes

$$\frac{\delta^{(4)}}{\delta^{(2)}} \sim \frac{1}{2}$$

These values of the parameters f and $\omega_{\max}$ are approximately the ones necessary to account for the behavior of the 33 phase shifts at low energy.

It remained for Low (1955) to deduce a set of relations between observable quantities which were exact consequences of the Chew model and which could be used to check the theory. These relations were independently derived by Wick (1955) and further explored and made the basis of a complete theory of low energy meson phenomena, including photo production, by Chew and Low (1956a, b). The papers by Chew and Low are landmarks in quantum field theory. Every serious student of modern field theory is urged to study them, for no review of the articles can do full justice to the clarity and simplicity of the Wick–Chew-Low approach.

The important observation of Wick (1955) was to realize that, in order to solve the problem of meson-nucleon scattering, the scattering solution must consist of a free meson incident on a physical "dressed" nucleon plus a scattered wave. If we denote by $|s, t\rangle_+$ the four eigenstates of H corresponding to the physical nucleon, the problem is then to find those solutions of

$$(H - E)\,|\,j\mathbf{k}; s, t\rangle_+ = 0 \tag{275a}$$

$$E = M + \omega_{\mathbf{k}} \tag{275b}$$

which are of the form

$$|j, \mathbf{k}; s, t\rangle_+ = a_{j\mathbf{k}}^*\,|\,s, t\rangle_+ + |\chi'\rangle_+ \tag{276}$$

where $|\chi'\rangle_+$ denotes a state representing an outgoing free meson. In a time-dependent formulation of the scattering problem, one verifies that the term $a_{j\mathbf{k}}^*\,|\,s, t\rangle_+$ corresponds at $t = -\infty$ to a free meson incident on the physical nucleon. The requirement that $|\mathbf{k}, j; s, t\rangle_+$ as given by Eq. (276) be a solution of (275) is that

$$\begin{aligned} 0 &= (H - M - \omega_{\mathbf{k}})\,(a_{j\mathbf{k}}^*\,|\,s, t\rangle_+ + |\chi'\rangle_+) \\ &= (a_{j\mathbf{k}}^* H + [H, a_{j\mathbf{k}}^*])\,|\,s, t\rangle_+ - (M + \omega_{\mathbf{k}})\,a_{j\mathbf{k}}^*\,|\,s, t\rangle_+ \\ &\quad + (H - M - \omega_{\mathbf{k}})\,|\,\chi'\rangle_+ \end{aligned} \tag{277}$$

Since

$$H = M + \sum_{j\mathbf{k}} \omega_{\mathbf{k}} a_{j\mathbf{k}}^* a_{j\mathbf{k}} + \sum_{j\mathbf{k}} (a_{j\mathbf{k}} V_{j\mathbf{k}}^{(0)} + a_{j\mathbf{k}}^* V_{j\mathbf{k}}^{(0)*}) - \Delta M \tag{278}$$

$$[H, a_{j\mathbf{k}}^*] = \omega_{\mathbf{k}} a_{j\mathbf{k}}^* + V_{j\mathbf{k}}^{(0)} \tag{279}$$

whence

$$(H - M - \omega_{\mathbf{k}}) \mid \chi' \rangle_+ = -V_{j\mathbf{k}}^{(0)} \mid s, t\rangle_+ \tag{280}$$

and

$$|\chi'\rangle_+ = \frac{1}{M + \omega_{\mathbf{k}} - H + i\epsilon} V_{j\mathbf{k}}^{(0)} \mid s, t\rangle_+ \tag{281}$$

That

$$|j, \mathbf{k}; s, t\rangle_+ = a_{j\mathbf{k}}{}^* \mid s, t\rangle_+ + \frac{1}{M + \omega_{\mathbf{k}} - H + i\epsilon} V_{j\mathbf{k}}^{(0)} \mid s, t\rangle_+ \tag{282}$$

is indeed the correct scattering state corresponding to an incident meson of momentum $\mathbf{k}$, isotopic spin j being scattered by a nucleon in the state $|s, t\rangle_+$, is verified by the fact that $|j, \mathbf{k}; s, t\rangle_+$ satisfies Eq. (275) and has the correct singular behavior to guarantee that only outgoing scattered waves exist at time $t = +\infty$. Similarly the solution $|j, \mathbf{k}; s, t\rangle_-$ which corresponds to the ingoing wave Lippmann-Schwinger solution is:

$$|j, \mathbf{k}; s, t\rangle_- = a_{j\mathbf{k}}{}^* \mid s, t\rangle_+ + \frac{1}{M + \omega_{\mathbf{k}} - H - i\epsilon} V_{j\mathbf{k}}^{(0)} \mid s, t\rangle_+ \tag{283}$$

Note that $|s, t\rangle_+ = |s, t\rangle_-$ since the one-nucleon state is stationary, so that $S \mid s, t\rangle_+ = |s, t\rangle_- = |s, t\rangle_+$.

Comparing (283) and (282), we deduce that

$$\begin{aligned} |j, \mathbf{k}; s, t\rangle_+ &= |j, \mathbf{k}; s, t\rangle_- \\ &\quad + \left[\frac{1}{M + \omega_{\mathbf{k}} - H + i\epsilon} - \frac{1}{M + \omega_{\mathbf{k}} - H - i\epsilon}\right] V_{j\mathbf{k}}^{(0)} \mid s, t\rangle_+ \\ &= |j, \mathbf{k}; s, t\rangle_- - 2\pi i \delta(M + \omega_{\mathbf{k}} - H)\, V_{j\mathbf{k}}^{(0)} \mid s, t\rangle_+ \end{aligned} \tag{284}$$

whence the S-matrix element for the scattering of a meson in an initial state $|j, \mathbf{k}\rangle$ by a nucleon in the state $|s, t\rangle_+$ to a final state, in which the meson has momentum $\mathbf{k}'$, isotopic spin j' and the nucleon is in the state $|s', t'\rangle_+$, is given by

$$\begin{aligned} {}_+\langle j'\mathbf{k}'; s', t' \mid S \mid j, \mathbf{k}; s, t\rangle_+ &= {}_-\langle j', \mathbf{k}'; s', t' \mid j, \mathbf{k}; s, t\rangle_+ \\ &= {}_-\langle j', \mathbf{k}'; s', t' \mid j, \mathbf{k}; s, t\rangle_- - 2\pi i\, {}_-\langle j', \mathbf{k}'; s', t' \mid \delta(M + \omega_{\mathbf{k}} - H)\, V_{j\mathbf{k}}^{(0)} \mid s, t\rangle_+ \\ &= \delta_{jj'}\delta_{ss'}\delta_{tt'}\delta^{(3)}(\mathbf{k} - \mathbf{k}') - 2\pi i \delta(\omega_{\mathbf{k}} - \omega_{\mathbf{k}'})\, {}_-\langle j', \mathbf{k}'; s', t' \mid V_{j\mathbf{k}}^{(0)} \mid s, t\rangle_+ \\ &= \delta_{jj'}\delta_{ss'}\delta_{tt'}\delta^{(3)}(\mathbf{k} - \mathbf{k}') - 2\pi i \delta(\omega_{\mathbf{k}'} - \omega_{\mathbf{k}})\, R_{j\mathbf{k}, st}(j'\mathbf{k}', s't') \end{aligned} \tag{285}$$

with[9]

$$R_{j\mathbf{k}, st}(j'\mathbf{k}'; s't') = {}_-\langle j', \mathbf{k}'; s', t' \mid V_{j\mathbf{k}}^{(0)} \mid s, t\rangle_+ \tag{286}$$

[9] Note that in the case that the theory is not of the Yukawa type (i.e., H_I is not linear in ϕ) the results are amended as follows: If $H = H_0 + H_I$, define

$$[H_I, a_{r\mathbf{k}}{}^*] = j_r(\mathbf{k})$$

then

$$R_{l\mathbf{k}, st}(l'\mathbf{k}'; s't') = {}_-\langle l', \mathbf{k}'; s', t' \mid j_l(\mathbf{k}) \mid s, t\rangle_+$$

For $\omega_{\mathbf{k}} = \omega_{\mathbf{k}'}$, the quantity $R_{j\mathbf{k},st}(j'\mathbf{k}', s't') = {}_-\langle j', \mathbf{k}'; s', t' \mid V_{j\mathbf{k}}^{(0)} \mid s, t\rangle_+$ is the conventional R matrix of scattering theory but is no longer so when $\omega_{\mathbf{k}} \neq \omega_{\mathbf{k}'}$. However, since

$$V_{j\mathbf{k}}^{(0)} = \frac{f_0}{\mu} \frac{iv(\mathbf{k}^2)}{\sqrt{2\omega_{\mathbf{k}}}} i\sigma \cdot \mathbf{k}\tau_j \tag{287}$$

it will be noted that the dependence of $R_{j\mathbf{k},st}(j'\mathbf{k}'; s't')$ on $\mathbf{k}$ is completely known: A factor $kv(\mathbf{k}^2)/\sqrt{2\omega_{\mathbf{k}}}$ can be taken outside the matrix element defining $R_{j\mathbf{k},st}(j'\mathbf{k}', s't')$ and the remaining matrix element is independent of k. This factorability is a special feature of the Chew model and has its origin in the neglect of nucleon recoil. It implies that $R_{j\mathbf{k},st}(j'\mathbf{k}', s', t)$ for $\omega_{\mathbf{k}} \neq \omega_{\mathbf{k}'}$ remains closely related to $R_{j\mathbf{k},st}(j'\mathbf{k}', s't')$ on the energy shell, i.e., to the scattering amplitude, since $R_{j\mathbf{k},st}(j'\mathbf{k}', s't')$ depends on the variable $\mathbf{k}$ and j in a known and factorable way. This fact proves to be of the utmost importance in what follows for it allows one to relate the $R_{j\mathbf{k},st}(j'\mathbf{k}', s't')$s on and off the energy shell in a trivial way. In contrast, the conventional R matrix depends on its indices in a nonfactorable way, and its values on and off the energy shell are not simply related. If in (286) we substitute for ${}_-\langle j', \mathbf{k}'; s, t'|$ Eq. (283), we then obtain

$$\begin{aligned} R_{j\mathbf{k},st}(j'\mathbf{k}'; s't') &= {}_+\langle s', t' \mid a_{j'\mathbf{k}'} V_{j\mathbf{k}}^{(0)} \mid s, t\rangle_+ \\ &\quad + {}_+\langle s', t' \mid V_{j'\mathbf{k}'}^{(0)*} \frac{1}{M + \omega_{\mathbf{k}'} - H + i\epsilon} V_{j\mathbf{k}}^{(0)} \mid s, t\rangle_+ \end{aligned} \tag{288}$$

The first term of (288) can be recast into the following form: Since

$$[H, a_{j\mathbf{k}}] = -\omega_{\mathbf{k}'} a_{j'\mathbf{k}'} - V_{j'\mathbf{k}'}^{(0)*} \tag{289}$$

we find that

$$\begin{aligned} Ha_{j'\mathbf{k}'} \mid s', t'\rangle_+ &= \{Ma_{j'\mathbf{k}'} + [H, a_{j'\mathbf{k}'}]\} \mid s', t'\rangle_+ \\ &= \{(M - \omega_{\mathbf{k}'})\, a_{j'\mathbf{k}'} - V_{j'\mathbf{k}'}^{(0)*}\} \mid s', t'\rangle_+ \end{aligned} \tag{290}$$

Since the spectrum of $H - M + \omega_{\mathbf{k}}$ is always positive, $H - M + \omega_{\mathbf{k}}$ has a well-defined inverse, and

$$a_{j'\mathbf{k}'} \mid s', t'\rangle_+ = -\frac{1}{H - M + \omega_{\mathbf{k}'}} V_{j'\mathbf{k}'}^{(0)*} \mid s', t'\rangle_+ \tag{291}$$

In the Yukawa theory V and a commute, so that Eq. (291) allows us to rewrite R in the following form:

$$\begin{aligned} R_{j\mathbf{k},st}(j'\mathbf{k}', s't') &= {}_+\langle s', t' \mid V_{j'\mathbf{k}'}^{(0)*} \frac{1}{\omega_{\mathbf{k}'} + M - H + i\epsilon} V_{j\mathbf{k}}^{(0)} \\ &\quad - V_{j\mathbf{k}}^{(0)} \frac{1}{\omega_{\mathbf{k}'} - M + H} V_{j'\mathbf{k}'}^{(0)*} \mid s, t\rangle_+ \end{aligned} \tag{292}$$

In a perturbation expansion the first term corresponds to the sum of all diagrams in which the incident and final meson lines do not cross, whereas the second term corresponds to the sum of all crossed diagrams. Upon introducing into the Eq. (292) the completeness relations for the incoming wave eigenstates of the total Hamiltonian $|n\rangle_-$

$$\sum_{|n\rangle_-} |n\rangle_- {}_-\langle n| = 1 \tag{293}$$

where the states with $n = 0$ correspond to the one-nucleon states $|0\rangle_- = |s, t\rangle_-$, those with $n = 1$, to the meson-nucleon scattering states $|j, \mathbf{k}; s, t\rangle_-$, those with $n = 2$ to the scattering states with two (plane wave) mesons at time $t = \infty$, etc., we obtain the equation

$$R_{j\mathbf{k},\,st}(j'\mathbf{k}', s't') = \sum_{|n\rangle_-} \left\{ \frac{{}_+\langle s', t' \mid V_{j'\mathbf{k}'}^{(0)*} \mid n\rangle_- {}_-\langle n \mid V_{j\mathbf{k}}^{(0)} \mid s, t\rangle_+}{M + \omega_{\mathbf{k}'} - E_n + i\epsilon} + \frac{{}_+\langle s't' \mid V_{j\mathbf{k}}^{(0)} \mid n\rangle_- {}_-\langle n \mid V_{j'\mathbf{k}'}^{(0)*} \mid s, t\rangle_+}{M - \omega_{\mathbf{k}'} - E_n} \right\} \tag{294}$$ [10]

Consider next the matrix element ${}_-\langle n \mid V_{j\mathbf{k}}^{(0)} \mid s, t\rangle_+$. We shall now show that when $E_n = \omega_{\mathbf{k}} + M$, this matrix element is precisely the matrix element of the R matrix for the process $\pi + \mathrm{N} \to n$. *Proof*: The S-matrix element for the scattering from the initial state $|j, \mathbf{k}; s, t\rangle_+$ to the final state $|n\rangle_-$ is given by

$${}_-\langle n \mid j, \mathbf{k}; s, t\rangle_+ = \delta(n; j\mathbf{k}, st) - 2\pi i \delta(E_n - \omega_{\mathbf{k}} - M)\, R_{j\mathbf{k},\,st}(n) \tag{295}$$

where when $E_n = \omega_{\mathbf{k}} + M$, $R_{j\mathbf{k},\,st}(n)$ is the R-matrix element for the transition and where

$$\delta(n; j\mathbf{k}, st) = {}_-\langle n \mid j, \mathbf{k}; s, t\rangle_- = {}_+\langle n \mid j, \mathbf{k}; s, t\rangle_+ \tag{296}$$

Substituting for $|j, \mathbf{k}; s, t\rangle_+$ in the left-hand side of Eq. (295) the expression (284), we deduce that

$$R_{j\mathbf{k},\,st}(n) = {}_-\langle n \mid V_{j\mathbf{k}}^{(0)} \mid s, t\rangle_+ \tag{297}$$

Note that once again the dependence of $R_{j\mathbf{k},\,st}(n)$ on $\mathbf{k}$ is factorable and known, so that ${}_-\langle n \mid V_{j\mathbf{k}}^{(0)} \mid s, t\rangle_+$ for $E_n \neq \omega_{\mathbf{k}} + M$ is still simply related to its energy shell value. Finally noting that

$$\begin{aligned} {}_+\langle s', t' \mid V_{j'\mathbf{k}'}^{(0)*} \mid n\rangle_- &= \overline{{}_-\langle n \mid V_{j'\mathbf{k}'}^{(0)} \mid s', t'\rangle_+} \\ &= \overline{R_{j'\mathbf{k}',\,s't'}(n)} \end{aligned} \tag{298}$$

and that

$$V_{j\mathbf{k}}^{(0)*} = -V_{j\mathbf{k}}^{(0)} \tag{299}$$

allows us to rewrite Eq. (294) in the form:

$$R_{j\mathbf{k},\,st}(j'\mathbf{k}', s't') = \sum_n \left\{ \frac{\overline{R_{j'\mathbf{k}',\,s't'}(n)}\, R_{j\mathbf{k},\,st}(n)}{M + \omega_{\mathbf{k}'} - E_n + i\epsilon} + \frac{\overline{R_{j\mathbf{k},\,s't'}(n)}\, R_{j'\mathbf{k}',\,st}(n)}{M - \omega_{\mathbf{k}'} - E_n} \right\} \tag{300}$$

[10] Note that the denominator of the second term in (294) never vanishes since the lowest value that E_n can assume is M. If it proved convenient, we could therefore add the term $-i\epsilon$ to this denominator since the limit $\epsilon \to 0$ will always be considered.

Equation (300), which is the Low equation for the Chew model, is a nonlinear equation relating the meson-nucleon scattering amplitude to all other scattering and production amplitudes starting from the same initial state or ending on the same final state. By deriving similar equations for these other amplitudes $R_{j\mathbf{k},\,st}(n)$, one can obtain a closed, infinite set of equations which are, in some sense to be specified later, equivalent to the original Hamiltonian formulation. Such a set of equations has, in fact, been derived and analyzed for a somewhat simpler model [a scalar field interacting with sources through an interaction of the form $H_I = = \sum_{i=1}^{N} \lambda_i \left(\int \rho(\mathbf{x})\,\phi(\mathbf{x})\,d^3x\right)^i$] by Norton and Klein [Norton (1958); see also Fukuda (1956)].

It will be noted that since the dependence of $R_{j\mathbf{k},\,st}(j'\mathbf{k}', s't')$ on $\mathbf{k}$ and $\omega_{\mathbf{k}}$ is completely determined by the dependence of $V_{j\mathbf{k}}^{(0)}$ on $\mathbf{k}$ and $\omega_{\mathbf{k}}$, the dependence of $\mathbf{k}$ and $\omega_{\mathbf{k}}$ drops out of the Low equation. This is the case because the same dependence appears on both the right- and left-hand side of Eq. (300). Furthermore, although for the determination of scattering amplitude we shall be principally interested in the solutions for $\omega_{\mathbf{k}'} = \omega_{\mathbf{k}}$, it will be noted that the Low equation, Eq. (300), is in fact valid for $\omega_{\mathbf{k}'} \neq \omega_{\mathbf{k}}$.

There are three important properties that the Low equation exhibits, namely

1. unitarity
2. crossing symmetry
3. analyticity properties

The unitary of the S matrix

$$SS^* = S^*S = 1 \tag{301}$$

where

$${}_+\langle n \mid S \mid j, \mathbf{k}; s, t\rangle_+ = \delta(n; j\mathbf{k}, st) - 2\pi i \delta(E_n - \omega_{\mathbf{k}} - M)\, R_{j\mathbf{k},\,st}(n) \tag{302}$$

implies that $R_{j\mathbf{k},\,st}(j'\mathbf{k}', s't')$ satisfies the equation

$$\overline{R_{j'\mathbf{k}',\,s't'}(j\mathbf{k}, st)} - R_{j\mathbf{k},\,st}(j'\mathbf{k}', s't') = 2\pi i \sum_n \overline{R_{j'\mathbf{k}',\,s't'}(n)}\, R_{j\mathbf{k},\,st}(n)\, \delta(E_n - M - \omega_{\mathbf{k}}) \tag{303}$$

where $\omega_{\mathbf{k}} = \omega_{\mathbf{k}'}$. (Incidentally, it should be noted that the no-meson states $n = 0$, $E_0 = M$, do not contribute to the right-hand side, since $\omega_{\mathbf{k}}$ never vanishes.) Now from Low's equation, Eq. (300), one verifies that

$$\overline{R_{j'\mathbf{k}',\,s't'}(j\mathbf{k}, st)} - R_{j\mathbf{k},\,st}(j'\mathbf{k}', s't') = \sum_n \left\{ \frac{R_{j\mathbf{k},\,st}(n)\, \overline{R_{j'\mathbf{k}',\,s't'}(n)}}{\omega_{\mathbf{k}} + M - E_n - i\epsilon} - \frac{R_{j\mathbf{k},\,st}(n)\, \overline{R_{j'\mathbf{k}',\,s't'}(n)}}{\omega_{\mathbf{k}'} + M - E_n + i\epsilon} \right\} \tag{304}$$

so that for $\omega_{\mathbf{k}} = \omega_{\mathbf{k}'}$, by making use of the identity

$$\frac{1}{x \pm i\epsilon} = \mathrm{P}\frac{1}{x} \mp i\pi\delta(x) \tag{305}$$

one verifies that any $R_{j\mathbf{k},st}(j'\mathbf{k}', s't')$ satisfying the Low equation automatically gives rise to a unitary S matrix.

To analyze the further symmetry properties of the scattering amplitude, it is convenient to define the function r of the complex variable z by

$$r_{j\mathbf{k},st\,|\,j'\mathbf{k}',s't'}(z) = \sum_n \left\{ \frac{\overline{R_{j'\mathbf{k}';s't'}(n)}\, R_{j\mathbf{k},st}(n)}{z + M - E_n} + \frac{\overline{R_{j\mathbf{k},s't'}(n)}\, R_{j'\mathbf{k}',st}(n)}{M - z - E_n} \right\} \tag{306}$$

with

$$\lim_{z \to \omega_{\mathbf{k}} + i\epsilon} r_{j\mathbf{k},st\,|\,j'\mathbf{k}',s't'}(z) = R_{j\mathbf{k},st}(j'\mathbf{k}', s't') \tag{307}$$

We note that the dependence of r on $\mathbf{k}$ and $\mathbf{k}'$ is known, and only the dependence on z is unknown. From its definition by the right-hand side of Eq. (306), it follows that

$$r_{j\mathbf{k},s't'\,|\,j'\mathbf{k}',st}(z) = r_{j'\mathbf{k}',s't'\,|\,j\mathbf{k},st}(-z) \tag{308}$$

Equation (308) is the mathematical statement of the "crossing theorem" due to Gell-Mann and Goldberger [Gell-Mann (1954)]. In terms of diagrams, one may express this symmetry by the statement that for any given diagram there exists another which is obtained from the former by the interchange of incoming and outgoing meson line. More specifically, the matrix element of the crossed diagram can be obtained from the uncrossed one by (*a*) interchanging $\mathbf{k}$ and $\mathbf{k}'$, (*b*) interchanging τ_j and $\tau_{j'}$ and (*c*) changing the sign of ω in the energy denominators. Since the R matrix is a sum of contributions from all diagrams, it must remain unchanged

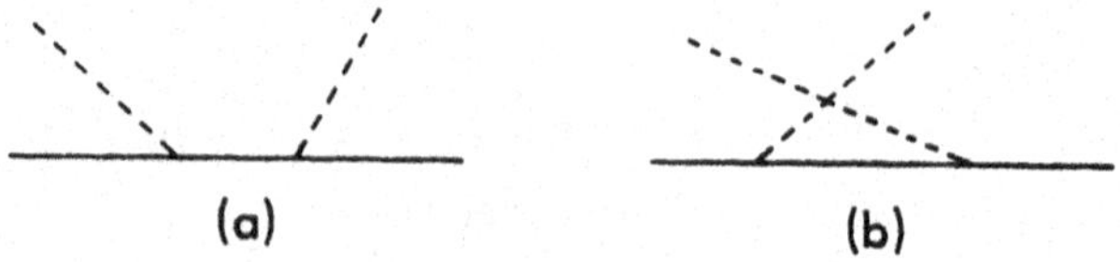

Fig. 12.9

under the above operations. This is easily verified for the second-order contribution arising from the diagrams 12.9*a* and 12.9*b*. Similarly, one notes that

$$\overline{r_{j\mathbf{k},s't'\,|\,j'\mathbf{k}',st}(\bar{z})} = r_{j\mathbf{k},st\,|\,j'\mathbf{k}',s't'}(-z) \tag{309a}$$

$$= r_{j'\mathbf{k}',st\,|\,j\mathbf{k},s't'}(z) \tag{309b}$$

where in obtaining the second line we have made use of crossing symmetry. It is sometimes convenient to consider $r_{j\mathbf{k},st\,|\,j'\mathbf{k}',s't'}(z)$ as the st, $s't'$ matrix element of the operator $r_{\mathbf{pq}}(z)$; i.e., $\langle s, t \,|\, r_{\mathbf{pq}}(z) \,|\, s', t' \rangle = r_{j\mathbf{k},st\,|\,j'\mathbf{k}',s't'}(z)$

where we have further abbreviated $j\mathbf{k}$ by $\mathbf{p}$ and $j'\mathbf{k}'$ by $\mathbf{q}$. The crossing symmetry relation (308) and the reality condition (309) can then be written as

$$r_{\mathbf{qp}}(z) = r_{\mathbf{pq}}(-z) \tag{310}$$

and

$$r_{\mathbf{qp}}^*(\bar{z}) = r_{\mathbf{pq}}(z) \tag{311}$$

Further properties of the function $r_{j\mathbf{k},\,st\,|\,j'\mathbf{k}',\,s't'}(z)$ can be deduced by using certain assumed properties of the energy spectrum of the intermediate states which enter in the definition of r, Eq. (306). We assume that H has a point eigenvalue at $E_n = M$ corresponding to the nucleon states and has a continuous spectrum starting at $E = M + \mu, M + 2\mu, \cdots$, corresponding to the states composed of nucleon + one incoming meson, nucleon + two incoming mesons, etc. We shall also assume that there are no point eigenvalues between M and $M + \mu$, which would correspond to a bound state of a nucleon and a meson.

Let us consider in detail the contribution to r of the one-nucleon states ($n = 0$), for which $E_n = M$. These states give rise to a pole in $r(z)$ at $z = 0$, the residue of which is proportional to

$$\sum_{s''t''} \{\overline{R_{j'\mathbf{k}',\,s't'}(s''t'')}\, R_{j\mathbf{k},\,st}(s''t'') - \overline{R_{j\mathbf{k},\,s't'}(s''t'')}\, R_{j'\mathbf{k}',\,st}(s''t'')\}$$

$$= \sum_{s''t''} \{\overline{{}_+\langle s'', t'' \mid V^{(0)}_{j'\mathbf{k}'} \mid s', t'\rangle_+}\, {}_+\langle s'', t'' \mid V^{(0)}_{j\mathbf{k}} \mid s, t\rangle_+$$

$$- \overline{{}_+\langle s'', t'' \mid V^{(0)}_{j\mathbf{k}} \mid s', t'\rangle_+}\, {}_+\langle s'', t'' \mid V^{(0)}_{j'\mathbf{k}'} \mid s, t\rangle_+\} \tag{312}$$

This residue clearly vanishes for forward scattering without charge exchange, i.e., when $(j'\mathbf{k}') = (j\mathbf{k})$. Now apart from the dependence on $\mathbf{k}$ which is known, the matrix element of $V^{(0)}_{j\mathbf{k}}$ between physical one-nucleon states is proportional to ${}_+\langle s', t' \mid \sigma_i\tau_j \mid s, t\rangle_+$. One next notes that the transformation properties of the states $|s, t\rangle_+$ and of the operators σ_i, τ_j under rotations in space and isotopic spin space imply that

$${}_+\langle s', t' \mid \sigma_i\tau_j \mid s, t\rangle_+ = Z\langle s', t' \mid \sigma_i\tau_j \mid s, t\rangle \tag{313}$$

i.e., that the ratio of the matrix elements of $\sigma_i\tau_j$ between the physical one-nucleon states and the matrix element of $\sigma_i\tau_j$ between the same bare one-nucleon states is a constant, Z, independent of $s't'$ and st. The proof uses the commutation rules between σ_i, τ_j and the total angular momentum and isotopic spin operators. From the commutation rules between $\mathbf{J}^4$ and σ_i, one derives [see, for example, Pake (1953)] that

$${}_+\langle s', t' \mid \sigma_i\tau_j \mid s, t\rangle_+ = \tfrac{3}{4}\,{}_+\langle \tfrac{1}{2}, t' \,\|\, \mathbf{J} \cdot \boldsymbol{\sigma}\tau_j \,\|\, \tfrac{1}{2}, t\rangle_+ \, {}'_+\langle \tfrac{1}{2}, s' \mid J_i \mid \tfrac{1}{2}, s\rangle_+{}' \tag{314a}$$

where $|\tfrac{1}{2}, s\rangle_+{}'$ denotes an eigenfunction of $\mathbf{J}^2$ (with $J = \tfrac{1}{2}$) and $J_3 = s$ and the notation ${}_+\langle \tfrac{1}{2}, t' \,\|\, \mathbf{J} \cdot \boldsymbol{\sigma}\tau_j \,\|\, \tfrac{1}{2}, t\rangle_+$ indicates that this matrix element is independent of the J_3 value of the one-nucleon state $|s, t\rangle_+$. Similarly, since τ_j has the same properties with respect to the total isotopic spin as σ_j with respect to $\mathbf{J}$, we deduce that

$$_+\langle s', t' \mid \sigma_i\tau_j \mid s, t\rangle_+ = \tfrac{9}{16}\,{}_+\langle \tfrac{1}{2}, \tfrac{1}{2} \,||\, \mathbf{J}\cdot\boldsymbol{\sigma}\,\mathbf{T}\cdot\boldsymbol{\tau} \,||\, \tfrac{1}{2}, \tfrac{1}{2}\rangle_+ \cdot {}'_+\langle \tfrac{1}{2}, s' \mid J_i \mid \tfrac{1}{2}, s\rangle_+' \; {}''_+\langle \tfrac{1}{2}, t' \mid T_j \mid \tfrac{1}{2}, t\rangle_+'' \quad (314b)$$

where $|\tfrac{1}{2}, t\rangle_+''$ is an eigenstate of $\mathbf{T}^2$ (with $T = \tfrac{1}{2}$) and $T_3 = t$, and the notation $_+\langle \tfrac{1}{2}, \tfrac{1}{2} \,||\, \mathbf{J}\cdot\boldsymbol{\sigma}\mathbf{T}\cdot\boldsymbol{\tau} \,||\, \tfrac{1}{2}, \tfrac{1}{2}\rangle_+$ denotes the fact that this matrix element is independent of s', t', s and t. Clearly, a similar expression holds for the matrix elements of $\sigma_j\tau_i$ between the bare states and, furthermore,

$$'_+\langle \tfrac{1}{2}, s' \mid J_i \mid \tfrac{1}{2}, s\rangle_+' = {}'\langle \tfrac{1}{2}, s' \mid J_i^{(0)} \mid \tfrac{1}{2}, s\rangle' \quad (314c)$$

$$''_+\langle \tfrac{1}{2}, t' \mid T_i \mid \tfrac{1}{2}, t\rangle_+'' = {}''\langle \tfrac{1}{2}, t' \mid T_i^{(0)} \mid \tfrac{1}{2}, t\rangle'' \quad (314d)$$

where the superscript (0) on $J_i^{(0)}$ and $T_i^{(0)}$ indicates that these are the operators for the noninteracting meson-nucleon system. Therefore

$$\frac{_+\langle s', t' \mid \sigma_i\tau_j \mid s, t\rangle_+}{\langle s', t' \mid \sigma_i\tau_j \mid s, t\rangle} = \frac{_+\langle \tfrac{1}{2}, \tfrac{1}{2} \,||\, \mathbf{J}\cdot\boldsymbol{\sigma}\,\mathbf{T}\cdot\boldsymbol{\tau} \,||\, \tfrac{1}{2}, \tfrac{1}{2}\rangle_+}{\langle \tfrac{1}{2}, \tfrac{1}{2} \,||\, \mathbf{J}^{(0)}\cdot\boldsymbol{\sigma}\,\mathbf{T}^{(0)}\cdot\boldsymbol{\tau} \,||\, \tfrac{1}{2}, \tfrac{1}{2}\rangle} \quad (315)$$

The right-hand side from the derivation is independent of s, t and of s' and t'. The ratio is not equal to one, reflecting the presence of the meson cloud surrounding the physical nucleon. This effect of the meson cloud can be described by a change in the strength of the coupling. We define the renormalized coupling constant f by

$$f_0\,{}_+\langle s', t' \mid \sigma_i\tau_j \mid s, t\rangle_+ = f\langle s', t' \mid \sigma_i\tau_j \mid s, t\rangle \quad (316a)$$

i.e.,

$$f = Zf_0 \quad (316b)$$

so that

$$R_{j'\mathbf{k}', s't'}(s''t'') = \langle s''t'' \mid V_{j\mathbf{k}} \mid st\rangle \quad (317)$$

where $V_{j\mathbf{k}}$ is obtained from $V_{j\mathbf{k}}^{(0)}$ by replacing f_0 by f:

$$V_{j\mathbf{k}} = i\,\frac{f}{\mu}\,\frac{v(\mathbf{k}^2)}{\sqrt{2\omega_{\mathbf{k}}}}\,\boldsymbol{\sigma}\cdot\mathbf{k}\tau_j \quad (318)$$

and $|s, t\rangle$ are "bare" nucleon state vectors. The function $r_{\mathbf{qp}}(z)$ is thus completely determined in the neighborhood of $z = 0$ by the renormalized coupling constant, and vice versa, since

$$\begin{aligned}\lim_{z\to 0} z r_{\mathbf{qp}}(z) &= \sum_{s''t''} \{_+\langle s', t' \mid V_{j'\mathbf{k}'}^{(0)*} \mid s'', t''\rangle_+ \, {}_+\langle s'', t'' \mid V_{j\mathbf{k}}^{(0)} \mid s, t\rangle_+ \\ &\quad - \langle s', t' \mid V_{j\mathbf{k}}^{(0)*} \mid s'', t''\rangle_+ \, {}_+\langle s'', t'' \mid V_{j'\mathbf{k}'}^{(0)} \mid s, t\rangle_+\} \\ &= \sum_{s''t''} \{\langle s', t' \mid V_{j'\mathbf{k}'}{}^* \mid s'', t''\rangle \langle s'', t'' \mid V_{j\mathbf{k}} \mid s, t\rangle \\ &\quad - \langle s', t' \mid V_{j\mathbf{k}}{}^* \mid s''t''\rangle \langle s'', t'' \mid V_{j'\mathbf{k}'} \mid s, t\rangle\} \\ &= \langle s', t' \mid [V_{j'\mathbf{k}'}{}^*, V_{j\mathbf{k}}] \mid s, t\rangle \end{aligned} \quad (319)$$

By Eq. (318) the zero energy behavior of the scattering is thus completely determined by the renormalized coupling constant f. Furthermore, we note that this zero energy behavior is identical with that in Born approximation, but where in the latter f_0 has been replaced by the f, the renormalized coupling constant. The requirement that the exact zero energy scattering amplitude coincide with the "renormalized" Born approximation scattering amplitude can be taken as a prescription for the renormalization of the coupling constant [Lee (1954)].

The meaning of the renormalized coupling constant is further elucidated by a consideration of the expectation value of the meson field in the one-nucleon state $|s, t\rangle_+$. Let us compute ${}_+\langle s, t \mid \phi_j(\mathbf{x}) \mid s, t\rangle_+$. Since $|s, t\rangle_+$ is an eigenstate of H, clearly

$$ {}_+\langle s, t \mid [H, a_{i\mathbf{k}}] \mid s, t\rangle_+ = {}_+\langle s, t \mid [H, a_{i\mathbf{k}}^*] \mid s, t\rangle_+ = 0 \tag{320} $$

From (279) and (289) it follows that

$$ {}_+\langle s, t \mid a_{j\mathbf{k}} \mid s, t\rangle_+ = +{}_+\langle s, t \mid \frac{1}{\omega_\mathbf{k}} V_{j\mathbf{k}}^{(0)} \mid s, t\rangle_+ \tag{321a} $$

$$ {}_+\langle s, t \mid a_{j\mathbf{k}}^* \mid s, t\rangle_+ = -{}_+\langle s, t \mid \frac{1}{\omega_\mathbf{k}} V_{j\mathbf{k}}^{(0)} \mid s, t\rangle_+ \tag{321b} $$

so that

$$ \begin{aligned} {}_+\langle s, t \mid \phi_j(\mathbf{x}) \mid s, t\rangle_+ &= \frac{1}{(2\pi)^{3/2}} \int \frac{d^3k}{\sqrt{2\omega_\mathbf{k}}}\, {}_+\langle s, t \mid a_{j\mathbf{k}}\, e^{i\mathbf{k}\cdot\mathbf{x}} + a_{j\mathbf{k}}^*\, e^{-i\mathbf{k}\cdot\mathbf{x}} \mid s, t\rangle_+ \\ &= \frac{1}{(2\pi)^{3/2}} \frac{f_0}{\mu} \partial_i \int d^3k\, v(\mathbf{k}^2)\, \frac{e^{i\mathbf{k}\cdot\mathbf{x}} + e^{-i\mathbf{k}\cdot\mathbf{x}}}{2\omega_\mathbf{k}^2}\, {}_+\langle s, t \mid \sigma_i\tau_j \mid s, t\rangle_+ \\ &= \frac{1}{(2\pi)^{3/2}} \frac{f}{\mu} \langle s, t \mid \tau_j \boldsymbol{\sigma} \cdot \nabla \mid s, t\rangle \int d^3k\, v(\mathbf{k}^2)\, \frac{e^{i\mathbf{k}\cdot\mathbf{x}}}{\mathbf{k}^2 + \mu^2} \end{aligned} \tag{322} $$

For $|\mathbf{x}|$ much larger than the radius R_0 of the source, only small values of $|\mathbf{k}|$ will contribute to the right-hand side of (322), and for these values of $|\mathbf{k}|$ $v(\mathbf{k}^2) \approx 1$. We therefore obtain

$$ \lim_{|\mathbf{x}| \gg R_0} {}_+\langle s, t \mid \phi_j(\mathbf{x}) \mid s, t\rangle_+ \sim \frac{1}{\sqrt{4\pi}} \frac{f}{\mu} \langle s, t \mid \tau_j \boldsymbol{\sigma} \cdot \nabla \mid s, t\rangle \frac{e^{-\mu r}}{r} \tag{323} $$

where $|\mathbf{x}| = r$. That is, at a distance large compared with the source radius the meson "potential" due to the nucleon is a Yukawa potential whose strength is determined by the *renormalized* coupling constant f. We further note that if we were to place a second nucleon (denoted by labels 2) at the position $\mathbf{x}$ where $|\mathbf{x}|$ is very much larger than R_0, so that (323) is valid and the presence of this second nucleon does not perturb appreciably the meson field around the nucleon 1 located at the origin, then the energy of this second nucleon in the field of the first is approximately given by the expectation value of the operator

$$\sum_j \frac{f}{\mu} \int d^3x' \, \rho_2(\mathbf{x}') \, \boldsymbol{\sigma}_2 \cdot \nabla_{\mathbf{x}'} \langle s_1, t_1 \mid \phi_j(\mathbf{x}') \mid s_1, t_1 \rangle \, \tau_{2j}$$

in the state $|s_2, t_2\rangle_+$ of nucleon 2. Here $\rho_2(\mathbf{x}')$ is the source function for nucleon 2 which is different from zero only around the point $\mathbf{x}$. Thus

$$E(r) \sim \frac{f^2}{\mu^2} \sum_j \int d^3x' \, \rho_2(\mathbf{x}') \langle s_2, t_2 \mid \tau_{2j}\boldsymbol{\sigma}_2 \cdot \nabla_{\mathbf{x}'} \mid s_2, t_2 \rangle \langle s_1, t_1 \mid \tau_{1j}\boldsymbol{\sigma}_1 \cdot \nabla_{\mathbf{x}'} \mid s_1, t_1 \rangle \frac{e^{-\mu|\mathbf{x}'|}}{|\mathbf{x}'|}$$

$$\sim \frac{f^2}{\mu^2} \sum_j \langle s_2, t_2 \mid \tau_{2j}\boldsymbol{\sigma}_2 \cdot \nabla \mid s_2, t_2 \rangle \langle s_1, t_1 \mid \tau_{1j}\boldsymbol{\sigma}_1 \cdot \nabla \mid s_1, t_1 \rangle \frac{e^{-\mu|\mathbf{x}|}}{|\mathbf{x}|}$$

$$= \frac{f^2}{\mu^2} \langle s_2t_2, s_1t_1 \mid \boldsymbol{\tau}_1 \cdot \boldsymbol{\tau}_2 \boldsymbol{\sigma}_1 \cdot \nabla \boldsymbol{\sigma}_2 \cdot \nabla \mid s_2t_2, s_1t_1 \rangle \frac{e^{-\mu|\mathbf{x}|}}{|\mathbf{x}|} \tag{324}$$

which is an approximate expression for the two-nucleon interaction potential in the adiabatic limit.

We now return to the problem at hand, the description of the scattering of mesons by nucleons, and in particular, to the properties of the function $r_{\mathbf{qp}}(z)$. As $z \to \infty$, $r_{\mathbf{qp}}(z)$ again behaves like $1/z$. This is seen from a consideration of Eq. (306). As z approaches infinity, the dependence of the denominator on the energy E_n can be neglected and closure may be applied to evaluate the sum over states. One finds

$$\lim_{z\to\infty} z r_{\mathbf{pq}}(z) = -{}_+\langle s', t' \mid [V_{\mathbf{p}}^{(0)}, V_{\mathbf{q}}^{(0)}] \mid s, t\rangle_+ \tag{325}$$

From (306) it is also clear that all the singularities of $r(z)$ lie along the real axis since E_n is real. The states $|n = 1\rangle_-$, i.e., the one incoming meson-nucleon scattering states have as their energy eigenvalues the positive reals starting at $M + \mu$ and extending to $+\infty$. Similarly, the states $|n = 2\rangle_-$ have for their energy spectrum the reals from $M + 2\mu$ to ∞, etc. Thus $r_{\mathbf{qp}}(z)$ has a branch point at $z = \mu$ and a branch cut along the real axis for $z \geqslant \mu$ from the contribution of the one-meson states. Similarly, due to the contribution of the n-meson states, it has branch points at $z = n\mu$ $(n = 2, 3, \cdots)$ and branch lines starting at $z = n\mu$ $(n = 2, 3, \cdots)$. By crossing symmetry, r also has branch cuts at $z = -n\mu$ $(n = 1, 2, 3, \cdots)$ and branch lines starting at $z = -n\mu$ $(n = 1, 2, 3, \cdots)$. Summarizing, $r(z)$ has a pole at $z = 0$, with a residue proportional to f^2, goes like $1/z$ at infinity and has branch point singularities with cuts running along the real axis for $z \geqslant \mu$ and $z \leqslant -\mu$; $r(z)$ has no other singularities besides the enumerated ones. Now the most general analytic function which has these properties must have the form

$$r_{\mathbf{qp}}(z) = \frac{\mathfrak{R}_{\mathbf{qp}}}{z} + \int_\mu^\infty dx' \left[\frac{F_{\mathbf{qp}}(x')}{x' - z} + \frac{G_{\mathbf{qp}}(x')}{x' + z}\right] \tag{326}$$

where $\mathfrak{R}_{\mathbf{qp}}$ is the residue of $r_{\mathbf{qp}}(z)$ at $z = 0$ and $F(x)$ and $G(x)$ are weighting functions defined for $x \geqslant \mu$. Since

$$\lim_{z\to x+i\epsilon} r_{\mathbf{qp}}(z) = \frac{\mathcal{R}_{\mathbf{qp}}}{x+i\epsilon} + \int_\mu^\infty dx' \left[\frac{F_{\mathbf{qp}}(x')}{x'-x-i\epsilon} + \frac{G_{\mathbf{qp}}(x')}{x'+x+i\epsilon}\right]$$

$$= \frac{\mathcal{R}_{\mathbf{qp}}}{x+i\epsilon} + \mathrm{P}\int_\mu^\infty dx' \left[\frac{F_{\mathbf{qp}}(x')}{x'-x} + \frac{G_{\mathbf{qp}}(x')}{x'+x}\right]$$

$$+ i\pi\theta(x-\mu)\, F_{\mathbf{qp}}(x) - i\pi\theta(-x-\mu)\, G_{\mathbf{qp}}(x) \quad (327a)$$

and

$$\lim_{z\to x-i\epsilon} r_{\mathbf{qp}}(z) = \frac{\mathcal{R}_{\mathbf{qp}}}{x-i\epsilon} + \mathrm{P}\int_\mu^\infty dx' \left[\frac{F_{\mathbf{qp}}(x')}{x'-x} + \frac{G_{\mathbf{qp}}(x')}{x'+x}\right]$$

$$- i\pi\theta(x-\mu)\, F_{\mathbf{qp}}(x) + i\pi\theta(-x-\mu)\, G_{\mathbf{qp}}(x) \quad (327b)$$

evidently

$$2\pi i F_{\mathbf{qp}}(x) = \lim_{z\to x+i\epsilon} r_{\mathbf{qp}}(z) - \lim_{z\to x-i\epsilon} r_{\mathbf{qp}}(z) \qquad (x > \mu) \quad (328a)$$

$$2\pi i G_{\mathbf{qp}}(x) = \lim_{z\to -x-i\epsilon} r_{\mathbf{qp}}(z) - \lim_{z\to -x+i\epsilon} r_{\mathbf{qp}}(z) \quad (x < -\mu) \quad (328b)$$

i.e., F and G are given by the jump in the function r in going across the cuts on the real axis in the right and left half planes respectively. Recall that the scattering amplitude $R_{\mathbf{q}}(\mathbf{p})$ (considered as an operator in the nucleon spin and isotopic spin space) is defined as the limit of $r_{\mathbf{qp}}(z)$ as $z = \omega_{\mathbf{p}} + i\epsilon$, so that if we impose the reality condition $r_{\mathbf{qp}}(z) = r_{\mathbf{pq}}{}^*(\bar{z})$, then (328a) leads to

$$2\pi i F_{\mathbf{qp}}(\omega_{\mathbf{p}}) = [R_{\mathbf{q}}(\mathbf{p}) - R_{\mathbf{p}}{}^*(\mathbf{q})]_{\omega_{\mathbf{p}}=\omega_{\mathbf{q}}} \quad (329a)$$

The crossing relation (310), allows us to rewrite (328b) as

$$2\pi i G_{\mathbf{qp}}(\omega_{\mathbf{p}}) = [R_{\mathbf{p}}(\mathbf{q}) - R_{\mathbf{q}}{}^*(\mathbf{p})]_{\omega_{\mathbf{p}}=\omega_{\mathbf{q}}} \quad (329b)$$

Finally, the unitarity condition (303) transforms Eq. (323), using (329a) and (329b), into the original Low equation (300) for the case $\omega_{\mathbf{p}} = \omega_{\mathbf{q}}$. This restriction is, however, of no consequence, since as noted previously, one can move off the energy shell at will.

Conversely, if it were possible to find a matrix function $r_{\mathbf{qp}}(z)$ which has a simple pole at the origin of residue $\mathcal{R}_{\mathbf{qp}}$, goes to zero at ∞ like $1/z$, and otherwise has only branch points and cuts along the real axis for $z > \mu$ and $z < -\mu$ and which satisfies unitarity and the crossing and reality relations (310) and (311), then one would have a solution of the Low equation. The unitary condition, however, involves two or more meson states and can therefore not be written down in terms of $r_{\mathbf{qp}}(z)$ only. However, if the multimeson states are neglected, the above conditions then form a practical basis for solving the scattering problem. The neglect of these states will be valid if the cross sections for inelastic processes are small compared to the elastic ones (for all values of the energy). This approximation is called the one-meson approximation. Let us consider it in detail. The Low equation in this approximation becomes

$$R_{j\mathbf{k},\, st}(j'\mathbf{k}', s't') =$$

$$= \frac{1}{\omega_{\mathbf{k}}} \sum_{|s'', t''\rangle_+} \{\overline{R_{j'\mathbf{k}', s't'}(s''t'')}\, R_{j\mathbf{k}, st}(s''t'') - \overline{R_{j\mathbf{k}, s't'}(s''t'')}\, R_{j'\mathbf{k}', st}(s''t'')\}$$

$$+ \sum_{|j'', \mathbf{k}''; s'', t''\rangle_-} \left\{ \frac{\overline{R_{j'\mathbf{k}', s't'}(j''\mathbf{k}'', s''t'')}\, R_{j\mathbf{k}, st}(j''\mathbf{k}'', s''t'')}{\omega_{\mathbf{k}'} - \omega_{\mathbf{k}''} + i\epsilon} - \frac{\overline{R_{j\mathbf{k}, s't'}(j''\mathbf{k}'', s''t'')}\, R_{j'\mathbf{k}', st}(j''\mathbf{k}'', s''t'')}{\omega_{\mathbf{k}'} + \omega_{\mathbf{k}''}} \right\} \quad (330)$$

It is an inhomogeneous, nonlinear integral equation for $R_{j\mathbf{k}, st}(j'\mathbf{k}', s''t')$. Note that the summation in Eq. (330) goes over all states $|j''\mathbf{k}''; s'', t''\rangle_-$, including states for which $\omega_{\mathbf{k}''} \neq \omega_{\mathbf{k}}$. Therefore, as it stands, Eq. (330) is not an equation involving only scattering amplitudes [i.e., $R_{j'\mathbf{k}', s't'}(j''\mathbf{k}'', s''t'')$ for $\omega_{\mathbf{k}'} = \omega_{\mathbf{k}''}$]. However, in the Chew model, as noted previously, it is possible to relate the matrix elements off the energy shell to ones on the energy shell. More precisely,

$$ {}_-\langle j', \mathbf{k}'; s', t' \mid V_{j\mathbf{k}}^{(0)} \mid s, t\rangle_+ = \frac{if_0}{\mu} \frac{v(\mathbf{k}^2)\, k_l}{\sqrt{2\omega_{\mathbf{k}}}}\, {}_-\langle j', \mathbf{k}'; s't' \mid \sigma_l \tau_j \mid s, t\rangle_+ $$

$$ = \frac{v(\mathbf{k}^2)\, k}{v(\mathbf{q}^2)\, q} \sqrt{\frac{\omega_{\mathbf{q}}}{\omega_{\mathbf{k}}}}\, {}_-\langle j', \mathbf{k}'; s', t' \mid V_{j\mathbf{q}}^{(0)} \mid s, t\rangle_+ \quad (331)$$

where $\mathbf{q}$ is an arbitrary momentum which can be so chosen that $\omega_{\mathbf{q}} = \omega_{\mathbf{k}'}$. In the static model the Low equation (in the one-meson approximation as well as in its exact form) can therefore be written in terms of observable quantities only, i.e., the scattering amplitudes ${}_-\langle n \mid V_{j\mathbf{q}} \mid s, t\rangle_+$ with $E_n = \omega_{\mathbf{q}} + M$. In relativistic theories this procedure is not possible and the Low equation is not as interesting since it is a relation between matrix elements which are off the energy shell. For relativistic theories a different set of equations, known as dispersion relations, can be derived containing only matrix elements on the energy shell. We shall consider these in Chapter 18, Section d.

In the one-meson approximation the unitarity condition, (303), becomes

$$\overline{R_{j'\mathbf{k}', s't'}(j\mathbf{k}, st)} - R_{j\mathbf{k}, st}(j'\mathbf{k}', s', t')$$
$$= 2\pi i \sum_{j''\mathbf{k}'', s'', t''} \delta(\omega_{\mathbf{k}} - \omega_{\mathbf{k}''})\, \overline{R_{j'\mathbf{k}', s't'}(j''\mathbf{k}'', s''t'')}\, R_{j\mathbf{k}, st}(j''\mathbf{k}'', s''t'') \quad (332)$$

where $\omega_{\mathbf{k}} = \omega_{\mathbf{k}'}$. The zero-meson states cannot have the same energy as any one-meson state and therefore do not contribute to the right-hand side of (303). Note also that, if one restricts one's attention to energies $\omega_{\mathbf{k}} < 2\mu$, then the states with two or more mesons cannot contribute to the right-hand side of Eq. (303), the general unitary condition, so that for the case $\omega_{\mathbf{k}} < 2\mu$ Eq. (332) is exact.

To further simplify our notation, we shall denote the nucleonic variables st by the single letter n where n runs from 1 to 4, and consider $R_{j\mathbf{k}, n}(j'\mathbf{k}', n')$ as the $n'n$ matrix element of the operator $R_{j\mathbf{k}}(j'\mathbf{k}')$, i.e.,

$$R_{j\mathbf{k},\,n}(j'\mathbf{k}', n') = \langle n' \mid R_{j\mathbf{k}}(j'\mathbf{k}') \mid n\rangle \tag{333}$$

With this notation Eq. (330), recalling Eq. (319), reads:

$$\langle n' \mid R_{j\mathbf{k}}(j'\mathbf{k}') \mid n\rangle = \frac{1}{\omega_{\mathbf{k}'}} \sum_{n''} \langle n' \mid [V_{j'\mathbf{k}'}{}^*, V_{j\mathbf{k}}] \mid n\rangle$$
$$- \sum_{j'', \mathbf{k}'', n''} \left\{ \frac{\overline{\langle n'' \mid R_{j'\mathbf{k}'}(j''\mathbf{k}'') \mid n'\rangle}\, \langle n'' \mid R_{j\mathbf{k}}(j''\mathbf{k}'') \mid n\rangle}{\omega_{\mathbf{k}''} - \omega_{\mathbf{k}'} - i\epsilon} \right.$$
$$\left. + \frac{\overline{\langle n'' \mid R_{j\mathbf{k}}(j''\mathbf{k}'') \mid n'\rangle}\, \langle n'' \mid R_{j'\mathbf{k}'}(j''\mathbf{k}'') \mid n\rangle}{\omega_{\mathbf{k}'} + \omega_{\mathbf{k}''}} \right\} \tag{334}$$

If, as above, we denote $\overline{\langle n'' \mid R_{j'\mathbf{k}'}(j''\mathbf{k}'') \mid n'\rangle}$ by $\langle n' \mid R_{j'\mathbf{k}'}{}^*(j''\mathbf{k}'') \mid n''\rangle$, then the sum over the four intermediate nucleonic states can be performed with the result that (334) can be written as:

$$R_{j\mathbf{k}}(j'\mathbf{k}') = \frac{1}{\omega_{\mathbf{k}}} [V_{j'\mathbf{k}'}{}^*, V_{j\mathbf{k}}]$$
$$- \sum_{j''\mathbf{k}''} \left\{ \frac{R_{j'\mathbf{k}'}{}^*(j''\mathbf{k}'')\, R_{j\mathbf{k}}(j''\mathbf{k}'')}{\omega_{\mathbf{k}''} - \omega_{\mathbf{k}'} - i\epsilon} + \frac{R_{j\mathbf{k}}{}^*(j''\mathbf{k}'')\, R_{j'\mathbf{k}'}(j''\mathbf{k}'')}{\omega_{\mathbf{k}''} + \omega_{\mathbf{k}'}} \right\} \tag{335}$$

In the following we shall again further abbreviate the meson indices $j\mathbf{k}$ by a single index, e.g., $(j\mathbf{k}) = \mathbf{p}$, $(j'\mathbf{k}') = \mathbf{q}$, and denote the operator $R_{j\mathbf{k}}(j'\mathbf{k}')$ considered as a function of $\omega_{\mathbf{k}'} = z$ by $r_{\mathbf{qp}}(z)$. We next decompose $r_{\mathbf{qp}}(z)$ into contributions corresponding to states of definite total angular momentum and isotopic spin, i.e., write

$$r_{\mathbf{qp}}(z) = -v(\mathbf{q}^2)\, v(\mathbf{p}^2)\, \frac{4\pi}{\sqrt{4\omega_{\mathbf{p}}\omega_{\mathbf{q}}}} \sum_{\alpha=1}^{4} P_\alpha(\mathbf{p}, \mathbf{q})\, h_\alpha(z) \tag{336}$$

where the P_α are the projection operators for the four states of total angular momentum and isotopic spin

$$P_{11} = \tfrac{1}{3}\tau_{\mathbf{p}}\tau_{\mathbf{q}}(\boldsymbol{\sigma}\cdot\mathbf{p})\,(\boldsymbol{\sigma}\cdot\mathbf{q}) = \mathfrak{I}_{1/2}(\mathbf{q}, \mathbf{p})\, \mathcal{J}_{1/2}(\mathbf{q}, \mathbf{p}) \tag{337}$$
$$P_{13} = \tfrac{1}{3}\tau_{\mathbf{p}}\tau_{\mathbf{q}}[3\mathbf{p}\cdot\mathbf{q} - (\boldsymbol{\sigma}\cdot\mathbf{p})\,(\boldsymbol{\sigma}\cdot\mathbf{q})] = \mathfrak{I}_{1/2}\mathcal{J}_{3/2} \tag{338}$$
$$P_{31} = (\delta_{\mathbf{pq}} - \tfrac{1}{3}\tau_{\mathbf{p}}\tau_{\mathbf{q}})\,(\boldsymbol{\sigma}\cdot\mathbf{p})\,(\boldsymbol{\sigma}\cdot\mathbf{q}) = \mathfrak{I}_{3/2}\mathcal{J}_{1/2} \tag{339}$$
$$P_{33} = (\delta_{\mathbf{pq}} - \tfrac{1}{3}\tau_{\mathbf{p}}\tau_{\mathbf{q}})\,[3\mathbf{p} - \mathbf{q} - (\boldsymbol{\sigma}\cdot\mathbf{p})\,(\boldsymbol{\sigma}\cdot\mathbf{q})] = \mathfrak{I}_{3/2}\mathcal{J}_{3/2} \tag{340}$$

The subscript $\alpha = (2T, 2J)$ where T is the total isotopic spin and J is the total angular momentum. If this decomposition for $r_{\mathbf{qp}}$ is substituted into the unitarity relation (332), we find that the latter becomes

$$\operatorname{Im} h_\alpha(\omega_{\mathbf{p}}) = v^2(\mathbf{p})\, p^3 \mid h_\alpha(\omega_{\mathbf{p}})\mid^2 \tag{341}$$

In arriving at (341), use has been made of the fact that $k^2\,dk\,d\Omega_{\mathbf{k}} = k\omega_{\mathbf{k}}\,d\omega_{\mathbf{k}}\,d\Omega_{\mathbf{k}}$ when carrying out the indicated summation in the right-hand side of the unitarity condition, Eq. (332), as well as of the identity

$$\int d\Omega_{\mathbf{k}}\, \boldsymbol{\sigma}\cdot\mathbf{k}(\boldsymbol{\sigma}\cdot\mathbf{p})\boldsymbol{\sigma}\cdot\mathbf{k} = -\frac{4\pi\mathbf{k}^2}{3}\boldsymbol{\sigma}\cdot\mathbf{p} \tag{342}$$

It follows from Eq. (341) that we can write

$$\lim_{z\to\omega_{\mathbf{p}}+i\epsilon} h_\alpha(z) = \frac{e^{i\delta_\alpha(p)}\sin\delta_\alpha(p)}{v^2(\mathbf{p}^2)\,p^3} \tag{343}$$

where $\delta_\alpha(p)$ are real and denote the phase shifts.

If the one-meson approximation is not made in the unitarity condition, then the $\delta_\alpha(p)$ are real and identical with the conventional phase shifts only for $\omega_{\mathbf{p}} \leqslant 2\mu$. Note that even for $\omega_{\mathbf{p}} > 2\mu$ the unitarity condition asserts that

$$\operatorname{Im} h_\alpha(\omega_{\mathbf{p}}) = \frac{\sigma_\alpha(\omega_{\mathbf{p}})}{4\pi v^2(\mathbf{p}^2)\,p} \tag{344}$$

where σ_α is the total cross section for the state α including all inelastic processes, normalized to $4\pi\sin^2\delta_\alpha/p^2$ in the elastic range, $\omega_{\mathbf{p}} \leqslant 2\mu$.

The condition (310) on $r_{\mathbf{pq}}$ when translated into a condition on $h_\alpha(z)$ reads:

$$\sum_{\alpha=1}^{4} P_\alpha(\mathbf{q},\mathbf{p})\,h_\alpha(z) = \sum_{\alpha=1} P_\alpha(\mathbf{p},\mathbf{q})\,h_\alpha(-z) \tag{345}$$

From the definitions (337–40) of the projection operators P_α, one verifies that

$$P_\beta(\mathbf{p},\mathbf{q}) = \sum_{\alpha=1}^{4} A_{\alpha\beta}P_\alpha(\mathbf{q},\mathbf{p}) \tag{346}$$

where the 4×4 "crossing-matrix" A is given by

$$A = \frac{1}{9}\begin{pmatrix} 1 & -4 & -4 & 16 \\ -2 & -1 & 8 & 4 \\ -2 & 8 & -1 & 4 \\ 4 & 2 & 2 & 1 \end{pmatrix} \tag{347}$$

Since A is a reflection matrix it has the property that

$$\sum_{\beta=1}^{4} A_{\alpha\beta}A_{\beta\gamma} = \delta_{\alpha\gamma} \tag{348}$$

Finally, using the orthogonality properties of the P_α, we obtain from (345) that

$$h_\alpha(z) = \sum_{\beta=1}^{4} A_{\alpha\beta}h_\beta(-z) \tag{349}$$

which are the crossing relations for the function $h_\alpha(z)$. The condition that $r_{\mathbf{qp}}(z)$ be a Hermitian matrix function of z implies that $h_\alpha(z)$ is a real function of z, in the sense that

$$h_\alpha(\bar{z}) = \overline{h_\alpha(z)} \tag{350}$$

Since the inhomogeneous term in (335), i.e., the Born approximation term, can be written as

$$-\frac{1}{\omega_{\mathbf{q}}}\sum_{\alpha}\lambda_\alpha P_\alpha(\mathbf{p},\mathbf{q})\cdot\frac{v(\mathbf{p}^2)\,v(\mathbf{q}^2)}{\sqrt{4\omega_{\mathbf{p}}\omega_{\mathbf{q}}}} \tag{351}$$

where

$$\lambda_\alpha = \frac{2}{3}\left(\frac{f}{\mu}\right)^2\begin{bmatrix}-4 \text{ for } \alpha = 11\\ -1 \text{ for } \alpha = 13\\ -1 \text{ for } \alpha = 31\\ 2 \text{ for } \alpha = 33\end{bmatrix} \tag{352}$$

the Low equation in terms of $h_\alpha(\omega)$ becomes:

$$h_\alpha(\omega) = \frac{\lambda_\alpha}{\omega} + \frac{1}{\pi}\int_\mu^\infty d\omega_{\mathbf{p}}\, p^3 v^2(\mathbf{p}^2)\left\{\frac{|h_\alpha(\omega_{\mathbf{p}})|^2}{\omega_{\mathbf{p}} - \omega - i\epsilon} + \sum_\beta A_{\alpha\beta}\frac{|h_\beta(\omega_{\mathbf{p}})|^2}{\omega_{\mathbf{p}} + \omega}\right\} \tag{353}$$

With the help of the unitarity condition (344) it can also be written in the following form:

$$h_\alpha(\omega) = \frac{\lambda_\alpha}{\omega} + \frac{1}{\pi}\int_\mu^\infty d\omega_{\mathbf{p}}\left\{\frac{\operatorname{Im} h_\alpha(\omega_{\mathbf{p}})}{\omega_{\mathbf{p}} - \omega - i\epsilon} + \sum_\beta A_{\alpha\beta}\frac{\operatorname{Im} h_\beta(\omega_{\mathbf{p}})}{\omega_{\mathbf{p}} + \omega}\right\} \tag{354}$$

$$= \frac{\lambda_\alpha}{\omega} + \frac{1}{4\pi^2}\int_\mu^\infty d\omega_{\mathbf{p}}\left\{\frac{\sigma_\alpha(\omega_{\mathbf{p}})}{\omega_{\mathbf{p}} - \omega - i\epsilon} + \sum_\beta A_{\alpha\beta}\frac{\sigma_\beta(\omega_{\mathbf{p}})}{\omega_{\mathbf{p}} + \omega}\right\}\frac{1}{pv^2(\mathbf{p}^2)} \tag{355}$$

Equations (355) and (354) are actually *exact* consequences of the Chew model and unitarity and do not depend on the one-meson approximation, as will be verified by combining Eqs. (300), (303), and (344) and (336). We shall return to this point at the end of the present section. Equation (354) is equivalent to the following properties of $h_\alpha(z)$:

(*a*) the singularities of $h_\alpha(z)$ are entirely on the real axis

(*b*) $h_\alpha(z)$ has a simple pole at the origin with residue λ_α

(*c*) $h_\alpha(z)$ has branch points at $z = \pm\mu$, with cuts along the real axis to $\pm\infty$

(*d*) $h_\alpha(z)$ behaves at infinity like $1/z$

(*e*) $h_\alpha(z)$ satisfies the crossing relation (349)

To solve Eq. (349), Chew and Low introduce the (real) function

$$g_\alpha(z) = \frac{\lambda_\alpha}{z}\frac{1}{h_\alpha(z)} \tag{356}$$

where $g_\alpha(0) = 1$, since $h_\alpha(z)$ has a pole at $z = 0$ with residue λ_α. Also, $g_\alpha(z)$ behaves like a constant at ∞. The reason for introducing the function $g_\alpha(z)$ is that its discontinuities across the cuts for $z > \mu$ and $z < -\mu$ are independent of $g_\alpha(z)$. For example, in the region $\omega_{\mathbf{p}} > \mu$, since $\overline{h_\alpha(z)} = h_\alpha(\bar{z})$,

$$\lim_{z\to\omega_{\mathbf{p}}+i\epsilon} g_\alpha(z) - \lim_{z\to\omega_{\mathbf{p}}-i\epsilon} g_\alpha(z) = \frac{\lambda_\alpha}{\omega_{\mathbf{p}}}\left[\frac{1}{h_\alpha(\omega_{\mathbf{p}}+i\epsilon)} - \frac{1}{\overline{h_\alpha(\omega_{\mathbf{p}}+i\epsilon)}}\right]$$

$$= -\frac{2i\lambda_\alpha}{\omega_{\mathbf{p}}}\frac{\operatorname{Im} h_\alpha(\omega_{\mathbf{p}}+i\epsilon)}{|h_\alpha(\omega_{\mathbf{p}}+i\epsilon)|^2}$$

$$= -\frac{\lambda_\alpha}{\omega_{\mathbf{p}}}\left[\frac{2ip^3v^2(\mathbf{p}^2)\,|\,h_\alpha(\omega_{\mathbf{p}})|^2}{|h_\alpha(\omega_{\mathbf{p}})|^2}\right]$$

$$= -2i\frac{\lambda_\alpha p^3}{\omega_{\mathbf{p}}}v^2(\mathbf{p}^2) \tag{357}$$

The crossing relation for g_α becomes

$$\sum_\beta B_{\alpha\beta}\frac{1}{g_\beta(z)} = \frac{1}{g_\alpha(-z)} \tag{358}$$

where

$$B_{\alpha\beta} = -\frac{1}{\lambda_\alpha}A_{\alpha\beta}\lambda_\beta \tag{359}$$

The singularities of $g_\alpha(z)$ are as follows: Just as for $h_\alpha(z)$, it clearly has branch points at $z = \pm\mu$ with cuts along the real axis to $\pm\infty$. If $h_\alpha(z)$ has no zeros, these will be the only singularities of $g_\alpha(z)$. Now from Eq. (353) it follows that $\operatorname{Im} h_\alpha(z) = c_\alpha \operatorname{Im} z$ where c_α is a constant. Therefore, unless c_α vanishes accidentally, the zeros of $h_\alpha(z)$ can only occur on the real axis. For $(f/\mu)^2$ sufficiently small, $h_\alpha(z)$ will have no zeros even on the real axis (since the leading term λ_α/z then dominates), so that in this case the boundary conditions on $g_\alpha(z)$, together with the nature of its singularities imply that $g_\alpha(z)$ has the following representation

$$g_\alpha(z) = 1 - \frac{z}{\pi}\int_\mu^\infty dx'\left[\frac{A_\alpha(x')}{x'-z} + \frac{B_\alpha(x')}{x'+z}\right] \tag{360}$$

where A_α and B_α are weight functions defined for $x \geqslant \mu$. The function $A_\alpha(x)$ gives the jump of $g_\alpha(z)$ in going across the real axis for $z \geqslant \mu$, so that by virtue of (357)

$$A_\alpha(\omega_{\mathbf{p}}) = \frac{\lambda_\alpha p^3}{\omega_{\mathbf{p}}^2}v^2(\mathbf{p}^2) \tag{361}$$

The crossing relation is just sufficient to determine the second weighting function, $B_\alpha(x)$, which gives the jump of $g_\alpha(z)$ in going across the real axis for $z \leqslant \mu$. Therefore, for sufficiently small f^2/μ^2

$$h_\alpha(z) = \frac{\lambda_\alpha/z}{1 - \frac{z}{\pi}\lambda_\alpha\int_\mu^\infty \frac{d\omega_{\mathbf{p}}}{\omega_{\mathbf{p}}^2}\frac{p^3v^2(\mathbf{p}^2)}{\omega_{\mathbf{p}}-z} - \frac{z}{\pi}\int_\mu^\infty \frac{d\omega_{\mathbf{p}}}{\omega_{\mathbf{p}}^2}\frac{p^3v^2(\mathbf{p}^2)}{\omega_{\mathbf{p}}+z}B'_\alpha(\omega_{\mathbf{p}})} \tag{362}$$

where $p = \sqrt{\omega_{\mathbf{p}}^2 - \mu^2}$ and the function B' defined by

$$\frac{p^3v^2(\mathbf{p}^2)}{\omega_{\mathbf{p}}^2}B'_\alpha(\omega_{\mathbf{p}}) = B_\alpha(\omega_{\mathbf{p}}) \tag{363}$$

must be so chosen as to make $h_\alpha(z)$ satisfy the crossing relation (349). [For numerical methods to accomplish this task, see Salzman (1955); also Chew (1956).] Equation (362) furnishes a solution of the Low equation in the one-meson approximation which has the property that it is the analytic continuation of the perturbation solution in $(f/\mu)^2$. This solution has been shown by Castillejo, Dalitz, and Dyson [Castillejo (1956); see also Klein (1956)] not to be unique but rather that it is only one of an infinite number of solutions. These other solutions are generated by giving $h_\alpha(z)$ an arbitrary number of zeros on the real axis and giving $g_\alpha(z)$ arbitrary positive residues at these points. The Chew-Low solution is thus distinguished from the others in that it has the fewest number of zeros in the scattering amplitude.

This nonuniqueness of the solutions is due to the fact that the Low equation does not manifest the full physical content of the theory [Dyson (1957), Norton and Klein (Norton 1958a, b)]. More specifically, the nonuniqueness is due to lack of information of the Low equation on the hidden structure of the system [Haag (1957)]. It has been shown by Fairlie and Polkinghorne [Fairlie (1958)] that the energy spectrum of the unperturbed Hamiltonian suffices to choose the correct "physical" solution, which is unique. In particular, the solution obtained by Chew and Low describes that theory of the infinite class characterized by the same Low equation which has a target (the nucleon) with a minimum of internal structure. All other solutions correspond to models for which the target (nucleon) has excited states which lead to additional resonances.

We next note that Eqs. (356) and (343), together, assert that

$$\operatorname{Re} g_\alpha(z)\Big|_{z\to\omega_{\mathbf{p}}+i\epsilon} = \frac{\lambda_\alpha p^3 v^2(\mathbf{p}^2)}{\omega_{\mathbf{p}}} \cot \delta_\alpha(\omega_{\mathbf{p}}) \tag{364}$$

and it follows from Eq. (362) that

$$\operatorname{Re} g_\alpha(\omega) = 1 - \omega \left\{ \frac{\lambda_\alpha}{\pi} \mathrm{P} \int_\mu^\infty \frac{d\omega_{\mathbf{p}}\, p^3 v^2(\mathbf{p}^2)}{\omega_{\mathbf{p}}{}^2(\omega_{\mathbf{p}} - \omega)} + \frac{1}{\pi} \int \frac{d\omega_{\mathbf{p}}}{\omega_{\mathbf{p}}{}^2}\, p^3 v^2(\mathbf{p}^2)\, \frac{B'_\alpha(\omega_{\mathbf{p}})}{\omega_{\mathbf{p}} + \omega} \right\} \tag{365}$$

Now in the absence of a cutoff, i.e., for $v(\mathbf{p}^2) = 1$, the integrals in Eq. (365) would be linearly divergent. The contribution to the integrals, therefore, come primarily from the region $\omega_{\mathbf{p}} \sim \omega_{\max}$, where $\omega_{\max}$ is the maximum energy effectively allowed by the cutoff factor $v(\mathbf{p}^2)$. Therefore, for ω small compared with $\omega_{\max}$, the dependence on ω occurring in the denominators of the integrand can be neglected. The error incurred by this neglect is expected to be of order $\omega/\omega_{\max}$ since

$$\frac{1}{\omega_{\mathbf{p}} - \omega} = \frac{1}{\omega_{\mathbf{p}}} + \frac{\omega}{\omega_{\mathbf{p}}} \frac{1}{\omega_{\mathbf{p}} - \omega} \tag{366}$$

With a cutoff in the neighborhood of 1 Bev, this approximation should be reasonable for meson energies less than 250 Mev. In general, Re $g_\alpha(\omega)$ can be written in the form:

$$\frac{\lambda_\alpha p^3 v^2(\mathbf{p}^2)}{\omega_\mathbf{p}} \cot \delta_\alpha(\omega_\mathbf{p}) = 1 - \omega_\mathbf{p} \mathfrak{r}_\alpha(\omega_\mathbf{p}) \tag{367}$$

where by the above arguments $\mathfrak{r}_\alpha(\omega_\mathbf{p})$ is almost a constant for small $\omega_\mathbf{p}$. The approximation consisting in the complete neglect of the energy dependence of $\mathfrak{r}_\alpha(\omega_\mathbf{p})$ for $\omega_\mathbf{p} \ll \omega_{\max}$ is called the effective range approximation.[11] We note from Eq. (365) that $\mathfrak{r}_\alpha$ is of order $\lambda_\alpha \omega_{\max}$. Crossing symmetry requires that $\mathfrak{r}_{11} = -\mathfrak{r}_{33} - \frac{1}{4}\mathfrak{r}_{13}$, or equivalently that

$$\mathfrak{r}_\alpha = \mathfrak{r}_{33} \begin{pmatrix} -1 + \frac{1}{4}x \\ -x \\ -x \\ 1 \end{pmatrix} \tag{368}$$

where $x = -\mathfrak{r}_{13}/\mathfrak{r}_{33}$. Furthermore, by comparing (365) and (355) for $|z| < \mu$, it can be shown that

$$\mathfrak{r}_{33} = \frac{\mu^2}{3f^2} \frac{1}{4\pi^2} \int_\mu^\infty d\omega_\mathbf{p} \frac{(\frac{8}{9}\sigma_3 + \sigma_1 + \sigma_2)}{\omega_\mathbf{p} p v^2(\mathbf{p}^2)} \tag{369}$$

i.e., that the 33 effective range is positive. With an appropriate choice of cutoff, there will then be a resonance in the 33 state, since it follows from Eq. (367) that to order $1/\omega_{\max}$

$$\frac{p^3 v^2(\mathbf{p}^2)}{\omega_\mathbf{p}} \cot \delta_{33}(\omega_\mathbf{p}) = \frac{3\mu^2}{4f^2}(1 - \mathfrak{r}_{33}\omega_\mathbf{p}) \tag{370}$$

The resonance will occur when $\omega_\mathbf{p} = 1/\mathfrak{r}_{33}$, at which point $\delta_{33} = 90°$. The low energy P-wave phase shifts are thus predicted in terms of two parameters, the coupling constant f^2 and the effective range $\mathfrak{r}_{33}$, where the latter is determined from the cutoff, $\omega_{\max}$ by $\mathfrak{r}_{33} \sim f^2\omega_{\max}$.

The coupling constant can be obtained by making a plot of $p^3 \cot \delta_{33}(\omega_\mathbf{p})/\omega_\mathbf{p}$ against $\omega_\mathbf{p}$. The effective range formula asserts that this should be straight line with the intercept at zero energy equal to $\frac{3}{4}(\mu^2/f^2)$. The expected linear dependence is indeed found and leads to a value of $f^2/\mu^2 = 0.08$ [see Bernardini (1956) for a complete discussion of the steps involved]. Given the experimental fact that a resonance occurs at $\omega_\mathbf{p} \sim 200$ Mev, the relation between resonance energy ω_0 and $\mathfrak{r}_{33}$, $\mathfrak{r}_{33} \approx 1/\omega_0$ together with the theoretical relation $\mathfrak{r}_{33} \sim \left[\left(\frac{f^2}{\mu^2}\right)\omega_{\max}\right] \Big/ \mu^2$ implies that $\omega_{\max} \sim 6\mu$; which a posteriori justifies the validity of the approximation that $1/\omega_{\max}$ is small.

The Chew model only gives rise to P-wave scattering and thus cannot

[11] Actually the power series $\mathfrak{r}_\alpha(z) = \mathfrak{r}_\alpha(0) + z\mathfrak{r}_\alpha'(0) + \cdots$ has a radius of convergence $|z| < \mu$, with $z = \mu$ just the beginning of the region of physical interest. Nevertheless for $\mu \lesssim \omega \ll \omega_{\max}$, $\mathfrak{r}_\alpha(\omega) \approx \mathfrak{r}_\alpha(0)$ [see Chew (1958)].

account for the (low energy) S-wave phase shifts. To extend the Chew-Low analysis for the low energy S-wave scattering Drell, Friedman, and Zachariasen [Drell (1956)] have extended the Chew model to include an S-wave meson-nucleon interaction by adopting as the interaction Hamiltonian

$$H_I = H_p + H_{s1} + H_{s2} \tag{371a}$$

$$H_p = \frac{f_0}{\mu} \sum_j \int \rho(\mathbf{x})\, (\boldsymbol{\sigma} \cdot \boldsymbol{\nabla})\, \phi_j(\mathbf{x})\, \tau_j\, d^3x \tag{371b}$$

$$H_{s1} = g_{01} \int \boldsymbol{\phi}(\mathbf{x}) \cdot \boldsymbol{\phi}(\mathbf{x}')\, \rho(\mathbf{x})\, \rho(\mathbf{x}')\, d^3x\, d^3x' \tag{371c}$$

$$H_{s2} = g_{02} \int \boldsymbol{\tau} \cdot (\boldsymbol{\phi}(\mathbf{x}) \times \boldsymbol{\pi}(\mathbf{x}'))\, \rho(\mathbf{x})\, \rho(\mathbf{x}')\, d^3x\, d^3x' \tag{371d}$$

This interaction Hamiltonian is the static limit of the Hamiltonian (177) with terms of order $1/M$ and G^2/M^2 included. The interaction $H_s = H_{s1} + H_{s2}$ give rise to S-wave scattering. The S and P-wave scattering amplitude which Drell *et al.* derive are then functions of the renormalized coupling constants:

$$f = \frac{{}_+\langle s, t \mid \sigma_i\tau_j \mid s, t\rangle_+}{\langle s, t \mid \sigma_i\tau_j \mid s, t\rangle} f_0 = Zf_0 \tag{372}$$

$$g_1 = \frac{{}_+\langle s, t \mid s, t\rangle_+}{\langle s, t \mid s, t\rangle} g_{01} = g_1 \tag{373}$$

and

$$g_2 = \frac{{}_+\langle s, t \mid \tau_i \mid s, t\rangle_+}{\langle s, t \mid \tau_i \mid s, t\rangle} g_{02} = Z' g_{02} \tag{374}$$

By taking g_1 and g_2 as adjustable parameters and f and ρ determined from the P-wave data, Drell *et al.* find they could obtain agreement with the experimental low energy S-wave phase shifts

$$\delta_{J=0,\, T=1/2} = \delta_1 = 0.16\, \frac{\hbar k}{\mu c} \tag{375a}$$

$$\delta_{J=0,\, T=3/2} = \delta_3 = -0.11\, \frac{\hbar k}{\mu c} \tag{375b}$$

by taking

$$g_1 = \frac{0.4}{\mu} \tag{376}$$

and

$$g_2 = \frac{0.4}{\mu^2} \tag{377}$$

[see also Bonnevay (1959)]. Bincer (1958) has calculated double S-wave pion production by a P-wave meson using the Wick–Chew-Low approach with the Hamiltonian (371) and obtained results in reasonable agreement with experiments. Calculations of meson-production in meson-nucleon collisions using the Chew-Low formalism have also been performed by Franklin (1957), Rodberg (1957), and Kazes (1957) [see also Barshay (1956), who starts with the relativistic pseudoscalar meson theory in the Heisenberg picture, obtains the meson production amplitude using the Low (1955) approach, and then reduces this expression to the static limit].

Because of space limitation, our discussion of the Chew model must be limited to its predictions concerning meson-nucleon scattering, and we cannot say very much about the inclusion of electromagnetic interactions into the model nor about its predictions concerning nuclear forces. This is unfortunate since the calculation of the photoproduction of mesons is the real success and, perhaps, the greatest achievement of the Chew-Low approach, for it interrelates the meson-nucleon results to the photoproduction problem. In fact, once the pion-nucleon scattering phase shifts are known, the corresponding photomeson cross sections are almost unambiguously predicted and are found to be in good agreement with experiment. The theory furthermore predicts that the zero energy limit of photomeson production depends only on the value of f^2/μ^2, the same renormalized coupling constant which appeared in the P-wave meson-nucleon scattering. The reader is referred to the second paper of Chew and Low [Chew (1956b); also Chew (1956c)] for full details.

Experimentally it is found that the cross section for photoproduction of charged pions behaves as a function of energy very differently from that for photoproduction of neutral mesons. Both $\sigma(\gamma + \mathrm{p} \to \pi^+ + \mathrm{n})$ and $\sigma(\gamma + \mathrm{p} \to \pi^0 + \mathrm{p})$ have a maximum at a photon energy of about 330 Mev which corresponds to the 33 resonance in πN scattering. The threshold behavior of $\sigma(\gamma + \mathrm{p} \to \pi^0 + \mathrm{p})$ indicates that the π^0 is produced in a p state, whereas the threshold behavior of $\sigma(\gamma + \mathrm{p} \to \pi^+ + \mathrm{n})$ indicates that the charged pions are produced mainly in an s state. This marked difference can be understood when one recalls that the gauge invariant introduction of electromagnetic interactions requires that the operator $\boldsymbol{\nabla}\boldsymbol{\phi}$ in H_I be replaced by $(\boldsymbol{\nabla} - et_3\mathbf{A})\,\boldsymbol{\phi}$, i.e., the $\boldsymbol{\nabla}$ operator be replaced by $\boldsymbol{\nabla} + ie\mathbf{A}$ when acting on the π^+ field operator. Therefore in the presence of electromagnetic effects an additional term of the form $\left(\frac{ef}{\mu}\,\boldsymbol{\sigma}\cdot\mathbf{A}\phi\tau_- + \text{h.a.}\right)$ occurs in the interaction and provides a mechanism for s state production of charged pions.

Other electromagnetic effects which have been computed with the Chew model and based on the Chew-Low approach are the photoproduction of a pair of πs (one S wave and one P wave) [Cutkosky and Zachariasen (Cutkosky 1956)], the charge distribution and anomalous magnetic moments of nucleons [Salzman (1955), Trieman and Sachs (Trieman 1956), Miyazawa (1956)], and the scattering of photons from a nucleon [Karzas *et al.* (Karzas 1958)].

Finally, Low (1957) has given a lucid and simple exposition of the connection between the static Chew model and the fully relativistic γ_5 theory and the basis for the success of the simple static model. Essentially the reason that so crude a theory proves so useful is the fact that invariance considerations (isotopic spin, rotational parity, etc.) together with the assumed energy spectrum of the states of the meson-nucleon dictate the low energy behavior of the theory: almost any theory with the same invariances and spectrum would have the same low energy P-wave behavior [Horwitz (1957)] as long as no direct meson-meson interaction terms are included. The success of the Chew model thus also reflects the relative unimportance of pion-pion interactions in low energy P-wave scattering.

We finally conclude this section on the Chew model with another derivation of the Low equation which is based on the connection between causality and "dispersion relations." [For this connection in the present application, see Toll (1956).] By the concept of causality we here mean the condition that the response of a system to a disturbance must vanish for times earlier than the application of the disturbance, i.e., "no output can occur before the input." In its application to the scattering problem, the causality condition asserts that there can be no scattered waves before the primary wave has hit any part of the scatterer. By a dispersion relation is meant a relation connecting the real and imaginary part of a function, where the latter is usually some physically observable quantity, e.g., the scattering amplitude.

A simple (nonrigorous) derivation of the connection between the causality principle and dispersion relations proceeds as follows: Let the input into a given physical system as a function of time be denoted by $I(t)$ and let the resulting output be denoted by $R(t)$. We shall assume the system to be linear, although dispersion relations also hold for somewhat more general systems. We shall also assume the system to be time-independent, i.e., to be such that a displacement in time of the whole input causes a corresponding displacement in time of the response. Under these assumptions, we can write the following relation between response and input:

$$R(t) = \frac{1}{\sqrt{2\pi}} \int_{-\infty}^{+\infty} T(t - t')\, I(t')\, dt' \tag{378}$$

where $T(t)$, the time delay distribution function, is the output corresponding to a δ-function input. If we denote by $r(\omega)$, $t(\omega)$, and $i(\omega)$ the Fourier transforms of R, T, and I respectively, i.e.,

$$G(t) = \frac{1}{\sqrt{2\pi}} \int_{-\infty}^{+\infty} g(\omega)\, e^{-i\omega t}\, d\omega \tag{379}$$

Eq. (378) can then be written as

$$r(\omega) = t(\omega)\, i(\omega) \tag{380}$$

or equivalently

$$R(t) = \frac{1}{\sqrt{2\pi}} \int_{-\infty}^{+\infty} d\omega\, e^{-i\omega t}\, t(\omega)\, i(\omega) \tag{381}$$

The causality principle can be translated into the statement that if $I(t) = 0$ for $t < 0$, then $R(t) = 0$ for $t < 0$. Since

$$i(\omega) = \frac{1}{\sqrt{2\pi}} \int_{-\infty}^{+\infty} e^{i\omega t}\, I(t)\, dt \tag{382}$$

if $I(t) = 0$ for $t < 0$, then

$$i(\omega) = \frac{1}{\sqrt{2\pi}} \int_{0}^{\infty} e^{i\omega t}\, I(t)\, dt \tag{383}$$

considered as function of the complex variable $\omega = \omega_1 + i\omega_2$ will be analytic for $\operatorname{Im} \omega = \omega_2 > 0$, i.e., in the upper half of the complex ω plane. [For simplicity, assume $I(t)$ bounded or square integrable, so that the integral (383) is then always convergent for $\omega_2 > 0$.]

It will next be noticed from Eq. (381) that $R(t)$ will also vanish for $t < 0$ if the function $t(\omega)\, i(\omega)$ is analytic in the upper half of the complex ω plane. Since $i(\omega)$ is analytic there (and this is true for the class of *all* inputs vanishing for $t < 0$), the system will be causal if the Fourier transform, $t(\omega)$, of the time delay distribution function, $T(t)$, is an analytic function in the upper half of the complex ω plane, or equivalently, if $T(t) = 0$ for $t < 0$, i.e., if $T(t) = \theta(t)\, T'(t)$ where $\theta(t) = 0$ for $t < 0$, and $= 1$ for $t > 0$. By Cauchy's theorem therefore, for any contour C lying entirely in the upper half of the complex ω plane and any ω with $\operatorname{Im} \omega > 0$, we can write

$$t(\omega) = \frac{1}{2\pi i} \int_C \frac{t(\zeta)}{\zeta - \omega}\, d\zeta \tag{384}$$

Let us suppose that the behavior at infinity of $t(\zeta)$ is such that the contour can be extended to a large semicircle, $\frown$, and that the contribution from the integration over the circumference of this semicircle vanishes in the limit as the radius R goes to ∞, then

$$t(\omega) = \frac{1}{2\pi i} \int_{-\infty}^{+\infty} \frac{t(\zeta)}{\zeta - \omega}\, d\zeta \tag{385}$$

Finally, since we are interested in real values of ω, in the limit as $\omega \to \omega_1 + i\epsilon$, we obtain

$$\begin{aligned} \lim_{\omega \to \omega_1 + i\epsilon} t(\omega) &= t(\omega_1) \\ &= \frac{1}{2\pi i} \lim_{\epsilon \to 0+} \int_{-\infty}^{+\infty} \frac{t(\zeta)}{\zeta - \omega_1 - i\epsilon}\, d\zeta \\ &= \frac{1}{2\pi i} \mathrm{P} \int_{-\infty}^{+\infty} \frac{t(\zeta)}{\zeta - \omega_1}\, d\zeta + \tfrac{1}{2} t(\omega_1) \end{aligned} \tag{386}$$

whence for real ωs

$$t(\omega) = \frac{1}{i\pi} \mathrm{P} \int_{-\infty}^{+\infty} \frac{t(\zeta)}{\zeta - \omega} d\zeta \tag{387}$$

It follows from (387) that

$$\mathrm{Re}\, t(\omega) = \frac{1}{\pi} \mathrm{P} \int_{-\infty}^{+\infty} \frac{\mathrm{Im}\, t(\omega')}{\omega' - \omega} d\omega' \tag{388a}$$

$$\mathrm{Im}\, t(\omega) = -\frac{1}{\pi} \mathrm{P} \int_{-\infty}^{+\infty} \frac{\mathrm{Re}\, t(\omega')}{\omega' - \omega} d\omega' \tag{388b}$$

Since

$$i\, \mathrm{Im}\, t(\omega) = \frac{1}{\pi} \int_{-\infty}^{+\infty} \mathrm{Im}\, t(\omega')\, i\pi\delta(\omega - \omega')\, d\omega' \tag{389}$$

adding (388a) and (389), we can also write

$$t(\omega) = \frac{1}{\pi} \lim_{\epsilon \to 0+} \int_{-\infty}^{+\infty} \frac{\mathrm{Im}\, t(\omega')}{\omega' - \omega - i\epsilon} d\omega' \tag{390}$$

Equations (388) and (390) are called dispersion relations and evidently are completely equivalent to the analyticity of $t(\omega)$ in $\mathrm{Im}\, \omega > 0$ and the statement that the behavior of $t(\omega)$ at infinity is such as to guarantee that $\int t(\omega)/\omega\, d\omega \to 0$ as $R \to \infty$.

With this preliminary out of the way, let us return to the description of meson-nucleon scattering. Since the scattering amplitude determines the response of the system when a meson is scattered by a nucleon (recall $|\rangle_- = S\,|\rangle_+$ and $R \propto S - 1$), we therefore expect the amplitude $R_{JT}(\omega)$ (supposed to be defined by analytic continuation for all values of ω in $\mathrm{Im}\, \omega > 0$) to obey dispersion relations. However, due to the finite extent of the nucleon, it is not the scattering amplitude but the function $h_\alpha(\omega)$ [i.e., the ratio of the amplitude to $v^2(\mathbf{k}^2)\, k^3$] which satisfies the dispersion relation:

$$h_\alpha(\omega) = \frac{1}{\pi} \lim_{\epsilon \to 0+} \int_{-\infty}^{+\infty} \frac{\mathrm{Im}\, h_\alpha(\omega')}{\omega' - \omega - i\epsilon} d\omega' \tag{391}$$

We shall now show that Eq. (391), when taken together with crossing symmetry and unitarity relations, is in fact the Low equation. Since the latter is an integral relation involving only quantities defined for $\omega > \mu$, i.e., physical ωs, we first of all wish to rewrite (391) in such a way that the integration also only involves ωs greater than μ. Now crossing symmetry asserts that

$$h_\alpha(-\omega) = \sum_\beta A_{\alpha\beta} h_\beta(\omega) \tag{392}$$

so that (391) can be rewritten as

$$h_\alpha(\omega) = \frac{1}{\pi}\int_{-\mu}^{\mu} d\omega' \frac{\operatorname{Im} h_\alpha(\omega')}{\omega' - \omega - i\epsilon} + \frac{1}{\pi}\int_{\mu}^{\infty} d\omega' \frac{\operatorname{Im} h_\alpha(\omega')}{\omega' - \omega - i\epsilon} + \frac{1}{\pi}\int_{\mu}^{\infty} d\omega' \sum_\beta A_{\alpha\beta} \frac{\operatorname{Im} h_\beta(\omega)}{\omega' + \omega} \tag{393}$$

To obtain the contributions from the range $|\omega| < \mu$, we use the unitarity relation which can be written in the form

$$\operatorname{Im} h_\alpha(\omega) \propto -\pi \sum_\beta h_{\alpha\beta}\bar{h}_{\beta\alpha}\delta(\omega + M - E_\beta) \tag{394}$$

where $h_{\alpha\beta}$ is the amplitude for the process $\alpha \to \beta$ (in this notation $h_\alpha = h_{\alpha\alpha}$). Now for ω in the range $|\omega| \leqslant \mu$, clearly only the one-nucleon states contribute in the summation in the right-hand side of (394), since the argument in the δ function will always be different from zero for all other states. Using these results, we obtain

$$\operatorname{Im} h_\alpha(\omega) = \pi\lambda_\alpha\delta(\omega) \quad |\omega| < \mu \tag{395}$$

so that finally

$$h_\alpha(\omega) = \frac{\lambda_\alpha}{\omega} + \frac{1}{\pi}\int_{\mu}^{\infty} d\omega' \frac{\operatorname{Im} h_\alpha(\omega')}{\omega' - \omega - i\epsilon} + \frac{1}{\pi}\int_{\mu}^{\infty} d\omega' \sum_\beta A_{\alpha\beta} \frac{\operatorname{Im} h_\beta(\omega')}{\omega' + \omega} \tag{396}$$

which is our previous equation (354).

13

Reduction of S Matrix

13a. Formal Introductions

Following our study in the last chapter of some "simple" and soluble models of field theories, we turn our attention to a discussion of the application of perturbation methods to the quantized theory of interacting relativistic fields. In particular, we shall first consider the application of perturbation theory to the description of the scattering of particles. Our discussion will be based on the previously derived formula for the S matrix in the Dirac picture. Recall that in this picture the field operators satisfy free-field equations and that the time dependence of the state vector, $|\Psi(t)\rangle$, is determined by the interaction part of the Hamiltonian operator, i.e., by $H_I(t)$

$$i\hbar\partial_t \mid \Psi(t)\rangle = H_I(t) \mid \Psi(t)\rangle \tag{1}$$

The S matrix in this picture is defined as:

$$S = U(\infty, -\infty) = \lim_{t\to+\infty} \lim_{t_0\to-\infty} U(t, t_0) \tag{2a}$$

$$= \sum_{n=0}^{\infty} \left(-\frac{i}{\hbar}\right)^n \frac{1}{n!} \int_{-\infty}^{+\infty} dt_1 \cdots \int_{-\infty}^{+\infty} dt_n\, P(H_I(t_1) \cdots H_I(t_n)) \tag{2b}$$

and determines the overall change of the state of the system due to the interaction. For field theoretic applications it is convenient to introduce explicitly into (2b) the Hamiltonian density $\mathcal{H}_I(x)$. Since

$$H_I(t) = \int d^3x\, \mathcal{H}_I(x) \tag{3}$$

the S matrix becomes

$$S = \sum_{n=0}^{\infty} \left(\frac{-i}{\hbar c}\right)^n \frac{1}{n!} \int_{-\infty}^{+\infty} d^4x_1 \cdots \int_{-\infty}^{+\infty} d^4x_n\, P(\mathcal{H}_I(x_1) \cdots \mathcal{H}_I(x_n)) \tag{4}$$

In all our applications, the interaction Hamiltonian density will have the property that

$$[\mathcal{H}_I(x), \mathcal{H}_I(x')] = 0 \quad \text{for } (x - x')^2 < 0 \tag{5}$$

so that the ordering of $\mathcal{H}_I(x)$ and $\mathcal{H}_I(x')$ when $x_0 = x_0'$ is immaterial, and the P operator completely specifies the ordering. The advantage of the expression (4) for S over (2b) is that it is manifestly covariant. This can be made more explicit by writing (4) in the form

$$S = P\left[\exp\left(-\frac{i}{\hbar c}\int_{-\infty}^{+\infty} \mathcal{H}_I(x)\, d^4x\right)\right] \tag{6}$$

Since $\mathcal{H}_I(x)$ is a scalar, and so is d^4x, the exponent is an invariant; by virtue of Eq. (5), the time-ordering operation P is likewise invariant. Q.E.D.

There are, however, several difficulties connected with the application of Eq. (4) to field theoretic problems. Firstly, a prescription must be given on how the limiting procedure $\lim t \to \infty$, $\lim t_0 \to -\infty$ is to be carried out on $U(t, t_0)$ to obtain the S matrix. In particular, there are $n!$ ways to go to the limit $t_0 \to -\infty$ for the nth order term. Furthermore, a prescription must be given on how to smooth out any transient behavior due to the assumption that at t_0 the state vector is assumed to be an eigenstate of H_0, yet its subsequent time dependence is determined by $H_I(t)$.

To avoid these difficulties Dyson (1951b, c, d, e) has introduced the "intermediate interaction picture" (or smoothed interaction picture) which essentially gives the interaction Hamiltonian a convergence factor of the form $\exp(-\lambda|t|)$. The limit $\lambda \to 0$ is to be taken after the calculation is performed. Such a convergence factor removes the ambiguity from the order in which the limit $t \to \pm\infty$ is performed (it is equivalent to averaging over the $n!$ orders of taking the limits) and removes transient effects.

Another difficulty encountered in the application of (4) to field theoretic problems is the following: As mentioned above, the application of the perturbation theory in the Dirac (interaction) picture usually assumes that initially the state of the system is in an eigenstate of H_0, the unperturbed Hamiltonian. Due to the presence of $H_I(t)$, the state vector is altered. Eventually, for large t, its behavior is again assumed to be determined by H_0 only.

The applicability of perturbation theory thus rests on the assumption that perturbed and unperturbed state vectors lie in the same Hilbert space. However, it is known that this is not the case for local point interactions. Van Hove (1951, 1952) has shown that for theories with trilinear point interactions (or theories with cutoff functions which do not decrease fast enough) there are no normalizable vectors (other than the vacuum, for those theories where the bare and physical vacuum coincide) which are in the common domain of both H and H_0.

To circumvent this difficulty, it has become fashionable to work in terms of operators, the so-called "in" and "out" operators [Yang and Feldman (Yang 1950c), Källén (1950); see also Schweber (1955)] which also satisfy free-field commutation rules and free-field equations, but whose domains are the eigenstates of H. To define the in-operators, let us assume that the time at which the Heisenberg, Schrödinger and Dirac pictures coincide is $t = 0$. We then define the Heisenberg operator $V_+(t)$ by

$$V_+(t) = e^{iHt}\,\Omega^{(+)}\,e^{-iHt} \tag{7}$$

where $\Omega^{(+)}$ is the Møller wave matrix whose properties, it will be recalled, are that

$$\Omega^{(+)} \mid \varphi_a\rangle = |\psi_a{}^+\rangle \tag{8a}$$

$$\Omega^{(+)} H_0 \Omega^{(+)-1} = H \tag{8b}$$

or

$$\Omega^{(+)} H_0 = H\Omega^{(+)} \tag{8c}$$

In the following[1] we shall assume that the Hamiltonians H_0 and H have the property that no bound states exist for either, in which case $\Omega^{(+)-1} = \Omega^{(+)*}$ and

$$\begin{aligned} V_+(t)^{-1} &= e^{iHt}\,\Omega^{(+)*}\,e^{-iHt} \\ &= e^{iHt}\,\Omega^{(+)-1}\,e^{-iHt} \end{aligned} \tag{9}$$

The time evolution of $V_+(t)$ is determined by the following differential equation:

$$\begin{aligned} i\partial_t V_+(t) &= e^{iHt}\,[\Omega^{(+)}, H]\,e^{-iHt} \\ &= e^{iHt}\,\Omega^{(+)}\,(H - H_0)\,e^{-iHt} \\ &= e^{iHt}\,\Omega^{(+)}\,e^{-iHt}\,e^{iHt}\,H_I(0)\,e^{-iHt} \\ &= V_+(t)\,H_{IH}(t) \end{aligned} \tag{10}$$

where the subscript H is used to denote a Heisenberg operator. We define the Heisenberg "in" operator by:

$$\begin{aligned} O_{\text{in}}(t) &= V_+(t)\,O_H(t)\,V_+(t)^{-1} \\ &= e^{iHt}\,\Omega^{(+)}\,e^{-iHt}\,O_H(t)\,e^{iHt}\,\Omega^{(+)-1}\,e^{-iHt} \\ &= e^{iHt}\,\Omega^{(+)}\,O_H(0)\,\Omega^{(+)-1}\,e^{-iHt} \\ &= e^{iHt}\,O_{\text{in}}(0)\,e^{-iHt} \end{aligned} \tag{11}$$

The equation of motion which $O_{\text{in}}(t)$ satisfies is easily derived from (11), and one deduces that

$$-i\partial_t O_{\text{in}}(t) = [H, O_{\text{in}}(t)] \tag{12a}$$

as required for a Heisenberg operator. Also,

$$\begin{aligned} -i\partial_t O_{\text{in}}(t) &= e^{iHt}\,\{H\Omega^{(+)}O_H(0)\,\Omega^{(+)-1} - \Omega^{(+)}O_H(0)\,\Omega^{(+)-1}H\}\,e^{-iHt} \\ &= e^{iHt}\,\Omega^{(+)}[H_0, O_H(0)]\,\Omega^{(+)-1}\,e^{-iHt} \\ &= e^{iHt}\,\Omega^{(+)}\,e^{-iHt}\,[H_{0H}(t), O_H(t)]\,e^{iHt}\,\Omega^{(+)-1}\,e^{-iHt} \\ &= [H_{0\,\text{in}}(t), O_{\text{in}}(t)] \end{aligned} \tag{12b}$$

[1] The operators H, H_0, H_I, etc. without any time label are Heisenberg operators at time $t = 0$. Their time dependence is determined by

$$O_H(t) = e^{iHt}\,O_H(0)\,e^{-iHt}$$

so that $H(t) = H_H(t) = H(0) = H$.

so that $O_{\text{in}}(t)$ obeys free-field equations as determined by $H_{0\,\text{in}}$. Note that $H_{0\,\text{in}}$ is given by the expression for H_0 in which the Heisenberg operators have been replaced by "in" operators. From (11) and (8b) it follows that

$$\begin{aligned} H_{0\,\text{in}}(t) &= e^{iHt}\,\Omega^{(+)} H_{0H}(0)\,\Omega^{(+)-1}\,e^{-iHt} \\ &= e^{iHt}\,H_H\,e^{-iHt} \\ &= H = H_{0\,\text{in}}(0) \end{aligned} \tag{13}$$

so that the eigenstates of $H_{0\,\text{in}}$ are also the eigenstates of H.

Upon inserting a factor $V_+(t)^{-1}\,V_+(t)$ in the right-hand side of Eq. (10), it can be rewritten in the form

$$i\partial_t V_+(t) = H_{I\,\text{in}}(t)\,V_+(t) \tag{14}$$

Further properties of $V_+(t)$ are obtained from a consideration of the matrix element:

$$\begin{aligned} \langle \psi_b{}^+ \mid V_+(t) \mid \psi_a{}^+ \rangle &= e^{-i(E_a - E_b)t}\,\langle \psi_b{}^+ \mid \Omega^{(+)} \mid \psi_a{}^+ \rangle \\ &= e^{-i(E_a - E_b)t}\,\langle \varphi_b \mid \psi_a{}^+ \rangle \\ &= e^{-i(E_a - E_b)t}\left\{\delta(b - a) - \frac{R_{ba}}{E_b - E_a - i\epsilon}\right\} \end{aligned} \tag{15}$$

Taking the limit $t \to \pm\infty$, we obtain

$$\lim_{t \to -\infty} \langle \psi_b{}^+ \mid V_+(t) \mid \psi_a{}^+ \rangle = \delta(b - a) \tag{16a}$$

$$\begin{aligned} \lim_{t \to +\infty} \langle \psi_b{}^+ \mid V_+(t) \mid \psi_a{}^+ \rangle &= \delta(b - a) - 2\pi i \delta(E_b - E_a)\,R_{ba} \\ &= S_{ba} \end{aligned} \tag{16b}$$

i.e., $\lim_{t \to +\infty} V_+(t) = S$, the S matrix in the Heisenberg picture, the matrix elements S_{ba} of which are defined in terms of the exact Heisenberg scattering states $|\psi^+\rangle$, which are eigenstates of H [and by (13) also of $H_{0\,\text{in}}$]. The S-matrix formalism in terms of the in-operators parallels the development given previously in terms of the interaction picture operators. For example, by virtue of Eq. (14) and the boundary condition $\lim_{t \to -\infty} V_+(t) = 1$, $V_+(t)$ satisfies the integral equation

$$V_+(t) = 1 - i\int_{-\infty}^{t} H_{I\,\text{in}}(t')\,V_+(t')\,dt' \tag{17}$$

whence

$$V_+(t) = \sum_{n=1}^{\infty} \frac{(-i)^n}{n!} \int_{-\infty}^{t} dt_1 \cdots \int_{-\infty}^{t} dt_n\,P(H_{I\,\text{in}}(t_1) \cdots H_{I\,\text{in}} \cdot (t_n)) \tag{18}$$

and $S = V_+(+\infty)$. The S-matrix elements are obtained by evaluating S between eigenstates of the unperturbed Hamiltonian (now $H_{0\,\text{in}}$!).

We again wish to emphasize at this point that the formalism is valid only in the absence of bound states. We shall later return to the problem of incorporating bound states.

An identical formalism can be constructed from the operator $V_-(t)$ defined by

$$V_-(t) = e^{iHt}\,\Omega^{(-)}\,e^{-iHt} \tag{19}$$

in terms of which the S matrix is given by

$$\lim_{t\to-\infty} \langle\psi_b^- \mid V_-(t) \mid \psi_a^-\rangle = S_{ba} \tag{20a}$$

$$\lim_{t\to+\infty} \langle\psi_b^- \mid V_-(t) \mid \psi_a^-\rangle = \delta(b-a) \tag{20b}$$

With the help of $V_-(t)$, we can define the Heisenberg out-operators:

$$O_{\text{out}}(t) = V_-(t)\,O_H(t)\,V_-(t)^{-1} \tag{21}$$

which again obey free-field equations and build up the out-eigenstates $|\psi_a^-\rangle$. The relation between in- and out-operators is easily established by recalling that the S matrix can also be defined through the equation $S \mid \psi_a^-\rangle = |\psi_a^+\rangle$ or $S\Omega^{(-)} \mid \varphi_a\rangle = \Omega^{(+)} \mid \varphi_a\rangle$, whence

$$S = \Omega^{(+)}\Omega^{(-)-1} \tag{22}$$

and

$$\begin{aligned} O_{\text{out}}(0) &= \Omega^{(-)}O_H(0)\,\Omega^{(-)-1} = \Omega^{(-)}\Omega^{(+)-1}O_{\text{in}}(0)\,\Omega^{(+)}\Omega^{(-)-1} \\ &= S^{-1}O_{\text{in}}(0)\,S \end{aligned} \tag{23a}$$

and since S and $\exp iHt$ commute, we have in general

$$O_{\text{out}}(t) = S^{-1}O_{\text{in}}(t)\,S \tag{23b}$$

The in-out formalism bypasses the difficulty of unperturbed and perturbed eigenstates not lying in the same Hilbert space. However, it should be noted that in a rigorous mathematical sense the operator $\Omega^{(\pm)}$ will not exist as a unitary operator whenever H_0 and H have no vectors in common. The original difficulty is therefore only shifted. We shall therefore, for the time being, continue our exposition in the Dirac picture.

Historically, the introduction of the interaction picture [Tomonaga (1946), Schwinger (1948b)] was responsible for many of the technical advances in field theory in the 1947–52 period. The advantage of the interaction picture stems from the fact that in this picture the field operators satisfy free-field equations, so that invariant commutation rules can be written down for all times. It also permits a generalization of Eq. (1) which makes the equation covariant. Equation (1), as it stands, is not covariant since it singles out a special Lorentz frame in defining the time derivative, ∂_t.

The generalization consists in the introduction of the concept of a general space-like surface instead of the "flat" surface $t =$ constant. The only condition which must be satisfied by such a surface is that the normal to it at any point x, $n_\mu(x)$ be time-like, i.e., $n_\mu(x)\,n^\mu(x) > 0$. This then requires that no two points on the surface can be connected by a light signal, or, alternatively, that any two points on the surface have a space-like separation. Call such a surface σ. Then to each point $\mathbf{x}$ on this surface we may assign a time $t(\mathbf{x})$, its local time. In the limit that the surface becomes plane, each point has the same time t, the co-ordinate of the plane t = constant. A natural generalization of $|\Psi(t)\rangle$ now affords itself, namely $|\Psi(t(\mathbf{x}))\rangle$. We may then consider the basic equation:

$$i\hbar\partial_t \mid \Psi(t)\rangle = H_I(t) \mid \Psi(t)\rangle \tag{24}$$

as the result of summing an infinite set of equations obtained by introduction of a local time for each point of a space-like surface. If we express the interaction Hamiltonian in the general case as a sum over small three-dimensional cells ΔV over the space-like surface σ,

$$H_I(t) = \sum_{\text{over } \sigma} \mathcal{H}_I(x)\, \Delta V \tag{25}$$

then the equation which holds in a little cell about the space-time point $\mathbf{x}$, $t(\mathbf{x})$ may be written as

$$i\hbar \frac{\partial \mid \Psi(t(\mathbf{x}))\rangle}{\partial t(\mathbf{x})} = \mathcal{H}_I(x)\, \Delta V \mid \Psi(t(\mathbf{x}))\rangle \tag{26}$$

This equation is the generalization of (24). For, if the variation of $|\Psi(t)\rangle$ corresponding to a rigid infinitesimal displacement of the whole surface t = constant is determined by $\int_t \mathcal{H}_I(x)\, d^3x$, then it is reasonable that the variation of $|\Psi(t(\mathbf{x}))\rangle$ about the point $\mathbf{x}$, $t(\mathbf{x})$ will be determined by the interaction energy density $\mathcal{H}_I(x)\, \Delta V$ in the infinitesimal region around x.

Since the combination $\Delta V t(\mathbf{x})$ is invariant, the following invariant differentiation procedure suggests itself: Consider the function of the space-like surface $|\Psi(t(\mathbf{x}))\rangle = |\Psi(\sigma)\rangle$. Let us compare the value of this function on two space-like surfaces, σ and σ', which differ from each other only by an infinitesimal amount in the neighborhood of the space-time point x (see Fig. 13.1). We then define the invariant operation $\delta/\delta\sigma(x)$ as

$$\begin{aligned} \frac{\delta}{\delta\sigma(x)} \mid \Psi(\sigma)\rangle &= \lim_{\Delta t \Delta V \to 0} \frac{|\Psi(t(\mathbf{x}) + \Delta t(\mathbf{x}))\rangle - |\Psi(t(\mathbf{x}))\rangle}{c \int_{\Delta V} d^3x\, \Delta t(\mathbf{x})} \\ &= \lim_{\Delta t \Delta V \to 0} \frac{|\Psi(\sigma')\rangle - |\Psi(\sigma)\rangle}{c\Delta t(\mathbf{x})\, \Delta V} \\ &= \lim_{\Omega(x) \to 0} \frac{|\Psi(\sigma')\rangle - |\Psi(\sigma)\rangle}{\Omega(x)} \end{aligned} \tag{27}$$

Here $\Omega(x)$ is the four-dimensional volume enclosed between σ and σ', σ' being the space-like surface obtained by a small deformation of σ about the point x. In the limit, $\Omega(x)$ goes over the point x. We can therefore rewrite Eq. (26) as

$$i\hbar c \frac{\delta \mid \Psi(\sigma)\rangle}{\delta\sigma(x)} = \mathcal{H}_I(x) \mid \Psi(\sigma)\rangle \tag{28}$$

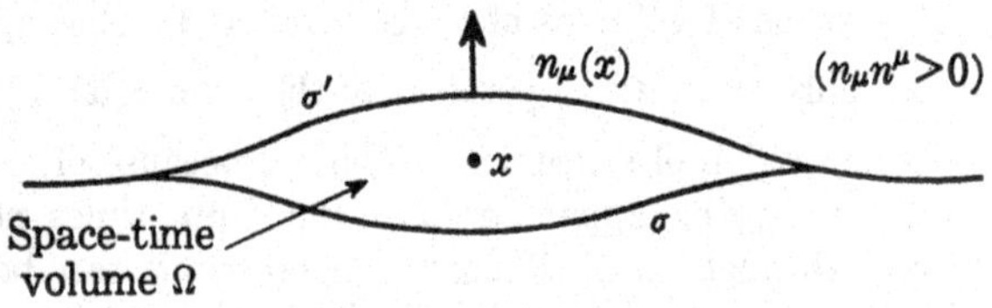

Fig. 13.1

This equation is now covariant since $\mathcal{H}_I$ is an invariant in the case of nonderivative coupling (see the remarks below for the case of derivative coupling) and the concept of a space-like surface does not presuppose any Lorentz frame for its definition. We have therefore rewritten the equation of motion of the system without reference to any particular co-ordinate frame. Equation (28) is usually called the Tomonaga-Schwinger equation.

In order that solutions to (28) exist, the following integrability condition must be satisfied:

$$\frac{\delta^2 \,|\, \Psi(\sigma)\rangle}{\delta\sigma(x)\, \delta\sigma(x')} - \frac{\delta^2 \,|\, \Psi(\sigma)\rangle}{\delta\sigma(x')\, \delta\sigma(x)} = 0 \tag{29}$$

which is essentially a restriction on the domain of variation of $\sigma(x)$. Equation (29) in turn implies that

$$[\mathcal{H}_I(x), \mathcal{H}_I(x')] = 0 \tag{30}$$

for x and x' on the space-like surface σ. The invariant commutation rules for the field operators are such that (30) is automatically satisfied for all interacting fields with local nonderivative couplings. When the interaction Lagrangian density involves the time derivatives of the field variables (such as is the case for a charged boson field interacting with the electromagnetic field), a slight extension of the above methods is needed. In this case, the interaction Hamiltonian density must explicitly depend on the surface in order that the integrability condition be satisfied [see, for example, Kroll (1949b), Matthews (1949a)]. The integrability condition, through the relation (30), implies that a solution exists only when σ is a space-like surface, for in general Eq. (30) does not vanish if x and x' are separated by a time-like distance. Furthermore, it guarantees that when the space-like surface is the plane t = constant, every solution of (28) is a solution of (24).

We shall not have occasion in our work to make explicit use of these surfaces. They do not simplify the theory, nor do they add to the physical content of the theory, although they are not more unphysical than plane space-like surfaces, t = constant. What is unphysical is the assumption that Hermitian field operators defined at precise space-time points exist and are measurable in the ordinary quantum mechanical sense. In fact, Bohr and Rosenfeld [Bohr (1933, 1950)] in their classic papers on the question of the measurability of electromagnetic fields have already shown that only averages over small volumes of space-time of the field operators are measurable, i.e., only Hermitian operators of the form $\varphi(R) = \int_R \varphi(x)\, d^4x$, where R is a small region of space-time. This is because of the finite size of the classical measuring apparatus and the finite times necessary to determine forces through their effects on macroscopic test bodies. The connection between these measurability limitations and the problem of the self-charge and self-energy divergences in the relativistic field theories has yet not been answered. It is, however, known [Heisenberg (1934c), Stückelberg (1951a, b), Corinaldesi (1951, 1953)] that further "boundary divergences" are introduced into the theory if the space-time region, in which a measurement of the fluctuations of a field or charge-current distribution is made, has sharp spatio-temporal boundaries.

To make more precise the remarks up to this point, let us illustrate them with a simple field theoretic model, namely that of a spinor field interacting with a neutral meson field. The unsymmetrized Hamiltonian of the theory in the Schrödinger picture is:

$$H = \int d^3x\, \tilde{\psi}(-i\gamma \cdot \partial + M_0)\, \psi(\mathbf{x}) + \tfrac{1}{2} \int d^3x\, \{\pi^2(\mathbf{x}) + \mu_0{}^2\phi^2(x) + (\partial\phi(\mathbf{x}))^2\} + G \int d^3x\, \tilde{\psi}(\mathbf{x})\, \Gamma\psi(\mathbf{x})\, \phi(\mathbf{x}) \quad (31)$$

In the Dirac picture the field operators satisfy the following equations of motion and commutation rules:

$$(-i\gamma \cdot \partial + M_0)\, \psi_D(x) = 0 \qquad [\psi_D(x), \tilde{\psi}_D(x')]_+ = -iS(x - x') \quad (32)$$

$$(\Box + \mu_0{}^2)\, \phi_D(x) = 0 \qquad [\phi_D(x), \phi_D(x')] = i\Delta(x - x') \quad (33)$$

and the Tomonaga-Schwinger equation for the state vector of the system reads:

$$i\, \frac{\delta \mid \Psi(\sigma)\rangle}{\delta\sigma(x)} = \mathcal{H}_{ID}(x) \mid \Psi(\sigma)\rangle \quad (34)$$

with

$$\mathcal{H}_{ID}(x) = G : \tilde{\psi}_D(x)\, \Gamma\psi_D(x) : \phi_D(x) \quad (35)$$

The equations satisfied by the in-operators $\phi_{\text{in}}(x)$, $\psi_{\text{in}}(x)$ are the same as (32) and (33), and the operator $V_+(t)$ satisfies the equation:

$$i\partial_t V_+(t) = G \int d^3x : \tilde{\psi}_{\text{in}}(x)\, \Gamma\psi_{\text{in}}(x) : \phi_{\text{in}}(x)\, V_+(t) \quad (36)$$

To apply the formalism developed in Chapter 11 to the scattering of a meson by a nucleon, or for that matter to any scattering process, it is necessary that the spectrum of H and H_0 be the same. This will not be the case if the split-up of the Hamiltonian (31) into an unperturbed and perturbed part is

$$H = H_0 + H_I(0) \quad (37)$$

with

$$H_I(0) = G \int d^3x : \tilde{\psi}(x)\, \Gamma\psi(x) : \phi(x) \Big|_{x_0 = 0} \quad (38)$$

The reason is twofold. Firstly, in relativistic field theories in general, and in the model under discussion in particular, the "bare" vacuum $|\Phi_0\rangle$ (which is an eigenstate of H_0) is not an eigenvector of H, because $H_I(0) \mid \Phi_0\rangle \neq 0$ due to the presence of the pair creation term $\bar{\psi}^{(-)}\Gamma\psi^{(-)}\phi^{(-)}$ in H_I. Due to this possibility of pair creation, the vacuum energy is shifted relative to the "bare" vacuum by an (infinite) amount E_0, the vacuum self-energy. To make the energy eigenvalue of the bare and physical vacuum be the same we add and subtract to H a term E_0, where E_0 is the level shift of the vacuum, and consider $H_0 + E_0$ as the unperturbed Hamiltonian and $H_I - E_0$ as the perturbation. Secondly, a similar phenomenon occurs with respect to the one-particle states: the energies of the bare one-particle states (eigenstates of H_0) and the corresponding energies of the physical one-particle states (eigenstates of H) of the same momentum are not the same. In a relativistically invariant theory the only effect that the interaction can have on a single (free) particle is to change its mass, say by δM for the nucleon, and $\delta\mu^2$ for the meson, so that the change in energy of the one-nucleon state is given by

$$
\begin{aligned}
E + \Delta E &= \sqrt{(M_0 + \delta M)^2 + \mathbf{p}^2} \\
&\approx \sqrt{M_0{}^2 + \mathbf{p}^2} + \frac{M_0 \delta M}{\sqrt{M_0{}^2 + \mathbf{p}^2}} + \cdots
\end{aligned} \tag{39}
$$

and to first order the level shift will be

$$
\Delta E = \frac{M_0 \delta M}{E} \tag{40}
$$

The adjustment of the spectrum of H_0 and H to take into account this level shift proceeds as in the simple models discussed in Chapter 12. We add to H_0 a term $\int d^3x \, \delta M : \bar{\psi}\psi :$ and subtract the same term from H_I and identify $M_0 + \delta M$ with the physical (observed) mass M of the nucleon. Similarly, we add and subtract a term $\frac{1}{2} \int d^3x \, \delta\mu^2 : \phi^2 :$ and identify $\delta\mu^2 + \mu_0{}^2$ with the square of the observed mass of the meson. The unperturbed and interaction terms of H therefore become

$$
\begin{aligned}
H_0' &= \int d^3x : \bar{\psi}(-i\gamma \cdot \partial + M)\,\psi : + \tfrac{1}{2} \int d^3x : \pi^2 + \mu^2\phi^2 + (\partial\phi)^2 : + E_0 \\
H_I' &= G \int d^3x : \bar{\psi}\Gamma\psi : \phi - E_0 - \int d^3x : \bar{\psi}\psi : \delta M - \tfrac{1}{2} \int d^3x \, \delta\mu^2 : \phi^2 :
\end{aligned} \tag{41}
$$

where the renormalization counterterms are to be so determined that the one-particle states of momentum $\mathbf{p}$ have energy $\sqrt{\mathbf{p}^2 + M^2}$ for the one-nucleon state, and $\sqrt{\mathbf{p}^2 + \mu^2}$ for the one-meson state. These counterterms then cancel the self-energy terms which arise in the course of calculations. Note again that this procedure introduces the parameter M and μ phenomenologically into the theory.

Let us next analyze the interaction term $G : \bar{\psi}\Gamma\psi\phi :$ in detail in the

Schrödinger picture. Consider first the spinor part of it: $\bar{\psi}(\mathbf{x})\,\Gamma\psi(\mathbf{x})$:. The substitution into this expression of the expansion of $\bar{\psi}(\mathbf{x})$ and $\psi(\mathbf{x})$ into creation and annihilation operators indicates that four types of interaction occur, identified with the terms:

1. $b_{\mathbf{p}'}{}^*b_{\mathbf{p}}$, corresponding to the destruction of a nucleon of momentum $\mathbf{p}$ and the subsequent creation of a nucleon of momentum $\mathbf{p}'$; i.e., corresponding to the scattering of the nucleon;
2. $d_{\mathbf{p}}{}^*d_{\mathbf{p}'}$, corresponding to the scattering of an antinucleon;
3. $b_{\mathbf{p}}d_{\mathbf{p}'}$, corresponding to the annihilation of a nucleon and an antinucleon, i.e., *pair destruction*;
4. $b_{\mathbf{p}}{}^*d_{\mathbf{p}'}{}^*$, corresponding to pair creation.

Similarly, the expansion of the meson field operator into creation and annihilation operators is

$$\phi(\mathbf{x}) = \frac{1}{(2\pi)^{3/2}} \int \frac{d^3k}{\sqrt{2\omega_{\mathbf{k}}}}\,(a_{\mathbf{k}}{}^* + a_{-\mathbf{k}})\, e^{-i\mathbf{k}\cdot\mathbf{x}} \tag{42}$$

It therefore follows that the interaction term H_I gives rise to eight elementary processes corresponding to:

(*a*) boson emission with,
   1. nucleon scattering
   2. antinucleon scattering
   3. pair annihilation
   4. pair creation

(*b*) boson absorption with the same nucleon-antinucleon processes, 1–4, occurring.

These elementary processes can be represented graphically by time-ordered Feynman diagrams. If we denote a boson by a dashed line, a nucleon by a directed line moving forward in time, and an antinucleon by a directed line moving backward in time, then process (*a*) 1, corresponding to boson emission with nucleon scattering, can be represented by Figure 13.2*a*. The direction of time is as indicated and the direction of the lines may not be topologically distorted. This is what is meant by the adjective "time-ordered." Similarly, process (*a*) 2 would be represented by Figure 13.2*b*, and the others by Figures 13.2*c*–13.2*h*.

As a result of the integration over d^3x in the interaction operator H_I, momentum is conserved at each of these vertices. Consider, for example, the process (*a*) 1, corresponding to Figure 13.2*a*. The part of H_I which gives rise to this process is

$$(H_I)_{a1} = \frac{G}{(2\pi)^{9/2}} \int d^3x \int d^3p' \int d^3p \int d^3k \sum_{r=1}^{2}\sum_{s=1}^{2} e^{-i(\mathbf{p}'+\mathbf{k}-\mathbf{p})\cdot\mathbf{x}}$$
$$\cdot \sqrt{\frac{M^2}{2\omega(\mathbf{k})\,E(\mathbf{p})\,E(\mathbf{p}')}}\;\bar{w}^r(\mathbf{p}')\,\Gamma w^s(\mathbf{p})\,b_r{}^*(\mathbf{p}')\,b_s(\mathbf{p})\,a^*(\mathbf{k}) \tag{43}$$

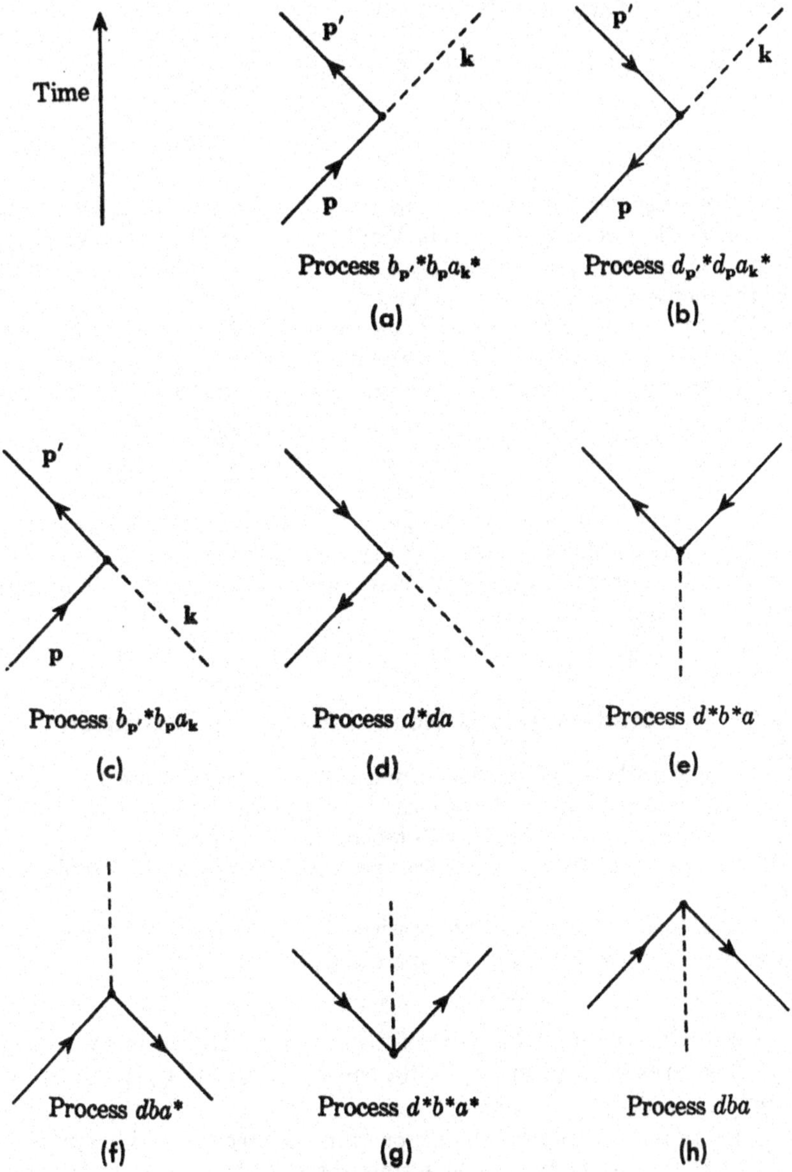
Time
p′
k
p
Process $b_{\mathbf{p}'}^*b_{\mathbf{p}}a_{\mathbf{k}}^*$
(a)
p′
k
p
Process $d_{\mathbf{p}'}^*d_{\mathbf{p}}a_{\mathbf{k}}^*$
(b)
p′
k
p
Process $b_{\mathbf{p}'}^*b_{\mathbf{p}}a_{\mathbf{k}}$
(c)
Process d^*da
(d)
Process d^*b^*a
(e)
Process dba^*
(f)
Process $d^*b^*a^*$
(g)
Process dba
(h)

Fig. 13.2

which upon integration over d^3x yields

$$(H_I)_{a1} = \frac{G}{(2\pi)^{3/2}} \int d^3p \int d^3k \int d^3p' \sqrt{\frac{M^2}{2\omega(\mathbf{k})\, E(\mathbf{p})\, E(\mathbf{p}')}}$$

$$\sum_{r=1}^{2} \sum_{s=1}^{2} \delta^{(3)}(\mathbf{p} - \mathbf{p}' - \mathbf{k})\, \tilde{w}^r(\mathbf{p}')\, \Gamma w^s(\mathbf{p})\, b_r^*(\mathbf{p}')\, b_s(\mathbf{p})\, a^*(\mathbf{k}) \quad (44)$$

the δ function corresponding to conservation of momentum at the vertex. Thus, if the incoming nucleon in Figure 13.2a has a momentum $\mathbf{p}$, the outgoing nucleon's momentum, $\mathbf{p}'$, is related to $\mathbf{p}$ by $\mathbf{p}' = \mathbf{p} - \mathbf{k}$ where $\mathbf{k}$ is the momentum of the emitted meson.

The application of perturbation theory then proceeds as in the case of ordinary quantum mechanics. If we limit ourselves to second-order processes,[2] the matrix element for a process $i \to f$ is given by

$$R_{fi}^{(2)} = (\Phi_f, R^{(2)}\Phi_i) = \sum_{\substack{m \\ E_m \neq E_i}} \frac{\langle \Phi_f, H_I \Phi_m \rangle \langle \Phi_m, H_I \Phi_i \rangle}{E_i - E_m + i\epsilon} \quad (45)$$

where $|\Phi_f\rangle$, $|\Phi_i\rangle$, and $|\Phi_m\rangle$ denote the final, initial, and intermediate state wave functions of the system. (They are eigenfunctions of H_0.) The summation is over all intermediate states which differ from the initial state (of energy E_i).

13b. The Scattering of a Neutral Meson by a Nucleon

As an illustration of the above remarks, we here calculate the lowest order matrix elements for the scattering of a neutral meson by a nucleon in pseudoscalar meson theory with pseudoscalar coupling.

The scattering of the two particles proceeds through states described as follows:

Initial state. The initial state consists of a meson of momentum $\mathbf{k}_1$, and a nucleon of momentum $\mathbf{p}_1$, and spin s_1, i.e.,

$$|\Phi_i\rangle = b_{s_1}^*(\mathbf{p}_1)\, a^*(\mathbf{k}_1) \mid \Phi_0\rangle \quad (46)$$

Intermediate states. As a result of the interaction H_I, the following four possibilities exist, which constitute the possible (second-order) intermediate states:

(a) The nucleon absorbs the meson (this corresponds to the term b^*ba in H_I) so that in the intermediate state there is present just one nucleon of momentum $\mathbf{p}_i = \mathbf{p}_1 + \mathbf{k}_1$ and energy $\sqrt{\mathbf{p}_i^2 + M^2}$.

(b) The nucleon emits the final meson of momentum $\mathbf{k}_2$ (interaction term

[2] No real first-order process can occur since energy and momentum cannot be conserved for these.

b^*ba^*) so that in the intermediate state there are present two mesons of momenta $\mathbf{k}_1$ and $\mathbf{k}_2$, and one nucleon of momentum $\mathbf{p}_1 - \mathbf{k}_2$.

(*c*) A nucleon-antinucleon pair and a meson are created ($d^*b^*a^*$) so that in the intermediate state there are present two nucleons, one antinucleon, and two mesons. The nucleon and meson which are created in the process are the final state ones.

(*d*) The initial meson creates a pair and is thus "annihilated" (d^*b^*a); hence, in the intermediate state there are present two nucleons and one antinucleon, and no mesons.

Final state. A second application of the operator H_I in Eq. (45) must bring us to the final state in which one meson of momentum $\mathbf{k}_2$, and a nucleon of spin s_2, and momentum $\mathbf{p}_2$, are present, i.e.,

$$|\Phi_f\rangle = b_{s_2}{}^*(\mathbf{p}_2)\, a^*(\mathbf{k}_2) \mid \Phi_0\rangle \tag{47}$$

This is accomplished as follows: In the process

(*a*) The nucleon in the intermediate state emits the final state meson and is scattered into its final state (interaction term b^*ba^*).

(*b*) The nucleon in the intermediate state absorbs the initial meson (interaction term b^*ba).

(*c*) The antinucleon created in the intermediate state annihilates the initial nucleon, and the initial meson is absorbed (dba).

(*d*) The antinucleon created in the intermediate state annihilates the initial nucleon and the final state meson is created (dba^*). These correspond to the time-ordered Feynman diagrams illustrated in Figure 13.3.

Consider first the contribution to $R_{fi}^{(2)}$ from process (*a*). It proves useful to rewrite the matrix element (45) for this case as

$$R_{fi}^{(2)a} = \langle \Phi_f, H_I{}^{(a2)} \frac{1}{E_i - H_0 + i\epsilon} H_I{}^{(a1)}\Phi_i\rangle \tag{48}$$

where $E_i = E(\mathbf{p}_1) + \omega(\mathbf{k}_1)$ is the energy of the initial state.[3] The operator $H_I{}^{(a1)}$ is given by

$$H_I{}^{(a1)} = \frac{G}{(2\pi)^{3/2}} \int d^3p \int d^3p' \int d^3k \left(\frac{M^2}{2E(\mathbf{p})\, E(\mathbf{p}')\, \omega(\mathbf{k})}\right)^{1/2}$$
$$\sum_{s,r=1}^{2} \delta^{(3)}(\mathbf{p}' - \mathbf{p} - \mathbf{k})\, b_r{}^*(\mathbf{p}')\, b_s(\mathbf{p})\, a(\mathbf{k})\, \bar{w}^r(\mathbf{p}')\, \gamma_5 w^s(\mathbf{p})$$

and $H_I{}^{(a2)}$ is given by Eq. (44) with $\Gamma = \gamma_5$. Now the intermediate state $|\Phi_m^{(a)}\rangle = H_I{}^{(a1)} \mid \Phi_i\rangle$ can easily be evaluated using Eqs. (46), (48), and the property of the vacuum: $0 = a(\mathbf{k}) \mid \Phi_0\rangle = b_s(\mathbf{p}) \mid \Phi_0\rangle = d_s(\mathbf{p}) \mid \Phi_0\rangle$. The state $|\Phi_m'^{(a)}\rangle$ defined by

$$|\Phi_m'^{(a)}\rangle = b_r{}^*(\mathbf{p}')\, b_s(\mathbf{p})\, a(\mathbf{k}) \mid \Phi_i\rangle \tag{49}$$

[3] Strictly speaking, we ought to include in (48) a factor $(1 - P_i)$ where P_i is a projection operator onto the initial state. However, $(\Phi_i, H_I\Phi_i) = 0$.

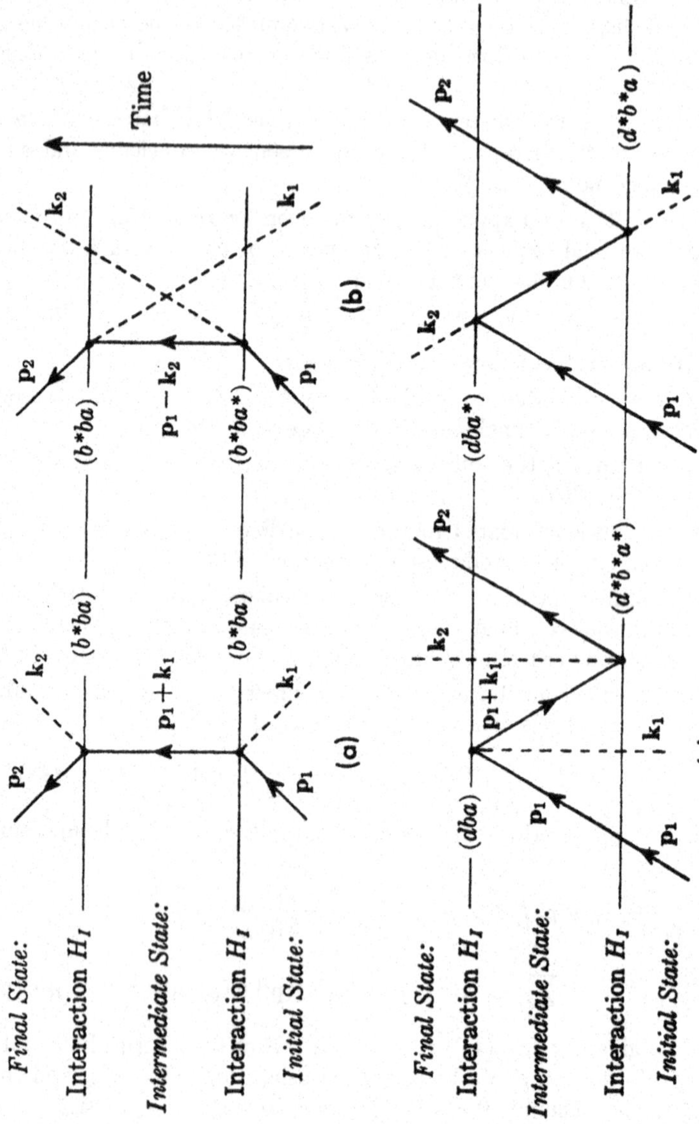
Final State:
Interaction H_I
Intermediate State:
Interaction H_I
Initial State:
Time
p_2
k_2
(b^*ba)
p_1+k_1
(b^*ba)
k_1
p_1
(a)
(b^*ba)
p_1-k_2
(b^*ba^*)
(b)
Final State:
Interaction H_I
Intermediate State:
Interaction H_I
Initial State:
(dba)
$(d^*b^*a^*)$
(c)
(dba^*)
(d^*b^*a)
(d)

Fig. 13.3

upon using the commutation rules for the b and a operators becomes

$$|\Phi_m'^{(a)}\rangle = b_r^*(\mathbf{p}')\, b_s(\mathbf{p})\, a(\mathbf{k})\, a^*(\mathbf{k}_1)\, b_{s_1}^*(\mathbf{p}_1) \mid \Phi_0\rangle$$
$$= \delta^{(3)}(\mathbf{k} - \mathbf{k}_1)\, \delta_{s_1s}\delta^{(3)}(\mathbf{p} - \mathbf{p}_1)\, b_r^*(\mathbf{p}') \mid \Phi_0\rangle \tag{50}$$

Hence

$$|\Phi_m^{(a)}\rangle = H_I^{(a1)} \mid \Phi_i\rangle$$
$$= \frac{G}{(2\pi)^{3/2}}\left(\frac{M^2}{2E(\mathbf{p}_1 + \mathbf{k}_1)\, E(\mathbf{p}_1)\, \omega(\mathbf{k}_1)}\right)^{1/2}$$
$$\sum_{r=1}^{2} \tilde{w}^r(\mathbf{p}_1 + \mathbf{k}_1)\, \gamma_5 w^{s_1}(\mathbf{p}_1)\, b_r^*(\mathbf{p}_1 + \mathbf{k}_1) \mid \Phi_0\rangle \tag{51}$$

and $|\Phi_m^{(a)}\rangle$ is a one-nucleon state with momentum $\mathbf{p}_1 + \mathbf{k}_1$. The operator H_0 operating on this state gives $E(\mathbf{p}_1 + \mathbf{k}_1) \mid \Phi_m^{(a)}\rangle$. Therefore, $(E_i - H_0 + i\epsilon)^{-1}$ operating on $|\Phi_m^{(a)}\rangle$ will yield $[E_i - E(\mathbf{p}_1 + \mathbf{k}_1)]^{-1} \mid \Phi_m^{(a)}\rangle$.

Next, $H_I^{(a2)}$ operating on this last expression, using the same methods as above, is readily shown to result in the expression

$$H_I^{(a2)}(E_i - H_0 + i\epsilon)^{-1} H_I^{(a1)} \mid \Phi_i\rangle$$
$$= \frac{G^2}{2(2\pi)^3}\sum_{r=1}^{2}\int d^3k\, \frac{M}{E(\mathbf{p}_1 + \mathbf{k}_1)}\left(\frac{M^2}{\omega(\mathbf{k})\,\omega(\mathbf{k}_1)\, E(\mathbf{p}_1 + \mathbf{k}_1 - \mathbf{k})\, E(\mathbf{p}_1)}\right)^{1/2}$$
$$\frac{[\tilde{w}^s(\mathbf{p}_1 + \mathbf{k}_1 - \mathbf{k})\,\gamma_5 w^r(\mathbf{p}_1 + \mathbf{k}_1)][\tilde{w}^r(\mathbf{p}_1 + \mathbf{k}_1)\,\gamma_5 w^{s_1}(\mathbf{p}_1)]}{E(\mathbf{p}_1) + \omega(\mathbf{k}_1) - E(\mathbf{p}_1 + \mathbf{k}_1) + i\epsilon}$$
$$\cdot\, b_s^*(\mathbf{p}_1 + \mathbf{k}_1 - \mathbf{k})\, a^*(\mathbf{k}) \mid \Phi_0\rangle \tag{52}$$

Finally, to obtain $R_{fi}^{(2)a}$, we have to evaluate the matrix element

$$(\Phi_f, b_s^*(\mathbf{p}_1 + \mathbf{k}_1 - \mathbf{k})\, a^*(\mathbf{k})\, \Phi_0)$$
$$= (b_{s_2}^*(\mathbf{p}_2)\, a^*(\mathbf{k}_2)\, \Phi_0, b^*(\mathbf{p}_1 + \mathbf{k}_1 - \mathbf{k})\, a^*(\mathbf{k})\, \Phi_0)$$
$$= \delta_{s_2s}\delta^{(3)}(\mathbf{p}_2 - \mathbf{p}_1 - \mathbf{k}_1 + \mathbf{k})\, \delta^{(3)}(\mathbf{k}_2 - \mathbf{k}) \tag{53}$$

so that,

$$R_{fi}^{(2)a} = \frac{G^2}{2(2\pi)^3}\frac{M}{E(\mathbf{p}_1 + \mathbf{k}_1)}\left(\frac{M^2}{\omega(\mathbf{k}_2)\,\omega(\mathbf{k}_1)\, E(\mathbf{p}_2)\, E(\mathbf{p}_1)}\right)^{1/2}$$
$$\delta^{(3)}(\mathbf{p}_2 - \mathbf{p}_1 - \mathbf{k}_1 + \mathbf{k}_2)\cdot\sum_{r=1}^{2}\frac{[\tilde{w}^{s_2}(\mathbf{p}_2)\,\gamma_5 w^r(\mathbf{p}_1 + \mathbf{k}_1)][\tilde{w}^r(\mathbf{p}_1 + \mathbf{k}_1)\,\gamma_5 w^{s_1}(\mathbf{p}_1)]}{E(\mathbf{p}_1) + \omega(\mathbf{k}_1) - E(\mathbf{p}_1 + \mathbf{k}_1) + i\epsilon} \tag{54}$$

The δ-function factor corresponds to the overall momentum conservation for the process.

We can interpret the resulting matrix element as follows: The factor

$$\frac{G}{\sqrt{2(2\pi)^9}}\left(\frac{M^2}{\omega(\mathbf{k}_1)\, E(\mathbf{p}_1)\, E(\mathbf{p}')}\right)^{1/2}\tilde{w}^r(\mathbf{p}')\,\gamma_5 w^{s_1}(\mathbf{p}_1) \tag{55}$$

corresponds to the absorption of a meson of momentum $\mathbf{k}_1$ [factor

$(2(2\pi)^3\,\omega(\mathbf{k}_1))^{-1/2}]$ by the initial nucleon of momentum $\mathbf{p}_1$, which then makes a transition to a state with momentum $\mathbf{p}'$

$$\left[\text{factor}\ \frac{G}{(2\pi)^3}\left(\frac{M^2}{E(\mathbf{p}_1)\,E(\mathbf{p}')}\right)^{1/2}\tilde{w}^r(\mathbf{p}')\,\gamma_5 w^{s_1}(\mathbf{p}_1)\right]$$

Since momentum is conserved in the process, this gives rise to a factor $(2\pi)^3\,\delta^{(3)}(\mathbf{p}'-\mathbf{p}_1-\mathbf{k}_1)$. Similarly, the factor,

$$\frac{G}{\sqrt{2(2\pi)^9}}\left(\frac{M^2}{\omega(\mathbf{k}_2)\,E(\mathbf{p}_2)\,E(\mathbf{p}')}\right)^{1/2}\tilde{w}^{s_2}(\mathbf{p}_2)\,\gamma_5 w^r(\mathbf{p}') \tag{56}$$

corresponds to the matrix element for the transition of a nucleon from the state $(\mathbf{p}'r)$ to the final state $(\mathbf{p}_2 s_2)$, with the emission of a meson of momentum $\mathbf{k}_2$. Again momentum is conserved, hence a factor, $(2\pi)^3\,\delta^{(3)}(\mathbf{p}_2+\mathbf{k}_2-\mathbf{p}')$. Finally, according to perturbation theory we must sum over all intermediate states $\left(\text{which here corresponds to } \int d^3p \sum_r^{1,2}\right)$ after having divided by $E_i - E_m + i\epsilon$, where, in our example, $E_m{}^a = E(\mathbf{p}')$. This procedure, indeed, gives back the expression (54).

Using the above reasoning, we can immediately write down the matrix element corresponding to Figure 13.3*b*. Here, the initial nucleon first emits the final meson of momentum $\mathbf{k}_2$, and hence makes a transition to the state with momentum $\mathbf{p}_1 - \mathbf{k}_2$. In the intermediate state, therefore, we have two mesons and the nucleon present, so that its energy is

$$E_m{}^b = \omega(\mathbf{k}_1) + \omega(\mathbf{k}_2) + E(\mathbf{p}_1 - \mathbf{k}_2) \tag{57}$$

The matrix element $R_{fi}^{(2)b}$ is therefore

$$R_{fi}^{(2)b} = \frac{G^2}{2(2\pi)^3}\left(\frac{M}{E(\mathbf{p}_1-\mathbf{k}_2)}\right)\left(\frac{M^2}{\omega(\mathbf{k}_1)\,\omega(\mathbf{k}_2)\,E(\mathbf{p}_1)\,E(\mathbf{p}_2)}\right)^{1/2}$$
$$\delta^{(3)}(\mathbf{p}_2+\mathbf{k}_2-\mathbf{p}_1-\mathbf{k}_1)\sum_{r=1}^{2}\frac{[\tilde{w}^{s_2}(\mathbf{p}_2)\,\gamma_5 w^r(\mathbf{p}_1-\mathbf{k}_2)]\,[\tilde{w}^r(\mathbf{p}_1-\mathbf{k}_2)\,\gamma_5 w^{s_1}(\mathbf{p}_1)]}{E(\mathbf{p}_1)-E(\mathbf{p}_1-\mathbf{k}_2)-\omega(\mathbf{k}_2)} \tag{58}$$

An analogous procedure to the one outlined above indicates that the factor corresponding to meson emission (of momentum $\mathbf{k}_2$) with pair creation, the nucleon having variables $\mathbf{p}_2 s_2$ and the antinucleon $\mathbf{p}'r$, is

$$\frac{G}{\sqrt{2(2\pi)^9}}\left(\frac{M^2}{\omega(\mathbf{k}_2)\,E(\mathbf{p}_2)\,E(\mathbf{p}')}\right)^{1/2}\tilde{w}^{s_2}(\mathbf{p}_2)\,\gamma_5 v^r(\mathbf{p}') \tag{59}$$

Similarly, the factor corresponding to pair annihilation (nucleon $\mathbf{p}_1 s_1$, antinucleon $\mathbf{p}'r$) with meson absorption ($\mathbf{k}_1$) is

$$\frac{G}{\sqrt{2(2\pi)^9}}\left(\frac{M^2}{\omega(\mathbf{k}_1)\,E(\mathbf{p}_1)\,E(\mathbf{p}')}\right)^{1/2}\tilde{v}^r(\mathbf{p}')\,\gamma_5 w^{s_1}(\mathbf{p}_1) \tag{60}$$

The matrix element corresponding to the time-ordered Feynman diagram of Figure 13.3c is therefore

$$R_{fi}^{(2)c} = -\frac{G^2}{2(2\pi)^3}\left(\frac{M}{E(\mathbf{p}_1+\mathbf{k}_1)}\right)\left(\frac{M^2}{\omega(\mathbf{k}_1)\,\omega(\mathbf{k}_2)\,E(\mathbf{p}_1)\,E(\mathbf{p}_2)}\right)^{1/2}$$
$$\delta^{(3)}(\mathbf{p}_2+\mathbf{k}_2-\mathbf{p}_1-\mathbf{k}_1)$$
$$\sum_{r=1}^{2}\frac{[\tilde{w}^{s_2}(\mathbf{p}_2)\,\gamma_5 v^r(\mathbf{p}_1+\mathbf{k}_1)]\,[\tilde{v}^r(\mathbf{p}_1+\mathbf{k}_1)\,\gamma_5 w^{s_1}(\mathbf{p}_1)]}{E_i - E(\mathbf{p}_1) - E(\mathbf{p}_2) - E(\mathbf{p}_1+\mathbf{k}_1) - \omega(\mathbf{k}_1) - \omega(\mathbf{k}_2)} \tag{61}$$

The minus sign in front of the matrix element is due to the Pauli principle. Similarly, we obtain, for Figure 13.3*d*,

$$R_{fi}^{(2)d} = -\frac{G^2}{2(2\pi)^3}\left(\frac{M}{E(\mathbf{p}_1-\mathbf{k}_2)}\right)\left(\frac{M^2}{\omega(\mathbf{k}_1)\,\omega(\mathbf{k}_2)\,E(\mathbf{p}_1)\,E(\mathbf{p}_2)}\right)^{1/2}$$
$$\delta^{(3)}(\mathbf{p}_2+\mathbf{k}_2-\mathbf{p}_1-\mathbf{k}_1)\sum_{r=1}^{2}\frac{[\tilde{w}^{s_2}(\mathbf{p}_2)\,\gamma_5 v^r(\mathbf{p}_1-\mathbf{k}_2)]\,[\tilde{v}(\mathbf{p}_1-\mathbf{k}_2)\,\gamma_5 w^{s_1}(\mathbf{p}_1)]}{E_i - E(\mathbf{p}_1) - E(\mathbf{p}_2) - E(\mathbf{p}_1-\mathbf{k}_2)} \tag{62}$$

These are the four lowest order matrix elements which give rise to meson-nucleon scattering. Other processes involving a meson and a nucleon are possible to the second order—for example, Figure 13.4, *a* and *b*—which give

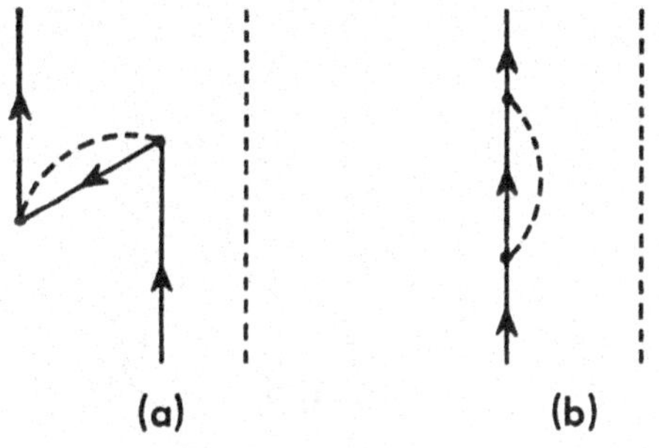

Fig. 13.4

rise to the self-energy of the nucleon. The contribution of these diagrams are exactly canceled by the $\delta M\bar{\psi}\psi$ term in the Hamiltonian upon suitable adjustment of the δM factor. Similarly, the diagrams *a* and *b* of Figure 13.5, which correspond to the second order of the meson self-energy, are canceled by the $\Delta\mu^2\phi^2$ in $\mathcal{H}_I(x)$. Finally, the diagram of Figure 13.6, which corresponds to the vacuum self-energy, is canceled by the E_0 term in H_I by a suitable adjustment of $E_0{}^{(2)}$, i.e., the value of E_0 to order G^2.

If we wished, we could now use Eqs. (4.143a and b) to carry out the sum over the two intermediate states and carry out the necessary quadratures to obtain the cross section for scattering. We shall not do so here. Instead, we shall consider the simplification in the matrix element that

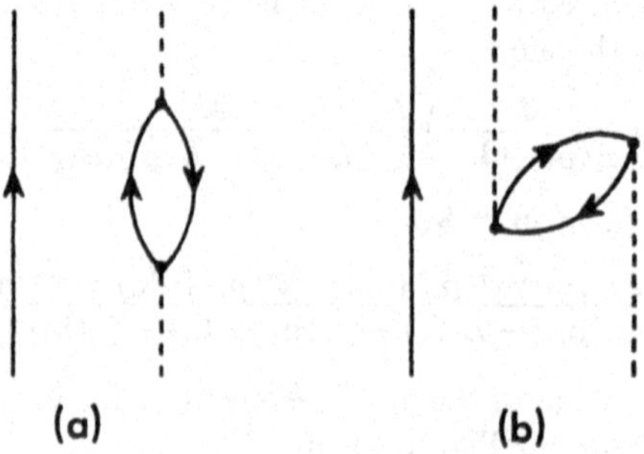

Fig. 13.5

obtains in the limit when the meson momentum is small compared to the nucleon mass in the center-of-momentum system ($\mathbf{k}_1 = -\mathbf{p}_1 = \mathbf{k}_i$; $\mathbf{k}_2 = -\mathbf{p}_2 = \mathbf{k}_f$). We shall also not concern ourselves with the lack of validity of the Born approximation in this situation [see Bethe (1955)].

We note that in the center-of-mass the nucleon's motion is nonrelativistic so that the matrix element $\tilde{w}^r(\mathbf{p}_1 + \mathbf{k}_1)\, \gamma_5 w^{s_1}(\mathbf{p}_1) = \tilde{w}^r(0)\, \gamma_5 w^{s_1}(-\mathbf{k}_1)$ is small, as it corresponds to the operator γ_5 between two positive energy

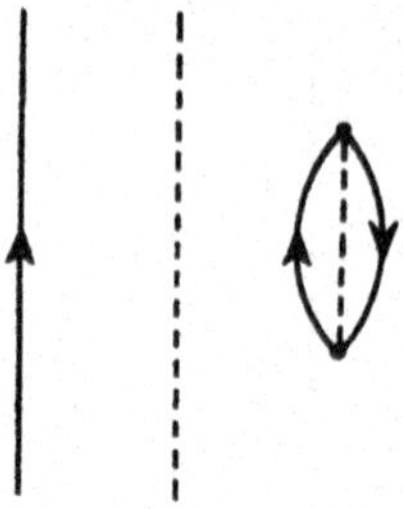

Fig. 13.6

amplitudes $w^r(0)$ and $w^{s_1}(-\mathbf{k}_1)$ whose momentum difference is small [recall the discussion in Chapter 4 of the operator γ_5, Eq. (4.130) ff.]. Using the representation of γ_5, and w, its magnitude is $i\boldsymbol{\sigma} \cdot \mathbf{k}_1/2M$ evaluated between Pauli two-component spinors $|r\rangle$ and $|s_1\rangle$ [$\boldsymbol{\sigma}$ is the nucleon spin, a 2×2 matrix now!]. Similarly, the matrix element $\tilde{w}^{s_2}(\mathbf{p}_2)\, \gamma_5 w^r(\mathbf{p}_1 + \mathbf{k}_1)$ in the center-of-mass system, in our approximation, is determined by $-i\boldsymbol{\sigma} \cdot \mathbf{k}_f/2M$. The form $i\boldsymbol{\sigma} \cdot \mathbf{k}/2M$ of the matrix element corresponding to the emission or absorption of a meson with the nucleon going from a positive to a positive energy state indicates some important features of the pseudoscalar coupling: first, that in such a transition the meson is absorbed or emitted in a P state, and secondly, that the interaction is weakened since it involves the "small" operator, γ_5. Thus, diagrams 13.3a and 13.3b lead predominantly to P-wave meson scattering. In the limit $k \ll M$, the energy denominator corresponding to (54) becomes (in the center-of-mass)

$$E(\mathbf{k}_i) + \omega(\mathbf{k}_i) - M \approx M + \omega(\mathbf{k}_i) - M$$
$$\approx \omega(\mathbf{k}_i) \tag{63}$$

so that our matrix element (54), omitting the δ-function factor, now is given by

$$r_{fi}^{(2)a} = \frac{G^2}{(2\pi)^3\,(2M)^2}\,\frac{1}{\sqrt{4\omega(\mathbf{k}_i)\,\omega(\mathbf{k}_f)}}\,\frac{\boldsymbol{\sigma}\cdot\mathbf{k}_f\boldsymbol{\sigma}\cdot\mathbf{k}_i}{\omega(\mathbf{k}_i)}$$
$$= \left(\frac{G\mu}{2M}\right)^2 \frac{1}{\mu^2\sqrt{4\omega(\mathbf{k}_i)\,\omega(\mathbf{k}_f)}}\,\frac{\boldsymbol{\sigma}\cdot\mathbf{k}_f\boldsymbol{\sigma}\cdot\mathbf{k}_i}{\omega(\mathbf{k}_i)} \tag{64}$$

The second form of the matrix element indicates the relation of the matrix element to the pseudovector coupling one, which was discussed in Chapter 10. Similarly, the matrix element for diagram 13.3*b* in this approximation becomes:

$$r_{fi}^{(2)a} = \left(\frac{G\mu}{2M}\right)^2 \frac{1}{\mu^2\sqrt{4\omega(\mathbf{k}_i)\,\omega(\mathbf{k}_f)}}\,\frac{\boldsymbol{\sigma}\cdot\mathbf{k}_i\boldsymbol{\sigma}\cdot\mathbf{k}_f}{\omega(\mathbf{k}_f)} \tag{65}$$

The matrix elements (64) and (65) will be recognized as the ones obtained in lowest approximation from the static Chew model (apart from the cutoff function), and thus give an insight into the domain of validity of this model.

The matrix elements (61) and (62), involving pair creation, are large. In our approximation, the matrix elements are spherically symmetric, so that the diagrams *c* and *d* in Figure 13.3 are primarily responsible for *S*-wave scattering. A more accurate treatment of the scattering problem not using the Born approximation actually indicates that the *S*-wave scattering is greatly reduced as compared with the Born approximation, while the *P*-wave scattering is enhanced [Bethe (1955)].

We close this section with a qualitative discussion as to why in the limit $k \ll M$ the diagrams *a* and *b* in Figure 13.3 lead predominantly to *P*-wave scattering, whereas the diagrams *c* and *d* in Figure 13.3 lead to *S*-wave scattering. The reason is essentially the following: For meson energies $\omega(\mathbf{k}) \ll M$ in the center-of-mass system the nucleon moves very slowly compared to the meson. As a first approximation, we may therefore neglect its motion completely. Consider, then, the application of angular momentum and parity conservation to the two different types of mesonic absorption processes, corresponding to Figure 13.7, diagrams *a* and *b*. If we define the intrinsic parity of a nucleon in an *S* state to be even, then that of the antinucleon in an *S* state is odd. Therefore, the intrinsic parity of the nucleon after absorption is even in the case of the scattering of the nucleon (Fig. 13.7*a*), whereas the intrinsic parity of the system is odd in the final state, in the case of nucleon-antinucleon production (Fig. 13.7*b*). The meson we are considering being pseudoscalar, the parity of

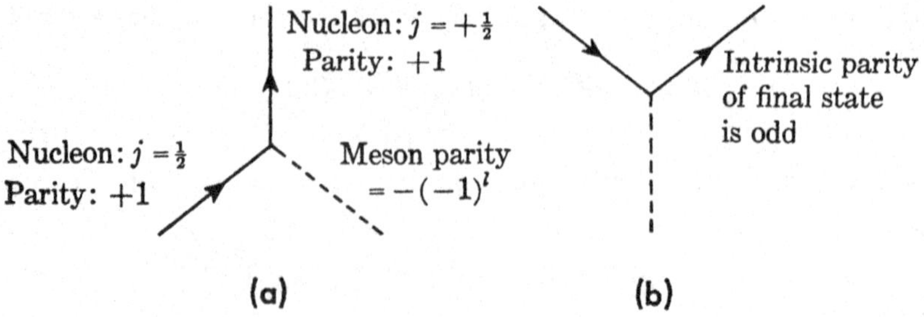

Fig. 13.7

the total system before the absorption is even or odd, according to whether the meson is in a P or S state, respectively. Therefore, a pseudoscalar meson can only be absorbed from an odd state of angular momentum in the diagram a in Figure 13.7 and only from even angular momentum states in the diagram b in Figure 13.7. Now since the nucleon is in an S state in diagram 13.7a before and after the absorption, the meson which was absorbed must have been a P-wave meson. Similarly, in diagram 13.7b the meson must have been absorbed in an S state (with odd parity).

Let us next consider the description of the scattering in the Dirac picture using the expression (4) for the S matrix. Suppose we are again interested in computing the scattering of a meson by a nucleon in the above theory. To lowest order, the amplitude for scattering from the initial state $|\mathbf{p}s;\mathbf{k}\rangle = b_s^*(\mathbf{p})\, a^*(\mathbf{k})\, |\, \Phi_0\rangle$ to the final state $|\mathbf{p}'s';\mathbf{k}'\rangle$ is then given by:

$$\langle \mathbf{p}'s';\mathbf{k}' \,|\, S \,|\, \mathbf{p}s;\mathbf{k}\rangle$$

$$= -\frac{1}{2!}\int d^4x_1 \int d^4x_2\, \langle \mathbf{p}'s';\mathbf{k}' \,|\, P(\mathcal{H}_I(x_1)\, \mathcal{H}_I(x_2) \,|\, \mathbf{p}s;\mathbf{k}\rangle$$

$$= -\frac{G^2}{2!}\int d^4x_1 \int d^4x_2$$

$$\langle \mathbf{p}'s';k' \,|\, P(\bar{\psi}(x_1)\, \Gamma\psi(x_1)\, \phi(x_1)\, \bar{\psi}(x_2)\, \Gamma\psi(x_2)\, \phi(x_2)) \,|\, \mathbf{p}s;\mathbf{k}\rangle$$

$$= -\frac{G^2}{2!}\int d^4x_1 \int d^4x_2\, \langle \mathbf{p}'s' \,|\, P(\bar{\psi}(x_1)\, \Gamma\psi(x_1)\, \bar{\psi}(x_2)\, \Gamma\psi(x_2)) \,|\, \mathbf{p}s\rangle$$

$$\cdot\, \langle \mathbf{k}' \,|\, P(\phi(x_1)\, \phi(x_2) \,|\, \mathbf{k}\rangle \qquad (66)$$

We could next evaluate analytically the matrix elements of the fermion and boson factors to obtain the S-matrix element for the scattering. However, since a general method exists for the reduction of the S matrix into parts which contribute to specific processes, we first present this general method before once again taking up the specific problem of meson-nucleon scattering.

13c. Wick's Theorem

Each term in the power series expansion (4) of the S matrix can give rise to a variety of processes, virtual as well as real. Dyson (1951a, b, c) and Wick (1950) have shown how to express $P(\mathcal{H}_I(x_1) \cdots \mathcal{H}_I(x_n))$ in a form in which all the virtual processes are represented explicitly. This is the so-called decomposition of the chronological product into *normal products*. It will be recalled that the latter type of product is defined as a product of free-particle creation and destruction operators in which all the creation operators stand to the left of all destruction operators. Then, given any initial and final state specified by definite numbers of free particles with specified spins and momenta, there exists one, and only one, normal product with a nonzero matrix element between these states. A decomposition of S into normal products is thus equivalent to the listing of all the matrix elements of S in a representation in which the free-particle occupation numbers are diagonal. A Feynman graph is then simply a concise way of representing a normal product. We shall here indicate the reduction of the S matrix to normal form, following the algebraic method of Wick (1950).

Consider first the case of a simple product of interaction picture operators, Q, such as arises in the expansion (4). The decomposition of such a product into normal products, i.e., products where all creation operators stand to the left of the destruction operators, is then derivable from the commutation rules between the factors of Q. It is an operator identity independent of the particular states in which we may be interested. Let us again introduce the operator N which, acting upon a product of creation and annihilation operators, rewrites the product in normal form, with the understanding that the rearrangement is to be made as if all commutation or anticommutation rules which the operators satisfy had a vanishing right-hand side. The operator N thus includes the change of sign which arises when the order of anticommuting field variables is changed. Recall that for a product of two boson factors, we have

$$N(\phi^{(-)}(x)\,\phi^{(+)}(y)) = N(\phi^{(+)}(y)\,\phi^{(-)}(x) = \phi^{(-)}(x)\,\phi^{(+)}(y) \qquad (67)$$

and similarly, for a product of two fermion factors, we have

$$\begin{aligned} N(\psi^{(+)}(x)\,\bar{\psi}^{(-)}(y)) &= -N(\bar{\psi}^{(-)}(y)\,\psi^{(+)}(x)) \\ &= -\bar{\psi}^{(-)}(y)\,\psi^{(+)}(x) \end{aligned} \qquad (68)$$

By definition the distributive law is valid for the normal product operation, e.g.,

$$\begin{aligned} N(\phi(x)\,\phi(y)) &= N[(\phi^{(+)}(x) + \phi^{(-)}(x))\,(\phi^{(+)}(y) + \phi^{(-)}(y))] \\ &= N(\phi^{(+)}(x)\,\phi^{(+)}(y)) + N(\phi^{(+)}(x)\,\phi^{(-)}(y)) \\ &\quad + N(\phi^{(-)}(x)\,\phi^{(+)}(y)) + N(\phi^{(-)}(x)\,\phi^{(-)}(y)) \end{aligned} \qquad (69)$$

It is then a consequence of the commutation rules that we can write:

$$\bar{\psi}_\alpha(x)\,\psi_\beta(y) = N(\bar{\psi}_\alpha(x)\,\psi_\beta(y)) - iS^{(-)}_{\beta\alpha}(y - x) \tag{70a}$$

$$\psi_\alpha(x)\,\bar{\psi}_\beta(y) = N(\psi_\alpha(x)\,\bar{\psi}_\beta(y)) - iS^{(+)}_{\alpha\beta}(x - y) \tag{70b}$$

$$\psi(x)\,\psi(y) = N(\psi(x)\,\psi(y)) \tag{70c}$$

$$\phi(x)\,\phi(y) = N(\phi(x)\,\phi(y)) + i\hbar c\Delta^{(+)}(x - y) \tag{70d}$$

$$\phi(x)\,\psi(y) = N(\phi(x)\,\psi(x)) \tag{70e}$$

We next introduce Wick's chronological T operator, which is defined in the same way as the Dyson chronological operator P, except that the T operator includes in its definition the sign of the permutation of the fermion

$$T(UV \cdots Z) = \delta_P XY \cdots \tag{71}$$

In other words, T operating on a product of time-labeled operators, $UV \cdots Z$, rewrites the product in chronological order $XY \cdots$ (the operator having the latest time standing furthest to the left), and the whole product is given the sign plus or minus (δ_P), according to whether the permutation of fermion factors in going from the left side to the right side of Eq. (71) is even or odd. That is, the T operator in rewriting the factors in chronological order operates as if all commutators and anticommutators were zero. For the case of two factors,

$$T(\psi(x)\,\psi(y)) = +\psi(x)\,\psi(y) \quad \text{if } x_0 > y_0 \tag{72a}$$

$$= -\psi(y)\,\psi(x) \quad \text{if } y_0 > x_0 \tag{72b}$$

By virtue of the commutation rules $0 = [\psi(x), \psi(y)]_+$, the right-hand side of (72) may in both cases be written as

$$T(\psi(x)\,\psi(y)) = \psi(x)\,\psi(y) \tag{73}$$

Similarly, we have

$$\begin{aligned} T(\psi(x)\,\bar{\psi}(y)) &= +\psi(x)\,\bar{\psi}(y) \quad \text{if } x_0 > y_0 \\ &= -\bar{\psi}(y)\,\psi(x) \quad \text{if } y_0 > x_0 \end{aligned} \tag{74}$$

$$\begin{aligned} T(\phi(x)\,\phi(y)) &= \phi(x)\,\phi(y) \quad \text{if } x_0 > y_0 \\ &= \phi(y)\,\phi(x) \quad \text{if } y_0 > x_0 \end{aligned} \tag{75}$$

We now turn to our basic problem, that of expressing a chronological product in terms of a sum of normal products. To this end, we introduce the notion of a contraction between two factors to represent the commutator or anticommutator which arises in going from a chronological product to the normal form. The contraction symbol will be denoted by appending a dot superscript to the two factors, and is defined by

$$T(UV) = U^{\cdot}V^{\cdot} + N(UV) \tag{76}$$

where $U^{\cdot}V^{\cdot}$ represents the contracted factors. In the interaction picture, the relevant commutator or anticommutator of two creation or annihilation operators are always c-numbers, so that the contracted factors are likewise always c-numbers. Furthermore, it follows from the definition of the operator N that the (bare) vacuum expectation value of a normal product of interaction picture operators vanishes, so that taking the vacuum expectation value of Eq. (76), we obtain

$$U^{\cdot}V^{\cdot} = (\Phi_0, T(UV)\,\Phi_0) \tag{77}$$

From the definitions of the T and N products, we have for $x_0 > y_0$

$$\begin{aligned}\phi^{(+)\cdot}(x)\,\phi^{(-)\cdot}(y) &= T(\phi^{(+)}(x)\,\phi^{(-)}(y)) - N(\phi^{(+)}(x)\,\phi^{(-)}(y))\\ &= \phi^{(+)}(x)\,\phi^{(-)}(y) - \phi^{(-)}(y)\,\phi^{(+)}(x)\\ &= [\phi^{(+)}(x), \phi^{(-)}(y)] = i\hbar c\Delta^{(+)}(x-y) \quad \text{for } x_0 > y_0\end{aligned} \tag{78}$$

Similarly, one verifies that for $x_0 < y_0$

$$\phi^{(+)\cdot}(x)\,\phi^{(-)\cdot}(y) = 0 \quad \text{for } y_0 > x_0 \tag{79}$$

which of course agrees with the definition of the contracted factors given by Eq. (77), namely

$$\begin{aligned}\phi^{(+)\cdot}(x)\,\phi^{(-)\cdot}(y) &= (\Phi_0, T(\phi^{(+)}(x)\,\phi^{(-)}(y))\,\Phi_0)\\ &= i\hbar c\Delta^{(+)}(x-y) \quad \text{for } x_0 > y_0\\ &= 0 \qquad\qquad\qquad\quad\ \text{for } y_0 > x_0\end{aligned} \tag{80}$$

More generally, we have

$$\begin{aligned}\phi^{\cdot}(x)\,\phi^{\cdot}(y) &= T(\phi(x)\,\phi(y)) - N(\phi(x)\,\phi(y))\\ &= (\Phi_0, T(\phi(x)\,\phi(y))\,\Phi_0)\\ &= \begin{cases} = +i\hbar c\Delta^{(+)}(x-y) & \text{for } x_0 > y_0\\ = -i\hbar c\Delta^{(-)}(x-y) & \text{for } y_0 > x_0\end{cases}\end{aligned} \tag{81a}$$

We shall, with Dyson, denote the function which has the properties indicated by the right-hand side of (81) by $\Delta_F(x-y)$, i.e.,

$$\tfrac{1}{2}\hbar c\Delta_F(x-y) = \begin{cases} i\hbar c\Delta^{(+)}(x-y) & \text{for } x_0 > y_0\\ -i\hbar c\Delta^{(-)}(x-y) & \text{for } y_0 > x_0\end{cases} \tag{81b}$$

Similarly for the fermion factors, we obtain using (76) and the anticommutation rules:

$$\begin{aligned}\psi_\alpha^{\cdot}(x)\,\tilde{\psi}_\beta^{\cdot}(y) &= (\Phi_0, T(\psi_\alpha(x)\,\tilde{\psi}_\beta(y))\,\Phi_0)\\ &= \begin{cases} -iS^{(+)}_{\alpha\beta}(x-y) & \text{for } x_0 > y_0\\ +iS^{(-)}_{\alpha\beta}(x-y) & \text{for } y_0 > x_0\end{cases}\\ &= -\tfrac{1}{2}S_{F\alpha\beta}(x-y) = K_{+\alpha\beta}(x-y)\end{aligned} \tag{82}$$

From the definition of normal and T products, it also follows that

$$\begin{aligned}\tilde{\psi}_\beta^{\cdot}(y)\,\psi_\alpha^{\cdot}(x) &= -\psi_\alpha^{\cdot}(x)\,\tilde{\psi}_\beta^{\cdot}(y)\\ &= +\tfrac{1}{2}S_{F\alpha\beta}(x-y)\end{aligned} \tag{83}$$

Thus, whenever the contracted product does not vanish, the order in it is important. Finally, mixed contracted products of boson and fermion factors vanish as does the contraction of two ψ or $\bar{\psi}$ factors:

$$\psi^{\cdot}(x)\,\phi^{\cdot}(y) = \bar{\psi}^{\cdot}(x)\,\phi^{\cdot}(y) = \psi^{\cdot}(x)\,\psi^{\cdot}(y) = \bar{\psi}^{\cdot}(x)\,\bar{\psi}^{\cdot}(y) = 0 \quad (84)$$

Summarizing, we have established that the nonvanishing contractions are

$$\begin{aligned} \psi_\alpha{}^{\cdot}(x)\,\bar{\psi}_\beta{}^{\cdot}(y) &= (\Phi_0,\, T(\psi_\alpha(x)\,\bar{\psi}_\beta(y))\,\Phi_0) \\ &= -\tfrac{1}{2}S_{F\alpha\beta}(x-y) \end{aligned} \quad (85)$$

and

$$\begin{aligned} \phi^{\cdot}(x)\,\phi^{\cdot}(y) &= (\Phi_0,\, T(\phi(x)\,\phi(y))\,\Phi_0) \\ &= +\tfrac{1}{2}\hbar c\Delta_F(x-y) \end{aligned} \quad (86)$$

The functions $-\frac{1}{2}S_F = K_+$ and Δ_F are also sometimes called S_c and Δ_c, or causal function, after Stückelberg (1949) who discovered them by noting that it is the Green's function which maintains the correct causal time sequence of events in the quantum theory. [See also Fierz (1950c).]

To facilitate the subsequent exposition, we next define a normal product with one or more contracted pairs of factors in it. In order to distinguish between different contracted pairs, we shall use different superscripts such as double dots, or triplet dots, etc. Then if $UV \cdots XYZ$ are creation or annihilation operators, we define

$$N(U^{\cdot}V^{\cdot\cdot} \cdots R \cdots X^{\cdot\cdot}YZ^{\cdot}) = \pm(U^{\cdot}Z^{\cdot})\,(V^{\cdot\cdot}X^{\cdot\cdot})\,N(\cdots R \cdots Y) \quad (87)$$

the $+$ or $-$ sign being given according to whether the permutation of the fermion factors, in going from the left-hand side to the right-hand side, is even or odd.

We are now in a position to prove an auxiliary theorem which will be very useful.

Theorem:

If Z is an operator labeled with a time which is earlier than $UV \cdots XY$, then

$$\begin{aligned} N(UV \cdots XY)\,Z = N(UV \cdots XY^{\cdot}Z^{\cdot}) + N(UV \cdots X^{\cdot}YZ^{\cdot}) \\ \cdots + N(U^{\cdot}V \cdots XYZ^{\cdot}) + N(UV \cdots XYZ) \end{aligned} \quad (88)$$

Since the distributive law holds for the N operation, it is clearly sufficient to prove Eq. (88), for the case where each factor $U, V, \cdots X, Y$ is either a creation or annihilation operator. Furthermore, according to the definition of a normal product, Eq. (88) is still true if we regroup the operators $UV \cdots XY$, as long as the same regrouping of the factors is done on both sides of the equation. We may therefore assume that the operators $UV \cdots XY$ are already in normal order with all creation operators standing to the left of all destruction operators.

We shall prove the theorem under the assumption that Z is a creation operator and that all the operators $UV \cdots XY$ are destruction operators.

This is clearly sufficient. If $UV \cdots XY$ are all destruction operators and Z is a creation operator, we may then add any number of creation operators to the left of all the factors on both sides of Eq. (88) within the N product without impairing the validity of our theorem, since the contraction between two creation operators gives zero. If, on the other hand, Z is a destruction operator and $UV \cdots XY$ are creation operators, then Eq. (88) reduces to a trivial identity, since the contraction between a creation and destruction operator, with the time label of the destruction operator earlier than that of the creation operator, also gives zero [see, for example, Eq. (80)]. In this case, therefore, only the last term of the right-hand side of (88) contributes, reducing it to an identity.

The proof of the theorem is by induction. It is clearly true for the case of one factor in N, since then (recall that Z is earlier than Y, by definition)

$$N(Y)\,Z = YZ = T(YZ) \tag{89}$$

which by the definition (76) is equal to

$$T(YZ) = Y^{\cdot}Z^{\cdot} + N(YZ) \tag{90}$$

Assume now that (88) is true for n factors, and multiply (88) by another destruction operator D on the left, D having a time label which is later than that of Z. Then

$$DN(UV \cdots XY)\,Z = N(DUV \cdots XY^{\cdot}Z^{\cdot}) + \cdots + N(DU^{\cdot}V \cdots XYZ^{\cdot}) + DN(UVX \cdots YZ) \tag{91}$$

since the factor D can be brought within the normal product in all the terms in which the Z operator is contracted. However, since by definition Z is a creation operator and all the $UV \cdots$ are destruction operators,

$$N(UV \cdots XYZ) = \delta_P ZUV \cdots XY \tag{92}$$

δ_P being the sign of the permutation of the fermion factors, so that

$$DN(UV \cdots XYZ) = \delta_P DZUV \cdots XY \tag{93}$$

Again, since Z is labeled by a time which is earlier than that of D, we have

$$\begin{aligned} DZ = T(DZ) &= D^{\cdot}Z^{\cdot} + N(DZ) \\ &= D^{\cdot}Z^{\cdot} + \delta_Q ZD \end{aligned} \tag{94}$$

and hence,

$$DZUV \cdots XY = D^{\cdot}Z^{\cdot}UV \cdots XY + \delta_Q ZDUV \cdots XY \tag{95}$$

Now, by virtue of our definition of a normal product with contracted factors in it,

$$D^{\cdot}Z^{\cdot}UV \cdots XY = \delta_P N(D^{\cdot}UV \cdots XYZ^{\cdot}) \tag{96}$$

Similarly,

$$ZDUV \cdots XY = \delta_P\delta_Q N(DUV \cdots XYZ) \tag{97}$$

so that, collecting results, we have

$$DN(UV \cdots XYZ) = \delta_P DZUV \cdots XY$$
$$= \delta_P{}^2 N(D^{\cdot}UV \cdots XYZ^{\cdot}) + \delta_P{}^2\delta_Q{}^2 N(DUV \cdots XYZ) \quad (98)$$

Since $\delta_Q{}^2 = \delta_P{}^2 = 1$, we have proved our theorem (24) for the case of $n + 1$ destruction operators and hence, by induction, the theorem is proved in general.

This theorem is immediately generalized. Multiply Eq. (88) on the left by a factor $R^{\cdot\cdot}S^{\cdot\cdot}$; then, by virtue of our definition (87), we can make Eq. (88) read as follows, for example:

$$N(UR^{\cdot\cdot}V \cdots S^{\cdot\cdot}XY)\,Z = N(UR^{\cdot\cdot}V \cdots S^{\cdot\cdot}XY^{\cdot}Z^{\cdot})$$
$$+ \cdots + N(U^{\cdot}R^{\cdot\cdot}V \cdots S^{\cdot\cdot}XYZ^{\cdot}) + N(UR^{\cdot\cdot}V \cdots S^{\cdot\cdot}XYZ) \quad (99)$$

Similarly, one can add several such contracted factors. The theorem (88) is thus seen to hold when any number of contractions (though, of course, the same in every term) are marked within $UV \cdots XY$.

Wick's theorem now states: A T product can be decomposed into a unique sum of normal products as follows:

$$T(UV \cdots XYZ) = N(UV \cdots XYZ) + N(U^{\cdot}V^{\cdot} \cdots XYZ) + \cdots$$
$$+ N(U^{\cdot}V^{\cdot\cdot} \cdots X^{\cdot\cdot}YZ^{\cdot}) + \cdots$$
$$+ N(U^{\cdot}V^{\cdot\cdot}W^{\cdot\cdot} \cdots X^{\cdot\cdot\cdot}Y^{\cdot}Z^{\cdot\cdot\cdot}) + \cdots \quad (100)$$

where the sum on the right-hand includes all possible sets of contractions between pairs of factors.

The proof is again by induction. It is true for one factor. For two factors it is again true, as it is a restatement of Eq. (76). Assume it to hold true for n factors. Then multiply Eq. (100) on the right by an operator Ω, belonging to a time earlier than any other factor, so that

$$T(UV \cdots XYZ)\,\Omega = T(UV \cdots XYZ\Omega) \quad (101)$$

The terms obtained from the right-hand side of (100) after multiplication by Ω are of the form $N(UV \cdots XYZ)\,\Omega$. These terms can be brought back into normal product form using our previous theorem (88). The theorem (100) is then seen to be true for $n + 1$ factors, if Ω is earlier than $U, V, \cdots X, Y$, and Z. However, the restriction due to the time label of Ω can easily be removed. Once Ω is inside the T and N products we may rearrange the order of the operators, since by the definition of the T and N products Eq. (100) is invariant under the same regrouping of factors on both sides of the equation. Hence, we have proved the theorem in general for $n + 1$ factors.

A similar rule to the above for the decomposition of a simple product of operators into a sum of normal products can be written down [Dyson (1951a, b)]. In this case the contraction symbol, denoted by a line joining the factors, is defined by Eqs. (70a)–(70d), with

$$\underbracket{\psi_\alpha(x)\,\bar{\psi}_\beta(x')} = -iS^{(+)}_{\alpha\beta}(x-x') = (\Phi_0,\, \psi_\alpha(x)\,\bar{\psi}_\beta(x')\,\Phi_0) \tag{102}$$

$$\underbracket{\bar{\psi}_\beta(x')\,\psi_\alpha(x)} = -iS^{(-)}_{\alpha\beta}(x-x') = (\Phi_0,\, \bar{\psi}_\beta(x')\,\psi_\alpha(x)\,\Phi_0) \tag{103}$$

$$\underbracket{\phi(x)\,\phi(x')} = i\Delta^{(+)}(x-x') = (\Phi_0,\, \phi(x)\,\phi(x')\,\Phi_0) \tag{104}$$

$$\underbracket{\psi(x)\,\psi(y)} = \underbracket{\bar{\psi}(x)\,\bar{\psi}(y)} = \underbracket{\phi(x)\,\psi(y)} = \underbracket{\phi(x)\,\bar{\psi}(y)} = 0 \tag{105}$$

The decomposition theorem, which is easily proved by induction, now states that

$$UV\cdots XYZ = N(UV\cdots XYZ) + N(\underbracket{UV}\cdots XYZ) + \cdots N(\underbracket{U\underbracket{V\cdots X}YZ}) + \cdots \tag{106}$$

where again the sum on the right-hand side includes all possible sets of contractions between pairs of factors.

The theorems we have just proved are not yet in a form in which they can be applied to the reduction of the S matrix for the following reasons: First, the S-matrix expansion was defined in terms of the Dyson chronological operator P, and not Wick's T operator. Secondly, the S matrix contains noncommuting factors which are labeled with the same time, and the T symbol does not prescribe their order.

In quantum electrodynamics, where the interaction term is $\mathcal{H}_I(x) = = j_\mu(x)\,A^\mu(x)$, the second difficulty is removed by recalling that the expression for the matter current density,

$$j_\mu(x) = -e\bar{\psi}(x)\,\gamma_\mu\psi(x) \tag{107}$$

is incorrect and must be redefined as

$$j'_\mu(x) = -\frac{e}{2}\,[\bar{\psi}(x)\,\gamma_\mu,\, \psi(x)] \tag{108}$$

The expression $j'_\mu(x)$ for the current has the property that

$$(\Phi_0,\, j'_\mu(x)\,\Phi_0) = 0 \tag{109}$$

and is obtained by symmetrizing the theory in particle and antiparticle variables. It can easily be shown that this symmetrized current (we now drop primes) can also be written as a normal product

$$j_\mu(x) = -eN(\bar{\psi}(x)\,\gamma_\mu\psi(x)) \tag{110}$$

so that when $j_\mu(x)$ occurs in a chronological product, the normal product then orders the factor of $j_\mu(x)$. The S matrix in quantum electrodynamics can therefore be written as

$$S = \sum_{n=0}^{\infty}\left(\frac{ie}{\hbar c}\right)^n \frac{1}{n!}\int d^4x_1 \int d^4x_2 \cdots \int d^4x_n \; T\{N(\bar{\psi}(x_1)\,\gamma^{\mu_1}\psi(x_1)\,A_{\mu_1}(x_1)) \cdots N(\bar{\psi}(x_n)\,\gamma^{\mu_n}\psi(x_n)\,A_{\mu_n}(x_n))\} \tag{111}$$

where we have included the A_μ factor in the normal product, since the order of the factors $\bar{\psi}\gamma^\mu\psi(x)$ and $A_\mu(x)$ relative to each other is irrelevant, because they commute. Furthermore, we have rewritten the Dyson expansion with the T operator replacing the P operator, as there is now no difference, since only *pairs* of fermion factors, which bear the same time label, are always involved.

A T product in which normal products occur is called a mixed product. Wick has extended theorem (100) to include such mixed T products. The theorem then states:

> A mixed T product can be decomposed as in (100), but with the omission of contractions between factors already in normal product form.

The proof of the theorem is based on the following device. Assume that we have used the distributive law to reduce a mixed T product, such as (111), into a sum of mixed T products, each of which contains only creation or annihilation operators X, Y, Z (i.e., only $A^{(+)}$, $A^{(-)}$, $\psi^{(+)}$, $\psi^{(-)}$, etc., and not A, ψ, $\bar{\psi}$). We may then consider the mixed T product $T(N(RST) \cdots N(XYZ))$ (where XYZ bear the same time label and similarly RST have the same time co-ordinate, etc.), as the limit of $T(RST \cdots XYZ)$ where the time label of the creation operators amongst $RST, \cdots, XYZ$, is assumed to be later than the destruction operator by an infinitesimal amount. The theorem (100) may now be applied. Since the contractions to be omitted in the theorem for mixed T products actually vanish [the destruction operator having a time label earlier than that of the creation operator], the theorem follows.

In pseudoscalar meson theory, where

$$\mathcal{H}_I(x) = \tfrac{1}{2}G[\bar{\psi}(x)\,\gamma_5,\,\psi(x)]\,\phi(x) = GN(\bar{\psi}(x)\,\gamma_5\psi(x)\,\phi(x)) \tag{112}$$

by Wick's theorem on mixed products it is again unnecessary to contract the factors $\bar{\psi}(x)$ and $\psi(x)$ in each $\mathcal{H}_I(x)$.

13*d*. The Representation of the Invariant Functions

We have defined the function Δ_F arising from the contraction of two boson factors as

$$\Delta_F(x) = \begin{cases} +2i\Delta^{(+)}(x) & \text{for } x_0 > 0 \\ -2i\Delta^{(-)}(x) & \text{for } x_0 < 0 \end{cases} \tag{113}$$

We recall that

$$\Delta(x) = \Delta^{(+)}(x) + \Delta^{(-)}(x) \tag{114}$$

and

$$\Delta^{(+)}(x) = \tfrac{1}{2}[\Delta(x) - i\Delta^{(1)}(x)] \tag{115}$$

$$\Delta^{(-)}(x) = \tfrac{1}{2}[\Delta(x) + i\Delta^{(1)}(x)] \tag{116}$$

the Δ, $\Delta^{(1)}$, and $\Delta^{(\pm)}$ singular functions having been previously defined in Chapter 7. We may thus re-express the Δ_F function as

$$\Delta_F(x) = \Delta^{(1)}(x) + i\epsilon(x)\,\Delta(x) \tag{117}$$

where $\epsilon(x) = x_0/|x_0|$ is the sign of the time. Since $\Delta^{(1)}(x)$ is an even function of x, and $\Delta(x)$ and $\epsilon(x)$ odd ones, $\Delta_F(x)$ is an even function of x. Although $\Delta(x)$ vanishes outside the light cone, $\Delta^{(1)}(x)$ does not, so that $\Delta_F(x)$ is nonzero outside the light cone (it is, in fact, equal to $\Delta^{(1)}(x)$ there).

In order to obtain the integral representation of $\Delta_F(x)$, we here recall those for $\Delta(x)$ and $\Delta^{(1)}(x)$, namely,

$$\Delta(x) = -\frac{i}{(2\pi)^3}\int_{-\infty}^{+\infty} d^4k\, e^{-ik\cdot x}\,\delta(k^2 - \mu^2)\,\epsilon(k) \tag{118}$$

$$\Delta^{(1)}(x) = \frac{1}{(2\pi)^3}\int_{-\infty}^{+\infty} d^4k\, e^{-ik\cdot x}\,\delta(k^2 - \mu^2) \tag{119}$$

Entirely equivalent definitions for these singular functions are given by the following expressions:

$$\Delta(x) = \frac{-1}{(2\pi)^4}\int_C \frac{e^{-ik\cdot x}}{k^2 - \mu^2}\,d^4k \tag{120}$$

$$\Delta^{(1)}(x) = \frac{i}{(2\pi)^4}\int_{C_1} \frac{e^{-ik\cdot x}}{k^2 - \mu^2}\,d^4k \tag{121}$$

In these, the integration over k_0 is to be performed first, and the contours C and C_1 in the complex k_0 plane are given in Figure 13.8, diagrams a and b,

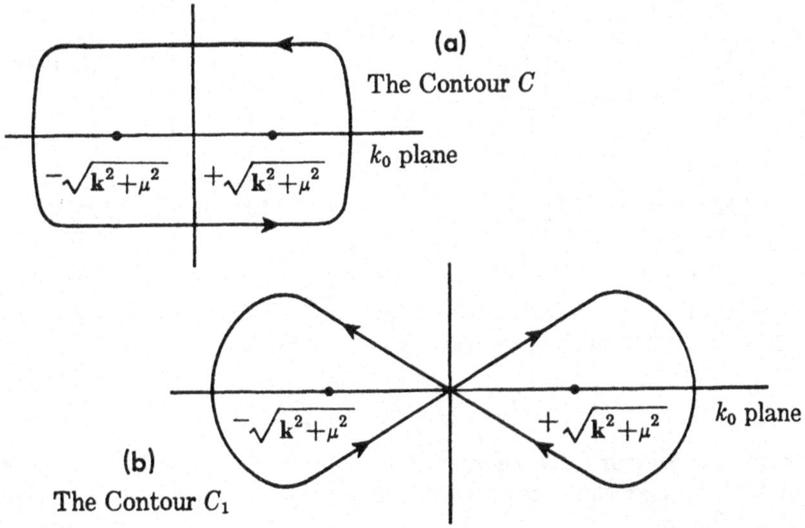

Fig. 13.8

respectively. Thereafter, the integration over the real variables k^1, k^2, and k^3 is to be carried out.

Similarly, since

$$\Delta^{(+)}(x) = \frac{-i}{(2\pi)^3} \int_{k_0>0} e^{-ik\cdot x}\, \delta(k^2 - \mu^2)\, d^4k \tag{122}$$

we may replace the form (122) by the following integral representation which holds for $x_0 > 0$:

$$\Delta^{(+)}(x) = \frac{-1}{(2\pi)^4} \int_{C_+} \frac{e^{-ik\cdot x}}{k^2 - \mu^2}\, d^4k \quad \text{for } x_0 > 0 \tag{123}$$

where the contour C_+ is from $+\infty$ to $-\infty$ along the real k_0 axis, going below the singularity at $k_0 = -\sqrt{\mathbf{k}^2 + \mu^2} = -\omega_{\mathbf{k}}$, and above the one at $k_0 = +\omega_{\mathbf{k}}$, and closed by a large semicircle in the negative imaginary half plane (Fig. 13.9). If $x_0 < 0$, then the following representation holds for $\Delta^{(-)}(x)$:

$$\Delta^{(-)}(x) = \frac{-1}{(2\pi)^4} \int_{C_-} \frac{e^{-ik\cdot x}}{k^2 - \mu^2}\, d^4k \quad \text{for } x_0 < 0 \tag{124}$$

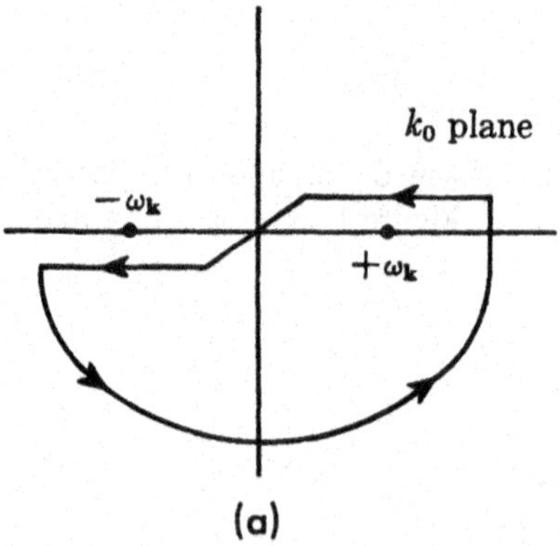

The Contour C_+ for $x_0 > 0$

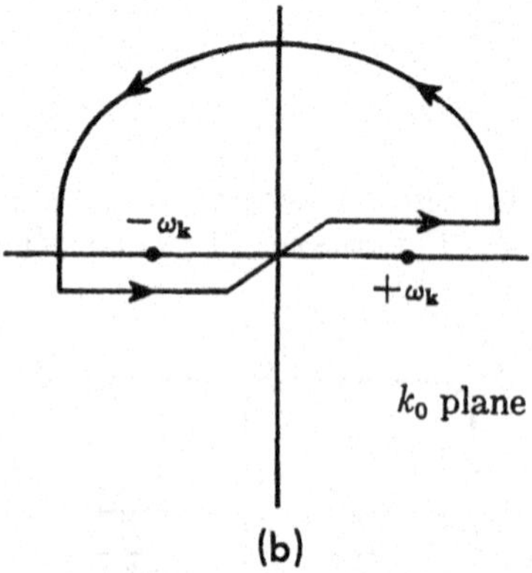

The Contour C_- for $x_0 < 0$

Fig. 13.9

with C_- the contour indicated in Figure 13.9*b*. Recalling the definition (113) of the Δ_F function, we may write

$$\Delta_F(x) = \frac{2i}{(2\pi)^4} \int_{C_F} \frac{d^4k}{k^2 - \mu^2}\, e^{-ik\cdot x} \tag{125}$$

where the contour C_F is the contour indicated in Figure 13.10. Indeed, if $x_0 > 0$, we may close the contour in the lower half plane, and therefore, according to (123)

$$\Delta_F(x) = 2i\Delta^{(+)}(x) \quad \text{for } x_0 > 0 \tag{126}$$

Similarly, if $x_0 < 0$, we close the contour in the upper half plane obtaining

$$\Delta_F(x) = -2i\Delta^{(-)}(x) \quad \text{for } x_0 < 0 \tag{127}$$

which agrees with our previous definition (113).

The same result as integrating over C_F may be obtained by integrating along the real k_0 axis from $-\infty$ to $+\infty$, but giving μ a small negative imaginary part, which is made to vanish after the integration is performed. In this case, our denominator in (125) becomes $k^2 - \mu^2 + i\epsilon$. The effect of this addition of a small negative imaginary part is to displace the poles. These now occur at

$$\begin{aligned} k_0 &= \pm\sqrt{\mathbf{k}^2 + (\mu - i\epsilon)^2} = \pm\sqrt{\mathbf{k}^2 + \mu^2 - i\epsilon'} \\ &= \pm\sqrt{k^2 + \mu^2} \mp i\epsilon'' \end{aligned} \tag{128}$$

so that the pole at $-\omega_\mathbf{k}$ is moved upward, whereas the one at $+\omega_\mathbf{k}$ is displaced downward, as should be the case.

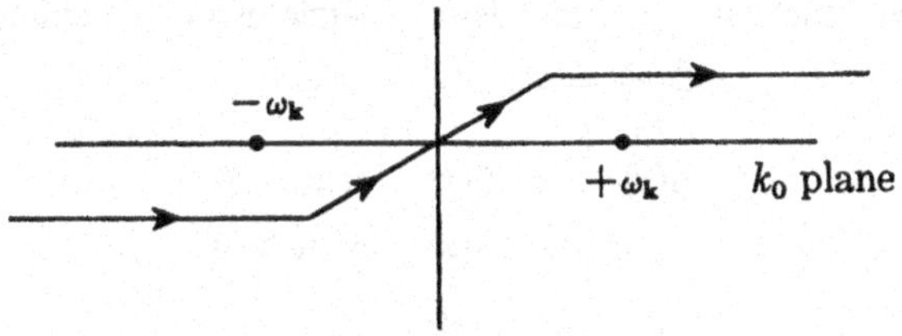

The Contour C_F

Fig. 13.10

For the sake of completeness, we here exhibit the advanced and retarded singular functions $\Delta_A(x)$ and $\Delta_R(x)$, which are defined by the following integral representations:

$$\Delta_A(x) = \frac{-1}{(2\pi)^4}\int_{C_A} \frac{e^{-ik\cdot x}}{k^2 - \mu^2}\, d^4k \tag{129}$$

$$\Delta_R(x) = \frac{-1}{(2\pi)^4}\int_{C_R} \frac{e^{-ik\cdot x}}{k^2 - \mu^2}\, d^4k \tag{130}$$

where the contours C_A and C_R are indicated in Figure 13.11. If $x_0 > 0$,

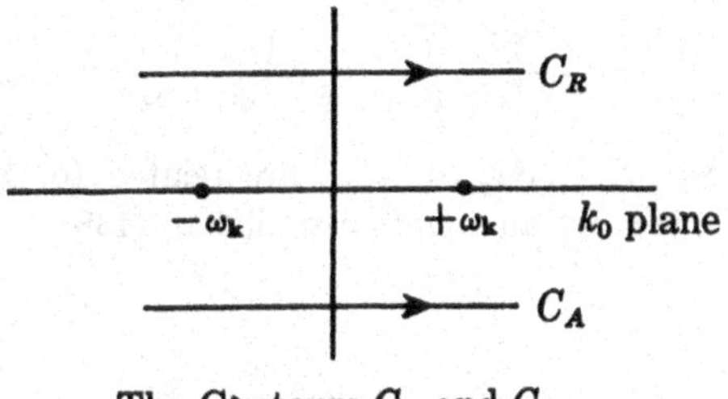

The Contours C_A and C_R

Fig. 13.11

we can close the contours with a large semicircle in the lower half plane so that we obtain

$$\Delta_A(x) = 0 \qquad \text{for } x_0 > 0$$
$$\Delta_R(x) = -\Delta(x) \quad \text{for } x_0 > 0 \tag{131}$$

Similarly, if $x_0 < 0$, the contours can be closed in the upper half planes so that

$$\Delta_A(x) = \Delta(x) \quad \text{for } x_0 < 0$$
$$\Delta_R(x) = 0 \qquad \text{for } x_0 < 0 \tag{132}$$

hence the names, retarded and advanced, for these singular functions. It follows from their definitions that

$$\Delta(x) = \Delta_A(x) - \Delta_R(x) \tag{133}$$
$$\Delta_A(x) = \Delta_R(-x) \tag{134}$$

The singular function $\bar{\Delta}(x)$ which is symmetric in time is defined as

$$\bar{\Delta}(x) = \tfrac{1}{2}(\Delta_A(x) + \Delta_R(x)) \tag{135}$$

so that

$$\bar{\Delta}(x) = +\tfrac{1}{2}\Delta(x) \quad \text{for } x_0 < 0 \tag{136a}$$
$$= -\tfrac{1}{2}\Delta(x) \quad \text{for } x_0 > 0 \tag{136b}$$

or

$$\bar{\Delta}(x) = -\tfrac{1}{2}\epsilon(x)\,\Delta(x) \tag{137}$$

The $\bar{\Delta}$ and $\Delta^{(1)}$ functions have been extensively used by Schwinger (1948a, c; 1949a, b).

The functions $\Delta_F(x)$, $\bar{\Delta}(x)$, $\Delta_A(x)$, $\Delta_R(x)$ satisfy the inhomogeneous Klein-Gordon equation with a $\delta^{(4)}(x)$ source function. On the other hand, the functions $\Delta^{(\pm)}(x)$, $\Delta(x)$, $\Delta^{(1)}(x)$ obey the homogeneous Klein-Gordon equation.

The relation of the S_F function to the Δ_F function is again given by

$$S_F(x) = -(i\gamma \cdot \partial + m)\,\Delta_F(x) \tag{138}$$

so that

$$S_F(x) = -\frac{2i}{(2\pi)^4}\int \frac{\gamma \cdot p + m}{p^2 - m^2 + i\epsilon}\, e^{-ip\cdot x}\, d^4p$$
$$= -\frac{2i}{(2\pi)^4}\int \frac{1}{\gamma \cdot p - m + i\epsilon}\, e^{-ip\cdot x}\, d^4p \tag{139}$$

The functions $S^{(1)}$, $S^{(\pm)}$, S_A, S_R, etc., are related to the corresponding $\Delta^{(1)}$, $\Delta^{(\pm)}$, Δ_A, Δ_R, etc., by an equation similar to (138).

14

Feynman Diagrams

In order to familiarize ourselves with the S-matrix formalism in the Dirac picture, and in order to be able to write down the matrix elements without direct recourse to Wick's theorem in each case, we consider a few examples from which we shall be able to draw certain rules.

14a. Interaction with External Electromagnetic Field

Let us first consider the case of the electron-positron field interacting with a prescribed external electromagnetic field. The interaction energy is then $\mathcal{H}_I(x) = j_\mu(x)\, A^{e\mu}(x)$, and the S matrix given by

$$S = 1 + \left(\frac{ie}{\hbar c}\right) \int d^4x_1\, N(\bar{\psi}(x_1)\, \gamma^\mu A^e{}_\mu(x_1)\, \psi(x_1))$$

$$+ \frac{1}{2!}\left(+\frac{ie}{\hbar c}\right)^2 \int d^4x_1 \int d^4x_2\, T(N(\bar{\psi}(x_1)\, \gamma^\mu A^e{}_\mu(x_1)\, \psi(x_1))$$

$$\cdot\, N(\bar{\psi}(x_2)\, \gamma^\mu A^e{}_\mu(x_2)\, \psi(x_2)) + \cdots \qquad (1)$$

Consider the first-order term $N(\bar{\psi}(x)\, \gamma^\mu \psi(x))\, A^e{}_\mu(x)$. The spinor factors have previously been written out in normal product form. We now represent the creation and destruction operators in the normal products by directed lines pointing upwards (forward in time) for electrons and pointing downwards (backward in time) for positrons. Thus:

1. The operator $\bar{\psi}^{(-)}(x)$, corresponding to the creation of an electron at the space-time point x, is represented by a directed line leaving the point x upward, as indicated in Figure 14.1*a*.

2. The operator $\psi^{(+)}(x)$ (destruction of an electron at x) by a directed line upward and toward the point x, as indicated in Figure 14.1*b*.

3. The operator $\bar{\psi}^{(+)}(x)$ (destruction of a positron at x) by a directed line downward and leaving the point x, as indicated in Figure 14.1*c*.

4. The operator $\psi^{(-)}(x)$ (creation of a positron at x) by a directed line downward and coming to the point x, as indicated in Figure 14.1*d*.

The direction of increasing time is supposed to be upward in a space-time diagram as in the ones shown in Figure 14.1. This arrow convention on positron factors is in agreement with the Feynman picture [Feynman (1949)] of looking upon the latter as electrons running backward in time.

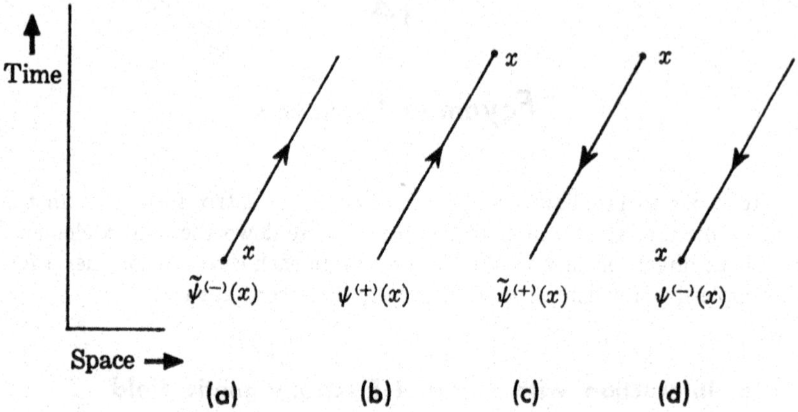

Fig. 14.1

It is seen that in such a description $\tilde{\psi}$ operators are represented by lines leaving the point x (upward or downward) and ψ operators by lines coming to the point x; also that positive frequency lines are below and negative frequency lines above x (later in time). With this convention, then, we

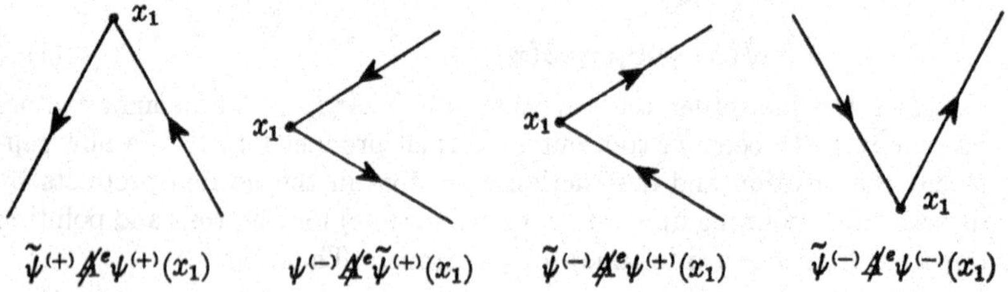

Fig. 14.2

may represent the first-order term in the S matrix by the four Feynman diagrams given in Figure 14.2. The external field is understood to be acting at the point x_1. This is sometimes explicitly exhibited by representing the external potential by a wavy line with a cross at the end. Using this convention, the diagrams indicated in Figure 14.2 are redrawn in Figure 14.3.

Consider next the second-order term. Using Wick's theorem, and writ-

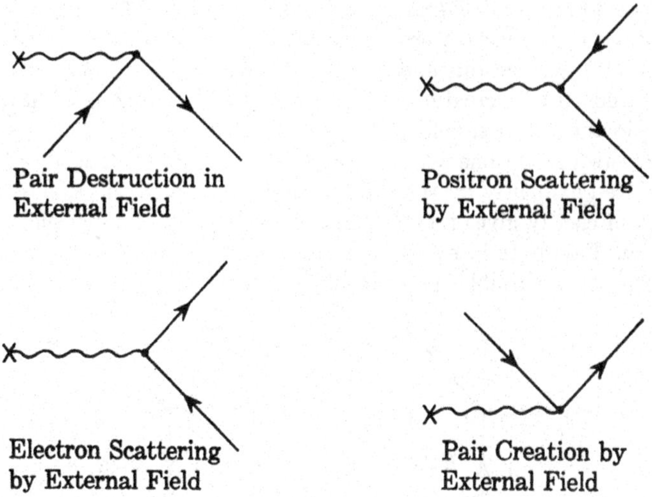
Pair Destruction in External Field
Positron Scattering by External Field
Electron Scattering by External Field
Pair Creation by External Field

Fig. 14.3

ing the contraction of two spinor factors in terms of the Feynman propagator $-\frac{1}{2}S_F = K_+$, we obtain (suppressing the $A\!\!\!/^e$ factors)

$$
\begin{aligned}
&T(N(\bar{\psi}_\alpha(x_1)\,\psi_\beta(x_1))\,N(\bar{\psi}_\rho(x_2)\,\psi_\sigma(x_2)))\\
&\quad = N(\bar{\psi}_\alpha(x_1)\,\psi_\beta(x_1)\,\bar{\psi}_\rho(x_2)\,\psi_\sigma(x_2)) + N(\bar{\psi}_\alpha(x_1)\,K_{+\beta\rho}(x_1 - x_2)\,\psi_\sigma(x_2))\\
&\qquad + N(\bar{\psi}_\rho(x_2)\,K_{+\sigma\alpha}(x_2 - x_1)\,\psi_\beta(x_1)) - K_{+\beta\rho}(x_1 - x_2)\,K_{+\sigma\alpha}(x_2 - x_1) \qquad (2)
\end{aligned}
$$

The first term in the expansion corresponds, in terms of Feynman diagrams, to the joint occurrence of any two of the elementary processes represented in Figure 14.2. We indicate a few of the possible processes in Figure 14.4.

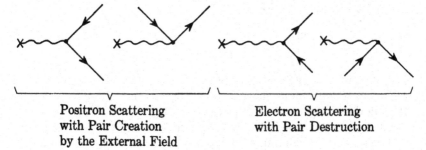

Positron Scattering with Pair Creation by the External Field
Electron Scattering with Pair Destruction

Fig. 14.4

If we represent the factor $K_+(x_1 - x_2) = -\frac{1}{2}S_F(x_1 - x_2)$ arising from the contraction of $\psi(x_1)$ and $\bar{\psi}(x_2)$ by an internal line directed from x_2 to x_1, then the term

$$N(\bar{\psi}(x_1)\,A\!\!\!/^e(x_1)\,K_+(x_1 - x_2)\,A\!\!\!/^e(x_2)\,\psi(x_2)) \qquad (3)$$

gives rise to the four diagrams indicated in Figure 14.5. These diagrams correspond to the second-order corrections (two interactions with the external field) to electron and positron scattering (Fig. 14.5, diagrams *a* and *b*) and to pair creation and pair annihilation (Fig. 14.5, diagrams *c* and *d*) in the external field. For each diagram indicated in Figure 14.5, there exists another one with the points x_1 and x_2 interchanged. This fact is exhibited explicitly by the third term of the right-hand side of Eq. (2). This term clearly gives rise to identical Feynman diagrams, such as those shown in Figure 14.5, except for the labeling of the space-time points. Since the x_1, x_2 variables are integration (dummy) variables, we can com-

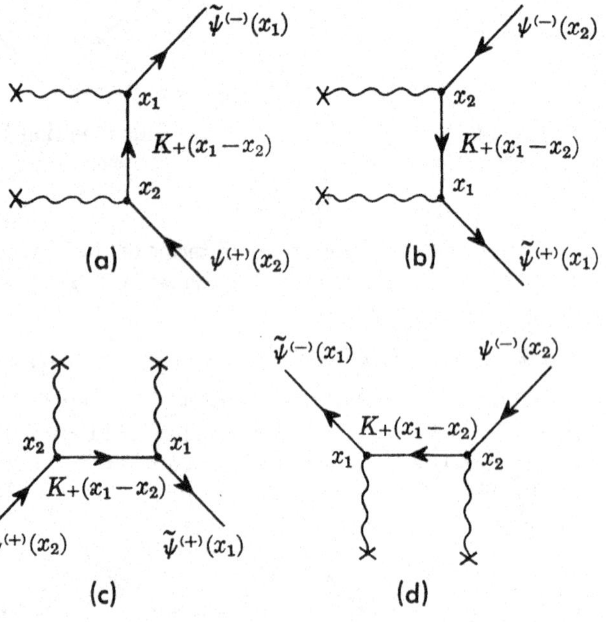

Fig. 14.5

bine the second and third terms in Eq. (2) into one. This then yields a factor 2!, which cancels the 1/2! occurring in front of the second-order term in the expansion of the S matrix, Eq. (1).

Finally, the last term in (2), Tr $(A\!\!\!/^e(x_1)\, K_+(x_1 - x_2)\, A\!\!\!/^e(x_2)\, K_+(x_2 - x_1))$, is represented by the diagram in Figure 14.6. This is a vacuum process. It corresponds to the creation of a pair and its subsequent annihilation. Its matrix element is given by

$$M_V^{(2)} = -\frac{1}{2!}\left(\frac{ie}{\hbar c}\right)^2 \int d^4x_1 \int d^4x_2$$

$$\sum_{\alpha\beta\rho\sigma} A^e{}_\mu(x_1)\, A^e{}_\nu(x_2)\, \{(\gamma^\mu)_{\alpha\beta}\, (\gamma^\nu)_{\rho\sigma}\, K_{+\beta\rho}(x_1 - x_2)\, K_{+\sigma\alpha}(x_2 - x_1)\}$$

$$= +\frac{1}{2!}\left(\frac{e}{\hbar c}\right)^2 \int d^4x_1 \int d^4x_2 \operatorname{Tr}\{A\!\!\!/^e(x_1)\, K_+(x_1 - x_2)\, A\!\!\!/^e(x_2)\, K_+(x_2 - x_1)\} \tag{4}$$

and it is to be interpreted as the probability amplitude for the vacuum to remain a vacuum, i.e., the probability that there be no particle initially and

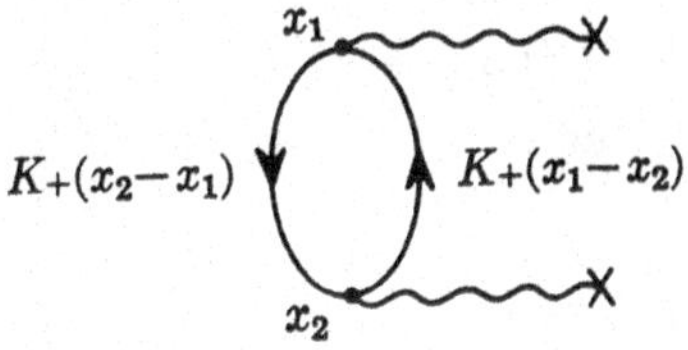

Fig. 14.6

finally. When evaluated, this matrix element has an infinite imaginary part independently of the field. We shall later analyze this difficulty in detail.

It is clear from these examples that there exists a one-to-one correspondence between the decomposition into normal products and Feynman graphs

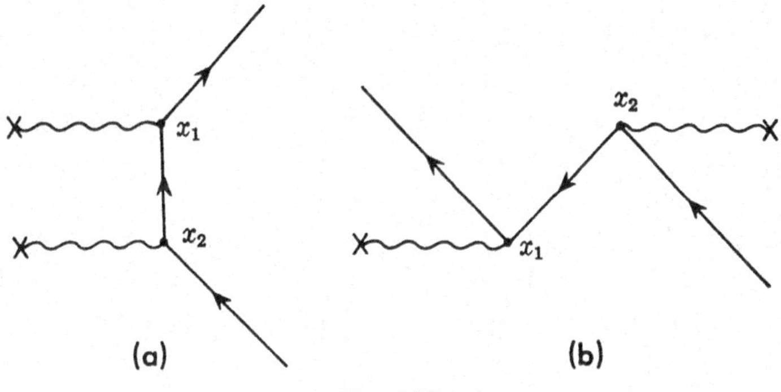

Fig. 14.7

with corners identified; that is, two graphs which differ only by an interchange of the label of their corners are considered distinct. However, care must be exercised in writing down the correct sign for the matrix element corresponding to each diagram, a fact which is automatically taken into account by Wick's decomposition rules. The great advantage of the Feynman approach lies in the fact that the diagrams *a* and *b* of Figure 14.7 are both included in the single term

$$\int d^4x_1 \int d^4x_2\, \bar{\psi}^{(-)}(x_1)\, A\!\!\!/^e(x_1)\, K_+(x_1 - x_2)\, A\!\!\!/^e(x_2)\, \psi^{(+)}(x_2)$$

i.e., the various time orderings of a diagram are included in a single matrix element. There are, therefore, far fewer diagrams than in the time-independent approach outlined in Section b of Chapter 13.

As an example, consider in lowest order the amplitude for positron scattering by the external field. Let the initial state be

$$|\Phi_i\rangle = d_{s_1}{}^*(\mathbf{p}_1) \mid \Phi_0\rangle \tag{5}$$

and we are interested in the scattering to the final state

$$|\Phi_f\rangle = d_{s_2}{}^*(\mathbf{p}_2) \mid \Phi_0\rangle \tag{6}$$

The probability amplitude for this event, to first order in the external potential, is

$$M = (\Phi_f, (1 + S^{(1)})\, \Phi_i) \tag{7}$$

Consider the term

$$\begin{aligned} M^{(1)} &= (\Phi_f, S^{(1)}\Phi_i) \\ &= \left(\frac{ie}{\hbar c}\right) \int d^4x_1\, (d_{s_2}{}^*(\mathbf{p}_2)\, \Phi_0, N(\bar{\psi}(x_1)\, A\!\!\!/^e(x_1)\, \psi(x_1))\, d_{s_1}{}^*(\mathbf{p}_1)\, \Phi_0) \end{aligned} \tag{8}$$

In the normal product expansion, only the term $-\psi_\beta{}^{(-)}(x_1)\, \bar{\psi}_\alpha{}^{(+)}(x_1)$ will contribute: The operator $\bar{\psi}_\alpha{}^{(+)}(x_1)$ destroys the positron in the initial state, and $\psi_\beta{}^{(-)}(x_1)$ creates the positron in the final state. Now, according to (8.45) and (8.47),

$$\begin{aligned} -\psi_\beta{}^{(-)}(x)\, \bar{\psi}_\alpha{}^{(+)}(x) &= -\frac{1}{(2\pi)^3} \int d^3p \int d^3p' \left(\frac{m^2}{E(\mathbf{p})\, E(\mathbf{p}')}\right)^{1/2} \\ &\qquad \sum_{r,s=1}^{2} v^r{}_\beta(\mathbf{p})\, \bar{v}^s{}_\alpha(\mathbf{p}')\, e^{i(p-p')\cdot x}\, d_r{}^*(\mathbf{p})\, d_s(\mathbf{p}') \end{aligned} \tag{9}$$

Hence

$$\begin{aligned} M^{(1)} &= \left(\frac{-ie}{\hbar c}\right) \frac{1}{(2\pi)^3} \int d^4x_1 \int d^3p \int d^3p' \left(\frac{m^2}{E(\mathbf{p})\, E(\mathbf{p}')}\right)^{1/2} e^{i(p-p')\cdot x_1} \\ &\qquad \sum_{r,s=1}^{2} \bar{v}^s(\mathbf{p}')\, A\!\!\!/^e(x_1)\, v^r(\mathbf{p})\, (\Phi_0, d_{s_2}(\mathbf{p}_2)\, d_r{}^*(\mathbf{p})\, d_s(\mathbf{p}')\, d_{s_1}{}^*(\mathbf{p}_1)\, \Phi_0) \\ &= \frac{1}{(2\pi)^3} \left(-\frac{ie}{\hbar c}\right) \int d^4x_1 \left(\frac{m^2}{E(\mathbf{p}_1)\, E(\mathbf{p}_2)}\right)^{1/2} e^{-i(p_1-p_2)\cdot x_1} \\ &\qquad \cdot\, \bar{v}_{s_1}(\mathbf{p}_1)\, A\!\!\!/^e(x_1)\, v_{s_2}(\mathbf{p}_2) \end{aligned} \tag{10}$$

If we call

$$a_\mu(q) = \frac{1}{(2\pi)^4} \int d^4x\, e^{iq\cdot x}\, A^e{}_\mu(x) \tag{11}$$

our matrix element then becomes

$$\begin{aligned} M^{(1)} &= -\frac{e}{\hbar c} \left(\frac{m^2}{E(\mathbf{p}_1)\, E(\mathbf{p}_2)}\right)^{1/2} 2\pi i \bar{v}_{s_1}(\mathbf{p}_1)\, a\!\!\!/(p_2 - p_1)\, v_{s_2}(\mathbf{p}_2) \\ &= -\frac{e}{\hbar c} \left(\frac{m^2}{E(\mathbf{p}_1)\, E(\mathbf{p}_2)}\right)^{1/2} 2\pi i \tilde{u}^c{}_{s_2}(\mathbf{p}_2)\, a\!\!\!/(p_2 - p_1)\, u^c{}_{s_1}(\mathbf{p}_1) \end{aligned} \tag{12}$$

where u^c denotes the charge conjugate spinor. Note the "$-$" sign in contradistinction to the "$+$" sign which would occur for electron scattering. The factors $(m/E(\mathbf{p}))^{1/2}$ are the normalization factors for the initial and final positonic wave functions. Equation (12) further indicates that

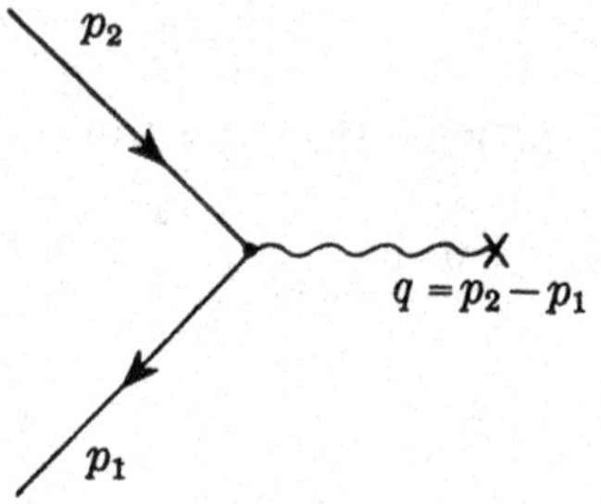

Fig. 14.8

if we draw a Feynman diagram in momentum space wherein each line carries a definite momentum, then the assignment of factors to each line and vertices is most easily accomplished in a formalism using charge conjugate spinors (rather than negative energy Dirac spinors) to describe the positons. (See Fig. 14.8.)

The matrix element, which describes to first order the amplitude for the scattering of an electron by an external potential from the state $b_{s_1}{}^*(p_1) \mid \Phi_0\rangle$ to the state $b_{s_2}{}^*(p_2) \mid \Phi_0\rangle$ is

$$M_e{}^{(1)} = \left(\frac{e}{\hbar c}\right)\left(\frac{m^2}{E(\mathbf{p}_1)\,E(\mathbf{p}_2)}\right)^{1/2} 2\pi i \tilde{u}_{s_2}(\mathbf{p}_2)\, \not{a}(p_2 - p_1)\, u_{s_1}(\mathbf{p}_1) \quad (13)$$

corresponding to the Feynman diagrams of Figure 14.2. If the external electromagnetic field is the Coulomb field of a nucleus of charge $+Ze$, so that

$$A^0 = \frac{Ze}{4\pi r} \quad (14)$$

then

$$\begin{aligned} \not{a}(q) &= \frac{1}{(2\pi)^4}\int e^{iq\cdot x}\, \gamma^\mu A_\mu(x)\, d^4x \\ &= \frac{1}{(2\pi)^3}\,\delta(q_0)\,\gamma^0 \int d^3x\, e^{-i\mathbf{q}\cdot\mathbf{x}}\, A_0(\mathbf{x}) \\ &= \frac{Ze}{4\pi}\,\frac{1}{2\pi^2\mathbf{q}^2}\,\delta(q_0)\,\gamma^0 \end{aligned} \quad (15)$$

The transition amplitude for this situation is therefore given by

$$M_e{}^{(1)} = \frac{Ze^2}{4\pi\hbar c}\,\frac{i}{\pi}\left(\frac{m^2}{E(\mathbf{p}_1)\,E(\mathbf{p}_2)}\right)^{1/2} \tilde{u}_{s_2}(\mathbf{p}_2)\,\frac{1}{|\mathbf{p}_2 - \mathbf{p}_1|^2}\,\gamma^0 u_{s_1}(\mathbf{p}_1)\,\delta(E_2 - E_1) \quad (16)$$

where $E_1 = E_2 = E$ is the energy of the electron. The transition probability is given by the absolute square of the transition amplitude and is

$$|M_e^{(1)}|^2 = Z^2\alpha^2 \frac{m^2}{E(\mathbf{p}_1)\,E(\mathbf{p}_2)} \frac{1}{\pi^2}\,\delta(E_2 - E_1)\,\delta(0) \cdot \left|\tilde{u}_{s_2}(\mathbf{p}_2)\,\frac{1}{|\mathbf{p}_2 - \mathbf{p}_1|^2}\,\gamma^0 u_{s_1}(\mathbf{p}_1)\right|^2 \tag{17}$$

where $\alpha = e^2/4\pi\hbar c \simeq (137)^{-1}$. The factor $\delta(0)$ is to be interpreted as

$$\delta(0) = \lim_{T\to\infty}\lim_{E_1\to E_2}\frac{1}{2\pi}\int_{-T/2}^{T/2} e^{i(E_1-E_2)t}\,dt = \lim_{T\to\infty}\frac{T}{2\pi} \tag{18}$$

with T the time during which the interaction takes place. Hence the transition probability per unit time, $w_{1\to2}$ to a single final state $u_{s_2}(\mathbf{p}_2)$ is given by

$$w_{1\to2} = \frac{Z^2\alpha^2}{2\pi^3}\left(\frac{m}{E}\right)^2 \left|\tilde{u}_{s_2}(\mathbf{p}_2)\,\frac{1}{|\mathbf{p}_2 - \mathbf{p}_1|^2}\,\gamma^0 u_{s_1}(\mathbf{p}_1)\right|^2 \delta(E_1 - E_2) \tag{19}$$

In the case that we are not interested in the spin of the scattered electron, and that the initial beam is unpolarized, we must sum $|M_e^{(1)}|^2$ over the final spin states and average over the two initial spin states. We are also only interested in the transition probability per unit time to a group of final states with density $\rho_f = dn_f/dE_f$. This transition rate is given by

$$w = \frac{4Z^2\alpha^2}{(2\pi)^3}\left(\frac{m}{E}\right)^2\left(\frac{dn_f}{dE_f}\right)_{E_f=E_i}\frac{1}{2}\,\mathrm{Tr}\left(\frac{\not{p}_2 + m}{2m}\,\gamma^0\,\frac{\not{p}_1 + m}{2m}\,\gamma^0\right)\frac{1}{|\mathbf{p}_2 - \mathbf{p}_1|^4} \tag{20}$$

The density of final states is obtained by noting that in the one-particle subspace the operator

$$\int dn_f\, b_s^*(\mathbf{p}_f)\,|\,\Phi_0\rangle\,\langle\Phi_0\,|\,b_s(\mathbf{p}_f)$$

must equal the identity operator. Equivalently this operator operating on a one-particle state $|\Phi\rangle = b_r^*(\mathbf{q})\,|\,\Phi_0\rangle$ must reproduce it. Hence $dn_f = d^3p_f$ and

$$\frac{dn_f}{dE} = \frac{p^2dpd\Omega}{dE} = Epd\Omega \tag{21}$$

since

$$\frac{dE}{dp} = \frac{d}{dp}\sqrt{\mathbf{p}^2 + m^2} = \frac{p}{E} \tag{22}$$

The trace in (20) can easily be evaluated using the properties of traces of γ matrices as stated in Section 5e. Recalling that the trace of an odd number of γ matrices vanishes, and using (4.172), we obtain

$$\frac{1}{2}\operatorname{Tr}\left(\frac{\not p_2 + m}{2m}\gamma^0 \frac{\not p_1 + m}{2m}\gamma^0\right) = \frac{1}{8m^2}\operatorname{Tr}\left(\not p_2\gamma^0\not p_1\gamma^0 + m^2\right)$$

$$= \frac{1}{2m^2}\left(2p_{20}p_{10} - p_2 \cdot p_1 + m^2\right)$$

$$= \frac{1}{2m^2}\left(E^2 + \mathbf{p}_2 \cdot \mathbf{p}_1 + m^2\right) \qquad (23)$$

Since the field is time-independent, the energy of the particle is conserved,

$$E_1 = E_2 = (\mathbf{p}_1^2 + m^2)^{1/2} = (\mathbf{p}_2^2 + m^2)^{1/2} \qquad (24)$$

so that the magnitude of the three-momentum of the particle is likewise conserved. If we call θ the scattering angle, then

$$\mathbf{p}_1 \cdot \mathbf{p}_2 = |\mathbf{p}_1|\,|\mathbf{p}_2| \cos\theta = p^2 \cos\theta$$

$$= (E^2 - m^2)\cos\theta \qquad (25)$$

where $|\mathbf{p}_1| = |\mathbf{p}_2| = p$. From relativistic mechanics we know that

$$\frac{m}{(1 - v^2)^{1/2}} = E \qquad \text{or} \qquad \frac{m^2}{E^2} = 1 - v^2 \qquad (26)$$

where v is the velocity of the particle, so that the expression (23) for the trace may be rewritten as follows:

$$\frac{1}{2}\operatorname{Tr}\left(\frac{\not p_2 + m}{2m}\gamma^0 \frac{\not p_1 + m}{2m}\gamma^0\right) = \frac{1}{m^2}\left[\tfrac{1}{2}(1 + \cos\theta)\,(E^2 - m^2) + m^2\right]$$

$$= \frac{E^2}{m^2}\left[v^2 \cos^2 \tfrac{1}{2}\theta + 1 - v^2\right]$$

$$= \frac{E^2}{m^2}\left(1 - v^2 \sin^2 \tfrac{1}{2}\theta\right) \qquad (27)$$

Similarly, since the magnitude of the momentum is conserved,

$$|\mathbf{p}_2 - \mathbf{p}_1| = 2p \sin \tfrac{1}{2}\theta \qquad (28)$$

We therefore obtain the following expression for w

$$w = \frac{Z^2\alpha^2}{4p^4 \sin^4 \frac{1}{2}\theta}\left(1 - v^2 \sin^2 \tfrac{1}{2}\theta\right)\frac{1}{(2\pi)^3}\left(\frac{dn_f}{dE}\right)_{E_f = E_i} \qquad (29)$$

To obtain the differential cross section, we must divide w by the incident flux. The density of the incoming particles can be obtained from a consideration of the charge-current density corresponding to the initial state vector describing the electron-positron system. Consider the expectation value of the charge-current operator for the initial state, $b_{s_1}{}^*(\mathbf{p}_1) \mid \Phi_0\rangle$ representing one electron of momentum $\mathbf{p}_1$ and spin s_1:

$$\langle j_\mu(x)\rangle = -\left(b_{s_1}{}^*(\mathbf{p}_1)\,\Phi_0, \frac{e}{2}\left[\tilde{\bar\psi}(x), \gamma_\mu\psi(x)\right] b_{s_1}{}^*(\mathbf{p}_1)\,\Phi_0\right) \qquad (30)$$

Clearly, in the expansion of the current operator in terms of the creation and annihilation operators for particles of definite momenta and spin, only the terms proportional to b^*b will contribute in the matrix element (30), whence

$$\begin{aligned}\langle j_\mu(x)\rangle &= -e\,(b_{s_1}{}^*(\mathbf{p}_1)\,\Phi_0, \widetilde{\psi^{(+)}}(x)\,\gamma_\mu\psi^{(+)}(x)\,b_{s_1}{}^*(\mathbf{p}_1)\,\Phi_0)\\ &= -\frac{e}{(2\pi)^3}\frac{m}{E_1}\,\bar{w}^{s_1}(\mathbf{p}_1)\,\gamma_\mu w^{s_1}(\mathbf{p}_1)\\ &= -\frac{e}{(2\pi)^3}\frac{p_{1\mu}}{E_1} = -\frac{1}{(2\pi)^3}\,ev_{1\mu} \qquad (31)\end{aligned}$$

where $v_{1\mu}$ is the velocity of the incident particle. Hence, the flux of incoming electrons is

$$\text{flux} = \frac{1}{(2\pi)^3}\,v = \frac{1}{(2\pi)^3}\frac{p}{E} \qquad (32)$$

The differential cross section is therefore given by

$$\frac{d\sigma}{d\Omega} = \frac{Z^2\alpha^2}{4p^2v^2\sin^4\dfrac{\theta}{2}}\,(1 - v^2\sin^2\tfrac{1}{2}\theta) \qquad (33)$$

which is the Rutherford formula, $Z^2\alpha^2/4p^2v^2\sin^4\frac{1}{2}\theta$, multiplied by $(1 - v^2\sin^2\frac{1}{2}\theta)$. The latter is the correction to the Rutherford formula due to the spin of the electron. It did indeed arise from the trace terms. The next order, i.e., the double scattering contribution, has been calculated by McKinley and Feshbach [McKinley (1948); see also Dalitz (1951a)]. A straightforward application of the perturbation expansion (1) will actually yield a divergent result for the second order (and in fact all higher orders) matrix element. This is connected with the infinite range of the Coulomb field. The solution of this puzzle has been given by Dalitz (1951a), who has shown that the scattering amplitude in a screened Coulomb field

$$A^0 = \frac{Ze}{4\pi}\frac{e^{-\lambda r}}{r} \qquad (34)$$

has the property in the nonrelativistic limit it reduces to the lowest order scattering amplitude (of order $Z\alpha$) times a phase factor which is infinite as $1/\lambda$, the range of the screening, goes to infinity (i.e., $\lambda \to 0$). It is precisely the (infinite) contribution of this phase factor which gives rise to divergences in the higher orders of the power series expansion (1).

For the description of the scattering of electrons by protons at very high energies, modifications to the Mott formula (33) will be required due to:

(*a*) the recoil of the proton

(*b*) the contribution to the scattering associated with the magnetic moment of the proton

(*c*) the finite extent of the proton charge and magnetic moment

(d) the radiative corrections to elastic scattering to account for effects of virtual photons and the emission of the real quanta permitted by the finite resolution of the experiment

The effects of a and b have been included in a calculation made by Rosenbluth (1950). The effect of the finite size of the nucleon is to multiply (33) by the square of the nuclear form factor. The form factor is a function of the momentum transfer $|\mathbf{p}_1 - \mathbf{p}_2|$ and is given by the Fourier transform of the charge density of the nucleon. The effects of d will be taken up in Chapter 15. For an excellent review of the field of high energy electron scattering, the reader is referred to the article of Hofstadter (1957).

We next turn to a brief discussion of the vacuum diagrams. We have noted above that to lowest order the probability amplitude that the vacuum remain a vacuum in the presence of the external field is given by

$$M_V^{(2)} = \frac{e^2}{2} \int d^4x_1 \int d^4x_2 \operatorname{Tr} \{A\!\!\!/^e(x_1) K_+(x_1 - x_2) A\!\!\!/^e(x_2) K_+(x_2 - x_1)\} \tag{35}$$

which diverges independently of the external field A^e. This is due to confluence of the singularities of $K_+(x_1 - x_2)$ and $K_+(x_2 - x_1)$ at $x_1 = x_2$. To exhibit the divergence, we introduce the Fourier transforms $\tilde{K}$ and a of K_+ and A^e into Eq. (35), which then reads:

$$M_V^{(2)} = \frac{e^2}{2(2\pi)^8} \int d^4k \, a_\mu(k) \, \Pi^{\mu\nu}(k) \, a_\nu(-k) \tag{36}$$

where

$$\begin{aligned} \Pi^{\mu\nu}(k) &= \int d^4q \operatorname{Tr} \{\gamma^\mu \tilde{K}_+(q + k) \, \gamma^\nu \tilde{K}_+(q)\} \\ &= \Pi^{\nu\mu}(k) \end{aligned} \tag{37}$$

Gauge invariance requires that if we replace $a_\mu(k)$ by $a_\mu(k) + k_\mu \Lambda(k)$ the matrix element $M_V^{(2)}$ remains invariant. Stated differently, if $a_\mu(k)$ is of the form k_μ times a function of k, then $M_V^{(2)}$ must vanish, hence

$$k_\mu \Pi^{\mu\nu}(k) = \Pi^{\mu\nu} k_\nu = 0 \tag{38}$$

Equation (38) and relativistic invariance imply that $\Pi^{\mu\nu}$ must be of the form

$$\Pi_{\mu\nu}(k) = (k_\mu k_\nu - g_{\mu\nu} k^2) \, \Pi(k^2) \tag{39}$$

The form (39) allows us to rewrite (36) in a manifestly gauge invariant form involving only the field strengths by noting that

$$(k_\mu k_\nu - g_{\mu\nu} k^2) \, a^\mu(-k) \, a^\nu(k) = -\tfrac{1}{2} \tilde{F}_{\mu\nu}(-k) \, \tilde{F}^{\mu\nu}(k) \tag{40}$$

Substituting this expression into (36) we obtain

$$M_V^{(2)} = \frac{e^2}{4} \int d^4x_1 \int d^4x_2 \, F_{\mu\nu}(x_1) \, \tilde{\Pi}(x_1 - x_2) \, F^{\mu\nu}(x_2) \tag{41}$$

where $\tilde{\Pi}(x)$ is the Fourier transform of $\Pi(k)$. An explicit expression for

$\tilde{\Pi}(x)$ has been given by many people, and we shall do so in Chapter 15. We here give an expression for $\tilde{\Pi}(x)$ due to Schwinger (1954):

$$\tilde{\Pi}(x - x') = \frac{i}{16\pi^2} \int_{4m^2}^{\infty} d\kappa^2 \left[\left(1 - \frac{4m^2}{\kappa^2}\right)^{1/2} - \frac{1}{3}\left(1 - \frac{4m^2}{\kappa^2}\right)^{3/2} \right] \Delta_F(x - x'; \kappa^2) \tag{42}$$

The divergent part of $\tilde{\Pi}(x)$ is separated by writing

$$\Delta_F(x - x'; \kappa^2) = -\frac{2i}{\kappa^2} \delta^{(4)}(x - x') - \frac{1}{\kappa^2} \Box_x \Delta_F(x - x'; \kappa^2) \tag{43}$$

which yields

$$\tilde{\Pi}(x - x') = C\delta^{(4)}(x - x') + \Box_x \tilde{\Pi}'(x - x') \tag{44}$$

where C is logarithmically divergent and has the form

$$C = \frac{1}{6\pi^2} \left(\log\left(\frac{K}{m}\right) - \frac{5}{6} \right)\Bigg|_{K\to\infty} \tag{45}$$

The kernel $\tilde{\Pi}'(x)$ will yield a convergent integral for most external fields that would be considered. In particular, it will give a convergent contribution to $M_V^{(2)}$ for a field whose worst singularities are of the form $|x - x_0|^{-(2-\gamma)}$ with $\gamma > 0$.

The vacuum expectation value of $S^{(3)}$, corresponding to the third-order contribution to the amplitude that the vacuum remain a vacuum under the influence of the external field, is given by

$$M_V^{(3)} = \left(\Phi_0, \left(\frac{ie}{\hbar c}\right)^3 \frac{1}{3!} \int d^4x_1 \int d^4x_2 \int d^4x_3 \right.$$
$$\left. T\{N(\bar{\psi} A^e \psi(x_1))\, N(\bar{\psi} A^e \psi(x_2))\, N(\bar{\psi} A^e \psi(x_3))\}\ \Phi_0 \right) \tag{46}$$

The only contribution to this matrix element will come from those normal products in the expansion of (35), in which all the fermion factors are contracted. There are two such normal products which correspond to the following two sets of contractions:

$$-\psi^{\cdot}(x_1)\, \bar{\psi}^{\cdot}(x_2)\, \psi^{\cdot\cdot}(x_2)\, \bar{\psi}^{\cdot\cdot}(x_3)\, \psi^{\cdot\cdot\cdot}(x_3)\, \bar{\psi}^{\cdot\cdot\cdot}(x_1) \tag{47a}$$

and

$$-\psi^{\cdot}(x_1)\, \bar{\psi}^{\cdot}(x_3)\, \psi^{\cdot\cdot}(x_3)\, \bar{\psi}^{\cdot\cdot}(x_2)\, \psi^{\cdot\cdot\cdot}(x_2)\, \bar{\psi}^{\cdot\cdot\cdot}(x_1) \tag{47b}$$

These factors correspond to the two diagrams indicated in Figure 14.9 in which the external electromagnetic field acts at the points x_1, x_2 and x_3.

Note the minus sign in front of the expressions (47). This is due to the fact that in contracting the first and last fermion factors we must always jump an odd number of ψ and $\bar{\psi}$ factors. This is the same minus sign as the one which occurred in the last term of the right-hand side of (2). In general, any closed-loop Feynman diagram will give rise to a factor of -1 in its matrix element.

The matrix element, corresponding to the diagrams of Figure 14.9, is therefore given by

$$M_{V^{(3)}} = -\frac{1}{3!}\left(\frac{ie}{\hbar c}\right)^3 \int d^4x_1 \int d^4x_2 \int d^4x_3$$
$$[\mathrm{Tr}\,\{A\!\!\!/^e(x_1)\, K_+(x_1 - x_2)\, A\!\!\!/^e(x_2)\, K_+(x_2 - x_3)\, A\!\!\!/^e(x_3)\, K_+(x_3 - x_1)\}$$
$$+ \mathrm{Tr}\,\{A\!\!\!/^e(x_1)\, K_+(x_1 - x_3)\, A\!\!\!/^e(x_3)\, K_+(x_3 - x_2)\, A\!\!\!/^e(x_2)\, K_+(x_2 - x_1)\}] \tag{48}$$

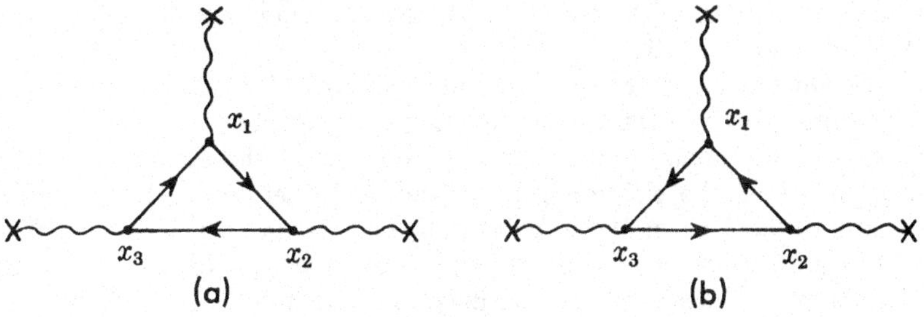

Fig. 14.9

Although the second term can, by a change of integration variables, be cast into a form identical to the first, we have nevertheless preferred to keep $M_{V^{(3)}}$ in the form given by (48), for reasons which will become clear shortly.

We shall now show that the matrix element (48) actually vanishes. The proof uses the fact that there exists a unitary skew-symmetric matrix C, with the property that

$$C^{-1}\gamma^\mu C = -\gamma^{\mu T} \tag{49a}$$

$$C^*C = CC^* = 1 \tag{49b}$$

$$C^T = -C \tag{49c}$$

If we insert in the first term of Eq. (48), the factor $C^{-1}C = I$, we obtain

$$\mathrm{Tr}\,\{\ \} = \mathrm{Tr}\,\{C^{-1}A\!\!\!/^e(x_1)\, CC^{-1}K_+(x_1 - x_2)\, CC^{-1}A\!\!\!/^e(x_2)\, CC^{-1}$$
$$\cdot\, K_+(x_2 - x_3)\, CC^{-1}A\!\!\!/^e(x_3)\, CC^{-1}K_+(x_3 - x_1)\, C\} \tag{50}$$

However, it follows from (49), that

$$C^{-1}K_+(x_1 - x_2)\, C = \frac{i}{(2\pi)^4}\int \frac{-p\!\!\!/^T + m}{p^2 - m^2}\, e^{-ip\cdot(x_1 - x_2)}\, d^4p$$
$$= \frac{i}{(2\pi)^4}\int \frac{p\!\!\!/^T + m}{p^2 - m^2}\, e^{-ip\cdot(x_2 - x_1)}\, d^4p$$
$$= K_+{}^T(x_2 - x_1) \tag{51}$$

We may therefore rewrite the trace (50) as follows:

$$\begin{aligned}\operatorname{Tr}\{\ \} &= (-1)^3 \\ &\operatorname{Tr}\{A^{eT}(x_1)\,K_+{}^T(x_2 - x_1)\,A^{eT}(x_2)\,K_+{}^T(x_3 - x_2)\,A^{eT}(x_3)\,K_+{}^T(x_1 - x_3)\} \\ &= -\operatorname{Tr}\{K_+(x_1 - x_3)\,A^e(x_3)\,K_+(x_3 - x_2)\,A^e(x_2)\,K_+(x_2 - x_1)\,A^e(x_1)\} \\ &= -\operatorname{Tr}\{A^e(x_1)\,K_+(x_1 - x_3)\,A^e(x_3)\,K_+(x_3 - x_2)\,A^e(x_2)\,K_+(x_2 - x_1)\}\end{aligned} \tag{52}$$

so that the two terms in (48) cancel and, indeed, $M_V{}^{(3)} = 0$. Actually, of course, each term vanishes singly, since by a change of variables we can transform (52) into its original form, the left-hand side of (50), so this trace equals its negative.

The cancellation of the two terms in (48) may be looked upon physically as follows: The first term, corresponding to Figure 14.9*a*, may be viewed as an electron going around one way, whereas in the second term, i.e., diagram 14.9*b*, the electron goes around the other way. However, reversing the motion of the electron makes it behave like a positon, thus changing the sign of the charge and hence the sign of each potential interaction. The sum of the two diagrams is therefore zero. Quite generally, a closed loop with an odd number of corners vanishes by a similar argument. This theorem is due to Furry (1937).

Let us call, with Feynman, the sum of all connected closed-loop diagrams $-L$, i.e.,

$$\sum_n M_V{}^{(n)} = -L \tag{53}$$

The minus sign exhibits the fact that we are dealing with closed loops. In addition to these single loops, there exists the possibility that two independent pairs may be created, and each pair may annihilate itself again. The contribution from such pairs of loops is $L^2/2!$ since in L^2 we count every pair of loops twice. The total vacuum-vacuum amplitude is then

$$(\Phi_0, S\Phi_0) = 1 - L + \frac{L^2}{2!} - \frac{L^3}{3!} + \cdots = e^{-L} \tag{54}$$

L has an infinite imaginary part, which corresponds to the vacuum self-energy; however, this infinity has no effect on the normalization constant, that is, the probability that the vacuum remain a vacuum, which is given by

$$|(\Phi_0, S\Phi_0)|^2 = \exp(-2 \operatorname{Re} L) \tag{55}$$

and the real part of L is generally finite. Alternatively, had we included in the interaction Hamiltonian a term of the form ΔE_0 corresponding to the level shift of the physical vacuum relative to the bare vacuum state, then for a static field $\Delta E_0 = \operatorname{Im} L/T$.[1] [$T$ is the (infinite) time interval be-

[1] That in the static external field case, when $A^e(\mathbf{x}) = A^e(x)$, L is proportional to T can be established by introducing in M_V the relative times $x_{10} - x_{20}$, $x_{20} - x_{30}$, etc. as new variables of integration. For example, in (4) introduce x_{10} and $x_{20} - x_{10}$ as variables of integration. The integration over x_{10} then yields a factor T.

tween initial and final states, i.e., $U(T, -\infty) = S$.] The ΔE_0 term is so adjusted that it cancels the divergence in Im L.

Consider now the scattering of an electron in third order. In addition to the triple interaction with the external field, Figure 14.10a, there exists also the possibility that a single scattering occurs along with a disconnected vacuum process as indicated in Figure 14.10b. Now the matrix element

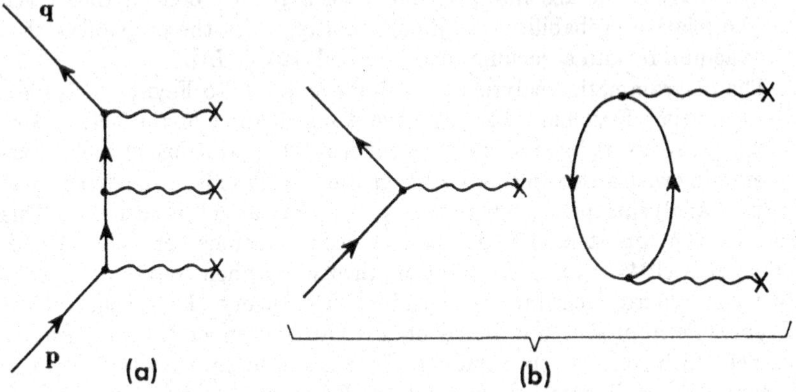

Fig. 14.10

corresponding to the transition from the state p to q for the Feynman diagram of Figure 14.10b may be written as follows:

$$(q \mid S^{(3b)} \mid p) = (q \mid R^{(1)} \mid p)\,(0 \mid S^{(2)} \mid 0) \tag{56}$$

where $(q \mid R^{(1)} \mid p)$ is the first Born approximation matrix element corresponding to a single scattering, i.e.,

$$(q \mid R^{(1)} \mid p) = 2\pi i \left(\frac{e}{\hbar c}\right) \left(\frac{m^2}{E(\mathbf{p})\,E(\mathbf{q})}\right)^{1/2} \tilde{u}^s(\mathbf{q})\, \not{a}(q - p)\, u^r(\mathbf{p}) \tag{57}$$

and $(0 \mid S^{(2)} \mid 0)$ is the vacuum expectation value of the second-order element of the S matrix. The disconnectedness of the two parts of the diagram of 14.10b is reflected in the matrix element (56) in the fact that the product of two factors occurs and that these factors have no variables in common.

Similarly, in higher orders we shall obtain the single scattering diagram occurring with every possible vacuum process. We may, therefore, write the total transition amplitude for going from the state p to q as

$$\begin{aligned}(q \mid S \mid p) &= (q \mid R \mid p)\,(0 \mid S \mid 0) \\ &= (q \mid R \mid p)\, e^{-L}\end{aligned} \tag{58}$$

where $(0 \mid S \mid 0)$ is the vacuum expectation value of the S matrix, which in terms of diagrams is represented by the vacuum Feynman diagrams, while $(q \mid R \mid p)$ is represented by the connected graphs for the scattering process only, that is, those graphs in which no disconnected vacuum process occurs. Feynman (1949a) calls $(q \mid S \mid p)$ the absolute probability amplitude for the transition, and $(q \mid R \mid p)$ the relative probability amplitude. In effect therefore, for the problem of electrons moving in an external field, we need only calculate L once and for all for that given field, and consider for the scattering matrix elements only connected diagrams [the R matrix of Eq. (32)]. The absolute probability for a given process is then given by the relative probability $|(p \mid R \mid q)|^2$ multiplied by the probability that the vacuum remain a vacuum [exp (-2 real part of L)].

The diagrammatic analysis outlined above is due to Feynman (1949a). In this paper Feynman not only gave a formulation of the hole theory which greatly simplified all calculations involving positrons and electrons interacting with an external electromagnetic field, but also indicated a novel way of analyzing perturbation theory in a diagrammatic fashion. This formulation was then extended to include interactions between particles [Feynman (1949b)], and the resulting theory was then proven equivalent to the standard procedure of quantized field theory [Feynman (1950)]. Feynman's approach was based on a consideration of the propagation kernel which connects the solutions of the Dirac equation at two different times. We shall here briefly establish Feynman's formulation starting from the quantized field theory. However, no review of Feynman's work can do full justice to the clarity, simplicity, and elegance of his original papers. The reader is therefore urged to study these papers.

The relative amplitude for arrival at the point x for an electron which initially was in the state $\bar{\psi}(x') \mid \Phi_0\rangle$ is, according to the previous considerations, given by

$$K_+{}^A(x, x') = \frac{\langle \Phi_0 \mid \psi(x)\, U(t, t')\, \bar{\psi}(x') \mid \Phi_0\rangle}{\langle \Phi_0 \mid U(t, t') \mid \Phi_0\rangle} \tag{59}$$

where $x_0 = t$ and $x'_0 = t'$ and $U(t, t')$ is given by

$$U(t, t') = \sum_{n=0}^{\infty} \left(\frac{ie}{\hbar c}\right)^n \frac{1}{n!} \int_{t'}^{t} d^4x_1 \int_{t'}^{t} d^4x_2 \cdots \int_{t'}^{t} d^4x_n$$
$$T\{N(\bar{\psi}(x_1) A\!\!\!/^e(x_1)\, \psi(x_1)) \cdots N(\bar{\psi}(x_n)\, A\!\!\!/^e(x_n)\, \psi(x_n))\} \tag{60}$$

We have here ignored the fact that the state $\bar{\psi}(x) \mid \Phi_0\rangle$ does not really correspond to an electron localized at x. An appropriate scalar product of both sides of Eq. (59) with localized Newton-Wigner amplitudes would, however, overcome this objection. Since these scalar products can be taken at any stage of the calculation, we shall not further concern ourselves with this point. Equation (59) can be further rewritten as

$$K_+{}^A(x, x') = \sum_{n=0}^{\infty} \left(\frac{ie}{\hbar c}\right)^n \frac{1}{n!} \int_{t'}^{t} d^4x_1 \int_{t'}^{t} d^4x_2 \cdots \int_{t'}^{t} d^4x_n$$
$$\cdot \langle \Phi_0 \mid T(\psi(x)\, N(\bar{\psi}(x_1)\, A\!\!\!/^e(x_1)\, \psi(x_1)) \cdots N(\bar{\psi}(x_n)\, A\!\!\!/^e(x_n)\, \psi(x_n))\, \bar{\psi}(x') \mid \Phi_0\rangle_C \tag{61}$$

where the subscript C denotes that only connected diagrams are to be considered. The inclusion of the factors $\psi(x)$ and $\bar{\psi}(x')$ inside the brackets at the positions indicated is possible since x is later, and x' is earlier, than all points $x_1, \cdots, x_n$.

The connected diagrams which contribute to (61) are those indicated in Figure 14.11, and they yield the following contribution:

$$K_+^A(x, x') = K_+(x - x') + ie \int K_+(x - x_1) \mathcal{A}^e(x_1) K_+(x_1 - x') d^4x_1$$
$$+ (ie)^2 \int d^4x_1 \int d^4x_2 K_+(x - x_1) \mathcal{A}^e(x_1) K_+(x_1 - x_2) \mathcal{A}^e(x_2) K_+(x_2 - x')$$
$$+ \cdots \tag{62a}$$
$$= K_+(x - x') + ie \int K_+(x - x_1) \mathcal{A}^e(x_1) K_+^A(x_1, x') d^4x_1 \tag{62b}$$

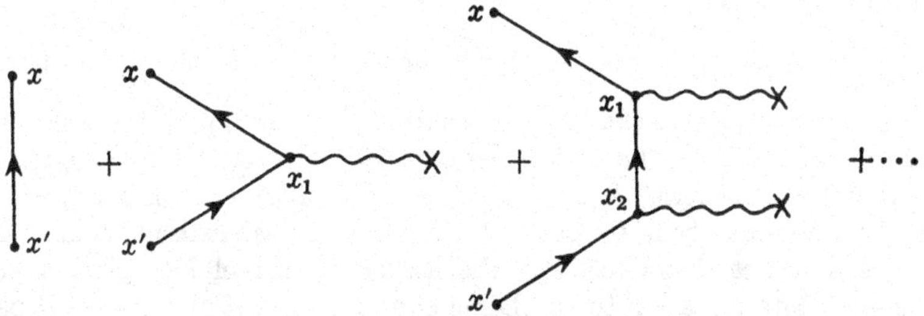

Fig. 14.11

In arriving at Eq. (62b) we noted that (62a) is the Neumann-Liouville expansion of Eq. (62b).

More generally, we infer from the above that if the initial one-electron state is given by

$$|\Phi_i\rangle = \int_{\sigma(x') = t'} \bar{\psi}(x') \gamma_\mu f(x') d\sigma^\mu(x') \mid \Phi_0\rangle \tag{63}$$

and the final state by

$$|\Phi_f\rangle = \int_{\sigma(x) = t} \bar{\psi}(x) \gamma_\mu g(x) d\sigma^\mu(x) \mid \Phi_0\rangle \tag{64}$$

then the transition amplitude is given by

$$R_{fi} = \int_{\sigma'} d\sigma^\mu(x) \int_\sigma d\sigma^\nu(x') \tilde{g}(x) \gamma_\mu K_+^A(x, x') \gamma_\nu f(x') \tag{65}$$

The kernel K_+^A can thus be considered as propagating the initial amplitude f to the surface σ'

$$f_\sigma(x) = \int_{\sigma'} d\sigma^\mu(x') K_+^A(x, x') \gamma_\mu f(x') \tag{66}$$

and the transition amplitude given by the scalar product

$$(g_\sigma, f_\sigma) = \int_\sigma d\sigma_\mu(x) \tilde{g}_\sigma(x) \gamma^\mu f_\sigma(x) \tag{67}$$

where $g_\sigma = g$. Feynman (1949a) was led to precisely this expression from a consideration of the Green's function $G^A(x, x')$ which connects solutions of the Dirac equation in the external field A^e at two different times. An argument based on hole theoretic considerations allowed him to infer that the correct Green's function must be such as to contain only positive frequencies for $t > t'$, which then yields $G^A = K_+{}^A$. By a Green's function for the Dirac equation in the external field A^e is meant a solution of

$$[i\gamma_\mu(\partial^\mu - ieA^{e\mu}(x)) - m]\, G^A(x, x') = i\delta^{(4)}(x - x') \qquad (68)$$

If ϕ is a solution of

$$[i\gamma_\mu(\partial^\mu - ieA^{e\mu}(x)) - m]\, \phi(x) = 0 \qquad (69)$$

then Gauss's theorem asserts that if x is a point inside the volume Ω which is bounded by the surface σ,

$$\phi(x) = \int_\sigma G^A(x, x')\, \gamma_\mu\phi(x')\, n^\mu(x')\, d\sigma(x') \qquad (70)$$

One readily verifies that $K_+{}^A$, as characterized by Eq. (62), satisfies the differential equation (68). Furthermore by virtue of the inhomogeneous term K_+ and the kernel K_+A^e appearing in (62), $K_+{}^A$ contains only positive frequencies for $x_0 > x_0'$ in the case where A^e is a weak external field. The above indicates that effectively the use of the kernel $K_+{}^A$ gives a hole theoretic interpretation to the Dirac equation, and allows one to bypass the field theoretic formalism in problems involving positons and negatons interacting with external fields.

The theory of negatons and positons interacting with an external field presented thus far was based on the perturbation expansion of the S matrix given by (1). However, for sufficiently weak time-independent external fields, an exact formal solution can be given [see, for example, Schwinger (1953b, 1954a)]. Let ϕ_n be the c-number solutions of the Dirac equation in the external field

$$[i\gamma_\mu(\partial^\mu - ieA^{e\mu}(x)) - m]\, \phi_n(x) = 0 \qquad (71)$$

where n specifies a complete set of one-particle commuting observables. We shall always assume in the following that the time-independent field is sufficiently weak so that a gap exists between positive and negative energy eigenvalues. If this is not the case, a stable vacuum will not exist.

The Heisenberg operator $\psi^e(x)$ which satisfies the (operator) equation of motion

$$[i\gamma_\mu(\partial^\mu - ieA^{e\mu}(x)) - m]\, \psi^e(x) = 0 \qquad (72)$$

can be expanded in terms of the $\phi_n(x)$s as follows:

$$\psi^e(x) = \sum_{(n', E+)} b^e{}_n\phi_n(x) + \sum_{(n', E-)} d^e{}_n{}^*\phi_n(x) \qquad (73)$$

where the first summation was only over the positive energy solutions and the second over only the negative energy solutions. From the canonical commutation rules:

$$[\psi^e(x), \tilde{\psi}^e(x')]_+\Big|_{x_0=x'_0} = \gamma^0\delta^{(3)}(\mathbf{x} - \mathbf{x}') \tag{74}$$

and the orthogonality properties of the solutions $\phi_n(x)$, one verifies that the operators $b^e{}_n$, $d^e{}_n \cdots$ satisfy anticommutation rules analogous to the free-field case, e.g.,

$$[b^e{}_n, b^e{}_{n'}{}^*]_+ = \delta_{nn'}$$

$$[b^e{}_n, d^e{}_{n'}]_+ = [b^e{}_n, d^e{}_{n'}{}^*]_+ = 0 \quad \text{etc.} \cdots \tag{75}$$

The current operator is given by

$$j^e{}_\mu(x) = -\frac{e}{2}\,[\tilde{\psi}^e(x)\,\gamma_\mu, \psi^e(x)] \tag{76}$$

and is divergence-free by virtue of the equations of motion. From the commutation rules of the total charge operator

$$Q = -\frac{e}{2}\int d^3x\,[\tilde{\psi}^e(x)\,\gamma^0, \psi^e(x)] \tag{77}$$

with the operators ψ^e and $\tilde{\psi}^e$, one obtains the interpretation of ψ^e as a destruction operator for an amount of charge $-e$ and a creation operator for a charge $+e$. The Hamiltonian operator, as well as the charge operator, are diagonal in the $b^e{}_n$ and $d^e{}_n$ operators. This allows the interpretation of the operators $b^e{}_n$ and $d^e{}_n$ as destruction operators for a negaton and positon in the state n, respectively. The vacuum state $|\Phi^e{}_0\rangle$ is defined as the state of lowest energy and satisfies

$$d^e{}_n \mid \Phi^e{}_0\rangle = b^e{}_n \mid \Phi^e{}_0\rangle = 0 \quad \text{for all } n \tag{78}$$

If $|\Phi^e{}_1\rangle$ is a one-electron state, then its only nonvanishing amplitude is

$$\chi(x) = \langle\Phi^e{}_0 \mid \psi^e(x) \mid \Phi^e{}_1\rangle \tag{79}$$

This amplitude satisfies the Dirac equation (71) in the presence of the external field since $\psi^e(x)$ does. Furthermore, using the expansion (73) and the fact that $\langle\Phi^e{}_0 \mid d^e{}_n{}^* = 0$, one notes that the amplitude χ is a superposition of positive energy solutions

$$\chi(x) = \sum_{n', E+} \langle\Phi^e{}_0 \mid b^e{}_n \mid \Phi^e{}_1\rangle\,\phi_{n', E+}(x) \tag{80}$$

The propagation kernel $K_+{}^A$ takes on its simplest form in terms of the Heisenberg operators ψ^e. In terms of these, it is given by

$$K_+{}^A(x, x') = (\Phi^e{}_0, T(\psi^e(x)\,\tilde{\psi}^e(x'))\,\Phi^e{}_0) \tag{81a}$$

$$= \begin{cases} \displaystyle\sum_{n', E+} \phi_n(x)\,\tilde{\phi}_n(x') & \text{for } x_0 > x'_0 \\ \displaystyle -\sum_{n', E-} \phi_n(x)\,\tilde{\phi}_n(x') & \text{for } x_0 < x'_0 \end{cases} \tag{81b}$$

14b. Feynman Diagrams for Interacting Fields

Consider next the theory of a neutral pseudoscalar field interacting with a spin $\frac{1}{2}$ field through a pseudoscalar coupling so that the interaction Lagrangian density is given by

$$\mathfrak{L}_I = -\tfrac{1}{2}G[\tilde{\psi}(x)\,\gamma_5,\,\psi(x)]\,\phi(x) \tag{82a}$$

The Hamiltonian density in the interaction picture can be written as

$$\mathcal{H}_I(x) = GN(\tilde{\psi}(x)\,\gamma_5\psi(x))\,\phi(x) \tag{82b}$$

In this case, the first-order matrix elements of the S matrix vanish, since energy and momentum cannot be conserved in the act of emission or absorption of a free meson by a free nucleon. Let us therefore consider the second-order terms,

$$S^{(2)} = \left(-\frac{iG}{\hbar c}\right)^2 \frac{1}{2!} \int d^4x_1 \int d^4x_2\, T\{N(\tilde{\psi}(x_1)\,\gamma_5\psi(x_1))\,N(\tilde{\psi}(x_2)\,\gamma_5\psi(x_2))\} \cdot T(\phi(x_1)\,\phi(x_2)) \tag{83}$$

We have written separate T products for the nucleon and meson factors, which is permissible since they commute with one another. We have previously shown that the contraction $\tilde{\psi}^{\cdot}(x)\,\gamma_5\psi^{\cdot}(x))$ vanishes. Hence, the expansion into normal products of T (fermion factors) is, in fact, identical to (2). We have therefore only to consider the meson factors. For these, using Wick's theorem, we obtain

$$T(\phi(x_1)\,\phi(x_2)) = N(\phi(x_1)\,\phi(x_2)) + \tfrac{1}{2}\hbar c\Delta_F(x_1 - x_2) \tag{84}$$

If we now represent the contraction of two boson factors, $\phi^{\cdot}(x_1)\,\phi^{\cdot}(x_2)$, by a dashed line joining x_1 to x_2, a meson creation operator $\phi^{(-)}(x)$, by a dashed line moving upward and away from x_1, and a meson destruction operator, $\phi^{(+)}(x)$, by a dashed line moving toward x_1, then the diagrams of Figures 14.12 through 14.19 represent the second-order matrix elements of S. Let us consider the diagrams and the terms from which they arise individually.

1. $N(\tilde{\psi}\psi\tilde{\psi}\psi)\,N(\phi\phi)$: These terms are equal to $(S^{(1)})^2$ and therefore, since energy and momentum cannot be conserved, they do not contribute to $S^{(2)}$.

2. $N(\tilde{\psi}\psi\tilde{\psi}\psi)\,\phi^{\cdot}\phi^{\cdot}$: These terms correspond to the exchange of a boson between the two fermions and hence are the analogue of Møller scattering for nucleons. For identical particle interaction, the two diagrams of Figure 14.12 occur. There are in addition to the diagrams of Figure 14.12 two more with the points x_1 and x_2 interchanged. This is true for every diagram in this section except for the diagram in Figure 14.18, which corresponds to a vacuum fluctuation. The matrix elements corresponding to these topologically identical diagrams, when added together, just cancel the factor 1/2! in the perturbation expansion. The exchange scattering

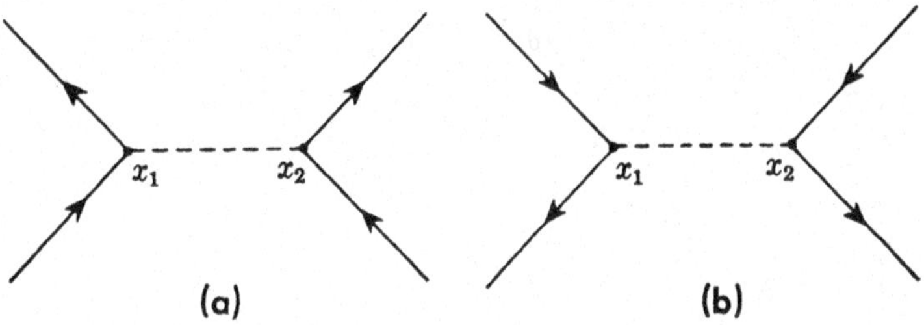

Fig. 14.12

will automatically be included when one evaluates the matrix element between initial and final states.

To nucleon-antinucleon scattering there correspond in second order the two diagrams given in Figure 14.13. Figure 14.13*b* arises from the possibility of the two particles annihilating each other and emitting a virtual boson, which then creates the particles in the final state.

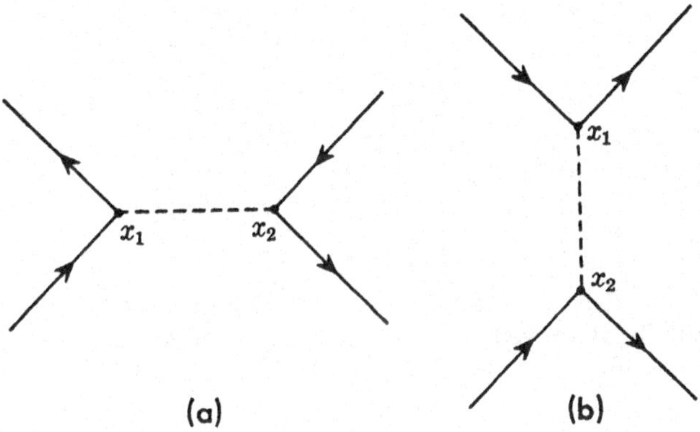

Fig. 14.13

The processes just discussed correspond to the presence of two fermions in the initial and final state.

3. $N(\bar{\psi}K_+\psi)\,N(\phi\phi)$: $S^{(2)}$ likewise contains a variety of processes involving one fermion and one boson initially and finally. The diagrams which correspond to these processes are those of Figure 14.14. The term $N(\bar{\psi}K_+\psi)\,N(\phi\phi)$, in addition, gives rise to the processes indicated in Figure 14.15.

4. $N(\bar{\psi}K_+\psi)\,\phi^{\cdot}\phi^{\cdot}$: The one-fermion, no-meson term $N(\bar{\psi}K_+\psi)\,\phi^{\cdot}\phi^{\cdot}$ gives rise to fermion self-energy diagrams which will be discussed in greater detail later. These are illustrated in Figure 14.16.

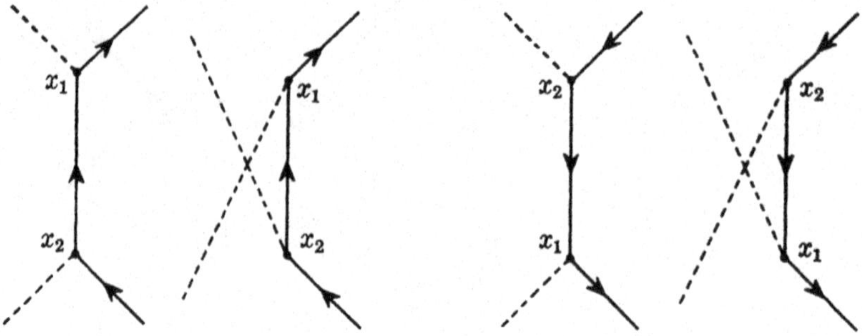

(a)

Scattering of a Meson by a Nucleon

(b)

Scattering of a Meson by an Antinucleon

Fig. 14.14

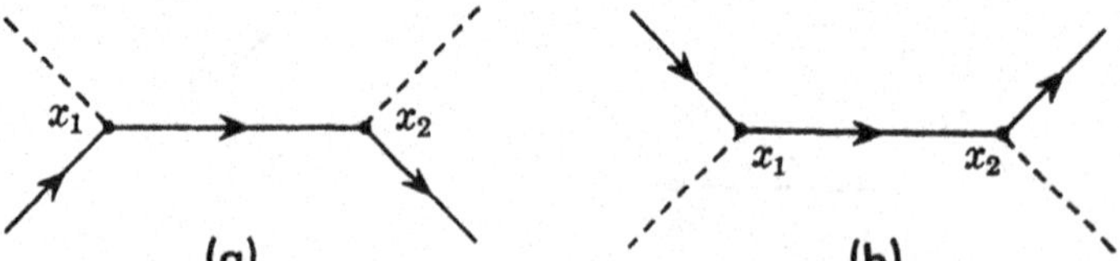

(a)

Nucleon-Antinucleon Annihilation with Emission of 2 Mesons

(b)

Nucleon Pair Production by 2 Interacting Mesons

Fig. 14.15

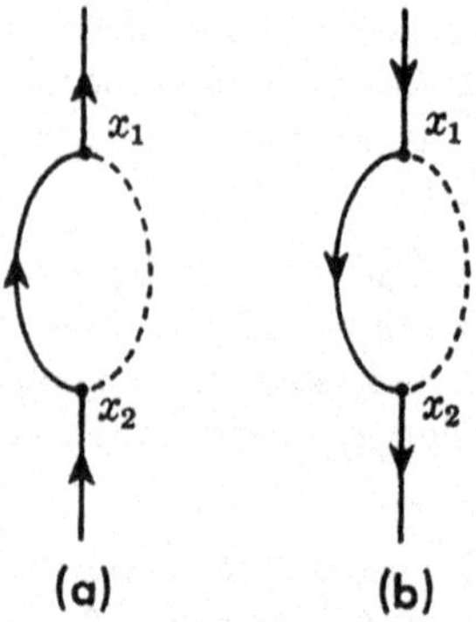

(a) (b)

Fig. 14.16

5. $N(\phi\phi)\,K_+K_+$: Similarly, the no-fermion, one-meson term, $K_+K_+N(\phi\phi)$, gives rise to a meson self-energy illustrated in Figure 14.17.

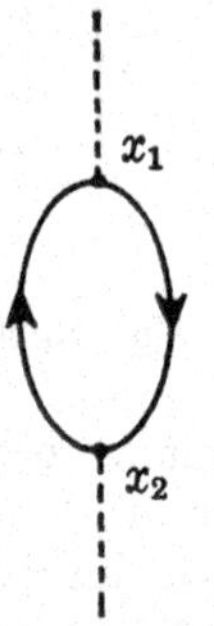

Fig. 14.17

6. Finally, there is the term $K_+K_+\Delta_F$, the Feynman diagram of which is given in Figure 14.18. This diagram corresponds to a vacuum fluctuation.

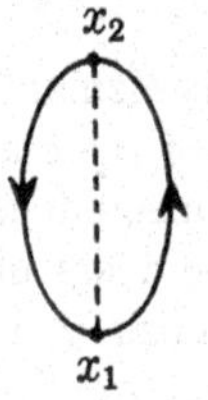

Fig. 14.18

It should, perhaps, be re-emphasized that the time sequence of the intermediate states is here irrelevant. Thus, the diagram of Figure 14.19 is a

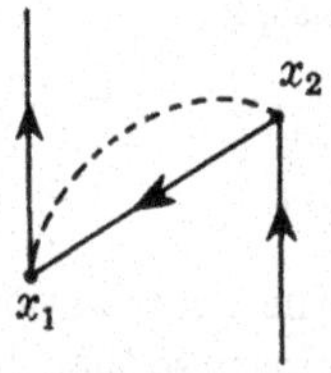

Fig. 14.19

special case of Figure 14.16*a* when $x_{10} < x_{20}$. In other words, we may deform a Feynman graph in any way we wish, as long as the correct arrow direction is kept for the external lines, i.e., for the initial and final states.

Examples of vacuum processes are indicated in Figure 14.20. These vacuum diagrams can once again be omitted from consideration. *Proof*:

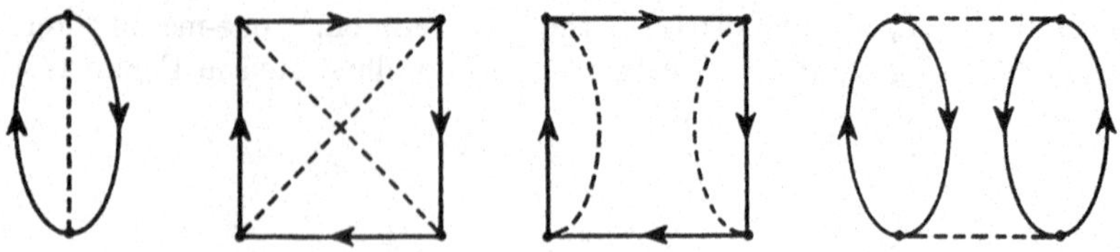

(Γ_1) (Γ_2) (Γ_3) (Γ_4)

Fig. 14.20

The total contribution of all vacuum processes is given by $(\Phi_0 \mid S \mid \Phi_0) = M_V$. The diagrams which contribute to M_V are those which have no external lines. In general, a vacuum diagram is composed of several disconnected pieces. If we denote by $\Gamma_1, \Gamma_2, \Gamma_3 \cdots$ the elementary connected vacuum-vacuum diagram, i.e., those diagrams which can be traced out by following internal lines only, then the most general vacuum fluctuation diagram is made up of n_1 diagrams of type Γ_1, n_2 of type Γ_2, etc., and the contribution to M_V of this diagram, $\Gamma = n_1\Gamma_1 + n_2\Gamma_2 + \cdots$, is given by

$$(\Phi_0 \mid S_\Gamma \mid \Phi_0) = (\Phi_0 \mid S_{\Gamma_1} \mid \Phi_0)^{n_1} (\Phi_0 \mid S_{\Gamma_2} \mid \Phi_0)^{n_2} \cdots \tag{85}$$

where $(\Phi_0 \mid S_{\Gamma_1} \mid \Phi_0)$ is the matrix element corresponding to the diagram Γ_1. The fact that the contribution of the diagrams $\Gamma_1, \cdots \Gamma_n$ is multiplicative is the reflection of their disconnectedness. If a diagram is composed of several identical pieces, the members of the family are not all distinct, since a permutation of their labels will give rise to the same diagrams; hence, the total vacuum-vacuum amplitude is given by

$$\begin{aligned} M_V &= \langle \Phi_0 \mid S \mid \Phi_0 \rangle \\ &= \sum_{n_1, n_2} \frac{1}{n_1! n_2! \cdots} \langle \Phi_0 \mid S_{\Gamma_1} \mid \Phi_0 \rangle^{n_1} \langle \Phi_0 \mid S_{\Gamma_2} \mid \Phi_0 \rangle^{n_2} \cdots \\ &= e^{\langle \Phi_0 | S_{\Gamma_1} | \Phi_0 \rangle} e^{\langle \Phi_0 | S_{\Gamma_2} | \Phi_0 \rangle} \cdots \\ &= e^{\langle \Phi_0 | S_{\Gamma_1} + S_{\Gamma_2} + \cdots | \Phi_0 \rangle} \\ &= e^{\langle \Phi_0 | S | \Phi_0 \rangle_C} = e^{-L} \end{aligned} \tag{86}$$

where $\langle \Phi_0 \mid S \mid \Phi_0 \rangle_C$ denotes the contribution of all connected diagrams. The quantity L is purely imaginary so that the probability that the vacuum remain a vacuum is one.

If we focus our attention on any real scattering process, for example, the process indicated in Figure 14.12 with matrix element M, then in higher orders this same diagram will appear together with unconnected vacuum processes. Summing over all vacuum processes, then, merely multiplies M by $\exp(-L)$, a phase factor of absolute magnitude one. We may therefore omit all disconnected graphs from consideration.

A procedure similar to the one outlined above for the second-order matrix elements can be used to establish the one-to-one correspondence between the nth-order elements of the S matrix in normal form and Feynman diagrams with n vertices, in which the space-time points $x_1, x_2, \cdots x_n$ are identified. Since the decomposition of a T product includes all possible sets of contractions, there exists, likewise, a one-to-one correspondence between all possible graphs with a given number, n, of vertices (with points identified) and the normal decomposition of S_n.

This latter correspondence is the basis of the usefulness of the concept of Feynman diagrams. Thus, in practice, one draws all possible topologically different[2] graphs consistent with the interaction energy term. If the latter is of the form $\bar{\psi}\Gamma\psi\phi$, then two fermion lines and one boson line must meet at every vertex (corner). The matrix element corresponding to any nth-order diagram can then be obtained by writing down the following factors:

1. A factor $(-i/\hbar c)^n$ for the diagram as a whole from the perturbation expansion.

2. A factor $(G\Gamma)_{\alpha\beta}$ for each vertex.

3. A factor $\frac{1}{2}\hbar c\Delta_F(x_j - x_l)$ for an internal boson line connecting the points x_j and x_l.

4. A factor $[-\frac{1}{2}S_F(x_l - x_j)]_{\alpha\beta} = [K_+(x_l - x_j)]_{\alpha\beta}$ for an internal fermion line directed from x_j to x_l. The factor $(G\Gamma)_{\sigma\alpha}$ is assumed to occur at the vertex labeled by x_l and the factor $(G\Gamma)_{\beta\delta}$ at the vertex labeled by x_j.

5. The correct creation or annihilation operator $\psi^{(\pm)}(x)$, $\bar{\psi}^{(\pm)}(x)$, $\phi^{(\pm)}(x)$ for each external free fermion or boson line leaving or arriving at x.

6. A factor (-1) for each internal closed fermion loop.

7. Integrate over $x_1, x_2, \cdots x_n$.

If the quantized Bose field is the electromagnetic field (henceforth represented by wavy lines), then rule 2 above becomes replaced by

2. A factor $(e\gamma^\mu)_{\alpha\beta}$ for each vertex, and rule 3 by

3. A factor $-\frac{1}{2}\hbar c D_F(x_j - x_l)\, g_{\mu\nu}$ for an internal photon line connecting the points x_j and x_l. It is assumed that the factors $(e\gamma^\mu)$ and $(e\gamma^\nu)$ act at these vertices. The summation convention is here used with respect to the polarization indices μ and ν.

It should be emphasized that the representation

$$(\Phi_0, T(A_\mu(x)\, A_\nu(x'))\, \Phi_0) = -\tfrac{1}{2}\hbar c D_F(x - x')\, g_{\mu\nu} \qquad (87a)$$

corresponds to a particular choice of gauge for the electromagnetic po-

[2] That is, a graph which differs from another merely by an interchange of the labels of the vertices is not considered different.

tentials. By a change of gauge it is possible to change the right-hand side of Eq. (87a) so that it reads

$$(\Phi'_0, T(A_\mu(x)\, A_\nu(x'))\, \Phi'_0) = -\tfrac{1}{2}(g_{\mu\nu} + \lambda \partial_\mu \partial_\nu \Box^{-2})\, D_F(x - x') \quad (87b)$$

where λ is an arbitrary constant. The particular form with $\lambda = -1$

$$(\Phi'_0, T(A_\mu(x)\, A_\nu(x'))\, \Phi_0) = -\tfrac{1}{2}(g_{\mu\nu} - \partial_\mu \partial_\nu \Box^{-2})\, D_F(x - x') \quad (87c)$$

is often convenient, since with this choice of gauge the right-hand side is transverse, i.e., $\partial^\mu(\Phi'_0, T(A_\mu(x)\, A_\nu(x'))\, \Phi'_0) = 0$. We shall call the gauge in which the photon propagator is given by the right-hand side of (87c) the transverse or Landau gauge.

Similarly, in rule 4, $G\gamma$ is replaced by $e\gamma^\mu$; and in rule 5, ϕ is replaced by A_μ, where the index μ is the same as that of the γ operating at that vertex. Strictly speaking, when dealing with the electromagnetic field, we must take into account the subsidiary condition. However, if we restrict ourselves to a consideration of scattering problems wherein the initial and final states are "bare" states specified at time $t = \pm\infty$, then the discussion and formalism presented in Chapter 8 can be taken over *in toto*, since the evolution of the system is described in terms of such "bare" states. Therefore, for scattering problems we need not worry about any modification of the subsidiary condition [see, in this connection, Coester and Jauch (Coester 1950)].

Note that the factor $1/n!$ is not included since we are dealing only with topologically different graphs. Clearly, there are $n!$ permutations of the points $x_1, \cdots x_n$ among themselves, which leave the graph topologically unchanged, i.e., they remain the same except for the labeling of the points. This last remark does not apply to vacuum fluctuation graphs, but as noted above vacuum diagrams need not be considered.

In the above consideration, we have not made use of the fact that in the absence of external fields the total energy and momentum of the field system is a constant of the collision. We should, therefore, expect considerable simplification to ensue if we go to momentum space.

14c. Momentum Space Considerations

To illustrate the methods involved in momentum space, we consider the scattering of two fermions. In configuration space, the operator corresponding to Figure 14.21 can easily be written down according to the rules stated in the previous section. It is given by

$$S^{(2)}_{NN} = \left(-\frac{i}{\hbar c}\right)^2 G^2 \int d^4x_1 \int d^4x_2$$

$$N(\widetilde{\psi^{(+)}}(x_1)\, \Gamma\psi^{(+)}(x_1)\, \frac{\hbar c}{2}\, \Delta_F(x_1 - x_2)\, \widetilde{\psi^{(+)}}(x_2)\, \Gamma\psi^{(+)}(x_1)) \quad (88)$$

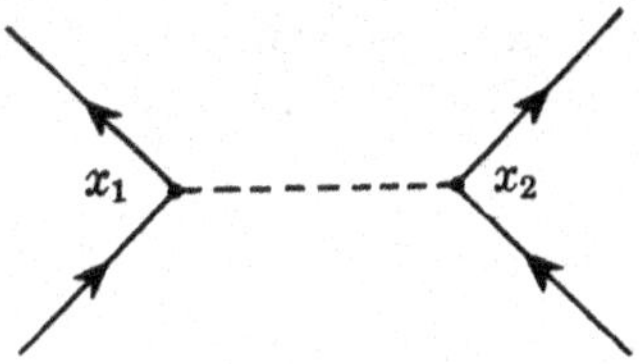

Fig. 14.21

Again note that the factor 1/2! does not occur in (88), since there are 2! diagrams of the type indicated, namely, Figure 14.21, and the same diagram with x_1 and x_2 interchanged. To obtain the probability amplitude for this process, we must next take the matrix element of $S_{NN}^{(2)}$ between the initial and final states, $|\Phi_i\rangle$ and $|\Phi_f\rangle$, respectively. Let these states be

$$|\Phi_i\rangle = b_{s_1}{}^*(\mathbf{p}_1)\, b_{r_1}{}^*(\mathbf{q}_1) \mid \Phi_0\rangle \tag{89a}$$

$$|\Phi_f\rangle = b_{s_2}{}^*(\mathbf{p}_2)\, b_{r_2}{}^*(\mathbf{q}_2) \mid \Phi_0\rangle \tag{89b}$$

where $\mathbf{p}_1 s_1$, $\mathbf{q}_1 r_1$ and $\mathbf{p}_2 s_2$, $\mathbf{q}_2 r_2$ are the momenta and spins of the nucleons before and after the collision, respectively. An analysis similar to Eqs. 7–12 then indicates that we may look upon the factor $\widetilde{\psi^{(+)}}(x_1)\, \Gamma\psi^{(+)}(x_1)$ as destroying the particle labeled by the variables $\mathbf{p}_1 s_1$ with an amplitude $w^{s_1}(\mathbf{p}_1)$ and creating particle $\mathbf{p}_2 s_2$ with an amplitude $\widetilde{w}^{s_2}(\mathbf{p}_2)$. The operator $\widetilde{\psi^{(+)}}(x_2)\, \Gamma\psi^{(+)}(x_2)$ does the same for the particle labeled with the $\mathbf{q}r$ variables. In addition, $S^{(2)}$ also gives rise to an exchange scattering matrix element in which the wave functions of the two particles in the final states are interchanged. The exact structure of both these terms is obtained by evaluating the matrix element $(\Phi_f, N(\widetilde{\psi^{(+)}}\Gamma\psi^{(+)}(x_1)\, \widetilde{\psi^{(+)}}\Gamma\psi^{(+)}(x_2))\, \Phi_i)$.

Alternatively, we can immediately replace in expression (88) the operators $\widetilde{\psi^{(+)}}(x)$ and $\psi^{(+)}(x)$ by the corresponding amplitudes of the free particles they create or destroy:

$$\widetilde{\psi^{(+)}}(x_1) \to \frac{1}{(2\pi)^{3/2}}\, \widetilde{w}^{s_2}(\mathbf{p}_2)\, e^{ip_2\cdot x_1} \left(\frac{M}{E(\mathbf{p}_2)}\right)^{1/2} \tag{90a}$$

$$\psi^{(+)}(x_1) \to \frac{1}{(2\pi)^{3/2}}\, w^{s_1}(\mathbf{p}_1)\, e^{-ip_1\cdot x_1} \left(\frac{M}{E(\mathbf{p}_1)}\right)^{1/2} \tag{90b}$$

with $p_1{}^2 = p_2{}^2 = M^2$ and $p_{0i} = E(\mathbf{p}_i) = (\mathbf{p}_i{}^2 + M^2)^{1/2}$.

The matrix element for the nonexchange scattering, including factors $\hbar$ and c, then becomes

$$R = \left(-\frac{iG}{\hbar c}\right)^2 \hbar c\, \frac{i}{(2\pi)^4}\, \frac{1}{(2\pi)^6} \int d^4x_1 \int d^4x_2 \int d^4k$$

$$\left(\frac{M^4}{E(\mathbf{p}_1)\, E(\mathbf{p}_2)\, E(\mathbf{q}_1)\, E(\mathbf{q}_2)}\right)^{1/2} e^{i(p_2-p_1)\cdot x_1}\, e^{i(q_2-q_1)\cdot x_2}\, e^{-ik\cdot(x_2-x_1)}$$

$$\cdot\, \frac{1}{k^2 - \mu^2 + i\epsilon}\, [\widetilde{w}^{s_2}(\mathbf{p}_2)\, \Gamma w^{s_1}(\mathbf{p}_1)]\, [\widetilde{w}^{r_2}(\mathbf{q}_2)\, \Gamma w^{r_1}(\mathbf{q}_1)] \tag{91}$$

The matrix element for exchange scattering, R_{ex}, is obtained by interchanging in $-R$ the wave function of the two particles in the final state, and is given by

$$R_{ex} = -\left(-\frac{iG}{\hbar c}\right)^2 \hbar c \frac{i}{(2\pi)^4}\frac{1}{(2\pi)^6} \int d^4x_1 \int d^4x_2 \int d^4k$$
$$\left(\frac{M^4}{E(\mathbf{p}_1)\,E(\mathbf{p}_2)\,E(\mathbf{q}_1)\,E(\mathbf{q}_2)}\right)^{1/2} e^{i(q_2-p_1)\cdot x_1}\, e^{i(p_2-q_1)\cdot x_2}\, e^{-ik\cdot(x_2-x_1)}$$
$$\cdot \frac{1}{k^2-\mu^2+i\epsilon}\,[\tilde{w}^{r_2}(\mathbf{q}_2)\,\Gamma w^{s_1}(\mathbf{p}_1)]\,[\tilde{w}^{s_2}(\mathbf{p}_2)\,\Gamma w^{r_1}(\mathbf{q}_1)] \tag{92}$$

Carrying out the indicated integrations over x_1 and x_2, we obtain for R the following expression:

$$R = -\frac{i}{\pi}\frac{G^2}{4\pi\hbar c}\int d^4k\, \delta^{(4)}(p_2-p_1-k)\,\delta^{(4)}(q_2-q_1-k)$$
$$[\tilde{w}^{s_2}(\mathbf{p}_2)\,\Gamma w^{s_1}(\mathbf{p}_1)]\,[\tilde{w}^{r_2}(\mathbf{q}_2)\,\Gamma w^{r_1}(\mathbf{q}_1)]$$
$$\left(\frac{M^4}{E(\mathbf{p}_1)\,E(\mathbf{p}_2)\,E(\mathbf{q}_1)\,E(\mathbf{q}_2)}\right)^{1/2}\frac{1}{k^2-\mu^2+i\epsilon} \tag{93}$$

We have purposely left our matrix element in the form (93), in which it can readily be interpreted. Thus we may imagine (see Fig. 14.22) that an

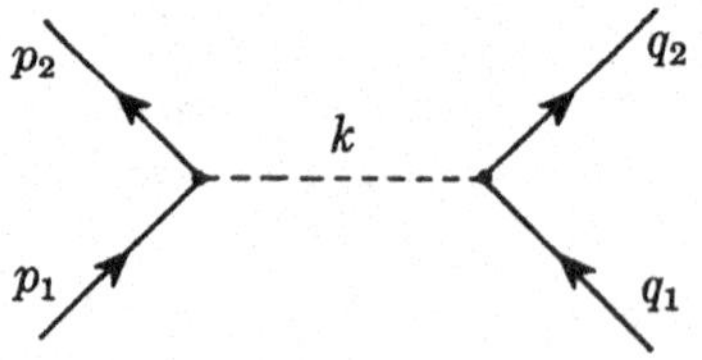

Fig. 14.22

incoming nucleon of momentum $\mathbf{p}_1$ and spin s_1 [wave function $w^{s_1}(\mathbf{p}_1)$] emits (factor $G\Gamma$) a Bose quantum of momentum k and moves on to its final state, with a momentum $\mathbf{p}_2$ and spin s_2 [wave function $w^{s_2}(\mathbf{p}_2)$].

The delta function $(2\pi)^4\,\delta^{(4)}(p_2-p_1+k)$, arising from the integration over x_1, now states that energy and momentum are conserved during the act of emission. The nucleon thus moves on with momentum $p_2 = p_1 - k$. Similarly, the second nucleon with momentum q_1 [wave function $w^{r_1}(\mathbf{q}_1)$], comes along and absorbs (factor $G\Gamma$) the quantum which was emitted by nucleon 1 and continues on to its final state [wave function $w^{r_2}(\mathbf{q}_2)$]. As indicated by the delta function, $(2\pi)^4\,\delta^{(4)}(q_2-q_1-k)$, energy and momentum are again conserved in the process.[3] The propa-

[3] In the old-fashioned perturbation theory, energy was not conserved in intermediate states, although the momentum and the mass were. In the Feynman-Dyson perturbation expansion, energy and momentum are conserved, but the mass (which is an invariant quantity) is not. This is the origin of the invariant character of the Feynman perturbation theory. See Umezawa and Kawabe [Umezawa (1949a, 1950)].

gation function for the Bose quantum is $\frac{i\hbar c}{(2\pi)^4}\frac{1}{k^2-\mu^2+i\epsilon}$ and one has to integrate over all virtual quanta, i.e., internal lines. The quantities $\left(\frac{1}{(2\pi)^{3/2}}\right)^4\left(\frac{M^4}{E(\mathbf{p}_1)\,E(\mathbf{p}_2)\,E(\mathbf{q}_1)\,E(\mathbf{q}_2)}\right)^{1/2}$ are the normalization factors for the nucleon wave functions. Finally, the expression is to be multiplied by $(-i/\hbar c)^2$ arising from the perturbation expansion.

In the present example we can integrate over k, thus obtaining the final form of our matrix element:

$$R = -\frac{i}{\pi}\frac{G^2}{4\pi\hbar c}\,\delta^{(4)}(p_2+q_2-p_1-q_1)\left(\frac{M^4}{E(\mathbf{p}_1)\,E(\mathbf{p}_2)\,E(\mathbf{q}_1)\,E(\mathbf{q}_2)}\right)^{1/2}$$
$$\frac{[\tilde{w}^{s_2}(\mathbf{p}_2)\,\Gamma w^{s_1}(\mathbf{p}_1)]\,[\tilde{w}^{r_2}(\mathbf{q}_2)\,\Gamma w^{r_1}(\mathbf{q}_1)]}{(q_2-q_1)^2-\mu^2} \tag{94}$$

The $\delta^{(4)}$-function factor here again expresses the overall conservation of energy and momentum in the process.

As a second illustrative example, consider the second-order self-energy of a nucleon corresponding to the Feynman diagram of Figure 14.23. The operator described by this diagram is given by

$$S_{\text{SE}}{}^{(2)} = -\frac{1}{2}\left(-\frac{iG}{\hbar c}\right)^2\int d^4x_1\int d^4x_2$$
$$N[\tilde{\psi}(x_1)\,\Gamma S_F(x_1-x_2)\,\Gamma\psi(x_2)]\,\frac{\hbar c}{2}\,\Delta_F(x_1-x_2) \tag{95}$$

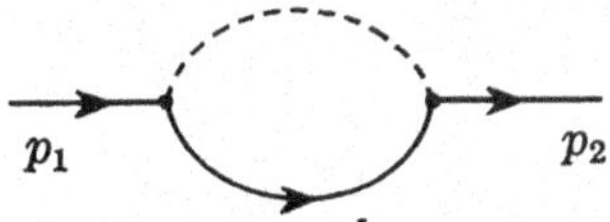

Fig. 14.23

The matrix element of $S_{\text{SE}}{}^{(2)}$ between one-nucleon states of momentum p_1 and p_2 is given by

$$R' = \left(-\frac{iG}{\hbar c}\right)^2\left(\frac{i}{(2\pi)^4}\right)^2\frac{\hbar c}{(2\pi)^3}\int d^4x_1\int d^4x_2\int d^4q\int d^4k\left(\frac{M^2}{E(\mathbf{p}_1)\,E(\mathbf{p}_2)}\right)^{1/2}$$
$$e^{ip_2\cdot x_1-iq\cdot(x_1-x_2)-ik\cdot(x_1-x_2)-ip_1\cdot x_2}$$
$$\tilde{w}^s(\mathbf{p}_2)\,\Gamma\frac{\gamma\cdot q+M}{q^2-M^2+i\epsilon'}\,\Gamma w^r(\mathbf{p}_1)\,\frac{1}{k^2-\mu^2+i\epsilon}$$
$$= \frac{1}{2\pi^2}\left(\frac{G^2}{4\pi\hbar c}\right)\int d^4k\int d^4q\left(\frac{m^2}{E(\mathbf{p}_1)\,E(\mathbf{p}_2)}\right)^{1/2}\delta^{(4)}(p_2-q-k)$$
$$\delta^{(4)}(p_1-k-q)\,\tilde{w}^s(\mathbf{p}_2)\,\Gamma\frac{\gamma\cdot q+M}{q^2-M^2+i\epsilon'}\,\Gamma w^r(\mathbf{p}_1)\,\frac{1}{k^2-\mu^2+i\epsilon} \tag{96}$$

This matrix element can again readily be interpreted in terms of a momentum space Feynman diagram (Fig. 14.23). A nucleon of momentum $\mathbf{p}_1$, spin r [factor $w^r(\mathbf{p}_1)$] emits a boson (factor $G\Gamma$) of momentum k. The delta function $\delta^{(4)}(p_1 - k - q)$ indicates that energy-momentum is conserved in the process, so that after the emission the nucleon travels on with a momentum $q = p_1 - k$. Its propagation function (the inverse Dirac operator) is $(\gamma \cdot q - M + i\epsilon)^{-1}$. The virtual meson of momentum k, which has been propagated by the function $(k^2 - \mu^2)^{-1}$ (inverse Klein-Gordon operator), is then reabsorbed by the nucleon (factor $G\Gamma$), energy and momentum again being conserved $[\delta^{(4)}(p_2 - q - k)]$. One is to integrate over all virtual quanta (internal lines), that is, over q and k. Carrying out the integration over q then yields

$$R' = \frac{1}{(2\pi)^2}\,\delta^{(4)}(p_1 - p_2)\left(\frac{M}{E(\mathbf{p}_1)}\right)\left(\frac{G^2}{4\pi\hbar c}\right)\int d^4k$$
$$\tilde{w}^s(\mathbf{p}_2)\,\Gamma\,\frac{\gamma\cdot(p_2 - k) + M}{(p_2 - k)^2 - M^2 + i\epsilon}\,\Gamma w^r(\mathbf{p}_1)\,\frac{1}{k^2 - \mu^2 + i\epsilon'} \qquad (97)$$

The $\delta^{(4)}(p_2 - p_1)$ again corresponds to overall energy-momentum conservation. The energy-momentum of the nucleon must be the same before and after, for otherwise the matrix element vanishes. There exists only a diagonal matrix element for $p_1 s \to p_1 s$. This, the only nonvanishing element, may be rewritten in slightly different form by noting that in the present situation the creation and annihilation operators create and destroy particles of the same energy and momentum, so that in (95)

$$\bar{\psi}(x_1) \to \frac{1}{(2\pi)^{3/2}}\left(\frac{M}{E(\mathbf{p})}\right)^{1/2}\tilde{w}^s(\mathbf{p})\,e^{ip\cdot x_1}$$
$$\psi(x_2) \to \frac{1}{(2\pi)^{3/2}}\left(\frac{M}{E(\mathbf{p})}\right)^{1/2} w^s(\mathbf{p})\,e^{-ip\cdot x_2} \qquad (98)$$

The integration over x_1 and x_2 in (96) can now be replaced by an integration over $x_1 - x_2$ and $x_1 + x_2$, the latter yielding a factor VT, the spatio-temporal volume over which the integral extends. We therefore obtain the following expression for our matrix element:

$$R' = \frac{1}{4\pi^3}\left(\frac{G^2}{4\pi\hbar c}\right)\frac{V}{(2\pi)^3}\,T\left(\frac{M}{E(\mathbf{p})}\right)$$
$$\int d^4k\,\tilde{w}^s(\mathbf{p})\,\Gamma\,\frac{\gamma\cdot(p - k) + M}{(p - k)^2 - M^2 + i\epsilon'}\,\Gamma w^s(\mathbf{p})\,\frac{1}{k^2 - \mu^2 + i\epsilon} \qquad (99)$$

The proportionality of R' to V is due to the fact that the spinor wave functions were normalized in the continuum. Had we normalized these to a volume V, the factor $V/(2\pi)^3$ would not appear, and the matrix element would simply be proportional to T. The significance of this dependence on T will be discussed in our discussion of self-energies in Chapter 15.

As a result of these examples it becomes clear that it is really unnecessary to first write down the matrix element in configuration space. It can be written down immediately in momentum space. Consider, for example, the more complicated Feynman diagram of Figure 14.24. Physically, this

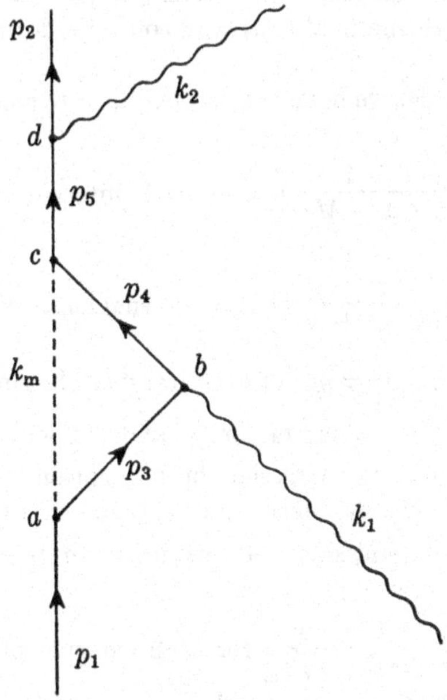

Fig. 14.24

diagram represents a correction to the Compton scattering of a photon by a proton due to the emission and reabsorption of a virtual neutral meson. The fermion (proton), which initially is in the free-particle state $w^{s_1}(\mathbf{p}_1)$, is scattered at a with the emission of a virtual meson, then absorbs the incident light quantum at b, reabsorbs the virtual meson at c, emits the final light quantum at d, and finally continues in the free-particle state $w^{s_2}(\mathbf{p}_2)$. The diagram thus has four external lines which do not end in the diagram. These correspond to the fermion and photon in the initial and final states. The diagram also has four internal lines, namely, ab, bc, cd, and ac, of which the first three are fermion propagation lines and the last one that of a virtual meson. The diagram has four vertices, so that it corresponds to a fourth-order process arising from $S^{(4)}$, which is given by

$$S^{(4)} = e^2 G^2 \left(\frac{-i}{\hbar c}\right)^4 \frac{1}{4!} \int d^4x_1 \int d^4x_2 \int d^4x_3 \int d^4x_4$$
$$T\{N(\bar{\psi}A\!\!\!/\psi(x_1))\, N(\bar{\psi}A\!\!\!/\psi(x_2))\, N(\bar{\psi}\Gamma\psi\phi(x_3))\, N(\bar{\psi}\Gamma\psi\phi(x_4))\} \quad (100)$$

In Figure 14.24 no attention is paid to the direction in time of the internal fermion or boson lines. However, in order to write down the matrix element, we do have to pay attention to the order in which the individual processes occur along the fermion line (which may not necessarily be the sequence in time). It is a matter of convenience from which end (in time) of the diagram one starts. It has, however, become customary to start with the final state and continue down along the fermion lines.

Recall next the rules we have derived by our previous examples. There is

(*a*) a factor $\frac{i}{(2\pi)^4}\frac{1}{\gamma \cdot p - M + i\epsilon}$ for each internal nucleon line of momentum p;

(*b*) a factor $\frac{i\hbar c}{(2\pi)^4}\frac{1}{k^2 - \mu^2 + i\epsilon}$ for each internal meson line of momentum k;

(*c*) a factor $-\frac{i\hbar c}{(2\pi)^4}\frac{1}{k^2 + i\epsilon}g_{\mu\nu}$ for each internal photon line of momentum k joining two vertices where factors γ^μ and γ^ν operate;

(*d*) a factor $+e\gamma^\mu$ for the emission or absorption of a virtual photon (internal photon line) by a fermion of charge e at a vertex;

(*e*) a factor $G\Gamma$ for the emission or absorption of an (internal) virtual meson by a nucleon at a vertex;

(*f*) a factor $\epsilon^{(\lambda)}{}_\mu(\mathbf{k})\frac{(\hbar c)^{1/2}}{(2\pi)^{3/2}}\frac{1}{\sqrt{2|k_0|}}$ for each external photon line which is emitted or absorbed at a vertex (where a factor γ^μ operates). $\epsilon^{(\lambda)}{}_\mu(\mathbf{k})$ is the polarization four-vector ($\epsilon_\mu\epsilon^\mu = -1$) for the emitted or absorbed photon of energy-momentum k and state of polarization λ ($k_0 = |\mathbf{k}|$; note also that $k_\mu\epsilon^{(\lambda)\mu}(\mathbf{k}) = 0$);

(*g*) a factor $\frac{1}{(2\pi)^{3/2}}\frac{(\hbar c)^{1/2}}{\sqrt{2\omega_{\mathbf{k}}}}$ for each external meson of energy $\omega_{\mathbf{k}} = (\mathbf{k}^2 + \mu^2)^{1/2}$ which is emitted or absorbed by a nucleon at a vertex (where a factor $G\Gamma$ operates);

(*h*) a factor $\frac{1}{(2\pi)^{3/2}}\sqrt{\frac{M}{E(\mathbf{p})}}\,\bar{w}^s(\mathbf{p})$ for each external nucleon line of momentum $\mathbf{p}$ and spin s leaving the diagram;

(*i*) a factor $\frac{1}{(2\pi)^{3/2}}\sqrt{\frac{M}{E(\mathbf{p})}}\,w^s(\mathbf{p})$ for each external nucleon line of momentum $\mathbf{p}$ and spin s entering the diagram;

(*j*) a factor $(2\pi)^4\,\delta^{(4)}(p - p' \pm k)$ for each vertex, corresponding to energy and momentum conservation for all the lines joining at that vertex; p and p' are the momenta of the fermion lines, and k that of the internal or external photon or meson line ending at the vertex;

(k) a factor (-1) for each closed nucleon loop;

(l) a factor $(-i/\hbar c)^n$ corresponding to the nth-order term in the perturbation expansion; and finally,

(m) one is to integrate over the momenta of all the internal lines.

As a result of these rules, we may immediately write down the matrix element corresponding to Figure 14.24. It is

$$R = \left(-\frac{i}{\hbar c}\right)^4 \int d^4k_m \int d^4p_3 \int d^4p_4 \int d^4p_5 \,[(2\pi)^4]^4\, \delta^{(4)}(p_1 - p_3 - k_m)$$

$$\delta^{(4)}(p_4 - p_3 - k_1)\, \delta^{(4)}(p_5 - p_4 - k_m)\, \delta^{(4)}(p_2 - p_5 + k_2)\, \frac{1}{(2\pi)^{3/2}} \left(\frac{M}{E(\mathbf{p}_2)}\right)^{1/2}$$

$$\tilde{w}^{s_2}(\mathbf{p}_2)\, \frac{(\hbar c)^{1/2}}{(2\pi)^{3/2}\,(2|\mathbf{k}_2|)^{1/2}}\, e\gamma^\nu \epsilon^{(\lambda_2)}{}_\nu(\mathbf{k}_2)\, \frac{i}{(2\pi)^4}\, \frac{1}{\gamma \cdot p_5 - M + i\epsilon}\, G\Gamma\, \frac{i}{(2\pi)^4}$$

$$\frac{1}{\gamma \cdot p_4 - M + i\epsilon}\, \frac{(\hbar c)^{1/2}}{(2\pi)^{3/2}\,(2|\mathbf{k}_1|)^{1/2}}\, e\gamma^\mu \epsilon^{(\lambda_1)}{}_\mu(\mathbf{k}_1)\, \frac{i}{(2\pi)^4}\, \frac{1}{\gamma \cdot p_3 - M + i\epsilon}$$

$$G\Gamma\, \frac{1}{(2\pi)^{3/2}} \left(\frac{M}{E(\mathbf{p}_1)}\right)^{1/2} w^{s_1}(\mathbf{p}_1)\, \frac{i\hbar c}{(2\pi)^4}\, \frac{1}{k_m{}^2 - \mu^2 + i\epsilon} \tag{101}$$

In writing down this matrix element, we have followed the nucleon line, writing down the appropriate factor for every process we encountered. Note that the emitted photon at d has been given a momentum $-k_2$. This is because the emission operator has the exponential factor $ik_2 \cdot x$, and integration over the appropriate x yields the term $-k_2$ in the δ function. This is in agreement with the intuitive picture that you take away a momentum k_2 from p_5 in order to give the final momentum p_2 to the nucleon. Similarly, the absorbed photon at b has momentum $+k_1$ (the absorption operator has a phase factor exp $-ik_1 \cdot x$). The sign of k_m does not matter as long as we are consistent as to its use in the δ functions. If k_m is assumed to be emitted at a, then its value is $+k_m$ at that corner, so that $p_3 + k_m = p_1$. It will then be absorbed at c, and hence $p_4 + k_m = p_5$.

In our example, it is clear that three of the integrations over internal momenta can be carried out by virtue of the δ-function factors. The remaining δ function then again expresses the conservation of overall energy-momentum involving the external lines.

Let us next consider the case of two nonidentical positively charged fermions interacting through the electromagnetic field as indicated in Figure 14.25. In writing down the matrix element corresponding to this diagram, the procedure is again to follow each fermion line from its final state to its initial one. It does not matter on which particle one starts, since the operators referring to the different particles commute. However, the fact that the two photon lines k_2 and k_3 cross is of importance. It means that we must be careful to put the vertex operators $e\gamma^\mu$ in their proper order, since we must sum over their polarization index. Applying

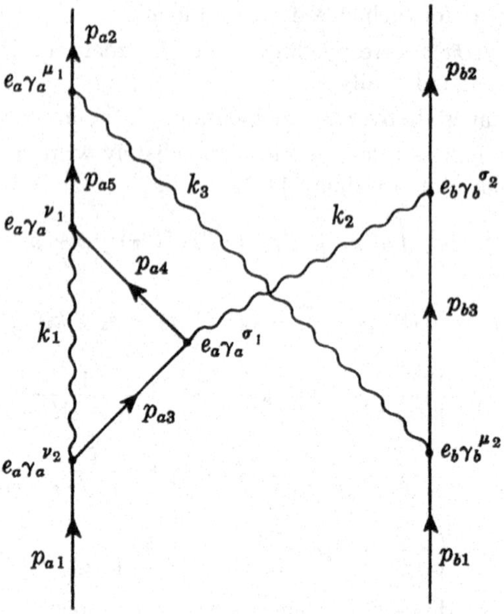

Fig. 14.25

our previous rules, but now differentiating between the operators (e.g., γ matrices) for the two particles by subscripts a and b, we obtain the following matrix element for the process:

$$
\begin{aligned}
R = \left(\frac{-i}{\hbar c}\right)^6 & \int d^4k_1 \int d^4k_2 \int d^4k_3 \int d^4p_{a3} \int d^4p_{a4} \int d^4p_{a5} \int d^4p_{b3} \\
& [(2\pi)^4]^6 \, \delta^{(4)}(k_1 + p_{a3} - p_{a1}) \, \delta^{(4)}(p_{a4} + k_2 - p_{a3}) \, \delta^{(4)}(p_{a5} - p_{a4} - k_1) \\
& \delta^{(4)}(p_{a2} - p_{a5} - k_3) \, \delta^{(4)}(p_{b2} - p_{b3} - k_2) \, \delta^{(4)}(p_{b3} + k_3 - p_{b1}) \\
& \left[\left(\frac{M_a}{(2\pi)^3 E(\mathbf{p}_{a2})}\right)^{1/2} \tilde{w}_a^{s_{a2}}(\mathbf{p}_{a2}) \, e_a\gamma_a^{\mu_1} \frac{i}{(2\pi)^4} \frac{1}{\gamma_a \cdot p_{a5} - M_a + i\epsilon} e_a\gamma_a^{\nu_1} \frac{i}{(2\pi)^4} \right. \\
& \frac{1}{\gamma_a \cdot p_{a4} - M_a + i\epsilon} e_a\gamma_a^{\sigma_1} \frac{i}{(2\pi)^4} \frac{1}{\gamma_a \cdot p_{a3} - M_a + i\epsilon} \left(\frac{M_a}{(2\pi)^3 E(\mathbf{p}_{a1})}\right)^{1/2} \\
& \left. e_a\gamma_a^{\nu_2} w^{s_{a1}}(\mathbf{p}_{a1}) \right] \cdot \left[\left(\frac{M_b}{(2\pi)^3 E(\mathbf{p}_{b2})}\right)^{1/2} \tilde{w}_b^{s_{b2}}(\mathbf{p}_{b2}) \, e_b\gamma_b^{\sigma_2} \frac{i}{(2\pi)^4} \right. \\
& \left. \frac{1}{\gamma_b \cdot p_{b3} - M_b + i\epsilon} e_b\gamma_b^{\mu_2} \cdot \left(\frac{M_b}{(2\pi)^3 E(\mathbf{p}_{b1})}\right)^{1/2} w_b^{s_{b1}}(\mathbf{p}_{b1}) \right] \\
& \cdot \left[\frac{-i\hbar c}{(2\pi)^4}\right]^3 g_{\mu_1\mu_2} \frac{1}{k_3^2 + i\delta} g_{\nu_1\nu_2} \frac{1}{k_1^2 + i\delta} g_{\sigma_1\sigma_2} \frac{1}{k_2^2 + i\delta} \qquad (102)
\end{aligned}
$$

The summation over repeated indices is here implied, and this amounts to summing over the polarization of the virtual photons. The sign we attach to k_1, k_2, k_3 is again not important, but we must be consistent in our treatment of momentum conservation by considering one end of the photon line as emission of the photon and the other as absorption.

The extension of the Feynman techniques to an interaction of the form $G : \bar{\psi}\Gamma\tau\psi \cdot \phi :$ is straightforward. The rules are that to a virtual meson of isotopic spin i emitted or absorbed at a vertex corresponds a factor $G\tau_i\Gamma$. The propagation function for a virtual meson between two vertices at which τ_i and τ_j operate is given by a factor $\delta_{ij}[i(2\pi)^{-4}]\,(k^2 - \mu^2 + i\epsilon)^{-1}$. The sum over i and j is to be performed. The τ factors automatically take into account charge conservation at each vertex. It is also possible in practice to simply keep track of whether a nucleon line is a proton or a neutron line. The charged and neutral mesons then are given a propagation factor $i(2\pi)^{-4}\,(k^2 - \mu^2 + i\epsilon)^{-1}$. However, if this procedure is followed, care must be exercised in giving the correct coupling constant to the pion-nucleon vertices: the emission or absorption of a charged meson by a nucleon has a coupling constant $\sqrt{2}\,G$, whereas the emission or absorption of a neutral meson by a proton has a coupling constant G and the emission or absorption of a neutral meson by a neutron has a coupling constant $-G$.

To obtain the rules for the Feynman diagrams describing charged mesons interacting with photons, we recall that the interaction of a charged boson field with the electromagnetic field is described by an interaction Lagrangian density

$$\mathcal{L}_I(x) = +\,ieN[\phi^*(x) \cdot \partial_\mu\phi(x) - \partial_\mu\phi^*(x) \cdot \phi(x)]\,A^\mu(x) + e^2\,N[A_\mu(x)\,A^\mu(x)\,\phi^*(x)\,\phi(x)] \qquad (103)$$

If, as in previous cases, we transform to the interaction picture and to the covariant Tomonaga-Schwinger equation, it turns out that the integrability condition is not satisfied for the interaction Hamiltonian density obtained by a straightforward application of the canonical formalism. Kanesawa and Tomonaga [Kanesawa (1948); see also Neuman and Furry (Neuman 1949) and Kinoshita (1950a, b)] have shown that the correct covariant generalization of the interaction Hamiltonian density is given by

$$\begin{aligned}\mathcal{H}_I(x;\sigma) &= N(-ie[\phi^*(x) \cdot \partial_\mu\phi(x) - \partial_\mu\phi^*(x) \cdot \phi(x)]\,A^\mu(x) \\ &\quad - e^2\,\phi^*(x)\,\phi(x)\,[A_\mu(x)\,A^\mu(x) - (n_\mu(x)\,A^\mu(x))^2]) \\ &= \mathcal{H}_I'(x) + \mathcal{H}_I''(x;\sigma) \qquad (104)\end{aligned}$$

where n_μ is the normal to the space-like surface at x. Note that this density depends explicitly on the surface through its dependence on $n_\mu(x)$. It reduces to the usual interaction Hamiltonian in the case of a flat surface t = constant, i.e., when $n_\mu = (1, 0, 0, 0)$. The dependence of $\mathcal{H}_I(x;\sigma)$ on the space-like surface is a feature of all derivative couplings.

We could next introduce the S matrix as defined in Chapter 13, using the expression (104) for $\mathcal{H}_I$. It can, however, be shown [Matthews

(1949b)] that when the S matrix is calculated for this theory, the effects of the normal-dependent parts are always exactly canceled by certain singular expressions arising from higher order effects of the parts of $\mathcal{H}_I$ not depending on n_μ. In fact, the correct procedure is to ignore the surface-dependent part entirely, and at the same time use the simple rules previously developed to represent the effects of the interaction. The Hamiltonian to be used is thus given by

$$\mathcal{H}'_I(x) = -ieN[\phi^*(x) \cdot \partial_\mu\phi(x) - \partial_\mu\phi^*(x) \cdot \phi(x)]\, A^\mu(x) \\ - e^2\, N[\phi^*(x)\, \phi(x)\, A_\mu(x)\, A^\mu(x)] \quad (105)$$

and we are to use[4] the following expressions for the contraction of two boson factors involving derivatives:

$$\frac{\partial\phi^{\cdot}(x)}{\partial x_\mu}\, \phi^{*\cdot}(x') = \left(\Phi_0,\, T\left(\frac{\partial\phi(x)}{\partial x_\mu}\, \phi^*(x')\right) \Phi_0\right) \\ \Rightarrow \frac{hc}{2}\, \frac{\partial\Delta_F(x - x')}{\partial x_\mu} \quad (106a)$$

$$\frac{\partial\phi^{\cdot}(x)}{\partial x_\mu}\, \frac{\partial\phi^{*\cdot}(x')}{\partial x'_\nu} = \left(\Phi_0,\, T\left(\frac{\partial\phi(x)}{\partial x_\mu}\, \frac{\partial\phi^*(x')}{\partial x'_\nu}\right) \Phi_0\right) \\ \Rightarrow \frac{hc}{2}\, \frac{\partial^2\Delta_F(x - x')}{\partial x_\mu \partial x'_\nu} \quad (106b)$$

Using these formulae, the graphical methods previously developed can be derived for the present case in a straightforward manner. Thus the term of order e in $\mathcal{H}'_I(x)$ will give rise to the interaction diagrams involving two meson lines and one photon at each vertex (the meson lines should now be directed dashed lines to indicate charge!) as indicated in Figure 14.26*a*.

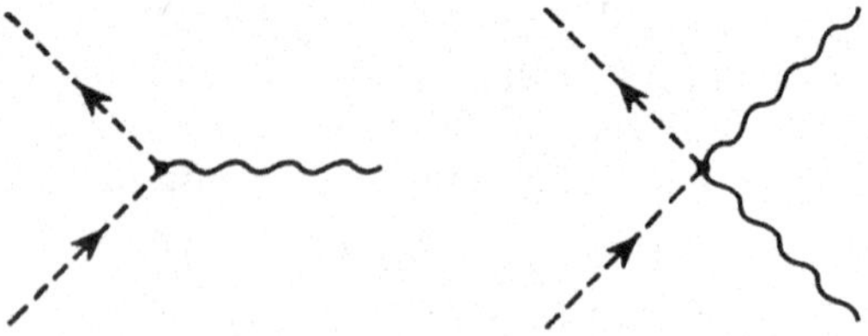

(a) (b)

Fig. 14.26

The interaction term : $\phi^*(x)\, \phi(x)\, A_\mu(x)\, A^\mu(x)$: gives rise to the simultaneous emission or absorption of two photons at a vertex as indicated in Figure 14.26*a*.

[4] These rules omit the singular expressions arising when differentiating singular functions. See Matthews (1949b) and Rohrlich (1950).

The rules for obtaining the matrix element from an nth-order diagram (that is, a diagram containing n corners) can again be derived by an application of Wick's theorem to the S-matrix expansion for the interaction Hamiltonian (105). [See Rohrlich (1950) where such an analysis is carried out in detail.] Note, however, that in the present situation the nth-order contribution to the S matrix will contain terms of all orders between e^n and e^{2n}. In momentum space the rules are the following:

There is a factor

(*a*) $+e(p_\mu + p'_\mu)$ at each single corner where p_μ and p'_μ are the momenta of the meson line coming in and going out of the corner, respectively;

(*b*) $(2\pi)^4\,\delta^{(4)}(p - p' \pm k)$ at each single corner, where p and p' are the momenta associated with the meson lines and k that of the internal or external photon line ending there (the $+$ or $-$ sign is used depending on whether the photon is emitted or absorbed);

(*c*) $e^2\,(2\pi)^4\,g_{\mu\nu}\delta^{(4)}(p - p' \pm k \pm k')$ for each double corner (i.e., a corner at which two photon lines of polarization μ and ν are absorbed or emitted). Each photon emitted or absorbed at the double corner is given the factor appropriate to that photon [see rules (*d*) and (*f*) below];

(*d*) $-\dfrac{i\hbar c}{(2\pi)^4}\dfrac{1}{k^2 + i\delta}\,g_{\mu\nu}$ for each internal photon line of momentum k joining vertices where factors $(p + p')^\mu$ and $(p + p')^\nu$ operate in the case of single corners, or factors $g^{\mu\rho}$ and $g^{\nu\sigma}$ in the case of double corners (ρ, σ refer to the second photon emitted or absorbed at the double corner), and factors $(p + p')^\mu$ and $g^{\nu\rho}$ in the mixed case;

(*e*) $\dfrac{i\hbar c}{(2\pi)^4}\dfrac{1}{p^2 - \mu^2 + i\epsilon}$ for each internal meson line of momentum p;

(*f*) $\sqrt{\dfrac{\hbar c}{(2\pi)^3\,2|\mathbf{k}|}}\;\epsilon^{(\lambda)}{}_\mu(\mathbf{k})$ for each external photon line which is emitted or absorbed at a single or double corner; $\epsilon^{(\lambda)}{}_\mu(\mathbf{k})$ is the polarization four-vector ($\epsilon_\mu\epsilon^\mu = -1$, $k_\mu\epsilon^\mu(\mathbf{k}) = 0$) for the emitted or absorbed photon of energy $k_0 = |\mathbf{k}|$, momentum $\mathbf{k}$, and state of polarization λ;

(*g*) $\sqrt{\dfrac{\hbar c}{(2\pi)^3\,2\omega_{\mathbf{k}}}}$ for each external meson line of momentum $\mathbf{k}$ and energy $\omega_{\mathbf{k}} = (\mathbf{k}^2 + \mu^2)^{1/2}$, which enters or leaves the diagram;

(*h*) $(-i/\hbar c)^n$ from the perturbation expansion; and finally,

(*i*) one is to integrate over all internal momenta and multiply the integral by a weight factor $w = 2^g$, where $g = d - b$; where d is the number of double corners in the diagram, and b is the number of pairs of double corners connected by two photon lines. The use of a normal product form for the interaction term implies, by Wick's theorem, that we are to ignore those Feynman diagrams in which two lines originating from the same vertex rejoin each other without any other interaction. Thus, diagrams

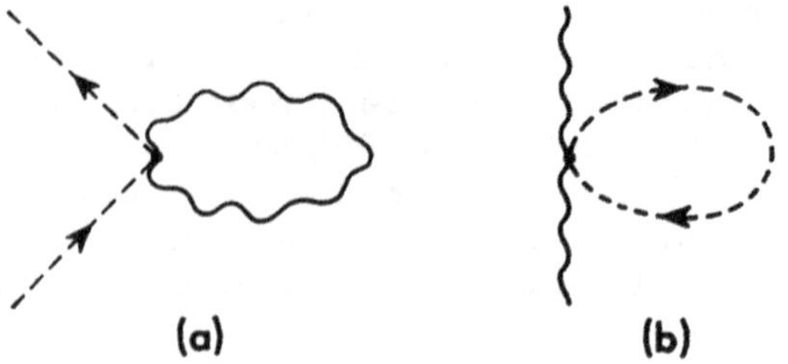

Fig. 14.27

such as those illustrated in Figure 14.27*a* and *b* are *not* to be considered.

A similar situation as the above occurs for any theory with derivative coupling, e.g., the case of the pseudoscalar meson theory with pseudovector coupling for which

$$\mathfrak{L}_I = \frac{f}{\mu} : \bar{\psi}(x)\, \gamma^\mu \gamma_5 \psi(x)\, \partial_\mu \phi(x) : \tag{107}$$

The momentum canonically conjugate to ϕ is now

$$\pi(x) = \partial_0 \phi(x) - \frac{f}{\mu} : \bar{\psi}(x)\, \gamma^0 \gamma_5 \psi(x) : \tag{108}$$

and the interaction Hamiltonian density is no longer the negative of $\mathfrak{L}_I$ but is given by

$$\mathfrak{H}_I(x;\sigma) = : \frac{f}{\mu} \bar{\psi}(x)\, \gamma_\mu \gamma_5 \psi(x)\, \partial^\mu \phi(x) - \frac{1}{2}\left(\frac{f}{\mu}\right)^2 (\bar{\psi} \gamma_\mu \gamma_5 \psi n^\mu(x))^2 : \tag{109}$$

and both these terms must be used in calculating the S matrix. A detailed analysis again indicates that the correct answer is obtained by adopting for the S matrix the expansion

$$S = \sum_{n=0}^{\infty} \left(\frac{+i}{\hbar c}\right)^n \int d^4x_1 \cdots \int d^4x_n\, P(\mathfrak{L}_I(x_1) \cdots \mathfrak{L}_I(x_n)) \tag{110}$$

and neglecting the additional singular terms obtained from the contraction of $\partial_\nu \phi$ and $\partial_\mu \phi$ for $\mu = \nu = 0$.

14*d*. Cross Sections

The rules previously derived in this chapter enable us to write down the matrix element corresponding to any given Feynman diagram, and therefore enable us to calculate the S matrix. Now the matrix element of the S matrix between specified initial and final states, $|\Phi_a\rangle$ and $|\Phi_b\rangle$, is the probability amplitudes for a transition from the initial state $|\Phi_a\rangle$ to the final state $|\Phi_b\rangle$. Let us write this transition amplitude as

$$R_{ba} = (2\pi)^4\, \delta^{(4)}\left(\sum_{b=1}^{n'} p_b' - \sum_{a=1}^{n} p_a\right) (b \mid M \mid a) \left[\frac{1}{(2\pi)^{3/2}}\right]^{n+n'} NN' \tag{111}$$

where we have explicitly exhibited the δ functions corresponding to the overall energy-momentum conservation for the process $\left(\sum_{a=1}^{n} p_a = \text{sum of}\right.$ four momenta of the n incoming particles, $\sum_{b=1}^{n'} p_b' =$ sum of four momenta of the n' outgoing particles$\left.\right)$; and N, N' denote the normalization factors for the incoming and outgoing particles [$(m/E)^{1/2}$ for fermions, $1/\sqrt{2\omega}$ for bosons] so that $M_{ba} = (b \mid M \mid a)$ is a relativistically invariant matrix element.

We have treated in Section a of this chapter the case of the scattering of a particle by an external field which is time-independent. In this case momentum is not conserved and the transition amplitude is of the form

$$R'_{ba} = +2\pi i \delta^{(1)}\left(\sum p_{a0} - \sum p_{b0}'\right)(b \mid M' \mid a) \tag{112}$$

The way to obtain cross sections in these cases was outlined in Section a, so we shall here only consider the situation with no external fields. Furthermore, since the most important applications will be to collisions in which the initial state is composed of two particles, we here also restrict ourselves to this case.

The probability for scattering into a final state in which the momenta of the final state particles lie in a set Ω with momenta between $\mathbf{p}_1'$ and $\mathbf{p}_1' + d\mathbf{p}_1'$, $\cdots$ $\mathbf{p}_n'$ and $\mathbf{p}_n' + d\mathbf{p}_n'$ and specified spins, is given by

$$dw = \int_\Omega |R_{ba}|^2 \, d\mathbf{p}_1' \cdots d\mathbf{p}_n' \tag{113}$$

This transition probability is infinite as it contains a factor $\delta^{(4)}(0)$, due to the fact that our state vectors were not normalizable. To remedy this situation, we should construct wave packets for the incoming states. Instead, with Lippmann and Schwinger [Lippmann (1950)], we can again interpret the $\delta^{(4)}(0)$ factor as corresponding to $(2\pi)^{-4}\, VT$ where V is the (large) volume and T the (long) time in which and during which the scattering process takes place. Now the quantity of physical interest is dw', the transition probability per unit time and unit volume when the incoming states are normalized to one particle per unit volume. Let the normalization of the incoming one-particle *states* be such that their density is η_i ($i = 1, 2$) so that

$$\begin{aligned} dw' &= \frac{dw}{\eta_1\eta_2 VT} \\ &= (2\pi)^4 \int_\Omega N^2 N'^2 \delta^{(4)}\left(\sum_{b=1}^{n'} p_b' - \sum_{a=1}^{n} p_a\right) \\ &\quad \cdot \frac{1}{(2\pi)^6\, \eta_1\eta_2} |(b \mid M \mid a)|^2 \frac{d\mathbf{p}_1'}{(2\pi)^3} \cdots \frac{d\mathbf{p}_n'}{(2\pi)^3} \end{aligned} \tag{114}$$

The cross section $d\sigma$ is the number of transitions per unit time and volume, divided by the incident flux and by the number of target particles per unit volume. Since by dividing by η_i we have normalized our incident states so as to contain one particle per unit volume, the flux of incident particles is $|\mathbf{v}_1 - \mathbf{v}_2| = v$, where $\mathbf{v}_1$ and $\mathbf{v}_2$ are the velocities of the two incident particles, which are supposed to be in the same direction. The cross section $d\sigma$ is therefore obtained by dividing dw' by the relative velocity $|\mathbf{v}_1 - \mathbf{v}_2|$ of the incident particles;

$$d\sigma = \frac{dw'}{|\mathbf{v}_1 - \mathbf{v}_2|} \tag{115}$$

The relativistic invariance of the cross section is made manifest by noting that the factor $E_1E_2|\mathbf{v}_1 - \mathbf{v}_2|$ can be replaced by the invariant expression $|(p_1 \cdot p_2)^2 - p_1{}^2p_2{}^2|^{1/2}$ and that d^3p/E is also invariant.

In the case of a reaction with only two products

$$p_1 + p_2 \rightarrow p_1' + p_2'$$

Eqs. (114)–(115) define an ordinary differential cross section by taking for Ω the set of final states in which the direction of the outgoing particle $\mathbf{p}_1'$ lies in a given small solid angle $d\Omega$. Given $d\Omega$ the magnitudes of $\mathbf{p}_1'$, $\mathbf{p}_2'$, $p_{10} = E_1$ and $p_{20} = E_2$ are then fixed by energy-momentum conservation. In the center-of-mass system $|\mathbf{p}_1'| = |\mathbf{p}_2'|$ and Eq. (115) becomes

$$\begin{aligned} d\sigma &= \frac{(2\pi)^{-8}}{v\eta_1\eta_2} \int N^2N'^2p_1'^2dp_1'd\Omega|M_{ba}|^2\,\delta(E_1 + E_2 - E_1' - E_2') \\ &= \frac{(2\pi)^{-8}}{\eta_1\eta_2 v} N^2N'^2p_1' \frac{E_1'E_2'}{E_1' + E_2'} |M_{ba}|^2\,d\Omega \end{aligned} \tag{116}$$

Since the relative velocity of two particles in the final state is

$$v' = \frac{p_1'(E_1' + E_2')}{E_1'E_2'}$$

the cross section in the case of a two-particle collision is given by

$$\frac{d\sigma}{d\Omega} = \frac{(2\pi)^{-8}}{\eta_1\eta_2 vv'} (p_1')^2\,|M_{ba}|^2\,N^2N'^2 \tag{117}$$

Consider next the case of the decay of an unstable particle of momentum $\mathbf{k}$ into two product particles of momentum $\mathbf{p}_1$ and $\mathbf{p}_2$ (e.g., the $\pi^0 \rightarrow 2\gamma$ decay or the $\mathrm{K} \rightarrow 2\pi$ decay). The decay rate (the inverse lifetime) is the total transition probability per unit time and is given by

$$w = \frac{1}{\tau} = (2\pi)^4 \int \delta^{(4)}(k - p_1 - p_2)\,\frac{d^3p_1}{(2\pi)^3}\,\frac{d^3p_2}{(2\pi)^3}\,|M|^2\,\frac{N^2N'^2}{\eta(2\pi)^3} \tag{118}$$

where the integration over the possible final states includes a summation over the spin states of the daughter particles (if the particles have spin).

In the rest frame of the decaying particle $k = (m_0, \mathbf{0})$ (m_0 is the mass of the decaying particle) and $|\mathbf{p}_1| = |\mathbf{p}_2|$, hence

$$\delta^{(4)}(k - p_1 - p_2) = \delta^{(1)}(m_0 - p_{10} - p_{20})\, \delta^{(3)}(\mathbf{p}_1 + \mathbf{p}_2) \tag{119}$$

so that

$$w = (2\pi)^4 \int \delta^{(1)}(m_0 - \sqrt{\mathbf{p}_1^2 + m_1^2} - \sqrt{\mathbf{p}_1^2 + m_2^2})\, |M_{ba}|^2 \frac{N^2 N'^2}{\eta(2\pi)^3} \frac{d^3 p_1}{(2\pi)^6} \tag{120}$$

If the decaying particle has no internal structure, i.e., is a spin zero particle, the decay must be isotropic and hence $|M_{ba}|^2$ is spherically symmetric in momentum space and

$$\frac{1}{\tau} = (2\pi)^4\, 4\pi \int_0^\infty \delta(m_0 - \sqrt{p^2 + m_1^2} - \sqrt{p^2 + m_2^2})\, |M_{ba}|^2 \cdot \frac{N^2 N'^2}{\eta(2\pi)^9}\, p^2 dp \tag{121}$$

14e. Examples

1. Compton Scattering

We shall illustrate the application of (114) by calculating the cross section for the scattering of a photon by free electrons (Compton scattering) in the lowest order of perturbation theory.

There are two Feynman diagrams which contribute in lowest order. These are shown in Figure 14.28. In the first, the electron in state $p_1 s_1$

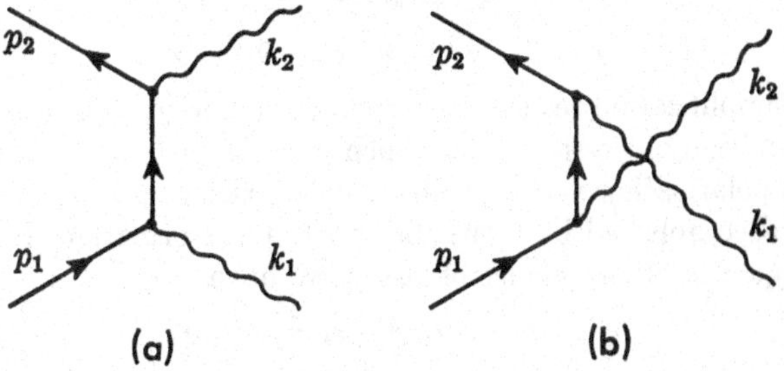

Fig. 14.28

first absorbs a photon of momentum k_1, polarization vector $\epsilon^{(\lambda_1)}(\mathbf{k}_1)$, and then emits a second photon of momentum k_2, polarization $\epsilon^{(\lambda_2)}(\mathbf{k}_2)$, to arrive in the final state of momentum p_2 and spin s_2. In the second process, the act of emission of the final photon precedes that of absorption of the initial photon. One readily verifies, using the rules of Section 14c, that the matrix element corresponding to these two Feynman diagrams is:

$$R_{ba} = \delta^{(4)}(p_2 + k_2 - p_1 - k_1) \frac{r_0}{2\pi i} \frac{m^2}{\sqrt{E_1 E_2 \omega_1 \omega_2}}$$

$$\bar{w}^{s_2}(\mathbf{p}_2) \left\{ \gamma \cdot \epsilon_2 \frac{\gamma \cdot (p_1 + k_1) + m}{(p_1 + k_1)^2 - m^2 + i\delta} \gamma \cdot \epsilon_1 + \gamma \cdot \epsilon_1 \frac{\gamma \cdot (p_1 - k_2) + m}{(p_1 - k_2)^2 - m^2 + i\delta} \gamma \cdot \epsilon_2 \right\} w^{s_1}(\mathbf{p}_1) \tag{122a}$$

$$= (2\pi)^4 \delta^{(4)}(p_2 + k_2 - p_1 - k_1) (p_2 s_2, k_2\epsilon_2 \mid M \mid p_1 s_1, k_1\epsilon_1) \frac{1}{(2\pi)^6} \sqrt{\frac{m^2}{4E(\mathbf{p}_1) E(\mathbf{p}_2) \omega(\mathbf{k}_1) \omega(\mathbf{k}_2)}} \tag{122b}$$

where, since no confusion can arise, we have written $\epsilon^{(\lambda_1)}(\mathbf{k}_1)$ as ϵ_1, and $\omega(\mathbf{k}_1)$ as ω_1; and similarly for ϵ_2, ω_2 and E_1, E_2. In Eq. (122), r_0 is the classical electron radius, $e^2/4\pi mc^2$. This factor is obtained as follows: The application of the rules of Section 14c will result in a numerical factor $\alpha/2\pi i$ in Eq. (122a). If we express the photon normalization factors $(2|\mathbf{k}|)^{-1/2}$ in terms of the photon energy, $\hbar c\mathbf{k}$, this will then give rise to an additional factor $\hbar c$ in the numerator, which in combination with a factor m changes α to r_0. This also explains the factor m^2 in the numerator: One m arises from the electron normalization factor, the other from changing α to r_0. Since we are working in a system of units in which $\hbar = c = 1$, $\hbar\omega$, the energy of the photon, is denoted by $\omega = |\mathbf{k}|$.

Let us call the factor within the braces in Eq. (122a) $\mathfrak{M}$. $\mathfrak{M}$ can be simplified by recalling that the electron and photon are free, so that

$$p_1^2 = p_2^2 = m^2 \tag{123}$$

and also

$$k_1^2 = k_2^2 = 0 \tag{124}$$

$$k_1 \cdot \epsilon_1 = k_2 \cdot \epsilon_2 = 0 \tag{125}$$

A further simplification results if we choose the Lorentz frame in which the electron is initially at rest, i.e., for which $p_1 = (m, 0, 0, 0)$. Also, we shall choose the polarization vectors to have only space components which can always be accomplished by a suitable gauge transformation (Section 9b). As a consequence of this choice of gauge, we have

$$p_1 \cdot \epsilon_1 = p_1 \cdot \epsilon_2 = 0 \tag{126}$$

since p_1 has only a time component and ϵ only space components. Finally, we note that $\mathfrak{M}$ operates to the right on a Dirac wave function which is an eigenfunction of $\gamma \cdot p_1$ with eigenvalue m. We use the commutation rules for the γ matrices:

$$\gamma_\mu \gamma_\nu + \gamma_\nu \gamma_\mu = 2g_{\mu\nu} \tag{127}$$

to write

$$(\gamma \cdot p_1)(\gamma \cdot \epsilon_1) = -(\gamma \cdot \epsilon_1)(\gamma \cdot p_1) + 2p_1 \cdot \epsilon_1 \tag{128}$$

The second term, however, vanishes by virtue of Eq. (126). Similarly, one finds that $\gamma \cdot p_1$ anticommutes with $\gamma \cdot \epsilon_2$. Using these results, we may eliminate the terms $\gamma \cdot p_1 + m$ occurring in $\mathfrak{M}$. Finally, using Eq. (125), we may write

$$R_{ba} = (2\pi)^4 \, \delta^{(4)}(p_2 + k_2 - p_1 - k_1) \frac{-ir_0 m^2}{(2\pi)^5 \sqrt{E_1 E_2 \omega_1 \omega_2}}$$

$$\tilde{w}^{s_2}(\mathbf{p}_2) \left\{ \gamma \cdot \epsilon_2 \frac{\gamma \cdot k_1}{2p_1 \cdot k_1} \gamma \cdot \epsilon_1 + \gamma \cdot \epsilon_1 \frac{\gamma \cdot k_2}{2p_1 \cdot k_2} \gamma \cdot \epsilon_2 \right\} w^{s_1}(\mathbf{p}_1) \quad (129)$$

To obtain the transition probability, we shall have to square R_{ba}. If we are not interested in the spin of the final electron, and if we restrict ourselves to the case of unpolarized initial electrons, then we have seen in Chapter 4 that summing $|\tilde{w}^{s_2}(\mathbf{p}_2) \, \mathfrak{M} w^{s_1}(\mathbf{p}_1)|^2$ over s_2 and averaging over s_1 corresponds to carrying out a certain trace operation. A somewhat laborious calculation then yields the result

$$F = \tfrac{1}{2} \sum_{s_1 s_2} |\tilde{w}^{s_2}(\mathbf{p}_2) \, \mathfrak{M} w^{s_1}(\mathbf{p}_1)|^2 \quad (130a)$$

$$= \frac{1}{8m^2} \operatorname{Tr} \{(\gamma \cdot p_2 + m) \, \mathfrak{M}(\gamma \cdot p_1 + m) \, \tilde{\mathfrak{M}}\} \quad (130b)$$

$$= \frac{1}{4m^2} \left[\frac{p_1 \cdot k_1}{p_1 \cdot k_2} + \frac{p_1 \cdot k_2}{p_1 \cdot k_1} - 2 + 4(\epsilon_1 \cdot \epsilon_2)^2 \right] \quad (130c)$$

$$= \frac{1}{4m^2} \left[\frac{\omega_1}{\omega_2} + \frac{\omega_2}{\omega_1} - 2 + 4 \cos^2 \Theta \right] \quad (130d)$$

where, in going from Eq. (130c) to (130d), we have used the fact that in the laboratory frame $p_1 = (m, 0, 0, 0)$ and Θ is the angle between the directions of polarization of the incident and emitted photon. The expression (130d) will be recognized as the principal factor in the Klein-Nishina formula [Klein and Nishina (Klein 1929), Nishina (1929), Tamm (1930); see also Heitler (1947)].

To compute the cross section for the process, we must know the density of incoming particles. This can be obtained from a consideration of the charge-current density of the electron-positron field and the Poynting vector of the electromagnetic field. The expectation value of the current operator in the initial electron state $|\Phi_i\rangle = b_{s_1}{}^*(\mathbf{p}_1) \mid \Phi_0\rangle$ is

$$\langle j_\mu(x)\rangle_i = -\tfrac{1}{2} e \langle \Phi_i, [\tilde{\psi}(x) \, \gamma_\mu, \psi(x)] \, \Phi_i\rangle$$

$$= -e\langle b_{s_1}{}^*(\mathbf{p}_1) \, \Phi_0, \widetilde{\psi^{(+)}}(x) \, \gamma_\mu \psi^{(+)}(x) \, b_{s_1}{}^*(\mathbf{p}_1) \, \Phi_0\rangle$$

$$= -\frac{e}{(2\pi)^3} \frac{m}{E_1} \tilde{w}^{s_1}(\mathbf{p}_1) \, \gamma_\mu w^{s_1}(\mathbf{p}_1)$$

$$= -\frac{e}{(2\pi)^3} \frac{p_{1\mu}}{E_1} = -\frac{e}{(2\pi)^3} v_{1\mu} \quad (131)$$

where $v_{1\mu}$ is the velocity of the incident electron. The charge density in the initial state is therefore

$$\langle \rho(x) \rangle_i = -\frac{e}{(2\pi)^3}\frac{m}{E_1} w_{p_1}{}^*(x)\, w_{p_1}(x) \tag{132}$$

so that the density of incoming electrons, η_1, is given by

$$\eta_1 = \frac{1}{(2\pi)^3} \tag{133}$$

This result is obtained from Eq. (132) by recalling that the Dirac spinors have been normalized not to unit volume but to the invariant volume m/E_1. Similarly, one readily establishes that the density of incoming photons is $1/(2\pi)^3$. Now $|R_{ba}|^2/VT$ is the transition probability per unit volume and unit time for the above density of incident particles. To obtain a probability per unit density of incoming particles, we must divide $|R_{ba}|^2/VT$ for the present situation by $\frac{1}{(2\pi)^6}$. Hence dw' for the present situation is given by

$$dw' = r_0^2 \int d^3k_2 \int d^3p_2\, \delta^{(4)}(p_2 + k_2 - p_1 - k_1)\, \frac{m^4}{E_1E_2\omega_1\omega_2} F \tag{134}$$

We may now eliminate the delta function in the momentum variables by carrying out the integration over d^3p_2, so that

$$dw' = r_0^2 \int d^3k_2\, \delta^{(1)}(E_f - E_i)\, \frac{m^4}{E_1E_2\omega_1\omega_2} F \tag{135}$$

where $E_f = p_{20} + k_{20}$ is the total energy of the system in the final state and $E_i = p_{10} + k_{10}$ is the energy of the initial state. Let us now write

$$d^3k_2 = (k_{20})^2\, dk_{20} d\Omega = (k_{20})^2 \left(\frac{dk_{20}}{dE_f}\right) d\Omega dE_f \tag{136}$$

We can then carry out the integration over dE_f and remove the final remaining δ function over the energy. We thus finally obtain

$$dw' = r_0^2 \frac{m^4}{E_1E_2} \left(\frac{\omega_2}{\omega_1}\right) \left(\frac{d\omega_2}{dE_f}\right)_{E_f = E_i} d\Omega \tag{137}$$

Note that the integration over d^3p_1 fixes the momenta of the final particles in terms of the initial ones by momentum conservation, and similarly the integration over dE_f fixes the total energy of the final particles by energy conservation.

Let us now carry out the differentiation indicated in Eq. (136). In the laboratory system, where $E_1 = m$, recalling that

$$\begin{aligned} E_f &= \omega_2 + E_2 = \omega_1 + E_1 = \omega_2 + (\mathbf{p}_2{}^2 + m^2)^{1/2} \\ &= \omega_2 + (\omega_1{}^2 - 2\omega_1\omega_2 \cos\phi + \omega_2{}^2 + m^2)^{1/2} \end{aligned} \tag{138}$$

where ϕ is the scattering angle

$$\mathbf{k}_1 \cdot \mathbf{k}_2 = \omega_1\omega_2 \cos \phi \tag{139}$$

and using the Compton relation

$$\omega_1\omega_2(1 - \cos \phi) = m(\omega_1 - \omega_2) \tag{140}$$

we obtain

$$\begin{aligned} \frac{dE_f}{d\omega_2} &= 1 + \frac{\omega_2 - \omega_1 \cos \phi}{E_2} = \frac{E_2 + \omega_2 - \omega_1 \cos \phi}{E_2} \\ &= \frac{m + \omega_1(1 - \cos \phi)}{E_2} \\ &= \frac{m}{E_2}\frac{\omega_1}{\omega_2} \end{aligned} \tag{141}$$

The transition probability for the scattered photon to emerge in the solid angle $d\Omega$ having energy ω_2, is therefore given by

$$dw' = r_0^2m^2 \left(\frac{\omega_1}{\omega_2}\right)^2 F d\Omega \tag{142}$$

The differential cross section, $d\sigma$, is defined as the transition probability divided by the incident flux ($c = 1$) and by the number of scatterers per unit volume ($= 1$). It is therefore given by

$$d\sigma = r_0^2 \left(\frac{\omega_1}{\omega_2}\right)^2 m^2 F d\Omega \tag{143}$$

and is the well-known Klein-Nishina formula.

We shall here only consider the nonrelativistic limit of the cross section, that is, the limit in which $\omega_1, \omega_2 \ll m$ and $p_1 = p_2 = (m, 0, 0, 0)$. In this limit, Eq. (130d) reduces to

$$F_{\text{nr}} = \frac{1}{m^2} \cos^2 \Theta \tag{144}$$

(since $\omega_1 = \omega_2$), and the differential scattering cross section becomes

$$d\sigma_{\text{nr}} = r_0^2 \cos^2 \Theta d\Omega \tag{145}$$

which is the classical Thomson formula for the scattering of low energy radiation by a static charge. The differential cross section for unpolarized light is obtained by summing over the two final states of polarization of the photon and averaging over the two initial states. This summation can easily be carried out,[5] and gives

[5] Recall that the three-dimensional vectors $\boldsymbol{\epsilon}^{(1)}(\mathbf{k})$, $\boldsymbol{\epsilon}^{(2)}(\mathbf{k})$, $\mathbf{k}/|\mathbf{k}|$ form an orthonormal set, so that the closure relation for them reads

$$\epsilon^{(1)}{}_r(\mathbf{k})\, \epsilon^{(1)}{}_s(\mathbf{k}) + \epsilon^{(2)}{}_r(\mathbf{k})\, \epsilon^{(2)}{}_s(\mathbf{k}) + \frac{k_r k_s}{|\mathbf{k}|^2} = \delta_{rs}$$

where the subscripts $r, s = 1, 2, 3$ denote the components of the vector. We may therefore write

$$d\bar{\sigma}_{nr} = \tfrac{1}{2}r_0^2(1 + \cos^2\phi)\, d\Omega \tag{146}$$

where ϕ is the angle between $\mathbf{k}_1$ and $\mathbf{k}_2$, i.e., the scattering angle. The total cross section is now obtained by integrating over $d\Omega$ and is

$$\sigma = \frac{8\pi}{3} r_0^2 \tag{147}$$

We shall see in Chapter 17 that Eq. (147), the Thomson scattering formula, plays an important role in understanding charge renormalization.

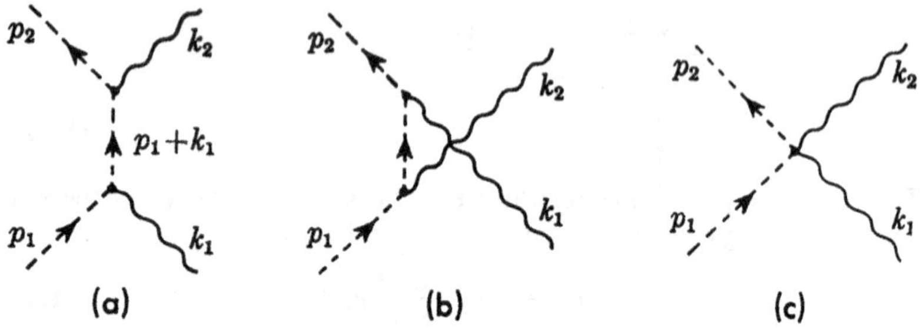

Fig. 14.29

It is interesting to compare the scattering of photons from spin $\frac{1}{2}$ particles to that from spin 0 particles. To this end let us compute the cross section for scattering of photons by pions in lowest order of perturbation theory. The three diagrams which contribute are indicated in Figure 14.29, and their contributions to the S matrix are:

$$\begin{aligned} T = &-\frac{\alpha}{4\pi\sqrt{p_{10}p_{20}\omega_1\omega_2}}\,\delta(p_1 + k_1 - p_2 - k_2) \\ &\left\{(p_2 + p_1 + k_1)\cdot\epsilon_2 \frac{1}{(p_1 + k_1)^2 - \mu^2}(2p_1 + k_1)\cdot\epsilon_1 \right. \\ &\left. + (p_2 + p_1 - k_2)\cdot\epsilon_1 \frac{1}{(p_1 - k_2)^2 - \mu^2}(2p_1 - k_2)\cdot\epsilon_2 - 2\epsilon_1\cdot\epsilon_2\right\} \end{aligned} \tag{148}$$

$$\sum_{\lambda=1}^{2} \epsilon^{(\lambda)}{}_r(\mathbf{k})\, \epsilon^{(\lambda)}{}_s(\mathbf{k}) = \delta_{rs} - \frac{k_r k_s}{|\mathbf{k}|^2}$$

In this fashion, one readily verifies that

$$\begin{aligned} \tfrac{1}{2}\sum_{\lambda_1\lambda_2} [\boldsymbol{\epsilon}^{(\lambda_1)}(\mathbf{k}_1)\cdot\boldsymbol{\epsilon}^{(\lambda_2)}(\mathbf{k}_2)]^2 &= \tfrac{1}{2}\sum_{\lambda_1,\lambda_2}\sum_{r,s} \epsilon^{(\lambda_1)}{}_r(\mathbf{k}_1)\,\epsilon^{(\lambda_2)}{}_r(\mathbf{k}_2)\cdot\epsilon^{(\lambda_1)}{}_s(\mathbf{k}_1)\,\epsilon^{(\lambda_2)}{}_s(\mathbf{k}_2) \\ &= \tfrac{1}{2}\sum_{rs}\left(\delta_{rs} - \frac{k_{1r}k_{1s}}{|\mathbf{k}_1|^2}\right)\left(\delta_{rs} - \frac{k_{2r}k_{2s}}{|\mathbf{k}_2|^2}\right) \\ &= \tfrac{1}{2}(1 + \cos^2\phi) \end{aligned}$$

where ϕ is the angle between $\mathbf{k}_1$ and $\mathbf{k}_2$, i.e., the scattering angle.

The factor 2 appears in the contribution from diagram 14.29c, since either of the As in the interaction term $: A_\mu(x)\, A^\mu(x)\, \phi^*\phi(x) :$ could have emitted photon 2 or destroyed photon 1. In the frame of reference where the initial pion is at rest, $\mathbf{p}_1 = 0$, and with the choice of a gauge such that ϵ_1 and ϵ_2 are pure space-like vectors (so that $p_1 \cdot \epsilon_1 = p_1 \cdot \epsilon_2 = 0$), only diagram c contributes. This is a consequence of the particular choice of gauge such that $\epsilon_{10} = \epsilon_{20} = 0$. It is interesting to note that the amplitude for each diagram a, b, and c of Figure 14.29 is not separately gauge invariant, but the sum is (as is easily verified by making a gauge transformation $\epsilon_\mu \rightarrow \epsilon'_\mu = \epsilon_\mu + \lambda k_\mu$). The cross section is given by

$$d\sigma = \alpha^2 \int d^3k_2 \int d^3p_2\, \delta^{(4)}(p_1 + k_1 - p_2 - k_2) \frac{|(\epsilon_1 \cdot \epsilon_2)|^2}{p_{10}p_{20}\omega_1\omega_2} \tag{149}$$

Carrying out the integration in a manner completely analogous to the spin $\frac{1}{2}$ case, we finally obtain

$$d\sigma = r_\pi^{\,2} \frac{\mu^2}{[\mu + \omega_1(1 - \cos\phi)]^2} |(\epsilon_1 \cdot \epsilon_2)|^2\, d\Omega \tag{150}$$

where $r_\pi = e^2/4\pi\mu$, the classical "pion" radius, and $\cos\phi\ \omega_1\omega_2 = \mathbf{k}_1 \cdot \mathbf{k}_2$. For $\omega_1 \ll \mu$, the extreme nonrelativistic limit, the cross section again reduces to the classical result

$$d\sigma = r_\pi^{\,2} |(\epsilon_1 \cdot \epsilon_2)|^2\, d\Omega \tag{151}$$

The scattering of a very low energy photon off a spin 0 particle is thus the same as off a spin $\frac{1}{2}$ particle. The explanation for this is that for very long wavelength the interaction takes place only through the total charge of the system from which the photon is scattered.

We close this section by noting that, by virtue of the unitarity of the S matrix, certain relations hold between the various order terms of the S matrix. Thus, if we develop the S matrix in a power series in the coupling constant, as is the case in the Dyson expansion,

$$S = 1 + eS_1 + e^2S_2 + \cdots \tag{152}$$

then the condition for the unitarity of S becomes

$$\begin{aligned} S^*S = 1 &+ e(S_1^* + S_1) + e^2(S_1^*S_1 + S_2 + S_2^*) \\ &+ e^3(S_1^*S_2 + S_2^*S_1 + S_3^* + S_3) \\ &+ e^4(S_1^*S_3 + S_2^*S_2 + S_3^*S_1 + S_4^* + S_4) + \cdots \end{aligned} \tag{153}$$

so that each of the expressions in parentheses must vanish. The relations between the S_ns thus obtained can be very useful in computational work [see, for example, Jost (1950b)].

2. *Pion Photoproduction*

As a second example, consider the photoproduction of positively-charged pions by the reaction

$$\gamma + \mathrm{p} \rightarrow \pi^+ + \mathrm{n}$$

In pseudoscalar meson theory, with nonderivative coupling, the lowest order diagrams which contribute to the amplitude for this process are given in Figure 14.30. Figure 14.30a is analogous to the photoelectric effect in

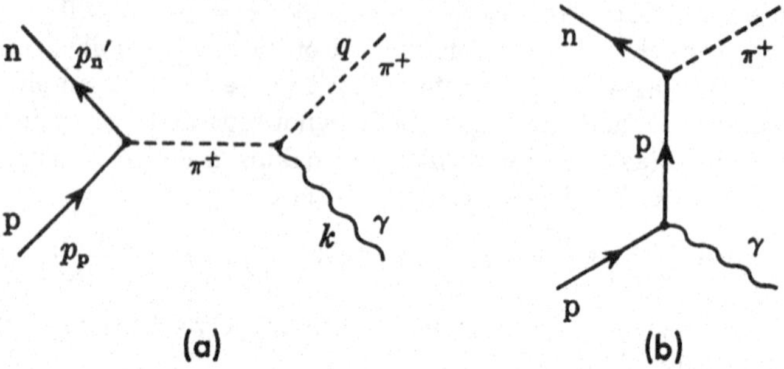

Fig. 14.30

the sense that a virtual meson in the nucleon cloud (for which $q^2 \neq \mu^2$) is made "real" (i.e., one for which $q^2 = \mu^2$) by absorbing the photon. We have endowed the nucleons only with a charge and have neglected the anomalous magnetic moment. The results of these lowest order calculations are not to be taken too seriously. They are in violent disagreement with experiment. They are here presented as illustrations of the Feynman techniques. Diagram 14.30a contributes

$$R^{(a)} = \frac{1}{(2\pi)^6} \frac{1}{\sqrt{4q_0k_0E_{\mathbf{p}}E_{\mathbf{p}'}}} (2\pi)^8 \delta^{(4)}(p' + q - p - k)$$

$$\bar{w}_{\mathbf{n}}(\mathbf{p}') \sqrt{2}G\gamma_5 w_{\mathbf{p}}(\mathbf{p}) \frac{1}{(q-k)^2 - \mu^2} \frac{i}{(2\pi)^4} [-e\epsilon^{(\lambda)}(\mathbf{k}) \cdot (2q - k)] \tag{154a}$$

$$= -\frac{i}{(2\pi)^2} \frac{\sqrt{2}eGM}{\sqrt{E_{\mathbf{p}'}E_{\mathbf{p}}4q_0k_0}} \bar{w}_{\mathbf{n}}(\mathbf{p}') \gamma_5 \frac{q \cdot \epsilon^{(\lambda)}(\mathbf{k})}{q \cdot k} w_{\mathbf{p}}(\mathbf{p}) \cdot \delta^{(4)}(p' + q - p - k) \tag{154b}$$

where to arrive at (154b) we have used the fact that $q^2 = \mu^2$, $k^2 = 0$, $\epsilon^{(\lambda)}(\mathbf{k}) \cdot k = 0$ and have assumed that the mass of the proton and neutron are the same and equal to M. Diagram 14.30b contributes a partial amplitude

$$R^{(b)} = \frac{e}{(2\pi)^6} \frac{M}{\sqrt{E_{\mathbf{p}}E_{\mathbf{p}'}4q_0k_0}} (2\pi)^8 \delta^{(4)}(p' + q - p - k)$$

$$\bar{w}_{\mathbf{n}}(\mathbf{p}') G\gamma_5 \frac{i}{(2\pi)^4} \frac{\gamma \cdot (p + k) + M}{(p + k)^2 - M^2} \gamma \cdot \epsilon^{(\lambda)}(\mathbf{k}) w_{\mathbf{p}}(\mathbf{p}) \tag{155}$$

With the choice of gauge $\epsilon^{(\lambda)}{}_0 = 0$, in both the laboratory and center-of-mass frames $p \cdot \epsilon = 0$ so that

$$R^{(b)} = \frac{i}{(2\pi)^2} \frac{eGM}{\sqrt{E_{\mathbf{p}'}E_{\mathbf{p}}4q_0k_0}} \tilde{w}_{\mathrm{n}}(\mathbf{p}')\, \gamma_5 \frac{\gamma \cdot k\gamma \cdot \epsilon^{(\lambda)}(\mathbf{k})}{2p \cdot k} w_{\mathrm{p}}(\mathbf{p})$$

$$\delta^{(4)}(p' + q - p - k) \quad (156)$$

so that together diagrams a and b contribute

$$R = -\frac{i}{(2\pi)^2} \frac{eGM}{\sqrt{E_{\mathbf{p}'}E_{\mathbf{p}}4q_0k_0}} \delta^{(4)}(p' + q - p - k)$$

$$\tilde{w}_{\mathrm{n}}(\mathbf{p}')\, \gamma_5 \left[\frac{q \cdot \epsilon^{(\lambda)}(\mathbf{k})}{q \cdot k} - \frac{\gamma \cdot k\gamma \cdot \epsilon^{(\lambda)}(\mathbf{k})}{2p \cdot k}\right] w_{\mathrm{p}}(\mathbf{p}) \quad (157)$$

To compute the differential cross section (independent of spin of proton) we must compute $\sum_{\text{spin}} |R|^2$. In the center-of-mass system

$$\mathrm{Tr}\left\{\left[\frac{q \cdot \epsilon}{q \cdot k} - \frac{\gamma \cdot k\gamma \cdot \epsilon}{2p \cdot k}\right] \gamma_5 \frac{\gamma \cdot p + M}{2M} \gamma_5 \left[\frac{q \cdot \epsilon}{q \cdot k} - \frac{\gamma \cdot k\gamma \cdot \epsilon}{2p \cdot k}\right] \frac{\gamma \cdot p' + M}{2M}\right\}$$

$$= \frac{1}{2M^2}\left(1 - \frac{\mu^2}{2} \frac{\mathbf{q}^2 \sin^2\theta}{\mathbf{k}^2(q_0 - |\mathbf{q}| \cos\theta)^2} - \frac{q_0 - |\mathbf{q}| \cos\theta}{E_{\mathbf{p}} + |\mathbf{k}|}\right) = \mathfrak{T} \quad (158)$$

where

$$\mathbf{q} \cdot \mathbf{k} = |\mathbf{q}|\, \omega \cos\theta \quad (159)$$

The differential cross section is obtained by applying (114)–(115) and in the center-of-mass is given by

$$\frac{d\sigma}{d\Omega} = \frac{e^2G^2}{(4\pi)^2} \frac{1}{2E_{\mathbf{p}}E_{\mathbf{p}'}} \frac{|\mathbf{q}|}{|\mathbf{k}|} \frac{1}{1 + (|\mathbf{k}|/E_{\mathbf{p}})} \frac{1}{1 + (q_0/E_{\mathbf{p}'})} \mathfrak{T} \quad (160)$$

3. *Pion Decay*

As another example of calculational techniques, consider the following phenomenological model to account for the decay of the charged pion into a muon and a neutrino

$$\mathfrak{K}_I = G_{\mathrm{eff}} : \tilde{\mu}\gamma_\mu \tfrac{1}{2}(1 - i\gamma_5)\, \nu : \partial^\mu \pi(x) + \text{h.a.} \quad (161)$$

where π denotes the meson operator and μ and ν the (four-component) spinor operators for the muon and neutrino, respectively. The neutrino wave functions are normalized so that

$$\sum_{r=1}^{2} u_{(\nu)r}(\mathbf{p})\, \tilde{u}_{(\nu)r}(\mathbf{p}) = \sum_{r=1}^{2} v_{(\nu)r}(\mathbf{p})\, \tilde{v}_{(\nu)r}(\mathbf{p}) = \frac{\gamma \cdot p}{2|\mathbf{p}|} \quad (162)$$

and the decomposition of the operator $\nu(x)$ in the interaction picture is

$$\nu(x) = \frac{1}{(2\pi)^{3/2}} \int d^3p \sum_{r=1}^{2} \{b_r(\mathbf{p})\, u_{(\nu)r}(\mathbf{p})\, e^{-ip\cdot x} + d_r^*(\mathbf{p})\, v_{(\nu)r}(\mathbf{p})\, e^{ip\cdot x} \quad (163)$$

with $p_0 = |\mathbf{p}|$; $b_r(\mathbf{p})$ and $d_r(\mathbf{p})$ are destruction operators for neutrinos and antineutrinos, respectively, satisfying the usual anticommutation rules $\{b_r(\mathbf{p}), b_{r'}^*(\mathbf{p}')\} = \delta_{rr'}\delta^{(3)}(\mathbf{p} - \mathbf{p}')$, etc., so that

$$\{\nu(x), \bar{\nu}(x')\} = -iS(x - x'; m = 0) \tag{164}$$

The interaction Hamiltonian (161) describes the elementary processes $\pi^- \to \mu^- + \bar{\nu}$ and $\pi^+ \to \mu^+ + \nu$. Due to the presence of the factor $\frac{1}{2}(1 - i\gamma_5)$, $\mathcal{H}_I$ is not P invariant. The Hamiltonian (161) can be considered as the result of the "elimination" of the strong interactions from the Hamiltonian describing the strong interaction between baryons and mesons and the weak Fermi interactions between baryons and leptons. In such a more complete theory the decay of the pion can proceed in lowest order by the process indicated in Figure 14.31a. To lowest order,

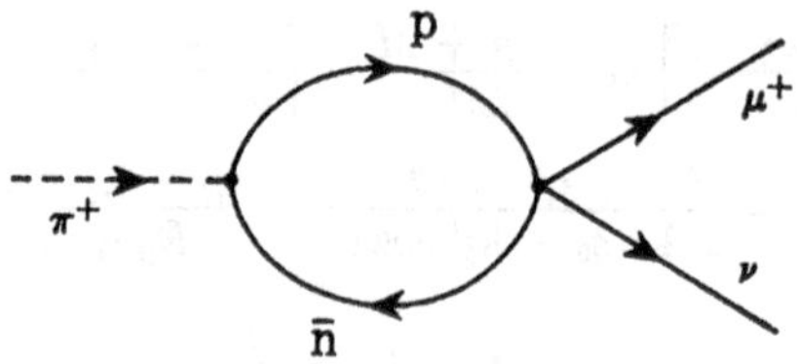

Fig. 14.31(a)

the Hamiltonian (161) gives rise to the diagram of Figure 14.31b for the decay of the meson which corresponds to a matrix element:

$$R = -i(2\pi)^4\,\delta^{(4)}(q - p - k)\,\frac{G_{\rm eff}}{(2\pi)^3}\sqrt{\frac{m_{(\mu)}}{E_{(\mu)}}}$$

$$\tilde{u}_{(\mu)}(\mathbf{p})\,\gamma^\rho\tfrac{1}{2}(1 - i\gamma_5)\,v_{(\nu)}(\mathbf{k})\,\frac{1}{(2\pi)^{3/2}}\,\frac{1}{\sqrt{2\omega_\pi}}\,q_\rho \tag{165}$$

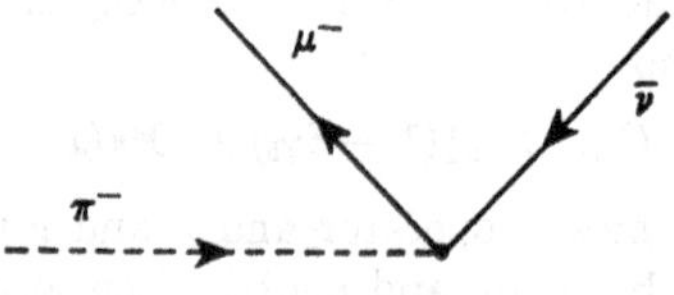

Fig. 14.31(b)

where q is the four momentum of the meson, and p and k those of the muon and antineutrino, respectively. Energy and momentum conservation asserts that $q = p + k$ so that $\gamma \cdot q = \gamma \cdot p + \gamma \cdot k$, which then allows us to simplify (165) since $\tilde{u}_{(\mu)}(\mathbf{p})\,\gamma \cdot p = m_{(\mu)}\tilde{u}(\mathbf{p})$ and $\gamma \cdot k v_{(\nu)}(\mathbf{k}) = 0$. Hence

$$R = -i(2\pi)^4\frac{G_{\rm eff}}{(2\pi)^3}\sqrt{\frac{m_{(\mu)}{}^3}{(2\pi)^3\,2\omega_\pi E_{(\mu)}}}\,\tilde{u}_{(\mu)}\tfrac{1}{2}(1 - i\gamma_5)\,v_{(\nu)}\delta^{(4)}(p + k - q) \tag{166}$$

and the transition probability per unit time and volume irrespective of the neutrino's spin is

$$\sum_{(\nu)} \frac{1}{VT} |R|^2 = (2\pi)^4 \frac{G_{\text{eff}}^2}{(2\pi)^9} \frac{m_{(\mu)}^3}{2\omega_\pi E_{(\mu)}}$$

$$\tilde{u}_{(\mu)}(\mathbf{p})\, \tfrac{1}{2}(1 + i\gamma_5) \frac{\gamma \cdot k}{2k_0} \tfrac{1}{2}(1 - i\gamma_5)\, u_{(\mu)}(\mathbf{p}) \cdot \delta^{(4)}(p + k - q) \qquad (167)$$

If we are only interested in the lifetime, then the only quantity of interest is the transition probability irrespective of the spin of the muon, which quantity is equal to

$$\frac{1}{VT} \sum_{\text{spin } \mu,\, \nu} |R|^2 = \frac{G_{\text{eff}}^2}{(2\pi)^5} \frac{m_{(\mu)}^2}{8\omega_\pi E_{(\mu)} k_0} \delta^{(4)}(p + k - q)$$

$$\cdot \operatorname{Tr}\left[\tfrac{1}{2}(1 + i\gamma_5)\, \gamma \cdot k \tfrac{1}{2}(1 - i\gamma_5)\, (\gamma \cdot p + m_{(\mu)})\right] \qquad (168)$$

The trace can be further simplified by recalling that $\frac{1}{4}(1 - i\gamma_5)^2 = \frac{1}{2}(1 - i\gamma_5)$ and that the trace of an odd number of γ matrices vanishes, so that

$$\begin{aligned} \operatorname{Tr}[\ \ ] &= \tfrac{1}{2} \operatorname{Tr}[\gamma \cdot k\, \gamma \cdot p + i\gamma \cdot k\, \gamma \cdot p\, \gamma_5] \\ &= 2\, k \cdot p \end{aligned} \qquad (169)$$

and

$$\frac{1}{VT} \sum_{\text{spin } \mu,\, \nu} |R|^2 = \frac{G_{\text{eff}}^2}{(2\pi)^5} \frac{m_{(\mu)}^2}{4\omega_\pi E_{(\mu)} k_0} k \cdot p \delta^{(4)}(p + k - q) \qquad (170)$$

Since the neutrino is massless, $k^2 = 0$, also $q^2 = m_\pi^2 = k^2 + 2p \cdot k + p^2 = m_{(\mu)}^2 + 2p \cdot k$, so that the quantity $p \cdot k$ is equal to

$$\begin{aligned} p \cdot k &= \tfrac{1}{2}(m_\pi^2 - m_{(\mu)}^2) = \frac{m_\pi^2}{2}\left(1 - \frac{m_{(\mu)}^2}{m_\pi^2}\right) \\ &= (q - k) \cdot k = q \cdot k \\ &= q_0 k_0 (1 - \mathbf{v}_\pi \cdot \hat{\mathbf{k}}) \end{aligned} \qquad (171)$$

where $\hat{\mathbf{k}}$ denotes a unit vector in the direction of the neutrino. To compute the total transition probability per unit time, Γ, it is convenient to specialize to the case of the decay of a pion at rest. We then find

$$\Gamma = \frac{G_{\text{eff}}^2}{(2\pi)^2} \int d^3p \int d^3k \frac{m_{(\mu)}^2}{4\omega_\pi E_{(\mu)} k_0} \delta^{(4)}(p + k - q)\, k \cdot p \qquad (172)$$

[a factor $(2\pi)^3$ has been introduced into (172) to normalize the initial state to one particle per unit volume]. In the rest frame of the pion $m_\pi = q_0 = \omega_\pi = (p_0 + k_0) = (\mathbf{p}^2 + m_{(\mu)}^2)^{1/2} + |\mathbf{p}|$, so that

$$|\mathbf{p}| = \frac{1}{2m_\pi}(m_\pi^2 - m_{(\mu)}^2) = \frac{m_\pi}{2}\left(1 - \frac{m_{(\mu)}^2}{m_\pi^2}\right) \qquad (173)$$

and

$$\Gamma = \frac{1}{4} \frac{G_{\text{eff}}^2}{4\pi} m_\pi m_{(\mu)}^2 \left(1 - \frac{m_{(\mu)}^2}{m_\pi^2}\right)^2 \qquad (174)$$

It is interesting to note that, if the assumption of a universal Fermi interaction is invoked, then the effective Hamiltonian describing the decay of a pion into an electron and an antineutrino is

$$\mathcal{H}_I' = G_{\text{eff}} : \tilde{e}\gamma_\mu \tfrac{1}{2}(1 - i\gamma_5)\, \nu : \partial^\mu \pi(x) + \text{h.a.} \tag{175}$$

where G_{eff} is the same coupling constant as in (161). The predicted branching ratio is given by

$$\frac{\Gamma_{\pi\to e}}{\Gamma_{\pi\to\mu}} = \frac{m_e^2(m_\pi^2 - m_e^2)^2}{m_{(\mu)}^2(m_\pi^2 - m_{(\mu)}^2)^2} = 1.3 \times 10^{-4} \tag{176}$$

which is consistent with the branching ratio found experimentally [Ashkin *et al.* (1959a)].

4. β *Decay of Neutron*

As a final example of the calculational techniques involved in quantum field theory, we here calculate some properties of the decay of the neutron in the β-decay theory proposed by Gell-Mann and Feynman and by Marshak and Sudarshan (see Chapter 10). Recall that the interaction Hamiltonian they proposed for the (parity-nonconserving) β interaction of nucleons is

$$\mathcal{H}_I = G : (\tilde{p}\gamma_\mu \tfrac{1}{2}(1 - i\gamma_5)\, n)\, (\tilde{e}\gamma^\mu \tfrac{1}{2}(1 - i\gamma_5)\, \nu) : + \text{h.a.} \tag{177}$$

where p, n, e, ν denote the (four-component spinor) operators for proton, neutron, electron and neutrino, respectively. The interaction Hamiltonian gives rise to various processes; the Feynman diagram illustrated in Figure 14.32 is the one responsible for the decay of the neutron and corresponds

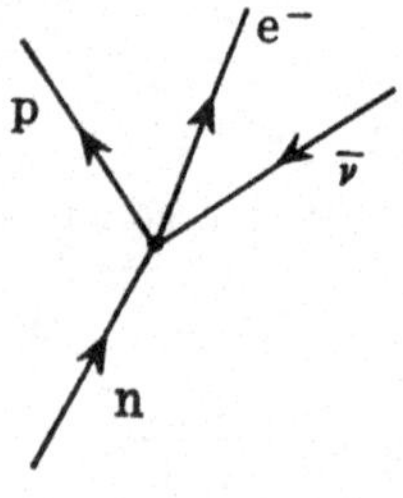

Fig. 14.32

to the elementary process $n \to p + e^- + \bar{\nu}$. One easily derives that the Feynman diagram of Figure 14.32 corresponds to a matrix element

$$R = \frac{G}{(2\pi)^6}\sqrt{\frac{M_n M_p m_e}{E_n E_p E_e}}\, (\tilde{u}_p \gamma_\mu \tfrac{1}{2}(1 - i\gamma_5)\, u_n)\, (\tilde{u}_e \gamma^\mu \tfrac{1}{2}(1 - i\gamma_5)\, v_\nu) \cdot (2\pi)^4\, \delta^{(4)}\left(\sum p_f - p_n\right) \tag{178}$$

We are to calculate $|R|^2$. For the sake of generality, let us consider the decay of a neutron polarized in some direction $\mathbf{n}$ in its rest frame. We therefore need the projection operator for a Dirac spinor corresponding to a particle polarized in the direction $\mathbf{n}$ in its rest frame. To this end we note that the operator $i\gamma_5\gamma \cdot n$ has its square equal to $+1$ if $n_\mu n^\mu = -1$ and commutes with $\gamma \cdot p + m$ if $p \cdot n = p_\mu n^\mu = 0$. Hence simultaneous eigenstates of these two operators can be found. In the rest frame of the particle when $p = (M, 0, 0, 0)$, since $p \cdot n = 0$, we must have $n_0 = 0$ ($\mathbf{n}$ can be arbitrary), so that $i\gamma_5\gamma \cdot n = -i\gamma_5\boldsymbol{\gamma} \cdot \mathbf{n} = i\gamma_5\gamma_0\boldsymbol{\gamma} \cdot \mathbf{n}\gamma_0 = = \boldsymbol{\Sigma} \cdot \mathbf{n}\gamma_0$, and a simultaneous eigenfunction of $\gamma \cdot p + m = \gamma_0 p_0 + m$ and $i\gamma_5\gamma \cdot n$ corresponds to the amplitude for a particle at rest with its spin pointing in the direction $\mathbf{n}$. The projection operator $\frac{1}{2}(1 - i\gamma \cdot n\gamma_5)$ is therefore the needed operator. Note, however, that it has a straightforward physical meaning only in the rest frame of the particle.[6]

If we ask for the probability for a neutron (polarized in the direction $\mathbf{n}_n$ in its rest frame) to decay into a proton, an antineutrino and an electron (polarized in the direction $\mathbf{n}_e$ in its rest frame) irrespective of the polarization of proton and neutrino, we must compute:

$$\frac{1}{VT} \sum_{\substack{\text{proton spin}\\ \text{neutrino spin}}} |R|^2 = (2\pi)^4 \, \delta^{(4)}\left(\sum p' - p\right) \frac{G^2}{(2\pi)^{12}} \frac{M_n M_p m_e}{E_n E_p E_e}$$

$$\mathrm{Tr}\left[\tfrac{1}{2}(1 + i\gamma_5)\,\gamma_\nu \frac{\gamma \cdot p_p + M_p}{2M_p} \gamma_\mu \tfrac{1}{2}(1 - i\gamma_5) \frac{\gamma \cdot p_n + M_n}{2M_n} \frac{1 - i\gamma \cdot n_n\gamma_5}{2}\right]$$

$$\mathrm{Tr}\left[\tfrac{1}{2}(1 + i\gamma_5)\,\gamma^\nu \frac{\gamma \cdot p_e + m_e}{2m_e} \tfrac{1}{2}(1 - i\gamma \cdot n_e\gamma_5)\,\tfrac{1}{2}(1 - i\gamma_5) \frac{\gamma \cdot p_{(\nu)}}{2|\mathbf{p}_{(\nu)}|}\right] \tag{179}$$

The first trace corresponds to the nucleon factors, the second to the lepton factors. The traces can be simplified by recalling that $\gamma_5{}^2 = -1$ so that $\gamma_5(1 \pm i\gamma_5) = \mp i(1 \pm i\gamma_5)$; also, $\frac{1}{4}(1 \pm i\gamma_5)^2 = \frac{1}{2}(1 \pm i\gamma_5)$ and $(1 \mp i\gamma_5)(1 \pm i\gamma_5) = 0$. Using these facts, the trace of the nucleon factors becomes

$$\mathrm{Tr}[\;\;]_{\text{nucleon}}$$

$$= \frac{1}{2^4 M_n M_p} \mathrm{Tr}\,[\gamma_\nu\gamma \cdot p_p\gamma_\mu(1 - i\gamma_5)(\gamma \cdot p_n + M_n)(1 - i\gamma \cdot n_n\gamma_5)]$$

$$= \frac{1}{2^4 M_n M_p} \mathrm{Tr}\,[\gamma_\nu\gamma \cdot p_p\gamma_\mu(\gamma \cdot p_n + M_n)(1 - \gamma \cdot n_n)(1 + i\gamma_5)] \tag{180}$$

Furthermore, since the trace of an odd number of γ matrices vanishes, we obtain

[6] Alternatively, we could introduce the noncovariant projection operator $\frac{1}{2}(1 - s\boldsymbol{\Sigma} \cdot \mathbf{p})$ corresponding to states of definite helicity [see also Michel and Wightman (Michel 1957)].

$$\text{Tr}\,[\;\;]_{\text{nucleon}} = \frac{1}{2^4 M_n M_p}\,[\gamma_\nu \gamma \cdot p_p \gamma_\mu \gamma \cdot p_n(1 + i\gamma_5) \\ - \gamma_\nu \gamma \cdot p_p \gamma_\mu M_n \gamma \cdot n_n - i\gamma_\nu \gamma \cdot p_p \gamma_\mu M_n \gamma \cdot n_n \gamma_5] \quad (181)$$

The traces are readily computed using the fact that

$$\text{Tr}\,(\gamma_\mu \gamma_\nu \gamma_\rho \gamma_\sigma) = 4(g_{\mu\sigma} g_{\nu\rho} - g_{\mu\rho} g_{\nu\sigma} + g_{\mu\nu} g_{\rho\sigma}) \quad (182a)$$

$$\text{Tr}\,(\gamma_\mu \gamma_\nu \gamma_\rho \gamma_\sigma \gamma_5) = -4\epsilon_{\mu\nu\rho\sigma} \quad (182b)$$

$$\text{Tr}\,(\gamma_5 \gamma_\mu \gamma_\nu) = 0 \quad (182c)$$

and one finds that the trace of the nucleon factors is equal to

$$4(g_{\nu\beta} g_{\alpha\mu} - g_{\nu\mu} g_{\alpha\beta} + g_{\nu\alpha} g_{\mu\beta} - i\epsilon_{\nu\alpha\mu\beta})\,(p_p{}^\alpha p_n{}^\beta - M_n p_p{}^\alpha n_n{}^\beta)$$

and, similarly, the trace of the leptonic factor is equal to

$$4(g^{\mu\beta} g^{\alpha\nu} - g^{\mu\nu} g^{\alpha\beta} + g^{\mu\alpha} g^{\nu\beta} - i\epsilon^{\mu\alpha\nu\beta})\,(p_{(\nu)\alpha} p_{e\beta} - m_e p_{(\nu)\alpha} n_{e\beta})$$

Using the fact that

$$\epsilon^{\alpha\beta\mu\nu}\epsilon_{\alpha\beta\rho\sigma} = -2!(g^\mu{}_\rho g^\nu{}_\sigma - g^\mu{}_\sigma g^\nu{}_\rho) \quad (183)$$

the product of the traces is found equal to

$$4[p_{(\nu)} \cdot (p_n - M_n n_n)]\,[(p_e - m_e n_e) \cdot p_p] \quad (184)$$

For an electron of definite helicity, in the rest frame of the electron we have $n_{e0} = 0$ and $\mathbf{n}_e = s_e \mathbf{p}_e / |\mathbf{p}_e|$ (where $s_e = \pm 1$). Since $n_{e\mu}$ transforms like a four-vector, in the moving system we will have

$$n_{e0} = \frac{E_e}{m_e}\,(0 + v_e s_e)$$

$$= s_e \frac{E_e}{m_e} v_e \quad (185)$$

so that the transition probability, in the nonrelativistic limit for the proton's motion, is proportional to

$$G^2 M E_e E_{(\nu)} (1 + \cos\theta_\nu)\,(1 - s_e v_e) \quad (186)$$

where θ_ν is the angle between the spin of the neutron and the direction of the emitted antineutrino. These predictions of the theory—in particular the neutrino angular correlation—have been confirmed by Telegdi *et al.* [Telegdi (1958)]. Equation (186) implies that for $v_e/c \sim 1$ the probability for finding electrons with helicity $+1$ is approximately 0, whereas for $v_e/c \sim 1$, the relative probability that the electron is emitted with helicity -1 is approximately one. Hence the electrons emitted in the β decay of the neutron are polarized (spinning to the left). For the neutron decay under consideration, they are emitted isotropically. This has also been verified by Telegdi.

The fact that electrons emitted in β decay should be strongly longitudinally polarized (of order v_e/c) has stimulated experimental and theoretical physicists to devise methods to detect such electrons. It has thus led to a recalculation of cross sections for

(*a*) the scattering of polarized electrons by nuclei (Mott scattering);
(*b*) the scattering of polarized electrons by polarized electrons (which can be realized by scattering polarized electrons from ferromagnetic 3*d* electrons of iron in a magnetic field);
(*c*) the circular polarization of bremstrahlung from polarized electrons;
(*d*) the annihilation of polarized positrons in ferromagnetic materials;
(*e*) the Compton scattering of polarized radiation.

For a review of the experimental and theoretical work in this connection, the reader is referred to the review article of Sternheimer (1959).

14f. Symmetry Principles and S Matrix

The invariance of the Lagrangian under certain symmetry operations, such as space reflections, Lorentz transformations, charge conjugations, etc., entails that the S matrix is likewise invariant under the same symmetry operation. This is expressed by the statement that

$$USU^{-1} = S \tag{187}$$

where U is the unitary (or antiunitary) operator which induces the symmetry operation on the Hilbert space of state vectors. Equation (187) can also be written as

$$[U, S] = 0 \tag{188}$$

Thus U can be called a constant of the collision. It is an observable only in the case that the symmetry operation is a discrete operation for which $U^2 = 1$, so that U is also Hermitian. However, it will be recalled that in the case of a continuous transformation U can be written in the form $\exp i\theta G$, where G is Hermitian and is the generator for an infinitesimal transformation. One therefore deduces from the infinitesimal situation that $[1 + i\epsilon G, S] = 0 = [G, S]$. Since G is Hermitian, the states of the system can be classified by the eigenvalues of G. The equation $[G, S] = 0$ then asserts that the transition amplitude, $\langle g'' \mid S \mid g' \rangle$, between two states with different eigenvalues of G, vanishes. Examples of operators G are the energy-momentum, angular momentum, isotopic spin, strangeness, baryon number, lepton number, etc. Many examples of the selection rules which follow from the fact that there exists a G such that $[S, G] = 0$ can be given. The most familiar are probably Laporte's rule in the emission of dipole radiation in atomic transitions, and the Fermi and Gamov-Teller selection rules in nuclear beta decay.

In the case that U corresponds to a discrete operation such that $U^* = U$ and $U^2 = 1$ (as, for example, for the case of space inversion and charge conjugation), U is then unitary and Hermitian.[7] The vanishing of the commutator $[S, U]$ together with the hermiticity of U then implies that the S matrix vanishes between states having different eigenvalues of the U operator.

We shall illustrate the consequences of the equation $[S, U] = 0$ in the discrete case, by deriving certain selection rules for the decay of positronium from the invariance of quantum electrodynamics under the operation of charge conjugation. Incidentally, it will be recalled that only states of zero charge can be eigenstates of U_c (since $U_cQ = -QU_c$), so that selection rules due to charge conjugation invariance can be derived only for neutral systems. That quantum electrodynamics is invariant under the unitary operation U_c under which

$$U_cA_\mu(x)\, U_c^{-1} = \eta_A A_\mu(x) \tag{189a}$$

$$U_c\psi(x)\, U_c^{-1} = \eta_\psi \tilde{\psi}(x)\, C \tag{189b}$$

$$U_c\tilde{\psi}(x)\, U_c^{-1} = -\bar{\eta}_\psi C^*\psi(x) \tag{189c}$$

$$C^T = -C; \quad C^{-1}\gamma^\mu C = -\gamma^{\mu T}; \quad C^* = C^{-1} \tag{189d}$$

follows from the fact that it is possible to choose phases η_A and η_ψ such that the field equations and commutation rules of the operators are preserved and such that the Hamiltonian $\mathcal{H}_I$ is invariant under U_c: $\mathcal{H}_I = U_c\mathcal{H}_I U_c^{-1}$. The invariance of the commutation rules implies that $|\eta_\psi|^2 = |\eta_A|^2 = +1$. The choice of phase

$$\eta_A = -1 \tag{190}$$

guarantees that $\mathcal{H}_I(x) = -e : \tilde{\psi}(x)\, \gamma^\mu\psi(x) : A_\mu(x) = j_\mu(x)\, A^\mu(x)$ is invariant under U_c. [Recall that $U_c j_\mu(x)\, U_c^{-1} = -j_\mu(x)$.] Furthermore, we can choose the phase of U_c such that $U_c \mid \Phi_0\rangle = 0$.

Consider next a system consisting of an electron and a positron whose state vector is

$$|\Phi\rangle = \sum_{t,\, s} \int d^3p \int d^3q\, \chi(\mathbf{p}s;\, \mathbf{q}t)\, b_s^*(\mathbf{p})\, d_t^*(\mathbf{q}) \mid \Phi_0\rangle \tag{191}$$

where χ corresponds (in the nonrelativistic limit) to the momentum space wave function of the system. Under U_c the state $|\Phi\rangle$ transforms into

$$\begin{aligned} U_c \mid \Phi\rangle &= \sum_{t,\, s} \int d^3p \int d^3q\, \chi(\mathbf{p}s;\, \mathbf{q}t)\, d_s^*(\mathbf{p})\, b_t^*(\mathbf{q}) \mid \Phi_0\rangle \\ &= -\sum_{t,\, s} \int d^3p \int d^3q\, \chi(\mathbf{p}s;\, \mathbf{q}t)\, b_t^*(\mathbf{q})\, d_s^*(\mathbf{p}) \mid \Phi_0\rangle \\ &= -\sum_{t,\, s} \int d^3p \int d^3q\, \chi(\mathbf{q}t;\, \mathbf{p}s)\, b_s^*(\mathbf{p})\, d_t^*(\mathbf{q}) \mid \Phi_0\rangle \end{aligned} \tag{192}$$

[7] Recall that the choice $U^2 = +1$ (rather than $e^{i\beta}$, say) implies a certain choice of phases for the spinor operators.

Therefore, if χ is symmetric under the exchange of negaton and positon, $|\Phi\rangle$ is an eigenfunction of U_c with eigenvalue -1 and if χ is antisymmetric, $|\Phi\rangle$ is an eigenfunction with eigenvalue $+1$.

Now one readily verifies that an n-photon state

$$|\Phi_n\rangle = \sum_{\lambda_1, \cdots \lambda_n}^{1,2} \int d^3k_1 \cdots \int d^3k_n \, \phi(\mathbf{k}_1\lambda_1, \cdots, \mathbf{k}_n\lambda_n) \, a_{\lambda_1}{}^*(\mathbf{k}_1) \cdots a_{\lambda_n}{}^*(\mathbf{k}_n) \mid \Phi_0\rangle \tag{193}$$

is an eigenstate of U_c with eigenvalue $(-1)^n$. Since $[S, U_c] = 0$, it follows that the transition amplitude $\langle\Phi_n \mid S \mid \Phi\rangle = \langle\Phi_n \mid U_c^{-1}SU_c \mid \Phi\rangle$ vanishes unless the charge conjugation quantum number of the initial and final state is the same. Now for a ${}^1S$ state, χ is antisymmetric under the exchange of the two particles, so that $|\Phi\rangle$ in this case has a charge conjugation eigenvalue $+1$. Such a state of the electron-positron system can therefore only annihilate into an even number of photons 2, 4, $\cdots$, etc. (the decay rate in 4 photons is of course a factor α^2 smaller than that for decay into 2 photons). Similarly, since the ${}^3S$ wave function is symmetric under exchange of the particles, the electron-positron system when in this state can only decay into an odd number of photons.

If C is a good quantum number for the theory describing the interaction of mesons and nucleons with one another and with the electromagnetic field, then, since the π^0 field is a charge self-conjugate field $U_c\pi^0U_c^{-1} = \eta_{\pi^0}\pi^0$ ($\eta_{\pi^0}{}^2 = 1$, η_{π^0} is real), the fact that the π^0 meson decays into two γ rays implies that $\eta_{\pi^0} = +1$. *Proof:* Since the final state of two photons has a charge conjugation parity of $+1$, the initial state of a π^0 meson must have the same parity. [In terms of Feynman diagrams, the mechanism for the decay is indicated in Fig. 14.33.] Note, incidentally, that the assignment

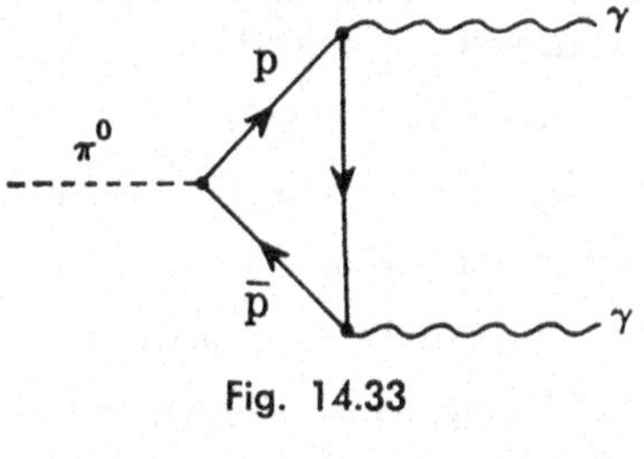

Fig. 14.33

$\eta_{\pi^0} = +1$ is also in accordance with the description of the interaction between the neutral (pseudoscalar) mesons and nucleons by a charge conjugation invariant Yukawa coupling of the form $\mathcal{H}_I = \sum_{i=1}^{3} :\bar{\psi}\gamma_5\tau_i\psi:\pi_i$. *Proof:* Under charge conjugation, $U_c:\bar{\psi}\gamma_5\tau_3\psi:U_c^{-1} = :\bar{\psi}\gamma_5\tau_3\psi:$ so that for the invariance of the 3-component under U_c, η_{π^0} must equal $+1$.

A further consequence of the invariance of a theory under the operation of charge conjugation is that the transition amplitude between two states $|a\rangle$ and $|b\rangle$ is the same as the transition amplitude between the states $|\bar{a}\rangle = U_c \mid a\rangle$ and $|\bar{b}\rangle = U_c \mid b\rangle$ where $|\bar{a}\rangle$, $|\bar{b}\rangle$ are the states obtained by replacing every particle in the states $|a\rangle$, $|b\rangle$ by its antiparticle. *Proof:* Since U_c is unitary and $[S, U_c] = 0$ by assumption,

$$\begin{aligned}\langle b \mid S \mid a\rangle &= \langle b \mid U_c{}^*U_cSU_c{}^*U_c \mid a\rangle \\ &= \langle \bar{b} \mid S \mid \bar{a}\rangle \end{aligned} \tag{194}$$

Another relation which follows from C invariance is that the charge and magnetic moment of particle and antiparticle have opposite signs. *Proof:* The magnetic moment of a particle is given by the expectation value of the operator $\int \mathbf{r} \times \mathbf{j}\, d^3x$ where $\mathbf{j}$ is the total current operator. If we call $|a\rangle$ the state of a particle and $|\bar{a}\rangle$ the state of the antiparticle with the same symmetry quantum number, then

$$\begin{aligned}\boldsymbol{\mu}_a &= \langle a \mid \int \mathbf{r} \times \mathbf{j}(x)\, d^3x \mid a\rangle = \langle a \mid U_c{}^{-1}U_c \int \mathbf{r} \times \mathbf{j}(x)\, d^3x\, U_c{}^{-1}U_c \mid a\rangle \\ &= -\langle \bar{a} \mid \int \mathbf{r} \times \mathbf{j}\, d^3x \mid \bar{a}\rangle \\ &= -\boldsymbol{\mu}_{\bar{a}} \end{aligned} \tag{195}$$

since $U_c j_\mu(x)\, U_c{}^{-1} = -j_\mu(x)$. Note therefore that a purely neutral (charge self-conjugate) particle must have zero magnetic moment. [For further consequences see Pais (1952b), Wolfenstein (1952), Michel (1953), Feinberg (1959).]

The consequences of the invariance of a field theory under the operation of time reversal are different from those of inversion and charge conjugation invariance due to the antiunitary character of the operator $U_T = T$ inducing the transformation. Recall that a theory is said to be invariant under time inversion if by a suitable adjustment of the phases η_{iT} of the field operators when transformed under U_T:

$$U_T\phi(x)\, U_T{}^{-1} = \eta_{\phi T}\phi(i_t x) \tag{196a}$$

$$U_T\psi(x)\, U_T{}^{-1} = \eta_{\psi T}C^{-1}\gamma_5\psi(i_t x) \tag{196b}$$

the field equations and commutation rules are invariant under T and

$$U_T\mathcal{H}_I(x)\, U_T{}^{-1} = \mathcal{H}_I(i_t x) \tag{197}$$

Consider the action of U_T on an incoming solution $|\Psi_a{}^+\rangle$ of the Lippmann-Schwinger which satisfies

$$|\Psi_a{}^+\rangle = |\Phi_a\rangle + \frac{1}{E_a - H_I + i\epsilon} H_I \mid \Psi_a{}^+\rangle \tag{198}$$

and which reduces to the plane wave state $|\Phi_a\rangle$ at $t = -\infty$. If the theory is T invariant, so that $TH_IT^{-1} = H_I$, we deduce that

$$U_T \mid \Psi_a{}^+\rangle = U_T \mid \Phi_a\rangle + \frac{1}{E_a - H_I - i\epsilon} H_I U_T \mid \Psi_a{}^+\rangle \tag{199}$$

By virtue of the antiunitary character of T the sign of $i\epsilon$ in the denominator has been changed, so that $U_T \mid \Psi_a{}^+\rangle$ is an outgoing solution of the Lippmann-Schwinger equation which corresponds to the plane wave state $U_T \mid \Phi_a\rangle$ at $t = +\infty$. Now we have previously determined (in Chapters 7, 8, and 9) that the operation U_T on $|\Phi_a\rangle$ reverses the velocity of all the particles, so that if, for example, $|\Phi_a\rangle$ is a two-particle state in which the particles have velocity $\mathbf{v}_1$ and $\mathbf{v}_2$, respectively, the state $U_T \mid \Phi_a\rangle$ corresponds to the state wherein the particles have velocities $-\mathbf{v}_1$ and $-\mathbf{v}_2$, respectively. We shall denote $U_T \mid \Phi_a\rangle$ by $|\Phi_{Ta}\rangle$, i.e.,

$$U_T \mid \Phi_a\rangle = |\Phi_{Ta}\rangle \tag{200}$$

Since the S matrix is defined by $S_{ba} = \langle \Psi_b{}^{(-)} \mid \Psi_a{}^{(+)}\rangle$, the invariance of the theory under time inversion implies that

$$\begin{aligned} S_{ba} &= (\Phi_b, S\Phi_a) = (\Psi_b{}^{(-)}, \Psi_a{}^{(+)}) \\ &= \overline{(U_T\Psi_b{}^{(-)}, U_T\Psi_a{}^{(+)})} = \overline{(\Psi_{Tb}{}^{(+)}, \Psi_{Ta}{}^{(-)})} \\ &= (\Psi_{Ta}{}^{(-)}, \Psi_{Tb}{}^{(+)}) \\ &= S_{TaTb} \end{aligned} \tag{201}$$

which is the "detailed balance" theorem. Under certain circumstances (when the theory is rotationally invariant, and the basis vectors $|\Phi_a\rangle$ are chosen to be eigenstates of the total angular momentum) Eq. (201) reduces to the statement that the S matrix is symmetric [Coester (1953)].

We have previously made reference to the fact that, if a field theory has the properties that

(*a*) it is invariant under proper Lorentz transformations,
(*b*) its Hamiltonian is Hermitian,
(*c*) the usual connection between spin and statistics is made,
(*d*) the field operators are local,

that it can then be shown [Lüders (1957)] that such a field theory is automatically invariant under the product operation TCP ($= U_TU_CU_P$) taken in any order. More precisely stated, it can then be shown that the phases η_{iP}, η_{iC}, η_{iT} for the various field operators ($\phi_i = \phi, \psi, A_\mu$ etc.) under the operation P, C, and T can be so chosen that $[TCP, H] = 0$. In particular, the choice $\eta_{iT}\eta_{iC}\eta_{iP} = 1$ guarantees that TCP will be conserved: this even if C, P, and T are not individually conserved. That the phases can always be chosen so that $\eta_{iT}\eta_{iC}\eta_{iP} = 1$ follows from the fact that, if T, C, and P are not individually conserved, their phases are arbitrary and can therefore be so chosen as to make their product one. If on the other hand, T, C, and P are conserved individually so that their phases are fixed, then η_{iC} and η_{iP} can be so chosen that P and C are conserved and η_{iT} so that $\eta_{iT}\eta_{iP}\eta_{iC} = +1$. We shall not here present the proof of this important theorem since several excellent review articles exist [Lüders (1957), Pauli (1955), Grawert (1959); see also Feinberg (1959)].

One of the consequences of the TPC theorem is that the mass of a particle is always equal to the mass of its antiparticle; also, that for an unstable particle of mass m, the lifetime of particle and antiparticle are the same [Lee, Oehme, and Yang (Lee 1957b)]. To prove this assertion, it is assumed that the particle is stable when only strong interactions are considered and that its instability is brought about by a weak interaction whose effects need only be considered to first order. The Hamiltonian H_{strong} of the strong interactions then defines the one-particle state $|\Psi_a\rangle$, such that, if the particle is at rest, $H_{\text{strong}} \mid \Psi_a\rangle = m \mid \Psi_a\rangle$. This particle then decays as a result of the action of H_{weak}. The end product is a state $|\Psi_b\rangle$ of the decay products. The total transition rate is then determined by the matrix element

$$\delta^{(4)}(P_b - P_a) \langle b \mid M \mid a\rangle = \int_{-\infty}^{+\infty} \langle \Psi_b \mid H_{\text{weak}}(t) \mid \Psi_a\rangle \, dt$$

and is proportional to

$$\Gamma_a \propto \frac{1}{m} \int |\langle b \mid M \mid a\rangle|^2 \, \delta^{(4)}(P_b - P_a) \, db \tag{202}$$

The integration is over all possible final states. It is clear that the insertion of the operator TPC into the matrix element $|\langle b \mid M \mid a\rangle|^2 = |(\Psi_b, H_{\text{weak}}(0)\Psi_a)|^2$ will change particle to antiparticle, from which one then infers that $\Gamma_a = \Gamma_{\bar{a}}$.

15

Quantum Electrodynamics

In the present chapter we begin our presentation of the methods that have been devised to circumvent the divergence difficulties inherent in all relativistic local field theories with nontrivial interaction terms. These methods, due principally to Feynman and Schwinger, have their most striking success in quantum electrodynamics where, due to the smallness of the coupling constant, a perturbation approach gives remarkable quantitative agreement between calculations and observations for the Lamb shift, the separation of the ground state doublet of positronium, the hyperfine structure of hydrogen, the line shape of emitted radiation in atomic transitions, and other atomic phenomena. The Feynman-Schwinger-Dyson methods circumvent the difficulties caused by the appearance of divergent expressions in the calculations of the radiative corrections, by absorbing them into parameters to which finite values are imposed a posteriori as the observed experimental value. This method of bypassing the divergence difficulties has become known as renormalization because *all* the infinities in the S-matrix expansion for certain important field theories can be traced to certain basic ones in the parameters of the theory (e.g., the correction to the mass and charge of the electron in quantum electrodynamics). It was shown by Dyson that in quantum electrodynamics if one replaces these infinite corrected parameters (of mass and charge) everywhere in the theory by their observed values in nature, then all the coefficients of the power series expansions of the S matrix are finite. In its original form, the renormalization procedure was applicable only to the power series expansion of the theory, and in particular of the S matrix. The restriction to power series expansion was not too severe a restriction in the case of quantum electrodynamics where the smallness of the coupling constant, α, made it hopeful that even if the power series did not converge, such expansions might be useful asymptotic expressions. In meson theories, however, the largeness of the coupling constants makes it doubtful that a perturbation expansion converges and that such series are useful representations of the predictions of the theory. Källén, Valatin, and others have given a formulation of the renormalization procedure which frees it from the restriction to power series.[1]

The question, however, still remains as to whether the ultraviolet divergences are due to inadequate mathematical method used or are intrinsic in the present formulation of the physics. The belief at present is that the second statement is probably the "truer." We shall return to these questions at the end of our discussion of the renormalization program.

In the present chapter we present these renormalization methods as applied to the perturbation expansion of the S matrix in quantum electrodynamics. In Section h we briefly take up the renormalization methods for pseudoscalar meson theory with nonderivative coupling.

15a. The Self-Energy of a Fermion

We have previously noted that in the decomposition of the S matrix describing a theory of interacting bosons and fermions $[\mathcal{L}_I = G : \bar{\psi}(x)\, \Gamma \psi(x) : \phi(x)]$, one of the possible processes in second order corresponded to the virtual emission and reabsorption of a boson quantum by a fermion. The Feynman diagram corresponding to this process is again

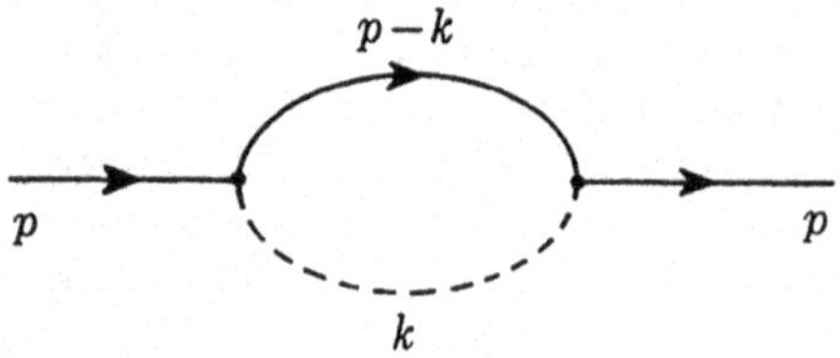

Fig. 15.1

redrawn in Figure 15.1. The matrix element for the case when the lines p are external lines has been previously written down: it is

$$R = \frac{G^2}{4\pi\hbar c}\frac{1}{4\pi^3}\, T\, \frac{M_0}{E_0(\mathbf{p})}\, \tilde{w}^s(\mathbf{p})$$

$$\int \Gamma \frac{\gamma \cdot (p-k) + M}{(p-k)^2 - M_0^2 + i\epsilon}\, \Gamma \frac{1}{k^2 - \mu_0^2 + i\epsilon'}\, w^s(\mathbf{p})\, d^4k$$

$$E_0(\mathbf{p}) = \sqrt{\mathbf{p}^2 + M_0^2} \qquad (1)$$

[1] For a masterly historical review of the developments of quantum electrodynamics, see the preface by J. Schwinger to the selected set of papers on quantum electrodynamics he has edited [Schwinger (1958)]. The reader is urged to acquaint himself with the content of that volume for a correct appreciation of the more recent developments in the light of the advances made during the 1930's.

where we have normalized our wave functions to a volume V. Hereafter we shall call $\Sigma^{(2)}(p)$ the operator

$$\Sigma^{(2)}(p) = \frac{G^2}{4\pi\hbar c} \int \Gamma \frac{1}{\gamma \cdot (p - k) - M_0 + i\epsilon} \Gamma \frac{d^4k}{k^2 - \mu_0{}^2 + i\epsilon'} \tag{2}$$

More generally it may happen that the diagram in Figure 15.1 is an internal part of a larger diagram, so that the two outer fermion lines correspond actually to fermion propagators. The factor corresponding to such an internal part is then given by

$$4\pi \left(\frac{i}{(2\pi)^4}\right) \frac{1}{\gamma \cdot p - M_0} \Sigma^{(2)}(p) \frac{1}{\gamma \cdot p - M_0}$$

It is important to realize that we can then look upon the diagram of Figure 15.1 as modifying the propagation kernel $(\gamma \cdot p - M_0)^{-1}$ (Fig. 15.2) which

Fig. 15.2

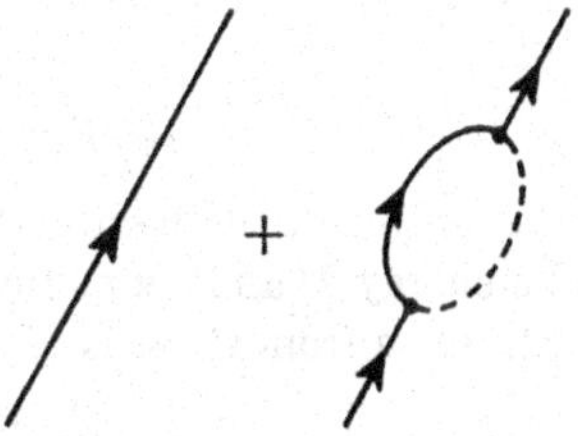

Fig. 15.3

arises from a lower order term in the perturbation expansion. The modified propagation kernel then corresponds to the diagram of Figure 15.3 and is given analytically by a factor

$$\frac{i}{(2\pi)^4} \left[\frac{1}{\gamma \cdot p - M_0} + \frac{1}{\gamma \cdot p - M_0} \frac{i}{4\pi^3} \Sigma^{(2)}(p) \frac{1}{\gamma \cdot p - M_0} \right]$$

Clearly this viewpoint can be extended to higher order fermion self-energy diagrams. Let us, however, first consider the case the lines p are external lines.

It will be recalled that in the interaction picture, the evolution in time of the state vector $|\Phi(t_0)\rangle = |\Phi(\mathbf{p}s)\rangle = b_s{}^*(\mathbf{p}) \mid \Phi_0\rangle$ describing at time t_0 (in the remote past) a free fermion, is determined by the equation

$$|\Phi(t)\rangle = U(t, t_0) \mid \Phi(t_0)\rangle \tag{3}$$

where $U(t, t_0)$ is the U matrix given by Eq. (11.141). As a result of the interaction of the fermion with the quantized boson field, the energy of the state $|\Phi(\mathbf{p}s)\rangle$ is changed and the shift in energy, ΔE, is given by

$$e^{-i\Delta E(t - t_0)} = \frac{\langle \Phi(\mathbf{p}s) \mid U(t, t_0) \mid \Phi(\mathbf{p}s)\rangle}{\langle \Phi_0 \mid U(t, t_0) \mid \Phi_0\rangle} \tag{4}$$

In (4) the time difference $t - t_0$ is to be taken as large and any oscillating dependence on t and t_0 is to be averaged out. Equation (4) states that the change in the energy of the state $|\Phi(\mathbf{p}s)\rangle$ as a result of the interaction manifests itself by a change of phase. [Alternatively, Eq. (11.189) for the level shift could be used; however, this formula does not involve the large time difference $T = t - t_0$ in as explicit a manner.] The factor $(\Phi_0 \mid U(t, t_0) \mid \Phi_0)$ cancels the infinite phase factor present in the numerator in the right-hand side of Eq. (4) due to the disconnected vacuum-vacuum diagrams. To lowest order in ΔE we thus obtain

$$-i\Delta ET = (iG)^2 \langle \Phi(\mathbf{p}s) \mid \int d^4x_1 \int d^4x_2$$

$$P\{ :\bar{\psi}(x_1)\, \Gamma\psi(x_1)\, \phi(x_1): \; :\bar{\psi}(x_2)\, \Gamma\psi(x_2)\, \phi(x_2): \}_C \mid \Phi(\mathbf{p}s)\rangle$$

$$= \frac{1}{4\pi^3}\, T\, \frac{M_0}{E_0(\mathbf{p})}\, \tilde{w}^s(\mathbf{p})\, \Sigma^{(2)}(p)\, w^s(\mathbf{p}) \tag{5a}$$

or

$$\Delta E = \frac{i}{4\pi^3} \frac{M_0}{E_0(\mathbf{p})}\, \tilde{w}^s(\mathbf{p})\, \Sigma^{(2)}(p)\, w^s(\mathbf{p}) \tag{5b}$$

Now the only measurements we can make on a free particle are to observe its energy E and its momentum $\mathbf{p}$. The only invariant quantity that we can form from these is $E^2 - \mathbf{p}^2$, the mass of the particle:

$$E^2 - \mathbf{p}^2 = M^2 \tag{6}$$

If we have a relativistically invariant interaction, then the only effect this interaction can have on a free particle is to change its mass, let us say, by δM. The change in the energy of the bare particle must therefore be given by

$$E_0 + \Delta E_0 = \sqrt{(M_0 + \delta M)^2 + \mathbf{p}^2} \tag{7}$$

Expanding (7) in δM, we find

$$E_0 + \Delta E_0 = \sqrt{M_0^2 + \mathbf{p}^2} + \frac{M_0 \delta M_0}{\sqrt{M_0^2 + \mathbf{p}^2}} + O((\delta M)^2) \tag{8}$$

and to first order

$$\Delta E_0 = \frac{M_0 \delta M}{E_0} \tag{9}$$

In other words, in an invariant theory the first-order self-energy must depend on the momentum of the particle in a special way, viz., it must be inversely proportional to $\sqrt{M_0^2 + \mathbf{p}^2}$. That the self-energy is inversely proportional to E_0 was already recognized by Weisskopf (1934a, b) in his perturbation theoretic treatment of the self-energy in the hole theory.

The fact that the self-energy of a free particle can only lead to a change of its mass is the basis of the modern solution of the divergence difficulties. The original suggestion was made by Kramers in 1947 [Kramers (1950)] in connection with classical electrodynamics. He pointed out that we can never experimentally observe M_0, the bare mass of the electron, but only $(M_0 + \delta M)$, the observed mass, which we denote as usual by M, i.e.,

$$M = M_0 + \delta M \tag{10}$$

Kramers therefore suggested, returning to Heisenberg's original view of quantum mechanics, that only the observable quantity M should play a role in the theory and not the separate quantities M_0 or δM. This principle is known as the principle of *mass renormalization.* Kramers accepts the fact that the change in mass may be infinite (however, hoping that a future theory might make it small and finite), but does not consider this of consequence since the physically observable quantities, such as energy levels and cross sections, are finite when expressed in terms of the observable mass, M. This principle has proved exceedingly powerful in practice and led to the successful development of quantum electrodynamics and pseudoscalar meson theory in the past decade.

The mass renormalization can be included explicitly into the theory by introducing a counterterm in the Lagrangian in a manner already explained in Chapter 12. The Lagrangian of the free-matter field contains a term $M_0 : \tilde{\psi}(x)\, \psi(x) :$ where M_0 is the mass of the bare quantum. This term can be written as $M : \tilde{\psi}(x)\, \psi(x) : - \, \delta M : \tilde{\psi}(x)\, \psi(x) :$ and the term $-\delta M : \tilde{\psi}(x)\, \psi(x) :$ considered as part of the perturbation energy along with the interaction $G : \tilde{\psi}(x)\, \Gamma\psi(x)\, \phi(x) :$. More precisely, the contribution of the term $-\delta M : \tilde{\psi}(x)\, \psi(x) :$ is always to be considered together with the contribution of the fermion self-energy diagrams (which they cancel in the case of a free particle). For example, we must consider together the diagrams of Figure 15.4*a* and *b*, the small circle in diagram *b*

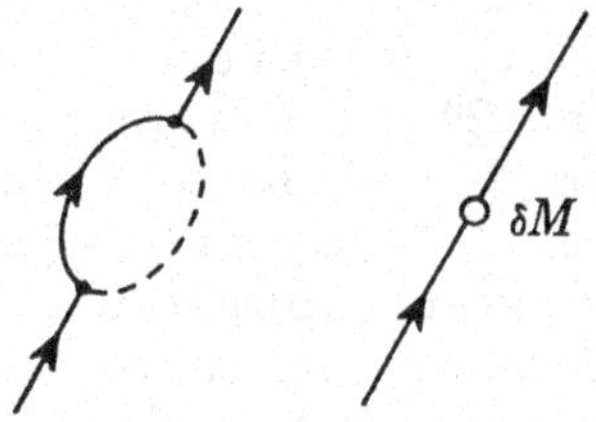

Fig. 15.4

corresponding to the contribution from the $\delta M : \tilde{\psi}\psi :$ term. The divergent quantity δM is then to be determined such that the contribution of the self-energy diagram *a* should be exactly canceled by the diagram *b* if the fermion is free, so that its mass is exactly M, the observed mass of the fermion. More generally, if we consider all the self-energy diagrams which contribute to the amplitude A that an electron in the state $|\Phi(E_{\mathbf{p}}, \mathbf{p}s)\rangle$,

$E_{\mathbf{p}} = (\mathbf{p}^2 + M^2)^{1/2}$, will remain in the state $|\Phi(E_{\mathbf{p}}, \mathbf{p}s)\rangle$, then, apart from numerical constants,

$$A = \bar{w}^s(\mathbf{p}) \left[1 + \Sigma'(p)\right] w^s(\mathbf{p}) \tag{11a}$$

where $\Sigma'(p)$ denotes the contributions of all fermion self-energy diagrams, some of which are illustrated in Figure 15.5. The requirement of mass

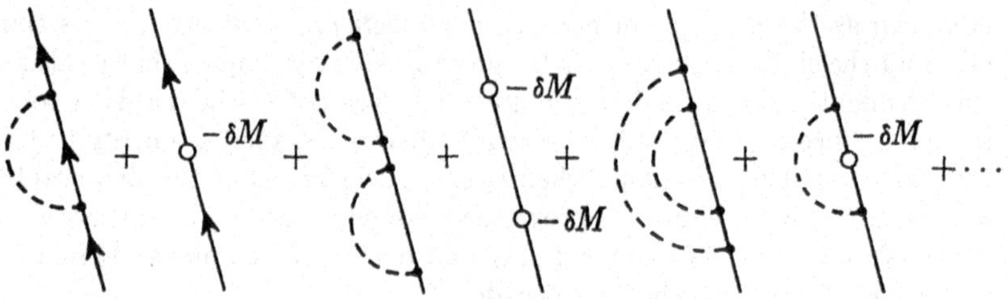

Fig. 15.5

renormalization is then that for $\not p = M$, $p^2 = M^2$, $\Sigma'(p)$ is to vanish (as we are including the contributions of the δM term in Σ'). In the case that the diagrams of Figure 15.5 correspond to the modification of an internal line, the propagator of the original internal line $S_F(p) = (\not p - M)^{-1}$ becomes replaced by

$$S_F'(p) = \frac{1}{\gamma \cdot p - M} + \frac{1}{\gamma \cdot p - M}\, \Sigma'(p)\, \frac{1}{\gamma \cdot p - M} \tag{11b}$$

The principle of mass renormalization can also be stated as the requirement that in the limit $\gamma \cdot p = M$ the pole of the modified propagator $S_F'(p)$, considered as a function of $\gamma \cdot p$, occur at the experimental mass of the particle. Note that these requirements fix δM. For example, since the diagram a of Figure 15.4 gives a contribution of order G^2, this determines δM to that order. Higher order contributions to δM can be determined in a similar fashion. A mass renormalization counter $\delta\mu^2 : \phi^2(x) :$ must also be introduced into the Lagrangian to take into account the meson self-energy. We return to this point later in Section h.

Let us next calculate to second order the electron's self-energy (or equivalently δm) due to its interaction with the quantized electromagnetic field. In that case (2) becomes

$$\Sigma^{(2)}(p) = \frac{e^2}{4\pi\hbar c} \int d^4k\, \gamma^\mu \frac{\gamma \cdot (p - k) + m}{(p - k)^2 - m^2 + i\epsilon}\, \gamma_\mu \frac{1}{k^2 + i\epsilon'} \tag{12}$$

and we need only to evaluate (12) for the case $\not p = m$, $p^2 = m^2$, i.e., the case of a free particle. Using the relations

$$\gamma^\mu \gamma_\mu = 4 \tag{13a}$$

$$\gamma^\mu \gamma^\nu \gamma_\mu = -2\gamma^\nu \tag{13b}$$

we obtain

$$\Sigma^{(2)}(\not{p} = m) = \frac{e^2}{4\pi\hbar c} 2 \int \frac{\gamma \cdot k + m}{-2p \cdot k + k^2} \frac{d^4k}{k^2 + i\epsilon} \tag{14}$$

For large k both denominators become equal to k^2. In this limit, the contribution of the term $\gamma \cdot k$ vanishes by symmetry and we are thus left with an integrand proportional to $k^3\,dk/k^4$ for large k. The integral is, therefore, logarithmically divergent. Although we have overlooked the fact that we are dealing with a hyperbolic metric rather than a Euclidian metric, we shall show below that a more careful analysis substantiates this result. That the self-energy of an electron in quantum electrodynamics would be infinite was recognized as long ago as 1929. Heisenberg and Pauli [Heisenberg (1929, 1930)] pointed this out at that time, and Waller (1930), Oppenheimer (1930a), Rosenfeld (1931c), and others calculated the self-energy of the electron in the Dirac one-electron theory and found a quadratic divergence. In 1934, Weisskopf (1934a, b) recalculated the self-energy of the electron using the Dirac hole theory and found that the divergence was only logarithmic. In the "old-fashioned" calculation of the electromagnetic self-energy, it was usual to consider separately the contributions from the static Coulomb field and those from the transverse modes [see, for example, Weisskopf (1939)]. This procedure was suggested by the nonrelativistic and correspondence limit where the static contribution predominates.

It is of interest to examine the relationship of the self-energy appearing in quantum electrodynamics to that appearing in the classical theory of radiation[2] and to understand the difference. The simplest way to introduce the self-energy in classical electrodynamics is to identify it with the electric and magnetic energy in the electron's own field. A second approach starts from the fact that the electron's field makes it more difficult to accelerate the electron in an external field and hence is equivalent to an additional mass of the electron [Lorentz (1916, 1937)]. This self-mass is given by an expression similar to the self-energy, but there are certain differences which made considerable trouble in the classical theory of self-energy. The self-energy is given classically (except for a numerical factor equal to $\frac{1}{2}$) by e^2/r_0, where r_0 is the "radius" of the electron. Thus a point electron leads to infinite self-energy.

If we ascribe a finite radius to the electron, then the interaction of the electromagnetic field with the electron will contain a form factor which will deviate appreciably from unity when the wavelength $\lambda/2\pi$ is approximately equal to r_0. Setting

$$r_0 = 1/k_{\max} \tag{15}$$

we see that the classical self-energy e^2/r_0 is proportional to $k_{\max}$, i.e., we

[2] See, for example, a discussion in Heitler (1944), pp. 29 ff.

have a linear divergence in k. The quantity k_{max} is usually spoken of as the cutoff.

One theory that used to be very popular ascribed the entire mass of the electron to the interaction energy [see Abraham (1933)]. This would make the electron radius r_0 equal to

$$r_0 = \frac{e^2}{mc^2} \tag{16}$$

a quantity commonly known as the "classical electron radius." Poincaré (1905, 1906, 1924) pointed out several defects of this model. For example, if one computes the momentum carried by the field when the electron is moving with a velocity v, one finds that it would correspond to a particle of rest mass different from that given by (16); in other words, the model does not give the right transformation properties nor the correct relationship between the energy and momentum of the system to ascribe these observables to a particle. Poincaré inferred from this that additional forces besides electromagnetic must contribute to the binding of the electron and that they would contribute an additional self-energy. The observations of Poincaré are related to the so-called self-stress problem which will be briefly discussed later in the present section.

In contrast to the classical theory, the self-energy calculated from the Dirac hole theory is proportional to

$$\frac{e^2}{4\pi\hbar c} mc^2 \log\left(\frac{\hbar k_{max}}{mc}\right) \tag{17}$$

i.e., it is only logarithmically divergent. Another important difference is that in quantum theory the self-energy does depend on m, the rest mass of the electron, whereas in classical theory m does not occur in the expression for the self-energy. These differences already indicate that the self-energy problem is of a rather different nature in quantum theory from that found in classical theory. The main point is that serious deviations from classical behavior occur already at distances of the order of the Compton wavelength $\hbar/mc$. The classical electron radius is 137 times smaller than the electron Compton wavelength. Therefore, as we consider smaller and smaller distances, quantum effects have to be taken into account before effects from the classical electron radius can possibly occur, and it thus seems unlikely that a solution of the classical self-energy problem is either necessary or sufficient for the solution of the quantum mechanical one.

If in the Dirac hole theory we wish to ascribe most of the mass of the electron to its self-energy, then according to (17) we would have to postulate a cutoff radius which is of the order

$$R_{max} = \frac{\hbar}{mc} e^{-137} \tag{18}$$

It should be borne in mind that at distances very much larger than (18), namely, at the meson Compton wavelength $\hbar/\mu_\pi c$, virtual mesons can be produced and these in turn may interact strongly with virtual nucleons. Thus, quantum electrodynamics may break down long before energies corresponding to (18) are reached. We note then that the reason for a breakdown of quantum electrodynamics may be that interactions with fields other than the electromagnetic and the electron-positron field play a role.

In spite of the differences between classical and quantum theory, numerous attempts have been made to reformulate the classical Maxwell-Lorentz theory in the presence of point charges so that it is free of divergences. It was hoped that, if a classical electrodynamics could be devised which would not contain the difficulties of infinite self-energy and finite self-stress (exploding electron) and if this theory could be quantized, then the problem of a self-consistent quantum electrodynamics would be solved. Many successful attempts have been made to produce such a classical theory, but none of them has particularly contributed to the solution of the quantum problem.

Thus, for example, the approach of Wentzel (1933, 1934, and 1949) made use of the fact that we may regard a charged point as the limiting case in which a space-like vector, the electron radius, tends to zero. Dirac, on the other hand [(1938, 1939); see also Eliezer (1947) for a review of the preceding works], makes use of the possibility of the retarded and advanced solutions of the Maxwell equations. He defines a "proper" and an "external" field as half the sum and half the difference, respectively, of the retarded and the advanced fields. This splitting of the field is an invariant one. He still retains the Maxwell-Lorentz theory to describe the field right up to the point electron, and shows that the terms which give rise to infinities can be subtracted out in a Lorentz invariant manner. This corresponds to subtracting the proper field from the total field to obtain the force acting on a given electron. The subtraction procedure does not violate any of the conservation laws, as the reaction of the radiation field on the motion of the electron is properly taken into account. However, in this classical Dirac theory, a free electron would undergo self-accelerated motions. In this respect, this theory is unsatisfactory.

An alternative possibility at the classical level, closely related to the above classical Dirac theory, is the action at a distance theory of Wheeler and Feynman [Wheeler (1945, 1949)]. In this theory, the electromagnetic field is again invariantly split up into a self-field (proper field) and an external field. However, it is postulated that the proper field of a given electron shall not act back on the particle which produces it. There is, therefore, no such thing as the self-energy of a particle in this theory. The formalism is developed as an action at a distance theory, in which the fields are merely subordinate entities with no degrees of freedom of their own. They are integrated out of the equation of motion of the particles, so that they no longer appear explicitly. This theory requires that the interaction between particles is by means of half-retarded plus half-advanced fields. Radiative reaction then appears, in this point of view, as a consequence of statistical mechanics (rather than pure electrodynamics) in a world in which all emitted radiation is eventually absorbed.

Another very interesting theory is a point electron model due to Stückelberg (1939, 1941a), which not only gives a finite self-energy but also zero self-stress. Because of Coulomb self-repulsion, a point charge tends to explode, thus leading to an infinite self-energy and a finite self-stress. Stückelberg's idea is that if there were, in addition to the electromagnetic interaction, another kind of interaction characterized by a "new" (e.g., mesic) charge, such that like charges attracted one another, then the effect of the new interaction might just cancel the Coulomb repulsion. He therefore postulates an interaction of the electron with a short-range scalar field, for which it is true that two like particles attract each other. The short range is necessary in order that there be no observable effect on the interaction of charges which are widely separated. If the strength of the coupling with the scalar field is f, Stückelberg then finds that if $e^2 = f^2$ the total classical self-energy is finite and the self-stress is zero.

Bopp (1940), and Landé and Thomas [Landé (1941a, b, 1944)] have developed a theory similar to that of Stückelberg, except that a vector field rather than a scalar is used. Although a vector field gives rise to a repulsion between particles of like charge, they suggest that one subtract the total energy of the vector field from that of the electron. This makes compensation possible, but the energy is no longer positive-definite.

Actually, the first relativistic gauge invariant theory giving finite self-stress and finite self-energy in the classical case was that of Born and Infeld [Born (1934a, b, 1937)]. In their theory, the Maxwell equations were replaced by a nonlinear set of equations, which in the limit of weak fields reduce to the Maxwell ones. Quantization of the theory becomes nearly impossible, due to the nonlinearity. Furthermore, there still exist singularities in the field quantities in the presence of point charges.

Nonlinear unitary field theories[3] have been proposed by Mie [see Pauli (1921) for a review of this work], Rosen (1939, 1942, 1952), Jehle (1947/48, unpublished), Finkelstein (1949, 1951), and Drell (1950). In a unitary field theory, particles appear not as singularities, but as small volumes in which energy and charge of the field are concentrated. In such a theory, the equations of motion of the particle follow from the field equations. It is in this respect that a unitary theory differs from a theory in which the particles are represented by singularities in the field. In the latter case, the field equations break down at the singularities and therefore cannot in general determine the motion of the particle.

[3] We do not include here the Einstein theory of general relativity nor his unified theories which are unitary theories par excellence.

A unitary theory is necessarily nonlinear, for one particle could not influence another if the equations were linear. Such a theory has also the property that once the charge of the particle is fixed, a discrete spectrum of masses follows uniquely. Rosen (1939, 1942) and Jehle (unpublished) have shown that some of the ordinary classical field theories, such as the Dirac field interacting with a Klein-Gordon field with scalar coupling, permit the existence of static and spherically symmetric solutions which can be interpreted as particles. These solutions are free from singularities everywhere and are quadratically integrable, so that the classical infinities never appear. The quantization of unitary field theories has been investigated formally by Finkelstein (1949).

Dirac (1951, 1952a, b, 1954) has developed a classical theory for continuous distributions of charges which move in accordance with the Lorentz equation of motion. The special feature of his theory is that only the ratio e/m of the charge to the mass appears, and not e and m separately. However, in the present version of the theory, only one ratio e to m can be introduced. It was Dirac's hope that a new method of quantization of this theory would be found which would introduce $\hbar$ and at the same time explain the fact that electric charge occurs only in multiples of e. This would then give a connection between e and $\hbar$ and thus fix the value of the fine structure constant $e^2/4\pi\hbar c$.

Many of the other classical theories have had as their aim the introduction of an electron radius in a relativistically invariant way. The difficulty which one encounters here is that of specifying the charge distribution in a way that is consistent with the relativistic interpretation of causality. Thus, consider, for example, the behavior of a perfectly rigid electron at which we direct a pulse of electromagnetic waves. As soon as the pulse strikes the edge of the charge distribution, the electron as a whole is set in motion. Thus an impulse is transmitted instantaneously, contradicting the relativistic law of causality. The attempt is therefore made to keep these acausal effects to space-time intervals which would not conflict with experience. [See, in this connection, Chrétien (1953).]

Linear extended source theories have been developed, and among these the best attempts at a classical theory are those of Bopp (1942), Peierls and McManus [McManus (1948)], and Feynman (1948a). McManus (1948) assumed a nonlocal interaction of the kind mentioned in Section 11c. Explicitly, the interaction Lagrangian he assumed was

$$\int \mathfrak{L}_I(x)\, d^4x = -\iint j^\mu(x)\, F[(x-x')^2]\, A_\mu(x')\, d^4x\, d^4x' \qquad (19)$$

corresponding to an averaging of the interaction over a region of space-time. The kernel is relativistically invariant, depending only on the invariant four-distance between the two points. The requirements on the function F are that it fall off sufficiently rapidly when $R^2 = (x - x')^2$ becomes large, so that F can satisfy

$$\int F(x^2)\, d^4x = 1 \tag{20}$$

which is necessary in order to make the total charge finite and equal to e. Such F functions can be obtained by considering the four-dimensional Fourier transform $g(k^2)$ of $F(R^2)$, defined by

$$F(R^2) = \int e^{-ik\cdot(x-x')}\, g(k^2)\, d^4k \tag{21}$$

One such function which gives rise to an $F(R^2)$ satisfying the above requirements is

$$g(k^2) = \frac{1}{(2\pi)^4}\frac{1}{1 + (k^2r_0^2)^2} \tag{22}$$

Thus a finite electron radius r_0 is introduced in a relativistically invariant way. This theory of McManus', which pictures the electron as a distribution of charge in both space and time, results in a finite self-energy of the electron.

Feynman (1948a) introduces the finite radius of the electron directly in the interaction between two electrons rather than in the interaction between electron and electromagnetic fields, since he formulates his theory in the language of action at a distance. The two theories of McManus and Feynman were subsequently shown to be equivalent for a class of F functions. The great advantage of these theories is that they can be quantized [Feynman (1948b)], and, in fact, yield a finite self-energy in lowest order of perturbation theory.

However, in quantum theory generally there is another divergence which does not occur in classical theory, namely, that due to vacuum polarization. This process corresponds to pair production by an electromagnetic field followed by the annihilation of the same pair. To remedy this situation by a nonlocal theory, we would have to consider an interaction Lagrangian of the form

$$-\int \bar{\psi}(x')\, \gamma^\mu \psi(x'')\, A_\mu(x''')\, F(x - x', x - x'', x - x''')\, d^4x'\, d^4x''\, d^4x''' \tag{23}$$

Thus we can no longer speak of a current j_μ at one space-time point, since the expression for the current now involves both x and x'. In fact, although the total charge is conserved, we no longer have a local continuity equation for electric charge, since the interaction (23) is not gauge invariant. Chrétien and Peierls [Chrétien (1954)] have attempted to formulate nonlocal electrodynamics in which the requirement of gauge invariance is satisfied. Although they succeeded in constructing such a theory, they were not able to remove the vacuum polarization divergences with it.

While the classical attempts to obtain finite self-energies have not been very successful in the larger framework of quantum field theory, they still provide a useful method for a formal elimination of divergences from the quantized theory. It is often convenient to introduce a cutoff, in the manner proposed by Feynman (1948b), which makes the self-energy finite but dependent on the cutoff parameter λ. This cutoff can be viewed as a special case of an interaction of the form (23), which then replaces the factor $\frac{1}{2}\Delta_F(x - y) = \phi^{\cdot}(x)\,\phi^{\cdot}(y)$ that appears in the usual local form of the field theory, by a new function,

$$\tfrac{1}{2}\Delta_{F\,\text{Reg}}(x - y) = \int d^4x' \int d^4y'\, F(x - x')\, F(y - y')\, \phi^{\cdot}(x')\, \phi^{\cdot}(y') \tag{24}$$

By a suitable choice of F functions, $\Delta_{F\,\text{Reg}}$ can be made a smooth function of $(x - y)^2$ having no singularities. Clearly, there are many possible choices for $\Delta_{F\,\text{Reg}}$, corresponding to different F functions. The simplest one is probably the one proposed by Feynman (1948b) himself, which replaces the boson propagation amplitude, $(k^2 - \mu^2)^{-1}$, by

$$\begin{aligned}\Delta_{F\,\text{Reg}}(k^2) &= \int_0^\infty \left[\frac{1}{k^2 - \mu^2} - \frac{1}{k^2 - \mu^2 - \lambda^2}\right] G(\lambda)\, d\lambda \\ &= \int_0^\infty \frac{1}{k^2 - \mu^2}\left(\frac{-\lambda^2}{k^2 - \mu^2 - \lambda^2}\right) G(\lambda)\, d\lambda \end{aligned} \tag{25}$$

The quantity G involves values of λ which are large compared to M and μ, and is so normalized that

$$\int_0^\infty G(\lambda)\, d\lambda = 1 \tag{26}$$

Hence, if one does not integrate over λ, the limit $\lambda \to \infty$ gives back the local theory. Every integral over an intermediate boson quantum which previously involved a factor $(k^2 - \mu^2)^{-1}$ is now supplied with a convergence factor $C(k^2)$

$$C(k^2) = -\int_0^\infty d\lambda\, G(\lambda)\, \frac{\lambda^2}{k^2 - \mu^2 - \lambda^2} \tag{27}$$

which is sufficient to make convergent all integrals over virtual Bose quanta. The poles are still defined in the Feynman sense, that is, μ has a small negative imaginary part. It should, however, be emphasized that the Feynman method of introducing a cutoff into the theory is a formal calculational device and not a consistent theory. This can be inferred from the fact that, for example, in quantum electrodynamics the photon propagator $1/k^2$ would be replaced by $(k^2)^{-1} - (k^2 - \lambda^2)^{-1}$, which can be looked upon as introducing an additional interaction of the electron-positron field with a vector field whose quanta have mass λ and whose propagators are $-(k^2 - \lambda^2)^{-1}$. The $-$ sign, however, indicates that these quanta are coupled to the charges by a factor $-e^2$, so that the interaction Lagrangian would have the coupling constant ie, which implies that $\mathfrak{L}_I$ is not Hermitian. Probability would therefore not be conserved. One indeed verifies that with the above modified propagators, violations of conservation of probabilities of order m^2/λ^2 are introduced into the theory.

As an illustration of the modern covariant techniques of integration [see

Feynman (1949b), especially the appendices], we shall here calculate the self-energy of a fermion for a theory for which $\mathfrak{L}_I = G : \bar{\psi}\Gamma\psi\phi :$ using the Feynman cutoff method. If we leave the integration over $d\lambda$ to the end, and note that

$$\frac{1}{k^2 - \mu^2} - \frac{1}{k^2 - \mu^2 - \lambda^2} = -\int_0^{\lambda^2} \frac{dL}{(k^2 - \mu^2 - L)^2} \tag{28}$$

the self-energy integral (5), (2) becomes

$$\Delta E = \frac{-i}{4\pi^3} \frac{G^2}{4\pi\hbar c} \frac{M}{E(\mathbf{p})}$$

$$\tilde{w}(\mathbf{p}) \int_0^{\lambda^2} dL \int d^4k \Gamma \frac{\gamma \cdot (p - k) + M}{(p - k)^2 - M^2 + i\epsilon} \Gamma \frac{1}{(k^2 - \mu^2 - L)^2} w(\mathbf{p}) \tag{29}$$

Let us consider the integral over k first. Due to the cutoff, it is now a convergent integral. Since we are dealing with a free particle, $p^2 = M^2$, the fermion denominator, $(p - k)^2 - M^2$, can be simplified to

$$p^2 - 2p \cdot k + k^2 - M^2 = k^2 - 2p \cdot k \tag{30}$$

Next, we combine denominators in order to make the denominator a function of k^2 only, by using the remarkable Feynman formula[4]

[4] This formula is obtained by noting that

$$\frac{1}{ab} = \frac{1}{b - a}\left(\frac{1}{a} - \frac{1}{b}\right) = \frac{1}{b - a}\int_a^b \frac{dx}{x^2}$$

and introducing in the last integral the new variable, $x = az + b(1 - z)$. Equation (31), as it stands, holds for all values of a and b. If, however, a and b are of the opposite sign, z is to be considered as a complex variable and the path of integration must deviate from the real axis so as to avoid the singularity at $z = b/b - a$. Any path joining $z = 0$ and $z = 1$ (but not passing through the singularity) may actually be chosen, since the residue of the integrand at the singularity vanishes. The Feynman integral (31) is actually a special case of the following general formula:

$$\frac{1}{a_1a_2a_3 \cdots a_n} = (n - 1)! \int_0^1 \cdots \int_0^1 \frac{dz_1dz_2dz_3 \cdots dz_n}{[a_1z_1 + a_2z_2 + \cdots + a_nz_n]^n}$$
$$\sum_{i=1}^n z_i = 1$$

To evaluate this last integral, see, for example, Goursat and Hedrick [Goursat (1904), Vol. I, p. 312], or introduce the delta function $\delta\left(\sum_{i=1}^n z_i - 1\right)$ in the integrand. Alternatively, we may write

$$\frac{1}{a_1a_2a_3 \cdots a_n} = (n - 1)! \int_0^1 dz_1 \int_0^{z_1} dz_2 \cdots \int_0^{z_{n-2}} dz_{n-1}$$
$$\cdot \frac{1}{[a_nz_{n-1} + a_{n-1}(z_{n-2} - z_{n-1}) + \cdots + a_1(1 - z_1)]^n}$$
$$= (n - 1)! \int_0^1 \zeta_1^{n-2}\, d\zeta_1 \int_0^1 \zeta_2^{n-3}\, d\zeta_2 \cdots \int_0^1 d\zeta_{n-1}$$
$$\cdot \frac{1}{[a_1\zeta_1\zeta_2 \cdots \zeta_{n-1} + a_2\zeta_1 \cdots \zeta_{n-2}(1 - \zeta_{n-1}) + \cdots + a_n(1 - \zeta_1)]^n}$$

$$\frac{1}{ab} = \int_0^1 \frac{dz}{[az + b(1 - z)]^2} \tag{31}$$

Since, in our example, the factor $(k^2 - \mu^2 - L)^2$ occurs, we shall use the formula

$$\frac{1}{a^2b} = \int_0^1 \frac{2z\,dz}{[az + b(1 - z)]^3} \tag{32}$$

obtained from (31) by differentiation with respect to a. Our combined denominator therefore becomes

$$\begin{aligned}(k^2 - \mu^2 - L)\,z + (k^2 - 2p \cdot k)\,(1 - z) \\ = k^2 - 2p \cdot k(1 - z) - (\mu^2 + L)\,z \\ = [k - p(1 - z)]^2 - M^2(1 - z)^2 - (\mu^2 + L)\,z \quad (33)\end{aligned}$$

where we have again used the fact that $p^2 = M^2$. We next introduce the variable $k' = k - p(1 - z)$, so that our integral becomes

$$2\int_0^1 dz\,z \int d^4k'\,\Gamma\,\frac{\gamma \cdot pz - \gamma \cdot k' + M}{[k'^2 - M^2(1 - z)^2 - (\mu^2 + L)\,z]^3}\,\Gamma \tag{34}$$

Dropping the prime, the denominator is now a function of k^2 only. Furthermore, an integral of the form $\int d^4k\,k_\mu f(k^2)$ vanishes by symmetry. In general, the symmetry implies

$$\int d^4k \text{ (odd number of } k_\mu \text{ factors) } f(k^2) = 0 \tag{35a}$$

$$\int d^4k\,k_\mu k_\nu f(k^2) = \tfrac{1}{4}g_{\mu\nu}\int d^4k\,k^2 f(k^2) \tag{35b}$$

The term $\gamma \cdot k$ in the integral (34) therefore vanishes and this is equivalent, in the absence of the cutoff, to the reduction of the integral from a linearly to a logarithmically divergent one.

We thus have to evaluate the following integral:

$$I = \int_{-\infty}^{+\infty} d^4k\,\frac{1}{(k^2 - C + i\epsilon)^3} \tag{36}$$

where $C = m^2(1 - z)^2 + (\mu^2 + L)\,z$ is a positive-definite quantity. As indicated, the path of integration is along the real k_0 axis, the poles having been displaced by the Feynman addition of $-i\epsilon$ to the masses of the particles. We may, however, rotate the path of integration by 90° in the complex plane, so that the integration is up along the imaginary axis from $-i\infty$ to $+i\infty$. This is permissible, since in rotating the contour we never cross any singularities, the latter being located above the negative real axis and below the positive real axis. We may, therefore, introduce a new integration variable, $k_0 = ik_4$, so that the integral (36) is now transformed into an integral over a Euclidian four-space

$$I = \int_{-\infty}^{+\infty} d^3k \int_{-i\infty}^{+i\infty} \frac{dk_0}{(k^2 - C + i\epsilon)^3} = -i \int \frac{dk_4\, d^3k}{(k_4{}^2 + k_1{}^2 + k_2{}^2 + k_3{}^2 + C)^3} \tag{37}$$

Introducing four-dimensional spherical polar co-ordinates, the volume element becomes $2\pi^2 k^3\, dk$, and the integral I finally reduces to

$$I = -2\pi^2 i \int_0^\infty \frac{k^3\, dk}{(k^2 + C)^3} = -i\pi^2 \int_0^\infty \frac{x\, dx}{(x + C)^3} = -\frac{i\pi^2}{2C} \tag{38}$$

Note that the value of similar integrals with different powers in the denominator can be obtained from I by integration or differentiation with respect to C.

The second-order self-energy is therefore equal to (omitting $\int d\lambda$)

$$\Delta E = -\frac{G^2}{4\pi\hbar c} \frac{M}{E(\mathbf{p})} \frac{1}{4\pi} \int_0^1 dz\, z \int_0^{\lambda^2} dL$$
$$\tilde{w}(\mathbf{p}) [\Gamma(\gamma \cdot pz + M)\, \Gamma]\, w(\mathbf{p}) \frac{1}{[M^2(1 - z)^2 + (\mu^2 + L)\, z]} \tag{39}$$

If we specialize to the case $\Gamma = \gamma_5$, and recall that $\gamma \cdot pw(\mathbf{p}) = Mw(\mathbf{p})$, we obtain

$$\Delta E = \frac{G^2}{4\pi\hbar c} \frac{M}{E(\mathbf{p})} \frac{1}{4\pi} \tilde{w}(\mathbf{p})\, w(\mathbf{p}) \int_0^1 dz \int_0^{\lambda^2} dL \frac{Mz(1 - z)}{M^2(1 - z)^2 + (\mu^2 + L)\, z} \tag{40a}$$

The integration over L now yields

$$\Delta E = \frac{M}{E(\mathbf{p})} \frac{G^2}{4\pi\hbar c} \frac{1}{4\pi} \tilde{w}(\mathbf{p})\, w(\mathbf{p})$$
$$M \int_0^1 dz\, (1 - z) \log \left[\frac{M^2(1 - z)^2 + (\mu^2 + \lambda^2)\, z}{M^2(1 - z)^2 + \mu^2 z} \right] \tag{40b}$$

In the limit $\lambda \to \infty$, we may neglect in the numerator of the logarithm the terms $\mu^2 z + M^2(1 - z)^2$ compared with $\lambda^2 z$. If, in addition, we neglect μ compared with M, we obtain

$$\Delta E = \frac{M}{E(\mathbf{p})} \frac{G^2}{4\pi\hbar c} \frac{M}{8\pi} \left[\log\left(\frac{\lambda^2}{M^2}\right) - \frac{1}{2} \right] \tilde{w}(\mathbf{p})\, w(\mathbf{p}) \tag{40c}$$

which expression diverges logarithmically in the limit $\lambda \to \infty$. The integral (40b) can be evaluated without any approximations to give

$$\Delta E = \frac{M}{E(\mathbf{p})} \frac{G^2}{4\pi\hbar c} \frac{M}{8\pi} \left[\log\left(\frac{\lambda^2}{M^2}\right) - \frac{1}{2} + \frac{\mu^2}{M^2} + 2\left(\frac{\mu}{M}\right)^2 \left(1 - \frac{\mu^2}{2M^2}\right) \log\frac{\mu}{M} \right.$$
$$\left. - 2\left(\frac{\mu}{M}\right)^3 \sqrt{1 - \frac{\mu^2}{2M^2}} \cos^{-1}\frac{\mu}{2M} \right] \tilde{w}(\mathbf{p})\, w(\mathbf{p}) \tag{40d}$$

The change in mass is therefore given by

$$\delta M = \frac{G^2}{4\pi\hbar c} \frac{M}{8\pi} \left[\log\left(\frac{\lambda^2}{M^2}\right) - \frac{1}{2} + O\left(\frac{\mu}{M}\right) \right] \tag{41}$$

In Feynman's covariant treatment, the fact that the self-energy corresponds to a change in mass is thus exhibited directly. The self-energy of a fermion interacting with a scalar boson field through a scalar coupling is also logarithmically divergent but of opposite sign from the pseudoscalar case with

$$\delta M \propto -M \frac{3g^2}{4\pi\hbar c} \log \frac{\lambda}{M}$$

The sign of the second-order self-energy of the electron in quantum electrodynamics is the same as in the pseudoscalar case with

$$\delta m \propto m \frac{3e^2}{4\pi} \left(\log \frac{\lambda^2}{m^2} + \frac{1}{2} \right)$$

With this choice for δm to order e^2, the contributions of diagrams a and b of Figure 15.4 will vanish for a *free* electron.

The Feynman cutoff method which we have just presented may be regarded as a special case of a general formal covariant method of handling divergent integrals due to Pauli and Villars [Pauli (1949)]. [See also Stückelberg (1948a, b) and Rivier (1949).] Their technique is known as the "regulator" method. It consists in first "regularizing" all singular expressions, and then performing a limiting process. Thus, if

$$I(M) = \int d^4k \, J(k, M) \tag{42a}$$

is a divergent integral over an internal line with mass M, the regularized integral $I_R(M)$ is defined by the formula

$$I_R(M) = \sum_{i=1}^{N} C_i I(M_i) = \int \sum_{i=1}^{N} C_i J(k, M_i) \, d^4k \tag{42b}$$

with $C_1 = 1$ and $M_1 = M$.

Certain restrictions are then imposed on the C_i and M_i sufficient to make the integral (42b) converge. In electrodynamics it is found that the conditions

$$\sum_{i=1}^{N} C_i = 0 \tag{43a}$$

$$\sum_{i=1}^{N} C_i M_i^2 = 0 \tag{43b}$$

are sufficient to achieve this purpose. The limiting process then consists in letting the auxiliary masses M_i ($i = 2, 3, \cdots$) tend to infinity.

The conditions (43) insure that the functions Δ and $\Delta^{(1)}$ are free of singularities on the light cone, unlike their unregularized counterparts. [Recall Eqs. (7.179) and (7.182) and note that Eq. (43) just eliminates the singularities.] The conditions (43) are also enough to lead to a convergent expression for the electron self-energy, provided one regularizes the expression as a whole and not in parts. Upon performing the final limiting process $M_i \to \infty$ $(i = 2, 3, \cdots)$, logarithmic divergences reappear.

The above procedure constitutes, in the terminology of Pauli and Villars, a "formalistic theory" wherein a purely formal recipe for the evaluation of divergent integrals is given, in contrast to a "realistic theory" in which the mathematical device which affects convergence is a consequence of interactions with other fields [see, for example, Pais (1947)].

It should be noted that the evaluation of the self-energy by the Feynman methods maintains the relativistic invariance of the theory at each step. This fact is of great importance because it allows an unambiguous separation of the self-energy effects and in addition implies that the self-stress $S(0)$ of the fermion vanishes. The quantity $S(0)$ is defined as the expectation value of the space-space component of the symmetrized energy-momentum tensor of the field system in the one-particle state of zero momentum

$$S(0) = \langle \Phi_1, T^i{}_i(x)\, \Phi_1 \rangle \quad \text{(not summed over } i) \tag{44}$$

Recall that the energy and momentum operators of the system, P^μ, are related to $T^{\mu\nu}$ by $P_\mu = \int d\sigma^\nu\, T_{\mu\nu}$, and that these operators transform like the components of a four-vector if $T_{\mu\nu}$ is a tensor of rank 2. If we now consider two different Lorentz frames O' and O, one (O) in which the particle is at rest and one (O') in which it has momentum $\mathbf{p}'$ and energy E', the question arises whether the values $\mathbf{p}'$ and E' (as calculated from $T'^{\mu\nu}(x')$) are connected by a Lorentz transformation to those calculated from $T^{\mu\nu}$. It turns out [Pais and Epstein (1949), Rohrlich (1950a, 1960), Villars (1950), Borowitz (1952), Takahashi and Umezawa (1952); see also Arnous and Heitler (Arnous 1955)] that necessary conditions for this to be the case is that the self-stress vanishes. The latter *will* vanish for any finite relativistically covariant theory. Therefore any covariant cutoff (such as Feynman's method) will automatically guarantee a vanishing self-stress because it assures the theory to remain covariant throughout the calculation.

15b. Mass Renormalization and the Nonrelativistic Lamb Shift

In the present section we shall illustrate the renormalization principle by means of the first problem historically to be solved thereby, namely, the Lamb shift. When Kramers made his suggestion in 1947, Lamb and Retherford [Lamb (1947)] had just measured the $2S$ level displacement from the $2P$ level in hydrogen. Stimulated by Kramers' ideas, Bethe (1947) interpreted Lamb's level displacement as an effect of the interaction of the electron with the radiation field. In fact, Bethe simply calculated the self-energy of an electron bound in the atom. But, according to

Kramers' idea, a large part of this self-energy has already been taken into account if we use the observed mass of the electron, m, in the calculation, rather than the bare mass, m_0. Therefore, the true level shift is the difference between the self-energy of a bound electron and that of a free electron.

In the nonrelativistic approximation which Bethe used in 1947, only the interaction of the electron with the transverse electromagnetic waves need be considered. The Hamiltonian describing the electron moving in some potential $V(\mathbf{x})$, and interacting with the radiation field, is taken to be

$$\begin{aligned} H &= H_{\text{rad}} + H_{\text{matter}} + H_{\text{int}} \\ &= H_0 + H_I \end{aligned} \tag{45}$$

with[5]

$$H_{\text{rad}} = \tfrac{1}{2} \int (\mathcal{E}^2 + \mathcal{H}^2)\, d^3x \tag{45a}$$

$$H_{\text{matter}} = \frac{\mathbf{p}^2}{2m_0} + V(\mathbf{x}) \tag{45b}$$

$$H_I = -\frac{e}{m_0 c}\, \mathbf{p} \cdot \mathbf{A}(\mathbf{x}) \tag{45c}$$

Here, the radiation gauge, div $\mathbf{A} = 0$, has been adopted so that only the vector potential $\mathbf{A}$ is used in the description of the radiation field and $\mathcal{E} = -\partial_0 \mathbf{A} = \frac{i}{\hbar} [\mathbf{A}, H]$ and $\mathcal{H} = \text{curl}\, \mathbf{A}$. In the Schrödinger picture that we are working in, the expansion of the quantized field operator $\mathbf{A}(\mathbf{x})$ in terms of creation and destruction operators for photons of momentum $\mathbf{k}$ and polarization λ is

$$\mathbf{A}(\mathbf{x}) = \sqrt{\frac{\hbar c}{(2\pi)^3}} \sum_{\lambda=1}^{2} \int \frac{d^3k}{\sqrt{2\omega_{\mathbf{k}}}}\, \boldsymbol{\epsilon}_\lambda(\mathbf{k})\, (c_\lambda(\mathbf{k})\, e^{i\mathbf{k}\cdot\mathbf{x}} + c_\lambda^*(\mathbf{k})\, e^{-i\mathbf{k}\cdot\mathbf{x}}) \tag{46}$$

Note that the summation over the polarization index λ here runs only from 1 to 2, the two transverse directions, and since div $\mathbf{A} = 0$, $\mathbf{k} \cdot \boldsymbol{\epsilon}_\lambda(\mathbf{k}) = 0$ $(\lambda = 1, 2)$. The commutation rules for the operators are the usual ones with

$$\begin{aligned} [c_\lambda(\mathbf{k}), c_{\lambda'}^*(\mathbf{k}')] &= \delta_{\lambda\lambda'} \delta^{(3)}(\mathbf{k} - \mathbf{k}') \\ [c_\lambda(\mathbf{k}), c_{\lambda'}(\mathbf{k}')] &= [c_\lambda^*(\mathbf{k}), c_{\lambda'}^*(\mathbf{k}')] = 0 \end{aligned} \tag{47}$$

[5] In the following we shall only be interested in the self-energy effects due to the interaction term, H_{int}. We have therefore omitted the $\mathbf{A}^2$ term which arises from the gauge invariant introduction of the interaction of the charged particle with the radiation field obtained from the replacement $\mathbf{p} \to \mathbf{p} - \frac{e}{c}\mathbf{A}$. The $\mathbf{A}^2$ term contributes an uninteresting constant contribution to the self-energy of the particle (i.e., it is independent of the momentum of the particle). It is, however, of great importance in problems dealing with the scattering of radiation from charged particles and, of course, for the gauge invariance of the theory.

The operator $N_\lambda(\mathbf{k}) = c_\lambda{}^*(\mathbf{k})\, c_\lambda(\mathbf{k})$ is again the number operator for photons of momentum $\mathbf{k}$ and polarization λ, and $H_{\text{rad}} = \hbar \sum_{\mathbf{k},\lambda} \omega_{\mathbf{k}} N_\lambda(\mathbf{k})$, where $\omega_{\mathbf{k}} = c|\mathbf{k}|$.

The Hamiltonian (45) corresponds to the interaction of a nonrelativistic point particle with the radiation field. However, due to the interaction term H_I, the charged particle may, for example, emit virtual photons of arbitrarily high momentum and thus acquire a recoil momentum so large that its motion is no longer nonrelativistic. Hence, the Hamiltonian (45) is, strictly speaking, not internally self-consistent. This difficulty may be bypassed, in part, if we consider the particle to have a spread-out charge, so that the interaction term is really of the form $-(e/m_0c) \int F(\mathbf{x} - \mathbf{x}')\, \mathbf{p} \cdot \mathbf{A}(\mathbf{x}')\, d\mathbf{x}'$ where $F(\mathbf{x}')$ describes the shape of the charge distribution. Now if $a \approx \hbar/mc$ is the radius of the charge distribution, then wavelengths smaller than a will not interact with the particle, or in other words, only those k for which $\hbar k < mc$ will contribute in the expansion (46).

Consider now a free electron: $V(\mathbf{x}) = 0$. If H_I were zero, then the eigenfunctions of H would be the eigenfunctions of H_0, i.e., products of an eigenfunction of $\mathbf{p}^2/2m_0$ and of an eigenfunction of H_{rad}. If we assume that the coupling between the particle and radiation field is small and, therefore, that perturbation theory is valid, then the term H_I will cause a shift in the unperturbed energy eigenvalues. Let us calculate this change in energy to second order for the unperturbed state in which there is just the electron with momentum $\mathbf{p}$ and there are no photons present, i.e., the state

$$|\Phi\rangle = \frac{1}{\sqrt{V}}\, e^{i\mathbf{p}\cdot\mathbf{x}}\, |\, \Phi_0\rangle \tag{48}$$

where V is the normalization volume for the electron wave function and $|\Phi_0\rangle$ is the vacuum state of the radiation field for which $c_\lambda(\mathbf{k})\, |\, \Phi_0\rangle = 0$. Note that the energy, E_0, of this state is $\mathbf{p}^2/2m_0$, the kinetic energy of the electron.

Proceeding in a manner similar to that of Section 12*d*, one readily establishes that the energy shift to second order[6] is due to the emission of a photon by the electron and its subsequent reabsorption, and is given by the following expression:

$$\Delta E^{(2)} = \sum_{i \neq 0} \frac{\langle \Phi \,|\, H_I \,|\, i\rangle \langle i \,|\, H_I \,|\, \Phi\rangle}{E_0 - E_i} \tag{49a}$$

$$= \left(\Phi, H_I \frac{1}{E_0 - H} H_I \Phi\right) \tag{49b}$$

[6] There is no first-order contribution from the term $\mathbf{p} \cdot \mathbf{A}$ since $(\Phi, \mathbf{p} \cdot \mathbf{A}\Phi) = 0$. The $\mathbf{A}^2$ term would, however, contribute a constant term to this order.

$$= \frac{e^2}{m_0{}^2c^2} \frac{\hbar c}{(2\pi)^3} \int \frac{d^3k}{\sqrt{2\omega_{\mathbf{k}}}} \int \frac{d^3k'}{\sqrt{2\omega_{\mathbf{k}'}}} \int d^3x \, \frac{1}{V} \, e^{-i\mathbf{p}\cdot\mathbf{x}} \, e^{i\mathbf{p}\cdot\mathbf{x}}$$

$$\sum_{\lambda,\lambda'=1}^{2} \frac{(\mathbf{p} \cdot \boldsymbol{\epsilon}_\lambda(\mathbf{k}) \, e^{i\mathbf{k}\cdot\mathbf{x}}) \, (\mathbf{p} \cdot \boldsymbol{\epsilon}_{\lambda'}(\mathbf{k}') \, e^{-i\mathbf{k}'\cdot\mathbf{x}})}{E_0 - \frac{1}{2m_0} (\mathbf{p} - \hbar\mathbf{k}')^2 - \hbar c|\mathbf{k}'|} F(\mathbf{k}) \, F^*(\mathbf{k}')$$

$$\cdot \, (\Phi_0, c_\lambda(\mathbf{k}) \, c_{\lambda'}{}^*(\mathbf{k}') \, \Phi_0) \tag{49c}$$

$$= -\frac{e^2}{m_0{}^2c^2} \frac{\hbar c}{(2\pi)^3} \int \frac{d^3k}{2\omega_{\mathbf{k}}} |\, F(\mathbf{k})|^2 \sum_{\lambda=1}^{2} \frac{|\mathbf{p} \cdot \boldsymbol{\epsilon}_\lambda(\mathbf{k})|^2}{-\frac{\hbar\mathbf{k} \cdot \mathbf{p}}{m_0} + \frac{\hbar^2\mathbf{k}^2}{2m_0} + \hbar c|\mathbf{k}|} \tag{49d}$$

Now $\mathbf{p}/m_0 = \mathbf{v}$ is the velocity of the bare particle. Hence, the first term in the denominator in (49d) is of order v/c compared with the term $\hbar c|\mathbf{k}|$. Since we are working in the nonrelativistic limit, we shall therefore neglect the $\hbar\mathbf{v} \cdot \mathbf{k}$ term. Furthermore, we know that $F(\mathbf{k})$ cuts off the large k contributions ($|\mathbf{k}| > 1/a$); we shall therefore also neglect $\hbar^2\mathbf{k}^2/2m_0$ in comparison to the $\hbar c|\mathbf{k}|$ term. Finally, recalling that $\mathbf{k}$ and $\boldsymbol{\epsilon}_\lambda(\mathbf{k})$ are orthogonal, so that

$$\sum_{\lambda=1}^{2} |\mathbf{p} \cdot \boldsymbol{\epsilon}_\lambda(\mathbf{k})|^2 = \mathbf{p}^2 - \frac{(\mathbf{p} \cdot \mathbf{k})^2}{\mathbf{k}^2} \tag{50}$$

we can then carry out the integration over angles in (49d) by choosing $\mathbf{p}$ as the polar axis and obtain

$$\Delta E^{(2)} \approx -\frac{\mathbf{p}^2}{2m_0} \left[\frac{e^2}{4\pi\hbar c} \frac{4}{3\pi} \frac{\hbar}{m_0c} \int_0^\infty dk \,|\, F(k)|^2 \right] \tag{51}$$

We therefore note that for a point charge, for which $F(k) = 1$, $\Delta E^{(2)}$ diverges linearly. Actually, the important thing to recognize with this self-energy contribution is the fact that it is proportional to $\mathbf{p}^2/2m_0$, the kinetic energy of the bare particle, so that the energy of the system to second order is given by

$$E = E_0 + \Delta E^{(2)} \tag{52a}$$

$$= \frac{\mathbf{p}^2}{2m_0} \left[1 - \frac{4\alpha}{3\pi} \frac{\hbar}{m_0c} \int_0^\infty dk \,|\, F(k)|^2 \right] \tag{52b}$$

$$= \frac{\mathbf{p}^2}{2m_0} \left[1 - \frac{\delta m}{m_0} \right] \tag{52c}$$

where

$$\delta m = \frac{4\alpha}{3\pi} \frac{\hbar}{c} \int_0^\infty dk \,|\, F(k)|^2 \tag{52d}$$

We now accept Kramers' viewpoint: We recognize δm as the electromagnetic mass which the particle has acquired (over and above its bare, mechanical mass) by virtue of its interaction with the radiation field. However, only $m_0 + \delta m$ is observable, since no experiment can differentiate between the part of the mass of a charged particle which is of electromagnetic origin and the one which is of mechanical origin. We therefore *identify* $m_0 + \delta m$ as the observed experimental mass, m. (There will be, of course, higher order electromagnetic contributions to m and, strictly speaking, it is the sum of all these contributions and m_0 which should be identified with m.) Since in our cutoff theory δm is small (of the order of α), we may view Eq. (52c) as the second-order term arising from the expansion of

$$E = \frac{\mathbf{p}^2}{2(m_0 + \delta m)} \approx \frac{\mathbf{p}^2}{2m_0}\left(1 - \frac{\delta m}{m_0} + \cdots\right) \tag{53}$$

Therefore, if we re-express the energy in terms of the observable quantity, m, then the result of perturbation theory makes sense: It yields the fact that the energy of the particle is indeed $E = \mathbf{p}^2/2m$. Note that it is irrelevant whether δm is finite or infinite for the mass renormalization to *formally* go through.

The theory presented thus far is still unsatisfactory in one respect, in that the bare mass appears in the original Hamiltonian. Let us again adopt Kramers' principle to rewrite it as follows:

$$\begin{aligned} H &= H_{\text{rad}} + \frac{\mathbf{p}^2}{2(m - \delta m)} - \frac{e}{(m - \delta m)c}\,\mathbf{p}\cdot\mathbf{A} \\ &\approx H_{\text{rad}} + \frac{\mathbf{p}^2}{2m} - \frac{e}{mc}\,\mathbf{p}\cdot\mathbf{A} + \frac{\delta m}{m}\frac{\mathbf{p}^2}{2m} \end{aligned} \tag{54}$$

where we have neglected terms of order e^3. In Eq. (54) we are to treat not only the term $(e/mc)\,\mathbf{p}\cdot\mathbf{A}$ as a perturbation, but also the term $(\delta m/m)\,(\mathbf{p}^2/2m)$ where δm to order α is chosen equal to

$$\delta m = \frac{4\alpha}{3\pi}\frac{\hbar}{c}\int_0^\infty dk\,|F(k)|^2 \tag{55}$$

If we do this, then to order e^2 the energy of the one-electron state will be rigorously equal to $\mathbf{p}^2/2m$. Conversely, we could have determined δm to order e^2 from the requirement that the energy of the physical one-electron state be $\mathbf{p}^2/2m$. The term $(\delta m/m)\,(\mathbf{p}^2/2m)$ is usually called the mass renormalization counterterm.

Consider next the case of an electron bound in a Bohr orbit around a proton, i.e., the case $V(\mathbf{x}) = -Ze^2/|\mathbf{x}|$. The Hamiltonian for this system is taken to be

$$H = H_{\text{rad}} + \frac{\mathbf{p}^2}{2m} + V(\mathbf{x}) - \frac{e}{mc}\,\mathbf{p}\cdot\mathbf{A} + \frac{\delta m}{m}\frac{\mathbf{p}^2}{2m} \tag{56}$$

where δm is given above by (55). A considerable simplification can now be made in this Hamiltonian, arising from the fact that we shall consider only the bound states of the electron. For these, the position co-ordinate of the electron, $\mathbf{x}$, will always be of the order of a Bohr radius. Hence, since $F(k)$ cuts off momenta larger than mc, we may now neglect the exponential factors in the expansion (46), since $\mathbf{k} \cdot \mathbf{x} \approx \alpha$, and write $\exp i\mathbf{k} \cdot \mathbf{x} \approx 1$. This approximation is known as the dipole approximation. It is equivalent to neglecting retardation and recoil effects in the act of emission and absorption of a photon by the electron. In other words, one considers only the interaction of the electron with such long wavelengths that the momentum transfer (factor $\exp i\mathbf{k} \cdot \mathbf{x}$) is negligible. In this approximation the expansion for $\mathbf{A}(\mathbf{x})$ is given by

$$\mathbf{A}(\mathbf{x}) = \sqrt{\frac{\hbar c}{(2\pi)^3}} \int \frac{d^3k}{\sqrt{2\omega_{\mathbf{k}}}} \sum_{\lambda=1}^{2} \epsilon_\lambda(\mathbf{k}) \, (c_\lambda(\mathbf{k}) + c_\lambda{}^*(\mathbf{k})) \tag{57}$$

Again, applying perturbation theory, but this time choosing the unperturbed eigenfunctions as those of $H_{\text{rad}} + (\mathbf{p}^2/2m) + V(\mathbf{x})$, we now find [Bethe (1947)] the level shift for the electron level n to be

$$\Delta E_n = -\frac{2\alpha}{3\pi c^2} \int_0^\infty dk \, k|F(k)|^2 \sum_r \frac{|\mathbf{v}_{rn}|^2}{E_r - E_n + k} + \frac{\delta m}{m} \frac{|\mathbf{p}^2|_{nn}}{2m} \tag{58}$$

where r runs over all the energy states of the electron and where $\mathbf{v}$ is the velocity operator of the electron. Using the identity

$$\frac{1}{E_r - E_n + k} = \frac{1}{k} - \frac{E_r - E_n}{k(E_r - E_n + k)} \tag{59}$$

we may rewrite the expression (58) as

$$\Delta E_n = -\frac{2\alpha}{3\pi c^2} \int_0^\infty dk \, |F(k)|^2 \sum_r |\mathbf{v}_{nr}|^2 + \frac{\delta m}{m} \frac{|\mathbf{p}^2|_{nn}}{2m}$$
$$+ \frac{2\alpha}{3\pi c^2} \int_0^\infty dk \, |F(k)|^2 \sum_r \frac{|\mathbf{v}_{nr}|^2 \, (E_r - E_n)}{E_r - E_n + k} \tag{60}$$

Now recalling that $\overline{\mathbf{v}_{nr}} = \mathbf{v}_{rn}$ and that

$$\sum_r \mathbf{v}_{nr} \cdot \mathbf{v}_{rn} = |\mathbf{v}^2|_{nn} = \frac{|\mathbf{p}^2|_{nn}}{m^2} \tag{61}$$

we note that the first term is just the self-energy of an electron having the kinetic energy distribution of the state n. This term is therefore exactly canceled by the mass renormalization counterterm, the second term in (60). The level shift is therefore finally given by

$$\Delta E_n = \frac{2\alpha}{3\pi c^2} \int_0^K dk \sum_r \frac{|\mathbf{v}_{nr}|^2 \, (E_r - E_n)}{E_r - E_n + k} \tag{62a}$$

$$= \frac{2\alpha}{3\pi c^2} \sum_r |\mathbf{v}_{nr}|^2 \, (E_r - E_n) \log \frac{K}{|E_r - E_n|} \tag{62b}$$

where K is the cutoff energy $K \sim mc^2$, which is assumed to be large compared with all energy differences $E_r - E_n$ in the atom.

Actually, in his original work, Bethe (1947) did not use a counterterm, nor did he consider the electron as having a finite size of the order of the Compton wavelength. Instead, he argued that renormalization requires that the correction to the kinetic energy due to the self-mass must be subtracted from the self-energy of the bound electron. This reduces the divergence of the self-energy of a bound electron from linear to logarithmic. [Let $K \to \infty$ in (62b).] Bethe further surmised that the same principle would lead to convergent results in the hole theory (in which the self-energy diverges only logarithmically) and therefore introduced a cutoff of the nonrelativistic theory at a photon energy of mc^2. With this assumption, the formula for the level shift which he obtained is Eq. (62b), which when evaluated for the $2S$ level [Bethe (1947), (1950b), and Harriman (1956)] yields a value of 1040 megacycles, in surprisingly close agreement with the most recent experimental value [Triebwasser *et al.* (1953)] of 1057.77 ± 0.10 megacycles.

The evaluation of Eq. (62b) proceeds as follows: Since the log in Eq. (62b) will be quite large in the nonrelativistic range, it is convenient to write

$$\sum_r (E_r - E_n)\,|\mathbf{p}_{rn}|^2 \log \frac{|E_r - E_n|}{\text{Ry}} = \left[\sum_r (E_r - E_n)\,|\mathbf{p}_{rn}|^2\right] \log \frac{(E_{\text{ave}} - E_n)}{\text{Ry}} \tag{63}$$

which is the definition of $E_{\text{ave}} - E_n$. For the $2S$ level of H, it was found [Bethe (1950b)] that $E_{\text{ave}} - E_n \approx 16.6\text{Ry}$. In the hydrogenic case with $H_{\text{matter}} = (\mathbf{p}^2/2m) + V(\mathbf{x})$; $V(\mathbf{x}) = -Ze^2/|\mathbf{x}|$, one finds

$$\sum_r (E_r - E_n)\,|\mathbf{p}_{rn}|^2 = \tfrac{1}{2} \sum_{i=1}^{3} ([p_i, [H_{\text{matter}}, p_i]])_{nn}$$
$$= \tfrac{1}{2} \int \bar{\psi}_n(\mathbf{x})\, \nabla^2 V(\mathbf{x})\, \psi_n(\mathbf{x})\, d^3x = 2\pi\hbar^2 e^2 Z\,|\psi_n(\mathbf{0})|^2 \tag{64}$$

To this approximation, there is a level shift only for S states. Since for S states with quantum number n, $|\psi_n(\mathbf{0})|^2 = (Z/n\pi^{1/3}a_0)^3$ where a_0 is the Bohr radius, the level shift becomes

$$\Delta E(S \text{ states}) = \frac{8}{3\pi} \frac{\alpha^3 Z^4}{n^3} \text{Ry} \log \frac{K}{(E - E_n)_{\text{ave}}} \tag{65}$$

which yields the value of 1040 megacycles for the $2S$ state.

Subsequent relativistic calculations confirmed that the level shift calculated by Kramers' renormalization principle indeed converges. These relativistic calculations [Schwinger and Weisskopf (Schwinger 1948b), French and Weisskopf (French 1949), Kroll and Lamb (Kroll 1949a), Feynman (1948b), Schwinger (1949c), and Bethe (1950a, b)] gave a level shift of 1052 megacycles. Further refinements [Baranger (1951a, b, 1953), Karplus (1952b), Fried (1960), and Layzer (1960)], especially in the treatment of the high momentum components of the electron in the $2S$ state, give a theoretical result within 0.1 megacycle of the observed value. [For a review of the present status of the theoretical knowledge of atomic energy levels in hydrogen-like atoms, see Petermann (1958) and Layzer (1960).]

15c. Radiative Corrections to Scattering

We shall now consider in detail the scattering of an electron by an external electromagnetic field, including radiative corrections, to illustrate the renormalization procedure in a relativistic field theory[7] [Feynman (1949)]. The interaction Hamiltonian density for this situation is given by

$$\mathcal{H}_I(x) = -eN[\tilde{\psi}\gamma^\mu\psi A_\mu(x)] - eN[\tilde{\psi}\gamma^\mu\psi A^e{}_\mu(x)] - \delta m N[\tilde{\psi}\psi(x)] \quad (66)$$

The first term represents the interaction of the electron-positron field with the quantized radiation field, the second term the interaction with the external potential $A^e{}_\mu(x)$ which, it is assumed, satisfies $\partial_\mu A^{e\mu}(x) = 0$, and the third is the counterterm effecting the mass renormalization. In the S-matrix expansion to order e^3, the diagrams indicated in Figure 15.6 contribute to the scattering. In Figure 15.6, the convention of indicating the action of the external potential by a wavy line ending with a cross has been used.

The matrix element corresponding to diagram a in Figure 15.6 has been investigated previously. This matrix element for a transition from a state $\mathbf{p}_1$ to the state $\mathbf{p}_2$ is given by

$$R^{(a)}(p_1, p_2) = 2\pi i\tilde{u}(\mathbf{p}_2)\, M^{(a)} u(\mathbf{p}_1) \quad (67a)$$

$$M^{(a)} = e\, \gamma^\mu a_\mu(p_2 - p_1) \quad (67b)$$

$$q^\mu a_\mu(q) = 0 \quad (67c)$$

We have here omitted the normalization factors $[m/E(\mathbf{p})]^{1/2}$. Using our previous rules, the operator $M^{(b)}$ corresponding to the Feynman diagram b in Figure 15.6 is given by

$$M^{(b)} = -\frac{i\alpha}{4\pi^3} \int d^4k\, \gamma^\mu \frac{1}{\gamma \cdot (p_2 - k) - m}\, e\gamma^\nu a_\nu(p_2 - p_1) \frac{1}{\gamma \cdot (p_1 - k) - m}\gamma_\mu \frac{1}{k^2 - \lambda_{\min}{}^2 + i\epsilon} \quad (68a)$$

$$= \Lambda^\nu(p_2, p_1)\, e a_\nu(p_2 - p_1) \quad (68b)$$

where $\alpha = e^2/4\pi\hbar c = 1/137$ (e is here in rational units). The diagram 15.6b is called a vertex graph, and $\Lambda^\nu(p_2, p_1)$ the vertex operator to order α. In (68a), we have replaced $(k^2)^{-1}$, the photon propagator, by $(k^2 - \lambda_{\min}{}^2)^{-1}$

[7] For a semiclassical treatment of radiative corrections which permits a simple physical picture, see Welton (1948), Koba (1949), and also the excellent review article of Weisskopf (1949). For the renormalization procedure carried out using "old-fashioned" perturbation theoretic method, consult Lewis (1948) and Epstein (1948).

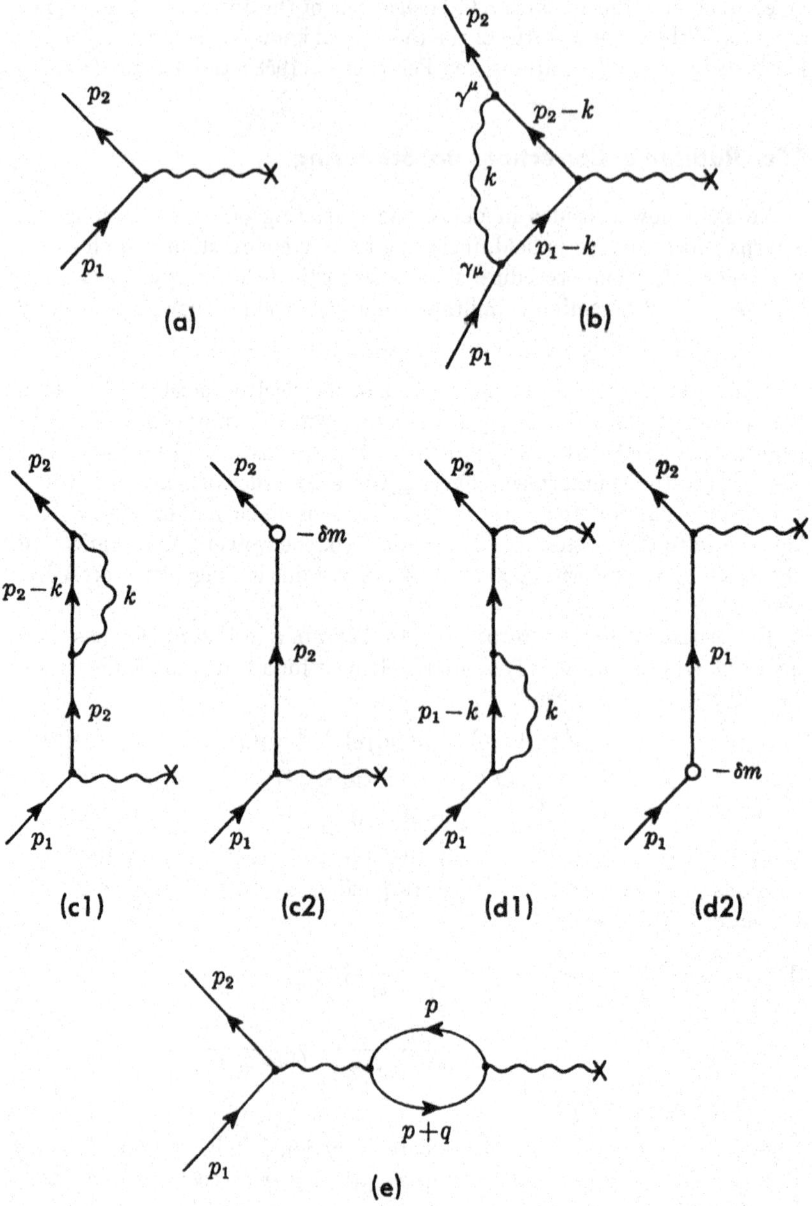

p2
p1
(a)
p2
γ^μ
p2 − k
k
p1 − k
γμ
p1
(b)
p2
p2 − k
k
p2
p1
(c1)
p2
− δm
p2
p1
(c2)
p2
p1 − k
k
p1
(d1)
p2
p1
− δm
p1
(d2)
p2
p
p + q
p1
(e)

Fig. 15.6

to avoid the divergence which is present for small k. The interpretation of this device will be considered later.

Similarly, we may write the operators corresponding to the diagrams $c1$ and $d1$ as

$$M^{(c1)} = -\frac{ie}{4\pi^3}\Sigma^{(2)}(p_2)\frac{1}{\gamma \cdot p_2 - m}\gamma \cdot a(p_2 - p_1) \tag{69}$$

$$M^{(d1)} = -\frac{ie}{4\pi^3}\gamma \cdot a(p_2 - p_1)\frac{1}{\gamma \cdot p_1 - m}\Sigma^{(2)}(p_1) \tag{70}$$

where $\Sigma^{(2)}(p)$, the self-energy operator to second order, is given by

$$\Sigma^{(2)}(p) = \alpha \int d^4k\, \gamma^\mu \frac{1}{\gamma \cdot (p - k) - m}\gamma_\mu \frac{1}{k^2 - \lambda_{\min}{}^2 + i\epsilon} \tag{71}$$

The mass correction diagrams $c2$ and $d2$ are given by

$$M^{(c2)} = -e\delta m \frac{1}{\gamma \cdot p_2 - m}\gamma \cdot a(p_2 - p_1) \tag{72}$$

$$M^{(d2)} = -e\delta m \gamma \cdot a(p_2 - p_1)\frac{1}{\gamma \cdot p_1 - m} \tag{73}$$

Finally, the contribution of diagram e which corresponds to a vacuum polarization effect, and which will be discussed in Section 15e, is given by

$$M^{(e)} = \frac{i\alpha}{4\pi^3}\gamma^\mu \frac{1}{(p_2 - p_1)^2}\Pi_{\mu\nu}a^\nu(p_2 - p_1) \tag{74}$$

where

$$\Pi_{\mu\nu}(q) = \int d^4p\, \mathrm{Tr}\left[\gamma_\mu \frac{1}{\gamma \cdot (p + q) - m}\gamma_\nu \frac{1}{\gamma \cdot p - m}\right] \tag{75}$$

and

$$q = p_2 - p_1 \tag{76}$$

It should be remarked that, although we have specialized to the case of an external field, we could just as easily have considered the situation where the external field is replaced by an internal photon line. In this case, all the factors $a_\mu(q)$ would be replaced by $q^{-2}e\gamma_\mu$ corresponding to the photon propagator coming to that vertex (there would also be a δ function corresponding to momentum conservation, $p_2 - p_1 = q$). The subsequent analysis will then still go through. In this situation, we could consider the diagrams of Figure 15.6 as a part of larger diagrams, for example, those indicated in Figure 15.7 corresponding to the interaction of an electron with a proton (denoted by double lines).

An analysis of the diagram 15.7a actually gives an insight as to the conditions under which a source of virtual photons can be considered an "external" potential. Let the amplitude for production by the source of a virtual photon of momentum q, polarization μ, be $a_\mu{}^{(1)}(q)$. Then the matrix element for the scattering of an electron by this source due to the

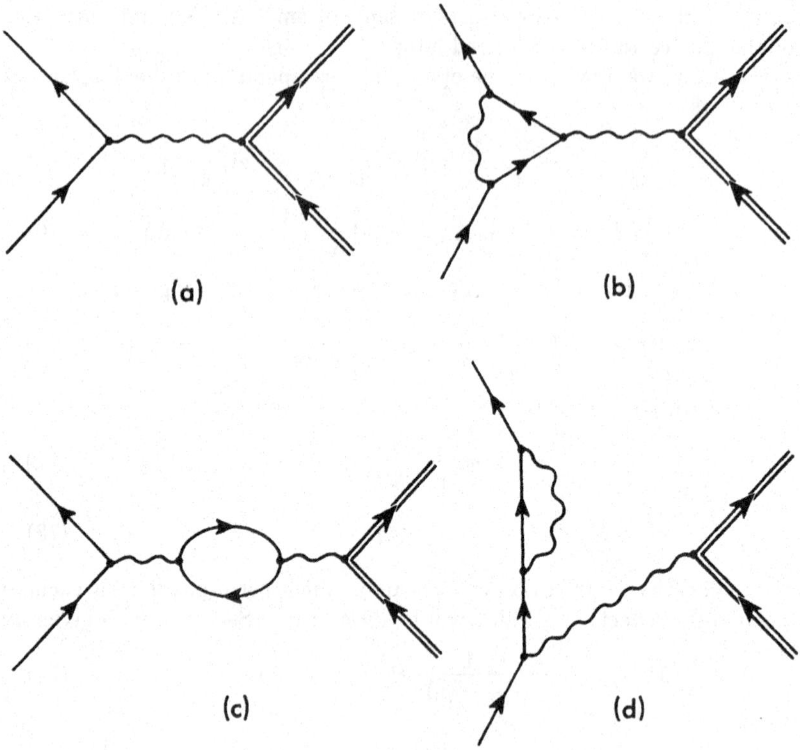

Fig. 15.7

absorption of a single photon (corresponding to the diagram of Figure 15.8) is proportional to

$$M^{(1)} = \int d^4q \, \frac{1}{\gamma \cdot (p_1 + q) - m} \gamma^\mu \frac{1}{\gamma \cdot p_1 - m} a_\mu{}^{(1)}(q) \qquad (77)$$

Similarly, if the amplitude for emission of two photons by the source is

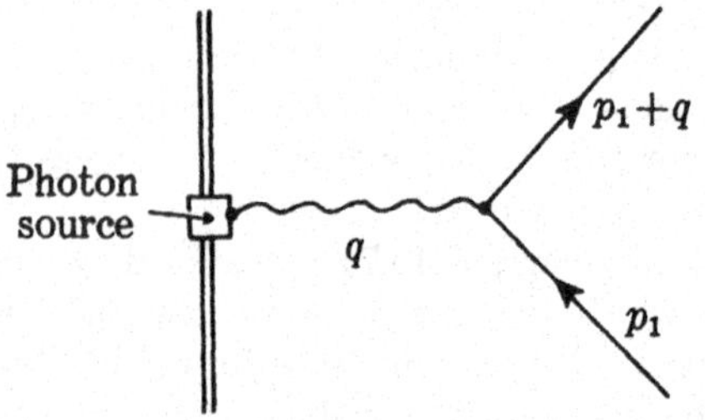

Fig. 15.8

$a^{(2)}_{\mu_1\mu_2}(q_1, q_2)$, the contribution to the scattering of the electron due to two-photon absorption will be proportional to

$$M^{(2)} = \int d^4q_1 \int d^4q_2 \frac{1}{\gamma \cdot (p_1 + q_1 + q_2) - m} \gamma^{\mu_1} \frac{1}{\gamma \cdot (p_1 + q_1) - m} \gamma^{\mu_2} \frac{1}{\gamma \cdot p_1 - m} a^{(2)}_{\mu_1\mu_2}(q_1, q_2) \quad (78)$$

If the source has the property that the emission of the first photon leaves its state essentially unchanged, so that the amplitude for the emission of a second photon is independent of whether a prior photon has been emitted or not, i.e., if

$$a^{(n)}_{\mu_1\mu_2\cdots\mu_n}(q_1, \cdots q_n) = a_{\mu_1}{}^{(1)}(q_1) \cdots a_{\mu_n}{}^{(1)}(q_1) \quad (79)$$

we then call this source an "external" potential. A very heavy charged particle has the property that its recoil in the act of emission of a photon of energy $\omega_{\mathbf{k}} \ll M$ is negligible. Such a "source" therefore behaves like an external potential. [For the description of the emission of radiation by classical systems, see Glauber (1951), Thirring (1951), Schwinger (1953b).] Similarly, the real emission of very soft photons by a charged particle can be described classically, to a very good approximation. This approximation which consists in neglecting the reaction of the electromagnetic field on the charge was first justified by Bloch and Nordsieck [Bloch (1937)].

To see this in somewhat greater detail in the case of virtual emission, consider again diagram *a* of Figure 15.7 on the assumption that the double line represents a heavy spin $\frac{1}{2}$ particle of mass $M \gg m$ and charge Ze. The amplitude for the scattering of an electron from the state p_1 to $p_1 + k = p_2$ is then proportional to

$$R' = Ze^2\delta^{(4)}(p_2 + q_2 - p_1 - q_1) \cdot \tilde{u}_M(\mathbf{q}_2)\, \gamma^\mu u_M(\mathbf{q}_1) \frac{1}{k^2} \tilde{u}_m(\mathbf{p}_2)\, \gamma_\mu u_m(\mathbf{p}_1) \quad (80)$$

We recall that

$$\tilde{u}_M(\mathbf{q}_2)\, \gamma_\mu u_M(\mathbf{q}_1) = \frac{1}{2M} \tilde{u}_M(\mathbf{q}_2)\, [\gamma \cdot q_2\gamma_\mu + \gamma_\mu\gamma \cdot q_1]\, u_M(\mathbf{q}_1) \quad (81a)$$

$$= \frac{1}{2M} (q_1 + q_2)_\mu\, \tilde{u}_M(\mathbf{q}_2)\, u_M(\mathbf{q}_1) + \frac{i}{2M} (q_2 - q_1)^\nu\, \tilde{u}_M(\mathbf{q}_2)\, \sigma_{\nu\mu} u_M(\mathbf{q}_1) \quad (81b)$$

In this decomposition of the (current) operator γ_μ, the first term corresponds to the contribution of the charge and the second is the contribution of the magnetic moment. The latter vanishes in the approximation that recoil is neglected, i.e., that $q_2 = q_1$. In the frame in which the heavy particle is at rest, the neglect of recoil is the statement that

$$q_{1\mu} \approx q_{2\mu} = M g_{0\mu} \quad (82)$$

in which case (81) becomes equal to 1 since $\tilde{u}_M(\mathbf{0})\, u_M(\mathbf{0}) = 1$. Since $q_2 = q_1 - k$,

$$q_2^2 = M^2 = (q_1 - k)^2 = M^2 - 2k \cdot q_1 + k^2 \tag{83}$$

The space-like vector $k^2 = 2k \cdot q_1$ is therefore equal to $2k_0M$ in the frame in which the massive particle is at rest; whence $k_0 \approx -\mathbf{k}^2/2M$ and

$$\frac{1}{k^2} \simeq -\frac{1}{\mathbf{k}^2}\left(\frac{1}{1 - \mathbf{k}^2/4M^2}\right) \approx -\frac{1}{\mathbf{k}^2} \tag{84}$$

which is just the Coulomb potential of the charge Ze. Therefore, in the approximation that the recoil can be neglected, Eq. (80) reduces to

$$R' \sim -\tilde{u}_m(\mathbf{p}_2)\, \gamma^0 \frac{Ze^2}{|\mathbf{p}_2 - \mathbf{p}_1|^2} u_m(\mathbf{p}_1)\, \delta^{(1)}(p_{20} - p_{10})\, \delta^{(3)}(\mathbf{p}_2 - \mathbf{p}_1 - \Delta\mathbf{q}) \tag{85}$$

which is the amplitude for the scattering of the electron with a momentum transfer $\Delta\mathbf{q}$ in the Coulomb field of the massive particle.

We now return to the principal consideration of this section, namely the analysis of the diagrams of Figure 15.6. Consider first diagram *b* which, as we have seen, corresponds to the operator $\Lambda^\nu(p_2, p_1)\, a_\nu(q)$. The integral for Λ

$$\Lambda_\nu(p_2, p_1) = -\frac{i\alpha}{4\pi^3}\int d^4k\, \gamma^\mu \frac{1}{\gamma \cdot (p_2 - k) - m}\gamma_\nu \frac{1}{\gamma \cdot (p_1 - k) - m}\gamma_\mu \cdot \frac{1}{k^2 - \lambda_{\min}{}^2 + i\epsilon} \tag{86}$$

is evidently logarithmically divergent, since for large k the denominator will contribute a factor k^4 whereas the numerator behaves as $d^4k \propto k^3\, dk$. The expression (86) can be evaluated[8] using the regulator method [see Feynman (1949b), Appendix B]. For this purpose one replaces the photon propagator $(k^2 - \lambda_{\min}{}^2)^{-1}$ in (86) by $(k^2 - \lambda_{\min}{}^2)^{-1} - (k^2 - M^2)^{-1}$; M is the large "regulator mass." The regulated expression can then be written in the form

$$\Lambda_{R\mu}(q) = \frac{\alpha}{2\pi}\left(\log\frac{M}{m} + \frac{9}{4} - 2\log\frac{m}{\lambda_{\min}}\right)\gamma_\mu + \Lambda_{C\mu}(q) \tag{87}$$

The quantity $\Lambda_{C\mu}(q)$ is represented by an integral which converges at large momenta k (although it would diverge in the limit $\lambda_{\min} \to 0$). It has been evaluated by Feynman for the case $q^2 \ll 4m^2$ and $p_1{}^2 = p_2{}^2 = m^2$, neglect-

[8] The three energy denominators can be combined by the repeated use of the Feynman formula (15.31) or by using the formula,

$$\frac{1}{abc} = 2\int_0^1 dx\,x \int_0^1 dy \frac{1}{[a(1 - x) + bxy + cx(1 - y)]^3}$$

ing terms which vanish in the limit $\lambda_{\min} \to 0$, $M \to \infty$.[9] Introducing θ by $\sin^2\theta = q^2/4m^2$, the expression for $\Lambda_{C\mu}$ under the stated assumptions is:

$$\Lambda_{C\mu}(q) = \frac{\alpha}{2\pi}\left\{\left[2\left(\log\frac{m}{\lambda_{\min}} - 1\right)\left(1 - \frac{2\theta}{\tan 2\theta}\right) + \theta\tan\theta + \frac{4}{\tan 2\theta}\int_0^\theta x\tan x\,dx\right]\gamma_\mu + \frac{\gamma\cdot q\gamma_\mu - \gamma_\mu\gamma\cdot q}{4m}\frac{2\theta}{\sin 2\theta}\right\} \quad (88)$$

For small q, $\Lambda_{C\mu}(q)$ is of the form

$$\Lambda_{C\mu}(q) = \frac{\alpha}{2\pi}\left[\frac{1}{4m}(\gamma\cdot q\gamma_\mu - \gamma_\mu\gamma\cdot q) + \frac{2q^2}{3m^2}\left(\log\frac{m}{\lambda_{\min}} - \frac{3}{8}\right)\gamma_\mu\right] \quad (89)$$

All these expressions are valid only if p_1 and p_2 are free-particle momenta, $p_1{}^2 = p_2{}^2 = m^2$.

It will be noted that (88) and (89) go to zero as the momentum change q goes to zero, and that on the other hand, the logarithmically divergent first term of (87) is independent of q. In other words, only the part of Λ_μ which remains when $q \to 0$, or when p_2 approaches p_1, contains a divergent integral. That this must be so can be seen as follows. Let us consider p_1 as fixed and expand Λ_μ, as defined by Eq. (86) in a series in powers of $p_2 - p_1$, or rather of the components of this four-vector. This involves the expansion of $[\gamma\cdot(p_2 - k) - m]^{-1}$ using the following general expansion valid for any operator A, B [Feynman (1949b)]:

$$\frac{1}{A+B} = \frac{1}{A} - \frac{1}{A}B\frac{1}{A} + \frac{1}{A}B\frac{1}{A}B\frac{1}{A} + \cdots \quad (90a)$$

obtained by iterating the identity (90b):

$$\frac{1}{A+B} = \frac{1}{A} - \frac{1}{A}B\frac{1}{A+B} \quad (90b)$$

$$= \frac{1}{A}(A+B)\frac{1}{A+B} - \frac{1}{A}B\frac{1}{A+B}$$

$$= \frac{1}{A}(A+B-B)\frac{1}{A+B} \quad (90c)$$

Equation (90a) applied to $[\gamma\cdot(p_2 - k) - m]^{-1}$ yields

$$\frac{1}{\gamma\cdot p_2 - \gamma\cdot k - m} = \frac{1}{\gamma\cdot p_1 + \gamma\cdot(p_2 - p_1) - \gamma\cdot k - m}$$

$$= \frac{1}{\gamma\cdot p_1 - \gamma\cdot k - m} - \frac{1}{\gamma\cdot p_1 - \gamma\cdot k - m}\gamma\cdot(p_2 - p_1)\frac{1}{\gamma\cdot p_1 - \gamma\cdot k - m} + \cdots \quad (91)$$

[9] In the case $q^2 > 4m^2$, the external potential can create real pairs and the denominators in the integration over the auxiliary variable xy has poles. The integral is, however, well defined since the $i\epsilon$ factors give an unambiguous prescription for its evaluation. The possibility of real pair production is, however, not of practical importance, and we therefore shall not consider the case $q^2 > 4m^2$. An evaluation of $\Lambda_\nu(p_1, p_2)$ when $p_1{}^2 \neq m^2$, $p_2{}^2 \neq m^2$ can be found in Karplus and Kroll [Karplus (1950)].

Clearly, the first term is of the order of $1/k$ for large k, the second of the order of $1/k^2$, etc. Thus, when the expansion (91) is inserted back into the integral (86), only the first term will give rise to a logarithmically divergent integral over k; the second term will give an integrand behaving as dk/k^2 and hence a convergent integral, and the further terms in the series will give even better convergence of the integrals, the term proportional to $(p_2 - p_1)^n$ being associated with an integral of the form dk/k^{n+1}. This proves our previous statement that only $\Lambda_\mu(p_1, p_1)$ is divergent [i.e., depends on the regulator mass M, as in Eq. (87)], but that

$$\Lambda_{C\mu}(p_1, p_2) = \Lambda_\mu(p_1, p_2) - \Lambda_\mu(p_1, p_1) \tag{92}$$

is finite (i.e., represented by a convergent integral over k, and independent of the regulator mass). Indeed, Eq. (92) may be considered as the definition of the corrected, finite part of Λ_μ which we have called $\Lambda_{C\mu}$ in (88) (at least in the case when both p_1 and p_2 represent free particles).

Let us consider the term $\Lambda_\mu(p_1, p_1)$ in somewhat greater detail. Using the identities

$$\gamma^\mu \not{k} \gamma_\mu = -2\not{k} \tag{93a}$$

$$\gamma^\mu \not{k} \not{p} \gamma_\mu = 2p \cdot k \tag{93b}$$

$$\gamma^\mu \not{k} \not{p} \not{q} \gamma_\mu = -2\not{q}\not{p}\not{k} \tag{93c}$$

one verifies that $\Lambda_\mu(p_1, p_1)$ can be written as follows:

$$\Lambda_\mu(p_1, p_1) = B\gamma_\mu + R_\mu(p_1) \tag{94a}$$

where B is a divergent constant

$$B = \frac{\alpha}{2\pi} \lim_{M\to\infty} \left(\log\frac{M}{m} + \frac{9}{4} - 2\log\frac{m}{\lambda_{\min}}\right) \tag{94b}$$

and $R_\mu(p_1)$ has the property that it is finite and that

$$\tilde{u}(\mathbf{p}_1)\, R_\mu(p_1)\, u(\mathbf{p}_1) = 0 \tag{94c}$$

$$R_\mu(p_1)\Big|_{\gamma\cdot p_1 = m} = 0 \tag{94d}$$

The complete vertex function can be written in the form

$$\begin{aligned}\Lambda_\mu(p_2, p_1) &= \Lambda_\mu(p_1, p_1) + \{\Lambda_\mu(p_2, p_1) - \Lambda_\mu(p_1, p_1)\} \\ &= B\gamma_\mu + R_\mu(p_1) + \Lambda_{C\mu}(p_2, p_1)\end{aligned} \tag{95}$$

The contributions of diagrams 15.6a and b can therefore be combined to give

$$\begin{aligned}M^{(a)} + M^{(b)} = {} & +2\pi i(1 + B)\, e\tilde{u}(\mathbf{p}_2)\, \gamma \cdot a(p_2 - p_1)\, u(\mathbf{p}_1) \\ & + 2\pi i\tilde{u}(\mathbf{p}_2)\, e\Lambda_{C\mu}(p_2, p_1)\, a^\mu(p_2 - p_1)\, u(\mathbf{p}_1)\end{aligned} \tag{96}$$

We shall see that in electrodynamics the contributions of diagrams 15.6c and d actually cancel the B term. It should be emphasized that the quantity B is gauge dependent. Thus in the Landau gauge in which the photon propagator has the form $D_{F\mu\nu}(k) = (g_{\mu\nu} - k_\mu k_\nu k^{-2})\, k^{-2}$, B is *convergent* to order e^2. In the Landau gauge the only divergence to order e^2 in the S-matrix expansion is that corresponding to a mass renormalization and is canceled by the δm counterterm [Fried and Yennie (Fried 1958)].

The principle of renormalization by subtraction, similar to Eq. (95), is used very generally in renormalization theory. One expands an operator, in the present case $\Lambda_\mu(p_1, p_2)$, in a power series of some quantity related to the external variables, in our case $p_2 - p_1$. Then only the first term (or the first few terms) of this expansion will contain divergent integrals over the internal variables (in our case k). These terms are then discarded in a relativistically invariant manner, by either a mass renormalization or by a charge renormalization as in Section 15*e*, or by noting that they cancel another divergent term, as will be found to be the case for our Λ_μ, and in a few cases by introducing special terms into the original Lagrangian as, for example, in cases discussed later in Chapter 16. The finite terms which remain after this procedure are then taken to be the renormalized expression for the operator. We shall encounter this procedure throughout the rest of this volume.

We next study the terms (69) and (72). The appearance of the factor $(\gamma \cdot p - m)^{-1}$ in these leads to difficulties, since the matrix elements of $M^{(c1)}$ and $M^{(c2)}$ are to be taken between two free-particle wave functions, thus giving rise to zero denominators. It is true that the sum of the numerators in $M^{(c1)}$ and $M^{(c2)}$ also cancels, but this only leads to the result 0/0, which is indeterminate. To obtain an unambiguous result, we must explicitly introduce the damping factors, which are necessary for the correct definition of the initial and final "bare" states (Chapter 11), because the above ambiguity is essentially related to the limiting process of letting the initial and final times of the $U(t, t_0)$ matrix go to $\pm\infty$. If care is not exercised, the limiting process destroys the unitarity of the S matrix. Feynman (1949b) and Dyson (1949b, 1951b) have shown how to get over this difficulty [see also Karplus and Kroll (Karplus 1950), p.542, and Lüders (1952b)]. If we explicitly introduce a damping function, $g(t)$, which adiabatically switches on and switches off the coupling between fields, the interaction Hamiltonian is replaced by[10]

$$\mathcal{H}_I(x) \to -eg(t)\, N[\bar{\psi}(x)\, \gamma_\mu \psi(x)]\, (A^\mu(x) + A^{e\mu}(x)) - \delta m[g(t)]^2\, N(\bar{\psi}(x)\, \psi(x)) \tag{97}$$

It is assumed that the time T over which $g(t)$ varies is very long compared to the duration of the scattering process. If we denote the Fourier transform of $g(t)$ by $G(\Gamma_0)$, then

[10] The self-energy term δm being a second-order radiative effect must be multiplied by $[g(t)]^2$.

$$g(t) = \int_{-\infty}^{+\infty} G(\Gamma_0)\, e^{-i\Gamma_0 t}\, d\Gamma_0$$
$$= \int_{-\infty}^{+\infty} G(\Gamma_0)\, e^{-i\Gamma\cdot x}\, d\Gamma_0 \tag{98}$$

where the vector Γ_μ is defined by $\Gamma = (\Gamma_0, 0, 0, 0)$. The normalization is such that

$$g(0) = \int_{-\infty}^{+\infty} G(\Gamma_0)\, d\Gamma_0 = 1 \tag{99}$$

and it is supposed that $G(\Gamma_0)$ is almost a delta function, being large only for values of Γ_0 in a range of about T^{-1}.

If we substitute the interaction Hamiltonian given by (97) in the S-matrix expansion, then, for example, Eq. (70) is replaced by

$$M^{(d1)} = -\frac{ie}{4\pi^3} \iint \gamma\cdot a(p_2 - p_1 - \Gamma - \Gamma') \frac{1}{\gamma^\mu(p_{1\mu} + \Gamma_\mu + \Gamma'_\mu) - m}$$
$$\Sigma^{(2)}(p_1 + \Gamma)\, G(\Gamma_0)\, G(\Gamma'_0)\, d\Gamma_0\, d\Gamma'_0 \tag{100}$$

As $T \to \infty$, and Γ_0 and $\Gamma'_0 \to 0$, the integrand may be expanded as follows: The electron propagator becomes

$$\frac{1}{\gamma^\mu(p_{1\mu} + \Gamma_\mu + \Gamma'_\mu) - m} = \frac{\gamma^\mu(p_{1\mu} + \Gamma_\mu + \Gamma'_\mu) + m}{2p_1\cdot(\Gamma + \Gamma') + (\Gamma + \Gamma')^2}$$
$$\approx \frac{\gamma\cdot p_1 + m}{2p_{10}(\Gamma_0 + \Gamma'_0)} \tag{101}$$

where we have used the free-particle relation $p_1^2 = m^2$. Hence, if we expand $\Sigma^{(2)}(p_1 + \Gamma)$, using the expansion

$$\frac{1}{A + B} = \frac{1}{A} - \frac{1}{A} B \frac{1}{A} + \frac{1}{A} B \frac{1}{A} B \frac{1}{A} + \cdots \tag{102}$$

we need only retain terms to order Γ in $\Sigma^{(2)}$

$$\Sigma^{(2)}(p_1 + \Gamma)$$
$$= \alpha \int d^4k\, \gamma^\mu \frac{1}{\gamma\cdot(p_1 + k + \Gamma) - m} \gamma_\mu \frac{1}{k^2 - \lambda_{\min}^2 + i\epsilon}$$
$$= \Sigma^{(2)}(p_1) - \alpha \int d^4k\, \gamma^\mu \frac{1}{\gamma\cdot(p_1 + k) - m} \gamma^\nu \frac{1}{\gamma\cdot(p_1 + k) - m} \gamma_\mu$$
$$\cdot \frac{1}{k^2 - \lambda_{\min}^2 + i\epsilon} \Gamma_\nu$$
$$= \Sigma^{(2)}(p_1) - I^\nu(p_1)\, \Gamma_\nu \tag{103a}$$

where $I_\nu(p)$ is a logarithmically divergent integral which is closely related to $\Lambda^\nu(p_1, p_1)$

$$I_\nu(p_1) = 4\pi^3 i \Lambda_\nu(p_1, p_1) \tag{103b}$$

Using (103), we may write the integrand of Eq. (100) as

$$\frac{1}{\gamma \cdot (p_1 + \Gamma + \Gamma') - m} [\Sigma^{(2)}(p_1) - I^\nu(p_1)\,\Gamma_\nu] \quad (104)$$

Now if we had likewise introduced the damping factor $g(t)$ in the expression (73) for $M^{(d2)}$, it would read

$$M^{(d2)} = -\iint e\delta m\gamma \cdot a(p_2 - p_1 - \Gamma - \Gamma') \frac{1}{\gamma \cdot (p_1 + \Gamma + \Gamma') - m} G(\Gamma_0)\, G(\Gamma'_0)\, d\Gamma_0\, d\Gamma'_0 \quad (105)$$

Since $\Sigma^{(2)}(p_1{}^2 = m^2)$ is apart from a constant factor $(-i/4\pi^3)$ defined as δm, it is clear that $M^{(d2)}$ cancels the part involving $\Sigma^{(2)}(p_1)$ in the matrix element $M^{(d1)}$. This, in fact, is the mass renormalization. We therefore obtain

$$M^{(d1)} + M^{(d2)} = -\frac{ie}{4\pi^3} \iint d\Gamma_0\, d\Gamma'_0\, G(\Gamma_0)\, G(\Gamma'_0) \gamma \cdot a(p_2 - p_1 - \Gamma - \Gamma') \frac{1}{\gamma \cdot (p_1 + \Gamma + \Gamma') - m} I^\nu(p_1)\,\Gamma_\nu \quad (106)$$

Now relativistic invariance requires that $I^\nu(p_1)$ transform like a four-vector under Lorentz transformation, and hence it must be of the form[11]

$$I^\nu(p_1) = I_{(1)}(p_1{}^2)\,\gamma^\nu + I_{(2)}(p_1{}^2)\,(\gamma \cdot p_1 - m)\,\gamma^\nu + I_{(3)}(p_1{}^2)\,\gamma^\nu(\gamma \cdot p_1 - m) + I_{(4)}(p_1{}^2)\,(\gamma \cdot p_1 - m)\,\gamma^\nu(\gamma \cdot p_1 - m) \quad (107)$$

where $I_{(1)}, I_{(2)}, \cdots$ are functions of the invariant p^2. Since $I^\nu(p_1)$ will operate on a free Dirac spinor of momentum p_1, only the first two terms will contribute. Moreover, the second term will clearly give a result of order Γ and can therefore be neglected. Hence, only the first term needs to be considered. For $p_1{}^2 = m^2$, we may then write

$$I^\nu(p_1{}^2) = I_{(1)}\gamma^\nu \quad (108)$$

where $I_{(1)} = I_{(1)}(m^2)$ is a constant which is equal to $+4\pi^3 iB$, B being defined by Eq. (94) (recall Eq. 103b). We next symmetrize the factor at the end of Eq. (106) by writing $\Gamma \to \frac{1}{2}(\Gamma + \Gamma')$. Also, recalling that Eq. (106) operates on the free-particle Dirac spinor, we may add to the factor $\frac{1}{2}(\Gamma + \Gamma')$ the operator $\frac{1}{2}(\gamma \cdot p - m)$; Eq. (106) then becomes

$$M^{(d1)} + M^{(d2)} = +\frac{ie\alpha}{4\pi^3} \iint d\Gamma_0\, d\Gamma'_0\, G(\Gamma_0)\, G(\Gamma'_0)\, \gamma \cdot a(p_2 - p_1 - \Gamma - \Gamma') \cdot \frac{1}{\gamma \cdot (p_1 + \Gamma + \Gamma') - m} I_{(1)}\tfrac{1}{2}[\gamma \cdot (p_1 + \Gamma + \Gamma') - m] \quad (109)$$

[11] Note that a term of the form $I'(p_1{}^2)\,p^\nu$ can always be written in the form $\frac{1}{2}I'(p_1{}^2)\,(\gamma^\nu \not p + \not p \gamma^\nu)$, so that Eq. (107) is indeed the most general form for $I_\nu(p_1)$.

In the limit $\Gamma_0 \to 0$, this becomes

$$M^{(d1)} + M^{(d2)} = +\frac{1}{2}\frac{ie}{4\pi^3} I_{(1)}\gamma \cdot a(p_2 - p_1)$$
$$= -\frac{1}{2} eB\gamma \cdot a(p_2 - p_1) \qquad (110)$$

where the factor $\frac{1}{2}$ is especially to be noted. A similar analysis can be carried out for the terms (69) and (72) which then yields

$$M^{(c1)} + M^{(c2)} = +\frac{1}{2}\frac{ie}{4\pi^3} I_{(1)}\gamma \cdot a(p_2 - p_1)$$
$$= -\frac{1}{2} eB\gamma \cdot a(p_2 - p_1) \qquad (111)$$

Now it was shown above, Eq. (95), that the divergent contribution to the matrix element corresponding to the diagram 15.6*b* can be written in the form

$$M^{(b)}{}_D = \Lambda^\nu(p_1, p_1)\, ea_\nu(p_2 - p_1)$$
$$= Be\gamma \cdot a(p_2 - p_1) \qquad (112)$$

Therefore, although the matrix elements corresponding to the diagrams *b*, *c*, and *d* of Figure 15.6 were individually divergent, their sum is finite and in fact equal to

$$M = +2\pi i\tilde{u}(\mathbf{p}_2)\, e\Lambda_{C\nu}(p_2, p_1)\, a^\nu(p_2 - p_1)\, u(\mathbf{p}_1) \qquad (113)$$

where $\Lambda_{C\nu}$ is given by Eq. (88). We shall show in Chapter 16 that this cancellation of divergences between the self-energy part and the vertex part occurs to all orders in quantum electrodynamics as a result of the requirements of gauge invariance [Ward (1950a, b)].

In order to understand the physical significance of the divergences in the sum of the two diagrams of Figures 15.6*c* and 15.6*d*, consider the following: We are calculating the effect of the interaction of an electron with an external field up to order e^3. To order e we had to consider the diagram of Figure 15.6*a* where the electron interacts with the external field and with nothing else. In order e^3, two changes must be made. First of all, there is a probability P that the electron may interact with the external field while a virtual photon is around. This is described by Figure 15.6*b*. Secondly, the probability that the electron interacts with the external field without a virtual photon around is now no longer 1, but is $1 - P$, since it is reduced by the probability of having a virtual photon around the electron. (These statements will automatically be true if S is unitary, since this condition guarantees conservation of total probability.) This probability P is given by

$$P = \int |b(\mathbf{k})|^2\, d^3k = -\frac{i\alpha}{4\pi^3} I_{(1)} \qquad (114)$$

where $b(\mathbf{k})$ is the probability amplitude that a virtual quantum of particular momentum $\mathbf{k}$ is present.

The reduction of the probability of finding the electron unaccompanied by virtual quanta is equivalent to a change in the normalization of the wave function of the electron. The diagrams c and d of Figure 15.6 are therefore often referred to as wave function renormalization. Figure 15.6d is the renormalization of the incident, Figure 15.6c that of the outgoing wave function. The wave function (i.e., the probability amplitude) should be renormalized by a factor $(1 - P)^{1/2}$, so that the first-order correction should be $\frac{1}{2}P$ which explains the factor $\frac{1}{2}$ in the result (111).

15*d*. The Anomalous Magnetic Moment and the Lamb Shift

In the last section we have shown that after mass renormalization, the sum of the contributions from the diagrams 15.6c and d is finite and that the matrix element corresponding to them is given by Eq. (113). Perhaps the easiest way to understand the meaning of these radiative corrections is to consider them in terms of an "equivalent" potential in which the electron moves. The scattering amplitude which includes the contributions of the diagrams 15.6a–d is given by

$$R = +2\pi ie\tilde{u}(\mathbf{p}_2)\,(\gamma_\nu + \Lambda_{C\nu}(p_2, p_1))\,u(\mathbf{p}_1)\,a^\nu(p_2 - p_1) \tag{115}$$

Equation (115) can also be written in the form

$$R = +2\pi ie\tilde{u}(\mathbf{p}_2)\left\{F(q^2)\,\gamma_\nu - \frac{i}{2m}\,G(q^2)\,q^\mu\sigma_{\nu\mu}\right\}u(p_1)\,a^\nu(p_2 - p_1) \tag{116}$$

which is actually the most general form of the matrix element and follows from the relativistic and gauge invariance of the S-matrix formalism and from the assumption that the external field is weak so that only terms linear in the external field need be considered [Salzman (1955)].[12] Including terms of order α, the quantities F and G are given by

$$F(q^2) = 1 + \frac{\alpha}{2\pi}\left[\left(2\log\frac{m}{\lambda_{\min}} - 1\right)\left(1 - \frac{2\theta}{\tan 2\theta}\right) + \theta\tan\theta + \frac{4}{\tan 2\theta}\int_0^\theta x\tan x\,dx\right] \tag{117a}$$

$$G(q^2) = \frac{\alpha}{2\pi}\,\frac{2\theta}{\sin 2\theta}; \quad \sin^2\theta = \frac{q^2}{4m^2} \tag{117b}$$

as is easily verified from the explicit form for $\Lambda_{C\nu}$, Eq. (88). Now to first order in the external field, the scattering matrix element that describes the elastic scattering of a Dirac particle whose electromagnetic properties are

[12] We shall actually derive the form (116) for the matrix element in Chapter 17.

characterized by Foldy coefficients ϵ_n and μ_n (ϵ_0 is the static charge and μ_0 the static anomalous magnetic moment; recall Section 4*h*) is:

$$
\begin{aligned}
(p_2 \mid s \mid p_1) &= -\frac{i}{(2\pi)^3} \int d^4x \, \tilde{u}(\mathbf{p}_2) \, e^{ip_2 \cdot x} \\
&\qquad \sum_{n=0} \{\epsilon_n \Box^n \gamma^\nu A^e{}_\nu(x) + \tfrac{1}{2}\mu_n \sigma_{\mu\nu} \Box^n F^{e\mu\nu}(x)\} \cdot u(\mathbf{p}_1) \, e^{-ip_1 \cdot x} \\
&= -2\pi i \tilde{u}(\mathbf{p}_2) \sum_{n=0} [\epsilon_n q^{2n} \gamma_\nu - i q^\mu \sigma_{\nu\mu} \mu_n q^{2n}] \, u(\mathbf{p}_1) \, a^\nu(p_2 - p_1)
\end{aligned}
\tag{118}
$$

Comparing Eqs. (118) and (116) when in the latter $F(q^2)$ and $G(q^2)$ are expanded in a power series in q^2, we see that the expansion coefficients $F^{(n)}$, $G^{(n)}$

$$F(q^2) = \sum_{n=0}^{\infty} F^{(n)} q^{2n} \tag{119a}$$

$$G(q^2) = \sum_{n=0}^{\infty} G^{(n)} q^{2n} \tag{119b}$$

are apart from constant factors the Foldy coefficients ϵ_n and μ_n which characterize phenomenologically the intrinsic electromagnetic structure of a spin $\frac{1}{2}$ particle. Thus, due to the radiative corrections, the electron behaves as if it had a charge distribution $-eF(q^2)$ and an anomalous magnetic moment distribution $\frac{-e}{2m} G(q^2)$. The function $F(q^2)$ and $G(q^2)$ are called the charge and magnetic moment form factors of the electron [Yennie *et al.* (1957)].

By expanding Eq. (117b) in powers of q^2, it will be noted that to second order, as a consequence of the radiative corrections, the electron in its interaction with slowly varying external electromagnetic field behaves as if it had a static magnetic moment of magnitude $\left(1 + \frac{\alpha}{2\pi}\right)$ Bohr magnetons since $G(0) = \frac{2m}{-e}\mu_0 = \frac{\alpha}{2\pi}$. The existence of this anomalous magnetic moment was first discovered experimentally by Nafe, Nelson, and Rabi [Nafe (1947)] and by Nagle, Julian, and Zacharias [Nagle (1947)]. Breit (1947, 1948a, b) was the first to suggest that the observed effect could be accounted for by a small additional electron spin magnetic moment. Schwinger (1948a) was the first to show that part of the radiative correction of quantum electrodynamics corresponds to an additional magnetic moment of magnitude $\alpha/2\pi$ associated with the electron spin [see also Luttinger (1948)]. The contribution to the anomalous magnetic moment, i.e., to $G(q^2)$, due to the fourth-order diagrams of Figure 15.9 has been computed by Karplus and Kroll [Karplus (1950a)]. In Figure 15.9 we have omitted the δm counterterms which go with diagram *d*.

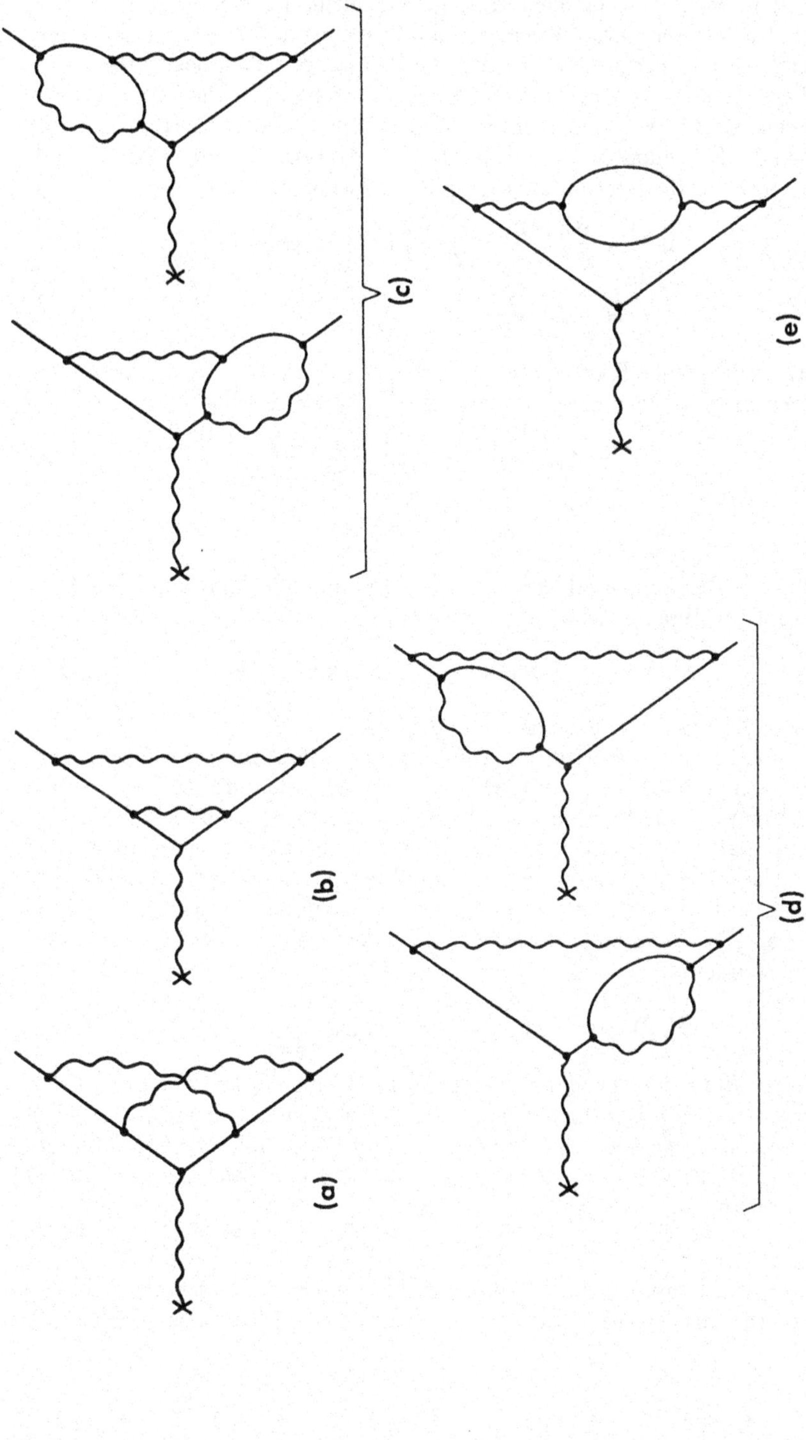

Fig. 15.9

The importance of the Karplus-Kroll calculation derives from the fact that it was the first α^2 calculation and gave explicit proof that the renormalization program circumvented all divergence difficulties to this order. Their results contained some numerical mistakes, and the calculation has been redone by Sommerfield (1957, 1958), by Petermann (1957), and by Kroll (1957, unpublished). The fourth-order contribution to the magnetic moment of the electron quoted by these authors is

$$\begin{aligned}\Delta\mu^{(4)} &= \frac{\alpha^2}{\pi^2}\left(\frac{197}{144} + \frac{\pi^2}{12} + \frac{3}{4}\zeta(3) - \frac{1}{2}\pi^2 \log 2\right)\frac{e\hbar}{2mc} \\ &= -0.328\,\frac{\alpha^2}{\pi^2}\,\frac{e\hbar}{2mc} \qquad (120)\end{aligned}$$

where $\zeta(3)$ is the Riemann zeta function of argument 3. The total anomalous magnetic moment of the electron to order α^2 is thus

$$\begin{aligned}\Delta\mu_{\text{theo}} &= \left(\frac{\alpha}{2\pi} - 0.328\,\frac{\alpha^2}{\pi^2}\right)\frac{e\hbar}{2mc} \\ &= 0.0011596\,\frac{e\hbar}{2mc} \qquad (121)\end{aligned}$$

The best experimental value of $\Delta\mu$ [Schupp, Pidd, and Crane (Schupp 1961)], as obtained from a measurement of the g factor of free electrons, is

$$\Delta\mu_{\text{exp}} = (0.0011609 \pm 0.0000024)\,\frac{e\hbar}{2mc} \qquad (122)$$

in good agreement with the theoretical value.

A comparison between Eq. (118) and Eq. (117a) further indicates that one can view the net effect of the radiative corrections due to diagrams *a–d* of Figure 15.6 as altering the potential so that the electron sees an effective potential $(\gamma_\mu + \Lambda_{C\mu})\,a^\mu$ (instead of $\gamma^\mu a_\mu$). Omitting the contribution due to the anomalous magnetic moment, Eqs. (117a) and (89) indicate that these radiative corrections spread out the potential over a distance of the order of a Compton wavelength. More precisely, the effective potential the charge sees is

$$F(q^2)\,a_\nu(q) \sim a_\nu(q) - \frac{2q^2}{3m^2}\,\frac{\alpha}{2\pi}\left(\log\frac{m}{\lambda_{\min}} - \frac{3}{8}\right)a_\nu(q) \qquad (123)$$

In an atom this change of the potential will modify the structure of the energy levels. For the case of the hydrogen atom assuming that the electron finds itself in a pure Coulomb field, Eq. (123) implies that the effective potential between proton and electron also has a term proportional to

$$-\frac{\alpha}{3\pi m^2}\,\boldsymbol{\nabla}^2\left(\frac{Ze^2}{r}\right) = 4\pi\,\frac{\alpha}{3\pi m^2}\,Ze^2\delta^{(3)}(\mathbf{r}) \qquad (124)$$

Due to the presence of the $\delta(\mathbf{r})$, this term affects primarily the S levels. For the case of the $2^2S_{1/2}$ level, it is found that the perturbation (124) shifts this level by about 1010 Mc/sec relative to the $2P_{1/2}$ level, with which it is degenerate in the Dirac theory. The anomalous magnetic moment contribution to the $2S_{1/2} - 2P_{1/2}$ level shift is approximately 68 Mc/sec. Vacuum polarization effects (the contribution of diagram 15.6*e*) give rise to a $2S_{1/2} - 2P_{1/2}$ shift of -27 Mc/sec, so that the predicted shift to order $\alpha(\alpha Z)^4$ is approximately 1052 Mc/sec. The experimental value for the shift is 1057.77 Mc/sec. The inclusion of $\alpha(\alpha Z)^5$ and $\alpha^2(\alpha Z)^4$ corrections, together with the finite size and finite mass corrections, bring the theoretical value for the $2S_{1/2} - 2P_{1/2}$ shift in hydrogen to 1057.95 ± 0.15 Mc/sec in good agreement with experiments. The inclusion of $\alpha(\alpha Z)^6 \log^2 (\alpha Z)$ and $\alpha(\alpha Z)^6 \log (\alpha Z)$ terms [Layzer (1960), Fried and Yennie (Fried 1960)] brings the agreement for hydrogen to within 0.07 ± 0.15 Mc/sec (see Table 15.1). The uncertainty of 0.15 Mc/sec in the theoretical value corresponds to uncalculable finite size effects (0.02 Mc/sec) and to the fact that the fourth-order contribution to $F(q^2)$ [Weneser, Bersohn, and Kroll (Weneser 1953)] could not be evaluated analytically but only estimated by rigorous upper and lower bounds.

TABLE 15.1

THEORETICAL AND EXPERIMENTAL VALUES OF THE LAMB SHIFT IN MC/SEC

	H	*D*	*He*
Theoretical	1057.70 ± 0.15	1058.96 ± 0.16	14046.3 ± 3.0
Experimental	1057.77 ± 0.10	1059.00 ± 0.10	14040.2 ± 4.5

An approximate value for the Lamb shift can thus be obtained from the expectation value of $\Lambda_{C\mu}(q)\, a^{\mu}(q)$ for the hydrogenic states under consideration. Strictly speaking, however, since in this situation the electron is no longer free (i.e., $p^2 \neq m^2$, it is bound to the proton!) the $\Lambda_{C\mu}(p_1, p_2)$ given by (89) is not applicable since this vertex part was derived under the assumption that $p_1{}^2 = p_2{}^2 = m^2$. However, for $|\mathbf{p}| \ll mc$, the deviation from (89) is small so that it is expected that these corrections can be neglected. There is a second difficulty connected with this procedure: the second term of (89) becomes infinite when the photon mass λ_{min} is allowed to go to zero. This infrared catastrophe is spurious and should not arise in the Lamb shift of a bound electron.

In the description of the scattering of free particles, the infrared divergences arise because the electron can emit and absorb soft photons without being displaced very far off the mass shell $p^2 = m^2$. This results in electron propagators which have very small denominators which, in combination with the photon propagator, lead to infrared divergences. In the Lamb shift calculation, on the other hand, no infrared divergence can arise because the electron's four-momentum is off the free particle mass shell.

Thus, even for very small photon energies, the electron propagator has a nonvanishing denominator equal to $m^2 - p^2$. The resulting vertex operator, however, then depends logarithmically on this quantity, i.e., the quantity $\log \lambda_{\min}/m$ in the free-particle scattering is effectively replaced by $\log (m^2 - p^2)/m^2$. Since for a bound hydrogenic state $p_0 = m - \epsilon_n$, where ϵ_n is the binding energy in the state n, and $|\mathbf{p}| \sim Z\alpha m$

$$\frac{1}{m}(m^2 - p^2) \sim 2\left[\frac{\mathbf{p}^2}{2m} + \epsilon_n\right]$$

so that the $\log \lambda_{\min}/m$ term for the bound state problem becomes replaced by $\log 2[(\mathbf{p}^2/2m) + \epsilon_n]$. The expectation value of this operator yields the major portion of the Bethe logarithm.

That no infrared divergences can arise in the Lamb shift can also be seen from the nonrelativistic treatment given in Section *b* of the present chapter. In that treatment, we assumed the light quanta to have zero rest mass and the electron to be bound. The expression (62a) for the Lamb shift has the resonance denominator $E_r - E_n + |\mathbf{k}|$ which does not vanish for light quanta of zero energy, $|\mathbf{k}| = 0$, because only states $r \neq n$ contribute, due to the factor $E_r - E_n$ in the numerator.

In the older Lamb shift calculations the following procedure was usually adopted: The nonrelativistic Lamb shift is calculated with zero rest mass for the light quanta, as given by Eq. (62a), and then again with rest mass $\lambda_{\min}$ for the light quanta. For the latter calculation, the k in the denominator of (62a) is to be replaced by

$$\omega_{\mathbf{k}} = (\mathbf{k}^2 + \lambda_{\min}{}^2)^{1/2} \tag{125}$$

The difference between these two calculations represents the difference between the Lamb shifts of a bound electron interacting with transverse quanta of zero and of finite rest mass, respectively; the difference is

$$\Delta E(0) - \Delta E(\lambda_{\min}) = \frac{2\alpha}{3\pi c^2}\int_0^K dk \sum_r |\mathbf{v}_{rn}|^2 (E_r - E_n) \left[\frac{1}{E_r - E_n + k} - \frac{1}{E_r - E_n + (k^2 + \lambda_{\min}{}^2)^{1/2}}\right] \tag{126}$$

The integral in (126) converges both at the lower limit, $k = 0$, and at the upper limit; K may therefore be replaced by ∞. Moreover, only values of $k \leqslant \lambda_{\min}$ contribute appreciably to (126); therefore, if we choose $\lambda_{\min} \ll mc^2$, the important intermediate states (r, k) in (126) will all be nonrelativistic so that the approximations of Section 15*b*, including the dipole approximation, are justified. One then adds to (126) the expectation value of (89) for the hydrogenic state under consideration, leaving the photon mass $\lambda_{\min}$ finite. Indeed, in order to justify the use of the free-electron Lamb shift operator (89) for our bound state problem, one has to choose $\lambda_{\min}$ large compared to the hydrogen binding energy, i.e., $\lambda_{\min} \gg$ $\gg 1$ Ry. This condition is clearly compatible with the requirement

$\lambda_{\min} \ll mc^2$. Evaluation then shows that the term $\log \lambda_{\min}$ cancels between (126) and the expectation value of (89), as it should: the result is independent of the fictitious photon mass. As French (1949) has pointed out, one must still take into account the contribution of longitudinal and timelike photons to the nonrelativistic part (126). When this is done, the result is equivalent to taking the sum of (62b) and the expectation value of (89) and setting

$$\log \lambda_{\min} = \log 2K - \tfrac{5}{6} \tag{127}$$

From the calculations just described, one obtains the previously mentioned result of 1052 megacycles for the Lamb shift (after inclusion of the vacuum polarization effects). For the refinements of these calculations and a discussion of the higher approximations, the reader is referred to the papers of Baranger, Bethe, and Feynman [Baranger (1953)], Karplus *et al.* [Karplus (1952b)], Mills and Kroll [Mills (1955)], Layzer (1960), and of Fried and Yennie [Fried (1958, 1960)]. [See also Kroll and Pollock (Kroll 1952) and Karplus and Klein (Karplus 1952a) for a calculation of the radiative corrections to the hyperfine structure.]

We now turn to a brief discussion of the significance of $\lambda_{\min}$ for problems involving the scattering of free electrons. The fact that the expressions (89) and (113) diverge logarithmically as $\lambda_{\min} \to 0$ is known as "infrared divergence." The procedure just outlined for bound electrons, which eliminates the infrared divergence in that case, is not applicable to free electrons. Still, it is clearly not permissible to calculate a scattering cross section directly from (89) and (113) with $\lambda_{\min} \to 0$. This difficulty is resolved if one notes that it is impossible to design an experiment which will guarantee that no photon is emitted by the electron in the scattering process. The best one can do in an experiment is to require that, if a photon is emitted, its energy should be less than some value k_0, determined by the accuracy of the measuring device. It turns out that the differential cross section for scattering with the emission of a photon (i.e., bremsstrahlung) of energy less than k_0 also contains an infrared divergence, and that this divergence just cancels the similar divergence in the radiative correction. More precisely, if, as in the above, we suppose the photon to have a very small mass $\lambda_{\min}$, then the amplitude for diagrams 15.6*b*–*d* contains a term $\log (m/\lambda_{\min})$. The probability for scattering with no photon being emitted, $|R^{(a)} + \cdots + R^{(e)}|^2$, thus contains a term proportional to $be^6 \log (m/\lambda_{\min})$. [There is also a term in $[\log (m/\lambda_{\min})]^2$ which is of higher order in α.] It is, however, found that the diagrams of Figure 15.10 corresponding to the emission of a real photon ($k^2 = 0$) yields a probability for scattering with the emission of a real photon with energy less than k_0 which to lowest order in α is equal to $-be^6 \log (k_0/\lambda_{\min})$. Therefore to order e^6, the sum of the probabilities no longer contains $\lambda_{\min}$ so that the infrared divergence is removed to this order. In general, however, the infrared divergences always cancel if one considers all the possible

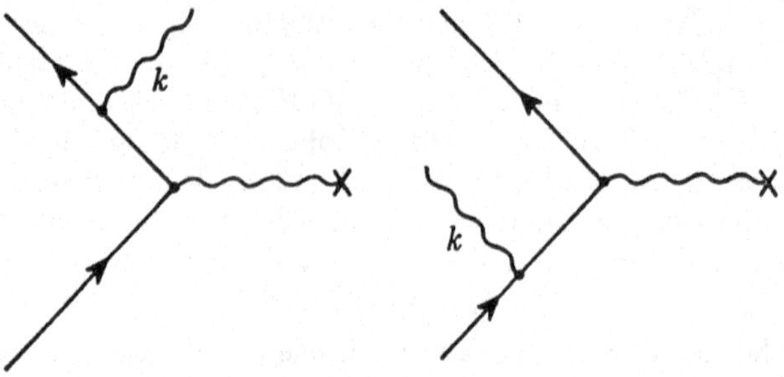

Fig. 15.10

processes (including real emission) to a given order. For a general discussion of this question the reader is referred to the papers by Bloch and Nordsieck [Bloch (1937)] who were responsible for the original solution of the problem of the infrared divergences; Pauli and Fierz [Pauli (1937)]; Bethe and Oppenheimer [Bethe (1946)]; Jost (1947); Glauber (1951); Brown and Feynman [L. Brown (1952)]; Jauch and Rohrlich [Jauch (1954, 1955)]; Lomon (1956); Prange (1957); Ascoli (1955a, b); Nakanishi (1958); and Yennie, Frautschi, and Suura [Yennie (1961)]. We shall return to this question in Section *f* of the present chapter.

15e. Vacuum Polarization

Thus far in our discussion we have not included the term corresponding to diagram 15.6*e* and Eq. (74). For the interpretation of this term, one may imagine the closed electron-positron loop to produce a current

$$j_\mu(q) = \frac{i\alpha}{4\pi^3} \Pi_{\mu\nu}(q)\, a^\nu(q) \tag{128}$$

which is the source of a photon of momentum q which then acts on the electron and scatters it. The invariance of this induced current with respect to a gauge transformation

$$a_\nu \to a_\nu + q_\nu \Lambda(q) \tag{129}$$

requires that

$$\Pi_{\mu\nu}(q)\, q^\nu = 0 \tag{130}$$

Similarly, in order that $j_\mu(q)$ be conserved, the continuity equation must be satisfied:

$$q^\mu j_\mu(q) = q^\mu \Pi_{\mu\nu}(q) = 0 \tag{131}$$

which is equivalent to Eq. (130), since $\Pi_{\mu\nu}$ is symmetrical. The operator $\Pi_{\mu\nu}(q)$ is usually called the polarization tensor. If we write $\Pi_{\mu\nu}(q)$ in the form

$$\Pi_{\mu\nu}(q) = \Pi_{(1)}(q^2)\, q_\mu q_\nu + q^2 \Pi_{(2)}(q^2)\, g_{\mu\nu} \tag{132}$$

where, because of relativistic invariance, $\Pi_{(1)}$ and $\Pi_{(2)}$ are functions of q^2 only, then (130) is equivalent to the condition

$$\Pi_{(1)}(q^2) = -\Pi_{(2)}(q^2) \tag{133a}$$

so that the current $j_\mu(q)$ given by Eq. (128) must be of the form

$$j_\mu(q) = \frac{i\alpha}{4\pi^3} (q_\mu q_\nu - g_{\mu\nu} q^2)\, \Pi_{(1)}(q^2)\, a^\nu(q) \tag{133b}$$

or

$$j_\mu(q) = \frac{i\alpha}{4\pi^3} \Pi_{(1)}(q^2)\, J^e{}_\mu(q) \tag{133c}$$

where $J^e{}_\mu(q)$ is the current producing the external field $a_\mu(q)$. In obtaining the form (133c) we have used Maxwell's equation for the external field namely,

$$\partial^\nu F^e{}_{\mu\nu}(x) = J^e{}_\mu(x) \tag{134a}$$

which relates the external field to the current producing it. In momentum space Eq. (134a) reads

$$(q_\mu q_\nu - g_{\mu\nu} q^2)\, a^\nu(q) = J^e{}_\mu(q) \tag{134b}$$

The expression $\Pi_{(1)}$ can be evaluated as follows:

$$\Pi_{\mu\nu}(q) = \int d^4p \frac{\mathrm{Tr}\,\{\gamma_\mu[\gamma \cdot (p+q) + m]\, \gamma_\nu[\gamma \cdot p + m]\}}{[(p+q)^2 - m^2]\,[p^2 - m^2]} \tag{135a}$$

$$= \int_0^1 dz \int d^4p \frac{\mathrm{Tr}\,\{\gamma_\mu(\not p + \not q + m)\, \gamma_\nu(\not p + m)\}}{[(p+qz)^2 + q^2(z - z^2) - m^2]^2} \tag{135b}$$

$$= \int_0^1 dz \int d^4p \frac{\mathrm{Tr}\,\{\gamma_\mu(\not p + \not q(1-z) + m)\, \gamma_\nu(\not p - \not q z + m)\}}{[p^2 + q^2(z - z^2) - m^2]^2} \tag{135c}$$

In obtaining Eq. (135c) from (135b), we have shifted the origin of integration from p to $p - qz$. Since the constant m has a small negative imaginary part, we may again rotate the contour of integration so that the integration path is along $-i\infty$ to $+i\infty$. Upon evaluating the trace term, dropping the terms which are odd in p [Eq. (135a)] and making use of the symmetry relationship of Eq. (135b), we obtain

$$\Pi_{\mu\nu}(q) = 4 \int_0^1 dz \int d^4p \frac{-(2q_\nu q_\mu - g_{\mu\nu} q^2)(z - z^2) - g_{\mu\nu}(\frac{1}{2}p^2 - m^2)}{[p^2 + q^2(z - z^2) - m^2]^2} \tag{136}$$

Thus the expression for $\Pi_{\mu\nu}(q)$ is quadratically divergent because there are six powers of integration variables in the numerator ($p^2\, d^4p$) and only four in the denominator. However, if we impose the requirement of gauge invariance, Eqs. (130) and (131), on the expression (136) we then obtain the result that

$$\int d^4p \frac{-q^2(z - z^2) - \frac{1}{2}p^2 + m^2}{[p^2 + q^2(z - z^2) - m^2]^2} = 0 \tag{137}$$

Although the integral is, strictly speaking, meaningless, since it is divergent, nonetheless we shall use this expression[13] (imposed by the physical requirement of gauge invariance) to rewrite Eq. (136) in the form

$$\Pi_{\mu\nu}(q) = -8(q_\mu q_\nu - g_{\mu\nu}q^2) \int_0^1 dz\,(z - z^2) \cdot \int d^4p\,[p^2 + q^2(z - z^2) - m^2]^{-2} \tag{138a}$$

$$= (q_\mu q_\nu - g_{\mu\nu}q^2)\,\Pi_{(1)}(q^2) \tag{138b}$$

with

$$\Pi_{(1)}(q^2) = \Pi(q^2)$$

$$= -8 \int dz\,(z - z^2) \int \frac{d^4p}{[p^2 + (z - z^2)\,q^2 - m^2]^2} \tag{138c}$$

If we[14] expand $\Pi(q^2)$ in powers of q^2

$$\Pi(q^2) = \Pi(0) + \left.\frac{\partial\Pi(q^2)}{\partial q^2}\right|_{q^2=0} q^2 + \cdots \tag{139}$$

then it is clear that $\Pi(0)$ diverges logarithmically but all other terms are finite. This may be seen as follows: An expansion in q^2 lowers the power of the integration variable p to the same extent as the external variable q is raised, as has been shown at the end of Section *c*. Thus, the term independent of q^2 in Π is multiplied by an integral over p whose integrand is of order zero (i.e., dp/p) giving a logarithmic divergence for $\Pi(0)$. The next term in Π is of order q^2, multiplied by an integral $\int dp/p^3$ which converges for large p; hence Π' and all higher terms in (139) will be convergent integrals over p. We may now use a procedure analogous to Eq. (92), namely, subtract from $\Pi(q^2)$ its value for $q^2 = 0$ to obtain a finite result. We shall show below that the idea of charge renormalization indicates that just this finite remainder, $\Pi(q^2) - \Pi(0)$ represents the observable effect of vacuum polarization.

Note that the expansion (139) is valid only when $q^2 < m^2$. When $q^2 > (2m)^2$, the external field can produce real pairs and $M^{(e)}$ will then have a real as well as an imaginary part. The imaginary part corresponds to a decrease in time for the probability amplitude for the occurrence of a pure scattering process (i.e., without real pair production occurring). On the other hand, the probability amplitude for a transition to a final state which includes some (real) electron-positron pairs will now correspondingly increase in time. The increase and corresponding decrease of these sets of matrix elements are such as to make the total transition probability out of the initial state unity, the S matrix being unitary [see, in the present

[13] A somewhat more satisfactory approach would consist in first "regulating" expression (136) over the electron mass [see Feynman (1949b), Appendix C, and Pauli and Villars (Pauli 1949)].

[14] Note that since $\Pi(q^2)$ is dimensionless, Π' has the dimensions of m^{-2}, Π'' of m^{-4}, etc.

connection, Dalitz (1951b)]. Since the case $q^2 < m^2$ is of greatest interest physically, we limit ourselves to this situation.

Using the methods previously illustrated in the covariant calculation of the electron self-energy, one can evaluate [see Feynman (1949b), Appendix C] the expression for $\Pi_{\mu\nu}$ with the result that

$$\Pi_{\mu\nu}(q) = 4\pi^2 i(q_\mu q_\nu - g_{\mu\nu}q^2)\left[-\frac{2}{3}\lim_{M\to\infty}\log\frac{M}{m} - \frac{4m^2+2q^2}{3q^2}\left(1-\frac{\theta}{\tan\theta}\right)+\frac{1}{9}\right] \tag{140}$$

with $q^2 = 4m^2\sin^2\theta$. The logarithmically divergent term $-(8\pi^2 i/3)\log(M/m)$ corresponds to the term, $\Pi(0)$.

Since we have adopted the Lorentz gauge for the external field

$$q^\mu a_\mu(q) = 0 \tag{141}$$

using Eqs. (138) and (140), we may rewrite the matrix element (74) as follows:

$$\begin{aligned} M^{(e)} &= \frac{ie\alpha}{4\pi^3}\Pi(q^2)\,\gamma^\mu a_\mu(q) \\ &= \frac{ie\alpha}{4\pi^3}[\Pi(0) + q^2\Pi'(0) + q^4\Pi''(0) + \cdots]\,\gamma^\mu a_\mu(q) \end{aligned} \tag{142}$$

Now, we can combine the first term of the right-hand side with the matrix element (67) corresponding to the diagram a in Figure 15.6. Then, although the term $(i\alpha e/4\pi^3)\,\Pi(0)$ is, in fact, divergent, we may view it as merely reducing the strength of the external field by a constant factor. It thus produces an unobservable external field renormalization. From Eq. (133c) we see that the divergent part of the induced charge $j_\mu(q)$ is exactly proportional to the inducing external charge, $J^e{}_\mu(q)$. The renormalization may, therefore, be regarded as one which renormalizes the external charge-current density $J^e{}_\mu(q)$, or through it the charge.

Note the similarity between charge and mass renormalization. In both cases a divergent effect is recognized as producing no observable phenomenon. Thus, in the mass renormalization the divergence δm only changed the mechanical mass m_0 of the electron. Since only $m_0 + \delta m = m$ is observable, the divergent effect disappears when the result is written in terms of the observed mass. The same is true in the case of the charge renormalization. It is experimentally impossible to separate the external charge-current $J^e{}_\mu$ from the induced charge-current proportional to it. For, just as the electron always has its cloud of virtual quanta around it (so that one will always measure m), similarly the external current will always polarize the vacuum, and in all measurements of the external current and charge at large distances therefrom the observed charge-current density is given by

$$J^e{}_{\mu R}(x) = \left[1 + \frac{i\alpha}{4\pi^3}\Pi(0)\right]J^e{}_\mu(x) \tag{143}$$

where $J^e_{\mu R}(x)$ is the "renormalized" charge-current density. Therefore, if we express the matrix element $M^{(a)} + M^{(e)}$ in terms of the renormalized external charge-current density $J^e_{\mu R}(x)$ (instead of J^e_μ), the results will be finite, i.e.,

$$M^{(a)} + M^{(e)} = e\gamma \cdot a_R(q) + \frac{ie\alpha}{4\pi^3} \Pi'(0)\, q^2\gamma \cdot a(q) + \cdots \tag{144}$$

We should also have labeled as renormalized the other terms in the series (144), since to the order we are working they are renormalized. Higher order diagrams will, however, renormalize these terms in a nontrivial fashion.

The above remark should make it sufficiently clear that even in the absence of infinities, we still would have to renormalize the theory. The origin of renormalization is due to the fact that we describe the state of the system in terms of unperturbed bare wave functions, whereas in the actual world we can never switch off the interaction between fields. Therefore, corrections to the bare mass and charge will occur. However, since only the bare mass (charge) plus the corrections to it can ever be observed, we always must express the observables in terms of the renormalized constants. In some sense, therefore, the questions of divergences and renormalization are separate ones. Nonetheless, since all local relativistic field theories with interactions are divergent, we shall use the term "renormalizability" to express the fact that when the observable quantities are re-expressed in terms of the renormalized charge and mass, no divergences appear.

The evaluation of $\Pi'(0)$ is straightforward and one finds

$$\left.\frac{\partial \Pi(q)}{\partial q^2}\right|_{q^2=0} = 16 \int_0^1 dz\, (z - z^2)^2 \int \frac{d^4p}{[p^2 - m^2]^3}$$

$$= \frac{16}{30} \frac{i\pi^2}{2m^2} \tag{145}$$

The effective potential seen by the electron as a result of vacuum polarization effects (to first order in α) is therefore

$$a_{\mu\,\text{eff}}(q) = \left(1 - \frac{\alpha}{15\pi m^2} q^2\right) a_\mu(q) \tag{146}$$

which in configuration space become

$$A_{\mu\,\text{eff}}(x) = \left(1 + \frac{\alpha}{15\pi m^2} \Box\right) A_\mu(x) \tag{147}$$

Thus the polarization of the vacuum spreads the potential out over a region whose radius is of the order of the Compton wavelength of the electron. In the hydrogen atom, in addition to the radiative corrections treated in Sections *c* and *d*, the Coulomb potential of the proton is also altered as a result of the vacuum polarization phenomenon and the potential seen by the electron (including only vacuum polarization effects) is $-Ze^2\left(\frac{1}{r} - \frac{\alpha}{15\pi m^2} 4\pi\delta(\mathbf{r})\right)$.

The fact that the electrostatic potential due to a point charge in the vacuum is no longer given exactly by Coulomb's law, but is modified due to the vacuum polarization is known as the Uehling effect [Uehling (1935)], after Uehling, who first calculated the deviation from the Coulomb law [i.e., the coefficient $\Pi'(0)$, etc.] shortly after the first discussions on the vacuum polarization by Dirac (1934a, b) and Heisenberg (1934a, b). [For a review of these works, see Weisskopf (1936); see also Valatin (1954a, b).] In the nonrelativistic approximation the Uehling effect affects only the hydrogenic S states, and in particular the $2S_{1/2}$ level is lowered by 27 Mc/sec relative to the $2P_{1/2}$ level. Since the agreement between theory and experiment is within $\frac{1}{4}$ Mc/sec, this constitutes direct proof that vacuum polarization effects are real.

As stated above, the net effect of the vacuum polarization is essentially to spread out the effective charge of point charges over distances of the order of $\hbar/mc$. Physically, what happens in the polarization phenomenon may be viewed as follows: A charge Q_0, as a result of its interaction with the electron-positron field, surrounds itself by a cloud of electrons and positrons. Some of these, with net charge δQ of the same sign as Q, escape to infinity [see Schwinger (1949b), Appendix] leaving a net charge, $-\delta Q$, in the part of the cloud which is closely bound to the test body, i.e., within a distance $\hbar/mc$. If one observes the charge of the body from a distance which is large compared with $\hbar/mc$, one sees an effective charge $Q = Q_0 - \delta Q$ (the renormalized charge). However, as one inspects within distances much less than $\hbar/mc$, the charge that will be seen is Q_0 (the bare unrenormalized charge!).

The higher order vacuum polarization effects arising from diagrams such as the one illustrated in Figure 15.11 have been analyzed by Källén (1949,

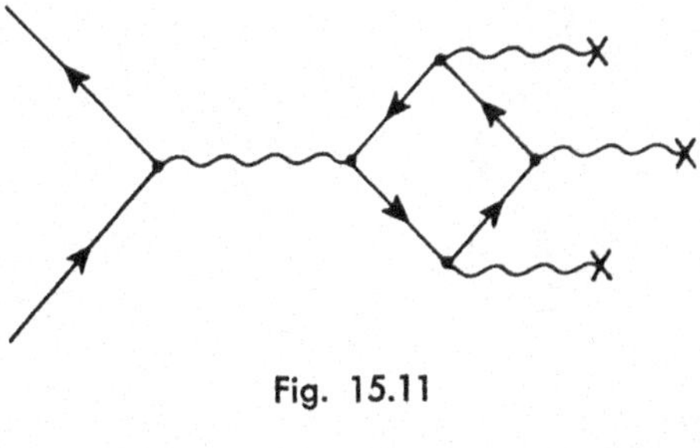

Fig. 15.11

1950), Furry (1951), and Umezawa (1951). In the case of a constant external field, the phenomenon of vacuum polarization has been discussed to all orders in the strength of the inducing field by Weisskopf (1936) and by Schwinger (1951a). For the case of a Coulomb field, the problem has been studied by Wichmann and Kroll [Wichmann (1956)].

The radiative corrections to the "bubble" diagram 15.6e such as the ones indicated in Figure 15.12 have been worked out by Baranger,

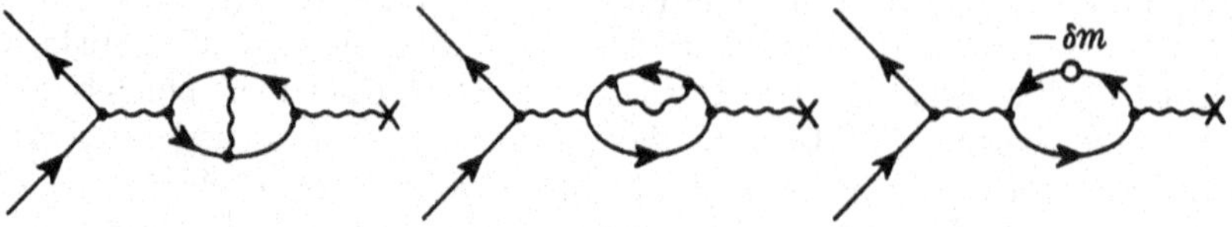

Fig. 15.12

Dyson, and Salpeter [Baranger (1953)] and by Källén and Sabry [Källén (1955)]. They are of order $\alpha^2(\alpha Z)^2$ and contribute -0.24 Mc/sec to the $2^2S_{1/2} - 2^2P_{1/2}$ shift in hydrogen.

A similar situation to the above obtains if, instead of considering the interaction with an external field, we had considered the interaction between two charges. To carry through this analysis, it is, however, necessary to consider the gauge invariance of the theory somewhat more closely. Up to this point we have been using for the photon propagator the expression

$$-\tfrac{1}{2}g_{\mu\nu}D_F(x-y) = (\Phi_0, T(A_\mu(x)\, A_\nu(y))\, \Phi_0) \qquad (148)$$

It will, however, be noted that this expression does not manifestly satisfy the subsidiary condition $\partial_\mu A^{\mu(+)}(x) = 0$. For the Lorentz condition to be manifestly satisfied we should replace the propagator $g_{\mu\nu}k^{-2}$ by $(g_{\mu\nu} - k_\mu k_\nu k^{-2})\, k^{-2}$ which corresponds to the propagator in a particular gauge, sometimes called the Landau gauge [Landau (1954); see also Bogoliubov (1959); Zumino (1960)]. It should be noted that in scattering problems the use of the Landau photon propagator, and in particular, the term $-k_\mu k_\nu(k^{-2})^2$, does not alter any of our previous considerations [Feynman (1949, pp. 780–81)]. Consider, for example, the case of the Møller

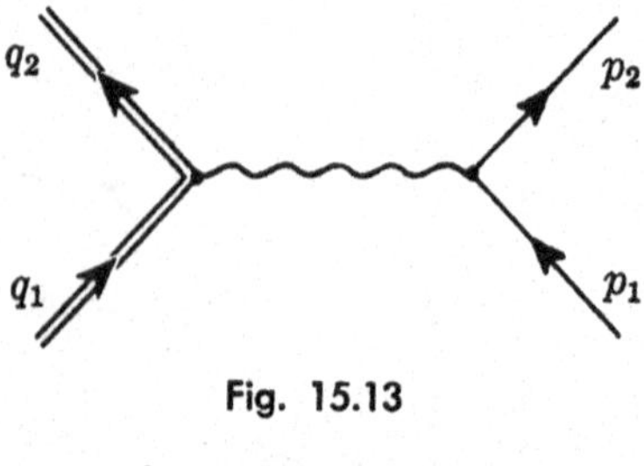

Fig. 15.13

interaction between two charge particles in lowest order (Fig. 15.13). This diagram gives rise to a matrix element proportional to

$$R_M = e^2\, \tilde{u}_M(\mathbf{q}_2)\, \gamma^\mu u_M(\mathbf{q}_2) \left(g_{\mu\nu} - \frac{k_\mu k_\nu}{k^2}\right) \frac{1}{k^2}\, \tilde{u}(\mathbf{p}_2)\, \gamma^\nu u(\mathbf{p}_1) \qquad (149)$$

where $q_1 - q_2 = k = p_2 - p_1$. The contribution to this matrix element from the $k_\mu k_\nu$ term in the photon propagator vanishes since

$$\begin{aligned}\tilde{u}(\mathbf{p}_2)\,\gamma \cdot k u(\mathbf{p}_1) &= \tilde{u}(\mathbf{p}_2)\,\gamma \cdot (p_2 - p_1)\,u(\mathbf{p}_1)\\ &= (m - m)\,\tilde{u}(\mathbf{p}_2)\,u(\mathbf{p}_1)\\ &= 0 \qquad (150)\end{aligned}$$

This will be true to all orders [Feynman (1949)]. The reason the $k_\mu k_\nu$ terms in the photon propagators do not contribute is that all the sources of the electromagnetic potential, namely the currents j_μ of the charged particles, obey the differential continuity equation $\partial^\mu j_\mu(x) = 0$. (This last fact is, of course, intimately connected with the gauge invariance of the theory.)

The lowest order vacuum polarization correction to the Møller interaction diagram 15.13 is illustrated in Figure 15.14. The matrix element cor-

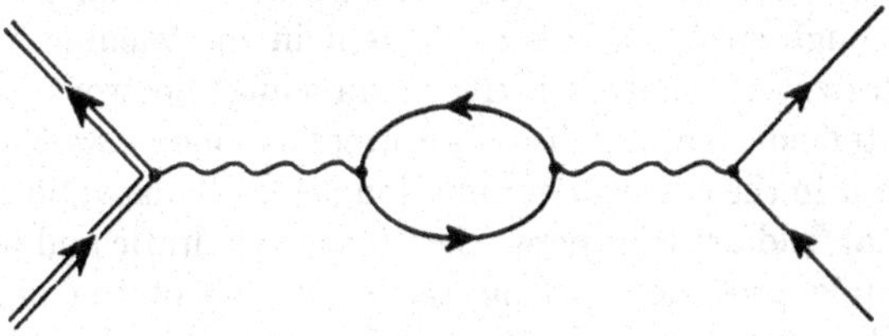

Fig. 15.14

responding to this diagram can be obtained by replacing the factor $(g_{\mu\nu} - k_\mu k_\nu k^{-2})\,k^{-2}$ in the Møller matrix element (149) by the contribution due to the insertion of the "bubble" which (apart from a constant) is

$$e^2(g_{\mu\rho} - k_\mu k_\rho k^{-2})\,k^{-2}\Pi^{\rho\sigma}(k^2)\,(g_{\sigma\nu} - k_\sigma k_\nu k^{-2})\,k^{-2} \qquad (151)$$

where $\Pi_{\rho\sigma}(k^2)$ is the polarization tensor given by Eqs. (75) and (138). We may again expand $\Pi_{\rho\sigma}$ as in (139), and the $\Pi(0)$ term in Eq. (151) then gives rise to a multiple of the Møller matrix element (149). We can therefore combine these two terms in the S matrix and view the factor $\left[1 + \frac{i\alpha}{4\pi^3}\Pi(0)\right]e^2$ which now multiplies the lowest order Møller matrix element as the *renormalized* charge e_R^2. The term in $\Pi'(0)\,k^2$ gives rise to the actual radiative correction of the lowest order matrix elements due to vacuum polarization effects. It corresponds to the modification of the interaction between the two charges due to the vacuum polarization phenomenon.

We also note that the polarization tensor $\Pi_{\mu\nu}(k^2)$ in the case $k^2 = 0$ describes the lowest order self-energy of the photon, the self-energy in this case arising from the virtual creation of an electron-positron pair and its subsequent annihilation, as illustrated in Figure 15.15. This diagram

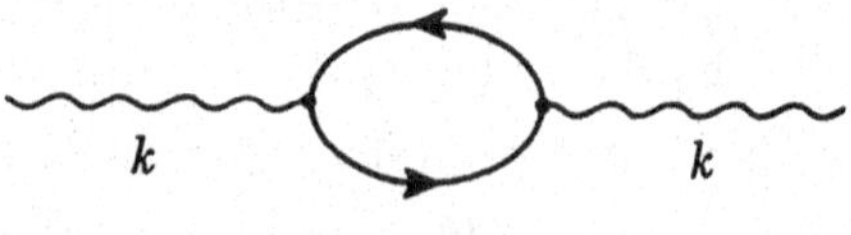

Fig. 15.15

contributes an amplitude proportional to $\epsilon^{(\lambda)}{}_\mu(\mathbf{k})\, \Pi^{\mu\nu}(k^2)\, \epsilon^{(\lambda)}{}_\nu(\mathbf{k})$ to the matrix element of the S matrix between a one-photon state of momentum $\mathbf{k}$ and polarization λ. Using Eq. (138), we obtain

$$\begin{aligned}\epsilon^{(\lambda)}{}_\mu(\mathbf{k})\, \Pi^{\mu\nu}(k^2)\, \epsilon^{(\lambda)}{}_\nu(\mathbf{k}) &= \epsilon^{(\lambda)}{}_\mu(\mathbf{k})\, (k^\mu k^\nu - g^{\mu\nu}k^2)\, \epsilon^{(\lambda)}{}_\nu(\mathbf{k}) \cdot \Pi(k^2) \\ &= -k^2 \epsilon^{(\lambda)}{}_\mu(\mathbf{k})\, \epsilon^{(\lambda)\mu}(\mathbf{k})\, \Pi(k^2) \\ &= 0 \quad \text{for } k^2 = 0 \end{aligned} \tag{152}$$

The vanishing of (152) is a consequence of the gauge invariance of the expression (138) for $\Pi_{\mu\nu}$ and implies that the mass of the photon is strictly zero. The fact that $\Pi^{\mu\nu}(k)\, \epsilon^{(\lambda)}{}_\nu(\mathbf{k})$ vanishes implies that a single photon traveling through space induces no current in the vacuum. There is no vacuum polarization effect for freely traveling electromagnetic waves. This last statement, however, is only correct as long as we consider effects to lowest order in the external or radiation field. Thus, if nonlinear effects in the external field are considered (i.e., terms quadratic and higher in A^e), then observable processes such as the scattering of the photon by the external field are possible. This is illustrated in Figure 15.16. Similarly,

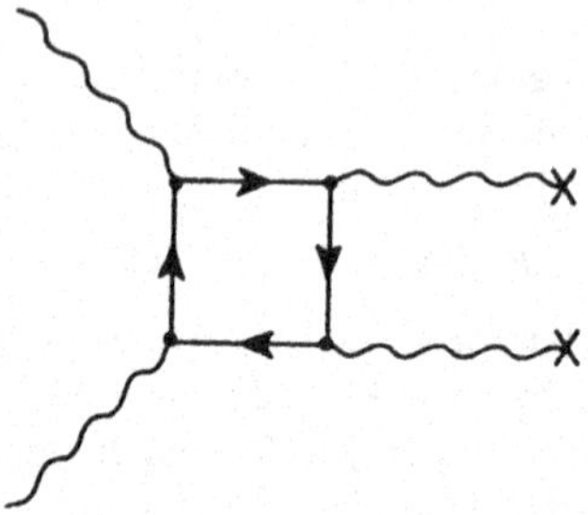

Fig. 15.16

in higher orders, the scattering of two photons by one another is possible as a result of higher vacuum polarization (closed-loop) processes (Fig. 15.17).

As noted above, gauge invariance implies that the self-energy of a photon vanishes identically and that the mass of a photon is identically zero. Gauge invariance thus makes unnecessary a "photon" mass renormalization counterterm.[15] We can state these facts somewhat differently in terms

[15] The ambiguities connected with the fact that one is dealing with divergent integrals should, however, not be overlooked; recall the discussion following Eq. (136).

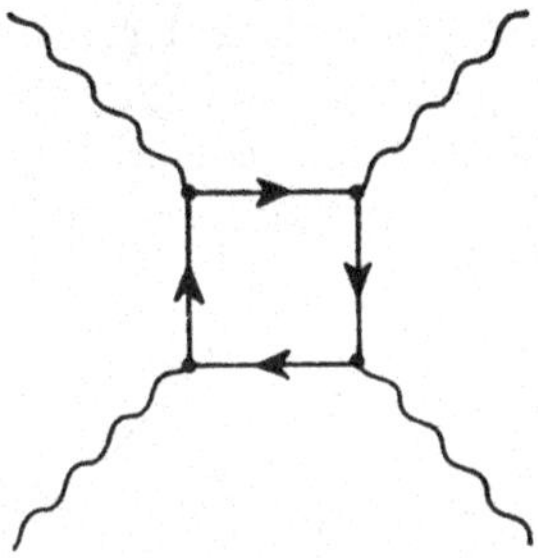

Fig. 15.17

of the photon propagator $D'_F(k^2)$ corresponding to the diagrams illustrated in Figure 15.18, i.e., to the bare propagator together with all the diagrams which give rise to a photon self-energy. The fact that the "real" photon has a vanishing rest mass then corresponds to the statement that the propagator $D'_F(k^2)$ (which includes all radiative corrections) still has its pole at $k^2 = 0$, the mass of the physical photon.

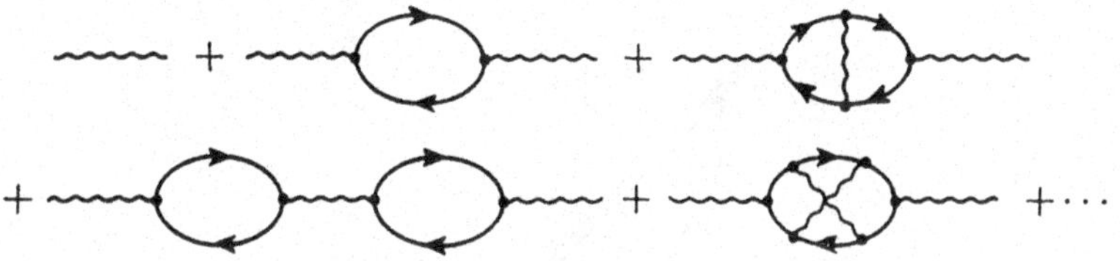

Fig. 15.18

Finally, we note that vacuum polarization effects will arise not only from the virtual creation of electron-positron pairs but also from charged meson pairs, nucleon pairs, etc. In fact every type of charged particle will contribute to the vacuum polarization and the contributions from the various types will simply add [Feldman (1949) and Umezawa (1949b, c, 1950, 1951)]. As a consequence, this type of charge renormalization will apply to the charge of any particle which may interact with the electromagnetic field. In other words, if we assume the *un*renormalized charges of all particles to be the same, then also the renormalized charges will be the same regardless of the particular interactions which contribute to the vacuum polarization.

That every type of charged particle contributes to the vacuum polarization phenomenon has indeed been verified in the level shift of μ-mesic and π-mesic atoms. The dominant contribution to the level shifts due to vacuum polarization effects is, in these cases, actually due to electron-positron pair creation, i.e., to the diagram of Figure 15.19, where the double line represents either a pion or a μ meson and single lines are electron lines. (Recall that the radiative effects due to the phenomena of vacuum polarization, to lowest order in the external field, are inversely proportional to the square of the mass of the particles making up the bubble.) The contributions of the diagram 15.20 to the level shift is therefore a factor $(m_e/m_\pi)^2$ or $(m_e/m_\mu)^2$ smaller than that arising from Figure 15.19.

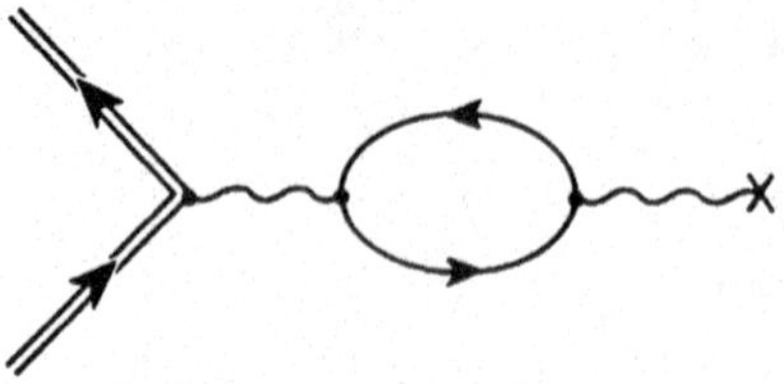

Fig. 15.19

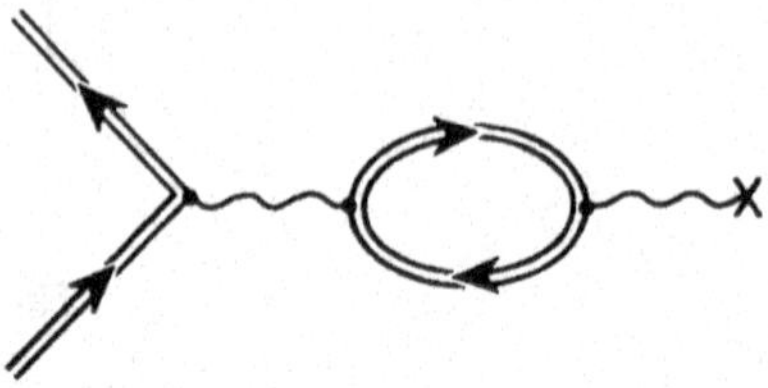

Fig. 15.20

Similarly in the fourth order radiative corrections of the magnetic moment of the μ meson, the contribution of the diagram 15.21 must be included. This contribution has been evaluated by Suura and Wichmann [Suura (1957)] and by Petermann (1957) who find that diagram 15.21

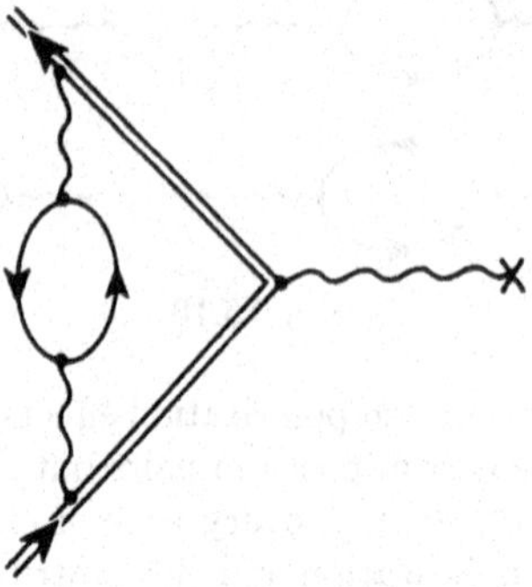

Fig. 15.21

contributes to the anomalous magnetic moment of the muon an amount

$$\begin{aligned}\Delta\mu_{(\mu)} &= \frac{\alpha^2}{\pi^2}\left[\frac{1}{6}\log\frac{m_{(\mu)}{}^2}{4m_e{}^2} + \frac{1}{3}\log 2 - \frac{25}{36} + O\left(\frac{m_e}{m_{(\mu)}}\right)\right]\frac{e\hbar}{2m_{(\mu)}c} \\ &\approx 1.08\,\frac{\alpha^2}{\pi^2}\frac{e\hbar}{2m_{(\mu)}c} \qquad (153)\end{aligned}$$

The calculated anomalous magnetic moment of the muon due to electromagnetic radiative corrections is, therefore, larger than the electronic one and is

$$\Delta\mu_{(\text{muon})} = \left(\frac{\alpha}{2\pi} + 0.75\,\frac{\alpha^2}{\pi^2}\right)\frac{e\hbar}{2m_{(\mu)}c} \qquad (154)$$

as compared to the electronic anomalous moment

$$\Delta\mu_{(\text{electron})} = \left(\frac{\alpha}{2\pi} - 0.328\frac{\alpha^2}{\pi^2}\right)\frac{e\hbar}{2m_e c} \tag{155}$$

No very accurate measurements of the μ-meson g factor have as yet been made. The best value at present is $g_{(\mu)} \simeq 2(1.0015 \pm 0.0006)$ so that (154) has as yet not been verified.

15f. Applications

In the preceding subsections, we have outlined the modern methods of handling the divergences that occur in the higher orders of a power series expansion of the solutions of the equations of motion of quantum electrodynamics. These methods have been applied to compute the lowest order radiative corrections to most of the elementary scattering processes involving electrons and photons.

The first such process (i.e., besides the Lamb shift, which is a bound state problem) for which the radiative corrections were calculated was the process for which we described the renormalization procedure, i.e., the Coulomb scattering of electrons. The lowest order radiative corrections to the Rutherford scattering formula were first derived by Schwinger (1949b). The calculation is a first-order Born approximation with respect to the Coulomb field as well as with respect to the radiation field. Its validity requires that $\alpha \ll 1$ and $Z\alpha \ll 1$, separately (where Z is the nuclear charge). The scattering cross section for the process (including radiative corrections) is obtained in the usual fashion from the square of the transition amplitude

$$R = +2\pi i\tilde{u}(\mathbf{p}_2)\, Mu(\mathbf{p}_1) \tag{156a}$$

where M is given by

$$M = \left\{\gamma_\nu + \Lambda_{C\nu}(p_2, p_1) - \frac{i\alpha}{4\pi^3}\left[\Pi(q^2) - \Pi(0)\right]\gamma_\nu\right\} ea^\nu(q)$$

$$q = p_2 - p_1 \tag{156b}$$

(recall the derivation of the Rutherford formula in Chapter 14). The use of Eq. (156b), however, cannot be entirely correct since the matrix element M of Eq. (156b) contains an infrared divergence in the limit $\lambda_{\min} \to 0$ [see Eq. (89)].

The resolution of this difficulty has been mentioned previously at the end of Section 15*d*. It hinges on the realization that, together with the radiationless scattering process, there exists a nonvanishing (and in fact infinite) probability for the electron to emit a single photon and be scattered. To second order, there are two such diagrams described in Figure 15.10*a* and *b*, one in which the electron is first scattered by the external potential and then emits a photon, and the other describing the reverse situation wherein a photon is first emitted and the electron is subsequently scattered. The matrix elements corresponding to these two situations are easily written down using the rules of Chapter 14. For a transition with the emission of a photon of energy-momentum k and polarization $\epsilon^{(\lambda)}(\mathbf{k})$ $[k \cdot \epsilon^{(\lambda)}(\mathbf{k}) = 0]$, these are

$$R_1 = -2\pi i \tilde{u}(\mathbf{p}_2) \left[\frac{e}{\sqrt{2(2\pi)^3 k_0}} \gamma \cdot \epsilon^{(\lambda)}(\mathbf{k}) \frac{\gamma \cdot (p_2 + k) + m}{(p_2 + k)^2 - m^2} \cdot e\gamma \cdot a(p_2 + k - p_1) \right] u(\mathbf{p}_1) \tag{157a}$$

$$= -2\pi i \tilde{u}(\mathbf{p}_2) M_1 u(\mathbf{p}_1) \tag{157b}$$

and

$$R_2 = -2\pi i \tilde{u}(\mathbf{p}_2) \left[e\gamma \cdot a(p_2 - p_1 + k) \frac{\gamma \cdot (p_1 - k) + m}{(p_1 - k)^2 + m^2} \frac{e}{\sqrt{2(2\pi)^3 k_0}} \gamma \cdot \epsilon^{(\lambda)}(\mathbf{k}) \right] u(\mathbf{p}_1) \tag{158a}$$

$$= -2\pi i \tilde{u}(\mathbf{p}_2) M_2 u(\mathbf{p}_1) \tag{158b}$$

Consider now the case where the emitted photon is very soft, i.e., $k_0 \ll m$ and $|\mathbf{k}| \ll |\mathbf{p}_1|, |\mathbf{p}_2|$. Recalling that

$$\gamma \cdot p \gamma \cdot q = -\gamma \cdot q \gamma \cdot p + 2p \cdot q \tag{159}$$

and the fact that in the problem under consideration

$$k^2 = 0 \tag{160a}$$

$$(\gamma \cdot p_1 - m)\, u(\mathbf{p}_1) = 0 \quad p_1^2 = m^2 \tag{160b}$$

$$\tilde{u}(\mathbf{p}_2)\, (\gamma \cdot p_2 - m) = 0 \quad p_2^2 = m^2 \tag{160c}$$

we obtain

$$R_1 + R_2 \simeq -2\pi i \frac{e}{\sqrt{2(2\pi)^3 k_0}} \tilde{u}(\mathbf{p}_2) \left[\left(\frac{p_2 \cdot \epsilon^{(\lambda)}(\mathbf{k})}{p_2 \cdot k} - \frac{p_1 \cdot \epsilon^{(\lambda)}(\mathbf{k})}{p_1 \cdot k} \right) e\gamma \cdot a(p_2 - p_1) \right] u(\mathbf{p}_1) \tag{161a}$$

$$\approx -\frac{e}{\sqrt{2(2\pi)^3 k_0}} \left[\frac{p_2 \cdot \epsilon^{(\lambda)}(\mathbf{k})}{p_2 \cdot k} - \frac{p_1 \cdot \epsilon^{(\lambda)}(\mathbf{k})}{p_1 \cdot k} \right] R^{(a)}(p_1, p_2) \tag{161b}$$

where $R^{(a)}(p_1, p_2)$ is given by Eq. (67). The probability that the electron be scattered from p_1 to p_2 with the emission of a single photon having energy between 0 and $\Delta E \ll m$, and arbitrary polarization, is therefore given by

$$|R'|^2 = \sum_{\lambda=0}^{3} \int_0^{\Delta E} d^3k\, |R_1 + R_2|^2 = \frac{e^2}{2(2\pi)^3} \sum_{\lambda=0}^{3} \int_0^{\Delta E} \frac{d^3k}{k_0} \cdot \left| \frac{p_2 \cdot \epsilon^{(\lambda)}(\mathbf{k})}{p_2 \cdot k} - \frac{p_1 \cdot \epsilon^{(\lambda)}(\mathbf{k})}{p_1 \cdot k} \right|^2 |R^{(a)}(p_1, p_2)|^2 \tag{162}$$

It is evident from (162) that this probability diverges logarithmically at the lower limit. If the change of momentum of the electron is small, $|\mathbf{q}| = |\mathbf{p}_1 - \mathbf{p}_2| \ll m$, it is possible to simplify this result further by transforming to a co-ordinate system in which the electron is substantially at rest both initially and finally; in this system

$$p_1 \cdot k = p_2 \cdot k \cong mk_0 \tag{163}$$

The factor in between the absolute signs in Eq. (162) then has a numerator $(p_2 - p_1) \cdot \epsilon^{(\lambda)} = q \cdot \epsilon^{(\lambda)}$. We now take into account only the transverse quanta which are the ones which can actually be emitted, and average over all directions of $\mathbf{k}$ to obtain

$$\langle \sum_{\lambda=1}^{2} |q \cdot \epsilon^{(\lambda)}(\mathbf{k})|^2 \rangle = -\tfrac{2}{3} q^2 \tag{164}$$

so that

$$|R'|^2 = -\frac{2\alpha}{3\pi}\frac{q^2}{m^2}|R^{(a)}(p_1, p_2)|^2 \log \frac{\Delta E}{k_{\min}} \tag{165}$$

if the lower limit of the integration over k in (162) is replaced by $k_{\min}$. The quantity ΔE now refers to the Lorentz system in which the electron is essentially at rest. Apart from this quantity, all other factors are relativistically invariant. [For an evaluation for arbitrary initial energy and momentum change of the electron, see Jost (1947), Schwinger (1949a), and also Bloch and Nordsieck (Bloch 1937).]

Now the probability that an electron is scattered between the states $u(\mathbf{p}_1)$ and $u(\mathbf{p}_2)$ without emitting a photon is given by the absolute value squared of Eq. (156), or more explicitly by[16]

$$\begin{aligned} |R|^2 &= |+2\pi i\tilde{u}(\mathbf{p}_2)\,[M^{(a)} + e\Lambda_{\nu C}(p_2, p_1)\,a^\nu(p_2 - p_1)]\,u(\mathbf{p}_1)|^2 \\ &= |R^{(a)}(p_1, p_2)|^2 + 2\pi i\overline{R^{(a)}(p_1, p_2)}\,\tilde{u}(\mathbf{p}_2)\,e\Lambda_{C\nu}(p_2, p_1)\,a^\nu(p_2 - p_1)\,u(\mathbf{p}_1) \\ &\quad - 2\pi i\overline{(\tilde{u}(\mathbf{p}_2)\,e\Lambda_{C\nu}(p_2, p_1)\,a^\nu(p_2 - p_1)\,u(\mathbf{p}_1))}\,R^{(a)}(p_2, p_1) \end{aligned} \tag{166}$$

where we have neglected quantities of the order of e^8. Now it follows from Eq. (88) that the infrared divergent part of $\Lambda_{\nu C}$, call it $\Lambda^i{}_{\nu C}$, contributes to (166) a factor

$$\begin{aligned} &+2\pi i\tilde{u}(\mathbf{p}_2)\,e\Lambda^i{}_{C\nu}(p_2, p_1)\,a^\nu(p_2 - p_1)\,u(\mathbf{p}_1) \\ &\qquad = \frac{\alpha}{3\pi}\frac{q^2}{m^2}\left(\log\frac{m}{\lambda_{\min}}\right)[+2\pi i\tilde{u}(\mathbf{p}_2)\,e\gamma \cdot a(p_2 - p_1)\,u(\mathbf{p}_1)] \\ &\qquad = \frac{\alpha}{3\pi}\frac{q^2}{m^2}\log\frac{m}{\lambda_{\min}}\,R^{(a)}(p_2, p_1) \end{aligned} \tag{167}$$

(provided again the momentum change $|\mathbf{q}|$ is small). Thus we may rewrite

[16] We have neglected the contribution of the vacuum polarization. The inclusion of this term would in no way alter the conclusions arrived at below since $\Pi(q) - \Pi(0)$ is not infrared divergent.

Eq. (166) exhibiting only the infrared divergent parts of the second and third term as follows:

$$|R|^2 = |R^{(a)}(p_1, p_2)|^2 + \frac{2\alpha}{3\pi}\frac{q^2}{m^2}\left(\log\frac{m}{\lambda_{\min}}\right)|R^{(a)}(p_1, p_2)|^2 \tag{168}$$

Comparing (168) and (165), we can see [cf. also Bethe and Oppenheimer (Bethe 1946)] that the sum of $|R|^2$ and $|R'|^2$, i.e., the total cross section for all possible processes to order e^6, is finite and free of divergences in the limit $k_{\min}$ and $\lambda_{\min} \to 0$. It does, however, depend on the energy ΔE which corresponds to the energy below which real photons will not be detected in the scattering experiment. The value of ΔE is determined by the energy resolution of the detector used in the experiment.

Schwinger (1949b) has evaluated the cross section for scattering with energy losses less than ΔE in the extreme relativistic as well as in the intermediate and nonrelativistic limit. The cross section in the nonrelativistic limit (including the contribution from the vacuum polarization due to the external field) is given by

$$\frac{d\sigma}{d\Omega}\Big/\left(\frac{d\sigma}{d\Omega}\right)_M = 1 - \frac{8\alpha}{3\pi}\left(\frac{19}{30} + \log\frac{m}{2\Delta E}\right)v^2 \sin^2 \tfrac{1}{2}\theta \tag{169}$$

where $(d\sigma/d\Omega)_M$ is the elastic Mott cross section given by

$$\left(\frac{d\sigma}{d\Omega}\right)_M = \frac{Z^2e^4}{4p^2v^2 \sin^4 \frac{1}{2}\theta}(1 - v^2 \sin^2 \tfrac{1}{2}\theta) \tag{170}$$

In the high energy limit, the cross section is

$$\frac{d\sigma}{d\Omega}\Big/\left(\frac{d\sigma}{d\Omega}\right)_M = 1 - \frac{4\alpha}{\pi}\log\left(\frac{2E \sin \frac{1}{2}\theta}{m}\right) \cdot \log\left(\frac{E}{\Delta E}\right) \tag{171}$$

Thus, both in the nonrelativistic and in the extreme relativistic limit the cross section can be written in the form

$$\left(\frac{d\sigma}{d\Omega}\right)\Big/\left(\frac{d\sigma}{d\Omega}\right)_M = 1 - \delta(E) \tag{172}$$

That the cross section can be expressed in the form (172) even if higher order contributions of the external potential are included has been shown by Suura (1955). Explicit calculations to second order in the external field, but including only the effects of a single virtual photon, have been performed by Newton (1953, 1955) and by Chrétien (1955), and confirm that the ratio of $d\sigma/d\Omega$ to $(d\sigma/d\Omega)_M$ can be written in the form given by Eq. (172); i.e., that the radiative correction result in fractional correction to the elastic cross section. It was also noted by Schwinger that, if the energy resolution of the detector were improved, i.e., ΔE decreased, δ would become large and (172) would lose its meaning. Under these circumstances, however, higher order radiative corrections would become important. On the basis of arguments patterned after those of Bloch and Nordsieck, Schwinger conjectured that in general

$$\frac{d\sigma}{d\Omega} \Big/ \left(\frac{d\sigma}{d\Omega}\right)_M = e^{-\delta} \tag{173}$$

Yennie and Suura [Yennie (1956); see also Yennie (1961)] have shown that Schwinger's conjecture is asymptotically true for small ΔE ($\Delta E \ll E$). [See also Gupta (1955a, 1955b); Tsai (1960).] Experiments by Tautfest and Panofski [Tautfest (1957)] on the radiative corrections to high energy electron scattering have confirmed the lowest order correction (171).

The fourth-order radiative corrections to electron scattering by an external field (i.e., the contributions to $F(q^2)$ from the diagrams including two virtual photons but only one interaction with the external field, Fig. 15.9) have been calculated by Weneser, Bersohn, and Kroll [Weneser (1953)].

Another problem for which the radiative corrections have been calculated is that of the Compton scattering of a photon by a free spin 0 and spin $\frac{1}{2}$ particle. Corinaldesi and Jost [Corinaldesi (1948)] have calculated the radiative correction for the spin 0 case. Brown and Feynman [L. Brown (1952)] have calculated all the lowest order radiative corrections to the Klein-Nishina formula and have exhibited the cross section to the order of e^6 in both the nonrelativistic and extreme relativistic cases. The reader is referred to their paper for a clear and thorough discussion of both the practical and computational techniques used in calculating radiative corrections in scattering problems. In Chapter 16 we shall show that in the limit of vanishing energy of the photon all the radiative corrections for the Compton scattering vanish. The relation of this important fact to the principle of charge renormalization [Thirring (1950)] will be discussed there.

We have already mentioned the calculation of Karplus and Kroll (1950a) of the fourth-order radiative correction to the magnetic moment of the electron. The reader is referred to their paper for the separation of the divergences in the self-energy and vertex operators in the case when $p^2 \neq m^2$, i.e., when all the lines are internal.

Another process for which the lowest order radiative corrections have been calculated is the internal pair creation [Dalitz (1951b)]. The radiative corrections for the process of bremsstrahlung have been computed by Mitra (1952, 1959).

In general, these electromagnetic radiative corrections for the scattering processes are quite small (1 per cent or less), so that even under the most favorable conditions they are hard to detect. Nonetheless, the success of the theory in predicting the correct level displacement in the Lamb shift leaves little doubt as to the reality of these effects.

15g. The Furry Picture

In Section d we had some difficulty in deriving the Lamb shift for bound electrons from the radiative corrections calculated for free electrons. In point of fact, all the results derived up to this point are, strictly speaking, only applicable to scattering problems wherein all the particles in the initial and final state are free. There exists, however, a large class of interesting scattering experiments in which the electron is bound in an atom (for example, Rayleigh scattering) to which the methods described thus far are not applicable.

Furry (1951) has indicated how the Feynman-Dyson formalism can be extended to those situations in which the electron is bound or where an expansion in $Z\alpha$ is not valid. It is clear that these situations could be encompassed in the Schrödinger picture by choosing for the unperturbed Hamiltonian of the electron-positron field the operator

$$H_0 = \int d^3x : \psi^*(\mathbf{x}) \{-i\boldsymbol{\alpha} \cdot \nabla + m\} \psi(\mathbf{x}) : \\ - e \int d^3x : \psi^*(\mathbf{x}) \gamma_0 \gamma_\mu A^{e\mu}(\mathbf{x}) \psi(\mathbf{x}) : \quad (174)$$

where $A^e{}_\mu(\mathbf{x})$ describes the static external potential in which the electron is bound. The interaction part of the Hamiltonian, describing the interaction between matter and radiation, is then given by

$$H_I = -e \int d^3x : \bar{\psi}(\mathbf{x}) \gamma^\mu \psi(\mathbf{x}) : A_\mu(\mathbf{x}) \quad (175)$$

where $A_\mu(\mathbf{x})$ is the quantized electromagnetic potential in the Schrödinger picture. The eigenfunctions of H_0 are then state vectors describing situations in which a definite number of (noninteracting) electrons and positrons are present in (one-particle) states described by the solution of the Dirac equation in the potential, $A^{e\mu}(x)$. Stated more precisely, one finds that the analysis of Chapter 8 for the quantization of the Dirac field goes through unchanged for the Hamiltonian H_0 of Eq. (174), if one expands the field operators $\psi^*(\mathbf{x})$, $\psi(\mathbf{x})$ occurring therein in terms of the complete set of solutions of the Dirac equation in the external potential $A^{e\mu}(\mathbf{x})$. Precisely such a procedure was used for the first relativistic calculations of the Lamb shift [see, for example, French (1949), Kroll (1949a)]. Such an approach is, however, noncovariant, and the separation of the divergences is far from being unambiguous [see, for example, Kroll (1949a)]. Furry's method accomplishes the above separation, i.e., the inclusion of the effects of the external potential in the field variables, within the framework of the covariant formalism.

Consider again the equation of motion of the state vector in the interaction picture

$$i\hbar c \frac{\partial \mid \Psi(\tau)\rangle}{\partial \tau} = \int_\tau d\sigma(x)\, \mathcal{H}_I(x) \mid \Psi(\tau)\rangle \quad (176)$$

where $\mathcal{H}_I(x)$ is the interaction energy density given by Eq. (66). Here we have written the Tomonaga-Schwinger equation for the case of a plane space-like surface whose equation is

$$n_\mu x^\mu = \tau \tag{177}$$

n_μ being the constant unit vector normal to τ. If $n_\mu = (1, 0, 0, 0)$, then τ is the hyperplane $t =$ constant, and $d\sigma = d^3x$, so that Eq. (176) reduces to the more familiar one

$$i\hbar \frac{\partial \mid \Psi(t)\rangle}{\partial t} = \int d^3x \left\{ -\frac{e}{2} [\bar{\psi}(x)\, \gamma^\mu, \psi(x)]\, (A_\mu(x) + A^e{}_\mu(x)) + \tfrac{1}{2}\delta m[\bar{\psi}(x), \psi(x)] \right\} \mid \Psi(t)\rangle \tag{178a}$$

$$= H_I(t) \mid \Psi(t)\rangle \tag{178b}$$

With Furry, we now make the following unitary transformation on $|\Psi(t)\rangle$:

$$|\Psi_F(t)\rangle = V^{-1} \mid \Psi(t)\rangle \tag{179}$$

The state vector $|\Psi_F(t)\rangle$ then satisfies the equation

$$i\hbar\partial_t \mid \Psi_F(t)\rangle = [V^{-1}(t)\, H_I(t)\, V(t) + i\hbar\partial_t V^{-1}(t) \cdot V(t)] \mid \Psi_F(t)\rangle \tag{180}$$

If we choose the unitary operator V, $V^* = V^{-1}$, so that it satisfies the equation

$$i\hbar\partial_t V^{-1}(t) \cdot V(t) = -i\hbar V^{-1}(t)\, \partial_t V(t)$$
$$= +V^{-1}(t) \int_{x_0=t} d^3x \frac{e}{2} [\bar{\psi}(x)\, \gamma^\mu, \psi(x)]\, A^e{}_\mu(x)\, V(t) \tag{181a}$$

or

$$i\hbar\partial_t V(t) = -\int_{x_0=t} d^3x \frac{e}{2} [\bar{\psi}(x)\, \gamma^\mu, \psi(x)]\, A^e{}_\mu(x)\, V(t) \tag{181b}$$

then $|\Psi_F(t)\rangle$ will obey the equation

$$i\hbar\partial_t \mid \Psi_F(t)\rangle = \int_{x_0=ct} d^3x \left\{ -\frac{e}{2} [\bar{\psi}_F(x)\, \gamma^\mu, \psi_F(x)]\, A_{F\mu}(x) - \tfrac{1}{2}\delta m[\bar{\psi}_F(x), \psi_F(x)] \right\} \mid \Psi_F(t)\rangle \tag{182}$$

where the transformed operators ψ_F and A_F are defined by

$$\psi_F(x) = V^{-1}(t)\, \psi(x)\, V(t) \qquad x_0 = ct \tag{183}$$

$$A_{F\mu}(x) = V^{-1}(t)\, A_\mu(x)\, V(t) \quad x_0 = ct \tag{184}$$

They satisfy the same equal-time commutation rules as ψ and A_μ.

Let us derive the equation of motion that $\psi_F(x)$ satisfies. To this end, we differentiate Eq. (183) with respect to $x_0 = ct$, and using (181), we obtain

$$\partial_0 \psi_F(x) = \partial_0 V^{-1}(t) \cdot \psi(x)\, V(t) + V^{-1}(t)\, \psi(x)\, \partial_t V(t) + V^{-1}(t)\, \partial_0 \psi(x) \cdot V(t)$$

$$= -\frac{i}{\hbar c} V^{-1}(t) \left[-\int_{x'_0 = t} d^3x' \, \frac{e}{2} [\bar{\psi}(x')\, \gamma^\mu, \psi(x')]\, A^e{}_\mu(x'), \psi(x) \right] V(t)$$

$$+ V^{-1}(t)\, \partial_0 \psi(x) \cdot V(t) \quad (185)$$

The evaluation of the commutator in (185) yields

$$\tfrac{1}{2} \int_{x'_0 = ct} d^3x' \, [[\bar{\psi}(x')\, \gamma^\mu, \psi(x')]\, \psi(x)]\, A^e{}_\mu(x')$$

$$= i \int d^3x' \, S(x - x')\, \gamma^\mu A^e{}_\mu(x')\, \psi(x')$$

$$= -\gamma^0 \gamma^\mu A^e{}_\mu(x)\, \psi(x) \quad (186)$$

since $-iS(x - x')$ for $t = t'$ is equal to $\gamma^0 \delta^{(3)}(\mathbf{x} - \mathbf{x}')$. Therefore, if we recall that $\psi(x)$ is an interaction picture operator satisfying the equation

$$\partial_0 \psi(x) = (-\gamma^0 \boldsymbol{\gamma} \cdot \boldsymbol{\partial} - i\gamma^0 m)\, \psi(x) \quad (187)$$

upon collecting terms we find that ψ_F obeys the following equation:

$$(i\gamma^\mu \partial_\mu - m)\, \psi_F(x) = \frac{e}{\hbar c} \gamma^\mu A^e{}_\mu(x)\, \psi_F(x) \quad (188)$$

that is, it satisfies the Dirac equation in the presence of the external electromagnetic field $A^e{}_\mu(x)$. Similarly, since $A_\mu(x)$ and $\psi(x)$ commute, we find that

$$A_{F\mu}(x) = A_\mu(x) \quad (189)$$

so that

$$\Box A_{F\mu}(x) = 0 \quad (190)$$

The picture characterized by Eqs. (182), (188), and (190) has become known as the Furry (bound state interaction) picture because bound state problems can be handled with it.

This can be seen as follows: Let $\{\phi_n(x)\}$ be the solutions of Eq. (188) in the nonoperator case, where n specifies the quantum number of the state (which may be a bound or a scattering state). If we expand the operator $\psi_F(x)$ in terms of the $\phi_n(x)$s, viz.,

$$\psi_F(x) = \sum_{n', E+} b^{(e)}{}_n \phi_n(x) + \sum_{n', E-} d^{(e)}{}_n{}^* \phi_n(x) \quad (191)$$

where the first term runs only over the positive energy ($E+$) solutions and the second over the negative energy solutions, then the operator $b^{(e)}{}_n$ will be a destruction operator for an electron in the state n, $d^{(e)}{}_n{}^{(*)}$ a creation operator for a positron, etc. This is easily established using the properties of the Hamiltonian $H_{0F} = V^{-1} H_{0D} V$, and in particular its commutation rules with $b^{(e)}{}_n{}^*$. Note that the expansion (191) is meaningful only in the case of sufficiently weak fields for which a gap exists between the positive and negative energy eigenvalues. If this is not the case, a stable vacuum, $|\Phi_{F0}\rangle$, will not exist. [See Snyder (1940), Furry (1951), and Salam (1953).]

Notice also that the time dependence of the amplitudes $\phi_n(x)$ is $\exp(-iE_n x^0)$, where E_n is the energy eigenvalue of $\phi_n(x)$.

The solution of Eq. (182) may again be written in the form

$$|\Psi_F(t_2)\rangle = U_F(t_2, t_1) \mid \Psi_F(t_1)\rangle \tag{192}$$

where $U_F(t_2, t_1)$ is given by the Dyson expansion

$$U_F(t_2, t_1) = \sum_{n=0}^{\infty} \left(-\frac{i}{\hbar c}\right)^n \frac{1}{n!} \int_{t_1}^{t_2} d^4x_1 \int_{t_1}^{t_2} d^4x_2 \cdots \int_{t_1}^{t_2} d^4x_n \; T(\mathcal{H}_{FI}(x_1)\, \mathcal{H}_{FI}(x_2) \cdots \mathcal{H}_{FI}(x_n)) \tag{193}$$

with $\mathcal{H}_{FI}$ given by

$$\mathcal{H}_{FI}(x) = -\tfrac{1}{2}e[\bar{\psi}_F(x)\, \gamma^\mu, \psi_F(x)]\, A_{F\mu}(x) - \tfrac{1}{2}\delta m[\bar{\psi}_F(x), \psi_F(x)] \tag{194}$$

Consider now a state vector $|\Phi_{Fa}\rangle = b^{(e)}{}_a{}^* \mid \Phi_{F0}\rangle$ of the uncoupled system which describes the situation in which a single "bare" electron is in some bound state a and no photons are present. The energy of this state is E_a, the energy eigenvalue of the eigenfunction ϕ_a of the Dirac equation in the external field. However, as a result of the interaction, $\mathcal{H}_{FI}$, the energy of this state will change. In fact, in complete analogy with the discussion of Section a, the change in the energy, ΔE_a, of the state $|\Phi_{Fa}\rangle$ will be given by

$$e^{-i\Delta E_a(t_2-t_1)} = \frac{(\Phi_{Fa}, U_F(t_2, t_1)\, \Phi_{Fa})}{(\Phi_{F0}, U_F(t_2, t_1)\, \Phi_{F0})} \tag{195}$$

where the time difference $t_2 - t_1$ is to be taken large, and all oscillating dependence on the limits t_2 and t_1 is to be averaged out. (Recall the discussion of Chapter 11 which is easily extended to include the present discussion for which H_0 has bound states.)

The quantity ΔE_a has a real part as well as an imaginary part. The real part corresponds to the level shift due to the interaction with the radiation field. The imaginary part of ΔE_a is negative and corresponds to the decay of the state $|\Phi_{Fa}\rangle$ arising from a transition of the electron to a lower energy state accompanied by the emission of one or more photons. It gives, therefore, the lifetime (or the line breadth) of the state a. The imaginary part of ΔE_a is zero for the ground state.

The operator $U_F(t_2, t_1)$ can again be analyzed by means of Feynman diagrams. The only difference in the present situation and that analyzed in the free interaction picture is that here the contraction of two fermion factors is given by

$$\begin{aligned} -\tfrac{1}{2}S^e{}_F(x_2, x_1) &= (\Phi_{F0}, T(\psi_F(x_2)\, \bar{\psi}_F(x_1))\, \Phi_{F0}) \\ &= +\sum_{n,\,E+} \phi_n(x_2)\, \bar{\phi}_n(x_1) \quad \text{for } x_{20} > x_{10} \\ &= -\sum_{n,\,E-} \phi_n(x_2)\, \bar{\phi}_n(x_1) \quad \text{for } x_{10} > x_{20} \end{aligned} \tag{196}$$

The function $S^e{}_F$ is no longer a function of the difference of the space-time co-ordinates, but in the case of a static field is a function of $\mathbf{x}_1$ and $\mathbf{x}_2$ separately and of $t_2 - t_1$.

The function $-\frac{1}{2}S^e{}_F(x, y)$ is identical to the Feynman propagator in the external field, $K^A{}_+$, that we discussed in Section 14a. It satisfies the following equations:

$$(i\gamma^\mu\partial_\mu - e\gamma^\mu A^e{}_\mu(x) - m)\, S^e{}_F(x, y) = -2i\delta^{(4)}(x - y) \quad (197a)$$

$$\frac{i\partial S^e{}_F(x, y)}{\partial y^\mu}\gamma^\mu + S^e{}_F(x, y)\,[e\gamma^\mu A^e{}_\mu(y) + m] = -2i\delta^{(4)}(x - y) \quad (197b)$$

by virtue of the equations of motion of the ψ_F, $\tilde{\psi}_F$ operators and the definition (196). The δ function on the right-hand side arises from the time differentiation of Eq. (196). For note that we may write

$$T(\psi_F(x)\,\tilde{\psi}_F(y)) = \tfrac{1}{2}[\psi_F(x), \tilde{\psi}_F(y)] + \tfrac{1}{2}\epsilon(x - y)\,[\psi_F(x), \tilde{\psi}_F(y)]_+ \quad (198)$$

so that $i\gamma^0\partial_0$ operating on this expression will receive a contribution from the $\epsilon(x - y)$ factor of $2i\gamma^0\delta^{(1)}(x_0 - y_0)$. Due to the presence of this δ function in the time, we may then use the equal-time canonical anticommutation rules for $\psi_F(x)$ and $\tilde{\psi}_F(y)$ which yields $\gamma^0\delta^{(3)}(\mathbf{x} - \mathbf{y})$, whence

$$\begin{aligned}(i\gamma^0\partial_0 + i\boldsymbol{\gamma}\cdot\boldsymbol{\partial} - e\gamma^\mu A^e{}_\mu(x) - m)\,&(\Phi_{0F}, T(\psi_F(x)\,\tilde{\psi}_F(y))\,\Phi_{0F})\\ &= -\tfrac{1}{2}(i\gamma^\mu\partial_\mu - e\gamma^\mu A^e{}_\mu(x) - m)\,S^e{}_F(x, y)\\ &= (\Phi_{0F}, T((i\gamma^\mu\partial_\mu - e\gamma^\mu A^e{}_\mu(x) - m)\,\psi_F(x)\,\tilde{\psi}_F(y))\,\Phi_{0F}) + i\delta^{(4)}(x - y)\\ &= i\delta^{(4)}(x - y) \qquad (199)\end{aligned}$$

where in obtaining the last line we have used the equation of motion for the operator $\psi_F(x)$. The derivation of Eq. (197b) proceeds in a similar manner.

With the introduction of the Fourier transform[17]

$$S^e{}_F(x, y) = \int d^4p_1 \int d^4p_2\, e^{-ip_1\cdot x}\, e^{ip_2\cdot y}\, S^e{}_F(p_1, p_2) \quad (200)$$

$$S_F(p) = \frac{-2i}{(2\pi)^4}\frac{1}{\gamma\cdot p - m} \quad (201)$$

Eqs. (197a) and (197b) are easily transcribed to momentum space. The iteration of these equations in combination then yields the following equation, which is still exact:

$$\begin{aligned}S^e{}_F(p_1, p_2) &= S_F(p_1)\,\delta^{(4)}(p_1 - p_2)\\ &\quad + \left(\frac{(2\pi)^4}{-2i}\right) S_F(p_1)\, e\gamma\cdot a(p_1 - p_2)\, S_F(p_2) + \left(\frac{(2\pi)^4}{-2i}\right)^2\\ S_F(p_1)&\left[\int d^4q \int d^4q'\, e\gamma\cdot a(q)\, S^e{}_F(p_1 - q, p_2 + q')\, e\gamma\cdot a(q')\right] S_F(p_2)\end{aligned} \quad (202)$$

[17] If the external field is time-independent, the situation of greatest interest for practical application, then in Eq. (200) $S^e{}_F(p_1, p_2)$ is replaced by

$$S^e{}_F(p_1, p_2) \to \delta^{(1)}(p_{10} - p_{20})\, S^e{}_F(p_1, p_2)$$

In the Furry bound state interaction picture, the vacuum expectation value of the current no longer vanishes, i.e.,

$$-\tfrac{1}{2}e(\Phi_{F0}, [\tilde{\psi}_F(x)\, \gamma_\mu, \psi_F(x)]\, \Phi_{F0}) = j_P(x) \neq 0 \qquad (203)$$

but corresponds to the current induced in the vacuum due to the presence of the external field, and hence contains all the vacuum polarization phenomena. The nonvanishing of the vacuum expectation value of the current operator now implies that in the decomposition of the T products in Eq. (193) into normal products we must include the contractions of the factor $\frac{1}{2}[\tilde{\psi}_F(x)\, \gamma_\mu, \psi_F(x)]$ which occurs in the expression for $\mathcal{H}_{FI}$. The normal constituents can be represented by Feynman diagrams in a manner entirely analogous to the corresponding problem for a free electron. The second-order radiative corrections are represented by the diagrams of Figure 15.22, where double lines are used to denote electron propagation in

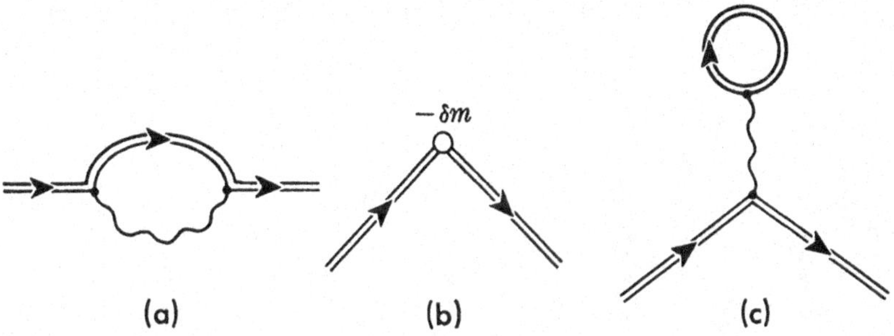

Fig. 15.22

the external field $A^e_\mu(x)$. The (in general complex) level shift in the stationary state ϕ_a is given, to first order, by

$$\begin{aligned}
-\frac{i}{\hbar} T\Delta E_a = & -\alpha\pi \int_{t_0}^{t} \tilde{\phi}_a(x_2)\, \gamma^\mu S^e_F(x_2, x_1)\, \gamma_\mu D_F(x_2 - x_1)\, \phi_a(x_1)\, d^4x_1\, d^4x_2 \\
& - \alpha\pi \int_{t_0}^{t} \tilde{\phi}_a(x_2)\, \gamma^\mu \phi_a(x_2)\, D_F(x_2 - x_1)\, \mathrm{Tr}\, [\gamma_\mu S^e_F(x_1, x_1)]\, d^4x_1\, d^4x_2 \\
& + i\delta m \int_{t_0}^{t} \tilde{\phi}_a(x_1)\, \phi_a(x_1)\, d^4x_1 \qquad (204)
\end{aligned}$$

In view of the fact that for a static external field $S^e_F(x_1, x_2)$ depends upon the time-like components of x_1 and x_2 only through their difference $(x_2 - x_1)_0$, the introduction in the first two terms of new variables of integration $(x_2 - x_1)_0$ and $\frac{1}{2}(x_2 + x_1)_0$ permits the integration over $(x_2 + x_1)_0$ to be performed. The integrand in the δm term in (204) is independent of x_1 so that these time integrations give rise to a factor $t - t_0 = T$, with the result that

$$\Delta E_a = -i\alpha\pi \int \bar{\phi}_a(x_2)\, \gamma^\mu S^e{}_F(x_2, x_1)\, \gamma_\mu \phi_a(x_1)\, D_F(x_2 - x_1)\, d^3x_1\, d^3x_2\, d(x_2 - x_1)_0$$
$$- i\alpha\pi \int \bar{\phi}_a(x_2)\, \gamma^\mu \phi_a(x_2)\, D_F(x_2 - x_1)\, \mathrm{Tr}\,[\gamma_\mu S^e{}_F(x_1, x_1)]\, d^3x_1\, d^3x_2\, d(x_2 - x_1)_0$$
$$- \delta m \int d^3x_1\, \bar{\phi}_a(x_1)\, \phi_a(x_1) \tag{205}$$

The shift in energy of the level a is given by the real part of ΔE_a. The first term of Eq. (205) contains an infinite self-energy term which must be isolated and canceled against the δm terms. Also, the second term contains an infinite term corresponding to a charge renormalization which must be identified and removed.

The isolation of these divergent terms can be carried out through the use of the relation (202), which corresponds to the beginning of the iterative solution of $S^e{}_F$ in powers of the external field. The result of iterating the bound state propagator can be pictorially represented as indicated in Figure 15.23. The matrix element corresponding to diagram a in Figure

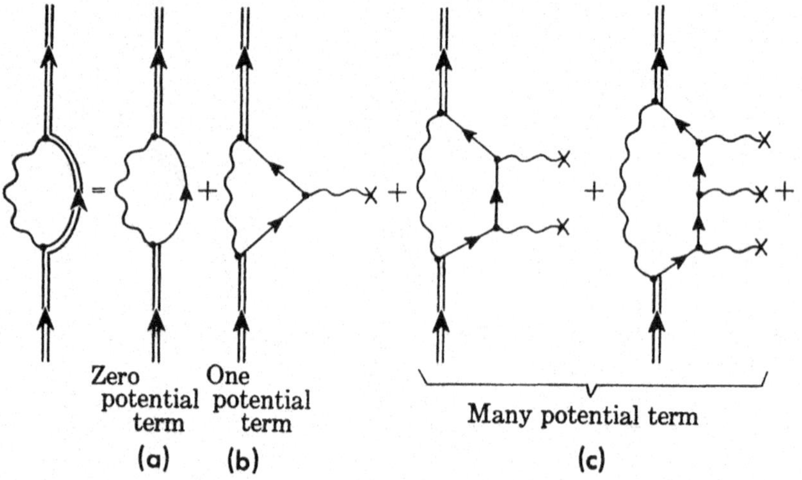

Fig. 15.23

15.23 contains a divergent self-energy term which cancels against the third term of Eq. (205), the δm term. More precisely, in the present situation diagram a corresponds to a term proportional to $\int \bar{\phi}_a(\mathbf{p})\Sigma^{(2)}(p)\, \phi_a(\mathbf{p})$ where $\Sigma^{(2)}(p)$ is given by Eq. (71). It can be expressed in the form

$$\begin{aligned}\Sigma^{(2)}(p) &= \Sigma^{(2)}(\not{p} = m) + A(\not{p} - m) + (\not{p} - m)\, \Sigma_f{}^{(2)}(p) \\ &= 2\pi i \delta m + A(\not{p} - m) + (\not{p} - m)\, \Sigma_f{}^{(2)}(p)\end{aligned} \tag{206}$$

where A is divergent and $\Sigma_f{}^{(2)}(p)$ is finite. Note that since $\phi_a(\mathbf{p})$ is not a free particle Dirac spinor but satisfies the equation

$$(\gamma \cdot p - m)\, \phi_a(\mathbf{p}) = -e \int d^3\mathbf{q}\, \gamma \cdot a(\mathbf{p} - \mathbf{q})\, \phi_a(\mathbf{q}) \tag{207}$$

the second term in (206) will not vanish. The first as mentioned above cancels against the third term in (205). The contribution due to the A term in (206) can be shown to cancel the divergence in the contribution from diagram 15.23*b*. Similarly, on inserting expression (202) for $S^e_F(x_1, x_2)$ into the trace of the second term of Eq. (205), one finds that the contribution from the first term, $S_F(x_1 - x_2)$, vanishes identically while that of the second is identical to the weak field vacuum polarization effect calculated in Section *d*. The divergence occurring therein can therefore be isolated and discarded.

The reader is referred to the very lucid papers of Kroll and Pollock [Kroll (1952)] and of Baranger, Bethe, and Feynman [Baranger (1953)] for the details of these calculations; in particular, for the covariant separation of the divergences (ultraviolet and infrared) in the bound state problem [see also Mills and Kroll (Mills 1955)]. Kroll and Pollock examined the effect of the electrodynamic radiative correction on the hyperfine structure separation. This is of great importance in view of the fact that the most accurate numerical value of the fine structure constant, α, is obtained from a determination of the hyperfine separation in hydrogen.[18] Baranger, Bethe, and Feynman calculated the level shift proper (Lamb shift) and obtained the high momentum corrections to the calculations of Bethe, French, and Weisskopf; Lamb and Kroll; and Feynman, which were mentioned previously.

The radiative corrections to the hyperfine structure and on the energy levels of a bound electron have also been calculated by Karplus, Klein, and Schwinger [Karplus (1952b, c)]. These more formal calculations make use of techniques developed by Schwinger (1951a, b) which, because of limitations of space, cannot be presented in this book. We have therefore referred the reader mainly to those papers which are closest in spirit to the presentation of field theory outlined in this volume.[19] The papers of Kroll and Pollock, and of Baranger *et al.*, using the Feynman-Dyson methods, have the advantage, perhaps, that the physical content of the calculations and approximations is clearer than in the more formal approach of Karplus, Klein, and Schwinger.

The calculations of Baranger *et al.*, and Karplus *et al.* include the radiative effects of a single virtual photon and the $Z\alpha$ correction to the lowest order level shift. When all other known corrections (e.g., recoil and finite size effects, α^2 radiative correction, $\alpha(\alpha Z)^6 \ln^2 (\alpha Z)$ corrections, etc.) are included, the theoretical results for the $2^2S_{1/2} - 2^2P_{1/2}$ level shift in hydrogen are in phenomenal agreement with experiment: 1057.70 ± 0.15 Mc/sec as compared with the experimental value of 1057.77 ± 0.10 Mc/sec.

[18] For a review of the corrections to the Fermi formula for the hyperfine structure due to radiative, recoil and structure effects, see Bethe and Salpeter [Bethe (1958)], also Iddings (1959a, 1959b).

[19] For an excellent introduction to the Schwinger techniques, see Sommerfield (1958).

(There exists at present a discrepancy of 6.1 ± 4.5 Mc/sec in the case of singly ionized helium.) [For a review of the state of knowledge of atomic energy level shifts as of 1959, see Petermann (1958) and Layzer (1960).]

The major difficulties encountered in carrying out the Lamb shift calculations are primarily those concerned with avoiding the infrared divergences. We noted in Section *d* that in the earliest calculations [Feynman (1948), Kroll (1949), French (1949)] the infrared divergence difficulty was avoided by introducing a photon mass and by using the (renormalized) vertex operator calculated for $|\mathbf{p}_1| \ll mc$, $|\mathbf{p}_2| \ll mc$, $|\mathbf{p}_1 - \mathbf{p}_2| \ll mc$ to obtain the level shift arising from the high energy photons. The contribution from the "low energy" photons was then calculated by using the nonrelativistic dipole approximation (Section *b*). When these two contributions were added [using the French connection formula (127)], the lowest order level shift is obtained which is independent of $\lambda_{\min}$, the assumed photon mass. In the calculation of Karplus *et al.*, and Baranger *et al.*, the infrared divergences are shown to cancel out in the final result.

Important progress in the treatment of the Lamb shift has been made by Fried and Yennie [Fried (1958)]. They noticed that it is possible to choose a gauge such that the zeroth order photon propagator $D_{F\mu\nu}(k)$ has the form

$$D_{F\mu\nu}(k) = \left(g_{\mu\nu} + 2\,\frac{k_\mu k_\nu}{k^2}\right)\frac{1}{k^2} \tag{208}$$

and that this gauge is particularly suited to the study of the infrared divergences. It turns out that in this gauge the major contribution to the Lamb shift comes from the first two diagrams *a* and *b* of Figure 15.23, corresponding to the first two terms in the expansion (202) of $S^e{}_F$. After a mass renormalization of the zero potential term, the remaining ultraviolet diagram is canceled by a corresponding divergence arising from the one-potential term. Furthermore, a natural prescription for the isolation of these divergent terms can be given which does not introduce any spurious infrared divergences. The reader is referred to their papers for this analysis, as well as for a simple derivation of the Lamb shift to order $\alpha(Z\alpha)^4\, m$.

The Furry picture has also been used in the calculation of Rayleigh scattering. This is the scattering of a photon by a bound electron. In its full generality the problem has proved very difficult even in second approximation. A calculational has been developed by Brown, Peierls, and Woodward [G. E. Brown (1952b, 1954)] who also give references to the earlier literature. The problem of the natural line shape has been discussed in the Furry picture by Low (1952).

This ends our brief survey of quantum electrodynamics, except for the remarks on renormalization in Chapter 16. The reader interested in a more complete presentation is referred to the treatises of Jauch and Rohrlich [Jauch (1955)], Akhiezer and Berezetski [Akhiezer (1957)], and

Källén (1957); and to Bogoliubov and Shirkov [Bogoliubov (1959)] for the more formal aspects of the theory [see also Thirring (1958)].

15h. Renormalization in Meson Theory

The renormalization techniques we have applied to electrodynamics can readily be carried over to meson theories, and in particular to the case of pseudoscalar mesons interacting with nucleons through a pseudoscalar coupling. We shall, however, see that in meson theory, in contradistinction to quantum electrodynamics, there exist several possible definitions of the renormalized coupling constant, depending on whether, for example, one defines

$$\lim_{p'-p\to 0} \tilde{u}(\mathbf{p}')\,\Lambda_5(p', p)\,u(\mathbf{p}) \to 0 \tag{209a}$$

or

$$\lim_{(p'-p)^2\to \mu^2} \tilde{u}(\mathbf{p}')\,\Lambda_5(p', p)\,u(\mathbf{p}) \to 0 \tag{209b}$$

i.e., whether the radiative corrections vanish in the limit of zero total or zero kinetic energy mesons. In quantum electrodynamics, these two limits are the same since the photon has zero rest mass due to the gauge invariance of the theory. It is then found that either of the following requirements:

(*a*) that Coulomb's law should hold for the interaction of two widely separated charges,

(*b*) that the amplitude for the scattering of a photon by a charged particle in the limit of zero photon energy be given by the Thomson amplitude,

leads to the same value of the renormalized charge.

For the sake of definiteness, we shall illustrate our remarks concerning renormalization in meson theory by considering the problem of the scattering of nucleons by nucleons in *PS-PS* theory, i.e., in pseudoscalar meson theory with pseudoscalar coupling. [Watson and Lepore (Watson 1949)] for which the interaction Hamiltonian is

$$\mathcal{H}_I = G : \tilde{\psi}(x)\,\tau_i\gamma_5\psi(x) : \phi_i(x) - \delta M : \tilde{\psi}(x)\,\psi(x) : - \delta\mu^2 : \phi_i\phi_i : \tag{210}$$

[In Eq. (210) a summation over repeated indices is implied.] The lowest order diagram is given in diagram *a* of Figure 15.24 and corresponds to a matrix element [omitting spinor normalization factors $\sqrt{M/E}$, a factor of $-2\pi i$, and the delta function corresponding to overall energy-momentum conservation $\delta^{(4)}(p_1 + q_1 - p_2 - q_2)$]:

$$T^{(a)} = \frac{G^2}{(2\pi)^3}\,\frac{\tilde{u}(\mathbf{p}_2)\,\tau_i\gamma_5 u(\mathbf{p}_1)\,\delta_{ij}\tilde{u}(\mathbf{q}_2)\,\tau_j\gamma_5 u(\mathbf{q}_1)}{k^2 - \mu^2 + i\epsilon}$$

$$k = q_2 - q_1 = p_1 - p_2 \tag{211}$$

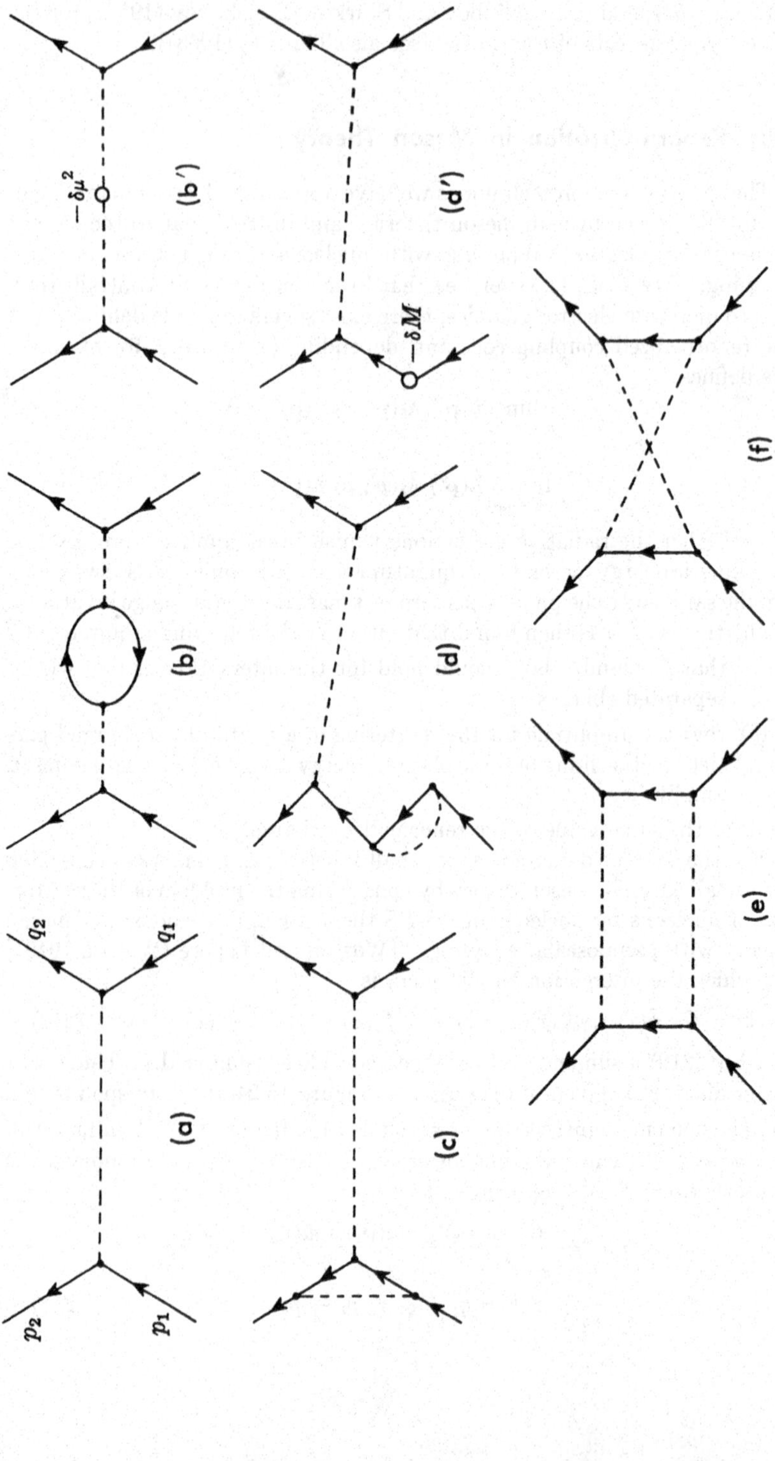

Fig. 15.24

In addition, since the two nucleons are considered identical, diagram a of Figure 15.24 gives rise to an exchange matrix element which can be obtained from $-T^{(a)}$ by interchanging p_2 and q_2. In the following we do not consider the exchange matrix elements which can always be obtained from the matrix element we shall write down by interchanging p_2 and q_2 and affixing a minus sign to the resulting expression. The types of diagrams which give rise to the lowest order radiative corrections to $T^{(a)}$ are illustrated by diagrams b to f of Figure 15.24.

Figure 15.24b is the analogue of the vacuum polarization of quantum electrodynamics. We can view diagram 15.24b as being obtained from diagram 15.24a by replacing in the latter the internal meson line by the diagram of Figure 15.25. Correspondingly, the matrix element $T^{(b)}$

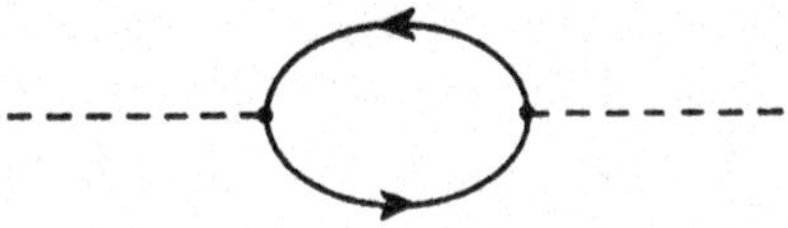

Fig. 15.25

determined by the diagram b of Figure 15.24 can be obtained from $T^{(a)}$ by replacing in the latter the factor $(i/(2\pi)^4)\,\delta_{ij}(k^2-\mu^2)^{-1}$ by $(i/(2\pi)^4)\,(k^2-\mu^2)^{-1}\,K_{ij}(k)\,(i/(2\pi)^4)\cdot(k^2-\mu^2)^{-1}$ where

$$K_{ij}(k) = -G^2\int d^4p\,\mathrm{Sp}\left\{\tau_i\gamma_5\frac{1}{\gamma\cdot(p+k)-M}\tau_j\gamma_5\frac{1}{\gamma\cdot p-M}\right\} \tag{212}$$

In Eq. (212) "Sp" denotes the fact that one is to take the trace of the Dirac matrices as well as of the isotopic spin matrices occurring in the braces. Since $\mathrm{Tr}\,\tau_i\tau_j = \delta_{ij}$, upon using the Feynman formula $(ab)^{-1} = \int_0^1 dz\,[az+b(1-z)]^{-1}$ to combine denominators, and regulating the expression with respect to the nucleon mass, one finds

$$K_{ij}(k) = \delta_{ij}K(k) \tag{213a}$$

$$K(k) = -G^2\int_0^1 dz\int d^4p\sum_i c_i\frac{\mathrm{Tr}\,[\gamma_5(\not p+\not k-M_i)\,\gamma_5(\not p+M_i)]}{(p^2+2zp\cdot k+zk^2-M_i^2+i\epsilon)^2} \tag{213b}$$

A shift of the origin of integration, together with the application of the symmetry relation $\int d^4p\,p_\mu f(p^2) = 0$, makes it evident that the integral is a function of k^2 only and is quadratically divergent:

$$K(k^2) = G^2\int_0^1 dz\int d^4p\sum_i c_i\frac{4[p^2+k^2(z^2-z)-M_i^2]}{[p^2+k^2(z-z^2)-M_i^2+i\epsilon]} \tag{214}$$

If we expand $K(k^2)$ about $k^2=\mu^2$, in a manner similar to the polarization tensor in the electrodynamic case, we obtain

$$K(k^2) = K(0)+(k^2-\mu^2)\,K'(0)+(k^2-\mu^2)^2\,K_R(k^2) \tag{215}$$

The Taylor expansion corresponds to a differentiation of the integrand and thus to an increasing of the power of the exponent of the denominator in the expression (214), so that $K'(0)$ is only logarithmically divergent, and K_R is finite. By a suitable adjustment of $\delta\mu^2$, the first term in the expansion (215) exactly cancels the contribution of diagram b of Figure 15.24. It thus corresponds to the meson self-energy. [For a free meson ($k^2 = \mu^2$) there are according to (215) no other effects but this mass shift.] The contributions of diagrams 15.24b and b' are thus given by

$$T^{(b)} + T^{(b')}$$
$$= \frac{G^2}{8\pi^3} \frac{\tilde{u}(\mathbf{p}_2)\, \tau_i\gamma_5 u(\mathbf{p}_1)\, \delta_{ij}\tilde{u}(\mathbf{q}_2)\, \tau_j\gamma_5 u(\mathbf{q}_1)}{k^2 - \mu^2} \frac{i}{(2\pi)^4} [K'(0) + (k^2 - \mu^2)\, K_R(k^2)] \tag{216}$$

It will be noted that the term involving $K'(0)$ is a constant multiple of the matrix element $T^{(a)}$, Eq. (211). We can therefore combine this term with $T^{(a)}$ and redefine $G^2\left[1 + \frac{i}{(2\pi)^4} K'(0)\right]$ as the renormalized coupling constant. The last term in (216) involving $K_R(k^2)$ is finite and constitutes a radiative correction to the diagram $T^{(a)}$.

Consider next diagram c of Figure 15.24 which corresponds to the insertion of a vertex part into the lowest order diagram 15.24a. An analysis similar to the quantum electrodynamic case indicates that the matrix element $T^{(c)}$ is obtained from $T^{(a)}$ by replacing $\gamma_5\tau_i$ inside the nucleon spinors labeled by p by

$$\tau_i\gamma_5 \to \frac{iG^2}{(2\pi)^4} \int \gamma_5\tau_j \frac{1}{\gamma\cdot(p_2 - l) - M} \gamma_5\tau_i \frac{1}{\gamma\cdot(p_1 - l) - M} \gamma_5\tau_j \frac{d^4l}{l^2 - \mu^2} \tag{217}$$

Since $\tau_j\tau_i\tau_j = -\tau_i$, this is equivalent to the replacement of γ_5 by $\Lambda_5^{(2)}(p_2, p_1)$, where

$$\Lambda_5^{(2)}(p_2, p_1) = \frac{iG^2}{(2\pi)^4} \int d^4l \frac{[\gamma\cdot(-p_2 + l) + M]\, \gamma_5[\gamma\cdot(-p_1 + l) + M]}{[(p_2 - l)^2 - M^2]\, [(p_1 - l)^2 - M^2]\, [l^2 - \mu^2]} \tag{218}$$

The operator $\Lambda_5^{(2)}(p_2, p_1)$ is logarithmically divergent. It can be further simplified by recalling that it is to be evaluated between free-particle spinors $u(\mathbf{p}_2)$ and $u(\mathbf{p}_1)$ so that we can set $\gamma\cdot p_1 = M$, $\gamma\cdot p_2 = M$. The quantity of interest is therefore given by the expression

$$\Lambda_5^{(2)}(p_2, p_1)\Big|_{\substack{\gamma\cdot p_1 = M\\ \gamma\cdot p_2 = M}} = \frac{iG^2}{(2\pi)^4} \gamma_5 \int \frac{l^2\, d^4l}{[(p_2 - l)^2 - M^2]\, [(p_1 - l)^2 - M^2]\, [l^2 - \mu^2]} \tag{219}$$

This quantity is relativistically invariant and hence can only be a function of p_1^2, p_2^2 and $(p_1 - p_2)^2$. Since p_1 and p_2 are the momenta of real particles $p_1^2 = p_2^2 = M^2$, Λ_5 is only a function of $(p_1 - p_2)^2$: $\Lambda_5^{(2)}(p_1, p_2) = \Lambda_5^{(2)}((p_1 - p_2)^2)$. The fact that (219) is logarithmically divergent implies that a single subtraction will render it finite. Therefore if we write

$$\Lambda_5^{(2)}((p_2 - p_1)^2) = \Lambda_5^{(2)}(\mu^2) + \Lambda_{5C}^{(2)}((p_2 - p_1)^2) \tag{220}$$

then $\Lambda_{5C}^{(2)}$ is finite and $\Lambda_5^{(2)}(\mu^2) = C\gamma_5$ where C is a logarithmically divergent constant.[20] The structure of this divergent term is again such that its contribution to diagram 15.24*c* can be combined with the lowest order matrix element, and the divergence absorbed in a redefinition of the coupling constant. Similarly, the contribution of diagram *d* of Figure 15.24 can be obtained from that of diagram 15.24*a* by the replacement of $u(\mathbf{p}_1)$ by the factor $(i/(2\pi)^4)\ (\gamma \cdot p_1 - M)^{-1}\ \Sigma^{(2)}(p_1)\ u(\mathbf{p}_1)$ where

$$\Sigma^{(2)}(p_1) = G^2 \int d^4l\ \tau_i\gamma_5 \frac{1}{\gamma \cdot (p - l) - M}\ \tau_i\gamma_5 \frac{1}{l^2 - \mu^2 + i\epsilon} \tag{221}$$

is a logarithmically divergent integral. It can be expanded about $\gamma \cdot p = M$ as follows:

$$\Sigma^{(2)}(p) = \delta M^{(2)} + B^{(2)}(\gamma \cdot p - M) + \Sigma_C^{(2)}(p) \tag{222}$$

where $\delta M^{(2)}$ and $B^{(2)}$ are divergent constants and $\Sigma_C^{(2)}$ is finite and vanishes for $\gamma \cdot p = M$. The first defines the mass renormalization and cancels the contribution of the δM term, Figure 15.24*d*′. The contribution of the second term (using techniques similar to those outlined in Section 15*c*) can be identified as a wave function renormalization. [In any case it is a constant (divergent) multiple of the matrix element $T^{(a)}$ and can therefore be absorbed into it.] The term $\Sigma_C^{(2)}(p)$ vanishes for $\gamma \cdot p = M$ and hence in the present context does not contribute since $\Sigma_C^{(2)}(p)$ is acting on $u(\mathbf{p})$.

Graphs 15.24*e* and *f* correspond to scattering processes wherein two mesons are exchanged by the nucleons. They do not contain any renormalization effects and are finite.

We leave it as an exercise to the reader to analyze the structure of these same diagrams in the case of *PS-PV* theory, i.e., a pseudoscalar meson theory with pseudovector coupling $\mathfrak{L}_I = \frac{F}{\mu} : \tilde{\psi}\tau_i\gamma_5\gamma_\mu\psi : \partial^\mu\phi$. We here note only that to lowest order the two couplings give identical matrix elements if

$$\frac{G}{2M} = \frac{F}{\mu} \tag{223}$$

Proof: The lowest order diagram 15.24*a* in *PS-PV* theory corresponds to a matrix element

$$T_{PV}^{(a)} = \frac{F^2}{\mu^2(2\pi)^3} \frac{\tilde{u}(\mathbf{p}_2)\ \tau_i\gamma_5\gamma \cdot (p_2 - p_1)\ u(\mathbf{p}_1)\ \tilde{u}(\mathbf{q}_2)\ \tau_i\gamma_5\gamma \cdot (q_2 - q_1)\ u(\mathbf{q}_1)}{(q_2 - q_1)^2 - \mu^2} \tag{224}$$

[20] Note that we could just as well define a finite $\Lambda_{5C}^{(2)}$ by subtracting $\Lambda_5^{(2)}(0)$ from $\Lambda_5^{(2)}((p_2 - p_1)^2)$. This would define a different renormalized coupling constant.

Since

$$\tilde{u}(\mathbf{p}_2)\,\gamma_5\gamma\cdot(p_2 - p_1)\,u(\mathbf{p}_1) = -2M\tilde{u}(\mathbf{p}_2)\,\gamma_5 u(\mathbf{p}_1) \qquad (225)$$

the matrix element $T_{PV}^{(a)}$ agrees with $T_{PS}^{(a)}$ if the coupling constants are related by Eq. (223).

We conclude this section with a brief inquiry into the possibility of describing the interaction between the nucleons in terms of a potential, at least in the low energy nonrelativistic region. The most obvious way to define a potential is to require that when inserted into a Schrödinger equation and the scattering computed from this Schrödinger equation, it shall reproduce the field theoretic S matrix in the nonrelativistic range. If we assume that in the nonrelativistic region the interaction between the particles can be described by a potential $V(|\mathbf{x}|)$ and by a Schrödinger equation

$$\left\{-\frac{\hbar^2}{2M}\nabla_1^2 - \frac{\hbar^2}{2M}\nabla_2^2 + V(|\mathbf{x}_1 - \mathbf{x}_2|)\right\}\psi(\mathbf{x}_1, \mathbf{x}_2; t) = i\hbar\partial_t\psi(\mathbf{x}_1, \mathbf{x}_2; t) \qquad (226)$$

their relative motion is then determined by the equation

$$\left\{-\frac{\hbar^2}{2M^*}\nabla^2 + V(|\mathbf{x}|)\right\}\varphi(\mathbf{x}) = W\varphi(\mathbf{x}) \qquad (227)$$

where the reduced mass M^* is equal to $\frac{1}{2}M$. In the center-of-mass system, $(\mathbf{p}_1 = -\mathbf{q}_1)$, the scattering from a state of relative momentum $\frac{1}{2}(\mathbf{p}_1 - \mathbf{q}_1) = \mathbf{k}_1$ to one with relative momentum $\mathbf{k}_2 = \frac{1}{2}(\mathbf{p}_2 - \mathbf{q}_2)$ is then determined by an S matrix[21]

$$(\mathbf{k}_2 \mid s \mid \mathbf{k}_1) = \delta^{(3)}(\mathbf{k}_2 - \mathbf{k}_1) - 2\pi i\delta(W_2 - W_1)\,(\mathbf{k}_2 \mid t \mid \mathbf{k}_1) \qquad (228)$$

where $W = \mathbf{k}^2/M$, $(|\mathbf{k}_1| = |\mathbf{k}_2| = |\mathbf{k}|)$. This T matrix is related to the potential by Eq. (10.88), namely

$$(\mathbf{k}_2 \mid t \mid \mathbf{k}_1) = (\mathbf{k}_2 \mid V \mid \mathbf{k}_1) - \int d^3k\,(\mathbf{k}_2 \mid V \mid \mathbf{k})\,\frac{(\mathbf{k} \mid t \mid \mathbf{k}_1)}{W - W_1} \qquad (229)$$

We shall only consider the solution of this integral equation in lowest order, in which case

$$(\mathbf{k}_2 \mid V \mid \mathbf{k}_1) \approx (\mathbf{k}_2 \mid T \mid \mathbf{k}_1) \qquad (230)$$

Any ambiguities connected with the fact that from a field theoretic model we know T only for $\mathbf{k}_2{}^2 = \mathbf{k}_1{}^2$ do not arise in this approximation. [Note, however, that in general values of $(\mathbf{k} \mid T \mid \mathbf{k}')$ for $\mathbf{k}_1{}^2 \neq \mathbf{k}_2{}^2$ will affect the definition of V.] Now the field theoretic expression for the S matrix is given by

$$\begin{aligned}(p_2q_2 \mid S \mid p_1q_1) &= \delta^{(3)}(\mathbf{p}_2 - \mathbf{p}_1)\,\delta^{(3)}(\mathbf{q}_1 - \mathbf{q}_2) \\ &\quad - 2\pi i\delta^{(4)}(p_2 + q_2 - p_1 - q_1)\,\frac{M^2}{E(\mathbf{p}_1)\,E(\mathbf{p}_2)\,E(\mathbf{q}_1)\,E(\mathbf{q}_2)}\sum_n T^{(n)} \qquad (231)\end{aligned}$$

[21] We are suppressing the spin indices of the nucleons.

where the $T^{(n)}$ are the contributions of the various diagrams 15.24*a*, 15.24*b*, *b'* ... etc. The reduced S matrix, s, from which the kinematics of the center-of-mass is factored out, is defined as

$$(p_2q_2 \mid S \mid p_1q_1) = \delta^{(3)}(\mathbf{P}_2 - \mathbf{P}_1)\, \delta(P_{20} - P_{10})\, (\mathbf{k}_2 \mid s \mid \mathbf{k}_1) \quad (232)$$

where $P_i = p_i + q_i$ $(i = 1, 2)$ is the total energy-momentum and $k_i = \frac{1}{2}(p_i - q_i)$ is the relative momentum of the particles. The corresponding reduced T matrix is therefore precisely given by Eq. (211) to second order. Comparing expressions (231) and (228) and taking into account Eq. (232), we deduce that in general in the center-of-mass system $(\mathbf{p}_1 + \mathbf{q}_1 = 0;\ q_{10} = q_{20} = p_{10} = p_{20} = E)$

$$(\mathbf{k}_2 \mid t \mid \mathbf{k}_1) = \frac{M^2}{E^2} \sum_n T^{(n)}(\mathbf{k}_1, \mathbf{k}_2) \quad (233)$$

Let us next see what effective nonrelativistic potential corresponds to the second-order matrix element $T^{(a)}$. As indicated above, it is to be expected that the notion of a potential is only valid in the nonrelativistic region, so that we need to specialize the matrix element to that case. We recall that

$$\begin{aligned}\tilde{u}(\mathbf{p}_2)\, \gamma_5 u(\mathbf{p}_1) &= \frac{1}{2M}\, \tilde{u}(\mathbf{p}_2)\, (\gamma \cdot p_2\gamma_5 + \gamma_5\gamma \cdot p_1)\, u(\mathbf{p}_1) \\ &= \frac{1}{2M}\, \tilde{u}(\mathbf{p}_2)\, \gamma^\mu\gamma_5 u(\mathbf{p}_1)\, (p_{2\mu} - p_{1\mu}) \quad (234)\end{aligned}$$

and that in the nonrelativistic limit $|\mathbf{p}_2|$, $|\mathbf{p}_1| \ll M$, this matrix element reduces to

$$\tilde{u}(\mathbf{p}_2)\, \gamma_5 u(\mathbf{p}_1) \rightarrow i\, \frac{\tilde{u}_2\sigma \cdot (\mathbf{p}_2 - \mathbf{p}_1)\, u_1}{2M} \quad (235)$$

where $\tilde{u}_2$ and u_1 are two-component Pauli spinors. In the nonrelativistic limit, $T^{(a)}$ therefore becomes equal to

$$(\mathbf{k}_2 \mid t^{(a)} \mid \mathbf{k}_1) = -\frac{1}{8\pi^3}\left(\frac{G^2}{4M^2}\right) \boldsymbol{\tau}_1 \cdot \boldsymbol{\tau}_2\, \frac{\boldsymbol{\sigma}_1 \cdot (\mathbf{k}_2 - \mathbf{k}_1)\, \boldsymbol{\sigma}_2 \cdot (\mathbf{k}_2 - \mathbf{k}_1)}{(\mathbf{k}_2 - \mathbf{k}_1)^2 + \mu^2} \quad (236)$$

(where we have used the fact that in the center-of-mass frame $q_{10} = q_{20} = p_{20} = p_{20}$). The potential in co-ordinate space is therefore equal to

$$\begin{aligned}V(|\mathbf{x}|) &= -\frac{1}{(2\pi)^3}\left(\frac{G^2}{4M^2}\right) \boldsymbol{\tau}_1 \cdot \boldsymbol{\tau}_2 \int e^{i\mathbf{k}\cdot\mathbf{x}}\, \frac{\boldsymbol{\sigma}_1 \cdot \mathbf{k}\boldsymbol{\sigma}_2 \cdot \mathbf{k}}{\mathbf{k}^2 + \mu^2} \quad (237\text{a}) \\ &= -\frac{1}{4\pi}\left(\frac{G^2}{4M^2}\right) \boldsymbol{\tau}_1 \cdot \boldsymbol{\tau}_2 (\boldsymbol{\sigma}_1 \cdot \boldsymbol{\nabla})\, (\boldsymbol{\sigma}_2 \cdot \boldsymbol{\nabla})\, \frac{e^{-\mu r}}{\mu r} \quad (237\text{b})\end{aligned}$$

It is interesting to note that, apart from a factor due to the cutoff function, (237a) is precisely the potential one would derive in the static Chew model in second order [Gartenhaus (1955)]. The Chew model is easily extended to the case of two nucleons, one of which is fixed at the origin and the other at the point $\mathbf{x}$, respectively. The potential between the nucleons is then the energy of the two-nucleon states (with no physical mesons present) minus their self-energies. The diagrams which contribute to lowest order are indicated in Figure 15.26 and their contribution is given by

$$V(|\mathbf{x}|) = \frac{f}{\mu} \int \frac{d^3k}{(2\pi)^3} \frac{v^2(\mathbf{k}^2)}{2\omega_{\mathbf{k}}} \tau_1 \cdot \tau_2 \frac{\sigma_1 \cdot \mathbf{k}\sigma_2 \cdot \mathbf{k}}{-\omega_{\mathbf{k}}} e^{i\mathbf{k}\cdot\mathbf{x}} \qquad (238)$$

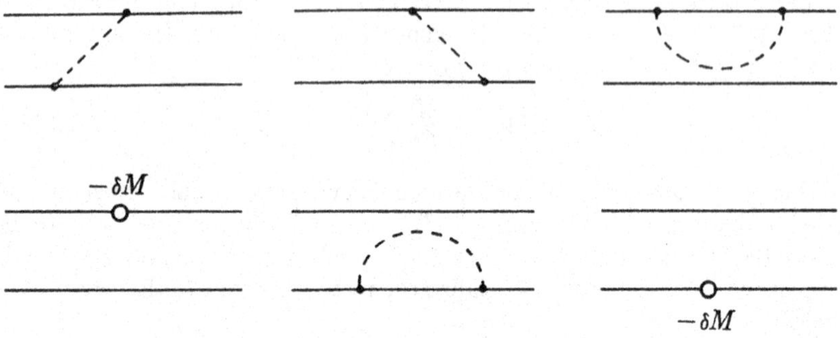

Fig. 15.26

which, apart from the factor $v^2(\mathbf{k}^2)$, is indeed the potential (237a). The cutoff can be viewed as effectively taking into account the (neglected) recoil effect. The potential (237a) should, of course, also have cut-effect arising from the neglected recoil corrections. These recoil corrections become important for momentum transfers of the order of M, the nucleon mass. The potential (237a) therefore becomes invalid for distances smaller than $1/M$, the Compton wavelength of the nucleon. Similarly, due to the presence of the (unknown) cutoff function, the static potential is valid only for $r > 1/M$. It should, however, be emphasized that the two-pion exchange contribution will invalidate the one-pion exchange potential already for distances $(\hbar/2\mu_\pi c) \leqslant r \leqslant (\hbar/2\mu_\pi c)$. It should also be recalled that historically Yukawa's original hypothesis of the existence of pions was put forth to account for the strong short-range forces between nucleons. The range of the force due to the exchange of one pion should be about $\hbar/\mu_\pi c$ where μ_π is the meson mass; that due to the exchange of two pions (as in Fig. 15.24e and f) approximately $\hbar/2\mu_\pi c$; that of n pions $\hbar/n\mu_\pi c$. This means that if one is only interested in the "tail" of the nuclear potential one need only consider the contribution corresponding to the exchange of a single pion at a time between nucleons.

The main feature of the one-pion potential (237) for large r (i.e., $r \gtrsim \hbar/\mu_\pi c$) is that it is attractive for both the $^1S_0$ and $^3S_1$ states, whereas experiments indicate that the $^3S_1$ force is somewhat stronger than the $^3S_0$ force. However, the tensor force can account for this difference; it will be attractive in the triplet state and does not act in the singlet state. Besides accounting for the singlet-triplet difference in S states, the tensor force is also essential in accounting for the deuteron's quadrupole moment. Thus the qualitative features of the one-pion exchange potential are in agreement with experiment. To obtain quantitative predictions it is necessary to know the potential at smaller internucleon distances and there exists at present no really satisfactory way to calculate the two-pion and higher pion number exchange potential.

There does exist, however, an interesting check of the one-pion exchange contribution to the nuclear forces through the analysis of high energy nucleon-nucleon scattering. The high l partial waves will be sensitive only to the tail end of the nuclear potential. (Recall that, classically, a particle of momentum p is scattered by a potential of range a only if its impact parameter, t, is less than a; quantum-mechanically we can translate this statement into the following: only the partial waves l for which $l\hbar \lesssim pa$ will experience scattering.) Thus if, in analyzing the proton-proton scattering data at 310 Mev, one assumes that the one-pion exchange contribution is correct at large distances and that this potential determines the phase shifts for $l \geqslant 5$ and one extracts the $l \leqslant 4$ phase shifts by then analyzing the data, one obtains a very good agreement with the experimental angular distribution. Furthermore, one finds that the coupling constant f when determined in this manner has the value $f \approx 0.06$ in fair agreement with the value obtained from pion-nucleon scattering [MacGregor, Moravscik, and Stapp (MacGregor 1959)]. The encouraging nature of these results is appreciated when one realizes that until 1958 no single quantity had been calculated and measured with sufficient accuracy to constitute a confirmation of the quantitative correctness of pseudoscalar meson theory.

An extensive literature has arisen dealing with the derivation of a potential between two nucleons. The reader is referred to the article of Phillips (1959) for a review of the problem, and to the articles of Gupta (1960) and Charap and Fubini [Charap (1959, 1960)] for more recent developments.

16

Quantitative Renormalization Theory

We have in Chapter 15 analyzed the divergences which occur in quantum electrodynamics and pseudoscalar meson theory when some of the higher order contributions of the S matrix were taken into account. We next investigate whether other types of diagrams besides the ones we have considered thus far are divergent. In Section a we discuss in a more general way which diagrams can give rise to infinities. We then outline Ward's proof for the renormalizability of the S matrix of quantum electrodynamics.

16a. Primitively Divergent Diagrams

When the S matrix is evaluated in the Feynman theory, there are three types of infinities that may arise and which have been classified by Dyson (1949a) as follows:

1. Singularities caused by the coincidence of two or more poles of the integrand;
2. Divergences at small momenta caused by a factor $1/k^2$ in the integrand;
3. Divergences at large momenta due to insufficiently rapid decrease of the whole integrand at infinity.

An example of a divergence of the first type arises when, for special values of the particle momenta, a many-particle scattering process can be divided into independent processes involving separate groups of particles [for a discussion of these divergences, see Eden (1952a)]. An example of a divergence of the second type is the so-called "infrared catastrophe" which was analyzed previously in Sections 15d and f. The third type of singularity arises from the confluence of singularities of $D_F(x)$, $\Delta_F(x)$ and $S_F(x)$ functions for small x. For example, the electron self-energy to lowest order is proportional to $\int d^4x_1 \int d^4x_2\, \bar{\psi}^{(+)}(x_1)\, \gamma^\mu S_F(x_1 - x_2)\, \gamma_\mu \psi^{(+)}(x_2)\, D_F(x_1 - x_2)$ and involves the product: $S_F(x)\, D_F(x)$. For small x^2 the function $\Delta_F(x)$ has the following singular behavior

$$\Delta_F(x) \sim \delta(x^2) - \frac{i}{\pi x^2} - \frac{\mu^2}{4}\theta(x^2) + \text{terms of O } (\log \mu\sqrt{x^2} \text{ and } \sqrt{x^2}\log \mu\sqrt{x^2})$$

so that to define $S_F(x)\, D_F(x)$ it is necessary to give meaning to products like $(\delta(x^2))^2$, $\delta(x^2)\, x^{-2}$, $\delta(x^2)\,\theta(x^2)$, etc. Renormalization theory can be regarded in part as an attempt to define the product of such singular (generalized) functions [see, for example, Bogoliubov (1959a)].

We shall not concern ourselves here with divergences of the first or second type, but only with the really troublesome ones (third type), examples of which have been discussed in previous sections for specific cases.

Consider now a general diagram consisting of external and internal boson and fermion lines. To obtain its contribution to the S matrix, we know that we must integrate over all internal momenta p_j, regardless of whether these are boson or fermion momenta. Thus we shall have to integrate over $d^4p_j = dp_j^0\, dp_j^1\, dp_j^2\, dp_j^3$. Dyson (1949b) has indicated that, for the purposes of examining possible divergences, we may rotate the contours of integration over p_j^0 so that the path of integration over p_j^0 is up along the imaginary axis in the complex p_j^0 plane (recall Section 15*b*). He further demonstrated [Dyson (1949b); see also (1951b)] that this is always possible without introducing divergences of a new type. This transformation has the advantage that all four components can then be considered on the same footing; for if we put $p_j^0 = ip_j^4$ where p_j^4 is *real*, then the usual (rationalized) denominators become

$$\begin{aligned} p \cdot p - m^2 = p^2 - m^2 &= p_0^2 - p_1^2 - p_2^2 - p_3^2 - m^2 \\ &= -(p_1^2 + p_2^2 + p_3^2 + p_4^2 + m^2) \qquad (1) \end{aligned}$$

These are then negative-definite and we can integrate over a Euclidean four-dimensional p space.

After this transformation, it is very easy to consider the integral corresponding to a certain Feynman diagram and to ascertain its degree of divergence. If the number of momentum factors in the numerator is lower than in the denominator, we get a convergent result. If, on the other hand, the number of momenta in the numerator is higher than or equal to that in the denominator, we may (but need not necessarily in all cases) get a divergent result. If we define the "dimension" of the integral as the total number of momenta in the numerator of the integral minus the total number of momenta in the denominator, and denote it by D, then a sufficient condition for convergence is that $D < 0$, and the condition for which divergences may occur is that $D \geqslant 0$.

We shall consider first the interaction of a spinor field with a boson field with a simple nonderivative coupling of the form $\mathfrak{L}_I = G{:}\bar{\psi}(x)\,\Gamma\psi(x){:}\phi(x)$. We shall discuss derivative couplings briefly later on. It follows from our previous discussion on direct interactions that at each "vertex" of any diagram two fermion lines and one boson line meet. Let us adopt the following notation to describe a diagram:

F = number of fermion lines
B = number of boson lines
C = number of vertices
F_e = number of external fermion lines
B_e = number of external boson lines
F_i = number of internal fermion lines
B_i = number of internal boson lines

Specifying the number and kind of external lines serves to describe the process. The specification of F_i and B_i tells us in what approximation we are calculating the process; we are then still free to choose the particular way the internal lines are arranged with respect to each other.

Now in a direct coupling theory, the operator Γ (where $\Gamma = \gamma_\mu, \gamma_5$ or I) does not contain the momentum of any of the particles involved. Also the external lines as such do not influence the divergence possibility, since we do not integrate over them. The factors which do influence the number of momenta occurring in numerator and denominator of the integrand are the following:

1. Each internal fermion or boson line contributes four momenta in the numerator by virtue of the integration d^4p. This is a contribution $4(F_i + B_i)$ to the quantity D.

2. Each integration over all four-space results in each vertex contributing a four-dimensional δ function. This is equivalent to the introduction of four momenta in the denominator. Recall, however, that one of these δ functions merely expresses overall momentum conservation between the initial and final states of the process, and is therefore not available for the reduction of the dimension of the integrand. The vertices therefore give a contribution $-4(C - 1)$ to D.

3. Each internal fermion line contributes one power of p in the denominator, due to its propagation kernel $1/(\gamma \cdot p - m)$. This is a contribution $-F_i$ to D.

4. Each internal boson line contributes two powers of p in the denominator, due to its propagation kernel $1/(p^2 - \mu^2)$. This is a contribution $-2B_i$ to D. Hence, D is given by

$$D = 3F_i + 2B_i - 4(C - 1) \tag{2}$$

We may re-express this in terms of the external lines by noting that the number of vertices is related both to the number of fermion lines and to the number of boson lines. In particular, since there are two fermion lines coming into each vertex, the number of fermion line endings is equal to twice the number of vertices But each internal fermion line has two endings, and each external line one, so that

$$2F_i + F_e = 2C \tag{3}$$

Similarly, there is one boson line ending for each vertex, and thus the number of boson line endings is equal to C, giving

$$2B_i + B_e = C \tag{4}$$

Substituting (3) and (4) into (2) yields

$$D = 4 - \tfrac{3}{2}F_e - B_e \tag{5}$$

An analysis of the aforementioned method of counting powers to find the degree of divergence of an integral reveals, however, that it is in general valid only for the case of matrix elements with one integration. In the case of multiple integrations it is not sufficient to count powers to ascertain whether the matrix element is convergent or divergent. Thus, in the case of multiple internal lines, it is possible that two or more of the δ functions turn out to have the same argument, in which case one cannot conclude that the vertices give a contribution of $-4(C - 1)$ to D. This point is made clear by noting that the diagrams a and b of Figure 16.1 both have

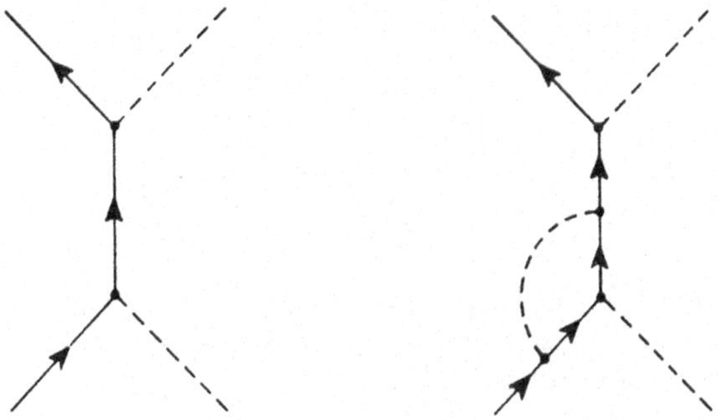

(a) (b)

Fig. 16.1

$D = -1$, since they have the same number of external lines. Diagram b, however, is known to diverge. The statement $D \geqslant 0$ or $D < 0$ for divergence or convergence is therefore immediately applicable only for the case of lowest order diagrams in which one integration occurs. If the diagram contains n internal lines, the condition $D < 0$ and $D \geqslant 0$ for convergence and divergence is valid only in the case that the result of carrying out any arbitrarily chosen $4(n - 1)$ subintegrations is finite; in other words, if the diagram is divergent, the divergence arises only in the last four-momentum integration.

The divergent graphs satisfying these restrictions are called *primitive divergents* [Dyson (1949b)]. They can be characterized by the fact that whenever any one of their internal lines is cut and replaced by two external lines, their matrix element converges. Therefore, since by assumption the integral converges when one of the integration variables is held fixed, the integral will diverge only if the degree of the numerator, $4(F_i + B_i - C + 1)$ is greater than that of the denominator, $2B_i + F_i$, i.e., if $4(F_i + B_i - C + 1) \geqslant 2B_i + F_i$ or, alternatively, if $D \geqslant 0$. Note that since D does not depend on C, the degree of divergence of a primitive divergent does not depend on the number of vertices in the diagram for a field theory with nonderivative coupling. The primitively divergent graphs constitute the basic radiative corrections from which all other divergent radiative corrections can be obtained by appropriate insertions.

We next enumerate these basic divergent diagrams. In this connection, we must remember that a fermion line can never come to an end so that F_e must always be even.

Case $F_e = 0$, $B_e = 0$

This corresponds to the diagram of Figure 16.2 and gives rise to a vacuum self-energy effect. The matrix element for this diagram is quad-

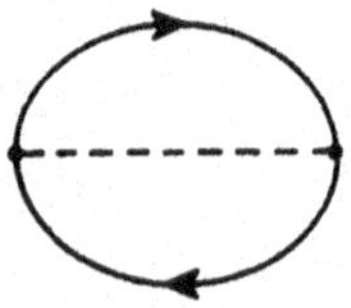

Fig. 16.2

ratically divergent. However, since we have already indicated that such vacuum-vacuum diagrams need never be considered, we shall omit them from further consideration.

Case $F_e = 2$, $B_e = 0$

The only primitively divergent fermion self-energy diagram is shown in Figure 16.3*a*. There are, of course, other (non-primitively divergent) self-

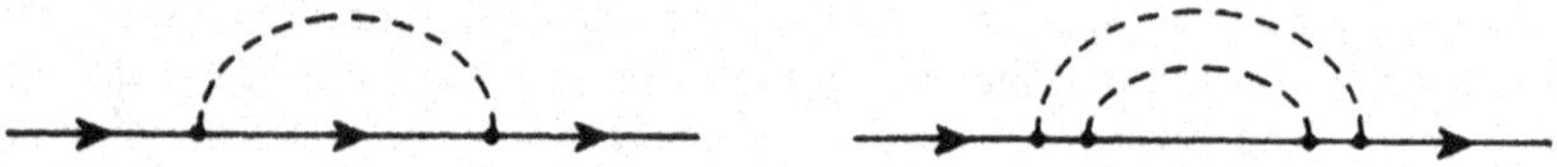

(a) (b)

Fig. 16.3

energy diagrams with more than one virtual boson, such as that shown in Figure 16.3*b*. From (4) we have $D = 1$ so that a linear divergence appears possible. Actually, we have seen in Section 15*a* that the divergence is only logarithmic, corresponding to $D = 0$. It is a general rule that, if D is odd, only the next lower even divergence compared to the maximum one actually occurs.

Case $F_e = 0$, $B_e = 2$

This corresponds to the boson self-energy, Figure 16.4, which we discussed in Sections 15*e* and 15*h*. From (4) we find $D = 2$. Indeed, this

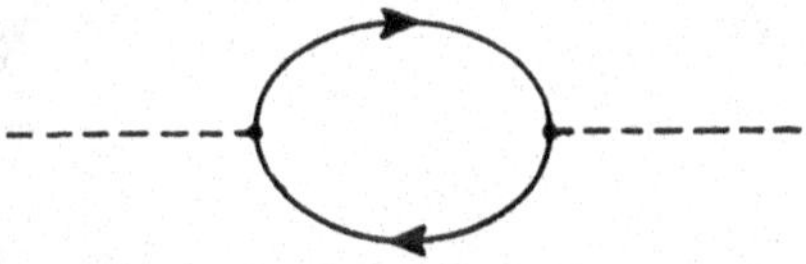

Fig. 16.4

quadratic divergence is present in the case of mesons; in the case of photons we have seen that the divergence is, however, only logarithmic due to gauge invariance.

Case $F_e = 2$, $B_e = 1$

The vertex diagram of Figure 16.5*a* is the simplest example of this kind. For this case $D = 0$, and a logarithmic divergence is, in principle, possible.

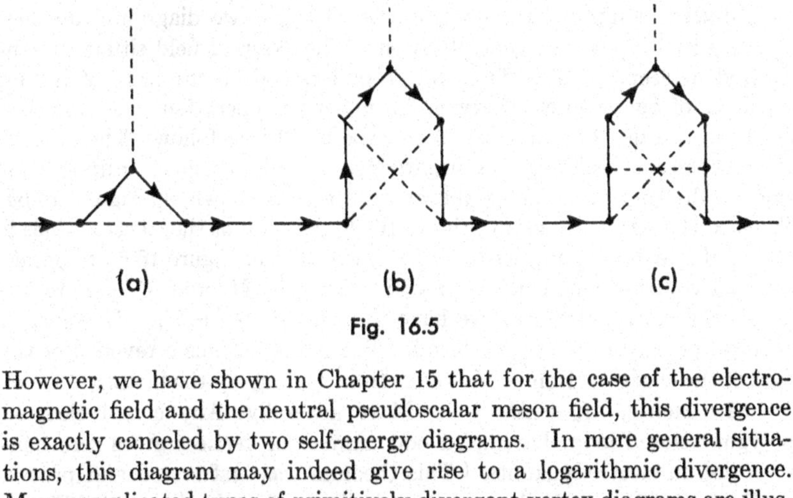

Fig. 16.5

However, we have shown in Chapter 15 that for the case of the electromagnetic field and the neutral pseudoscalar meson field, this divergence is exactly canceled by two self-energy diagrams. In more general situations, this diagram may indeed give rise to a logarithmic divergence. More complicated types of primitively divergent vertex diagrams are illustrated in Figure 16.5*b*, *c*; these will be discussed in Section *b*.

Case $F_e = 0$, $B_e = 3$

For this case $D = 1$, and we may have a linear divergence. A typical diagram is that of Figure 16.6*a*. For obvious reasons, it is often referred to as a triangle diagram. Many variations of this diagram are possible, depending on the particular nature of the fermion triangle and of the boson

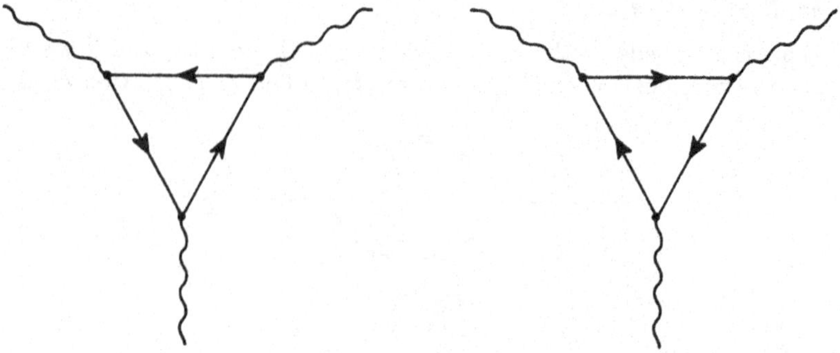

Fig. 16.6

lines joining the vertices. Consider first the case when the fermion field interacts with only one boson field, and, in particular, let us first specialize to the case this field is the electromagnetic field, i.e., to the case of the diagrams of Figures 16.6*a* and *b*. On evaluation, it turns out that instead of a divergent integral the contribution of these two diagrams vanishes. The present case is the generalization of the external field situation considered in Section 14*c*.[1] This can be understood on the basis of the invariance of quantum electrodynamics under the operation of charge conjugation. A qualitative way of seeing the result is as follows: The fermion lines represent positively- or negatively-charged electrons. Suppose for one of the matrix elements the charges are as shown in Figure 16.6*a*. Then in the second diagram, Figure 16.6*b*, the path of the electron lines is reversed. We must add its contribution to that of Figure 16.6*a* to obtain the full contribution to the S matrix. Thus, if M_a and M_b denote the respective matrix elements, we have to evaluate $M_a + M_b$. However, a reversal of path of a charged fermion line merely means a reversal of the sign of the charge. Hence, M_b differs from M_a only by the factor $(-1)^3 = -1$. Hence, $M_a + M_b = 0$. More generally: Any closed electron polygon (closed loop) having an odd number of vertices gives zero contribution in electrodynamics. This theorem was first stated and proved by Furry (1937). A similar argument indicates that the diagram of Figure 16.7 corresponding to the case $F_e = 0$, $B_e = 1$ vanishes. Were it not for Furry's theorem (and also conservation of angular momentum), this diagram would give rise to a cubically divergent integral since $D = 3$.

If we still consider the case of a single boson field interacting with a spinor field, and assume it to be a pseudoscalar meson field, the diagrams

[1] However, the contributions of the individual diagrams do *not*, in the present case, vanish. They vanished in the external field case due to the fact that an integration was carried out over the external fields acting at the three vertices.

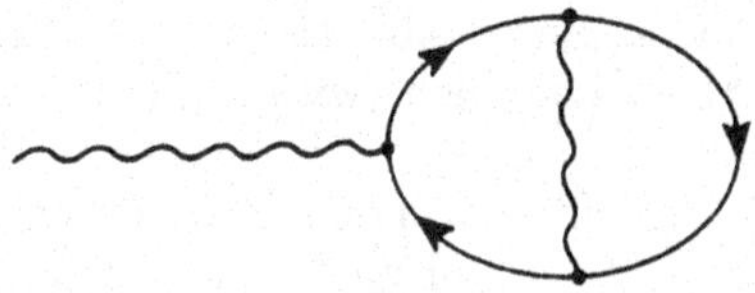

Fig. 16.7

of Figure 16.6 still yield zero because of the nature of the interaction operator γ_5 [Salam (1951a, b)]. This may be seen as follows: Suppose that the three internal fermion momenta are p_1, p_2, and p_3. Then the matrix element is proportional to

$$\gamma_5 \frac{1}{\gamma \cdot p_1 - M} \gamma_5 \frac{1}{\gamma \cdot p_2 - M} \gamma_5 \frac{1}{\gamma \cdot p_3 - M} \tag{6}$$

On rationalizing the denominator and remembering that γ_5 and any γ_μ occurring in $\gamma \cdot p$ anticommute, we find the numerator of (6) to be

$$-\gamma_5(\gamma \cdot p_1 + M)\,(-\gamma \cdot p_2 + M)\,(\gamma \cdot p_3 + M) \tag{7}$$

Now to obtain the S-matrix element, we must take the trace of (6) since the momenta and spins of the fermions are arbitrary. For this purpose, recall that

$$\mathrm{Tr}\,(\gamma_\mu) = \mathrm{Tr}\,\gamma_\mu\gamma_5 = \mathrm{Tr}\,\gamma_5\gamma_\mu\gamma_\nu = \mathrm{Tr}\,\gamma_5 = 0 \tag{8}$$

since $\gamma_5\gamma_\mu\gamma_\nu = -\sum_{\rho\sigma} \epsilon_{\mu\nu\rho\sigma}\gamma^\rho\gamma^\sigma$. Similarly,

$$\mathrm{Tr}\,(\gamma_5\gamma_\mu\gamma_\nu\gamma_\rho) = -\mathrm{Tr}\,(\gamma_\mu\gamma_\nu\gamma_\rho\gamma_5) = -\mathrm{Tr}\,(\gamma_5\gamma_\mu\gamma_\nu\gamma_\rho) = 0 \tag{9}$$

so that, in fact, the trace of (7) is zero and the triangle diagram for the pseudoscalar case does not contribute.[2] An immediate generalization of this is that a heavy pseudoscalar meson cannot disintegrate into two (or, in general, an even number of) lighter pseudoscalar mesons.

Next, consider the boson field to be a scalar meson field. In this case the three vertices each have a factor $\Gamma = 1$, and the equation corresponding to (7) is now

$$(\gamma \cdot p_1 + M)\,(\gamma \cdot p_2 + M)\,(\gamma \cdot p_3 + M) \tag{10}$$

Again, when the trace is taken, those factors of (10) which contain one or three γs vanish, but this time there is also a nonvanishing term with no γs or two γs, so that indeed we get a nonzero element. In actuality, since D is odd, we find the divergence to be logarithmic, i.e., of the order of

[2] This statement can be proved generally for an odd-cornered closed loop for the pseudoscalar case using the invariance of the S matrix under a space inversion. For general selection rules which are consequences of theorems analogous to Furry's for processes involving mesons and nucleons, see Fukuda, Hayakawa, and Miyamoto [Fukuda (1950a,b)]; Pais and Jost [Pais (1952)]; and Wolfenstein and Ravenhall [Wolfenstein (1952)].

$D = 0$ and not the maximum possible one, $D = 1$. The introduction of a counterterm $a\phi^3$ in the Lagrangian (with a infinite) enables one to compensate this divergence.

Let us next admit two different boson fields; for example, let two of the interactions be with photons and one with a meson, as in Figure 16.8.

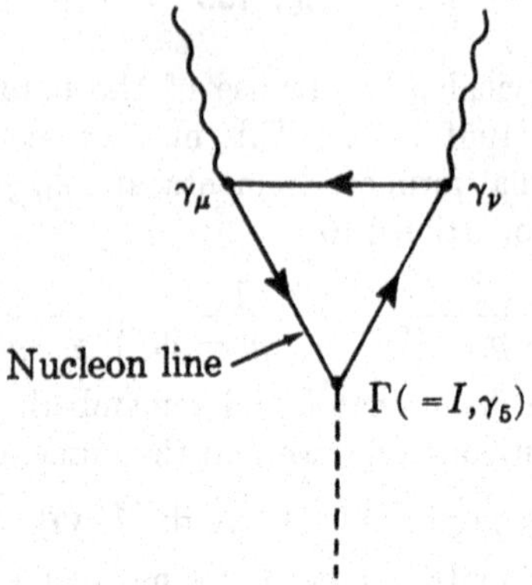

Fig. 16.8

Physically, this diagram describes to lowest order the disintegration of a neutral meson into two photons [Yang (1950), Steinberger (1949)]. This time the coupling constant e appears only twice, i.e., an even number of times, and the charge conjugation theorem of Furry does not force the matrix element to vanish. Furthermore, the argument concerning the vanishing of the trace no longer holds, regardless of whether $\Gamma = \gamma_5$ or $\Gamma = 1$. In particular, the matrix element now is proportional to

$$\Gamma(\gamma \cdot p_1 + M)\, \gamma_\nu(\gamma \cdot p_2 + M)\, \gamma_\mu(\gamma \cdot p_3 + M) \tag{11}$$

When $\Gamma = 1$, we have a nonvanishing contribution from terms that involve an even number of γs. When $\Gamma = \gamma_5$, we have more than four operators involving γ in addition to γ_5, so that another factor, γ_5, can be formed, giving $\gamma_5^2 = -1$. Therefore, the pseudoscalar case also gives a nonvanishing result.

If the incoming meson in Figure 16.8 is a scalar one, then a judicious use of the gauge condition [Fukuda (1949a, b), Schwinger 1951a)], shows that while $D = 1$ only the term in the expansion which has dimensions $D = -1$ actually contributes, and the result therefore converges. For the case of a pseudoscalar meson in Figure 16.8, the result is also gauge invariant and convergent. The disintegration of the neutral pseudoscalar π meson into two γ rays is experimentally observed with a lifetime $<10^{-15}$ second.

Lastly, we must examine the case where the two boson fields are both meson fields. One then finds a nonzero result for a scalar meson disintegrating into two pseudoscalar mesons.

Case $B_e = 4$, $F_e = 0$

For this case from (4) we find $D = 0$. We consider first the scattering of light by light. The Feynman diagram for this process is given in Figure 16.9. The result of the process is that the two incident photons, k_1 and k_2,

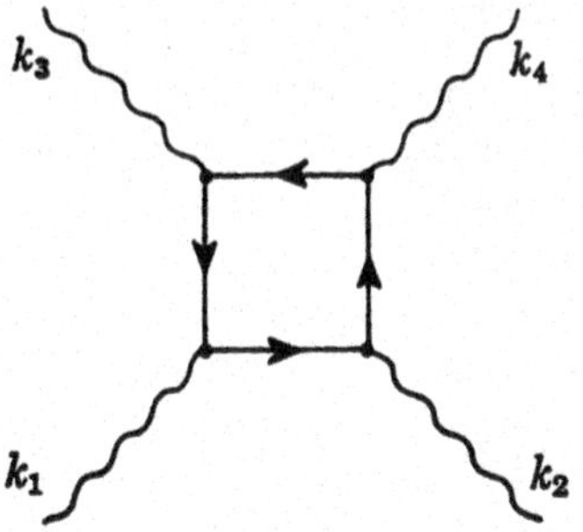

Fig. 16.9

scatter each other and give rise to two outgoing photons, k_3 and k_4. The diagram is actually not logarithmically divergent. Only the term with $D = -4$ contributes, i.e., the diagram is strongly convergent [Feynman (1949), Karplus (1950b); see also Ward (1950a)]. This again is due to gauge invariance and can be understood as follows: the S-matrix element $(k_3, k_4 \mid S \mid k_1, k_2)$ which gives rise to the scattering can be written in the form

$$(k_3, k_4 \mid S \mid k_1, k_2) = (k_3, k_4 \mid -i \int \mathcal{L}_{I\,\text{eff}} \mid k_1, k_2) \tag{12}$$

where $\mathcal{L}_{I\,\text{eff}}$ must be a relativistically and gauge invariant function of a product of four A_μ operators; that is, it must be expressible in terms of $(\frac{1}{2}F_{\mu\nu}F^{\mu\nu})^2$ and $(\frac{1}{4}\epsilon_{\mu\nu\rho\sigma}F^{\mu\nu}F^{\rho\sigma})^2$, where $\frac{1}{2}F_{\mu\nu}F^{\mu\nu} = \mathcal{H}^2 - \mathcal{E}^2$ and $\frac{1}{4}\epsilon_{\mu\nu\rho\sigma}F^{\mu\nu}F^{\rho\sigma} = \mathcal{H} \cdot \mathcal{E}$. Since the electromagnetic field operators $F_{\mu\nu}$ have dimensionality kA, the gauge invariant matrix element corresponding to diagram (16.9) must be proportional to a factor involving the product of k_1, k_2, k_3 and k_4. This requirement reduces the highest power of the electron momentum which occurs inside the integral corresponding to the left-hand side of Eq. (12) by four units, and only a term with $D = -4$ contributes.

To lowest order (α^2), $\mathcal{L}_{I\,\text{eff}}$ for spinor electrodynamics is given by

$$\mathcal{L}_{I\,\text{eff}} = \frac{2\alpha^2}{45m^2}[(\mathcal{H}^2 - \mathcal{E}^2)^2 + 7(\mathcal{H} \cdot \mathcal{E})^2] \tag{13}$$

That virtual pair production in the vacuum gives rise to interactions between photons was recognized as long ago as 1934 [Halpern (1934)]. Euler (1935, 1936) and Heisenberg (1936) considered the problem in 1936, but, due to the difficulties of calculating a fourth-order process in the old theory, they were able to carry through the computation only in the limit of low photon energy. The cross section for the scattering of two photons of equal and very low energy ($\omega \ll m$) was found by Euler (1936) to increase as the sixth power of the energy. However, a subsequent calculation by Akhiezer (1937a) showed that in the extreme relativistic case the cross section goes down as $\frac{\alpha^2}{\pi^2} r_0^2 \left(\frac{m}{\omega}\right)^2 \left(\ln \frac{\omega}{m}\right)^4$. It was not until the advent of modern quantum electrodynamics that the problem of bridging these energy regions was solved by Karplus and Neuman [Karplus (1950b, 1951)]. When Heisenberg discussed this process in 1936, he suggested that perhaps the scattering of light by light might be observable in stars. Actually, the calculations of Karplus and Neuman showed that even at the most favorable photon energy of $2mc^2$, or about 1 Mev, when there is just enough energy to create a real pair, the cross section[3] is only about 3×10^{-30} cm^2. It will be difficult to observe this effect, since even in interstellar space the scattering of radiation from interstellar hydrogen (and dust) is much greater.

Another electromagnetic process involving $B_e = 4$, $F_e = 0$, may, however, be subject to experimental verification. It is the scattering of light by an electrostatic field in second order.[4] Indicating the static interaction by crosses in the Feynman diagram, we represent the process in Figure 16.10. In particular, the process may be observable in strong Coulomb

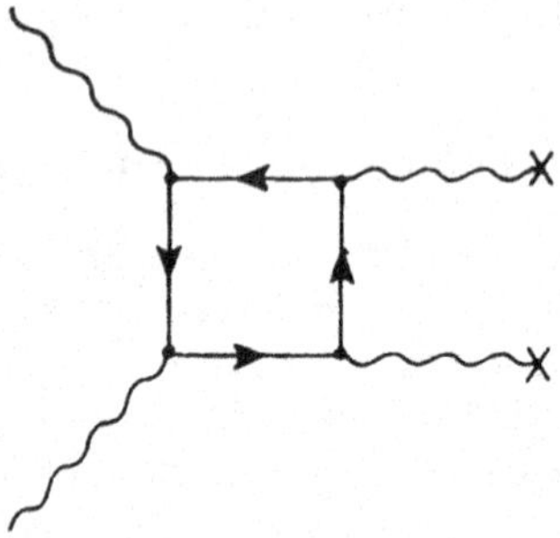

Fig. 16.10

fields of heavy nuclei of charge Z. The interaction indicated by a cross is characterized by Ze^2 instead of e. This gives in the cross section a factor $(Z^2e^2/4\pi\hbar c)^2$ compared with the scattering of light by light. Thus, for

[3] The order of magnitude is evident since there are four interactions e, each contributing $e/\sqrt{4\pi\hbar c}$ to the matrix element. This means the cross section should be of the order of $(e^2/4\pi\hbar c)^4 (\hbar/mc)^2 \sim 4 \times 10^{-30}$ cm^2.

[4] Scattering in first order by the static potential gives a zero contribution because it would be described by a triangle diagram and therefore the Furry rule operates; in fact, Furry's theorem was derived for just this purpose.

Z of the order of 82, the cross section is about $\frac{1}{2} \times 10^4 \times 3 \times 10^{-30} = 1.5 \times 10^{-26}$ cm^2, or 15 millibarns.

This type of scattering was first suggested by Delbrück (1933) and is therefore often referred to as "Delbrück scattering." It was calculated in some limiting cases before the advent of modern quantum electrodynamics by Akhiezer and Pomeranchuk [Akhiezer (1937b)], and by Kemmer and Ludwig [Kemmer (1937a, b)]. Delbrück scattering for the simple case of forward scattering of the photon, where its momentum is practically unchanged, was calculated by Rohrlich and Gluckstern [Rohrlich (1952)] who found that the differential cross section for Delbrück scattering in the forward direction continues to increase for higher and higher energy of the γ ray. It is one of the few processes that show such a continued increase.

A less accurate calculation than that of Rohrlich and Gluckstern was carried out by Bethe and Rohrlich [Bethe (1952)] to obtain the angular distribution of Delbrück scattering for small angles of the order of $\theta \approx mc^2/\hbar\omega$ and less. Unfortunately, no calculations have as yet been made for Delbrück scattering at angles and energies of experimental interest, namely, large angles and energies of a few Mev. [See, however, Claesson (1957) for an effort in this direction.] Mention should also be made of the investigations on Delbrück scattering using dispersion relations and the optical theorem, which relate the Delbrück scattering amplitude to the associated absorptive process, pair production. This method has been used by Rohrlich and Gluckstern, and by Toll (1952). The results of this approach are in agreement with the perturbation theoretic approach. Present experimental indications [Bernstein and Mann (Bernstein 1958) and Moffat and Stringfellow (Moffat 1960)] are that Delbrück scattering is more likely to exist than not. Experimental verification whether such a closed-loop process exists would still be desirable but is now less urgent than previously, since it has been demonstrated that the -27 megacycles vacuum polarization effect in the Lamb shift is real.

Thus far we have discussed only the case of the electromagnetic field as the boson field for the $B_e = 4$, $F_e = 0$ diagram. We can also consider the scattering of mesons by mesons by this type of diagram [Salam (1951a, b)]. In this case, gauge invariance is not available to reduce the order of divergence, and in fact both scalar and pseudoscalar mesons lead to a divergent result. This is the first (and only) divergent process which is not eliminated by a renormalization of mass or coupling constant in pseudoscalar theory. In the spirit of our renormalization program, it is necessary to remove the divergence of meson-meson scattering also by a suitable renormalization. For this purpose it is necessary to introduce a term $\lambda\phi^4$ into the original Lagrangian with a suitably chosen (infinite) coefficient λ. It should be noted that the constant λ is dimensionless. [In the charge-symmetric pseudoscalar theory the counterterm would be of the form $\frac{1}{4}\lambda : (\boldsymbol{\phi} \cdot \boldsymbol{\phi})^2 : = \frac{1}{4}\lambda : \phi_i\phi_i\phi_j\phi_j :$.]

The term with ϕ^4 is not wholly foreign to us. In the interaction of scalar or pseudoscalar mesons with the electromagnetic field, we have already encountered terms of the type $:\phi^*\phi A_\mu A^\mu:$ in the interaction Lagrangian, i.e., of fourth order in boson operators. Of course, the fact that the coefficient λ has to be infinite to compensate for the infinity introduced by the fourth-order meson-meson scattering is something new. It is not known whether it is permissible to give λ a finite as well as an infinite part, i.e., whether the combination of $\lambda\phi^4$ and the diagram in Figure 16.9 may still give a finite effect.

The fact that various models of pion-nucleon scattering (e.g., Chew-Low) agree quite well with experiment without including a π-π coupling has been suggested as evidence that the specific meson-meson coupling $\lambda(\boldsymbol{\phi} \cdot \boldsymbol{\phi})^2$ is weak. However, it is certainly possible that even a strong meson-meson interaction could have but little effect on low energy meson-nucleon scattering. An answer to the problem of whether a finite term of the form $\delta\lambda :\boldsymbol{\phi} \cdot \boldsymbol{\phi}:^2$ should occur in the description of meson-nucleonic phenomena must await further experimental data and better computational methods.

Meson-meson scattering will also occur as a part of more complicated diagrams. That this does not lead to intrinsic difficulties in the theory has been shown by Salam (1951a, b).

We have now completed the descriptions of diagrams that can lead to divergences in the interaction of fermions and bosons when described by nonderivative couplings. Most of these divergences can be removed by a renormalization of mass or coupling constant. More specifically:

For electrodynamics, mass and coupling constant renormalization removes all divergences.

For pseudoscalar mesodynamics, mass and coupling constant renormalization removes all divergences except those due to the square diagram which necessitates a term $\lambda\phi^4$ in the Lagrangian.

For scalar mesodynamics, both the triangle and square diagrams lead to divergences which cannot be removed by mass and coupling constant renormalization. This necessitates the introduction of counterterms $a\phi^3$ and $b\phi^4$ in the Lagrangian. (Note that the constant a will *not* be dimensionless.)

It follows from this summary that from the standpoint of field theory pseudoscalar mesodynamics is slightly simpler than scalar mesodynamics. The two are, however, intrinsically different since for the scalar case a new dimensional constant has to be introduced into the theory to make it meaningful.

We next turn to the discussion of the primitively divergent diagrams for the case of the interaction of two boson fields, and in particular to the case of charged bosons of spin 0 (i.e., charged mesons) interacting with the electromagnetic field (photons), i.e., scalar electrodynamics. Let C denote the number of vertices, P the number of boson particle lines, and Q the number of photon lines, where the subscripts i and e will again distinguish internal and external lines. We obtain the equation for D in a manner analogous to that used above. Before doing so, we shall make one specialization for the present; we shall consider only the interactions due to the $:(\phi^*\partial_\mu\phi - \partial_\mu\phi^* \cdot \phi):A^\mu(x)$ term. This corresponds to permitting only Figure 16.11a and not Figure 16.11b at a vertex. With this restriction, momenta are introduced into the integrand as follows:

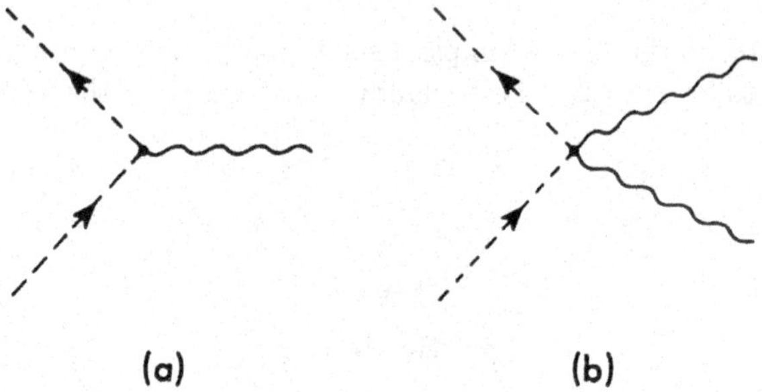

Fig. 16.11

1. Each internal line contributes d^4p in the numerator. This is a contribution $+4(P_i + Q_i)$ to D.

2. Each vertex contributes a $\delta^{(4)}$. This is equivalent to the introduction of four momenta in the denominator. One $\delta^{(4)}$ function expresses overall momentum conservation. The contribution from the vertices is therefore $-4(C - 1)$ to D. In addition, at each vertex an interaction takes place. For the moment since we admit only the interaction $(p + p')^\mu A_\mu$, there is an additional contribution $+C$ to D.

3. Each internal meson line contributes two powers of p in the denominator, due to its propagation kernel $1/(p^2 - \mu^2)$. This is a contribution $-2P_i$ to D.

4. Each internal photon line contributes two powers of p in the denominator, due to its propagation kernel $1/p^2$. This is a contribution $-2Q_i$ to D.

Hence, D is given by

$$D = 2P_i + 2Q_i - 3C + 4 \qquad (14)$$

We may re-express this expression for D in terms of the number of external lines, since

$$2Q_i + Q_e = C \qquad (15a)$$

$$2P_i + P_e = 2C \qquad (15b)$$

Substituting (9) and (10) into (8) then gives

$$D = 4 - P_e - Q_e \qquad (16)$$

As in the case of the interaction between fermions and bosons, the coefficient of the term C turns out to be zero, that is, the degree of divergence does not depend on the number of vertices.

We shall now enumerate all the primitively divergent diagrams for scalar electrodynamics. In this connection, we must remember that a meson line can never come to an end (due to charge conservation), so that P_e must always be even.

Case $P_e = 2, Q_e = 0$

Figure 16.12 shows the simplest such case. The quantity D is found from (11) to be $D = 2$. This quadratic divergence of the self-energy of a

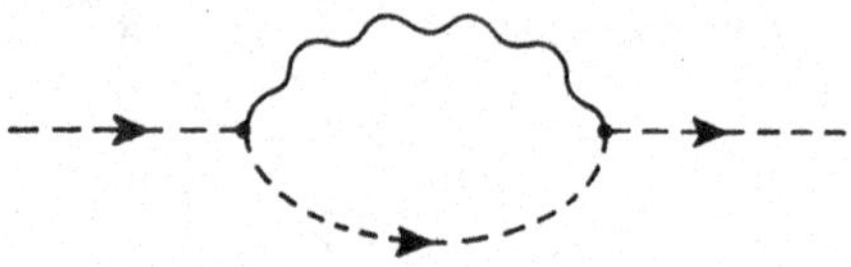

Fig. 16.12

spin 0 boson actually occurs. Recall that for the fermion the self-energy due to interaction with photons was logarithmically divergent, whereas in classical theory the self-energy of the electron was linearly divergent. This is a clear indication that any classical estimate of the self-energy either for a fermion or a boson cannot be trusted.

Case $P_e = 2, Q_e = 1$

The lowest order single photon vertex diagram is illustrated in Figure 16.13*a*. It gives rise to the Lamb shift for a charged meson. As for the

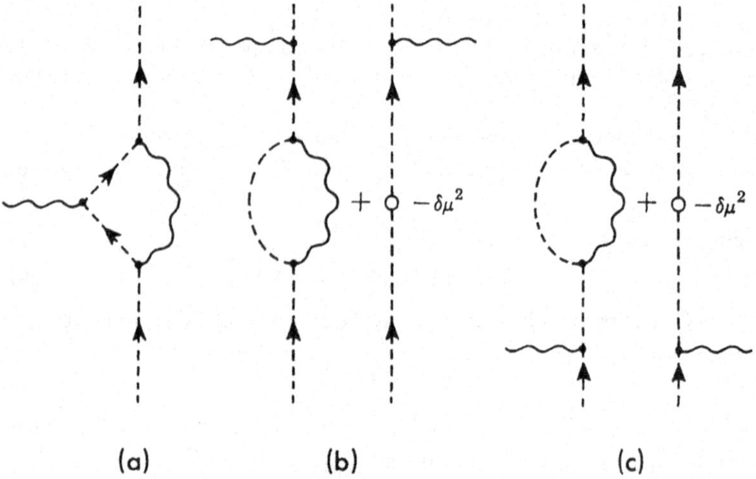

Fig. 16.13

spin $\frac{1}{2}$ case, the three diagrams, Figure 16.13*a*, *b*, and *c*, exist. Each separate diagram diverges, with $D = 1$ giving rise to a linear divergence, but the sum of all three is finite.

Case $P_e = 2$, $Q_e = 2$

As Figure 16.14 shows, this case gives a radiative correction to the Compton effect, i.e., to the scattering of a photon by a spin 0 boson. From

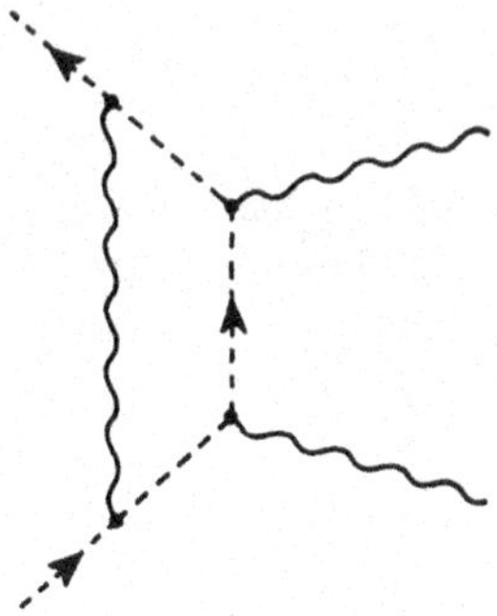

Fig. 16.14

Eq. (11), we have $D = 0$. Actually, however, due to the requirements of gauge invariance, only the next term with $D = -2$ has a nonzero coefficient. Hence, we obtain a convergent result.

Case $P_e = 0$, $Q_e = 2$

This corresponds to the photon self-energy and gives rise to another contribution to the vacuum polarization due to the production of a virtual pair of charged mesons, as shown in Figure 16.15. Here $D = 2$. The

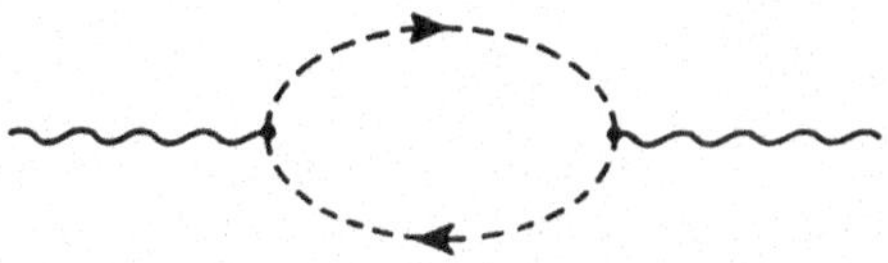

Fig. 16.15

quadratically divergent term actually does not occur due to gauge invariance. This leaves in the expansion of the polarization tensor the next term with $D = 0$, i.e., the logarithmically divergent term and the higher finite terms. These finite terms are of physical significance. They have different numerical factors as compared to those due to a virtual electron pair [see e.g., Schwinger (1951a)], but the contribution to the finite terms is still inversely proportional to the square of the mass of the virtual particle. We have already mentioned, in Section 15e, that the vacuum polarization effects due to virtual electron-positron pairs contribute −27 megacycles to the Lamb shift. The contribution to the Lamb shift from π-mesonic vacuum polarization effects which is smaller by a factor $(275)^2$ is therefore only 0.3 kilocycles and is unobservable. Since theoretical and experimental numbers for the Lamb effect now agree to within .25 megacycle, one may conclude that there do not exist any charged particles which can be produced in pairs and which have a mass less than ten times that of the electron.

Case $P_e = 0, Q_e = 4$

This is the scattering of light by light where, in this square diagram, the internal line is that of a spin 0 boson (Figure 16.16*a*). While $D = 0$ this

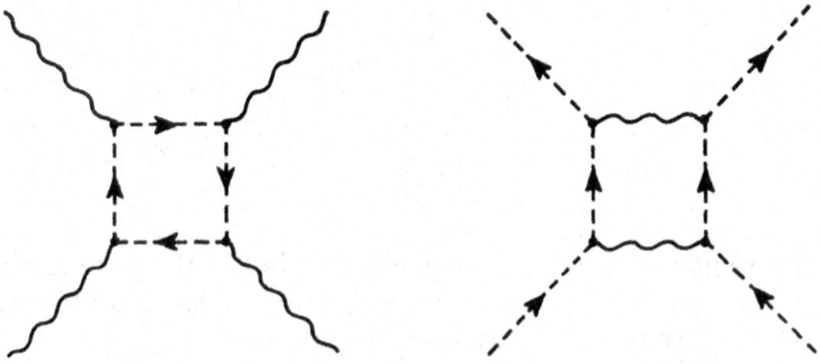

(a) (b)

Fig. 16.16

diagram converges, since gauge invariance assures that only the next term with $D = -4$ and higher convergent terms contribute.

Case $P_e = 4, Q_e = 0$

This case describes the scattering of two Bose particles by each other by virtue of their electromagnetic interaction, as shown in Figure 16.16*b*. Here $D = 0$, and since no gauge condition reduces this divergence, one actually finds a logarithmic divergence. As in the case of meson-meson scattering via virtual nucleon pair production, the present divergence can be compensated by introducing a $\lambda_{em}(\phi^*\phi)^2$ term into the original Lagrangian (λ_{em} is again dimensionless!).

Let us next turn to the analysis of diagrams in which we allow the $:\phi^*\phi A_\mu A^\mu:$ term to operate. We now differentiate between vertices at which *one* and those at which *two* photons interact. Let C_1 be the number of vertices arising from the interaction, $:\phi^*\overleftrightarrow{\partial_\mu}\phi : A^\mu$, and C_2 the number of those stemming from the interaction $:\phi^*\phi A^\mu A_\mu:$. Then, by a now familiar argument, one finds that

$$D = -4(C_1 + C_2 - 1) + 4(P_i + Q_i) + C_1 - 2P_i - 2Q_i \quad (17)$$

Since the number of photon line endings is $2C_2 + C_1$,

$$2Q_i + Q_e = 2C_2 + C_1 \quad (18a)$$

and since there are two boson lines coming into each vertex

$$2P_i + P_e = 2(C_1 + C_2) \quad (18b)$$

upon substituting Eqs. (18a) and (18b) into Eq. (17), we again deduce that

$$D = 4 - P_e - Q_e \quad (19)$$

The condition for D is identical to (16) so that the two-photon interaction does not cause additional difficulties. It does necessitate, however, the inclusion of the diagram of Figure 16.17c in the definition of counterterm

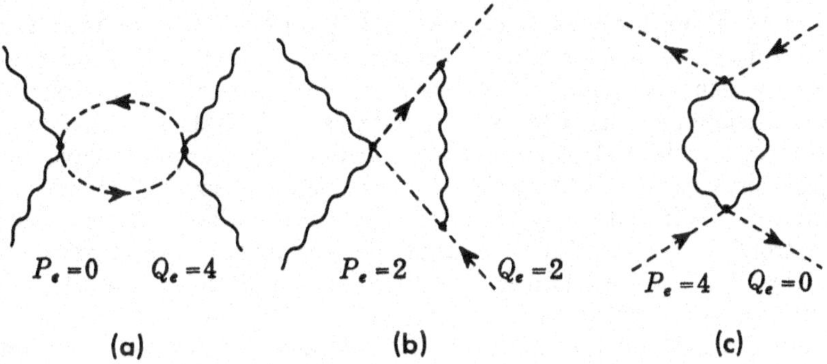

Fig. 16.17

λ_{em}. [For further details, see Rohrlich (1950b).] Gauge invariance again implies that the contributions of diagrams 16.17a and b are actually finite. (Recall also that, due to the normal ordering prescription, no photon line can be reabsorbed at the vertex at which it was emitted, so that these divergent tadpole diagrams do not contribute to the meson self-energy nor to other processes.)

In concluding this section, we shall briefly examine the types of primitively divergent diagrams which occur in field theories with derivative couplings and, more generally, field theories for which the coupling constant is not dimensionless. Consider first the case of a *PS-PV* or *S-V* theory. The interaction at a vertex in these cases is determined by $\mathcal{L}_I = (F/\mu) : \bar{\psi}(x)\, \Gamma\gamma_\mu\psi(x) : \partial^\mu\phi(x)$ ($\Gamma = \gamma_5$ or I). Due to the fact that the derivative of the meson operator appears in $\mathcal{L}_I$, there is a factor $\Gamma\gamma \cdot k$ for each vertex, where k_μ is the momentum of the meson absorbed or emitted at the vertex. In determining D for the primitively divergent diagrams, each vertex therefore contributes an additional momentum factor, so that in the present case

$$D = 4 - \tfrac{3}{2}F_e - B_e + C \tag{20}$$

In contrast to all the theories discussed thus far, the value of D depends on C, so that there are a denumerable infinity of primitively divergent graphs.

Thus, if we were to apply the renormalization principle to these interactions, an infinite number of renormalization constants would be needed to make the theory finite. Since renormalization theory as presently formulated requires one experiment to determine each renormalization constant, an infinite number of such experiments would be needed to make the theory meaningful. The renormalization program, as presently formulated in terms of a power series expansion, is therefore inadequate to cope with such derivative coupling theories. One calls such a theory nonrenormalizable.

Although we have not considered field theories of finite rest mass particles with spin greater than $\frac{1}{2}$, we here mention that such theories are usually nonrenormalizable. An exception is the case of a neutral spin 1 boson field interacting with spin $\frac{1}{2}$ fermion field. This theory is analogous to quantum electrodynamics except that the theory does not have any gauge invariance. Field theories of neutral bosons of spin higher than one interacting with spin $\frac{1}{2}$ fermions are nonrenormalizable. The quantum electrodynamics of charged bosons of spin 1 or higher are also nonrenormalizable. The same is true for the interaction of such charged bosons with spin $\frac{1}{2}$ fermions.[5]

Similarly, the weak Fermi interactions of the form $(\tilde{p}O_i n)\,(\tilde{\nu}O^i e)$ are not renormalizable. Consider, for example, the interaction $(\tilde{p}n)\,(\tilde{\nu}e)$. If we denote by N the number of nucleon lines and by L the number of lepton lines in any diagram, then

$$D = -4C + 4 + 3N_i + 3L_i \tag{21}$$

Moreover, since

$$2N_i + N_e = 2C \tag{22a}$$

and

$$2L_i + L_e = 2C \tag{22b}$$

D can be re-expressed as

$$D = 2C + 4 - \tfrac{3}{2}N_e - \tfrac{3}{2}L_e \tag{23}$$

so that the number of primitively divergent graphs is again denumerably infinite and the theory is not renormalizable.

[5] An exception seems to be the mixed theory of Beard and Bethe [Beard (1951)], in which both pseudoscalar and pseudovector mesons are coupled to the nucleon by pseudovector coupling. As far as the theory was carried in this paper, no divergences appeared. A general investigation has not been made to date. The interaction of spin 1 charged mesons with the electromagnetic field is, however, not renormalizable.

When an invariant cutoff or regulator method is used to define an (otherwise divergent) theory, the renormalizability of the theory implies that after mass and charge renormalization (and possibly the renormalization of a finite number of other parameters) the predictions of the theory (at least when represented by a power series) are independent of the cutoff. This is not the case for the so-called nonrenormalizable theories. In such theories even after mass and charge renormalization, the theory is still strongly cutoff dependent. Whether this is an argument for casting such theories aside is doubtful, since it is a widely held belief that the effects of those interactions which have been neglected (e.g., gravitational;[6] or the strange particles effects) could be such as to introduce a cutoff.

An intuitive way to make this notion somewhat more precise is as follows: Qualitatively one of the effects of the interaction between fields is to "dress" each particle with a cloud of quanta of *all* the fields, so that the particles acquire a *structure*. Present-day theories are at a loss to account for many of the structural features of the particles, e.g., their mass, and one might argue that a more correct field theoretic description of the interaction between the "fundamental" particles would ascribe a nonlocal character to the theory, or at least embody a new fundamental constant, of the dimension of a length. In the case of nonlocal interactions, the theory would automatically contain in it a constant with the dimension of a length to define the nonlocal region (this constant ℓ would probably be of the order of 10^{-13} cm or smaller). In general, this length, ℓ, could be either a coupling constant of a nonlinear interaction term [Heisenberg (1957)] or the radius of a nonlocal domain. Let us further assume that the theory can be described in terms of a Lagrangian $\mathcal{L}(\ell)$. The expansion of this Lagrangian $\mathcal{L}(\ell)$ in powers of ℓ

$$\mathcal{L}(\ell) = \mathcal{L}_{(0)} + \ell\mathcal{L}_{(1)} + \ell^2\mathcal{L}_{(2)} + \cdots \tag{24}$$

then classifies the interactions into various terms $\mathcal{L}_{(0)}$, $\mathcal{L}_{(1)}$, $\cdots$ which have decreasing strengths, since ℓ is a small constant. Notice that in the current classification of interactions according to their strength (strong, electromagnetic weak; recall the discussion in Chapter 10), the weakest (the Fermi interactions) are characterized by a coupling constant which has the dimension of a length squared.

This heuristic argument makes it plausible that the fact that the weak interactions are not renormalizable, when isolated from the effects of other interactions and written in the form $\ell^2\mathcal{L}_{(2)}$, is not an argument against using this form in whatever domain it is valid. It is probably the case that an expansion such as (24) is valid only in a certain energy range and that the Fermi interactions which are responsible for the weak decays may not be representable in the form $\ell^2\mathcal{L}_{(2)} \sim G_\beta(\bar{p}n)\ (\bar{\nu}e)$ at high energies (say,

[6] For example, the combination of quantum theory and gravitation defines a length $\sqrt{\gamma\hbar/c^3} \approx 10^{-32}$ cm, where γ is the ordinary gravitational constant.

energies comparable to the nucleon mass). In fact, it is possible that the Fermi interactions are mediated by a heavy charged boson X which couples the heavy particle current to the leptonic current in such a way that the neutron decay proceeds via the two-step reaction $n \to p + X^- \to p + e^- + \bar{\nu}$ as illustrated in Figure 16.18. If this were the case, the β-decay

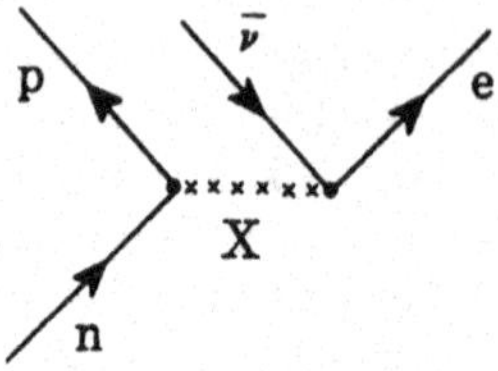

Fig. 16.18

amplitude $G_\beta(\bar{p}n)\,(\bar{\nu}e)$ would really be the low energy representation of an amplitude $-g^2(\bar{p}n)\,(k^2 - \mu_X{}^2)^{-1}\,(\bar{\nu}e)$ corresponding to the exchange of an X boson. Due to the small energy and momentum transfers involved in nuclear β decay $\mu_X{}^2 \gg k^2$, so that the two amplitudes are identical in the low energy region if $G_\beta = g^2/\mu_X{}^2$. [Actually if one wants to obtain the V-A theory in the local, low energy limit, the heavy boson must have spin 1 and its propagator be $(k_X{}^2 - \mu_X{}^2)^{-1}\,(g_{\mu\nu} - k_\mu k_\nu/\mu_X{}^2)$.]

A similar example is the use of the nonrenormalizable interaction $-e : \bar{\psi}\gamma_\mu\psi : A^\mu + \frac{1}{2}\Delta\mu : \bar{\psi}\sigma_{\mu\nu}\psi : F^{\mu\nu}$ to describe the interaction of a proton (which has an anomalous magnetic moment $\Delta\mu = 1.7896$) with the electromagnetic field. This phenomenological representation of the magnetic properties of the proton is valid only in slowly varying electromagnetic fields and for nonrelativistic motion of the particle. It is not permissible to use this description to inquire about the radiative corrections induced by high energy fluctuations in the quantized electromagnetic field. Actually, the extra moment arises from mesonic radiative corrections (such as those illustrated in Figure 16.19) in the complete theory which takes into account meson-nucleon and electromagnetic interactions. The term $\frac{1}{2}\Delta\mu : \bar{\psi}\sigma_{\mu\nu}\psi : F^{\mu\nu}$ is therefore not to be regarded as part of the fundamental interaction between the proton and the electromagnetic field but only as a phenomenological representation of this interaction at low energies.

If the point of view outlined above has some validity, the renormalizability or nonrenormalizability of a given theory is not necessarily a fundamental distinction. Rather, it becomes of great interest to inquire up to what energies is the present description of electromagnetic phenomena in terms of a Lagrangian characterized by a dimensionless coupling constant $e/\sqrt{\hbar c}$ valid. Similarly, what does the fact that quantum electrodynamics has no cutoff dependence (after mass and charge renormalization) imply about the Lagrangian $\mathcal{L}(\ell)$? Also, what does the fact that field theories whose coupling constants have the dimension of a length raised to some

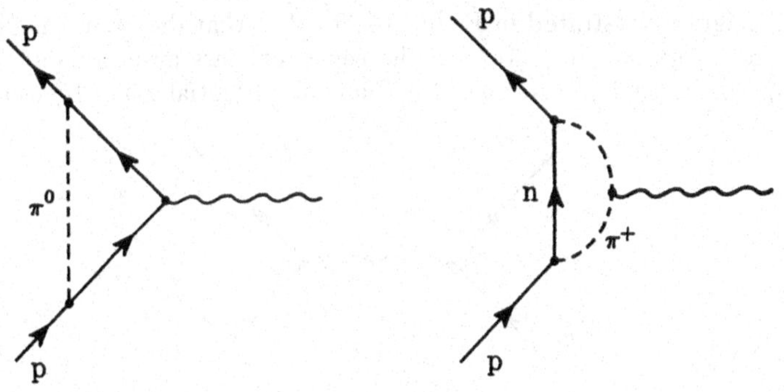

Diagrams Contributing to Anomalous Proton Magnetic Moment in Second Order

Fig. 16.19

positive power are *non*renormalizable whereas those with dimensionless coupling constant are renormalizable reflect? [Interaction with $G \propto [L]^\eta$ are called of the first kind if $\eta \leqslant 0$ and of the second kind if $\eta > 0$; see Umezawa (1959).]

Experiments on the charge distribution of the neutron and proton, the magnetic moment of the muon, the Lamb shift and the magnetic form factor of the proton have indicated [Drell (1958)] that there exist no discrepancies between the predictions of quantum electrodynamics and the observed data for energies up to 600 Mev, or equivalently for distances greater than 0.33×10^{-13} cm.[7] Experiments to test quantum electrodynamics at small distances have been suggested [Drell (1958), Frautschi (1958)] but require either very high energies or extreme accuracies. For example, the interpretation of an electron-electron scattering experiment requires the computation of the matrix element corresponding to the lowest

[7] The inference that a breakdown of quantum electrodynamics occurs for distances less than 0.3×10^{-13} can be made if one assumes that the neutron-proton mass difference ($\sim 2.53m_e$) is of electromagnetic origin [Feynman and Speisman (Feynman 1955)]. On this point of view, the mass difference $M_n - M_p$ is the result of competition between the electrostatic repulsion (which is of order $1/R$ in the leading term) and a magnetic moment self-energy (which is of order $1/R^2$). The magnetic moment interaction may decrease M_p relative to M_n. Since $M_n > M_p$, the magnetic effects are important. From the observed mass difference $M_n - M_p \sim 2.53m_e$, Feynman and Speisman deduced an rms proton radius of 0.4×10^{-13} cm, which is half the radius suggested by electron-proton scattering experiment. These experiments [Hofstadter (1957)] have given an apparent rms radius of 0.8×10^{-13} cm for the proton and essentially a zero charge radius for the neutron. The interpretation of this data assumes the validity of quantum electrodynamics and that all deviations from theoretical expectations are due to nucleon structure effects. The discrepancy between the Feynman-Speisman theoretical rms radius and the observed radius can be explained by a cutoff at about 0.3×10^{-13} cm on electrodynamics.

order diagram illustrated in Figure 16.20. For that diagram, the photon propagator, $q^{-2} = (q_2 - q_1)^{-2}$, in the center-of-mass system simplifies to $-1/\mathbf{q}^2$ the Fourier transform of the Coulomb potential $1/r$. If electrody-

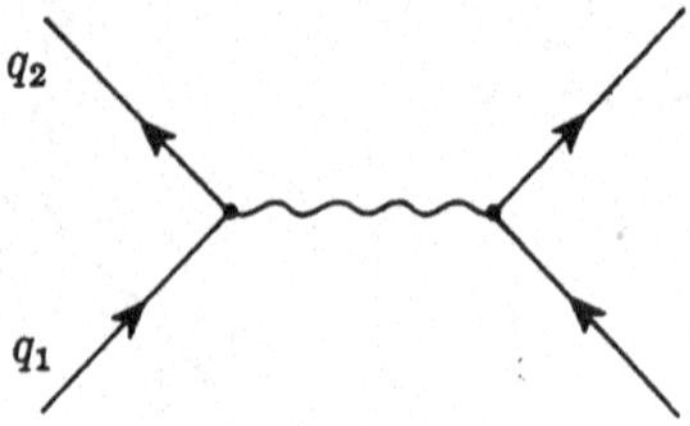

Fig. 16.20

namics is altered at small distance by replacing $1/r$ by $1/r \cdot (1 - e^{-\Lambda r})$, the photon propagator is altered from $1/q^2$ to $\frac{1}{q^2} - \frac{1}{q^2 - \hbar^2\Lambda^2}$. For such an alteration of quantum electrodynamics, the Møller scattering cross section is reduced by the factor $1 + (2q^2/\hbar^2\Lambda)$. An experiment with a 10% accuracy, in which electrons of 8 Bev are scattered by stationary target electrons, could just detect a cutoff length $1/\Lambda$ of $\sim 0.5 \times 10^{-13}$ cm. In due time, the answer to the question: "What is the limit of applicability of quantum electrodynamics?" will therefore be experimentally answered.

Although we have argued against the discarding of nonrenormalizable theories, we shall in this book only concern ourselves with quantum field theories of the renormalizable variety. This, firstly, because there is evidence that the PS-PS theory can account for low energy meson-nucleon phenomena and the fact that quantum electrodynamics has proven itself correct in those areas where it has been tested. Secondly, because the structure of derivative coupling theories is much more complex than those with direct coupling [see e.g., Arnowitt (1955), Cooper (1955), Guttinger (1958a, b)], and very little is known about them in general. In particular, we do not know how to calculate the observable consequences of such a theory.

In conclusion, it is to be emphasized that the success of the renormalization program lies in its *circumvention* of the divergence problem: it does *not* solve the problem. Renormalization is therefore *not* the final answer to the divergence problems which beset local relativistic quantized field theories. The fact that the renormalized quantum electrodynamics is in excellent agreement with experiments over a significant energy range is the primary reason for its detailed study, since this fact suggests that the renormalized theory embodies some of the features of the much heralded "future correct theory." This "correct theory" should not only contain no divergences but, in addition, should account for the mass spectrum of the elementary particles as well as the magnitude of the coupling constants, such as the fine structure constant [8] (quantities which at present are fitted empirically). Undoubtedly, some aspects of renormalization theory will be embodied (or be present) in the future correct theory. It is also very probable that in the "future theory" the interrelation and simultaneous coexistence of all the "elementary" particles (such as the electrons, nucleons, etc.) will be an integral part of the theory. The hope is that, by studying the present forms of renormalized quantum electrodynamics and of the renormalizable meson theories, an insight will be obtained into those parts of the present-day theories which are dispensable and into those which are consequences of very general principles, such as relativistic invariance and causality, and which, therefore, may also be present in the "future theory." Such a study has, in fact, yielded a different formulation in which one deals only with finite renormalized quantities and whose structure sheds some light on the points enumerated above. [Lehmann, Symanzik, and Zimmermann (Lehmann 1955a, 1957); Wightman (1956).] We shall take up these formulations in Chapter 18.

16b. The Renormalizability of Quantum Electrodynamics

We shall now show that the ideas of charge and mass renormalization are, in fact, sufficient to remove all the ultraviolet divergences in the S-matrix expansion in quantum electrodynamics.

We have seen in Section *a* that the integral corresponding to a primitively divergent graph diverges only if

$$\tfrac{3}{2}F_e + B_e < 5$$

where, in the present context, F_e and B_e are the number of external electron and photon lines, respectively. This inequality, as we have seen, implies that there are only a finite number of primitively divergent graphs that can introduce divergences into the theory. The reason for this is the fact that the degree of divergence of the integrals corresponding to these graphs does not depend on the order of the graphs. The renormalization program now consists of a subtraction procedure which removes these infinities and the theoretical justification for these subtractions. In our presentation of the renormalization of quantum electrodynamics, we shall follow closely

[8] There was some hope in the early days of renormalization theory that the magnitude of the fine structure constant might be obtained from the requirement that the divergences occurring in the successive orders of perturbation theory cancel one another. This idea was suggested by Pauli. It was, however, shown by Jost and Luttinger [Jost (1950a)] that such a compensation cannot occur. Thus, the contribution to the vacuum polarization divergence has the same sign in second and fourth order. A somewhat related idea concerning the finiteness and convergence of the self-energy contributions was formulated by Racah (1946), but has since been disproved by Frank (1951) by an explicit calculation of the fourth-order contribution to the self-energy of a fermion. [See also Gell-Mann and Low (Gell-Mann 1956).]

Dyson's two articles of 1949 [Dyson (1949a, b)], and the proof of renormalizability follows that given by Ward (1951).

We first introduce the concepts of connected, reducible, and irreducible graphs, and of vertex and self-energy parts [Dyson (1949b)].

A graph is called *connected* if it is such that any part of it is joined to the remainder by at least two lines. A *vertex part* is defined as a connected part (consisting only of vertices and internal lines) joined to the rest of the graph by just two fermion lines and one photon line. Figure 16.21*a*

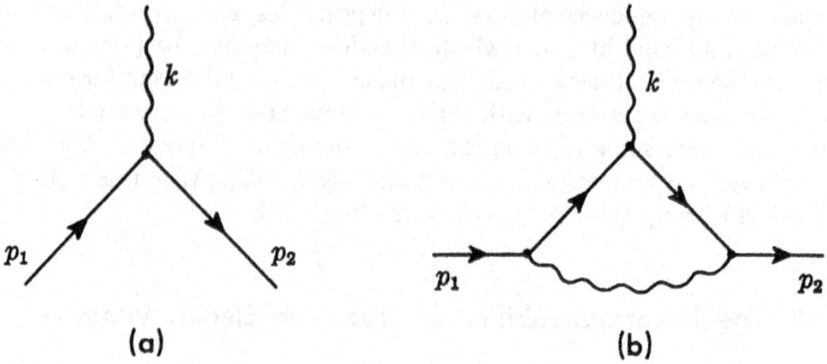

Fig. 16.21

illustrates the simple vertex part. We can also have a vertex part, V, of higher order (for example, Figure 16.21*b*) which can be substituted for any vertex of a graph G, and the modified graph G' will have a definite physical meaning. The matrix element for G' may be obtained from that of G by replacing the factor γ_μ, for the particular vertex under consideration, by an operator $\Lambda_\mu(V, p_1, p_2)$, which depends only on the structure of V and on the two fermion lines by which V is joined to the rest of the graph, i.e., on the propagation vectors p_1 and p_2 in Figure 16.21.

Similar , a *self-energy part, W*, of a graph G is a connected part (consisting only of vertices and internal lines) which can be inserted in the middle of a line in G (electron or photon line) such that the modified graph is consistent with the rules for building Feynman diagrams. Figures 16.23*a* and *b* illustrate the simplest electron and photon self-energy parts possible. Figure 16.22*a* illustrates the insertion of an electron self-energy part into the electron line p_1 of Figure 16.21*b*, and Figure 16.22*b* the insertion of a photon self-energy part in the photon line k of Figure 16.21*b*. These lines p_1 or k may be external, or they may be internal, for the graph as a whole. As a special case, a self-energy part may consist of only a point, corresponding to the term $-\delta m : \bar{\psi}(x)\, \psi(x) :$ in $\mathcal{H}_I(x)$.

Let us define, with Dyson, the momentum transform of the ψ and A operators as follows:

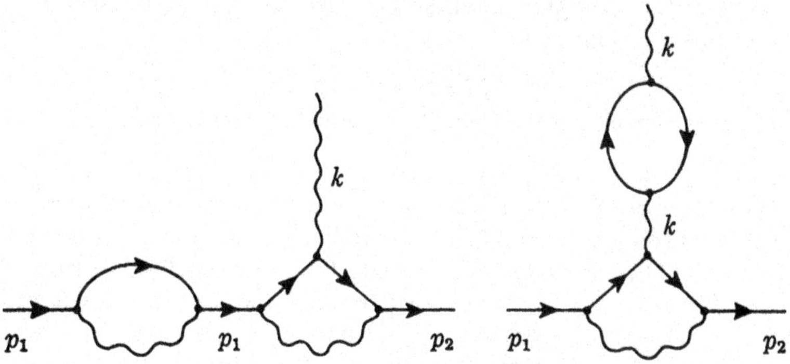

(a) (b)

Fig. 16.22

$$\psi(x) = \int d^4p\, e^{-ip\cdot x}\, \psi(p) \tag{25}$$

$$A_\mu(x) = \int d^4k\, e^{-ik\cdot x}\, A_\mu(k) \tag{26}$$

and, throughout this section, we define $S_F(p)$ and $D_F(k)$ as essentially the Fourier transforms of $S_F(x)$ and $D_F(x)$, respectively. Explicitly,

$$D_F(k) = \frac{1}{2\pi}\frac{1}{k^2} \tag{27a}$$

$$S_F(p) = \frac{1}{2\pi}\frac{1}{\gamma\cdot p - m} \tag{27b}$$

If now the modified graph due to any one of the above insertions in G is denoted by G', then our previous rules for writing down matrix elements show that the matrix for G' may be obtained from that of G by making the following modifications:

(*a*) Insertion of an electron self-energy part, W, into an external electron line of momentum p_1

$$\psi(p_1) \to S_F(p_1)\,[\Sigma(W, p_1) - 2\pi i\delta m]\,\psi(p_1) \tag{28}$$

$$\bar{\psi}(p_1) \to \bar{\psi}(p_1)\,[\Sigma(W, p_1) - 2\pi i\delta m]\,S_F(p_1) \tag{29}$$

(*b*) Insertion of an electron self-energy part, W, into an internal electron line of momentum p_1

$$S_F(p_1) \to S_F(p_1)\,[\Sigma(W, p_1) - 2\pi i\delta m]\,S_F(p_1) \tag{30}$$

(*c*) Insertion of a photon self-energy part, W', into an external photon line of momentum k

$$A_\mu(k) \to A_\mu(k)\,\Pi(W', k)\,D_F(k) \tag{31}$$

(*d*) Insertion of a photon self-energy part, W', into an internal photon line of momentum k

$$D_F(k) \rightarrow D_F(k)\, \Pi(W', k)\, D_F(k) \tag{32}$$

(*e*) Insertion of a vertex part, V, for simple vertex operator γ_μ

$$\gamma_\mu \rightarrow \Lambda_\mu(V, p_1, p_2) \tag{33}$$

Here the operator Σ (or Π) depends only on the structure of the graph W (or W'), and on p_1 (or k), and not on the rest of the diagram. In Eqs. (28), (29), and (30), the term $-2\pi i\delta m$, which is the contribution from the $-\delta m : \bar{\psi}\psi(x) :$ term, has been added to effect the mass renormalization. For the photon self-energy operator Π no such counterterm is necessary, since the photon self-energy vanishes by the requirements of gauge invariance. (For mesons, however, a mass renormalization term would again be necessary. Recall Section 15*h*). Similarly, the vertex part $\Lambda_\mu(V, p_1, p_2)$ depends only on the structure of graph V and on the momenta associated with the two fermion lines, p_1 and p_2. Note that the operators Λ_μ, Σ, and Π can be calculated without reference to the rest of the graph and the results will be of a general nature. These may then be used for various particular applications.

If from a graph G we omit all self-energy and vertex parts, we obtain a reduced graph G_0, which is called the *skeleton* of G. A graph which is its own skeleton is called *irreducible*.* The second-order self-energy parts W and W' of Figure 16.23 are the only examples of irreducible self-energy parts. Higher order self-energy parts are *reducible*, i.e., not irreducible, and may be obtained by inserting self-energy and vertex parts in the lines and vertices of W and W' of Figure 16.23. However, vertex parts of

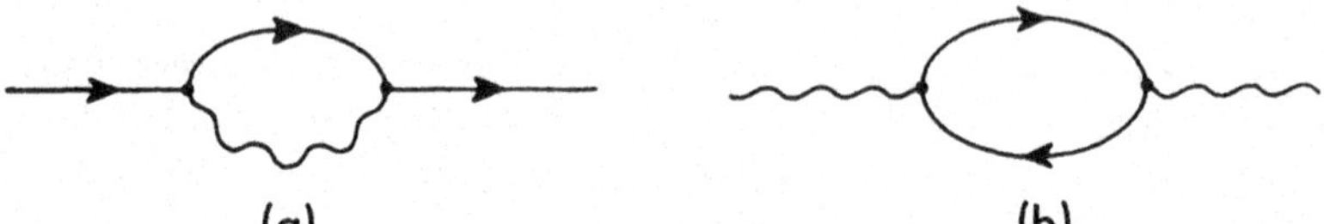

(a) (b)

Fig. 16.23

higher order than the second may be reducible or irreducible. A vertex or self-energy part is called *proper* if it cannot be divided into two parts joined by a single line; otherwise, it is called *improper*. Figure 16.24 is an example of an improper fermion self-energy part.

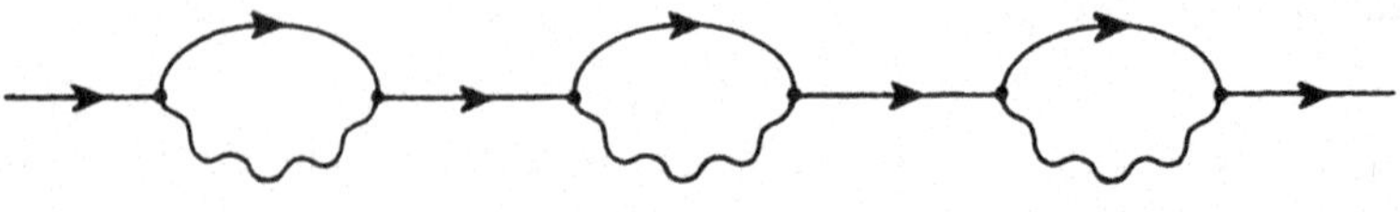

Fig. 16.24

(*) Conversely, an arbitrary graph is said to be *reducible* if some part of the diagram, but not the entire diagram, may be recognized as a vertex part or self-energy part. An *irreducible* diagram is one which is not *reducible*.

Now we can obtain a reducible graph G of a certain order, say n, by inserting independently and in all possible ways self-energy and vertex parts of requisite orders in the various lines and vertices of an irreducible graph G_0 of lower order. These graphs G form a well-defined class L, and it is thus possible to enumerate them accurately.

Suppose that M_0 is the matrix element of G_0. Then every matrix element M in L will yield additional contributions to the process corresponding to the matrix element M_0 of the S matrix. Let the sum of *all* such contributions, including M_0, be denoted by M. In view of the independence of the insertions made at the different vertices and lines of G, the sum M will be obtained from M_0 by making the following modifications in the factors of M_0:

For every internal electron line of G_0, a factor $S_F(p)$ of M_0 is replaced by

$$S'_F(p) = S_F(p) + S_F(p)\,\Sigma^*(p)\,S'_F(p) \tag{34}$$

(proof will be given below). Similarly, for every external fermion line, the operator ψ or $\bar{\psi}$ is replaced by

$$\psi(p) \rightarrow \psi'(p) = \psi(p) + S'_F(p)\,\Sigma^*(p)\,\psi(p) \tag{35a}$$

$$\bar{\psi}(p) \rightarrow \bar{\psi}'(p) = \bar{\psi}(p) + \bar{\psi}(p)\,\Sigma^*(p)\,S'_F(p) \tag{35b}$$

For photon lines, the replacements are

$$D_F(k) \rightarrow D'_F(k) = D_F(k) + D_F(k)\,\Pi^*(k)\,D'_F(k) \tag{36a}$$

$$A_\mu(k) \rightarrow A'_\mu(k) = A_\mu(k) + A_\mu(k)\,\Pi^*(k)\,D'_F(k) \tag{36b}$$

Finally, the vertex operator γ_μ is to be replaced by

$$\Gamma_\mu(p_1, p_2) = \gamma_\mu + \Lambda_\mu(p_1, p_2) \tag{37}$$

Here $\Sigma^*(p)$ and $\Pi^*(p)$ are the sum operators $\Sigma(W, p)$ and $\Pi(W', k)$ over all *proper* self-energy parts. It is assumed that in Σ^* we have included the contribution of the mass renormalization counterterm.

Actually, somewhat greater care must be exercised with regard to the gauge invariance of the theory in order to obtain a consistent renormalization procedure [see in this connection Bogoliubov and Shirkov (Bogoliubov 1956), Bogoliubov (1959), Umezawa (1956a, b), and Fried (1960)]. Thus Eq. (36a) should be replaced by

$$D'_{F\mu\nu}(k) = D_{F\mu\nu}(k) + D_{F\mu\lambda}(k)\,\Pi^{*\lambda\sigma}(k)\,D'_{F\sigma\nu}(k) \tag{36c}$$

where $\Pi^*_{\lambda\sigma}(k)$ is the sum over all proper self-energy diagram parts and is assumed to have the gauge invariant form

$$\Pi^*_{\mu\nu}(k) = \left(g_{\mu\nu} - \frac{k_\mu k_\nu}{k^2}\right)\Pi^*(k) \tag{36d}$$

If one takes for the representation of the bare propagator the form

$$D_{F\mu\nu}(k) = \frac{1}{2\pi}\left(g_{\mu\nu} - \frac{k_\mu k_\nu}{k^2}\right)\frac{1}{k^2} \tag{38}$$

then

$$D'_{F\mu\nu}(k) = \left(g_{\mu\nu} - \frac{k_\mu k_\nu}{k^2}\right) D'_F(k) \tag{39}$$

and Eq. (36a) follows, upon inserting Eqs. (38), (39), and (36d) into Eq. (36c).

Similarly, $\Lambda_\mu(p_1, p_2)$ in Eq. (37) is the sum of all $\Lambda_\mu(V, p_1, p_2)$ over all *proper* vertex parts since a vertex part which is not proper is a proper vertex part plus one or more self-energy parts, and the latter have already been included in S'_F and D'_F.

The improper parts being nonoverlapping combinations of proper parts, these have already been accounted for in the implicit definition of S'_F, D'_F given above. Thus, it is easy to see that, for example, the improper diagrams of the type given in Figure 16.24 are included in the definition of

$$S'_F(p, W) = S_F(p) + S_F(p)\,\Sigma(p, W)\,S'_F(p, W) \tag{40}$$

with W corresponding to the fermion self-energy operator of Figure 16.23*a*. In fact, if we solve for S'_F, we obtain

$$\begin{aligned} S'_F(p, W) &= \frac{1}{1 - S_F(p)\,\Sigma(p, W)}\,S_F(p) \\ &= S_F(p) + S_F(p)\,\Sigma(p, W)\,S_F(p) \\ &\quad + S_F(p)\,\Sigma(p, W)\,S_F(p)\,\Sigma(p, W)\,S_F(p) + \cdots \end{aligned} \tag{41}$$

which, in graphical representation, corresponds to the diagrams in Figure 16.25.

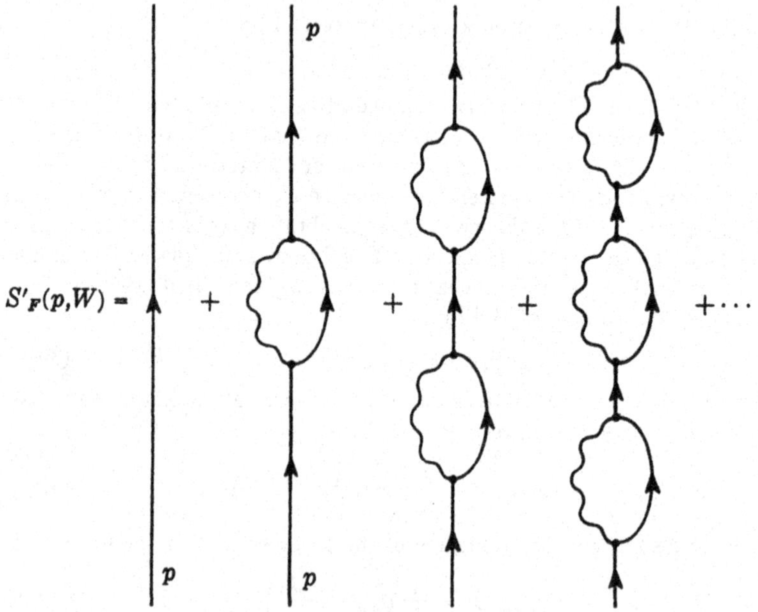

Fig. 16.25

The matrix elements of the S matrix will be correctly calculated if one includes only contributions from irreducible graphs, after making in each matrix element corresponding to a skeleton (irreducible) graph the replacements (34)–(39). These equations are integral equations. In co-ordinate space, Eq. (34), for example, would read

$$S'_F(x - y) = S_F(x - y) + \int d^4x' \int d^4y' \, S_F(x - x') \, \Sigma^*(x' - y') \, S'_F(y' - y) \tag{42}$$

Similarly, Eq. (39) corresponds to a replacement in co-ordinate space of

$$\gamma_\mu \delta^{(4)}(x - x') \, \delta^{(4)}(x' - x'') \to \Gamma_\mu(x - x'; x' - x'') \tag{43}$$

To calculate the operators Λ_μ, Σ^*, and Π^*, it is necessary to write down explicitly the integrals corresponding to every self-energy part W and vertex part V. For higher order effects, the parts W and V will themselves often be reducible (they always are for W and W'), containing self-energy and vertex parts in their interior. In this case again, it will be convenient to omit such reducible V and W and to include their effects by making substitutions (34)–(39) in the integral corresponding to irreducible V and W. One thus obtains integral equations for Λ_μ, Σ^*, and Π^*.

It should be noted, however, that reducible self-energy parts are to be enumerated by inserting vertex parts at only one and not both of the vertices of the irreducible self-energy parts, Figure 16.23a and b; otherwise, the same self-energy part would appear more than once in the enumeration. With these statements in mind, we can, in fact, immediately write down the integral equations that Σ^* and Π^* satisfy (α is the fine structure constant):

$$\Sigma^*(x - x') = -2\pi\alpha i \int d^4y \int d^4y' \, \gamma^\mu S'_F(x - y) \, \Gamma_\mu(y - x; y - y') \, D'_F(y' - x) \tag{44}$$

$$g_{\mu\nu}\Pi^*(x - x') = 2\pi\alpha \int d^4y \int d^4y' \, \mathrm{Tr}\,[\gamma_\mu S'_F(x - y) \, \Gamma_\nu(y - x'; y - y') \, S'_F(y' - x)] \tag{45}$$

On the other hand, no closed expression in terms of Γ_ν, S'_F, and D'_F can be written down for Λ_μ. We can, however, write down a power series integral equation for the vertex operator Γ_μ, which in momentum space reads [see Schwinger (1951b) and Edwards (1953)]:

$$\begin{aligned}\Gamma_\mu(p_1, p_2) &\equiv \gamma_\mu + \Lambda_\mu(p_1, p_2) \\ &= \gamma_\mu - 2\alpha i \int \Gamma^\nu(p_1; p_1 - k) \, S'_F(p_1 - k) \, \Gamma_\mu(p_1 - k, p_2 - k) \\ &\qquad S'_F(p_2 - k) \, \Gamma_\nu(p_2 - k, p_2) \, D'_F(k) \, d^4k + \cdots \end{aligned} \tag{46}$$

In other words, to obtain the integral equation for Γ_μ, draw all the irreducible vertex diagrams and substitute in each, S'_F for S_F, Γ_μ for γ_μ, and D'_F for D_F. The complete series for Γ_μ is then generated.

One should, of course, view with reservations the above formal manipulations since one is dealing with divergent quantities. For practical applications, it is always wise to use regulators, so that all integrals become finite, though cutoff dependent. The quantities which would be finite in the absence of a cutoff remain unchanged, while logarithmically divergent quantities become finite logarithmic functions of the regulator masses.

We next turn to the task of identifying the divergences of the theory and to a prescription for removing these.

Recall that we have called a divergent matrix element M *primitive* if, whenever any one of the integration four-vector variables is held fixed, the resulting integral over the remaining variables is convergent. Holding an integration four-variable, p^j_μ, fixed means replacing an internal line in a connected graph by two external lines. Let G be a primitive divergent graph with F_e external fermion lines and B_e external photon lines; then, as we have seen in Section a, the graph may diverge if $\frac{3}{2}F_e + B_e < 5$.

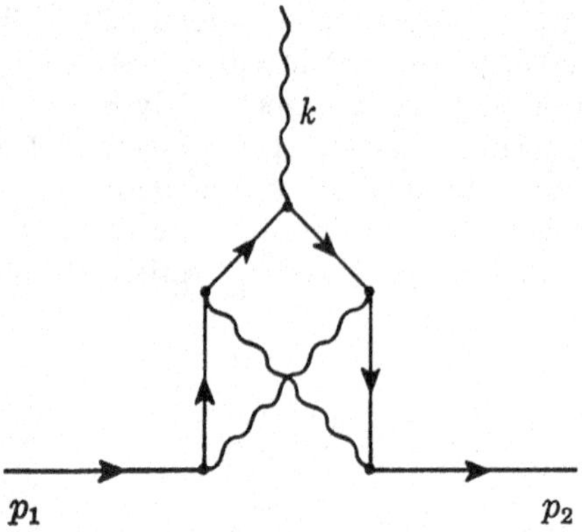

Fig. 16.26

The primitively divergent graphs have been previously enumerated in Section a. They are

1. the electron self-energy part ($F_e = 2$, $B_e = 0$),
2. the photon self-energy part ($B_e = 2$, $F_e = 0$),
3. the vertex part ($F_e = 2$, $B_e = 1$), and
4. the scattering of light by light ($B_e = 4$, $F_e = 0$).

The triangle diagrams have not been considered because of Furry's theorem. Also, as stated previously, the scattering of light by light is actually convergent because of the requirements of gauge invariance.

Again, note that most of the diagrams do not satisfy the condition that their integrals converge if any one of their internal momenta is held fixed. In fact, by the definition of a reducible graph, a reducible divergent graph cannot be primitively divergent. Hence, the primitive divergents correspond to the irreducible (i.e., skeleton) graphs of the enumeration (1)–(4) above. There is only one irreducible self-energy part for the electron and photon, given in Figure 16.23*a* and *b*; however, there can be irreducible vertex parts of any order. Thus in a higher order, a possible irreducible vertex part is illustrated in Figure 16.26.

16c. The Separation of Divergences from Irreducible Graphs

Dyson (1949b) has given an unambiguous method for the separation of a primitively divergent integral into an infinite and a finite part. This is accomplished in each case by a Taylor expansion of the integrand corresponding to the primitively divergent graph in terms of the external momenta [recall Eq. (15.90) and, especially, the discussion following Eq. (15.89)]. Thus, if $R(p, t)$ is the integrand with p and t the external and internal momenta, respectively, then we write

$$R(p, t) = R(0, t) + p_\mu \left(\frac{\partial R(p, t)}{\partial p_\mu}\right)_{p=0} + \tfrac{1}{2} p_\mu p_\nu \left(\frac{\partial^2 R(p, t)}{\partial p_\mu \partial p_\nu}\right)_{p=0} + \cdots \quad (47)$$

Now, each differentiation increases the difference of powers of t of the numerator and denominator by one. Therefore, a logarithmically, linearly, or quadratically divergent integral will have an expansion of R in which only the first one, two, or three terms are divergent. The form of the divergent terms then follows from relativistic invariance. We shall now consider the separation of the divergences from the irreducible divergent diagrams.

Electron Self-Energy Part

The only irreducible electron self-energy part may be represented by the integral

$$\Sigma(W, p) = \int R_e(p, k)\, d^4k \quad (48)$$

which is linearly divergent [$R_e(p, k) \propto e^2\, \gamma^\mu S_F(p - k)\, \gamma_\mu D_F(k)$]. The Taylor expansion is now given by

$$R_e(p, k) = R_e(0, k) + p_\mu \left(\frac{\partial R_e(p, k)}{\partial p_\mu}\right)_{p=0} + R_{eC}(p, k) \quad (49)$$

Since Σ is linearly divergent, R_{eC} gives a convergent integral since it contains two extra powers of k in the denominator as compared with $R_e(p, k)$. Therefore, we may write (48) as

$$\Sigma(W, p) = A'(W) + B'_\mu(W)\, p^\mu + \Sigma_C(W, p) \quad (50)$$

where $A'(W)$ and $B'_\mu(W)$ are divergent matrix operators which are independent of p. Now, since $\Sigma(W, p)$ is a 4×4 scalar matrix, it can be expanded in terms of the sixteen linearly independent Γ matrices as follows:

$$\Sigma(p) = \Sigma_S(p) + \Sigma_{V\mu}(p)\,\gamma^\mu + \tfrac{1}{2}\Sigma_{T\mu\nu}(p)\,\sigma^{\mu\nu} + \Sigma_P(p)\,\gamma_5 + \Sigma_{A\mu}(p)\,\gamma^\mu\gamma_5 \quad (51)$$

Relativistic covariance requires that $\Sigma_S(p)$ be a scalar valued function of p^2, $\Sigma_{V\mu}(p)$ be a vector valued function of the invariant p^2, $\Sigma_{T\mu\nu}$ a tensor valued function of p^2, and that $\Sigma_{A\mu}(p)$ and $\Sigma_P(p)$ vanish. Since p_μ is the only vector at our disposal, $\Sigma_{V\mu}(p^2) = p_\mu\Sigma_V'(p^2)$, $\Sigma_{T\mu\nu}(p) = p_\mu p_\nu\Sigma_T'(p^2)$ so that, since $\sigma_{\mu\nu}p^\mu p^\nu = 0$,

$$\Sigma(W, p) = \Sigma_S(W, p^2) + \gamma^\mu p_\mu\Sigma_V(W, p^2) \quad (52)$$

where we have dropped the prime on Σ_V'. Comparing Eqs. (52) and (50), we obtain

$$A'(W) = \Sigma_S(W, 0) \quad (53)$$

$$B'_\mu(W) = \gamma_\mu\Sigma_V(W, 0) = \gamma_\mu B'(W) \quad (54)$$

Furthermore, since Σ_C is convergent, it can be written uniquely in the form

$$\Sigma_C(W, p^2) = A''(W) + B''(W)\,(\gamma \cdot p - m) + (\gamma \cdot p - m)\,S_C(W, p^2) \quad (55)$$

where $S_C(W, p^2)$ vanishes for a free electron, i.e., one satisfying $\gamma \cdot p = m$ and $p^2 = m^2$. Making use of Eqs. (54) and (55), Eq. (50) can be then rewritten in the form

$$\Sigma(W, p) = A''(W) + A'(W) + B'(W)\,\gamma \cdot p + B''(W)(\gamma \cdot p - m) + (\gamma \cdot p - m)\,S_C(W, p^2) \quad (56a)$$

so that if we set

$$A'(W) + A''(W) + mB'(W) = A(W) \quad (56b)$$

$$B'(W) + B''(W) = B(W) \quad (56c)$$

and include the contribution of the $-\delta m : \bar\psi\psi :$ term, we obtain after summing over all irreducible self-energy diagrams (there is only one!)[9]

$$\Sigma_I{}^*(p) = A_I - 2\pi i\delta m_I + B_I(\gamma \cdot p - m) + (\gamma \cdot p - m)\,S_C(I, p^2) \quad (57)$$

which is the contribution of the irreducible (I) self-energy diagram to $\Sigma^*(p)$. In (57), we have written

$$\sum_{\text{irreducible } W_i} A(W_i) = A(W) = A_I \quad (58)$$

Similarly, $B_I = B(W)$, etc. Now the mass renormalization is defined such that for $\gamma \cdot p = m$, that is, for a free electron, the mass should be exactly given by m. Therefore,

[9] In practice, this separation is accomplished by expanding $S_F(p - k)$ in R_e according to the formula: $(A + B)^{-1} = A^{-1} - A^{-1}BA^{-1} + A^{-1}B(A + B)^{-1}\,BA^{-1}$ with $A = \gamma \cdot k$ and $B = \gamma \cdot p - m$.

$$\bar{\psi}(p)\, \Sigma_I^*(W, p)\, \psi(p) = 0 \quad \text{if } \gamma \cdot p = m \tag{59}$$

from which it follows that

$$A_I = 2\pi i \delta m_I \tag{60}$$

Photon Self-Energy Part

The irreducible photon self-energy part W', given by Figure 16.23*b* may be represented in the form

$$\Pi(W', k) = \int R_p(k, q)\, d^4q \tag{61}$$

where Π is a scalar which is quadratically divergent. In this case, the subtraction procedure is defined by

$$R_p(k, q) = R_p(0, q) + k_\mu \left(\frac{\partial R_p(k, q)}{\partial k_\mu}\right)_{k=0} + \tfrac{1}{2} k_\mu k_\nu \left(\frac{\partial^2 R_p(k, q)}{\partial k_\mu \partial k_\nu}\right)_{k=0} + R_{pC}(k, q) \tag{62}$$

Since $R_{pC}(k, q)$ contains three extra powers of p in the denominator, as compared with $R_p(k, q)$, it is a convergent integral over p. Proceeding in an analogous fashion to the electron self-energy, we obtain, on grounds of covariance,

$$\Pi(W', k) = A_1(W') + B_{1\mu}(W')\, k^\mu + C_{\mu\nu}(W')\, k^\mu k^\nu + \Pi_C(W', k) \tag{63}$$

The A_1, $B_{1\mu}$, $C_{\mu\nu}$ are constant tensors with the same components in all Lorentz frames, whence it follows that

$$B_{1\mu}(W') = 0 \tag{64}$$

$$C_{\mu\nu}(W') = C(W')\, g_{\mu\nu} \tag{65}$$

Furthermore, gauge invariance forbids the appearance of the constant $A_1(W)$, so that this term is to be set equal to zero. Therefore, since $C_{\mu\nu}(W')\, k^\mu k^\nu = C(W')\, k^2$ we may write

$$\Pi(W', k) = k^2 D_C(W', k) + C'(W')\, k^2 \tag{66a}$$

where

$$D_C(W', k) = 0 \quad \text{for } k^2 = 0 \tag{66b}$$

and where $C(W')$ has been absorbed in $C'(W')$. Finally, since there is only one irreducible photon self-energy part, we obtain

$$\Pi^*_I(k^2) = k^2 C_I + k^2 D_C(I, k^2) \tag{67}$$

where $C'(W')$ has been denoted by C_I. Again $\Pi^*_I(k^2)$ is the contribution to Π^* of all the irreducible photon self-energy diagrams.

Vertex Part

The contribution from vertex parts can at most be logarithmically divergent. In this case, therefore, no derivative term is required in the Taylor expansion of the integrand of Λ_μ, so that

$$\Lambda_\mu(V, p_1, p_2) = L_\mu(V) + \Lambda_{\mu C}(V, p_1, p_2) \tag{68}$$

where $L_\mu(V)$ is a constant divergent matrix operator, and $\Lambda_{\mu C}$ is convergent and zero for $p_1 = p_2 = 0$. Covariance requires that

$$L_\mu(V) = L(V)\, \gamma_\mu \tag{69}$$

Also, if $p_1 = p_2$ satisfies $\gamma \cdot p_1 = m$, then $\Lambda_{\mu C}$ reduces to a constant multiple of γ_μ which can be included in the term $L(V)\, \gamma_\mu$. Therefore, it may be supposed that $\Lambda_{\mu C}$ in (68) is zero not for $p_1 = p_2 = 0$, but for $\gamma \cdot p_1 = = \gamma \cdot p_2 = m$, and $p_1 = p_2$. Physically, the meaning of this is as follows: The quantity $\Lambda_{\mu C}$ now gives zero contribution in a constant electromagnetic potential, since such a potential transfers zero momentum. This is as it should be since a constant electromagnetic potential can always be removed by a gauge transformation. Furthermore, since $\bar{\psi}(p)\, \gamma_\mu \psi(p)$ is the charge-current four-vector of an electron without radiative corrections, $\bar{\psi}(p)\, \Gamma_\mu \psi(p)$ may be interpreted as the current four-vector of an electron including radiative corrections. Therefore, Eq. (68) implies that the whole static charge is now included in the term $L\gamma_\mu$, since for a free electron when $\gamma \cdot p = m$, $\Lambda_{\mu C}(p, p, V) = 0$, and hence $\bar{\psi}(p)\, \Gamma_\mu \psi(p) = L\bar{\psi}(p)\, \gamma_\mu \psi(p)$.

Summing (68) over all irreducible vertex parts gives

$$\sum_{\text{all irreducible } V_i} \Lambda_\mu(V^i, p_1, p_2) = \Lambda_{\mu I}(p_1, p_2)$$
$$= L_I \gamma_\mu + \Lambda_{\mu IC}(p_2, p_2) \tag{70}$$

where L_I is a divergent constant.

Equations (57), (67), and (70) form the basis of the separation of the divergent terms in a matrix element corresponding to an irreducible graph. The reader is referred to the paper of Karplus and Kroll [Karplus (1950)], where the separation is carried out explicitly for the lowest order self-energy and vertex diagram. The above method is identical to the one presented in Chapter 15, where we discussed the renormalization of the lowest order radiative corrections (using, however, regularized expressions to carry out the separation).

We must next establish formulae similar to (57), (67), and (70) for the contribution to Σ^*, Π^*, and Γ_μ from the reducible proper diagrams, which again separate the finite and infinite parts in a relativistically invariant and unambiguous fashion. It is clear that from invariance considerations, it follows that these functions must be of the form

$$\Sigma^*(p) = (A - 2\pi i \delta m) + 2\pi B(\gamma \cdot p - m) + S_C(p)\,(\gamma \cdot p - m) \tag{71}$$

$$\Pi^*(k) = 2\pi C k^2 + k^2 D_C(k^2) \tag{72}$$

$$\Lambda_\mu(p_1, p_2) = L\gamma_\mu + \Lambda_{\mu C}(p_1, p_2) \tag{73}$$

where

$$S_C(p) = 0 \quad \text{if } \gamma \cdot p = m$$

$$\Lambda_{\mu C}(p, p) = 0 \quad \text{if } \gamma \cdot p = m \tag{74}$$

and

$$D_C(k^2) = 0 \quad \text{if } k^2 = 0 \tag{75}$$

and A, B, C, and L are divergent constants.

However, Eqs. (71), (72), and (73) are only the final form for these expressions. We must first show, given a reducible graph, how the separation of divergent factors can be effected so as to leave a unique covariant and convergent result.

16d. The Separation of Divergences from Reducible Graphs

In the last section, we considered Dyson's invariant method of separation of divergences for the irreducible graphs. To show the possibility of renormalization of the theory as a whole, we must consider graphs of all possible types and of all orders, reducible as well as irreducible ones. We must therefore show how to separate the divergences for the reducible diagrams.

Now the graphs G, obtained from an irreducible graph G_0 by making insertions, V, W, W' in the various lines and vertices of G_0, may be broadly divided into three classes:

1. Graphs with completely nonoverlapping insertions;
2. Graphs in which any two insertions are such that one is completely contained within the other; and
3. Graphs with overlapping insertions.

For the first two classes the removal of divergences can be accomplished by means of Dyson's method of invariant separation, as explained in the previous section. For nonoverlapping insertions, the separation can be made independently for each insertion. For the second type of insertion, we may *successively* remove the divergences by the above method, starting from the *innermost*, and ending with the divergence arising from G itself. For example, in the diagram of Figure 16.27 the divergence corresponding

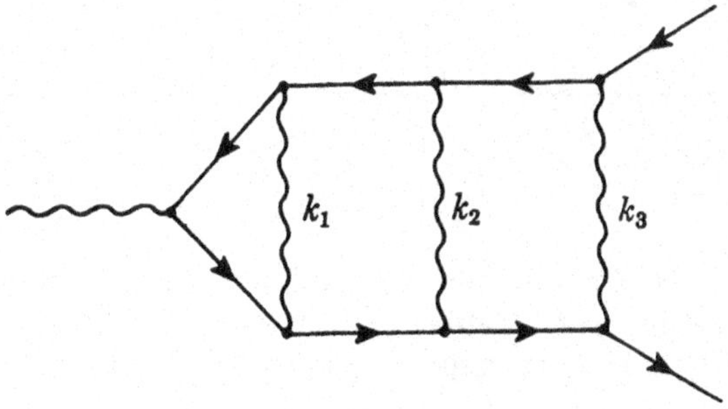

Fig. 16.27

to the integration over k_1 is first isolated, then the one arising from the integration over k_2, and finally that arising from k_3. Similarly, the combination of the first two types of divergences presents no difficulty. Thus, in the diagram of Figure 16.28, one first eliminates the divergence in the

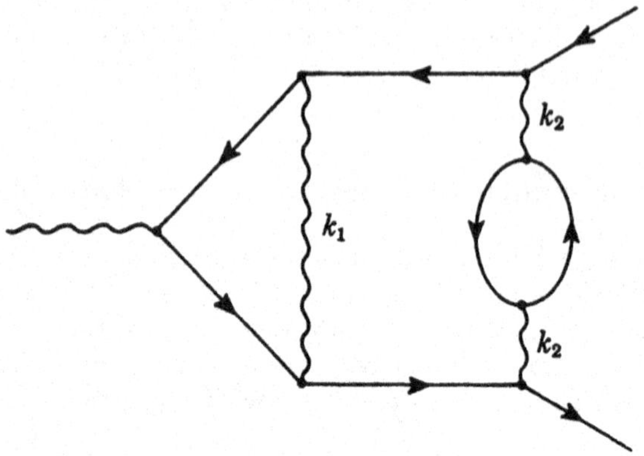

Fig. 16.28

k_1 integration, then that due to the photon self-energy part in the line k_2, and finally one removes the divergence corresponding to the subsequent k_2 integration.

For overlapping insertions, however, the Dyson method is not sufficient, since in such cases we encounter simultaneous integrals over more than one variable which diverge. They are divergent even if the integration over any one of the variables is performed holding the other variables fixed [see, e.g., Eq. (76a) below]. This is in contrast to the graphs falling under 1 and 2 for which one has to deal with only one integration variable at a time, so in that case Eq. (47) does separate out the divergences.

An illustration of such an overlapping divergence is given in Figure 16.29.

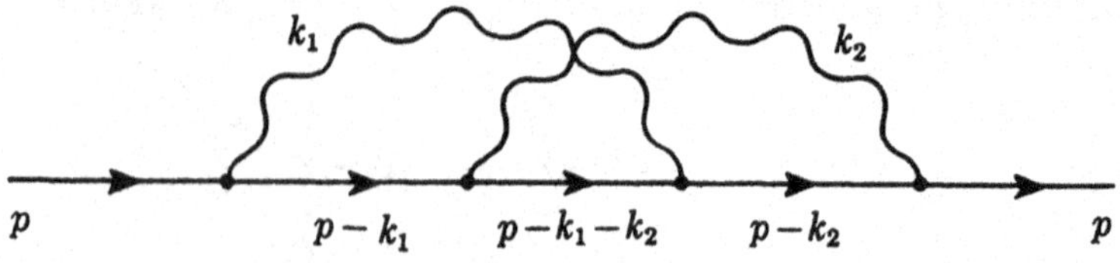

Fig. 16.29

It represents an electron self-energy graph which can be obtained by the insertion of a vertex part at either the vertex a or b of Figure 16.30. The diagram of Figure 16.29 corresponds to a matrix element

$$\Sigma(W_2, p) = e^4 \int d^4k_1 \int d^4k_2\, F^\mu(p, k)\, G_{\nu\mu}(p, k_1, k_2)\, H^\nu(p, k) \quad (76a)$$

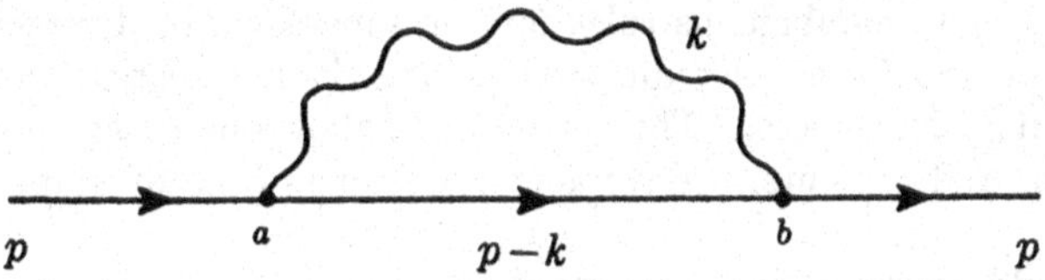

Fig. 16.30

where

$$F^{\mu}(p, k_1) = \frac{1}{k_1^2} \gamma^{\mu} \frac{\gamma \cdot (p - k_1) + m}{(p - k_1)^2 - m^2} \tag{76b}$$

$$G^{\nu\mu}(p, k_1, k_2) = \gamma^{\nu} \frac{\gamma \cdot (p - k_1 - k_2) + m}{(p - k_1 - k_2)^2 - m^2} \gamma^{\mu} \tag{76c}$$

$$H^{\nu}(p, k_2) = \frac{1}{k_2^2} \frac{\gamma \cdot (p - k_2) + m}{(p - k_2)^2 - m^2} \gamma^{\nu} \tag{76d}$$

Not only is the double integral over k_1k_2 linearly divergent, but it also diverges logarithmically if the integration is performed over either k_1 or k_2 while the other variable is held fixed.

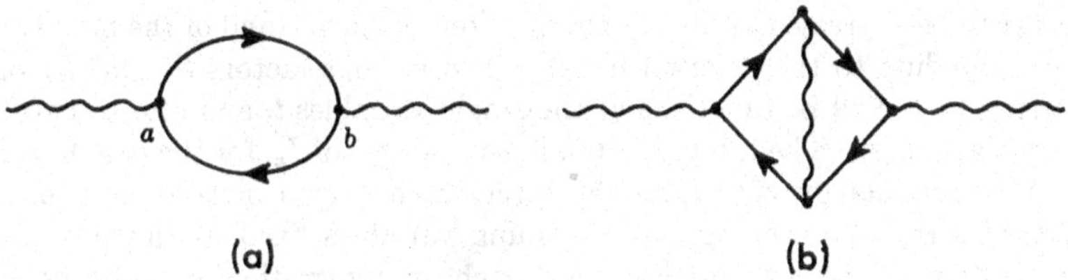

Fig. 16.31

Similarly, consider the insertion of a vertex part at a in the second-order photon self-energy graph of Figure 16.31a. This insertion then appears simultaneously as an insertion at the other vertex b (Fig. 16.31b). In general, the contribution to Π^* (or Σ^*) arising from a reducible diagram, such as Figure 16.31b, is an integral which involves divergences corresponding to each of the ways in which the diagram might have been built up by insertions of vertex parts at either or both vertices of the original irreducible diagram. Dyson has called these "b divergences."

In other words, these b divergences give rise to difficulties because it is possible to regard a reducible self-energy part as constructed from irreducible components in many different ways, according to whether vertex parts are considered to be inserted at one vertex or the other of the irreducible second-order graph, and each possible way, in fact, contributes separately to the resulting divergences.

To deal with these cases, Dyson (1949, unpublished) and Salam (1951a,

b) have given unambiguous rules for the separation of divergences from overlapping graphs, which reduce to the rules given in Section *c* for the non-overlapping divergences. These rules then allow one to separate unique, covariant, and absolutely convergent remainders from overlapping divergent parts.

It should, perhaps, be emphasized that the difficulty in obtaining the correct prescription for the separation of divergences from overlapping graphs is only concerned with the *correct* enumeration of *all* the divergent parts such that the subtractions can properly be interpreted as renormalization effects and a unique finite part be left over. This proved to be a difficult task because one had to analyze extremely complicated multiple integrals for which no adequate notation existed. (As defined by Salam, an adequate notation is one which is "both concise and intelligible to at least two persons of whom one may be the author.") Such a notation was presented for the first time by Salam (1951a) in an important paper which put the whole discussion of renormalization on a clearer and more rigorous basis. We shall now briefly describe Salam's method without going into details.

Let us consider an n-fold integral I_n arising from a graph with n "basic" momentum four-vectors t_i such that the propagation vectors of the other lines can be expressed as linear combinations of the t_is (and of the momenta corresponding to the external lines). The various factors S_F and D_F occurring in I_n will be functions of the single variables t_i and also of two or more variables. The integral (76a) is an example of I_n, for the case $n = 2$. In the terminology of Salam, the integration over a smaller number of variables than n, keeping the remaining variables fixed, is called a subintegration. The convergence over each subintegration may be determined by counting the powers of t_j (which occur in the subintegration) in the numerator and denominator of the integrand. In counting powers, each integration variable is to be counted on the same footing. This is how we arrived at the conclusion that the double integral over k_1k_2 in (76a) was linearly divergent, for there are eleven powers of k in the numerator and only ten in the denominator. For the convergence of the integral I_n as a whole, not only should the final integration over the variables $t_1 \cdots t_n$ be convergent, but also the various subintegrations over a smaller number of the basic variables, i.e., over $t_1, t_it_j, \cdots, t_it_j \cdots t_{n-1}$, these basic variables being chosen in all possible ways. It is possible that a certain integral is convergent (by counting the powers of t_i) but some of the subintegrations are not. In such a case, the integral is said to be superficially convergent.

Now to make I_n convergent, a series of divergent terms have to be subtracted from the integrand. First, those corresponding to all possible subintegrations over the variables from 1 to $(n - 1)$, and then those corresponding to the final integration over all the n variables. The rules according to which the divergent terms are subtracted are as follows:

First, hold fixed all variables in the integrand except t_1, and subtract the divergent terms corresponding to the subintegration over t_1, following the rules given by Dyson [Eq. (47)]. In these subtraction terms, leave unchanged those factors which are free from t_1 and give free-particle values to the variables $t_2, t_3, \cdots t_n$, and to the external momenta in those factors which involve t_1. Similar subtractions have to be made corresponding to subintegration over each of the other single variables $t_2, \cdots t_n$. In the new integrand thus obtained, one now holds fixed all variables except t_1 and t_2, and subtracts divergent terms corresponding to the subintegration over $t_1 t_2$, these terms being obtained from the new integrand by leaving unchanged those factors which are free from both t_1 and t_2, and giving free-particle values to the variables $t_3, t_4, \cdots t_n$ (and the external momenta) in those factors which involve t_1, t_2, or both. Similar subtractions have to be made for all other possible combinations $t_i t_j$ of the variables, taken two at a time. In the same manner one is to subtract the divergent terms corresponding to the subintegration over $t_i t_j t_k, \cdots, t_1 t_2 \cdots t_{n-1}$, remembering that at each stage, before subtracting the divergent terms corresponding to any subintegration, the integrand has to be modified by the subtraction of divergent terms corresponding to all possible subintegrations of the next lower order. The "true divergence" over a subintegration may now be defined as the divergent part to be subtracted from the "modified" integrand (as just defined), in order to make that subintegration convergent.

After having subtracted the true divergences over all the subintegrations up to the $(n-1)$th order in this same manner, a final subtraction is to be made in order to separate out the true divergence from the integration over all the n variables. Finally, if, during this subtraction procedure we find that any particular subintegration is *superficially convergent*, then the corresponding subtraction does not have to be made.

The entire procedure may be represented mathematically as follows:

$$I_n = \sum_{i=1}^{n} D(t_i)\, R(t_1, \cdots t_{i-1}, t_{i+1}, \cdots t_n)$$

$$+ \sum_{ij}^{n} D(t_i, t_j)\, R(t_1, \cdots t_{i-1}, t_{i+1}, \cdots t_{j-1}, t_{j+1}, \cdots t_n)$$

$$+ \cdots + D(t_1, \cdots, t_n) + I_C(t_1, \cdots t_n) \tag{77}$$

Here, $D(t_i, t_j \cdots)$ is the "true divergence" corresponding to the $t_i t_j \cdots$ subintegration; and the corresponding "R" multiplying it is called the "reduced integral"; $D(t_1, t_2 \cdots t_n)$ is the "true divergence" for the final integration over all the n variables. It has been shown by Salam (1951b) [see also Weinberg (1960)] that this subtraction procedure in a renormalizable field theory does, in fact, lead to an *absolutely convergent* and *unique* remainder, I_C.

Note that if I_n does not contain any overlap factor, we can write I_n as

$$I_n = \left(\int F_1(t_1)\, d^4t_1\right)\left(\int F_2(t_2)\, d^4t_2\right) \cdots \left(\int F_n(t_n)\, d^4t_n\right) \tag{78}$$

In this case, we may apply Dyson's method of separation to each variable separately, and we find that the convergent part of (78) is

$$I_C = \prod_{r=1}^{n} \left(\int \{F_r(t_r) - D_r(t_r)\}\, d^4t_r\right) \tag{79}$$

where $D_r(t_r)$ is the term giving rise to true divergences in the t_r integration.

If we now write the analogous definitions of true divergences, D, and reduced integrals, R, in the case of no overlaps, we find from (78) and (79) that

$$\begin{aligned} I_n = &\sum_{i=1}^{n} D(t_i)\, R(t_1, \cdots t_{i-1}, t_{i+1}, \cdots t_n) \\ &- \sum_{ij}^{n} D(t_i, t_j)\, R(t_1, \cdots t_{i-1}, t_{i+1}, \cdots t_{j-1}, t_{j+1}, \cdots t_n) \\ &+ (-1)^{n-1} D(t_1, t_2, \cdots t_n) + I_C(t_1, t_2, \cdots t_n) \end{aligned} \tag{80}$$

Thus, unlike (77), the terms in (80) are alternatively of positive and negative sign.

One can now apply Eqs. (77) and (80) to construct the Eqs. (71), (72), and (73). Then if S'_{F1}, D'_{F1}, and $\Gamma_{\mu 1}$ are the finite, physically significant parts of S'_F, D'_F, and Γ_μ obtained from the latter by *dropping* the divergent parts in Σ^*, Π^*, and Γ_μ, one can show that, after a suitable mass renormalization, these finite and corresponding infinite functions are related as follows:

$$\begin{aligned} S'_F(e) &= Z_2 S'_{F1}(e_1) \\ D'_F(e) &= Z_3 D'_{F1}(e_1) \\ \Gamma_\mu(e) &= Z_1^{-1}\Gamma_{\mu 1}(e_1) \end{aligned} \tag{81}$$

where Z_1, Z_2, Z_3 are divergent constants and $e_1 = Z_1^{-1}Z_2Z_3^{1/2}e$ is the renormalized charge.

We shall not here reproduce these constructions. They are contained in Salam's papers (1951a, b) which include a proof of the renormalizability of meson theories with nonderivative couplings [see also Takeda (1952)]. Instead, we shall follow the modification of the subtraction procedure due to Ward (1951b) which avoids the construction of reducible self-energy parts. It makes use of a formal identity, derived by him [Ward (1950b)], which we present in the next section.

It should be pointed out in concluding this section that, when considering reducible self-energy parts, the Salam rules presented above are necessary if the subtractions are to be correctly interpreted as multiplicative Z factors. Only if these (unique) rules for the separation of the divergences are adopted can one prove the renormalizability of the theory along the lines indicated by Dyson (1949b). However, as far as practical applications are concerned, the correct finite parts are obtained by the following more "naïve" approach which is illustrated for the overlap diagram in Figure 16.29.

This diagram may be viewed as arising from an insertion of a vertex part $\Lambda_\mu(p - k, p)$ at the vertex b (or a) in Figure 16.30. Now this operator may be separated into a convergent and a divergent part, according to (70), as follows:

$$\Lambda_\mu(p - k, p) = L\gamma_\mu + \Lambda_{\mu C}(p - k, p) \tag{82}$$

The divergent factor $L\gamma_\mu$ may be absorbed as a charge renormalization of the diagram of Figure 16.30, when the matrix elements corresponding to Figures 16.30 and 16.29 are combined. Alternatively, we may drop the divergent terms with the tacit understanding that one can always look upon these divergent contributions as adding a constant multiple to a lower order graph which redefines the charge (coupling constant) multiplying the matrix element. If we now insert $\Lambda_{\mu C}(p - k, p)$ for γ_μ at the vertex b (or a) of Figure 16.30, then the *finite* part of Σ corresponding to Figure 16.29, calculated by subtracting divergent parts by the rule of Eq. (57), is the same as the finite part obtained by Salam's rules. However, the divergent parts (if these are kept) are not the same in the two procedures. In fact, if this "naïve" procedure were adopted, one would not be able to prove the renormalizability of the theory, since these divergences could not be properly interpreted as Z factors which renormalize the charge e the *same* way in every matrix element.

16e. The Ward Identity

If we differentiate the identity $S_F(p)\, S_F^{-1}(p) = 1$ with respect to p_μ, we obtain, after rearrangement,

$$\frac{\partial S_F(p)}{\partial p^\mu} = -S_F(p)\, \frac{\partial S_F^{-1}(p)}{\partial p^\mu}\, S_F(p) \tag{83a}$$

Upon explicitly inserting the expression $2\pi(\gamma \cdot p - m)$ for $S_F^{-1}(p)$, Eq. (83a) becomes

$$\frac{\partial S_F(p)}{\partial p^\mu} = -2\pi S_F(p)\, \gamma_\mu S_F(p) \tag{83b}$$

The right-hand side of Eq. (83b) has the correct factors to correspond to the vertex diagram of Figure 16.32, in which the momentum of the photon line is zero. Formally, therefore, differentiating the propagation function $S_F(p)$ with respect to p^μ corresponds to the insertion of a zero momentum photon line into the electron line.

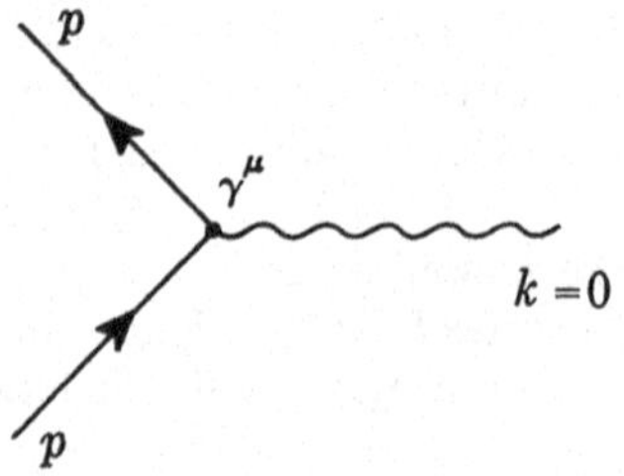

Fig. 16.32

Consider next the self-energy diagram W of Figure 16.23a. It corresponds to an operator

$$\Sigma(p, W) = -\frac{ie^2}{2\pi}\int d^4k\, \gamma^\nu S_F(p-k)\,\gamma_\nu D_F(k) \tag{84}$$

If we differentiate $\Sigma(p, W)$ with respect to p_μ, we obtain, using (83a),

$$\frac{\partial\Sigma(p, W)}{\partial p^\mu} = ie^2\int d^4k\, \gamma^\nu S_F(p-k)\,\gamma_\mu S(p-k)\,\gamma_\nu D_F(k) \tag{85}$$

But this, apart from a factor (-2π), is precisely the operator $\Lambda_\mu(V, p, p)$ corresponding to the lowest order vertex diagram indicated in Figure 16.21b, in which the photon momentum k is equal to zero.[10]

Now it is evident that a proper vertex diagram $\Lambda(V)$ can be obtained from a proper self-energy diagram $\Sigma(W, p)$ by the insertion of an external photon line into any of the internal electron lines in W. However, as we have seen, the insertion of a zero momentum photon line into an electron line formally corresponds to differentiating the propagator of this line with respect to p^μ, the external momentum it carries. Thus, differentiation of $\Sigma(W, p)$ with respect to p_μ inserts a photon line into *every*[11] electron line

[10] The numerical factor, $-2\alpha i$, in Eq. (84) is obtained by recalling that $\Sigma(p, W)$, according to Eq. (30), is defined as $i/(2\pi)^3$ times the matrix element corresponding to W. Similarly, note that $\Lambda_\mu(p_1, p_2, V)$ is defined as that operator which replaces γ_μ in the simple vertex diagram of Fig. 16.21a corresponding to the operator:

$$\left(-\frac{i}{\hbar c}\right)(2\pi)^4\,\delta^{(4)}(p_1 - p_2 \pm k)\,\frac{i}{(2\pi)^3}S_F(p_1)\,\gamma_\mu\,\frac{i}{(2\pi)^3}S_F(p_2)\,\frac{i}{(2\pi)^3}D_F(k)$$

so that the new vertex operator corresponds to V. Hence, the Λ_μ corresponding to Fig. 16.21b is given by

$$\Lambda_\mu(p_1, p_2, V) = -2\alpha i\int d^4k\gamma_\nu S_F(p_1 - k)\,\gamma_\mu S_F(p_2 - k)\,\gamma^\nu D_F(k)$$

[11] A comment concerning proper self-energy parts which contain closed loops. Consider, for example, the diagram of Fig. 16.33. In that diagram the momenta p_2 and p_3 can be chosen to be independent of p. However, one can always add a momentum p to these momenta without changing the value of the integrand since p_2 and p_3 are variables of integration. If this is done, differentiating with respect to p will also introduce a zero photon momentum photon line into the closed loop so that these proper vertex diagrams are also automatically included. Actually, they will give zero con-

of W which carries the momentum p. Therefore, if we sum over all proper self-energy diagrams, and then differentiate with respect to p, we obtain *all* proper vertex diagrams, whence

$$-\frac{1}{2\pi}\frac{\partial\Sigma^*(p)}{\partial p^\mu} = \Lambda_\mu(p, p) \tag{86}$$

This relation between Σ^* and Λ_μ was first derived by Ward (1950).

The identity (86) enables one to obtain the following relation between S'_F and Γ: From (34) it follows that

$$\begin{aligned}(S'_F(p))^{-1} &= S_F^{-1}(p) - \Sigma^*(p)\\ &= 2\pi\left(\gamma\cdot p - m - \frac{1}{2\pi}\Sigma^*(p)\right)\end{aligned} \tag{87}$$

hence,

$$\begin{aligned}\frac{1}{2\pi}\frac{\partial}{\partial p^\mu}(S'_F(p))^{-1} &= \frac{1}{2\pi}\frac{\partial}{\partial p^\mu}[S_F^{-1}(p) - \Sigma^*(p)]\\ &= \gamma_\mu + \Lambda_\mu(p, p) = \Gamma_\mu(p, p)\end{aligned} \tag{88}$$

This is an alternate form of the Ward identity.

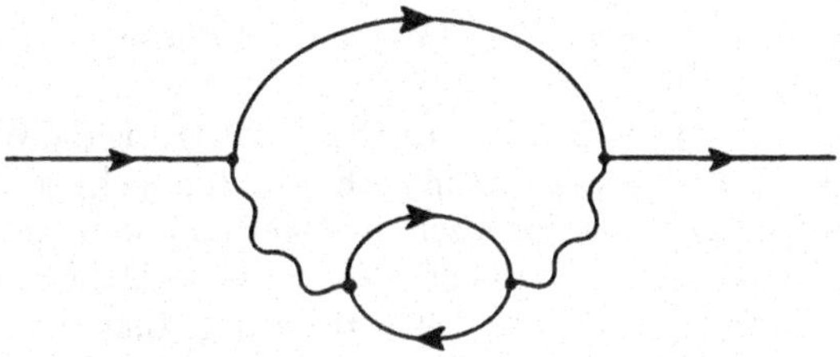

Fig. 16.33

A slightly different derivation, which indicates the connection of Ward's identity with the requirement of gauge invariance, uses the fact that a constant external electromagnetic *potential* transfers zero momentum. Consider the proper self-energy operator $\Sigma^*(p)$ (Fig. 16.34), in the presence

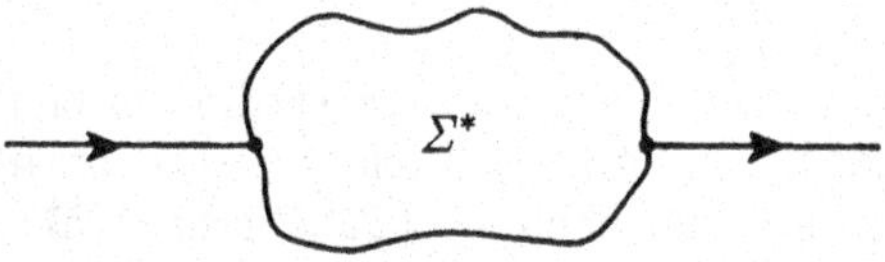

Fig. 16.34

tribution by Furry's theorem since the closed loop will now contain an odd number of photon lines and there exists a complementary diagram in which the electron lines in the closed loop have the opposite sense. Alternatively, in the case of a closed loop we may take as the variable of integration, t_μ, the momentum common to all the electron lines of the loop. The insertion of a zero momentum photon line in such a loop is then obtained by differentiating the integrand with respect to t_μ. Carrying out the integration of such an integrand over the integration variable t_μ will clearly give zero [Ward (1950a)].

of a constant potential, a_μ, and denote it by $\Sigma_a^*(p)$ [p is unchanged!]. Now expand $\Sigma_a^*(p)$ in a series according to the number of times the external potential acts. Clearly, one has

$$\Sigma_a^*(p) = \Sigma(p) + 2\pi e a^\mu \Lambda_\mu(p, p) + \frac{(2\pi e)^2}{2!} a^\mu a^\nu \Xi_{\mu\nu}(p, p) + \cdots \tag{89}$$

However, due to the requirements of gauge invariance, $\Sigma_a^*(p)$ must also be equal to $\Sigma^*(p - ea)$ so that

$$\begin{aligned}\Sigma_a^*(p) &= \Sigma^*(p - ea) \\ &= \Sigma^*(p) - ea^\mu \left(\frac{\partial \Sigma^*(p)}{\partial p^\mu}\right)_{a=0} + \frac{1}{2!} e^2 a^\mu a^\nu \left(\frac{\partial \Sigma^*(p)}{\partial p^\mu \partial p^\nu}\right)_{a=0} + \cdots\end{aligned} \tag{90}$$

hence, comparing (89) and (90), we obtain

$$-\frac{1}{2\pi}\frac{\partial \Sigma^*(p)}{\partial p^\mu} = \Lambda_\mu(p, p) \tag{91}$$

$$\frac{1}{(2\pi)^2}\frac{\partial^2 \Sigma^*(p)}{\partial p^\mu \partial p^\nu} = \Xi_{\mu\nu}(p, p) \tag{92}$$

The operator $\Xi_{\mu\nu}$ is a part of the Compton scattering operator for zero energy photons.

The connection between these two methods of deriving the Ward identity is made evident by the remark that in both cases it is a charge carrying line which is altered either by having a zero momentum photon inserted into it or having its momentum p replaced by $p - ea$ where e is the charge that the line carries. That it is possible to follow the charge that enters and leaves the diagram in continuous and connected fashion through the diagram is, of course, a consequence of charge conservation. It also guarantees that in differentiating $\Sigma^*(p)$ with respect to p_μ, a zero momentum photon line is inserted in every possible way such that the resulting diagram corresponds to a proper vertex part. Conversely, charge conservation may be viewed as a consequence of the invariance of the theory under the gauge transformation $\psi \to \exp iea_\mu x^\mu \psi$ under which $p\psi \to (p - ea)\,\psi$, and from which Eq. (90) follows.

These observations actually yield a generalization of the above Ward identity and of Eqs. (91)–(92), which relate any Feynman diagram to one containing one less external photon. [Takahashi (1957); see also Kazes (1959).] This generalized Ward's identity reads

$$2\pi\,(p'_\mu - p_\mu)\,\Gamma^\mu(p', p) = S'_F{}^{-1}(p') - S'_F{}^{-1}(p) \tag{93}$$

We shall not have occasion to use this generalized Ward's identity in the following.

We conclude this section with the observation that an immediate consequence of the Ward identity is that in the expressions (71) and (73)

$$B + L = 0 \tag{94}$$

thus proving the cancellation of vertex divergence with that of the wave function renormalization to all orders in quantum electrodynamics.

16f. Proof of Renormalizability

The proof of the renormalizability of quantum electrodynamics which we now present is due to Ward (1951b). It bypasses the problem of the overlap divergences by reducing the problem of the electron proper self-energy diagrams to that of proper vertex parts, and that of the photon proper self-energy diagrams to a certain other class of diagrams in which overlapping divergences do not occur.

As a simple illustration of the method in the case of electron self-energy diagrams, consider the self-energy operator $\Sigma^*(p, W_2)$ corresponding to the diagram of Figure 16.35. As the analysis in the last section indicated,

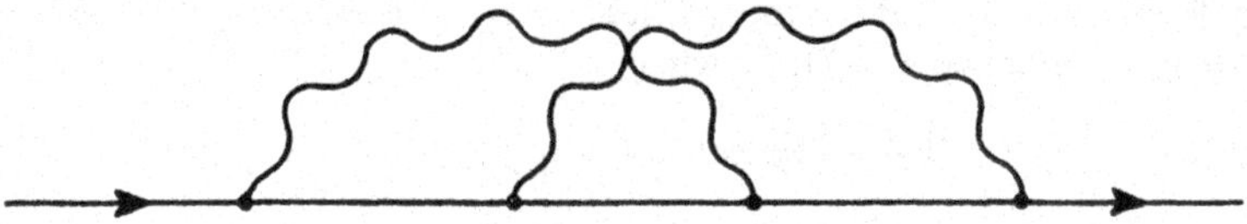

Fig. 16.35

$\Sigma^*(p, W_2)$ contains overlap divergences. Upon performing the operation $\partial/\partial p_\mu(\Sigma^*(p, W_2))$, we obtain three terms corresponding to the three diagrams indicated in Figure 16.36 arising from the insertion of a zero momentum external photon line into the three internal electron lines of diagram

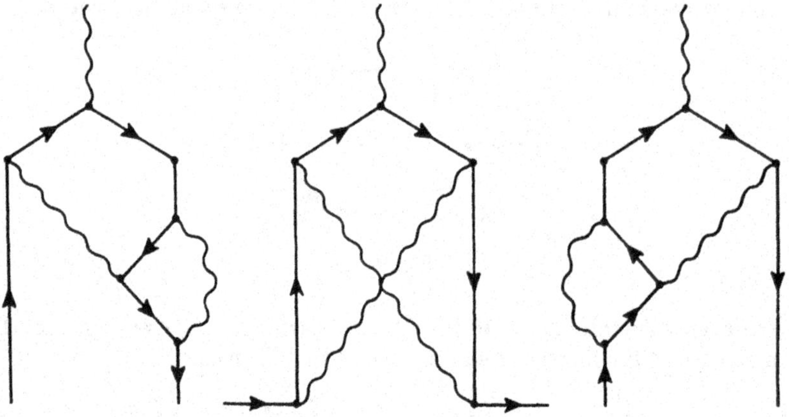

(a) (b) (c)

Fig. 16.36

16.35. We shall denote the operators corresponding to these three diagrams by $\Lambda(p_1, p_2, V^{(2)}{}_i)$, $i = 1, 2, 3$. Note that the introduction of the external photon line first of all reduces the degree of divergence (the resulting diagrams 16.36 are logarithmically divergent as compared to the linear divergence of diagram 16.35) and secondly, and more importantly for our purposes, the divergent parts of these proper vertex diagrams can be isolated without ambiguities. Let us therefore assume that these divergences have been isolated from $\Lambda_\mu(p, p, V^{(2)}{}_i)$ according to the rules outlined above. Then since

$$-\frac{1}{2\pi}\frac{\partial\Sigma^*(p, W_2)}{\partial p_\mu} = \sum_{i=1}^{3} \Lambda_\mu(p, p, V^{(2)}{}_i) = \Lambda_\mu(p, p, V_2) \tag{95}$$

$\Sigma^*(p, W_2)$ can be recovered from $\Lambda_\mu(p, p, V_2)$ by integrating Eq. (95)

$$\Sigma^*(p, W_2) - \Sigma^*(p', W_2) = \int_{p'}^{p} \Lambda_\mu(p'', p'', V_2)\, dp''^\mu \tag{96}$$

To determine the limits of integration, we recall that $\Sigma^*(p, W_2)$ for $\not p = m$, $p^2 = m^2$ is equal to a $\delta m(W_2)$, whence

$$\Sigma^*(p, W_2) - 2\pi i\delta m(W_2) = \int_{p'}^{p} \Lambda_\mu(p'', p'', V_2)\, dp''^\mu \tag{97a}$$

$$\gamma \cdot p' = m, \qquad p'^2 = m^2 \tag{97b}$$

Thus, by separating the divergences in $\Lambda(p'', p''; V_2)$ (which corresponds to a *proper* vertex part), an expression for $\Sigma^*(p, W_2)$ is obtained in which the divergences have been isolated and classified without having to concern ourselves with the overlap problem.

A similar procedure can be devised for the overlap divergences arising in photon self-energy parts. Thus consider, for example, the proper self-energy diagram illustrated in Figure 16.37, corresponding to a photon

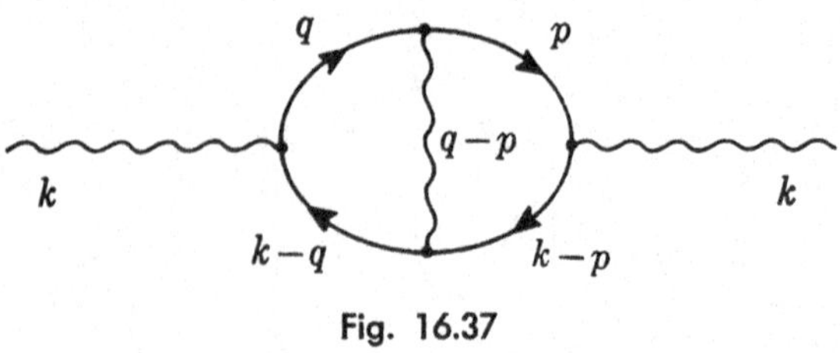

Fig. 16.37

self-energy operator $\Pi^*(k, W'_2)$. Note, in particular, the assignment of the momenta of the internal electron lines in that diagram. Since

$$\frac{\partial}{\partial k_\mu}\frac{1}{\gamma \cdot (k - q) - m} = -\frac{1}{\gamma \cdot (k - q) - m}\gamma^\mu\frac{1}{\gamma \cdot (k - q) - m} \tag{98}$$

the operator $\Delta_\mu(k, k; W'_2)$ resulting from the differentiation of $\Pi^*(k, W'_2)$ with respect to k_μ can be obtained from the diagrams 16.38a and b which

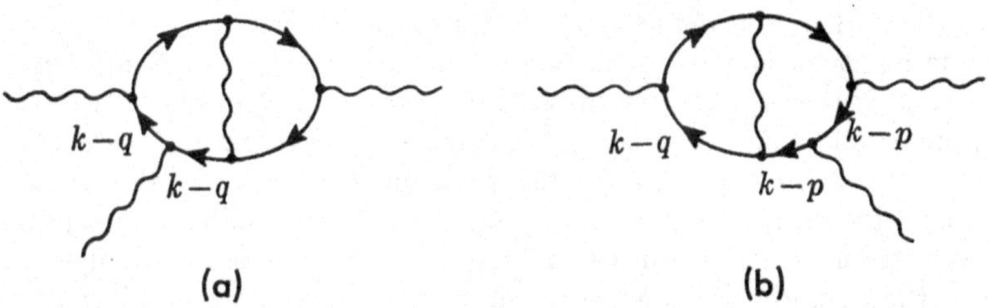

Fig. 16.38

arise from the insertion of an external zero momentum photon line into those electron lines of diagram 16.37 which carry the external momentum.

Again the divergence of the resulting diagrams is reduced as compared to that of diagram 16.37. The diagrams corresponding to $\Delta_\mu(k, k; W'_2)$ are, in fact, only logarithmically divergent (linear divergences do not arise due to relativistic covariance). Moreover, the divergences can be unambiguously isolated. Once the divergences have been separated Π^* can again be recovered by an integration over a finite interval. In higher orders the situation is somewhat more complicated since there exist photon self-energy diagrams where some of the internal photons must carry the external photon momentum (see, for example, Fig. 16.39). In that case,

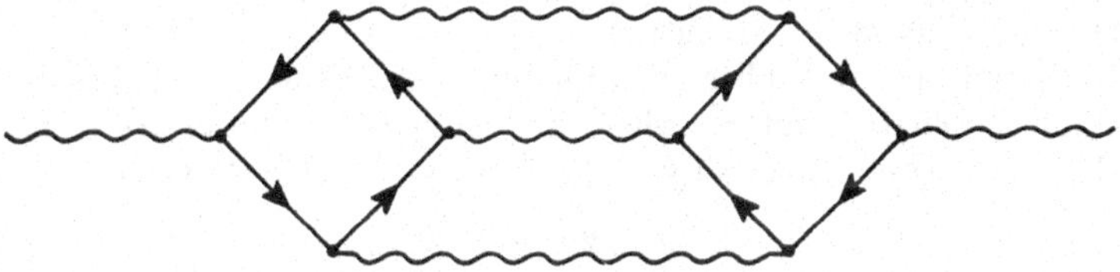

Fig. 16.39

differentiating with respect to $\partial/\partial k_\mu$ will not give rise to a factor γ_μ, but rather to a factor k_μ. It will also raise the power of the denominator, thus again rendering the integration more convergent. In any case, the differentiation $\partial/\partial k_\mu$ always removes the overlap divergences, since by inserting a zero momentum photon line into a part which had an overlapping divergence, the part which has the photon line inserted is no longer divergent: only one of the two overlapping divergences remain.

Let us see how the method works in general. For the consideration of the electron self-energy parts, we integrate Eq. (86) to obtain

$$\Sigma^*(p) - \Sigma^*(p') = -2\pi \int_{p'}^{p} dq^\mu \, \Lambda_\mu(q, q) \tag{99a}$$

$$= -2\pi \int_0^1 dx \, (p^\mu - p'^\mu) \, \Lambda_\mu(p^x, p^x) \tag{99b}$$

The second form, Eq. (99b), is obtained after a change of variable

$$q = p^x = px + (1 - x)\, p'$$

is made. Here p' is defined to be a free-electron momentum, that is, after the integration on x, p'^2 is to be set equal to m^2, and $\gamma \cdot p'$ equal to m, where m is the *experimental* mass of the electron. The term $\Sigma^*(p')$ then automatically takes into account the mass renormalization, since $\Sigma^*(p')$ for $\gamma \cdot p' = m$, $p'^2 = m^2$, is by definition equal to δm.[12] In the following we shall not carry the term $\Sigma^*(p')$ along, thus assuming that the effect of the δm term is included in the definition of $\Sigma^*(p)$. As indicated by Eq. (46), the operator Λ_μ is correctly obtained if in each irreducible vertex diagram we replace each γ_μ by Γ_μ, each S_F by S'_F, and each D_F by D'_F, and then sum the resulting series. There are therefore no overlaps in Λ_μ, since it is built up from irreducible diagrams, so that the finite parts can be separated out unambiguously.

Similarly, we shall define the operator Δ_μ by the equation

$$-\frac{1}{2\pi}\frac{\partial \Pi^*(k)}{\partial k^\mu} = \Delta_\mu(k, k) \tag{100}$$

so that

$$\begin{aligned}\Pi^*(k) &= -2\pi \int_0^k dq^\mu \, \Delta_\mu(q, q) \\ &= -2\pi \int_0^1 dy \, k^\mu \Delta_\mu(ky, ky)\end{aligned} \tag{101}$$

where the photon self-energy $\Pi^*(0)$ has been set equal to zero by the requirements of gauge invariance.

It is important to realize that the above definitions by differentiation are really implicit definitions when the functions Π^* and Δ_μ are calculated by the use of irreducible graphs. In fact, if we define the function

$$W_\mu(k) = 2k_\mu + \Delta_\mu(k, k) \tag{102}$$

which is then related to D'_F by the equation

$$W_\mu(k) = \frac{1}{2\pi}\frac{\partial}{\partial k^\mu}[D'_F(k)]^{-1} \tag{103}$$

then

$$[D'_F(k)]^{-1} = 2\pi \int_0^1 dy \, k^\mu W_\mu(ky) \tag{104}$$

Equation (104) indicates that this divergent function W_μ will occur in the integral equations, defining D'_F, Π^*, etc.

In terms of Feynman diagrams, W_μ is the contribution of all the irreducible diagrams obtained from photon self-energy diagrams by the insertion of an external photon line into an electron line carrying the external photon variable (Fig. 16.39). Again, the replacements $\gamma_\mu \to \Gamma_\mu$, $D_F \to D'_F$ and $S_F \to S'_F$ are to be made in reading off the matrix element

[12] This is sufficient, since p' can only appear in the covariant combination $\gamma \cdot p'$ or $p' \cdot p' = p'^2$.

from the original Feynman diagrams. The first two terms, corresponding to $2k_\mu$ and the simplest bubble diagram, are given by

$$W_\mu(k) = 2k_\mu - \tfrac{2}{3}\alpha \int \mathrm{Tr}\,\{\Gamma^\nu(p, p + k)\, S'_F(p + k)$$
$$\cdot\, \Gamma_\mu(p + k, p + k)\, S'_F(p + k)\, \Gamma_\nu(p + k, p)\, S'_F(p)\}\, d^4p + \cdots \quad (105)$$

There will also be higher order terms. The next one corresponds to a bubble term with two crossed photon lines inside, etc. (see Fig. 16.40b). Thus, both Λ_μ and W_μ are defined sums of contributions of irreducible diagrams which are, at worst, logarithmically divergent.

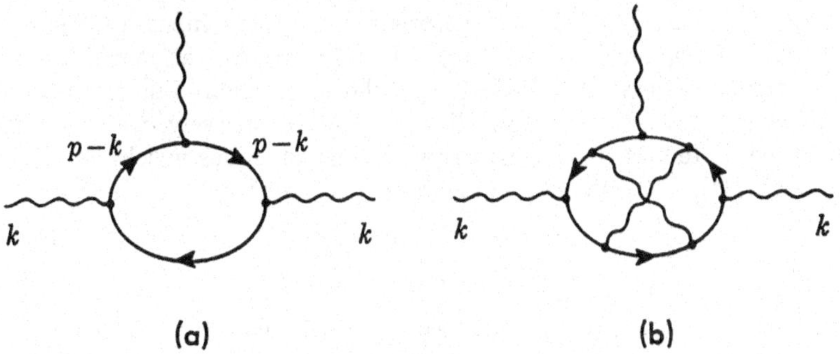

Fig. 16.40

It is now possible to define a scheme for constructing, by repeated approximations, the finite parts of the functions Γ_μ, S'_F, and D'_F which are to be inserted in the skeleton graphs in order to calculate the sum of all the contributions to the matrix element.

In order to do so, recall that Γ_μ is logarithmically divergent, so that the operator $\Gamma_{\mu 1}$, defined by

$$\Gamma_{\mu 1}(p_1, p_2) = \gamma_\mu + \Lambda_\mu(p_1, p_2) - \Lambda_\mu(p_1', p_1') \quad (106)$$

where p_1' is again a free-electron four-momentum (i.e., $\gamma \cdot p_1' = m$, $p_1'^2 = m^2$), will be finite. Similarly, recall that Δ_μ is linearly divergent, since Π^* was quadratically divergent. [Actually, covariance will reduce this divergence to a logarithmic one, since there is no invariant four-vector, i.e., $\Delta_\mu(0, 0)$ of (107) below is zero.] Therefore, the quantity

$$\Delta_\mu(yk, yk) - \Delta_\mu(0, 0) - (yk)_\nu \left(\frac{\partial \Delta_\mu(ky, ky)}{\partial (yk)_\nu}\right)_{k=0} \quad (107)$$

is convergent. Let us now, with Ward, introduce a subtraction procedure which makes the kernels in Eqs. (34)–(39) finite. The solutions of these new equations will then be finite functions S'_{F1}, D'_{F1}, $\Gamma_{\mu 1}$. Thus, we write

$$S'_{F1}(p) = S_F(p) - 2\pi S_F(p) \int_0^1 dx\, (p^\mu - p'^\mu)\, \{\Lambda_\mu(p^x, p^x) - \Lambda_\mu(p', p')\}\, S'_{F1}(p) \quad (108)$$

$$D'_{F1}(k) = D_F(k) - 2\pi D_F(k) \int_0^1 dy\, k^\mu \cdot \left\{ \Delta_\mu(ky, ky) - \Delta_\mu(0, 0) - (yk_\nu) \left(\frac{\partial \Delta_\mu(ky, ky)}{\partial (ky)_\nu} \right)_{k=0} \right\} D'_{F1}(k) \quad (109)$$

$$\Gamma_{\mu 1}(p_1, p_2) = \gamma_\mu + \Lambda_\mu(p_1, p_2) - \Lambda_\mu(p_1', p_1') \quad (110)$$

The operators S'_{F1}, D'_{F1}, $\Gamma_{\mu 1}$, defined by these integral equations, are now finite. In these modified equations, let us everywhere replace the coupling constant, e, by another, e_1. We shall then show that if e_1 is a properly chosen function of e and m, the modified functions $S'_{F1}(p, e_1)$, $D_{F1}(k, e_1)$, $\Gamma_{\mu 1}(p_1, p_2, e_1)$ are multiples of the original divergent functions $S'_{F1}(p, e)$, $D'_F(k, e)$, $\Gamma_\mu(p_1, p_2; e)$, respectively. In other words, the modification of the original integral Eqs. (34)–(39) by the subtraction of infinite terms is equivalent to a charge renormalization. More precisely, we shall show that the finite and the corresponding infinite functions which are defined by Eqs. (34), (37), and (39) are related as follows:

$$S'_F(p; e) = Z_2(e_1)\, S'_{F1}(p; e_1) \quad (111)$$

$$D'_F(k; e) = Z_3(e_1)\, D'_{F1}(k; e_1) \quad (112)$$

$$\Gamma_\mu(p_1, p_2; e) = Z_1^{-1}(e_1)\, \Gamma_{\mu 1}(p_1, p_2; e_1) \quad (113)$$

where Z_1, Z_2, Z_3 are (infinite) constants to be determined, and

$$e_1 = Z_1^{-1} Z_2 Z_3^{1/2} e \quad (114)$$

These equations, first written down by Dyson (1949b), thus interpret the *subtraction* of infinite constants as the *extraction* of infinite constant multipliers.

The proof of the renormalizability of quantum electrodynamics thus consists of two steps. Step 1 is the specification of a subtraction (or regularization) procedure which yields finite results for an otherwise divergent integral. Step 2 consists in showing that this subtraction or regularization procedure acts as a renormalization, that is, that the suppressed divergent terms amount only to a modification of the parameters (such as, for example, m_0, e) in terms of which the original unrenormalized theory was written.

That Eqs. (111), (112), (113) are in fact true can be seen by substituting the expressions (111)–(113) for S'_F, D'_F, and Γ_μ into the original set of integral equations which defined these quantities, namely,

$$S'_F = S_F + S_F \Sigma^* S'_F \quad (115)$$

$$D'_F = D_F + D_F \Pi^* D'_F \quad (116)$$

$$\Gamma_\mu = \gamma_\mu + \sum_{\text{all irreducible } V^i} \Lambda_\mu(V^i) \quad (117)$$

where Σ^* and Π^* are given by (44) and (45), respectively. Now, in the expression (46) for Γ_μ, each term with coefficient $(e^2)^n$ contains exactly

$n\,D'_F$ functions, $2n\,S'_F$ functions, and $(2n+1)\,\Gamma$ functions. Hence, if we replace in these equations e by e_1 as given by (114) and D'_F, S'_F, and Γ_μ by their finite counterparts to which they are related by (111), (112), and (113), we shall obtain

$$\Lambda_\mu(e, \Gamma_\mu, S'_F, D'_F) = Z_1^{-1}\Lambda_\mu(e_1; \Gamma_{\mu 1}, S'_{F1}, D'_{F1}) \tag{118}$$

Similarly, in the expression for Δ_μ, Eq. (105), each term with coefficient $(e^2)^n$ contains $(2n+1)\,\Gamma$ and S'_F functions and $(n-1)\,D_F$ functions, so that if again we replace e by e_1, S'_F by S'_{F1}, D'_F by D'_{F1}, and Γ_μ by $\Gamma_{\mu 1}$,

$$\Delta_\mu(e, \Gamma_\mu, S'_F, D'_F) = Z_1^{-1}Z_3^{-1}Z_2\Delta_\mu(e_1, \Gamma_{\mu 1}, S'_{F1}, D'_{F1}) \tag{119a}$$

or

$$W_\mu(e, \Gamma_\mu, S'_F, D'_F) = Z_1^{-1}Z_2Z_3^{-1}W_\mu(e_1, \Gamma_{\mu 1}, S'_{F1}, D'_{F1}) \tag{119b}$$

Therefore, using (99) and (101), we obtain

$$\Sigma^*(p; e_1) = -2\pi Z_1^{-1} \int_0^1 dx\,(p^\mu - p'^\mu)\,\Lambda_\mu(p^x, p^x; e_1, \Gamma_1, S'_{F1}, D'_{F1}) \tag{120}$$

and

$$\Pi^*(k; e_1) = -2\pi Z_1^{-1}Z_2Z_3^{-1} \int_0^1 dy\; k^\mu\Delta_\mu(yk, yk; e_1, \Gamma_1, S'_{F1}, D'_{F1}) \tag{121}$$

The integral equations (115), (116), and (117), when expressed in terms of (120) and (121), become

$$Z_2S'_{F1}(e_1) = S_F - 2\pi S_F \cdot Z_1^{-1}Z_2 \int_0^1 dx\,(p^\mu - p'^\mu)\,\Lambda_\mu(p^x, p^x; e_1, \Gamma_1, S'_{F1}, D'_{F1}) \cdot S'_{F1}(e_1) \tag{122}$$

$$Z_3D'_{F1}(e_1) = D_F - 2\pi D_F Z_1^{-1}Z_2 \int_0^1 dy\; k^\mu\Delta_\mu(yk, yk; e_1, \Gamma_1, S'_{F1}, D'_{F1}) \cdot D'_{F1}(e_1) \tag{123}$$

$$Z_1^{-1}\Gamma_{\mu 1}(e_1) = \gamma_\mu + Z_1^{-1}\Lambda_\mu(e_1, \Gamma_1, S'_{F1}, D'_{F1}) \tag{124}$$

These equations have to be identical to the integral equations (108), (109), and (110) for the finite operators S'_{F1}, D'_{F1}, and $\Gamma_{\mu 1}$. The two sets of equations will, in fact, be the same if in (110) and (124)

$$Z_1(e_1)\,\gamma_\mu = \gamma_\mu - \Lambda_\mu(p', p'; e_1) \tag{125a}$$

and therefore, comparing with Eqs. (73) and (94),

$$Z_1(e_1) = 1 + B(e_1) = 1 - L(e_1) \tag{125b}$$

If we now substitute this result into (108), we find that (122) and (108) will be identical if

$$Z_1 = Z_2 \tag{126}$$

and, finally, (109) and (123) are the same if

$$(Z_3 - 1)\,k^2 = ik^\mu k^\nu \int_0^1 dy\; y \left(\frac{\partial\Delta_\mu(ky, ky)}{\partial(ky)^\nu}\right)_{k=0} \tag{127}$$

or, therefore, if

$$Z_3 = 1 + k^{-2}k^\mu k^\nu C_{\mu\nu} \tag{128a}$$

where

$$C_{\mu\nu} = i\int_0^1 dy\, y \left(\frac{\partial\Delta_\mu(ky, ky)}{\partial(ky)^\nu}\right)_{k=0} \tag{128b}$$

In obtaining (127), we have made use of the fact that covariance requires that $\Delta_\mu(0, 0) = 0$ in (109). Finally, covariance again requires that $C_{\mu\nu} = Cg_{\mu\nu}$ so that

$$Z_3 = 1 + C \tag{129}$$

with $Cg_{\mu\nu}$ given by (128b).

We have therefore shown that the relations (111), (112), and (113) are true if e_1 is given by (114) and if the Z factors are determined by Eqs. (125), (126) and (128).

To obtain a further characterization of the renormalization procedure, we recall that the self-energy operator after mass renormalization can be written in the form

$$\Sigma^*(p) = 2\pi B(\gamma\cdot p - m) + S_C(p)\,(\gamma\cdot p - m) \tag{130}$$

[see Eq. (71)], with $S_C(p)$ vanishing for $\gamma\cdot p = m$. Equations (130) and (87) allow us to infer that $S'_F(p)$ can be written in the form

$$\begin{aligned} S'_F(p) &= \frac{1}{S_F^{-1}(p) + \Sigma^*(p)} \\ &= \frac{1}{2\pi}\frac{1+B}{\gamma\cdot p - m} + (\text{terms which are regular at } \gamma\cdot p = m) \\ &= \frac{1}{2\pi}\frac{Z_2}{\gamma\cdot p - m} + (\text{terms which are regular at } \gamma\cdot p = m) \end{aligned} \tag{131}$$

so that

$$\lim_{\gamma\cdot p\to m} 2\pi(\gamma\cdot p - m)\, S'_{F1}(p; e_1) = 1 \tag{132}$$

Similarly one verifies that

$$\Gamma_{\mu 1}(p_1, p_2)\Big|_{\gamma\cdot p_1 = \gamma\cdot p_2 = m} = \gamma_\mu \tag{133}$$

and that

$$\lim_{k^2\to 0} 2\pi k^2 D'_{F1}(k) = 1 \tag{134}$$

These equations state that there are no observable radiative corrections to the motion of a free electron or photon to all orders in the coupling constant. In fact, they may be considered as the definition of mass and charge renormalization, requiring that a free (physical) electron have a definite mass: m (the experimental mass of the electron) and a definite charge: e_1 (the experimental electronic charge), and that a free photon travel unperturbed through free space with zero rest mass.

We next turn to Eq. (35). Recalling that p in that equation corresponds to a free-electron momentum, and the fact that for $\gamma \cdot p = m$

$$\Sigma^*(p) = 2\pi(Z_2 - 1)\,(\gamma \cdot p - m) \tag{135}$$

we can rewrite Eq. (35) as follows:

$$\psi'(p) = \psi(p) + 2\pi(Z_2 - 1)\, S_F(p)\,(\gamma \cdot p - m)\,\psi(p) \tag{136}$$

The expression (136) is indeterminate, since $\gamma \cdot p - m$ operating on $\psi(p)$ gives zero, while operating on $S_F(p)$ it gives the constant $1/2\pi$. Thus, according to the order in which the factors are evaluated, (136) gives for $\psi'(p)$ either $\psi(p)$ or $Z_2\psi(p)$. We have seen in our discussion of Section 15c that this ambiguity is removed if greater care is taken in evaluating the S matrix such that its unitarity is guaranteed at all stages of the calculations. In fact, we there indicated that the correct wave function renormalization is given by

$$\psi'(p) = Z_2^{1/2}\psi(p) \tag{137}$$

[see also Karplus and Kroll (Karplus 1950)]. Similar considerations indicate that

$$\tilde{\psi}'(p) = Z_2^{1/2}\tilde{\psi}(p) \tag{138}$$

and that

$$A'_\mu(k) = Z_3^{1/2}A_\mu(k) \tag{139}$$

Consider now an irreducible scattering matrix element M. Let the structure of the graph of M be such that there appears in it F_e external electron lines, B_e external photon lines, and n vertices. Then, according to the analysis of the present section, M will contain n factors Γ_μ, F_e factors ψ or $\tilde{\psi}$, B_e factors A, F_i factors S'_F corresponding to the F_i internal electron lines, and B_i factors D'_F for the internal photon lines. Recall that the number of internal and external lines are related by

$$n = \tfrac{1}{2}F_e + F_i = B_e + 2B_i \tag{140}$$

Hence, if in M we replace Γ_μ, S'_F, and D'_F by their finite counterparts $\Gamma_{\mu 1}$, S'_{F1}, and D'_{F1} to which they are related by (111), (112), and (113), then the Z factors will occur in M only in the combination

$$(Z_1 Z_2^{-1} Z_3^{1/2})^n$$

This multiplier is now exactly correct to convert the factor e^n, multiplying M into the factor e_1^n. Thereby, both e and the Z factors disappear from M, leaving only the finite operators $\Gamma_{\mu 1}(e_1)$, $D'_{F1}(e_1)$, $S'_{F1}(e_1)$, and e_1. If now e_1 is identified with the finite observed electronic charge, there no longer appear any divergent expressions in M. Since M is completely general, the complete elimination of all ultraviolet divergences from the S matrix has been accomplished.[13]

[13] The infrared divergences and their cancellation within the renormalization program are discussed in Jauch (1954) and Yennie (1961).

We conclude this section with the remark that Eq. (126), which states that $Z_1 = Z_2$, indicates again that the wave function renormalization divergence is exactly canceled by the vertex part divergence to all orders. It is a direct consequence of Ward's identity and hence of gauge invariance. Therefore, the only true divergence in quantum electrodynamics is connected with the photon self-energy parts, i.e., with the vacuum polarization phenomenon, and is embodied in the divergent constant Z_3.

16g. The Meaning of Charge Renormalization

In order to clarify the meaning of charge renormalization, let us consider the Compton scattering of a very low frequency photon by a free electron [Thirring (1950)]. There are two classes of irreducible graphs which will contribute. The first is exhausted by the two diagrams in Figure 16.41.

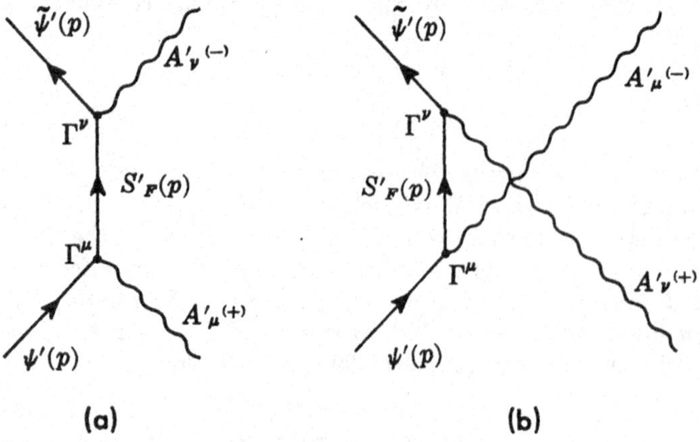

Fig. 16.41

The second class of diagrams is illustrated by Figure 16.42. This second class of diagrams can be obtained by the insertion into Σ^* of two external photon lines, as indicated in Figure 16.43. If we consider the limiting situation of zero frequency photons, then Figure 16.41 gives rise to the following operator (which includes all radiative corrections):

$$M_a = (2\pi)^2 e^2 \tilde{\psi}'(p) \{\Gamma^\mu(p, p) S'_F(p) \Gamma^\nu(p, p) + \Gamma^\nu(p, p) S'_F(p) \Gamma^\mu(p, p)\} \psi'(p) A'_\mu{}^{(+)} A'_\nu{}^{(-)} \quad (141)$$

where p is the electron momentum and $A'^{(+)}$ $A'^{(-)}$ are the creation and annihilation operators for the zero frequency photons including radiative corrections. Recalling our discussion of Section e [especially Eq. (92)],

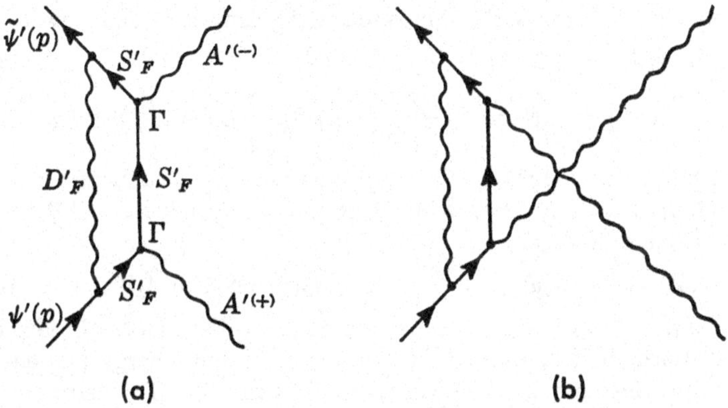

Fig. 16.42

the diagram of Figure 16.43 corresponds to the following matrix element:

$$M_b = e^2\,\tilde{\psi}'(p)\,\frac{\partial^2\Sigma^*(p)}{\partial p_\mu \partial p_\nu}\,\psi'(p)\,A'_\mu{}^{(+)}A'_\nu{}^{(-)} \tag{142}$$

Using the identity

$$\frac{\partial S'_F(p)}{\partial p_\mu} = -S'_F(p)\,\frac{\partial S'_F{}^{-1}(p)}{\partial p_\mu}\,S'_F(p) \tag{143}$$

and Eq. (88), one readily establishes that

$$\begin{aligned}\frac{\partial^2 S'_F(p)}{\partial p_\mu \partial p_\nu} = {} & (2\pi)^2\,\{S'_F(p)\,\Gamma^\nu(p,p)\,S'_F(p)\,\Gamma^\mu(p,p)\,S'_F(p) \\ & + S'_F(p)\,\Gamma^\mu(p,p)\,S'_F(p)\,\Gamma^\nu(p,p)\,S'_F(p)\} \\ & + S'_F(p)\,\frac{\partial^2\Sigma^*(p)}{\partial p_\mu \partial p_\nu}\,S'_F(p)\end{aligned} \tag{144}$$

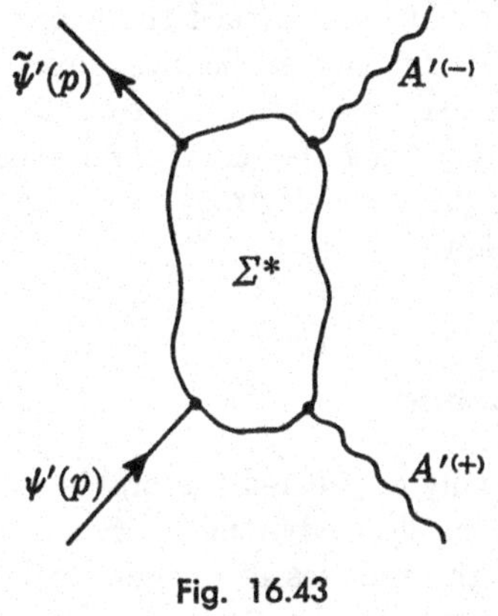

Fig. 16.43

Hence, we may rewrite the complete matrix element M as follows:

$$M = M_a + M_b$$
$$= e^2\,\bar{\psi}'(p)\left\{S'_F{}^{-1}(p)\,\frac{\partial^2 S'_F(p)}{\partial p_\mu \partial p_\nu}\,S'_F{}^{-1}(p)\right\}\psi'(p)\,A'_\mu{}^{(+)}A'_\nu{}^{(-)} \quad (145)$$

Upon introducing (131), (137), (138), and (139), and carrying out the indicated operations within the braces of (145), one finds that all the factors Z_2 cancel and M becomes

$$M = Z_3\,e^2\,\bar{\psi}(p)\,\{\gamma^\mu S_F(p)\,\gamma^\nu + \gamma^\nu S_F(p)\,\gamma^\mu\}\,\psi(p)\,A_\mu{}^{(+)}A_\nu{}^{(-)} \quad (146)$$

Setting $Z_3\,e^2 = e_1{}^2$, M is then exactly equal to the renormalized second-order (Born matrix element for scattering for zero energy photons. In this limit, therefore, all radiative corrections vanish. The matrix element M gives rise to the well-known Thomson formula,

$$\sigma = \frac{8\pi}{3}\left(\frac{e_1{}^2}{4\pi mc^2}\right) \quad (147)$$

[for a derivation, see Sec. 14e and L. Brown (1952) where radiative corrections are also considered].

We therefore see that charge renormalization can be regarded as being experimentally defined such that the low frequency Compton scattering is given by the Thomson formula. This defines the experimental charge e_1 of the electron.

Physically, these statements can be made because Thomson scattering is a purely classical effect and depends only on the total charge. In this situation, the photon wavelength is so long that no details of the charge distribution of the scatterer are relevant. Since, by virtue of the charge conservation law, the radiative corrections cannot change the total charge, we can in fact use the Thomson formula to define the experimental charge e_1 of the electron.

Alternatively, we could have defined the charge from the requirement that the Coulomb law for two static electrons hold at large distances. This definition would lead to the same value of e_1. In fact, it can be shown [Källén (1953b, 1954)] that the charge renormalization is essentially uniquely defined in quantum electrodynamics as a consequence of the charge conservation law.

16*h*. General Remarks

We have thus far only considered the problem of the renormalizability of the spinor quantum electrodynamics. Salam (1952) has proved the renormalizability of the theories of charged scalar or pseudoscalar spin zero bosons interacting with the electromagnetic field. In this case, in addition to the renormalization of mass and charge, a counterterm

$\lambda : (\phi^*\phi)^2 :$ (λ infinite) has to be added to the Lagrangian, in order to cancel consistently all divergences arising from the Møller scattering of one spinless particle by another. The proof of the renormalizability, i.e., of equations like (111)–(114), is in this case far more difficult than in the spinor electrodynamics because of the more complicated overlaps. Ward (1951b) has extended the method presented in Section 16*f* to cover this case. However, just as in spinor electrodynamics, Salam's method is perhaps more transparent and gives a clearer physical picture of the method of renormalization than the simpler but more formal and implicit approach of Ward.

The proof of the renormalizability of the field theories of mesons interacting with nucleons with *non*derivative coupling has also been given by Salam (1951a, b). For the case of pseudoscalar mesons, again, an infinite counterterm, $\lambda' : (\phi^*\phi)^2 :$ for charged mesons and $\lambda'' : \phi^4 :$ for neutral mesons, has to be added to the Hamiltonian. For scalar mesons, as indicated in Section 16*a*, in addition to the ϕ^4 terms, a term $\lambda''' : \phi^3 :$ has to be added in order to cancel the divergences arising from parts with three external meson lines. It has been found that one can renormalize the combined interactions of mesons, nucleons, photons, and electron-positron fields without introducing any essentially new complications [Matthews (1950, 1951a, b), Salam (1950), Ward (1951b)].

The subtraction methods we have presented in this section were based on the use of the ordinary interaction representation and considered only the removal of divergences from the *S* matrix. There exist alternative methods for removing these divergences. For example, one can introduce a further counterterm (like the mass renormalization term) in the Lagrangian, so as to cancel out the effects of charge renormalization. One can then work directly in terms of the renormalized charge (like the renormalized mass) from the very beginning. Such a formulation has been given by Gupta (1951). (The Gupta formalism actually describes charge renormalization as a renormalization of the unit in which the electromagnetic field is measured, rather than as a renormalization of the coupling constant.) Takeda (1952) has used a variant of Salam's method to derive the required counterterms. An excellent exposition of these methods is given in an article by Matthews and Salam [Matthews (1954b)], which likewise treats the problems of meson interactions in a clear and simple manner. All these methods are equivalent in the sense that they are related to one another by (unbounded) "unitary" transformations.

To illustrate this approach, consider the case of a neutral pseudoscalar meson field interacting with a fermion field. The Lagrangian for the theory is

$$\begin{aligned}\mathfrak{L} = &-\tfrac{1}{2} : \mu^2\phi^2(x) + \partial_\nu\phi\partial^\nu\phi(x) : - \tfrac{1}{4}[\bar{\psi}(x), (-i\gamma\cdot\partial + M)\,\psi(x)] \\ &- \tfrac{1}{4}[i\partial_\mu\bar{\psi}(x)\,\gamma^\mu + M\bar{\psi}(x), \psi(x)] + \tfrac{1}{2}\delta M[\bar{\psi}(x), \psi(x)] \\ &- \tfrac{1}{2}G[\bar{\psi}(x)\,\gamma_5, \psi(x)]\,\phi(x) - \tfrac{1}{2}\delta\mu^2 : \phi^2(x) : - \delta\lambda : \phi^4(x) : \qquad (148)\end{aligned}$$

where $\delta\mu^2$, δM are the mass renormalization counterterms and $\delta\lambda : \phi^4(x) :$ the meson-meson divergence counterterm. We then define the "renormalized" operators ϕ_R and ψ_R and the renormalized coupling constant G_R as follows:

$$\phi(x) = Z_3^{1/2}\phi_R(x) \tag{149}$$

$$\psi(x) = Z_2^{1/2}\psi_R(x) \tag{150}$$

$$G = Z_2^{-1}Z_3^{-1/2}Z_1G_R \tag{151}$$

We shall write the renormalization constant Z_1, Z_2, Z_3 in the form

$$Z_1 = 1 - L \tag{152}$$

$$Z_2 = 1 + B \tag{153}$$

$$Z_3 = 1 + C \tag{154}$$

where L, B, and C are divergent constants [analogous to the quantities $L(e_1)$, $B(e_1)$, and $C(e_1)$] encountered in electrodynamics. (In the present case, $L \neq B$ since there is no Ward's identity.)

The Lagrangian (148), when re-expressed in terms of these renormalized quantities, can be written in the form

$$\mathfrak{L} = \mathfrak{L}_{0R} + \mathfrak{L}_{IR} \tag{155a}$$

with

$$\begin{aligned}\mathfrak{L}_0 = &-\tfrac{1}{2} : \mu^2\phi_R^2(x) + \partial_\nu\phi_R\partial^\nu\phi_R(x) : - \tfrac{1}{4}[\bar{\psi}_R(x), (-i\gamma\cdot\partial + M)\,\psi_R(x)] \\ &- \tfrac{1}{4}[i\partial_\mu\bar{\psi}_R(x)\,\gamma^\mu + M\bar{\psi}_R(x), \psi_R(x)] \end{aligned}\tag{155b}$$

$$\begin{aligned}\mathfrak{L}_I = &-\tfrac{1}{2}C : (\mu^2\phi_R + \partial_\nu\phi_R\partial^\nu\phi_R) : - \tfrac{1}{4}B[\psi_R(x), (-i\gamma\cdot\partial + M)\,\psi_R(x)] \\ &- \tfrac{1}{4}B[i\gamma_\mu\bar{\psi}_R(x)\,\gamma^\mu + M\bar{\psi}_R(x), \psi_R(x)] - \tfrac{1}{2}\delta\mu^2Z_3 : \phi_R^2(x) : \\ &- \tfrac{1}{2}\delta MZ_2[\bar{\psi}_R(x), \psi_R(x)] - \delta\lambda Z_3^2 : \phi_R^4(x) : \\ &- Z_1G_R\tfrac{1}{2}[\bar{\psi}_R(x)\,\gamma_5, \psi_R(x)]\,\phi_R(x) \end{aligned}\tag{155c}$$

The Lagrangian $\mathfrak{L}_0$ defines a new "renormalized" interaction picture wherein the renormalized operators ϕ_R and ψ_R satisfy free-field equations of motion and commutation rules. The new counterterms in $\mathfrak{L}_I$ are such that they precisely cancel all the divergences arising from the interaction.[14]

Let us next briefly state the limitations of the renormalization theory presented thus far. The first and most important is that all the above considerations (e.g., the Dyson analysis of the divergences into primitive divergents) depend in an essential manner on the expansion of the S matrix in powers of the coupling constant. Furthermore it was tacitly assumed that the S-matrix series expansion was convergent after the renormaliza-

[14] For a discussion of the consequences of the fact that it is always possible to add a finite part to Z_1, Z_2, Z_3 without altering any observable effects by a suitable redefinition of the coupling constant, see Stückelberg (1953), Gell-Mann (1954), and Bogoliubov (1956 and 1959); in particular Chapter VIII of this last reference.

tions were carried out. It is known [Ferretti (1951), Nishijima (1951)] that the series after renormalization will diverge in a trivial way if there exists the possibility of the particles being captured into permanently bound states. In this situation a perturbation expansion of the scattering amplitude is known to diverge even in nonrelativistic quantum mechanics [Jost and Pais (Jost 1951)] and in a relativistic field theory the series diverges for the same reason. Ferretti (1951) has made a detailed investigation of this problem. He analyzed the unitary operator

$$V(t) = \prod_{n=0}^{\infty} \left\{1 - \frac{i}{\hbar} \int_{t_{n+1}}^{t_n} \lambda(t')\, H_I(t')\, dt'\right\} \tag{156}$$

where $t, t_1, \cdots t_n$ are sequences of times extending from $-\infty$ to t and becoming, in the limit, infinitely close to each other. The operator $V(t)$ diagonalizes the complete Hamiltonian $H_0 + \lambda H_I$. That is, $V(t)$ is an operator such that

$$E(\lambda) = V^{-1}(t)\, (H_0 + \lambda H_I)\, (t)\, V(t) \tag{157}$$

is diagonal in the representation in which H_0 is diagonal. In (156) Ferretti has explicitly introduced the adiabatic hypothesis by letting the coupling constant λ be a function of the time, $\lambda = \lambda(t)$. This function $\lambda(t)$ is supposed to increase very slowly from the value $\lambda = 0$ at $t = -\infty$ to the value $\lambda = 1$ at $t' = t$. This, as previously stated, is necessary in order to make the operator $V(t)$ well defined.

Now, if the product $V(t)$ is expanded formally into a power series in the coupling constant, the result is

$$S(t) = \sum_{n=0}^{\infty} \left(\frac{-i}{\hbar}\right)^n \int_{-\infty}^{t} dt_1 \int_{-\infty}^{t_1} dt_2 \cdots \int_{-\infty}^{t_{n-1}} dt_n\, \lambda(t_1)\, H_I(t_1) \cdots \lambda(t_n)\, H_I(t_n) \tag{158}$$

Ferretti then analyzes under what condition is $V(t)$ equal to $S(t)$. He shows quite generally that this identity is not valid when the total Hamiltonian $H_0 + \lambda H_I$ has discrete eigenvalues. This is true even when the discrete eigenvalues can be calculated by a perturbation expansion. In fact, Ferretti was able to show that the series $S(t)$ actually diverges in the case of discrete eigenvalues [see also Glaser and Zimmermann (Glaser 1952)]. Therefore, even leaving aside the question of convergence, the renormalization of the S matrix as described in this chapter is applicable, strictly speaking, only to pure scattering processes between free particles not involving bound states.

Caianiello (1956) and Yennie and Gartenhaus [Yennie (1958)] have investigated the convergence of the $U(t, t')$ matrix elements when expanded in powers of the coupling constant. The latter authors have shown that for a nonlocal theory of the form

$$H_I = G \int d^4x \int d^4x' \int d^4x''\, K(x - x', x - x'') : \bar{\psi}(x')\, \psi(x'') : \phi(x)$$

the transition amplitude for any given process over a finite time interval is bounded term by term by an exponential series. They further demonstrated that for such an interaction the elements of the U matrix considered as power series in the unrenormalized coupling constant G converges with an infinite radius (the convergence taking place because of cancellations between Feynman diagrams). However, no definite results were obtained in the limiting case that the time interval $t - t' \to \infty$. Certain similar inferences were deduced by Yennie and Gartenhaus but were based on the assumption that an adiabatic switching of the interaction is valid, so that no transitions are induced from the vacuum to other states. This assumption is, however, open to question if one believes certain conjectures of Dyson (1952).

Dyson (1952) has given arguments suggesting that the S-matrix expansion in quantum electrodynamics should only be viewed, at best, as an asymptotic expansion. His line of reasoning is as follows: Suppose one were to calculate a physical observable in a power series in the coupling constant, e^2. If this series is convergent for some positive value of e^2, it must converge in some circle of radius e^2 centered around the origin in the complex $Z = e^2$ plane. The series must therefore also converge for $e = 0$ and, moreover, must also converge on some interval of the negative real axis, i.e., for e^2 negative. Now e^2 negative corresponds to a world in which like charges would attract one another and opposite charges would repel each other.[15] However, if e^2 is negative, then a state which contains a large number, N, of electron-positron pairs in which the electrons are clustered together in a region V_1 of space and the positrons are clustered together in another region V_2 of space far from V_1, would have an energy lower than that of the vacuum for N large enough. Stated differently, if the forces between like charges are attractive and of long range then the binding energy of a large collection of particles could exceed the energy necessary to create them. The vacuum state is therefore unstable relative to such states. Since the larger the number of pairs is, the more pronounced the effect becomes, the higher order terms in the power series expansion must become more and more important so that the series cannot converge.

The series can therefore at best only be an asymptotic expansion. The divergence of the series will only become noticeable when terms of a very high order in the power series expansion are considered. Dyson estimates in quantum electrodynamics the terms of the series will decrease to a minimum and then increase again without limit, and that the index of the minimum term is of the order of magnitude $1/\alpha \sim 137$.

Arguments have been put forward suggesting that the propagators D'_F, S'_F actually have an essential singularity in the region of vanishing coupling

[15] The argument is perhaps not entirely convincing since a negative means an imaginary coupling constant in the original Lagrangian, which in turn implies that $\mathcal{L}$ is non-Hermitian and also implies that probabilities are not necessarily conserved.

constant [Redmond (1960)]. Under these circumstances, any expansion in powers of the coupling constant is clearly at best asymptotic. We shall return to these questions in Chapter 17 after obtaining closed expressions for the propagators and renormalization constants Z_1, Z_2, Z_3 in terms of Heisenberg operators.

The mathematically inclined reader undoubtedly by now will have had serious misgivings about the validity and meaningfulness of the renormalization program, since this program has as its point of departure a set of meaningless equations which it then proceeds to manipulate according to rules which are outside the bounds of conventional mathematics to obtain (presumably) finite results (not to mention the fact these prescriptions, as outlined in the present chapter, are applicable only to the power series expansion of the "meaningless equations," which power series expansion in all probability does not converge!).

Actually a generalization of the renormalization method which frees it from the restriction to power series has been given by Källén (1952) and Valatin (1954). Also important contributions have been made by Caianiello (1953, 1954a, b, 1955, 1956, 1957, 1958, 1959a, b) towards the solution of the problem of giving a more precise mathematical meaning to the manipulations involved in the renormalization program. Caianiello has stressed the fact that since one is dealing with a system of hyperbolic differential equation the integral involved in expressing the solution of these equations must be redefined. (It is known that the straightforward application of the ordinary Riemann integral to the solution of hyperbolic equations may generate divergences, which disappear if a redefinition of the concept of the integrals [such as, for example, the "partie finie" of Hadamard (1932)] is made. Caianiello concludes that, if a suitable redefinition of the integral is made, the problem of the renormalization of ultraviolet divergences can be completely bypassed.

However, even with these improved methods, no answer has as yet been given to the central problem of present-day quantum field theory: "Do solutions exist of the renormalized equations?" and if solutions exist, what are their analytic properties as functions of the renormalized coupling constant? We shall present the conjectures that have been advanced as answers to these questions in the next chapter.

Part Four

FORMAL DEVELOPMENTS

17

The Heisenberg Picture

The properties of relativistic field theories deduced in Chapters 14–16 were obtained in the interaction picture and were based on perturbation theory. In the present chapter we shall deduce certain important features of relativistic field theories making use of the Heisenberg picture for the description of the field system. In these considerations the relativistic invariance of the theory will play a key role. This explains the preferred role of the Heisenberg picture, for it is in the Heisenberg picture that the expression of the relativistic invariance of a quantum mechanical theory is most easily stated. The reason for this is principally that the Schrödinger and Dirac pictures deal only with experiments which are instantaneous in the inertial frame which they use. Such experiments are, in general, difficult to express in terms of similar experiments in different inertial systems. Furthermore it should be noted that, in order to define an instantaneous state, it must be possible to define a complete set of commuting observables which correspond to compatible experiments at a single instant of time. Since it is not clear that such a set exists for relativistic field systems, it might be the case that instantaneous states cannot be defined for such systems.

Also, up to this point, we have concerned ourselves mostly with the description of pure scattering processes between initially and finally free particles. It was to deal with these situations that the S matrix was developed. This viewpoint corresponded to the description of the results of measurements on the asymptotic behavior of the system. There exists, however, another class of observable situations which cannot be described if only the S matrix for the scattering of (initially and finally) free particles is known. Thus, for example, the knowledge of the S matrix does not determine the bound states of the system [Jost (1947b)] nor does it include the results of possible local measurements, such as the measurements of the charge-current density or field strengths in a region of space-time [Bohr and Rosenfeld (Bohr 1933, 1950)]. To describe these situations, the formalism presented thus far must be extended. We shall see that the Heisenberg picture is well suited for just these situations.

17a. Vacuum Expectation Values of Heisenberg Operators

The Heisenberg picture is defined by the fact that in this picture the state vector of the system is constant in time, whereas the operators carry the full interaction. In the case of quantum electrodynamics, the field operators[1] satisfy the following equations of motion in the Heisenberg picture:

$$(i\gamma_\mu\partial^\mu - m)\,\psi(x) = -\delta m\psi(x) - e\gamma_\mu \mathbf{A}^\mu(x)\,\psi(x) \tag{1}$$

$$\Box \mathbf{A}_\mu(x) = \mathbf{j}_\mu(x) = -\frac{e}{2}\,[\bar{\psi}(x)\,\gamma_\mu, \psi(x)] \tag{2}$$

The Heisenberg state vector of the system $|\boldsymbol{\Psi}\rangle$, however, is time-independent:

$$\partial_t\,|\,\boldsymbol{\Psi}\rangle = 0 \tag{3a}$$

In addition, the physically realizable states of the system satisfy the subsidiary condition

$$(\partial_\mu \mathbf{A}^\mu(x))^{(+)}\,|\,\boldsymbol{\Psi}\rangle = 0 \tag{3b}$$

The subsidiary condition (3b) is relativistically invariant since the operator $\partial_\mu \mathbf{A}^\mu(x)$ satisfies, by virtue of the equations of motion (1) and (2), $\Box(\partial_\mu \mathbf{A}^\mu(x)) = 0$. Hence a decomposition into only positive and negative frequencies is meaningful and invariant. The equations of motion (1) and (2) are derivable from the Lagrangian density:

$$\begin{aligned}\mathfrak{L}(x) &= -\frac{1}{2}\frac{\partial \mathbf{A}_\mu(x)}{\partial x_\nu}\frac{\partial \mathbf{A}^\mu(x)}{\partial x^\nu} - \frac{1}{4}\,[\bar{\psi}(x), (-i\gamma^\mu\partial_\mu + m)\,\psi(x)] \\ &\quad - \frac{1}{4}\,[i\partial_\mu\bar{\psi}(x)\,\gamma^\mu + m\bar{\psi}(x), \psi(x)] + \frac{e}{2}\,[\bar{\psi}(x)\,\gamma^\mu, \psi(x)]\,\mathbf{A}_\mu(x) \\ &\quad + \frac{1}{2}\,\delta m[\bar{\psi}(x), \psi(x)] \\ &= \mathfrak{L}_{EM} + \mathfrak{L}_D + \mathfrak{L}_I \end{aligned} \tag{4}$$

It is now impossible to write down the commutation rules for the Heisenberg operators which are valid for all times. This would require the knowledge of the solutions of the equations of motion (1) and (2) for all times, which is, in fact, the problem to be solved. However, from our Lagrangian we can still obtain the canonical commutation rules which are valid for the equal time operators.[2] These are

[1] Throughout this chapter boldface letters will be used to denote the Heisenberg operators and state vectors.

[2] For more modern methods for obtaining commutation rules, see Peierls (1952), Schwinger (1951c, 1953a, d), Burton (1953a, b), and Cini (1953).

$$[\psi(x), \bar{\psi}(x')]_+\Big|_{x_0 = x'_0} = \gamma^0 \delta^{(3)}(\mathbf{x} - \mathbf{x}') \tag{5a}$$

and

$$[\mathbf{A}_\mu(x), \partial'_0 \mathbf{A}_\nu(x')]_{x_0 = x'_0} = -i\hbar c g_{\mu\nu} \delta^{(3)}(\mathbf{x} - \mathbf{x}') \tag{5b}$$

with all other commutators (anticommutators) vanishing. These commutation rules can be generalized to arbitrary space-like separations [Schwinger (1948c)] and then read

$$[\psi(x), \bar{\psi}(x')]_+ = -iS(x - x') \tag{6a}$$

$$[\mathbf{A}_\mu(x), \mathbf{A}_\nu(x')] = -i\hbar c g_{\mu\nu} D(x - x') \tag{6b}$$

$$[\psi(x), \psi(x')]_+ = [\bar{\psi}(x), \bar{\psi}(x')]_+ = 0 \tag{6c}$$

$$[\psi(x), \mathbf{A}_\mu(x')] = [\bar{\psi}(x), \mathbf{A}_\mu(x')] = 0 \tag{6d}$$

$$\text{for } (x - x')^2 < 0$$

The Lagrangian also enables us to construct an energy-momentum tensor $\mathbf{T}_{\mu\nu}$ from which the total energy-momentum four-vector $\mathbf{P}_\mu$, and the angular momentum tensor $\mathbf{M}_{\mu\nu}$, of the field can be defined. The operators $\mathbf{P}_\mu$ and $\mathbf{M}_{\mu\nu}$ are the generators for infinitesimal translations and Lorentz transformations, respectively: $\mathbf{U}(a, 1) = \exp ia_\mu \mathbf{P}^\mu$, $\mathbf{U}(0, \Lambda) = \exp \frac{1}{2}\Lambda_{\mu\nu}\mathbf{M}^{\mu\nu}$. The time component $\mathbf{P}_0$ of the translation operators $\mathbf{P}_\mu$ is the Hamiltonian of the system. Under a Lorentz transformation, $\{a, \Lambda\}$, the field operators $\psi(x)$ and $\mathbf{A}_\mu(x)$, by definition, satisfy the following transformation laws:

$$\mathbf{U}(a, \Lambda)\, \mathbf{A}_\mu(x)\, \mathbf{U}(a, \Lambda)^{-1} = \sum_{\nu=0}^{3} \Lambda^{-1}{}_{\mu\nu} \mathbf{A}^\nu(\Lambda x + a) \tag{7a}$$

$$\mathbf{U}(a, \Lambda)\, \psi_\alpha(x)\, \mathbf{U}(a, \Lambda)^{-1} = \sum_{\beta=1}^{4} S^{-1}{}_{\alpha\beta}(\Lambda)\, \psi_\beta(\Lambda x + a) \tag{7b}$$

where $\Lambda \to S(\Lambda)$ is the usual up to a factor four-by-four spinor representation of the homogeneous Lorentz group, and $\mathbf{U}(a, \Lambda)$ is a unitary (or antiunitary in the case $\Lambda^0{}_0 < -1$) representation of the Lorentz group in the Hilbert space of the physically realizable states of the system. Equations (7a) and (7b) characterize the operators $\psi(x)$ and $\mathbf{A}_\mu(x)$ as a spinor field and vector field, respectively. The commutation rules satisfied by the generators for infinitesimal translation and rotations are

$$[\mathbf{P}_\mu, \mathbf{P}_\nu] = 0 \tag{8a}$$

$$[\mathbf{P}_\mu, \mathbf{M}_{\kappa\lambda}] = i(g_{\mu\kappa}\mathbf{P}_\lambda - g_{\mu\lambda}\mathbf{P}_\kappa) \tag{8b}$$

$$[\mathbf{M}_{\kappa\lambda}, \mathbf{M}_{\mu\nu}] = i(g_{\lambda\mu}\mathbf{M}_{\kappa\nu} + g_{\kappa\nu}\mathbf{M}_{\lambda\mu} - g_{\kappa\mu}\mathbf{M}_{\lambda\nu} - g_{\lambda\nu}\mathbf{M}_{\kappa\mu}) \tag{8c}$$

which are identical with the structure relations for the inhomogeneous Lorentz group, thus proving the Lorentz covariance of the theory. Equation (8a) expresses the conservation of the total energy and momentum of the field system. The operators $\mathbf{P}_\mu$ are the space-time displacement operators, and induce the following change in an operator $\mathbf{F}(x)$ under an infinitesimal translation:

$$i\delta x^\mu[\mathbf{P}_\mu, \mathbf{F}(x)] = \delta\mathbf{F}(x) \tag{9a}$$

or

$$i[\mathbf{P}_\mu, \mathbf{F}(x)] = \partial_\mu\mathbf{F}(x) \tag{9b}$$

In this last equation, $\mathbf{F}(x)$ is an arbitrary Heisenberg operator which depends on x only through the operators $\mathbf{A}_\nu(x)$ and $\psi(x)$.

Since all the operators $\mathbf{P}_\mu$ commute with one another, we can choose a representation in which every basis vector is an eigenfunction of all the $\mathbf{P}_\mu$s, with eigenvalue p_μ:

$$\mathbf{P}_\mu \mid \Psi_a\rangle = p^{(a)}{}_\mu \mid \Psi_a\rangle \tag{10}$$

Every such $|\Psi_a\rangle$ describes a stationary state and will have a definite energy and momentum ascribed to it. It should, however, be noted that the specification of an energy and a momentum does *not* uniquely characterize a state. The specification of a state entails, besides the characterization of its total energy and momentum, the specification of other quantum numbers such as, for example, the total charge. We shall denote these other eigenvalues, which together with $p^{(a)}{}_\mu$ form a complete set of observables, by α. We shall often use the notation $|\mathbf{p}, \boldsymbol{\alpha}\rangle$ to denote the states of the system: here p stands for the momentum of the state, $\mathbf{P}_\mu \mid \mathbf{p}, \boldsymbol{\alpha}\rangle = p_\mu \mid \mathbf{p}, \boldsymbol{\alpha}\rangle$ and α for the other quantum numbers necessary to specify the state.

In the momentum representation, the matrix element of the commutator of $\mathbf{P}_\mu$ and $\mathbf{F}(x)$ can, using (9b) and (10), be written in the form

$$\begin{aligned} i(\Psi_a, [\mathbf{F}(x), \mathbf{P}_\mu]\, \Psi_b) &= i(p^{(b)}{}_\mu - p^{(a)}{}_\mu)\, (\Psi_a, \mathbf{F}(x)\, \Psi_b) \\ &= -\partial_\mu(\Psi_a, \mathbf{F}(x)\, \Psi_b) \end{aligned} \tag{11}$$

whence

$$(\Psi_a, \mathbf{F}(x)\, \Psi_b) = (\Psi_a, \mathbf{F}(0)\, \Psi_b)\, e^{-i(p^{(b)} - p^{(a)})\cdot x} \tag{12}$$

In Eq. (12), $\mathbf{F}(0)$ is the operator $\mathbf{F}(x)$ evaluated at the point $x = 0$. Equation (12) gives the x dependence of the matrix element of an arbitrary Heisenberg operator in the representation in which the $\mathbf{P}_\mu$s are diagonal. We shall work in this representation throughout this section.

In order to proceed with the development, the following three physical assumptions concerning the states of the theory are usually made [Källén (1952), Wightman (1952, unpublished)]:

I. (*a*) There exists a unique vacuum state, $|\Psi_0\rangle$, which is invariant under all Lorentz transformation

$$\mathbf{U}(a, \Lambda) \mid \Psi_0\rangle = |\Psi_0\rangle \tag{13}$$

(*b*) This state is the state of lowest energy ($p^{(0)}{}_0 = 0$).

II. The states of the system are such that they contain only time-like or light-light momenta of non-negative energy; i.e., no state $|\Psi_n\rangle$

exists whose momentum four-vector $p^{(n)}{}_\mu$ satisfies $p^{(n)}{}_\mu p^{(n)\mu} < 0$ or $p^{(n)}{}_\mu p^{(n)\mu} > 0$ with $p^{(n)}{}_0 < 0$.

III. There exists a set of states $|\mathbf{p}^{(n)}, \boldsymbol{\alpha}\rangle$ with $p^{(n)}{}_\mu p^{(n)\mu} \geqslant 0$, $p^{(n)0} \geqslant 0$ which is complete within the Hilbert space of realizable (physical) states of the field system.

The following remarks can be made concerning assumption I*a*. If one considers the manifold $\mathfrak{M}_0$ of states satisfying $\mathbf{U}(a, \Lambda) \mid \Psi_0\rangle = |\Psi_0\rangle$ then the multiplication law of the $\mathbf{U}(a, \Lambda)$s

$$\mathbf{U}(a, \Lambda)\, \mathbf{U}(b, N) = \mathbf{U}(\Lambda b + a, \Lambda N) \tag{14}$$

implies that $\mathfrak{M}_0$ is invariant under the transformation $\{a, \Lambda\}$. From the continuity of $\mathbf{U}(a, \Lambda)$, this manifold can be shown to be a closed linear manifold. Within this manifold $\mathfrak{M}_0$, the $\mathbf{U}(a, \Lambda)$ generate a unitary representation up to a factor of the inhomogeneous Lorentz group in which the translation operators are trivial. Since it is unitary, any such representation is either equivalent to the trivial one $\mathbf{U}(a, \Lambda) \to 1$ or is infinite dimensional (Chapter 2). Thus if any vacuum states exist at all, they are of two kinds—those invariant under Lorentz transformation and those belonging to infinite families, whose elements are transformed into each other by Lorentz transformation. Assumption I*a* states that the vacuum is that state which transforms according to the one-dimensional unitary representation of the inhomogeneous Lorentz group.

The further assumption that if there exists a state of lowest energy that it be the vacuum is self-evident. A field theory for which no state of lowest energy exists, i.e., for which the energy spectrum has no lower bound, would not be a consistent one, for under such conditions the system would undergo radiative collapse. Several examples of relativistic field theories for which a ground state does not exist are known [Fierz (1955), Baym (1960)]. The Ward theory, a scalar field with a $\lambda\phi^3$ self-interaction, is one such example. The Hamiltonian for this model is

$$\mathbf{H} = \tfrac{1}{2} \int d^3x : (\phi^0)^2 + (\phi^k)^2 + \mu^2\phi^2 + \lambda\phi^3 :$$

$$\phi^k = \partial^k\phi, \qquad \phi^0 = \partial^0\phi \tag{15}$$

The reason no ground state exists for this model is that a Bose field may be given arbitrarily high excitation so that the nonpositive-definite cubic term $\lambda\phi^3$ in $\mathbf{H}$ will, for large field excitations, in general, dominate the positive-definite quadratic terms. The Ward theory is therefore not a suitable model of an interacting field.

Concerning assumption II, the argument against the existence of states with negative energy or space-like momenta is that no one has ever seen such a state in the laboratory. With regard to the existence of such states in a theory, one may argue as follows: A necessary and sufficient condition that such states cause no conflict with our present knowledge is that

1. they be orthogonal to all physically reasonable states,
2. they should not come into existence as a result of interaction.

The first condition is satisfied in any relativistically invariant theory, but the second cannot be satisfied unless one excludes all external perturbations which cause transitions into the unphysical states. If condition 2 were satisfied, one might be able to exclude the undesired states from the theory completely. Under what circumstances 2 is satisfied is not known, so that we assume outright that every state contains only time-like momenta and positive energies.

Relativistic invariance asserts that the energy-momentum four-vector of the vacuum state is zero. Since, by assumption, it is unique and is the state of lowest energy, it follows that the eigenvalue of $\mathbf{P}_0$ for every other state is such that ${p^{(n)}}_0 > 0$ for all n.

In quantum electrodynamics, the requirement that ${p^{(0)}}_\mu = 0$ is not sufficient to characterize the vacuum state uniquely, due to the freedom of making gauge transformations. There are an infinity of vacuum states which differ only by a choice of gauge. In what follows, we shall adopt the gauge [Källén (1952)] which makes $(\boldsymbol{\Psi}_0, \eta \mathbf{A}_\mu(x)\, \boldsymbol{\Psi}_0) = 0$. It is assumed that (13) does not contradict the second postulate.

We next turn to the main concern of the present section, namely the study of the vacuum expectation values of time-ordered products of Heisenberg operators

$$G_{\alpha_1 \cdots \alpha_m \beta_1 \cdots \beta_m \mu_1 \cdots \mu_n}(x_1, x_2, \cdots x_m; y_1, \cdots y_m; z_1, \cdots z_n)$$
$$= (\boldsymbol{\Psi}_0, T(\psi_{\alpha_1}(x_1) \cdots \psi_{\alpha_m}(x_m)\, \bar{\psi}_{\beta_1}(y_1) \cdots \bar{\psi}_{\beta_m}(y_m)\, \mathbf{A}_{\mu_1}(z_1) \cdots \mathbf{A}_{\mu_n}(z_n)), \boldsymbol{\Psi}_0) \tag{16}$$

The relevance of such Green's functions in the description of the particle aspect of quantum field theory was first emphasized by Gell-Mann and Low [Gell-Mann (1951)] and by Schwinger (1951b). These authors established the connection between these Green's functions and the Feynman propagators of multiparticle systems. Schwinger has made them the basis of his formulation of quantum field theory. To elucidate their meaning, we shall first of all relate them to interaction picture variables.

We define the Heisenberg state vectors to be the same as the interaction picture ones for the time $t = 0$. If the interaction picture state vectors $|\Psi(t)\rangle$ are the solutions of

$$i\partial_t \mid \Psi(t)\rangle = H_I(t) \mid \Psi(t)\rangle \tag{17}$$

then the corresponding Heisenberg state vectors $|\boldsymbol{\Psi}\rangle$ are defined as follows:

$$U(t, 0) \mid \boldsymbol{\Psi}\rangle = |\Psi(t)\rangle \tag{18}$$

where the unitary operator $U(t, 0)$ satisfies the equation

$$i\partial_t U(t, 0) = H_I(t)\, U(t, 0) \tag{19a}$$

and the boundary condition

$$U(0, 0) = 1 \tag{19b}$$

The relation between interaction picture operators and the corresponding operators in the Heisenberg picture is obtained from the requirement that their expectation value in terms of their respective state vectors be the same. That is, by definition

$$(\boldsymbol{\Psi}, \mathbf{F}(x)\, \boldsymbol{\Psi}) = (\Psi(t), F(x)\, \Psi(t)) \tag{20}$$

where the time t of the state vector $|\Psi(t)\rangle$ is the same as the time x_0 of the operator $F(x)$. Hence, using (18), we obtain

$$\begin{aligned}(\boldsymbol{\Psi}, \mathbf{F}(x)\, \boldsymbol{\Psi}) &= (U(t, 0)\, \boldsymbol{\Psi}, F(x)\, U(t, 0)\, \boldsymbol{\Psi}) \\ &= (\boldsymbol{\Psi}, U^{-1}(t, 0)\, F(x)\, U(t, 0)\, \boldsymbol{\Psi})\end{aligned} \tag{21a}$$

or

$$\mathbf{F}(x) = U^{-1}(t, 0)\, F(x)\, U(t, 0) \tag{21b}$$

where $t = x_0$.

An explicit representation of the operator U can be obtained as follows: Recall that we may transform from the Schrödinger picture, in which the state vector $|\Psi_S(t)\rangle$ satisfies

$$i\partial_t \mid \Psi_S(t)\rangle = (H_{0S} + H_{IS}) \mid \Psi_S(t)\rangle \tag{22}$$

(with H_{0S} and H_{IS} time-independent operators), to the interaction picture by the unitary transformation

$$|\Psi(t)\rangle = e^{iH_0 t} \mid \Psi_S(t)\rangle \tag{23}$$

The two pictures coincide for the time $t = 0$. Now, we can also transform directly from the Schrödinger to the Heisenberg picture, remembering that the Hamiltonian $H = H_{0S} + H_{IS} = \mathbf{H}$ is time-independent, by the unitary transformation

$$|\boldsymbol{\Psi}\rangle = e^{iHt} \mid \Psi_S(t)\rangle \tag{24}$$

One readily establishes that this $|\boldsymbol{\Psi}\rangle$ is indeed time-independent: $\partial_t \mid \boldsymbol{\Psi}\rangle = 0$. Upon introducing into (24) $|\Psi_S(t)\rangle$ as given by (23), we obtain

$$|\boldsymbol{\Psi}\rangle = e^{iHt}\, e^{-iH_0 t} \mid \Psi(t)\rangle \tag{25}$$

Comparing Eqs. (18) and (25), we obtain the following representation for the operator $U(t, 0)$:

$$U(t, 0) = e^{iH_0 t}\, e^{-iHt} \tag{26}$$

It will prove useful in the following to be able to express the "physical" (or "true") vacuum $|\boldsymbol{\Psi}_0\rangle$ (the eigenstate of H with eigenvalue 0) in terms of the vacuum of the bare quanta $|\Phi_0\rangle$ ($H_0 \mid \Phi_0\rangle = 0$). The desired relation has been derived previously in Chapter 11 and is given by

$$\lambda \mid \Psi_0(0)\rangle = \lambda \mid \boldsymbol{\Psi}_0\rangle = \frac{U(0, \pm\infty) \mid \Phi_0\rangle}{(\Phi_0, U(0, \pm\infty)\, \Phi_0)} \tag{27}$$

In Eq. (27), λ is a normalization constant. It should be noted that the (infinite) phase factors (corresponding to the disconnected vacuum-vacuum diagrams) which are present in both numerator and denominator in the right-hand side of Eq. (27) cancel out in the quotient.

With these formal preliminaries out of the way, let us consider next the vacuum expectation value of the time-ordered product of two Heisenberg operators. For the sake of simplicity, let us first consider the vacuum expectation value of two boson operators in a meson theory for which $H_I(t) = G \int d\sigma(x)\, N(\bar{\psi}(x)\, \gamma\psi(x))\, \phi(x)$ with $\gamma = 1$ or γ_5. Let us denote this matrix element by

$$R(x_1, x_2) = (\Psi_0, T(\phi(x_1)\, \phi(x_2))\, \Psi_0) \tag{28}$$

and consider first the following time ordering: $x_{10} > x_{20}$. If we express the Heisenberg operators and state vector on the right-hand side of (28) in terms of interaction picture variables using (18), (21), and (27), we obtain the following expression for R when $x_{10} > x_{20}$:

$$R(x_1, x_2) = (U(0, +\infty)\, \Phi_0,\, U^{-1}(t_1, 0)\, \phi(x_1)\, U(t_1, 0) \cdot U^{-1}(t_2, 0)\, \phi(x_2)\, U(t_2, 0)\, U(0, -\infty)\, \Phi_0) \tag{29}$$

We have omitted the denominator of Eq. (27), which then implies that if we calculate (29) in terms of Feynman diagrams, we are to omit all disconnected closed-loop diagrams. Now the U operator is unitary and obeys the group property, so that

$$U(t_1, t_3)\, U(t_3, t_2) = U(t_1, t_2) \tag{30}$$

and

$$U^{-1}(t_1, t_2) = U^*(t_1, t_2) = U(t_2, t_1) \tag{31}$$

Hence,

$$U^{-1}(t_1, t_2)\, U(t_1, t_3) = U(t_2, t_3) \tag{32}$$

where $U(t, t')$ satisfies the equation

$$i\partial_t U(t, t') = H_I(t)\, U(t, t') \tag{33a}$$

and the boundary condition

$$U(t, t) = 1 \tag{33b}$$

Recall, incidentally, that the solution of Eq. (33) can be expressed in the form

$$U(t, t') = \sum_{n=0}^{\infty} (-i)^n \frac{1}{n!} \int_{t'}^{t} dt_1 \cdots \int_{t'}^{t} dt_n\, T(H_I(t_1) \cdots H_I(t_n)) \tag{34}$$

We may therefore rewrite R as follows:

$$R(x_1, x_2) = (\Phi_0, U^{-1}(0, +\infty)\, U^{-1}(t_1, 0)\, \phi(x_1)\, U(t_1, t_2)\, \phi(x_2)\, U(t_2, -\infty)\, \Phi_0) \tag{35a}$$

$$= (\Phi_0, U(\infty, t_1)\, \phi(x_1)\, U(t_1, t_2)\, \phi(x_2)\, U(t_2, -\infty)\, \Phi_0) \tag{35b}$$

$$= (\Phi_0, T(U(+\infty, t_1)\, \phi(x_1)\, U(t_1, t_2)\, \phi(x_2)\, U(t_2, -\infty))\, \Phi_0) \tag{35c}$$

$$= (\Phi_0, T(U(+\infty, -\infty)\, \phi(x_1)\, \phi(x_2))\, \Phi_0) \tag{35d}$$

The insertion of the chronological operator in Eq. (35b) is evident. Wick's theorem then permits us to rearrange factors so as to yield (35d). The operator factor in the latter equation can explicitly be written as

$$T(U(+\infty, -\infty)\, \phi(x_1)\, \phi(x_2))$$

$$= \sum_{n=0}^{\infty} \frac{(-i)^n}{n!} \int_{-\infty}^{+\infty} d^4y_1 \cdots \int_{-\infty}^{+\infty} d^4y_n\, T(\mathcal{H}_I(y_1) \cdots \mathcal{H}_I(y_n)\, \phi(x_1)\, \phi(x_2))_C \tag{36}$$

where the subscript C denotes that only connected diagrams are to be considered. One easily verifies that Eq. (36) is actually valid for arbitrary times x_{10} and x_{20}. $R(x_1, x_2)$ can therefore be evaluated in a power series expansion using Wick's theorem. The first few terms of the series are given by

$$R(x_1, x_2) = \tfrac{1}{2}\Delta_F(x_1 - x_2) + \tfrac{1}{16}G^2 \int d^4y_1 \int d^4y_2\, \Delta_F(x_1 - y_1)$$
$$\mathrm{Tr}\, (\gamma S_F(y_1 - y_2)\, \gamma S_F(y_2 - y_1))\, \Delta_F(y_2 - x_2) + \cdots \tag{37}$$

It is clear that in terms of Feynman diagrams, $R(x_1, x_2)$ corresponds to a simple boson line and all the possible self-energy insertions into this line

$$\tfrac{1}{2}\Delta'_F(x_1, x_2) = R(x_1, x_2)$$

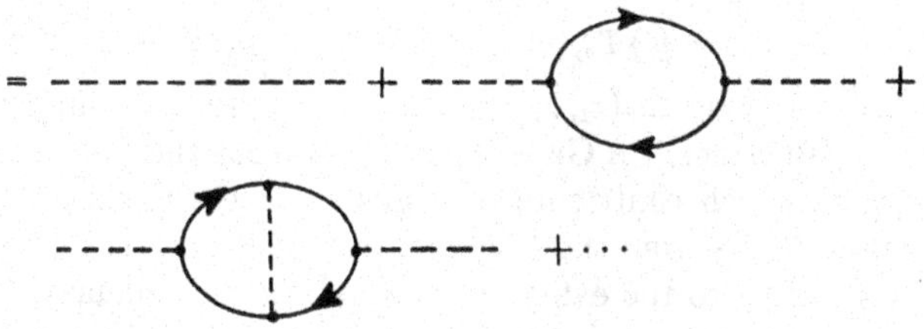

Fig. 17.1

(see Fig. 17.1). It is thus identical to what we called the primed boson propagation function in Chapter 16,

$$R(x_1, x_2) = \tfrac{1}{2}\Delta'_F(x_1, x_2)$$
$$= (\Psi_0, T(\phi(x_1)\, \phi(x_2))\, \Psi_0) \tag{38}$$

Using the same methods as outlined above for the boson operators, one readily establishes that

$$-\tfrac{1}{2}S'_F(x_1, x_2) = (\Psi_0, T(\psi(x_1)\,\tilde{\psi}(x_2))\,\Psi_0) \tag{39a}$$

$$= (\Phi_0, T(S\psi(x_1)\,\tilde{\psi}(x_2))\,\Phi_0) \tag{39b}$$

$$S = U(+\infty, -\infty) \tag{39c}$$

Similarly one verifies by a procedure identical to the above that the vacuum expectation value of the operator $T(\psi(x)\,\tilde{\psi}(y)\,\phi(z))$ corresponds to the Feynman diagrams illustrated in Figure 17.2. This vacuum expectation

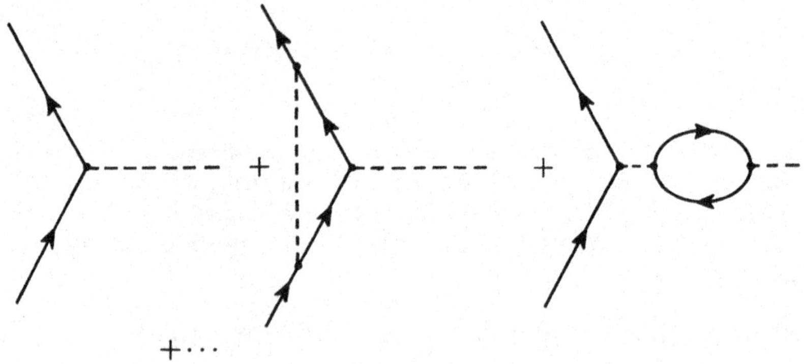

Fig. 17.2

value can therefore be expressed in terms of the functions S'_F, D'_F and Γ as follows:

$$\langle\Psi_0 \mid T(\tilde{\psi}_\beta(x)\,\psi_\alpha(y)\,\phi(z)) \mid \Psi_0\rangle = -\tfrac{1}{8}\int d^4x' \int d^4y' \int d^4z'$$

$$S'_{F\alpha\sigma}(y - y')\,\Gamma_{\sigma\rho}(y' - z', x' - z')\,S'_{F\rho\beta}(x' - x)\,\Delta'_F(z' - z) \tag{40}$$

The functions $\Delta'_F(x_1, x_2)$ and $S'_F(x_1, x_2)$ are the simplest examples of Green's functions. A Green's function is thus the sum of all the Feynman diagrams which contribute to a given process to all orders. A Green's function differs from the S matrix in that it does not contain the factors corresponding to the external lines and that it is defined for all values of the momenta of the external lines, not merely their mass shell values. The conservation laws such as charge and momentum are, however, implicit in the definition of Green's functions. For example, in momentum space, the Green's function for Compton scattering is defined as that function $G_{\mu_1\mu_2}(p'k'; pk)$ such that

$$\cdot \sqrt{\frac{m^2}{8E'E\omega\omega'}}\,\frac{1}{(2\pi)^6}\,\tilde{u}(\mathbf{p}')\,\epsilon^{(\lambda')}{}_{\mu'}(\mathbf{k}')\,G^{\mu'\mu}(p'k'; pk)\,u(\mathbf{p})\,\epsilon^{(\lambda)}{}_{\mu}(\mathbf{k})\ \cdot$$

is the S matrix describing the process to all orders. It should be noted that the same Green's function G describes the process of pair annihilation into two photons, since the same Feynman diagrams (now lying on their "sides") enter into this process. This is illustrated in Figure 17.3. However, in the case of pair annihilation, it is the function $G(-p', k'; p, -k)$

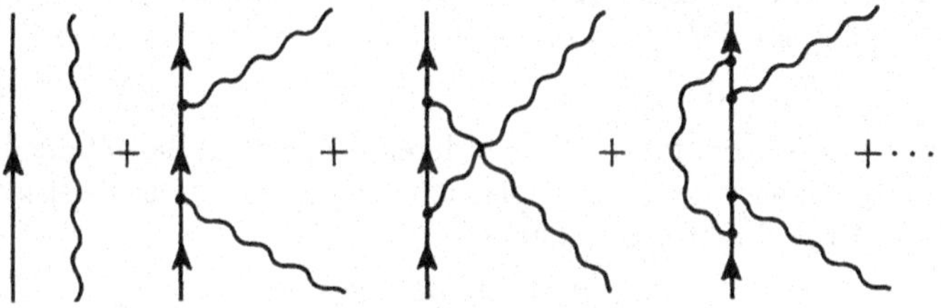

(a) Compton Scattering

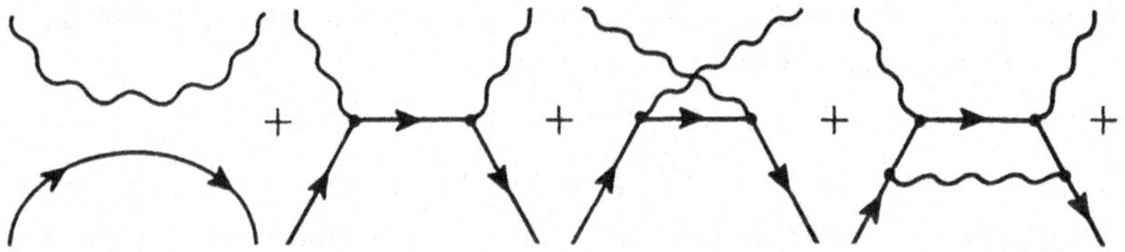

(b) Pair Annihilation

Diagrams Contributing to $G_{\mu\mu'}(p', k'; p, k)$

Fig. 17.3

which will enter into the S-matrix element if the initial electron and positron have momenta p, p' and the final photons have momenta k, k'.

17b. The Lehmann Spectral Representation

The assumptions

(*a*) of relativistic invariance,

(*b*) spectral conditions (existence of unique vacuum and only states with $p_\mu p^\mu \geqslant 0$, $p_0 \geqslant 0$)

(*c*) that the physical states span a Hilbert space endowed with a Hermitian scalar product so that every state has a positive norm

allow us to infer general properties concerning the structure of the vacuum expectation value of two Heisenberg operators, i.e., about the Δ'_F and S'_F functions. These general representations were first used by Källén (1952), and first derived in a systematic fashion from the general principles of quantum field theory by Wightman (1953, unpublished) and independently by Lehmann (1954). They have become known as the Lehmann form for the "twofold."

We shall first derive the Lehmann form for the vacuum expectation value of the time-ordered product of two neutral pseudoscalar boson operators which we have established to correspond to the Δ'_F function

$$\tfrac{1}{2}\Delta'_F(x, y) = (\boldsymbol{\Psi}_0, T(\boldsymbol{\phi}(x)\,\boldsymbol{\phi}(y))\,\boldsymbol{\Psi}_0) \tag{37}$$

Since the spectral form for the time-ordered product can be obtained from that of the ordinary product of the two operators, we shall study the properties of the Wightman function:

$$W^{(2)}(x, y) = (\boldsymbol{\Psi}_0, \boldsymbol{\phi}(x)\,\boldsymbol{\phi}(y)\,\boldsymbol{\Psi}_0) \tag{41}$$

The (generalized) functions defined by the vacuum expectation values of products of Heisenberg operators are known as Wightman functions, after Wightman who first extensively studied their properties [Wightman (1956)]. The Lorentz covariance of the theory, i.e., the fact that under an inhomogeneous Lorentz transformation

$$\mathbf{U}(a, \Lambda)\,\boldsymbol{\phi}(x)\,\mathbf{U}(a, \Lambda)^{-1} = \boldsymbol{\phi}(\Lambda x + a) \tag{42a}$$

$$\mathbf{U}(a, \Lambda)\,|\,\boldsymbol{\Psi}_0\rangle = |\boldsymbol{\Psi}_0\rangle \tag{42b}$$

where $\mathbf{U}(a, \Lambda)$ is a unitary operator (time inversions will not be considered), implies that

$$\begin{aligned}(\boldsymbol{\Psi}_0, \boldsymbol{\phi}(x)\,\boldsymbol{\phi}(y)\,\boldsymbol{\Psi}_0) &= (\mathbf{U}(a, \Lambda)\,\boldsymbol{\Psi}_0, \mathbf{U}(a, \Lambda)\,\boldsymbol{\phi}(x)\,\boldsymbol{\phi}(y)\,\boldsymbol{\Psi}_0)\\ &= (\boldsymbol{\Psi}_0, \boldsymbol{\phi}(\Lambda x + a)\,\boldsymbol{\phi}(\Lambda y + a)\,\boldsymbol{\Psi}_0)\end{aligned} \tag{43}$$

If we specialize to the case of pure translations $\Lambda = I$, then (43) asserts that for arbitrary displacements

$$W^{(2)}(x, y) = W^{(2)}(x + a, y + a) \tag{44a}$$

The function $W^{(2)}$ is therefore only a function of the difference of the co-ordinates x, y, i.e.,

$$W^{(2)}(x, y) = W^{(2)}(x - y) \tag{44b}$$

The invariance under proper homogeneous transformations then asserts that

$$W^{(2)}(x - y) = W^{(2)}(\Lambda(x - y)) \tag{44c}$$

whence $W^{(2)}$ is a function of only $(x - y)^2$ for space-like separations and of $(x - y)^2$ and $\epsilon(x - y)$ for time-like and null separations.

We next turn to the properties of $W^{(2)}$ which follow from the assumption that the states of the systems have only positive energy and time-like momenta. Using the complete set of states $|\mathbf{p}^{(n)}, \boldsymbol{\alpha}\rangle$, $\mathbf{P}_\mu\,|\,\mathbf{p}, \boldsymbol{\alpha}\rangle = p_\mu\,|\,\mathbf{p}, \boldsymbol{\alpha}\rangle$, where α denotes the other observables characterizing the basis vectors, we can rewrite Eq. (41) as follows:

$$(\Psi_0, \phi(x)\,\phi(y)\,\Psi_0) = \sum_{|\mathbf{p}^{(n)},\,\alpha\rangle} \langle \Psi_0 \mid \phi(x) \mid \mathbf{p}^{(n)}, \alpha\rangle \langle \mathbf{p}^{(n)}, \alpha \mid \phi(y) \mid \Psi_0\rangle \tag{45}$$

Next, upon using Eq. (12) to factor out the spatial dependence of the matrix elements, we obtain

$$\begin{aligned}(\Psi_0, \phi(x)\,\phi(y)\,\Psi_0) &= \sum_{|\mathbf{p}^{(n)},\,\alpha\rangle} \langle \Psi_0 \mid \phi(0) \mid \mathbf{p}^{(n)}, \alpha\rangle \langle \mathbf{p}^{(n)}, \alpha \mid \phi(0) \mid \Psi_0\rangle \cdot e^{-ip^{(n)}\cdot(x-y)} \\ &= \sum_{|\mathbf{p}^{(n)},\,\alpha\rangle} |\langle \Psi_0 \mid \phi(0) \mid \mathbf{p}^{(n)}, \alpha\rangle|^2\, e^{-ip^{(n)}\cdot(x-y)} \end{aligned} \tag{46}$$

The sum in (46) runs over all physically distinct states (α) and over all values of their total energy-momentum ($p^{(n)}$). The total energy-momentum $p^{(n)}{}_\mu$ of the state $|\mathbf{p}^{(n)}, \alpha\rangle$ defines the mass $M^{(n)}$ of the system through the relation

$$M^{(n)2} = p^{(n)}{}_\mu p^{(n)\mu} \tag{47}$$

This mass $M^{(n)}$ corresponds to the total energy of the state in the rest frame of the system.

Let us next introduce the *positive-definite* quantity $\rho(p^{(n)})$ defined by

$$\rho(p^{(n)}) = (2\pi)^3 \sum_\alpha |\langle \Psi_0 \mid \phi(0) \mid \mathbf{p}^{(n)}, \alpha\rangle|^2 \tag{48}$$

where the sum runs over all states $|\mathbf{p}^{(n)}, \alpha\rangle$ for fixed $p^{(n)}$. It is easily verified that there are only a finite number of states for any given value of the vector $p^{(n)}$, so that the sum defining $\rho(p^{(n)})$ can never diverge. Two facts should be noted about $\rho(p^{(n)})$. First that it is only defined for timelike momenta $p^{(n)}{}_\mu$ with $p^{(n)}{}_0 \geqslant 0$ since $p^{(n)}{}_\mu$ is the energy-momentum of a physical state. Secondly, that the transformation properties of the matrix element $\langle \Psi_0 \mid \phi(0) \mid \mathbf{p}^{(n)}, \alpha\rangle$ under homogeneous Lorentz transformations

$$\begin{aligned}\langle \Psi_0 \mid \phi(0) \mid \mathbf{p}^{(n)}, \alpha\rangle &= \langle \Psi_0 \mid \phi(0)\, \mathrm{U}(\Lambda^{-1}, 0) \mid \mathbf{p}^{(n)}, \alpha\rangle \\ &= \langle \Psi_0 \mid \phi(0) \mid \Lambda\mathbf{p}^{(n)}, \alpha'\rangle \end{aligned} \tag{49}$$

asserts that $\rho(p^{(n)}) = \rho(\Lambda p^{(n)})$ is a function of $p^{(n)2}$ only. We summarize these statements by writing

$$\theta(p^2)\,\theta(p_0)\,\rho(p^2) = (2\pi)^3 \sum_\alpha |\langle \Psi_0 \mid \phi(0) \mid \mathbf{p}, \alpha\rangle|^2 \tag{50}$$

We can therefore rewrite Eq. (47) as follows:

$$(\Psi_0, \phi(x)\,\phi(y)\,\Psi_0) = \frac{1}{(2\pi)^3} \int d^4p\, \theta(p^2)\,\theta(p_0)\,\rho(p^2)\, e^{-ip\cdot(x-y)} \tag{51}$$

where we now have indicated the sum over the vectors $p^{(n)}$ by an integration over d^4p. Since we can write $\theta(p^2)$ as

$$\theta(p^2) = \int_0^\infty dm^2\, \delta(p^2 - m^2) \tag{52}$$

Eq. (51) can finally be rewritten in the form

$$(\Psi_0, \phi(x)\, \phi(y)\, \Psi_0) = i \int_0^\infty dm^2\, \rho(m^2) \frac{-i}{(2\pi)^3} \int d^4p\, \theta(p_0)\, \delta(p^2 - m^2)\, e^{-ip\cdot(x-y)} \tag{53}$$

The integral over p will be recognized as the (unprimed) singular function $\Delta^{(+)}(x - y; m^2)$ for the mass m, so that

$$(\Psi_0, \phi(x)\, \phi(y)\, \Psi_0) = i \int_0^\infty dm^2\, \rho(m^2)\, \Delta^{(+)}(x - y; m^2) \tag{54}$$

To obtain the complete representation for the time-ordered product, we use the fact that for $x_0 > y_0$

$$\begin{aligned}(\Psi_0, T(\phi(x)\, \phi(y))\, \Psi_0) &= (\Psi_0, \phi(x)\, \phi(y)\, \Psi_0) && \text{for } x_0 > y_0 \\ &= i \int_0^\infty dm^2\, \rho(m^2)\, \Delta^{(+)}(x - y; m^2) && \text{for } x_0 > y_0\end{aligned} \tag{55}$$

and that for $y_0 > x_0$

$$\begin{aligned}(\Psi_0, T(\phi(x)\, \phi(y))\, \Psi_0) &= (\Psi_0, \phi(y)\, \phi(x)\, \Psi_0) && \text{for } y_0 > x_0 \\ &= i \int_0^\infty dm^2\, \rho(m^2)\, \Delta^{(+)}(y - x; m^2) && \text{for } y_0 > x_0 \\ &= -i \int_0^\infty dm^2\, \rho(m^2)\, \Delta^{(-)}(x - y; m^2) && \text{for } y_0 > x_0\end{aligned} \tag{56}$$

so that combining Eqs. (55) and (56), we obtain

$$\begin{aligned}(\Psi_0, T(\phi(x)\, \phi(y))\, \Psi_0) &= \tfrac{1}{2}\Delta'_F(x - y) \\ &= \tfrac{1}{2} \int_0^\infty dm^2\, \rho(m^2)\, \Delta_F(x - y; m^2)\end{aligned} \tag{57}$$

The relation (54) also allows us to deduce the spectral representation for the vacuum expectation value of the commutator

$$\begin{aligned}(\Psi_0, [\phi(x), \phi(y)]\, \Psi_0) &= i \int_0^\infty dm^2\, \rho(m^2)\, (\Delta^{(+)}(x - y; m^2) + \Delta^{(-)}(x - y; m^2)) \\ &= i \int_0^\infty dm^2\, \rho(m^2)\, \Delta(x - y; m^2)\end{aligned} \tag{58}$$

Note that in the derivation of these spectral forms no dynamical assumptions have been made, beyond assuming that the complete set of states $|\mathbf{p}, \alpha\rangle$ have only time-like or null momenta ($p_\mu p^\mu \geqslant 0$ with $p_0 \geqslant 0$). It will also be noted that without any assumptions concerning the commutation rules of the operators $\phi(x)$ and $\phi(y)$, we have deduced the fact that the vacuum expectation value of the commutator $[\phi(x), \phi(y)]$ vanishes for space-like separations since $\Delta(x - y; m^2)$ has this property for all m^2.

Let us next study the structure of the spectral function $\rho(p^2)$ somewhat more fully. To do this, we shall have to specify the spectrum of the physical states somewhat more precisely than was necessary up to now.

Since we are dealing with a theory of neutral pseudoscalar bosons interacting with spin $\frac{1}{2}$ fermions, we assume that the energy-momentum operator $\mathbf{P}_\mu$ has the following *mass* spectrum:

(*a*) a discrete point at $p^2 = 0$ corresponding to the vacuum state;

(*b*) a discrete point at $p^2 = \mu^2$, corresponding to the stable one-meson states; μ^2 is the physical (observed) mass of the meson;

(*c*) a continuum starting at $p^2 = (2\mu)^2$ corresponding to the two-meson states; similarly a continuum starting at $p^2 = (n\mu)^2$, $n = 3, 4, \cdots$ corresponding to n-meson states;

(*d*) a discrete point at $p^2 = M^2$, corresponding to the stable one-nucleon states; M is the mass of the physical nucleon;

(*e*) a continuum starting at $p^2 = (nM + m\mu)$, $n = 1, 2, \cdots, m = 0, 1, 2, \cdots$ corresponding to meson-nucleon states.

The energy-momentum spectrum of the physical states is illustrated in Figure 17.4.

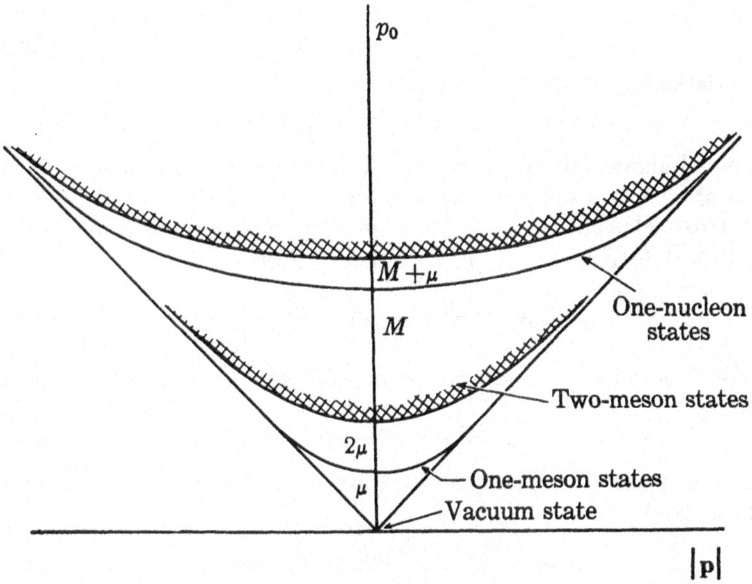

Fig. 17.4

These detailed assumptions concerning the energy-momentum spectrum of the stable physical states allow us to write $\rho(p^2)$ in the form

$$\rho(p^2) = (2\pi)^3 \sum_\alpha |\langle\Psi_0 \mid \phi(0) \mid \mathbf{p}, \alpha\rangle|^2 \tag{59a}$$

$$\begin{aligned} &= (2\pi)^3 |\langle\Psi_0 \mid \phi(0) \mid \mathbf{p}; p^2 = \mu^2\rangle|^2 \, \delta(p^2 - \mu^2) \\ &\quad + (2\pi)^3 \sum_{\alpha'} |\langle\Psi_0 \mid \phi(0) \mid \mathbf{p}, \alpha'\rangle|^2 \end{aligned} \tag{59b}$$

which decomposes ρ into the contribution from the one-meson state $|\mathbf{p}; p^2 = \mu^2\rangle$ of momentum p_μ and that from all other states of momentum p_μ. In Eq. (59a) the sum over α does not include the vacuum state since the vacuum expectation value of $\phi(x)$ vanishes for a pseudoscalar field. [*Proof:* By translation invariance $\langle \Psi_0 \mid \phi(x) \mid \Psi_0\rangle$ must be a constant. If the theory is invariant under spatial inversions

$$\mathrm{U}(i_s)\,\phi(x)\,\mathrm{U}(i_s)^{-1} = -\phi(i_s x) \tag{60a}$$

$$\mathrm{U}(i_s) \mid \Psi_0\rangle = |\Psi_0\rangle \tag{60b}$$

the value of this constant must be zero.] Our next task is to analyze the contribution of the one-meson states in detail. To that end consider the matrix element $\langle \Psi_0 \mid \phi(x) \mid \mathbf{p}; p^2 = \mu^2\rangle$ where $|\mathbf{p}; p^2 = \mu^2\rangle$, as above, denotes the one-meson eigenstate of $\mathbf{P}_\mu$, i.e., the state for which

$$\mathbf{P}_\mu \mid \mathbf{p}; p^2 = \mu^2\rangle = p_\mu \mid \mathbf{p}; p^2 = \mu^2\rangle \tag{61a}$$

with

$$p_\mu p^\mu = p^2 = \mu^2 \tag{61b}$$

Translation invariance asserts that

$$\langle \Psi_0 \mid \phi(x) \mid \mathbf{p}; p^2 = \mu^2\rangle = e^{-ip\cdot x}\langle \Psi_0 \mid \phi(0) \mid \mathbf{p}; p^2 = \mu^2\rangle \tag{62}$$

As noted above, relativistic invariance further asserts that $\langle \Psi_0 \mid \phi(0) \mid \mathbf{p}; p^2 = \mu^2\rangle$ is only a function of p^2. Since for the one-meson state $p^2 = \mu^2$, the matrix element $\langle \Psi_0 \mid \phi(0) \mid \mathbf{p}; p^2 = \mu^2\rangle$ is therefore a constant, call it $\sqrt{Z_3/(2\pi)^3}$, independent of p_μ. We therefore write

$$\langle \Psi_0 \mid \phi(x) \mid \mathbf{p}; p^2 = \mu^2\rangle = \frac{Z_3^{1/2}}{(2\pi)^{3/2}}\, e^{-ip\cdot x} \tag{63}$$

where Z_3 is some real constant. Using (63), we can rewrite Eq. (59b) as follows:

$$\rho(m^2) = Z_3\delta(m^2 - \mu^2) + \sigma(m^2) \tag{64}$$

where the positive-definite quantity σ is the contribution to ρ of the states with $p^2 > \mu^2$:

$$\sigma(p^2) = (2\pi)^3 \sum_{\alpha'} |\langle \Psi_0 \mid \phi(0) \mid \mathbf{p}, \alpha'\rangle|^2 \tag{65}$$

Using the result (64), we can finally rewrite Eq. (57) as follows:

$$\Delta'_F(x - y) = Z_3\Delta_F(x - y; \mu^2) + \int_{(3\mu)^2}^{\infty} dm^2\, \sigma(m^2)\, \Delta_F(x - y; m^2) \tag{66}$$

The limit of integration in the integral in (66) starts at $(3\mu)^2$ by virtue of our assumption that ϕ is a pseudoscalar field operator. In this case the matrix elements $\langle \Psi_0 \mid \phi(0) \mid \mathbf{p}; \alpha'\rangle$ with α' a two-meson state vanish and the first states which contribute to σ are the three-meson states whose mass spectrum starts at 3μ.

More precise statements can be made about the normalization constant Z_3 if it is assumed that the ϕs are local fields, i.e., satisfy

$$[\phi(x), \phi(y)] = i\Delta(x - y) \quad \text{for } (x - y)^2 < 0 \tag{67}$$

for space-like separations, $(x - y)^2 < 0$. We recall that for $(x - y)^2 < 0$, $\Delta(x - y; m^2)$ is independent of the mass m and is proportional to $\frac{1}{4\pi}\,\epsilon(x - y)\,\delta((x - y)^2)$. For local fields Eq. (58) therefore reads

$$i\Delta(x - y) = i\int_0^\infty dm^2\,\rho(m^2)\,\Delta(x - y; m^2) \quad \text{for } (x - y)^2 < 0 \tag{68}$$

or since $\Delta(x - y; m^2)$ is independent of m for $(x - y)^2 < 0$

$$\int_0^\infty \rho(m^2)\,dm^2 = 1 \tag{69a}$$

and equivalently

$$Z_3 + \int_{(3\mu)^2}^\infty \sigma(m^2)\,dm^2 = 1 \tag{69b}$$

Since $\sigma(m^2) \geqslant 0$, it follows from (69b) that $0 \leqslant Z_3 \leqslant 1$.

Objections might be raised to the above derivation since, in fact, $\Delta(x - y; m^2)$ vanishes for $(x - y)^2 < 0$. The same results are also obtained under the stronger assumption that $\phi(x)$ obeys canonical commutation rules, which will be the case if the field theory is derived from a local Lagrangian. If

$$[\partial_0\phi(x), \phi(y)]_{x_0=y_0} = -i\delta^{(3)}(\mathbf{x} - \mathbf{y}) \tag{70}$$

Eq. (69) then follows by differentiating both sides of Eq. (58) with respect to x_0, setting $x_0 = y_0$ and recalling that $\partial_0\Delta(x)$ for $x_0 = 0$ is equal $-\delta^{(3)}(\mathbf{x})$. Further important properties of the Green's functions $\Delta'_F(x)$ are most easily expressed in terms of its Fourier transform $\Delta'_F(k^2)$

$$\Delta'_F(k^2) = \frac{Z_3}{k^2 - \mu^2 + i\epsilon} + \int_{(3\mu)^2}^\infty dm^2\,\frac{\sigma(m^2)}{k^2 - m^2 + i\epsilon} \tag{71}$$

Consider the function $\Delta'_F(\zeta)$ of the complex variable ζ defined by

$$\Delta'_F(\zeta) = \frac{Z_3}{\zeta - \mu^2} + \int_{(3\mu)^2}^\infty dm^2\,\frac{\sigma(m^2)}{\zeta - m^2} \tag{72a}$$

where since Z_3 and $\sigma(m^2)$ are real

$$\overline{\Delta'_F(\zeta)} = \Delta'_F(\bar{\zeta}) \tag{72b}$$

The quantity of physical interest $\Delta'_F(k^2)$ can be obtained from $\Delta'_F(\zeta)$ by taking the limit $\zeta \to k^2 + i\epsilon$

$$\lim_{\zeta \to k^2 + i\epsilon} \Delta'_F(\zeta) = \Delta'_F(k^2) \tag{73}$$

It is evident from the representation (72) that in the complex ζ plane, $\Delta'_F(\zeta)$, has a pole on the real axis at $\zeta = \mu^2$, a branch point at $\zeta = (3\mu)^2$

with a branch line extending along the positive real axis from $(3\mu)^2$ to $+\infty$, and that $\Delta'_F(\zeta)$ is analytic everywhere else. We also note, writing $\zeta = \xi + i\eta$ that

$$\Delta'_F(\zeta) = \frac{Z_3(\xi - \mu^2)}{(\xi - \mu^2)^2 + \eta^2} + \int_{(3\mu)^2}^{\infty} dm^2 \frac{\sigma(m^2)\,(\xi - m^2)}{(\xi - m^2)^2 + \eta^2}$$
$$- i\left[\frac{Z_3\eta}{(\xi - \mu^2)^2 + \eta^2} + \int_{(3\mu)^2}^{\infty} dm^2 \frac{\sigma(m^2)\,\eta}{(\xi - m^2)^2 + \eta^2}\right] \quad (74)$$

so that $\operatorname{Im} \Delta'_F(\zeta)$ has the opposite sign as $\operatorname{Im} \zeta$ and is different from zero for $\operatorname{Im} \zeta \neq 0$. The function $\Delta'_F(\zeta)$ therefore has no complex zeros. Furthermore it can have zeros along the real axis only for $\xi > 0$. The discontinuity of $\Delta'_F(\zeta)$ across the branch line is given by

$$\Delta'_F(\xi + i\epsilon) - \Delta'_F(\xi - i\epsilon)$$
$$= -2\pi i Z_3 \delta(\xi - \mu^2) - 2\pi i \int_{(3\mu)^2}^{\infty} dm^2\, \sigma(m^2)\, \delta(\xi - m^2) \quad (75a)$$

or

$$\operatorname{Im} \Delta'_F(\xi + i\epsilon) = -\pi Z_3 \delta(\xi - \mu^2) - \pi\theta(\xi - 9\mu^2)\, \sigma(\xi) \quad (75b)$$

Using (75b), we can rewrite (72a) in the form

$$\Delta'_F(k^2) = \frac{1}{\pi} \int_{-\infty}^{+\infty} \frac{\operatorname{Im} \Delta'_F(k'^2)}{k'^2 - k^2 - i\epsilon}\, dk'^2 \quad (76)$$

which is a dispersion relation. The importance of this dispersion relation lies in the fact that it is a relation which is independent of perturbation theory and is an exact consequence of field theory, resulting from the spectral assumptions and from the relativistic invariance of the theory. Note that Eq. (76) does not determine $\Delta'_F(k^2)$ (all boson propagators of "reasonable" field theories will satisfy it) but is merely a shorthand notation for expressing the analytic properties of $\Delta'_F(k^2)$. The latter, however, are consequences of physical assumptions.

The results obtained thus far for the boson propagator have been obtained independently of any specific dynamical assumptions. We shall next obtain certain relations which are consequences of specific dynamical assumptions. The first is a relation between the bare and physical mass of the boson. The particular theory we shall consider is again that of a neutral pseudoscalar meson theory described by operators ψ and ϕ satisfying

$$(\Box + \mu_0^2)\, \phi(x) = \tfrac{1}{2} G[\bar{\psi}(x)\, \gamma_5, \psi(x)] \equiv \mathbf{J}(x) \quad (77)$$
$$(-i\gamma \cdot \partial + M_0)\, \psi(x) = G\gamma_5 \phi(x)\, \psi(x) \quad (78)$$

where μ_0^2 and M_0 are the *bare* masses of the quanta. The equal-time commutation rules of the operators ψ and ϕ are the ones noted previously, Eqs. (6a), (6c), (67), and (70). We here also recall that for equal times ψ

and ϕ commute. If we apply $\square_x + \mu_0^2$ on the vacuum expectation value of the commutator $[\phi(x), \phi(y)]$, we obtain

$$\begin{aligned}(\square_x + \mu_0^2)\,(\Psi_0, [\phi(x), \phi(y)]\,\Psi_0) &= (\Psi_0, [\mathbf{J}(x), \phi(y)]\,\Psi_0) \\ &= i(\square_x + \mu_0^2) \int_0^\infty dm^2\,\rho(m^2)\,\Delta(x-y; m^2) \\ &= i \int_0^\infty dm^2\,\rho(m^2)\,(\mu_0^2 - m^2)\,\Delta(x-y; m^2) \end{aligned} \tag{79}$$

since $\square_x\Delta(x-y; m^2) = -m^2\Delta(x-y; m^2)$. Taking the derivative of Eq. (79) with respect to x_0, and setting $x_0 = y_0$, and noting that for equal times $\phi(y)$ commutes with $\mathbf{J}(x)$ and $\partial_0\mathbf{J}(x)$, i.e., with $\psi(x)$ and $\partial_0\psi(x)$ [this is consistent with Eq. (78)], we obtain

$$\mu_0^2 = \frac{\int_0^\infty dm^2\,\rho(m^2)\,m^2}{\int_0^\infty dm^2\,\rho(m^2)} = Z_3\mu^2 + \int_{(3\mu)^2}^\infty m^2\sigma(m^2)\,dm^2 \tag{80a}$$

so that

$$\mu_0^2 - \mu^2 = \int_0^\infty dm^2\,\rho(m^2)\,(m^2 - \mu^2) \tag{80b}$$

$$= \int_{\mu^2}^\infty dm^2\,\rho(m^2)\,(m^2 - \mu^2) \tag{80c}$$

Since $\rho(m^2)$ is positive, Eq. (80c) indicates that the bare mass is larger than the physical mass.[3]

Let us next consider the asymptotic behavior of $\Delta'_F(k^2)$ for large momenta, or what is equivalent the small space-time values of $\Delta'_F(x)$. From (71) it follows that

$$\begin{aligned}\lim_{k^2\to\infty} \Delta'_F(k^2) &= \lim_{k^2\to\infty} \int_0^\infty dm^2 \frac{\rho(m^2)}{k^2\left(1 - \frac{m^2}{k^2} + i\epsilon\right)} \\ &\simeq \lim_{k^2\to\infty} \int_0^\infty dm^2\,\rho(m^2)\,\frac{1}{k^2}\left(1 + \frac{m^2}{k^2} \pm \cdots\right) \\ &\simeq \frac{1}{k^2}\left(1 + \frac{\mu_0^2}{k^2} \pm \cdots\right) \simeq \frac{1}{k^2 - \mu_0^2 + i\epsilon} + O\left(\frac{1}{k^6}\right)\end{aligned} \tag{81}$$

so that the large momentum or small space-time behavior of the boson propagator is determined by the *bare* mass of the Bose quantum. Con-

[3] Equation (80) is valid in the presence of a $\lambda\phi^4$ counterterm in the Lagrangian only if a suitable normal ordering is prescribed. If in the presence of such a counterterm the equation of motion of the boson field is

$$(\square + \mu_0^2)\,\phi(x) = \tfrac{1}{2}G[\bar\psi(x)\,\gamma_5, \psi(x)] - \lambda[\phi^3(x) - 3\phi(x)\,\langle\phi^2(x)\rangle_0]$$

where $\langle\phi^2(x)\rangle_0 = \langle\Psi_0 \mid \phi^2(x) \mid \Psi_0\rangle$, then (80) follows.

versely, the behavior for small momenta or for large time and spatial distance is dominated by the pole, so that

$$\lim_{k^2\to\mu^2} \Delta'_F(k^2) \simeq \frac{Z_3}{k^2-\mu^2+i\epsilon} \tag{82}$$

Since $\Delta'_F(x-x')$ is essentially the boson propagator which takes into account all radiative corrections, Eq. (81), which when Fourier transformed to configuration space gives the propagator in the limit as $x - x' \to 0$, allows us to calculate the amplitude that in an idealized, almost instantaneous, measurement a Bose quantum located at time x'^0 at $\mathbf{x}'$ be found at $\mathbf{x}$ at a very short time x^0 later. What Eq. (81) says is that for such an instantaneous experiment the coupling has no time to take effect and that the quantum therefore propagates according to its bare mass. On the other hand, for large-time intervals (corresponding to realistic measurements) the propagator which determines the outcome is that with the physical (observed) mass of the particle.

Similar results to the above hold for the photon propagator in quantum electrodynamics. The Lehmann form for the function $D'_{F\mu\nu}(k^2)$ is

$$D'_{F\mu\nu}(k^2) = \left(g_{\mu\nu} - \frac{k_\mu k_\nu}{k^2}\right) D'_F(k^2) \tag{83a}$$

$$D'_F(k^2) = \frac{Z_3}{k^2} + \int_0^\infty dM^2 \frac{\sigma(M^2)}{k^2-M^2+i\epsilon} \tag{83b}$$

and one again proves that

$$Z_3 + \int_0^\infty \sigma(M^2)\, dM^2 = 1 \tag{84}$$

The proof of the positive-definiteness of $\sigma(M^2)$ and Z_3 are somewhat more difficult to carry out, due to the nonpositive-definiteness of the Gupta scalar product. To second order in the coupling constant e^2, D'_F is given by

$$D'_F(k^2) \approx \frac{Z_3{}^{(2)}}{k^2} + \int_0^\infty dM^2 \frac{\sigma^{(2)}(M^2)}{k^2-M^2+i\epsilon} \tag{85a}$$

with

$$Z_3{}^{(2)} \simeq 1 - \frac{e^2}{12\pi^2}\log\frac{\lambda^2}{m^2} \tag{85b}$$

$$\sigma^{(2)}(M^2) = \frac{e^2}{12\pi^2}\frac{1}{M^2}\left(1+\frac{2m^2}{M^2}\right)\left(1-\frac{4m^2}{M^2}\right)^{1/2}\theta(M^2-4m^2) \tag{85c}$$

The contribution to $\sigma^{(2)}$ arises from a bubble diagram (electron-positron pair creation and annihilation). The computation of $\sigma^{(2)}$ is straightforward. We recall that in the transverse gauge by definition

$$\begin{aligned} -\frac{1}{2} D'_{F\mu\nu}(x-y) &= -\frac{1}{2}\left(g_{\mu\nu} - \frac{\partial_\mu\partial_\nu}{\Box}\right) D'_F(x-y) \\ &= (\mathbf{\Psi}_0, T(\mathbf{A}_\mu(x)\, \mathbf{A}_\nu(y)\, \mathbf{\Psi}_0) \\ &= (\Phi_0, T(A_\mu(x)\, A_\nu(y)\, S)_C\, \Phi_0) \end{aligned} \tag{86}$$

where the subscript C denotes that only connected diagrams are to be considered, so that to order e^2

$$\tfrac{1}{2} D'_{F\mu\nu}(x - y) = \tfrac{1}{2} D_{F\mu\nu}(x - y) + \frac{e^2}{16} \int d^4z_1 \int d^4z_2$$

$$\cdot \operatorname{Tr} \{\gamma^{\mu_1} S_F(z_2 - z_1) \gamma^{\nu_1} S_F(z_1 - z_2)\} D_{F\mu\mu_1}(x - z_1) D_{F\nu_1\nu}(y - z_2) \quad (87)$$

where with our choice of gauge

$$D_{F\mu\nu}(x - y) = \frac{2i}{(2\pi)^4} \int d^4k\, e^{-ik\cdot(x-y)} \left(\frac{k_\mu k_\nu}{k^2} - g_{\mu\nu}\right) \frac{1}{k^2 + i\epsilon} \quad (88)$$

The Fourier transform of the second term in (87) is

$$\frac{e^2}{(2\pi)^8} \int d^4k\, e^{-ik\cdot(x-y)} \frac{k^{-2}k_\mu k_{\mu_1} - g_{\mu\mu_1}}{k^2 + i\epsilon} \cdot \Pi^{\mu_1\nu_1(2)}(k^2) \cdot \frac{k^{-2}k_{\nu_1\nu} - g_{\nu_1\nu}}{k^2 + i\epsilon} \quad (89a)$$

where

$$\Pi^{(2)}_{\mu\nu}(k^2) = -8(k_\mu k_\nu - g_{\mu\nu}k^2)\, \Pi^{(2)}(k^2) \quad (89b)$$

is the polarization tensor previously evaluated in Chapter 15 [Eqs. (138a)–(138c)]. The divergent part of $\Pi^{(2)}(k^2)$, $\Pi^{(2)}(0)$ contributes a term proportional to $D_{F\mu\nu}(x - y)$

$$\frac{e^2}{(2\pi)^8} \Pi^{(2)}(0) \int d^4k\, e^{-ik\cdot(x-y)} \left(\frac{k_\mu k_\nu}{k^2} - g_{\mu\nu}\right) \frac{1}{k^2 + i\epsilon}$$

$$= -\frac{e^2}{12\pi^2} \log \frac{\lambda^2}{m^2} \frac{1}{2} D_{F\mu\nu}(x - y) \quad (90)$$

which combines with the first term of the right-hand side of (87). Comparing (87) and the general form

$$D'_{F\mu\nu}(x - y) = Z_3 D_{F\mu\nu}(x - y)$$

$$+ \int_0^\infty dM^2\, \sigma(M^2) \left(\frac{\partial_\mu \partial_\nu}{\Box} - g_{\mu\nu}\right) \Delta_F(x - y; M^2) \quad (91)$$

we obtain

$$Z_3 = 1 - \frac{e^2}{12\pi^2} \log \frac{\lambda^2}{M^2} + \cdots \quad (92)$$

The convergent contribution, $\Pi^{(2)}(k^2) - \Pi^{(2)}(0)$, can be written in the form

$$\Pi^{(2)}(k^2) - \Pi^{(2)}(0) = i\pi^2 \int_0^1 dz\, (z - z^2) \log \left|\frac{m^2 - (z - z^2)\, k^2}{m^2}\right| \quad (93a)$$

$$= -\tfrac{1}{2} i\pi^2 k^2 \int_0^1 \frac{z^2\, dz \left(1 - \frac{z^2}{3}\right)}{4m^2 - k^2(1 - z^2) - i\epsilon} \quad (93b)$$

In going from (93a) to (93b) the change of variable $z \to 1 - 2z$ has been made and an integration by part performed. Finally making the change of variables $4m^2/(1 - z^2) = M^2$, we obtain

$$\Pi^{(2)}(k^2) - \Pi^{(2)}(0) = \frac{i\pi^2 k^2}{6} \int_{4m^2}^{\infty} dM^2 \frac{\left(1 + \frac{2m^2}{M^2}\right)\sqrt{1 - \frac{4m^2}{M^2}}}{M^2(k^2 - M^2 + i\epsilon)} \tag{94}$$

Note that in this form k^2 appears in the integrand only in the denominator and in the form $[k^2 - M^2 + i\epsilon]^{-1}$, corresponding to a propagator for a mass M boson. Combining (94), (89), and (87), we see that the finite contribution to (91) is given by a term which can be written in the form

$$\int_0^{\infty} dM^2\, \sigma^{(2)}(M^2)\, (\partial^{-2}\partial_\mu\partial_\nu - g_{\mu\nu})\, \Delta_F(x - y; M^2)$$

with $\sigma^{(2)}(M^2)$ given by (85c).

The behavior of $D'_F(k^2)$ for large k is given by

$$\lim_{k^2\to\infty} D'_F(k^2) \simeq \frac{1}{k^2}\left(Z_3 + \int_0^{\infty} \sigma(M^2)\, dM^2 + O\left(\frac{1}{k^6}\right)\right)$$

$$\simeq \frac{1}{k^2} \tag{95a}$$

whereas that for small k^2 is given by

$$\lim_{k^2\to 0} D'_F(k^2) \simeq \frac{Z_3}{k^2} \tag{95b}$$

To obtain a physical interpretation for Eqs. (95a) and (95b), recall that the potential energy between two heavy point test particles, with unrenormalized charges q_0 and q'_0, separated by a distance r is given by

$$V(r) = \frac{q_0 q'_0}{(2\pi)^3} \int d^3k\, e^{i\mathbf{k}\cdot\mathbf{r}}\, D'_F(k^2) \tag{96}$$

For small r ($r \ll 1/m$), the potential will be determined by the large k behavior of $D'_F(k)$, whence

$$\lim_{r\to 0} V(r) \sim \frac{q_0 q'_0}{4\pi r} + \cdots \tag{97a}$$

whereas for large r ($r \gg 1/m$), the potential is determined by the behavior of $D'_F(k)$ for small k values, so that

$$\lim_{r\to\infty} V(r) \simeq \frac{Z_3 q_0 q'_0}{4\pi r} + \cdots$$

$$\simeq \frac{q q'}{4\pi r} + \cdots \tag{97b}$$

where q and q' are the *renormalized* charges of the sources: $q = Z_3^{1/2} q_0$; $q' = Z_3^{1/2} q'_0$. Thus for small distances it is the unrenormalized charge which determines the character of the interaction, whereas for large separation it is the renormalized charge which determines the interaction between particles. The fact that $0 \leqslant Z_3 \leqslant 1$ implies that the renormalized charge is less than the bare charge. This was made plausible by our discussion in Section 15e where it was indicated that the phenomenon of vacuum polarization could be pictured as follows: The bare charge surrounds itself with a cloud of (virtual) quanta of the opposite charge as itself (whereas the quanta of the same charge are repelled to infinity.) These quanta are the members of virtual pairs constantly created by the vacuum fluctuations. Thus at large distances one sees the bare charge minus the charge of the cloud. This by definition is the renormalized charge. Its magnitude therefore is smaller than the bare charge.

The calculations carried out above for Z_3 in electrodynamics (and similar ones for mesodynamics) indicate that at least when calculated in perturbation theory Z_3^{-1} diverges, so that the matrix element of the boson operator ϕ between a physical one-meson state and the vacuum vanishes:

$$\langle \Psi_0 \mid \phi(0) \mid \mathbf{p}; p^2 = \mu^2 \rangle = \frac{Z_3^{1/2}}{(2\pi)^{3/2}} \tag{98}$$

One therefore defines a new, renormalized operator,

$$\phi_R(x) = Z_3^{-1/2}\phi(x) \tag{99}$$

which has the property that its matrix element between the physical one-meson state and the vacuum is finite

$$\begin{aligned}\langle \Psi_0 \mid \phi_R(x) \mid \mathbf{p}; p^2 = \mu^2 \rangle &= \langle \Psi_0 \mid Z_3^{-1/2}\phi(x) \mid \mathbf{p}; p^2 = \mu^2 \rangle \\ &= \frac{1}{(2\pi)^{3/2}} e^{-ip\cdot x} \end{aligned} \tag{100a}$$

More generally, the matrix element of $\phi_R(x)$ between arbitrary physical states will also be finite. The renormalized weight function

$$\rho_R(p^2) = (2\pi)^3 \sum_\alpha |\langle \Psi_0 \mid \phi_R(0) \mid \mathbf{p}; \alpha \rangle|^2 \tag{100b}$$

is related to the unrenormalized one by

$$\rho_R(p^2) = Z_3^{-1}\rho(p^2) \tag{101}$$

Since $\phi_R(x)$ is the renormalized field operator the computation of $\rho_R(p^2)$ can be carried through without encountering any divergences. In terms of renormalized quantities Eq. (69) reads

$$1 + \int_{(3\mu)^2}^\infty \sigma_R(m^2)\, dm^2 = Z_3^{-1} \tag{102}$$

We also note that the renormalized operators satisfy the following equal-time commutation rules:

$$[\phi_R(x), \phi_R(y)] = 0 \quad \text{for } (x - y)^2 < 0 \tag{103a}$$

$$[\partial_0\phi_R(x), \phi_R(y)]_{x_0 = y_0} = -iZ_3^{-1}\delta^{(3)}(\mathbf{x} - \mathbf{y}) \tag{103b}$$

Techniques similar to the ones outlined for the twofold vacuum expectation value of boson fields can be carried out for the spinor case [Wightman (1953, unpublished); Lehmann (1954); Gell-Mann and Low (Gell-Mann

1954)]. Let us denote the vacuum expectation value of two spinor fields, the fermion twofold Wightman function, by

$$W^{\psi}_{\alpha\beta}(x - y) = (\Psi_0, \psi_\alpha(x)\, \tilde{\psi}_\beta(y)\, \Psi_0) \tag{104}$$

where use has already been made of the translational invariance of the theory to write W^{ψ} as a function of $x - y$. Since under a homogeneous Lorentz transformation

$$\mathbf{U}(0, \Lambda) \mid \Psi_0\rangle = |\Psi_0\rangle \tag{105a}$$

$$\mathbf{U}(0, \Lambda)\, \psi_\alpha(x)\, \mathbf{U}(0, \Lambda)^{-1} = \sum_{\beta=1}^{4} S^{-1}{}_{\alpha\beta}(\Lambda)\, \psi_\beta(\Lambda x) \tag{105b}$$

it follows that

$$W^{\psi}_{\alpha\rho}(x - y) = \sum_{\delta\beta} S^{-1}{}_{\alpha\beta}(\Lambda)\, W^{\psi}{}_{\beta\delta}(\Lambda(x - y))\, S_{\delta\rho}(\Lambda) \tag{106}$$

A further reduction of W^{ψ} can be made by expanding it in terms of the sixteen linearly independent matrices $\Gamma^{(i)}$ formed of the γ matrices and their product:

$$W^{\psi}_{\alpha\beta}(x) = \sum_{i=1}^{16} [\Gamma^{(i)}]_{\alpha\beta}\, W^{\psi}{}_{(i)}(x) \tag{107}$$

Dividing the $\Gamma^{(i)}$ into sets with well-defined tensor transformation properties, we have

$$W^{\psi}_{\alpha\beta}(x) = \delta_{\alpha\beta} W^{\psi}{}_S(x) + (\gamma^\mu)_{\alpha\beta}\, W^{\psi}{}_{V_\mu}(x) + (\gamma^5)_{\alpha\beta}\, W^{\psi}{}_P(x)$$
$$+ \tfrac{1}{2}(\sigma^{\mu\nu})_{\alpha\beta}\, W^{\psi}{}_{T_{\mu\nu}}(x) + i(\gamma^5\gamma^\mu)_{\alpha\beta}\, W^{\psi}{}_{A_\mu}(x) \tag{108}$$

Using the fact that $S(\Lambda)^{-1}\, \gamma^\mu S(\Lambda) = \Lambda^\mu{}_\nu \gamma^\nu$, upon substituting the expansion (108) into (106) and taking appropriate traces, we find that

$$W^{\psi}{}_S(x) = W^{\psi}{}_S(\Lambda x) \tag{109a}$$

$$W^{\psi}{}_{V_\mu}(x) = \Lambda^{-1}{}_\mu{}^\nu W^{\psi}{}_{V_\nu}(\Lambda x) \tag{109b}$$

$$W^{\psi}{}_{T_{\mu\nu}}(x) = \Lambda^{-1}{}_\mu{}^\rho \Lambda^{-1}{}_\nu{}^\lambda W^{\psi}{}_{T_{\rho\lambda}}(\Lambda x) \tag{109c}$$

$$W^{\psi}{}_{A_\mu}(x) = (\det \Lambda)\, \Lambda^{-1}{}_\mu{}^\nu W^{\psi}{}_{A_\nu}(\Lambda x) \tag{109d}$$

$$W^{\psi}{}_P(x) = \det \Lambda \cdot W^{\psi}{}_P(\Lambda x) \tag{109e}$$

which justifies the labels S (scalar), V (vector), T (tensor), A (axial vector), and P (pseudoscalar). By taking the complex conjugate of Eq. (104), we find that

$$\overline{W^{\psi}_{\alpha\beta}(x - y)} = \overline{(\Psi_0, \psi_\alpha(x)\, \tilde{\psi}_\beta(y)\, \Psi_0)} = (\Psi_0, (\psi_\alpha(x)\, \tilde{\psi}_\beta(y))^*\, \Psi_0)$$
$$= \sum_{\rho\lambda} \gamma^0{}_{\beta\rho}(\Psi_0, \psi_\rho(y)\, \tilde{\psi}_\lambda(x)\, \Psi_0)\, \gamma^0{}_{\lambda\alpha}$$
$$= \sum_{\rho\lambda} \gamma^0{}_{\beta\rho} W^{\psi}{}_{\rho\lambda}(y - x)\, \gamma^0{}_{\lambda\alpha} \tag{110a}$$

or

$$W^{\psi}(x)^* = \gamma^0 W^{\psi}(-x)\,\gamma^0 \tag{110b}$$

Furthermore, since $\gamma^0[\Gamma^{(i)}]^*\,\gamma^0 = \Gamma^{(i)}$, we have

$$W^{\psi}{}_{(i)}(x) = \overline{W_{(i)}(-x)} \tag{111}$$

The application of PT invariance and T invariance, when joined with Eqs. (105)–(111), indicates that $W^{\psi}{}_P$, $W^{\psi}{}_T$ and $W^{\psi}{}_A$ vanish so that

$$W^{\psi}{}_{\alpha\beta}(x) = \delta_{\alpha\beta}W^{\psi}{}_S(x) + (\gamma^{\mu})_{\alpha\beta}W^{\psi}{}_{V\mu}(x) \tag{112}$$

We next make use of the spectral conditions. Upon introducing the complete set of states $|\mathbf{p}, \boldsymbol{\alpha}\rangle$ into the defining equation (100), we find that

$$W^{\psi}{}_{\rho\sigma}(x-y) = \sum_{|\mathbf{p},\boldsymbol{\alpha}\rangle} \langle \boldsymbol{\Psi}_0 \mid \psi_\rho(0) \mid \mathbf{p}, \boldsymbol{\alpha}\rangle \langle \mathbf{p}, \boldsymbol{\alpha} \mid \tilde{\psi}_\sigma(0) \mid \boldsymbol{\Psi}_0\rangle\, e^{-ip\cdot(x-y)} \tag{113}$$

Denoting the sum over the distinct states $|\mathbf{p}, \boldsymbol{\alpha}\rangle$ with fixed energy-momentum p_μ by $w_{\rho\sigma}(p)$

$$\theta(p^2)\,\theta(p_0)\,w_{\rho\sigma}(p) = (2\pi)^3 \sum_{\alpha} \langle \boldsymbol{\Psi}_0 \mid \psi_\rho(0) \mid \mathbf{p}, \boldsymbol{\alpha}\rangle \langle \mathbf{p}, \boldsymbol{\alpha} \mid \tilde{\psi}_\sigma(0) \mid \boldsymbol{\Psi}_0\rangle \tag{114}$$

we may then write

$$W^{\psi}{}_{\rho\sigma}(x) = \frac{1}{(2\pi)^3} \int d^4p\, e^{-ip\cdot(x-y)}\,\theta(p_0)\,\theta(p^2)\,w_{\rho\sigma}(p) \tag{115a}$$

where

$$w_{\rho\sigma}(p) = \delta_{\rho\sigma}w_S(p) + (\gamma^{\mu})_{\rho\sigma}\,w_{V\mu}(p) \tag{115b}$$

The covariance statements [Eqs. (109a–e)], when translated in terms of the transforms $w_{V\mu}(p)$ and $w_S(p)$, assert that

$$w_S(p) = w_S(\Lambda p) \tag{116a}$$

$$w_{V\mu}(p) = \Lambda^{-1}{}_{\mu}{}^{\nu} w_{V\nu}(\Lambda p) \tag{116b}$$

From (116a) it follows that $w_S(p) = w_{(1)}(p^2)$. Equation (116b) asserts that $w_{V\mu}(p)$ can be computed from $w_{V\mu}(\lambda, \mathbf{0})$ for all p on the same hyperboloid $p^2 = \lambda^2$. Furthermore, $w_{V\mu}(\lambda, \mathbf{0})$ is a four-vector which is invariant under all Λ which leaves $(\lambda, \mathbf{0})$ invariant, i.e., under all rotations. Therefore $w_{Vi}(\lambda, \mathbf{0}) = 0$ for $i = 1, 2, 3$ and

$$w_{V\mu}(p) = p_\mu w_{(2)}(p^2) \tag{117}$$

so that finally we can express $W^{\psi}(x)$ in the form

$$\begin{aligned} W^{\psi}(x-y) &= \frac{1}{(2\pi)^3} \int d^4p\, e^{-ip\cdot(x-y)}\,\theta(p_0)\,\theta(p^2)\,(w_{(1)}(p^2) + \gamma\cdot p\,w_{(2)}(p^2)) \\ &= i\int_0^\infty dm^2\,\{w_{(1)}(m^2) + w_{(2)}(m^2)\,i\gamma\cdot\partial\}\,\Delta^{(+)}(x-y; m^2) \end{aligned} \tag{118}$$

By a redefinition of the weight functions

$$w_{(1)} - mw_{(2)} = \rho_{(2)} \tag{119a}$$

$$w_{(2)} = \rho_{(1)} \tag{119b}$$

we can write

$$\begin{aligned} W^{\psi}(x - y) &= (\boldsymbol{\Psi}_0, \psi(x)\, \tilde{\psi}(y)\, \boldsymbol{\Psi}_0) \\ &= \int_0^\infty dm^2\, \{\rho_{(1)}(m^2)\,(-iS^{(+)}(x - y; m)) + \rho_{(2)}(m^2)\, i\Delta^{(+)}(x - y; m^2)\} \end{aligned} \tag{120}$$

By comparing Eqs. (114), (115b), and (119), we can infer that the weight function $\rho_{(1)}$ and $\rho_{(2)}$ are real and that in meson theory (where the norm of any state is positive) these weight functions satisfy the following positive-definiteness relations: $\rho_{(1)}(m^2) \geqslant 0$, $2m\rho_{(1)}(m^2) \geqslant \rho_{(2)}(m^2) \geqslant 0$. In quantum electrodynamics, the nonpositive-definite nature of the Gupta scalar product does not allow us to draw these inferences.

If the theory is invariant under charge conjugation, so that the spectral representation for $(\boldsymbol{\Psi}_0, \tilde{\psi}(y)\, \psi(x)\, \boldsymbol{\Psi}_0)$ can be related to the one for $(\boldsymbol{\Psi}_0, \psi(x)\, \tilde{\psi}(y)\, \boldsymbol{\Psi}_0)$, we can deduce that

$$(\boldsymbol{\Psi}_0, T(\psi(x)\, \tilde{\psi}(y))\, \boldsymbol{\Psi}_0)$$

$$= -\tfrac{1}{2}S'_F(x - y) \tag{121a}$$

$$= \int_0^\infty dm^2\, \{w_{(1)}(m^2) + w_{(2)}(m^2) i\gamma^\mu\partial_\mu\}\, \Delta_F(x - y; m^2) \tag{121b}$$

Denoting the Fourier transform of $S'_F(x - y)$ by $S'_F(p)$, we can write

$$-\tfrac{1}{2}S'_F(p) = i\int_0^\infty dm^2 \frac{1}{p^2 - m^2 + i\epsilon}\,(w_{(1)}(m^2) + \gamma \cdot p\, w_{(2)}(m^2)) \tag{122}$$

It is convenient to introduce two new functions $h^{(1)}(m^2)$, $h^{(2)}(m^2)$ defined by

$$w_{(1)}(m^2) = h^{(1)}(m^2) - h^{(2)}(m^2) \tag{123a}$$

$$w_{(2)}(m^2) = \frac{1}{m}\,(h^{(1)}(m^2) + h^{(2)}(m^2)) \tag{123b}$$

in which case

$$-\tfrac{1}{2}S'_F(p) = i\int_0^\infty \frac{dm^2}{m}\left[\frac{h^{(1)}(m^2)}{\gamma \cdot p - m + i\epsilon} + \frac{h^{(2)}(m^2)}{\gamma \cdot p + m - i\epsilon}\right] \tag{124}$$

It is of interest to consider the form of S'_F when the contribution of the one-fermion states is explicitly separated out. To this end, let us investigate the structure of the functions $w_{(1)}$ and $w_{(2)}$. From Eqs. (114) and (115b), it follows that

$$w_{(1)}(p^2) = \frac{(2\pi)^3}{4}\sum_{\rho=1}^{4}\sum_{|\mathbf{p}\alpha\rangle} \langle \boldsymbol{\Psi}_0 \mid \psi_\rho(0) \mid \mathbf{p}, \boldsymbol{\alpha}\rangle \langle \mathbf{p}, \boldsymbol{\alpha} \mid \tilde{\psi}_\rho(0) \mid \boldsymbol{\Psi}_0\rangle \tag{125}$$

whence, separating off the contribution of the one-fermion states, we obtain

$$w_{(1)}(p^2) = \frac{(2\pi)^3}{4} \sum_{\rho=1}^{4} \sum_{s=1}^{2} \langle \Psi_0 \mid \psi_\rho(0) \mid \mathbf{p}, \mathbf{s}; p^2 = M^2 \rangle$$

$$\langle \mathbf{p}, \ \mathbf{s}; \ \mathbf{p}^2 = M^2 \, | \tilde{\psi}_\rho(0) \mid \Psi > + \, w_{(1)}'(p^2) \tag{126}$$

where $w_{(1)}'(p^2)$ is the contribution from all other states $|\mathbf{p}, \boldsymbol{\alpha}'\rangle$. Relativistic invariance implies that

$$\langle \Psi_0 \mid \psi_\rho(x) \mid \mathbf{p}, \mathbf{s}; p^2 = M^2 \rangle = \frac{Z_2^{1/2}}{(2\pi)^{3/2}} e^{-ip\cdot x} \sqrt{2M} \, u^s{}_\rho(\mathbf{p}) \tag{127}$$

where $(\gamma \cdot p - M) \, u^s(\mathbf{p}) = 0$ and Z_2 is a normalization constant. Substituting the expression (127) into Eq. (126), we see that the two one-fermion states ($s = 1, 2$) contribute to $w_{(1)}(p^2)$ a term

$$\tfrac{1}{2} Z_2 M \operatorname{Tr} \left(\sum_{s=1}^{2} u^s(\mathbf{p}) \, \tilde{u}^s(\mathbf{p}) \right) = \tfrac{1}{2} Z_2 M \operatorname{Tr} \left(\frac{\gamma \cdot p + M}{2M} \right)$$

$$= Z_2 M \tag{128}$$

whence

$$w_{(1)}(p^2) = Z_2 M \delta(p^2 - M^2) + w_{(1)}'(p^2) \tag{129}$$

Similarly [since by (115b) Tr $(\gamma^0 w) = 4p^0 w_{(2)}$], one readily computes that

$$w_{(2)}(p^2) = Z_2 \delta(p^2 - M^2) + w_{(2)}'(p^2) \tag{130}$$

It therefore follows from Eq. (123) that

$$h^{(1)}(m^2) = M Z_2 \delta(m^2 - M^2) + \tfrac{1}{2}[w_{(1)}'(m^2) + m w_{(2)}'(m^2)] \tag{131}$$

$$h^{(2)}(m^2) = \tfrac{1}{2}[m w_{(2)}'(m^2) - w_{(1)}'(m^2)] \tag{132}$$

so that Eq. (124) reads

$$-\tfrac{1}{2} S'_F(p) = i \frac{Z_2}{\gamma \cdot p - M + i\epsilon}$$

$$+ \tfrac{1}{2} \int_{M+\mu}^{\infty} \left\{ \frac{w_{(1)}'(m^2) + m w_{(2)}'(m^2)}{\gamma \cdot p - m + i\epsilon} - \frac{w_{(1)}'(m^2) - m w_{(2)}'(m^2)}{\gamma \cdot p + m - i\epsilon} \right\} \frac{dm^2}{m} \tag{133}$$

For a local field theory wherein the spinor operators obey the canonical equal-time commutation rules:

$$[\psi(x), \tilde{\psi}(x')]_+ \Big|_{x_0 = x'_0} = \gamma^0 \delta^{(3)}(\mathbf{x} - \mathbf{x}') \tag{134}$$

a further condition on the weight function $\rho_{(1)}$ can be deduced by comparing the Lehmann form for $(\Psi_0, [\psi(x), \tilde{\psi}(x')]_+ \Psi_0)$ when $x_0 = x'_0$, with the vacuum expectation of Eq. (126). By invoking the assumed C-invariance of the theory (which enables us to relate $(\Psi_0, \psi(x) \, \tilde{\psi}(y) \, \Psi_0)$ to $(\Psi_0, \tilde{\psi}(x) \, \psi(y) \, \Psi_0)$), we find that the vacuum expectation value of the anticommutator of $\psi(x)$ and $\tilde{\psi}(y)$ has the following spectral representation:

$$(\Psi_0, [\psi(x), \bar{\psi}(x')]_+ \Psi_0)$$
$$= \int_0^\infty dm^2 \, \{\rho_{(1)}(m^2) \, (-iS(x - x'; m) + \rho_{(2)}(m^2) \, i\Delta(x - x'; m^2)\} \quad (135)$$

Comparing this equation for $x_0 = x'_0$ with the vacuum expectation value Eq. (134), we deduce that

$$\int_0^\infty dm^2 \, \rho_{(1)}(m^2) = 1 = Z_2 + \int_0^\infty dm^2 \, w_{(2)}'(m^2) \quad (136)$$

since $\Delta(\mathbf{x}, 0; m^2) = 0$.

Finally, if the equation of motion for the ψ operator is

$$(i\gamma_\mu \partial^\mu - M) \, \psi(x) = G\gamma_5 \phi(x) \, \psi(x) - \delta M \psi(x) \quad (137)$$

a procedure similar to that followed in the boson case indicates that

$$M - M_0 = \delta M = \frac{\int_0^\infty [(M - m) \, \rho_{(1)}(m^2) + \rho_{(2)}(m^2)] \, dm^2}{\int_0^\infty \rho_{(1)}(m^2) \, dm^2} \quad (138)$$

Equation (138) is obtained by operating with $i\gamma \cdot \partial - M$ on

$$(\Psi_0, [\bar{\psi}(x), \psi(x')]_+ \Psi_0)$$

and making use of the equation of motion (137), the representation (135) and thereafter going to the equal-time limit.

When calculated in perturbation theory, Z_2 is divergent. One therefore defines a renormalized Heisenberg operator $\psi_R(x)$ by the equation

$$\psi_R(x) = Z_2^{-1/2} \psi(x) \quad (139a)$$
$$\bar{\psi}_R(x) = Z_2^{-1/2} \bar{\psi}(x) \quad (139b)$$

The renormalized operator $\psi_R(x)$ has the property that its matrix element between the physical one-nucleon state and the vacuum is finite. The renormalized operator satisfies the commutation rules:

$$[\psi_R(x), \bar{\psi}_R(x')]_+ \Big|_{x_0 = x'_0} = -iZ_2^{-1} S(x - x') \quad (140)$$

The renormalized operators have a spectral representation for the twofold vacuum expectation value in terms of weight functions $\rho_{(1)R}$ and $\rho_{(2)R}$, which is identical to that presented above. It is, in fact, clear that the renormalized weight functions are related to the unrenormalized one by a multiplicative factor: $Z_2 \rho_{(i)R} = \rho_{(i)}$ $(i = 1, 2)$ whence, by Eq. (136),

$$Z_2^{-1} = \int_0^\infty \rho_{(1)R}(m^2) \, dm^2 \quad (141)$$

Upon defining the renormalized charge G_R, by the relation

$$G_R = Z_1 Z_2^{-1} Z_3^{1/2} G \quad (142)$$

the equation of motion for the renormalized Heisenberg operators for neutral pseudoscalar meson theory reads:

$$(i\gamma \cdot \partial - M)\,\psi_R(x) = -\delta M \psi_R(x) + G_R Z_1 Z_2^{-1} \gamma_5 \phi_R(x)\,\psi_R(x)$$
$$(\square + \mu^2)\,\phi_R(x) = \tfrac{1}{2} G_R Z_1 Z_3^{-1} [\bar{\psi}_R(x)\,\gamma_5, \psi_R(x)] + \tfrac{1}{2}\delta\mu_0^2 \phi_R(x) - \delta\lambda' \phi_R^3(x) \tag{143}$$

When the predictions of the theory are calculated in perturbation theory in terms of these renormalized operators (with Z_1, Z_2, Z_3 defined as above) all cross sections and level shifts are finite. However, there still exist divergences connected with such local quantities as $(\boldsymbol{\Psi}_0, \mathbf{J}_R(x)\,\mathbf{J}_R(x')\,\boldsymbol{\Psi}_0)$ or $(\boldsymbol{\Psi}_0, \phi_R(x)\,\phi_R(x')\,\boldsymbol{\Psi}_0)$. This is a reflection of the fact that a local operator such as $\mathbf{J}_R(x)$ or $\phi_R(x)$ cannot be an observable since it would correspond to the possibility of making measurement at a single space-time point which is, however, impossible by the uncertainty principle. An analysis of possible measurements indicates that only weighted averages over finite space-time regions are observable, i.e., only quantities of the type $\int_\Omega d^4x\, \mathbf{J}(x)\, f(x)$ where $f(x)$ is some weight function.

The spectral representations for the Green's function $(\boldsymbol{\Psi}_0, T(\psi(x)\,\bar{\psi}(y))\,\boldsymbol{\Psi}_0)$ and $(\boldsymbol{\Psi}_0, T(\phi(x)\,\phi(y))\,\boldsymbol{\Psi}_0)$ which we have discussed in this section are general and only depend on the relativistic invariance of the theory and the certain properties of the energy-momentum spectrum of the physical states of the theory. Clearly, one criterion that must be demanded of any perturbation theoretic approach to the problem of finding solutions of a quantized field theory is that it yield approximate solutions which have the same analytic properties as those the exact solutions are known to possess. Attempts to incorporate in perturbation theory the analytic properties of the Δ'_F and S'_F have been formulated by Redmond (1958) and Falk (1958). For this, and other reasons, it is of great interest to have similar representations for the higher order Green's functions, and a great deal of attention has recently been given to this problem. We refer the reader to the papers of Chisholm (1952), Nambu (1957, 1958), Nakanishi (1957), Symanzik (1958), Mathews (1959), Landau (1959), Okun (1960), and Björken (1959) for an introduction to these problems. In Chapter 18 we shall derive a representation for the matrix element $\langle \mathbf{p}, \boldsymbol{\alpha} \,|\, [\phi(x), \phi(y)] \,|\, \mathbf{q}, \boldsymbol{\beta} \rangle$ on the assumption that ϕ is a local field.

17c. The Magnitude of the Renormalization Constants

The formal considerations of the last section indicated that the renormalization constant Z_3 had the property that $0 \leqslant Z_3 \leqslant 1$, so that three possibilities must be discussed: (*a*) $0 < Z_3 < 1$, (*b*) $Z_3 = 0$, (*c*) $Z_3 = 1$. A fourth possibility: (*d*) $Z_3 < 0$, actually occurs for some theory (recall the Lee model), so that this phenomenon must also be considered. To see what these various possibilities mean, we shall discuss as a representative theory the case of quantum electrodynamics. Recall that the renormalized propagator $D'_{FR\mu\nu}(x - y)$ can be expressed in terms of the renormalized operators, $\mathbf{A}_{R\mu}(x) = Z_3^{-1/2}\mathbf{A}_\mu(x)$, as follows:

$$D'_{FR\mu\nu}(x - y) = (\Psi_0, T(\mathbf{A}_{R\mu}(x)\, \mathbf{A}_{R\nu}(y))\, \Psi_0) \tag{144a}$$

and has the following spectral representation:

$$D'_{FR\mu\nu}(k^2) = (g_{\mu\nu} - k_\mu k_\nu k^{-2})\, D_{FR}(k^2) \tag{144b}$$

$$D'_{FR}(k^2) = \frac{1}{k^2 + i\epsilon} + \int_0^\infty dM^2 \frac{\sigma_R(M^2)}{k^2 - M^2 + i\epsilon} \tag{144c}$$

Recall also that Z_3 is given in terms of $\sigma_R(M^2)$ by the expression

$$Z_3^{-1} = 1 + \int_0^\infty \sigma_R(M^2)\, dM^2 \tag{145}$$

It is convenient for the discussion which follows to introduce the function

$$P(k^2) = k^2 D'_{FR}(k^2) \tag{146}$$

which has the properties

$$\lim_{k^2 \to 0} P(k^2) = P(0) = 1 \tag{147}$$

and

$$\begin{aligned}\lim_{k^2 \to \infty} P(k^2) &= \lim_{k^2 \to \infty} \left(1 + \int_0^\infty \frac{dM^2\, \sigma_R(M^2)}{\left(1 - \dfrac{M^2}{k^2} + i\epsilon\right)}\right) \\ &\approx 1 + \int_0^\infty \sigma_R(M^2)\, dM^2 + O\left(\frac{1}{k^2}\right) \\ &= Z_3^{-1}\end{aligned} \tag{148}$$

As noted previously, perturbation theory to lowest order in e^2, gives the result that asymptotically

$$\lim_{M^2 \to \infty} \sigma_R(M^2) \simeq \frac{e_R^2}{3\pi} \frac{1}{M^2} \tag{149}$$

so that

$$Z_3^{-1} \approx 1 + \int^\infty \frac{e_R^2}{3\pi} \frac{dM^2}{M^2} \tag{150a}$$

$$\approx \infty \tag{150b}$$

i.e., that $Z_3 = 0$. The implication of this result is that the bare charge $e_0^2 = Z_3^{-1}\, e_R^2$ is infinite. This result can be stated in a somewhat less drastic fashion by noting that $Z_3^{-1}\, e_R^2$ is the coefficient of $1/r$ in the electrostatic potential between two charges at *close* distances. The above perturbation theoretic result can alternatively be interpreted as saying that Coulomb law, e_R^2/r, which for large distances defines the (renormalized) electric charges of the particles, is actually more singular than $1/r$ at close distances. Stated more precisely, we have noted previously that the potential energy between two heavy point test bodies separated by a distance

r (whose renormalized charges are q_R and q'_R respectively) is given by

$$V(r) = \frac{q_R q'_R}{(2\pi)^3} \int d^3p \, e^{i\mathbf{p}\cdot\mathbf{r}} \, D_{FR}(p^2, e_R^2)$$

$$\approx \frac{q_R q'_R}{4\pi r}\left[1 + \frac{\alpha_R}{3\pi}\int_{4m^2}^{\infty} dM^2 \, e^{-r/(\hbar^2/M^2c^2)^{1/2}} \cdot \left(1 + \frac{2m^2}{M^2}\right)\left(1 - \frac{4m^2}{M^2}\right)^{1/2} \frac{1}{M^2}\right] \quad (151)$$

which for $r \ll \hbar/mc$ takes the asymptotic form [Gell-Mann and Low (Gell-Mann 1955)]

$$V(r) \approx \frac{q_R q'_R}{4\pi r}\left[1 + \frac{2\alpha_R}{3\pi}\left\{\log\left(\frac{\hbar}{mcr}\right) - \frac{5}{6} - \log\gamma\right\} + O(\alpha_R^2)\right] \quad (152)$$

where $\gamma = 1.781 \cdots$. We have also previously indicated that as $r \to 0$, $V(r)$ must approach $qq'/4\pi r$ where q, q' are the bare charges, so that we may write

$$q_0 q'_0 = q_R q'_R \lim_{r\to 0}\left\{1 + \frac{2\alpha_R}{3\pi}\log\frac{(\hbar/mc)}{r} + \cdots\right\} \quad (153)$$

Hence, instead of saying that the bare charges become infinite, we can say that, when vacuum polarization effects are included, the potential between charged particles is more singular than $1/r$ near $r = 0$, the potential actually varying as $1/r \log 1/r$. Under these circumstances, it is of course not very useful to speak in terms of the bare charge e_0. The theory is defined in terms of its large distance behavior (or equivalently in terms of its low energy behavior). This behavior of Z_3, namely that $Z_3 = 0$ is a distinct possibility and would, apart from renormalizations, yield a consistent physical picture. The conjecture that $Z_3 = 0$ is based on the result of a second-order perturbation theoretic calculation for the computation of $\sigma_R(M^2)$. In this calculation only the contribution of the single closed loop illustrated in Figure 17.5 was included in σ_R.

It is interesting to consider the form of $D'_{FR}(k^2)$, or equivalently of $P(k^2)$, when the contribution of all the diagrams of the form illustrated in Figure 17.6 are included. We have seen in Chapter 16, Section b that the effect of the inclusion of such improper diagrams is to replace $D'_F{}^{(2)}$ by

$$D'_F(k) = \frac{D_F(k)}{1 - \Pi^{(2)}(k^2)} \sim \frac{1}{(1 + C^{(2)})\,k^2 - (\Pi^{(2)}(k^2) - \Pi^{(2)}(0))\,k^2} \quad (154)$$

with $(Z_3{}^{(2)})^{-1} = 1 + C^{(2)}$. To this order $D'_{FR}(k^2)$ in the $\lim k^2 \to \infty$ is therefore given by

$$\lim_{k^2\to\infty} D'_{FR}(k^2) \sim \frac{1}{k^2\left(1 - \frac{\alpha_R}{3\pi}\log\frac{k^2}{4m^2} + \cdots\right)} \quad (155)$$

or equivalently

$$\lim_{k^2\to\infty} P(k^2) \sim \frac{1}{1 - \frac{\alpha_R}{3\pi}\log\frac{k^2}{4m^2} + \cdots} \tag{156}$$

If this result is accepted, then Z_3^{-1}, which is given by $\lim_{k^2\to\infty} P(k^2)$, in the absence of a cutoff can become negative. This contradicts the general result $0 \leqslant Z_3 \leqslant 1$ derived on the assumption that every state has a positive

Fig. 17.5

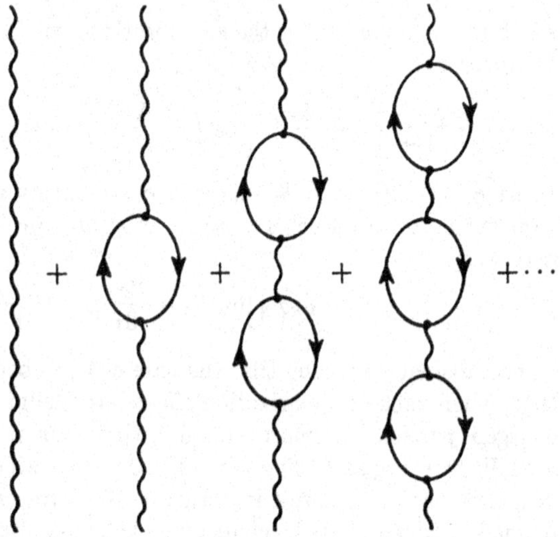

Fig. 17.6

norm in Hilbert space, so that if the theory is to be mathematically consistent, states with negative norm ("ghost states") must exist. The S matrix is therefore no longer unitary and the situation is, in fact, very similar to the problem encountered in the Lee model (Chap. 12, Sec. b) when the critical value of the coupling constant was exceeded [Källén (1955), Ford (1957), Umezawa (1956a, b), Landau (1955)].

This result for the function D'_F was obtained by summing a particular class of Feynman diagrams without regard to whether the approximation satisfies the general requirements of the Lehmann representation, or equivalently Eq. (76). Let us investigate what the result is of forcing Eq. (76) to be satisfied to this approximation [Redmond (1958), Bogoliubov (1959)]. To this end we write the contribution of the nth-order term to $P(k^2)$ for large k^2 in the form

$$P^{(n)}(k^2) \simeq -(F(k^2, m^2))^n \tag{157a}$$

$$F(k^2, m^2) = \frac{\alpha_R}{3\pi}\log\frac{k^2 - 4m^2}{4m^2} \tag{157b}$$

where the term $4m^2$ has been introduced into the argument of the logarithm in order to correctly represent the imaginary part of F, namely,

$$\operatorname{Im} F(k^2, m^2) = -\frac{\alpha_R}{3} \theta(k^2 - 4m^2) \tag{157c}$$

and at the same time retaining its normalization $F(0, m^2) = 0$. A direct summation of $P^{(n)}(k^2)$ yields as we have seen

$$\lim_{k^2 \to \infty} P(k^2) = \sum_{n=0}^{\infty} P^{(n)}(k^2) = \frac{1}{1 - \frac{\alpha_R}{3\pi} \log \frac{k^2 - 4m^2}{4m^2}} \tag{158}$$

Let us, however, proceed in such a way that Eq. (76) be satisfied. To this end we note that the nth order contribution to the propagator, $D'_{FR}{}^{(n)}(k^2)$, is given by

$$D'_{FR}{}^{(n)}(k^2) = \frac{1}{k^2} P^{(n)}(k^2) \approx \frac{1}{k^2} \left(\frac{\alpha_R}{3\pi} \log \frac{k^2 - 4m^2}{4m^2} \right)^n \tag{159a}$$

It can be represented in the Lehmann spectral form by writing

$$\begin{aligned} D'_{FR}{}^{(n)}(k^2) &= \frac{1}{k^2} \left(\frac{\alpha_R}{3\pi} \log \frac{k^2 - 4m^2}{4m^2} \right)^n \\ &= \int_{4m^2}^{\infty} \frac{I_n(z)}{z - k^2} dz \end{aligned} \tag{159b}$$

with

$$\pi I_n(z) = \operatorname{Im} D'_{FR}{}^{(n)}(k^2) \tag{159c}$$

Now

$$\pi I_n(z) = \operatorname{Im} \frac{1}{z} \left(\frac{\alpha_R}{3\pi} \log \frac{z - 4m^2}{4m^2} \right)^n \tag{160}$$

so that

$$\begin{aligned} I(z) &= \sum_{n=0}^{\infty} I_n(z) \\ &= \frac{\alpha_R}{3\pi z} \left[\left(1 - \frac{\alpha_R}{3\pi} \log \frac{(z - 4m^2)}{4m^2} \right)^2 + \frac{\alpha_R}{9} \right]^{-1} \theta(z - 4m^2) \end{aligned} \tag{161}$$

and to this approximation the photon Green's function which satisfies (76) is given by

$$\begin{aligned} D'_{FR}(k^2) &= \frac{1}{k^2} + \frac{\alpha_R}{3\pi} \int_{4m^2}^{\infty} \frac{dz}{z - k^2 - i\epsilon} \left[\left(1 - \frac{\alpha_R}{3\pi} \log \left(\frac{z - 4m^2}{4m^2} \right) \right)^2 + \frac{\alpha_R}{9} \right]^{-1} \\ &= \frac{1}{k^2} \left(1 - \frac{\alpha_R}{3\pi} \log \frac{k^2 - 4m^2}{4m^2} \right)^{-1} \\ &\quad - \frac{3\pi}{\alpha_R} \frac{1}{\left[1 - 4e^{-\frac{3\pi}{\alpha_R}} \right] \left[k^2 - 4m^2 + 4m^2 e^{\frac{3\pi}{\alpha_R}} \right]} \end{aligned} \tag{162a}$$

which for $k^2 \gg m^2$ and $\alpha_R < 1$ becomes

$$D'_{FR}(k^2) \approx \frac{1}{k^2} \frac{1}{1 - \frac{\alpha_R}{3\pi} \log \frac{-k^2}{4m^2}} - \frac{3\pi}{\alpha_R} \frac{1}{k^2 + 4m^2 e^{3\pi/\alpha_R}} \tag{162b}$$

The propagator given by Eq. (162b) has the following properties:

1. it does not have a logarithmic pole;
2. in the neighborhood of $\alpha_R = 0$, $D'_{FR}(k^2)$ has an essential singularity of the type $\exp(-3\pi/\alpha_R)$. Nonetheless in the neighborhood of the point $\alpha_R = 0$, $D_{FR}(k^2)$ has an asymptotic expansion that coincides with the ordinary perturbation theoretic expansion, the nth-order term of which can be represented in the form (159a);
3. from the fact that $Z_3^{-1} = \lim_{k^2 \to \infty} k^2 D'_{FR}(k^2)$, we deduce that

$$Z_3 = \frac{\alpha_R}{3\pi} \tag{163}$$

so that in this method of calculation the bare charge $\alpha_0 = Z_3^{-1}\alpha_R = 3\pi$ is independent of α_R. [This possibility was already suggested by Gell-Mann and Low (Gell-Mann 1954) who indicated that, if Z_3 is finite, then the bare charge would be independent of the renormalized one.]

The possibility that the solutions of relativistic field theories have the property that they have an essential singularity for vanishing value of the coupling constant has been especially stressed by Redmond (1958a, b, 1959). In particular, Redmond and Uretski [Redmond (1958)] have conjectured that all nontrivial field theories are characterized by spectral functions $\rho(M^2)$ which have essential singularities for $M^2 = 0$, so that if one expands these spectral functions in powers of the coupling constant, all the coefficients diverge as functions of M^2 as $M^2 \to \infty$, even though the sum of the series goes to zero.[4] Redmond and Uretski have further conjectured that the difference between a renormalizable and a nonrenormalizable theory is that the former permits an asymptotic expansion in the neighborhood of zero coupling constant while the latter does not.

We have thus indicated how the possibility that $0 < Z_3 < 1$ can arise. Which of the above possibilities is the one which a correct solution will exhibit is at present not known. This remains the outstanding mathematical problem in the present formulation of quantum field theory.

Källén (1954, 1958) has attempted to give a proof that the renormaliza-

[4] Consider the function $\left(\frac{M^2}{4m^2}\right)^{-\alpha_R/3\pi}$, which for large M^2 goes to zero. However, its powers series expansion

$$\left(\frac{M^2}{4m^2}\right)^{-\alpha_R/3\pi} = e^{-\frac{\alpha_R}{3\pi}\log\frac{M^2}{4m^2}} = 1 - \frac{\alpha_R}{3\pi}\log\frac{M^2}{4m^2} + \frac{1}{2}\left(\frac{\alpha_R}{3\pi}\log\frac{M^2}{4m^2}\right)^2 + \cdots$$

has the property that every coefficient diverges as $M^2 \to \infty$.

tion constants in quantum electrodynamics are infinite without explicit use of perturbation theory. The main steps of his proof can be summarized as follows: Under the assumption that electrodynamics is mathematically consistent and that the multiplicative renormalization constants Z_1, $Z_2 = Z_1$, and Z_3 are finite, Källén establishes a formula for the asymptotic behavior of the renormalized vertex function $\Gamma_{R\mu}(p, p') = \Gamma_{1\mu}(p, p') = \Gamma_{1\mu}(p^2, p'^2, (p - p')^2)$ when the momentum transfer $(p - p')^2$ tends to infinity and the two momenta p, p' are on the mass shell. Källén's formula asserts that

$$\lim_{(p-p')^2 \to \infty} \Gamma_{1\mu}(p^2 = p'^2 = m^2; (p - p')^2) = Z_1 \gamma_\mu \tag{164}$$

i.e., that in this limit $\Gamma_{1\mu}$ is given by the Born approximation multiplied by Z_1. One then shows by using a result of Lehmann, Symanzik, and Zimmermann [Lehmann (1955a)] generalized to quantum electrodynamics by Evans (1960) and by Drell and Zachariasen [Drell (1960)] that the vertex operator actually vanishes in the above limit, so that one concludes $Z_1 = 0$ or equivalently $Z_1^{-1} = \infty$. Källén's derivation has been criticized on two scores. One concerns the gauge invariance of the result (164). Thus Johnson (1958) has observed that, while $\Gamma_{1\mu}$ is gauge invariant and directly connected with experimentally observable quantities, the renormalization constant Z_1 is not gauge invariant. In fact, we have noted that in second order in the gauge in which $D_{F\mu\nu}(k^2) = (g_{\mu\nu} - k_\mu k_\nu k^{-2})\, k^{-2}$, Z_1 and Z_2 are finite, whereas in the gauge in which $D_{F\mu\nu}(k^2) = g_{\mu\nu} k^{-2}$, Z_1^{-1} is divergent. Johnson's remark, therefore, indicates that (164) cannot be true, unless it happens to be accidentally so in a particular gauge. Zumino (1960) has verified that Eq. (164) is indeed inconsistent with the gauge invariance of quantum electrodynamics and, furthermore, that it is very unlikely that it holds in any particular gauge.

A second objection to Källén's proof has been raised by Gasiorowicz, Yennie, and Suura [Gasiorowicz (1959)], who have pointed out that Källén's enumeration of the possible forms of the dispersion relations for the vertex function is incomplete, and that because of this his conclusion cannot be drawn.

17*d*. The *S* Matrix in the Heisenberg Picture

In Section *a* of the present chapter we have indicated how to express the Green's function for a given process in terms of the vacuum expectation value of the Heisenberg operators associated with the ingoing and outgoing particles partaking in the process. In the present section we show how to express S-matrix elements entirely in terms of Heisenberg states and operators. The way this could be done was already indicated in Chapter 13 where we demonstrated that it was possible to formally define Heisenberg "in" operators by

$$\boldsymbol{\phi}_{\text{in}}(x) = \mathbf{V}_+(t)\, \boldsymbol{\phi}(x)\, \mathbf{V}_+(t)^{-1}; \; x_0 = t \tag{165a}$$

where

$$\mathbf{V}_+(t) = e^{i\mathbf{H}t}\, \Omega^{(+)}\, e^{-i\mathbf{H}t} \tag{165b}$$

and $\Omega^{(+)} = U(0, -\infty)$ is the Møller wave matrix which has the property that it transforms a positive energy eigenstate of $\mathbf{H}_0 = \mathbf{H}^{(0)}(0)$ into an ingoing (scattering) eigenstate of $\mathbf{H}$ of the same energy. Specifically, the Møller wave matrix $\Omega^{(+)}$ has the property that

$$\begin{aligned} \mathbf{H}(0)\,(1 - \Lambda) = \mathbf{H}(1 - \Lambda) &= \Omega^{(+)}\mathbf{H}^{(0)}(0)\,\Omega^{(+)*} \\ &= \mathbf{H}^{(0)}_{\text{in}}(0) = \mathbf{H}^{(0)}_{\text{in}}(t) = \mathbf{H}^{(0)}_{\text{in}} \end{aligned} \tag{166}$$

where Λ is the projection operator on the bound states of $\mathbf{H}$. We there also showed that the matrix elements of $\mathbf{V}_+(+\infty)$ evaluated between the true scattering states were the elements of the S matrix. Let us study the properties of the "in" operators in field theory in somewhat greater detail. For the present we shall assume that the theory contains no bound states, so that $\Omega^{(+)*} = \Omega^{(+)-1}$.

It is an immediate consequence of Eqs. (165) and (166) that $\boldsymbol{\phi}_{\text{in}}(x)$ obeys the equation of motion:

$$i\partial_0\boldsymbol{\phi}_{\text{in}}(x) = [\boldsymbol{\phi}_{\text{in}}(x), \mathbf{H}^{(0)}_{\text{in}}] \tag{167}$$

and that the "in" operators satisfy free-field commutation rules:

$$[\boldsymbol{\phi}_{\text{in}}(x), \boldsymbol{\phi}_{\text{in}}(y)] = i\Delta(x - y) \tag{168}$$

for all times. For the Bose field under consideration, the repeated use of Eq. (167), i.e.,

$$-\partial_0{}^2\boldsymbol{\phi}_{\text{in}}(x) = [[\boldsymbol{\phi}_{\text{in}}(x), \mathbf{H}^{(0)}_{\text{in}}], \mathbf{H}^{(0)}_{\text{in}}] \tag{169}$$

yields the Klein-Gordon equation for

$$-\partial_0{}^2\boldsymbol{\phi}_{\text{in}}(x) = (-\boldsymbol{\partial}^2 + \mu^2)\, \boldsymbol{\phi}_{\text{in}}(x) \tag{170}$$

It is to be emphasized, however, that although $\boldsymbol{\phi}_{\text{in}}(x)$ satisfies a free-field equation, nonetheless $\boldsymbol{\phi}_{\text{in}}(x)$ is a Heisenberg field, i.e., its time dependence is determined by $\mathbf{H}$ as indicated by Eq. (171a) below:

$$\begin{aligned} \boldsymbol{\phi}_{\text{in}}(x) &= e^{i\mathbf{H}t}\, \Omega^{(+)}\, e^{-i\mathbf{H}t}\, \boldsymbol{\phi}(x)\, e^{i\mathbf{H}t}\, \Omega^{(+)*}\, e^{-i\mathbf{H}t} \\ &= e^{i\mathbf{H}t}\, \Omega^{(+)}\boldsymbol{\phi}(\mathbf{x}, 0)\, \Omega^{(+)*}\, e^{-i\mathbf{H}t} \\ &= e^{i\mathbf{H}t}\, \boldsymbol{\phi}_{\text{in}}(\mathbf{x}, 0)\, e^{-i\mathbf{H}t} \qquad\qquad (171\text{a}) \\ &= e^{i\mathbf{H}^{(0)}_{\text{in}}t}\, \boldsymbol{\phi}_{\text{in}}(\mathbf{x}, 0)\, e^{-i\mathbf{H}^{(0)}_{\text{in}}t} \qquad\qquad (171\text{b}) \end{aligned}$$

It is possible to establish an explicit relation between $\boldsymbol{\phi}_{\text{in}}(x)$ and $\boldsymbol{\phi}(x)$ which does not involve $\mathbf{V}_+(t)$. To establish this correspondence, it is easiest to relate both these operators to the interaction picture operator $\phi(x)$. The latter, it will be recalled, is related to the Heisenberg operator $\boldsymbol{\phi}(x)$ by

$$U(t, 0)^{-1}\, \phi(x)\, U(t, 0) = \boldsymbol{\phi}(x) \tag{172}$$

We have taken the Dirac, Heisenberg, and Schrödinger pictures to coincide at $t = 0$, so that

$$\phi(\mathbf{x}, 0) = \boldsymbol{\phi}(\mathbf{x}, 0) \tag{173}$$

Since the time dependence of the in-operators has been established above, it suffices to obtain an explicit relation between $\boldsymbol{\phi}_{\text{in}}(\mathbf{x}, 0)$ and $\boldsymbol{\phi}(\mathbf{x}, 0)$ at time $t = 0$. To do so we write, using Eqs. (165a) and (172),

$$\begin{aligned}\boldsymbol{\phi}_{\text{in}}(\mathbf{x}, 0) &= \mathbf{V}_+(0)\, \boldsymbol{\phi}(\mathbf{x}, 0)\, \mathbf{V}_+(0)^{-1} \\ &= U(0, -\infty)\, \phi(\mathbf{x}, 0)\, U(0, -\infty)^{-1} \\ &= \boldsymbol{\phi}(\mathbf{x}, 0) + [U(0, -\infty), \phi(\mathbf{x}, 0)]\, U(0, -\infty)^{-1}\end{aligned} \tag{174a}$$

where

$$[U(0, -\infty), \phi(\mathbf{x}, 0)] = \sum_{n=0}^{\infty} \frac{(-i)^n}{(n-1)!} \int_{-\infty}^{0} d^4x_1 \cdots \int_{-\infty}^{0} d^4x_n$$
$$P([\mathcal{H}_I(x_1), \phi(\mathbf{x}, 0)]\, \mathcal{H}_I(x_2) \cdots \mathcal{H}_I(x_n)) \tag{174b}$$

As a specific example, we shall consider a theory for which

$$\mathcal{H}_I(x) = J(x)\, \phi(x)$$

where J only involves fermion factors (e.g., $J(x) = G : \bar{\psi}(x)\, \gamma\psi(x) :$), in which case the Heisenberg equation of motion for $\boldsymbol{\phi}$ is:

$$(\Box + \mu^2)\, \boldsymbol{\phi}(x) = \mathbf{J}(x) \tag{175}$$

For such a theory, Eq. (174b) can easily be evaluated, and the result is

$$\begin{aligned}&[U(0, -\infty), \phi(\mathbf{x}, 0)] \\ &\quad = +\sum_{n=0}^{\infty} \frac{(-i)^{n-1}}{(n-1)!} \int_{-\infty}^{0} d^4y \int_{-\infty}^{0} d^4x_1 \int_{-\infty}^{0} d^4x_2 \cdots \int_{-\infty}^{0} d^4x_{n-1} \\ &\qquad \cdot \Delta(\mathbf{x} - \mathbf{y}; -y_0)\, P(J(y)\, \mathcal{H}_I(x_1) \cdots \mathcal{H}_I(x_{n-1})) \\ &\quad = +\int_{-\infty}^{0} d^4y\, \Delta(\mathbf{x} - \mathbf{y}, -y_0)\, P(J(y)\, U(0, -\infty))\end{aligned} \tag{176}$$

Using the fact that $y_0 < 0$ and the properties of the chronological operator, the expression $P(J(y)\, U(0, -\infty))$ can be rewritten as follows:

$$\begin{aligned}P(J(y)\, U(0, -\infty)) &= P(U(0, y_0)\, J(y)\, U(y_0, -\infty)) \\ &= U(0, y_0)\, J(y)\, U(y_0, -\infty) \\ &= U(y_0, 0)^{-1}\, J(y)\, U(y_0, 0)\, U(0, y_0)\, U(y_0, -\infty) \\ &= \mathbf{J}(y)\, U(0, -\infty)\end{aligned} \tag{177}$$

Collecting these results, we derive that

$$\boldsymbol{\phi}_{\text{in}}(x, 0) = \boldsymbol{\phi}(x, 0) + \int_{-\infty}^{0} d^4y\, \Delta(x - y, -y_0)\, \mathbf{J}(y) \tag{178}$$

Since all the operators appearing in Eq. (178) are Heisenberg ones, the equation relating $\phi_{in}(x)$ to $\phi(x)$ for arbitrary x_0 is obtained by displacing the entire equation in time,

$$\begin{aligned}\phi_{in}(\mathbf{x}, t) &= \phi_{in}(x) = e^{-i\mathbf{H}t}\,\phi_{in}(\mathbf{x}, 0)\,e^{i\mathbf{H}t} \quad (t = x_0)\\ &= \phi(x) + \int_{-\infty}^{0} d^4y\,\Delta(\mathbf{x} - \mathbf{y}, -y_0)\,\mathbf{J}(y_0 + t, \mathbf{y})\\ &= \phi(x) + \int_{-\infty}^{t} d^4y\,\Delta(\mathbf{x} - \mathbf{y}, x_0 - y_0)\,\mathbf{J}(y_0, \mathbf{y}) \qquad (179)\end{aligned}$$

or with the four-dimensional notation $x = (\mathbf{x}, t)$

$$\begin{aligned}\phi(x) &= \phi_{in}(x) - \int_{-\infty}^{t} d^4y\,\Delta(x - y)\,\mathbf{J}(y)\\ &= \phi_{in}(x) + \int_{-\infty}^{+\infty} d^4y\,\Delta_R(x - y)\,\mathbf{J}(y) \qquad (180)\end{aligned}$$

where $\Delta_R(x)$ is the retarded singular function, $\Delta_R(x) = -\theta(x)\,\Delta(x)$. Equation (180) can be viewed as the integrated form of the equation of motion $(\Box + \mu^2)\,\phi(x) = \mathbf{J}(x)$ for which the solution satisfies the initial condition

$$\lim_{x_0 \to -\infty} \phi(x) = \phi_{in}(x) \qquad (181)$$

A heuristic and mathematically "vague" proof of this limiting procedure can be established by allowing the interaction to be adiabatically switched off, i.e., by letting $G \to G \exp(-\alpha|x_0|)$; see e.g., Källén (1953c). By "vague" we mean the following: since $\phi(x)$ is an operator, the sense in which $\phi(x)$ converges to $\phi_{in}(x)$ must be made more precise.

It was Haag (1955) who first stressed that the type of operator convergence to be used in Eq. (181) must be specified. Haag required that the limiting procedure $\phi(x) \to \phi_{in}(x)$ for $x_0 \to -\infty$ is to be understood in the sense of strong convergence, that is

$$\lim_{x_0 \to -\infty} ||(\phi(x) - \phi_{in}(x))\,\Psi|| = 0 \qquad (182)$$

for arbitrary but fixed vector $|\Psi\rangle$ in the domain of $\phi(x)$ and $\phi_{in}(x)$. In (182), $||\Phi||$ denotes the norm of the vector $|\Phi\rangle$, i.e., $||\Phi|| = (\Phi, \Phi)^{1/2}$. It was shown, however, by Lehmann, Symanzik, and Zimmermann [Lehmann (1955a)] that there exist two difficulties with the requirement that the convergence be that indicated by Eq. (182). The first difficulty is connected with the fact the states $\phi(x)\,|\,\Psi\rangle$ and $\phi_{in}(x)\,|\,\Psi\rangle$ do not have a finite norm. This difficulty can, however, be overcome by considering the operator

$$\phi^f(t) = -i\int_t d\sigma^\mu(x)\,\phi(x)\,\overleftrightarrow{\partial_\mu}f(x) \qquad (183)$$

and similarly the operator $\phi_{in}{}^f(t)$, where $f(x)$ is a normalized positive energy wave-packet solution of the Klein-Gordon equation. [Note that $\phi_{in}{}^f(t) = \phi_{in}{}^f$ is actually time-independent because both f and ϕ_{in} satisfy the Klein-Gordon equation.] The second objection is concerned with the fact that, if one requires that

$$\lim_{t \to -\infty} \|(\phi^f(t) - \phi_{\text{in}}{}^f)\, \Psi\| = 0 \tag{184}$$

then with the choice $|\Psi\rangle = |\Psi_0\rangle$, Eq. (184) implies that the twofold vacuum expectation of the Heisenberg field is the same as that of the free field ϕ_{in}, in contradistinction to perturbation theoretic calculations. Lehmann, Symanzik, and Zimmermann therefore concluded that the convergence cannot be a strong convergence and that one can only demand that it be a weak convergence. In the present context, the operator $\phi^f(t)$ is said to converge weakly to $\phi_{\text{in}}{}^f$ if

$$\lim_{t \to -\infty} |(\Psi, (\phi^f(t) - \phi_{\text{in}}{}^f)\, \Phi)| = 0 \tag{185}$$

for all pairs of vectors $|\Psi\rangle$, $|\Phi\rangle$ in the domain of $\phi^f(t)$ and $\phi_{\text{in}}{}^f$. It is in this sense that Eq. (181) is to be understood.[5] The conditions under which an operator $\phi_{\text{in}}{}^f$ exists as a weak limit in *local* relativistic field theories have been investigated by Greenberg and Wightman. They have shown that even for theories with infinite wave function renormalization constants (i.e., Z_2^{-1}, $Z_3^{-1} = \infty$) in-operators exist as weak limits of the Heisenberg operators in the sense of Eq. (185). They have also shown that these operators have relativistic transformation properties of free fields of a given mass, and furthermore that these in-operators obey free-field commutation rules [see also Zimmermann (1959)]. We shall return to these points in Chapter 18.

We can similarly define the out-operator

$$\phi_{\text{out}}(x) = \mathbf{V}_-(t)\, \phi(x)\, \mathbf{V}_-(t)^{-1} \tag{186a}$$

where

$$\mathbf{V}_-(t) = e^{i\mathbf{H}t}\, \Omega^{(-)}\, e^{-i\mathbf{H}t} \tag{186b}$$

with $\Omega^{(-)} = U(0, +\infty)$ the outgoing Møller wave matrix. One then verifies that the Heisenberg out-operator has the following properties

$$[\Box + \mu^2]\, \phi_{\text{out}}(x) = 0 \tag{187}$$

$$[\phi_{\text{out}}(x), \phi_{\text{out}}(y)] = i\Delta(x - y; \mu) \tag{188}$$

$$\phi(x) = \phi_{\text{out}}(x) + \int_{-\infty}^{+\infty} \Delta_A(x - y)\, \mathbf{J}(y)\, d^4y \tag{189}$$

Furthermore, from the definition of in- and out-operators, it follows that

$$\begin{aligned} \phi_{\text{out}}(x) &= \mathbf{V}_-(t)\, \mathbf{V}_+(t)^{-1}\, \phi_{\text{in}}(x)\, [\mathbf{V}_-(t)\, \mathbf{V}_+(t)^{-1}]^{-1} \\ &= \mathbf{S}^{-1} \phi_{\text{in}}(x)\, \mathbf{S} \end{aligned} \tag{190}$$

i.e., the S matrix is the matrix relating in- and out-fields.

Equations (180) and (189) were first written down by Yang and Feldman [Yang (1950)] and Källén (1950) who obtained them by integrating the

[5] The situation is actually somewhat more complex if $Z_3 = 0$. In this case a time averaging of $\phi^f(t)$ has also to be made, since if $Z_3 = 0$ the vector $\phi^f(t)\,|\,\Psi\rangle$ may not exist [Greenberg (1957, unpublished); see Wightman (1959)].

Heisenberg equations of motion using the retarded and advanced Green's functions Δ_R and Δ_A. Since the operators $\boldsymbol{\phi}_{\text{in}}$ and $\boldsymbol{\phi}_{\text{out}}$ satisfy free-field equations, they postulated free-field commutation rules for them.[6] From the fact that the operators $\boldsymbol{\phi}_{\text{in}}$ and $\boldsymbol{\phi}_{\text{out}}$ obey the same commutation rules, they inferred that they must be related by a unitary transformation, i.e., $\boldsymbol{\phi}_{\text{out}} = \mathbf{S}^{-1}\boldsymbol{\phi}_{\text{in}}\mathbf{S}$. (This inference involves the extra assumption that $\boldsymbol{\phi}_{\text{in}}$ and $\boldsymbol{\phi}_{\text{out}}$ have the same no-particle state.) Yang and Feldman then showed the so-defined unitary matrix S is identical with the S matrix as usually defined in the interaction picture.

Similar results hold for the nucleon operator. For a theory of neutral mesons interacting with nucleons through an interaction $\mathfrak{L}_I = G[\tilde{\boldsymbol{\psi}}(x)\,\gamma, \boldsymbol{\psi}(x)]\,\boldsymbol{\phi}(x)$, the Yang-Feldman equation for the nucleon operator takes the form

$$\boldsymbol{\psi}(x) = \boldsymbol{\psi}_{\text{in}}(x) + \int S_R(x-y)\,\boldsymbol{\phi}(y)\,\gamma\boldsymbol{\psi}(y)\,d^4y \tag{191}$$

$$= \boldsymbol{\psi}_{\text{out}}(x) - \int S_A(x-y)\,\boldsymbol{\phi}(y)\,\gamma\boldsymbol{\psi}(y)\,d^4y \tag{192}$$

$$(-i\gamma\cdot\partial + M)\,\boldsymbol{\psi}_{\text{in}}(x) = (-i\gamma\cdot\partial + M)\,\boldsymbol{\psi}_{\text{out}}(x) = 0 \tag{193}$$

$$[\boldsymbol{\psi}_{\text{out}}(x), \tilde{\boldsymbol{\psi}}_{\text{out}}(y)]_+ = [\boldsymbol{\psi}_{\text{in}}(x), \tilde{\boldsymbol{\psi}}_{\text{in}}(y)]_+ = -iS(x-y; M) \tag{194}$$

$$\boldsymbol{\psi}_{\text{out}}(x) = \mathbf{S}^{-1}\boldsymbol{\psi}_{\text{in}}(x)\,\mathbf{S} \tag{195}$$

Since the in- and out-operators satisfy free-field equations, they can be decomposed into positive and negative frequency parts in an invariant manner valid for all times. That is we can write, for example,

$$\boldsymbol{\phi}_{\substack{\text{in}\\ \text{out}}}(x) = \frac{1}{\sqrt{(2\pi)^3}} \int \frac{d^3k}{\sqrt{2\omega_{\mathbf{k}}}} \left(e^{-ik\cdot x}\,\mathbf{a}_{\substack{\text{in}\\ \text{out}}}(\mathbf{k}) + e^{ik\cdot x}\,\mathbf{a}_{\substack{\text{in}\\ \text{out}}}{}^*(\mathbf{k})\right)$$

$$k_0 = \omega_{\mathbf{k}} = \sqrt{\mathbf{k}^2 + \mu^2} \tag{196a}$$

$$= \boldsymbol{\phi}^{(+)}_{\substack{\text{in}\\ \text{out}}}(x) + \boldsymbol{\phi}^{(-)}_{\substack{\text{in}\\ \text{out}}}(x) \tag{196b}$$

Similar equations hold for the decomposition of the nucleon in- and out-operators. The physical vacuum state $|\boldsymbol{\Psi}_0\rangle$ is that eigenstate of $\mathbf{H} = \mathbf{H}_{\text{in}}^{(0)}$ with eigenvalue 0. It can be characterized by the equations

$$\boldsymbol{\phi}^{(+)}_{\substack{\text{in}\\ \text{out}}}(x)\,|\,\boldsymbol{\Psi}_0\rangle = \boldsymbol{\psi}^{(+)}_{\substack{\text{in}\\ \text{out}}}(x)\,|\,\boldsymbol{\Psi}_0\rangle = \tilde{\boldsymbol{\psi}}^{(+)}_{\substack{\text{in}\\ \text{out}}}(x)\,|\,\boldsymbol{\Psi}_0\rangle = 0 \tag{197}$$

The eigenstates of

$$\mathbf{H}_{\text{in}}^{(0)} = \int d^3x : \tilde{\boldsymbol{\psi}}_{\text{in}}(x)\,(-i\gamma\cdot\partial + M)\,\boldsymbol{\psi}_{\text{in}}(x) : + \tfrac{1}{2}\int d^3x : (\partial^0\boldsymbol{\phi}_{\text{in}})^2 + (\boldsymbol{\partial}\boldsymbol{\phi}_{\text{in}})^2 + \mu^2\boldsymbol{\phi}_{\text{in}}{}^2 : \tag{198}$$

[6] This, of course, need not be the case. For example, the field $\boldsymbol{\phi}_{\text{sym}}(x) = \frac{1}{2}(\boldsymbol{\phi}_{\text{in}}(x) + \boldsymbol{\phi}_{\text{out}}(x))$ obeys free-field equations of motion but does not satisfy free-field commutation rules.

are constructed in the same way as in the free-particle case considered in Chapters 7 and 8, since ϕ_{in} and ψ_{in} have identical properties with respect to $|\Psi_0\rangle$ as those the free-particle operators in Chapters 7 and 8 had with respect to $|\Phi_0\rangle$. Thus the n-incoming meson state

$$\frac{1}{\sqrt{n!}} \mathbf{a}_{\text{in}}^*(\mathbf{k}_1) \cdots \mathbf{a}_{\text{in}}^*(\mathbf{k}_n) \mid \Psi_0\rangle$$

is an eigenstate of $\mathbf{H}_{\text{in}}^{(0)}$ with eigenvalue $\sum_{i=1}^{n} \omega(\mathbf{k}_i)$, and an eigenfunction of the total linear momentum $\mathbf{P}_i$ with eigenvalue $\sum_{i=1}^{n} \mathbf{k}_i$. It is, however, important to realize that since $\mathbf{H}_{\text{in}}^{(0)} = \mathbf{H}$, this n-incoming meson state is also an eigenstate of the total Hamiltonian. In particular, for example, the state $\mathbf{a}_{\text{in}}^*(\mathbf{k}_1)\, \mathbf{a}_{\text{in}}^*(\mathbf{k}_2) \mid \Psi_0\rangle$ is a stationary state describing two beams of free mesons of momentum $\mathbf{k}_1$ and $\mathbf{k}_2$ colliding and producing outgoing waves of collision products (e.g., mesons, nucleon pairs, etc.) consistent with the conservation laws. The incoming beams consist of dressed (physical) particles. Similarly, the states $\phi_{\text{in}}^{(-)}(x) \mid \Psi_0\rangle$, $\psi_{\text{in}}^{(-)}(x) \mid \Psi_0\rangle$, and $\bar{\psi}_{\text{in}}^{(-)}(x) \mid \Psi_0\rangle$ are physical one-particle states. To make these last statements more precise, consider the state vector

$$|\mathbf{p}s\rangle_{\text{in}} = \frac{1}{(2\pi)^{3/2}} \int d\sigma^\mu(x)\, \bar{\psi}_{\text{in}}(x)\, \gamma_\mu w^s(\mathbf{p})\, e^{-ip\cdot x} \mid \Psi_0\rangle \qquad (199)$$

where $w^s(\mathbf{p})$ is a positive energy Dirac free-particle spinor, so that $p^2 = M^2$, $p_0 > 0$. If we operate on both sides of Eq. (199) with the operator $\mathbf{P}_\mu$, make use of Eq. (96) which is valid for any Heisenberg operator and recall that $\mathbf{P}_\mu \mid \Psi_0\rangle = 0$, we deduce that

$$\mathbf{P}_\nu \mid \mathbf{p}s\rangle_{\text{in}} = \frac{1}{(2\pi)^{3/2}} \int [\mathbf{P}_\nu, \bar{\psi}_{\text{in}}(x)]\, \gamma_\mu w^s(\mathbf{p})\, e^{-ip\cdot x}\, d\sigma^\mu(x) \mid \Psi_0\rangle \qquad (200a)$$

$$= -\frac{i}{(2\pi)^{3/2}} \int \partial_\nu \bar{\psi}_{\text{in}}(x)\, \gamma_\mu w^s(\mathbf{p})\, e^{-ip\cdot x}\, d\sigma^\mu(x) \mid \Psi_0\rangle \qquad (200b)$$

$$= p_\nu \mid \mathbf{p}s\rangle_{\text{in}} \qquad (200c)$$

In going from Eq. (200b) to (200c), an integration by part has been performed and the surface terms neglected.[7] Since $p_\mu = (p_0 = \sqrt{\mathbf{p}^2 + M^2}, \mathbf{p})$, $|\mathbf{p}s\rangle_{\text{in}}$ is an eigenstate of the total energy-momentum operator $\mathbf{P}_\mu$ with eigenvalue p_μ and mass $M = \sqrt{p_\mu p^\mu}$. The state $|\mathbf{p}s\rangle_{\text{in}}$ is therefore a one-nucleon state of momentum $\mathbf{p}$. An identical procedure indicates that

[7] To justify the neglect of these terms, consider the spinor $f(x) = w(\mathbf{p}) \exp(-ip \cdot x)$ replaced by a wave packet $\hat{f}(x)$ and the limiting procedure $\hat{f} \to f$ carried out at the end. The ∂_0 term will vanish by virtue of the constancy in time of the right-hand side of Eq. (199) because both $w(\mathbf{p}) \exp(-ip \cdot x)$ and $\psi_{\text{in}}(x)$ satisfy the free-particle Dirac equation.

$\boldsymbol{\phi}_{\text{in}}^{(-)}(x) \mid \boldsymbol{\Psi}_0\rangle$ is a one-meson state. Note also that, since the one-particle states are stationary,

$$\mathbf{S}\mathbf{b}_{\text{in}}^*(\mathbf{p}s) \mid \boldsymbol{\Psi}_0\rangle = \lambda_{\mathbf{p}s}\mathbf{b}_{\text{in}}^*(\mathbf{p}s) \mid \boldsymbol{\Psi}_0\rangle = \mathbf{b}_{\text{out}}^*(\mathbf{p}s) \mid \boldsymbol{\Psi}_0\rangle \tag{201}$$

where $|\lambda_{\mathbf{p}s}|^2 = +1$. Similarly for the one-antinucleon and one-meson state, so that quite generally by an appropriate choice of phase factors

$$|1 \text{ particle}\rangle_{\text{in}} = |1 \text{ particle}\rangle_{\text{out}} \tag{202}$$

The outgoing states $|\mathbf{p}_1s_1, \cdots \mathbf{p}_l, s_l; \mathbf{q}_1, t_1 \cdots \mathbf{q}_m, t_m; \mathbf{k}_1 \cdots \mathbf{k}_n\rangle_{\text{out}}$ are similarly constructed:

$$|\mathbf{p}_1s_1, \cdots \mathbf{p}_ls_l; \mathbf{q}_1t_1 \cdots \mathbf{q}_mt_m; \mathbf{k}_1 \cdots \mathbf{k}_n\rangle_{\text{out}} = \frac{1}{\sqrt{l!m!n!}}\mathbf{b}_{\text{out}}^*(\mathbf{p}_1s_1) \cdots \mathbf{a}_{\text{out}}^*(\mathbf{k}_n) \mid \boldsymbol{\Psi}_0\rangle \tag{203}$$

These states are also eigenfunctions of $\mathbf{H}$ corresponding to outgoing wave boundary conditions. Finally, the S matrix can be expressed as

$$_{\text{out}}\langle\mathbf{b} \mid \mathbf{a}\rangle_{\text{in}} = S_{ba} \tag{204}$$

The formal developments outlined up to this point allow us to re-express matrix elements defined by interaction picture quantities in terms of Heisenberg variables. Of particular interest will be matrix elements of the form $\langle a \mid T(\phi(x_1) \cdots \phi(x_n)\, S) \mid b\rangle$, where $|a\rangle$ and $|b\rangle$ are Dirac picture state vectors. In this connection, recall that, if $|a\rangle$ and $|b\rangle$ are (bare) interaction picture eigenstates of H_0, we have deduced previously that

$$U(0, -\infty) \mid a\rangle = |\mathbf{a}\rangle_{\text{in}} \tag{205a}$$

$$U(0, +\infty) \mid b\rangle = |\mathbf{b}\rangle_{\text{out}} \tag{205b}$$

More explicitly, for example, if

$$|a\rangle = a^*(\mathbf{k}_1) \cdots a^*(\mathbf{k}_n) \mid \Phi_0\rangle \tag{206}$$

then

$$\begin{aligned} U(0, -\infty) \mid a\rangle &= U(0, -\infty)\, a^*(\mathbf{k}_1)\, U(0, -\infty)^{-1}\, U(0, -\infty) \cdots U(0, -\infty) \mid \Phi_0\rangle \\ &= \mathbf{a}_{\text{in}}^*(\mathbf{k}_1) \cdots \mathbf{a}_{\text{in}}^*(\mathbf{k}_n) \mid \boldsymbol{\Psi}_0\rangle \\ &= |\mathbf{a}\rangle_{\text{in}} \end{aligned} \tag{207}$$

The following relation is now easily verified

$$\langle a \mid T(\phi(x_1) \cdots \phi(x_n)\, S) \mid b\rangle = {}_{\text{out}}\langle\mathbf{a} \mid T(\boldsymbol{\phi}(x_1) \cdots \boldsymbol{\phi}(x_n)) \mid \mathbf{b}\rangle_{\text{in}} \tag{208}$$

The proof is similar to the one outlined previously. Assume the time ordering of $x_1, \cdots x_n$ is such that, for example $x_{10} > x_{20} \cdots > x_{n0}$, then

$$\begin{aligned} T(\phi(x_1) \cdots \phi(x_n)\, S) &= U(\infty, 0)\, U(0, x_{10})\, \phi(x_1)\, U(x_{10}, 0)\, U(0, x_{10}) \\ &\quad \cdot U(x_{10}, x_{20}) \cdots \phi(x_n)\, U(x_{n0}, 0)\, U(0, -\infty) \\ &= U(\infty, 0)\, \boldsymbol{\phi}(x_1) \cdots \boldsymbol{\phi}(x_n)\, U(0, -\infty) \end{aligned} \tag{209}$$

For an arbitrary time ordering it is evident that the following relation holds:

$$T(\phi(x_1) \cdots \phi(x_n)\, S) = U(\infty, 0)\, T(\boldsymbol{\phi}(x_1) \cdots \boldsymbol{\phi}(x_n))\, U(0, -\infty) \tag{210}$$

from which Eq. (208) follows immediately.

As an application of Eq. (208), we shall derive a closed expression for the meson-nucleon scattering amplitude in terms of Heisenberg operators in neutral pseudoscalar meson theory [Low (1955)]. To this end recall that in the Dirac picture the scattering amplitude for a transition from an initial state $|\mathbf{p}s; \mathbf{k}\rangle$ to a final one $|\mathbf{p}'s'; \mathbf{k}'\rangle$ is given by

$$\langle \mathbf{p}'s'; \mathbf{k}' \mid S \mid \mathbf{p}s; \mathbf{k}\rangle = \langle \mathbf{p}'s' \mid a(\mathbf{k}')\, Sa^*(\mathbf{k}) \mid \mathbf{p}s\rangle \tag{211}$$

We shall in the following make extensive use of the fact that we can write the creation and annihilation operators for a meson in the one-particle Klein-Gordon state f_α, $a_\alpha{}^*$ and a_α, in the form

$$a_\alpha{}^* = i \int \phi(x)\, \overleftrightarrow{\partial_0} f_\alpha(x)\, d^3x \tag{212a}$$

and

$$a_\alpha = i \int_t \bar{f}_\alpha(x)\, \overleftrightarrow{\partial_0} \phi(x)\, d^3x \tag{212b}$$

respectively, where f_α is a normalizable *positive* energy wave-packet solution of the Klein-Gordon equation. The advantage gained by considering wave packets lies in the fact that the state $a_\alpha{}^* \mid \Phi_0\rangle$ has a finite norm, whereas the state $a_{\mathbf{k}}{}^* \mid \Phi_0\rangle$ does not. In the limit of a plane wave

$$f_\alpha(x) \to f_{\mathbf{k}}(x) = \frac{1}{\sqrt{2(2\pi)^3\, \omega_{\mathbf{k}}}}\, e^{-ik\cdot x} \tag{213}$$

and in this limit $a_\alpha \to a_{\mathbf{k}}$ and $a_\alpha{}^* = a_{\mathbf{k}}{}^*$. Note that, since both f and ϕ satisfy the Klein-Gordon equation, the time t at which the integration is carried out in Eqs. (212a) and (212b) is arbitrary. We can therefore rewrite Eq. (211) as follows:

$$\langle \mathbf{p}'s'; \mathbf{k}' \mid S \mid \mathbf{p}s; \mathbf{k}\rangle$$
$$= -\int_{t'} d^3x' \int_t d^3x\, \overline{f_{\mathbf{k}'}(x')}\, \frac{\overleftrightarrow{\partial}}{\partial x'_0} \langle \mathbf{p}'s' \mid \phi(x')\, S\phi(x) \mid \mathbf{p}s\rangle \frac{\overleftrightarrow{\partial}}{\partial x_0} f_{\mathbf{k}}(x) \tag{214}$$

In the following we shall suppose $f_{\mathbf{k}}$ and $f_{\mathbf{k}'}$ to correspond to *normalizable* solutions of the Klein-Gordon equation with momenta nearly equal to $\mathbf{k}$ and $\mathbf{k}'$, respectively. We shall carry out the limit: $f_{\mathbf{k}} \to$ plane wave of momentum $\mathbf{k}$ only at the end of the calculation. It should be recalled that a correct description of an actual scattering experiment requires the use of such wave packets, and that the amplitude $\langle \mathbf{p}'s'; \mathbf{k}' \mid S \mid \mathbf{p}s; \mathbf{k}\rangle$ is in fact an idealized limit.

Since the times t and t' in (214) are arbitrary, we shall conveniently take t' in the infinite past and t in the infinite future so as to enable us to rewrite (214) in the form:

$$\langle \mathbf{p}'s'; \mathbf{k}' \mid S \mid \mathbf{p}s; \mathbf{k}\rangle$$

$$= -\int_{t'=+\infty} d^3x' \int_{t=-\infty} d^3x\, \overline{f_{\mathbf{k}'}(x')}\, \frac{\overleftrightarrow{\partial}}{\partial x'^0} \langle \mathbf{p}'s' \mid T(\phi(x')\, S\phi(x)) \mid \mathbf{p}s\rangle \frac{\overleftrightarrow{\partial}}{\partial x_0} f_{\mathbf{k}}(x) \tag{215}$$

Using the lemma (210), we can rewrite the matrix element in (215) as follows:

$$\langle \mathbf{p}'s' \mid T(S\phi(x')\, \phi(x)) \mid \mathbf{p}s\rangle = {}_{\text{out}}\langle \mathbf{p}'s' \mid T(\boldsymbol{\phi}(x')\, \boldsymbol{\phi}(x) \mid \mathbf{p}s\rangle_{\text{in}} \tag{216}$$

so that finally

$$\langle \mathbf{p}'s'; \mathbf{k}' \mid S \mid \mathbf{p}s, \mathbf{k}\rangle = -\lim_{t'\to+\infty} \lim_{t\to-\infty} \int_{t'} d^3x' \int_t d^3x$$

$$\overline{f_{\mathbf{k}'}(x')}\, \frac{\overleftrightarrow{\partial}}{\partial x'^0}\, {}_{\text{out}}\langle \mathbf{p}'s' \mid T(\boldsymbol{\phi}(x')\, \boldsymbol{\phi}(x)) \mid \mathbf{p}s\rangle_{\text{in}} \frac{\overleftrightarrow{\partial}}{\partial x^0} f_{\mathbf{k}}(x) \tag{217}$$

This matrix element can be further reduced if the asymptotic condition is satisfied. The asymptotic condition is the requirement that for arbitrary (but fixed) normalizable state vectors $|\boldsymbol{\Psi}\rangle$ and $|\boldsymbol{\Phi}\rangle$, $\lim_{x_0\to\pm\infty} \langle\boldsymbol{\Psi}|\, \phi'(x_0)\, |\boldsymbol{\Phi}\rangle$ exist, and more precisely that

$$\lim_{x_0\to\pm\infty} \langle \boldsymbol{\Psi} \mid i \int_{x_0} d^3x\, \overline{f_{\mathbf{k}}(x)}\, \overleftrightarrow{\partial_0} \boldsymbol{\phi}(x) \mid \boldsymbol{\Phi}\rangle$$

$$= \langle \boldsymbol{\Psi} \mid i \int_{x_0} d^3x\, \overline{f_{\mathbf{k}}(x)}\, \overleftrightarrow{\partial_0} \boldsymbol{\phi}_{\substack{\text{out}\\ \text{in}}}(x) \mid \boldsymbol{\Phi}\rangle \tag{218}$$

In other words, the asymptotic condition is the requirement that the limit defined by Eq. (182) and its extension to the definition of the out-field operator, Eq. (218), be meaningful. More generally, the asymptotic condition is the requirement that a field theory have an interpretation in terms of asymptotic particle observables, i.e., in terms of observables for stationary incoming and outgoing states which have the same transformation, statistical, and orthogonality properties as free particles of a given mass. These asymptotic states are constructed from the in- and out-operators, and differ from the free-particle states in that they contain particles of different momenta from the ingoing (or outcoming) particles corresponding to the outcoming (or incoming) scattered particles. The asymptotic condition requires that these in- and out-fields be obtainable from the Heisenberg operators by means of a well-defined limit: the asymptotic limit, Eq. (218). In the following we assume that the field theory under consideration is such that the asymptotic conditions are satisfied.

Now, in the limit that $f_{\mathbf{k}}$ is a plane wave, Eq. (218) becomes

$$\lim_{f\to f_{\mathbf{k}}} \lim_{x_0\to\pm\infty} \langle \boldsymbol{\Psi} \mid i \int d^3x\, \boldsymbol{\phi}(x)\, \overleftrightarrow{\partial_0} f_{\mathbf{k}}(x) \mid \boldsymbol{\Phi}\rangle = \langle \boldsymbol{\Psi} \mid \mathbf{a}^*_{\substack{\text{out}\\ \text{in}}}(\mathbf{k}) \mid \boldsymbol{\Phi}\rangle \tag{219}$$

The scattering amplitude can, therefore, be rewritten in the form:

$$\begin{aligned}\langle \mathbf{p}'s', \mathbf{k}' \mid S \mid \mathbf{p}s, k\rangle &= {}_{\text{out}}\langle \mathbf{p}'s' \mid \mathbf{a}_{\text{out}}(\mathbf{k}')\, \mathbf{a}_{\text{in}}^*(\mathbf{k}) \mid \mathbf{p}s\rangle_{\text{in}} \\ &= {}_{\text{out}}\langle \mathbf{p}'s'; \mathbf{k}' \mid \mathbf{a}_{\text{in}}^*(\mathbf{k}) \mid \mathbf{p}s\rangle_{\text{in}} \\ &= \lim_{t\to-\infty} {}_{\text{out}}\langle \mathbf{p}'s', \mathbf{k}' \mid i \int_t d^3x\, \boldsymbol{\phi}(x) \overleftrightarrow{\partial_0} f_{\mathbf{k}}(x) \mid \mathbf{p}s\rangle_{\text{in}} \end{aligned} \tag{220}$$

We next use the identity

$$\int d^3x \int_{-\infty}^{+\infty} dx^0\, \partial_0 F(x) = \int_{t=+\infty} d^3x\, F(x) - \int_{t=-\infty} d^3x\, F(x) \tag{221}$$

to cast Eq. (220) into the form

$$\begin{aligned} i \int_{t=-\infty} d^3x\, {}_{\text{out}}\langle \mathbf{p}'s'; \mathbf{k}' \mid \boldsymbol{\phi}(x) \mid \mathbf{p}s\rangle_{\text{in}} \overleftrightarrow{\partial_0} f_{\mathbf{k}}(x) \\ = i \int_{t=+\infty} d^3x\, {}_{\text{out}}\langle \mathbf{p}'s'; \mathbf{k}' \mid \boldsymbol{\phi}(x) \mid \mathbf{p}s\rangle_{\text{in}} \overleftrightarrow{\partial_0} f_{\mathbf{k}}(x) \\ - i \int d^3x \int_{-\infty}^{+\infty} dx^0\, \partial_0[{}_{\text{out}}\langle \mathbf{p}'s'; \mathbf{k}' \mid \boldsymbol{\phi}(x) \mid \mathbf{p}s\rangle_{\text{in}} \overleftrightarrow{\partial_0} f_{\mathbf{k}}(x)] \\ = {}_{\text{out}}\langle \mathbf{p}'s'; \mathbf{k}' \mid \mathbf{a}_{\text{out}}^*(\mathbf{k}) \mid \mathbf{p}s\rangle_{\text{in}} \\ - \int d^4x\, (\Box_x + \mu^2)\, {}_{\text{out}}\langle \mathbf{p}'s', \mathbf{k}' \mid \boldsymbol{\phi}(x) \mid \mathbf{p}s\rangle_{\text{in}} \cdot f_{\mathbf{k}}(x) \end{aligned} \tag{222}$$

To arrive at the last line of Eq. (222), we have made use of the asymptotic condition (219) to write $\mathbf{a}_{\text{out}}^*(\mathbf{k})$ in the first term. We have also used the fact that $f_{\mathbf{k}}(x)$ satisfies the Klein-Gordon equation, so that

$$\partial_0^2 f_{\mathbf{k}}(x) = (\boldsymbol{\partial}^2 - \mu^2)\, f_{\mathbf{k}}(x) \tag{223}$$

and thereafter we have performed a partial integration over the spatial variables in the second term. The first term can be further simplified by recalling that the one-particle state has the property that $|\mathbf{p}s\rangle_{\text{in}} = |\mathbf{p}s\rangle_{\text{out}}$ so that

$$\begin{aligned} {}_{\text{out}}\langle \mathbf{p}'s', \mathbf{k} \mid \mathbf{a}_{\text{out}}^*(\mathbf{k}) \mid \mathbf{p}s\rangle_{\text{in}} &= {}_{\text{out}}\langle \mathbf{p}'s' \mid \mathbf{a}_{\text{out}}(\mathbf{k}')\, \mathbf{a}_{\text{out}}^*(\mathbf{k}) \mid \mathbf{p}s\rangle_{\text{out}} \\ &= \delta^{(3)}(\mathbf{k} - \mathbf{k}')\, {}_{\text{out}}\langle \mathbf{p}'s' \mid \mathbf{p}s\rangle_{\text{out}} \\ &= \delta^{(3)}(\mathbf{k} - \mathbf{k}')\, \delta^{(3)}(\mathbf{p} - \mathbf{p}')\, \delta_{ss'} \end{aligned} \tag{224}$$

since

$$[\mathbf{a}_{\substack{\text{out}\\\text{in}}}(\mathbf{k}), \mathbf{a}_{\substack{\text{out}\\\text{in}}}^*(\mathbf{k}')] = \delta^{(3)}(\mathbf{k} - \mathbf{k}') \tag{225a}$$

$$[\mathbf{a}_{\text{out}}(\mathbf{k}), \mathbf{b}_{\text{out}}(\mathbf{p}s)] = 0 \tag{225b}$$

and

$$\langle \boldsymbol{\Psi}_0 \mid \mathbf{a}_{\text{out}}^*(\mathbf{k}) = 0 \tag{226}$$

The scattering matrix element is therefore given by the expression

$$\begin{aligned}\langle \mathbf{p}'s', \mathbf{k}' \mid S \mid \mathbf{p}s, \mathbf{k}\rangle &= \delta_{ss'}\delta^{(3)}(\mathbf{p} - \mathbf{p}')\, \delta^{(3)}(\mathbf{k} - \mathbf{k}') \\ &\quad - i \int d^4x\, (\Box_x + \mu^2)\, {}_{\text{out}}\langle \mathbf{p}'s'; \mathbf{k}' \mid \boldsymbol{\phi}(x) \mid \mathbf{p}s\rangle_{\text{in}} \cdot f_{\mathbf{k}}(x) \end{aligned} \tag{227}$$

A similar procedure as the above can be used to write

$$ {}_{\text{out}}\langle \mathbf{p}'s', k' \mid \phi(x) \mid \mathbf{p}s\rangle_{\text{in}} = {}_{\text{out}}\langle \mathbf{p}'s' \mid \mathbf{a}_{\text{out}}(\mathbf{k})\, \phi(x) \mid \mathbf{p}s\rangle_{\text{in}} $$

$$ = i \int_{t\to+\infty} \overline{f_{\mathbf{k}'}(x')}\, \frac{\overleftrightarrow{\partial}}{\partial x'^0}\, {}_{\text{out}}\langle \mathbf{p}'s' \mid \phi(x')\, \phi(x) \mid \mathbf{p}s\rangle_{\text{in}} \tag{228} $$

Since, in this last expression, x'^0 is in the infinite future, we shall write

$$ \phi(x')\, \phi(x) = T(\phi(x')\, \phi(x)) \tag{229} $$

The application of Eq. (221) now yields

$$ i \int_{t'=+\infty} d^3x' \overline{f_{\mathbf{k}'}(x')}\, \frac{\overleftrightarrow{\partial}}{\partial x'^0}\, {}_{\text{out}}\langle \mathbf{p}'s' \mid T(\phi(x')\, \phi(x) \mid \mathbf{p}s\rangle_{\text{in}} $$

$$ = i \int_{t'=-\infty} d^3x'\, \overline{f_{\mathbf{k}'}(x')}\, \frac{\overleftrightarrow{\partial}}{\partial x'^0}\, {}_{\text{out}}\langle \mathbf{p}'s' \mid T(\phi(x')\, \phi(x)) \mid \mathbf{p}s\rangle_{\text{in}} $$

$$ + i \int d^4x'\, \frac{\partial}{\partial x'^0} \left[\overline{f_{\mathbf{k}'}(x')}\, \frac{\overleftrightarrow{\partial}}{\partial x'^0}\, {}_{\text{out}}\langle \mathbf{p}'s' \mid T(\phi(x')\, \phi(x)) \mid \mathbf{p}s\rangle_{\text{in}} \right] \tag{230} $$

Since the surface integration in the first term is now carried out at $x'^0 = -\infty$, the T product rearranges the factors as follows:

$$ T(\phi(x')\, \phi(x)) = \phi(x)\, \phi(x') \quad \text{for } x'^0 \to -\infty \tag{231} $$

The first term on the right-hand side of Eq. (230) therefore vanishes, since

$$ i \int_{t'=-\infty} d^3x'\, \overline{f_{\mathbf{k}'}(x')}\, \frac{\overleftrightarrow{\partial}}{\partial x'^0}\, {}_{\text{out}}\langle \mathbf{p}'s' \mid \phi(x)\, \phi(x') \mid \mathbf{p}s\rangle_{\text{in}} $$

$$ = {}_{\text{out}}\langle \mathbf{p}'s' \mid \phi(x)\, \mathbf{a}_{\text{in}}(\mathbf{k}') \mid \mathbf{p}s\rangle_{\text{in}} \tag{232} $$

since $\mathbf{a}_{\text{in}}(\mathbf{k}) \mid \mathbf{p}s\rangle_{\text{in}} = \mathbf{b}_{\text{in}}^*(\mathbf{p}s)\, \mathbf{a}_{\text{in}}(\mathbf{k}) \mid \Psi_0\rangle = 0$. This is the motivation for introducing the T product. Carrying out the indicated differentiation and performing an integration by part over the spatial variables finally allows us to rewrite (230) as follows:

$$ \langle \mathbf{p}'s', \mathbf{k}' \mid S \mid \mathbf{p}s, \mathbf{k}\rangle = {}_{\text{in}}\langle \mathbf{p}'s', \mathbf{k}' \mid \mathbf{p}s, \mathbf{k}\rangle_{\text{in}} + \int d^4x \int d^4x' $$

$$ \overline{f_{\mathbf{k}'}(x')}\, f_{\mathbf{k}}(x)\, (\Box_x + \mu^2)\, (\Box_{x'} + \mu^2)\, {}_{\text{out}}\langle \mathbf{p}'s' \mid T(\phi(x')\, \phi(x) \mid \mathbf{p}s\rangle_{\text{in}} \tag{233} $$

In the limiting case of plane waves, Eq. (233) becomes

$$ \langle \mathbf{p}'s', \mathbf{k}' \mid S \mid \mathbf{p}s, \mathbf{k}\rangle = \delta^{(3)}(\mathbf{p} - \mathbf{p}')\, \delta_{ss'} \delta^{(3)}(\mathbf{k} - \mathbf{k}') + \frac{1}{(2\pi)^3 \sqrt{4\omega_{\mathbf{k}}\omega_{\mathbf{k}'}}} $$

$$ \int d^4x \int d^4x'\, e^{i(k'\cdot x' - k\cdot x)}\, (\Box_x + \mu^2)\, (\Box_{x'} + \mu^2)\, {}_{\text{out}}\langle \mathbf{p}'s' \mid T(\phi(x')\, \phi(x)) \mid \mathbf{p}s\rangle_{\text{in}} \tag{234} $$

Before proceeding further with the analysis of the meson-nucleon scattering amplitude as given by Eq. (234), we note that a slightly different expression could have been obtained by rewriting Eq. (228) as follows:

$$i \int_{t\to\infty} d^3x' \, \overline{f_{\mathbf{k}'}(x')} \frac{\overleftrightarrow{\partial}}{\partial x'^0} {}_{\text{out}}\langle \mathbf{p}'s' \mid \phi(x') \, \phi(x) \mid \mathbf{p}s\rangle_{\text{out}}$$

$$= i \int_{t\to\infty} d^3x' \, \overline{f_{\mathbf{k}'}(x')} \frac{\overleftrightarrow{\partial}}{\partial x'^0} {}_{\text{out}}\langle \mathbf{p}'s' \mid \theta(x' - x) \, [\phi(x'), \phi(x)] \mid \mathbf{p}s\rangle_{\text{out}} \quad (235)$$

where we have made use of the fact that the one-particle states are steady, to write $|\mathbf{p}s\rangle_{\text{in}} = |\mathbf{p}s\rangle_{\text{out}}$. In introducing the retarded commutator

$$R(\phi(x') \, \phi(x)) = -i \, \theta(x' - x) \, [\phi(x'), \phi(x)] \quad (236a)$$

$$\theta(x' - x) = 1 \quad \text{if } x'^0 > x^0$$
$$= 0 \quad \text{if } x'^0 < x^0 \quad (236b)$$

we have made use of the fact that x'^0 is in the infinite future, so that $\theta(x' - x) = 1$ and that the contribution of the term

$$i \int_{t\to\infty} d^3x' \, \overline{f_{\mathbf{k}'}(x')} \frac{\overleftrightarrow{\partial}}{\partial x'^0} {}_{\text{out}}\langle \mathbf{p}'s' \mid \phi(x) \, \phi(x') \mid \mathbf{p}s\rangle_{\text{out}}$$
$$= {}_{\text{out}}\langle \mathbf{p}'s' \mid \phi(x) \, \mathbf{a}_{\text{out}}(\mathbf{k}) \mid \mathbf{p}s\rangle_{\text{out}} \quad (237)$$

vanishes since $\mathbf{a}_{\text{out}}(\mathbf{k}) \mid \mathbf{p}s\rangle_{\text{out}} = 0$.

An analysis similar to that which led to Eq. (234) now yields

$$i \int_{t\to\infty} d^3x' \, \overline{f_{\mathbf{k}'}(x')} \frac{\overleftrightarrow{\partial}}{\partial x'^0} {}_{\text{out}}\langle \mathbf{p}'s' \mid \theta(x' - x) \, [\phi(x'), \phi(x)] \mid \mathbf{p}s\rangle_{\text{out}}$$

$$= -\int_{t=-\infty} d^3x' \, \overline{f_{\mathbf{k}'}(x')} \frac{\overleftrightarrow{\partial}}{\partial x'^0} {}_{\text{out}}\langle \mathbf{p}'s' \mid R(\phi(x') \, \phi(x)) \mid \mathbf{p}s\rangle_{\text{out}}$$

$$- \int d^4x' \, \partial_0 \left[\overline{f_{\mathbf{k}'}(x')} \frac{\overleftrightarrow{\partial}}{\partial x'^0} {}_{\text{out}}\langle \mathbf{p}'s' \mid R(\phi(x') \, \phi(x)) \mid \mathbf{p}s\rangle_{\text{out}} \right] \quad (238)$$

The surface term at $t = -\infty$ does not contribute due to the presence of the $\theta(x' - x)$ factor which vanishes for $x'^0 = -\infty$, so that the S-matrix element can also be written as

$$\langle \mathbf{p}'s', \mathbf{k}' \mid S \mid \mathbf{p}s, \mathbf{k}\rangle = \delta_{ss'}\delta^{(3)}(\mathbf{p} - \mathbf{p}') \, \delta^{(3)}(\mathbf{k} - \mathbf{k}') + \frac{1}{(2\pi)^3} \frac{1}{\sqrt{4\omega_{\mathbf{k}}\omega_{\mathbf{k}'}}}$$

$$i \int d^4x \int d^4x' \, e^{i(k' \cdot x' - k \cdot x)} \, (\Box_x + \mu^2) \, (\Box_{x'} + \mu^2) \, {}_{\text{out}}\langle \mathbf{p}'s' \mid R(\phi(x') \, \phi(x)) \mid \mathbf{p}s\rangle_{\text{in}} \quad (239)$$

It should be stressed that the representations (234) and (239) for the S-matrix element represent the same function only for $k^2 = k'^2 = \mu^2$ and

$p^2 = p'^2 = M^2$, i.e., only on the mass shell. Off the mass shell they will, in general, correspond to different functions.

Returning to the expression (234) for the scattering matrix element, we note that the right-hand side of this equation can be further reduced by recalling that

$$T(\phi(x)\,\phi(y)) = \tfrac{1}{2}\epsilon(x-y)\,[\phi(x), \phi(y)] + \tfrac{1}{2}[\phi(x), \phi(y)]_+ \tag{240}$$

so that

$$\begin{aligned}\frac{\partial}{\partial x^0} T(\phi(x)\,\phi(y)) &= \tfrac{1}{2}\epsilon(x-y)\,[\partial_0\phi(x), \phi(y)] + \tfrac{1}{2}[\partial_0\phi(x), \phi(y)]_+ \\ &\quad + \delta^{(1)}(x_0 - y_0)\,[\phi(x), \phi(y)] \\ &= T(\partial_0\phi(x)\,\phi(y))\end{aligned} \tag{241}$$

since the equal-time commutator of $\phi(x)$ and $\phi(y)$ vanishes. The second derivative is similarly computed, and we obtain

$$\begin{aligned}\partial_0{}^2 T(\phi(x)\,\phi(y)) &= \delta^{(1)}(x^0 - y^0)\,[\partial_0\phi(x), \phi(y)] + T(\partial_0{}^2\phi(x)\,\phi(y)) \\ &= -i\delta^{(4)}(x-y) + T(\partial_0{}^2\phi(x)\,\phi(y))\end{aligned} \tag{242}$$

so that

$$(\Box_x + \mu^2)\,T(\phi(x)\,\phi(y)) = -i\delta^{(4)}(x-y) + T(\mathbf{J}(x)\,\phi(y)) \tag{243}$$

where

$$(\Box + \mu^2)\,\phi(x) = \mathbf{J}(x) \tag{244}$$

Similarly, we deduce that

$$(\Box_y + \mu^2)\,T(\mathbf{J}(x)\,\phi(y)) = T(\mathbf{J}(x)\,\mathbf{J}(y)) - \delta^{(1)}(x^0 - y^0)\,[\mathbf{J}(x), \partial^0\phi(y)] \tag{245}$$

where we have assumed that for equal times $x_0 = y_0$, $\phi(y)$ and $\mathbf{J}(x)$ commute. This is an application of the causality requirement which states that the commutator of two local observables (Hermitian operators) taken at space-like points must vanish. We shall discuss this requirement in greater detail in Chapter 18. Combining all the above results finally permits us to write the scattering amplitude in the form

$$\begin{aligned}\langle \mathbf{p}'s', \mathbf{k}' \mid S \mid \mathbf{p}s, \mathbf{k}\rangle &= \delta_{ss'}\delta^{(3)}(\mathbf{p} - \mathbf{p}')\,\delta^{(3)}(\mathbf{k} - \mathbf{k}') \\ &\quad + \frac{1}{(2\pi)^3}\frac{1}{\sqrt{4\omega_{\mathbf{k}}\omega_{\mathbf{k}'}}}\int d^4x \int d^4x'\, e^{i(k'\cdot x' - k\cdot x)} \\ &\{{}_{\text{out}}\langle \mathbf{p}'s' \mid T(\mathbf{J}(x')\,\mathbf{J}(x)) + \delta^{(1)}(x^0 - x'^0)\,[\mathbf{J}(x'), \partial^0\phi(x)] \mid \mathbf{p}s\rangle_{\text{in}}\}\end{aligned} \tag{246}$$

which is the desired expression.

17e. Low Energy Theorems

As an application of the expression (246) for the scattering amplitude of a boson by a fermion, we investigate the scattering of very low frequency photons by a charged spin $\frac{1}{2}$ system. We have previously noted in Chapter 16 that the scattering of zero-energy photons by a charged system is described by the Thomson formula and is independent of the structure of the system. This fact was used to define the renormalized charge. We here shall show that not only the zero frequency limit of the scattering amplitude but also its first derivative with respect to photon frequency is determined by the static properties of the system, i.e., by the charge, mass, and static magnetic moment of the spin $\frac{1}{2}$ fermion the photon is scattered off [Low (1954), Gell-Mann (1954)]. Throughout the following analysis we shall always assume that the photon wavelength is very large compared to the dimensions of the system. For the scattering of photons by nucleons the length with respect to which the photon wavelength is large is the pion Compton wavelength, i.e., $k \ll \mu_\pi c$.

For definiteness let us consider the scattering of a photon by a nucleon in pseudoscalar meson theory. The Lagrangian describing this coupled meson-nucleon-electromagnetic field system is given by Eq. (10.92). The amplitude for the scattering of a photon by a nucleon from an initial state wherein the photon has momentum $\mathbf{k}$, energy $\omega = |\mathbf{k}|$, polarization $\boldsymbol{\epsilon}^{(\lambda)}(\mathbf{k})$, $\mathbf{k} \cdot \boldsymbol{\epsilon}^{(\lambda)}(\mathbf{k}) = 0$, and the nucleon momentum $\mathbf{p}$, spin s, to a final state wherein the photon has quantum numbers $\mathbf{k}'$, ω', $\boldsymbol{\epsilon}'$, $\boldsymbol{\epsilon}' \cdot \mathbf{k}' = 0$, and the nucleon has a momentum $\mathbf{p}'$ and a spin s' is given by[8]

$$\begin{aligned}
&\langle \mathbf{p}'s', \mathbf{k}'\boldsymbol{\epsilon}' \mid S \mid \mathbf{p}s, \mathbf{k}\boldsymbol{\epsilon}\rangle \\
&\quad = {}_{\text{out}}\langle \mathbf{p}'s', \mathbf{k}'\boldsymbol{\epsilon}' \mid \mathbf{p}s, \mathbf{k}\boldsymbol{\epsilon}\rangle_{\text{in}} = \delta^{(3)}(\mathbf{k} - \mathbf{k}')\, \delta_{\lambda\lambda'}\delta^{(3)}(\mathbf{p} - \mathbf{p}')\, \delta_{ss'} \\
&\qquad + \frac{1}{(2\pi)^3} \frac{1}{\sqrt{4\omega\omega'}} \int d^4x \int d^4x'\, e^{i(k'\cdot x' - k\cdot x)}\, \{{}_{\text{out}}\langle \mathbf{p}'s' \mid T(\boldsymbol{\epsilon}' \cdot \mathbf{j}(x')\, \boldsymbol{\epsilon} \cdot \mathbf{j}(x)) \\
&\qquad + \delta(x_0 - x'_0)\, [\boldsymbol{\epsilon}' \cdot \mathbf{j}(x'), \boldsymbol{\epsilon} \cdot \partial^0 \mathbf{A}(x)] \mid \mathbf{p}s\rangle_{\text{in}}\}
\end{aligned} \tag{247}$$

where $\mathbf{j}(x)$ is the spatial part[9] of the charge-current four-vector $\mathbf{j}_\mu(x)$, the source of the electromagnetic field:

$$\Box \mathbf{A}_\mu(x) = \mathbf{j}_\mu(x) \tag{248}$$

The expression for the current for the theory under discussion is given by

$$\begin{aligned}
\mathbf{j}_\mu(x) = -ie\left[\left(\frac{\partial \boldsymbol{\phi}^*}{\partial x^\mu} - ie\mathbf{A}_\mu \boldsymbol{\phi}^*\right)\boldsymbol{\phi} - \boldsymbol{\phi}^*\left(\frac{\partial \boldsymbol{\phi}}{\partial x^\mu} + ie\mathbf{A}_\mu \boldsymbol{\phi}\right)\right] \\
+ \tfrac{1}{2}e[\tilde{\boldsymbol{\psi}}(x)\, \gamma_\mu \tfrac{1}{2}(1 + \tau_3), \boldsymbol{\psi}(x)]
\end{aligned} \tag{249}$$

In Eq. (249) $\boldsymbol{\phi}$, $\boldsymbol{\phi}^*$ are the Heisenberg operators corresponding to the charged mesons. Using this expression for $\mathbf{j}_\mu$, the equal-time commutator in Eq. (247) is readily evaluated:

[8] We suppress the eigenvalue of T_3 for the nucleon.

[9] In the present instance the boldface setting for $\mathbf{j}$ denotes both the fact that $\mathbf{j}$ is a (renormalized) Heisenberg operator and that it is a three-vector.

$$\delta(x_0 - x'_0)\,[\mathbf{j}_i(x'), \partial^0\mathbf{A}_l(x)] = -2e^2\delta(x_0 - x'_0)\,\boldsymbol{\phi}^*(x')\,\boldsymbol{\phi}(x')\,[\mathbf{A}_i(x'), \partial^0\mathbf{A}_l(x)]$$
$$= +2ie^2\delta(x_0 - x'_0)\,\delta^{(3)}(\mathbf{x} - \mathbf{x}')\,g_{il}\boldsymbol{\phi}^*(x')\,\boldsymbol{\phi}(x') \qquad (250)$$

so that the S matrix is given by

$$(\mathbf{p}'s', \mathbf{k}'\epsilon' \mid S \mid \mathbf{p}s, \mathbf{k}\epsilon) = \delta^{(3)}(\mathbf{k} - \mathbf{k}')\,\delta_{\lambda\lambda'}\delta^{(3)}(\mathbf{p} - \mathbf{p}')\,\delta_{ss'}$$
$$+ \frac{2ie^2}{(2\pi)^3}\frac{1}{\sqrt{4\omega\omega'}}\int d^4x\, e^{i(k'-k)\cdot x}\,{}_{\text{out}}\langle \mathbf{p}'s' \mid \boldsymbol{\phi}^*(x)\,\boldsymbol{\phi}(x) \mid \mathbf{p}s\rangle_{\text{in}}\,\boldsymbol{\epsilon}' \cdot \boldsymbol{\epsilon}$$
$$+ \frac{1}{(2\pi)^3}\frac{1}{\sqrt{4\omega\omega'}}\int d^4x \int d^4x'\, e^{ik'\cdot x' - ik\cdot x}$$
$$\cdot\ {}_{\text{out}}\langle \mathbf{p}'s' \mid T(\boldsymbol{\epsilon}' \cdot \mathbf{j}(x')\,\boldsymbol{\epsilon} \cdot \mathbf{j}(x)) \mid \mathbf{p}s\rangle_{\text{in}} \qquad (251)$$

If we write $S = 1 - 2\pi iT$, then T can be expressed as

$$(\mathbf{p}'s', \mathbf{k}'\epsilon' \mid T \mid \mathbf{p}s, \mathbf{k}\epsilon) = \frac{i}{(2\pi)^4\sqrt{4\omega\omega'}}\sum_{l,m=1}^{3} \epsilon_l t_{lm}\epsilon_m \qquad (252a)$$

where

$$t_{lm} = 2ie^2\int d^4x\, e^{i(k'-k)\cdot x}\,{}_{\text{out}}\langle \mathbf{p}'s' \mid \boldsymbol{\phi}^*(x)\,\boldsymbol{\phi}(x) \mid \mathbf{p}s\rangle_{\text{in}}\,\delta_{lm}$$
$$+ \int d^4x \int d^4x'\, e^{i(k'\cdot x' - k\cdot x)}\,{}_{\text{out}}\langle \mathbf{p}'s' \mid T(\mathbf{j}_l(x')\,\mathbf{j}_m(x)) \mid \mathbf{p}s\rangle_{\text{in}} \qquad (252b)$$

We next note that

$$\sum_{l,m=1}^{3} k'_l t_{lm} k_m = \omega'\omega \int d^4x \int d^4x'\, e^{ik'\cdot x'}\,e^{-ik\cdot x} \cdot {}_{\text{out}}\langle \mathbf{p}'s' \mid T(\boldsymbol{\rho}(x')\,\boldsymbol{\rho}(x)) \mid \mathbf{p}s\rangle_{\text{in}} \qquad (253)$$

where $\boldsymbol{\rho}(x) = \mathbf{j}_0(x)$ is the charge density. *Proof*: Let us first compute $\sum_{l=1}^{3} k'_l t_{lm}$:

$$\sum_{l=1}^{3} k'_l t_{lm} = 2ie^2\int d^4x\, e^{i(k'-k)\cdot x}\,{}_{\text{out}}\langle \mathbf{p}'s' \mid \boldsymbol{\phi}^*(x)\,\boldsymbol{\phi}(x) \mid \mathbf{p}s\rangle_{\text{in}}\,k'_m$$
$$+ i\int d^4x \int d^4x'\, e^{ik'\cdot x'}\,e^{-ik\cdot x}\,{}_{\text{out}}\langle \mathbf{p}'s' \mid T(\boldsymbol{\partial}' \cdot \mathbf{j}(x')\,\mathbf{j}_m(x)) \mid \mathbf{p}s\rangle_{\text{in}} \qquad (254)$$

Using the equation of continuity

$$\sum_{i=1}^{3} \frac{\partial \mathbf{j}_i}{\partial x_i} = -\frac{\partial \boldsymbol{\rho}}{\partial x_0} \qquad (255)$$

we can transform the second term on the right-hand side of (254) into the form ${}_{\text{out}}\langle \mathbf{p}'s' \mid T(\partial'_0\boldsymbol{\rho}(x')\,\mathbf{j}_m(x)) \mid \mathbf{p}s\rangle_{\text{in}}$. Since

$$T\left(\frac{\partial \boldsymbol{\rho}(x')}{\partial x'^0}\,\mathbf{j}_m(x)\right) = \frac{\partial}{\partial x'^0}\,T(\boldsymbol{\rho}(x')\,\mathbf{j}_m(x)) - \delta(x_0 - x'_0)\,[\boldsymbol{\rho}(x'), \mathbf{j}_m(x)] \qquad (256)$$

upon introducing Eq. (256) into Eq. (254), we find after an integration by part that

$$\sum_{l=1}^{3} k'_l t_{lm} = 2ie^2 \int d^4x \, e^{i(k'-k)\cdot x} \, {}_{\text{out}}\langle \mathbf{p}'s' \mid \boldsymbol{\phi}^*(x) \, \boldsymbol{\phi}(x) \mid \mathbf{p}s\rangle_{\text{in}} \, k'_m$$
$$- i \int d^4x \int d^4x' \, e^{ik'\cdot x'} \, e^{-ik\cdot x} \, {}_{\text{out}}\langle \mathbf{p}'s' \mid [\boldsymbol{\rho}(x'), \mathbf{j}_m(x)] \mid \mathbf{p}s\rangle_{\text{in}} \, \delta(x_0 - x'_0)$$
$$+ \int d^4x \int d^4x' \, \omega' \, e^{ik'\cdot x'} \, e^{-ik\cdot x} \, {}_{\text{out}}\langle \mathbf{p}'s' \mid T(\boldsymbol{\rho}(x') \, \mathbf{j}_m(x)) \mid \mathbf{p}s\rangle_{\text{in}} \tag{257}$$

By explicitly computing the equal-time commutator $[\boldsymbol{\rho}(x'), \mathbf{j}_m(x)]$, using the representation (249) of the current four-vector and the equal-time canonical commutation rules, one verifies that the first two terms in Eq. (257) cancel one another. Proceeding as above, one then obtains the result, shown in Eq. (253) and repeated here, that

$$\sum_{l,m=1}^{3} k'_l t_{lm} k_m = \omega'\omega \int d^4x \int d^4x' \, e^{ik\cdot x} \, e^{-ik'\cdot x'} \, {}_{\text{out}}\langle \mathbf{p}'s' \mid T(\boldsymbol{\rho}(x') \, \boldsymbol{\rho}(x) \mid \mathbf{p}s\rangle_{\text{in}}$$

if use is made of the fact that the equal-time commutator of $\boldsymbol{\rho}(x)$ and $\boldsymbol{\rho}(y)$ vanishes. Equation (253) is a consequence of the gauge invariance of the theory. If enough information regarding the tensorial character of t_{lm} can be obtained from other sources, then Eq. (253) can be used to calculate t_{lm} itself. The advantage of being able to use Eq. (253) to compute the scattering lies in the fact that if we are interested in the scattering amplitude only to order k, then, since $\int d^3x \, \boldsymbol{\rho}(x) = \mathbf{Q}$ is the total charge operator and $\mathbf{Q} \mid \mathbf{p}s\rangle_{\text{in}} = e_R \mid \mathbf{p}s\rangle_{\text{in}}$ or $0 \mid \mathbf{p}s\rangle_{\text{in}}$, considerable simplification is achieved. We shall, in fact, show that the scattering amplitude to order k can be expressed in terms of matrix elements of the charge density $\boldsymbol{\rho}(x)$ rather than in terms of the current density $\mathbf{j}(x)$. Equation (252b), which defines t_{lm}, indicates that the first term (which is proportional to $\boldsymbol{\phi}^*\boldsymbol{\phi}$) is to order k^2 independent of the spin and is proportional to δ_{lm}. The second term can be rewritten in terms of a sum over intermediate states by inserting a complete set of states between the factors $\mathbf{j}_l(x')$ and $\mathbf{j}_m(x)$. The contribution, $t_{lm}^{(0)}$, to the sum which arises from the one-nucleon intermediate states (unexcited states) can readily be calculated to order k, and an explicit computation then shows that $\sum_{lm} k'_l t_{lm}^{(0)} k_m$ vanishes. The contribution, $t_{lm}^{(e)}$, arising from the excited states can be shown to order k to be proportional to δ_{lm} and to matrix elements of σ_{lm}. In other words, to order k

$$t_{lm} = A\delta_{lm} + B\sigma_{lm} + t_{lm}^{(0)} \tag{258}$$

so that, by substituting Eq. (258) into (253), we find that

$$\mathbf{k}' \cdot \mathbf{k}A + \boldsymbol{\sigma} \cdot \mathbf{k}' \times \mathbf{k}B = \omega'\omega C \tag{259a}$$

where

$$C = \int d^4x \int d^4x' \, e^{ik'\cdot x'} \, e^{-ik\cdot x} \, {}_{\text{out}}\langle \mathbf{p}'s' \mid T(\boldsymbol{\rho}(x') \, \boldsymbol{\rho}(x)) \mid \mathbf{p}s\rangle_{\text{in}} \tag{259b}$$

In Eqs. (258) and (259a), the spinor factors $\tilde{u}_{s'}(\mathbf{p}')$ and $u_s(\mathbf{p})$ have been suppressed. Let us see how C is calculated. The translational invariance of the theory allows us to simplify Eq. (259b). Since the (Hermitian) operator $\mathbf{P}_\mu$, corresponding to the total energy-momentum of the system, is the generator for space-time translations, we have

$$e^{i\mathbf{P}_\mu a^\mu} \boldsymbol{\rho}(x)\, e^{-i\mathbf{P}_\mu a^\mu} = \boldsymbol{\rho}(x + a) \tag{260}$$

Using the fact that $\mathbf{U}(1, a) = \exp i\mathbf{P} \cdot a$ is a unitary operator, we can write

$$\begin{aligned} {}_{\text{out}}\langle \mathbf{p}'s' \mid T(\boldsymbol{\rho}(x')\, \boldsymbol{\rho}(x)) \mid \mathbf{p}s\rangle_{\text{in}} &= {}_{\text{out}}\langle \mathbf{p}'s' \mid e^{-i\mathbf{P}\cdot a} e^{i\mathbf{P}\cdot a} \cdot T(\boldsymbol{\rho}(x')\, \boldsymbol{\rho}(x))\, e^{-i\mathbf{P}\cdot a} e^{i\mathbf{P}\cdot a} \mid \mathbf{p}s\rangle_{\text{in}} \\ &= e^{i(p-p')\cdot a}\, {}_{\text{out}}\langle \mathbf{p}'s' \mid T(\boldsymbol{\rho}(x' + a)\, \boldsymbol{\rho}(x + a)) \mid \mathbf{p}s\rangle_{\text{in}} \end{aligned} \tag{261}$$

Equation (261) is valid for arbitrary displacement a. The choice $a = -x$ permits us to rewrite Eq. (253) as follows:

$$\begin{aligned} \sum_{l,\, m=1}^{3} k'_l t_{lm} k_m &= \omega'\omega \int d^4x \int d^4x'\, e^{-i(p+k-p'-k')\cdot x}\, e^{ik'\cdot(x'-x)} \\ &\qquad {}_{\text{out}}\langle \mathbf{p}'s' \mid T(\boldsymbol{\rho}(x' - x)\, \boldsymbol{\rho}(0)) \mid \mathbf{p}s\rangle_{\text{in}} \\ &= (2\pi)^4\, \omega'\omega\delta^{(4)}(p + k - p' - k') \int d^4y\, e^{ik'\cdot y}\, {}_{\text{out}}\langle \mathbf{p}'s' \mid T(\boldsymbol{\rho}(y)\, \boldsymbol{\rho}(0)) \mid \mathbf{p}s\rangle_{\text{in}} \end{aligned} \tag{262}$$

The δ-function factor corresponds to the overall conservation of energy-momentum in the process. Upon writing

$$T(\boldsymbol{\rho}(y)\, \boldsymbol{\rho}(0)) = \theta(y)\, \boldsymbol{\rho}(y)\, \boldsymbol{\rho}(0) + \theta(-y)\, \boldsymbol{\rho}(0)\, \boldsymbol{\rho}(y) \tag{263}$$

and inserting between $\boldsymbol{\rho}(0)$ and $\boldsymbol{\rho}(y)$ the completeness relation for the states $|\mathbf{q}, \boldsymbol{\alpha}\rangle_{\text{in}}$ which are eigenfunctions of $\mathbf{P}_\mu$ with eigenvalue q_μ, we obtain

$$\begin{aligned} &\int d^4y\, e^{ik'\cdot y}\, {}_{\text{out}}\langle \mathbf{p}'s' \mid T(\boldsymbol{\rho}(y)\, \boldsymbol{\rho}(0) \mid \mathbf{p}s\rangle_{\text{in}} \\ &= \sum_{|\mathbf{q},\alpha\rangle_{\text{in}}} \Big\{ \int_0^\infty dy_0 \int d^3y\, e^{ik\cdot y}\, {}_{\text{out}}\langle \mathbf{p}'s' \mid \boldsymbol{\rho}(y) \mid \mathbf{q}, \boldsymbol{\alpha}\rangle_{\text{in}}\, {}_{\text{in}}\langle \mathbf{q}, \boldsymbol{\alpha} \mid \boldsymbol{\rho}(0) \mid \mathbf{p}s\rangle_{\text{in}} \\ &\qquad + \int_{-\infty}^0 dy_0 \int d^3y\, e^{ik'\cdot y}\, {}_{\text{out}}\langle \mathbf{p}'s' \mid \boldsymbol{\rho}(0) \mid \mathbf{q}, \boldsymbol{\alpha}\rangle_{\text{in}}\, {}_{\text{in}}\langle \mathbf{q}, \boldsymbol{\alpha} \mid \boldsymbol{\rho}(y) \mid \mathbf{p}s\rangle_{\text{in}} \Big\} \\ &= \sum_{|\mathbf{q},\alpha\rangle_{\text{in}}} \Big\{ \int_0^\infty dy_0 \int d^3y\, e^{i(k'+p'-q)\cdot y}\, {}_{-}\langle \mathbf{p}'s' \mid \boldsymbol{\rho}(0) \mid \mathbf{q}, \boldsymbol{\alpha}\rangle_{+}\, {}_{+}\langle \mathbf{q}, \boldsymbol{\alpha} \mid \boldsymbol{\rho}(0) \mid \mathbf{p}s\rangle_{+} \\ &\qquad + \int_{-\infty}^0 dy_0 \int d^3y\, e^{i(k-p+q)\cdot y}\, {}_{-}\langle \mathbf{p}'s' \mid \boldsymbol{\rho}(0) \mid \mathbf{q}, \boldsymbol{\alpha}\rangle_{+}\, {}_{+}\langle \mathbf{q}, \boldsymbol{\alpha} \mid \boldsymbol{\rho}(0) \mid \mathbf{p}s\rangle_{+} \Big\} \\ &= i \sum_{q_0,\, \alpha} \Big\{ \frac{{}_{-}\langle \mathbf{p}'s' \mid \boldsymbol{\rho}(0) \mid \mathbf{p}' + \mathbf{k}', q_0; \boldsymbol{\alpha}\rangle_{+}\, {}_{+}\langle q_0, \mathbf{p}' + \mathbf{k}'; \boldsymbol{\alpha}' \mid \boldsymbol{\rho}(0) \mid \mathbf{p}, s\rangle_{+}}{k'_0 + p'_0 - q_0 + i\epsilon} \\ &\qquad - \frac{{}_{-}\langle \mathbf{p}'s' \mid \boldsymbol{\rho}(0) \mid \mathbf{p} - \mathbf{k}', q_0; \boldsymbol{\alpha}\rangle_{+}\, {}_{+}\langle q_0, \mathbf{p} - \mathbf{k}'; \boldsymbol{\alpha} \mid \boldsymbol{\rho}(0) \mid \mathbf{p}s\rangle_{+}}{k'_0 - p_0 - q_0 - i\epsilon} \Big\} \end{aligned} \tag{264}$$

In (264), the summation is over all states $|\alpha\rangle$ which can be connected to the one-nucleon state by the charge operator $\boldsymbol{\rho}(0)$. Since $\mathbf{j}_\mu(x)$ commutes with the nucleon number operator N (the number of nucleons minus the number of antinucleons), the states $|\alpha\rangle$ must be eigenstates of N with eigenvalue $+1$; they must clearly also have the same charge as the state $|\mathbf{p}s\rangle_\pm$. The states $|\alpha\rangle$ which contribute are therefore the one-nucleon states, the one-nucleon plus one-meson states,[10] etc. It is, however, clear that the terms arising from excited states (i.e., the states other than the one-nucleon intermediate states) contribute of order k^2, since $\int d^3x\, \boldsymbol{\rho}(x)$ is the total charge operator, which when operating on $|\mathbf{p}s\rangle_\pm$ or $|\mathbf{p}'s'\rangle_\pm$ gives a c-number factor, the total charge of that state. Therefore, to calculate C to order k, it is only necessary to obtain the contribution to Eq. (264) from the one-nucleon intermediate states. If we separate off the contribution of the nucleon states in (262), we obtain

$$\int d^4y\, e^{ik'\cdot y}\, {}_{\text{out}}\langle \mathbf{p}'s' \mid T(\boldsymbol{\rho}(y)\, \boldsymbol{\rho}(0) \mid \mathbf{p}s\rangle_{\text{in}} = i \sum_{s''}$$
$$\left\{ \frac{{}_-\langle \mathbf{p}'s' \mid \boldsymbol{\rho}(0) \mid \mathbf{p}'+\mathbf{k}', E(\mathbf{p}'+\mathbf{k}'); s''\rangle_{+}\, {}_+\langle s'', \mathbf{p}'+\mathbf{k}', E(\mathbf{p}'+\mathbf{k}') \mid \boldsymbol{\rho}(0) \mid \mathbf{p}s\rangle_+}{\omega' + E(\mathbf{p}') - E(\mathbf{p}'+\mathbf{k}') + i\epsilon} \right.$$
$$\left. - \frac{{}_-\langle \mathbf{p}'s' \mid \boldsymbol{\rho}(0) \mid \mathbf{p}-\mathbf{k}', E(\mathbf{p}-\mathbf{k}'), s''\rangle_{+}\, {}_+\langle s'', \mathbf{p}-\mathbf{k}', E(\mathbf{p}-\mathbf{k}') \mid \boldsymbol{\rho}(0) \mid \mathbf{p}s\rangle_+}{\omega' - E(\mathbf{p}) + E(\mathbf{p}-\mathbf{k}')} \right\}$$
$$+ O(k^2) \qquad (265)$$

where $E(\mathbf{p}) = \sqrt{\mathbf{p}^2 + M^2}$.

To obtain an explicit form for the contribution of the one-nucleon states, we next determine the general form of the matrix element ${}_+\langle \mathbf{p}'s' \mid \mathbf{j}_\mu(x) \mid |\mathbf{p}s\rangle_-$. When expressed in terms of interaction picture operators, this matrix element is equal to

$$\begin{aligned} {}_+\langle \mathbf{p}'s' \mid \mathbf{j}_\mu(0) \mid \mathbf{p}s\rangle_+ &= -(p'-p)^2\, {}_+\langle \mathbf{p}'s' \mid \mathbf{A}_\mu(0) \mid \mathbf{p}s\rangle_+ \\ &= \langle \mathbf{p}'s' \mid U(0,\infty)\, j_\mu(0)\, U(0,-\infty) \mid \mathbf{p}s\rangle \\ &= \langle \mathbf{p}'s' \mid T(Sj_\mu(0)) \mid \mathbf{p}s\rangle \qquad (266a) \end{aligned}$$

The diagrammatic representation of the right-hand side of Eq. (266a) is given in Figure 17.7. The matrix element ${}_-\langle \mathbf{p}'s' \mid \mathbf{j}_{R\mu}(0) \mid \mathbf{p}s\rangle_+$ is thus equal to[11]

[10] The intermediate states containing one nucleon plus photons will give contributions of order α compared with the lowest order contribution. These contributions are therefore usually neglected compared to those coming from intermediate states containing mesons, which are of order G^2.

[11] We have omitted in the right-hand side of Eq. (266b) a kinematical factor $(M^2/p_0p_0')^{1/2}$. Since our spinors are normalized according to $\tilde{u}(\mathbf{p})\, u(\mathbf{p}) = +1$ (for a positive energy spinor), the factor (M^2/p_0p_0') is necessary to guarantee the Lorentz covariance of the matrix element. We have also appended a subscript R to the operators to remind the reader that we are dealing in the present section with renormalized Heisenberg operators.

$$_+\langle \mathbf{p}'s' \mid \mathbf{j}_{R\mu}(0) \mid \mathbf{p}s\rangle_- = (p' - p)^2 D'_{F1}(p' - p)\, \tilde{u}^{s'}(\mathbf{p}')\, \Gamma_{1\mu}(p', p)\, u^s(\mathbf{p}) \tag{266b}$$

where $\Gamma_{1\mu}$ is the (renormalized) vertex operator. To construct the operator

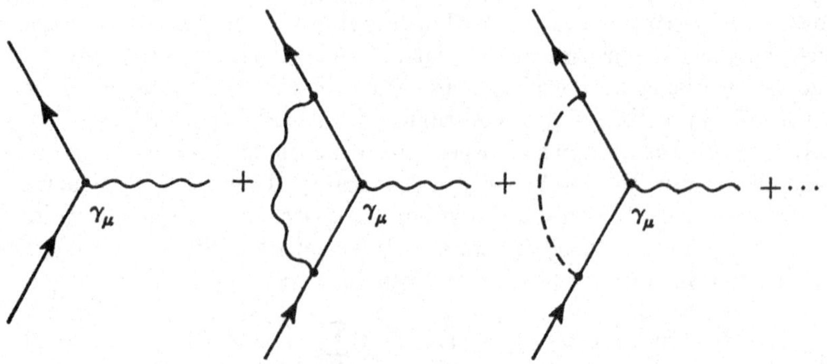

Fig. 17.7

$\Gamma_{1\mu}$, we note that, due to relativistic invariance, it must be of the form

$$\Gamma_{1\mu}(p', p)\Big|_{p^2 = p'^2 = M^2} = \gamma_\mu F_1(q^2) + i\sigma_{\mu\nu} p^\nu G_1(q^2) + i\sigma_{\mu\nu} p'^\nu G_2(q^2)$$

$$q = p' - p \tag{267}$$

where F_1, G_1 and G_2 are invariant functions. Furthermore, the requirement that $\partial^\mu {}_+\langle \mathbf{p}'s' \mid \mathbf{j}_\mu(x) \mid \mathbf{p}s\rangle_- = 0$, by virtue of the continuity equation $\partial^\mu \mathbf{j}_\mu(x) = 0$, implies that

$$(p' - p)^\mu\, \tilde{u}^{s'}(\mathbf{p}')\, \Gamma_{1\mu}(p', p)\, u^s(\mathbf{p}) = 0 \tag{268}$$

Now the term $\tilde{u}(\mathbf{p}')\, \gamma_\mu (p' - p)^\mu\, u(\mathbf{p})\, F_1(q^2)$ vanishes by virtue of the fact that $\tilde{u}$ and u obey the Dirac equation. The condition that $(p' - p)^\mu\, \tilde{u}(\mathbf{p}')\, \{\sigma_{\mu\nu} p^\nu G_1(q^2) + \sigma_{\mu\nu} p'^\nu G_2(q^2)\}\, u(\mathbf{p})$ vanishes is that

$$-G_1(q^2) = G_2(q^2) = F_2(q^2) \tag{269}$$

so that

$$\Gamma_{1\mu}(p', p) = \gamma_\mu F_1(q^2) + i\sigma_{\mu\nu}(p' - p)^\nu\, F_2(q^2) \tag{270}$$

The form factors $F_1(q^2)$ and $F_2(q^2)$ describe the distribution of charge and magnetization in the nucleon. They are the generalization of the electromagnetic form factors previously considered in describing the interaction of an electron with an electromagnetic field. Since the total charge is determined by the matrix element $_+\langle \mathbf{p}s \mid \int d^3x\, \mathbf{j}_0(x) \mid \mathbf{p}s\rangle_+$, i.e., by the matrix element for $q = 0$, we have

$$F_1(0) = \tfrac{1}{2}(1 + \tau_3)\, e_R \tag{271}$$

or $F_1{}^{p}(0) = e_R$ and $F_1{}^{n}(0) = 0$ where the superscript n, p corresponds to neutron and proton, respectively. The magnetic moment of the nucleon is obtained by calculating the expectation value of the operator $\frac{1}{2}\int d^3r\, \mathbf{r} \times \mathbf{j}$ in the state $|\Psi\rangle$ in which the nucleon is at rest. This expectation value is

$$\langle \Psi \mid \tfrac{1}{2} \int d^3r\, \mathbf{r} \times \mathbf{j}(x) \mid \Psi \rangle = \left[\frac{F_1(0)}{2M} + F_2(0) \right] \langle \tilde{u}(\mathbf{0})\, \boldsymbol{\sigma} u(\mathbf{0}) \rangle \quad (272)$$

where the right-hand side denotes the expectation value of $\boldsymbol{\sigma}$ for the Dirac spinor corresponding to the particle at rest ($\mathbf{p} = 0$). Since $F_1(0)/2M$ by virtue of Eq. (271) is the Dirac moment, we see that $F_2(0) = \Delta\mu$ is the anomalous moment. Higher moments of F_1, F_2 characterize the form of the charge and magnetization distribution. (One can think of the form factors as the Fourier transforms of the spatial charge and magnetization distributions.) The mean square radius of the spatial distribution is related to the derivative of $F_{1,2}$ evaluated at $q^2 = 0$. The experimental data [Hofstadter (1958, 1960)] on electron-nucleon and electron-deuteron scattering indicates that $F_{1p} \simeq F_{2p} \simeq F_{2n}$ (at least for the values of q^2 at which the experiments are carried out) and $F_{1n} \simeq 0$.

Summarizing, we have established that

$$_{-}\langle \mathbf{p}'s' \mid \mathbf{j}_\mu(0) \mid \mathbf{p}s \rangle_{+} = \tilde{u}^{s'}(\mathbf{p}') \left[\gamma_\mu F_1(q^2) + i\sigma_{\mu\nu} q^\nu F_2(q^2) \right] u^s(\mathbf{p}) \quad (q = p' - p) \quad (273)$$

so that $_{+}\langle \mathbf{p}'s' \mid \boldsymbol{\rho}(0) \mid \mathbf{p}s \rangle_{+}$ is given by

$$\begin{aligned} &_{+}\langle \mathbf{p}'s' \mid \boldsymbol{\rho}_R(0) \mid \mathbf{p}s \rangle_{-} \\ &\quad = q^2 D'_{F1}(q)\, \{ u^{s'*}(\mathbf{p}')\, u^s(\mathbf{p})\, F_1(q^2) + F_2(q^2)\, u^{s'*}(\mathbf{p}')\, \beta \boldsymbol{\alpha} \cdot \mathbf{q} u^s(\mathbf{p}) \} \end{aligned} \quad (274)$$

and $_{+}\langle \mathbf{p}'s' \mid \mathbf{j}_R(0) \mid \mathbf{p}s \rangle_{-}$ by

$$\begin{aligned} &_{+}\langle \mathbf{p}'s' \mid \mathbf{j}_R(0) \mid \mathbf{p}s \rangle_{-} \\ &\quad = q^2 D'_{F1}(q) u^{s'*}(\mathbf{p}')\, [\boldsymbol{\alpha} F_1(q^2) - i\mathbf{q} \times \boldsymbol{\sigma} F_2(q^2)]\, u^s(\mathbf{p}) \qquad (275a) \\ &\quad \cong u^{s'*}(\mathbf{p}') \left[\frac{F_1(0)}{2M} (\mathbf{p} + \mathbf{p}') + i \left(\frac{F_1(0)}{2M} + \Delta\mu \right) \boldsymbol{\sigma} \times \mathbf{q} \right] u^s(\mathbf{p}) + O(q^2) \\ &\quad \cong u^{s'*}(\mathbf{p}') \left[\frac{F_1(0)}{2M} (\mathbf{p} + \mathbf{p}') + i\mu \boldsymbol{\sigma} \times \mathbf{q} \right] u^s(\mathbf{p}) + O(q^2) \qquad (275b) \end{aligned}$$

Equation (275b) corresponds to the current of a nonrelativistic particle characterized by a charge $F_1(0)$ and a total magnetic moment μ. Such a particle would interact with a magnetic field $\mathcal{H}$ according to the Hamiltonian

$$H \sim -\frac{F_1(0)}{2M} (\mathbf{p} \cdot \mathbf{A} + \mathbf{A} \cdot \mathbf{p}) - \mu \boldsymbol{\sigma} \cdot \mathcal{H} \quad (276)$$

Substituting the expressions (274) and (275) into Eq. (265), one finds that to order k for the scattering of photons by protons

$$i \int d^4y\, e^{ik'\cdot y}\, {}_{\text{in}}\langle \mathbf{p}'s' \mid T(\boldsymbol{\rho}(y)\, \boldsymbol{\rho}(0)) \mid \mathbf{p}s\rangle_{\text{in}}$$

$$= u^{s'*}\left[\frac{e_R^2}{M}\left(\frac{\mathbf{k}\cdot\mathbf{k}'}{\omega\omega'}\right) + i\left(\frac{\boldsymbol{\sigma}\cdot\mathbf{k}'\times\mathbf{k}}{\omega}\right)\frac{e_R}{M}\left(2\mu - \frac{e_R}{2M}\right)\right]u^s + O(k^2) \tag{277}$$

where μ is the total magnetic moment of the nucleon being considered. The right-hand side of Eq. (277) is evaluated between the two-component Pauli spinors $u^{s'}$ and u^s, corresponding to the spin s and s', respectively. In other words, in Eq. (277) the reduction to the large components of $u^{s'}(\mathbf{p})$ and $u^s(\mathbf{p})$ has been carried out. Equation (277), which is proportional to $\sum_{lm} k'_l t_{lm} k_m$ [recall Eqs. (259a) and (262)], allows us to compute t_{lm}, which in turn permits the computation of the scattering amplitude to order k [see Eq. (252a)]. To order k the scattering amplitude is given by

$$\langle \mathbf{p}'s', \mathbf{k}'\boldsymbol{\epsilon}' \mid S \mid \mathbf{p}s, \mathbf{k}\boldsymbol{\epsilon}\rangle = \frac{-i(2\pi)^4}{\sqrt{4\omega\omega'}} \cdot \delta^{(4)}(p' + k' - p - k)$$

$$\times u^{s'*}\left[\frac{e_R^2}{M}\boldsymbol{\epsilon}'\cdot\boldsymbol{\epsilon} - i\frac{e_R}{M}\omega\boldsymbol{\sigma}\cdot(\boldsymbol{\epsilon}'\times\boldsymbol{\epsilon})\left(2\mu - \frac{e_R}{2M}\right) - \frac{2i\mu^2}{\omega}\boldsymbol{\sigma}\cdot(\boldsymbol{\epsilon}\times\mathbf{k})\right.$$

$$\left.\times(\boldsymbol{\epsilon}'\times\mathbf{k}') - \frac{e_R\mu i}{M\omega}((\boldsymbol{\epsilon}\cdot\mathbf{k}')\,\boldsymbol{\epsilon}\cdot(\boldsymbol{\sigma}\times\mathbf{k}') - \boldsymbol{\epsilon}'\cdot\mathbf{k}\boldsymbol{\epsilon}\cdot\boldsymbol{\sigma}\,(\times\mathbf{k}))\right]u^s$$

$$+ O(k^2) \tag{278}$$

For forward scattering $\mathbf{k} = \mathbf{k}'$, $\mathbf{p} = \mathbf{p}'$, the scattering amplitude becomes

$$\langle \mathbf{p}s', \mathbf{k}\boldsymbol{\epsilon}' \mid S \mid \mathbf{p}s, \mathbf{k}\boldsymbol{\epsilon}\rangle = \frac{-i(2\pi)^4}{\sqrt{4\omega\omega'}}$$

$$\cdot\,\delta^{(4)}(p' + k' - p - k)\, u^{s'*}\left[\frac{e_R^2}{M}\boldsymbol{\epsilon}\cdot\boldsymbol{\epsilon}' + i\omega(\Delta\mu)^2\,\boldsymbol{\sigma}\cdot(\boldsymbol{\epsilon}'\times\boldsymbol{\epsilon})\right]u^s \tag{279}$$

[For a neutron the e_R^2/M term is, of course, absent since $F_1^n(0) = 0$.] Equations (278) and (279) indicate that up to terms linear in the frequency of the photon, the scattering amplitude depends only on the charge, mass, and magnetic moment of the scatterer. Low (1958) has shown that the bremsstrahlung of low energy photons, up to terms linear in the frequency of the photon, can be related to the scattering amplitude without photon emission and that the expansion coefficients are again determined by the static electromagnetic properties of the system.

Other important low energy theorems for meson-nucleon scattering and photomeson production have been given by Kroll and Ruderman (1954), and by Klein (1955b) [see also Kazes (1959)]. The importance of these theorems lies in the fact that they provide a means of measuring parameters which are logically defined by other experiments. Thus the charge and magnetic moment of the proton is, in principle, measured in the scattering of the proton by a weak, slowly varying external electromagnetic field (weak enough so that only effects linear in the field need be considered, and sufficiently slowly varying so that any space-time variation of the field strength may be negleted). The above low energy theorem for Compton scattering then indicates how the so-defined parameters enter in the scattering of photons by protons.

17f. The Bound State Problem

In the present section, we outline some of the methods which have been used in treating bound state problems in relativistic field theories. The bound state problem can be considered from two closely related viewpoints. One consists in generalizing the notion of Feynman propagators to a many-particle system. The second approach works directly with amplitudes which are the analogues of "wave functions" in the nonrelativistic case [Feynman (1949a, b); Schwinger (1951b); Salpeter and Bethe (Salpeter 1951); Gell-Mann and Low (1951); Nishijima (1953a); and Freese (1953, 1954).]

Consider first the problem of describing a one-electron system in a weak static external field when one includes the effect of the quantized radiation and electron-positron field. We know that, if we neglect the effect of the quantized radiation field, the operators $\psi^e(x)$ describing the quantized electron-positron field satisfy the equation of motion

$$(i\gamma_\mu\partial^\mu + e\gamma_\mu A^{e\mu}(x) - m)\,\psi^e(x) = 0 \tag{280}$$

Furthermore, for sufficiently weak fields a stable vacuum $|\Phi^e{}_0\rangle$ exists and the state

$$|\Phi^e{}_n\rangle = \int_\sigma d\sigma_\mu(x)\,\tilde{\psi}^e(x)\,\gamma^\mu\phi_n(x)\mid\Phi^e{}_0\rangle \tag{281}$$

where $\phi_n(x)$ is a positive energy solution of the Dirac equation in the external field $A^{e\mu}(x)$, is a one-electron state and is independent of σ. This state $|\Phi^e{}_n\rangle$ is an eigenstate of $\mathbf{P}_0$, the total energy of the field system, with eigenvalue E_n, where E_n is the energy eigenvalue of ϕ_n. Conversely, the equal-time commutation rules:

$$\{\psi^e(x), \tilde{\psi}^e(x')\}_{x_0=x'_0} = \gamma^0\delta^{(3)}(\mathbf{x}-\mathbf{x}') \tag{282}$$

and the property of the vacuum that it is the state of lowest energy allows us to infer that

$$\langle\Phi^e{}_0\mid\psi^e(x)\mid\Phi^e{}_n\rangle = \phi_n(x) \tag{283}$$

This amplitude, $\phi_n(x)$, will satisfy the equation:

$$(i\gamma_\mu\partial^\mu + e\gamma_\mu A^{e\mu}(x) - m)\,\phi_n(x) = 0 \tag{284}$$

since $\psi^e(x)$ does, and also the equation:

$$i\partial_0\phi_n(x) = E_n\phi_n(x) \tag{285}$$

if $|\Phi^e{}_n\rangle$ is an eigenstate of $\mathbf{P}_0$ with eigenvalue E_0. *Proof:*

$$\begin{aligned}\langle\Phi^e{}_0 \mid [\mathbf{P}_0, \psi^e(x)] \mid \Phi^e{}_n\rangle &= i\partial_0\langle\Phi^e{}_0 \mid \psi^e(x) \mid \Phi^e{}_n\rangle \\ &= \langle\Phi^e{}_0 \mid \psi^e(x)\, E_n \mid \Phi^e{}_n\rangle\end{aligned} \tag{286}$$

When the effects of the quantized radiation field are included, the state vector of the system in the Furry picture satisfies the equation:

$$\begin{aligned}i\hbar\partial_t \mid \Psi_F(t)\rangle &= H_{IF}(t) \mid \Psi_F(t)\rangle \\ &= -e\int d\sigma(x) : \bar{\psi}^e(x)\, \gamma^\mu\psi^e(x) : A^e{}_\mu(x) \mid \Psi_F(t)\rangle\end{aligned} \tag{287}$$

where $\psi^e(x)$ are the Furry picture fermion operators which satisfy Eq. (280). The remarks in Section *a* of the present chapter suggest that we consider the vectors

$$|\boldsymbol{\Psi}_n\rangle = \lim_{t_0\to-\infty} \frac{U^e(0, t_0) \mid \Phi^e{}_n\rangle}{(\Phi^e{}_0, U^e(0, t_0)\, \Phi^e{}_0)} \tag{288}$$

as the "physical" eigenstates describing the system when the effects of the radiation field are taken into account. However, only in the case of the vacuum state $|\Phi^e{}_0\rangle$ and the state of one particle in the lowest bound state are the resulting states $|\boldsymbol{\Psi}\rangle$ eigenstates of $\mathbf{P}_0$ with discrete eigenvalues. What we are asserting is that the energy spectrum of the combined system electron-positron and radiation field has a discrete eigenvalue at $p_0 = 0$, corresponding to the vacuum; a continuous spectrum starting at $p_0 = 0$, corresponding to one, two, $\cdots$ photon states; a discrete eigenvalue at $m - \epsilon_0$ (where ϵ_0 is the binding energy of the 1*s* state including all radiative corrections); and a continuous spectrum starting at this point corresponding to states of one electron with one, two, $\cdots$ photons. The discrete eigenvalues of $\mathbf{P}_0{}^{(0)}$ which occurred at $m - E_n$, when the effect of the quantized radiation field was neglected, and which corresponded to the bound states ϕ_n are, as a result of the interaction with the radiation field, given a negative imaginary part and therefore no longer lie on the real axis in the complex energy plane. Physically, this is just the statement that the higher excited states are unstable against transitions to lower states with the emission of photons. (The state containing one electron in the lowest state is, of course, stable due to charge conservation.)

By analogy to the situation in the absence of the radiation field, let us describe the one-electron system whose Heisenberg state vector is $|\boldsymbol{\Psi}\rangle$ by the amplitude

$$f(x) = (\boldsymbol{\Psi}_0, \psi(x)\, \boldsymbol{\Psi}) \tag{289}$$

where $\psi(x)$ is the Heisenberg field describing the electron-positron field interacting with the external and radiation field and satisfying the equation:

$$(i\gamma^\mu\partial_\mu - m)\,\psi(x) = -e\gamma_\mu(A^{e\mu}(x) + \mathbf{A}^\mu(x))\,\psi(x) \tag{290}$$

Since $|\Psi\rangle$, by assumption, is a one-electron state (so that $\mathbf{Q}\,|\,\Psi\rangle = -e\,|\,\Psi\rangle$), the amplitudes

$$f(x_1, \cdots x_n) = (\Psi_0, T(\psi(x_1) \cdots \psi(x_n)\,\Psi) \tag{291}$$

vanish, but the amplitudes

$$f^{(n,m)}_{\mu_1\cdots\mu_n}(x, x_1, \cdots x_n; y_1 \cdots y_n; z_1 \cdots z_n)$$
$$= (\Psi_0, T(\psi(x)\,\psi(x_1) \cdots \psi(x_n)\,\tilde{\psi}(y_1) \cdots \tilde{\psi}(y_n)\,\mathbf{A}\mu_1(z_1) \cdots \mathbf{A}_{\mu_n}(z_n))\,\Psi_0) \tag{292}$$

are different from zero. Their nonzero value relates to the possibility of the external field creating pairs and to the possibility of photon emission by the one-electron system.

Let us consider "one-particle" amplitude $f(x)$ defined by Eq. (289) in somewhat greater detail. Using the fact that we can write

$$|\Psi\rangle = \frac{U^e(0, -\infty)\,|\,\Phi^e\rangle}{(\Phi^e{}_0, U^e(0, -\infty)\,\Phi^e{}_0)} \tag{293}$$

we can re-express the amplitude in terms of Furry picture variables as follows:

$$f(x) = \langle\Phi^e{}_0\,|\,U^{e*}(0, \infty)\,U^e(t, 0)^{-1}\,\psi^e(x)\,U^e(t, 0)\,U^e(0, -\infty)\,|\,\Phi^e\rangle_C$$
$$= \langle\Phi^e{}_0\,|\,T(S^e\psi^e(x))\,|\,\Phi\rangle_C \tag{294a}$$
$$S^e = U^e(+\infty, -\infty) \tag{294b}$$

Let us, in particular, consider the case that

$$|\Phi^e\rangle = |\Phi^e{}_n\rangle = \int_\sigma d\sigma_\mu(x)\,\tilde{\psi}^e\gamma^\mu\phi_n(x)\,|\,\Phi^e{}_0\rangle \tag{295}$$

where, since σ is arbitrary, we choose it to be a space-like surface at $t = -\infty$. With this choice of σ, Eq. (294) can be written in the form

$$f_n(x) = \int \langle\Phi^e{}_0, T(S^e\psi^e(x)\,\tilde{\psi}^e(x'))\,\Phi^e{}_0\rangle\,\gamma^\mu\phi_n(x')\,d\sigma_\mu(x') \tag{296a}$$
$$S^e = U^e(\infty, -\infty) \tag{296b}$$

or

$$f_n(x) = -\tfrac{1}{2}\int_\sigma S^{e\prime}{}_F(x, x')\,\gamma^\mu\phi_n(x')\,d\sigma_\mu(x') \tag{297}$$

where

$$-\tfrac{1}{2}S^{e\prime}{}_F(x, x') = \langle\Phi^e{}_0, T(S^e\psi^e(x)\,\tilde{\psi}^e(x'))\,\Phi^e{}_0\rangle \tag{298}$$

is the Feynman propagator for a particle in the external field including all radiative corrections. It is a function of $\mathbf{x}$, $\mathbf{x}'$ and $x_0 - x'_0$ and satisfies the integral equation:

$$-\tfrac{1}{2}S^{e\prime}_F(x, x') = -\tfrac{1}{2}S^e_F(x, x') + \int -\tfrac{1}{2}S^e_F(x, x'')\, \Sigma_e^*(x'', x''') \cdot \frac{-1}{2} S^{e\prime}_F(x''', x')\, d^4x'''\, d^4x'' \quad (299)$$

where S^e_F is given by

$$-\tfrac{1}{2}S^e_F(x, x') = (\Phi^e_0, T(\psi^e(x)\, \bar{\psi}^e(x'))\, \Phi^e_0) \quad (300)$$

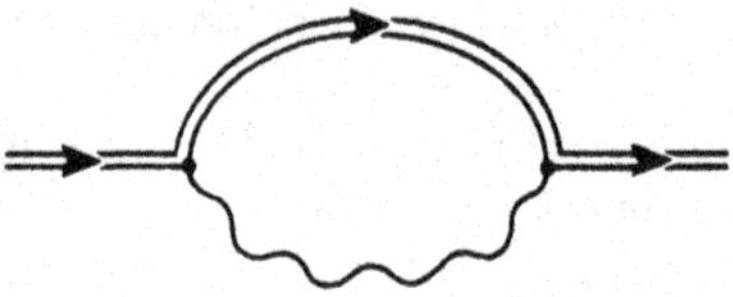

Fig. 17.8

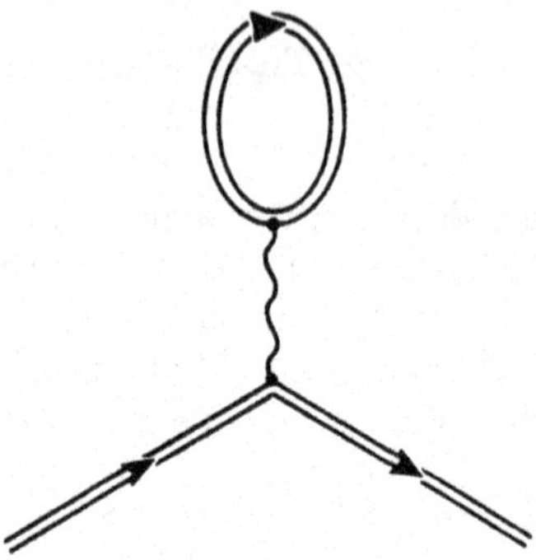

Fig. 17.9

The proper self-energy diagrams contributing to Σ_e^* are the same as those contributing to Σ^*, e.g., Figure 17.8. In addition, however, Σ_e^* will contain a whole series of diagrams arising from vacuum polarization by the external field, of which the lowest order is illustrated in Figure 17.9. This particular diagram, it will be recalled, contributes after renormalization a term

$$\Sigma_e^{*(2)}(x, x') \simeq \frac{e^2}{15\pi m^2}\, \delta^{(4)}(x - x')\, \gamma^\mu \Box A^e_\mu(x) \quad (301)$$

Substituting Eq. (299) into (297), we obtain the following integral equation for $f_n(x)$:

$$f_n(x) = \phi_n(x) + \int d^4x' \int d^4x'' \frac{-1}{2} S^e_F(x, x'')\, \Sigma_e^*(x'', x')\, f_n(x') \quad (302)$$

Alternatively, since $\phi_n(x)$ satisfies the Dirac equation in the external field $A^e(x)$ and $S^e_F(x, x')$ satisfies the equation

$$(i\gamma_\mu \partial^\mu + e\gamma_\mu A^{e\mu}(x) - m)\, S^e_F(x, x') = -2i\delta^{(4)}(x - x') \quad (303)$$

we find that $f_n(x)$ satisfies the differential equation:

$$(i\gamma_\mu \partial^\mu + e\gamma_\mu A^{e\mu}(x) - m)\, f_n(x) + \int d^4x'\, \Sigma_e^*(x, x')\, f_n(x') = 0 \quad (304)$$

This equation for the amplitude $f(x)$ was first derived by Schwinger (1951). It is the generalization of the Dirac equation which includes radiative corrections. As mentioned above, Eq. (304) will have solutions of the form

$$f(x) = e^{-iEx_0}\, g(\mathbf{x}) \tag{305}$$

only for the lowest state. For the other states, an approximate solution of the form

$$f(x) \sim e^{-i\left(E - \frac{i}{2}\Gamma\right)x_0}\, g(\mathbf{x}) \tag{306}$$

will exist where Γ is essentially the lifetime of the state in question. [For a method of solving Eqs. (299) and (304), see Low (1952); also see Mills and Kroll (Mills 1955).]

The description of a one-particle state in terms of amplitudes $f(x)$, $f(x, x_1; y_1) \cdots$, when no external fields are present, is as follows: Let the one-particle Heisenberg state vector which is an eigenstate of $\mathbf{P}_\mu$ with eigenvalue p_μ, with $p_\mu p^\mu = m^2$ be denoted by $|\boldsymbol{\Psi}_{\mathbf{p}s}\rangle$. The one-particle amplitude $f(x)$ is then defined by

$$f(x) = (\boldsymbol{\Psi}_0, \psi(x)\, \boldsymbol{\Psi}_{\mathbf{p}s}) \tag{307}$$

where $\psi(x)$ is the Heisenberg field operator which satisfies Eq. (1). Again expressing the right-hand side of Eq. (307) in terms of interaction picture operators, we have

$$f(x) = (\Phi_0, U(\infty, 0)\, U^{-1}(t, 0)\, \psi(x)\, U(t, 0)\, U(0, -\infty)\, \Phi_{\mathbf{p}s})_C \tag{308}$$

where

$$|\boldsymbol{\Psi}_{\mathbf{p}s}\rangle = U(0, -\infty)\, |\, \Phi_{\mathbf{p}s}\rangle_C \tag{309}$$

and

$$|\Phi_{\mathbf{p}s}\rangle = b_s{}^*(\mathbf{p})\, |\, \Phi_0\rangle = \int d\sigma^\mu(x)\, \bar{\psi}(x)\, \gamma_\mu w^s(\mathbf{p})\, e^{-ip\cdot x}\, |\, \Phi_0\rangle \tag{310}$$

We can therefore write

$$\begin{aligned} f(x) &= \int (\Phi_0, T(S\psi(x)\, \bar{\psi}(x'))\, \Phi_0)\, \gamma^\mu w(x')\, d\sigma_\mu(x') \\ &= \int d\sigma_\mu(x') \cdot \frac{-1}{2} S'_F(x - x')\, \gamma^\mu w(x') \end{aligned} \tag{311}$$

where $-\frac{1}{2}S'_F(x - x')$ is the renormalized propagation function (previously denoted by S'_{F1}) which satisfies the integral equation

$$\begin{aligned} -\tfrac{1}{2}S'_F(x - x') &= -\tfrac{1}{2}S_F(x - x') \\ &\quad + \int d^4x'\, \frac{-1}{2} S_F(x - x'')\, \Sigma^*(x'' - x''')\, \frac{-1}{2} S'_F(x''' - x') \end{aligned} \tag{312}$$

In Eq. (312) Σ^* is the contribution of all proper self-energy diagrams. Again, substituting this equation for S'_F into (311), we find that $f(x)$ satisfies the integral equation:

$$f(x) = w_{\mathbf{p}}(x) - \int d^4x'\, \Sigma^*(x - x')\, f(x') \tag{313}$$

which incorporates the boundary condition that $f(x')$ is to describe a particle of momentum p_μ since the inhomogeneous term is $w_{\mathbf{p}}(x)$. In fact,

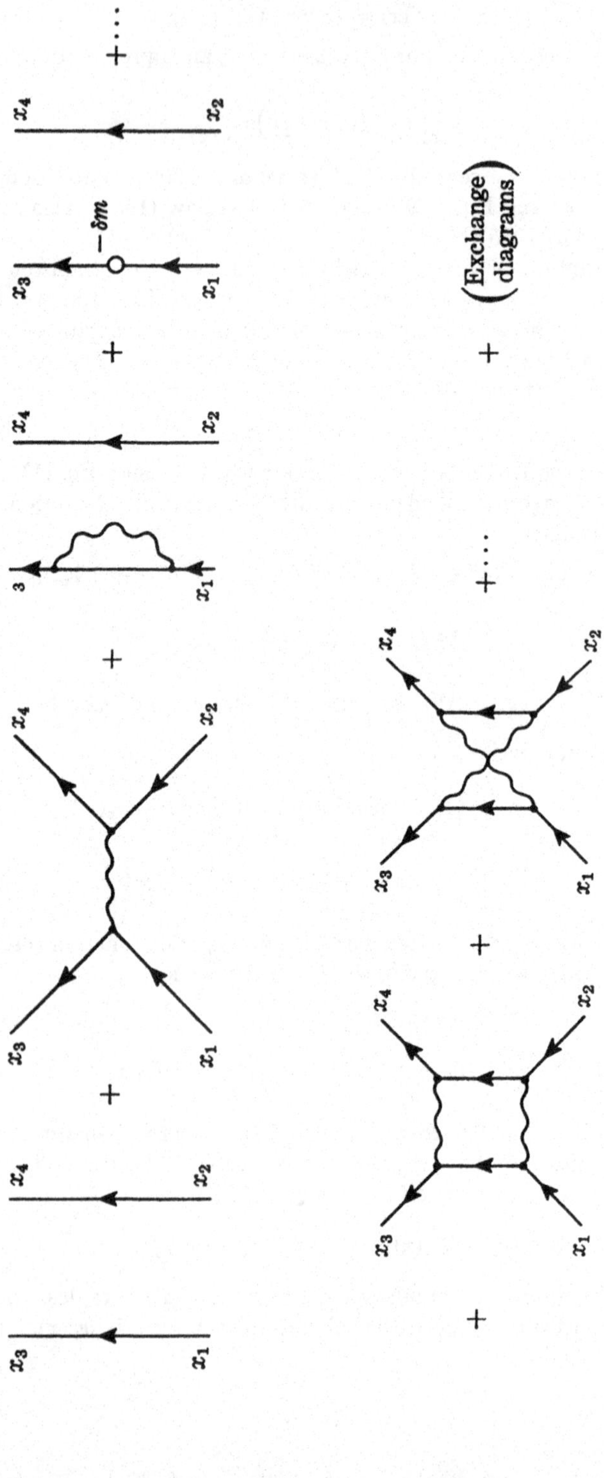

Fig. 17.10

upon iterating the above equation, since $\int d^4x'\, \Sigma^*(x - x')\, w_{\mathbf{p}}(x') = 0$, we have $f(x) = w_{\mathbf{p}}(x)$.

The above examples indicate that a study of the Feynman propagation kernels $K'_{A+} = -\frac{1}{2}S^{e\prime}{}_F$ or $K'_+ = -\frac{1}{2}S'_F$ is sufficient to determine the properties of the one-particle states. To extend such an analysis to a two-body system, we define the propagation kernel for the two-particle system as

$$K^{(2)\prime}{}_+(x_3, x_4; x_1, x_2) = \frac{(\Phi_0, T(S\psi(x_3)\, \psi(x_4)\, \bar{\psi}(x_1)\, \bar{\psi}(x_2))\, \Phi_0)}{(\Phi_0, S\Phi_0)}$$

$$= (\Psi_0, T(\psi(x_3)\, \psi(x_4)\, \bar{\psi}(x_1)\, \bar{\psi}(x_2))\, \Psi_0) \quad (314)$$

A power series expansion of (314) indicates that

$$K^{(2)\prime}{}_+(x_3, x_4; x_1, x_2) = K_+(x_3 - x_1)\, K_+(x_4 - x_2) - K_+(x_4 - x_1)\, K_+(x_3 - x_2)$$

$$- \frac{e^2}{2} \int d^4x' \int d^4x''\, K_+(x_3 - x')\, \gamma^\mu K_+(x' - x_1)\, D_F(x' - x'')$$

$$\cdot\, K_+(x_4 - x'')\, \gamma_\mu K_+(x'' - x_2) - (\text{exchange term with } x_3 \leftrightarrow x_4) + \cdots$$

$$(315)$$

which series can be represented by the diagrams in Figure 17.10.

Salpeter and Bethe [Salpeter (1951)] have analyzed the structure of the series (315) in a manner similar to that outlined in Chapter 16. They call a graph reducible if it can be split into two unconnected parts by drawing a line which cuts no boson line at all and each of the two fermion lines only once. Examples of such reducible diagrams are given in Figure 17.11.

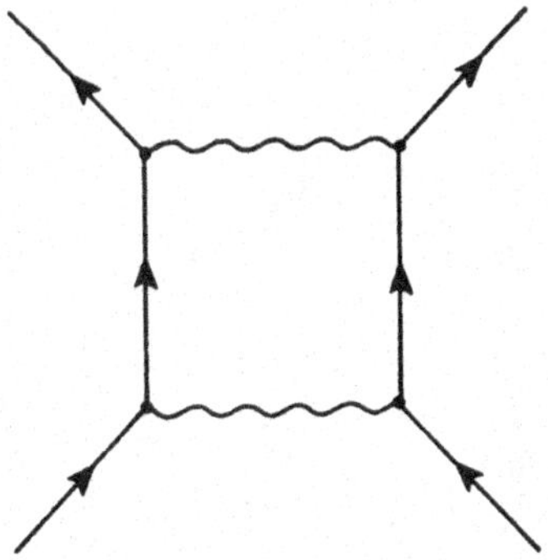
(a)

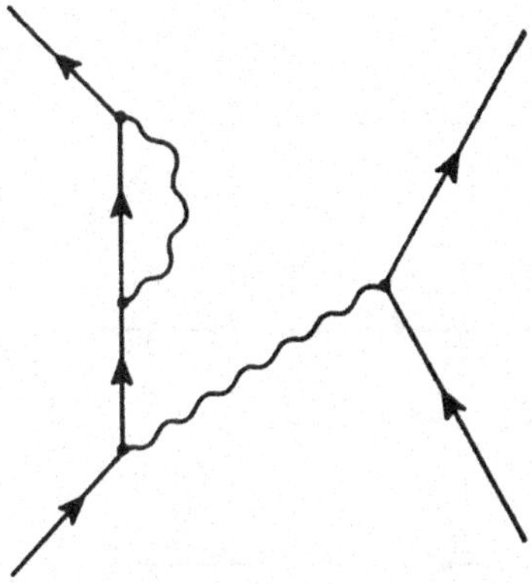
(b)

Fig. 17.11

The irreducible diagrams up to fourth order are given in Figure 17.12. Clearly, reducible diagrams can always be broken down into irreducible parts. Moreover, the irreducible graphs can be ordered according to the

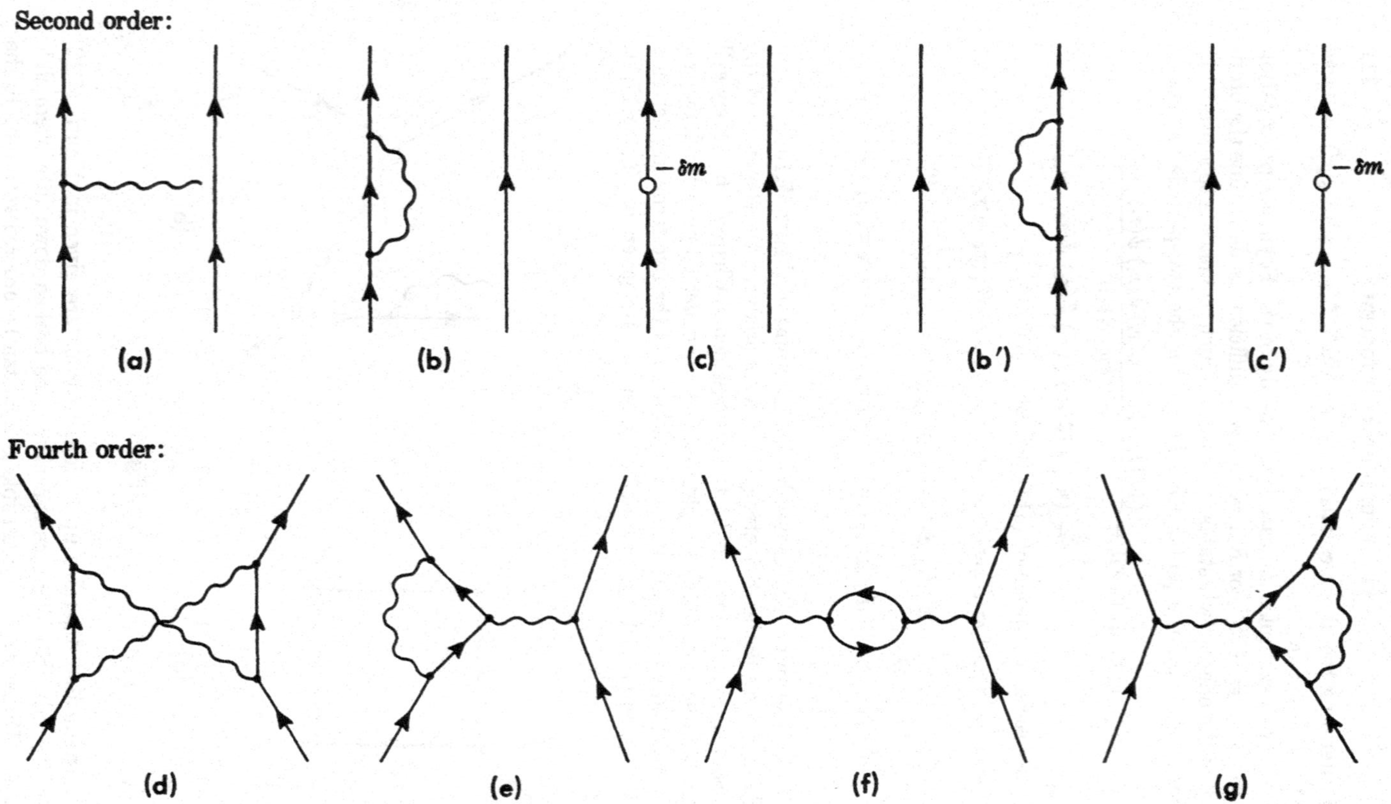

Fig. 17.12

power of the coupling constant occurring in the matrix element corresponding to the graph.

If we call $\sum_m G^{(m)}(x_1, x_2; x_3, x_4) = G(x_1, x_2; x_3x_4)$ the sum of all the irreducible graphs, the two-particle propagation kernel $K^{(2)\prime}{}_+(x_3, x_4; x_1, x_2)$ (omitting the exchange term) can then be written as [Salpeter (1951)]:

$$K^{(2)\prime}{}_+(x_3, x_4; x_1x_2) - K'_+(x_3 - x_1)\, K'_+(x_4 - x_2)$$
$$= i \iiiint d^4x_5\, d^4x_6\, d^4x_7\, d^4x_8\, K'_+(x_3 - x_5)\, K'_+(x_4 - x_6)$$
$$\sum_{(m)} G^{(m)}(x_5, x_6; x_7, x_8)\, K^{(2)\prime}{}_+(x_7, x_8; x_1, x_2) \qquad (316)$$

Although this integral equation contains the infinite series $\sum_{(m)} G^{(m)}$, it has clearly far fewer terms than Eq. (315).[12] From a Neumann-Liouville iteration of Eq. (316), it is evident that the reducible diagrams are already properly included by virtue of the integral equation, and that, in fact, such an iteration yields precisely all the terms in the power series expansion of Eq. (314). It should also be noted that in the form (316) we may omit from consideration all fermion self-energy parts, since their effect is already included by giving in $G^{(m)}$ each fermion line a factor $-\frac{1}{2}S'_F = K'_+$. The fact that all reducible diagrams are already included in (316) can be a decided advantage in theories where the coupling constant is small, and the expansion $\sum_{(m)} G^{(m)}$ may be considered as an asymptotic series. For example, most of the binding energy between a proton and an electron already comes from the repeated action of the diagram in Figure 17.12a, i.e., from the diagrams indicated in Figure 17.13. The advantage which is

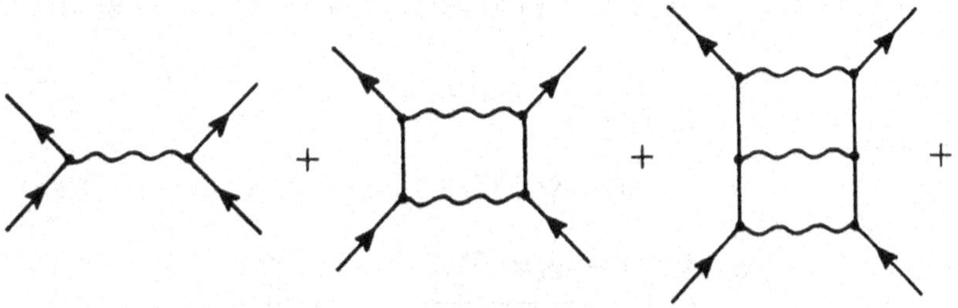

Fig. 17.13

gained by using (316) with $G^{(1)}$ is then equivalent to that of solving the Schrödinger equation with a potential, over that of treating the potential as a perturbation.

[12] The indications are, however, that this series does not converge [Hurst (1952a)] unless important cancellations occur between irreducible diagrams in a given order.

In lowest order, the interaction function $G^{(1)}$ in electrodynamics is given by

$$G^{(1)}(x_5, x_6; x_7, x_8) = e^2 D_F(x_5 - x_6)\, \delta^{(4)}(x_5 - x_7)\, \delta^{(4)}(x_6 - x_8)\, \gamma_a{}^{\mu}\gamma_{b\mu} \quad (317a)$$

where the subscripts a, b refer to the two fermions which are now assumed to be distinguishable. To this approximation, which is known as the ladder approximation since it includes all the diagrams of the form indicated in Figure 17.13, the propagator is given as the solution of the integral equation:

$$K'_{+(L)}(x_1, x_2; y_1, y_2) = \tfrac{1}{4}S_F(x_1 - y_1)\, S_F(x_2 - y_2) + \tfrac{1}{2}\pi\alpha \int d^4x_3 \int d^4x_4$$

$$S_F(x_1 - x_3)\, S_F(x_2 - x_4)\, \gamma^{(1)}{}_{\mu}\gamma^{(2)\mu} D_F(x_3 - x_4)\, K'_{+(L)}(x_3, x_4; y_1, y_2) \quad (317b)$$

where again, for the sake of simplicity, we have assumed the particles distinguishable.

The connection of the propagation kernel $K'_+(x_1, x_2; x_3, x_4)$ with the properties of the stationary states of the field system has been established by Gell-Mann and Low [Gell-Mann (1951)] and by Eden (1953a). Let us denote the eigenstates of the total energy-momentum four-vector of the field system by $|\mathbf{p}; \boldsymbol{\alpha}\rangle$. Then, if the theory is such that it admits of bound states of two fermions, these bound states will be characterized, in part, by their binding energy, or equivalently by their rest mass $M^{(\alpha)}$, i.e., if $p^{(\alpha)}{}_{\mu}$ is the total energy-momentum of the bound state system, then $p^{(\alpha)}{}_{\mu}p^{(\alpha)\mu} = M^{(\alpha)2}$. Now, when $t_1, t_2 > t_3, t_4$

$$T(\psi(x_1)\, \psi(x_2)\, \tilde{\psi}(x_3)\, \tilde{\psi}(x_4)) = T(\psi(x_1)\, \psi(x_2))\, T(\tilde{\psi}(x_3)\, \tilde{\psi}(x_4)) \quad (318)$$

and since it is assumed that the set of states $|\mathbf{p}, \boldsymbol{\alpha}\rangle$ are complete, we can write the two-body propagator in the following way:

$$K'_+(x_1, x_2; x_3, x_4) = \sum_{|\mathbf{p}, \alpha\rangle} \langle \boldsymbol{\Psi}_0 \mid T(\psi(x_1)\, \psi(x_2)) \mid \mathbf{p}, \boldsymbol{\alpha}\rangle \langle \mathbf{p}, \boldsymbol{\alpha} \mid T(\tilde{\psi}(x_3)\, \tilde{\psi}(x_4)) \mid \boldsymbol{\Psi}_0\rangle$$

$$= \sum_{p, \alpha} \chi_{p\alpha}(x_1, x_2)\, \chi_{p\alpha}{}^{\dagger}(x_3, x_4) \quad (319)$$

where

$$\chi_{p\alpha}(x_1, x_2) = \langle \boldsymbol{\Psi}_0 \mid T(\psi(x_1)\, \psi(x_2)) \mid \mathbf{p}, \boldsymbol{\alpha}\rangle \quad (320a)$$

and

$$\chi_{p\alpha}{}^{\dagger}(x_3, x_4) = \langle \mathbf{p}, \boldsymbol{\alpha} \mid T(\tilde{\psi}(x_3)\, \tilde{\psi}(x_4)) \mid \boldsymbol{\Psi}_0\rangle$$

$$= \overline{\langle \boldsymbol{\Psi}_0, T^{\dagger}(\tilde{\psi}(x_3)\, \tilde{\psi}(x_4)) \mid \mathbf{p}, \boldsymbol{\alpha}\rangle}\, \gamma^0\gamma^0 \quad (320b)$$

In (320b) $T^{\dagger}$ orders the operators in an antichronological fashion, i.e., the operators labeled with the later time standing to the right of those labeled with an earlier time.

Let us next determine the properties of $\chi_{p\alpha}(x_1, x_2)$ which follow from the translation invariance of the theory. Since, under a translation through a

$$\mathbf{U}(a) \mid \mathbf{p}, \boldsymbol{\alpha}\rangle = e^{i\mathbf{P}\cdot a} \mid \mathbf{p}, \boldsymbol{\alpha}\rangle = e^{ip\cdot a} \mid \mathbf{p}, \boldsymbol{\alpha}\rangle \quad (321)$$

we have

$$\begin{aligned}\chi_{p\alpha}(x_1, x_2) &= \langle \boldsymbol{\Psi}_0, e^{-i\mathbf{P}\cdot a} e^{i\mathbf{P}\cdot a} T(\psi(x_1)\,\psi(x_2))\, e^{-i\mathbf{P}\cdot a} e^{i\mathbf{P}\cdot a} \mid \mathbf{p}, \boldsymbol{\alpha}\rangle \\ &= \langle \boldsymbol{\Psi}_0, T(\psi(x_1 + a)\,\psi(x_2 + a)) \mid \mathbf{p}, \boldsymbol{\alpha}\rangle\, e^{ip\cdot a} \\ &= e^{ip\cdot a}\, \chi_{p\alpha}(x_1 + a, x_2 + a) \end{aligned} \tag{322}$$

Equation (322) is true for arbitrary space-time displacements and therefore in particular for the choice $a = -X$, where

$$X = \frac{m_1 x_1 + m_2 x_2}{m_1 + m_2}$$

X is the center-of-mass co-ordinate. With this choice for a, $\chi_{p\alpha}(x_1, x_2)$ can be written in the form

$$\chi_{p\alpha}(x_1, x_2) = e^{-ip\cdot X} f_{p\alpha}(x_1 - x_2) \tag{323}$$

which separates $\chi_{p\alpha}(x_1, x_2)$ into a dependence on the center-of-mass co-ordinate, $\exp(-ip \cdot X)$, and a dependence on the relative co-ordinate $x_1 - x_2$, $f_{p\alpha}(x_1 - x_2)$. Note that, as p varies, $f_{p\alpha}(x_1 - x_2)$ will vary.

If the field system under consideration allows for the existence of a (possibly degenerate) two-fermion bound state $|\mathbf{p}, \mathbf{b}\rangle$ of mass M_b, then by isolating the contribution of these terms in Eq. (319), the propagator for $X_0 > Y_0$ can be written as follows:

$$K'_+(x_1, x_2; y_1, y_2) = \sum_b \int d^4p\, e^{-ip\cdot(X-Y)}\, \delta(p^2 - M_b^2)\, \theta(p_0)\, f_{pb}(x)\, f_{pb}^\dagger(y) + \\ + \sum_{\alpha \neq b} \int d^4p\, \chi_{p\alpha}(x_1, x_2)\, \chi_{p\alpha}^\dagger(y_1, y_2) \tag{324}$$

where, in the first term, the summation over b is over all the degenerate states of rest mass M_b. Equation (324) allows us to write for $X_0 > Y_0$

$$K'_+(x_1, x_2; y_1, y_2) \\ = \frac{1}{2} \int d^3p\, e^{i\mathbf{p}\cdot(\mathbf{X}-\mathbf{Y})} \int_0^\infty \frac{dp_0}{p_0}\, e^{-ip_0(X_0 - Y_0)}\, \delta(p_0 - \sqrt{\mathbf{p}^2 + M_b^2}) \\ \sum_b f_{pb}(x)\, f_{pb}^\dagger(y) + \text{terms which are regular when } p_0 = \sqrt{\mathbf{p}^2 + M_b^2} \tag{325}$$

Note that, although $f_{pb}(x)$ varies as p takes on values for which $p^2 = M_b^2$, once f_{pb} is known for one $\mathbf{p}$, it can be obtained for any other $\mathbf{p}$ from a consideration of the Lorentz covariance of the theory. It is, therefore, usually convenient to choose that amplitude f_{pb} for which $\mathbf{p} = 0$ as the representative, in which case the degeneracy can be classified according to the eigenvalue of the angular momentum operator. Note also that for fixed relative co-ordinates, the contribution of the state $|\mathbf{p}, \mathbf{b}\rangle$ to $K'_+(x_1, x_2; y_1, y_2)$ when $Y_0 > X_0$ vanishes since $\langle \boldsymbol{\Psi}_0 \mid T(\bar{\psi}(x_1)\,\bar{\psi}(x_2)) \mid \mathbf{p}, \mathbf{b}\rangle$ vanishes owing to conservation laws (e.g., charge or baryon conservation). Hence the dependence of the propagator on $X - Y$ (for arbitrary X_0, Y_0) can be written in the form:

$$K'_+(X - Y; x, y) = \sum_b \int d^4p\, e^{-ip\cdot(X-Y)} \frac{f_{pb}(x)\, f_{pb}^\dagger(y)}{p_0 - \sqrt{\mathbf{p}^2 + (M_b - i\epsilon)^2}}$$
$$+ \text{ terms which are regular when } p_0 = \sqrt{\mathbf{p}^2 + M_b^2} \quad (326)$$

We shall next insert expression (326) for K'_+ into Eq. (316). Before doing so, it will be noted that Eq. (316), which can be abbreviated by $K'_+ = K_0 + K_0GK'_+$, can by virtue of the translation invariance of the theory be written in the form:

$$\int d^4p\, e^{-ip\cdot(X-Y)}\, K'_+(p; x, y)$$
$$= \int d^4p\, e^{-ip\cdot(X-Y)} \{K_0(p; x, y) + (K_0GK'_+)\,(p; x, y)\} \quad (327)$$

The integration over d^4p can therefore be dropped. If now we multiply the resulting expression by $p_0 - \sqrt{\mathbf{p}^2 + M_b^2}$ and take the limit $p_0 \to \to \sqrt{\mathbf{p}^2 + M_b^2}$, the inhomogeneous term $K_0(p; x, y)$ of Eq. (327) drops out, as well as those parts of the propagator which do not have a singularity at $p_0 = \sqrt{\mathbf{p}^2 + M_b^2}$, so that $\chi_{pb}(x_1, x_2)$ satisfies the equations:

$$\chi_{pb}(x_1, x_2) = -i \int d^4x_5 \int d^4x_6 \int d^4x_7 \int d^4x_8$$
$$K'_+(x_1 - x_5)\, K'_+(x_2 - x_6)\, G(x_5, x_6; x_7, x_8)\, \chi_{pb}(x_7, x_8) \quad (328)$$

Precisely this equation was derived by Salpeter and Bethe using somewhat more intuitive arguments based on an adiabatic switching-on of the interaction. For the scattering problem, the left-hand side of (328) becomes replaced by $\chi_{p\alpha}(x_1, x_2) - \phi_{p\alpha}(x_1, x_2)$, where $\phi_{p\alpha}$ is a product of two Dirac free-particle wave functions; $\phi_{p\alpha}$ may then be viewed as the incoming plane wave.

A differential equation which is equally applicable to the discussion of scattering and to the properties of bound states can be obtained if one recalls that $K'_+ = -\frac{1}{2}S'_F$ satisfies the equation:

$$-\tfrac{1}{2}S'_F(x - x') = -\tfrac{1}{2}S_F(x - x')$$
$$+ \int d^4y \int d^4y'\, \tfrac{1}{2}S_F(x - y)\, \Sigma^*(y - y')\, \tfrac{1}{2}S'_F(y' - x') \quad (329)$$

Hence, upon applying the Dirac operator to (329), we obtain

$$(-i\gamma_\mu\partial^\mu + m)\, S'_F(x - x') = 2i\delta^{(4)}(x - x') + \int d^4y\, \Sigma^*(x - y)\, \tfrac{1}{2}S'_F(y - x') \quad (330)$$

If we now apply the differential operator $-i\gamma_\mu\partial^\mu + m$ for each particle to Eq. (328), we obtain

$$\left(-i\gamma^{a\mu}\frac{\partial}{\partial x_1{}^\mu}+m_a\right)\left(-i\gamma^{b\mu}\frac{\partial}{\partial x_2{}^\mu}+m_b\right)\chi_{pa}(x_1, x_2)$$
$$= -\int d^4x_7 \int d^4x_8\, G(x_1, x_2; x_7, x_8)\,\chi_{pa}(x_7, x_8)$$
$$- i\int d^4x_5 \int d^4x_6 \int d^4x_7 \int d^4x_8\, d^4y \int d^4y'$$
$$[\Sigma^*(x_1 - y)\,\tfrac{1}{2}S'_F(y - x_5)]_a\,[\Sigma^*(x_2 - y')\,\tfrac{1}{2}S'_F(y' - x_6)]_b$$
$$G(x_5, x_6; x_7, x_8)\,\chi_{pa}(x_7, x_8) \tag{331}$$

If, in Eq. (331), we omit fermion self-energy parts, it takes on a particularly simple form, since then the last factor on the right-hand side of (331) no longer appears. If, in addition, we restrict ourselves to the lowest order interaction, i.e., the ladder approximation, the amplitude χ_{pa} then satisfies the equation:

$$(-i\gamma\cdot\partial_1 + m)_a\,(-i\gamma\cdot\partial_2 + m)_b\,\chi_{pa}(x_1, x_2)$$
$$= -ie^2\gamma^{a\mu}D_F(x_1 - x_2)\,\gamma^b{}_\mu\chi_{pa}(x_1, x_2) \tag{332}$$

The essential progress which has been achieved by Eq. (328) is a completely relativistic wave equation for the two-body system. Furthermore, it allows the graphical procedures of Feynman and Dyson, together with their renormalization prescription, to be applied to calculation of the binding energy of a bound state. This, however, is achieved at some cost in the physical interpretation of the amplitude. In particular, the dependence of the amplitude on the relative time $x_{10} - x_{20}$ is not readily interpreted physically.

Wick (1954) and Cutkosky (1954) have made decisive progress in clarifying some of these unfamiliar features of the Bethe-Salpeter equation. Wick has pointed out that an additional condition for the B-S amplitude is obtained from its definition (322), supplemented by simple stability requirements. We here briefly outline Wick's argument for the case of the hydrogen atom.

It follows from (323) that, if the system is at rest, i.e., $\mathbf{p}^{(a)} = 0$, the two-particle amplitude $\chi_{pa}(x_1, x_2)$ has the form

$$\chi_{pa}(x_1, x_2) = e^{-iET} f_{pa}(x) \tag{333}$$

where E is the total energy of the system and T the time co-ordinate of the center-of-mass co-ordinate. Here again, x is the relative co-ordinate of the two particles, $x = x_1 - x_2$. We next introduce the binding energy, B, of the bound state, by $E = m_e + M_p - B$, and it should be noted that

$$E = m_e + M_p - B < m_e + M_p \tag{334}$$

where m_e and M_p are the rest masses of the electron and proton.

Upon introducing into Eq. (320a), which defines $\chi_{pa}(x_1, x_2)$, a complete set of states when the time sequence is such that $x_{10} > x_{20}$, we obtain

$$\chi_{p\alpha}(x_1, x_2) = \sum_{|\mathbf{n}\rangle} \langle \Psi_0 \mid \psi_e(x_1) \mid \mathbf{n} \rangle \langle \mathbf{n} \mid \psi_p(x_2) \mid \mathbf{p}\alpha \rangle \tag{335}$$

The states $|\mathbf{n}\rangle$ giving a nonvanishing contribution to $\chi_{p\alpha}$ will belong to a special class. They clearly must be eigenfunctions of the total charge with eigenvalue $-e$. Also, in the frame of reference in which the total momentum is zero, they must be eigenfunctions of the total angular momentum with eigenvalue $-\frac{1}{2}\hbar$. Now all the states known in nature which satisfy the condition that their total charge is $-e$ and their angular momentum in their rest frame is $\frac{1}{2}\hbar$ also satisfy the inequality,

$$E_n^2 - \mathbf{p}_n^2 \geqslant m_e^2 \tag{336a}$$

where E_n and $\mathbf{p}$ are the total energy and momentum of the state $|\mathbf{n}\rangle$. Furthermore, the equality sign holds true only for the physical one-electron state, i.e., the state in which there is present one and only one physical electron. Wick calls this the stability condition for an electron. He then assumes that the physical one-electron state [for which the equality sign holds in (336)] is the state of lowest energy which will contribute to the sum in (335).

In a similar way, when the relative time x_0 is negative, one deduces that the states $|\mathbf{n}'\rangle$ which contribute to the amplitude are characterized, in part, by having energy and momentum eigenvalues such that

$$E_{n'}^2 - \mathbf{p}_{n'}^2 \geqslant M_p^2 \tag{336b}$$

which Wick calls the stability condition for the proton. The three inequalities (334), (336a), and (336b) form the basis of the following discussion. From (335) and (323) one readily establishes that $f_{p\alpha}(x)$ for $x_0 > 0$ has the following structure:

$$f_{p\alpha}(x) = \sum_{|\mathbf{n}\rangle} \langle \Psi_0 \mid \psi_e(0) \mid \mathbf{n} \rangle \langle \mathbf{n} \mid \psi_p(0) \mid \mathbf{p}\alpha \rangle \, e^{-i\left(p^{(n)} - \frac{m_e}{M_p + m_e} p\right)\cdot x} \tag{337}$$

where p is the energy-momentum of the state $|\mathbf{p}, \alpha\rangle$. Now, using the inequalities (334), (336a), and (336b), one finds that

$$p^{(n)0} - \frac{m_e}{m_e + M_p} p^0 \geqslant \frac{m_e}{M_p + m_e} B > 0 \tag{338}$$

Hence, for $x_0 > 0$, $f_{p\alpha}(x)$ has only positive frequencies. Similarly, it follows that for $x_0 < 0$, $f_{p\alpha}(x)$ is a superposition of negative frequency terms only; $f_{p\alpha}(x)$ has, therefore, properties similar to those of a Feynman propagator.

We may thus write for $x_0 > 0$

$$f_{p\alpha}(x) = \int d^3q \int_{\omega_{\min}}^{\infty} d\omega \, g_{p\alpha}(\mathbf{q}, \omega) \, e^{i(\mathbf{q}\cdot\mathbf{x} - \omega x_0)} \tag{339}$$

where

$$\omega_{\min} = \frac{m_e}{m_e + M_p} B + (m_e^2 + \mathbf{p}^2)^{1/2} - m_e \tag{340}$$

Considered as a function of the complex variable x_0, $f_{p\alpha}(x)$ can now be analytically continued in the lower half plane, in the region $0 \geqslant \arg x_0 > -\pi$. Similarly, starting from the negative real axis, $f_{p\alpha}(x)$ can be continued analytically in the upper half plane in the region $\pi \geqslant \arg x_0 > 0$. Furthermore, since $B > 0$, $f_{p\alpha}(x)$ goes to zero when $x_0 \to \infty$ in any direction in the lower or upper half plane different from the real axis.

In momentum space, if we write $f(x) = f_1(x) + f_2(x)$, where $f_1(x) = 0$ for $x_0 < 0$ and $f_2(x) = 0$ for $x_0 > 0$, then with

$$\phi_1(\mathbf{q}, q_0) = \frac{1}{(2\pi)^4} \int d^3x \, e^{-i\mathbf{q}\cdot\mathbf{x}} \int_0^\infty dx_0 \, e^{iq_0x_0} f(x) \tag{341}$$

$$\phi_2(\mathbf{q}, q_0) = \frac{1}{(2\pi)^4} \int d^3x \, e^{-i\mathbf{q}\cdot\mathbf{x}} \int_{-\infty}^0 dx_0 \, e^{iq_0x_0} f(x) \tag{342}$$

using (338), one readily finds that

$$\phi_1(\mathbf{q}, q_0) = \frac{1}{2\pi i} \int_{\omega_{\min}}^\infty \frac{g(\mathbf{q}, \omega)}{\omega - q_0 - i\epsilon} \, d\omega \tag{343}$$

where ϵ is an infinitesimal positive constant. It then follows from this representation that $\phi_1(\mathbf{q}, q_0)$ is an analytic function of q_0 in the whole complex plane in the region

$$2\pi > \arg (q_0 - \omega_{\min}) \geqslant 0 \tag{344}$$

Similar statements can be made about ϕ_2, and it can be shown that $\phi(q) = \phi(\mathbf{q}, q_0) = \phi_1(q) + \phi_2(q)$ is defined in the complex q_0 plane with two cuts from $\omega_{\min}$ to $+\infty$ and from $-\infty_{\min}$ to $\omega_{\max}$, where

$$-\omega_{\max} = \frac{M_p}{M_p + m_e} B + (M_p^2 + \mathbf{p}^2)^{1/2} - M_p \tag{345}$$

One can therefore continue the function analytically from the upper to the lower half plane through the gap between the two cuts, the existence of which depends on the fact that $B > 0$.

These analytic properties of the wave functions can now be used to transform the Bethe-Salpeter equation, by a rotation of the path of integration in the complex plane, to an equation in which q_0 is purely imaginary (or the relative time is purely imaginary). One then operates in a Euclidian space, and considerable simplifications in the mathematics ensue. Kemmer and Salam [Kemmer (1955)] have shown that this analytic continuation is also permissible in the case when the two-body equation is applied to scattering problems. Using this transformation to a Euclidean space, Wick and Cutkosky have been able to obtain the complete set of bound state solutions for the ladder approximation Bethe-Salpeter equation in the case of two scalar bosons interacting through scalar photons. The reader is referred to their papers for this analysis. [See also Hayashi and Munakata (Hayashi 1952), Goldstein (1953), Scarf (1955), and Green (1957).]

An extensive literature exists dealing with the properties of the Bethe-Salpeter and with the applications of this equation to various two-body problems. We here only mention that Mandelstam (1955) has shown how to express the matrix element of any dynamical variable between two bound states in terms of Bethe-Salpeter amplitudes. This allows him to find the normalization and orthogonality properties of these amplitudes, which in turn lead to the conditions which must be imposed on their singularities at the origin ($x = 0$). [See also Nishijima (1953a, 1954) and Allcock (1956, 1958).]

Salpeter (1952a, b) and Arnowitt (1953) have used the Bethe-Salpeter equation to calculate the mass corrections to the fine and hyperfine structure of hydrogen [see also Iddings (1959)]. Karplus and Klein [Karplus (1952c)] and Fulton and Martin [Fulton (1954b)] have used it for a description of the positronium atom. The work of Eden (1952, 1953, a, b, c) which extends the Bethe-Salpeter equation to include the possibility of decay from an excited state, enables one to calculate the lifetime of excited states within this framework.

The application of the covariant two-body equation to the problem of nuclear forces was initiated by Lévy (1952a, b), Klein (1953a, 1954), and Macke (1953a, b). The literature on this subject can be traced from the paper of Klein and McCormick [Klein (1958)].

18

The Axiomatic Formulation

Although insights into the structure of quantum field theories have been obtained by using perturbation methods and special simple models, and although general properties of relativistic field theories have been deduced based on the relativistic invariance of the theory and on spectral assumptions (Chapter 17), nonetheless it has proved extremely difficult to give any convincing answer to the central question of relativistic quantum theory: Do solutions of the renormalized equations of quantum electrodynamics or any meson theories exist? To answer this question, even if the equations were inherently simpler, is made almost impossible due to the fact that the renormalization method lies almost wholly outside the bounds of conventional mathematics. One has therefore no way of really asserting on the basis of the considerations of the previous chapters that the present description of the interactions between elementary particles based on the notions of local relativistic field theories is correct.

It is this inability to give convincing and mathematically rigorous proofs for many of the assertions made on the basis of the quantum theory of fields as presented thus far in this book which has been the principal motivation for the important investigations of the general structure of local field theories which have been carried out in recent years. In the present chapter we outline some of these investigations, principally those of Wightman (1956) and of Lehmann, Symanzik, and Zimmermann [Lehmann (1955a, 1957)], hereafter referred to as LSZ. Both of these approaches are based on certain general postulates such as Lorentz invariance, spectral conditions and locality, and do not use the notion of a Lagrangian (or Hamiltonian). More specifically, no assumptions are made about the form of the field equations or interactions. Their principal aim is to discover whether any local relativistic field theories exist. Their approach has mainly consisted in studying the consequences for observable quantities of the locality assumption, i.e., that the commutator (for integer spin fields) or anticommutator (for odd half-integer spin fields) of two local field operators vanishes for space-like separation.

All of the investigations which have been carried out make the following assumptions about the theory (usually taken for simplicity to be a neutral scalar field):

I. The usual postulates of quantum mechanics are valid, i.e., that the states of the systems are represented by vectors in a Hilbert space, $\mathfrak{H}$, and that the observables of the system can be represented by self-adjoint operators on $\mathfrak{H}$.

II. The theory is invariant under inhomogeneous Lorentz transformations.

III. The energy-momentum spectrum of the states is reasonable.

The requirement of Lorentz invariance implies the existence of a representation of the orthochronous inhomogeneous Lorentz group by unitary operators, $U(a, \Lambda)$, under which $\mathfrak{H}$ is invariant. From this follows that there exist Hermitian displacement operators[1] P_μ with the properties:

$$[P_\mu, P_\nu] = 0 \tag{1a}$$

$$[P_\mu, F(x)] = i\partial_\mu F(x) \tag{1b}$$

where $F(x)$ is any Heisenberg operator which does not depend explicitly on the space-time co-ordinates. In a representation in which the operators P_μ ($\mu = 0, 1, 2, 3$) are diagonal, and $\mathfrak{H}$ is spanned by basis vectors $|p, \alpha\rangle$ with

$$P_\mu \mid p, \alpha\rangle = p_\mu \mid p, \alpha\rangle \tag{2}$$

Postulate III can be more precisely formulated. It requires that

III(a). There exists a unique invariant vacuum state $|\Psi_0\rangle$, characterized by

$$U(a, \Lambda) \mid \Psi_0\rangle = |\Psi_0\rangle \tag{3a}$$

III(b). For all other states $|p, \alpha\rangle$ the eigenvalue p_μ is time-like or light-like with a positive value of the time component p_0.

By considering infinitesimal transformation, we derive from (3a) that

$$P_\mu \mid \Psi_0\rangle = M_{\mu\nu} \mid \Psi_0\rangle = 0 \tag{3b}$$

Finally, these investigations also usually assume that

IV. The theory is local, i.e., that a field observable at a point x commutes with a field observable at x', if the distance between x and x' is space-like, i.e., if $(x - x')^2 < 0$.

More precisely, if the theory is that of a neutral scalar field described by a field operator $\phi(x)$ which transforms as a scalar under the inhomogeneous Lorentz group

$$\phi^*(x) = \phi(x) \tag{4}$$

$$U(a, \Lambda)\, \phi(x)\, U(a, \Lambda)^{-1} = \phi(\Lambda x + a) \tag{5}$$

then it is assumed that

$$[\phi(x), \phi(x')] = 0 \quad \text{for } (x - x')^2 < 0 \tag{6}$$

[1] In the present chapters all operators and state vectors are (renormalized) Heisenberg operators and vectors, and will be denoted by light-face symbols since no confusion can arise.

This requirement, sometimes called the microscopic causality principle, is the mathematical statement of the fact that no signal can be exchanged between two points separated by a space-like interval and therefore that measurements at such points cannot interfere. Whether physical particles and their interactions can be accounted for by a description in terms of local operators remains at present an open question. In Section *d* of the present chapter, we shall examine in detail the analytical consequence for scattering amplitudes of the assumption of locality.

We next turn to a brief presentation of the Wightman and of the Lehmann, Symanzik, Zimmermann (LSZ) formulation of field theory. The approaches of Wightman (1956) and LSZ [Lehmann (1955a, 1957)] differ in that Wightman chooses to study the vacuum expectation of products of operators whereas LSZ analyze the theory in terms of vacuum expectation values of time-ordered products of operators [Lehmann (1955a)] and in terms of retarded products of operators [Lehmann (1957)]. The axioms of the theory, principally those of relativistic invariance, spectral conditions and locality, can be translated into certain properties of these vacuum expectation values and into certain relations among them. More precisely, when the theory is analyzed in terms of the vacuum expectation values of either products of field operators (Wightman functions) or retarded products of field operators (*r* functions), the requirements of relativistic invariance and locality as well as the spectral conditions are linear relations on these functions. On the other hand the positive-definiteness relations for the Wightman functions and the unitarity condition for the *r* functions are nonlinear relations among these functions. This division into a linear and a nonlinear part is an important feature of the axiomatic approach. In contrast to the Lagrangian approach in which any linear theory was trivial, and any interesting theory had necessarily nonlinear field equations whose complexity prevented any rigorous statements to be made about their solutions, in the axiomatic approach the linear relations have nontrivial content.

In the first two sections of the present chapter we outline these formulations. Section *c* then presents the Dyson-Jost-Lehmann representation for the matrix element of a causal commutator. This representation allows us to discuss the analytic properties of the two-particle elastic scattering amplitude and to introduce the notion of dispersion relations in Section *d*. Section *e* attempts to give a summary and to discern the direction of future work in the field.

18a. Wightman Formulation

One of the most important results in the axiomatic approach to quantum field theory is due to Wightman (1956) who has shown that a field theory can be reformulated in terms of the vacuum expectation values of products of field operators:

$$W^{(n)}(x_1, x_2, \cdots, x_n) = (\Psi_0, \phi(x_1)\,\phi(x_2) \cdots \phi(x_n)\,\Psi_0) \tag{7}$$

whose properties are directly related to the postulates I–IV enumerated above. A rigorous mathematical meaning can be assigned to these functions and they can be studied in an unambiguous and precise fashion. The importance of these Wightman functions, $W^{(n)}$, derives from the fact that Wightman has shown that, given a set of functions $W^{(n)}(x_1, \cdots x_n)$, $n = 1, 2, \cdots$, which satisfy certain specified properties, one can then reconstruct the field theory of which the $W^{(n)}$s are the vacuum expectation values. In the study of relativistic field theories, one can therefore restrict one's attention to these vacuum expectation values. We here briefly outline the Wightman program for the case of a neutral scalar field $\phi(x)$ interacting with itself. Although our presentation is for a scalar spin 0 field, the method is easily extended to the case of arbitrary fields. The same remarks apply to Wightman's result that a theory is uniquely determined by its vacuum expectation values.

The vacuum expectation value of products of field operators define the Wightman functions by Eq. (7). These Wightman functions are to be understood as linear functionals,[2] which assign a complex number $W^{(n)}[f] = \int d^4x_1 \cdots \int d^4x_n\, W^{(n)}(x_1, \cdots x_n)\, f(x_1, \cdots x_n)$ to each infinitely differentiable function $f(x_1, \cdots x_n)$ vanishing outside a bounded region of the $4n$ dimensional space-time continuum. This can be viewed as the mathematical reflection of the fact that only space-time averages of the field operator $\phi(x)$ are observable, i.e., only operators of the form

$$\phi(f) = \int d^4x\, f(x)\,\phi(x) \tag{8}$$

The class of functions $f(x)$ (usually called testing functions) for which $\phi(f)$ is assumed defined is taken as the class $\mathfrak{D}$ of all infinitely differentiable functions of compact support in space-time[3] to reflect the possibility of making field measurements in finite spatio-temporal regions. The quantity $W_\Psi[f] = (\Psi, \phi(f)\,\Psi)$ is then a linear functional with respect to f. To make more precise the mathematical meaning of the functional $W_\Psi[f]$, one further requires that, if $\{f_n\}$ is a sequence of test functions which converges to zero, then $W_\Psi[f_n] \to 0$. A sequence is said to converge to zero if all the f_ns have their support contained in a compact set and if

[2] A functional T on some space $\mathcal{S}$ is an operation which associates with every element f of $\mathcal{S}$ a complex number denoted by $T[f]$. A functional on $\mathcal{S}$ is said to be linear if
(a) $T[f_1 + f_2] = T[f_1] + T[f_2]$ for f_1, f_2 in $\mathcal{S}$
(b) $T[\alpha f] = \alpha T[f]$ for every f in $\mathcal{S}$ and every complex number α.

[3] The support of f is the complement of the largest open set on which f vanishes, i.e., essentially it is the set on which f is different from zero. For our purposes, a set is said to be compact if it is closed and bounded.

the f_ns, as well as their derivatives, converge uniformly to zero.[4] A linear functional T on $\mathfrak{D}$ which is endowed with the continuity property that $T[f_n]$ converges to zero for any sequence $\{f_n\}$ that converges to zero as defined above, is called a distribution [Schwartz (1950, 1951), Gärding (1959)]. The expectation value

$$W_\Psi[f] = (\Psi, \phi(f)\,\Psi) \tag{9}$$

is therefore a distribution on $\mathfrak{D}$, the space of all infinitely differentiable functions of compact support. More generally, the n-fold vacuum expectation value

$$W^{(n)}(x_1, \cdots x_n) = (\Psi_0, \phi(x_1) \cdots \phi(x_n)\,\Psi_0) \tag{10}$$

is a distribution in each variable x_i separately:

$$W^{(n)}[f_1, \cdots f_n] = (\Psi_0, \phi(f_1) \cdots \phi(f_n)\,\Psi_0) \tag{11}$$

It has a unique extension to become a distribution on $4n$ space, thus giving a precise meaning to $W^{(n)}(x_1, \cdots x_n)$ in terms of (infinitely differentiable) test functions $f(x_1, \cdots x_n)$ vanishing outside a bounded region of $4n$ space. Although it is only the distributions $W^{(n)}[f_1, \cdots f_n]$ or more generally $W^{(n)}[f]$ which are well-defined mathematical objects, we shall nonetheless continue to work with the Wightman function $W^{(n)}(x_1, \cdots x_n)$. For a more rigorous treatment, the reader is referred to Wightman (1958, 1959). [See also Schmidt (1956), Gärding (1959).]

The Lorentz invariance of the theory

$$U(a, \Lambda) \mid \Psi_0\rangle = |\Psi_0\rangle \tag{12a}$$

$$U(a, \Lambda)\,\phi(x)\,U(a, \Lambda)^{-1} = \phi(\Lambda x + a) \tag{12b}$$

implies that under orthochronous Lorentz transformations the Wightman functions have the property that

$$\begin{aligned} W^{(n)}&(x_1, x_2, \cdots, x_n) \\ &= (\Psi_0, \phi(x_1)\,\phi(x_2) \cdots \phi(x_n)\,\Psi_0) \\ &= (U(a, \Lambda)\,\Psi_0, U(a, \Lambda)\,\phi(x_1)\,U(a, \Lambda)^{-1}\,U(a, \Lambda) \cdots \\ &\qquad \cdots U(a, \Lambda)\,\phi(x_n)\,U(a, \Lambda)^{-1}\,U(a, \Lambda)\,\Psi_0) \\ &= W^{(n)}(\Lambda x_1 + a, \Lambda x_2 + a, \cdots, \Lambda x_n + a) \quad \text{(no time inversion)} \end{aligned} \tag{13}$$

If Λ includes a time inversion so that U is antiunitary, then

$$W^{(n)}(x_1, \cdots x_n) = \overline{W^{(n)}(\Lambda x_1 + a, \Lambda x_2 + a, \cdots \Lambda x_n + a)} \quad \text{(time inversion)} \tag{14}$$

From translation invariance it follows that $W^{(n)}(x_1, \cdots x_n)$ is a function only of the difference of co-ordinates:

$$W^{(n)}(x_1, \cdots, x_n) = W^{(n)}(\xi_1, \cdots \xi_{n-1}) \tag{15a}$$

[4] Uniform convergence is required only for each *fixed* order of the derivatives, not for all orders collectively.

where

$$\begin{aligned} \xi_1 &= x_1 - x_2 \\ &\cdot \\ &\cdot \\ \xi_j &= x_j - x_{j+1} \\ &\cdot \\ &\cdot \\ \xi_{n-1} &= x_{n-1} - x_n \end{aligned} \tag{15b}$$

The invariance under homogeneous Lorentz transformation then asserts that

$$W^{(n)}(\xi_1, \cdots \xi_{n-1}) = W^{(n)}(\Lambda\xi_1, \cdots \Lambda\xi_{n-1}) \tag{16}$$

From the assumed hermiticity of the field operator, $\phi(x) = \phi^*(x)$, it follows that

$$\begin{aligned} \overline{(\Psi_0, \phi(x_1) \cdots \phi(x_n) \Psi_0)} &= (\Psi_0, \phi^*(x_n) \cdots \phi^*(x_1) \Psi_0) \\ &= (\Psi_0, \phi(x_n) \cdots \phi(x_1) \Psi_0) \end{aligned} \tag{17}$$

which, when expressed in terms of Wightman functions, states that

$$\overline{W^{(n)}(x_1, \cdots x_n)} = W^{(n)}(x_n, \cdots, x_1) \tag{18}$$

If we assume that the $W^{(n)}$s have Fourier transforms

$$W^{(n)}(\xi_1, \cdots \xi_{n-1}) = \int d^4p_1 \cdots \int d^4p_{n-1}\, e^{-i\sum_{j=1}^{n-1} p_j\cdot\xi_j}\, \tilde{W}^{(n)}(p_1, \cdots p_{n-1}) \tag{19}$$

(this requires the $W^{(n)}$ functions to behave at most as a polynomial at infinity), it then follows from the fact that the spectrum of the physical states is such that no states with space-like momenta or negative energy exist, that $\tilde{W}^{(n)}(p_1, \cdots, p_{n-1})$ vanishes unless $p_i^2 \geqslant 0$ and $p_i \geqslant 0$, $i = 1, 2, \cdots n - 1$. *Proof:* By assumption there exists a *complete* set of states $|p, \alpha\rangle$ which are eigenstates of the energy-momentum operator P_μ with eigenvalues p_μ such that $p^2 \geqslant 0$, $p_0 \geqslant 0$. Hence upon inserting the completeness relation $\sum_{|p\alpha\rangle} |p\alpha\rangle \langle p\alpha| = 1$ into the defining relation for $W^{(n)}$, we obtain:

$$\begin{aligned} &W^{(n)}(x_1, \cdots x_n) \\ &= \sum_{|p_1\alpha_1\rangle\cdots|p_{n-1}\alpha_{n-1}\rangle} \langle\Psi_0 | \phi(x_1) | p_1\alpha_1\rangle \langle p_1\alpha_1 | \phi(x_2) | p_2\alpha_2\rangle \cdots \langle p_{n-1}\alpha_{n-1} | \phi(x_n) | \Psi_0\rangle \\ &= \sum_{|p_1\alpha_1\rangle\cdots|p_{n-1}\alpha_{n-1}\rangle} \langle\Psi_0 | \phi(0) | p_1\alpha_1\rangle \cdots \langle p_{n-1}\alpha_{n-1} | \phi(0) | \Psi_0\rangle \\ &\qquad e^{-ip_1\cdot(x_1-x_2)}\, e^{-ip_2\cdot(x_2-x_3)} \cdots e^{-ip_{n-1}\cdot(x_{n-1}-x_n)} \end{aligned} \tag{20a}$$

so that

$$\begin{aligned} &\tilde{W}^{(n)}(p_1, \cdots p_{n-1}) \\ &= \sum_{\substack{\alpha_1 \cdots \alpha_{n-1} \\ \text{for fixed } p_1, \cdots p_{n-1}}} \langle\Psi_0 | \phi(0) | p_1\alpha_1\rangle \langle p_1\alpha_1 | \phi(0) | p_2\alpha_2\rangle \cdots \langle p_{n-1}\alpha_{n-1} | \phi(0) | \Psi_0\rangle \end{aligned} \tag{20b}$$

From Eq. (20b) one deduces that $\tilde{W}^{(n)}(p_1, \cdots p_{n-1}) = 0$ unless $p_{i0} \geqslant 0$ and $p_i^2 \geqslant 0$, since in (20b) p_i is the energy-momentum of a physical state. It now follows from this that the $W^{(n)}(\xi_1, \cdots \xi_{n-1})$s are boundary values of analytic functions. This assertion is proved as follows: Write

$$\zeta_j = \xi_j - i\eta_j \qquad j = 1, 2, \cdots n - 1 \tag{21}$$

where the four-vectors η_j are restricted to lie in the future light cone. The $8(n-1)$ dimensional open region thus defined in the $8(n-1)$ dimensional space of the $\xi_1, \cdots \xi_{n-1}, \eta_1, \cdots \eta_{n-1}$ is called the future tube, T_n. The function $W^{(n)}(\zeta_1, \cdots \zeta_{n-1})$ defined by

$$\begin{aligned} W^{(n)}(\zeta_1, \cdots \zeta_{n-1}) &= W^{(n)}(\xi_1 - i\eta_1, \cdots \xi_{n-1} - i\eta_{n-1}) \\ &= \int d^4p_1 \cdots \int d^4p_n \, e^{-i\sum_{j=1}^{n-1} p_j \cdot (\xi_j - i\eta_j)} \tilde{W}^{(n)}(p_1, \cdots p_{n-1}) \end{aligned} \tag{22}$$

is clearly an analytic function in each variable $\zeta_{j\mu}$ (the factor $p_j \cdot \eta_j$ is always greater than zero and guarantees the convergence of the defining integral for a large class of $\tilde{W}^{(n)}$ functions). Thus $W^{(n)}(\xi_1, \cdots \xi_{n-1})$ is the boundary value of a function analytic in the future tube.

The local commutation rules, $[\phi(x), \phi(x')] = 0$ when $(x - x')^2 < 0$, imply that

$$W^{(n)}(x_1, \cdots x_j, x_{j+1}, \cdots x_n) = W^{(n)}(x_1, \cdots x_{j+1}, x_j, \cdots x_n) \tag{23}$$

as long as x_j and x_{j+1} are space-like separated points. These relations can be extended by analytic continuation to relations of the analytic function $W^{(n)}(\zeta_j)$.

Finally, since the length of any vector in Hilbert space is greater than or equal to zero, one has

$$\begin{aligned} ||\alpha_0 f_0 \Psi_0 + \alpha_1 \int d^4x_1 f_1(x_1)\, \phi(x_1)\, \Psi_0 + \cdots \\ + \alpha_n \int d^4x_1 \cdots \int d^4x_n f_n(x_1, \cdots x_n)\, \phi(x_1) \cdots \phi(x_n)\, \Psi_0 + \cdots ||^2 \geqslant 0 \end{aligned} \tag{24}$$

for all $\alpha_0, \alpha_1, \cdots \alpha_n, \cdots$ and all testing functions $f_1(x_1), \cdots f_n(x_1, \cdots, x_n), \cdots$. From Eq. (24), one deduces the infinite sequence of inequalities

$$\begin{aligned} \sum_{j,i} \bar{\alpha}_i \alpha_j \int d^4x_1 \cdots \int d^4x_i \int d^4y_1 \cdots \int d^4y_j \\ \bar{f}_i(x_1, x_2, \cdots x_i)\, W^{(i+j)}(x_i, x_{i-1}, \cdots x_1, y_1, \cdots y_j)\, f_j(y_1, \cdots y_j) \geqslant 0 \end{aligned} \tag{25}$$

Wightman calls Eq. (25) the positive-definiteness condition.

The importance of the above formulation of a relativistic field theory derives from the following theorem [Wightman (1956)]:

Let $W^{(n)}(x_1, \cdots x_n)$, $n = 0, 1, 2 \cdots$ be a sequence of distributions on $4n$ space which satisfy Eqs. (13), (14) [relativistic invariance], Eq. (18)

[hermiticity], Eq. (23) [local commutativity], and Eq. (25) [positive-definiteness]. There then exists a Hilbert space $\mathfrak{H}$, a representation of the inhomogeneous Lorentz group, $U(a, \Lambda)$, a vacuum state $|\Psi_0\rangle$ and a neutral scalar field $\phi(x)$, such that the n-fold vacuum expectation value of $\phi(x)$ is $W^{(n)}(x_1, x_2, \cdots x_n)$.

In other words, this theorem asserts that, if one is given a set of distributions satisfying (13), (14), (18) and (25), it is then possible to reconstruct a field theory which has these vacuum expectation values. One can therefore just study the properties of the vacuum expectation values. The advantage in being able to do this is that the Wightman functions $W^{(n)}$ are well-defined mathematical objects, and moreover are c-numbers.

We next note that Eqs. (13), together with Eq. (19), which defines the Fourier transform $\tilde{W}^{(n)}(p_1, \cdots p_{n-1})$ of $W^{(n)}$, assert that $\tilde{W}^{(n)}$ has the following transformation properties under Lorentz transformations

$$\tilde{W}^{(n)}(p_1, \cdots p_{n-1}) = \tilde{W}^{(n)}(\Lambda p_1, \cdots \Lambda p_{n-1}) \tag{26a}$$

for Λ without time inversion, and

$$\tilde{W}^{(n)}(p_1, \cdots, p_{n-1}) = \overline{W^{(n)}(-\Lambda p_1, \cdots -\Lambda p_n)} \tag{26b}$$

for Λ with time inversion. Equation (26a), together with Eq. (22), allow one to assert that $W^{(n)}(\zeta)$ has a single-valued analytic continuation to a domain which is called the extended tube [Hall and Wightman (Hall 1957)]. The extended tube, T_n', consists of all points of the form $\Lambda\zeta_1, \cdots \Lambda\zeta_{n-1}$ where Λ is an arbitrary complex proper Lorentz transformation[5] and

[5] The group of complex proper Lorentz transformations, $L_+(C)$, is the set of all complex matrices Λ (i.e., whose elements are complex numbers) such that $\Lambda^T g \Lambda = g$ and $\det \Lambda = +1$, where $(g)_{\mu\nu} = g_{\mu\nu}$ is the metric tensor.

We here also add a word as to how this extension of the domain of analyticity comes about. Equation (26a) together with Eq. (22) asserts that

$$W^{(n)}(\zeta_1, \cdots \zeta_{n-1}) = W^{(n)}(\Lambda\zeta_1, \cdots \Lambda\zeta_{n-1}) \tag{26c}$$

for ζ_i, $i = 1, 2, \cdots n-1$, in the future tube and Λ a real Lorentz transformation. Now $W^{(n)}(\zeta_1, \cdots \zeta_{n-1})$ is analytic in the future tube. If $\Lambda\zeta_1, \cdots \Lambda\zeta_{n-1}$ lies outside the future tube, Eq. (26c) defines an analytic continuation of the function $W^{(n)}$. One next uses the fact that in the neighborhood of any point of the group manifold of the Lorentz group, 6 real analytic co-ordinates can be introduced so that the Lorentz group is a Lie group. These co-ordinates are such that the matrix elements of Λ are analytic in these co-ordinates. In particular, in a neighborhood, N, of the identity of L_+, 6 real analytic co-ordinates $\lambda_1, \cdots \lambda_6$ can be introduced such that the matrix elements of Λ in this neighborhood are expressible as power series in $\lambda_1, \cdots \lambda_6$. Since an analytic function $W^{(n)}$ of analytic functions $\Lambda\zeta_i$, $i = 1, \cdots n-1$, is again analytic, $W^{(n)}(\Lambda\zeta_1, \cdots \Lambda\zeta_{n-1})$ is an analytic function of Λ in the neighborhood of the identity. When $\lambda_1, \cdots \lambda_6$ are extended to complex variables, the function $W^{(n)}(\Lambda\zeta_1, \cdots \Lambda\zeta_{n-1})$ of Λ remains constant equal to $W^{(n)}(\zeta_1, \cdots \zeta_{n-1})$, since the function $W^{(n)}(\Lambda\zeta_1, \cdots \Lambda\zeta_{n-1}) - W^{(n)}(\zeta_1, \cdots \zeta_{n-1})$ equals zero in a *real* environment $N \subset L_+$ and by analytic continuation must remain zero wherever both of the functions are analytic. One can next show that there exists in fact a *single-valued* analytic continuation for Λ any element of the complex Lorentz group [Hall (1957)].

$\zeta_1, \cdots \zeta_n$ lie in the future tube. Furthermore, this analytically continued function satisfies for $\zeta_1, \cdots \zeta_{n-1}$ in the extended tube

$$W^{(n)}(\zeta_1, \cdots \zeta_{n-1}) = W^{(n)}(\Lambda\zeta_1, \cdots \Lambda\zeta_{n-1}) \tag{27}$$

for real as well as complex Λ with det $\Lambda = +1$, i.e., it is invariant under the complex proper Lorentz group $L_+(C)$.

The future tube, i.e., the set ζ_i $(i = 1, \cdots, n-1)$ with $-\infty \leqslant \xi_i \leqslant +\infty$, $\eta_{i0} > 0$, $\eta_i^2 > 0$ contains by definition no real points. However, the extended tube, i.e., the set of points $\Lambda\zeta_i$ with Λ an element of the complex Lorentz group $L_+(C)$ and ζ_i in the future tube, does contain real points. The real points of the extended tube have been determined by Jost (1957a) and are characterized by the following theorem:

Theorem: The *real* points $\rho_1, \cdots \rho_{n-1}$ lie in the extended tube if, and only if, the convex hull of $\rho_1, \cdots \rho_{n-1}$ only contains space-like vectors.

By the convex hull of the points $(\rho_1, \cdots \rho_{n-1})$ is meant the set of all vectors of the form $\lambda_1\rho_1 + \cdots + \lambda_{n-1}\rho_{n-1}$ as the λ_i take on all *real* values with $\lambda_i \geqslant 0$ and $\sum_{i=1}^{n-1} \lambda_i = 1$. The real points of the extended tube are thus exactly those for which an arbitrary convex linear combination $\sum_{i=1}^{n-1} \lambda_i\rho_i$ $\left(\lambda_i \geqslant 0, \sum_{i=1}^{n-1} = 1\lambda_i\right)$ is always space-like. The necessity of this assertion is easily demonstrated.

Proof: Since $(\rho_1, \cdots, \rho_{n-1})$ is in the extended tube, there exists a complex Lorentz transformation Λ and a set of points $(\zeta_1 = \xi_1 - i\eta_1, \zeta_2 = \xi_2 - i\eta_2$ $\cdots, \zeta_{n-1} = \xi_{n-1} - i\eta_{n-1})$ in the future tube such that $\Lambda\zeta_i = \rho_i$. Now for real λ_i with $\lambda_i \geqslant 0, \sum_i \lambda_i = 1$

$$\left(\sum_i \lambda_i\rho_i\right)^2 = \left(\sum_i \lambda_i\Lambda\zeta_i\right)^2 = \left(\sum_i \lambda_i\zeta_i\right)^2$$
$$= \left(\sum_i \lambda_i\xi_i\right)^2 - \left(\sum_i \lambda_i\eta_i\right)^2 - 2i\left(\sum_j \lambda_j\xi_j\right)\cdot\left(\sum_l \lambda_l\eta_l\right) \tag{28}$$

Since by assumption the left-hand side of (29) is real,

$$\left(\sum_i \lambda_i\xi_i\right)\cdot\left(\sum_j \lambda_j\eta_j\right) = 0 \tag{29}$$

From Eq. (29) and the fact that the future cone is convex (i.e., if η_j is a time-like vector with $\eta_{j0} > 0$ then $\sum_j \lambda_j\eta_j$ is also time-like for all $\lambda_j > 0$, and $\left(\sum_j \lambda_j\eta_j\right)^2 > 0$) it follows that, since the η_js lie in the future cone, the vector $\sum_i \lambda_j \xi_j$ must be space-like $\left[\left(\sum_i \lambda_j \xi_j\right)^2 < 0\right]$. Therefore

$$\left(\sum_i \lambda_i \rho_i\right)^2 = \left(\sum_i \lambda_i \xi_i\right)^2 - \left(\sum_i \lambda_i \eta_i\right)^2 < 0 \tag{30}$$

i.e., $\sum_j \lambda_i \rho_i$ must be space-like. For a proof of the sufficiency, the reader is referred to Jost's paper [Jost (1957a)]. We shall call real points in the extended tube "Jost points." It can further be shown that each Jost point has a *real* neighborhood in the extended tube and that this neighborhood forms a real environment in the space of $(n - 1)$ complex four-vector. This means that, if there are two functions $f_1(\zeta_1, \cdots \zeta_{n-1})$ and $f_2(\zeta_1, \cdots \zeta_{n-1})$ which are *both* analytic in the extended tube and which coincide for a real neighborhood in the extended tube, that is

$$f_1(\xi'_1, \cdots \xi'_{n-1}) = f_2(\xi'_1, \cdots \xi'_{n-1}) \tag{31a}$$

for $\xi'_1, \cdots \xi'_{n-1}$ in a real neighborhood of a Jost point, then

$$f_1(\zeta_1, \cdots \zeta_{n-1}) = f_2(\zeta_1, \cdots \zeta_{n-1}) \tag{31b}$$

for all $\zeta_1, \cdots \zeta_{n-1}$ in the extended tube [see Bochner (1948), page 34]; in particular, going to the boundary values

$$f_1(\xi_1, \cdots \xi_{n-1}) = f_2(\xi_1, \cdots \xi_{n-1}) \quad \text{for } all \text{ real } \xi_1, \cdots \xi_{n-1} \tag{31c}$$

The importance of the determination of the real (Jost) points of the extended tube can now be understood. This determination indicates that there exists a domain of points $\rho_1, \cdots \rho_{n-1}$ with $\rho_1, \cdots \rho_{n-1}$ real and space-like, which lie in the interior of the extended tube but which lie on the boundary of the future tube. It follows that for $x_1 - x_2 = \rho_1, \cdots x_{n-1} - x_n = \rho_{n-1}$ the Wightman function $W^{(n)}(x_1, \cdots x_n)$ is an analytic function of $\rho_1, \cdots \rho_{n-1}$ (i.e., expandable in a convergent power series in these variables). By the above, this in turn implies that a knowledge of the vacuum expectation value in the neighborhood of a Jost point (i.e., where the arguments are separated by a space-like interval) uniquely determines it for *all* values of its argument. Since for $n = 1, 2, 3, 4$ the sets of real points $\rho_1, \cdots \rho_{n-1}$ with $\rho_{i0} = 0$, i.e., the equal-time point set, are real environments of the extended tube, an even stronger result holds in these cases: the knowledge of $W^{(1)}$, $W^{(2)}$, $W^{(3)}$, and $W^{(4)}$ for a neighborhood in which the arguments all have the same fixed time component determines $W^{(1)}$, $W^{(2)}$, $W^{(3)}$ and $W^{(4)}$ everywhere.

Hall and Wightman [Hall (1957)] have further proved the following important theorem:

A function f of n four-vector variables $\zeta_1, \cdots \zeta_n$ analytic in the future tube defined by

$$\zeta_j = \xi_j - i\eta_j \quad -\infty < \operatorname{Re} \zeta_{j\mu} < +\infty,\ \mu = 0, 1, 2, 3$$
$$(\operatorname{Im} \zeta_j)^2 > 0,\ (\operatorname{Im} \zeta_{j0}) < 0$$

and invariant under the orthochronous Lorentz group,

$$f(\zeta_1, \cdots \zeta_n) = f(\Lambda\zeta_1, \cdots, \Lambda\zeta_n)$$

is a function of the scalar products $\zeta_j \cdot \zeta_k = \zeta_{j\mu}\zeta_k{}^{\mu}$. It is analytic on a set $\mathfrak{M}_n$ over which the scalar products vary when the vectors $\zeta_1, \cdots \zeta_n$ vary over the future tube.

This theorem of Hall and Wightman states that as a result of the basic invariance properties given by Eq. (27), $W^{(n+1)}(\zeta_1, \cdots \zeta_n)$, which is an analytic function of $4n$ complex variables, is expressible as a function of scalar products

$$W^{(n+1)}(\zeta_1, \cdots \zeta_n) = W^{(n+1)}(\zeta_1{}^2, \zeta_1 \cdot \zeta_2, \cdots \zeta_n{}^2, \cdots) \tag{32}$$

analytic in a domain $\mathfrak{M}_n$ of the scalar products $\zeta_i \cdot \zeta_j$ which is the map of the future tube T_n'. The reduction in the number of variables is to be noted. Thus for $n = 2$, $W^{(2)}(\zeta)$ is expressible as an analytic function of one complex variable $z = \zeta^2$ (instead of 4); for $n = 3$, $W^{(3)}$ is expressible as an analytic function of 3 complex variables (instead of 8). Apart from the obvious simplification, this form makes explicit the invariance properties of $W^{(n)}$ and makes possible a succinct expression of the consequences of the local commutativity of the field ϕ. However, not all the scalar products are independent. It may at times be advantageous to work with the variables $\zeta_1, \cdots \zeta_n$ rather than the scalar products.

The above theorems are sufficient to give a concise proof [Jost (1957a)] of the *TPC* theorem of Pauli, Lüders, and Schwinger. The *TPC* theorem, loosely speaking, asserts that for a local field theory which is invariant under proper Lorentz transformations there exists an antiunitary operator $V = TPC$, with

$$V\phi(x)\,V^{-1} = \eta\phi^*(-x) \quad |\eta|^2 = 1 \tag{33}$$

and a choice of phase η such that the theory is invariant under this operation. In terms of vacuum expectation values, the *TPC* theorem therefore asserts the equality[6]

$$\begin{aligned}(\Psi_0, \phi_1(x_1) \cdots \phi_n(x_n)\,\Psi_0) &= \overline{(\Psi_0, \phi_1{}^*(-x_1) \cdots \phi_n{}^*(-x_n)\,\Psi_0)} \\ &= (\Psi_0, \phi_n(-x_n) \cdots \phi_1(-x_1)\,\Psi_0)\end{aligned} \tag{34}$$

We shall now show that Eq. (34) is, in fact, a consequence of the *locality* of the theory, and of the previously enumerated spectral assumptions as well as of the invariance of the theory under proper orthochronous inhomogeneous Lorentz transformations. Call $W^{\phi_1\cdots\phi_n}(x_1, \cdots x_n)$ the vacuum expectation value of the product $\phi_1(x_1), \cdots \phi_n(x_n)$

$$W^{\phi_1\cdots\phi_n}(x_1, \cdots x_n) = (\Psi_0, \phi_1(x_1) \cdots \phi_n(x_n)\,\Psi_0) \tag{35}$$

[6] Note that we are here not restricting ourselves to a single field, nor to the case of a Hermitian field. For simplicity, however, we have restricted ourselves to scalar fields. The generalization to include spinor fields is straightforward. [See Jost (1957a).]

Then by translation invariance $W^{\phi_1 \cdots \phi_n}$ is a function only of the difference of the $x_1, \cdots x_n$, i.e., $W^{\phi_1 \cdots \phi_n} = W^{\phi_1 \cdots \phi_n}(\xi_1, \cdots \xi_{n-1})$ and by virtue of the spectral conditions $W^{\phi_1 \cdots \phi_n}(\xi_1, \cdots \xi_{n-1})$ is the boundary value of an analytic function $W^{\phi_1 \cdots \phi_n}(\zeta_1, \cdots \zeta_{n-1})$ when $\zeta_j = \xi_j - i\eta_j$ lies in the future tube (i.e., when $-\infty < \xi_j < +\infty$, η_j in future cone). Furthermore, by Lorentz invariance $W^\phi(\zeta)$ has a uniform analytic continuation in the extended tube (i.e., for the set of points $\zeta_1, \cdots \zeta_{n-1}$ such that $\zeta_i = \Lambda \zeta'_i$ with ζ'_i in the future tube and Λ an element of the complex Lorentz group with $\det \Lambda = +1$).

Let now $\rho_1, \cdots \rho_{n-1}$ be a Jost point (i.e., a real point in the extended tube). From the theorem of Hall and Wightman, namely that $W^{\phi_1 \cdots \phi_n} = W^{\phi_1 \cdots \phi_n}(\zeta_1, \cdots \zeta_{n-1})$ is a function of the scalar products $\zeta_i \cdot \zeta_j$ which is analytic in the complex manifold over which the scalar products vary when $\zeta_1, \cdots \zeta_{n-1}$ vary over the extended tube, it follows that

$$W^{\phi_1 \cdots \phi_n}(\zeta_1, \cdots \zeta_{n-1}) = W^{\phi_1 \cdots \phi_n}(-\zeta_1, \cdots, -\zeta_{n-1}) \tag{36a}$$

[7]whence

$$W^{\phi_1 \cdots \phi_n}(\rho_1, \cdots \rho_{n-1}) = W^{\phi_1 \cdots \phi_n}(-\rho_1, \cdots -\rho_{n-1}) \tag{36b}$$

From the hermiticity condition

$$\begin{aligned}(\Psi_0, \phi_1(x_1) \cdots \phi_n(x_n) \Psi_0) &= \overline{(\phi_1(x_1) \cdots \phi_n(x_n) \Psi_0, \Psi_0)} \\ &= \overline{(\Psi_0, \phi_n^*(x_n) \cdots \phi_1^*(x_1) \Psi_0)}\end{aligned} \tag{37}$$

one deduces that

$$W^{\phi_1 \cdots \phi_n}(\xi_1, \cdots \xi_{n-1}) = \overline{W^{\phi_n^* \cdots \phi_1^*}(-\xi_{n-1}, \cdots, -\xi_1)} \tag{38}$$

for all real $\xi_1, \cdots \xi_{n-1}$. In particular, this relation holds in a real neighborhood of the extended tube, which allows one to deduce that the following relation holds in the entire extended tube

$$W^{\phi_1 \cdots \phi_n}(\zeta_1, \cdots, \zeta_{n-1}) = \overline{W^{\phi_n^* \cdots \phi_1^*}(-\bar{\zeta}_{n-1}, \cdots -\bar{\zeta}_1)} \tag{39}$$

By virtue of Eq. (39) we can therefore rewrite Eq. (36b) as follows:

$$W^{\phi_1 \cdots \phi_n}(\rho_1, \cdots \rho_{n-1}) = \overline{W^{\phi_n^* \cdots \phi_1^*}(\rho_{n-1}, \cdots, \rho_1)} \tag{40}$$

We are to compare this last statement with the assertion of the *TPC* theorem, Eq. (34), which, in terms of Wightman functions, would read

$$\begin{aligned}W^{\phi_1 \cdots \phi_n}(\xi_1, \cdots \xi_{n-1}) &= W^{\phi_n^* \cdots \phi_1^*}(+\xi_{n-1}, \cdots, +\xi_1) \\ &= \overline{W^{\phi_1^* \cdots \phi_n^*}(-\xi_1, \cdots, -\xi_{n-1})}\end{aligned}$$

$$\text{for all real } \xi_1, \cdots \xi_{n-1} \tag{41}$$

If in (41) we choose the points $x_1, \cdots x_n$ such that $x_k - x_{k+1} = \rho_k$, we

[7] This can also be inferred from the fact that the Lorentz transformation $\Lambda = -I$ is an element of the complex Lorentz group with $\det \Lambda = +1$; hence, we can deduce Eq. (36) from the Lorentz invariance of the Wightman functions: In Eq. (28) set $\Lambda = -I$ and note that the set of real points for which the complex Lorentz transformation can be chosen such that $\Lambda\rho = -\rho$ is precisely the set of Jost points.

deduce that a necessary condition for the *TPC* theorem to hold is that

$$\overline{W^{\phi_1^* \cdots \phi_n^*}(\rho_1, \cdots \rho_{n-1})} = \overline{W^{\phi_n^* \cdots \phi_1^*}(-\rho_{n-1}, \cdots, -\rho_1)} \tag{42}$$

or equivalently that

$$(\Psi_0, \phi_1(x_1) \cdots \phi_n(x_n) \Psi_0) = (\Psi_0, \phi_n(x_n) \cdots \phi_1(x_1) \Psi_0) \tag{43}$$

for all points $x_1, \cdots x_n$ such that $\rho_1, \cdots \rho_{n-1}$ are in the extended tube.

A set of fields which satisfy the condition (43) for $x_1, \cdots x_n$ such that $\left(\sum_i^{n-1} \lambda_i \xi_i\right)^2 < 0$, $\sum_{i=1}^{n-1} \lambda_i = 1$, $\lambda_i \geqslant 0$, $\xi_i = x_i - x_{i+1}$, are said to be *weakly local* with respect to each other. Note that a set of fields which are *local* with respect to each other, i.e., which are such that they commute for space-like separation, $[\phi_j(x), \phi_l(x')] = 0$ for $(x - x')^2 < 0$, are certainly weakly local with respect to each other since, from locality, follows:

$$W^{\phi_1 \cdots \phi_j \phi_{j+1} \cdots \phi_n}(x_1, \cdots x_n)$$
$$= W^{\phi_1 \cdots \phi_{j+1} \phi_j \cdots \phi_n}(x_1, \cdots x_{j+1}, x_j, \cdots x_n) \quad \text{for } (x_j - x_{j+1})^2 < 0 \tag{44}$$

A set of field operators $(\phi_1, \cdots \phi_n)$ will be said to have the property of weak local commutativity (*WLC*) at set of real points $(\xi'_1, \cdots \xi'_{n-1})$ if

$$(\Psi_0, \phi_1(x_1) \cdots \phi_n(x_n) \Psi_0) = (\Psi_0, \phi_n(x_n) \cdots \phi_1(x_1) \Psi_0) \tag{45a}$$

or equivalently if

$$W^{\phi_1 \cdots \phi_n}(\xi_1, \cdots \xi_{n-1}) = W^{\phi_n \cdots \phi_1}(-\xi_{n-1}, \cdots -\xi_1) \tag{45b}$$

holds for all $x_1, \cdots x_n$ such that the differences $\xi_1, \cdots \xi_{n-1}$, $\xi_j = x_j - x_{j+1}$ lie within some real neighborhood of $\xi'_1, \cdots \xi'_{n-1}$.

Jost has shown that the converse of the above theorem is also true, namely that, if a set of fields have the property of weak local commutativity (*WLC*) in the neighborhood of a real point of the extended tube, i.e., in the neighborhood of a Jost point, then the *TPC* theorem is valid and *WLC* [Eq. (45b)] is satisfied at all real points in the extended tube.

Proof: We can write the weak local commutativity (*WCL*) condition at $\xi_1, \cdots \xi_{n-1}$ as follows:

$$(\Psi_0, \phi_1(x_1) \cdots \phi_n(x_n) \Psi_0) = (\Psi_0, \phi_n(x_n) \cdots \phi_1(x_1) \Psi_0)$$
$$= \overline{(\Psi_0, \phi_1^*(x_1) \cdots \phi_n^*(x_n) \Psi_0)} \tag{46a}$$

or equivalently in terms of Wightman functions

$$W^{\phi_1 \cdots \phi_n}(\xi_1, \cdots \xi_{n-1}) = \overline{W^{\phi_1^* \cdots \phi_n^*}(\xi_1, \cdots \xi_{n-1})} \tag{46b}$$

By assumption, Eq. (46b) holds in a real neighborhood of a Jost point. Using Eqs. (46b) and (36a), we deduce that

$$W^{\phi_1^* \cdots \phi_n^*}(\zeta_1, \cdots \zeta_{n-1}) = \overline{W^{\phi_1 \cdots \phi_n}(-\bar{\zeta}_1, \cdots, -\bar{\zeta}_{n-1})} \tag{47}$$

for $(\zeta_1, \cdots \zeta_{n-1})$ in a real neighborhood of the Jost point $(\rho_1, \cdots \rho_{n-1})$. Now the Wightman function

$$W^{\phi_1^* \cdots \phi_n^*}(\xi_1, \cdots \xi_{n-1}) = (\Psi_0, \phi_1^*(x_1) \cdots \phi_n^*(x_n) \Psi_0) \tag{48}$$

has clearly the same analyticity properties as $W^{\phi_1\cdots\phi_n}(\xi_1, \cdots \xi_{n-1})$ by virtue of relativistic invariance and spectral conditions. Equation (47) thus has an analytic continuation into the extended tube, since, if $(\bar{\zeta}_1, \cdots \bar{\zeta}_{n-1})$ is in the future tube, then so is $(-\bar{\zeta}_1, \cdots, -\bar{\zeta}_{n-1})$. Equation (47), therefore, results in a relation for the function $W^{\phi_1^*\cdots\phi_n^*}$ and $\overline{W^{\phi_1\cdots\phi_n}}$ in the entire extended tube. Since the future tube is a subset of the extended tube, Eq. (47) also holds in the future tube. In particular, on the boundary points of the future tube, the following relation holds:

$$W^{\phi_1^*\cdots\phi_n^*}(\xi_1, \cdots \xi_{n-1}) = \overline{W^{\phi_1\cdots\phi_n}(-\xi_1, \cdots -\xi_{n-1})} \tag{49}$$

or equivalently

$$\overline{(\Psi_0, \phi_1^*(-x_1) \cdots \phi_n^*(-x_n)\, \Psi_0)} = (\Psi_0, \phi_1(x_1) \cdots \phi_n(x_n)\, \Psi_0) \tag{50}$$

which is the *TPC* theorem in terms of vacuum expectation values. *QED.*

Dyson (1958) has proved the following corollary to the *TPC* theorem: If *TPC* is valid, then a Wightman function will be analytic and one valued at a real set of space-time points, if and only if, the field possess the property of weak local commutativity at this point set.

The above is an indication of the connection which exists between the requirement of local commutativity and that of analyticity in a certain domain. In fact, Steinman and Jost [Steinman (1960)] have shown that local commutativity for all space-like separations follows from local commutativity in the neighborhood of any Jost point. More particularly, in the case of $W^{(2)}$, $W^{(3)}$, and $W^{(4)}$ the requirement that $[\phi(x), \phi(y)] = 0$ for $x_0 = y_0$ for a neighborhood in three-dimensional space is sufficient to imply local commutativity for *all* space-like separations.

Before proceeding further with the exposition of the Wightman formalism, let us illustrate all the previous remarks with a study of the twofold Wightman function W^{AB} for the case of two scalar fields, $A(x)$ and $B(x)$. The ensuing analysis constitutes a slight generalization of the Lehmann one previously considered in Chapter 17. The vacuum expectation value in question is given by:

$$W^{AB}(x - y) = \langle \Psi_0 \mid A(x)\, B(y) \mid \Psi_0 \rangle \tag{51}$$

where we have already used the translation invariance of the theory to write $W^{AB}(x, y) = W^{AB}(x - y)$. Lorentz invariance further asserts that

$$W^{AB}(\xi) = W^{AB}(\Lambda\xi); \quad \xi = x - y \tag{52}$$

By a now familiar argument, the absence of states of negative energy and space-like momenta implies that the Fourier transform $\tilde{W}^{AB}(p)$ of $W^{AB}(\xi)$

$$\begin{aligned} W^{AB}(\xi) &= \sum_{|p\alpha\rangle} \langle \Psi_0 \mid A(0) \mid p\alpha \rangle \langle p\alpha \mid B(0) \mid \Psi_0 \rangle\, e^{-ip\cdot(x-y)} \\ &= \int d^4p\, e^{-ip\cdot(x-y)}\, \tilde{W}^{AB}(p) \end{aligned} \tag{53a}$$

$$\tilde{W}^{AB}(p) = \sum_{\substack{\alpha \\ \text{for fixed } p}} \langle \Psi_0 \mid A(0) \mid p\alpha \rangle \langle p\alpha \mid B(0) \mid \Psi_0 \rangle \tag{53b}$$

has the property that $\tilde{W}^{AB}(p) = 0$ unless $p_0 \geqslant 0$ and $p^2 > 0$. Under Lorentz transformations

$$\tilde{W}^{AB}(p) = \tilde{W}^{AB}(\Lambda p) \tag{54}$$

so that $\tilde{W}^{AB}$ is a function of p^2 only. The Wightman function $W^{AB}(\xi)$ is the boundary value of an analytic function $W^{AB}(\zeta)$ defined by:

$$W^{AB}(\zeta) = \int d^4p \, e^{-ip\cdot(\xi - i\eta)} \, \tilde{W}^{AB}(p)$$

$$\zeta = \xi - i\eta \tag{55}$$

where η is a time-like vector in the forward light cone. By Lorentz invariance $W^{AB}(\zeta) = W^{AB}(\Lambda\zeta)$, so that in fact $W^{AB}(\xi)$ is the boundary value of an analytic function $W^{AB}(z)$ regular for all complex numbers z that can be written in the form

$$z = (\xi - i\eta)^2 \tag{56}$$

with $\eta_0 > 0$, $\eta^2 > 0$. This is the content of the Hall-Wightman theorem for the case of an invariant function of a single variable. Since $z = \xi^2 - 2i\eta \cdot \xi - \eta^2$ and therefore

$$\text{Re}\, z = \xi^2 - \eta^2 \tag{57a}$$

$$\text{Im}\, z = -2\xi \cdot \eta \tag{57b}$$

every point that does not lie on the positive real axis can be represented in the form (56). The set of point z when ξ varies over all space-time and η is in the future light cone, thus fills the entire complex z plane except for the positive real axis and the origin. In the limit $\eta \to 0$ Figure 18.1 illustrates this set with the labeled points indicating values of z for typical positions of ξ. The Wightman function $W^{AB}(z)$ is thus an analytic function in the cut plane, the cut extending from $z = 0$ to ∞ along to positive real axis.

The vacuum expectation value

$$W^{BA}(x' - x) = \langle \Psi_0 \mid B(x')\, A(x) \mid \Psi_0 \rangle \tag{58}$$

clearly defines in an analogous way another analytic function $W^{BA}(z)$, which again is regular for all points that can be written in the form

$$z = (-\xi - i\eta)^2 \quad (\eta \text{ in future cone}) \tag{59a}$$

$$\xi = x - x' \tag{59b}$$

Since Eq. (59) and Eq. (56) define the same set, $W^{BA}(z)$ and $W^{AB}(z)$ have the same domain of analyticity. If the theory is local, that is if $[A(x), B(x')] = 0$ for $(x - x')^2 < 0$, then

$$\langle \Psi_0 \mid A(x)\, B(x') \mid \Psi_0 \rangle = \langle \Psi_0 \mid B(x')\, A(x) \mid \Psi_0 \rangle \quad \text{for } (x - x')^2 < 0 \tag{60}$$

so that $W^{AB}(z) = W^{BA}(z)$ for z on the negative real z axis. As both $W^{AB}(z)$

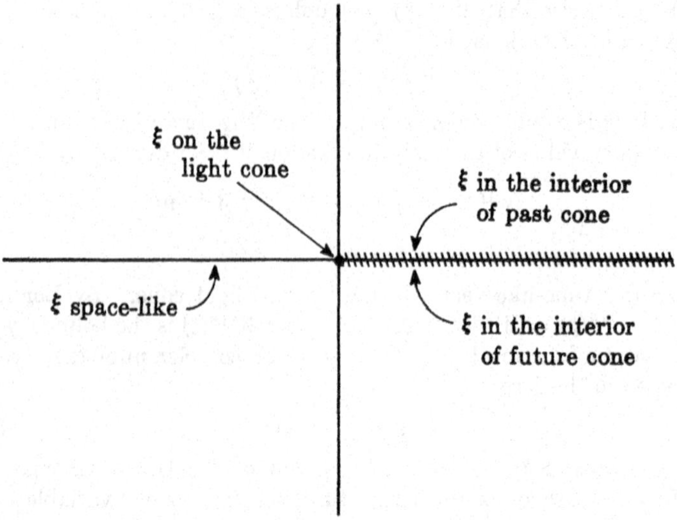

Complex $z = (\xi - i\eta)^2$ plane in the limit $\eta \to 0$

Fig. 18.1

and $W^{BA}(z)$ are analytic functions, $W^{AB}(z) = W^{BA}(z)$ for all points in their domain of analyticity, i.e.,

$$W^{AB}(z) = W^{BA}(z) \quad \text{in the cut plane.} \tag{61}$$

The above remarks, when specialized to the case $A(x) = \phi(x)$ and $B(x) = \phi^*(x)$, are sufficient to give a proof that such a scalar spin 0 field cannot have the wrong connection of spin with statistics, i.e., one cannot have $[\phi(x), \phi^*(x')]_+ = 0$ for space-like separations. [Burgoyne (1958), Lüders and Zumino (Lüders 1958)]. For assume that these "wrong" commutation rules actually were true, so that instead of (60) the following relation would hold

$$W^{\phi\phi^*}(\xi) + W^{\phi^*\phi}(-\xi) = 0 \quad \text{for } \xi^2 < 0 \tag{62a}$$

and by analyticity

$$W^{\phi\phi^*}(z) = -W^{\phi^*\phi}(z) \tag{62b}$$

everywhere in their domain of analyticity. Consider now the vectors $\phi(f) \mid \Psi_0\rangle$ and $\phi^*(g) \mid \Psi_0\rangle$. The sum of the norms of these vectors is

$$||\phi(f)\ \Psi_0||^2 + ||\phi^*(g)\ \Psi_0||^2$$

$$= \int d^4x \int d^4y\ \{\bar{f}(x)\ (\phi(x)\ \Psi_0, \phi(y)\ \Psi_0)\ f(y) + \bar{g}(x)\ (\phi^*(x)\ \Psi_0, \phi^*(y)\ \Psi_0)\ g(y)\}$$

$$= \int d^4x \int d^4y\ \{\bar{f}(x)\ W^{\phi^*\phi}(x - y)\ f(y) + \bar{g}(-x)\ W^{\phi\phi^*}(y - x)\ g(-y)\} \tag{63}$$

By choosing $f(x) = g(-x)$ and using the fact that $W^{\phi\phi^*}(\xi) + W^{\phi^*\phi}(-\xi) = 0$ for all ξ if the wrong connection between spin and statistics is assumed, we deduce from (63) that[8]

$$\phi(f) \mid \Psi_0\rangle = \phi^*(f) \mid \Psi_0\rangle = 0 \quad \text{for all } f \tag{64}$$

Consequently, all vacuum expectation values of products of ϕ and ϕ^* operators vanish, and from Wightman's work it then follows that such a field is identically zero, so that one must have the right connection between spin and statistics. The proof for higher spin fields (integer or half-integer) can be carried out in a similar manner [Burgoyne (1958)]. Note that this proof of the connection between spin and statistics makes no reference to the particular form of the equation of motion that the field ϕ obeys. This is in contrast to the Pauli (1940) proof which assumed in an essential fashion the linearity of the field equations. It should also be noticed that this proof based on the axiomatic formulation of field theory is a consequence of only the linear relations on the Wightman functions.

Returning to our discussion of the Wightman function $W^{AB}(\xi)$ and $W^{BA}(\xi)$, we note that if the fields are local, from (61) follows the fact that

$$\tilde{W}^{AB}(p) = \tilde{W}^{BA}(p) \tag{65}$$

The vacuum expectation value of the commutator is given by

$$\langle\Psi_0 \mid [A(x), B(x')] \mid \Psi_0\rangle = \lim_{\eta\to 0} \{W^{AB}((x - x' - i\eta)^2) - W^{BA}((x - x' + i\eta)^2) \tag{66}$$

If we make explicit use of the fact that the Fourier transform of $W^{AB}(\xi)$, $\tilde{W}^{AB}(p)$, vanishes for $p_0 < 0$ as well as of its transformation properties under Lorentz transformations by writing

$$\tilde{W}^{AB}(p) = \rho^{AB}(p^2)\,\theta(p) \tag{67}$$

the vacuum expectation value of the commutator of $A(x)$, $B(x')$ then has the following representation

$$\langle\Psi_0 \mid [A(x), B(x')] \mid \Psi_0\rangle = \int d^4p\, \rho^{AB}(p^2)\, \epsilon(p)\, e^{-ip\cdot\xi}$$
$$\xi = x - x' \tag{68}$$

The representation of the vacuum expectation value of the retarded product

$$R(A(x)\,B(x')) = -i\theta(x - x')\,[A(x), B(x')] \tag{69}$$

can now easily be derived from (68) by using the following integral representation for the θ function

$$\theta(x) = \frac{1}{2\pi i} \lim_{\delta\to 0} \int_{-\infty}^{+\infty} \frac{e^{-i\tau x_0}}{\tau + i\delta}\, d\tau \tag{70}$$

[8] Note that this conclusion can only be drawn if the norm in the vector space of the physical states is positive-definite, i.e., if the states are represented by vectors in a Hilbert space equipped with a Hermitian scalar product.

We shall denote by $r^{AB}(x; x')$ the expectation value of the retarded product of $A(x)$ and $B(x')$

$$r^{AB}(x; x') = \langle \Psi_0 \mid R(A(x)\, B(x')) \mid \Psi_0 \rangle \tag{71}$$

Since by assumption $A(x)$ and $B(x')$ commute for space-like separations, the retarded product $R(A(x)\, B(x'))$ vanishes except when $x - x'$ lies in the forward light cone. Under an orthochronous Lorentz transformation the retarded product therefore transforms as follows

$$\begin{aligned} U(a, \Lambda)\, R(A(x)\, B(x'))\, U(a, \Lambda)^{-1} \\ &= -i\theta(\Lambda(x - x'))\, [A(\Lambda x + a), B(\Lambda x' + a)] \\ &= R(A(\Lambda x + a)\, B(\Lambda x + a)) \end{aligned} \tag{72}$$

since under an orthochronous Lorentz transformation the sign of the time is invariant and a time-like vector remains time-like. These transformation properties have the following consequences for the vacuum expectation values

$$r^{AB}(x; x') = r^{AB}(x - x') \tag{73a}$$

$$r^{AB}(\xi) = r^{AB}(\Lambda\xi) \tag{73b}$$

One finds, using Eq. (70), that r^{AB} has the following representation:

$$\begin{aligned} r^{AB}(\xi) &= -\int d^4p\, e^{-ip_0\xi_0}\, e^{i\mathbf{p}\cdot\boldsymbol{\xi}}\, \rho^{AB}(p^2)\, \epsilon(p) \lim_{\delta\to 0} \frac{1}{2\pi} \int_{-\infty}^{+\infty} d\tau\, e^{-i\tau\xi_0} \frac{1}{\tau + i\delta} \\ &= -\int d^4p\, e^{-ip\cdot\xi} \lim_{\delta\to 0+} \frac{1}{2\pi} \int_{-\infty}^{+\infty} \frac{\rho^{AB}((p_0 - \tau)^2 - \mathbf{p}^2)}{\tau + i\delta}\, \epsilon(p_0 - \tau) \\ &= \int d^4p\, e^{-ip\cdot\xi} \lim_{\delta\to 0+} \tilde{r}^{AB}(\mathbf{p}^2, p_0, \delta) \end{aligned} \tag{74}$$

where

$$\begin{aligned} \tilde{r}^{AB}(\mathbf{p}^2, p_0, \delta) &= -\frac{1}{2\pi} \int_{-\infty}^{+\infty} d\tau\, \frac{\rho^{AB}((p_0 - \tau)^2 - \mathbf{p}^2)}{\tau + i\delta}\, \epsilon(p_0 - \tau) \\ &= -\frac{1}{2\pi} \int_{-\infty}^{+\infty} d\tau\, \frac{\rho^{AB}(\tau^2 - \mathbf{p}^2)}{p_0 - \tau + i\delta}\, \epsilon(\tau) \\ &= -\frac{1}{2\pi} \int_0^{\infty} d\tau\, \rho^{AB}(\tau^2 - \mathbf{p}^2) \left[\frac{1}{-\tau + p_0 + i\delta} - \frac{1}{\tau + p_0 + i\delta} \right] \\ &= \frac{1}{2\pi} \int_0^{\infty} d\tau^2\, \rho^{AB}(\tau^2 - \mathbf{p}^2)\, \frac{1}{\tau^2 - (p_0 + i\delta)^2} \\ &= \frac{1}{2\pi} \int_0^{\infty} d\tau^2\, \rho^{AB}(\tau^2)\, \frac{1}{\tau^2 + \mathbf{p}^2 - (p_0 + i\delta)^2} \end{aligned} \tag{75}$$

Equation (75) indicates that the Fourier transform, $\tilde{r}^{AB}(p^2)$, of the vacuum expectation value of $R(A(x)\, B(x'))$

$$\langle \Psi_0 \mid R(A(x')\, B(x') \mid \Psi_0 \rangle = \int d^4p\, \tilde{r}^{AB}(p^2)\, e^{-ip\cdot\xi} \tag{76}$$

is the boundary value of the analytic function

$$\tilde{r}^{AB}(z) = \frac{1}{2\pi}\int_0^\infty d\tau^2 \frac{\rho^{AB}(\tau^2)}{\tau^2 - z} = \frac{1}{2\pi}\int_0^\infty \frac{da\, \rho^{AB}(a)}{a - z} \tag{77}$$

which is regular in the cut z plane (the cut extending from $z = 0$ to $z = +\infty$ along the real positive axis).

An analogous calculation for the time-ordered product $T(A(x)\, B(x'))$ yields the result that the vacuum expectation value of $T(A(x)\, B(x'))$ admits of the following representation

$$\begin{aligned}\tau^{AB}(x - x') &= \langle \Psi_0 \mid T(A(x)\, B(x')) \mid \Psi_0\rangle \\ &= \lim_{\delta\to 0+} \int d^4p\, e^{-ip\cdot\xi} \frac{1}{2\pi}\int_0^\infty \frac{da\, \rho^{AB}(a)}{a - p^2 - i\delta}\end{aligned} \tag{78}$$

so that the Fourier transform of the time-ordered product is a different boundary value of the same analytic function $\tilde{r}^{AB}(z)$. As already noted in Chapter 17, the assumption that $W^{AB}(\xi)$ is not too singular for $\xi = 0$ is equivalent to the assumption that $\rho^{AB}(a)$ behaves in such a way at infinity that the integral over a appearing in the definition of the function $\tilde{r}^{AB}(z)$, Eq. (77), converges. In a renormalizable field theory, even if the integral (77) does not converge, it is assumed that either

$$\tilde{r}^{AB}(z) - \tilde{r}^{AB}(0) = \frac{z}{2\pi}\int_0^\infty \frac{da}{a}\frac{\rho^{AB}(a)}{a - z} \tag{79}$$

or more generally, that some further subtracted form of $\tilde{r}^{AB}(z)$ [and hence, with even higher power of a in the denominator of (79)] converges. The exact number of subtraction terms depends on the number of renormalization terms that enter in the theory.

The determination of the general analytic properties of the threefold vacuum expectation (which is related to the vertex operators) and more generally of the analytic properties of n-fold vacuum expectation values (related to scattering amplitudes) has been the object of intense studies in recent years [see in particular Källén (1958, 1959); Streater (1960a)]. Källén and Wightman [Källén (1958)] have made a detailed study of the analytic properties of the threefold Wightman function

$$\begin{aligned}\langle \Psi_0 \mid A(x_1)\, B(x_2)\, C(x_3) \mid \Psi_0\rangle &= W^{ABC}(x_1, x_2, x_3) = W^{ABC}(x_1 - x_2, x_2 - x_3) \\ &= \int e^{-ip_1\cdot(x_1 - x_2) - ip_2\cdot(x_2 - x_3)}\, \tilde{W}^{ABC}(p_1, p_2)\, d^4p_1\, d^4p_2\end{aligned} \tag{80}$$

where the Fourier transform $\tilde{W}^{ABC}(p_1, p_2)$ vanishes unless both vectors p_1, p_2 lie in the forward light cone. As noted previously, it follows from this fact that $W^{ABC}(x_1 - x_2, x_2 - x_3)$ is the boundary value of an analytic function of the two complex vectors $\zeta_1 = x_1 - x_2 - i\eta_1 = \xi_1 - i\eta_1$ and $\zeta_2 = x_2 - x_3 - i\eta_2 = \xi_2 - i\eta_2$, where η_1 and η_2 vary independently in the forward light cone. The Hall-Wightman theorem now asserts that the analytic function $W^{ABC}(\zeta_1, \zeta_2)$ depends only on the three Lorentz invariant variables

$$z_1 = (x_1 - x_2 - i\eta_1)^2 = (\xi_1 - i\eta_1)^2 = \zeta_1^2 \tag{81a}$$

$$z_2 = (x_2 - x_3 - i\eta_2)^2 = (\xi_2 - i\eta_2)^2 = \zeta_2^2 \tag{81b}$$

$$z_3 = (x_1 - x_3 - i(\eta_1 + \eta_2))^2 = (\zeta_1 + \zeta_2)^2 \tag{81c}$$

Now as ζ_1 and ζ_2 vary over their future tube, each variable z_1, z_2, z_3 varies in an open set of the complex z_i plane. This domain is called $\mathfrak{M}_2^{ABC}$. Since the z_is are not independent, it is fairly difficult and elaborate to determine the domain $\mathfrak{M}_2^{ABC}$. The boundary of the domain $\mathfrak{M}_2^{ABC}$ has been shown by Källén and Wightman to be composed of four so-called "analytic hypersurfaces."[9] They have further shown that, as a consequence of local commutativity, the domain of analyticity of W^{ABC} can be further extended. This can be seen as follows: The above has indicated that $W^{ABC}(\xi_1, \xi_2)$ is the boundary value of an analytic function $W^{ABC}(z_1, z_2, z_3)$ with z_1, z_2, z_3 defined by Eqs. (81a)–(81c). An analogous procedure indicates that

$$W^{BAC}(x_2, x_1, x_3) = \langle \Psi_0 \mid B(x_2)\, A(x_1)\, C(x_3) \mid \Psi_0 \rangle \tag{82}$$

is the boundary value of an analytic function $W^{BAC}(z_1, z_2, z_3)$ in a domain $\mathfrak{M}_2^{BAC}$. The domain $\mathfrak{M}_2^{BAC}$ can, in fact, be obtained from $\mathfrak{M}_2^{ABC}$ by a permutation of the variables z_2 and z_3 in (81). Now if the fields A and B are local with respect to each other, i.e., if $[A(x), B(x')] = 0$ for $(x - x')^2 < 0$, then

$$W^{BAC}(x_2, x_1, x_3) = W^{ABC}(x_1, x_2, x_3) \quad \text{for } \xi_1 = (x_1 - x_2)^2 < 0 \tag{83}$$

Now the set $\xi_1^2 < 0$ lies in the interior of the extended tube and constitutes a real environment for the analytic functions $W^{ABC}(z_1, z_2, z_3)$ and $W^{BAC}(z_1, z_2, z_3)$. The two analytic functions $W^{ABC}(z_1, z_2, z_3)$ and $W^{BAC}(z_1, z_2, z_3)$ since they coincide in this real neighborhood, coincide everywhere in the union of their domains of analyticity. The function W^{ABC} has thus an analytic continuation to the union of the domain $\mathfrak{M}_2^{ABC}$ and $\mathfrak{M}_2^{BAC}$. In a similar fashion one establishes that the function $W^{ACB}(z_3, z_2, z_1)$ is equal to $W^{ABC}(z_1, z_2, z_3)$ wherever both are defined if $[B(x_2), C(x_3)] = 0$ for $(x_2 - x_3)^2 < 0$. The function $W^{ABC}(z_1, z_2, z_3)$ is thus also analytic in the domain $\mathfrak{M}_2^{ACB}$, which can be obtained from $\mathfrak{M}_2^{ABC}$ by permuting z_1 and z_3 in (81). Similarly, for $W^{CBA}(z_2, z_1, z_3)$, etc. Local commutativity thus asserts that all six vacuum expectation values $W^{ABC}(\xi_1, \xi_2), \cdots W^{CBA}(-\xi_2, -\xi_1)$ are boundary values of the single analytic function $W^{ABC}(z_1, z_2, z_3)$ which is analytic in a domain $\mathfrak{M}_2$ which is the union of the domains defined by Eqs. (81a), (81b), and (81c), as ζ_1 and ζ_2

[9] A surface determined by an equation of the form $F(z_k, r) = 0$, where F is an analytic function of the complex variables z_k and also depends on one real parameter r, has been dubbed an "analytic hypersurface" by Källén and Wightman.

vary over the future tube and the two domains obtained from it by permuting z_1 with z_2 and z_3.

The Wightman function W^{ABC} can thus be shown to be an analytic function of three complex variables z_1, z_2, z_3 in a certain domain $\mathfrak{M}_2$. Now in the theory of a single complex variable z, it is well known that for every domain $\mathfrak{D}$ there exists a function $f(z)$ such that $\mathfrak{D}$ is the natural boundary of the function, i.e., f is analytic in $\mathfrak{D}$ and has singularities at every point of the boundary so that it cannot be continued outside $\mathfrak{D}$. However, in the theory of analytic functions of more than one complex variable, it turns out that an arbitrary domain $\mathfrak{D}_n$ in the $2n$-dimensional space of n complex variable cannot in general be a domain of analyticity for an analytic function. There exists, in fact, the possibility of continuing *every* analytic function regular in $\mathfrak{D}_n$ into a somewhat larger domain, called the envelope of holomorphy of $\mathfrak{D}_n$. A domain that is equal to its envelope of holomorphy is called a natural domain of holomorphy.

Källén and Wightman have shown that the union of the permuted domain $\mathfrak{M}_2^{ABC} \cup \mathfrak{M}_2^{BAC} \cup \cdots$ is not a domain of holomorphy. They have, however, computed this domain, which is the largest domain in which *every* function $W(z_1, z_2, z_3)$ satisfying the requirement of relativistic invariance, spectrum and locality is analytic.

There exists in the theory of analytic functions of several complex variables a theorem which states that it is possible to express this most general function $W(z_1, z_2, z_3)$ anywhere in the interior of the domain as an integral over certain low-dimensional subsets of the boundary of a certain kernel times the value of the function. For physical applications one would hope that these subsets could be chosen so as to correspond to physical points, i.e., sets of values of the arguments which can be obtained by using real vectors ξ_1, ξ_2. This has indeed been shown by Källén and Toll [Källén (1960)].

A start in the general problem of determining the domain of analyticity of the Wightman function $W^{(n)}(\zeta_1, \cdots \zeta_{n-1})$ has been made by Källén and Willhelmson [Källén (1959); see also Streater (1960)]. Some specific results for the case $n = 4, 5$ have been obtained by Bremmerman, Oehme, and Taylor [Bremmerman (1958)] and Kleitman (1959). The importance of these investigations derives from the close connection between Wightman functions and scattering amplitudes, form factors, and other observables. [Thus $W^{(3)}$ is related to the vertex function, $W^{(4)}$ to the two-particle elastic scattering amplitude, etc.] A representation of $W^{(n)}$ which exhibits in a transparent way its singularities as a function of the invariants $\zeta_i \cdot \zeta_j$ as allowed by locality, spectrum, and relativistic invariance could also serve as a first step in constructing a nontrivial, consistent local field theory.

Besides the remarks of Wightman (1956, 1959a), very little progress has been made in translating the positive-definiteness conditions (i.e., the nonlinear conditions) into properties of Wightman functions. For the twofold they impose the condition that $\tilde{W}^{(2)}$ be a positive measure of not too fast increase. For the threefold LSZ [Lehmann (1955b), Drell (1959)] have obtained certain restrictions on the behavior of form factors and vertex functions which are essentially consequences of the positive-definiteness inequality.

The positive-definiteness conditions are nonlinear properties which relate all W functions together, and therein lies the difficulty in handling them. It turns out that the consequences of these nonlinear conditions are somewhat easier to study in terms of another set of functions which also characterize the field theory, the retarded functions. The latter form the basis of the LSZ formulation of field theory which is the subject which we now turn to.

18*b*. The LSZ Formulation of Field Theory

A set of Wightman functions $W^{(n)}(\xi_1, \cdots \xi_{n-1})$ which satisfy the relations (13), (14), (18), and (25), define a local field theory. However, if such a field theory is to describe physical phenomena, it must, in addition to the field observables $\phi(x)$, contain particle observables. As a minimum requirement, the theory must contain those predictions of the results of collision experiments which are embodied in the usual formulation in the S matrix. At present the standard way to do this is to impose the asymptotic conditions on the field $\phi(x)$. The asymptotic condition is the requirement that a field theory have an interpretation in terms of asymptotic observables corresponding to particles of definite mass and charge. Expressed mathematically, the asymptotic condition requires that the limits

$$\lim_{t \to \pm\infty} (\Phi, \phi^f(t)\, \Psi) = (\Phi, \phi^f_{\substack{\text{out}\\ \text{in}}} \Psi) \tag{84}$$

exist. In Eq. (84) $|\Phi\rangle$, $|\Psi\rangle$ are any vectors in the domain of the operators, and

$$\phi^f(t) = i \int \left\{ \phi(x) \frac{\partial \bar{f}(x)}{\partial x^0} - \frac{\partial \phi(x)}{\partial x^0} \bar{f}(x) \right\} d^3x \tag{85}$$

where f is any normalizable solution of the Klein-Gordon equation with mass μ

$$(\Box + \mu^2) f(x) = 0 \tag{86}$$

It is also tacitly assumed that the limit, $\lim_{t \to \pm\infty} (\Phi, \phi^{f'}(t)\, \Psi)$, where $f'(x)$ is any normalizable solution of the Klein-Gordon equation with mass $\mu' \neq \mu$ vanishes. The asymptotic condition further requires that these fields, $\phi_{\text{in}}(x)$ and $\phi_{\text{out}}(x)$, satisfy the differential equations and commutation relations of the free-field theory for particles of mass μ:

$$(\Box + \mu^2)\, \phi_{\text{in}}(x) = (\Box + \mu^2)\, \phi_{\text{out}}(x) = 0 \tag{87}$$

$$[\phi_{\text{in}}(x), \phi_{\text{in}}(x')] = [\phi_{\text{out}}(x), \phi_{\text{out}}(x')] = i\Delta(x - x'; \mu) \tag{88}$$

μ being the mass of the particle described by these asymptotic fields.[10]

If the fields $\phi_{in}(x)$ and $\phi_{out}(x)$ are irreducible, and the vectors $|\Psi_0\rangle$, $\phi_{\substack{in\\out}}{}^{f} \mid \Psi_0\rangle, \cdots \phi_{\substack{in\\out}}{}^{f_1} \phi_{\substack{in\\out}}{}^{f_2} \cdots \phi_{\substack{in\\out}}{}^{f_n} \mid \Psi_0\rangle, \cdots$ span the entire Hilbert space $\mathfrak{H}$ of physical states, there then exists by virtue of (88) a unitary operator S, the S matrix, such that

$$\phi_{out}(x) = S^{-1}\phi_{in}(x)\, S \tag{89}$$

$$S^*S = SS^* = 1 \tag{90}$$

Lehmann, Symanzik, and Zimmermann [Lehmann 1955a, 1957)] have analyzed relativistic field theories, which obey Axioms I–IV together with the asymptotic conditions in terms of the vacuum expectation value of time-ordered product and retarded products of field operators. The reason that such time-ordered and retarded products are more convenient objects for the study of the consequences of the asymptotic conditions is that the latter imply certain recursion relations for the matrix elements of these products. Relativistic invariance, spectrum, and locality as in the case of Wightman functions imply certain properties for these vacuum expectation values. In addition, a set of coupled equations for these vacuum expectation values (the analogue of the positive-definiteness relations in the Wightman formulation) can again be derived. In deriving these coupled equations, explicit use is made of the asymptotic conditions and of the Hilbert space character of the set of physical states.

The Lehmann-Symanzik-Zimmermann (hereafter referred to as LSZ) formalism can be presented in either of two ways. On the one hand, one

[10] The problem of determining what additional properties a field must have in order to satisfy the asymptotic condition has been investigated in terms of Wightman functions by Greenberg (1956). Roughly speaking, Greenberg (1956) has shown that if the Wightman functions $\tilde{W}^{(n)}(p_1, \cdots p_{n-1})$ have at worst principal value and δ-function singularities near the free-particle masses, $p_i^2 = \mu^2$, and in addition the fields satisfy local commutation rules, i.e., the commutator $[\phi(x), \phi(y)]$ for $x_0 = y_0$ is proportional to $\delta^{(3)}(\mathbf{x} - \mathbf{y})$, then the field will satisfy the asymptotic conditions [see also Zimmermann (1958); Redmond and Uretski (Redmond 1959)].

The asymptotic condition for $t \to \pm\infty$ can be viewed as the requirement for the vanishing of the effective interaction between two particles as their spatial separation $R \to \infty$. One can therefore translate the asymptotic condition for $t \to \pm\infty$ into asymptotic conditions for large R at finite times [Haag (1958, 1960)]. Araki (1960) has given a proof that, with the usual assumptions (including locality) of the axiomatic approach, the truncated Wightman functions for equal time tend to zero exponentially as the largest distance R between points tends to infinity with an exponent mR, m being the lowest mass in the theory [see also Dell'Antonio and Gulmanelli (Dell'Antonio 1959)]. The truncated Wightman functions $W_T{}^{(n)}(x_1, \cdots x_n)$ are vacuum expectation values of products of field operators where the contribution from the vacuum intermediate states has been subtracted out in a symmetric manner:

$$W^{(n)}(x_1, \cdots x_n) = W_T{}^{(n)}(x_1, \cdots x_n) + \Sigma\, W_T{}^{(k)}(x_1, \cdots x_k) \cdots W_T{}^{(n-k)}(\cdots, x_n)$$

In this recursive definition the sum extends over all ways of dividing n points $x_1, \cdots x_n$ into more than one group and the order of the point $x_1, \cdots x_n$ on the right-hand side is the same as on the left-hand side.

can adopt the axioms of the Wightman formalism (i.e., existence of Hilbert space $\mathfrak{H}$, transformation properties of field operators and locality) together with the (postulate or) assumption that the field operators satisfy the asymptotic conditions, and investigate the consequences of these axioms for the functions

$$\tau^{(n)}(x_1, \cdots x_n) = (\Psi_0, T(\phi(x_1) \cdots \phi(x_n)\,\Psi_0) \tag{91}$$

called τ or time-ordered functions, and the functions

$$r^{(n+1)}(x; x_1, \cdots x_n) = (-i)^n \sum_{\text{permutations}} \theta(x - x_1) \cdots$$
$$\theta(x_{n-1} - x_n)\,(\Psi_0, [\cdots [\phi(x), \phi(x_1)], \cdots \phi(x_n)]\,\Psi_0) \tag{92}$$

called retarded or r functions. In (92) the sum is over all permutations of the set $\{1, \cdots n\}$. In particular one can try to see whether a theorem similar to that proved by Wightman for W functions holds for the τ and r functions, i.e., whether given a set of r and τ functions satisfying specified conditions, the field theory of which they are the vacuum expectation value of retarded and time-ordered products of field operators, can be reconstructed. This is the viewpoint adopted in LSZ I [Lehmann (1955a)]. It takes the Heisenberg field operator as the fundamental quantity. Alternatively, one can try [following Heisenberg's (1953) original suggestion] to base the formalism entirely on requirements imposed on the S matrix, i.e., regard the in- and out-fields as given and the Heisenberg fields as derived notions. This is the viewpoint adopted in LSZ II [Lehmann (1957); see also Bogoliubov (1958, 1959)].

We shall here briefly outline this second viewpoint. In this approach, the theory attempts only to describe in a relativistically invariant fashion the results of scattering experiments in which the particles initially and finally do not interact. These initial and final states are therefore represented by vectors corresponding to freely moving particles. The S matrix is the unitary operator which maps the Hilbert space of incoming states, $|\Psi\rangle_{\text{in}}$, on that of the outgoing states $|\Psi\rangle_{\text{out}}$. It is the operator from which the transition probability is computed in the usual manner. For the sake of simplicity, we present the LSZ formulation in terms of a model containing only one kind of stable, neutral particle of mass μ, spin 0. [Note that the mass and spin are given parameters!] We also assume that no bound states exist. In a mathematical characterization of the theory, two fields $\phi_{\text{in}}(x) = \phi_{\text{in}}^*(x)$ and $\phi_{\text{out}}(x) = \phi_{\text{out}}^*(x)$ are introduced satisfying the free-field equations

$$(\Box + \mu^2)\,\phi_{\text{in}}(x) = (\Box + \mu^2)\,\phi_{\text{out}}(x) = 0 \tag{93a}$$

and the commutation relations

$$[\phi_{\text{in}}(x), \phi_{\text{in}}(y)] = [\phi_{\text{out}}(x), \phi_{\text{out}}(y)] = i\Delta(x - y; \mu) \tag{93b}$$

These fields describe the asymptotically free ingoing and outgoing particles.

The in- and out-fields, by virtue of being free fields satisfying (93), have the following Fourier decomposition

$$\phi_{\substack{\text{in}\\\text{out}}}(x) = \frac{1}{(2\pi)^{3/2}} \int d^4k\, e^{-ik\cdot x}\, \delta(k^2 - \mu^2)\, \tilde{\phi}_{\substack{\text{in}\\\text{out}}}(k) \tag{94}$$

where due to the assumed hermiticity of $\phi_{\substack{\text{in}\\\text{out}}}(x)$

$$\tilde{\phi}_{\substack{\text{in}\\\text{out}}}{}^*(k) = \tilde{\phi}_{\substack{\text{in}\\\text{out}}}(-k) \tag{95}$$

With the help of these operators the complete set of ingoing and outgoing states can be constructed. In particular, an incoming state of n particles with momenta $k_1 = (\sqrt{\mathbf{k}_1^2 + \mu^2}, \mathbf{k}_1), k_2, \cdots k_n$ is represented in terms of the in-operators and the vacuum state $|\Psi_0\rangle$ by

$$|k_1, \cdots k_n\rangle_{\text{in}} = \frac{1}{\sqrt{n!}}\, a_{\text{in}}{}^*(k_1) \cdots a_{\text{in}}{}^*(k_n)\, |\, \Psi_0\rangle \tag{96}$$

The operator $a_{\text{in}}{}^*(k)$ is defined by the following invariant Fourier decomposition of the field $\phi_{\text{in}}(x)$:

$$\phi_{\text{in}}(x) = \frac{1}{(2\pi)^{3/2}} \int \frac{d^3k}{2k_0} \{a_{\text{in}}(k)\, e^{-ik\cdot x} + a_{\text{in}}{}^*(k)\, e^{+ik\cdot x}\} \tag{97}$$

Comparing Eq. (97) with the decomposition (94), we find

$$a_{\text{in}}(k) = \tilde{\phi}_{\text{in}}(k) \qquad k_0 > 0 \tag{98a}$$

$$a_{\text{in}}{}^*(k) = \tilde{\phi}_{\text{in}}(-k) \quad k_0 > 0 \tag{98b}$$

The commutation rules of the a_{in} operators are the same as the free-field ones with

$$[a_{\text{in}}(k), a_{\text{in}}{}^*(k')] = 2k_0\delta^{(3)}(\mathbf{k} - \mathbf{k}') \tag{99a}$$

$$[a_{\text{in}}(k), a_{\text{in}}(k')] = [a_{\text{in}}{}^*(k), a_{\text{in}}{}^*(k')] = 0 \tag{99b}$$

Similar relations hold for the outgoing states in terms of the out-operators a_{out}. The plane wave decomposition (97) results in non-normalizable state vectors $|k_1, \cdots, k_n\rangle_{\text{in}}$. It is often convenient, and for mathematical (and physical!) precision necessary, to replace the continuous system of plane wave amplitudes in the decomposition (97) by a discrete system, $\{f_\alpha\}$, of positive energy solutions of the Klein-Gordon equation. The properties of this set $\{f_\alpha\}$ are as follows:

(*a*) orthogonality:

$$-i \int d^3x \left\{f_\alpha(x)\, \frac{\partial \overline{f_\beta(x)}}{\partial x^0} - \frac{\partial f_\alpha(x)}{\partial x^0}\, \overline{f_\beta(x)}\right\} = \delta_{\alpha\beta} \tag{100}$$

(*b*) completeness:

$$\sum_{\alpha=1}^{\infty} f_\alpha(x)\, \overline{f_\alpha}(x') = i\Delta^{(+)}(x - x'; \mu) \tag{101}$$

The expansion of an arbitrary Hermitian operator $A(x)$ in terms of the functions $f_\alpha(x)$ is then given by

$$A(x) = \sum_\alpha \{f_\alpha(x)\, A_\alpha(x_0) + \bar{f}_\alpha(x)\, A^{\alpha *}(x_0)\} \tag{102}$$

with

$$A^\alpha(x_0) = i \int d^3x\, A(x) \frac{\overleftrightarrow{\partial}}{\partial x^0} \bar{f}_\alpha(x) \tag{103}$$

For the case $A(x) = \phi_{\substack{\text{in}\\\text{out}}}(x)$, the operator $\phi_{\substack{\text{in}\\\text{out}}}{}^\alpha(x_0)$ is time-independent and will be denoted by $\phi_{\substack{\text{in}\\\text{out}}}{}^\alpha$. These operators create ingoing (outgoing) particles in the state f_α. We shall denote the complete set of states obtained by the application of these operators on the vacuum by

$$|\alpha_1, \alpha_2 \cdots, \alpha_n\rangle_{\substack{\text{in}\\\text{out}}} = \frac{1}{\sqrt{n!}} \phi_{\substack{\text{in}\\\text{out}}}{}^{\alpha_1 *} \phi_{\substack{\text{in}\\\text{out}}}{}^{\alpha_2 *} \cdots \phi_{\substack{\text{in}\\\text{out}}}{}^{\alpha_n *} \mid \Psi_0\rangle \tag{104}$$

The S matrix is then the unitary operator which connects the ingoing and outgoing states

$$|\alpha_1, \cdots \alpha_n\rangle_{\text{in}} = S^* \mid \alpha_1, \cdots \alpha_n\rangle_{\text{out}} \tag{105}$$

or equivalently

$$\phi_{\text{out}}(x) = S^{-1}\phi_{\text{in}}(x)\, S \tag{106}$$

with

$$S^*S = SS^* = 1 \tag{107}$$

The theory is phenomenological in the sense that one is given the S matrix. For mathematical convenience we shall assume being given the operator η, with

$$S = e^{i\eta} \tag{108}$$

This operator η, the phase shift operator, is Hermitian (so that S is unitary). The precise specification of the S matrix is by the use of c-number functions $h_n(x_1, \cdots x_n)$, in terms of which η is given as

$$\eta = \sum_{n=4}^\infty \frac{1}{n!} \int d^4x_1 \cdots \int d^4x_n\, h_n(x_1, \cdots x_n) : \phi_{\substack{\text{in}\\\text{out}}}(x_1) \cdots \phi_{\substack{\text{in}\\\text{out}}}(x_n) : \tag{109}$$

The sum in (109) starts with $n = 4$ since the first real scattering process the S matrix is to describe is the elastic scattering between pairs of particles. We shall assume that the functions $h_n(x_1, \cdots x_n)$ are sufficiently well behaved so that they have Fourier transforms: $\tilde{h}'_n(k_1, \cdots k_n)$ defined by

$$h_n(x_1, \cdots x_n) = \frac{1}{(2\pi)^{5n/2}} \int d^4k_1 \cdots \int d^4k_n\, e^{i\sum_{j=1}^n k_j \cdot x_j}\, \tilde{h}'_n(k_1, \cdots k_n) \tag{110}$$

If we substitute into (109) this Fourier representation for h_n as well as that for the operators $\phi_{\substack{\text{in}\\\text{out}}}$, Eq. (94), the η matrix can then be written in the following form:

$$\eta = \sum_{n=4}^{\infty} \frac{1}{n!} \int d^4k_1 \cdots \int d^4k_n$$
$$\tilde{h}'_n(k_1, \cdots k_n)\, \delta(k_1{}^2 - \mu^2) \cdots \delta(k_n{}^2 - \mu^2) : \tilde{\phi}_{\text{in}}(k_1) \cdots \tilde{\phi}_{\text{in}}(k_n) : \quad (111)$$

Equation (111) indicates the important fact that only the mass shell values ($k_i{}^2 = \mu^2$) of $\tilde{h}'_n$ enter in the definition of the S matrix. Off the mass shell these functions can be chosen in arbitrary manner. In other words, if S is given, the functions $h_n(x_1, \cdots x_n)$ are not uniquely determined.

The assumed invariance of the S matrix under arbitrary inhomogeneous Lorentz transformation

$$U(a, \Lambda)\, SU(a, \Lambda)^{-1} = S \quad (112)$$

implies that

$$\tilde{h}'_n(k_1, \cdots k_n) = \delta(k_1 + \cdots + k_n)\, \tilde{h}_n(k_1, \cdots k_n) \quad \text{(translation invariance)} \quad (113a)$$

$$\tilde{h}_n(k_1, \cdots k_n) = \tilde{h}_n(\Lambda k_1, \cdots, \Lambda k_n) \quad (113b)$$

$\tilde{h}_n$ is therefore only a function of the scalar products $k_i \cdot k_j$. Furthermore, it is evident that, due to the hermiticity of the phase shift operator η and of the field ϕ,

$$\overline{\tilde{h}_n(k_1, \cdots, k_n)} = \tilde{h}_n(-k_1, \cdots, -k_n) \quad (114)$$

and also that due to the symmetry property of the normal product $\tilde{h}_n(k_1, \cdots k_n)$ is a symmetric function

$$\tilde{h}_n(k_1, \cdots k_n) = \mathcal{P}\tilde{h}_n(k_1, \cdots, k_n) \quad (115)$$

where $\mathcal{P}$ is an arbitrary permutation of the set $\{1, \cdots n\}$. If the theory is TPC invariant, the function $\tilde{h}_n$ will be real so that by virtue of Eq. (114)

$$\tilde{h}_n(k_1, \cdots k_n) = \tilde{h}_n(-k_1, \cdots -k_n) \quad (116)$$

Finally, in order that observable quantities such as scattering cross sections and reaction cross sections be finite $\tilde{h}_n(k_1, \cdots k_n)$ must not contain any four-dimensional delta function; $\tilde{h}_n(k_1, \cdots k_n)$ must also be a continuous function of its invariant variables. The above requirements on the functions $\tilde{h}_n$ are on the mass shell, i.e., when $k_i{}^2 = \mu^2$. However, since the extrapolation off the mass shell is thus far arbitrary, one can require that these symmetry and reality properties also hold off the mass shell.

Given the in-field ϕ_{in} and the S matrix, which incidentally is also expressible in the form

$$S = 1 + \sum_{n=1}^{\infty} \frac{1}{n!} \int d^4x_1 \cdots \int d^4x_n\, \sigma_n(x_1, \cdots x_n) : \phi_{\text{in}}(x_1) \cdots \phi_{\text{in}}(x_n) : \quad (117)$$

with the functions σ_n having properties similar to those of h_n discussed above,[11] the outfield $\phi_{out}(x)$ is then related to $\phi_{in}(x)$ by

$$\phi_{out}(x) = S^{-1}\phi_{in}(x)\, S \tag{118}$$

This equation can also be recast in the form:

$$\phi_{out}(x) = \phi_{in}(x) + S^{-1}[\phi_{in}(x), S] \tag{119}$$

By virtue of the commutation rules $[\phi_{in}(x), \phi_{in}(x')] = i\Delta(x - x'; \mu)$, we can formally write

$$[\phi_{in}(x), O] = i \int d^4x'\, \Delta(x - x'; \mu) \frac{\delta}{\delta\phi_{in}(x')} O \tag{120}$$

for any operator function O of ϕ_{in}. [Recall the definition of functional derivative in Chapter 7, Section *e*.] For the particular case that $O = \phi_{in}(y)$ since

$$\frac{\delta}{\delta\phi_{in}(x')} \phi_{in}(y) = \delta^{(4)}(x' - y) \tag{121}$$

Eq. (120) reduces to the correct expression for the commutator. Using this notion of functional derivative, we can rewrite Eq. (119) as follows:

$$\phi_{out}(x) = \phi_{in}(x) + i \int_{-\infty}^{+\infty} d^4x'\, \Delta(x - x')\, S^{-1} \frac{\delta S}{\delta\phi_{in}(x')} \tag{122a}$$

$$= \phi_{in}(x) - \int_{-\infty}^{+\infty} d^4x'\, \Delta(x - x')\, j(x') \tag{122b}$$

where the current operator $j(x)$ is introduced by the equality

$$\int_{-\infty}^{+\infty} d^4x'\, \Delta(x - x')\, j(x) = -i \int_{-\infty}^{+\infty} d^4x'\, \Delta(x - x')\, S^{-1} \frac{\delta S}{\delta\phi_{in}(x')} \tag{123}$$

It should be noted that, in order to define the current operator $j(x)$ by

$$j(x) = -iS^{-1} \frac{\delta S}{\delta\phi_{in}(x)} \tag{124}$$

an extrapolation of $\tilde{h}_n$ or $\tilde{\sigma}_n$ off the mass shell must be specified (whereas $\int \Delta(x - x')\, S^{-1}\, \delta S/\delta\phi_{in}(x')$ is uniquely determined if S is given).

Proof: Using the identity

$$\int_0^1 d\lambda \frac{d}{d\lambda} [e^{-i\lambda\eta} \phi_{in}(x)\, e^{i\lambda\eta}] = i \int_0^1 d\lambda\, e^{-i\lambda\eta} [\phi_{in}(x), \eta]\, e^{i\lambda\eta}$$

$$= \phi_{out}(x) - \phi_{in}(x) \tag{125}$$

Eq. (122b) can also be written in the form

$$\phi_{out}(x) = \phi_{in}(x) - \int_0^1 d\lambda \int \Delta(x - x')\, e^{-i\lambda\eta}\, \vartheta(x')\, e^{i\lambda\eta}\, d^4x' \tag{126}$$

[11] The requirement of unitarity cannot however be simply expressed in terms of the σ_ns. It becomes a nonlinear relation among all the σ_ns.

with

$$\vartheta(x') = \sum_{n=0}^{\infty} \frac{1}{n!} \int d^4x_1 \cdots \int d^4x_n \, h_{n+1}(x', x_1, \cdots x_n) : \phi_{\text{in}}(x_1) \cdots \phi_{\text{in}}(x_n) : \tag{127}$$

It is clear from this last expression[12] that $\tilde{h}_{n+1}(k, k_1, \cdots k_n)$ $(n = 0, 1, \cdots)$ has to be determined for $k^2 \neq \mu^2$ to determine $j(x)$ for all x. This extrapolation off the mass shell is for the present arbitrary except for the requirement that all the symmetry relations satisfied on the mass shell be fulfilled everywhere.

Lehmann, Symanzik, and Zimmermann next define the field operator $\phi(x)$ by

$$\phi(x) = \phi_{\text{in}}(x) + \int d^4x' \, \Delta_R(x - x') \, j(x') \tag{128a}$$

$$j(x) = (\Box + \mu^2) \, \phi(x) \tag{128b}$$

This operator is presumably well defined since none of the procedures involved in its construction from S and ϕ_{in} involve divergent steps. If the symmetry relations (113)–(116) are satisfied by $\tilde{h}_n(k_1, \cdots k_n)$ off the mass shell, the field $\phi(x)$ will then transform like a scalar field under Lorentz transformation and will be Hermitian. Moreover, due to the continuity conditions imposed on $\tilde{h}_n(k_1, \cdots k_n)$ as well as on the extrapolation procedure, $\phi(x)$ will satisfy the asymptotic condition

$$\lim_{t \to \pm\infty} (\Phi, \phi^f(t) \, \Psi) = (\Phi, \phi^f{}_{\substack{\text{out}\\\text{in}}} \Psi) \tag{129a}$$

where

$$\phi^f(t) = i \int d^3x \left\{ \phi(x) \frac{\partial \bar{f}(x)}{\partial x_0} - \frac{\partial \phi(x)}{\partial x_0} \bar{f}(x) \right\} \tag{129b}$$

with $f(x)$ an arbitrary normalizable solution of the Klein-Gordon equation with mass μ. That this limit exists is now a simple consequence of the Riemann-Lebesgue lemma on Fourier integrals and follows from the assumed continuity of $\tilde{h}_n(k_1, \cdots k_n)$ near the mass shell. Note, however, that given an S matrix the field $\phi(x)$ defined by (128) is not unique. This arbitrariness is a reflection of the arbitrariness which exists in the extrapolation of $\tilde{h}_n$ off the mass shell. Given an S matrix, there exist many interpolating interacting fields which satisfy the weak asymptotic limit $\lim_{t \to \pm\infty} \phi^f(t) = \phi^f{}_{\substack{\text{out}\\\text{in}}}$. The requirements for this to be true are that $\tilde{h}_n(k_1, \cdots k_n)$ be finite and continuous near $k_i^2 = \mu^2$. These interpolated fields will, however, not be local in general. The further condition on $\tilde{h}_n(k_1, \cdots, k_n)$ and on the extrapolation off the mass shell in order that

$$[\phi(x), \phi(x')] = 0 \quad \text{for } (x - x')^2 < 0 \tag{130}$$

[12] Even though the mass shell value of $\tilde{h}_j(k_1, \cdot, k_j)$ for $j = 1, 2, 3$ vanishes, it is, of course, not necessary that their extrapolated values also be identically zero. This accounts for the fact that the sum in Eq. (127) starts with $n = 0$.

is a very stringent one. As mentioned earlier it is known [Zimmermann (1958), Redmond (1959), Kashlun (1959, 1960); see also Greenberg (1956)] that for a Heisenberg field which is local and obeys canonical commutation rules, i.e., $[\phi(x), \dot{\phi}(x')]$ for $x_0 = x'_0$ is a c-number proportional to $\delta^{(3)}(\mathbf{x} - \mathbf{x}')$, the in-field defined by

$$\phi_{\text{in}}(x) = \phi(x) - \int d^4x' \, \Delta_R(x - x') \, (\Box_{x'} + \mu^2) \, \phi(x') \qquad (131)$$

will satisfy free-field commutation rules. (By construction this in-field clearly satisfies free-field equations.) Locality and canonical commutation rules are thus sufficient conditions for the existence of asymptotic fields. These conditions are however not necessary since it is known (at least to the lowest nontrivial order in perturbation theory) that an asymptotic limit exists also in nonlocal field theories [Kristensen and Møller (Kristensen 1952)].

A scattering matrix will be said to be causal if there exists at least one extrapolation such that the interpolated field $\phi(x)$ is local. Is it not known at present whether any scattering matrix exists yielding a local field theory described by an operator $\phi(x)$, besides the trivial one: $S = 1$ and $\phi(x) = \phi_{\text{in}}(x) = \phi_{\text{out}}(x)$. It is, however, possible to study the consequences for observable quantities of the requirement that the interpolated field $\phi(x)$ be local. In fact, although we have presented the "phenomenological" version of the LSZ formalism which takes the in- and out-fields as given and interpolates from them the field $\phi(x)$, an approach based on the Heisenberg field as the basic quantity in the theory is more fundamental and constitutes a more satisfactory attack on the problem of describing the interactions between elementary particles. In such an approach the most natural kind of theory to study initially is one in which the field operators are assumed to be local. Let us therefore, with LSZ, study the general properties of relativistically invariant local field theories which satisfy the asymptotic conditions. The important observation of LSZ was to realize that as a consequence of the asymptotic conditions, so-called reduction formulae can be derived for R products and T products of the Heisenberg field.

The retarded product of $n + 1$ operators, which is the generalization of the retarded product $R(x; y)$ of two operators

$$\begin{aligned} R(x; y) &= R(\phi(x) \, \phi(y)) \\ &= -i\theta(x - y) \, [\phi(x), \phi(y)] \end{aligned} \qquad (132)$$

is defined by[13]

[13] The retarded product for different Bose fields is defined as

$$R^{\phi_1\phi_2\cdots\phi_{n+1}}(x_1; x_2, \cdots x_{n+1}) = (-i)^n \sum_P \theta(x_1 - x_{i_2}) \cdots \theta(x_{i_n} - x_{i_{n+1}})$$

$$[\cdots [\phi_1(x_1), \phi_{i_2}(x_{i_2})], \cdots \phi_{i_{n+1}}(x_{i_{n+1}})]$$

where the summation is over all permutations P of $2, \cdots n + 1$. Note that the field ϕ_1 is again singled out.

$$n = 0 \quad R(x) = \phi(x)$$

$$n \geqslant 1 \quad R(x; x_1, \cdots x_n) = (-i)^n \sum_P \theta(x - x_1) \cdots \theta(x_{n-1} - x_n)$$

$$[\cdots [\phi(x), \phi(x_1)], \phi(x_2)], \cdots \phi(x_n)] \quad (133)$$

where in (133) the permutation is over all the n co-ordinates x_i. The vacuum expectation value of $R(x; x_1, \cdots x_n)$ will be denoted by $r(x; x_1, \cdots x_n)$:

$$r(x; x_1, \cdots x_n) = (\Psi_0, R(x; x_1, \cdots x_n) \Psi_0) \quad (134)$$

The properties of the R functions which are immediate consequences of their definition are the following:

(a) *Retardation property*: $R(x; x_1, \cdots x_n)$ vanishes if any one of the times x_{i0} is later than x_0.

(b) *Symmetry*: $R(x; x_1, \cdots x_n)$ is a symmetric function of $x_1, \cdots x_n$.

(c) *Hermiticity*: The R product of Hermitian operators is Hermitian. Furthermore, the retarded product of Lorentz covariant and *local* field operators $\phi(x)$ is covariant under Lorentz transformation.

Proof: For simplicity, consider first the retarded product of two operators, $R(x; y)$. By assumption

$$U(a, \Lambda) \phi(x) U(a, \Lambda)^{-1} = \phi(\Lambda x + a) \quad (135)$$

Clearly under translations since $\theta(x - y) = \theta(x + a - y - a)$

$$U(a, 1) R(x; y) U(a, 1)^{-1} = R(x + a; y + a) \quad (136)$$

so that, for example, $r(x; y) = r(x - y)$. In general, due to translation invariance, $r(x; x_1, \cdots x_n)$ will only be a function of the co-ordinate differences $x - x_1, x - x_2, \cdots x - x_n$

$$r(x; x_1, \cdots x_n) = r(x - x_1, x - x_2, \cdots x - x_n) \quad (137)$$

Now under a homogeneous Lorentz transformation a space-like vector remains space-like and a time-like vector time-like. Furthermore, for orthochronous transformations, if $x - y$ is time-like, $\theta(x - y) = \theta(\Lambda(x - y))$. Since $R(x; y)$ is different from zero only when $x - y$ is a time-like or null vector due to the locality of the operator $\phi(x)$ ($[\phi(x), \phi(y)] = 0$ for $(x - y)^2 < 0$), it follows that:

$$\begin{aligned} U(\Lambda) R(x; y) U(\Lambda)^{-1} &= -i\theta(x - y) [\phi(\Lambda x), \phi(\Lambda y)] \\ &= -i\theta(\Lambda(x - y)) [\phi(\Lambda x), \phi(\Lambda y)] \\ &= R(\Lambda x; \Lambda y) \end{aligned} \quad (138)$$

so that

$$r(x; y) = r(\Lambda x; \Lambda y) = r(\Lambda(x - y)) \quad (139)$$

The proof of the covariance under homogeneous Lorentz transformation of retarded products of more than two operators is nontrivial and requires the repeated use of the Jacobi identities. Actually, a convenient tool for the demonstration of the relativistic invariance of the retarded product of more than two operators, as well as the study of the formal aspects of the entire theory, is the use of generating functionals [Schwinger (1951b, 1954b), Symanzik (1954, 1960), Bogoliubov (1959a)]. These generating functionals are, in general, not supposed to exist in any other than a formal sense. The generating functional for the retarded product of operators can be obtained from the unitary functional

$$\begin{aligned}\mathfrak{T}\{J\} &= T\exp\left(i\int d^4x\, J(x)\,\phi(x)\right)\\ &= \sum_{n=0}^{\infty}\frac{i^n}{n!}\int d^4x_1\cdots\int d^4x_n\, J(x_1)\cdots J(x_n)\, T(\phi(x_1)\cdots\phi(x_n))\end{aligned} \tag{140}$$

where $J(x)$ is a source function which plays a purely algebraic role [see, however, Schwinger (1951, 1954)], and T is the Wick time-ordering operator. Recall incidentally, that the time-ordering operation is a relativistically invariant operation only if $[\phi(x), \phi(x')] = 0$ for $(x - x')^2 < 0$. The use of $\mathfrak{T}\{J\}$ presupposes that the products

$$T(\phi(x_1)\cdots\phi(x_n)) = \sum_{\text{permutations}}\theta(x_1 - x_2)\cdots\theta(x_{n-1} - x_n)\,\phi(x_1)\cdots\phi(x_n) \tag{141}$$

are well defined. It turns out to be sufficient for these products to be well-defined that the vacuum expectation values of products of operators, i.e. the Wightman function be well defined, and more precisely that the latter be tempered distributions [Zimmermann (1958), Araki (1960b), Symanzik (1960), Steinmann (1960)].

The retarded product of operators $R(\phi(x_1)\cdots\phi(x_n)$ can now be obtained from the functional

$$\mathfrak{R}(x;J) = -i\mathfrak{T}\{J\}^*\frac{\delta}{\delta J(x)}\mathfrak{T}\{J\} \tag{142}$$

by functional differentiation:

$$R(x;x_1,\cdots x_n) = \left.\frac{\delta^n\mathfrak{R}(x;J)}{\delta J(x_1)\cdots\delta J(x_n)}\right|_{J=0} \tag{143}$$

To illustrate these remarks consider first the case $n = 0$; since

$$\frac{\delta}{\delta J(x)}\mathfrak{T}\{J\} = iT(\phi(x)\,\mathfrak{T}\{J\}) \tag{144}$$

we have

$$\begin{aligned}R(x) &= \mathfrak{R}(x;J)|_{J=0}\\ &= -i\mathfrak{T}\{J\}^*\, iT(\phi(x)\,\mathfrak{T}\{J\})|_{J=0}\\ &= \phi(x)\end{aligned} \tag{145}$$

Similarly for $n = 1$, since

$$\frac{\delta\mathfrak{R}(x;J)}{\delta J(x_1)} = -i\left(\frac{\mathfrak{T}\{J\}^*}{\delta J(x_1)}\frac{\delta\mathfrak{T}\{J\}}{\delta J(x)} + \mathfrak{T}\{J\}^*\frac{\delta^2\mathfrak{T}\{J\}}{\delta J(x_1)\,\delta J(x)}\right) \tag{146}$$

we find that

$$\left.\frac{\delta\mathfrak{R}(x;J)}{\delta J(x_1)}\right|_{J=0} = -i[\phi(x_1)\,\phi(x) - T(\phi(x_1)\,\phi(x))]$$

$$= -i\theta(x - x_1)\,[\phi(x), \phi(x_1)] \tag{147}$$

etc. From the transformation properties of the functional $\mathfrak{T}\{J\}$

$$U(a,\Lambda)\,\mathfrak{T}\{J\}\,U(a,\Lambda)^{-1} = T\left(i\int d^4x\,J(\Lambda^{-1}(x-a))\,\phi(x)\right) \tag{148}$$

one immediately deduces that

$$U(\Lambda)\,R(x;x_1,\cdots x_n)\,U(\Lambda)^{-1} = R(\Lambda x;\Lambda x_1,\cdots \Lambda x_n) \tag{149a}$$

so that

$$r(x;x_1,\cdots x_n) = r(\Lambda x;\Lambda x_1,\cdots \Lambda x_n) \tag{149b}$$

The functions $r(x;x_1,\cdots x_n)$ are therefore Lorentz invariant distributions. Furthermore, by virtue of their retardation property, the r functions $r(x;x_1,\cdots x_n) = r(x - x_1,\cdots x - x_n)$ vanish unless the vectors $\xi_1 = x - x_1, \cdots, \xi_n = x - x_n$ lie in the forward light cone. The Fourier transform $\tilde{r}(p_1,\cdots p_n)$

$$\tilde{r}(p_1,\cdots p_n) = \frac{1}{(2\pi)^{4n}}\int d^4\xi_1\cdots\int d^4\xi_n\,e^{i\sum_j p_j\cdot\xi_j}\,r(\xi_1,\cdots \xi_n) \tag{150}$$

is, by virtue of the support properties of r in space-time, the boundary value of an analytic function whose domain of analyticity in the scalar product z_{ij} of the *complex* vectors $p_1,\cdots p_n$ is at least as large as the domain of analyticity of the Wightman function $W^{(n+1)}(\xi_1,\cdots \xi_n)$ in terms of the variables $z_{ij} = \zeta_i\cdot\zeta_j$. This has been explicitly demonstrated by Källén and Wightman [Källén (1958)] for the case $n = 3$, and by Kleitman (1959) for $n = 4$ and 5. Araki (1960b) has made a systematic investigation for the general case and has also obtained the necessary and sufficient conditions for retarded functions to be obtainable from Wightman functions. [See also Zimmermann (1960), Steinman (1960).]

In addition to the above (linear) properties, the retarded functions satisfy a system of nonlinear equations which couples the different functions and which are a consequence of the operator identity

$$R(x;y,x_1,\cdots x_n) - R(y;x,x_1,\cdots x_n)$$

$$= \sum_{i_1\cdots i_n}\sum_{k=0}^{n}\frac{1}{k!(n-k)!}\,[R(x;x_{i_1},\cdots x_{i_k}), R(y;x_{i_{k+1}},\cdots x_{i_n})] \tag{151}$$

Equation (151) is verified by noting that, by virtue of its definition, Eq. (142), $\mathfrak{R}(x;J)$ satisfies the following functional equation:

$$\frac{\delta\mathfrak{R}(x;J)}{\delta J(y)} - \frac{\delta\mathfrak{R}(y;J)}{\delta J(x)} = -[\mathfrak{R}(x;J), \mathfrak{R}(y;J)] \tag{152}$$

Proof: Recall Eq. (146) and note that $\mathfrak{T}\{J\}$ is unitary, $\mathfrak{T}\{J\}^* = \mathfrak{T}\{J\}^{-1}$, so that

$$\frac{\delta}{\delta J(x)} (\mathfrak{T}\{J\}^* \mathfrak{T}\{J\}) = 0 \tag{153}$$

Before continuing further with the characterization of the retarded functions, we next present the LSZ recursion relations which follow from the asymptotic conditions. We first prove the relation

$$[R(x; x_1, \cdots x_n), \phi_{\text{in}}(z)] = i \int d^4z' \, \Delta(z - z') \, K_{z'} R(x; x_1, \cdots x_n, z') \tag{154}$$

where $K_x = \square + \mu^2$.

Proof: Consider the matrix element

$$(\Phi, [R(x; x_1, \cdots x_n), \phi_{\text{in}}{}^{\alpha *}] \, \Psi)$$
$$= i \int d^3z \, (\Phi, [R(x; x_1, \cdots x_n), \phi_{\text{in}}(z)] \, \Psi) \overset{\leftrightarrow}{\frac{\partial}{\partial z_0}} f_\alpha(z) \tag{155}$$

The asymptotic conditions allow us to rewrite the right-hand side of (155) as follows:

$$i \int (\Phi, [R(x; x_1, \cdots x_n), \phi_{\text{in}}(z)] \, \Psi) \overset{\leftrightarrow}{\frac{\partial}{\partial z_0}} f_\alpha(z) \, d^3z$$
$$= i \lim_{z_0 \to -\infty} \int d^3z \, (\Phi, [R(x; x_1, \cdots x_n), \phi(z)] \, \Psi) \overset{\leftrightarrow}{\frac{\partial}{\partial z_0}} f_\alpha(z) \tag{156a}$$
$$= - \lim_{z_0 \to -\infty} \int d^3z \, (\Phi, R(x; x_1, \cdots x_n, z) \, \Psi) \overset{\leftrightarrow}{\frac{\partial}{\partial z_0}} f_\alpha(z) \tag{156b}$$

In writing down Eq. (156b), we have made use of the fact that z_0 is at $-\infty$ and hence earlier than $x_0, \cdots, x_{0n}$ so that a factor $\theta(x - z)$ can be supplied to the integrand on the right-hand side of (156a). The other terms required in order to be able to write $R(x; x_1, \cdots x_n, z)$ for $[R(x; x_1, \cdots x_n), \phi(z)]$ (corresponding to permutations of $x_1, \cdots x_n, z$) vanish since z_0 is earlier than $x_0, x_{10}, \cdots x_{n0}$. By a procedure already used in Chapter 17, Section *d*, we can now write

$$\lim_{z_0 \to -\infty} \int d^3z \, (\Phi, R(x; x_1, \cdots x_n, z) \, \Psi) \overset{\leftrightarrow}{\frac{\partial}{\partial z_0}} f_\alpha(z)$$
$$= \lim_{z_0 \to +\infty} \int d^3z \, (\Phi, R(x; x_1, \cdots x_n, z) \, \Psi) \overset{\leftrightarrow}{\frac{\partial}{\partial z_0}} f_\alpha(z)$$
$$- \int d^3z \int_{-\infty}^{+\infty} dz_0 \frac{\partial}{\partial z_0} \left\{ (\Phi, R(x; x_1, \cdots x_n, z) \, \Psi) \overset{\leftrightarrow}{\frac{\partial}{\partial z_0}} f_\alpha(z) \right\} \tag{157}$$

The first term on the right-hand side of Eq. (157) does not contribute since z_0 is now at $+\infty$ and the retarded product $R(x; x_1, \cdots x_n, z)$ vanishes whenever any one of the times $x_{10}, \cdots x_{n0}, z_0$ is later than x_0. Hence carrying out the indicated differentiation, we obtain

$$(\Phi, [R(x; x_1, \cdots, x_n), \phi_{\text{in}}^{\alpha *}] \Psi) = \int d^4z\, f_\alpha(z)\, K_z(\Phi, R(x; x_1, \cdots x_n, z)\, \Psi) \quad (158)$$

In arriving at (158) an integration by part over the spatial variables has been performed, after substituting for $\partial^2 f_\alpha(z)/\partial z_0{}^2$ the expression

$$\frac{\partial^2}{\partial z_0{}^2} f_\alpha(z) = (\nabla^2 - \mu^2)\, f_\alpha(z) \quad (159)$$

since $f_\alpha(z)$ satisfies the Klein-Gordon equation. By multiplying Eq. (158) by $\bar{f}_\alpha(x)$ and summing over α and adding to the resulting expression its complex conjugate, we obtain finally Eq. (154). By iterating Eq. (154), we derive

$$[\cdots[[R(x; x_1, \cdots x_m), \phi_{\text{in}}(z_1)], \phi_{\text{in}}(z_2)], \cdots \phi_{\text{in}}(z_n)] = i^n \int d^4z'_1 \cdots \int d^4z'_n$$
$$\Delta(z_1 - z'_1) \cdots \Delta(z_n - z'_n)\, K_{z'_1} \cdots K_{z'_n} R(x; x_1, \cdots x_m, z'_1, \cdots z'_n) \quad (160)$$

One should note that these recursion formulae involve differentiation of the retarded products which brings in δ functions. Therefore even if it is the case that the vacuum expectation values of R products define distributions, it is not at all clear that the differentiated R products define distributions. [For renormalizable theories, it is known that Eq. (160) is to be regarded as symbolic and must be supplemented by some renormalization prescription before it becomes mathematically well defined.]

The Yang-Feldman equation relating in- and out-field can be regarded as the result of the simplest application of the procedure which led to Eq. (154). From the asymptotic conditions, one deduces that

$$\begin{aligned}(\Phi, \phi_{\text{in}}{}^f \Psi) &= \lim_{x_0 \to -\infty} (\Phi, \phi^f(t)\, \Psi) \\ &= \lim_{x_0 \to -\infty} i \int d^3x\, (\Phi, \phi(x)\, \Psi) \frac{\overleftrightarrow{\partial}}{\partial x_0} \bar{f}(x) \\ &= \lim_{x_0 \to +\infty} i \int d^3x\, (\Phi, \phi(x)\, \Psi) \frac{\partial}{\partial x_0} \bar{f}(x) \\ &\quad - i \int_{-\infty}^{+\infty} d^4x \frac{\partial}{\partial x_0} \left\{ (\Phi, \phi(x)\, \Psi) \frac{\overleftrightarrow{\partial}}{\partial x_0} \bar{f}(x) \right\} \\ &= (\Phi, \phi_{\text{out}}{}^f \Psi) + i \int_{-\infty}^{+\infty} d^4x\, \bar{f}(x)\, K_x(\Phi, \phi(x)\, \Psi) \quad (161)\end{aligned}$$

whence, by an argument similar to the one which led to Eq. (154), we find that

$$\phi_{out}(x) = \phi_{in}(x) - \int \Delta(x - x')\, K_{x'}\phi(x')\, d^4x' \tag{162}$$

The commutation rules between in- and out-field are now easily derived by combining Eqs. (162) and (154). Thus, by taking the commutator of Eq. (162) with $\phi_{in}(y)$, we obtain

$$[\phi_{out}(x), \phi_{in}(y)] = i\Delta(x - y) - \int d^4x'\, \Delta(x - x')\, K_{x'}[\phi(x'), \phi_{in}(y)] \tag{163}$$

Now, Eq. (154) for the case $R(x) = \phi(x)$ asserts that

$$[\phi(x'), \phi_{in}(y)] = i \int d^4y'\, \Delta(y - y')\, K_{y'}R(x'; y') \tag{164}$$

whence

$$[\phi_{out}(x), \phi_{in}(y)] = i\Delta(x - y) - i \int d^4x' \int d^4y'\, \Delta(x - x')\, \Delta(y - y')\, K_{x'}K_{y'}R(x'; y') \tag{165}$$

In passing, we mention that Zimmermann (1958) has proved that the order of integration in (165) can be interchanged if, and only if, the incoming and outgoing field satisfy the same commutation rules. By inserting into Eq. (165) the expansion (97) for ϕ_{in} in terms of a_{in} and a_{in}^*, one deduces that

$$[a_{out}(k), a_{in}^*(k')] = 2k_0\delta^{(3)}(\mathbf{k} - \mathbf{k}') - \frac{i}{(2\pi)^3} \int d^4x \int d^4x'\, e^{-i(k'\cdot x' - k\cdot x)}\, K_xK_{x'}\, R(x; x') \tag{166a}$$

$$[a_{out}(k), a_{in}(k')] = \frac{i}{(2\pi)^3} \int d^4x \int d^4x'\, e^{+i(k\cdot x - k'\cdot x')}\, K_xK_{x'}R(x; x') \tag{166b}$$

We are now in a position to obtain the set of coupled equations which, together with the properties of retardation, reality, symmetry, and covariance, characterize the retarded functions $r(x; x_1, \cdots x_n)$. By taking the vacuum expectation value of Eq. (151), one obtains

$$\begin{aligned} r(x; y, x_1, \cdots x_n) - r(y; x, x_1, \cdots x_n) &= \sum_{i_1\cdots i_n} \sum_{k=0}^{n} \frac{1}{k!(n-k)!} \sum_{l=0}^{\infty} \sum_{\alpha_1\cdots\alpha_l} \\ &(\Psi_0, R(x; x_{i_1}, \cdots x_{i_k})\, \Phi_{in}^{\alpha_1\cdots\alpha_l})\, (\Phi_{in}^{\alpha_1\cdots\alpha_l} R(y; x_{i_{k+1}} \cdots x_{i_n})\, \Psi_0) \\ &- (\text{term with } x \leftrightarrow y) \end{aligned} \tag{167}$$

where a sum over the complete set of states

$$|\Phi_{in}^{\alpha_1\cdots\alpha_l}\rangle = \frac{1}{\sqrt{n!}} \phi_{in}^{\alpha_1 *} \cdots \phi_{in}^{\alpha_n *} \mid \Psi_0\rangle \tag{168}$$

has been introduced. Using the recursion relation

$$\begin{aligned} &(\Psi_0, R(x; x_1 \cdots x_n)\, \phi_{in}^{\alpha_1 *} \cdots \phi_{in}^{\alpha_n *}\Psi_0) \\ &= \int d^4z_1 \cdots \int d^4z_n\, f_{\alpha_1}(z_1) \cdots f_{\alpha_n}(z_n)\, K_{z_1} \cdots K_{z_n}R(x; x_1 \cdots x_m, z_1 \cdots z_n) \end{aligned} \tag{169}$$

one finally obtains the following set of equations:

$$r(x; y, x_1, \cdots x_n) - r(y; x, x_1, \cdots x_n) = \sum_{i_1 \cdots i_n} \sum_{k=0}^{n} \sum_{l=1}^{\infty} \frac{i^l}{k!(n-k)!l!}$$
$$\int d^4u_1 \cdots \int d^4u_l \int d^4v_1 \cdots \int d^4v_l \, K_{u_1} \cdots K_{u_l}$$
$$r(x; x_{i_1} \cdots x_{i_k}, u_1, \cdots u_l) \, \Delta^{(+)}(u_1 - v_1) \cdots \Delta^{(+)}(u_l - v_l)$$
$$K_{v_1} \cdots K_{v_l} r(y; x_{i_{k+1}}, \cdots x_{i_n} v_1 \cdots v_l) - (x \leftrightarrow y) \tag{170}$$

These coupled nonlinear integral equations[14] for the r functions are the analogue of the positive-definiteness inequalities for the Wightman functions. It should be noted that in their derivation explicit use has been made of the completeness relation for the in-fields, and hence the equations are not valid as they stand in the case that bound states exist [see, however, Baumann (1958)]. These equations express the absorptive part of r functions as a sum of bilinear terms arising from various intermediate states and can be viewed as "generalized unitary conditions." We shall see in Section d that the relations (170) play an important part in the proof of dispersion relations for scattering amplitudes [Jost (1958)].

In terms of r functions, the free field is characterized by

$$r(x; x_1, \cdots x_n) = 0 \quad \text{for } n > 1 \tag{171}$$

The equation which $r(x; x_1) = r(\xi_1)$, $\xi_1 = x - x_1$, satisfies is

$$\begin{aligned} r(x-y) - r(y-x) &= \int d^4u \int d^4v \, K_u r(x-u) \, \Delta^{(+)}(u-v) \, K_v r(y-v) \\ &\quad - \int d^4u \int d^4v \, K_u r(y-u) \, \Delta^{(+)}(u-v) \, K_v r(x-v) \\ &= \int d^4u \int d^4v \, K_u r(x-u) \, \Delta(u-v) \, K_v r(y-v) \end{aligned} \tag{172}$$

where $\Delta(x) = \Delta^{(+)}(x) - \Delta^{(+)}(-x) = \Delta^{(+)}(x) + \Delta^{(-)}(x)$. The only solution of this equation satisfying the three boundary conditions (symmetry, reality, retardation) is

$$\begin{aligned} r(x; y) &= -\Delta_R(x-y) \\ &= \theta(x-y) \, \Delta(x-y) \end{aligned} \tag{173}$$

If one now attempts to solve the coupled equations (170) as a power series in some perturbation parameter starting from the above free-field r function, one is led in a unique fashion to the renormalized perturbation theoretical expansions compatible with the assumed type of stable particle

[14] Nishijima (1958) looks upon these equations as self-consistency requirements on the recursion relations. The existence of solution of (170) is equivalent to the inner consistency of the axioms and the asymptotic conditions.

[Haag (1955), Nishijima (1960), Zimmermann (1960)]. What corresponds to the different types of interaction terms in a Lagrangian formulation here corresponds to r functions, $r(x; y, x_1, \cdots x_n)$, which differ from one another by terms of the form

$$P\left(\frac{\partial}{\partial y}, \frac{\partial}{\partial x_1}, \cdots \frac{\partial}{\partial x_n}\right) \delta(x-y)\, \delta(x-x_1)\, \delta(x-x_2) \cdots \delta(x-x_n)$$
$$= c_0 \delta(x-y)\, \delta(x-x_1) \cdots \delta(x-x_n)$$
$$+ c_1 \sum_i^n \Box_i \delta(x-y)\, \delta(x-x_1) \cdots \delta(x-x_n) + \cdots \quad (174)$$

where P is an invariant symmetric polynomial of differential operators. Apart from numerical factors, the constants $c_0, c_1, \cdots$ are related to the coupling constants for the different interactions. Conversely, the different constants of integration occurring in the solution of the equations (170) correspond to the coupling constants in a Lagrangian approach.

Finally, we note that in the case that the incoming and outgoing operators form an irreducible representation of the free-field commutation rules, i.e., in the case the vectors $|\alpha_1, \cdots \alpha_n\rangle_{\text{in}}$ and $|\alpha_1, \cdots \alpha_n\rangle_{\text{out}}$ $(n = 0, 1, 2 \cdots)$ span the entire Hilbert space, an arbitrary operator L can be expanded in terms of the in- (or out-) operators as follows:

$$L = \sum_{n=0}^{\infty} \frac{1}{n!} \int d^4x_1 \cdots \int d^4x_n\, \mathfrak{L}^{(n)}(x_1, \cdots x_n) : \phi_{\text{in}}(x_1) \cdots \phi_{\text{in}}(x_n) : \quad (175)$$

The expansion coefficients $\mathfrak{L}^{(n)}$ can be determined by recalling that the vacuum expectation value of a normal product of operators vanishes. Now

$$[L, \phi_{\text{in}}(y_1)] = i \int d^4y'_1\, \Delta(y - y'_1) \frac{\delta L}{\delta\phi(y'_1)}$$
$$= \sum_{n=0}^{\infty} \frac{1}{n!} \int d^4x_1 \cdots \int d^4x_n \int d^4y'_1$$
$$\mathfrak{L}^{(n+1)}(x_1, \cdots x_n; y'_1)\, i\Delta(y_1 - y'_1) : \phi_{\text{in}}(x_1) \cdots \phi_{\text{in}}(x_n) : \quad (176)$$

and by induction

$$[[\cdots [L, \phi_{\text{in}}(y_1)] \cdots \phi_{\text{in}}(y_m)]$$
$$= \sum_{n=0}^{\infty} \frac{1}{n!} \int d^4x_1 \cdots \int d^4x_n \int d^4y'_1 \cdots \int d^4y'_m\, i^m \Delta(y_1 - y'_1) \cdots$$
$$\Delta(y_m - y'_m)\, \mathfrak{L}^{(n+m)}(x_1, \cdots x_n, y'_1, \cdots y'_m) : \phi_{\text{in}}(x_1) \cdots \phi_{\text{in}}(x_n) : \quad (177)$$

so that

$$(\Psi_0, [\cdots [L, \phi_{\rm in}(y_1)], \cdots \phi_{\rm in}(y)] \Psi_0)$$
$$= i^m \int d^4y'_1 \cdots \int d^4y'_m \, \Delta(y_1 - y'_1) \cdots \Delta(y_m - y'_m) \, \mathfrak{L}^{(m)}(y'_1, \cdots y'_m) \tag{178}$$

By taking the Fourier transform of (178), with

$$\mathfrak{L}^{(n)}(x_1, \cdots x_n) = \frac{1}{(2\pi)^{5n/2}} \int d^4k_1 \cdots \int d^4k_n \, e^{-i \sum_{j=1}^{n} k_j \cdot x_j} \tilde{\mathfrak{L}}^{(n)}(k_1, \cdots k_n) \tag{179}$$

one deduces that

$$\tilde{\mathfrak{L}}^{(n)}(k_1, \cdots k_n) = \epsilon(k_1) \cdots \epsilon(k_n) \, (\Psi_0, [\cdots [L, \tilde{\phi}_{\rm in}(k_1)] \cdots \tilde{\phi}_{\rm in}(k_n)] \Psi_0) \tag{180}$$

so that

$$L = \sum_{n=0}^{\infty} \frac{1}{n!} \int d^4k_1 \cdots \int d^4k_n \, (\Psi_0, [\cdots [L, \tilde{\phi}_{\rm in}(-k_1)] \cdots \tilde{\phi}_{\rm in}(-k_n)] \Psi_0)$$
$$\epsilon(k_1) \cdots \epsilon(k_n) \, \delta(k_1{}^2 - \mu^2) \cdots \delta(k_n{}^2 - \mu^2) : \tilde{\phi}_{\rm in}(k_1) \cdots \tilde{\phi}_{\rm in}(k_n) : \tag{181}$$

In the particular case that L is the Heisenberg field operator $\phi(x)$, using the fact that the recursion relation (160) can be made to read

$$\epsilon(k_1) \cdots \epsilon(k_n) \, (\Psi_0, [\cdots [\phi(x), \tilde{\phi}_{\rm in}(-k_1)], \cdots \tilde{\phi}_{\rm in}(-k_n)] \Psi_0)$$
$$= \frac{1}{(2\pi)^{3n/2}} \int d^4k_1 \cdots \int d^4k_n \, e^{+i \sum_j k_j \cdot x_j} K_{x_1} \cdots K_{x_n} r(x; x_1, \cdots x_n) \tag{182}$$

we obtain

$$\phi(x) = \phi_{\rm in}(x) + \sum_{n=2}^{\infty} \frac{1}{n!} \int d^4x_1 \cdots \int d^4x_n$$
$$K_{x_1} \cdots K_{x_n} r(x; x_1, \cdots x_n) : \phi_{\rm in}(x_1) \cdots \phi_{\rm in}(x_n) : \tag{183}$$

In arriving at the representation (183), use has been made of the following identities

$$(\Psi_0, \phi(x) \Psi_0) = 0 \tag{184a}$$

$$\int d^4k \, (\Psi_0, [\phi(x), \tilde{\phi}_{\rm in}(-k)] \Psi_0) \, \epsilon(k) \, \delta(k^2 - \mu^2) \, \tilde{\phi}_{\rm in}(k) = \phi_{\rm in}(x) \tag{184b}$$

The proof of (184b) is easily established. We note that by virtue of relativistic invariance and the spectral conditions, the vacuum expectation value of the commutator of $\phi(x)$ and $\phi_{\rm in}(y)$ has the following Lehmann representation:

$$(\Psi_0, [\phi(x), \phi_{\rm in}(y)] \Psi_0) = i \int_0^\infty d\kappa^2 \, \rho(\kappa^2) \, \Delta(x - y; \kappa^2) \tag{185}$$

Since in addition $(\Box + \mu^2)\,\phi_{\text{in}}(x) = 0$, $\rho(\kappa^2) = a\delta(\kappa^2 - \mu^2)$ and

$$(\Psi_0, [\phi(x), \phi_{\text{in}}(y)]\,\Psi_0) = i\Delta(x - y; \mu^2) \qquad (186)$$

where the asymptotic condition has been used to fix the value of the constant a to be equal to $+1$.

With the help of the expansion (183), Glaser, Lehmann, and Zimmermann [Glaser (1957)] have proved two theorems which are essentially statements of the necessary and sufficient conditions that the retarded functions $r(x; x_1, \cdots x_n)$ must satisfy in order that they define a local field theory which satisfies the asymptotic conditions. They have shown that, given a set of arbitrary functions $r(x; x_1, \cdots x_n)$, $n = 0, 1, \cdots$, which have the property that

(*a*) they are real, symmetric and invariant functions of the variables $\xi_i = x - x_i$, $i = 1, 2, \cdots n$;

(*b*) they are retarded, i.e., $r(\xi_1, \cdots \xi_n)$ vanishes unless all the ξ_is are in the forward light cone;

(*c*) they satisfy the equations (170);

(*d*) if $\tilde{f}(k_1, \cdots k_n)$ denotes the Fourier transform of

$$f(x - x_1, x - x_2, \cdots x - x_n) = K_x K_{x_1} \cdots K_{x_n} r(x - x_1, \cdots x - x_n)$$

that $\tilde{f}(k_1, \cdots k_n)$ is finite on the mass shell $k_i^2 = \mu^2$, $\left(\sum_i^n k_i\right)^2 = \mu^2$ and does not depend on the order in which the limit $k_i^2 \to \mu^2$, $\left(\sum k_i\right)^2 \to \mu^2$ are taken;

then if $\phi(x)$ is defined by Eq. (183), the functions $r(x; x_1, \cdots x_n)$ are the vacuum expectation values of retarded products of this operator, and furthermore the so-defined field operator $\phi(x)$ satisfies the asymptotic condition, is local and is Lorentz covariant. This theorem is the analogue of the Wightman theorem which asserts that a set of Wightman functions which have certain specified properties defines a local field theory.

A very similar treatment of field theory to the above can be given in terms of expectation values of time-ordered products of operators [LSZ I, Lehmann (1955a)], the time-ordered product being defined as

$$\begin{aligned} T(\phi(x_1) \cdots \phi(x_n)) &= T(x_1, \cdots x_n) \\ &= \sum_P \theta(x_1 - x_{i_1})\,\theta(x_2 - x_{i_2}) \cdots \theta(x_n - x_{i_n})\,\phi(x_{i_1}) \cdots \phi(x_{i_n}) \qquad (187) \end{aligned}$$

The sum in (187) is over all permutations $i_1, \cdots i_n$ of $1, 2, \cdots n$. The vacuum expectation values of T products are usually called τ functions

$$\tau^{(n)}(x_1, \cdots x_n) = (\Psi_0, T(\phi(x_1) \cdots \phi(x_n)\,\Psi_0) \qquad (188)$$

It is possible to express the S matrix in terms of these τ functions. To show this, we first derive the following reduction formula:

$$[S \cdot T(\phi(x_1) \cdots \phi(x_n), \phi_{\text{in}}(z)] = -\int d^4z' \, \Delta(z - z') \, K_{z'} S \cdot T(\phi(x_1) \cdots \phi(x_n) \, \phi(z')) \quad (189)$$

Proof: Consider the matrix element of the commutator $[S \cdot T(x_1, \cdots x_n), \phi_{\text{in}}^{\alpha *}]$ between arbitrary normalizable states $|\Psi\rangle$ and $|\Phi\rangle$. As a consequence of the asymptotic conditions we again deduce that

$$(\Psi, [S \cdot T(x_1, \cdots x_n), \phi_{\text{in}}^{\alpha *}] \, \Phi) = \lim_{z_0 \to -\infty} (\Psi, [S \cdot T(x_1, x_2, \cdots x_n), \phi^{\alpha *}(z_0)] \, \Phi) \quad (190a)$$

$$= i \lim_{z_0 \to -\infty} \int (\Psi, [S \cdot T(x_1, \cdots x_n), \phi(z)] \, \Phi) \overleftrightarrow{\frac{\partial}{\partial z_0}} f_\alpha(z) \, d^3z \quad (190b)$$

$$= i \lim_{z_0 \to -\infty} \int d^3z \, \{(\Psi, S \cdot T(x_1, \cdots x_n, z) \, \Phi) - (\Psi, S\phi_{\text{out}}(z) \, T(x_1, \cdots x_n) \, \Phi)\} \overleftrightarrow{\frac{\partial}{\partial z_0}} f_\alpha(z) \quad (190c)$$

In arriving at Eq. (190c), we have made use of the fact that z_0 is earlier than the times $x_{10}, \cdots x_{n0}$ so that $\phi(z)$ can be inserted inside the T product. We have also used the fact that $\phi_{\text{in}}(z) \, S = S\phi_{\text{out}}(z)$. Continuing the derivation by again writing

$$\int_{-\infty}^{+\infty} dz_0 \frac{\partial}{\partial z_0} F(z) = \lim_{z_0 \to +\infty} F(z) - \lim_{z_0 \to -\infty} F(z) \quad (191)$$

we obtain

$$(\Psi, [S \cdot T(x_1, \cdots x_n), \phi_{\text{in}}^{\alpha *}] \, \Phi) = -i \int d^4z \frac{\partial}{\partial z_0} \left\{ (\Psi_0, S \cdot T(x_1, \cdots x_n, z) \, \Phi) \overleftrightarrow{\frac{\partial}{\partial z_0}} f_\alpha(z) \right\} \quad (192a)$$

$$= +i \int d^4z \, f_\alpha(z) \, K_z(\Psi, S \cdot T(x_1, \cdots x_n, z) \, \Phi) \quad (192b)$$

QED.

In (192a) the terms $\lim_{z_0 \to +\infty} S \cdot T(x_1, \cdots x_n, z)$ and $\lim_{z_0 \to +\infty} S\phi_{\text{out}}(z) \, T(x_1, \cdots x_n)$ cancel by virtue of the definition of the T product and the asymptotic limit. The term $\frac{\partial}{\partial z_0}\left\{ S\phi_{\text{out}}(z) \, T(x_1, \cdots x_n) \overleftrightarrow{\frac{\partial}{\partial z_0}} f_\alpha(z) \right\}$ likewise does not contribute since both $\phi_{\text{out}}(z)$ and $f_\alpha(z)$ satisfy the Klein-Gordon equation. By a now familiar procedure, one deduces Eq. (189) from (192b).

Again, if the in-fields are irreducible, we can expand the S matrix with respect to these operators as follows:

$$S = \sum_{n=0}^{\infty} \frac{(-i)^n}{n!} \int d^4x_1 \cdots \int d^4x_n\, \sigma_n(x_1, \cdots x_n) : \phi_{\text{in}}(x_1) \cdots \phi_{\text{in}}(x_n) : \tag{193a}$$

or in momentum space

$$S = \sum_{n=0}^{\infty} \frac{(-i)^n}{n!} \int d^4k_1 \cdots \int d^4k_n\, \tilde{\sigma}(k_1, \cdots k_n)$$
$$\delta(k_1{}^2 - \mu^2)\, \delta(k_2{}^2 - \mu^2) \cdots \delta(k_n{}^2 - \mu^2) : \tilde{\phi}_{\text{in}}(k_1) \cdots \tilde{\phi}_{\text{in}}(k_n) : \tag{193b}$$

By taking repeated commutators of S with $\phi_{\text{in}}(z_1), \cdots, \phi_{\text{in}}(z_n)$, etc. and taking vacuum expectation values, we deduce that

$$\tilde{\sigma}(k_1, \cdots k_n) = \epsilon(k_1) \cdots \epsilon(k_n)\, (\Psi_0, [\cdots [S, \tilde{\phi}_{\text{in}}{}^*(k_1)] \cdots \tilde{\phi}_{\text{in}}{}^*(k_n)]\, \Psi_0) \tag{194}$$

Using Eq. (192b), and transforming back to co-ordinate space, we finally get

$$S = \sum_{n=0}^{\infty} \frac{(-i)^n}{n!} \int d^4x_1 \cdots \int d^4x_n$$
$$K_{x_1} \cdots K_{x_n} (\Psi_0, T(x_1, \cdots x_n)\, \Psi_0) : \phi_{\text{in}}(x_1) \cdots \phi_{\text{in}}(x_n) : \tag{195}$$

so that

$$\sigma_n(x_1, \cdots x_n) = K_{x_1} \cdots K_{x_n} \tau(x_1, \cdots x_n) \tag{196}$$

The relation which is the analogue of Eq. (151) for the retarded product is here expressed in terms of the chronological and antichronological product. The antichronological operator $T^\dagger$ can be defined by

$$T^\dagger(\phi(x_1) \cdots \phi(x_n))$$
$$= \sum_P \theta(x_{i_1} - x_1)\, \theta(x_{i_2} - x_2) \cdots \theta(x_{i_n} - x_n)\, \phi(x_{i_1})\, \phi(x_{i_2}) \cdots \phi(x_{i_n}) \tag{197}$$

The desired relation is then

$$\sum_{k=0}^{n} \frac{(-i)^n (-1)^k}{k!(n-k)!} T^\dagger(x_1, \cdots x_k)\, T(x_{k+1}, \cdots x_n) = 0 \tag{198}$$

By taking the vacuum expectation value of Eq. (198), we obtain the coupled integral equations which characterize the τ and $\tau^\dagger$ functions

$$0 = \tau(x_1, \cdots x_n) + \tau^\dagger(x_1, \cdots x_n) + \sum_{i_1 \cdots i_n} \sum_{k=1}^{n-1} \sum_{l=0}^{\infty} \frac{i^l}{l!k!(n-k)!}$$
$$\int d^4u_1 \cdots \int d^4u_l \int d^4v_1 \cdots \int d^4v_l\, K_{u_1} K_{u_2} \cdots K_{u_l} \tau(x_1, \cdots x_k, u_1, \cdots u_l)$$
$$\Delta^{(+)}(u_1 - v_1) \cdots \Delta^{(+)}(u_l - v_l)\, K_{v_1} \cdots K_{v_l} \tau^\dagger(x_{k+1}, \cdots x_n, v_1, \cdots v_l) \tag{199}$$

These coupled equations are the "generalized unitary relations" for the τ functions. Their connection with the unitarity of the S matrix is made apparent by substituting into the equations $SS^* = S^*S = 1$ the expansion (195) for S and introducing into the resulting equations the completeness relation

$$\sum_{|\alpha>} |\alpha>_{\text{in in}}<\alpha| = 1$$

Similar remarks as those made in connection with the r functions can be made in connection with the τ functions [Nishijima (1960)]. The r functions are somewhat more convenient since the locality requirement on the operators is translated into a simple property of the r function: it is an invariant retarded function. The locality property is not so easily translatable into a linear property of the τ function. Nonetheless their close connection to the S matrix has made them the subject of intensive study. In perturbation theory, for a renormalizable field theory, the Fourier transform $\tilde{\tau}$ of the τ functions

$$\tilde{\tau}(p_1, \cdots p_n) = \frac{1}{(2\pi)^{4n}} \int d^4x_1 \cdots \int d^4x_n\, e^{-i\sum_{j=1}^{n} p_j \cdot x_j} \tau(x_1, \cdots x_n) \quad (200)$$

contains singularities of the form $\delta(p)$, $\delta(p^2 - m^2)$ and $P(p^2 - m^2)^{-1}$ in any variable $p = \sum_{\nu}^{r \leq n} p_{i_\nu}$. The singularities $\delta(p)$ arise from intermediate vacuum states; the singularities $\delta(p^2 - m^2)$ and $P(1/p^2 - m^2)$ from one-particle states of discrete mass m. We have seen that these one-particle singularities play an important role in scattering theory and that their residues are related to the coupling constants [Chew and Low (Chew 1959)]. The one-particle singularities of τ functions have been investigated in detail within the framework of the LSZ formulation of field theory by Zimmermann (1959). While the one-particle poles occur frequently in the nonphysical region of the momentum variables [e.g., in the π-N case the pole occurs at $(p \pm k)^2 = M^2$], they can correspond to physical conditions for physical momenta in the case of many-particle processes. The occurrence of these one-particle poles is then related to the requirements imposed by causality on such processes.

Finally, we conclude this section by noting that the formalism outlined in the present section can be extended to physical situations where one has stable bound states or composite particles [Zimmermann (1958), Nishijima (1958), Baumann (1958)]. In such an extension one defines a field operator for each composite particle or bound state. The masses and other characteristics of these composite systems enter the theory as given phenomenological constants. The operator for each composite system can be expressed explicitly in terms of the original Heisenberg field operator and satisfies an asymptotic condition appropriate to the particle it describes.

Reduction formulae which are the generalization of Eqs. (154) and (189) above can be derived and an S matrix obtained which includes in its description the production and annihilation of composite or bound systems.

18c. Integral Representations of a Causal Commutator

We now briefly interrupt our presentation of the axiomatic formulation of relativistic field theories to present the work of Jost and Lehmann [Jost (1957)] and of Dyson (1958) on the representation of the expectation value of a causal commutator between arbitrary physical states. This constitutes the generalization of the Lehmann representation for the vacuum expectation value of the twofold presented in Chapter 17 and in Section a of the present chapter. This representation will prove to be very useful in the analysis of the analytic properties of scattering amplitudes to be presented in Section d of the present chapter.

We are interested in obtaining a representation of the distribution

$$F_{PQ}(x, x') = \langle P\alpha \mid [A(x), B(x')] \mid Q\beta\rangle \tag{201}$$

which encompasses as many of the properties of the matrix element which follow from relativistic invariance, locality, and spectral conditions as possible. The result we will derive is a general one and will not depend on any particular Lagrangian, nor on any approximation method (such as perturbation theory). For simplicity, we take the fields A and B to be scalar fields

$$U(a, \Lambda)\, A(x)\, U(a, \Lambda)^{-1} = A(\Lambda x + a) \tag{202a}$$

$$U(a, \Lambda)\, B(x)\, U(a, \Lambda)^{-1} = B(\Lambda x + a) \tag{202b}$$

and by assumption the fields are local with respect to each other, that is

$$[A(x), B(x')] = 0 \quad \text{when } (x - x')^2 < 0 \tag{203}$$

so that $F_{PQ}(x, x')$ vanishes for $(x - x')^2 < 0$. Incidentally, the vectors P, Q which characterize in part the states $|P\alpha\rangle$ and $|Q\beta\rangle$ are the momenta of physical states and hence lie in the forward light cone. By translation invariance, we derive that

$$\begin{aligned}\langle P, \alpha \mid [A(x), B(x')] \mid Q\beta\rangle &\\ &= \langle P\alpha \mid U^{-1}(a, I)\, U(a, I)\, [A(x), B(x')]\, U^{-1}(a, I)\, U(a, I) \mid Q\beta\rangle \\ &= e^{-i(Q-P)\cdot a} \langle P\alpha \mid [A(x + a), B(x' + a)] \mid Q\beta\rangle \end{aligned} \tag{204}$$

so that

$$F_{PQ}(x, x') = e^{-i(Q-P)\cdot a} F_{PQ}(x + a, x' + a) \tag{205}$$

for arbitrary a. With the choice $a = -(x + x')/2$, one finds that

$$F_{PQ}(x, x') = e^{+i(Q-P)\cdot\frac{x+x'}{2}} F_{PQ}\left(\frac{x - x'}{2}, -\frac{x - x'}{2}\right) \tag{206}$$

It is therefore sufficient to consider the matrix element

$$f_{PQ}(x) = \langle P\alpha \,|\, \left[A\left(\frac{x}{2}\right), B\left(-\frac{x}{2}\right)\right] \,|\, Q\beta\rangle \tag{207}$$

which has the property that

$$f_{PQ}(x) = 0 \quad \text{for } x^2 < 0 \tag{208}$$

due to the locality of the operators. In the following we shall abbreviate f_{PQ} by f. We can further decompose f as follows:

$$f(x) = f_1(x) - f_2(x) \tag{209}$$

with

$$\begin{aligned} f_1(x) &= \langle P\alpha \,|\, A\left(\frac{x}{2}\right) B\left(-\frac{x}{2}\right) \,|\, Q\beta\rangle \\ &= \sum_{|K\delta\rangle} \langle P\alpha \,|\, A(0) \,|\, K\delta\rangle \langle K\delta \,|\, B(0) \,|\, Q\beta\rangle\, e^{-iK\cdot x}\, e^{i(Q+P)\cdot\frac{x}{2}} \end{aligned} \tag{210}$$

and

$$\begin{aligned} f_2(x) &= \langle P\alpha \,|\, B\left(-\frac{x}{2}\right) A\left(\frac{x}{2}\right) \,|\, Q\beta\rangle \\ &= \sum_{|K\delta\rangle} \langle P\alpha \,|\, B(0) \,|\, K\delta\rangle \langle K\delta \,|\, A(0) \,|\, Q\beta\rangle\, e^{+iK\cdot x}\, e^{-i(Q+P)\cdot\frac{x}{2}} \end{aligned} \tag{211}$$

where $|K\delta\rangle$ denotes a complete set of physical states. By carrying out the sum over all states with fixed momentum K, and introducing the notation

$$G^{(1)}_{PQ}(K) = \sum_{\delta} \langle P\alpha \,|\, A(0) \,|\, K\delta\rangle \langle K\delta \,|\, B(0) \,|\, Q\beta\rangle \tag{212a}$$

$$G^{(2)}_{PQ}(K) = \sum_{\delta} \langle P\alpha \,|\, B(0) \,|\, K\delta\rangle \langle K\delta \,|\, A(0) \,|\, Q\beta\rangle \tag{212b}$$

we can then write f_1 and f_2 as follows:

$$\begin{aligned} f_1(x) &= \int d^4K\, e^{-iK\cdot x}\, e^{i(Q+P)\cdot\frac{x}{2}}\, G^{(1)}_{PQ}(K) \\ &= \int d^4K\, e^{-iK\cdot x}\, G^{(1)}_{PQ}(K + \tfrac{1}{2}(Q+P)) \end{aligned} \tag{213a}$$

$$f_2(x) = \int d^4K\, e^{-iK\cdot x}\, G^{(2)}_{PQ}(-K + \tfrac{1}{2}(Q+P)) \tag{213b}$$

Due to the spectral conditions on the physical states, $G^{(1)}(K)$ and $G^{(2)}(K)$ vanish unless K lies in the forward cone. If we introduce the Fourier transforms $\tilde{f}_1(q)$ and $\tilde{f}_2(q)$ of $f_1(x)$ and $f_2(x)$,

$$f_1(x) = \int e^{-iq\cdot x} \tilde{f}_1(q)\, d^4q \tag{214a}$$

$$f_2(x) = \int e^{-iq\cdot x} \tilde{f}_2(q)\, d^4q \tag{214b}$$

then by comparing Eq. (214a) with (213a), we deduce that $f_1(q)$ vanishes unless $\frac{1}{2}(Q + P) + q$ is the momentum of a physical state, i.e., unless $\frac{1}{2}(Q + P) + q$ lies in the forward cone, and similarly, comparing (214b) with (213b), that $f_2(q)$ vanishes unless $\frac{1}{2}(Q + P) - q$ lies in the forward cone. More specifically $f_1(q)$ will vanish unless $\frac{1}{2}(P + Q) + q$ is the energy-momentum vector of a state $|n_1\rangle$, for which

$$\langle P\alpha \mid A(0) \mid n_1\rangle \neq 0, \qquad \langle n_1 \mid B(0) \mid Q\beta\rangle \neq 0 \tag{215}$$

and $f_2(q) = 0$ unless $\frac{1}{2}(P + Q) - q$ is the momentum vector of a state $|n_2\rangle$ with

$$\langle P\alpha \mid B(0) \mid n_2\rangle \neq 0, \qquad \langle n_2 \mid A(0) \mid Q\beta\rangle \neq 0 \tag{216}$$

Suppose that the least massive states for which Eq. (215) holds have a mass m_1 and those for which Eq. (216) holds have a mass not less than m_2, then in the co-ordinate system in which

$$\tfrac{1}{2}(P + Q) = (a, 0, 0, 0) \tag{217}$$

$f_1(q) \neq 0$ for all $q = (q_0, \mathbf{q})$ satisfying

$$(q_0 + a) \geqslant 0 \tag{218a}$$

$$(q_0 + a)^2 - \mathbf{q}^2 \geqslant m_1^2 \tag{218b}$$

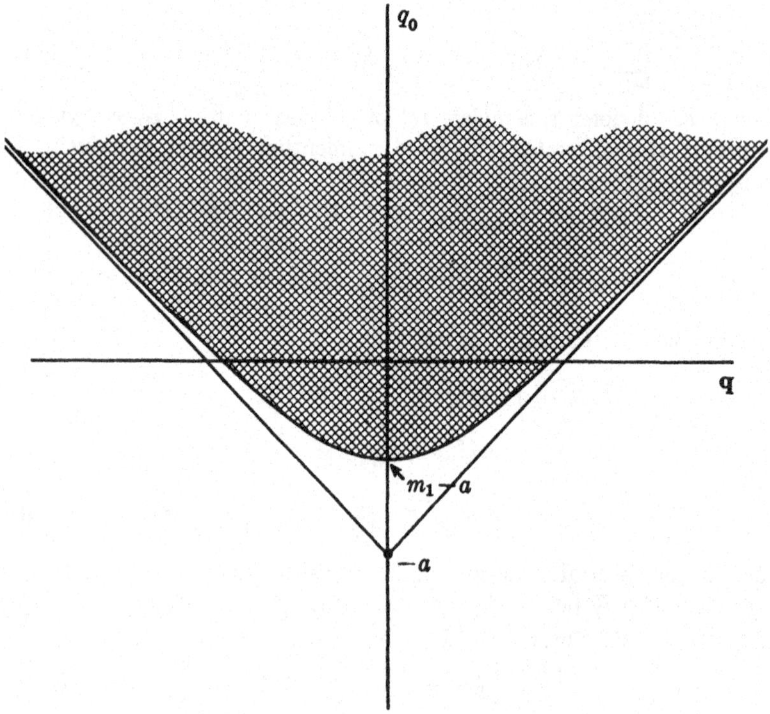

Fig. 18.2

i.e., inside the crosshatched hyperboloid of Figure 18.2. Alternatively, $\tilde{f}_1(q) = 0$ for all $q = (q_0, \mathbf{q})$ satisfying $q_0 < -a + \sqrt{\mathbf{q}^2 + m_1^2}$. Similarly, $\tilde{f}_2 \neq 0$ for all q satisfying

$$a - q_0 \geqslant 0 \tag{219a}$$

$$(a - q_0)^2 - \mathbf{q}^2 \geqslant m_2^2 \tag{219b}$$

or equivalently $\tilde{f}_2 = 0$ for those q satisfying $q_0 > a - \sqrt{\mathbf{q}^2 + m_2^2}$. Thus due to the spectral conditions, the Fourier transform $\tilde{f}(q) = \tilde{f}_1(q) - \tilde{f}_2(q)$ vanishes for all q satisfying

$$a - \sqrt{\mathbf{q}^2 + m_2^2} < q_0 < -a + \sqrt{\mathbf{q}^2 + m_1^2} \tag{220}$$

that is outside the two hyperboloids indicated in Figure 18.3. These hy-

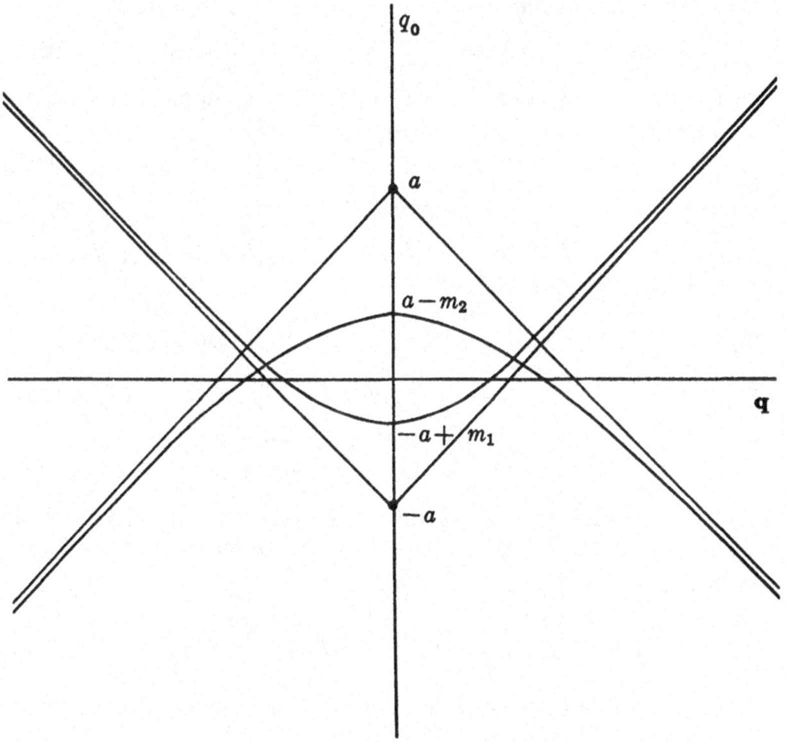

Fig. 18.3

perboloids will overlap if $a > \dfrac{m_1 + m_2}{2}$. (We have assumed $a > \frac{1}{2}|m_1 - m_2|$ which is satisfied in applications.)

The solution of the mathematical problem of finding the most general function $f(x)$ which is zero outside the light cone $x^2 \geqslant 0$, while its Fourier transform vanishes in some set R of momentum space, has been given by Dyson (1958). We next present his solution, which is based on the idea of going to a six-dimensional hyperbolic space and finding a connection between functions $f(x)$ which vanish for $x^2 < 0$ in four-space and functions defined on the light cone in six-space.[15] The representation obtained by Dyson is a nontrivial generalization of a representation by Jost and Lehmann [Jost (1957)] for a function vanishing outside the light cone but whose support in momentum space is symmetric about q_0, i.e., in the case $m_1 = m_2$.

Let z denote the six-vector ($z_0 = x_0$, $z_1 = x_1$, $z_2 = x_2$, $z_3 = x_3$, $z_4 = y_1$, $z_5 = y_2$) and r the six-vector $r_0 = q_0, r_1 = q_1, r_2 = q_2, r_3 = q_3, r_4 = p_1, r_5 = p_2$. The metric in the six-dimensional z space is defined so that

$$z \cdot z = z^2 = x^2 - y^2 = x_0{}^2 - x_1{}^2 - x_2{}^2 - x_3{}^2 - y_1{}^2 - y_2{}^2 \tag{221}$$

Consider now the function $F(z)$ defined from a four-dimensional function $f(x)$ as follows:

$$F(z) = 4\pi f(x)\, \delta(x^2 - y^2) \tag{222a}$$

$$= 4\pi f(x)\, \delta(z^2) \tag{222b}$$

As indicated by Eq. (222b), $F(z)$ is only defined on the light cone of the six-dimensional z space. Since

$$\iint_{-\infty}^{+\infty} dy_1\, dy_2\, F(z) = 4\pi f(x) \cdot 2\pi \int_0^\infty dy^2\, \tfrac{1}{2}\delta(x^2 - y^2)$$

$$= 4\pi^2 f(x)\, \theta(x^2) \tag{223a}$$

$$= \begin{cases} 4\pi^2 f(x) & \text{for } x^2 \geqslant 0 \\ 0 & \text{for } x^2 < 0 \end{cases} \tag{223b}$$

we note that for a function $f(x)$ which vanishes for $x^2 < 0$, the knowledge of $F(z)$ and $f(x)$ are equivalent since $f(x)$ can be recovered from $F(z)$ by integration:

$$f(x) = \frac{1}{4\pi^2} \int_{-\infty}^{+\infty} F(z)\, dy_1\, dy_2 = \frac{1}{4\pi} \int_0^\infty F(z)\, dy^2 \tag{224}$$

In the following we shall denote the special six-vector $(q_0, q_1, q_2, q_3, 0, 0)$ by $\hat{q}$ and the Fourier transform of $F(z)$ by $\tilde{F}(r)$

$$\tilde{F}(r) = \frac{1}{(2\pi)^6} \int e^{ir \cdot z}\, F(z)\, d^6 z \tag{225}$$

[15] Actually the number of dimensions that are added is immaterial, but the existence of a Huygens principle in spaces with an odd number of spatial dimensions makes the addition of two more spatial co-ordinates attractive.

Upon introducing into (225) the form (222b) for $F(z)$ we obtain

$$\begin{aligned}\tilde{F}(r) &= \frac{4\pi}{(2\pi)^6}\int d^6z\, e^{ir\cdot z}\,\delta(z^2)\int d^4q\, e^{-i\hat{q}\cdot z}\tilde{f}(q)\\ &= \frac{4\pi}{(2\pi)^6}\iint d^6z\, e^{i(r-\hat{q})\cdot z}\,\delta(z^2)\,\tilde{f}(q)\,d^4q\\ &= \int D^{(1)}(r-\hat{q})\,\tilde{f}(q)\,d^4q \end{aligned}\tag{226}$$

where

$$D^{(1)}(r) = D^{(1)}(r^2) = \frac{2}{(2\pi)^5}\int e^{-ir\cdot z}\,\delta(z^2)\,d^6z \tag{227a}$$

$$= \frac{1}{\pi^3}\,\mathrm{P}\,\frac{1}{(r^2)^2} \quad \text{(P denotes principal value)} \tag{227b}$$

is the even invariant function in six dimensions. Combining (227b) and (226), we finally deduce that

$$\begin{aligned}\tilde{F}(r) &= \frac{1}{\pi^3}\int d^4q\,\frac{\tilde{f}(q)}{[(r-\hat{q})^2]^2} = \frac{1}{\pi^3}\int d^4q\,\frac{\tilde{f}(q)}{[(u-q)^2-s]^2}\\ &= \tilde{F}(u,s)\end{aligned}\tag{228}$$

where $r = (r_0, r_1, r_2, r_3, r_4, r_5) = (u_0, u_1, u_2, u_3, p_4, p_5)$ and $s = p_1^2 + p_2^2$. Note, therefore, that the Fourier transform $\tilde{F}(r)$ of a function $F(z) = f(x)\,\delta(z^2)$ whose support is on the light cone in z space, is rotationally invariant in the r_4-r_5 plane (i.e., depends only on $s = r_4^2 + r_5^2$). Furthermore since $D^{(1)}(r)$ satisfies the wave equation in r space

$$\Box_6 D^{(1)}(r) = 0 \tag{229a}$$

$$\Box_6 = \frac{\partial^2}{\partial r_0^2} - \sum_{j=1}^{5}\frac{\partial^2}{\partial r_j^2} \tag{229b}$$

[as is readily seen from the representation (227a), by virtue of Eq. (226)], $\tilde{F}(r)$ clearly also satisfies the wave equation in six dimensions

$$\Box_6 \tilde{F}(r) = 0 \tag{230}$$

Finally we note that, if $f(x)$ vanishes for $x^2 < 0$, then $\tilde{f}(q)$ is the boundary value on the $s = 0$ plane of $\tilde{F}(r)$, i.e., $\tilde{F}(\hat{q}) = \tilde{f}(q)$.

Proof:

$$\tilde{F}(\hat{q}) = \frac{1}{(2\pi)^6}\int e^{i\hat{q}\cdot z}\,4\pi\delta(x^2 - y^2)\,f(x)\,d^6z \tag{231a}$$

$$= \frac{2\pi}{(2\pi)^6}\cdot\frac{1}{2}\int e^{iq\cdot x}\,4\pi\theta(x^2)\,f(x)\,d^4x \tag{231b}$$

and the right-hand side of (231b) is equal to $\tilde{f}(q)$ if, and only if, $f(x)$ vanishes for $x^2 < 0$. Thus the solutions of the integral equation

$$\tilde{f}(q) = \int d^4q' \frac{\tilde{f}(q')}{[(q - q')^2]^2} \tag{232}$$

constitute the class C of functions whose Fourier transforms $f(x)$ vanish for $x^2 < 0$.

We have therefore shown that a necessary condition for $f(x)$ to vanish outside the light cone is that $\tilde{f}(q)$ be the boundary value on the $s = 0$ plane of a solution $\tilde{F}(q, s)$ of the wave equation $\Box_6 \tilde{F}(r) = 0$ which is rotationally symmetric in the 4-5 plane. It should be noted, incidentally, that the $s = 0$ plane is a *time-like* surface, so that the boundary value of a solution of the hyperbolic equation $\Box_6 \tilde{F}(r) = 0$ is not arbitrary on this surface.

Conversely, consider a function $\tilde{F}(r)$ which satisfies the wave equation $\Box_6 \tilde{F}(r) = 0$ and which is rotationally invariant in the 4-5 plane. Its Fourier transform

$$F(z) = \int e^{-ir\cdot z} \tilde{F}(r)\, d^6r \tag{233}$$

then has the following properties: By virtue of the fact that $\Box_6 \tilde{F}(r) = 0$, $F(z) = \delta(z^2)\, G(z)$, i.e., $F(z)$ has its support on the light cone in z space. Furthermore, due to the assumed symmetry of $\tilde{F}(r)$ in the 4-5 plane

$$\begin{aligned} F(z) &= \int e^{-ir\cdot z} \tilde{F}(u, |p|)\, d^6r \\ &= \int d^4u\, e^{-iu\cdot x} \int_0^\infty dp\, p \int_0^{2\pi} d\theta\, e^{ip|y|\cos\theta} \tilde{F}(u, p) \\ &= 2\pi \int d^4u\, e^{-iu\cdot x} \int_0^\infty ds\, J_0(\sqrt{s}\,|y|)\, \tilde{F}(u, s) \end{aligned} \tag{234}$$

Since

$$J_0(\sqrt{s}\,|y|) = \sum_{n=0}^{\infty} \frac{(sy^2)^n}{(n!)^2} \tag{235}$$

is a function of y^2 only, it follows that $F(z)$ is a function only of x and y^2, i.e.,

$$F(z) = \delta(z^2)\, G_1(x, y^2) = \delta(x^2 - y^2)\, G_1(x, y^2) \tag{236}$$

Since $F(z)$ has its support on the light cone, $y^2 = x^2$, it follows that $F(z)$ must be of the form

$$\begin{aligned} F(z) &= \delta(x^2 - y^2)\, f(x) \\ &= \delta(z^2)\, f(x) \end{aligned} \tag{237}$$

[$f(x)$, however, need not vanish for $x^2 < 0$]. If one further requires that $\tilde{F}(\hat{q}) = \tilde{f}(q)$, then $f(x)$ will vanish outside the light cone.

Hence a necessary and sufficient condition for a function $f(x)$ to vanish outside the light cone is that $\tilde{f}(q)$ be the boundary value on the surface

$s = 0$ of a solution of $\square_6 \tilde{F}(r) = 0$ which is rotationally invariant in the 4-5 plane. Now a solution of the wave equation $\square_6 \tilde{F}(r) = 0$ can be written in terms of its boundary value and normal derivative on a *space-like* surface with the help of the singular function $D(r)$ which satisfies the homogeneous wave equations in six dimensions

$$\square_6 D(r) = 0 \tag{238}$$

and the initial conditions

$$D(r_0 = 0, r_1, \cdots r_5) = 0 \tag{239a}$$

$$\left.\frac{\partial D(r)}{\partial r_0}\right|_{r_0=0} = \prod_{i=1}^{5} \delta(r_i) \tag{239b}$$

Explicitly,

$$\begin{aligned} D(r) &= \frac{1}{(2\pi)^6} \int d^6z\, e^{-ir\cdot z}\, \epsilon(z)\, \delta(z^2) \\ &= \frac{-2i}{(2\pi)^3} \left(\frac{1}{\rho^3} - \frac{1}{\rho^2}\frac{d}{d\rho}\right) \int_0^\infty d\zeta \sin r_0\zeta \sin \rho\zeta \\ &= \frac{1}{2\pi^2}\, \epsilon(r_0)\, \delta'(r^2) \end{aligned} \tag{240}$$

If Σ is the space-like surface on which the initial data is prescribed, then a solution of the wave equation $\square_6 \tilde{F}(r) = 0$ which takes on the prescribed values $\tilde{F}(r')$ and $(\partial F(r')/\partial r'_\alpha)\, n^\alpha(r')$ on Σ, where n^α is the normal to Σ is given by

$$\tilde{F}(r) = \int_\Sigma d\Sigma_\alpha \left[\tilde{F}(r'), \frac{\partial}{\partial r'_\alpha} D(r' - r)\right] \tag{241}$$

where

$$\left[\tilde{F}, \frac{\partial}{\partial r'_\alpha} D\right] = F \frac{\partial D}{\partial r'_\alpha} - \frac{\partial D}{\partial r'_\alpha} F \tag{242}$$

and $d\Sigma_\alpha$ is the surface element ($d\Sigma_\alpha$ is a six-vector normal to the space-like surface Σ). If any function $\tilde{F}(r')$ is inserted on the right-hand side of (241), one obtains a solution of the wave equation on the left-hand side whose value and normal derivative on Σ are $\tilde{F}(r')$ and $(n^\alpha \partial/\partial r'^\alpha)\, \tilde{F}(r')$. A solution $\tilde{F}(r)$ with specified symmetry properties can therefore be obtained by suitably choosing Σ and the boundary values in (241). By setting $\tilde{F}(\hat{q}) = \tilde{f}(q)$ we then obtain the following integral representation for the Fourier transform $\tilde{f}(q)$ of a function $f(x)$ which vanishes outside the light cone:

$$\begin{aligned} \tilde{f}(q) &= \int_\Sigma d\Sigma'_\alpha \left[\tilde{F}(r'), \frac{\partial}{\partial r'_\alpha} D(r' - \hat{q})\right] \\ &= \frac{1}{2\pi^2} \int_\Sigma d\Sigma_\alpha \left[\tilde{F}(r), \frac{\partial}{\partial r_\alpha} \cdot \epsilon(u_0 - q_0)\, \delta'((u - q)^2 - s)\right] \end{aligned} \tag{243}$$

which is the basis of the discussion which follows. The representation (243) is unique in the sense that if a function $\tilde{f}(q)$ [for which $f(x) = 0$ for $x^2 < 0$] and a surface Σ are given, then if $\tilde{f}(q)$ has a representation (243) with any function $\tilde{F}(r) = \tilde{F}(u, s)$ which depends on r_4, r_5 only in the combination $s = r_4^2 + r_5^2$ and which satisfies $\Box_6 \tilde{F}(r) = 0$, then $\tilde{F}(r)$ is identical with the function defined by (228), i.e.,

$$\tilde{F}(r) = \int D^{(1)}(r - \hat{q})\, \tilde{f}(q)\, d^4q \tag{244}$$

We have therefore established the one-to-one correspondence between functions $\tilde{f}(q)$ of the class C and solutions $\tilde{F}(r)$ of the wave equation in six dimensions with rotational symmetry in the 4-5 plane.

The problem is next to apply the above representation (243) [which guarantees that $f(x) = 0$ for $x^2 < 0$] so that by an appropriate choice of $\tilde{F}(u, s)$ and Σ we can obtain a representation for a function $\tilde{f}(q)$ with specified support conditions in momentum space.

Suppose that $\tilde{f}(q) = 0$ in a region R of q space which is bounded by two space-like surfaces σ_1 and σ_2. More specifically, let the region R be defined by

$$R : s_1(\mathbf{q}) < q_0 < s_2(\mathbf{q}) \tag{245}$$

where s_1 and s_2 are functions satisfying the inequalities

$$|s_1(\mathbf{q}) - s_1(\mathbf{q}')| < |\mathbf{q} - \mathbf{q}'| \tag{246a}$$

$$|s_2(\mathbf{q}) - s_2(\mathbf{q}')| < |\mathbf{q} - \mathbf{q}'| \tag{246b}$$

which guarantee that the surfaces $q^0 = s_1(\mathbf{q})$ and $q^0 = s_2(\mathbf{q})$ are space-like. We denote by C_R the class of functions such that $f(x) = 0$ for $x^2 < 0$ and such that $\tilde{f}(q) = 0$ for q in R.

With Dyson, we shall call a hyperboloid

$$(q - u)^2 - s = 0 \tag{247}$$

in q space *admissible*, if its upper sheet does not come below σ_2 and its lower sheet above σ_1. Such hyperboloids correspond to points $r = (u_0, u_1, u_2, u_3, p_1, p_2)$, $s = p_1^2 + p_2^2$ lying in a certain region S of r space. Now, since for every r in S but q in R the function $D(r - \hat{q}) = (2\pi^2)^{-1}\epsilon(u_0 - q_0)\, \delta'((u - q)^2 - s)$ vanishes, this suggests that one should look for a representation of an $\tilde{f}(q)$ in C_R by a representation (243) with the points r confined to S. Every $\tilde{f}(q)$ defined by (243) with the points r on Σ in S only, clearly belongs to C_R. Dyson has shown that the converse is also true, namely that every $\tilde{f}(q)$ in C_R has a representation using only admissible hyperboloids. Now to be admissible the hyperboloid $(q - u)^2 - s = 0$ must not cross the surface defined by $q_0 = s_1(\mathbf{q})$ and $q_0 = s_2(\mathbf{q})$. Consider the upper sheet of the hyperboloid, that is, the branch $q_0 = u_0 + \sqrt{(\mathbf{q} - \mathbf{u})^2 + s}$. It will not cross σ_2 if for a fixed $\mathbf{q}$

$$u_0 + \sqrt{(\mathbf{q} - \mathbf{u})^2 + s} \geqslant s_2(\mathbf{q}) \tag{248}$$

and in general if

$$u_0 \geqslant \operatorname*{Max}_{\mathbf{q}} \{s_2(\mathbf{q}) - \sqrt{(\mathbf{q} - \mathbf{u})^2 + s}\} = m(\mathbf{u}, s) \tag{249}$$

Similarly the lower sheet $q_0 = u_0 - \sqrt{(\mathbf{q} - \mathbf{u})^2 + s}$ will not cross σ_1 if

$$u_0 \leqslant \operatorname*{Min}_{\mathbf{q}} \{s_1(\mathbf{q}) + \sqrt{(\mathbf{q} - \mathbf{u})^2 + s}\} = M(\mathbf{u}, s) \tag{250}$$

The region S in r space is therefore defined by

$$S : m(\mathbf{u}, s) \leqslant u_0 \leqslant M(\mathbf{u}, s) \tag{251}$$

and is bounded by two surfaces Σ_1 and Σ_2 in r space. Since these surfaces are envelopes of two families of hyperboloids, they are also space-like. Let us call T the complement of S, i.e., set of points in r space such that

$$M(\mathbf{u}, s) < u_0 < m(\mathbf{u}, s) \tag{252}$$

In order that the representation (243) yield an $\tilde{f}(q)$ in C_R, Dyson (1958) has shown that $\tilde{F}(r)$ must vanish for each r in T.

Suppose that in (243) one chooses for Σ a space-like surface lying between Σ_1 and Σ_2, e.g., the surface

$$u_0 = \tfrac{1}{2}[m(\mathbf{u}, s) + M(\mathbf{u}, s)] \tag{253}$$

Since, by virtue of Eqs. (252) and (253), every point of Σ belongs either to S or T, but since $\tilde{F}(r)$ must vanish for r in T, a function $\tilde{f}(q)$ belongs to C_R if, and only if, it admits the unique representation

$$\tilde{f}(q) = \int_{\Sigma(S)} d\Sigma_\alpha \left[\tilde{F}(r), \frac{\partial}{\partial r_\alpha} D(r - \hat{q}) \right] \tag{254}$$

with the integral extending only over those points r of the space-like surface Σ which belong to the set S, the set S being defined by Eqs. (249), (250) and (251). If one does not care about the uniqueness of the representation (i.e., two different $\tilde{F}(r)$ may give rise to the same $\tilde{f}(q)$), which is the case in most applications, then the following theorem answers the general problem posed at the beginning of this section:

Theorem: For a function $\tilde{f}(q)$ to vanish in the region

$$s_1(\mathbf{q}) < q_0 < s_2(\mathbf{q})$$

and to have a Fourier transform $f(x)$ vanishing outside the light cone, it is necessary and sufficient that it have a representation

$$\tilde{f}(q) = \int d^4u \int_0^\infty ds\, \epsilon(q_0 - u_0)\, \delta[(q - u)^2 - s]\, \Phi(u, s) \tag{255}$$

with $\Phi(u, s)$ vanishing outside the region S defined by Eqs. (249), (250), and (251), but arbitrary otherwise.

In other words, putting into the representation (255) a $\Phi(u, s)$ which vanishes unless the points u, s are such that the qs determined by $(u - q)^2 = s$ are entirely in R, but $\Phi(u, s)$ is arbitrary otherwise, reproduces on the left-hand side of (255) a function $\tilde{f}(q)$ with the desired support properties in both momentum and configuration space. An insight into the representation (255) is obtained by noting that its Fourier transform is

$$f(x) = \int_0^\infty ds\, \Delta(x; s)\, \tilde{\Phi}(x; s) \tag{256}$$

where $\tilde{\Phi}(x; s)$ is the Fourier transform of $\Phi(u, s)$ with respect to u, and $\Delta(x; s)$ is the odd invariant function for the mass $\sqrt{s}$ which vanishes for $x^2 < 0$ and which is equal to $J_0(\sqrt{s(x_0^2 - \mathbf{x}^2)})$ for $x_0 > |\mathbf{x}|$. Equation (256) is then the expansion of the function $f(x)$ in terms of the $\Delta(x; s)$s. (These functions are complete in the time-like direction by integrating over the masses s.) Such an expansion automatically guarantees that $f(x) = 0$ for $x^2 < 0$.

In the particular case that the space-like surfaces σ_1 and σ_2 are those determined by Eq. (220), i.e.,

$$s_1(\mathbf{q}) = a - \sqrt{\mathbf{q}^2 + m_2^2} \tag{257a}$$

$$s_2(\mathbf{q}) = -a + \sqrt{\mathbf{q}^2 + m_1^2} \tag{257b}$$

the region S is defined by Eq. (251) with

$$m(\mathbf{u}, s) = \operatorname*{Max}_{\mathbf{q}} \{\sqrt{\mathbf{q}^2 + m_1^2} - a - \sqrt{(\mathbf{q} - \mathbf{u})^2 + s}\} \tag{258}$$

$$M(\mathbf{u}, s) = \operatorname*{Min}_{\mathbf{q}} \{a - \sqrt{\mathbf{q}^2 + m_2^2} + \sqrt{(\mathbf{q} - \mathbf{u})^2 + s}\} \tag{259}$$

The bracketed expression in Eq. (258) has an extremum when each component of its gradient with respect to $\mathbf{q}$ vanishes. This occurs at

$$\mathbf{q} = \frac{m_1\mathbf{u}}{m_1 - \kappa} \tag{260}$$

where we have denoted $\sqrt{s}$ by κ

$$\kappa = \sqrt{s} \geqslant 0 \tag{261}$$

If $m_1 > \kappa$, $\mathbf{q}$ is parallel to $\mathbf{u}$ and (260) then corresponds to a maximum, so that

$$m(u, \kappa^2) = [\mathbf{u}^2 + (m_1 - \kappa)^2]^{1/2} - a \quad \text{for } m_1 > \kappa > 0 \tag{262}$$

If $\kappa > m_1$, (260) corresponds to a minimum. There are no other extremal points for finite $\mathbf{q}$. A second extremum of the bracketed expression in (258) occurs at $\mathbf{q} = \lambda\mathbf{u}$ as $\lambda \to \infty$. In this limit the bracketed expression in (258) assumes the value $|\mathbf{u}| - a$ which is a maximum provided $\kappa > m_1$. Hence,

$$m(\mathbf{u}, \kappa^2) = |\mathbf{u}| - a \quad \text{for } \kappa > m_1 \tag{263}$$

A similar procedure for Eq. (259) then indicates that for the case that $m_1 < m_2$ the region S is defined by

$$\begin{aligned} \sqrt{\mathbf{u}^2 + (m_1 - \kappa)^2} - a < u_0 < a - \sqrt{\mathbf{u}^2 + (m_2 - \kappa)^2} \quad & \text{for } 0 < \kappa < m_1 \\ |\mathbf{u}| - a < u_0 < a - \sqrt{\mathbf{u}^2 + (m_2 - \kappa)^2} \quad & \text{for } m_1 < \kappa < m_2 \\ |\mathbf{u}| - a < u_0 < a - |\mathbf{u}| \quad & \text{for } \kappa > m_2 \end{aligned} \tag{264}$$

A more useful specification of the region S is one in which the integration limits of u_0 and $|\mathbf{u}|$ are independent of κ and is given by

$$|\mathbf{u}| - a < u_0 < a - |\mathbf{u}|$$
$$|\mathbf{u}| < a$$
$$\kappa \geqslant \text{Max}\ \{0, m_2 - \sqrt{(u_0 - a)^2 - \mathbf{u}^2}, m_1 - \sqrt{(u_0 + a)^2 - \mathbf{u}^2}\} \tag{265}$$

This specification of S is most easily obtained by finding the maximum value of $s = \kappa^2$ for particular $(u_0, \mathbf{u})$ such that the hyperboloid $(q - u)^2 = s$ just touches the space-like surfaces $q_0 = s_1(\mathbf{q})$ and $q_0 = s_2(\mathbf{q})$. In the case s_1 and s_2 are given by Eqs. (257a) and (257b) upon maximizing the expression $(q - u)^2 + \lambda[(q_0 - a)^2 - \mathbf{q} - m_2]$ (λ is a Lagrangian parameter), one finds

$$\begin{aligned} \kappa_{\max} &= \text{Max}\ \{m_2 - \sqrt{(u_0 - a)^2 - \mathbf{u}^2}, 0\} \quad \text{for } |u_0| + |\mathbf{u}| < a \\ &= \infty \quad \text{for } |u_0| + |\mathbf{u}| > a \end{aligned} \tag{266}$$

with $\tilde{F}(r) = 0$ in the region where $\kappa < \kappa_{\max}$. Similarly for s_2; $\tilde{F}(r)$ is then different from zero for $\kappa < \kappa_{\max}$, whence (265). The Lorentz covariant specification of S is clearly given by

$$\left.\begin{aligned} &\left(\frac{P+Q}{2} + u\right) \epsilon\, L^+ \\ &\left(\frac{P+Q}{2} - u\right) \epsilon\, L^+ \\ &\kappa \geqslant \text{Max}\left\{0, m_1 - \sqrt{\left(\frac{P+Q}{2} + u\right)^2}, m_2 - \sqrt{\left(\frac{P+Q}{2} - u\right)^2}\right\} \end{aligned}\right\} \tag{267}$$

where L^+ denotes the forward light cone.

The representation for the retarded product, i.e., for the function

$$f_R(x) = \theta(x)\, f(x) \tag{268}$$

with $f(x) = 0$ for $x^2 < 0$ and $\tilde{f}(q) = 0$ for q in R, is obtained by formally using the representation (70) for $\theta(x)$ and proceeding as in the derivation of Eq. (75). In a formal manner we therefore obtain the following representation for $\tilde{f}_R(q)$, the Fourier transform of $f_R(x)$:

$$\tilde{f}_R(q) = \frac{1}{2\pi i} \int \frac{dq'_0\, \tilde{F}(q'_0, \mathbf{q})}{q'_0 - q_0}; \qquad \text{Im } q_0 > 0 \tag{269}$$

and using (255)

$$\tilde{f}_R(q) = -\frac{1}{2\pi} \int d^4u \int_0^\infty d\kappa^2\, \frac{\Phi(u, \kappa^2)}{(q - u)^2 - \kappa^2} \tag{270}$$

If the integral (270) exists, i.e., if Φ is sufficiently bounded in κ^2, Eq. (270) is the desired representation for $\tilde{f}_R(q)$. In general, however, the product of $\theta(x) f(x)$ is not well defined (as it involves the product of two distributions), and one may have to divide $\Phi(u, \kappa^2)$ by a polynomial in κ^2 of finite degree n to obtain a convergent integral. In that case $\tilde{f}_R(q)$ is only defined up to an arbitrary polynomial in κ^2 of degree n [Jost and Lehmann (Jost 1957)].

18*d*. Dispersion Relations

One of the main tasks of the axiomatic approach to quantum field theory is to derive observable consequences of the locality assumption, i.e., of the assumption that local field observables commute (or anticommute) for space-like separations. More generally the problem is to obtain spectral representations for all observable quantities (such as the scattering matrix elements) based on the general axioms and to confront these representations and their consequences with experimental data. We shall, in this section, primarily concern ourselves with the energy dependence of two-particle scattering amplitude. Here, although a certain representation has been conjectured by Mandelstam (1958) which exhibits the analytic properties of the scattering amplitude as a function of its invariant variables (essentially the energy and momentum transfer), this representation has not as yet been proved on the basis of the axioms. [See, however, Eden (1960) for a proof that under certain conditions each term in the perturbative expansion satisfies such a representation.] It has, however, been shown by Bogoliubov and others that a proof of the analyticity of the scattering amplitude as a function of (essentially) the energy—for fixed momentum transfer—can be given and that dispersion relations between the real and imaginary parts of this scattering amplitude (for fixed momentum transfer) can be rigorously derived. These dispersion relations have been used in pion physics for phenomenological and semi-phenomenological analysis of experimental data and to calculate specific particle properties (such as the electromagnetic form factors) in terms of the pion-nucleon and pion-pion phase shifts. Dispersion relations have also been used to give a general indication of the kind of answers any quantum field theory satisfying the basic axioms of locality, relativistic invariance and spectral conditions will give for processes involving strong interactions.

The theory of dispersion relations is primarily the work of Goldberger and his collaborators. Goldberger (1955a) was the first to give a heuristic proof of dispersion relations for boson-fermion scattering and to compare these dispersion relations with the experimental data in pion-nucleon scattering [Goldberger (1955b)]. He was also responsible for the application of dispersion theoretic methods to the computation of the properties of systems of strongly interacting particles (e.g., nucleon form factors, β-decay form factors, lifetimes of π^0 meson, etc.).

Let us briefly illustrate the above remarks with the deduction of the analytic properties of the forward scattering amplitude of a photon by a proton. This, in fact, is the problem for which dispersion relations were first derived using field theoretic techniques [Gell-Mann, Goldberger, and Thirring (Gell-Mann 1955); for a review of the Kramers-Kronig dispersion relations, see Toll (1952, 1958)].

In Chapter 17, Section *e*, we obtained a closed form for the desired scattering amplitude. For forward scattering (i.e., when $k' = k$, $p' = p$) the amplitude is related to the quantity

$$\begin{aligned} T_{\nu\mu}(k) &= i \int d^4x \, e^{ik\cdot x} \langle ps' \mid \{[j_\mu(x), j_\nu(0)]\, \theta(x) - \delta(x_0)\, [j_\mu(x), \dot{A}_\nu(0)]\} \mid ps\rangle \\ &= T_{\nu\mu}(k; s's) \end{aligned} \tag{271a}$$

The S-matrix element for the scattering of a photon of initial polarization $\epsilon^{(\lambda)}(k)$ to one of polarization $\epsilon^{(\lambda')}(k)$ $[\epsilon^{(\lambda)}{}_\mu(k)\, k^\mu = \epsilon^{(\lambda')}{}_\mu(k)\, k^\mu = 0]$ while the scatterer (proton) goes from the state $|ps\rangle$ to $|ps'\rangle$ is then given by $\epsilon^{(\lambda')}{}_\mu T^{\mu\nu}(k; s's)\, \epsilon^{(\lambda)}{}_\nu$. Two things are to be noted in the representation (271) for the forward scattering amplitude: Firstly, that the entire k dependence appears in the exponent

$$e^{ik\cdot x} = e^{i\omega(t - \mathbf{e}\cdot\mathbf{x})}; \quad \mathbf{e} = \frac{\mathbf{k}}{|\mathbf{k}|} \tag{271b}$$

and, secondly, that the quantity $\theta(x)\, [j_\mu(x), j_\nu(0)]$ vanishes not only for $x_0 < 0$ but also for $x^2 < 0$, according to the causality requirement. $T_{\mu\nu}(\omega)$ is therefore the Fourier transform of a function vanishing outside the forward light cone and therefore has certain analyticity properties as a function of the complex variable $\omega = \omega_1 + i\omega_2$.

It is convenient at this point to choose a definite Lorentz frame in which to evaluate the quantities appearing in Eq. (271a). This entails no loss of generality since T is effectively (i.e., apart from kinematic factors) an invariant. We shall choose the laboratory frame in which $\mathbf{p} = 0$. We shall also assume that the state $|ps\rangle$ has a definite parity. With this assumption, we easily derive the following symmetry property of T:

$$T_{\mu\nu}(\omega, \mathbf{k}; s's) = T_{\mu\nu}(\omega, -\mathbf{k}; s's) \tag{272}$$

where $k_0 = \omega$ and, of course, $\omega = |\mathbf{k}|$ since we are treating the case of the scattering of a physical photon. Thus, $T_{\mu\nu}$ is actually an even function of $\mathbf{k}$, and we could replace the factor $\exp(-i\mathbf{k}\cdot\mathbf{x})$ by $\cos \mathbf{k}\cdot\mathbf{x}$. The amplitude $T_{\mu\nu}$ therefore depends on $\mathbf{k}$ only through $\mathbf{k}^2 = \omega^2$ (since $\cos \mathbf{k}\cdot\mathbf{x} = 1 - \frac{1}{2}(\mathbf{k}\cdot\mathbf{x})^2 \pm \cdots$) and we can regard $T_{\mu\nu}$ as a function of ω only.

For future reference we here note that the ω dependence of the $\delta(x_0)\, [j_\mu(x), \dot{A}_\nu(0)]$ factor is easily ascertained. If we write

$$\int d^4x\, e^{+ik\cdot x}\, \delta(x_0)\, \langle ps' \mid [j_\mu(x), \dot{A}_\nu(0)] \mid ps \rangle = \int d^4x\, e^{+ik\cdot x}\, G^{s's}_{\mu\nu}(p, x) \quad (273)$$

locality asserts that $G^{s's}_{\mu\nu}(p, x)$ vanishes for $\mathbf{x} \neq 0$ and that it must be of the form

$$G^{s's}_{\mu\nu}(p, x) = \sum_{l=1}^{n} g^{s's}_{\mu\nu}(p; l)\, \delta^{(l)}(x) \quad (274)$$

with n finite[16] so that

$$\int d^4x\, e^{+ik\cdot x}\, \delta(x_0)\, \langle ps' \mid [j_\mu(x), \dot{A}_\nu(0)] \mid ps \rangle = \sum_{l=1}^{n} g^{s's}_{\mu\nu}(p; l)\, \omega^l \quad (275)$$

i.e., the contribution of (275) to $T_{\mu\nu}(\omega)$ is a polynomial in ω of finite degree. Since we are interested only in the analytic properties of T as a function of ω, we shall omit the $\delta(x_0)$ term in the following as its dependence on ω is now known. We also note the relation

$$\overline{T_{\mu\nu}(\omega; ss')} = T_{\mu\nu}(-\omega; s's) \quad (276)$$

so that, if we consider $T_{\mu\nu}$ as a matrix in the s, s' labels (i.e., in the nucleon spin space), changing ω into $-\omega$ causes $T_{\mu\nu}$ to go over into its Hermitian conjugate.

Now, due to the retarded property of $\theta(x)\, [j_\mu(x), j_\nu(0)]$, the integrand of the first term in the right-hand side of Eq. (271a) is different from zero only for $t > |\mathbf{x}|$, so that the right-hand side of (271a) defines a function $T_{\mu\nu}(\omega)$ of the complex variable $\omega = \omega_1 + i\omega_2$ which is analytic in the upper half of the complex ω plane. If we assume that the matrix element of the commutator is continuous inside the light cone, the asymptotic behavior of $T(\omega)$ for $\omega \to \infty$ is then, by the Riemann-Lebesgue lemma, governed by the singularities of the commutator on the light cone. Ignoring the possible singularities at $\omega = \infty$ arising from the polynomial dependence on ω of T, these analytic properties are summarized in the Hilbert relation

$$T_{\mu\nu}(\omega) = \frac{1}{\pi} \int_{-\infty}^{+\infty} \frac{d\omega' \operatorname{Im} T_{\mu\nu}(\omega')}{\omega' - \omega - i\epsilon} \quad (277a)$$

If we assume that in the limit $\omega \to \infty$, $\operatorname{Im} T_{\mu\nu}(\omega')$ behaves like a constant, then the following subtracted Hilbert relation is valid:

$$T_{\mu\nu}(\omega) - T_{\mu\nu}(0) = \frac{\omega}{\pi} \int_{-\infty}^{+\infty} \frac{d\omega' \operatorname{Im} T_{\mu\nu}(\omega')}{\omega'(\omega' - \omega - i\epsilon)} \quad (277b)$$

In order not to mar the inherent simplicity of the derivation of dispersion relations for photon-nucleon scattering with the algebraic problem connected with the spinorial and vector indices, we shall restrict ourselves in the following to the case of the completely coherent scattering ampli-

[16] The occurrence of only derivatives of finite order can be regarded as a substantial supplement to the definition of a local field.

tude, i.e., to the case $k = k'$, $p = p'$, $s = s'$, and $\lambda = \lambda'$. We shall denote the scattering matrix element $\epsilon^{(\lambda)}{}_\mu T^{\mu\nu}(\omega)\,\epsilon^{(\lambda)}{}_\nu$ by $M^{ss}_{\lambda\lambda}(\omega)$. In order to rewrite Eq. (277b) as a dispersion relation which relates the real and imaginary part of M for *physical* values of ω, we have to obtain some further properties of Im $T^{ss}_{\mu\nu}(\omega)$, in particular its even or odd character under the change of sign of ω. We first of all note that when $s = s'$, Eq. (276) asserts that

$$\begin{aligned}\text{Im } T^{ss}_{\mu\nu}(\omega) &= \frac{1}{2i}\left(T^{ss}_{\mu\nu}(\omega) - \overline{T^{ss}_{\mu\nu}(\omega)}\right) \\ &= \frac{1}{2i}\left(T^{ss}_{\mu\nu}(\omega) - T^{ss}_{\mu\nu}(-\omega)\right) \\ &= -\text{Im } T^{ss}_{\mu\nu}(-\omega) \end{aligned} \tag{278}$$

We also note that Im $T^{ss}_{\mu\nu}(\omega)$ can be written as follows:

$$\begin{aligned}\text{Im } T^{ss}_{\mu\nu}(\omega) = \tfrac{1}{2}\int d^4x\, e^{ik\cdot x}\, \{&\langle ps \mid [j_\mu(x), j_\nu(0)] \mid ps\rangle\,\theta(x) \\ &- \langle ps \mid [j_\mu(-x), j_\nu(0)] \mid ps\rangle\,\theta(-x)\}\end{aligned} \tag{279}$$

By translation invariance

$$\langle ps \mid [j_\mu(-x), j_\nu(0)] \mid ps\rangle = \langle ps \mid [j_\mu(0), j_\nu(x)] \mid ps\rangle \tag{280}$$

so that, since $\theta(x) + \theta(-x) = 1$,

$$\text{Im } M^{ss}_{\lambda\lambda}(\omega) = -\text{Im } M^{ss}_{\lambda\lambda}(-\omega) \tag{281a}$$

$$= \tfrac{1}{2}\int d^4x\, e^{ik\cdot x}\,\langle ps \mid [\epsilon^{(\lambda)}\cdot j(x), \epsilon^{(\lambda)}\cdot j(0)] \mid ps\rangle \tag{281b}$$

Note that, if we write $\theta(x) = \frac{1}{2}(\epsilon(x) + 1)$ in (271), then the decomposition

$$M^{ss}_{\lambda\lambda}(\omega) = D^{ss}_{\lambda\lambda}(\omega) + iA^{ss}_{\lambda\lambda}(\omega) \tag{282a}$$

with

$$A^{ss}_{\lambda\lambda}(\omega) = \text{Im } M^{ss}_{\lambda\lambda}(\omega) \tag{282b}$$

$$D^{ss}_{\lambda\lambda}(\omega) = \frac{i}{2}\int d^4x\, e^{ik\cdot x}\,\epsilon(x)\,\langle ps \mid [\epsilon^{(\lambda)}\cdot j(x), \epsilon^{(\lambda)}\cdot j(0)] \mid ps\rangle \tag{282c}$$

corresponds to the decomposition of the scattering amplitude into an absorptive and dispersive part. The absorptive part for $\omega > 0$

$$A^{ss}_{\lambda\lambda}(\omega) = \text{Im } M^{ss}_{\lambda\lambda}(\omega)$$

$$= \tfrac{1}{2}\sum_{|p_n\alpha\rangle}\int d^4x\, e^{ik\cdot x}\,|\langle ps \mid \epsilon^{(\lambda)}\cdot j(0) \mid p_n\alpha\rangle|^2\, e^{i(p-p_n)\cdot x} \tag{283a}$$

$$= \pi\sum_{p_{n0},\,\alpha} |\langle ps \mid \epsilon^{(\lambda)}\cdot j(0) \mid p_{n0}, \mathbf{p} + \mathbf{k};\alpha\rangle|^2\,\delta(p_{n0} - \omega - E_{\mathbf{p}}) \tag{283b}$$

receives its contributions from real intermediate states (real because of the δ function guaranteeing energy and momentum conservation). All the states corresponding to possible elastic and inelastic processes between the incoming particles contribute to the sum in (283b). The contribution to D, on the other hand, comes from virtual (non-energy conserving) states. It should be remarked that the dispersive and absorptive parts are the real and imaginary parts of the amplitude only in the coherent case. The situation is somewhat more complicated in the noncoherent case. The absorptive and dispersive parts are still defined by the decomposition $\theta(x) = \frac{1}{2}(\epsilon(x) + 1)$, but, due to the presence of proton spin and photon polarization variables, the functions A and D are no longer real.

If we now use Eq. (281a) to rewrite Eq. (277a) over only positive ω, we obtain

$$M^{ss}_{\lambda\lambda}(\omega) = \frac{1}{\pi}\int_0^\infty d\omega' \,\mathrm{Im}\, M^{ss}_{\lambda\lambda}(\omega')\left[\frac{1}{\omega' - \omega - i\epsilon} + \frac{1}{\omega + \omega'}\right] \quad (284a)$$

In the limit as $\epsilon \to 0$ the once subtracted dispersion relations thus read

$$\begin{aligned} D^{ss}_{\lambda\lambda}(\omega) - D^{ss}_{\lambda\lambda}(0) &= \mathrm{Re}\, M^{ss}_{\lambda\lambda}(\omega) - \mathrm{Re}\, M^{ss}_{\lambda\lambda}(0) \\ &= \frac{2\omega^2}{\pi}\,\mathrm{P}\int_0^\infty \frac{A^{ss}_{\lambda\lambda}(\omega')\,d\omega'}{\omega'(\omega'^2 - \omega^2)} \end{aligned} \quad (284b)$$

The optical theorem (Eq. 11.97), which is a consequence of the unitarity of the S matrix, next states that for $\omega > 0$

$$\begin{aligned} \mathrm{Im}\, M^{ss}_{\lambda\lambda}(\omega) &= A^{ss}_{\lambda\lambda}(\omega) \\ &= \frac{\omega}{4\pi}\,\sigma^{ss}_{\lambda\lambda}(\omega) \end{aligned} \quad (285)$$

so that finally the dispersion relations read

$$D^{ss}_{\lambda\lambda}(\omega) - D^{ss}_{\lambda\lambda}(0) = \frac{\omega^2}{2\pi^2}\,\mathrm{P}\int_0^\infty \frac{\sigma^{ss}_{\lambda\lambda}(\omega')\,d\omega'}{\omega'^2 - \omega^2} \quad (286a)$$

We have shown in Chapter 17, Section e, that

$$D^{ss}_{\lambda\lambda}(0) = -\frac{e^2}{M} \quad (286b)$$

so that (286a) can be rewritten as follows:

$$D^{ss}_{\lambda\lambda}(\omega) = -\frac{e^2}{M} + \frac{\omega^2}{2\pi^2}\,\mathrm{P}\int_0^\infty d\omega'\,\frac{\sigma_{\lambda\lambda}(\omega')}{\omega'^2 - \omega^2} \quad (286c)$$

A similar dispersion relation is also valid for the scattering of a photon by a spin 0 zero system. In the case of Compton scattering from protons, the states $|p_n\alpha\rangle$ which contribute to $A^{ss}_{\lambda\lambda}(\omega)$ in (285) are states containing photons and electron-positron pairs in addition to the proton, as well as states with mesons and nucleon-antinucleons present.

The states with two or more photons or with electron-positron pairs are of order e^4 so that, if one neglects terms of order e^4, the first contribution to $A^{ss}_{\lambda\lambda}$ comes from a meson-nucleon state, when $\omega > \mu$. The absorptive amplitude $A^{ss}_{\lambda\lambda}$ vanishes for $\omega < \mu$ (neglecting terms of order e^4), and one would expect that the cross section is well approximated by the photoproduction cross section [for a comparison of these dispersion relations with experiments, see Capps (1957a, b), Mathews (1957)].

The simplicity of the derivation of dispersion relations for processes involving photons is related to the fact that for photons $|\mathbf{k}| = \omega$, and by virtue of the causality assumption, the factor $\exp(i\omega t)$ in (271b) always dominates the factor $\exp(-i\omega\mathbf{e}\cdot\mathbf{x})$. For a particle with mass $[\omega = \sqrt{\mathbf{k}^2 + \mu^2}]$ considerable difficulties are encountered for the region $0 < \omega < \mu$. However, before turning to the case of the scattering of a particle with nonzero rest mass, let us indicate that dispersion relations can also be derived for observable quantities other than scattering amplitudes.

Consider the quantity

$$M_\mu(qi, kj) = \langle 0 \mid j_\mu(0) \mid qi, kj\rangle_{\text{in}} \tag{287}$$

where q, k characterize the momenta and i, j the isotopic spin variables of the two-meson state $|qi, kj\rangle_{\text{in}}$ and $j_\mu(x)$ is the electromagnetic charge-current operator. Equation (287) defines the meson electromagnetic form factor. [We could have just as easily considered the matrix element ${}_+\langle qi \mid j_\mu(0) \mid kj\rangle_+$ to which (287) is closely related.] Invariance under Lorentz transformation asserts that

$$M_\mu(q, i; k, j) = \Lambda_\mu{}^\nu M_\nu(\Lambda q, i; \Lambda k, j) \tag{288}$$

so that

$$M_\mu(qi; kj) = (aq_\mu + bk_\mu)\, M_{ij}((q+k)^2) \tag{289}$$

(We have suppressed the dependence of M_{ij} on k^2 and q^2 since k and q correspond to the momenta of physical mesons so that $k^2 = q^2 = \mu^2$.) Gauge invariance, the requirement that $\partial^\mu j_\mu(x) = 0$, implies that

$$(q+k)^\mu \langle 0 \mid j_\mu(0) \mid qi, kj\rangle_{\text{in}} = 0 \tag{290}$$

so that

$$M_\mu(qi, kj) = (q-k)_\mu\, M_{ij}((q+k)^2) \tag{291}$$

since $(q-k)\cdot(q+k) = q^2 - k^2 = 0$. Finally, the invariance of the matrix element under arbitrary rotations about the 3-axis in isotopic spin space implies that

$$M_{ij}((q+k)^2) = \epsilon_{3ij} M((q+k)^2) \tag{292}$$

Hence, combining the above symmetry requirements, we can write the following representation for $M_\mu(qi; kj)$:

$$M_\mu(qi, kj) = i\,\frac{e}{\sqrt{2}}\,\epsilon_{3ij}(q-k)_\mu\, M((q+k)^2) \tag{293}$$

Using the asymptotic conditions, Eq. (287) can be reduced to the following form:

$$\begin{aligned} M_\mu(qi, kj) &= \langle 0 \mid j_\mu(0)\, a_{j\text{in}}{}^*(k) \mid qi\rangle \\ &= \lim_{t\to-\infty} i \int d^3x\, \langle 0 \mid j_\mu(0)\, \phi_j(x) \mid qi\rangle\, \overleftrightarrow{\partial_0} f_k(x) \\ &= -i \int d^4x\, f_k(x)\, K_x \langle 0 \mid T(j_\mu(0)\, \phi_j(x)) \mid qi\rangle \end{aligned} \tag{294}$$

When the operation of taking K_x of the T product is carried out, there again appear terms coming from the equal-time commutator which, on invariance grounds and from the assumed locality of the theory, must have the form $(q-k)_\mu$ times polynomials in $(q+k)^2$. We shall omit these terms for the time being, and simply write

$$M_\mu(qi, kj) = i \int d^4x\, e^{-ik\cdot x} \langle 0 \mid T(j_\mu(0)\, J_j(x)) \mid qi\rangle \qquad (295)$$

where $J_j(x) = K_x\phi_j(x)$. Upon writing

$$T(j_\mu(0)\, J_j(x)) = \theta(-x)\,[j_\mu(0), J_j(x)] + J_j(x)\, j_\mu(0),$$

we note that the term

$$\int d^4x\, e^{-ik\cdot x} \langle 0 \mid J_j(x)\, j_\mu(0) \mid qi\rangle$$
$$= (2\pi)^4 \sum_{|p_n\alpha\rangle} \delta^{(4)}(k+p_n) \langle 0 \mid J_j(0) \mid p_n\alpha\rangle \langle p_n\alpha \mid j_\mu(0) \mid qi\rangle \qquad (296)$$

vanishes for $k_0 > \mu$ since p_{n0} must be the energy of a physical state. We shall therefore study the representation

$$M_\mu(qi, kj) = i \int d^4x\, e^{-ik\cdot x} \langle 0 \mid [j_\mu(0), J_j(x)] \mid qi\rangle\, \theta(-x) \qquad (297)$$

for the form factor, which for $q^2 = k^2 = \mu^2$ coincides with (295). Causality asserts that $[j_\mu(0), J_j(x)] = 0$ for space-like x, so that M_μ is the Fourier transform of a function that vanishes everywhere except in the past light cone. Again on the assumption that the state $|qi\rangle$ has a definite parity under spatial inversions ($J_j(x)$ having the same parity phase factor), one deduces that $M_\mu(qi; \omega, \mathbf{k}, j) = M_\mu(qi; \omega, -\mathbf{k}, j)$, so that one can again replace the factor $\exp i\mathbf{k}\cdot\mathbf{x}$ by $\cos \mathbf{k}\cdot\mathbf{x}$, and consider $M_\mu(qi; \omega, \mathbf{k}, j)$ for fixed q as a function of ω only (since $\mathbf{k}^2 = \omega^2 - \mu^2$). Note that once again the entire ω dependence of M_μ is exhibited in the exponent

$$\exp i(\omega t - \sqrt{\omega^2 - \mu^2}\,\mathbf{e}\cdot\mathbf{x})\,; \quad \mathbf{e} = \mathbf{k}/|\mathbf{k}|$$

All these facts suggest that it should be possible to derive a dispersion relation for M_μ, or more properly for the scalar function $M((q+k)^2)$. The latter can be isolated from M_μ by multiplying the expression (297) by $\epsilon_{3ij}(q-k)^\mu$ and summing over i, j and μ. The result is

$$[4\mu^2 - (q+k)^2]\, M((q+k)^2)$$
$$= \frac{i}{2} \sum_{ij} \int d^4x\, e^{-ik\cdot x}\, \epsilon_{3ij}(q-k)^\mu \langle 0 \mid [j_\mu(0), J_j(x)] \mid qi\rangle\, \theta(-x) \qquad (298)$$

since, due to gauge invariance $q_\mu M^\mu = -k_\mu M^\mu$ so that the factor $(q-k)_\mu$ can be replaced by $2q_\mu$ and this factor in turn by $-2i\partial^\mu$ when operating on the matrix element $\langle 0 \mid [j_\mu(0), J_j(x)] \mid qi\rangle$. Now the matrix element $\sum_{ij} \partial^\mu \langle 0 \mid [j_\mu(0), J_j(x)] \mid qi\rangle$ in the Lorentz frame in which $\mathbf{q} = 0$, is a scalar function of $x^2 = x_0{}^2 - r^2$ and of x_0.[17] We can therefore carry out the angular integration in the space integral for M. Thus

$$i\,[4\mu^2 - (q+k)^2]\, M((q+k)^2)$$

$$= \int_0^\infty 2\pi r\, dr \frac{\sin \sqrt{\omega^2 - \mu^2}\, r}{\sqrt{\omega^2 - \mu^2}} \int_{-\infty}^0 dt\, e^{-i\omega t} \sum_{ij} \epsilon_{3ij} \partial^\mu \langle 0 \mid [j_\mu(0), J_j(x)] \mid qi\rangle \tag{299a}$$

$$= \int_0^\infty 2\pi r\, dr\, M'_r(\omega) \tag{299b}$$

where in the expression for $M'_r(\omega)$

$$M'_r(\omega) = \frac{\sin \sqrt{\omega^2 - \mu^2}\, r}{\sqrt{\omega^2 - \mu^2}} \int_{-\infty}^0 e^{-i\omega t} \sum_{ij} \epsilon_{3ij} \partial^\mu \langle 0 \mid [j_\mu(0), J_j(x)] \mid qi\rangle\, dt \tag{300}$$

the right-hand side is to be understood as an angular average. $M'_r(\omega)$ is now a function which is analytic in the upper half of the complex ω plane. [The fact that $[j_\mu(0), J_j(x)] = 0$ for $r > t$ insures that $\sin kr\, e^{-i\omega t}$ $(t < 0)$ does not blow up for $\omega = i\infty$; furthermore, since $\sin \sqrt{\omega^2 - \mu^2} r / \sqrt{\omega^2 - \mu^2}$ is an even function of $\sqrt{\omega^2 - \mu^2}$, there are no branch points at $\omega = \pm\mu$]. Ignoring possible singularities at infinity arising from the discarded equal-time terms in (294) (or its analogous expression with T replaced by R) and possible δ-function singularities of the retarded commutator on the light cone for fixed r, we can therefore write the following Hilbert relation for $M_r(\omega)$:

$$M'_r(\omega) = \frac{1}{\pi} \int_{-\infty}^{+\infty} \frac{d\omega' \operatorname{Im} M'_r(\omega')}{\omega' - \omega - i\epsilon} \tag{301}$$

To obtain a dispersion relation for M we must integrate (301) over r. In order to perform this r integration, we split the integral into two parts corresponding to $|\omega| > \mu$ and $|\omega| < \mu$. We shall see that $\operatorname{Im} M'_r(\omega)$ also contains a factor $\sin \sqrt{\omega^2 - \mu^2}\, r$. In the range $|\omega| < \mu$, the factor $\sqrt{\omega^2 - \mu^2}$ is imaginary and $\sin \sqrt{\omega^2 - \mu^2}\, r$ therefore blows up for large r so that one does not have the right to interchange the order of the r and ω' integration. (Because for a photon $\omega = |\mathbf{k}|$, this difficulty does not arise in the derivation of the forward scattering dispersion relations for photons and this makes their derivation much simpler.) We therefore write

[17] On invariance ground the matrix element is a function only of $q^2 = \mu^2$, x^2 and $q \cdot x$. In the rest frame $\mathbf{q} = 0$ the above conclusion follows.

$$M(\omega) = \int_0^\infty 2\pi r\, dr \frac{1}{\pi} \int_{-\infty}^{+\infty} d\omega' \frac{\operatorname{Im} M'_r(\omega')}{\omega' - \omega - i\epsilon} \tag{302a}$$

$$= \int_0^\infty 2\pi r\, dr \frac{1}{\pi} \left\{ \int_{-\mu}^{+\mu} \frac{d\omega' \operatorname{Im} M'_r(\omega')}{\omega' - \omega - i\epsilon} + \left(\int_\mu^\infty + \int_{-\infty}^{\mu} \right) \frac{d\omega' \operatorname{Im} M'_r(\omega)}{\omega' - \omega - i\epsilon} \right\} \tag{302b}$$

It follows from P invariance [Oehme (1955)] that the imaginary or absorptive part of M is obtained by replacing in Eq. (297) $i\theta(-x)$ by $\frac{1}{2}$:

$$\operatorname{Im} M_\mu(qi; kj) = \tfrac{1}{2} \int d^4x\, e^{-ik\cdot x} \langle 0 \mid [j_\mu(0), J_j(x)] \mid qi \rangle \tag{303}$$

A somewhat lengthier and more explicit proof is obtained by inserting a complete set of states into the definition (300) for $M'_r(\omega)$ and carrying out the time integration (we also insert a factor exp ϵt to give meaning to the integration at $t = -\infty$). In the frame wherein $\mathbf{q} = 0$ we then obtain the following expression for $M'_r(\omega)$:

$$M'_r(\omega) = \frac{\sin \sqrt{\omega^2 - \mu^2}\, r}{\sqrt{\omega^2 - \mu^2}} \sum_{|n\rangle} \sum_{ij} \epsilon_{3ij} \frac{\sin |\mathbf{p}_n|\, r}{|\mathbf{p}_n|\, r}$$
$$\left\{ \langle 0 \mid j_0(0) \mid n \rangle \langle n \mid J_j(0) \mid qi \rangle \left[\mathrm{P} \frac{1}{\mu - p_{n0} + \omega} - i\pi\delta(\mu - p_{n0} + \omega) \right] \right.$$
$$\left. - \langle 0 \mid J_j(0) \mid n \rangle \langle n \mid j_0(0) \mid qi \rangle \left[\mathrm{P} \frac{1}{p_{n0} + \omega} - i\pi\delta(p_{n0} + \omega) \right] \right\} \tag{304}$$

In Eq. (304), in order to preserve the proper reality conditions, we understand the states $|n\rangle$ to be one-half the sum of "in" and "out" states. *TP* invariance then allows us to infer that the matrix elements occurring in (304) are real. Note also that we have replaced the factors exp $(\pm i\mathbf{p}_n \cdot \mathbf{x})$ by their angular averages, exp $(\pm i\mathbf{p}_n \cdot \mathbf{x}) \to \sin |\mathbf{p}_n|\, r/|\mathbf{p}_n|\, r$ as implied by our former procedure of integrating over angles.

The imaginary part of $M'_r(\omega)$ is thus the contribution of the δ-function factors and is given by

$$\operatorname{Im} M'_r(\omega) = \pi \frac{\sin \sqrt{\omega^2 - \mu^2}\, r}{\sqrt{\omega^2 - \mu^2}} \sum_{|n\rangle} \sum_{ij} \epsilon_{3ij} \frac{\sin |\mathbf{p}_n|\, r}{|\mathbf{p}_n|\, r}$$
$$[\langle 0 \mid j_0(0) \mid n \rangle \langle n \mid J_j(0) \mid qi \rangle\, \delta(\mu - p_{n0} + \omega)$$
$$- \langle 0 \mid J_j(0) \mid n \rangle \langle n \mid j_0(0) \mid qi \rangle\, \delta(p_{n0} + \omega)] \tag{305}$$

The states $|n\rangle$ which contribute to Im $M_r(\omega)$ must have the heavy particle quantum number equal to zero. The lowest states which can contribute are the one-meson states, with $p_{n0} \geqslant \mu$. (We are assuming that parity is a good quantum number in the description of our system and that mesons are pseudoscalar.) We therefore note that by virtue of the δ functions in (305)

$$\text{Im } M'_r(\omega) = 0 \quad \text{for } |\omega| < \mu \tag{306}$$

so that the troublesome unphysical region $|\omega| < \mu$ does in fact not contribute to the dispersion integral in (302b). Furthermore, if we carry out the r integration, conservation of momentum dictates that only states with momentum $-\mathbf{k}$ contribute to the second group of terms in (305), i.e., to the terms of the form $\langle 0 \mid J_j(0) \mid n\rangle \langle n \mid j_0(0) \mid qi\rangle$. These states must thus have $\mathbf{p}_n{}^2 = \mathbf{k}^2 = \omega^2 - \mu^2$ and also $p_{0n}{}^2 = \omega^2$, due to the δ-function factor $\delta(p_{0n} + \omega)$. Their mass is therefore μ^2 so that they are one-meson states. But these states do not contribute since

$$\begin{aligned} \langle 0 \mid J_j(0) \mid n\rangle &= (\Box + \mu^2) \langle 0 \mid \phi_j(x) \mid n\rangle\Big|_{x=0} \\ &= (\Box + \mu^2) \langle 0 \mid \phi_{j\,\text{in}}(x) \mid n\rangle\Big|_{x=0} \\ &= 0 \end{aligned} \tag{307}$$

if $|n\rangle$ is a one-meson state.

Summarizing: we need only consider that part of $\text{Im } M'_r(\omega)$ which is given by

$$\text{Im } M_r(\omega) = \pi \frac{\sin \sqrt{\omega^2 - \mu^2}\, r}{\sqrt{\omega^2 - \mu^2}} \sum_{|n\rangle} \sum_{ij} \epsilon_{3ij} \frac{\sin |\mathbf{p}_n|\, r}{|\mathbf{p}_n|\, r} \langle 0 \mid j_0(0) \mid n\rangle \langle n \mid J_j(0) \mid qi\rangle\, \delta(\mu - p_{n0} + \omega) \tag{308}$$

and which has the property that

$$\text{Im } M_r(\omega) = 0 \quad \omega < 2\mu \tag{309}$$

since by parity arguments the least massive states which contribute to the sum in Eq. (308) are two-meson states. The dispersion relation therefore reads

$$M(\omega) = \frac{1}{\pi} \int_{2\mu}^{\infty} d\omega' \frac{\text{Im } M(\omega')}{\omega' - \omega - i\epsilon} \tag{310}$$

Reverting to invariant variables $(q + k)^2 = 2\mu^2 + 2\mu\omega$, we can rewrite the dispersion relation for M as

$$M((q + k)^2) = \frac{1}{\pi} \int_{4\mu^2}^{\infty} d\zeta \frac{\text{Im } M(\zeta)}{\zeta - (q + k)^2 - i\epsilon} \tag{311}$$

In (311) we have assumed that M has no pole at $(q + k)^2 = 4\mu^2$. If the previously discarded equal-time terms are such that one subtraction is required (i.e., $\text{Im } M(\omega)/\omega \to 0$ as $\omega \to \infty$), the once subtracted dispersion relations read

$$M((q + k)^2) = M(0) + \frac{(q + k)^2}{\pi} \int_{4\mu^2}^{\infty} d\zeta \frac{\text{Im } M(\zeta)}{\zeta(\zeta - (q + k)^2 - i\epsilon)} \tag{312}$$

The dispersion relation (312) expresses the analytic properties of the meson form factor which follow from causality and spectrum.

It is of importance to note that dispersion theory can also be used to formulate an approximation method to calculate $M((q+k)^2)$. We have seen that the absorptive part Im M is related to

$$A_\mu = \pi \sum_{|n\rangle} \langle 0 \mid j_\mu(0) \mid n\rangle \langle n \mid J_j(0) \mid qi\rangle \, \delta(p_n - k - q) \qquad (313)$$

where the least massive states $|n\rangle$ which can contribute are the two-pion states. Suppose one were to limit one's attention to these two-pion intermediate states. Then the right-hand side of Eq. (313) becomes equal to $\sum_{|q'i', k'j'\rangle} \langle 0 \mid j_\mu(0) \mid q'i', k'j'\rangle \langle q'i', k'j' \mid J_j(0) \mid qi\rangle \, \delta(q' + k' - k - q)$ which again contains M_μ. In this approximation Eq. (312) therefore becomes an integral equation for the form factor $M((q+k)^2)$, the kernel $\langle q'i', k'j' \mid J_j(0) \mid qi\rangle$ of which is proportional to the scattering amplitude for pion-pion scattering and can be expressed in terms of the pion-pion scattering phase shifts [Federbush (1958)].

Similar attempts to the above have been made to derive dispersion relations for the nucleon form factors $F_{1,2}$ which are defined to lowest order in the electric charge [recall Eq. (17.266)] by

$$\langle p' \mid j_\mu(0) \mid p\rangle = \left(\frac{M^2}{E_{\mathbf{p}}E_{\mathbf{p}'}}\right)^{1/2} \cdot \bar{u}(p') \left[F_1(q^2)\, \gamma_\mu + iF_2(q^2)\, \sigma_{\mu\nu}(p' - p)^\nu\right] u(p)$$

$$q = (p' - p) \qquad (314)$$

where $|p'\rangle$, $|p\rangle$ are one-nucleon states. Considered as operators in the nucleon isotopic spin space, the functions $F_{1,2}$ may be subdivided into an isotopic scalar and vector component by writing

$$F_1 = F_1{}^S + \tau_3 F_1{}^V \qquad (315a)$$

$$F_2 = F_2{}^S + \tau_3 F_2{}^V \qquad (315b)$$

We have seen that

$$F_1{}^S(0) = F_1{}^V(0) = e/2 \qquad (316a)$$

$$F_2{}^S(0) = (\Delta\mu_{\mathrm{p}} + \Delta\mu_{\mathrm{n}})/2 \qquad (316b)$$

$$F_2{}^V(0) = (\Delta\mu_{\mathrm{p}} - \Delta\mu_{\mathrm{n}})/2 \qquad (316c)$$

where e is the proton charge and $\Delta\mu_{\mathrm{n}}$ and $\Delta\mu_{\mathrm{p}}$ are the static anomalous magnetic moments of neutron and proton respectively.

To give rigorous derivations for dispersion relations such as

$$F_1{}^{S,V}(q^2) = \frac{e}{2} + \frac{q^2}{2}\int_{m_{S,V}{}^2}^{\infty} d\sigma^2 \, \frac{\rho_1{}^{S,V}(\sigma^2)}{\sigma^2(\sigma^2 - q^2 - i\epsilon)} \qquad (317)$$

$$F_2{}^{S,V}(q^2) = \frac{1}{\pi}\int_{m_{S,V}{}^2}^{\infty} d\sigma^2 \, \frac{\rho_2{}^{S,V}(\sigma^2)}{\sigma^2 - q^2 - i\epsilon} \qquad (318)$$

where

$$\rho_1(\sigma^2) = \text{Im } F_1(\sigma^2) \tag{319a}$$

$$\rho_2(\sigma^2) = \text{Im } F_2(\sigma^2) \tag{319b}$$

and

$$m_S^2 = (3\mu)^2, \qquad m_V^2 = (2\mu)^2 \tag{319c}$$

is much more difficult than for the meson form factor. This problem is related to the problem of giving rigorous proof for dispersion relations for the vertex function [Oehme (1959)]. One can establish the dispersion relations (317), (318), i.e., one can establish that the form factors are analytic functions of q^2 in the cut q^2 plane [with the cut running from $(2\mu)^2$ to ∞], only if $\mu/M > \sqrt{2} - 1$. It has on the other hand been shown by Nambu (1958) that analyticity in the cut plane holds for *each* term in the perturbative expansion. However, in order to be able to use this result to make assertions independent of perturbation theory, one would have to prove the uniform convergence of the perturbative series.

Usually what is done is to assume the validity of dispersion relations such as the above Eqs. (317) and (318) in which the imaginary part is expressed as a sum over contributions from real intermediate states [Chew (1958), Federbush (1958), Bincer (1959)]. One then attempts to calculate the contributions of the least massive such states, hoping that no other states are important [for a review of the problem of nucleon form factors, see Drell (1961)].

As stated earlier, our main concern in the present section is the deduction of sufficient analyticity properties of the two-particle elastic scattering amplitude to be able to write down dispersion relations for this amplitude. We now turn to the problem of the proof of these dispersion relations.

The commutation rules between in- and out-operators derived in Section *b* of the present chapter by using the asymptotic conditions allow us to derive in a simple fashion a closed expression for the elastic scattering amplitude between two particles. We first write the scattering amplitude as follows:

$$\begin{aligned} {}_{\text{out}}\langle p'k' \mid pk\rangle_{\text{in}} &= {}_{\text{out}}\langle p' \mid a_{\text{out}}(k')\, a_{\text{in}}^*(k) \mid p\rangle_{\text{in}} \\ &= {}_{\text{out}}\langle p' \mid a_{\text{in}}^*(k)\, a_{\text{out}}(k') \mid p\rangle_{\text{in}} \\ &\qquad + {}_{\text{out}}\langle p' \mid [a_{\text{out}}(k'), a_{\text{in}}^*(k)] \mid p\rangle_{\text{in}} \end{aligned} \tag{320}$$

Now it will be recalled that in the two orthonormal sets $|k_1, \cdots k_n\rangle_{\text{in}}$, $|k_1, \cdots, k_n\rangle_{\text{out}}$, the vacuum and the one-particle states are identical (stability of the vacuum and the one-particle states):

$$|p\rangle_{\text{out}} = |p\rangle_{\text{in}} \tag{321}$$

$$|\Psi_0\rangle_{\text{out}} = |\Psi\rangle_{\text{in}} = |\Psi_0\rangle = |0\rangle \tag{322}$$

so that the first term on the right-hand side of Eq. (320) vanishes since

$a_{\text{in}}(k) \mid p\rangle_{\text{in}} = 0$. Substituting into Eq. (320) the expression (166a) for the commutator, we obtain

$$
{}_{\text{out}}\langle p'k' \mid pk\rangle_{\text{in}} = 4k_0 p_0 \delta^{(3)}(\mathbf{p} - \mathbf{p}')\, \delta^{(3)}(\mathbf{k} - \mathbf{k})
- \frac{i}{(2\pi)^3} \int d^4x \int d^4x'\, e^{-i(k\cdot x - k'\cdot x')} \cdot K_x K_{x'}\, {}_{\text{out}}\langle p' \mid R(x'; x) \mid p\rangle_{\text{in}} \quad (323a)
$$

where

$$
R(x; x') = -i\theta(x - x')\, [\phi(x), \phi(x')] \quad (323b)
$$

In the expression (323a) for the scattering amplitude no assumption concerning the nature of the target particle has yet been made, that is, no specification of the one-particle states $|p\rangle$, $|p'\rangle$ beyond the requirement that $a_{\text{in}}(k) \mid p\rangle_{\text{in}} = 0$ has been used. The scattering amplitude (323a) is still quite general and describes the interactions between pairs of bosons if the states $|p\rangle$ and $|p'\rangle$ are one-boson states with $|p\rangle_{\text{in}} = a_{\text{in}}^*(p) \mid 0\rangle$. It describes the scattering of mesons by nucleons if the states $|p\rangle$, $|p'\rangle$ correspond to nucleon states of momentum p, p', mass M, $p^2 = p'^2 = M^2$.

It should also be mentioned that the scattering matrix element can be further reduced so as to involve only vacuum expectation values of retarded products of field operators or of time-ordered products of operators. To see this, it will be recalled that the scattering matrix element can alternatively be written in the form

$$
{}_{\text{out}}\langle p'k' \mid pk\rangle_{\text{in}} = 4p_0 k_0 \delta^{(3)}(\mathbf{p} - \mathbf{p}')\, \delta^{(3)}(\mathbf{k} - \mathbf{k}')
+ \frac{1}{(2\pi)^3} \int d^4x \int d^4x'\, e^{-i(k\cdot x - k'\cdot x')} K_x K_{x'}\, {}_{\text{out}}\langle p' \mid T(\phi(x)\, \phi(x')) \mid p\rangle_{\text{in}} \quad (324)
$$

and by reducing out the particles in the initial and final states (here assumed to be "scalar" nucleons of mass M), we obtain

$$
{}_{\text{out}}\langle p'k' \mid pk\rangle_{\text{in}} = {}_{\text{out}}\langle p'k' \mid pk\rangle_{\text{out}}
+ \frac{1}{(2\pi)^6} \int d^4x \int d^4x' \int d^4y \int d^4y'\, e^{-i(k\cdot x + p\cdot y - k'\cdot x' - p'\cdot y')}
D_y D_{y'} K_x K_{x'} \langle 0 \mid T(\phi(x)\, \phi(x')\, \psi(y)\, \psi^*(y')) \mid 0\rangle \quad (325a)
$$

where

$$
D_y = \Box_y + M^2 \quad (325b)
$$

For the derivation of analyticity properties of the scattering amplitude ${}_{\text{out}}\langle p'k' \mid pk\rangle$ the form (323) is more convenient to use than (324). This is because the locality postulate when imposed on the retarded product of operators gives support properties in space-time to the integrand in the right-hand side of Eq. (323) from which analyticity properties in momentum space can be inferred for the left-hand side of Eq. (323). The same is not immediately true for the form (324). Similar statements can be made about the consequences of the locality assumption for the expression analogous to (325a) involving the retarded product of operators in its relation to the expression (325a).

In the subsequent analysis which concerns itself with the analyticity properties of the scattering amplitude as a function of its kinematic variables, it turns out that the isotopic spin and spin variables of the particles are not essential ones. Their inclusion introduces complications which are primarily of an indicial nature and will be omitted for the time being. For simplicity we shall hereafter consider the scattering of neutral "scalar" meson of momentum k, mass μ by a "heavy" scalar nucleon of momentum p, mass M. It will be assumed that a heavy particle conservation law operates for the nucleons. The problem is, then, to obtain as much information about the analyticity properties of the amplitudes ${}_{\text{out}}\langle p'k' \mid pk\rangle_{\text{in}}$ as is possible using only the axioms of relativistic invariance, spectrum and locality. The most natural approach would be to study the vacuum expectation value of the retarded products of four operators to which the scattering amplitude is related.[18] However, the problem of obtaining a representation for the four-fold vacuum expectation value which incorporates the above general properties of the theory has not yet been solved [see in this connection Bremmerman (1958); Screaton (1960)], and most of the knowledge about ${}_{\text{out}}\langle p'k' \mid pk\rangle_{\text{in}}$ obtained up to the present has been based on the representation (323) for the scattering amplitude.

If, in that amplitude, we carry out the indicated differentiations, we obtain

$$
{}_{\text{out}}\langle p'k' \mid pk\rangle_{\text{in}} = {}_{\text{in}}\langle pk \mid pk\rangle_{\text{in}} - \frac{1}{(2\pi)^3}\int d^4x' \int d^4x\, e^{-i(k\cdot x - k'\cdot x')}
$$

$$
\langle p' \mid \theta(x'-x)\,[j(x'), j(x)] + \delta^{(1)}(x_0 - x'_0)\,[j(x'), \partial_0\phi(x)] \mid p\rangle \qquad (326a)
$$

with

$$
K_x\phi(x) = j(x) \qquad (326b)
$$

where we have used the locality of the field $\phi(x)$ to set equal to zero terms of the form $\delta^{(1)}(x_0 - x'_0)\,[\phi(x), \phi(x')]$ and $\delta^{(1)}(x_0 - x'_0)\,[j(x), \phi(x')]$.

In the following work, we shall neglect the complications which arise from the fact that

$$
K_xK_yT(\phi(x')\,\phi(x)) \neq T(j(x')\,j(x)) \qquad (327a)
$$

or

$$
K_xK_yR(\phi(x')\,\phi(x)) \neq R(j(x')\,j(x)) \qquad (327b)
$$

[18] It should be mentioned in this connection that it is, however, unlikely that one would obtain all the results of perturbation theory in such an analysis because no use has been made of the unitarity conditions (the "nonlinear" part of the theory). This is known to be definitely the case for the vertex function [Jost (1958)]. Jost has shown that dispersion relations cannot be derived for the π-N vertex function if the mass ratio μ/M is less than $\frac{2}{\sqrt{3}} - 1$, if only causality, Lorentz invariance and the spectral conditions are used, i.e., if only the "linear" part of the theory is used.

This is because we shall be primarily concerned with the derivation of the analyticity properties of the scattering amplitudes and the difference between the right- and left-hand side of Eq. (327a) [or Eq. (327b)] contributes terms whose dependence on the variables k, k', p and p' is easily established. Locality, together with invariance, asserts that they are of the form $\sum_i^n g_i[(p - p')^2]\,(k + k')^{2i}$ with n finite, i.e., that this term contributes to the scattering amplitude a polynomial in $(k + k')^2$ of finite degree. We therefore once again omit from consideration the contribution of the term $\delta(x_0 - x'_0)\,[j(x'), \partial_0\phi(x)]$. Making use of the translation invariance of the theory, we can write

$$\begin{aligned}\langle p' \mid [j(x'), j(x)] \mid p\rangle &= \langle p' \mid U^*(a)\,U(a)\,[j(x'), j(x)]\,U^*(a)\,U(a) \mid p\rangle \\ &= e^{-i(p'-p)\cdot a}\,\langle p' \mid [j(x' + a), j(x + a)] \mid p\rangle \qquad (328)\end{aligned}$$

The choice $a = -x$ allows us to rewrite the scattering matrix in the form

$$\begin{aligned}{}_{\text{out}}\langle p'k' \mid pk\rangle_{\text{in}} - {}_{\text{in}}\langle p'k' \mid pk\rangle_{\text{in}} &= -\frac{1}{(2\pi)^3}\int d^4\xi \int d^4\eta \\ e^{-i(k+p-k'-p')\cdot\eta}\, e^{-i(-k-k'+p'-p)\cdot\frac{\xi}{2}}\,&\theta(\xi)\,\langle p' \mid [j(\xi), j(0)] \mid p\rangle \qquad (329)\end{aligned}$$

after the co-ordinates $x' - x = \xi$ and $\frac{1}{2}(x + x') = \eta$ are introduced. Carrying out the integration over η then yields

$$\begin{aligned}{}_{\text{out}}\langle p'k' \mid pk\rangle_{\text{in}} &- {}_{\text{in}}\langle p'k' \mid pk\rangle_{\text{in}} \\ &= 2\pi i\delta^{(4)}(p + k - p' - k')\,R(p', k'; p, k) \qquad (330)\end{aligned}$$

where

$$R(p'k'; pk) = i\int d^4x\, e^{ik'\cdot x}\,\theta(x)\,\langle p' \mid [j(x), j(0)] \mid p\rangle \qquad (331a)$$

$$= i\int d^4x\, e^{i(k+k')\cdot\frac{x}{2}}\,\theta(x)\,\langle p' \left| \left[j\left(\frac{x}{2}\right), j\left(-\frac{x}{2}\right)\right] \right| p\rangle \qquad (331b)$$

The expression (331a) or (331b) for R is defined only for $p + k = p' + k'$. If we would have adopted the form (324) for the scattering amplitude, the expression for R would then read

$$R(p'k'; pk) = i\int d^4x\, e^{ik'\cdot x}\,\langle p' \mid T(j(x)\,j(0)) \mid p\rangle \qquad (332)$$

The equality of these two forms for the mass shell values of the momenta, i.e., for $p^2 = p'^2 = M^2$, $k^2 = k'^2 = \mu^2$, is easily ascertained by noting that

$$T(j(x)\,j(0)) = \theta(x)\,[j(x), j(0)] + j(0)\,j(x) \qquad (333)$$

and that $j(0)\,j(x)$ contributes to R a term of the form

$$\begin{aligned}\int d^4x\, e^{ik'\cdot x}\,&\langle p' \mid j(0)\,j(x) \mid p\rangle \\ &= \sum_{|p_n\alpha\rangle}\int d^4x\, e^{-i(k'+p_n-p)\cdot x}\,\langle p' \mid j(0) \mid p_n\alpha\rangle\,\langle p_n\alpha \mid j(0) \mid p\rangle \\ &= (2\pi)^4\sum_{|p_n\alpha\rangle}\delta(k' + p_n - p)\,\langle p' \mid j(0) \mid p_n\alpha\rangle\,\langle p_n\alpha \mid j(0) \mid p\rangle \qquad (334)\end{aligned}$$

Since $|p'\rangle$ is a one-nucleon state and $j(0)$ conserves the heavy particle quantum number, the states $|p_n\alpha\rangle$ must likewise be states with heavy particle quantum number +1. Since p, p' and k, k' are the energy-momentum vectors of one-particle states, the square of the mass of the intermediate state $|p_n\alpha\rangle$ is by virtue of the delta function equal to

$$M_n^2 = p_n^2 = (p - k')^2 = M^2 + \mu^2 - 2p \cdot k' \tag{335}$$

In the co-ordinate system in which $p = (M, 0, 0, 0)$ (i.e., the laboratory system), we have

$$\begin{aligned} M_n^2 &= M^2 + \mu^2 - 2M\omega_{\mathbf{k}'} \\ &= (M - \mu)^2 - 2M(\omega_{\mathbf{k}'} - \mu) \\ &< M^2 \end{aligned} \tag{336}$$

But, since there is no state with heavy particle quantum number equal to +1 whose mass is less than M, the term $j(0)\,j(x)$ does not contribute to R and the forms (332) and (331a) are identical for $k^2 = k'^2 = \mu^2$ and $p'^2 = p^2 = M^2$. Off the mass shell the right-hand sides of (331) and (332) define two functions which, in general, are different from one another. Since only the mass shell representation has physical meaning, we can adopt the most convenient form should one need to make use of these functions off the mass shell. It turns out that, to obtain analyticity properties of the scattering amplitude, the form (331a) is most convenient and will be the one used in the following investigation.

Relativistic invariance now asserts that R is an invariant function of the (invariant) scalar products formed from the kinematical variables p, p', k, k' specifying the scattering. Since R is only defined for $p + k = p' + k'$, only three of these four vectors are linearly independent. These can be chosen to be p, k, $p'(k' = p + k - p')$. There are therefore six distinct scalar products: $p \cdot k$, $p \cdot p'$, $k \cdot p'$, p^2, k^2, p'^2. It is convenient to use the identity $k'^2 = (k + p - p')^2$ to re-express $k \cdot p'$ in terms of k'^2 and the other five scalar products so that the six scalar products can be taken as $k \cdot p$, $p \cdot p'$, k^2, k'^2, p^2, and p'^2. Since for physical applications $p^2 = p'^2$ and $k^2 = k'^2$, the number of independent scalar product is four. Actually we shall only be interested in R for fixed $p^2 = p'^2$ on the mass shell, i.e., for $p^2 = p'^2 = M^2$. It will also be noted that the form (331b) actually defines R for arbitrary vectors k, k'. (These variables appear in the exponential and can be assigned values at will.) We shall find it convenient in the following to let k and k' be arbitrary real vectors which are only restricted by the constraints $k + p = k' + p'$, $p^2 = M^2$, $p'^2 = M^2$ and $k^2 = k'^2 = \zeta$. The scattering amplitude is then obtained when $\zeta = \mu^2$. The number of independent scalar products in the problem is thus three, namely, $k \cdot p$, $p \cdot p'$, and $k^2 = k'^2$.[19] Alternatively, we can use the set

[19] Incidentally note that, if $k'^2 = k^2$, the quantity $(k + k') \cdot (p - p') = 0$.

$$W^2 = (p + k)^2 = (p + k) \cdot (p' + k') \tag{337a}$$

$$\Delta^2 = -\left(\frac{k - k'}{2}\right)^2 = -\left(\frac{p' - p}{2}\right)^2 \tag{337b}$$

and

$$\zeta = k^2 = k'^2 \tag{337c}$$

We shall also have occasion to use the variables

$$\omega = \frac{(k + k') \cdot (p + p')}{2\sqrt{(p + p')^2}} \tag{338}$$

and

$$K^2 = \frac{(W^2 + M^2 - \zeta)^2 - 4M^2W^2}{4W^2} \tag{339}$$

Although ω can be expressed in terms of W^2, Δ^2, and ζ, it is convenient to have a separate symbol for it. The relation between ω and these other variables is as follows: since

$$(p + p')^2 = (p - p')^2 + 4p \cdot p' \tag{340}$$

so that

$$2p \cdot p' = 4\Delta^2 + 2M^2 \tag{341a}$$

and similarly

$$2k \cdot k' = 4\Delta^2 + 2\zeta \tag{341b}$$

whence

$$(p + p')^2 = 4(\Delta^2 + M^2) \tag{342a}$$

$$(k + k')^2 = 4(\Delta^2 + \zeta) \tag{342b}$$

we thus find that

$$\begin{aligned} W^2 &= (p + k) \cdot (p + k) = \tfrac{1}{4}(p + k + p' + k') \cdot (p + k + p' + k') \\ &= \tfrac{1}{2}(p + p') \cdot (k + k') + \tfrac{1}{4}(p + p')^2 + \tfrac{1}{4}(k + k')^2 \\ &= 2\omega\sqrt{\Delta^2 + M^2} + 2\Delta^2 + M^2 + \zeta \end{aligned} \tag{343}$$

The function R defined by (331b) can now be considered a function of ω, Δ^2, and ζ with the scattering amplitude given by $R(\omega, \Delta^2; \zeta = \mu^2)$. It will be noted that R is invariant under the substitution $p \to p'$, $p' \to p$, since ω and Δ^2 are invariant under this transformation. The physical meaning of the above variables is easily established. For example, in the center-of-mass system in which $\mathbf{p} = -\mathbf{k}$, $\mathbf{p}' = -\mathbf{k}'$, $W = p_0 + k_0$ and is therefore the total energy. In the center-of-mass system $p_0^2 - k_0^2 = = M^2 - \zeta$, so that

$$\begin{aligned} K^2 &= \frac{(W^2 + M^2 - \zeta)^2 - 4M^2W^2}{4W^2} \\ &= \frac{[(p_0 + k_0)^2 + (p_0 + k_0)\,(p_0 - k_0)]^2 - 4M^2(p_0 + k_0)^2}{4(p_0 + k_0)^2} \\ &= p_0^2 - M^2 = \mathbf{p}^2 \end{aligned} \tag{344}$$

and hence K is the particle momentum while $1 - (2\Delta^2/K^2)$ is related to the scattering angle θ by

$$\cos\theta = 1 - \frac{2\Delta^2}{K^2} \tag{345}$$

For a physical scattering process, the range of these variables is

$$\left.\begin{aligned} &0 < \Delta^2 < \infty \\ &\omega > \sqrt{\Delta^2 + \mu^2} \\ &K^2 > \Delta^2 \end{aligned}\right\} \text{physical region} \tag{346}$$

corresponding to positive energies of the particles and real scattering angles.

As stated above, our aim is to obtain analyticity properties of the amplitude $R(\omega, \Delta^2, \zeta = \mu^2)$ as a function of ω, Δ^2, and ζ. These properties will be obtained from the representation

$$R(\omega, \Delta^2, \zeta) = i \int e^{i(k+k')\cdot\frac{x}{2}}\,\theta(x)\,\langle p' \,|\left[j\left(\frac{x}{2}\right), j\left(-\frac{x}{2}\right)\right]|\, p\rangle\, d^4x \tag{347}$$

where ζ is arbitrary. In addition to this expression for R, we shall also use an expression for $\operatorname{Im} R$ containing information related to unitarity. The desired expression is

$$\begin{aligned} \operatorname{Im} R(\omega, \Delta^2, \zeta) &= \frac{1}{2i}(R - \overline{R}) \\ &= \frac{1}{2}\int d^4x\, e^{i(k+k')\cdot\frac{x}{2}}\langle p' \,|\left[j\left(\frac{x}{2}\right), j\left(-\frac{x}{2}\right)\right]|\, p\rangle \end{aligned} \tag{348}$$

where use has been made of the fact that $\overline{R(\omega, \Delta^2, \zeta)}$ is invariant under the interchange $p \to p'$, $p' \to p$, and the fact that $\theta(x) + \theta(-x) = 1$. If we define the real quantity $M(\omega, \Delta^2, \zeta)$ for arbitrary real vectors k, k' such that $k + p = p' + k'$, $k^2 = k'^2 = \zeta$, by the expression

$$\begin{aligned} M(\omega, \Delta^2, \zeta) &= \int d^4x\, e^{i(k+k')\cdot\frac{x}{2}}\langle p' \,|\, j\left(\frac{x}{2}\right) j\left(-\frac{x}{2}\right) |\, p\rangle \\ &= \overline{M(\omega, \Delta^2, \zeta)} \end{aligned} \tag{349}$$

then $\operatorname{Im} R$ can be written in terms of $M(\omega, \Delta^2, \zeta)$ as follows:

$$\operatorname{Im} R(\omega, \Delta^2, \zeta) = \tfrac{1}{2}[M(\omega, \Delta^2, \zeta) - M(-\omega, \Delta^2, \zeta)] \tag{350a}$$

$$= -\operatorname{Im} R(-\omega, \Delta^2, \zeta) \tag{350b}$$

This oddness property of $\operatorname{Im} R$ as a function of ω is a consequence of the assumed scalar and neutral character of the particles described by $\phi(x)$ (both in space-time and isotopic spin space). If the particles had additional degrees of freedom, the two terms in the commutator would in general refer to different processes, but this does not change the mathematical problem in an essential way. Incidentally, it will be noted that, if we write

$$\begin{aligned} R &= \operatorname{Re} R + i \operatorname{Im} R \\ &= D + iA \end{aligned} \tag{351}$$

then, as indicated above, $A = \operatorname{Im} R$ is given by Eq. (350), whereas

$$\operatorname{Re} R = D = \int e^{i(k+k')\cdot\frac{x}{2}} \epsilon(x) \langle p' \mid \left[j\left(\frac{x}{2}\right), j\left(-\frac{x}{2}\right)\right] \mid p\rangle \tag{352}$$

The quantity $\operatorname{Re} R = D$ is the dispersive part of the amplitude since, if a complete set of states is introduced into Eq. (352) and the integration over x is carried out, the contribution to D comes from non-energy–momentum conserving virtual processes; the absorptive part $A = \operatorname{Im} R$ on the other hand, is a sum over states which can be reached from the initial state by energy conserving real processes.

The quantity $M(\omega, \Delta^2, \zeta)$ can be further reduced by using the asymptotic conditions. Let us first introduce a complete set of states $|p_n\alpha\rangle_{\text{in}}$ into the right-hand side and use translation invariance to carry out the integration over x. We then obtain

$$\begin{aligned} M(\omega, \Delta^2, \zeta) &= (2\pi)^4 \sum_{|p_n\alpha\rangle_{\text{in}}} \delta(p_n - p - k) \langle p' \mid j(0) \mid p_n\alpha\rangle_{\text{in}}\ {}_{\text{in}}\langle p_n\alpha \mid j(0) \mid p\rangle \\ &= (2\pi)^4 \sum_{\alpha} \langle p' \mid j(0) \mid p + k, \alpha\rangle_{\text{in}}\ {}_{\text{in}}\langle p + k, \alpha \mid j(0) \mid p\rangle \end{aligned} \tag{353}$$

Let now $\psi_{\text{in}}(p')$ be the annihilation operator for an incoming nucleon of momentum p', then

$$\begin{aligned} &\langle p' \mid j(0) \mid p + k, \alpha\rangle_{\text{in}} \\ &\quad = \langle 0 \mid \psi_{\text{in}}(p')\, j(0) \mid p + k, \alpha\rangle_{\text{in}} && (354\text{a}) \\ &\quad = \langle 0 \mid [\psi_{\text{in}}(p'), j(0)] \mid p + k, \alpha\rangle_{\text{in}} \quad \text{if } (p + k - p')^2 = k'^2 < m_1^2 && (354\text{b}) \end{aligned}$$

where m_1 is the mass of the least massive state with total momentum k' for which $\langle 0 \mid j(0) \mid k'\rangle \neq 0$. (In a pseudoscalar meson theory m_1 would be equal to 3μ.) In arriving at (354b) we have used the fact $\psi_{\text{in}}(p')$ destroys an amount of momentum p' so that the state $\psi_{\text{in}}(p') \mid p + k, \alpha\rangle_{\text{in}}$ is a state of momentum $p + k - p'$ which, by energy-momentum conservation, is equal to k'. Similarly, the second factor in (353) can be written as

$$\begin{aligned} &{}_{\text{in}}\langle p + k, \alpha \mid j(0) \mid p\rangle \\ &\qquad = {}_{\text{in}}\langle p + k, \alpha \mid j(0)\, \psi_{\text{in}}^*(p) \mid 0\rangle \\ &\qquad = {}_{\text{in}}\langle p + k, \alpha \mid [j(0), \psi_{\text{in}}^*(p)] \mid 0\rangle \quad \text{if } k^2 < m_1^2 \end{aligned} \tag{355}$$

Using the recursion relation (154) which, when generalized to the case of two fields, reads as follows:

$$[\psi_{\text{in}}(p), j(0)] = \frac{-i}{(2\pi)^{3/2}} \int d^4x\, e^{-ip'\cdot x}\, \theta(-x)\, [j(0), f(x)] \tag{356}$$

where $f(x)$ is the source term in the equation of motion for the ψ field, $(\Box + M^2)\,\psi = f$, we finally obtain if $k^2, k'^2 < m_1^2$

$$M(\omega, \Delta^2, \zeta) = 2\pi \int d^4x_1 \int d^4x_2\, e^{-i(k'-p')\cdot\frac{x_1}{2} + i(k-p)\cdot\frac{x_2}{2}}$$

$$\cdot \sum_\alpha \langle 0 \mid R\left(j\left(\frac{x_1}{2}\right) f\left(-\frac{x_1}{2}\right)\right) \mid p + k, \alpha\rangle_{\text{in}}$$

$$_{\text{in}}\langle p + k, \alpha \mid R\left(j\left(\frac{x_2}{2}\right) f^*\left(-\frac{x_2}{2}\right)\right) \mid 0\rangle \quad (357)$$

The additional usable information which the expression (357) for Im R contains, in contrast to the representation (331b) for R, is whether the field ψ and ϕ interact in a causal manner, i.e., if ψ and ϕ are local with respect to each other. This information is not made use of in the form (331) where the presence of the nucleons is only indicated by the states vectors $|p\rangle$, $|p'\rangle$. Whether an interaction is local or not is the property of the commutators of the Heisenberg fields and to make use of this information we must, so to say, transfer the particles from the state vectors to the operators. This property would, of course, be incorporated in R if the expression

$$2\pi i \delta^{(4)}(p + k - p' - k')\, R = \frac{1}{(2\pi)^6} \int d^4x \int d^4x' \int d^4y \int d^4y'$$

$$e^{-i(k\cdot x + p\cdot y - k'\cdot x' - p'\cdot y')}\, K_x{}^\mu K_y{}^M K_{x'}{}^\mu K_{y'}{}^M \langle 0 \mid R(\phi(x)\, \phi(x')\, \psi(y)\, \psi^*(y')) \mid 0\rangle \quad (358)$$

were used for it.

The form (350)–(357) for Im $R(\omega, \Delta^2, \zeta)$ also embodies what is called a "generalized unitary condition." [20] To see this, recall that, if we write

$$S = 1 + i\mathbf{R} \quad (359)$$

$$_{\text{in}}\langle n \mid \mathbf{R} \mid m\rangle_{\text{in}} = 2\pi\delta^{(4)}(p_n - p_m)\, R(n; m) \quad (360)$$

the unitary condition $S^*S = SS^* = 1$ implies that

$$i(\mathbf{R}^* - \mathbf{R}) + \mathbf{R}^*\mathbf{R} = 0 \quad (361)$$

In particular, for the elastic scattering amplitude, the relation (361) asserts that

$$_{\text{in}}\langle p'k' \mid \mathbf{R}^* - \mathbf{R} \mid pk\rangle_{\text{in}} = i \sum_{|n\rangle} {}_{\text{in}}\langle p'k' \mid \mathbf{R}^* \mid n\rangle \langle n \mid \mathbf{R} \mid pk\rangle_{\text{in}} \quad (362)$$

or equivalently

$$\overline{R(pk; p'k')} - R(p'k'; pk) = 2\pi i \sum_{|n\rangle} \delta(p' + k' - p_n)\, \overline{R(n; p'k')}\, R(n; pk) \quad (363)$$

[20] Compare Eq. (348) when combined with Eqs. (350) and (357) to the two-field generalization of Eq. (170). Note also that the relation (359) for **R** in terms of S differs by a minus sign from the one given previously in Chapter 10.

For the coherent forward scattering case, i.e., $p' = p$ and $k' = k$, Eq. (363) reduces to the optical theorem

$$\text{Im}\, R(pk, pk) = \frac{|\mathbf{k}|}{4\pi}\sigma \tag{364}$$

We next note that, had we carried out only a single reduction on the scattering amplitude ${}_{\text{out}}\langle p'k'; pk\rangle_{\text{in}}$, it would read

$$\begin{aligned}(2\pi)\,\delta(p + k - p' - k')\,R(p'k'; pk)&\\ &= \frac{1}{(2\pi)^{3/2}}\int d^4x\, e^{-ik\cdot x}\, K_x\, {}_{\text{out}}\langle p'k' \mid \phi(x) \mid p\rangle\\ &= (2\pi)^{5/2}\, {}_{\text{out}}\langle p'k' \mid j(0) \mid p\rangle\, \delta(p' + k' - k - p)\end{aligned} \tag{365}$$

More generally, the scattering amplitude for the process $p + k \to n$ is given by

$$\delta(p_n - p - k)\,R(n; pk) = (2\pi)^{3/2}\, {}_{\text{out}}\langle n \mid j(0) \mid p\rangle\, \delta(p_n - p - k) \tag{366}$$

Using the result of Eq. (366) and using translation invariance to carry out the x integration, Eq. (348) can be written in the form

$$\begin{aligned}R(p'k'; pk) - \overline{R(pk; p'k')} = 2\pi i \sum_{|n\rangle} \{&\delta(p' + k' - p_n)\,R(p'k'; n)\,\overline{R(pk; n)}\\ &- \delta(p - k' - p_n)\,R(p', -k; n)\,\overline{R(p, -k'; n)}\}\end{aligned} \tag{367}$$

The first term in the right-hand side of Eq. (367) is clearly identical with the right-hand side of Eq. (363). Let us therefore consider the contribution of the other terms in the right-hand side of (367) for the case the particle momenta are on the mass shell. It will be recalled that Eq. (363) was derived from the condition $SS^* = 1$ which is *only* assumed to hold on the mass shell, i.e., when $p^2 = p'^2 = M^2$, $k^2 = k'^2 = \mu^2$. By virtue of the δ function, only those states $|n\rangle$ will contribute to the second group of terms for which $p_n = p - k'$. The masses of these states therefore are given by

$$M_n^2 = p_n^2 = (p - k')^2 \tag{368a}$$

In the laboratory system the square of this (invariant) mass is equal to

$$\begin{aligned}M_n^2 &= M^2 + \mu^2 - 2Mk_0\\ &= (M - \mu)^2 - 2M(k_0 - \mu)\end{aligned} \tag{368b}$$

so that M_n is always less than $M - \mu$. But, by assumption, the theory contains a heavy particle conservation law so that the intermediate states must have a heavy particle quantum number equal to plus one since $j(x)$ conserves heavy particles. The contribution of these states therefore vanishes for $p^2 = p'^2 = M^2$ and $k^2 = k'^2 = \mu^2$ since, by assumption (stability condition on the nucleon), the least massive state with heavy particle number one is the one-nucleon state which has a mass M. On the other hand, if we consider the reaction $p + (-k') \rightarrow p' + (-k)$ $(-k_0, -k_0' > 0)$, in which particles p and k' are incoming, then this second term will be the only term contributing on the mass shell and the first term will then not contribute.

If we represent the elastic two-particle scattering matrix element under consideration by the Feynman diagram of Figure 18.4, then only those

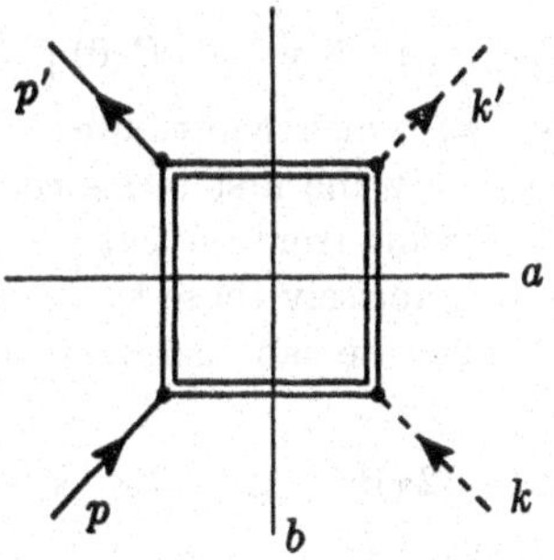

Fig. 18.4

intermediate states to be found by the cut indicated by line a will contribute to $\operatorname{Im} R(p'k'; pk)$ (for $p^2 = p'^2 = M^2$, $k^2 = k'^2 = \mu^2$) when p and k represent the incoming particle. These states are precisely those contributing to the sum in the first term of the right-hand side of (367). Similarly, if p and k' are the incoming particles (so that $-k_0' > 0$), then only the second term on the right-hand side of (367) contributes to the mass shell value of $\operatorname{Im} R$ in the physical range. Finally, if p and p' are the incoming particles (when $-p_0' > 0$), we would expect that only those states to be found if the cut b is made in the diagram 18.4 will contribute to $\operatorname{Im} R$. This is indeed the case. If we suitably contract the amplitude ${}_{\text{out}}\langle p'k' \mid pk\rangle_{\text{in}}$ with respect to the outgoing and ingoing variables, we can obtain the following two expressions:

$$R(p'k'; pk) = i \int d^4x \, e^{i(p'+k)\cdot\frac{x}{2}} \theta(x) \langle k' \mid \left[f\left(\frac{x}{2}\right), j\left(-\frac{x}{2}\right)\right] \mid p\rangle \quad \text{(369a)}$$

and

$$R(p'k'; pk) = i \int d^4x \, e^{i(p+k')\cdot\frac{x}{2}} \theta(x) \langle p' \mid \left[j\left(\frac{x}{2}\right), f^*\left(-\frac{x}{2}\right)\right] \mid k\rangle \quad \text{(369b)}$$

The first formula can also be written as

$$\overline{R(pk; p'k')} = -i \int d^4x \, e^{i(p+k')\cdot\frac{x}{2}} \theta(-x) \langle p' \mid \left[j\left(\frac{x}{2}\right), f^*\left(-\frac{x}{2}\right)\right] \mid k\rangle \quad \text{(370)}$$

so that

$$R(p'k';pk) - \overline{R(pk, p'k')}$$

$$= i \int d^4x\, e^{i(p+k')\cdot\frac{x}{2}} \langle p' \mid \left[j\left(\frac{x}{2}\right), f^*\left(-\frac{x}{2}\right)\right] \mid k\rangle$$

$$= (2\pi)^4 i \sum_{|n\rangle} \{\delta(p_n - p - k) \langle p' \mid j(0) \mid n\rangle \langle n \mid f^*(0) \mid k\rangle$$

$$- \delta(p_n - p' + p) \langle p' \mid f^*(0) \mid n\rangle \langle n \mid j(0) \mid k\rangle\} \quad (371)$$

The intermediate states $|n\rangle$ which contribute to the sum (371) are as follows: for p, k incoming only the first terms contribute, and the states $|n\rangle$ are the same as those arising from the cut a. For p and p' incoming, the intermediate states are precisely those to be expected from the cut b. Note, incidentally, that, since we can also write the scattering amplitude in the form

$$R(n;pk)\,\delta(p_n - p - k) = (2\pi)^{3/2}\, {}_{\text{out}}\langle n \mid f^*(0) \mid k\rangle\, \delta(p_n - p - k) \quad (372)$$

the unitarity relation (371) reads:

$$R(p'k';pk) - \overline{R(pk, p'k')} = 2\pi i \sum_{|n\rangle} \{\delta(p_n - p - k)\, R(p'k';n)\, \overline{R(pk;n)}$$

$$- \delta(p - p' + p_n)\, R(p', -p; n)\, \overline{R(k, -k'; n)}\} \quad (373)$$

The above considerations play an important role in the "derivation" of a Mandelstam representation for the amplitude ${}_{\text{out}}\langle p'k'; pk\rangle_{\text{in}}$. We shall return to this point in Section e of the present chapter.

Let us summarize what we have done up to now in our consideration of the two-particle amplitude $R(\omega, \Delta^2, \zeta)$. We have shown that it has a representation given by Eq. (347) and that Im $R(\omega, \Delta^2, \zeta)$ has a representation given by Eqs. (350) and (357) which embodies what we have called a generalized unitary relation. We are now ready to analyze the dependence of $R(\omega, \Delta^2, \zeta)$ on ω for fixed Δ^2 and ζ. However, before considering the derivation of the dispersion relations when $\Delta^2 \neq 0$, let us briefly sketch the derivation of dispersion relations for forward scattering, i.e., for the case $\Delta^2 = 0$. In this case the amplitude for meson-nucleon scattering can be written as

$$R(pk;pk) = \int d^4x\, e^{ip\cdot x} (\Box + M^2)^2 \langle k \mid R(\psi(x)\, \psi^*(0) \mid k\rangle$$

$$p^2 = M^2;\ k^2 = \mu^2 \quad (374)$$

where p and k denote the momenta of the nucleon and meson, respectively. Note that we have contracted with respect to the nucleon variables [Symanzik (1957)]. In the Lorentz frame in which the meson is at rest $\mathbf{k} = 0$, $k_0 = \mu$, the scattering amplitude has the following representation:

$$R(pk, pk) = i \int_0^\infty dt \int d^3x\, e^{ip_0x_0 - i\sqrt{p_0^2 - M^2}\mathbf{e}\cdot\mathbf{x}} f(\mathbf{x}^2, x_0) \quad (375)$$

where $f(\mathbf{x}^2, x_0)$ vanishes unless $x_0 > |\mathbf{x}|$. The angular integration can therefore again be carried out with the result that

$$R(pk;pk) = 4\pi i \int_0^\infty dr\, r^2 \int_r^\infty e^{iwt} \frac{\sin\sqrt{w^2 - M^2}\, r}{\sqrt{w^2 - M^2}\, r} \cdot \langle k \,|\, [f(x), f^*(0)] \,|\, k\rangle \tag{376a}$$

$$= \int_0^\infty dr\, F(w, r) \tag{376b}$$

with

$$F(w, r) = 4\pi i r \frac{\sin\sqrt{w^2 - M^2}\, r}{\sqrt{w^2 - M^2}} e^{iwr} \int_r^\infty dt\, e^{iw(t-r)} \langle k \,|\, [f(x), f^*(0)] \,|\, k\rangle \tag{377a}$$

$$w = p \cdot k/\mu \tag{377b}$$

Note that the quantity $F(w, r)$ is defined by (377a) for all values of w such that $\operatorname{Im} w \geqslant 0$. In fact, the right-hand side of Eq. (377a) defines an analytic function in $\operatorname{Im} w \geqslant 0$. [There are no branch points at $w = \pm M$ since $\sin\sqrt{w^2 - M^2}\, r/r\sqrt{w^2 - M^2}$ is an even function of $\sqrt{w^2 - M^2}\, r$. Also, when $w < M$, the factor $\exp(iwr)$ compensates for the growth of $\sin\sqrt{w^2 - M^2}\, r$.] Since, by assumption, the matrix element of the commutator is continuous *inside* the light cone, by the Riemann-Lebesgue lemma the asymptotic behavior of $F(w, r)$ for $w \to \infty$ is governed by the singularities of the commutator *on* the light cone. If $\langle k \,|\, [f(x), f^*(0)] \,|\, k\rangle$ behaves on the light cone like a derivative of order n ($n \geqslant 0$) of $\delta(x^2)$, the w dependence of $F(w, r)$ for large w will be like w^{n-1}. As indicated above in a rigorous derivation, we would then study the function $F(w, r)/\prod_{j=1} [(w + ia_j)^2 - b_j^2]$. But, since this is eventually equivalent to making a certain number of subtraction on the Hilbert relations for $F(w, r)$, we shall proceed as if $F(r, w)$ vanished sufficiently rapidly for $w \to \infty$ to be able to write

$$F(w, r) = \frac{1}{\pi} \int_{-\infty}^{+\infty} \frac{\operatorname{Im} F(w', r)}{w' - w - i\epsilon}\, dw' \tag{378}$$

Using the evenness property of $F(w, r)$,

$$F(w, r) = \overline{F(-w, r)} \tag{379}$$

which is readily verified from the representation (377a), it follows that

$$\operatorname{Im} F(w, r) = -\operatorname{Im} F(-w, r) \tag{380}$$

so that we can rewrite Eq. (378) in the form

$$F(w, r) = \frac{1}{\pi} \int_0^\infty dw' \operatorname{Im} F(w', r) \left\{ \frac{1}{w' - w - i\epsilon} + \frac{1}{w' + w + i\epsilon} \right\} \tag{381a}$$

$$= \frac{2}{\pi} \int_0^M dw' \operatorname{Im} F(w', r) \frac{w'}{w'^2 - (w + i\epsilon)^2} + \frac{2}{\pi} \int_M^\infty dw' \operatorname{Im} F(w', r) \frac{w'}{w'^2 - (w + i\epsilon)^2} \tag{381b}$$

To obtain a dispersion relation for the scattering amplitude $R(pk; pk)$, we now have to integrate Eq. (381b) over r. Let us therefore investigate the structure of Im $F(w', r)$ to see whether the interchange of the order of the r and w integration is permissible. From Eq. (374), it follows that

$$\operatorname{Im} R(pk, pk) = \tfrac{1}{2} \int d^4x \, e^{ip\cdot x} \langle k \mid [f(x), f^*(0)] \mid k \rangle \qquad (382)$$

so that again carrying out the angular integrations, we obtain the following representation for Im R:

$$\operatorname{Im} R(pk; pk) = \frac{1}{2} \int_0^\infty dr \, 4\pi r \, \frac{\sin \sqrt{w^2 - M^2}\, r}{\sqrt{w^2 - M^2}} \cdot \int_{-\infty}^{+\infty} dt \, e^{iwt} \langle k \mid [f(x), f^*(0)] \mid k \rangle \qquad (383)$$

from which, upon inserting a complete set and carrying out the t integration, we infer that

$$\operatorname{Im} F(w, r) = 2\pi r \frac{\sin \sqrt{w^2 - M^2}\, r}{\sqrt{w^2 - M^2}} \cdot \sum_{|n\rangle} \frac{\sin |\mathbf{p}_n| \, r}{|\mathbf{p}_n| \, r} |\langle k \mid f(0) \mid n \rangle|^2 \{\delta(w + p_{n0} - \mu) - \delta(w + \mu - p_{n0})\} \qquad (384)$$

[In Eq. (384) we have replaced the factors exp $(\pm i\mathbf{p}_n \cdot \mathbf{x})$ by their angular average.] From the representation (384) we see that, in the range $|w| < M$, we cannot interchange the order of the r and w integration since in this range $\sqrt{w^2 - M^2}\, r$ is imaginary and $\sin \sqrt{w^2 - M^2}\, r$ diverges for $r \to \infty$. Note, however, that only those states which have heavy particle quantum number equal to $+1$ contributes to the sum in (384). More particularly, in the range $0 \leqslant w < M$ only the one-nucleon states contribute.

By carrying out both the space and time integration in (382), we can also write Im $R(pk; pk)$ in the following form:

$$\operatorname{Im} R(pk; pk) = \tfrac{1}{2}(2\pi)^4 \sum_{|n\rangle} |\langle k \mid f(0) \mid n \rangle|^2 \cdot \{\delta^{(4)}(p + k - p_n) - \delta^{(4)}(p - k + p_n)\} \qquad (385)$$

If we now knew the analytic properties of $\langle k \mid f(0) \mid q \rangle = \phi((k - q)^2)$ as a function of $q \cdot k/\mu$ (where $|q\rangle$ is a one-nucleon state, $q^2 = M^2$), we could then try in Eq. (381b) to evaluate the w integration in the range by suitably deforming the path of integration. This procedure was, in fact, carried out by Symanzik (1957) and can be used to justify the interchange of the r and w integrations. One could also adopt the procedure used by Lehmann (1958) to prove dispersion relations for $\Delta^2 \neq 0$. This approach proves that Im R is analytic in M^2 in a strip about the real axis extending to $-\infty$. [This is demonstrated by using the representations (350a) and (357) for

Im R and the Dyson representation for the retarded commutators appearing in Eq. (357).] We can therefore consider Im R for M^2 large and negative, in which case no difficulty is encountered in interchanging the w and r integration. Thereafter we can continue the function to the mass shell value. Alternatively, with Bogoliubov, we could analyze instead of the function $R(pk; pk)$, the function

$$R_1(pk; pk) = ((k + p)^2 - M^2)\,((k - p)^2 - M^2)\,R(pk; pk) \tag{386}$$

It is evident from the structure of Im $R(pk; pk)$ as given by Eq. (385) that multiplying R by $[(k + p)^2 - M^2]$ and $[(k - p)^2 - M^2]$ removes the one-particle singularities so that R_1 has the property that

$$\text{Im}\, R_1(pk; pk) = 0 \quad \text{for } |w| < M \tag{387}$$

Clearly, proceeding as above, $R_1(pk; pk)$ can be shown to have an analytic continuation in the upper half of the complex w plane. Therefore, for R_1 considered as a function of $w = w_1 + iw_2$, we can write down twice-subtracted dispersion relations. [There is now nothing to prevent the interchange of the r and w integration by virtue of (387)!] The subtractions are necessary since, from the form (386), it is clear that R_1 has two more powers of w than R. The analytic properties of R can now be deduced from (386) and the analytic properties of R_1. Thus, besides the branch line singularities of R_1 extending from $-M$ to $-\infty$ and from M to $+\infty$, R has two simple poles at $w = \pm\mu/2$. Thus, one can write

$$\text{Re}\, R(w) = \frac{c}{\left(w - \frac{\mu}{2}\right)\left(w + \frac{\mu}{2}\right)} + \frac{2}{\pi}\,\text{P}\int_M^\infty dw'\,\frac{w'\,\text{Im}\, R(w')}{w'^2 - w^2} \tag{388}$$

By comparing the form (388) and (385), one can easily derive the value of the constant c [see, e. g., Zimmermann (1959)]. This constant c is closely related to the renormalized coupling constant in the Lagrangian formulation. We have thus derived dispersion relations for $R(w)$. It turns out that the same method also works for the case that we contract the meson variable, i.e., for the representation

$$R(pk, pk) = \int d^4x\, e^{ik\cdot x}\,\langle p \mid R(j(x)\, j(0)) \mid p\rangle \tag{389}$$

In this case one derives that in the laboratory system, $\mathbf{p} = 0$, the amplitude $R(\omega)$, where $\omega = p \cdot k/M = (\mu/M)\, w$, obeys the following dispersion relation:

$$\text{Re}\, R(\omega) = \frac{a}{\left(\omega + \frac{\mu}{2M}\right)\left(\omega - \frac{\mu}{2M}\right)} + \frac{2}{\pi}\,\text{P}\int_\mu^\infty d\omega'\,\frac{\omega'\,\text{Im}\, R(\omega')}{\omega'^2 - \omega^2} \tag{390}$$

(P denotes the fact that the principal value of the integral is to be taken at $\omega' = \omega$.) The constant a now is related to the matrix element $|\langle p \mid j(0) \mid q\rangle|^2$ for $(p - q)^2 = \mu^2$, i.e., to the Watson-Lepore coupling constant. The meson field of a nucleon at large distances is given in terms of this coupling constant by a Yukawa potential with a strength proportional to a. We shall return to a discussion of this coupling constant at the end of this section. Let us now return to the problem of obtaining dispersion relations for $R(\omega, \Delta^2, \zeta)$ for fixed Δ^2.

To do this we want to obtain some of the analyticity properties of $R(\omega, \Delta^2, \zeta)$ as a function of ω for fixed Δ^2. In general, the way this is done is to pick a co-ordinate system in which the exponent in the representation of the scattering amplitude contains (at least formally) all the dependence on the variable one wants to obtain dispersion relations in. Thus, the form (347) is most convenient for obtaining information about the ω dependence, whereas the representation

$$R = i \int d^4x\, e^{i(k'-p')\cdot\frac{x}{2}}\, \theta(x)\, \langle 0 \,|\left[j\left(\frac{x}{2}\right), f\left(-\frac{x}{2}\right)\right]|\, p, k\rangle_{\text{in}} \quad (391)$$

is most convenient for obtaining information about the momentum transfer since in the center-of-mass this variable appears only in the exponent.

If one chooses the Lorentz frame in which $\mathbf{p} + \mathbf{p}' = 0$, the so-called brick wall or Breit system, then the kinematics are as follows: since $\mathbf{p} = -\mathbf{p}'$, $p_0 = p'_0 = E(\mathbf{p}) = E(\mathbf{p}') = \sqrt{\mathbf{p}^2 + M^2} = \sqrt{\mathbf{p}'^2 + M^2}$ and by energy conservation $k_0 = k'_0 = \sqrt{\mathbf{k}^2 + \zeta} = \sqrt{\mathbf{k}'^2 + \zeta}$ from which it also follows that $|\mathbf{k}| = |\mathbf{k}'|$. Momentum conservation implies that $|\mathbf{p} + \mathbf{k}| = |\mathbf{p}' + \mathbf{k}'|$ so that $\mathbf{p} \cdot \mathbf{k} = \mathbf{p}' \cdot \mathbf{k}'$ and hence since $\mathbf{p}^2 = \mathbf{p}'^2$,

$$\begin{aligned}(\mathbf{p} + \mathbf{k}) \cdot \mathbf{p} &= -(\mathbf{p} + \mathbf{k}) \cdot \mathbf{p}' = -(\mathbf{p}' + \mathbf{k}') \cdot \mathbf{p}' \\ &= -(\mathbf{p} + \mathbf{k}) \cdot \mathbf{p} = 0 \end{aligned} \quad (392)$$

so that $\mathbf{p} + \mathbf{k}$ is orthogonal to $\mathbf{p}$ (Fig. 18.5). If we denote by $\mathbf{e}$ a unit

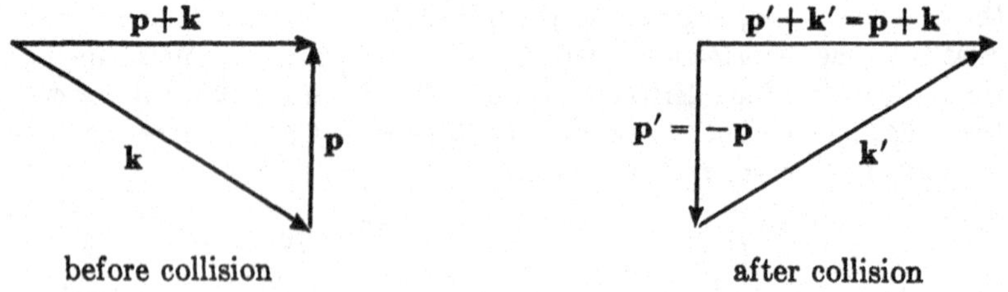

Kinematics in Breit Frame

Fig. 18.5

vector perpendicular to $\mathbf{p}$, $\mathbf{e} \cdot \mathbf{p} = \mathbf{e} \cdot \mathbf{p}' = 0$, and by $\boldsymbol{\Delta}$ the momentum transfer

$$\boldsymbol{\Delta} = \tfrac{1}{2}(\mathbf{p}' - \mathbf{p}) = \mathbf{p}' \quad (393)$$

then

$$\mathbf{k} = \rho\mathbf{e} + \boldsymbol{\Delta} \quad (394a)$$

$$\mathbf{k}' = \rho\mathbf{e} - \boldsymbol{\Delta} \quad (394b)$$

where ρ is the magnitude of the total momentum of the system in the Breit frame: since $\mathbf{e} \cdot \boldsymbol{\Delta} = 0$ and $(\rho\mathbf{e} + \boldsymbol{\Delta})^2 = \mathbf{k}^2$, one has

$$\rho^2 + \boldsymbol{\Delta}^2 = \mathbf{k}^2 \tag{395}$$

Summarizing, in the Breit frame

$$\left.\begin{aligned} p &= \{E_{\boldsymbol{\Delta}}, -\boldsymbol{\Delta}\} = p_{-\Delta}; \quad E_{\boldsymbol{\Delta}}{}^2 = \boldsymbol{\Delta}^2 + M^2 \\ p' &= \{E_{\boldsymbol{\Delta}}, \boldsymbol{\Delta}\} = p_{\Delta} \end{aligned}\right\} \tag{396a}$$

$$k = \{\sqrt{\mathbf{k}^2 + \zeta}, \rho\mathbf{e} + \boldsymbol{\Delta}\}$$

$$k' = \{\sqrt{\mathbf{k}^2 + \zeta}, \rho\mathbf{e} - \boldsymbol{\Delta}\} \tag{396b}$$

with $\mathbf{e} \cdot \boldsymbol{\Delta} = 0$, $\mathbf{e}^2 = 1$, $\rho^2 = \mathbf{k}^2 - \boldsymbol{\Delta}^2$, so that

$$\tfrac{1}{2}(p + p') = \{E_{\boldsymbol{\Delta}}, 0\} \tag{397a}$$

$$\tfrac{1}{2}(k + k') = Q = \{\sqrt{\mathbf{k}^2 + \zeta}, \rho\mathbf{e}\} \tag{397b}$$

and also

$$\Delta^2 = \boldsymbol{\Delta}^2 \tag{398}$$

$$\omega = \frac{(p + p') \cdot (k + k')}{2\sqrt{(p + p')^2}} = \sqrt{\mathbf{k}^2 + \zeta} \tag{399}$$

Note, therefore, that, when $\zeta = \mu^2$, ω reduces to the meson energy. In the Breit frame, the scattering amplitude is given by

$$R(\omega, \boldsymbol{\Delta}^2, \zeta) = i \int d^3x \int_0^\infty dt\, e^{i\omega t - i\sqrt{\omega^2 - \Delta^2 - \zeta}\,\mathbf{e}\cdot\mathbf{x}} \cdot \langle E_{\boldsymbol{\Delta}}, \boldsymbol{\Delta} | \left[j\left(\frac{x}{2}\right), j\left(-\frac{x}{2}\right)\right] | E_{\boldsymbol{\Delta}}, -\boldsymbol{\Delta}\rangle \tag{400a}$$

$$= i \int d^3x \int_0^\infty dt\, e^{i\omega t - i\sqrt{\omega^2 - \Delta^2 - \zeta}\,\mathbf{e}\cdot\mathbf{x}} f(p \cdot p', p \cdot x, p' \cdot x, x^2) \tag{400b}$$

The important consequence of this choice of the co-ordinate system is that the entire dependence of R on ω is contained in the exponential. Causality asserts that the commutator vanishes unless $t^2 > \mathbf{x}^2$ and furthermore, due to the $\theta(x_0)$ factor, the integration over t only extends over positive t values for $t > |\mathbf{x}|$. Let us define $R(\omega, \boldsymbol{\Delta}^2, \zeta)$ for complex $\omega = \omega_1 + i\omega_2$ by the right-hand side of (400a). Since for arbitrary real positive ϵ

$$\text{Im}\sqrt{\omega^2 - \epsilon} > \text{Im}\,\omega \tag{401}$$

the so-defined expression $R(\omega, \Delta^2, \zeta)$ will not necessarily be analytic in the upper half of the complex ω plane for ζ positive and real—since the integral may diverge. If, however, we let ζ likewise become complex $\zeta = \zeta_1 + i\zeta_2$, then (400a) defines an analytic function $R'(\omega, \Delta^2, \zeta)$ of ω and ζ regular in the domain $\mathcal{R}$

$$\{\mathcal{R}\} : \text{Im}\,\omega > |\text{Im}\sqrt{\omega^2 - \Delta^2 - \zeta}|$$

$$\text{Im}\,\omega > 0 \tag{402}$$

Explicitly, this domain is defined by

$$\omega_2 > 0 \qquad \omega_1^2 > \zeta_1 + \Delta^2 \tag{403a}$$

$$2\omega_2(\omega_1 - \sqrt{\omega_1^2 - \zeta_1 - \Delta^2}) < \zeta_2 < 2\omega_2(\omega_1 + \sqrt{\omega_1^2 - \zeta_1 - \Delta^2}) \tag{403b}$$

If we take ζ real, with $\zeta = \zeta_1 < -\Delta^2$, then R is regular in $\operatorname{Im} \omega > 0$. If now R is sufficiently bounded at infinity, we can then write

$$R(\omega, \Delta^2, \zeta_1) = \frac{1}{\pi} \int_{-\infty}^{+\infty} d\omega' \, \frac{\operatorname{Im} R(\omega', \Delta^2, \zeta_1)}{\omega' - \omega - i\epsilon} \tag{404}$$

In the case R does not have the requisite properties for large ω for (404) to hold, we can always divide R by a polynomial in ω which has no zeros in the upper half plane. Equivalently we can write

$$R(\omega, \Delta^2, \zeta_1) - R(\omega_0, \Delta^2, \zeta_1) = \frac{\omega - \omega_0}{\pi} \int_{-\infty}^{+\infty} d\omega' \, \frac{\operatorname{Im} R(\omega', \Delta^2, \zeta_1)}{(\omega' - \omega)(\omega' - \omega_0)} \tag{405}$$

where ω_0 is a fixed point. If necessary, further subtractions can be made to improve the convergence of the right-hand side.[21] Equation (404) can be rewritten as follows:

$$R(\omega, \Delta^2, \zeta_1) = \frac{1}{2\pi} \int_0^{\infty} d\omega' \, M(\omega', \Delta^2, \zeta_1) \left[\frac{1}{\omega' - \omega - i\epsilon} + \frac{1}{\omega' + \omega} \right] \tag{406}$$

using the representation (350a) for $\operatorname{Im} R$. More specifically, in the Breit frame, calling $\frac{1}{2}(k + k') = Q = \{\sqrt{\mathbf{k}^2 + \zeta_1}, \rho \mathbf{e}\}$, $M(\omega, \Delta^2, \zeta_1)$ has the following representation

$$\begin{aligned} M(\omega, \Delta^2, \zeta_1) &= \int d^4x \, e^{i(k+k') \cdot \frac{x}{2}} \langle p_\Delta | j\left(\frac{x}{2}\right) j\left(-\frac{x}{2}\right) | p_{-\Delta} \rangle \\ &= (2\pi)^4 \sum_{|p_n\alpha\rangle} \langle p_\Delta | j(0) | p_n\alpha \rangle \langle p_n\alpha | j(0) | p_{-\Delta} \rangle \, \delta\left(-p_n + \frac{k + k'}{2} + \frac{p_\Delta + p_{-\Delta}}{2}\right) \\ &= (2\pi) \sum_{p_{n0}, \alpha} \langle p_\Delta | j(0) | p_{n0}, \mathbf{Q}, \alpha \rangle \langle p_{n0}, \mathbf{Q}, \alpha | j(0) | p_{-\Delta} \rangle \cdot \delta(p_{n0} - E_\Delta - \omega) \end{aligned} \tag{407}$$

Now, since $p_{n0} \geqslant 0$,

$$\delta(p_{n0} - E_\Delta - \omega) = 2p_{n0}\delta(p_{n0}^2 - (E_\Delta + \omega)^2) \tag{408}$$

so that, if we call $\sigma^2 = p_{n0}^2 - \mathbf{p}_n^2 = p_{n0}^2 - \rho^2$, then

$$\delta(p_{n0}^2 - (E_\Delta + \omega)^2) = \delta(\sigma^2 + \rho^2 - (E_\Delta + \omega)^2) \tag{409}$$

Furthermore, since $\rho^2 = \mathbf{k}^2 - \Delta^2$, this delta function can also be rewritten as follows:

[21] The behavior of R for large ω has been investigated by Symanzik (1957) in the case of one-dimensional scattering. He has shown that in this case R behaves as ω^n (n finite) for large ω.

$$\delta(p_{n0}{}^2 - (E_\Delta + \omega)^2) = \delta(\sigma^2 + \mathbf{k}^2 - \Delta^2 - (E_\Delta + \omega)^2)$$
$$= \delta(\sigma^2 - M^2 - \zeta_1 - 2\Delta^2 - 2E_\Delta\omega) \quad (410)$$

so that finally, since $\sum_{p_{n0}} 2p_{n0} \to \int_0^\infty dp_{n0}\, 2p_{n0} = \int_0^\infty d\sigma^2$, we obtain the following representation for $M(\omega, \Delta^2, \zeta)$

$$M(\omega, \Delta^2, \zeta_1) = \int_0^\infty d\sigma^2\, \mu(\sigma^2, \Delta^2, \zeta_1)\, \delta(\sigma^2 - M^2 - \zeta_1 - 2\Delta^2 - 2E_\Delta\omega) \quad (411)$$

where

$$\mu(\sigma^2, \Delta^2, \zeta_1) = 2\pi \sum_\alpha \langle p_\Delta \mid j(0) \mid \sigma^2, \rho_\Delta(\zeta_1), \alpha\rangle \langle \sigma^2, \rho_\Delta(\zeta_1), \alpha \mid j(0) \mid p_{-\Delta}\rangle \quad (412)$$

and

$$\rho_\Delta(\zeta_1) = \sqrt{\omega_\Delta(\sigma)^2 - \zeta_1 - \Delta^2} \quad (413a)$$

$$\omega_\Delta(\sigma^2) = \frac{1}{2E_\Delta}(\sigma^2 - \Delta^2 - E_\Delta{}^2 - \zeta_1) \quad (413b)$$

If in the sum over α we separate the contribution from the one-nucleon state, for which $\sigma^2 = M^2$, then

$$\mu(\sigma^2, \Delta^2, \zeta_1) = \delta(\sigma^2 - M^2)\, g(\zeta_1)\, \overline{g(\zeta_1)} + \Theta(\sigma^2, \Delta^2, \zeta_1)\, \theta(\sigma^2 - (M + \mu)^2) \quad (414a)$$

where

$$g(\zeta_1) = \langle p_\Delta \mid j(0) \mid \sigma^2 = M^2, \rho_\Delta \mathbf{e}\rangle \quad (414b)$$

is the matrix element of the current operator between one-nucleon states of momentum Δ and $\rho_\Delta\mathbf{e}$, with ρ_Δ given by (413a, b) with $\sigma^2 = M^2$. By virtue of relativistic invariance g is a function only of $p_\Delta{}^2 = M^2$, $p_n{}^2 = M^2$ [where $p_n = \{p_{n0}, \rho_\Delta\mathbf{e}\}$; $p_{n0}{}^2 - \rho_\Delta{}^2 = M^2$] and of $(p_\Delta - p_n)^2$. But

$$(p_\Delta - p_n)^2 = (E_\Delta - p_{n0})^2 - (\Delta - \rho_\Delta\mathbf{e})^2$$
$$= (E_\Delta - (E_\Delta + \omega))^2 - \Delta^2 - \rho_\Delta{}^2$$
$$= \omega_\Delta{}^2 - \Delta^2 - \rho_\Delta{}^2 = \zeta_1 \quad (415)$$

whence $g = g(\zeta_1)$. The structure of $M(\omega, \Delta^2, \zeta_1)$ is therefore as follows:

$$M(\omega, \Delta^2, \zeta_1) = \frac{|g(\zeta_1)|^2}{2E_\Delta}\, \delta\left(\omega + \frac{\zeta_1 + 2\Delta^2}{2E_\Delta}\right)$$
$$+ \int_{(M+\mu)^2}^\infty d\sigma^2\, \Theta(\sigma^2, \Delta^2, \zeta_1)\, \delta(\sigma^2 - M^2 - \zeta_1 - 2\Delta^2 - 2E_\Delta\omega) \quad (416a)$$

so that the dependence of $M(\omega, \Delta^2, \zeta_1)$ on ω is such that $M(\omega, \Delta^2, \zeta_1)$ is different from zero only for values of ω such that

$$\omega = -\frac{\zeta_1 + 2\Delta^2}{2E_\Delta} \quad (416b)$$

and

$$\omega \geqslant \frac{(M+\mu)^2 - \zeta_1 - M^2 - 2\Delta^2}{2E_\Delta} \tag{416c}$$

If $\zeta_1 < -2\Delta^2$, then the nonvanishing values of $M(\omega, \Delta^2, \zeta)$ are as indicated in Figure 18.6. Note the existence of the gap between the discrete ω value

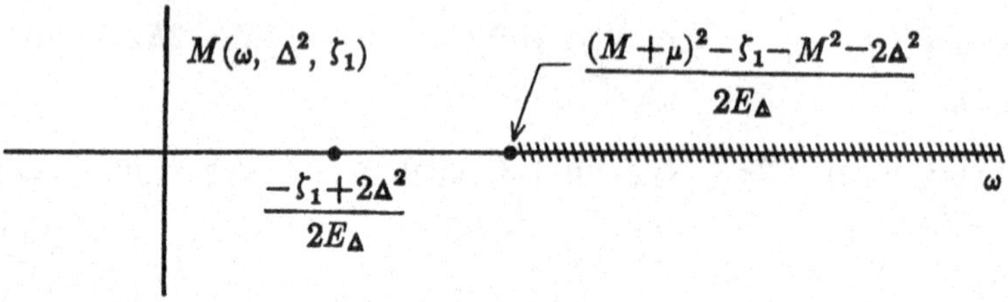

Fig. 18.6

(the contribution to M of the one-nucleon state) and the contributions of the higher mass states. Substituting the expression (415) into (406), we obtain

$$R(\omega, \Delta^2, \zeta_1) = \frac{|g(\zeta_1)|^2}{2\pi}\left[-\frac{1}{\zeta_1 + 2\Delta^2 + 2E_\Delta\omega} + \frac{1}{\zeta_1 + 2\Delta^2 - 2\omega E_\Delta}\right] + \frac{1}{2\pi}\int_{(M+\mu)^2}^{\infty} d\sigma^2\, \Theta(\sigma^2, \Delta^2, \zeta_1)$$
$$\left[\frac{1}{\sigma^2 - M^2 - \zeta_1 - 2\Delta^2 - 2\omega E_\Delta} + \frac{1}{\sigma^2 - M^2 - \zeta_1 - 2\Delta^2 + 2\omega E_\Delta}\right] \tag{417a}$$

$$= \frac{1}{(2\pi)}\int_{M^2}^{\infty} d\sigma^2\, M(\sigma^2, \Delta^2, \zeta_1)$$
$$\left[\frac{1}{\sigma^2 - 2\Delta^2 - M^2 - \zeta_1 - 2\omega E_\Delta} + \frac{1}{\sigma^2 - 2\Delta^2 - M^2 - \zeta_1 + 2\omega E_\Delta}\right] \tag{417b}$$

Introducing the variables $W^2 = 2\omega E_\Delta + 2\Delta^2 + M^2 + \zeta_1$, Eq. (417b) can also be written as

$$R(W, \Delta^2, \zeta_1) = \frac{|g(\zeta_1)|^2}{2\pi}\left[\frac{1}{W^2 - 4\Delta^2 - M^2 - 2\zeta_1} - \frac{1}{W^2 - M^2}\right] + \frac{1}{2\pi}\int_{(M+\mu)^2}^{\infty} dW'^2\, M(W'^2, \Delta^2, \zeta_1)$$
$$\left[\frac{1}{W'^2 - W^2} + \frac{1}{W'^2 - 2M^2 - 2\zeta_1 - 4\Delta^2 + W^2}\right] \tag{418}$$

The real part of (418) would be a dispersion relation but for unphysical, negative values of ζ. In order to obtain dispersion relations for $\zeta = \mu^2$, one must show that $R(\omega, \Delta^2, \zeta)$, as defined by the right-hand side of Eq. (417b), is analytic in ζ in a strip along the real axis, so that it can be continued to values of ζ near μ^2. One must then further show that the function thus obtained by analytic continuation in ζ is the same as the original scattering amplitude for real and positive ω.

Bogoliubov (1959) indeed has shown that a dispersion relation for the scattering amplitude $R(\omega, \Delta^2, \mu^2)$ follows from (417b) by analytic continuation [see also Vladimirov and Logunov (Vladimirov 1959)]. Using the theory of several complex variables, he showed that $M(\sigma^2, \Delta^2, \zeta)$ for $\Delta^2 \geqslant 0$ and $\sigma^2 \geqslant M^2$ is an analytic function of ζ regular for $\operatorname{Re} \zeta = \zeta_1 \leqslant \mu^2$ in a neighborhood of the real axis, $|\operatorname{Im} \zeta| < \delta$. If this is true, then (417b) defines an analytic function $R''(\omega, \Delta^2, \zeta)$ of ω and ζ regular in a region $\mathcal{R}''$ consisting of those points such that $\zeta_1 \leqslant \mu^2$, $|\zeta_2| < \delta$, $|\zeta_2| < 2E_\Delta \operatorname{Im} \omega$. For ζ real and $\zeta_1 < -\Delta^2$

$$R'(\omega, \Delta^2, \zeta_1) = R''(\omega, \Delta^2, \zeta_1) \tag{419}$$

Furthermore, the boundary value of R'' as $\omega_2 = \operatorname{Im} \omega \to 0$ for $\zeta = \mu^2$ and $\omega_1 > \sqrt{\Delta^2 + \mu^2}$ is the physical scattering amplitude $R(\omega, \Delta^2)$ defined by Eq. (331) because this boundary can be reached along a path which lies in the intersection of $\mathcal{R}'$ and $\mathcal{R}''$. [For a precise demonstration of these last two points, see Bogoliubov (1958); Bremmerman, Oehme, and Taylor (Bremmerman 1958); and Minguzzi and Streater (Minguzzi 1960).]

Lehmann (1958, 1959) has given a proof of the analyticity properties of $M(\omega, \Delta^2, \zeta)$ as a function of ζ required to deduce dispersion relations for $\zeta = \mu^2$ which does not make use of the theory of several complex variables, but instead uses the Dyson representation for the causal commutators appearing in Eq. (357) which defines $M(\omega, \Delta^2, \zeta)$ for $\zeta < m_1^2$. We here outline his proof.

To each retarded commutator in Eq. (357) there corresponds a weight function $\Phi_\alpha(u, \kappa^2, p + k)$ which vanishes unless

$$\begin{cases} \tfrac{1}{2}(p+k) + u \in L^+;\ \tfrac{1}{2}(p+k) - u \in L^+ \\ \kappa \geqslant \operatorname{Max}\left\{0; m_1 - \sqrt{\left(\dfrac{p+k}{2} + u\right)^2}; m_2 - \sqrt{\left(\dfrac{p+k}{2} - u\right)^2}\right\} \end{cases} \tag{420}$$

where for the π-N case $m_1 = 3\mu$ and $m_2 = M + \mu$. These weight functions are multiplied together and summed over α to give an overall weight function

$$\Phi(u_1, u_2, \kappa_1, \kappa_2; p + k) = \sum_\alpha \Phi_\alpha(u_1, \kappa_1; p + k)\, \overline{\Phi_\alpha(u_2, \kappa_2; p + k)} \tag{421}$$

which satisfies the support conditions (420) in each pair of variables u_i, κ_i ($i = 1, 2$) separately. This weight function is a real invariant function of the vectors u_1, u_2, $p + k$ and hence only a function of their scalar products. It is of the utmost importance in the following that Φ is only a function of $W^2 = (p + k)^2$ and does not independently depend on ζ. The representation of $M(W^2, \Delta^2, \zeta)$ which follows from inserting the Dyson representation for the matrix elements of the retarded commutators appearing in Eq. (357), after carrying out the summation (421), is

$$M(W^2, \Delta^2, \zeta) = \frac{1}{2\pi} \int \frac{d^4u_1\, d^4u_2\, d\kappa_1^2\, d\kappa_2^2\, \Phi(u_1, u_2, \kappa_1, \kappa_2; p + k)}{\left[\left(\frac{k' - p'}{2} - u_1\right)^2 - \kappa_1^2\right]\left[\left(\frac{k - p}{2} - u_2\right)^2 - \kappa_2^2\right]} \tag{422}$$

Since Φ depends only on the kinematic variable W, the entire Δ^2 and ζ dependence is contained in the denominator of Eq. (422). It is easiest to evaluate the expression (422) in the center-of-mass system, $\mathbf{p} + \mathbf{k} = 0$, in which case the weight function Φ becomes a function of u_1, u_2, u_1^2, u_2^2, u_{10}, u_{20}, κ_1^2, κ_2^2 and of W. With the following choice of co-ordinate systems,

$$\begin{aligned}
\mathbf{k} &= K(1, 0, 0)\\
\mathbf{k}' &= K(\cos\theta, \sin\theta, 0)\\
\mathbf{u}_1 &= u_1(\cos\varphi_1 \sin\vartheta_1, \sin\varphi_1 \sin\vartheta_1, \cos\vartheta_1)\\
\mathbf{u}_2 &= u_2(\cos\varphi_2 \sin\varphi_2, \sin\varphi_2 \sin\vartheta_2, \cos\vartheta_2)
\end{aligned} \tag{423}$$

and the use of $\chi = \varphi_1$ and $\alpha = \varphi_1 - \varphi_2$ as integration variables, some of the integration can be carried out [Lehmann (1958, 1959)]. In these integrations it is useful to consider the weight function Φ as a function of u_{10}, u_{20}, $\mathbf{u}_1^2$, $\mathbf{u}_2^2$, $\mathbf{u}_1 \cdot \mathbf{u}_2/|\mathbf{u}_1|\,|\mathbf{u}_2|$, κ_1^2, κ_2^2 and of W. With the introduction of the co-ordinates (423), $M(W^2, \Delta^2, \zeta)$ becomes equal to

$$M(W^2, \Delta^2, \zeta) = \frac{1}{8\pi K^2} \int du_{10} \int du_{20} \int du_1 \int du_2 \int d\kappa_1^2 \int d\kappa_2^2 \int_0^{2\pi} d\alpha$$
$$\int_0^{\pi} d\vartheta_1 \int_0^{\pi} d\vartheta_2 \int_0^{2\pi} d\chi\, \frac{\Phi(u_{i0}, u_i^2, \kappa_i^2, \cos\alpha \sin\vartheta_1 \sin\vartheta_2 + \cos\vartheta_1 \cos\vartheta_2, W)}{[x_1(\zeta) - \cos(\theta - \chi)]\,[x_2(\zeta) - \cos(\chi - \alpha)]} \tag{424}$$

where

$$x_i(\zeta) = \frac{K^2 + u_i^2 + \kappa_i^2 - \left(u_{i0} + \dfrac{M^2 - \zeta}{2W}\right)}{2Ku_i \sin\vartheta_i} \tag{425a}$$

and

$$K^2 = \frac{(W^2 + M^2 - \zeta)^2 - 4M^2W^2}{4W^2} \tag{425b}$$

$$\cos\theta = 1 - \frac{2\Delta^2}{K^2} \tag{425c}$$

Upon introducing the new variables of integrations u'_i and $\kappa_i'^2$ by

$$u'_i = u_i \sin \vartheta_i \tag{426a}$$

$$u'^2_i + \kappa'^2_i = u_i^2 + \kappa_i^2 \tag{426b}$$

[to eliminate the ϑ_i dependence of the denominators in (424)] and noting that

$$\int_0^{2\pi} \frac{d\chi}{[x_1 - \cos(\theta - \chi)]\,[x_2 - \cos(\chi - \alpha)]} = 2\pi \frac{\dfrac{x_1}{\sqrt{x_1^2 - 1}} + \dfrac{x_2}{\sqrt{x_2^2 - 1}}}{x_1 x_2 + \sqrt{x_1^2 - 1}\,\sqrt{x_2^2 - 1} - \cos(\theta - \alpha)} \tag{427}$$

we can then carry out the ϑ_i and χ integrations. The ϑ_i variables after the change of variables (426) appear only in the weight function Φ, so that integrating over them merely results in a new weight function $\Phi'(u_{i0}, u'_i, \kappa'^2_i, \cos\alpha, W)$. The range of the u'_i, κ'_i variables is, incidentally, the same as those of the u_i, κ_i. Carrying out the integration then yields (dropping primes)

$$M(W, \Delta^2, \zeta) = \int \cdots \int du_{10}\, du_1\, du_{20}\, du_2\, d\kappa_1^2\, d\kappa_2^2 \int_0^{2\pi} d\alpha\, \Phi'(u_{10}, u_1, u_{20}, u_2, \kappa_1^2, \kappa_2^2, \cos\alpha, W) \cdot \frac{\dfrac{y_1}{\sqrt{y_1^2 - K^2}} + \dfrac{y_2}{\sqrt{y_2^2 - K^2}}}{y_1 y_2 + \sqrt{y_1^2 - K^2}\,\sqrt{y_2^2 - K^2} - K^2 \cos(\theta - \alpha)} \tag{428a}$$

where

$$y_i = \frac{K^2 + u_i^2 + \kappa_i^2 - \left(u_{i0} - \dfrac{M^2 - \zeta}{2W}\right)^2}{2u_i} \tag{428b}$$

and the variables u_{i0}, u_i and κ_i vary over the region

$$\begin{cases} 0 \leqslant u_i \leqslant \dfrac{W}{2}, \quad |u_{i0}| \leqslant \dfrac{W}{2} - u_i \\ \kappa_i \geqslant \operatorname{Max}\left\{0; m_1 - \sqrt{\left(\dfrac{W}{2} + u_{i0}\right)^2 - u_i^2}; m_2 - \sqrt{\left(\dfrac{W}{2} - u_{i0}\right)^2 - u_i^2}\right\} \end{cases} \tag{428c}$$

We again wish to emphasize that the importance of the representation (428) for $M(W, \Delta^2, \zeta)$ lies in the fact that its entire dependence on θ (or alternatively Δ^2) and ζ is explicitly exhibited in the third line of Eq. (428a). The quantity $M(W, \Delta^2, \zeta)$ will have singularities in these variables if either

$$y_i^2 - K^2 = 0$$

and/or

$$y_1y_2 - \sqrt{y_1^2 - K^2}\sqrt{y_2^2 - K^2} - K^2\cos(\theta - \alpha) = 0$$

as u_i, u_{i0}, κ_i vary over the region (429). One verifies, however, that since for $K^2 \geqslant 0$, and $\zeta \leqslant \mu^2$, the minimum value of $y_i^2 - K^2$ as u_{i0}, u_i, κ_i^2 vary over the region (428c) is

$$\text{Min}\,\{y_i^2 - K^2\} = \frac{(m_1^2 - \mu^2)(m_2^2 - M^2)}{W^2 - (m_1 - m_2)^2} \tag{429}$$

$y_i^2 - K^2$ will never vanish if $m_1 > \mu$, $m_2 > M$ and $W^2 > (m_2 - m_1)^2$ which is always satisfied in applications. The second possibility when $\zeta^2 \leqslant \mu^2$ likewise will not cause any singularities if $\Delta^2 \leqslant \Delta_{\max}^2$ where

$$\Delta_{\max}^2 = \underset{W}{\text{Min}}\left\{K^2 + \frac{(m_1^2 - \mu^2)(m_2^2 - M^2)}{W^2 - (m_1 - m_2)^2}\right\} \tag{430}$$

[For the details of these assertions, see Lehmann (1958, 1959).] For π-N scattering[22]

$$\Delta_{\max}^2 = \frac{8}{3}\frac{2M + \mu}{2M - \mu}\mu^2 \sim 3\mu^2 \tag{431}$$

For Δ^2 less than this $\Delta_{\max}^2$, $M(W, \Delta^2, \zeta)$ is analytic in ζ in a strip about $|\text{Im}\,\zeta| < \delta$, $\text{Re}\,\zeta = \zeta_1 \leqslant \mu^2$. We can therefore analytically continue the right-hand side of Eq. (406) to the mass shell value $\zeta_1 = \mu^2$ [see, however, Minguzzi (1960) for a rigorous demonstration]. The same procedure can be used to show that the function $g(\zeta)$ defined by Eq. (414b) has an analytic continuation to the point $\zeta = \mu^2$. The quantity $g(\mu^2)$ plays the role of a coupling constant [Oehme and Taylor (Oehme 1959)].

Our task is however not complete yet since in the dispersion relations (417) or (418) when $\Delta^2 \neq 0$ (i.e., in the non-forward scattering case) there always exists a nonphysical region in the dispersion integral corresponding to values of ω' such that $\omega < \sqrt{\Delta^2 + \mu^2}$ in (417a), or $W^2 < (\sqrt{\Delta^2 + M^2} + \sqrt{\Delta^2 + \mu^2})^2$ in (418). In this nonphysical region the integrand $M(W, \Delta^2) = 2\,\text{Im}\,R(W, \Delta^2)$ has not yet been expressed in terms

[22] For nucleon-nucleon scattering, in which case $m_1 = m_2 = M + \mu$ and $W \geqslant 2\mu$,

$$\Delta_{\max}^2 = \mu(2M + \mu) - M^2$$

which is positive only if

$$\frac{\mu}{M} > \sqrt{2} - 1$$

Only if the ratio μ/M satisfies this condition can one *prove* a dispersion relation for forward scattering in the N-N case using the above methods. Symanzik (1958), on the other hand, has proved that each term in the perturbation theory expansion of $R(p'q'; pq)$, the N-N scattering amplitude, satisfies dispersion relations in W for fixed Δ^2 if $\Delta \leqslant \mu/2$ [see also Taylor (1959)]. For a discussion of the analytic properties of scattering amplitudes and vertex functions in perturbation theory, see, e.g., Nambu (1957); Nakanishi (1957); Symanzik (1957); Karplus (1959); Tarski (1960); Björken (1959); Landau (1959); and particularly Eden (1960).

of experimental quantities. For π-N elastic scattering, Lehmann (1958) has shown that $\text{Im}\, R(W, \Delta^2)$ is uniquely determined in the unphysical region by the nearly forward amplitudes at the same energy, and is, in fact, determined by analytic continuation in the momentum transfer variable Δ^2. The same situation holds whenever there is no unphysical region in forward scattering. More precisely, Lehmann shows that on the mass shell, $\zeta = \mu^2$, $M(W, \Delta^2)$ has the following representation:

$$M(W, \Delta^2) = \int_{2z_0^2-1}^{\infty} dz \int_{-1}^{+1} d(\cos\alpha) \frac{\Phi(W, \cos\alpha, z)}{z - \cos(\theta - \alpha)} \tag{432a}$$

$$= 2\, \text{Im}\, R(W, \Delta^2) \tag{432b}$$

as follows by introducing the new integration variable

$$z = (y_1 y_2 - \sqrt{y_1^2 - K^2}\, \sqrt{y_2^2 - K^2})/K^2 \tag{433}$$

into the representation (428). In Eq. (432a)

$$z_0 = 1 - \frac{2(m_1^2 - \mu^2)\,(m_2^2 - m^2)}{K^2[W^2 - (m_1 - m_2)^2]} \tag{434}$$

$M(W, \Delta^2)$ will therefore have certain analyticity properties as a function of $\cos\theta$, where $\cos\theta$ is the scattering angle. These properties clearly follow from Eq. (432a) since θ enters only in the kernel $1/(z - \cos(\theta - \alpha))$. More precisely, $M(W, \Delta^2)$ is an analytic function of $\cos\theta$, regular in an ellipse with foci at ± 1 and with axes $2z_0^2 - 1$ and $2z_0\sqrt{z_0^2 - 1}$ with z_0 given by (434). This ellipse includes the physical region $-1 \leqslant \cos\theta \leqslant +1$ and the region in which the dispersion relations are proved. This, of course, is a satisfactory result.

Actually already by using the representation (391) for $R(W, \Delta^2, \mu^2)$ one can prove that $R(W, \Delta^2, \mu^2)$ is an analytic function of $\cos\theta$, regular inside an ellipse in the $\cos\theta$ plane with center at the origin and with axes z_0, $\sqrt{z_0^2 - 1}$. This result follows from inserting the Dyson representation for the retarded commutator appearing in (391) [Lehmann (1958)], in which case Eq. (391) reads

$$R(W, \Delta^2, \mu^2) = \frac{1}{2\pi} \int \frac{d^4u\, d\kappa^2\, \Psi(u, \kappa^2, p, k)}{\left[\dfrac{k' - p'}{2} - u\right]^2 - \kappa^2} \tag{435}$$

where the weight function Ψ is an invariant function of the vectors u, p, k and vanishes outside the region

$$0 < u \leqslant \frac{W}{2};\ -\frac{W}{2} + u < u_0 < \frac{W}{2} - u$$

$$\kappa \geqslant \text{Max}\left\{0; m_1 - \sqrt{\left(\frac{W}{2} + u_0\right)^2 - u^2}; m_2 - \sqrt{\left(\frac{W}{2} - u_0\right)^2 - u^2}\right\} \tag{436}$$

In the center-of-mass system, $\mathbf{p} + \mathbf{k} = 0$, by a suitable choice of integration variables, the above representation can be reduced to the form

$$R(W, \Delta^2, \mu^2) = \int_{z_0}^{\infty} dz \int_0^{2\pi} d\alpha \frac{\Psi(z, \cos \alpha)}{z - \cos(\theta - \alpha)} \tag{437a}$$

If we expand the denominator of Eq. (437a), we obtain $z - \cos\theta\cos\alpha - \sqrt{1 - \cos^2\theta}\sin\alpha$. It would therefore appear that we must make a cut in the $\cos\theta$ plane due to the presence of the square-root factor. However, by separating the range of integration of the α variable from 0 to π and from π to 2π and making the change of variable $\alpha = 2\pi - \alpha$ in the range π to 2π, we can recast Eq. (437a) in the form

$$R(W, \Delta^2, \mu^2) = \int_{z_0}^{\infty} dz \int_0^{\pi} d\alpha \frac{\Psi(z, \cos\alpha)\,(z - \cos\theta\cos\alpha)}{(z - \cos\theta\cos x)^2 - \sin^2\theta\sin^2\alpha} \tag{437b}$$

This form makes it apparent that, in fact, R is a function of $\cos\theta$ only and that no branches are introduced by the $\sin\theta$ factor. Singularities of R as a function of θ occur only if the denominator vanishes, i.e., when $\cos\theta = z\cos\alpha \pm i\sin\alpha\sqrt{z^2 - 1}$. If we write $\cos\theta = u + iv$, then the singularities lie on or outside the ellipse

$$\frac{u^2}{z^2} + \frac{v^2}{z^2 - 1} = 1; \quad z_0 \leqslant z \leqslant \infty \tag{438}$$

The result of this investigation [Lehmann (1958)] is therefore that $R(W, \Delta^2)$, and hence also Re (W, Δ^2) and Im $R(W, \Delta^2)$, are analytic functions regular inside an ellipse in the $\cos\theta$ plane with center at the origin and with axes z_0 and $\sqrt{z_0^2 - 1}$, with z_0 given by (434).

Note, incidentally, that the representation (435), although yielding information about the θ or Δ^2 dependence of $R(W, \Delta^2)$, does not yield any information regarding the analyticity region of R as a function of ζ, since the weight function depends in an unknown fashion on $k^2 = \zeta$. Note also that the representation (428) for Im $R(W, \Delta^2)$ yielded the fact that Im $R(W, \Delta^2)$ was analytic in a larger region in the $\cos\theta$ plane than was implied by the representation (391).

The above results imply that if one expands $R(W^2, \Delta^2)$ into partial waves,

$$R(W^2, \Delta^2) = \frac{1}{\pi^2}\frac{W}{K}\sum_{l=0}^{\infty}(2l + 1)\,C_l(W)\,P_l\left(1 - \frac{2\Delta^2}{K^2}\right) \tag{439a}$$

$$C_l(W) = \frac{\pi^2 K}{2W}\int_{-1}^{+1} d(\cos\theta)\,R(W^2, \cos\theta)\,P_l(\cos\theta) \tag{439b}$$

then this expansion converges inside the ellipse with axes z_0, $\sqrt{z_0^2 - 1}$. One can therefore first obtain the coefficient $C_l(W)$ from the physical amplitudes and then define Im R in the unphysical region by analytic continuation of the imaginary part of (439a). Collecting all the above results, the unsubtracted physical dispersion relations are therefore given by

$$\text{Re}\, R(W, \Delta^2) = \frac{g^2(\mu^2)}{2\pi}\left[-\frac{1}{W^2 - M^2} + \frac{1}{W^2 - 4\Delta^2 - M^2 - 2\mu^2}\right] + \frac{1}{\pi} \text{P} \int_{(M+\mu)^2}^{\infty} dW'^2 \,\text{Im}\, R(W'^2, \Delta^2) \cdot \left\{\frac{1}{W'^2 - W^2} + \frac{1}{W'^2 - 4\Delta^2 - 2(M^2 + \mu^2) + W^2}\right\} \tag{440}$$

The decomposition (439a) of the scattering amplitude is, of course, essentially a phase shift analysis. If one were to attempt to see whether the non-forward dispersion relations (440) (or their straightforward generalization to more realistic situations involving charged particles with spin) are satisfied by the experimental data for a given physical system, these phase shifts would presumably be obtained from the differential cross section. Dyson [unpublished; quoted in Thirring (1959)] has, however, shown that the use of experimentally obtained phase shifts will not allow one to ascertain whether or not these dispersion relations are satisfied by the experimental data. This because any error in the determination of the phase shift in the physical region allows too wide an ambiguity in the continuation to the unphysical region [see also Allcock (1959)]. In fact, one can always choose a possible continuation so that the dispersion relations are satisfied [even though the exact experimental data (if it were known!) might not satisfy such relations]. A mathematical statement of the Dyson theorem is as follows: Given two arbitrary functions $A(k)$, $D(k)$ and given an interval d and a small positive number δ, it is always possible to find functions $A'(k)$, $D'(k)$ which satisfy dispersion relations and which are an approximation to $A(k)$ and $D(k)$ in the interval d with an accuracy δ, i.e., which have the property that

$$|A'(k) - A(k)| < \delta$$
$$|D'(k) - D(k)| < \delta \qquad \text{for } k \text{ in } d \tag{441}$$

Therefore, in order to test dispersion relations with experimental data, we should consider the case of forward scattering, and more particularly dispersion relations for forward scattering for which there is no unphysical region, e.g., the case of π-N scattering.

Having indicated how a rigorous proof of dispersion relations can be given, we next briefly outline a heuristic derivation of the dispersion relation for forward scattering of mesons by nucleons [Goldberger (1955a, b)], with particular reference to isotopic spin considerations. The amplitude for forward scattering of a meson of four-momentum k, isotopic spin α, by a nucleon of momentum p, isotopic spin i, to a final state of isotopic spin β and j is

$$R^{ji}_{\beta\alpha}(pk; pk) = +i \int d^4x\, e^{ik\cdot x}\, \theta(x)\, \langle pj \mid [j_\beta(x), j_\alpha(0)] \mid pi\rangle$$
$$p^2 = M^2;\ k^2 = \mu^2 \tag{442}$$

where $j_\beta(x) = (\Box + \mu^2)\,\phi_\beta(x)$ and we have again omitted the equal-time contributions. Recall that the ϕ_αs ($\alpha = 1, 2, 3$) are Hermitian operators with $\frac{1}{\sqrt{2}}(\phi_1 + i\phi_2)$ creating a π^+ meson, $\frac{1}{\sqrt{2}}(\phi_1 - i\phi_2)$ creating a π^- meson, and ϕ_3 describing the π^0 meson. We shall explicitly assume that the theory is invariant under spatial inversions and also under arbitrary rotations in isotopic space (charge independence). By virtue of the assumed charge independence of the theory, it is sufficient to consider the elastic scattering amplitude of charged mesons off protons, since from these amplitudes one can recover the $T = \frac{3}{2}$ and the $T = \frac{1}{2}$ amplitudes. We shall, therefore, assume that the nucleon charge state does not change, i.e., $j = i$, and more specifically that this quantum number corresponds to a proton. In the following we shall drop the subscript i,j on R and on the initial and final nucleon state vectors: $|pi\rangle$, $|pj\rangle$.

By going to the laboratory frame ($\mathbf{p} = 0$, $p_0 = M$), the angular integration in the space integral can once again be performed since $\langle p \mid [j_\beta(x), j_\alpha(0)] \mid p\rangle$ is a scalar function of $\mathbf{x}^2$ and x_0. Note that, considered as a spin matrix acting on the nucleon spinor amplitudes, $\langle p \mid [j_\beta(x), j_\alpha(0)] \mid p\rangle$ does not give rise to a term of the form $\boldsymbol{\sigma}\cdot\mathbf{x}$ since such a term is a *pseudoscalar* under inversions. One thus obtains

$$R_{\beta\alpha}(\omega) = \int_0^\infty dr\, 2\pi r R_{\beta\alpha}(\omega, r) \tag{443}$$

where

$$\omega = p\cdot k/M = k_0; \quad |\mathbf{k}| = \sqrt{\omega^2 - \mu^2} \tag{444}$$

and

$$R_{\beta\alpha}(\omega, r) = +i\,\frac{\sin\sqrt{\omega^2 - \mu^2}\,r}{\sqrt{\omega^2 - \mu^2}}\int_r^\infty e^{i\omega t}\,\langle p \mid [j_\beta(x), j_\alpha(0)] \mid p\rangle \tag{445}$$

In Eq. (445) $\langle p \mid [j_\beta(x), j_\alpha(0)] \mid p\rangle$ has again to be understood as its angular average since it is only a function of $r = |\mathbf{x}|$. By a now familiar argument, the representation (445) allows us to infer that $R_{\beta\alpha}(\omega, r)$ is an analytic function in $\mathrm{Im}\,\omega \geqslant 0$, so that it satisfies a Hilbert relation, from which we deduce that

$$\mathrm{Re}\,R_{\beta\alpha}(\omega) = \int_0^\infty 2\pi r\, dr\, \frac{\mathrm{P}}{\pi}\int_{-\infty}^{+\infty} \frac{d\omega'\,\mathrm{Im}\,R_{\beta\alpha}(\omega', r)}{\omega' - \omega} \tag{446a}$$

$$= \int_0^\infty 2\pi r\, dr\, \frac{\mathrm{P}}{\pi}\left\{\int_{-\mu}^{+\mu} \frac{d\omega'\,\mathrm{Im}\,R_{\beta\alpha}(\omega', r)}{\omega' - \omega} + \left(\int_\mu^\infty + \int_{-\infty}^{\mu}\right)\frac{d\omega'\,\mathrm{Im}\,R_{\beta\alpha}(\omega', r)}{\omega' - \omega}\right\} \tag{446b}$$

Before proceeding with the evaluation of the right-hand side of Eq. (446b), let us note the symmetry properties of $R_{\beta\alpha}$. From the representation

$$R_{\beta\alpha}(\omega) = +i \int d^3x \int_0^\infty dt\, e^{i\omega t - i\sqrt{\omega^2-\mu^2}\,\mathbf{e}\cdot\mathbf{x}} \cdot \langle pj \mid [j_\beta(x), j_\alpha(0)] \mid pi\rangle \quad (447)$$

upon noting that, due to the P invariance, the factor $\exp i\sqrt{\omega^2 - \mu^2}\,\mathbf{e}\cdot\mathbf{x}$ can be replaced by $\cos\sqrt{\omega^2-\mu^2}\,\mathbf{e}\cdot\mathbf{x}$, we infer that

$$\overline{R^{ji}_{\beta\alpha}(\omega)} = R^{ij}_{\beta\alpha}(-\omega) \quad (448)$$

Hence, if we consider $R^{ji}_{\beta\alpha}(\omega)$ as the ji matrix element of the operator $R_{\beta\alpha}(\omega)$ in the nucleon isotopic spin space, we can write Eq. (448) as

$$R_{\beta\alpha}(\omega)^* = R_{\beta\alpha}(-\omega) \quad (449)$$

Furthermore, by virtue of charge invariance, we can write

$$R_{\alpha\beta}(\omega) = T_1(\omega)\,\delta_{\alpha\beta} + \tfrac{1}{2}T_2(\omega)\,[\tau_\alpha, \tau_\beta] \quad (450)$$

where $T_1(\omega)$ and $T_2(\omega)$, by virtue of Eq. (449), satisfy

$$T_1(\omega) = \overline{T_1(-\omega)} \quad (451a)$$

$$T_2(\omega) = -\overline{T_2(-\omega)} \quad (451b)$$

The relation (451b) for $T_2(\omega)$ follows from the fact that $[\tau_\alpha, \tau_\beta]$ is skew-Hermitian. These amplitudes can also be expressed in terms of the amplitudes pertaining to states of definite total isotopic spin:

$$T_1(\omega) = \tfrac{1}{3}[T_{1/2}(\omega) + 2T_{3/2}(\omega)] = \tfrac{1}{2}[T_+(\omega) + T_-(\omega)] \quad (452a)$$

$$T_2(\omega) = \tfrac{1}{3}[T_{1/2}(\omega) - T_{3/2}(\omega)] = \tfrac{1}{2}[-T_+(\omega) + T_-(\omega)] \quad (452b)$$

where $T_{1/2}(\omega)$, $T_{3/2}(\omega)$ denote the scattering amplitude for the $\frac{3}{2}$ and $\frac{1}{2}$ states of total isotopic spin, and T_+ and T_- denote the coherent scattering amplitudes for π^+p and π^-p, respectively.

Proof: Write

$$R_{\alpha\beta}(\omega) = \sum_{t=1/2,\,3/2} T_t(\omega)\,[P_t]_{\alpha\beta} \quad (453)$$

where the P_ts are the projection operators on the $\frac{3}{2}$ and $\frac{1}{2}$ state. Since

$$P_{1/2} = \tfrac{1}{3}(1 - \boldsymbol{\tau}\cdot\mathbf{t}) \quad (454a)$$

$$P_{3/2} = \tfrac{1}{3}(2 + \boldsymbol{\tau}\cdot\mathbf{t}) \quad (454b)$$

($\mathbf{t}$ is the isotopic spin of the meson), Eq. (453) can be written as

$$R_{\alpha\beta}(\omega) = \tfrac{1}{3}(T_{1/2}(\omega) + 2T_{3/2}(\omega))\,\delta_{\alpha\beta} + \tfrac{1}{3}(T_{3/2}(\omega) - T_{1/2}(\omega))\,(\boldsymbol{\tau}\cdot\mathbf{t})_{\alpha\beta} \quad (455)$$

and since

$$(\boldsymbol{\tau}\cdot\mathbf{t})_{\alpha\beta} = -i\epsilon_{\alpha\beta\delta}\tau_\delta = -\tfrac{1}{2}[\tau_\alpha, \tau_\beta] \quad (456)$$

Equations (452a) and (452b) follow. Note also that

$$T_1(\omega)\,\delta_{\alpha\beta} = \tfrac{1}{2}(R_{\alpha\beta}(\omega) + R_{\beta\alpha}(\omega)) \quad (457)$$

$$\tfrac{1}{2}T_2(\omega)\,[\tau_\alpha, \tau_\beta] = \tfrac{1}{2}(R_{\alpha\beta}(\omega) - R_{\beta\alpha}(\omega)) \quad (458)$$

Let us next obtain the dispersion relation for $T_1(\omega)$. By virtue of Eq. (457), $T_1(\omega) = R_{\alpha\alpha}(\omega)$ (no summation over index α), so that

$$\operatorname{Re} T_1(\omega) = \int_0^\infty 2\pi r\,dr\,\frac{P}{\pi}\left\{\int_{-\mu}^{+\mu} \frac{d\omega' \operatorname{Im} R_{\alpha\alpha}(\omega', r)}{\omega' - \omega} + \left(\int_{-\infty}^{-\mu} + \int_{\mu}^{\infty}\right) d\omega' \frac{\operatorname{Im} R_{\alpha\alpha}(\omega', r)}{\omega' - \omega}\right\} \tag{459}$$

Making use of the time translation invariance to carry out the time integration in the representation (445) for $R_{\alpha\alpha}(\omega, r)$, and noting that $|\langle p \mid j_\alpha(0) \mid n\rangle|^2$ is obviously real, we easily deduce that

$$\operatorname{Im} R_{\alpha\alpha}(\omega, r) = \pi \frac{\sin \sqrt{\omega^2 - \mu^2}\, r}{\sqrt{\omega^2 - \mu^2}} \sum_n \frac{\sin |\mathbf{p}_n|\, r}{|\mathbf{p}_n|\, r} |\langle p \mid j_\alpha(0) \mid n\rangle|^2 \{\delta(\omega + E_n - M) - \delta(\omega - E_n + M)\} \tag{460}$$

so that $\operatorname{Im} R_{\alpha\alpha}$ satisfies the oddness relation

$$\operatorname{Im} R_{\alpha\alpha}(\omega, r) = -\operatorname{Im} R_{\alpha\alpha}(-\omega, r) \tag{461}$$

The decomposition (460) has been made in the laboratory system ($\mathbf{p} = 0$, $p_0 = M$). It is again evident that for $|\omega'| > \mu$ the interchange of the r and ω integration is valid. In the range $|\omega| < \mu$ the only states $|n\rangle$ which contribute to the sum in (460) are the one-nucleon states with momentum $p_N{}^2 = \omega(\omega + 2M)$. States with one nucleon and n mesons, $n = 1, 2, \cdots$, do not contribute to this range ($|\omega| < \mu$) since for these states $|E_n - M| > \mu$ which is forbidden by the δ function. (We have here assumed that there exist no bound states of the meson-nucleon system with mass less than $M + \mu$.) Thus for $|\omega| < \mu$

$$\operatorname{Im} R_{\alpha\alpha}(\omega) = \pi \frac{\sin \sqrt{\omega^2 - \mu^2}\, r \sin |\mathbf{p}_N|\, r}{\sqrt{\omega^2 - \mu^2} \quad |\mathbf{p}_N|\, r} \cdot \sum_{p_N, \gamma} |\langle p \mid j_\alpha(0) \mid p_N, \gamma\rangle|^2 \{\delta(\omega + E_N - M) - \delta(\omega - E_N + M)\} \quad \text{for } |\omega| < \mu \tag{462}$$

The interchange of the r and ω integration in Eq. (459) is therefore not *a priori* valid. We can now either make use of the Lehmann arguments to justify the interchange of the order of integration or alternatively consider (as we did earlier in this section) the amplitude

$$R'_{\alpha\alpha}(\omega) = \left[\omega^2 - \left(\frac{\mu}{2M}\right)^2\right] R_{\alpha\alpha}(\omega) \tag{463}$$

From the representation

$$\begin{aligned}\operatorname{Im} T_1(\omega) &= \tfrac{1}{2} \int d^4x\, e^{ik\cdot x} \langle p \mid [j_\alpha(x), j_\alpha(0)] \mid p\rangle \\ &= \frac{(2\pi)^4}{2} \sum_{|n\rangle} |\langle p \mid j_\alpha(0) \mid n\rangle|^2 \cdot \{\delta^{(4)}(k + p - p_n) - \delta^{(4)}(k + p_n - p)\}\end{aligned} \tag{464}$$

it follows that in the range $\mu > \omega \geqslant 0$ only the second δ function in the bracket need be considered and, as remarked earlier, only the one-nucleon states $|N\rangle$ contribute to the sum. Carrying out the integration over $\mathbf{p}_N$, we obtain that in the laboratory frame $\text{Im } T_1(\omega)$ for $0 \leqslant \omega < \mu$ has the following structure

$$\text{Im } T_1(\omega) = -\pi \sum_{\gamma = \binom{\text{spin}}{\text{isospin}}} |\langle p \mid j_\alpha(0) \mid -\mathbf{k} = \mathbf{e}\sqrt{\omega^2 - \mu^2}, \gamma\rangle|^2 \cdot \delta(\omega + E_{\mathbf{k}} - M) \quad \text{for } 0 \leqslant \omega < \mu \tag{465}$$

where $|-\mathbf{k}, \gamma\rangle$ denotes a one-nucleon state of momentum $-\mathbf{k}$, energy $E_{\mathbf{k}} = \sqrt{\mathbf{k}^2 + M^2} = \sqrt{M^2 - (\mu^2 - \omega^2)}$, and spin and isospin quantum number specified by γ. Note, incidentally, that the momentum $\mathbf{k}$ of the nucleon is imaginary. This is a manifestation of the fact that we are in the unphysical region. The factor $\delta(\omega + \sqrt{\omega^2 - \mu^2 + M^2} - M)$ in (465) can be rewritten as $\frac{1}{M}\left(M - \frac{\mu^2}{2M}\right)\delta\left(\omega - \frac{\mu^2}{2M}\right)$. Similarly, for the range $-\mu < \omega < 0$ the contribution to $\text{Im } T_1(\omega)$ will be of the form: *constant* times $\delta\left(\omega + \frac{\mu^2}{2M}\right)$ where, by virtue of Eq. (451a), the constant is the same as the one multiplying $-\delta\left(\omega - \frac{\mu^2}{2M}\right)$ for $\text{Im } T_1(\omega)$ in the range $\mu > \omega \geqslant 0$. In the range $-\mu < \omega < \mu$ we therefore have the following representation for $\text{Im } T_1(\omega)$

$$\text{Im } T_1(\omega) = -\pi \sum_{\gamma} |\langle p \mid j_\alpha(0) \mid \mathbf{k} = \mathbf{e}\sqrt{\omega^2 - \mu^2}, \gamma\rangle|^2 \cdot \left\{\delta\left(\omega - \frac{\mu^2}{2M}\right) - \delta\left(\omega + \frac{\mu^2}{2M}\right)\right\} \quad \text{for } -\mu < \omega < \mu \tag{466}$$

It is clear from this representation that the amplitude $R'_{\alpha\alpha}(\omega)$ as defined by Eq. (463)

$$T'_1(\omega) = R'_{\alpha\alpha}(\omega) = \left(\omega + \frac{\mu^2}{2M}\right)\left(\omega - \frac{\mu^2}{2M}\right) R_{\alpha\alpha}(\omega)$$

has the same analyticity properties as $R_{\alpha\alpha}(\omega)$, but in addition has the desirable property that

$$\text{Im } T'_1(\omega) = 0 \quad \text{for } |\omega| < \mu \tag{467}$$

Since $R'_{\alpha\alpha}(\omega)$ has the same properties under the interchange $\omega \to -\omega$ as did $R_{\alpha\alpha}(\omega)$, we can immediately infer (apart from subtractions) the following dispersion relations for $T'_1(\omega)$

$$\text{Re } T'_1(\omega) = \frac{2}{\pi} \text{P} \int_{\mu}^{\infty} d\omega' \, \frac{\omega' \text{ Im } T_1(\omega')}{\omega'^2 - \omega^2} \tag{468}$$

(since there are now no problems with interchanging the r and ω integration). Now $T_1(\omega)$ besides having the singularities of $T'_1(\omega)$ [which, according to (468) and (451a), consist of branch lines from ∞ to $-\mu$ and from $+\mu$

to ∞] also has simple poles at $\omega = \pm\mu^2/2M$. Thus [assuming that $\omega T_1(\omega) \to 0$ as $\omega \to \infty$], we can write

$$\operatorname{Re} T_1(\omega) = \frac{a}{\omega - \dfrac{\mu^2}{2M}} + \frac{b}{\omega + \dfrac{\mu^2}{2M}} + \frac{2}{\pi} \mathrm{P} \int_\mu^\infty d\omega' \, \frac{\omega' \operatorname{Im} T_1(\omega')}{\omega'^2 - \omega^2} \quad (469)$$

The constant a in (469) being the residue of the pole at $\omega = \mu^2/2M$ is clearly equal to

$$a = \lim_{\omega \to \mu^2/2M} \left(\omega - \frac{\mu^2}{2M}\right) \operatorname{Re} T_1(\omega) \quad (470)$$

Similarly,

$$b = \lim_{\omega \to -\mu^2/2M} \left(\omega + \frac{\mu^2}{2M}\right) \operatorname{Re} T_2(\omega) \quad (471)$$

In order to evaluate these constants, we revert to the defining equation for $T_1(\omega)$, Eq. (442), and note that

$$\begin{aligned} \operatorname{Re} T_1(\omega) &= +i \int d^4x \, e^{i\omega t - \mathbf{e}\cdot\mathbf{x}\sqrt{\omega^2-\mu^2}} \, \epsilon(x) \, \langle p \mid [j_\alpha(x), j_\alpha(0)] \mid p \rangle \\ &= \frac{\sum_\gamma |\langle p \mid j_\alpha(0) \mid \mathbf{k} = \mathbf{e}\sqrt{\omega^2 - \mu^2}, \gamma\rangle|^2}{E_\mathbf{k} - M - \omega} \\ &\quad - \frac{\sum_\gamma |\langle p \mid j_\alpha(0) \mid -\mathbf{k} = -\mathbf{e}\sqrt{\omega^2 - \mu^2}, \gamma\rangle|^2}{M - \omega - E_\mathbf{k}} + \cdots \end{aligned} \quad (472a)$$

where the dots indicate the contribution from the higher mass states ($M_n \gg M + \mu$). These higher mass states cannot give rise to a pole for $\omega < |\mu|$.

Applying the definitions (470) and (471) to the representation (472a), we find that

$$a = \frac{M - \dfrac{\mu^2}{2M}}{M} \sum_\gamma |\langle p \mid j_\alpha(0) \mid -\mathbf{k} = -\mathbf{e}\sqrt{\omega^2 - \mu^2}, \gamma\rangle|^2 \quad (472b)$$

$$b = -\frac{M - \dfrac{\mu^2}{2M}}{M} \sum_\gamma |\langle p \mid j_\alpha(0) \mid \mathbf{k} = \mathbf{e}\sqrt{\omega^2 - \mu^2}, \gamma\rangle|^2 \quad (472c)$$

Now from the tensorial properties of $j_\alpha(0)$, namely that it transforms like a vector in isotopic spin space and like a pseudoscalar in space-time, we deduce that the matrix element $\langle p \mid j_\alpha(0) \mid p' \rangle$ where $|p\rangle$ and $|p'\rangle$ are one-nucleon states, $p^2 = p'^2 = M^2$, has the following structure:

$$\langle p \mid j_\alpha(0) \mid p' \rangle = i \sqrt{\frac{M^2}{E_\mathbf{p} E_{\mathbf{p}'}}} \, g\tilde{u}(p) \, \tau_\alpha \gamma_5 u(p') \, K(p^2, p'^2, (p - p')^2) \quad (473)$$

where K is related to the renormalized vertex operator $\Gamma_5(p^2, p'^2, (p-p')^2)$ introduced in Chapter 16 as follows:

$$K(p^2 = p'^2 = M^2, (p-p')^2) = [(p-p')^2 - \mu^2]\,\Delta_{F1}(p-p')\,\Gamma_5(p^2 = p'^2 = M^2, (p-p')^2) \quad (474)$$

For the case under discussion, since $p = (M, 0)$, $p' = \{\sqrt{\omega^2 + M^2 - \mu^2}, \pm \mathbf{e}\sqrt{\omega^2 - \mu^2}\}$ with $\omega^2 = (\mu^2/2M)^2$, we find that

$$(p-p')^2 = (M - \sqrt{\omega^2 - \mu^2 + M^2})^2 - \omega^2 - \mu^2\Big|_{\omega^2 = \left(\frac{\mu^2}{2M}\right)^2} = \mu^2 \quad (475)$$

If we normalize K such that $K(M^2, M^2, \mu^2) = 1$, g is then the renormalized Watson-Lepore coupling constant. Now

$$\sum_{\gamma=(s,i)} \tilde{u}(p)\,\tau_\alpha\gamma_5 u^s(p')\,\tilde{u}^s(p')\,\tau_\alpha\gamma_5 u(p)\Big|_{p=(M,0)} \quad (476a)$$

$$= \tilde{u}(p)\,\gamma_5\,\frac{\gamma\cdot p' + M}{2M}\,\gamma_5 u(p)\Big|_{p=(M,0)} \quad (476b)$$

$$= \frac{1}{2M}(p'_0 - M) = -\frac{\mu^2}{(2M)^2} \quad (476c)$$

where we have suppressed the isotopic spin wave function $|i\rangle$ of the nucleons. In going from (476a) to (476b), we have used the fact that $\sum_{i=1}^{2} |i\rangle\langle i| = 1$ and that $(\tau_\alpha)^2 = 1$. Collecting terms, we have

$$a = -g^2\left(\frac{\mu}{2M}\right)^2 \quad (477)$$

$$b = +g^2\left(\frac{\mu}{2M}\right)^2 \quad (478)$$

neglecting terms of order (μ^2/M^2) higher than those kept in (477) and (478). These derivations are to be taken *cum grano salis* [e.g., Eq. (476) seems to assert that $\sum |\quad|^2$ is a negative quantity]. They nonetheless yield the same answer as a more rigorous approach which would study the analyticity properties of $\Gamma_5(M^2, M^2, \zeta)$ as a function of ζ [Oehme (1959)]. Finally, substituting (477) and (478) into (469), we obtain

$$\operatorname{Re} T_1(\omega) = \frac{f^2}{\omega + \dfrac{\mu^2}{2M}} - \frac{f^2}{\omega - \dfrac{\mu^2}{2M}} + \frac{2}{\pi}\,\mathrm{P}\int_\mu^\infty d\omega'\,\omega'\,\frac{\operatorname{Im} T_1(\omega')}{\omega'^2 - \omega} \quad (479)$$

where $f^2 = g^2(\mu/2M)^2$ is the so-called renormalized pseudovector coupling constant. The once-subtracted dispersion relations read

$$\text{Re } T_1(\omega) - \text{Re } T_1(\mu) = \frac{1}{M} \frac{f^2\mathbf{k}^2}{\left(\omega^2 - \frac{\mu^2}{4M^2}\right)\left(1 - \frac{\mu^2}{4M^2}\right)} + \frac{2\mathbf{k}^2}{\pi} \text{P} \int_\mu^\infty d\omega'\, \omega' \frac{\text{Im } T_1(\omega')}{(\omega'^2 - \omega^2)(\omega'^2 - \mu^2)} \tag{480}$$

The same method can be applied to derive the dispersion relation for $T_2(\omega)$ except that, in this case, $T_2(\omega) = -\overline{T_2(-\omega)}$. The dispersion relation is now

$$\text{Re } T_2(\omega) = \frac{2\omega f^2}{\left(\omega + \frac{\mu^2}{2M}\right)\left(\omega - \frac{\mu^2}{2M}\right)} + \frac{2\omega P}{\pi} \text{P} \int_\mu^\infty d\omega' \frac{\text{Im } T_2(\omega')}{\omega'^2 - \omega^2} \tag{481}$$

which, it will be noticed, contains a more convergent integral so that one might hope that no subtractions are necessary to insure convergence. In any case, the once-subtracted dispersion relations read

$$\text{Re } T_2(\omega) - \frac{\omega}{\mu} \text{Re } T_2(\mu) = -\frac{2f^2\omega\mathbf{k}^2}{\left(\omega^2 - \frac{\mu^2}{4M^2}\right)\left(1 - \frac{\mu^2}{4M^2}\right)} + \frac{2\mathbf{k}^2\omega}{\pi} \text{P} \int_\mu^\infty d\omega' \frac{\text{Im } T_2(\omega')}{(\omega'^2 - \mu^2)(\omega'^2 - \omega^2)} \tag{482}$$

We can obtain the subtracted dispersion relation for the amplitudes $T_+(\omega)$ and $T_-(\omega)$ by substituting into Eqs. (480) and (482) the expressions (452a) and (452b) for $T_{1,2}(\omega)$ in terms of $T_\pm(\omega)$. Furthermore, if we denote by $\sigma_+(\omega)$ and $\sigma_-(\omega)$ the total cross sections for all processes originating from positive and negative pions incident on protons, then the unitarity relation asserts that, for $\omega \geqslant \mu$

$$\sigma_\pm(\omega) = \frac{4\pi}{|\mathbf{k}|} \text{Im } T_\pm(\omega) \tag{483}$$

where $|\mathbf{k}| = \sqrt{\omega^2 - \mu^2}$. The dispersion relations for $T_\pm(\omega)$ can, therefore, be written as follows:

$$D_+(\omega) - \frac{1}{2}\left(1 + \frac{\omega}{\mu}\right) D_+(\mu) - \frac{1}{2}\left(1 - \frac{\omega}{\mu}\right) D_-(\mu) = 2\frac{f^2}{\mu^2} \frac{\mathbf{k}^2}{\omega - \frac{\mu^2}{2M}} + \frac{\mathbf{k}^2}{4\pi^2} \int_\mu^\infty \frac{d\omega'}{|\mathbf{k}'|} \left\{\frac{\sigma_+(\omega')}{\omega' - \omega} + \frac{\sigma_-(\omega')}{\omega' + \omega}\right\} \tag{484}$$

and

$$D_-(\omega) - \frac{1}{2}\left(1 + \frac{\omega}{\mu}\right) D_-(\mu) - \frac{1}{2}\left(1 - \frac{\omega}{\mu}\right) D_+(\mu) = -2\frac{f^2}{\mu^2} \frac{\mathbf{k}^2}{\omega + \frac{\mu^2}{2M}} + \frac{\mathbf{k}^2}{4\pi^2} \int_\mu^\infty \frac{d\omega'}{|\mathbf{k}'|} \left\{\frac{\sigma_-(\omega')}{\omega' - \omega} + \frac{\sigma_+(\omega')}{\omega' + \omega}\right\} \tag{485}$$

where we have neglected terms of order (μ^2/M^2) and have adopted the notation usually encountered in the literature

$$\text{Re } T_{\pm}(\omega) = D_{\pm}(\omega) \tag{486}$$

The dispersion relations (484) and (485) have been used by Haber-Schaim (1956) to determine the value of f^2 ($f^2 \sim 0.08$). By considering the scattering angle θ infinitesimal, dispersion relations can also be obtained for the spin-flip amplitudes[23] in a manner similar to the above derivations [Oehme (1956a, b)]. These dispersion relations were used by Davidon and Goldberger [Davidon (1956)] and by Gilbert and Screaton [Gilbert (1956)] to discard the Yang set of phase shifts in low energy meson-nucleon scattering.

Puppi and Stanghellini [Puppi (1957)] have compared the dispersion relations (484) and (485) directly with the experimental pion-nucleon scattering data. Their conclusion, based on the pion-nucleon data available until 1957, was that the forward scattering amplitude dispersion relations were in apparent disagreement with experiment. However, subsequent work has indicated that some of the experimental data used by them was not accurate and that, if more accurate data is used [see, e.g., Schnitzer (1958, 1959), Noyes (1960)], the discrepancy is almost totally removed. A complete test, however, must await further theoretical work on the energy dependence of the S-wave phase shifts, but the indication at present seems to be that the dispersion relations are satisfied by the experimental data.

Finally, the dispersion relations (484) and (485), together with certain plausible assumptions concerning the very high energy behavior of the total cross sections, have been used by Pomeranchuk (1958) [see also Amati (1960)] to show that the difference of the cross sections for a particle and its charge conjugate on the same target tends to vanish at infinity. More explicitly, if $\sigma^+(E)$ and $\sigma^-(E)$ are the total cross section for the scattering of π^+ and π^- mesons of energy E on protons in the laboratory system, and if

$$\lim_{E\to\infty} \sigma^+(E) = \sigma^+(\infty) = \text{constant} \tag{487a}$$

$$\lim_{E\to\infty} \sigma^-(E) = \sigma^-(\infty) = \text{constant} \tag{487b}$$

[23] Recall that the scattering matrix in the center-of-mass system on invariance ground must have the following form:

$$R_{\beta\alpha}(\omega, \theta) = A_{\beta\alpha}(\omega, \theta) + iB_{\beta\alpha}(\omega, \theta)\, \boldsymbol{\sigma} \cdot \mathbf{n}$$

where θ is the scattering angle $\cos\theta = \mathbf{k}_1 \cdot \mathbf{k}_2/|\mathbf{k}_1|\,|\mathbf{k}_2|$ and $\mathbf{n}$ is some *axial* vector, as required from invariance under spatial inversions. The only axial vector is $\mathbf{k}_1 \times \mathbf{k}_2$, so that we may take $\mathbf{n}$ as $\mathbf{k}_1 \times \mathbf{k}_2$, which is a vector perpendicular to the plane of scattering. If the direction $\mathbf{k}_1$ is chosen as the z axis and as the axis of quantization for the proton spin, then $\boldsymbol{\sigma} \cdot \mathbf{n}$ only involves σ_1 and σ_2 and hence flips the spin. One therefore speaks of $B_{\beta\alpha}$ as the spin-flip amplitude. Note that $\boldsymbol{\sigma} \cdot (\mathbf{k}_1 \times \mathbf{k}_2)\, B_{\alpha\beta}(\omega, \theta)$ vanishes in the forward direction, i.e., when $\mathbf{k}_1 = \mathbf{k}_2$.

then Pomeranchuk shows that

$$\sigma^{+}(\infty) = \sigma^{-}(\infty) \tag{488}$$

The most thorough discussion of the non-forward dispersion relations for pion-nucleon scattering is that of Chew, Goldberger, Low, and Nambu [Chew (1957)]. By assuming that the high energy contribution in the dispersion integrals are small, that there exists a resonance in the 3-3 state which exhausts the dispersion integrals, and that S- and D-wave phase shifts are small (smaller than the small P-wave phase shift), Chew, Goldberger, Low, and Nambu have shown that the non-forward dispersion relations can be made to yield a set of dynamical equations which in the limit as $M \to \infty$ reduce to the Chew-Low equations for the static Chew model.

Their attempt to base a dynamical theory of meson-nucleon scattering on a single variable dispersion relation was, however, frustrated by their inability to specify the momentum transfer dependence of the scattering amplitudes. It was, however, observed by Mandelstam (1958) that in the related process $\pi + \pi \to \mathrm{N} + \overline{\mathrm{N}}$ the momentum transfer variable, Δ, played the role of an $(\text{energy})^2$, and that the structure of the matrix element for this scattering amplitude suggested that a dispersion relation in Δ might also exist. By using the analyticity properties contained in these dispersion relations, Mandelstam (1958, 1959) proposed a "double dispersion" relation for the meson-nucleon scattering amplitude. The content of this double dispersion relation can be summarized by saying that the π-N scattering amplitude is the boundary value of a function of two complex variables which is regular except for cuts along parts of the real axes. It is the hope of Mandelstam (1958) that this representation and its generalization to multiparticle processes, combined with the unitarity relation, will replace the more usual equations of field theory and could be used to calculate all observable quantities in terms of a finite number of coupling constants and masses without specific introduction of a Lagrangian [Mandelstam (1959a, b)]. [For an excellent introduction to the "philosophy" of this program as well as to its mathematical formulation, see Chew (1959, 1960a, b).]

To illustrate the Mandelstam representation for the two-particle scattering amplitude, consider a scattering process involving two ingoing and outgoing particles. We assign four-momenta p_1, p_2, p_3, p_4 to the particles, and these momenta for convenience all formally correspond to ingoing particles. Two of these momenta will therefore always be negative time-like ($p_0 = -$) representing outgoing particles, and the other two are positive time-like ($p_0 = +$) corresponding to the incoming particles. Energy-momentum conservation states that

$$p_1 + p_2 + p_3 + p_4 = 0 \tag{489}$$

The fact that the p_is correspond to the momenta of physical particles implies that

$$p_i^2 = m_i^2 \quad i = 1, 2, 3, 4 \tag{490}$$

The scattering amplitude $T(p_1, p_2, p_3, p_4)$ will in general refer to different processes depending on the assignments of the variables $p_1, \cdots p_4$ to incoming or outgoing variables. Without further knowledge, this scattering amplitude would consist of a number of independent functions referring to different processes. We have, however, seen that causality asserts that these functions are actually closely related in the sense that they are boundary values of a single analytic function of complex variables. We shall also call this function $T(p_1, p_2, p_3, p_4)$.

The relativistic invariance of the scattering amplitude is made manifest when it is considered as a function of the three invariant kinematic variables:

$$s_1 = (p_1 + p_4)^2 = (p_2 + p_3)^2 \tag{491}$$

$$s_2 = (p_2 + p_4)^2 = (p_1 + p_3)^2 \tag{492}$$

$$s_3 = (p_3 + p_4)^2 = (p_1 + p_2)^2 \tag{493}$$

Each of these variables is the square of the total energy in the center-of-mass system for a particular pairing of incoming and outgoing particles. For example, if p_1 and p_2 are incoming and p_3 and p_4 outgoing, then s_3 is the total energy in the center-of-mass system, and s_1 and s_2 are the square of four-momentum transfers. These three variables are not independent, but due to the constraints (489) and (490) satisfy

$$s_1 + s_2 + s_3 = \sum_{i=1}^{4} m_i^2 \tag{494}$$

The scattering amplitude T can now be considered as a function of two of the three variables s_1, s_2, s_3. We shall nonetheless write $T = T(s_1, s_2, s_3)$ and consider the third variable fixed in terms of the others through the relation (494).

We have previously noted that the scattering amplitude will have singularities[24] for physical, as well as unphysical, values of the variables s_1, s_2, s_3, which are associated with the possible real intermediate states into which the scattering amplitude can be expanded. Moreover, the residues at the poles are connected to the coupling constants. The "rule" which locates the poles of the scattering amplitude in the case of two ingoing, two outgoing particles is as follows: If the two incoming and two outgoing particles can be "connected" by a stable single particle of mass m_0, then there will be a pole when the s variable corresponding to the square of the total four-momentum for this process is equal to m_0^2. By "connected" it is meant

[24] The theory of dispersion relations can, in fact, be defined as the study of the location and nature of the singularities of the scattering amplitude.

that the initial and final two-particle states can both have the same quantum numbers as the single-particle intermediate state. From the stability requirement on the intermediate particle, it follows that such single-particle poles, although on the real axis, are never in the physical region for the s variable under consideration. (If they were, the single particle could disintegrate into the two particles making up initial or final states!) For example, the π-N scattering amplitude, $T(s_1, s_2, s_3)$, must be of the form

$$T(s_1, s_2, s_3) = \frac{1}{s_1 - M^2}\frac{1}{s_3 - M^2} \times T_1(s_1, s_2, s_3) \tag{495}$$

This is established as follows: for π-N scattering, $s_3 = (p_1 + p_2)^2$ is the square of the incoming four-momentum. A possible one-particle intermediate state is illustrated in Figure 18.7b. Hence s_3 has a pole at $s_3 = M^2$.

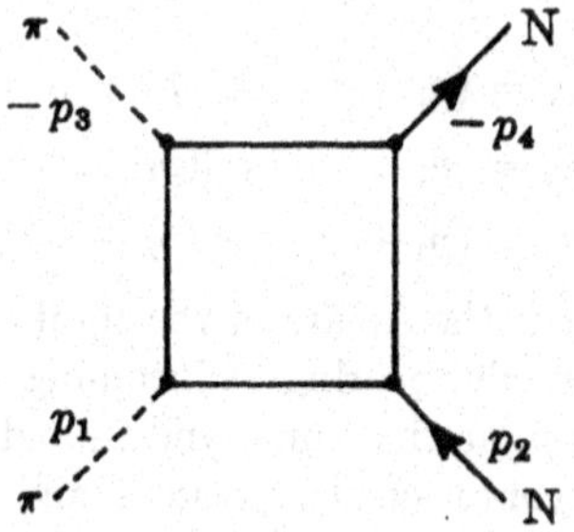

Fig. 18.7(a)

Note that the process $N + \overline{N} \to 2\pi$ is also described by the diagram 18.7*a*. In that case $s_2 = (p_2 + p_4)^2$ represents the square of the incoming four-momentum but, since there are no one-particle intermediate states "connected" with the initial or final states, s_2 does not have a pole. There is

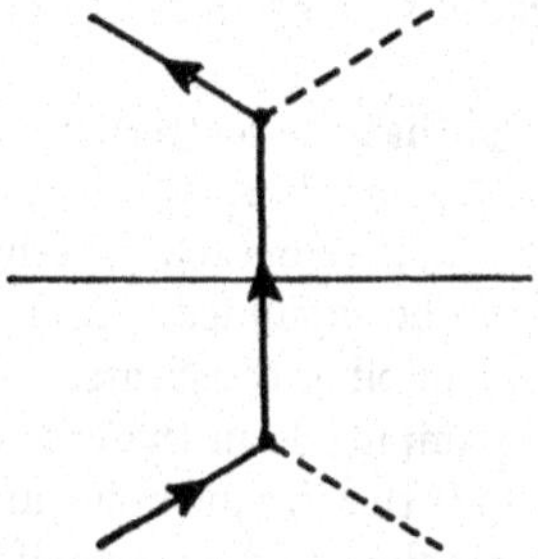

Fig. 18.7(b)

again a pole in the s_1 dependence, at $s_1 = M^2$, corresponding to a single nucleon connecting the initial and final π-N states. In (495) the function $T_1(s_1, s_2, s_3)$ is regular at $s_1 = s_3 = M^2$.

Earlier in this section we have indicated that the amplitude $T(s_1, s_2, s_3)$, obeys a dispersion relation in the variables s_3 with s_2 fixed $\left[s_1 = \sum_{i=1}^{4} m_i^2 - - s_2 - s_1\right]$. Similarly, by considering the processes in which s_1 and s_2 correspond to the total incoming total four-momentum, dispersion relations can be conjectured (although they cannot be proved) for T considered as function of s_1 and s_2 [Mandelstam (1958)]. These dispersion relations make it plausible that the scattering amplitude considered as a function of the complex variables s_1, s_2, s_3 is analytic in the cut plane of each variable. The question now arises what analyticity properties does it have as simultaneous function of two variables. Mandelstam (1958) conjectured that the scattering amplitude (except for subtractions necessary if the integrals do not converge) can be expressed as follows:

$$T(s_1, s_2, s_3) = \frac{1}{\pi^2} \iint ds_1'\, ds_2' \frac{\rho_{12}(s_1', s_2')}{(s_1' - s_1)(s_2' - s_2)}$$

$$+ \frac{1}{\pi^2} \iint ds_1'\, ds_3' \frac{\rho_{13}(s_1', s_3')}{(s_1' - s_3)(s_3' - s_3)} + \frac{1}{\pi^2} \iint ds_2'\, ds_3' \frac{\rho_{32}(s_3', s_2')}{(s_2' - s_2)(s_3' - s_3)} \tag{496}$$

where the weight functions ρ_{ij} are real and the integrations in each s' variable go over a region over the positive real axis extending to infinity. More precisely, the ρ_{ij}s fail to vanish only when an argument is equal to the square of the mass of an actual physical system that has the same quantum numbers as the corresponding channel. The representation also embodies what can be called generalized "crossing symmetry" relations, which are the generalization of the relations such as (350).

The Mandelstam representation is a generalization of a representation first suggested by Nambu (1955), which was based on perturbation theoretic considerations. This Nambu representation was, however, incorrect in that it was not compatible with unitarity. The Mandelstam representation does not suffer from this deficiency. Moreover, a proof that each term in the perturbation expansion of the scattering amplitude satisfies the Mandelstam representation has been given by Eden (1960). [The proof applies to any system that does not have so-called "anomalous thresholds" (Karplus 1958).] Some of the consequences of the Mandelstam representation are as follows:

1. Dispersion relations in the energy variables hold for arbitrary values of the momentum transfer;
2. The absorptive parts of the scattering amplitude satisfy dispersion relations as functions of the momentum transfer for fixed energy, i.e., Im T is analytic as a function of the momentum transfer variables except for cuts along the real axis;

3. Dispersion relations for fixed scattering angle, or rather fixed $\cos\theta$, can be derived [Cini, Fubini, and Stanghellini (Cini 1959)];
4. Analyticity properties of partial wave amplitudes can also be derived [see, e.g., MacDowell (1959)].

By combining the representation with the two-particle approximation to the unitarity conditions, Chew and Mandelstam proposed to make the representation the basis of a dynamical theory of strong interactions. For a review of this work, the reader is referred to the proceedings of the 1960 Rochester Conference on High Energy Physics.

With these brief remarks on the Mandelstam representation, we conclude our presentation of the theory of dispersion relations. Further details on the Mandelstam representation, its applications and limitations, can be obtained from the excellent review articles of Chew (1959, 1960a, b) and that of Gasiorowicz (1960). A complete account of the theory of dispersion relations is to be found in Goldberger and Oehme [Goldberger (1962, proposed publication); see also Goldberger (1961)].

18e. Outlook

We have come to the end of our exposition of modern field theory. Despite its difficulties, limitations, and *ad hoc* nature, clearly there is some "truth" in its approach and, moreover, in some sense it must be a crude approximation to a correct physical description of "elementary" particle interactions. Thus, the success of quantum electrodynamics cannot be fortuitous nor can the fact that the same coupling constant can account for low energy pion-nucleon and nucleon-nucleon phenomena be so.

It is, of course, tempting to speculate what the future formulation of field theory will be and to inquire into its possible relation to the dispersion theoretic approach based on the axiomatic formulation. Let us take an optimistic view and for the sake of argument assume that the Mandelstam program can be carried out. That is, let us assume that one can obtain spectral representations for all the scattering amplitudes which exhibit their analyticity properties as functions of the basic kinematics invariants and that these amplitudes satisfy generalized dispersion relations and crossing symmetry relations [see, e.g., Nishijima (1960)]. Let us furthermore assume that these dispersion relations together with unitarity can in fact replace the usual equations of field theory and can be used to calculate all observable quantities in terms of a finite number of coupling constants and masses [Gell-Mann (1956)]. It is then quite possible [see Mandelstam (1959, 1960)] that the observable consequences are identical with those derivable from a local, renormalizable field theory based on a Lagrangian approach. This can be made plausible by noting that even though dispersion relations use only the general principles of quantum field theory and do not make any assumptions about the Lagrangian other than locality and Lorentz invariance, the requirement of renormalizability, viewed as a self-consistency requirement, does in fact specify the Lagrangian to within a small number of coupling constants.

The success of such a program would, however, still seem to leave unexplained the observed mass spectrum of the "fundamental" particles, as well as the symmetries and strengths that the interactions among them are observed to possess.

Undoubtedly, further efforts will be concentrated in the future on trying to establish criteria for distinguishing "elementary" particles from bound states and to see whether some of the hyperons are "bound states" of other elementary particles. Probably, it will be necessary in this context to clarify the connection between the "microcausality condition" (the vanishing of commutators of field "observables" for space-like separated space-time points) and actual realistic measurements.

In the final analysis, however, it will probably be the new information that will be obtained from the high energy machines and colliding beam machines to go into operation in the next few years which will help unravel the puzzle of the elementary particles and their interactions. In particular, we may discover whether the notions of space and time upon which present-day field theories are based are in fact valid.

Problems and Suggested Further Reading

Chapter 1

1. The superposition principle asserts that every linear combination of states is again a state.

(**a**) Discuss this assumption in detail within the framework of nonrelativistic quantum mechanics with particular emphasis on its experimental realization.

(**b**) Given two states $|\Psi_1\rangle$ and $|\Psi_2\rangle$ of the hydrogen atom, can you devise an experiment which would realize the state $\alpha \mid \Psi_1\rangle + \beta \mid \Psi_2\rangle$?

2. Feynman (1948) has given a formulation of quantum mechanics which makes explicit the notion that quantum theory basically describes *processes*. It gives a rule for the calculation of the probability amplitude, a, for a given process in terms of an "integration over paths." The probability for the process is then given by $|a|^2$.

(**a**) Show that this formulation of quantum mechanics can be made to correspond to the usual one in terms of vectors in Hilbert space.

(**b**) Discuss whether the Feynman formulation is more general and whether, in fact, the correspondence between it and the usual one in terms of Hilbert space is necessary.

(**c**) Starting from Feynman's formulation, how could nonrelativistic quantum mechanics be altered if measurements of time intervals smaller than $\Delta\tau$ were impossible? if measurements of events separated by spatial distances smaller than $|\Delta\mathbf{l}|$ were impossible? What restrictions would Galilean invariance impose on such a formulation?

3. Consider the operators $\mathbf{T}$, N, and S, which obey the commutation relations

$$[T_l, T_m] = i\epsilon_{lmk}T_k$$

$$[N, \mathbf{T}] = [N, S] = [S, \mathbf{T}] = 0$$

Obtain the lowest dimensional faithful representation of this algebra. What other representations can you obtain?

For the applications of group theory to quantum mechanics, see, e.g., Heine (1960), besides the volumes noted in the text.

For a sketch of the content of the forthcoming book of Bargmann, Wigner, and Wightman on Relativistic Invariance and Quantum Mechanics, see Wightman (1959b).

For a review of the literature on symmetry principles and elementary particles, see Melvin (1960) and Roman (1960).

Chapter 2

1. Show that the sign of the scalar product of two real time-like four-vectors $x \cdot y = g_{\mu\nu}x^\mu y^\nu$ is determined by the product $x_0 y_0$ of their time components.

2. (**a**) Show that, if $v^\mu w_\mu = 0$ and $v^2 > 0$, then $w^2 < 0$.

(**b**) Show that, if $v^\mu w_\mu = 0$ and $v^2 = 0$, then w is either space-like, or parallel to v, i.e., either $w^2 < 0$ or $w = cv$, $c =$ constant.

3. Show that, if four vectors $v^{(1)}{}_\mu$, $v^{(2)}{}_\mu$, $v^{(3)}{}_\mu$, $v^{(4)}{}_\mu$ are pairwise orthogonal and linearly independent, one of them is time-like and three are space-like.

4. Discuss as thoroughly as possible the eigenfunctions and eigenvalues of a Lorentz transformation Λ.

5. Note that if one associates with the co-ordinates x_μ of a space-time point a matrix

$$X = \begin{pmatrix} x^0 + x^3 & x^1 - ix^2 \\ x^1 + ix^2 & x^0 - x^3 \end{pmatrix}$$

then $\det X = D(X)$ has the value $D(X) = g_{\mu\nu}x^\mu x^\nu$.

(**a**) If the matrix

$$W = \begin{pmatrix} \alpha & \beta \\ \gamma & \delta \end{pmatrix}$$

with complex elements α, β, γ, δ has determinant 1 and

$$W^* = \begin{pmatrix} \bar\alpha & \bar\gamma \\ \bar\beta & \bar\delta \end{pmatrix}$$

is its Hermitian adjoint, show that

$$X' = WXW^*$$

defines a proper Lorentz transformation $x'_\mu = \Lambda_\mu{}^\nu x_\nu$ [i.e., show that $\det \Lambda = +1$, $\Lambda^0{}_0 \geqslant 1$, that the $\Lambda_{\mu\nu}$s are real and that $D(X) = D(X')$].

(**b**) Discuss the transformations generated by Ws of the form

(1) $$W = \begin{pmatrix} \alpha & \beta \\ \bar{\beta} & \bar{\alpha} \end{pmatrix}$$

$$\det W = 1$$

(2) $$W = \begin{pmatrix} \alpha & \beta \\ -\bar{\beta} & \bar{\alpha} \end{pmatrix}$$

$$\det W = 1$$

6. Prove that, if $x_1, \cdots x_n$ and $y_1, \cdots y_n$ are two sets of real four-vectors, such that

(*a*) $x_i \cdot x_j = y_i \cdot y_j$, $j = 1, 2, \cdots n$

(*b*) each x_i and y_i, $i = 1, 2, \cdots n$, lies in or on the forward light cone

(*c*) if all x_i are light-like, not all are collinear

then there exists a Lorentz transformation Λ, $\Lambda \epsilon L^{\uparrow}$ such that

$$\Lambda x_i = y_i \quad i = 1, 2, \cdots n$$

7. Discuss in detail the properties that

(**a**) particles of negative mass and negative energy

(**b**) particles of positive mass but space-like momenta

would have if they existed.

For an interesting departure from the usual assignment of irreducible representations with elementary particles which takes into account the existence of an antiparticle with each particle, see Ekstein (1960).

Chapter 3

1. Obtain an explicit representation for the configuration space Newton-Wigner wave function corresponding to a spin zero particle localized at **y** at time y_0.

2. (**a**) Compute the probability of finding a Klein-Gordon particle at **y**, y_0 if it is known that at time $t = 0$ it was localized at the origin. Obtain explicit forms for the case $|\mathbf{y}| \gg y_0$ and the case $|\mathbf{y}| < y_0$.

(**b**) Discuss your result for the case $|\mathbf{y}| \gg y_0$ in the light of the relativistic causality principle which asserts that no signal can be propagated with a speed greater than the speed of light.

3. (**a**) Obtain the solutions of the second-order Klein-Gordon equation in a homogeneous constant external electromagnetic field for which only one component of $F_{\mu\nu}$ is different from zero.

(**b**) Discuss as far as possible the solution of the "square-root" equation for these cases.

Chapter 4

1. (**a**) Show that the set of Dirac matrices $i\gamma^\mu$, with $[\gamma^\mu, \gamma^\nu]_+ = 2g^{\mu\nu}$ can be so chosen as to make each $i\gamma^\mu$ ($\mu = 0, 1, 2, 3$) real. [Majorana (1937).]

(**b**) Obtain an explicit representation for these matrices.

(**c**) The matrices

$$\gamma'^\mu = (a - ib\gamma_5)\,\gamma^\mu$$

with

$$a^2 - b^2 = 1$$

obey the commutation rules $[\gamma'^\mu, \gamma'^\nu]_+ = 2g^{\mu\nu}$. Obtain the explicit representation of the similarity transformation S with relates γ'^μ and γ^μ:

$$\gamma'^\mu = S^{-1}\gamma^\mu S$$

2. (**a**) Show that the solutions to the Dirac equation for positive energies which are normalized to $\tilde{u}u = 1$ may be written in the form

$$u = \sqrt{\frac{E+M}{2M}}\left[1 - \frac{\gamma_0\boldsymbol{\alpha}\cdot\mathbf{p}}{E+M}\right]u_0$$

where u_0 is either $\begin{pmatrix}\alpha_+\\0\end{pmatrix}$ or $\begin{pmatrix}\beta_-\\0\end{pmatrix}$, with α_+ and β_- the usual two-component Pauli spinors corresponding to spin-up or spin-down.

(**b**) Obtain explicit values for the expectation value of the operators $O_i = \gamma^\mu, \gamma^\mu\gamma^5, \sigma^{\mu\nu}, \gamma^5$ in the states $u^s(p)$ and in the states $v^s(p)$ (i.e., compute $\tilde{u}^sO_iu^s$ and $\tilde{v}^sO_iv^s$). Analyze your answers for the case that $\mathbf{p} = (0, 0, p)$.

3. Obtain explicit representations for the matrix element of the operator $O_i = \gamma^\mu, \gamma^\mu\gamma^5, \sigma^{\mu\nu}, \gamma^5$ between spinors $u^s(p_2)$ and $u^r(p_1)$. Analyze in detail the case that

$$p_1 = (E, 0, 0, p)$$
$$p_2 = (E, 0, 0, -p)$$

i.e., compute $\tilde{u}^s(p_2)\,O_iu^r(p_1)$.

4. In a given Lorentz frame an electron undergoes a scattering from a state with momentum p_1 to one with momentum p_2. Show that there exists a Lorentz frame in which this scattering appears as a pure reversal of motion.

5. (**a**) Compute the following traces

$$\mathrm{Tr}\,\{\gamma^\alpha\gamma^\beta\gamma^5\}$$
$$\mathrm{Tr}\,\{\gamma^\alpha\gamma^\beta\gamma^\rho\gamma^5\}$$
$$\mathrm{Tr}\,\{\gamma^\alpha\gamma^\beta\gamma^\delta\gamma^\epsilon\gamma^5\}$$

(b) Show that

$$\mathrm{Tr}\,\{\not{a}_1\not{a}_2 \cdots \not{a}_n\} = 4\sum_P \delta_p(a_{p_1}\cdot a_{p_2})\,(a_{p_3}\cdot a_{p_4})\cdots(a_{p_{n-1}}\cdot a_{p_n})$$

where the sum is taken over all different ways one can dot the vectors into each other, subject to the restriction $p_1 < p_2$, $p_3 < p_4$, $\cdots$ and $\delta_p = \pm 1$ depending on whether $(p_1p_2 \cdots p_n)$ is an even or odd permutation of $(123 \cdots n)$. Show that there are $(2m)!/2^m m!$ terms.

6. The large and small components of the solution of the Dirac equation $(\not{p} - m)\,\psi = V\psi$ are defined as $\psi_\pm = \frac{1}{2}(1 \pm \gamma_0)\,\psi$. (We choose a representation in which $\gamma_0 = \beta$ is diagonal.)

(a) Obtain the equation for ψ_+ and show that when the kinetic energy, T, of the particle $(p_0 = m + T)$ is such that $T \ll m$, that ψ_+ reduces to the nonrelativistic Schrödinger wave function. (Note, however, that the "Hamiltonian" in the exact equation for ψ_+ is not necessarily Hermitian. Why?)

(b) Obtain the nonrelativistic limit of the equation that ψ_+ satisfies in the case that

(1) $V = \gamma_\mu A^\mu$ (2) $V = \gamma_5 U$ (3) $V = i\gamma_5\gamma_\mu U^\mu$

(4) $V = \sigma_{\mu\nu}F^{\mu\nu}$ (5) $V = \gamma_5\sigma_{\mu\nu}F^{\mu\nu}$ (6) $V = i\gamma_5\gamma_\mu\partial^\mu U$

7. (a) Obtain the solutions of the Dirac equation for the case of an electron in a Coulomb field by noting that $\mathbf{J} = \mathbf{L} + \frac{1}{2}\mathbf{\Sigma}$ and the parity operator $P = \beta R$ (where R is an operator which changes the sign of the spatial co-ordinates x_1, x_2, x_3) are constants of the motion.

(b) Discuss the relation of the operator $K = \beta(\mathbf{\Sigma}\cdot\mathbf{L} + 1)$ to P.

(c) Obtain a classification of the states in terms of the operators $\mathbf{J}^2$, J_3 and K.

(d) Discuss in detail the $S_{1/2}$, $P_{1/2}$, $P_{3/2}$, and $D_{3/2}$ levels and obtain explicitly the Dirac spinors for these states.

8. Obtain the solutions of the Dirac equation in the presence of

(a) a homogeneous time-independent electric field;

(b) a homogeneous time-independent magnetic field. Show that in this case there exist states with energy $E = m$.

9. Obtain the projection operators for states of definite helicity and momentum for the free-particle Dirac equation.

For the relativistic description of the polarization of a spin $\frac{1}{2}$ particle, see Michel (1959) and Bouchiat (1958).

10. Obtain the representation of the operators γ_5, γ^μ, $\sigma^{\mu\nu}$, $\gamma_5\gamma^\mu$, $\mathbf{\Sigma}\cdot\mathbf{p}$ in both the Foldy-Wouthuysen and the Cini-Touschek representations.

11. Discuss the relativistic invariance of the Dirac equation in an external field. What does "relativistic invariance" mean in this case since in effect the external field singles out a special Lorentz frame. [Eddington (1939, 1942); Dirac (1942b).]

Chapter 5

For an alternative formulation of the two-component neutrino theory, see Case (1957), McLennan (1957), and Theis (1959).

1. In a given Lorentz frame, L, a photon traveling in the z direction and polarized in the x direction can also be described by a four-potential

$$A_\mu(x) = \epsilon_\mu^{(1)} e^{-ik_\mu x^\mu}$$

$$k_\mu \epsilon^{\mu(1)} = 0$$

with

$$\epsilon^{\mu(1)} = (0, 1, 0, 0)$$

$$k^\mu = (\omega, 0, 0, \omega)$$

Obtain the description of this photon as seen by an observer traveling in the x direction with velocity v relative to L.

Chapter 6

1. Consider the operators $a_\mathbf{k}$, $a_\mathbf{k}^*$ which satisfy the commutation rules

$$[a_\mathbf{k}, a_{\mathbf{k}'}^*] = \delta_{\mathbf{k}\mathbf{k}'}$$

$$[a_\mathbf{k}^*, a_{\mathbf{k}'}^*] = [a_\mathbf{k}, a_{\mathbf{k}'}] = 0$$

and introduce the set of operators

$$b_\mathbf{k} = \cosh\lambda\, a_\mathbf{k} + \sinh\lambda\, a_\mathbf{k}^*$$

$$b_\mathbf{k}^* = \sinh\lambda\, a_\mathbf{k} + \cosh\lambda\, a_\mathbf{k}^*$$

(a) Obtain the commutation rules of the b operators.

(b) Show that, if the set of operators $a_\mathbf{k}$ is finite (i.e., $k = 1, \cdots n$), a unitary transformation

$$V_{(n)} = e^{iT_{(n)}}$$

exists such that

$$V_{(n)} a_\mathbf{k} V_{(n)}^* = b_\mathbf{k}$$

Obtain an explicit form for $T_{(n)}$ in terms of the a operators.

(c) Show that in the limit $n \to \infty$

$$\lim_{n\to\infty} \langle\Psi \mid V_{(n)} \mid \Phi\rangle \to 0$$

where $|\Psi\rangle$ and $|\Phi\rangle$ are any states of the form $\prod a_i^* \mid 0\rangle$.

2. Consider the second quantized formulation of the theory of spinless nonrelativistic particles interacting pairwise through a potential $V = V(|\mathbf{x} - \mathbf{y}|)$ which falls off faster than $1/r$, $(|\mathbf{x} - \mathbf{y}| = r)$.

If $f_\alpha(\mathbf{x}, t)$ is a normalizable (wave-packet) solution of the Schrödinger equation

$$\left(\frac{-h^2}{2m}\nabla^2 - i\hbar\frac{\partial}{\partial t}\right) f_\alpha(\mathbf{x}, t) = 0$$

form the operator

$$\psi_\alpha^*(t) = \int d^3x\, f_\alpha(\mathbf{x}, t)\, \psi^*(\mathbf{x}, t)$$

(a) Show that the state

$$|\alpha\rangle = \psi_\alpha^*(t)\,|\,0\rangle$$

is time-independent and, moreover, that if the set $\{f_\alpha(\mathbf{x}, t)\}$ is orthonormal then

$$\langle\alpha\,|\,\beta\rangle = \delta_{\alpha\beta}$$

(b) Assume that the potential V is such that it can bind a two-particle system. If the Schrödinger wave functions of these two-particle bound states are denoted by $\phi_b(\mathbf{x}_1, \mathbf{x}_2, t)$, obtain the equation of motion of the two-particle operator

$$\psi^B{}_b{}^*(t) = \frac{1}{\sqrt{2!}}\int d^3x \int d^3y\, \phi(\mathbf{x}, \mathbf{y}; t)\, \psi^*(\mathbf{x}, t)\, \psi^*(\mathbf{y}, t)$$

(c) Show that the state

$$|b\rangle = \psi^B{}_b{}^*(t)\,|\,0\rangle$$

is time-independent.

(d) Show that, if the potential has the asymptotic behavior assumed above, then the limit operators

$$\lim_{t\to\pm\infty} \psi_\alpha^*(t) = \psi^*_{\alpha\,\substack{\text{in}\\\text{out}}}$$

$$\lim_{t\to\pm\infty} \psi^B{}_b{}^*(t) = \psi^{B*}_{b\,\substack{\text{in}\\\text{out}}}$$

exist.

(e) Compute explicitly

(1) $\langle 0\,|\,\psi_{\alpha\,\text{out}}\psi_{\beta\,\text{in}}^*\,|\,0\rangle$

(2) $\langle 0\,|\,\psi_{\alpha\,\text{out}}\psi_{\beta\,\text{out}}\psi^B{}_b{}^*\,|\,0\rangle$

The theory of interacting nonrelativistic particles has been the subject of much interest recently and has seen important advances in the past few years. The interested student is referred to the 1958 notes of the Summer School at les Houches [deWitt (1958)], as well as to the Brandeis 1959 Summer Institute notes for a good introduction to many topics in this field.

For a more formal approach to this field, see Martin and Schwinger [Martin (1959)].

Chapters 7 and 8

1. Verify that the angular momentum operator for the scalar field is given by

$$\mathbf{M} = -\tfrac{1}{2} \int d^3x\, [\pi(\mathbf{x}, t), \mathbf{x} \times \nabla\phi(\mathbf{x}, t)]_+$$

and that it satisfies the commutation rules

$$[M_l, M_j] = i\epsilon_{ljm} M_m$$

For a discussion of the properties of a system of N mesons, see Pais (1960).

2. Obtain the angular momentum operator for the Dirac field. Obtain its eigenvalues for a one-particle state whose three-momentum is zero.

3. (**a**) Obtain explicit representations for the operators $U(i_s)$, $U(i_t)$ and U_c for the quantized field theory of spin $\frac{1}{2}$ particles.

(**b**) Discuss the transformation properties of the Dirac operator $\psi(x)$ under U_c in the Majorana representation of the γ matrices.

4. Obtain the Fock space treatment of the two-component theory of the neutrino in momentum space [Theis (1959)].

5. Work out the configuration (Fock) space treatment of a system of spin $\frac{1}{2}$ nucleons described by operators $\psi_\tau(x)$. (τ is the isotopic spin index.)

For a formulation of field theory using the Feynman principle [Feynman (1948)], see, e.g., Polkinghorne (1955) and Matthews and Salam (1955).

For an introduction to the Schwinger formulation of field theory starting from the action principle, see the Brandeis Summer Institute lecture notes of 1959 and 1960.

Chapter 9

1. Obtain the angular momentum operator for the photon field and discuss its eigenvalues in the one-photon state.

2. Discuss the properties of a two-photon state under T, C, P and TCP.

For the formulation of the quantized theory of the electromagnetic field in the Coulomb gauge, see, e.g., Schwinger (1953), Valatin (1951), Ozaki (1955).

For a presentation of the quantized field theory of spin 1 particle, see Pauli (1941) and Wentzel (1949).

Chapter 10

For a possible connection between the strong and weak interactions, see the paper of Gell-Mann and Lévy [Gell-Mann (1960a, b)].

For a theory of strong interactions based on the Yang-Mills philosophy, see the interesting article of Sakurai (1960).

Chapter 11

1. Rederive the Lippmann-Schwinger and general scattering formalism in the case that H_0 has bound states.

2. Prove that in the scattering of a nonrelativistic particle by a potential V

$$f(\theta, \varphi) = -\frac{m}{2\pi} T_{k\mathbf{n}, \mathbf{k}}$$

where $\mathbf{n} = \mathbf{r}/r$ is the direction of observation.

3. Show that by virtue of the unitarity of S the reduced S matrix, s, defined by

$$\begin{aligned}(E_b, \gamma_b|S|E_a, \gamma_a) &= \delta(E_a - E_b)s_{\gamma_b\gamma_a}(E_a)\\ &= \delta(E_a - E_b)(\gamma_b|s(E_a)|\gamma_a)\end{aligned}$$

can be brought to diagonal form. If we call Γ the representation in which this is possible, we define δ_Γ by

$$(\Gamma'|s|\Gamma) = e^{2i\delta_\Gamma}\delta_{\Gamma\Gamma'}$$

(**a**) Show that the reduced R matrix then has the following form:

$$(\Gamma'|t|\Gamma) = -\frac{1}{\pi} e^{i\delta_\Gamma} \sin \delta_\Gamma \, \delta_{\Gamma\Gamma'}$$

(**b**) Show that in the case of the scattering of a spinless nonrelativistic particle by a spherically symmetric potential V the δ_l are the phase shifts. Show that these phase shifts depend only on the asymptotic properties of $\psi^+_{klm}(r) = \langle rlm|\psi_\mathbf{k}^+\rangle$ by showing that

$$\begin{aligned}-\frac{1}{\pi} e^{i\delta_l} \sin \delta_l = \lim_{r\to\infty} \frac{\hbar^2}{2m} r^2 \Bigg\{&\Phi_{klm}(r) \frac{\partial\psi^+_{klm}(r)}{\partial r}\\ &- \frac{\partial\Phi_{klm}(r)}{\partial r} \psi^+_{klm}(r)\Bigg\}\end{aligned}$$

4. Prove that, if $\delta(p)$ is the scattering phase shift for a given angular momentum state, m and m_0 the number of bound states of H and H_0 for the same angular momentum, then

$$\delta(\infty) - \delta(0) = (m_0 - m)\,\pi$$

[Jauch (1951).]

5. Prove that

$$\frac{1}{2}\int_{-\infty}^{+\infty} \epsilon(t - t') \exp\left[-i(E_a - E_b)\,t'\right] dt' = \exp\left[-i(E_a - E_b)\,t\right] \mathrm{P}\frac{i}{E_a - E_b}$$

where $\epsilon(t) = t/|t|$.

For a comprehensive review of scattering theory, see the notes on lectures delivered by Low at the Brandeis University Summer Institute in Theoretical Physics (Summer 1959); see also Glauber (1959), and Brenig (1959).

Chapter 12

1. Show that in the neutral scalar theory the probability p_n of finding n (virtual) mesons around a nucleon is given by a Poisson distribution:

$$p_n = \int d^3k_1 \cdots \int d^3k_n \,|\langle k_1, \cdots k_n \mid \Psi_{(1)}\rangle|^2$$

Show that p_n vanishes in the limit of a point nucleon.

2. Compute the expectation value of the meson field operator $\phi(x)$ in the physical one-nucleon state for the neutral scalar theory.

3. Obtain the Heisenberg equations of motion for the field operators in the neutral scalar field model. Compute the vacuum expectation value of the product of Heisenberg operators $\psi(\mathbf{p}, t)$ and $\psi^*(\mathbf{p}', t')$

$$\langle 0 \mid \psi(\mathbf{p}, t)\, \psi^*(\mathbf{p}', t') \mid 0\rangle = G(\mathbf{p}, \mathbf{p}'; t - t') \quad \text{for } t > t'$$

Discuss the analytic properties of $\tilde{G}(\mathbf{p}, \mathbf{p}'; \omega)$, the Fourier transform with respect to $t - t'$ of G, in the complex ω plane. Similarly compute and discuss

$$\langle 0 \mid \phi(\mathbf{k}, t)\, \phi(\mathbf{k}', t') \mid 0\rangle = \mathcal{G}(\mathbf{k}, \mathbf{k}'; t - t')$$

4. Compute the action of $U_\alpha(0, -\infty)$ on the two-nucleon state $\psi^*(\mathbf{p}_1)\, \psi^*(\mathbf{p}_2) \mid 0\rangle$ in the neutral scalar theory.

5. (a) Show that the renormalized Heisenberg operator

$$\psi_R(t) = \sqrt{Z}\, \psi(t)$$

for the nucleon field in the scalar model, has finite matrix elements between physical states, i.e., eigenstates of H, even in the limit $v(\mathbf{k}^2) \to 1$ for all $\mathbf{k}$.

(b) Show that in the limit of a point source, $\psi_R^*(t)$ for $t \neq 0$ when operating on the vacuum state does not lead to a vector in the Hilbert space.

(c) Compute $||\psi_R^*(t)\,\Psi||^2$ and $||\psi_R(t)\,\Psi||^2$ where $|\Psi\rangle$ is an arbitrary state. Show that one of these two is proportional to Z^{-1}, hence infinite in the limit $f \to 1$.

6. Discuss as fully as possible the Lee model in which the θ particle has only 1 degree of freedom and whose Hamiltonian is

$$H_0 = (m_V + \delta m)\,V^*V + m_N N^*N + \mu\theta^*\theta$$

$$H_I = g(V^*N\theta + N^*V\theta^*)$$

The commutation rules of the operators are

$$[V^*, V]_+ = [N^*, N]_+ = [\theta, \theta^*] = 1$$

$$[\theta, \theta] = [V, N]_+ = [N, N]_+ = \cdots [V^*, N^*]_+ = 0$$

7. Consider the Lee model for a single V and N particle

$$H = m_V V^*V + m_N N^*N + \int d^3k\,\omega_k a^*(\mathbf{k})\,a(\mathbf{k}) + g_0 \int (V^*Na(\mathbf{k}) + N^*Va^*(\mathbf{k}))\,f(\mathbf{k})\,d^3k$$

Show that only θ particles with angular momentum zero interact with the V and N particles and that all other mesons are uncoupled. Write the Hamiltonian in terms of creation and annihilation operators for $l = 0$ mesons.

8. The amplitude $\varphi_{\mathbf{p}'\mathbf{k}'}(\mathbf{k}; \mathbf{p}) = \langle 0 \mid a_{\mathbf{k}} N(\mathbf{p}) \mid \mathbf{p}', \mathbf{k}'\rangle_+$ can be considered the "wave function" for the N-θ system undergoing scattering. Obtain the equation of motion of this amplitude from a consideration of the matrix element $\langle 0 \mid [H, a_{\mathbf{k}}N(\mathbf{p})] \mid \mathbf{p}', \mathbf{k}'\rangle_+$. Solve the scattering problem in this fashion.

9. Consider the Hamiltonian

$$H = \int d^3k\,\omega_{\mathbf{k}}(a_{\mathbf{k}}^*a_{\mathbf{k}} + b_{\mathbf{k}}^*b_{\mathbf{k}}) - g \int d^3k\,f(\mathbf{k})\,[(a_{\mathbf{k}} + b_{\mathbf{k}}^*)\,\tau_+ + (a_{\mathbf{k}}^* + b_{\mathbf{k}})\,\tau_-]$$

corresponding to the interaction of charged scalar mesons with an (infinitely heavy) nucleon.

(a) Obtain the equation satisfied by the Fock space amplitudes

$$\langle \mathbf{k}_1, \cdots \mathbf{k}_n; \mathbf{q}_1 \cdots \mathbf{q}_m \mid \ \rangle = \frac{1}{\sqrt{n!m!}} \langle 0 \mid a_{\mathbf{k}_1} \cdots a_{\mathbf{k}_n} b_{\mathbf{q}_1} \cdots b_{\mathbf{q}_m} \mid \ \rangle$$

when $|\ \ \rangle$ represents

(1) the physical nucleon state $|t\rangle_+$

(2) a meson-nucleon scattering state $|\mathbf{k}; t\rangle_+$ or $|\mathbf{q}; t\rangle_+$.

Two approximation methods immediately suggest themselves:

I. One assumes that $\langle \mathbf{k}_1, \cdots \mathbf{k}_n; \mathbf{q}_1, \cdots \mathbf{q}_m |\ \ \rangle$ can be approximated by

$$\langle \mathbf{k}_1, \cdots \mathbf{k}_n; \mathbf{q}_1, \cdots \mathbf{q}_m |\ \ \rangle = c_{nm} \prod_{i=1}^{n} \prod_{j=1}^{m} \phi(\mathbf{k}_i)\, \psi(\mathbf{q}_j) \quad n, m = 0, 1, 2, \cdots$$

with ϕ and ψ given. The coefficient c_{nm} are then determined so as to minimize the expectation value of the Hamiltonian. This is known as the Tomonaga intermediate coupling method [Tomonaga (1947); see also the Supplement No. 2 to the Progress of Theoretical Physics].

II. One assumes that only the amplitudes $\langle \mathbf{k}_1, \cdots \mathbf{k}_n; \mathbf{q}_1 \cdots \mathbf{q}_m | \rangle$ with $n \leqslant N$ and $m \leqslant M$ are important and sets the amplitudes with $n > N$ and $m > M$ equal zero. This is known as the Tamm-Dancoff method [see, e.g., Bethe (1955) and Silin and Tamm [Silin (1957)].

(**b**) Formulate these approximation methods for the Chew model.

(**c**) Show that in the general case the Tomonaga method is equivalent to assuming that just a few one-particle meson states are occupied but that these states can be populated by an arbitrary number of mesons. Show that the approximation can be made directly on the Hamiltonian by defining new operators

$$A_{il}{}^* = \int d^3k\, a_i{}^*(\mathbf{k})\, \phi_l(\mathbf{k})$$

where the index i refers to the charge quantum number and the $\phi_l(\mathbf{k})$ correspond to the wave function of the one-particle states which are assumed occupied [Maki (1953)].

(**d**) Obtain the equations for c_{nm} in the case of the charged scalar theory.

(**e**) For the Chew model solve the meson-nucleon scattering problem in the Tamm-Dancoff approximation in which only the amplitudes containing up to two mesons are kept. Show that this approximation violates crossing symmetry.

10. Prove that

$$|s, t\rangle_\pm = |s, t\rangle_+ = \Omega^\pm\, |\, s, t\rangle$$

where $\Omega^\pm$ is the Møller matrix and $|s, t\rangle$ is the bare nucleon state, by explicitly computing $\Omega^\pm$.

11. Prove that the projection operators $\mathcal{I}_J(\mathbf{k}_2, \mathbf{k}_1)$ defined by Eqs. (258) and (259) satisfy

$$\left[\frac{\boldsymbol{\sigma}}{2} + \mathbf{l}_2\right] \mathcal{I}_J(\mathbf{k}_2, \mathbf{k}_1) = J(J+1)\, \mathcal{I}_J(\mathbf{k}_2, \mathbf{k}_1)$$

where $\mathbf{l}_2 = -i(\mathbf{k}_2 \times \nabla_{\mathbf{k}_2})$ is the orbital angular momentum associated with $\mathbf{k}_2$.

12. (a) Recoil can be included in the Chew model when the nucleon is described nonrelativistically by considering the following Hamiltonian

$$H = \frac{\mathbf{p}^2}{2M} + \sum_{i=1}^{3} \int d^3k\, a_{i\mathbf{k}}{}^* a_{i\mathbf{k}} \omega_{\mathbf{k}} + H_I$$

By noting that the total momentum

$$\mathbf{P} = \mathbf{p} + \sum_{i=1}^{3} \int d^3k\, a_{i\mathbf{k}}{}^* a_{i\mathbf{k}} \mathbf{k}$$

commutes with H, show that in the center-of-mass system the effective Hamiltonian is

$$H' = \int d^3k \sum_{i=1}^{3} a_{i\mathbf{k}}{}^* a_{i\mathbf{k}} \left(\omega_{\mathbf{k}} + \frac{\mathbf{k}^2}{2M}\right) + H_I$$
$$+ \frac{1}{2M} \int d^3k \int d^3k' \sum_{ij=1}^{3} a_{i\mathbf{k}}{}^* a_{j\mathbf{k}'} a_{i\mathbf{k}} a_{j\mathbf{k}'} \mathbf{k} \cdot \mathbf{k}'$$

(b) Is this theory Galilean invariant?

(c) Estimate the magnitude of the recoil corrections for the 33 phase shift.

13. Obtain the modified Chew-Low equation in the case that a $(\boldsymbol{\phi} \cdot \boldsymbol{\phi})^4$ interaction (corresponding to a π-π interaction) is added to the Hamiltonian [Sugawara (1959)].

For an alternative approach to low energy meson-nucleon scattering in which one computes the components of the meson-nucleon scattering state $|\Psi\rangle_+$ on the state consisting of one physical nucleon and one free meson, $a_{\mathbf{k}i}{}^* \mid s, t\rangle_+$, see Bosco *et al.* [Bosco (1959)].

14. Derive and solve the Low equations for the scalar field interacting with fixed sources.

15. (a) Derive the Low equations for the charged scalar field interacting with a fixed nucleon.

(b) Solve the Low equation for meson scattering in the one-meson approximation.

16. Derive the Low equations for the Lee model.

17. Consider the class of Hamiltonian having the form

$$H = H_0 + H_I$$

$$H_0 = \int d\mathbf{k}\, a^*(\mathbf{k})\, \omega(\mathbf{k})\, a(\mathbf{k}) + \sum_{i=1}^{N} \psi_i^* M_i \psi_i$$

$$H_I = g \sum_{j=1}^{3} \sum_{u=1}^{N} \psi_u^* \sigma \tau_j \psi_1 \cdot \int \nabla\varphi_j(\mathbf{x})\, \rho(\mathbf{x})\, d\mathbf{x}$$

It represents the interaction of a "nucleon" capable of existing in N excited state with a meson field.

(a) Show that, if $M_i - M_1$ are sufficiently large compared to μ, the excited states of the unperturbed nucleon do not introduce any additional bound states in the spectrum of the complete Hamiltonian H.

(b) Show that the Low equation is the same for each member of the above class of Hamiltonian.

Chapter 13

1. Obtain the explicit representation of each of the singular functions D, S, and Δ in terms of known functions (Bessel, Hankel, and Neumann).

2. Derive the homogeneous or inhomogeneous equations obeyed by each of the singular functions.

3. Obtain the value of the contraction symbols for

(a) the nucleon operators $\psi_\tau(x)$

(b) the meson operator $\phi_i(x)$

(c) the hyperon operators in the theories of strong interactions discussed in Chapter 10.

Chapter 14

1. (a) Show that to order $(A^e)^2$ there will be a difference in the scattering of a positon and a negaton by an external field and trace the origin of the mechanism responsible for this.

(b) Discuss the degree of divergence of the higher order disconnected bubble diagrams in the theory of the electron-positron field interacting with an external field.

(c) Discuss the relation between the negative energy components of the amplitude $\langle 0 \mid \psi^e(x) \mid 1\ electron\rangle = \chi(x)$ when it is expanded in terms of the solution of the free Dirac equation and the probability for finding pairs.

2. (**a**) Prove that to lowest order the number of boson pairs produced per unit space-time by an external electric field $\boldsymbol{\mathcal{E}} = (\mathcal{E}_0 \cos \omega t, 0, 0)$ is

$$\frac{e\mathcal{E}_0^2}{48\pi} \frac{(\omega^2 - 4\mu^2)^{3/2}}{\omega^3} \theta(\omega - 2\mu)$$

(**b**) Obtain an expression for the total energy transformed and relate this quantity to the imaginary part of the (complex) dielectric constant of the "vacuum" of the charged boson field.

(**c**) Compute the real part assuming that the dielectric constant obeys a dispersion relation. [Euwema and Wheeler (Euwema 1956); Toll (1956).]

(**d**) Repeat these calculations for the electron-positron case.

3. Derive the Feynman rules for

(**a**) pseudoscalar meson theory (with ψ an isotopic spinor and ϕ an isotopic vector);

(**b**) the Fermi theory of β decay;

(**c**) the theories of strong interactions discussed in Chapter 10;

(**d**) the theory of nonrelativistic (Schrödinger) fermions interacting through a two-body potential [see, e.g., Dubois (1959a, b)];

(**e**) nonrelativistic bosons interacting through a two-body potential.

Discuss in detail the effect of closed-loop disconnected diagrams in each case. Show that in a relativistically invariant theory $\langle \Phi_0 \mid S \mid \Phi_0 \rangle_C$ is imaginary.

4. (**a**) Show that, if the S-matrix element for Compton scattering is written in the form

$$S_{fi} = \delta_{fi} - 2\pi i \delta(E_f - E_i) \left(\frac{M^2}{4\omega_1 \omega_2 E_1 E_2} \right)^{1/2} \tilde{u} T_{fi} u$$

then $\tilde{u} T_{fi} u$ is Lorentz invariant.

(**b**) Obtain the relation between the differential cross section for Compton scattering in the laboratory and center-of-momentum frame.

5. (**a**) Compute to lowest order the amplitude and cross section for the scattering of a photon off a spin $\frac{1}{2}$ particle having charge e, mass M, and anomalous magnetic moment λ. (The vertex operator in this case becomes

$$\Gamma_\mu = \gamma_\mu - \frac{\lambda}{2M} i\sigma_{\mu\nu} q^\nu$$

where q is the four-momentum transfer at the vertex.) [Powell (1949).]

(**b**) Obtain the low energy limit of this scattering amplitude and cross section.

6. Compute to lowest order the scattering of an electron off a spin $\frac{1}{2}$ particle having charge e, mass M, and anomalous magnetic moment λ. [Rosenbluth (1950).]

7. Obtain the lowest order scattering amplitudes and cross sections for π^+-π^- and $\pi^\pm$-$\pi^\pm$ scattering taking into account only electromagnetic interactions.

8. Obtain the amplitude for

(**a**) $e^+ + e^- \to 2\gamma$ annihilation (em interactions only)

(**b**) $\pi^+ + \pi^- \to 2\gamma$ annihilation (em interactions only)

(**c**) $p + \bar{p} \to 2\pi$ annihilation (in ps − ps theory)

to lowest order. Compute the lifetime in each case.

9. Discuss the relation between the amplitudes for

$$\pi^+ + \pi^- \to 2\gamma$$

and

$$\pi^\pm + \gamma \to \pi^\pm + \gamma$$

(*ibid.* for $e^+ + e^- \to 2\gamma$ and $e^\pm + \gamma \to e^\pm + \gamma$)

10. Obtain the lowest order amplitudes for π-N scattering in ps-ps theory. Compare these amplitudes with those for ps-pv theory.

11. Compute to lowest order the cross section for the production of

(**a**) an electron-positron pair;

(**b**) a μ^+-μ^- pair;

(**c**) a π^+-π^- pair;

in the scattering of

(**a**) a photon;

(**b**) an electron;

(**c**) a μ meson

by a lead nucleus. Consider the nucleus as infinitely heavy and take into account only electromagnetic interactions.

If sufficient fortitude is available, discuss the cross sections as a function of the polarization of the incident particle.

12. Compute the lifetime of a neutral pseudoscalar π^0 meson against decay into 2 γ rays on the assumption that the coupling is of the form

$$\mathcal{L}_{\text{eff}} = \frac{G}{M}\frac{\alpha}{m}\phi\mathcal{E}\cdot\mathcal{H}$$

where ϕ is the π^0 meson operator and $\mathcal{E}$ and $\mathcal{H}$ are the quantized electric and magnetic field intensity operators. Show that the photons are perpendicularly polarized.

13. Show that, if the interaction Hamiltonian

$$G\bar{\nu}\gamma_\alpha \tfrac{1}{2}(1 - i\gamma_5)\, \mu \bar{e}\gamma^\alpha \tfrac{1}{2}(1 - i\gamma_5)\, \nu + \text{h.a.}$$

is used to account for the $\mu^\pm \to e^\pm + \nu + \bar{\nu}$ decay, it predicts that in the decay of a polarized μ^+, the positon and antineutrino tend to come off in opposite directions and that the neutrino tends to be emitted in a direction opposite to that of the spin of the μ^+.

14. **(a)** Show that in general the invariance of a Hamiltonian under a symmetry transformation leads to the invariance of the transition rates under certain "kinematic" transformations of the quantum numbers in the initial and final states. (Example: invariance of H under space reflection implies invariance of the transition rates under the change of sign of all momenta in the initial and final states.)

(b) Discuss the effect on the transition rate of the invariance of H under T and TPC (i.e., antiunitary operators).

15. Show that the decay of the $^3S_1$ state of positronium into 2γs is forbidden by angular momentum conservation.

16. **(a)** Discuss the properties of the $\pi^0 \to 2\gamma$ amplitude which follow from Lorentz, TPC, PC, TP, T, P, C invariance. [Bernstein and Michel (Bernstein 1960).]

(b) Discuss the restriction imposed by time-reversal invariance on the following decay modes:

(1) $K_2^0 \to 2\pi$ [Weinberg (1958)]

(2) $K^+_{\mu 3} \to \mu^+ + \pi^0 + \nu$ [Sakurai (1958)]

17. Starting from the interaction Lagrangian density

$$\mathfrak{L}_I(x) = \sum_{\alpha\beta\gamma\delta} C_{\alpha\beta\gamma\delta}(\bar{\psi}_p(x))\,(\psi_n(x))_\beta\,(\bar{\psi}_e(x))_\gamma\,(\psi_\nu(x))_\delta + \text{h.a.}$$

to describe the β decay of nucleons [$C_{\alpha\beta\gamma\delta}$ is a (possibly complex) constant] systematically investigate the form of $C_{\alpha\beta\gamma\delta}$ if $\mathfrak{L}_I$ is to be invariant under

(a) homogeneous proper Lorentz transformations; and in addition

(b) TPC

(c) T or PC

18. Let $a_{t,t_3}{}^*(\mathbf{k})$ be a creation operator for a particle with total isotopic spin $t(t+1)$, *3* component of isotopic spin t_3 and momentum $\mathbf{k}$, and $a_{t,t_3}(\mathbf{k}_3)$ the creation operator for the corresponding antiparticles. These operators obey the following commutation rules

$$[T_{\pm}, a_{t,t_3}{}^*(\mathbf{k})] = \sqrt{(t \mp t_3)(t \pm t_3 + 1)}\, a_{t,t_3\pm 1}{}^*(\mathbf{k})$$

$$[T_3, a_{t,t_3}{}^*(\mathbf{k})] = t_3 a_{t,t_3}{}^*(\mathbf{k})$$

and the relation between particle and antiparticles operators is

$$a_{t,t_3}{}^*(\mathbf{k}) = C a_{t,t_3}(\mathbf{k})\, C^{-1}$$

(apart from a phase factor). (C is the unitary charge conjugation operator.)

(a) From the above relations, obtain the transformation properties of T under charge conjugation.

(b) Obtain the transformation properties of a_{t,t_3}, $a_{t,t_3}{}^*$ and T under the operation

$$R = e^{i\pi T_2}$$

corresponding to a rotation of 180° about the 2-axis in isotopic spin space.

(c) Discuss the transformation property of the operators under the operation

$$G = CR$$

Show that $CR = RC$.

(d) Discuss the relations of the eigenstates $|T', T'_3\rangle$ of T and T_3 to the states $G\,|\,T', T'_3\rangle$.

(e) Obtain the eigenvalues of R^2 and G^2 for the states $|T', T'_3\rangle$.

19. (a) Obtain explicit representations for the operators G, C, R and their squares for a system of

1. free nucleons
2. free pions
3. interacting pions and nucleons

(b) Discuss the consequences of the demanding that a theory of strong interactions be invariant under the operation G.

20. Obtain the selection rules for $N\bar{N}$ annihilation into pions for a G invariant theory.

21. Show that, if T is the antiunitary time-reversal operator, and $|j, m\rangle$ are eigenfunctions of the total angular momentum $\mathbf{J}^2$ and J_3,

$$T\,|\,j, m\rangle = \eta(-1)^m\,|\,j, -m\rangle$$

where η is a phase factor. (Recall that apart of a phase factor $T\mathbf{J}T^{-1} = -\mathbf{J}$.)

Chapter 15

1. (a) Using the Hamiltonian (45), compute the lifetime of a hydrogen atom in a $2P$ state.

(b) Using the Hamiltonian (56) in dipole approximation but including a spin orbit coupling in V, obtain an expression for the lifetime of a hydrogen atom in the $2s$ state against two-photon decay.

(c) Estimate this lifetime and discuss the influence of the cascade emission $2s \rightarrow 2p_{1/2} \rightarrow 1s$.

2. Discuss the convergence of the Bethe method of calculating the Lamb shift when the Hamiltonian (56) is used but the dipole approximation is *not* made.

3. **(a)** Using the Hamiltonian (56) in the dipole approximation, obtain the differential cross section for the elastic scattering of a photon by a hydrogen atom in its ground state.

(b) Show that in the "high energy limit" (but still $k < m$) the total cross section averaged over polarization is the Thompson cross section: $8\pi r_0^2/3$.

(c) Show that in the low energy limit the total cross section is given by the Rayleigh scattering formula, with $\sigma_{\text{total}} \propto k^4$.

(d) Formulate a Tamm-Dancoff method to discuss the case of resonant scattering paying particular attention to the relation of the width of the resonance and the self-energy of the resonant state.

4. Show that to lowest order in electromagnetic effects the mass shift for the proton and neutron can be written in the form

$$M = M_0[A + B\mu + C\mu^2]$$

where μ is the (static) anomalous magnetic moment and the factors A, B, C are numerical constants depending on the theory used. Show that the coefficients A and B are zero for the neutron. [Feynman and Speisman (Feynman 1955).]

5. In the Gell-Mann's theory of strong interactions, let M be the mass of the baryons when K interactions are switched off. Show that to second order in K-interactions (but taking into account pion interactions exactly)

$$M_N = M + (g_{N\Lambda K}^2 + 3g_{N\Sigma K}^2)\,\Delta$$
$$M_\Lambda = M + 2(g_{N\Lambda K}^2 + g_{\Xi\Lambda K}^2)\,\Delta$$
$$M_\Sigma = M + 2(g_{N\Sigma K}^2 + g_{\Xi\Sigma K}^2)\,\Delta$$
$$M_\Xi = M + (g_{\Xi\Lambda K}^2 + 3g_{\Xi\Sigma K}^2)\,\Delta$$

where Δ is the same integral in each case, and therefore that whatever the strength of the coupling of kaons to baryons

$$\frac{M_N + M_\Xi}{2} = \frac{M_\Lambda + 3M_\Sigma}{4}$$

For a discussion Σ^+-Σ^- mass difference, see, e.g., Kato and Takeda (1959).

6. (**a**) Obtain the unitary transformation which takes one from the Feynman gauge in which the photon propagator is given by $g_{\mu\nu}k^{-2}$ to the Landau gauge in which the photon propagator is $(g_{\mu\nu} - k_\mu k_\nu k^{-2})\, k^{-2}$.

(**b**) Show that the charge renormalization can be absorbed into the redefinition of A^e by a canonical transformation.

7. (**a**) Compute to lowest order the radiative corrections to the one- and two-photon vertex for charged spin 0 particles.

(**b**) Compute to lowest order the vacuum polarization effect due to a π^+-π^- bubble.

8. By considering the scattering of a proton and neutron by an external electromagnetic field, compute the static magnetic moment of the neutron and proton taking into account the lowest order mesonic effects in

(**a**) scalar meson theory with scalar coupling

(**b**) ps-ps theory

(**c**) ps-pv theory

9. (**a**) Calculate explicitly the charge and magnetic moment form factors of the electron to order α.

(**b**) Obtain the corresponding distributions in configuration space.

10. Calculate the charge and magnetic moment form factors of the neutron and proton to lowest order in mesonic effects in ps-ps theory.

11. Discuss the electromagnetic corrections to the $\mu^\pm \rightarrow e^\pm + \nu + \bar{\nu}$ decay. [Berman (1959).]

For a discussion of radiative effects in high energy electron-electron scattering, see Tsai (1960). A review of the subject of high energy electron scattering is to be found in the Proceedings of the High Energy Conference held in Kiev in 1959.

12. (**a**) Obtain $S_F{}^e$ for the case of

(1) a homogeneous constant magnetic field

(2) a homogeneous constant electric field

(**b**) Using these results, compute the self-energy of an electron in a homogeneous time-independent magnetic field to lowest order, and obtain an expression for its magnetic moment. [Geheniau (1950, 1951).]

(**c**) How would you use this approach to obtain the magnetic moment of the neutron and proton?

For a review of the nuclear force problem up to about 1956, see the Supplement of the Progress of Theoretical Physics, No. 3 (1956), in particular Nishijima's article [Nishijima (1956)]. For development until 1959, see Phillips. The dispersion theoretic approach is reviewed in Goldberger (1960).

Chapter 16

1. Discuss whether a Ward's identity holds for the nucleon-photon vertex. (Consider the ps-ps meson-nucleon system in the presence of electromagnetic interactions.)

2. Obtain Ward's identity in the theory of charged mesons interacting with the electromagnetic field.

3. Obtain Takahashi's (the generalized Ward) identity for the above cases.

Chapter 17

1. Calculate the spectral function $\rho(m^2)$ for the meson "twofold" to lowest order in the coupling constant for a neutral scalar and neutral pseudoscalar meson theory with scalar and pseudovector coupling.

2. (a) Generalize the Lehmann analysis to the case of meson operators which are isotopic vectors. Assume that the theory is invariant under arbitrary rotations in isotopic spin space and under charge conjugation.

(b) Do the same for the analysis of the nucleon "twofold."

3. Formulate a perturbation theory for the spectral functions in the case of ps-ps meson theory.

4. (a) Show that the Redmond-Bogoliubov method of summing weight functions does not alter content of theory if the theory has no ghost states in an exact solution.

(b) If the theory does have "ghost states," show that the method can be viewed as modifying the original Lagrangian in a nonlocal fashion.

5. Show that by virtue of the Riemann-Lebesgue lemma that

$$\lim_{x_0 \to \pm\infty} (\Psi, \phi(x)\, \Phi) \to 0$$

thus giving another reason for considering the operators $\phi^f(t)$ in the definition of the asymptotic limit.

6. Obtain an expression for the low energy limit of Compton scattering if only *TCP* invariance is assumed.

7. Obtain the Bethe-Salpeter equation for the p-$\bar{p}$ and p-$\bar{n}$ systems. Discuss the ladder approximation for these systems. [Okubo and Feldman (Okubo 1960).]

8. Does the exact Bethe-Salpeter equation determine in principle the higher excited states of the hydrogen atom?

9. Obtain an expression for the elastic scattering of a photon from a hydrogen atom in its ground state. Relate this quantity to an expression involving the Bethe-Salpeter amplitude for the ground state.

Chapter 18

1. Obtain the Wightman functions for the free-field theory of spin $\frac{1}{2}$ particles.

2. Show that if the twofold Wightman function $W(\xi)$ for a scalar field $\phi(x)$ coincides with the free-field twofold Wightman function $W_{(0)}(\xi)$ for $\xi_0 = 0$, then $\phi(x)$ is the free field. [This is Haag's theorem; see Haag (1955), Hall (1958), Greenberg (1959), Federbush (1960a, b).]

3. Consider the twofold Wightman function $W^{\psi\tilde{\psi}}_{\alpha_1\alpha_2}(x_1, x_2)$ for the spinor field $\psi(x)$.

(a) Obtain the requirements of relativistic invariance and show that

$$W^{\psi\tilde{\psi}}(-\xi) = -W^{\tilde{\psi}\psi}(\xi)$$

(b) Give a proof of the connection between spin and statistics for the spin $\frac{1}{2}$ case.

4. What does the *TPC* theorem assert regarding the equal-time commutation rules of *different* fermion fields?

5. **(a)** Show that the Wightman function for the vacuum expectation value of a product of three operators can be written in the form

$$W^{ABC}(x - x', x' - x'') = \frac{1}{(2\pi)^6} \int d^4p \int d^4p' \, e^{-ip\cdot(x-x') - ip'\cdot(x'-x'')}$$

$$\int_0^\infty da \int_0^\infty db \int_{\sqrt{ab}}^\infty \delta(p^2 - a)\, \delta(p'^2 - b)\, \delta(p \cdot p' - c)\, \theta(p)\, \theta(p')\, G^{ABC}(a, b, c)$$

by making explicit use of the spectral conditions and of relativistic invariance.

(**b**) Carry out the p integrations and obtain the explicit dependence on $x - x'$ and $x' - x''$ of

$$\Delta^{(+)}(x - x', x' - x'', a, b, c) = \frac{1}{(2\pi)^6} \int d^4p \int d^4p' \, e^{-ip\cdot(x-x')-ip'\cdot(x'-x'')}$$
$$\theta(p)\,\theta(p')\,\delta(p^2 - a)\,\delta(p'^2 - b)\,\delta(p \cdot p' - c)$$

(**c**) Obtain $G^{ABC}(a, b, c)$ in the case that $A(x) = B(x) = C(x) =$ free scalar field.

6. In the expansion (117) for the S matrix obtain the restrictions on σ_n so that the stability of the vacuum and one-particle state is guaranteed.

7. What is the characterization the scalar free-field theory in terms of τ functions? Starting from this free-field set, obtain an approximate perturbation solution of Eq. (199). What is the relation of these solutions to perturbation theory ones for renormalized field theories with interaction Lagrangian of the form $C_1\phi^4(x) + C_2\phi^6(x) + \cdots$ [Lehmann (1955a); Nishijima (1960)].

8. Formulate the Dyson method of obtaining the representation of a causal function by going to a five-dimensional hyperbolic space.

9. Show that the Heisenberg operator $\phi(x)$ whose Fourier transform $\tilde{\phi}(k)$ satisfies the commutation rule

$$[\tilde{\phi}(k), \tilde{\phi}(q)] = \int_0^\infty d\kappa^2 \int d^4u \, \delta([u + \tfrac{1}{2}(k - q)^2]^2 - \kappa^2)\, \epsilon(u - \tfrac{1}{2}(k - q))$$
$$\tilde{\Phi}(\kappa^2, u + \tfrac{1}{2}(k + q), -u + \tfrac{1}{2}(k + q))$$

[where $\tilde{\Phi}$ is an operator function of κ^2, u and $(k + q)$] is local.

10. (**a**) Obtain the reduction formulae for the T and R products of spin $\frac{1}{2}$ particle operators. In particular, use these formulae to deduce a closed expression for the scattering amplitude for the following processes:

(1) $N + \pi \rightarrow N + \pi$
(2) $N + N \rightarrow N + N$
(3) $N + \overline{N} \rightarrow N + \overline{N}$
(4) $N + \overline{N} \rightarrow \pi + \pi$
(5) $\overline{N} + \pi \rightarrow \overline{N} + \pi$
(6) $\pi + N \rightarrow 2\pi + N$

in which the nucleons (N) are treated as spin $\frac{1}{2}$ particles (in both spin and isotopic spin space) and the pions as spin 0 and isotopic spin 1 particles.

(**b**) Obtain an expression for the absorptive part of the scattering amplitude for each of the above processes.

(c) Discuss the relation of Eq. (348) when combined with Eqs. (350) and (357) to the two-field generalization of Eq. (170). Obtain the corresponding equations for each of the above processes.

The theory of dispersion relations for nonrelativistic Schrödinger scattering amplitudes which was not touched upon in the text can be traced from the papers of Khuri (1957), Klein and Zemach [Klein (1960)], and Blankenbecler (1960).

A fairly complete compendium of the literature on dispersion relation as well as a presentation of the theory of dispersion relations until about 1958 is to be found in Taylor (1958).

11. Assume that there are three fields ϕ_1, ϕ_2, ϕ_3 coupled to intermediate fields ϕ_a, ϕ_b, ϕ_c as follows:

$$\mathcal{L}_{in} = g_1\phi_1\phi_b\phi_c + g_2\phi_2\phi_a\phi_c + g_3\phi_3\phi_a\phi_b$$

The masses of the corresponding quanta are m_1, m_2, m_2 and m_a, m_b, m_c where the stability criteria are

$$m_2 < m_a + m_c$$

$$m_3 < m_a + m_b$$

(a) Show that the Fourier transform $T(p_1^2, p_2^2, p_3^2)$ of the vertex function $\langle 0 \mid T(\phi_1(x_1)\,\phi_2(x_2)\,\phi_3(x_3) \mid 0\rangle$ to lowest order of perturbation theory has the following representation:

$$T(p_1^2, p_2^2, p_3^2) = \text{constant} \int_0^1 \frac{\delta(1 - \alpha - \beta - \gamma)d\alpha d\beta d\gamma}{[p_1^2\beta\alpha + p_2^2\gamma\alpha + p_3^2\alpha\beta - m_a^2\alpha - m_b^2\beta - m_c^2\gamma]}$$

(b) Discuss the analyticity properties of T as a function of p_1^2 when p_2^2 and p_3^2 are on their mass shells. Express these analyticity properties in terms of a dispersion relation.

12. (a) Show that, if one restricts oneself to the contribution of one meson-nucleon states in the computation of

$$\text{Im}\, T(\omega, \Delta^2) = M(\omega, \Delta^2) \quad (\omega > 0)$$

by applying the reduction formulae to this expression, a nonlinear integral equation for T can be obtained. Discuss the relation of this equation to the Low equations derived for the Chew model.

(b) Obtain the Low equations for the meson-nucleon amplitude and discuss its relation to dispersion theory [Omnes (1957)].

13. Obtain the solutions of the singular integral equation

$$u(x) = f(x) + \frac{1}{\pi}\int_a^\infty \frac{\overline{h}(x')\,u(x')\,dx'}{x' - x - i\epsilon}$$

[where $f(x)$ and $h(x)$ are given] by considering the analyticity properties of the function

$$U(z) = f(z) + \frac{1}{\pi} \int_a^\infty \frac{\bar{h}(x')\, u(x')\, dx'}{x' - z}$$

[It is assumed that $f(x)$ can be analytically continued to a function $f(z)$ regular along the cut from a to $+\infty$ on the real axis.] [Omnes (1958); Caribbo (1959).]

For a detailed analysis of the minimization problem involved in the determination of the region of analyticity of $T(W, \Delta^2, \zeta)$, see Salam (1958); also Streater (1959).

14. Obtain the maximum momentum transfer, Δ_{max}, for which dispersion relations for the reaction $\pi + \pi \to \pi + \pi$ can be proved. Discuss the feasibility of proving dispersion relations for the processes [Goldberger (1958)].

$$K + N \to K + N$$

$$\pi + D \to \pi + D$$

For an application of dispersion theoretic methods to the computation of the electron self-energies and wave function renormalization constants, see De Celles (1961). The difficulties connected with the asymptotic conditions in quantum electrodynamics are discussed in Evans and Fulton [Evans (1961)].

References

NOTE: The numbers in parentheses appearing at the end of the reference indicate the Chapter sections in this book.

Abraham, M. (1933). (With R. Becker.) *Theorie der Electrizität*, 6th ed., Teubner, Leipzig (1933). (Sec. 15*a*)

Abrikosov, A. A. See Landau, L. D. (1954).

Akhiezer, A. (1937). (With I. Pomeranchuk.) Physik. Z. Sowjetunion, **11**, 478 (1937). (Sec. 16*a*)

——— (1953). (With V. B. Berezetski.) *Quantum Electrodynamics*, State Technico-Theoretical Literature Press, Moscow (1953). (Secs. 4*h*, 15*g*)

Akiba, T. (1958). (With I. Sato.) Prog. Theor. Phys. (Japan), **19**, 93 (1958). (Sec. 18*d*)

Allcock, G. R. (1956). Phys. Rev., **104**, 1799 (1956). (Sec. 17*f*)

——— (1958). (With D. J. Hooton.) Nuovo Cimento, **8**, 590 (1958). (Sec. 17*f*)

——— (1959/60). Nuclear Physics, **14**, 177 (1959/60). (Sec. 18*d*)

Amati, D. (1959). (With B. Vitale.) Fortschritte der Physik, **7**, 375 (1959). (Sec. 10*e*)

——— (1960). (With M. Fierz and V. Glaser.) Phys. Rev. Letters, **4**, 89 (1960). (Sec. 18*d*)

Ambler, E. See Wu, C. S. (1957).

Anderson, C. D. (1932). Phys. Rev., **41**, 405, 1932. (Sec. 4*g*)

Araki, H. (1957). (With Y. Munakata, M. Kawaguchi, and T. Goto.) Prog. Theor. Phys. (Japan), **17**, 419 (1957). (Sec. 12*b*)

——— (1960). Annals of Physics (New York), **11**, 260 (1960). (Sec. 18*b*)

——— (1961). Jour. Math. Physics (1961). (Sec. 18*b*)

Arnous, E. (1953). (With W. Heitler.) Proc. Roy. Soc. (London), **A220**, 290 (1953). (Sec. 15*a*)

——— (1955). (With W. Heitler.) Nuovo Cimento, **2**, 1282 (1955). (Sec. 15*a*)

Arnowitt, R. (1955). (With S. Deser.) Phys. Rev., **100**, 349 (1955). (Sec. 16*a*)

Ascoli, R. (1955a). Nuovo Cimento, **2**, 413 (1955). (Sec. 15*d*)

——— (1955b). Nuovo Cimento, **2**, 1 (1955). (Sec. 15*d*)

——— (1958/59). (With E. Minardi.) Nuclear Physics, **9**, 242 (1958/59). (Sec. 12*b*)

Ashkin, J. (1959a). (With T. Fazzini, G. Fidecaro, A. W. Merrison, H. Paul, and A. V. Tollestrup.) Nuovo Cimento, **13,** 1240 (1959). (Sec. 14*e*)

——— (1959b). Suppl. No. 2, Nuovo Cimento, **14,** pp. 221 and 310 (1959). (Sec. 10*a*). See also Ashkin (1960).

——— (1960). *Proc. of Tenth Annual High Energy Conference at Rochester,* Interscience Publishers, Ltd., New York (1960). (Sec. 10*a*)

Auerbach, L. B. See Yamagata, T. L. (1956).

Bade, W. L. (1953). (With H. Jehle.) Rev. Mod. Phys., **25,** 714 (1953). (Sec. 4*c*)

Baranger, M. (1951a). Doctoral Dissertation, Cornell University (1951). (Sec. 15*b*)

——— (1951b). Phys. Rev., **84,** 866 (1951). (Sec. 15*b*)

——— (1953). (With H. A. Bethe and R. P. Feynman.) Phys. Rev., **92,** 482 (1953). (Secs. 15*b*, 15*d*, 15*g*)

——— (1953). (With F. J. Dyson and E. E. Salpeter.) Phys. Rev., **88,** 680 (1953). (Sec. 15*e*)

Bardeen, J. (1936). Phys. Rev., **49,** 653 (1936). (Sec. 6*k*)

Bargmann, V. (1936). Zeits. für Phys., **99,** 576 (1936). (Sec. 1*f*)

——— (1947). Annals of Math., **48,** 568 (1947). (Secs. 2*b*, 2*c*)

——— (1948). (With E. P. Wigner.) Proc. Nat. Acad. Sci. U.S.A., **34,** 211 (1948). (Secs. 1*d*, 2*c*)

——— (1951). Mimeographed lecture notes on Advanced Quantum Mechanics, Princeton (1951). (Sec. 6)

——— (1953). (With E. P. Wigner and A. S. Wightman.) Ms. for a book on the representations of the Lorentz group and their relevance for the quantum theory of fields (to be published). (Secs. 1*a*, 1*d*)

——— (1954). Annals of Math., **59,** 1 (1954). (Sec. 1*e*)

Barshay, S. (1956). Phys. Rev., **103,** 1102 (1956). (Sec. 12*d*)

Baumann, K. (1958). Zeits. für Phys., **152,** 448 (1958). (Sec. 18*b*)

——— See also Schmidt, K. (1956).

Baym, G. (1960). Phys. Rev., **117,** 886 (1960). (Sec. 17*a*)

Becker, R. (1945). Göttingen Nachrichten (1945). (Sec. 4*f*)

——— (1946). (With G. Liebfried.) Phys. Rev., **69,** 34 (1946). (Sec. 6)

——— (1948). (With G. Liebfried.) Zeits. für Phys., **125,** 347 (1948). (Sec. 6)

——— See also Abraham, M. (1933).

Belinfante, F. J. (1939). Physica, **6,** 887 (1939). (Secs. 7*d*, 7*g*)

——— (1940). Physica, **7,** 305 (1940). (Sec. 7*d*)

——— See also Pauli, W. (1940b).

Berger, J. M. (1952). (With L. L. Foldy and R. K. Osborn.) Phys. Rev., **87,** 1061 (1952). (Sec. 10*g*)

Berezetski, V. B. See Akhiezer, A. I. (1953).

Berman, S. M. (1958). Phys. Rev., **112,** 267 (1958). (Problems)
Bernadini, G. (1956). *Encyclopedia of Physics,* 2d ed., Vol. 43, Lange and Springer, Berlin (1956). (Sec. 12*d*)
——— See also Yamagata, T. L. (1956).
Bernstein, A. M. (1958). (With A. K. Mann.) Phys. Rev., **110,** 805 (1958). (Sec. 16*a*)
Bernstein, J. (1960). (With L. Michel.) Phys. Rev., **118,** 871 (1960). (Problems)
Bersohn, R. See Weneser, J. (1953).
Bethe, H. A. (1946). (With J. R. Oppenheimer.) Phys. Rev., **70,** 451 (1946). (Secs. 15*d*, 15*f*)
——— (1947). Phys. Rev., **72,** 339 (1947). (Sec. 15*b*)
——— (1950a). *Rapports du 8e Conseil Solvay 1948,* R. Stoops, Brussels (1950). (Sec. 15*b*)
——— (1950b). (With L. M. Brown and J. R. Stehn.) Phys. Rev., **77,** 370 (1950). (Sec. 15*b*)
——— (1952). (With F. Rohrlich.) Phys. Rev., **86,** 10 (1952). (Sec. 16*a*)
——— (1955). (With F. de Hoffmann.) *Mesons and Fields,* Volume II: *Mesons,* Row, Peterson and Co., Evanston (1955). (Secs. 7*h*, 8*d*, 10*d*, 12*d*, 13*b*, Problems)
——— (1956). (With J. Hamilton.) Nuovo Cimento, **4,** 1 (1956).
——— (1957). (With E. E. Salpeter.) *Handbuch der Physik,* Volume XXXV/1. Springer-Verlag, Berlin (1957). (Sec. 4*h*)
——— See also Baranger, M. (1953); Salpeter, E. E. (1951).
Björken, J. D. (1959). Doctoral Dissertation, Stanford University (1959). (Secs. 17*b*, 18*d*)
Bincer, A. M. (1957). Phys. Rev., **105,** 1399 (1957). (Sec. 12*d*)
——— (1960). Phys. Rev., **118,** 855 (1960). (Sec. 18*d*)
Biswas, S. N. See Green, H. S. (1957).
Blankenbecler, R. (1960). (With M. L. Goldberger, N. N. Khuri, and S. B. Treiman.) Annals of Physics (New York), **10,** 62 (1960). (Problems)
Bleuler, K. (1950). Helv. Phys. Acta, **23,** 567 (1950). (Sec. 9*b*)
Bloch, F. (1937). (With A. Nordsieck.) Phys. Rev., **52,** 54 (1937). (Secs. 15*c*, 15*d*, 15*f*)
Bochner, S. (1948). (With W. T. Martin.) *Several Complex Variables,* Princeton University Press, Princeton (1948). (Sec. 18*a*)
Bogoliubov, N. N. (1956). (With D. V. Shirkov.) Nuovo Cimento, **3,** 77 (1956). (Secs. 16*b*, 16*h*)
——— (1958a). (With B. V. Medvedev and M. K. Polivanov.) *Problems in the Theory of Dispersion Relations,* Institute for Advanced Studies, Princeton (1958). (Sec. 18*b*)

Bogoliubov, N. N. (1958b). (With B. V. Medvedev and M. K. Polivanov.) Fortschritte der Physik, **6,** 169 (1958). (Sec. 18*b*)
——— (1959a). (With D. V. Shirkov.) *Introduction to the Theory of Quantized Fields*, Interscience Publishers, Ltd., New York (1959). (Secs. 15*e*, 15*g*, 16*b*, 16*h*, 18*b*)
——— (1959b). (With A. A. Logunov and D. V. Shirkov.) J. Exptl. Theor. Phys. (USSR), **37,** 805 (1959); Soviet Physics JETP (New York), **37** (10), No. 3, 574 (1960). (Sec. 17*c*)
Bohr, N. (1933). (With L. Rosenfeld.) Kgl. Danske Vidensk. Selsk. Mat.-Fys. Medd., **12,** No. 8 (1933). (Secs. 9*b*, 13*a*, 17)
——— (1950). (With L. Rosenfeld.) Phys. Rev., **78,** 794 (1950). (Secs. 7*f*, 9*b*, 13*a*, 17)
Bonnevay, G. (1959). Nuovo Cimento, **14,** 593 (1959). (Sec. 12*d*)
Booth, E. T. See Landé, K. (1956).
Bopp, F. (1940). Ann. der Phys. (Leipzig), **38,** 345 (1940). (Sec. 15*a*)
——— (1942/43). Ann. der Phys. (Leipzig), **42,** 473 (1942/43). (Sec. 15*a*)
Born, M. (1934a). Proc. Roy. Soc. (London), **A143,** 410 (1934). (Sec. 15*a*)
——— (1934b). (With W. Infeld.) Proc. Roy. Soc. (London), **A144,** 425 (1934). (Sec. 15*a*)
——— (1937). Ann. de l'Institut Henri Poincaré, **7,** 155 (1937). (Sec. 15*a*)
Borowitz, S. (1952). (With W. Kohn.) Phys. Rev., **86,** 985 (1952). (Sec. 15*a*)
Bosco, B. (1959). (With S. Fubini and A. Stanghellini.) Nuclear Physics, **10,** 663 (1959). (Problems)
Bose, S. K. (1959). (With A. Gamba and E. C. G. Sudarshan.) Phys. Rev., **113,** 1661 (1959). (Sec. 4*f*)
Bouchiat, C. (1958). (With L. Michel.) Nuclear Physics, **5,** 416 (1958). (Problems)
Breit, G. (1947). Phys. Rev., **72,** 984(L) (1947). (Sec. 15*d*)
——— (1948a). Phys. Rev., **73,** 1410(L) (1948). (Sec. 15*d*)
——— (1948b). Phys. Rev., **74,** 656 (1948). (Sec. 15*d*)
Bremerman, H. (1958). (With R. Oehme and J. G. Taylor.) Phys. Rev., **109,** 2178 (1958). (Secs. 18*a*, 18*d*)
Brenig, W. (1959). (With R. Haag.) Fortschritte der Physik, **7,** 183 (1959). (Secs. 11*a*, 11*c*, Problems)
Brenner, S. (1953). (With G. E. Brown.) Proc. Roy. Soc. (London), **A218,** 422 (1953). (Sec. 4*h*)
——— (1954). (With G. E. Brown and J. B. Woodward.) Proc. Roy. Soc. (London), **A227,** 59 (1954). (Sec. 4*h*)
Brown, G. E. (1955). (With D. F. Meyers.) Proc. Roy. Soc. (London), **A234,** 387 (1955). (Sec. 4*h*)

Brown, G. E. (1957). (With D. F. Meyers.) Proc. Roy. Soc. (London), **A242,** 89 (1957). (Sec. 4*h*)
——— (1959). (With J. S. Langer and G. W. Schaffer.) Proc. Roy. Soc. (London), **A251,** 92 (1959); see also *ibid.*, 105 (1959). (Sec. 4*h*)
——— See also Brenner, S. (1953), (1954).
Brown, L. M. (1952). (With R. P. Feynman.) Phys. Rev., **85,** 231 (1952). (Secs. 15*d*, 15*f*)
——— (1958). Phys. Rev., **111,** 462 (1958). (Sec. 4*e*)
——— See also Bethe, H. A. (1950b).
Brueckner, K. See Gell-Mann, M. (1957a).
Buccafurri, A. See Caianiello, E. R. (1958).
Bumiller, F. See Hofstadter, R. (1958).
Burgoyne, N. (1958). Nuovo Cimento, **8,** 604 (1958). (Sec. 18*a*)
Burgy, M. T. See Telegdi, V. L. (1958).
Burton, W. K. (1953a). (With B. F. Touschek.) Phil. Mag., **44,** 161 (1953). (Sec. 17*a*)

Caianiello, E. R. (1952). (With S. Fubini.) Nuovo Cimento, **9,** 1218 (1952). (Sec. 4*e*)
——— (1953). Nuovo Cimento, **10,** 1634 (1953). (Sec. 16*h*)
——— (1954a). Nuovo Cimento, **11,** 492 (1954). (Sec. 16*h*)
——— (1954b). Nuovo Cimento, **12,** 561 (1954). (Sec. 16*h*)
——— (1955). Nuovo Cimento, **2,** 186 (1955). (Sec. 16*h*)
——— (1956). Nuovo Cimento, **3,** 223 (1956). (Sec. 16*h*)
——— (1957). Nuovo Cimento, **5,** 739 (1957). (Sec. 16*h*)
——— (1958). (With A. Buccafurri.) Nuovo Cimento, **8,** 170 (1958). (Sec. 16*h*)
——— (1959a). Nuovo Cimento, **13,** 637 (1959). (Sec. 16*h*)
——— (1959b). Nuovo Cimento, **14,** 185 (1959). (Sec. 16*h*)
Cap, F. (1955). Fortschritte der Physik, **2,** Heft 5, 207 (1955). (Sec. 4*c*)
Capps, R. H. (1957a). Phys. Rev., **106,** 1031 (1957). (Sec. 18*d*)
——— (1957b). Phys. Rev., **108,** 1032 (1957). (Sec. 18*d*)
Caribbo, N. (1959). (With R. Gatto.) Nuovo Cimento, **13,** 1086 (1959). (Problems)
Cartan, E. (1938). *Leçons sur la Théorie des Spineurs* I, II, Hermann & Companie, Editeurs, Paris (1938). (Sec. 1*e*)
Case, K. M. (1949a). Phys. Rev., **76,** 1 (1949). (Sec. 10*g*)
——— (1949b). Phys. Rev., **76,** 14 (1949). (Sec. 10*g*)
——— (1954). Phys. Rev., **95,** 1323 (1954). (Sec. 3*d*)
——— (1957). Phys. Rev., **107,** 307 (1957). (Problems)
——— See also Moldauer, P. (1953).
Cassen, B. (1936). (With E. U. Condon.) Phys. Rev., **50,** 846 (1936). (Sec. 10*c*)

Castillejo, L. (1956). (With R. H. Dalitz and F. J. Dyson.) Phys. Rev., **101**, 453 (1956). (Sec. 12*d*)
Charap, J. M. (1959). (With S. F. Fubini.) Nuovo Cimento, **14**, 540 (1959). (Sec. 15*h*)
——— (1960a). (With S. F. Fubini.) Nuovo Cimento, **15**, 73 (1960). (Sec. 15*h*)
——— (1960b). (With M. Tausner.) Nuovo Cimento, **18**, 316 (1960). (Sec. 15*h*)
Chevalier, A. (1958). (With G. Rideau.) Nuovo Cimento, **10**, 228 (1958). (Sec. 12*c*)
Chew, G. F. (1954). Phys. Rev., **94**, 1748 (1954). (Sec. 12*d*)
——— (1956a). (With F. E. Low.) Phys. Rev., **101**, 1571 (1956). (Sec. 12*d*)
——— (1956b). (With F. E. Low.) Phys. Rev., **101**, 1579 (1956). (Sec. 12*d*)
——— (1956c). *Encyclopedia of Physics*, 2d ed., Vol. 43. Lange and Springer, Berlin (1956). (Sec. 12*d*)
——— (1957). (With M. L. Goldberger, F. E. Low, and Y. Nambu.) Phys. Rev., **106**, 1337 (1957). (Sec. 18*d*)
——— (1958). (With S. Gasiorowicz, R. Karplus, and F. Zachariasen.) Phys. Rev., **110**, 265 (1958). (Sec. 18*d*)
——— (1959). Ann. Rev. Nuclear Sci., **9**, 29 (1959). (Sec. 18*d*)
——— (1960a). (With S. Mandelstam.) Phys. Rev., **119**, 467 (1960). (Sec. 18*d*)
——— (1960b). Double Dispersion Relations and Unitarity as the Basis for a Dynamical Theory of Strong Interactions. UCRL-9289. (Sec. 18*d*)
Chinowski, W. See Landé, K. (1956).
Chisholm, J. S. R. (1952). Proc. Cambridge Phil. Soc. **48**, 300 (1952); for corrigendum see *ibid.* **48**, 518 (1952). (Sec. 17*b*)
Chrétien, M. (1953). (With R. E. Peierls.) Nuovo Cimento, **10**, 668 (1953). (Sec. 15*a*)
——— (1954). (With R. E. Peierls.) Proc. Roy. Soc. (London), **A223**, 468 (1954). (Sec. 15*a*)
——— (1955). Phys. Rev., **98**, 1515 (1955). (Sec. 15*f*)
Christy, R. F. (1957). *Proc. of Seventh Annual Rochester Conference on High Energy Physics*, Interscience Publishers, Ltd., New York (1957). (Sec. 10*a*)
Cini, M. (1953). Nuovo Cimento, **9**, 1025(L) (1953). (Sec. 17*a*)
——— (1958). (With B. Touschek.) Nuovo Cimento, **7**, 422 (1958). (Sec. 4*f*)
——— (1959). (With S. Fubini and A. Stanghellini.) Phys. Rev., **114**, 1633 (1959). (Sec. 18*a*)

Cini, M. (1960). (With S. Fubini.) Annals of Physics (New York), **3,** 352 (1960). (Sec. 18*d*)
Claesson, A. (1957). Kgl. Fysiograf. Sällskap Lund Forh., **27,** No. 1 (1957). (Sec. 16*a*)
Coester, F. (1950). (With J. M. Jauch.) Phys. Rev., **78,** 149 (1950). (Sec. 14*b*)
——— (1953). Phys. Rev., **89,** 619 (1953). (Sec. 14*f*)
Cohen, S. (1954). Doctoral Dissertation, Cornell University (1954). (Sec. 4*h*)
——— (1960). Phys. Rev., **118,** 489 (1960). (Sec. 4*h*)
Condon, E. U. See Cassen, B. (1936).
Corinaldesi, E. (1948). (With R. Jost.) Helv. Phys. Acta, **21,** 183 (1948). (Sec. 15*f*)
——— (1951). Nuovo Cimento, **8,** 494 (1951). (Sec. 13*a*)
——— (1953). Suppl. No. 2 to Nuovo Cimento, **10** (1953). (Secs. 9*b*, 13*a*)
Costa, G. (1955). (With N. Dellaporta.) Nuovo Cimento, **2,** 519 (1955). (Sec. 10*f*)
Crane, H. R. See Schupp, A. A. (1961).
Cutkosky, R. E. (1954). Phys. Rev., **96,** 1135 (1954). (Sec. 17*f*)
——— (1956). (With F. Zachariasen.) Phys. Rev., **103,** 1108 (1956). (Sec. 12*d*)

Dalitz, R. H. (1951a). Proc. Roy. Soc. (London), **A206,** 509 (1951). (Sec. 14*a*)
——— (1951b). Proc. Roy. Soc. (London), **A206,** 521 (1951). (Secs. 15*e*, 15*f*)
——— (1957). Reports on Progress in Physics, **20,** 163 (1957). (Sec. 10*a*)
——— (1959). Proc. of Ninth Annual International Conference on High Energy Physics, CERN, Geneva (1959). (Secs. 10*e*, 12*e*)
——— See also Castillejo, L. (1956).
Darwin, C. G. (1928). Proc. Roy. Soc. (London), **A118,** 654 (1928). (Sec. 4*h*)
Davidon, W. C. (1956). (With M. L. Goldberger.) Phys. Rev., **104,** 119 (1956). (Sec. 18*d*)
Dayhoff, E. S. See Triebwasser, S. (1953).
de Benedetti, S. (1956). Nuovo Cimento, **4,** Suppl. 3, 1234 (1956). (Sec. 4*g*)
de Celles, P. (1961). Phys. Rev. (1961). (Problems)
de Donder, The. (1926). (With H. van Dungen.) Comptes Rendus (July 1926). (Sec. 3*a*)
de Hoffmann, F. See Bethe, H. A. (1955).
Delbrück, M. (1933). Zeits. für Phys., **84,** 144 (1933). (Sec. 16*a*)

Dell'Antonio, G. F. (1959). (With P. Gulmanelli.) Nuovo Cimento, **12,** 38 (1959). (Sec. 18*b*)

Dellaporta, N. See Costa, G. (1955).

Demeur, M. See Géhéniau, J. (1951).

Deser, S. See Arnowitt, R. (1955).

d'Espagnat, B. (1956). (With J. Prentki.) Nuclear Physics, **1,** 33 (1956). (Sec. 10*e*)

——— (1958). (With J. Prentki.) Volume IV, *Elementary Particles and Cosmic Ray Physics*, ed. J. G. Wilson and S. A. Wouthuysen, North Holland Publishing Company, Amsterdam (1958). (Sec. 10*e*)

Deutsch, M. (1951a). Phys. Rev., **82,** 455L (1951). (Sec. 4*g*)

——— (1951b). (With E. Dulit.) Phys. Rev., **84,** 601L (1951). (Sec. 4*g*)

de Witt, B. S. (1955). Phys. Rev., **100,** 905 (1955). (Sec. 11*b*)

de Witt, C. (1958). *The Many Body Problem,* C. de Witt, editor. Lecture notes on courses given at les Houches. John Wiley and Sons, New York (1959). (Problems)

Dicke, R. H. See Pond, T. A. (1952).

Dirac, P. A. M. (1926). Proc. Roy. Soc. (London), **A112,** 661 (1926). (Sec. 11*c*)

——— (1927). Proc. Roy. Soc. (London), **A114,** 243 (1927). (Sec. 11*c*)

——— (1928). Proc. Roy. Soc. (London), **A117,** 610 (1928). (Secs. 4*a*, 4*h*)

——— (1930). Proc. Cambridge Phil. Soc., **26,** 376 (1930). (Sec. 4*g*)

——— (1934a). *Rapports du 7e Conseil Solvay, 1933*, R. Stoops, Brussels (1934). (Sec. 15*e*)

——— (1934b). Proc. Cambridge Phil. Soc., **30,** 150 (1934). (Sec. 15*e*)

——— (1938a). Proc. Roy. Soc. (London), **A167,** 148 (1938). (Sec. 15*a*)

——— (1938b). Ann. de l'Institut Henri Poincaré, **9,** 13 (1938). (Sec. 15*a*)

——— (1942). (With R. E. Peierls and M. H. L. Pryce.) Proc. Cambridge Phil. Soc., **38,** 193 (1942). (Problems)

——— (1947). *The Principles of Quantum Mechanics,* 3d ed., Oxford University Press, Oxford (1947). (Sec. 11*b*)

——— (1951). Proc. Roy. Soc. (London), **A209,** 291 (1951). (Sec. 15*a*)

——— (1952a). Ann. de l'Institut Henri Poincaré, **13,** 1 (1952). (Sec. 15*a*)

——— (1952b). Proc. Roy. Soc. (London), **A212,** 330 (1952). (Sec. 15*a*)

——— (1954). Proc. Roy. Soc. (London), **A223,** 438 (1954). (Sec. 15*a*)

——— (1958). *The Principles of Quantum Mechanics,* 4th ed., Clarendon Press, Oxford (1958). (Secs. 1*a*, 1*b*)

Drell, S. D. (1950). Phys. Rev., **79,** 220 (1950). (Sec. 15*a*)

——— (1952). (With E. M. Henley.) Phys. Rev., **88,** 1053 (1952). (Secs. 10*f*, 10*g*, 12*d*)

Drell, S. D. (1956). (With M. Friedman and F. Zachariasen.) Phys. Rev., **104**, 236 (1956). (Sec. 12*d*)
——— (1958). Annals of Physics (New York), **4**, 75 (1958). (Sec. 16*a*)
——— (1959). (With F. Zachariasen.) Phys. Rev., **111**, 1727 (1959). (Secs. 17*c*, 18*a*)
——— (1960). (With F. Zachariasen.) Phys. Rev., **119**, 463 (1960). (Sec. 17*c*)
——— (1961). (With F. Zachariasen.) *Electromagnetic Structure of Nucleons*, Oxford University Press, Oxford (1961). (Sec. 18*d*)
DuBois, D. F. (1959a). Annals of Physics (New York), **7**, 174 (1959). (Problems)
——— (1959b). Annals of Physics (New York), **8**, 24 (1959). (Problems)
Duffin, R. (1950). Phys. Rev., **77**, 242 (1950). (Sec. 4*e*)
Dulit, E. See Deutsch, M. (1951).
Dyson, F. J. (1948). Phys. Rev., **73**, 929 (1948). (Sec. 10*g*)
——— (1949a). Phys. Rev., **75**, 486 (1949). (Sec. 11*f*)
——— (1949b). Phys. Rev., **75**, 1736 (1949). (Secs. 11*f*, 15*c*, 16*a*, 16*b*, 16*c*, 16*f*)
——— (1951a). *Advanced Quantum Mechanics*, lithoprinted notes, Cornell University (1951). (Secs. 13*c*, 15*c*)
——— (1951b). Phys. Rev., **82**, 428 (1951). (Secs. 13*a*, 13*c*, 16*a*)
——— (1951c). Phys. Rev., **83**, 608 (1951). (Sec. 13*a*)
——— (1951d). Proc. Roy. Soc. (London), **A207**, 395 (1951). (Sec. 13*a*)
——— (1951e). Phys. Rev., **83**, 1207 (1951). (Sec. 13*a*)
——— (1952). Phys. Rev., **85**, 631 (1952). (Sec. 16*h*)
——— (1953). Phys. Rev., **91**, 1543 (1953). (Sec. 13*a*)
——— (1957). Phys. Rev., **106**, 157 (1957). (Sec. 12*d*)
——— (1958a). Phys. Rev., **110**, 579 (1958). (Sec. 18*a*)
——— (1958b). Phys. Rev., **110**, 1460 (1958). (Sec. 18*c*)
——— See also Baranger, M. (1953); Castillejo, L. (1956).

Eddington, A. (1939). Proc. Cambridge Phil. Soc., **35**, 186 (1939). (Problems)
——— (1942). Proc. Cambridge Phil. Soc., **38**, 201 (1942). (Problems)
Eden, R. J. (1952a). Proc. Roy. Soc. (London), **A210**, 388 (1952). (Secs. 16*a*, 17*f*)
——— (1952b). Proc. Roy. Soc. (London), **A215**, 133 (1952). (Sec. 17*f*)
——— (1953a). Proc. Roy. Soc. (London), **A217**, 390 (1953). (Sec. 17*f*)
——— (1953b). (With G. Rickayzen.) Proc. Roy. Soc. (London), **A219**, 109 (1953). (Sec. 17*f*)
——— (1953c). Proc. Roy. Soc. (London), **A219**, 516 (1953). (Sec. 17*f*)
——— (1954). Proc. Cambridge Phil. Soc., **50**, 592 (1954). (Sec. 17*f*)
——— (1960a). Phys. Rev., **119**, 1763 (1960). (Sec. 17*c*)
——— (1960b). Phys. Rev. Letters, **5**, 213 (1960). (Sec. 18*d*)

Eden, R. J. (1960c). *Proc. of Tenth Annual High Energy Conference at Rochester*, Interscience Publishers, Ltd., New York (1960). (Sec. 18*d*)

Edmonds, A. R. (1957). *Angular Momentum in Quantum Mechanics*, Princeton University Press, Princeton (1957). (Sec. 12*d*)

Edwards, D. N. See Noyes, H. P. (1960).

Edwards, S. F. (1953). Phys. Rev., **90,** 284 (1953). (Sec. 16*b*)

Eisenbud, L. (1956). J. of the Franklin Institute, **261,** 409 (1956). (Sec. 11*a*)

Ekstein, H. (1960). Phys. Rev., **120,** 1917 (1960). (Problems)

Eliezer, C. J. (1947). Rev. Mod. Phys., **19,** 147 (1947). (Sec. 15*a*)

Enz, C. P. (1956). Suppl. No. 3 to Vol. **3** of the Nuovo Cimento, p. 363 (1956). (Sec. 12*c*)

Epstein, S. T. (1948). Phys. Rev., **73,** 177L (1948); Erratum *73*, 630L (1948). (Sec. 15*c*)

——— See also Pais, A. (1949).

Eriksen, E. See Foldy, L. L. (1955).

Euler, E. (1936). Ann. der Phys. (Leipzig), **26,** 398 (1936). (Sec. 16*a*)

——— See also Heisenberg, W. (1936).

Euwema, R. W. (1956). (With J. A. Wheeler.) Phys. Rev., **103,** 803 (1956). (Problems)

Evans, P. (1961). (With T. Fulton.) Phys. Rev. (1961). (Problems)

Fainberg, G. B. See Silin, W. P. (1956).

Fairlie, D. B. (1958a). (With J. C. Polkinghorne.) Nuovo Cimento, **8,** 345 (1958). (Sec. 12*d*)

——— (1958b). (With J. C. Polkinghorne.) Nuovo Cimento, **8,** 555 (1958). (Sec. 12*d*)

Falk, D. S. (1959). Phys. Rev., **115,** 1069 (1959). (Sec. 17*b*)

Fazzini, T. See Ashkin, J. (1959).

Federbush, P. G. (1958a). (With M. T. Grisaru.) Nuovo Cimento, **9,** 1058 (1958). (Sec. 7*c*)

——— (1958b). (With M. L. Goldberger and S. B. Treiman.) Phys. Rev., **112,** 642 (1958). (Sec. 18*d*)

——— (1960a). Nuovo Cimento, **15,** 932 (1960). (Problems)

——— (1960b). (With K. A. Johnson.) Phys. Rev., **120,** 1926 (1960). (Problems)

Feinberg, G. (1957). Phys. Rev., **108,** 878 (1957). (Sec. 10*b*)

——— (1959). (With S. Weinberg.) Nuovo Cimento, **14,** 571 (1959). (Secs. 10*e*, 14*f*)

Feld, B. (1959). Annals of Physics (New York), **7,** 323 (1959). (Sec. 10*a*)

Feldman, D. (1949). Phys. Rev., **76,** 1369 (1949). (Sec. 15*e*)

——— (1956). Phys. Rev., **103,** 254 (1956). (Sec. 10*a*)

Feldman, D. See also Okubo, S. (1960); Yang, C. N. (1950b).
Fermi, E. (1929). Atti. Accad. Lincei, **9,** 881 (1929). (Sec. 9*a*)
——— (1930). Atti. Accad. Lincei, **12,** 431 (1930). (Sec. 9*a*)
——— (1932). Rev. Mod. Phys., **4,** 87 (1932). (Sec. 9*a*)
——— (1934). Zeits. für Phys., **88,** 161 (1934). (Sec. 10*f*)
——— (1949). (With C. N. Yang.) Phys. Rev., **76,** 1739 (1949). (Sec. 10*a*)
Ferretti, B. (1951). Nuovo Cimento, **8,** 108 (1951). (Sec. 16*h*)
——— (1959). Nuovo Cimento, **12,** 393 (1959). (Sec. 12*b*)
Feshbach, H. (1938). (With F. Villars.) Rev. Mod. Phys., **30,** 24 (1958). (Sec. 3*d*)
——— See also McKinley, W. A. (1948).
Feynman, R. P. (1948a). Phys. Rev., **74,** 1430 (1948). (Sec. 15*a*)
——— (1948b). Phys. Rev., **76,** 939 (1948). (Secs. 15*a*, 15*b*, 15*g*)
——— (1948c). Rev. Mod. Phys., **20,** 367 (1948). (Sec. 4*h*, and also Problems)
——— (1949a). Phys. Rev., **76,** 749 (1949). (Secs. 4*a*, 14*a*, 17*f*)
——— (1949b). Phys. Rev., **76,** 769 (1949). (Secs. 14*a*, 15*c*, 15*e*, 16*a*, 17*f*)
——— (1950). Phys. Rev., **80,** 440 (1950). (Sec. 14*a*)
——— (1951). Phys. Rev., **84,** 108 (1951). (Secs. 15*c*, 15*e*)
——— (1954). (With G. Speisman.) Phys. Rev., **94,** 500 (1954). (Secs. 8*d*, 16*a*, Problems)
——— (1958). (With M. Gell-Mann.) Phys. Rev., **109,** 193 (1958). (Secs. 4*h*, 10*f*)
——— See also Baranger, M. (1953); Brown, L. M. (1952); Wheeler, J. A. (1945 and 1949).
Fidecaro, G. See Ashkin, J. (1959).
Fierz, M. (1950). Helv. Phys. Acta, **23,** 731 (1950). (Sec. 13*c*)
——— (1955). *Proc. of Fifth Annual Rochester Conference on High Energy Physics*, p. 67, Interscience Publishers, Ltd., New York (1955). (Sec. 17*a*)
——— See also Amati, D. (1960); Pauli, W. (1937).
Filosofo, L. See Yamagata, T. L. (1956).
Finkelstein, R. J. (1949). Phys. Rev., **75,** 1079 (1949). (Sec. 15*a*)
——— (1951). (With R. LeLévier and M. Ruderman.) Phys. Rev., **83,** 326 (1951). (Sec. 15*a*)
Fock, V. (1926a). Zeits. für Phys., **38,** 242 (1926). (Sec. 3*a*)
——— (1926b). Zeits. für Phys., **39,** 226 (1926). (Sec. 3*a*)
——— (1930). Zeits. für Phys., **61,** 126 (1930). (Sec. 6*k*)
——— (1932). Zeits. für Phys., **75,** 622 (1932). (Sec. 6)
——— (1934). Physik Z. Sowjetunion, **6,** 425 (1934). (Sec. 7*e*)
——— (1936). Zeits. für Phys., **98,** 145 (1936). (Sec. 1*f*)

Foldy, L. L. (1950). (With S. A. Wouthuysen.) Phys. Rev., **78,** 29 (1950). (Secs. 4*f*, 4*h*)
——— (1951). Phys. Rev., **84,** 168 (1951). Sec. 10*g*)
——— (1952). Phys. Rev., **87,** 688 (1952). (Sec. 4*h*)
——— (1955). (With E. Eriksen.) Phys. Rev., **98,** 775 (1955). (Sec. 4*g*)
——— (1956). Phys. Rev., **102,** 568 (1956). (Sec. 2*c*)
——— See also Berger, J. M. (1952).
Ford, K. W. (1955). (With D. L. Hill.) Ann. Rev. Nuclear Sci., **5,** 25–72 (1955). (Sec. 4*h*)
——— (1957). Phys. Rev., **105,** 320 (1957). (Secs. 12*b*, 17*c*)
Frank, R. M. (1951). Phys. Rev., **83,** 1189 (1951). (Sec. 16*a*)
Franklin, J. (1957). Phys. Rev., **105,** 1101 (1957). (Sec. 12*d*)
Franzinetti, C. (1957). (With G. Morpugo.) Suppl to Vol. **6** of Nuovo Cimento, 469 (1957). (Sec. 10*a*)
Frautschi, S. C. (1958). Suppl. to Prog. of Theor. Phys. (Japan), No. **8,** 21 (1958). (Sec. 16*a*)
——— See also Yennie, D. R. (1961).
Freese, E. (1953). Zeits. für Naturforshung, **8a,** 775 (1953). (Sec. 17*f*)
——— (1954). Acta Physica Austriaca, **8,** 289 (1954). (Sec. 17*f*)
——— (1955). Nuovo Cimento, **2,** 50 (1955). (Sec. 17*f*)
French, J. B. (1949). (With V. Weisskopf.) Phys. Rev., **75,** 1240 (1949). (Secs. 15*b*, 15*d*, 15*g*)
Fried, B. D. (1952). Phys. Rev., **88,** 1142 (1952). (Problems)
Fried, H. M. (1958). (With D. R. Yennie.) Phys. Rev., **112,** 1391 (1958). (Secs. 15*c*, 15*d*, 15*g*)
——— (1959). Phys. Rev., **115,** 220 (1959). (Sec. 16*b*)
——— (1960). (With D. R. Yennie.) Phys. Rev. Letters, **4,** 580 (1960). (Secs. 15*b*, 15*d*)
Friedman, M. See Drell, S. (1956).
Friedrichs, K. (1953). *Mathematical Aspects of the Quantum Theory of Fields*, Interscience Publishers, Ltd., New York (1953). (Sec. 7*a*)
Fronsdal, C. (1959). Phys. Rev., **113,** 1367 (1959). (Secs. 2*c*, 5*a*)
Fubini, S. See Caianiello, E. R. (1952).
——— See also Bosco, B. (1959); Charap, J. M. (1959 and 1960); Cini, M. (1959 and 1960).
Fukuda, H. (1950a). (With S. Hayakawa and Y. Miyamoto.) Prog. Theor. Phys. (Japan), **5,** 283 (1950). (Sec. 16*a*)
——— (1950b). (With S. Hayakawa and Y. Miyamoto.) Prog. Theor. Phys. (Japan), **5,** 352 (1950). (Sec. 16*a*)
——— (1956). (With J. S. Kovacs.) Phys. Rev., **104,** 1784 (1956). (Sec. 12*d*)
Fulton, T. (1954). (With P. C. Martin.) Phys. Rev., **93,** 903 (1954). (Sec. 17*f*)

Fulton, T. See also Evans, P. (1961).
Furry, W. H. (1937). Phys. Rev., **51,** 125 (1937). (Sec. 14*a*)
——— (1951). Phys. Rev., **81,** 115 (1951). (Secs. 15*e*, 15*g*)

Gamba, A. See Bose, S. K. (1959).
Gärding, L. (1954). (With A. S. Wightman.) Proc. Nat. Acad. Sci. (USA), **40,** 612–26 (1954). (Sec. 7*a*)
——— (1959). (With J. L. Lions.) *Functional Analysis*, Suppl. No. 1 to Vol. **14,** Series X of the Nuovo Cimento (1959). (Sec. 18*a*)
Gartenhaus, S. (1955). Phys. Rev., **100,** 900 (1955). (Sec. 15*h*)
——— See also Yennie, D. R. (1958), (1961).
Gasiorowicz, S. G. (1959). (With D. R. Yennie and H. Suura.) Phys. Rev. Letters, **2,** No. 12, 513 (1959). (Sec. 17*c*)
——— (1960). Fortschritte der Physik, preprint (1960). (Sec. 18*d*)
——— See also Chew, G. F. (1958).
Gatto, R. See Caribbo, N. (1959).
Géhéniau, J. (1950). Physica, **16,** 822 (1950). (Problems)
——— (1951). (With M. Demeur.) Physica **17,** 71 (1951). (Problems)
Gel'fand, I. M. (1956). (With Z. Ya. Šapiro.) *The representations of the group of rotations in three-dimensional space and their applications,* Am. Math. Soc. Trans., Series 2, Vol. 2 (1956). (Sec. 1*e*)
Gell-Mann, M. (1951). (With F. Low.) Phys. Rev., **84,** 350 (1951). (Secs. 11*f*, 17*a*, 17*f*)
——— (1953a). (With M. L. Goldberger.) Phys. Rev., **91,** 398 (1953). (Secs. 11*b*, 11*c*, 11*e*)
——— (1953b). Phys. Rev., **92,** 833 (1953). (Secs. 10*d*, 10*e*)
——— (1954a). (With F. Low.) Phys. Rev., **95,** 1300 (1954). (Secs. 17*b*, 17*c*)
——— (1954b). (With M. L. Goldberger and W. Thirring.) Phys. Rev., **95,** 1612 (1954). (Secs. 12*d*, 18*d*)
——— (1954c). (With M. L. Goldberger.) Phys. Rev., **96,** 1433 (1954). (Sec. 17*e*)
——— (1955a). (With A. Pais.) *Proceedings of the 1954 Glascow Conference on Nuclear and Meson Physics*, Pergamon Press, London (1955). (Sec. 10*d*)
——— (1955b). (With A. Pais.) Phys. Rev., **97,** 1387 (1955). (Sec. 10*e*)
——— (1956a). Nuovo Cimento, **4,** Suppl. 2, 848 (1956). (Secs. 10*a*, 10*c*, 10*d*, 10*f*)
——— (1956b). *Proc. of Sixth Annual Rochester High Energy Conference 1956*, Interscience Publishers, Ltd., New York (1956). (Secs. 12*f*, 18*e*)
——— (1957a). (With K. Brueckner.) Phys. Rev., **106,** 364 (1957). (Sec. 6*k*)

Gell-Mann, M. (1957b). Phys. Rev., **106,** 367 (1957). (Sec. 6*k*)
——— (1957c). Phys. Rev., **106,** 1296 (1957). (Secs. 10*a*, 10*d*, 10*e*)
——— (1957d). (With A. H. Rosenfeld.) Ann. Rev. Nuclear Sci., **7,** 407 (1957). (Secs. 10*a*, 10*d*)
——— (1960). (With M. Lévy.) Nuovo Cimento, **16,** 705 (1960). (Problems)
——— See also Feynman, R. P. (1958).
Gilbert, W. (1956). (With G. R. Screaton.) Phys. Rev., **104,** 1158 (1956). (Sec. 18*d*)
Glaser, V. (1952). (With W. Zimmermann.) Zeits. für Physik, **134,** 346 (1952). (Sec. 16*h*)
——— (1956/57). (With G. Källén.) Nuclear Physics, **2,** 706 (1956/57). (Sec. 12*b*)
——— (1957). (With H. Lehmann and W. Zimmermann.) Nuovo Cimento, **6,** 1122 (1957). (Sec. 18*b*)
——— See also Amati, D. (1960).
Glauber, R. J. (1951). Phys. Rev., **84,** 395 (1951). (Secs. 15*c*, 15*d*)
——— (1959). Boulder Summer School Lecture Notes. Volume I of *Lectures in Theoretical Physics*, W. E. Brittin and L. G. Dunham, Editors, Interscience Publishers, Ltd., New York (1959). (Problems)
Gluckstern, R. L. See Rohrlich, F. (1952).
Goldberger, M. L. (1955a). Phys. Rev., **97,** 508 (1955). (Sec. 18*d*)
——— (1955b). Phys. Rev., **99,** 979 (1955). (Sec. 18*d*)
——— (1957). (With Y. Nambu and R. Oehme.) Annals of Physics (New York), **2,** 226 (1957). (Sec. 18*d*)
——— (1958). Proc. of Eighth Annual High Energy Nuclear Physics Conference at CERN, pp. 207–9, CERN, Geneva (1958). (Sec. 10*f*)
——— Proc. of the Midwest Conference on Theoretical Physics, p. 60, Purdue University (1960). (Problems)
——— (1961). Lectures given at les Houches, Summer (1960). (Sec. 18*d*)
——— (1962). (With R. Oehme.) *The Theory and Applications of Dispersion Relations* (to be published). (Sec. 18*d*)
——— See also Blankenbecler, R. (1960); Chew, G. F. (1957); Davidon, W. C. (1956); Federbush, P. (1958b); Gell-Mann, M. (1953 and 1954b, c).
Goldhaber, M. (1956). Phys. Rev., **101,** 433 (1956). (Sec. 10*a*)
Goldstein, J. S. (1953). Phys. Rev., **91,** 1516 (1953). (Sec. 17*f*)
Good, R. H. (1955). Rev. Mod. Phys. **27,** 187 (1955). (Sec. 4*b*)
Gordon, W. (1926a). Zeits. für Phys., **40,** 117 (1926). (Sec. 3*a*)
——— (1926b). Zeits. für Phys., **40,** 121 (1926). (Sec. 3*a*)
——— (1928). Zeits. für Phys., **48,** 11 (1928). (Sec. 4*h*)

Goto, T. See Araki, H. (1957).

Goursat, E. (1904). (With E. R. Hedrick.) *A Course in Mathematical Analysis*. Vol. I, Ginn and Co., Boston (1904). (Sec. 15*a*)

Grawert, G. (1959). (With G. Lüders and H. Rollnik.) Fortschritte der Physik, **7,** 291 (1959). (Sec. 14*f*)

Green, H. S. (1957). (With S. N. Biswas.) Prog. Theor. Phys. (Japan), **18,** 121 (1957). (Sec. 17*f*)

Green, T. A. See Stückelberg, E. C. G. (1951b).

Greenberg, O. W. (1956). Doctoral Dissertation, Princeton University (1956). (Sec. 18*b*)

——— (1958). (With S. S. Schweber.) Nuovo Cimento, **8,** 378 (1958). (Secs. 12*a*, 12*c*)

——— (1959). Phys. Rev., **115,** 706 (1959). (Problems)

Grisaru, M. T. See Federbush, P. G. (1958a).

Gulmanelli, P. See Dell'Antonio, G. F. (1959).

Gupta, S. N. (1950a). Proc. Phys. Soc. (London), **A63,** 681 (1950). (Secs. 9*a*, 9*b*)

——— (1950b). Phys. Rev., **77,** 294L (1950). (Sec. 15*a*)

——— (1951). Proc. Phys. Soc. (London), **A64,** 426 (1951). (Sec. 16*h*)

——— (1960). Phys. Rev., **117,** 1146 (1960). (Sec. 15*h*)

Güttinger, W. (1958a). Nuovo Cimento, **10,** 1 (1958). (Sec. 16*a*)

——— (1958b). Nuclear Physics, **9,** 429 (1958). (Sec. 16*a*)

Haag, R. (1955). Kgl. Danske Vidensk. Selsk. Mat.-Fys. Medd., **29,** No. 12 (1955). (Secs. 2*c*, 7*a*, 18*b*, Problems)

——— (1957). Nuovo Cimento, **5,** 203 (1957). (Sec. 12*d*)

——— (1958). Phys. Rev., **112,** 669 (1958). (Sec. 18*b*)

——— (1959a). (With G. Luzzatto.) Nuovo Cimento, **13,** 415 (1959). (Sec. 12*d*)

——— (1959b). Nuovo Cimento Suppl. to Vol. **14,** No. 1, 131 (1959). (Sec. 18*b*)

——— See also Brenig, W. (1959).

Haber-Schaim, U. (1956). Phys. Rev., **104,** 1113 (1956). (Sec. 18*d*)

Hack, M. N. (1954). Phys. Rev., **96,** 196 (1954). (Sec. 11*c*)

Hagedorn, R. (1959). Nuovo Cimento Suppl. to Vol. **12,** Series X, 73 (1959). (Sec. 1*d*)

Hall, D. (1957). (With A. S. Wightman.) Kgl. Danske Vidensk. Selsk. Mat.-Fys. Medd., **31,** No. 5 (1957). (Sec. 18*a*, Problems)

Halpern, O. (1934). Phys. Rev., **44,** 855 (1934). (Sec. 16*a*)

Hamilton, J. See Bethe, H. A. (1956).

Hansen, A. O. See Yamagata, T. L. (1956).

Harriman, J. M. (1956). Phys. Rev., **101,** 594 (1956). (Sec. 15*b*)

Hartree, D. R. (1928). Proc. Cambridge Phil. Soc., **24,** 89 (1928). (Sec. 6*k*)

Hayakawa, S. See Fukuda, H. (1950a, b).
Hayashi, C. (1952). (With Y. Munakata.) Prog. Theor. Phys. (Japan), **7,** 481 (1952). See *ibid.* **8,** 142 (1952) for correction. (Sec. 17*f*)
Hayward, R. W. See Wu, C. S. (1957).
Hedrick, E. R. See Goursat, E. (1904).
Heine, V. (1957). Phys. Rev., **107,** 620 (1957). (Sec. 2*b*)
——— (1960). *Group Theory in Quantum Mechanics,* Pergamon Press, Ltd., London (1960). (Problems)
Heisenberg, W. (1929). (With W. Pauli.) Zeits. für Phys., **56,** 1 (1929). (Sec. 15*a*)
——— (1930a). (With W. Pauli.) Zeits. für Phys., **59,** 168 (1930). (Secs. 6, 15*a*)
——— (1930b). Zeits. für Phys., **65,** 4 (1930). (Sec. 11*c*)
——— (1932). Zeits. für Phys. **77,** 1 (1932). (Sec. 10*d*)
——— (1934a). Zeits. für Phys., **90,** 209 (1934). (Secs. 4*g*, 15*e*)
——— (1934b). Zeits. für Phys., **92,** 692 (1934). (Secs. 4*g*, 15*e*). See also Heisenberg (1936), where some corrections are given.
——— (1934c). Leipziger Berichten, **86,** 317 (1934). (Sec. 15*e*)
——— (1936). (With E. Euler.) Zeits. für Phys., **98,** 714 (1936). (Secs. 15*e*, 16*a*)
——— (1938). Ann. der Physik (Leipzig), **32,** 20 (1938). (Sec. 11*c*)
——— (1943a). Zeits. für Phys., **120,** 513 (1943). (Secs. 11*c*, 13*a*)
——— (1943b). Zeits. für Phys., **120,** 673 (1943). (Secs. 11*c*, 13*a*)
——— (1957a). Rev. Mod. Phys., **29,** No. 3, 269 (1957). (Sec. 10*a*)
——— (1957b). Nuclear Physics, **4,** 532 (1957). (Sec. 12*b*)
Heitler, W. (1943). Proc. Roy. Irish Acad., **49,** 1 (1943). (Sec. 3*d*)
——— (1944). *The Quantum Theory of Radiation,* 2d ed., Oxford University Press (1944). (Secs. 14*e*, 15*a*)
——— See also Arnous, E. (1953), (1955).
Henley, E. M. See Drell, S. D. (1952).
Hill, D. L. See Ford, K. W. (1955).
Hill, E. L. (1951). Rev. Mod. Phys., **23,** 253 (1951). (Sec. 7*g*)
Hofstadter, R. (1957). Ann. Rev. Nuclear Sci., **7,** 231 (1957). (Secs. 2*c*, 4*h*, 14*a*, 16*a*)
——— (1958). (With F. Bumiller and M. R. Yearian.) Rev. Mod. Phys., **30,** 482 (1958). (Sec. 17*e*)
——— (1960). *Proc. of Tenth Annual Rochester High Energy Conference,* Interscience Publishers, Ltd., New York (1960). (Sec. 17*e*)
Höhler, G. (1958). Zeits. für Phys., **152,** 546 (1958). (Sec. 12*b*)
Hooton, D. J. See Allcock, G. R. (1958).
Hoppes, D. D. See Wu, C. S. (1957).
Horwitz, L. (1957). Phys. Rev., **108,** 886 (1957). (Sec. 12*d*)
Hudson, R. P. See Wu, C. S. (1957).
Huff, L. D. (1931). Phys. Rev., **38,** 501 (1931). (Sec. 4*h*)

Hurst, C. A. (1952a). Phys. Rev., **85,** 920L (1952). (Sec. 17*f*)
——— (1952b). Proc. Roy. Soc. (London), **A214,** 44 (1952). (Sec. 17*f*)
——— (1952c). Proc. Cambridge Phil. Soc., **48,** 625 (1952). (Sec. 17*f*)
Hylleraas, E. A. (1955). Zeits. für Phys., **140,** 626 (1955). (Sec. 4*h*)

Imamura, T. See Utiyama, R. (1952).
Impeduglia, J. See Landé, K. (1956).
Infeld, W. See Born, M. (1934b).
Isaev, P. S. See Logunov, A. A. (1958).

Jackson, J. D. (1958). *The Physics of Elementary Particles,* Princeton University Press (1958). (Sec. 10)
Jahn, H. (1959). Fortschritte der Physik, **7,** 451 (1959). (Problems)
Jauch, J. M. (1954). (With F. Rohrlich.) Helv. Phys. Acta, **27,** 613 (1954). (Sec. 16*f*)
——— (1955a). (With F. Rohrlich.) Phys. Rev., **98,** 181 (1955). (Sec. 15*d*)
——— (1955b). (With F. Rohrlich.) *The Theory of Photons and Electrons*, Addison-Wesley, Cambridge (1955). (Sec. 15*g*)
——— (1957). Helv. Phys. Acta, **30,** 143 (1957). (Problems)
——— See also Coester, F. (1950).
Jean, M. (1953). Annales de Physique (Paris), **8,** 338 (1953). (Sec. 7)
Jehle, H. See Bade, W. L. (1953).
Johnson, K. (1958). Phys. Rev., **112,** 1367 (1958). (Sec. 17*c*)
Johnson, M. H. (1949). (With B. A. Lippmann.) Phys. Rev., **76,** 828 (1949). (Sec. 4*h*)
Jordan, P. (1927). (With O. Klein.) Zeits. für Phys., **45,** 751 (1927). (Sec. 6)
——— (1928a). (With W. Pauli.) Zeits. für Phys., **47,** 151 (1928). (Sec. 9*a*)
——— (1928b). (With E. P. Wigner.) Zeits. für Phys., **47,** 631 (1928). (Secs. 6, 8*a*)
Jost, R. (1947a). Phys. Rev., **72,** 815 (1947). (Secs. 15*d*, 15*f*)
——— (1950a). (With J. M. Luttinger.) Helv. Phys. Acta, **23,** 201 (1950). (Secs. 16*a*, 16*b*)
——— (1950b). (With J. M. Luttinger and M. Slotnick.) Phys. Rev., **80,** 189 (1950). (Sec. 14*e*)
——— (1951). (With A. Pais.) Phys. Rev., **82,** 840 (1951). (Sec. 16*h*)
——— (1957a). Helv. Phys. Acta, **30,** 409 (1957). (Sec. 18*a*)
——— (1957b). (With H. Lehmann.) Nuovo Cimento, **5,** 1598 (1957). (Sec. 18*c*)
——— (1958). Helv. Phys. Acta, **31,** 263 (1958). (Secs. 18, 18*b*, 18*d*)
——— See also Corinaldesi, E. (1948); Pais, A. (1952).
Julian, R. S. See Nagel, D. E. (1947).

Källén, G. (1949). Helv. Phys. Acta, **22**, 637 (1949). (Sec. 15*e*)
——— (1950). Arkiv f. Physik, **2**, No. 37, 371 (1950). (Secs. 13*a*, 15*e*, 17*d*)
——— (1952). Helv. Phys. Acta, **25**, 417 (1952). (Secs. 16*h*, 17*a*, 17*b*)
——— (1953a). Kgl. Danske Vidensk. Selsk. Mat.-Fys. Medd., **27**, No. 12 (1953). (Sec. 16*g*)
——— (1953b). Helv. Phys. Acta, **26**, 755 (1953). (Sec. 16*g*)
——— (1953c). Physica, **19**, 850 (1953). (Sec. 17*d*)
——— (1954). Kgl. Danske Vidensk. Selsk. Mat.-Fys. Medd., **27**, No. 12 (1954). (Sec. 17*c*)
——— (1955a). (With A. Sabry.) Kgl. Danske Vidensk. Selsk. Mat.-Fys. Medd., **29**, No. 17 (1955). (Sec. 15*e*)
——— (1955b). (With W. Pauli.) Kgl. Danske Vidensk. Selsk. Mat.-Fys. Medd., **30**, No. 7 (1955). (Sec. 12*b*)
——— (1957). CERN Theoretical Division Tech. Report 57-43. (Sec. 12*b*)
——— (1958). (With A. S. Wightman.) Kgl. Danske Vidensk. Selsk. Mat.-Fys. Skr., **1**, No. 6 (1958). (Secs. 18*a*, 18*b*)
——— (1959). (With H. Wilhelmsson.) Kgl. Danske Vidensk. Selsk. Mat.-Fys. Skr., **1**, No. 9 (1959). (Sec. 18*a*)
——— (1960). (With J. S. Toll.) Helv. Phys. Acta, **23**, 753 (1960). (Sec. 18*a*)
——— See also Glaser, V. (1956).
Kamefuchi, S. See Umezawa, H. (1951, 1956a, b).
Kanesawa, A. See Sugawara, M. (1959).
Kanesawa, S. (1948a). (With S. Tomonaga.) Prog. Theor. Phys. (Japan), **3**, 1 (1948). (Sec. 14*c*)
——— (1948b). (With S. Tomonaga.) Prog. Theor. Phys. (Japan), **3**, 101 (1948). (Sec. 14*c*)
Karplus, R. (1950a). (With N. M. Kroll.) Phys. Rev., **77**, 536 (1950). (Secs. 15*c*, 15*d*, 15*f*, 16*c*, 16*e*)
——— (1950b). (With M. Neuman.) Phys. Rev., **80**, 380 (1950). (Sec. 16*a*)
——— (1950c). (With M. Neuman.) Phys. Rev., **83**, 776 (1950). (Sec. 16*a*)
——— (1952a). (With A. Klein.) Phys. Rev., **85**, 972 (1952). (Sec. 15*g*)
——— (1952b). (With A. Klein and J. Schwinger.) Phys. Rev., **86**, 288 (1952). (Secs. 15*b*, 15*d*, 15*g*)
——— (1952c). (With A. Klein.) Phys. Rev., **87**, 848 (1952). (Secs. 15, 17*f*)
——— (1955). (With M. Ruderman.) Phys. Rev., **98**, 771 (1955). (Sec. 18*d*)

Karplus, R. (1958). (With C. Sommerfield and E. H. Wichmann.) Phys. Rev., **111**, 1187 (1958). (Sec. 18*d*)

——— (1959). (With C. M. Sommerfield and E. H. Wichmann.) Phys. Rev., **114**, 376 (1959). (Sec. 18*d*)

——— See also Chew, G. F. (1958); Fulton, T. (1954a).

Karzas, W. J. (1958). (With W. K. R. Watson and F. Zachariasen.) Phys. Rev., **110**, 253 (1958). (Sec. 12*d*)

Kashlun, F. (1959). Nuovo Cimento, **12**, 541 (1959). (Sec. 18*b*)

——— (1960). Dubna Joint Institute for Nuclear Research, Preprint No. E-485 (1960). (Sec. 18*b*)

Kato, M. (1959). (With G. Takeda.) Suppl., Prog. Theor. Phys. (Japan), No. 7, 35 (1959). (Problems)

Kawabe, R. See Umezawa, H. (1949a, 1949b, 1949c, and 1950).

Kawaguchi, M. See Araki, H. (1957).

Kazes, E. (1957). Phys. Rev., **107**, 1131 (1957). (Sec. 12*d*)

——— (1959). Nuovo Cimento, **13**, 1226 (1959). (Secs. 16*e*, 17*e*)

Kelly, E. J. (1950). Phys. Rev., **79**, 399 (1950). (Sec. 10*g*)

Kemmer, N. (1937a). Helv. Phys. Acta, **10**, 112 (1937). (Sec. 16*a*)

——— (1937b). (With G. Ludwig.) Helv. Phys. Acta, **10**, 182 (1937). (Sec. 16*a*)

——— (1955). (With A. Salam.) Proc. Roy. Soc., **A230**, 266 (1955). (Sec. 17*f*)

——— (1959). (With D. L. Pursey and J. C. Polkinghorne.) Reports on Progress in Physics, **22**, 368 (1959), London. (Secs. 10*a*, 10*e*)

Khalatnikov, I. M. See Landau, L. D. (1954 and 1955a).

Khuri, N. N. (1957). Phys. Rev., **107**, 1148 (1957). (Problems)

——— See also Blankenbecler, R. (1960).

Kibble, T. W. B. (1958). (With J. C. Polkinghorne.) Nuovo Cimento, **8**, 74 (1958). (Sec. 4*h*)

Kinoshita, T. (1950a). Prog. Theor. Phys. (Japan), **5**, 473 (1950). (Sec. 14*c*)

——— (1950b). (With Y. Nambu.) Prog. Theor. Phys. (Japan), **5**, 749 (1950). (Sec. 14*c*)

Klein, A. (1953a). Phys. Rev., **90**, 1101 (1953). (Sec. 17*f*)

——— (1953b). Phys. Rev., **92**, 1017 (1953). (Sec. 17*f*)

——— (1955a). (With B. McCormick.) Phys. Rev., **98**, 1428 (1955). (Sec. 12*c*)

——— (1955b). Phys. Rev., **99**, 998 (1955). (Sec. 17*e*)

——— (1956). Phys. Rev., **104**, 1131 (1956). See also *ibid.*, **104**, 1136 (1956). (Sec. 12*d*)

——— (1958a). Prog. Theor. Phys. (Japan), **20**, 257 (1958). (Sec. 17*f*)

——— (1958b). (With B. McCormick.) Prog. Theor. Phys. (Japan), **20**, 876 (1958). (Sec. 17*f*)

Klein, A. (1960). Jour. Math. Physics, **1,** 41 (1960). (Problems)
——— See also Karplus, R. (1952a, b, c); Norton, R. E. (1958a, b).
Klein, O. (1926). Z. Phys., **37,** 895 (1926). (Sec. 3*a*)
——— (1929). (With Y. Nishina.) Zeits. für Phys., **52,** 853 (1929). (Sec. 14*e*)
——— (1948). Nature, **161,** 897 (1948). (Sec. 10*f*)
——— See also Jordan, P. (1927).
Kleitman, D. (1959). Nuclear Physics, **11,** 459 (1959). (Secs. 18*a*, 18*b*)
Kleppner, A. (1958). (With H. V. McIntosh.) *The theory of the three-dimensional rotation group*, Rias Technical Reports 58-5, 6, 7. (Sec. 1*f*)
Koba, Z. (1949). Prog. Theor. Phys. (Japan), **4,** 319 (1949). (Sec. 15*c*)
Kohn, W. See Borowitz, S. (1952).
Konopinski, E. J. (1959). Ann. Rev. Nuclear Sci. **9,** Annual Review, Stanford (1959). (Sec. 10*f*)
Konuma, M. See Umezawa, H. (1956a, c).
Kovacs, J. S. See Fukuda, N. (1956).
Kramers, H. A. (1937). Proc. Kgl. Ned. Acad. Wet., **40,** 814 (1937). (Sec. 4*g*)
——— (1950). *Rapports du 8e Conseil Solvay 1948*, p. 241, R. Stoops, Brussels (1950). (Sec. 15*a*)
Krohn, V. E. See Telegdi, V. L. (1958).
Kristensen, P. (1952). (With C. Møller.) Kgl. Danske Vidensk. Selsk. Mat.-Fys. Medd., **27,** No. 7 (1952). (Sec. 18*b*)
Kroll, N. M. (1949a). (With W. E. Lamb.) Phys. Rev., **75,** 388 (1949). (Secs. 15*b*, 15*g*)
——— (1949b). Phys. Rev., **75,** 1321A (1949). (Sec. 13*a*)
——— (1952). (With F. Pollock.) Phys. Rev., **86,** 876 (1952). (Secs. 15*d*, 15*g*)
——— (1954). (With M. A. Ruderman.) Phys. Rev., **93,** 233 (1954). (Sec. 17*e*)
——— See also Karplus, R. (1950a); Mills, R. L. (1955); Weneser, J. (1953); Wichmann, E. H. (1956).
Kudar, J. (1926). Ann. der Phys. (Leipzig), **81,** 632 (1926). (Sec. 3*a*)

Lamb, W. E. (1947). (With R. C. Retherford.) Phys. Rev., **72,** 241 (1947). (Sec. 15*b*)
——— (1951). Reports on Progress in Physics, **14,** 23 (1951), London. (Sec. 15*b*)
——— See also Kroll, N. M. (1949a); Triebwasser, S. (1953).
Landau, L. D. (1930). (With R. Peierls.) Zeits. für Phys., **62,** 188 (1930). (Sec. 6)
——— (1954). (With A. A. Abrikosov and I. M. Khalatnikov.) Doklady Akad. Nauk USSR, **95,** 773 (1954); **96,** 261 (1954). (Sec. 15*e*)

Landau, L. D. (1955a). (With I. M. Khalatnikov.) J. Exptl. Theor. Phys. (USSR), **29,** 89 (1955), English translation in Soviet Physics JEPT (New York), **2,** 69 (1956). (Sec. 15*e*)

——— (1955). *Niels Bohr and the Development of Physics*, McGraw-Hill, New York (1955). (Sec. 17*c*)

——— (1957). Nuclear Physics, **3,** 127 (1957). (Secs. 5*a*, 10*a*, 10*b*, 10*f*)

——— (1959). Nuclear Physics, **13,** 181 (1959). (Secs. 17*b*, 18*d*)

Landé, A. (1941a). (With L. Thomas.) Phys. Rev., **60,** 121 (1941). (Sec. 15*a*)

——— (1941b). (With L. Thomas.) Phys. Rev., **60,** 514 (1941). (Sec. 15*a*)

——— (1944). (With L. Thomas.) Phys. Rev., **65,** 175 (1944). (Sec. 15*a*)

Landé, K. (1956). (With E. T. Booth, J. Impeduglia, L. M. Ledermann, and W. Chinowski.) Phys. Rev., **103,** 1901 (1956); also Phys. Rev., **105,** 1925 (1957). (Sec. 10*e*)

Langer, J. S. See Brown, G. E. (1959).

Layzer, A. J. (1960). Phys. Rev. Letters, **4,** 580 (1960). (Secs. 15*b*, 15*d*, 15*g*)

Ledermann, L. M. (1956). See Landé, K. (1956).

Lee, T. D. (1954). Phys. Rev., **95,** 1329 (1954). (Secs. 12*b*, 12*d*)

——— (1955). Phys. Rev., **99,** 337 (1955). (Sec. 10*a*)

——— (1956a). (With C. N. Yang.) Phys. Rev., **102,** 290 (1956). (Secs. 5*a*, 10*c*)

——— (1956b). (With C. N. Yang.) Phys. Rev., **104,** 254 (1956). (Secs. 5*a*, 10*c*)

——— (1956c). (With C. N. Yang.) Phys. Rev., **104,** 822 (1956). (Secs. 5*a*, 10*c*)

——— (1956d). (With C. N. Yang.) Nuovo Cimento, **3,** 749 (1956). (Problems)

——— (1957a). (With C. N. Yang.) Phys. Rev., **105,** 1671 (1957). (Secs. 5*a*, 10*b*, 10*c*).

——— (1957b). (With R. Oehme and C. N. Yang.) Phys. Rev., **106,** 340 (1957). (Sec. 14*f*)

Lehmann, H. (1954). Nuovo Cimento, **11,** 342 (1954). (Sec. 17*b*)

——— (1955a). (With K. Symanzik and W. Zimmermann.) Nuovo Cimento, **1,** 1425 (1955). (Secs. 16*a*, 17*d*, 18, 18*b*, Problems)

——— (1955b). (With K. Symanzik and W. Zimmermann.) Nuovo Cimento, **2,** 425 (1955). (Secs. 17*c*, 18*a*)

——— (1957). (With K. Symanzik and W. Zimmermann.) Nuovo Cimento, **6,** 319 (1957). (Secs. 11*c*, 18, 18*b*)

——— (1958). Nuovo Cimento, **10,** 579 (1958). (Sec. 18*d*)

——— (1959). Suppl. No. 1 to Nuovo Cimento, **14,** 153 (1959). (Sec. 18*d*)

Lehmann, H. See also Glaser, V. (1957); Jost, R. (1957b).

LeLévier, R. See Finkelstein, R. J. (1951).

Lepore, J. V. (1952). Phys. Rev., **88**, 750 (1952). (Sec. 10*g*)

——— See also Watson, K. M. (1949).

Lévy, M. M. (1952a). Phys. Rev., **88**, 72 (1952). (Sec. 17*f*)

——— (1952b). Phys. Rev., **88**, 725 (1952). (Sec. 17*f*)

——— (1954). (With R. E. Marshak.) Nuovo Cimento, **11**, 366 (1954). (Sec. 10*a*)

——— (1959a). Nuovo Cimento, **13**, 115 (1959). (Sec. 12*b*)

——— (1959b). Nuovo Cimento, **14**, 612 (1959). (Sec. 12*b*)

——— See also Gell-Mann, M. (1960).

Lewis, H. W. (1948). Phys. Rev., **73**, 173 (1948). (Sec. 15*c*)

Liebfried, G. See Becker, R. (1946 and 1948).

Lions, J. L. See Gärding, L. (1959).

Lippmann, B. A. (1950). (With J. Schwinger.) Phys. Rev., **79**, 469 (1950). (Secs. 11*b*, 14*d*)

——— See also Johnson, M. H. (1949).

Logunov, A. A. (1958). (With P. S. Isaev.) Nuovo Cimento, **10**, 917 (1958). (Sec. 18*d*)

——— See also Bogoliubov, N. N. (1949b); Vladimirov, V. S. (1959).

Lomon, E. L. (1956). Nuclear Physics, **1**, 101 (1956). (Sec. 15*d*)

——— (1959). Phys. Rev., **113**, 726 (1959). (Sec. 15*d*)

Lopuszanski, J. (1959). Physica, **25**, 745 (1959). (Sec. 12*c*)

Lorentz, H. A. (1916). *The Theory of Electrons*, Teubner (1916). (Sec. 15*a*)

——— (1937). *The Collected Papers of Lorentz*, M. Nyhoff, The Hague (1937). (Sec. 15*a*)

Low, F. (1952). Phys. Rev., **88**, 53 (1952). (Secs. 15*g*, 17*f*)

——— (1954). Phys. Rev., **96**, 1428 (1954). (Sec. 17*e*)

——— (1955). Phys. Rev., **97**, 1392 (1955). (Secs. 11*d*, 12*d*, 17*d*)

——— (1957). Rev. Mod. Phys., **29**, 216 (1957). (Sec. 12*d*)

——— (1958). Phys. Rev., **110**, 974 (1958). (Sec. 17*e*)

——— See also Chew, G. F. (1956a, b and 1957); Gell-Mann, M. (1951 and 1954a).

Lüders, G. (1952a). Zeits. für Phys., **133**, 325 (1952). (Secs. 7*f*, 8*c*)

——— (1952b). Z. Naturforschung, **7a**, 206 (1952). (Sec. 15*c*)

——— (1954). Kgl. Danske Vidensk. Selsk. Mat.-Fys. Medd., **28**, No. 5 (1954). (Secs. 7*f*, 8*c*, 10*b*, 14*f*)

——— (1957). Annals of Physics (New York), **2**, 1 (1957). (Secs. 7*f*, 8*c*, 10*b*, 14*f*)

——— (1958). (With B. Zumino.) Phys. Rev., **110**, 1450 (1958). (Sec. 18*a*)

——— See also Grawert, G. (1959).

Ludwig, G. See Kemmer, N. (1937b).

Luttinger, J. M. (1948). Phys. Rev., **74**, 893 (1948). (Sec. 15*d*)
——— See also Jost, R. (1950).
Luzzatto, G. See Haag, R. (1959a).

McCormick, B. See Klein, A. (1955 and 1958b).
MacDowell, S. W. (1959). Phys. Rev., **116**, 774 (1959). (Sec. 18*d*)
MacGregor, M. H. (1959). (With M. J. Moravscik and H. P. Stapp.) Phys. Rev., **116**, 1248 (1959). (Sec. 15*h*)
McIntosh, H. V. See Kleppner, A. (1958).
Macke, W. (1953). Zeits. für Naturforschung, **8a**, 599 (1953); **8a**, 615 (1953). (Sec. 17*f*)
McKinley, W. A. (1948). (With H. Feshbach.) Phys. Rev., **74**, 1759 (1948). (Sec. 14*a*)
McLennan, J. A. (1957). Phys. Rev., **106**, 821 (1957). (Problems)
McManus, H. (1948). Proc. Roy. Soc. (London), **A195**, 323 (1948). (Sec. 15*a*)
Majorana, E. (1937). Nuovo Cimento, **14**, 171 (1937). (Secs. 4*h*, 8*c*, Problems)
Maki, Z. (1953). (With M. Sato and S. Tomonaga.) Prog. Theor. Phys. (Japan), **9**, 607 (1953). (Problems)
Malenka, B. J. (1957). (With H. Primakoff.) Phys. Rev., **105**, 338 (1957). (Sec. 8*d*)
Mandelstam, S. (1955). Proc. Roy. Soc. (London), **A233**, 248 (1955). (Sec. 17*f*)
——— (1958). Phys. Rev., **112**, 1344 (1958). (Sec. 18*d*)
——— (1959a). Phys. Rev., **115**, 1741 (1959). (Sec. 18*d*)
——— (1959b). Phys. Rev., **115**, 1752 (1959). (Sec. 18*d*)
——— See also Chew, G. F. (1960).
Mann, A. K. See Bernstein, A. (1958).
Markov, M. A. (1956). Report to the Sixth Annual Rochester Conference on High Energy Nuclear Physics (1956), Interscience Publishers, Ltd., New York (1956). (Sec. 10*a*)
Marshak, R. E. See Lévy, M. (1954).
——— See also Sudarshan, E. G. (1958).
Martin, P. C. (1959). (With J. Schwinger.) Phys. Rev., **115**, 1342 (1959). (Problems)
——— See also Fulton, T. (1954b).
Martin, W. T. See Bochner, S. (1948).
Mathews, J. (1957). Doctoral Dissertation, California Institute of Tech. (1957). (Sec. 18*d*)
——— (1959). Phys. Rev., **113**, 381 (1959). (Sec. 17*b*)
Matthews, P. T. (1949a). Phys. Rev., **75**, 1270 (1949). (Sec. 13*a*)
——— (1949b). Phys. Rev., **76**, 684L (1949); Erratum **76**, 1489(L) (1949). (Sec. 14*c*)

Matthews, P. T. (1950). Phys. Rev., **80,** 292 (1950). (Sec. 16*h*)
——— (1951a). Phil. Mag., **42,** 221 (1951). (Sec. 16*h*)
——— (1951b). Phys. Rev., **81,** 936 (1951). (Sec. 16*h*)
——— (1952). (With A. Salam.) Phys. Rev., **86,** 715 (1952). (Secs. 15*g*, 16*h*)
——— (1954a). (With A. Salam.) Proc. Roy. Soc. (London), **A221,** 128 (1954). (Sec. 16*h*)
——— (1954b). (With A. Salam.) Phys. Rev., **94,** 185 (1954). (Sec. 16*h*)
——— (1955). (With A. Salam.) Nuovo Cimento, **2,** 120 (1955). (Problems)
——— (1957). Nuovo Cimento, **6,** 642 (1957). (Sec. 10*e*)
——— See also Salam, A. (1953).
Medvedev, N. N. See Bogoliubov, N. N. (1958a).
Melvin, M. A. (1960). Rev. Mod. Phys. **32,** No. 3 (1960). (Problems)
Menius, A. See Rosen, N. (1942).
Merrison, A. W. See Ashkin, J. (1959).
Meyers, D. F. See Brown, G. E. (1955 and 1957).
Michel, L. (1953). Nuovo Cimento, **10,** 319 (1953). (Sec. 14*f*)
——— (1955). (With A. S. Wightman.) Phys. Rev., **98,** 1190 (1955). (Sec. 14*e*)
——— (1959). Suppl. No. 1 to Nuovo Cimento, **14,** 95 (1959). (Problems)
——— See also Bernstein, J. (1960); Bouchiat, C. (1958).
Mie, G. (1928). Ann. der Phys. (Leipzig), **85,** 711 (1928). (Sec. 7*d*)
Mills, R. L. (1955). (With N. M. Kroll.) Phys. Rev., **98,** 1489 (1955). (Secs. 15*d*, 15*g*)
——— See also Yang, C. N. (1954).
Minardi, E. See Ascoli, R. (1958/59).
Minguzzi, A. (1960). (With R. F. Streater.) Nuovo Cimento, **17,** 946 (1960). (Sec. 18*d*)
Mitra, A. (1952). Nature, **169,** 1009 (1952). (Sec. 15*f*)
——— (1959). (With P. Narayanaswamy and L. K. Pande.) Nuclear Physics, **10,** 629 (1959). (Sec. 15*f*)
Miyamoto, Y. See Fukuda, H. (1950a, b).
Miyazawa, H. (1956). Phys. Rev., **101,** 1564 (1956). (Sec. 12*d*)
Moffat, J. (1960). (With M. W. Stringfellow.) Proc. Roy. Soc. (London), **A254,** 242 (1960). (Sec. 16*a*)
Moldauer, P. (1953). (With K. M. Case.) Phys. Rev., **91,** 459A (1953). (Sec. 10*g*)
Møller, C. (1945). Kgl. Danske Vidensk. Selsk. Mat.-Fys. Medd., **23,** No. 1 (1945). (Secs. 11*c*, 11*d*)
——— (1946). Kgl. Danske Vidensk. Selsk. Mat.-Fys. Medd., **26,** No. 19 (1946). (Secs. 11*c*, 11*d*)

Møller, C. (1949a). Ann. de l'Institut Henri Poincaré, **11**, 251 (1949). (Secs. 3*c*, 7*d*, 7*g*)

——— (1949b). Commun. Dublin Inst. Advanced Stud., **4**, No. 5 (1949). (Secs. 3*c*, 7*d*, 7*g*)

——— (1952). *The Theory of Relativity*, Oxford University Press (1952). (Secs. 3*c*, 7*d*, 7*g*)

——— See Kristensen, P. (1952).

Moravscik, M. J. See MacGregor, M. H. (1959).

Morpugo, G. See Franzinetti, C. (1957).

Mott, N. F. (1929). Proc. Roy. Soc. (London), **A124**, 425 (1929). (Sec. 4*h*)

——— (1949). (With H. S. W. Massey.) *The Theory of Atomic Collisions*, 2d ed., Oxford (1949). (Sec. 4*h*)

Munakata, Y. See Hayashi, C. (1952).

——— See also Araki, H. (1957).

Nafe, J. E. (1947). (With E. B. Nelson and I. I. Rabi.) Phys. Rev., **71**, 914 (1947). (Sec. 15*d*)

Nakanishi, N. (1957). Prog. Theor. Phys. (Japan), **17**, 401 (1957). (Secs. 17*b*, 18*a*, 18*d*)

——— (1958). Prog. Theor. Phys. (Japan), **19**, 159 (1958). (Sec. 15*d*)

Naĭmark, M. A. (1957). *Linear representations of the Lorentz group*, American Math. Society Translations, Series 2, Vol. 6. (Secs. 2*b*, 2*c*)

Nambu, Y. (1955). Phys. Rev., **100**, 394 (1955). (Sec. 18*d*)

——— (1957). Nuovo Cimento, **6**, 1064 (1957). (Sec. 17*b*)

——— (1958). Nuovo Cimento, **9**, 610 (1958). (Sec. 17*b*)

——— See also Chew, G. F. (1957); Goldberger, M. F. (1957); Kinoshita, T. (1950b).

Narayanaswamy, P. See Mitra, A. (1959).

Nelson, E. B. See Nafe, J. E. (1947).

Neuman, M. See Karplus, R. (1950b, c).

Newton, R. (1955a). Phys. Rev., **97**, 1162 (1955). (Sec. 15*f*)

——— (1955b). Phys. Rev., **98**, 1514 (1955). (Sec. 15*f*)

Newton, T. D. (1949). (With E. P. Wigner.) Rev. Mod. Phys., **21**, 400 (1949). (Secs. 2*c*, 3*c*, 4*f*)

Nishijima, K. (1951). Prog. Theor. Phys. (Japan), **6**, 37 (1951). (Sec. 16*h*)

——— (1953a). Prog. Theor. Phys. (Japan), **10**, 549 (1953). (Sec. 17*f*)

——— (1953b). (With T. Nakano.) Prog. Theor. Phys. (Japan), **10**, 581 (1953). (Sec. 10*d*)

——— (1954). Prog. Theor. Phys. (Japan), **12**, 107 (1954). (Sec. 10*d*)

——— (1955). Prog. Theor. Phys. (Japan), **13**, 285 (1955); see also *ibid.*, **14**, 527 (1955). (Sec. 10*d*)

Nishijima, K. (1956a). Fortschritte der Physik, **4**, 519 (1956). (Sec. 10*a*)
——— (1956b). Suppl. to Prog. Theor. Phys., No. **3**, 138 (1956). (Problems)
——— (1957). Prog. Theor. Phys. (Japan), **17**, 765 (1957). (Sec. 18*b*)
——— (1958). Prog. Theor. Phys. (Japan), **11**, 995 (1958). (Sec. 18*b*)
——— (1960). Phys. Rev., **119**, 485 (1960). (Sec. 18*b*, 18*d*, Problems)
Nishina, Y. (1929). Zeits. für Phys., **52**, 869 (1929). (Sec. 14*e*)
——— See also Klein, O. (1929).
Nordsieck, A. (1937). See Bloch, F. (1937).
Norton, R. E. (1958a). (With A. Klein.) Phys. Rev., **109**, 584 (1958). (Sec. 12*d*)
——— (1958b). (With A. Klein.) Phys. Rev., **109**, 991 (1958). (Sec. 12*d*)
Novey, T. B. See Telegdi, V. L. (1958).
Novozilov, J. V. (1958). (With A. V. Tolub.) Fortschritte für Physik, **6**, 50 (1958). (Sec. 7*e*)
Noyes, H. P. (1960). (With D. N. Edwards.) Phys. Rev., **118**, 1409 (1960). (Sec. 18*d*)

Odian, A. C. See Yamagata, T. L. (1956).
Oehme, R. (1955a). Phys. Rev., **102**, 1174 (1955). (Sec. 18*d*)
——— (1955b). Phys. Rev., **100**, 1503 (1955). (Sec. 18*d*)
——— (1956). Nuovo Cimento, **10**, 1316 (1956). (Sec. 18*d*)
——— (1958). Nuovo Cimento, **13**, 778 (1958). (Sec. 18*d*)
——— (1959). (With J. G. Taylor.) Phys. Rev., **113**, 37 (1959). (Sec. 18*d*)
——— See also Bremerman, H. (1958); Goldberger, M. L. (1957 and 1962); Lee, T. D. (1957).
Okubo, S. (1954). Prog. Theor. Phys. (Japan), **11**, 80 (1954). (Sec. 10*b*)
——— (1960). (With D. Feldman.) Phys. Rev., **117**, 279 and 292 (1960). (Problems)
Okun, L. B. (1958). Report to the 1958 Ann. International Conf. on High Energy Physics at CERN (Geneva, Switzerland, 1958). (Sec. 10*a*)
——— (1959). Ann. Rev. of Nuclear Science, Vol. **9**, Annual Reviews, Palo Alto, Calif. (1959). (Sec. 10*a*)
——— (1960). (With A. P. Rudik.) Nuclear Physics, **15**, 261 (1960). (Secs. 17*b*, 18*d*)
Omnes, R. (1957). Nuovo Cimento, **5**, 983 (1957). (Problems)
——— (1958). Nuovo Cimento, **8**, 316 (1958). (Problems)
Oppenheimer, J. R. (1930a). Phys. Rev., **35**, 461 (1930). (Sec. 15*a*)
——— (1930b). Phys. Rev., **35**, 939 (1930). (Sec. 4*g*)
——— See also Bethe, H. A. (1946).

Osborn, R. K. (1952). Phys. Rev., **86**, 340 (1952). (Sec. 12*d*)
——— See also Berger, J. M. (1952).
Ozaki, S. (1955). Prog. Theor. Phys. (Japan), **14**, 511 (1955). (Problems)

Pac, Pong Y. (1959). Prog. Theor. Phys. (Japan), **21**, 640 (1959). (Sec. 4*f*)
Pais, A. (1947). *On the Theory of Elementary Particles*, Verh, Kgl. Aca., Amsterdam, Vol. **19** (1947). (Sec. 15*a*)
——— (1949). (With S. T. Epstein.) Rev. Mod. Phys., **21**, 445 (1949). (Sec. 15*a*)
——— (1952a). Phys. Rev., **86**, 633 (1952). (Secs. 10*a*, 10*d*, 10*e*, 14*f*)
——— (1952b). (With R. Jost.) Phys. Rev., **87**, 871 (1952). (Secs. 14*f*, 16*a*)
——— (1955). (With O. Piccioni.) Phys. Rev., **100**, 1487 (1955). (Sec. 10*e*)
——— (1958a). Phys. Rev., **110**, 574 (1958). (Secs. 10*e*, 12*e*)
——— (1958b). Phys. Rev., **110**, 1480 (1958). (Secs. 10*e*, 12*e*)
——— (1960). Annals of Physics (New York), **9**, 548 (1960). (Problems)
——— See also Gell-Mann, M. (1955a, b); Jost, R. (1951).
Pake, G. E. (1953). (With E. Feenberg.) *Notes on the Theory of Angular Momentum*, Addison-Wesley, Cambridge (1953). (Sec. 12*d*)
Pande, L. K. See Mitra, A. (1959).
Pandit, L. K. (1959). Nuovo Cimento, Suppl. Volume **11**, Series 10 (1959). (Secs. 9*b*, 12*b*)
Papapetrou, A. (1939). Acad. Athens, **14**, 540 (1939). (Sec. 3*c*)
Paul, H. See Ashkin, J. (1959).
Pauli, W. (1921). *Relativitatstheorie*, Enz. der Math. Wiss, **5**, 539 (1921). (Sec. 15*a*)
——— (1927). Zeits. für Phys., **43**, 601 (1927). (Sec. 4*f*)
——— (1933). *Handbuch der Physik*, 2d ed., Vol. **24/1**, J. Springer, Berlin (1933). (Sec. 8*a*)
——— (1934). (With V. Weisskopf.) Helv. Phys. Acta, **7**, 709 (1934). (Secs. 3*a*, 4*a*, 7)
——— (1935). *Zeeman Verhandelingen*, p. 31, Martinus Nijhoff, Haag (1935). (Secs. 4*b*, 4*h*)
——— (1936). Ann. de l'Institut Henri Poincaré, **6**, 137 (1936). (Sec. 4*b*)
——— (1937). (With M. Fierz.) Nuovo Cimento, **15**, 167 (1938). (Sec. 15*d*)
——— (1940a). Phys. Rev., **58**, 716 (1940). (Secs. 8*a*, 18*a*)
——— (1940b). (With F. J. Belinfante.) Physica, **7**, 177 (1940). (Secs. 8*a*, 18*a*)
——— (1941). Rev. Mod. Phys., **13**, 203 (1941). (Secs. 4*h*, 7*g*, Problems)

Pauli, W. (1943). Rev. Mod. Phys., **15**, 175 (1943). (Sec. 12*b*)
——— (1949). (With F. Villars.) Rev. Mod. Phys., **21**, 434 (1949). (Secs. 15*a*, 15*e*)
——— (1955). In *Niels Bohr and the Development of Physics*, McGraw-Hill, New York (1955). (Secs. 10*b*, 14*f*)
——— (1956). *Continuous Groups in Quantum Mechanics*, CERN Report 56-31, Geneva. (Secs. 1*e*, 1*f*)
——— (1958). *Handbuch der Physik*, Vol. **V**, Part 1, S. Flügge, editor, Springer-Verlag, Berlin (1958). (Sec. 1*a*)
——— See also Heisenberg, W. (1929 and 1930); Jordan, P. (1928a); Källén, G. (1955b).
Peierls, R. E. (1948). (With H. McManus.) Phys. Rev., **70**, 795A (1946). (Sec. 15*a*)
——— (1952). Proc. Roy. Soc. (London), **A214**, 143 (1952). (Sec. 17*a*)
——— See also Chrétien, M. (1953 and 1954); Dirac, P. A. M. (1942); Landau, L. (1930).
Petermann, A. (1957). Helv. Phys. Acta, **30**, 407 (1957). (Secs. 15*d*, 15*e*)
——— (1958). Fortschritte der Physik, **6**, 505 (1958). (Secs. 15*b*, 15*g*)
Phillips, R. J. N. (1959). Reports on Progress in Physics, **XXII**, 562 (1959). (Sec. 15*h*, Problems)
Piccioni, O. See Pais, A. (1955).
Pidd, R. W. See Schupp, A. A. (1961).
Pines, D. (1955). *Solid State Physics*, Vol. **1**, 368 (1955), F. Seitz and D. Turnbull, Editors, Academic Press, Inc., Publishers, New York.
Pirenne, J. (1949). Physica, **15**, 1023 (1949). (Sec. 6)
Poincaré, H. (1905). Comptes Rendus (Paris), **40**, 1504 (1905). (Sec. 15*a*)
——— (1906). Rend. Palmero, **21** (1906). (Sec. 15*a*)
——— (1924). *La Méchanique Nouvelle*, Gauthier-Villars, Paris (1924). (Sec. 15*a*)
Polivanov, M. K. See Bogoliubov, N. N. (1958*a*).
Polkinghorne, J. C. (1955). Proc. Roy. Soc. (London), **A230**, 272 (1955). (Problems)
——— See also Fairlie, D. B. (1958); Kemmer, N. (1959); Kibble, T. W. B. (1958); Salam, A. (1955).
Pollock, F. See Kroll, N. M. (1952).
Pomeranchuk, I. (1958). J.E.T.P. (USSR), **34**, 725 (1958). (Sec. 18*d*)
——— See also Akhiezer, A. (1937).
Pond, T. A. (1952). (With R. H. Dicke.) Phys. Rev., **85**, 489L (1952). (Sec. 4*g*)
Pontrjagin, L. (1946). *Topological Groups*, Princeton University Press (1946). (Sec. 1*e*)
Powell, J. L. (1949). Phys. Rev., **75**, 32 (1949). (Problems)

Prange, R. (1958). Phys. Rev., **110,** 240 (1958). (Sec. 15*d*)
Prentki, J. See d'Espagnat (1956 and 1958).
Primakoff, H. See Malenka, B. J. (1957).
Prosperi, G. M. (1959). (With A. Scotti.) Nuclear Physics, **13,** 140 (1959). (Problems)
Pryce, M. H. L. (1948). Proc. Roy. Soc. (London), **195A,** 62 (1948). (Secs. 3*c*, 7*d*, 7*g*)
——— See also Dirac, P. A. M. (1942).
Puppi, G. (1948). Nuovo Cimento, **5,** 505 (1948). (Sec. 12*f*)
——— (1949). Nuovo Cimento, **6,** Series 9, 194 (1949). (Sec. 10*f*)
——— (1957). (With A. Stanghellini.) Nuovo Cimento, **5,** 1305 (1957). (Sec. 18*d*)
Pursey, D. L. See Kemmer, N. (1959).

Rabi, I. I. (1928). Zeits. für Phys., **49,** 7 (1928). (Sec. 4*h*)
——— See also Nafe, J. E. (1947).
Racah, G. (1937). Nuovo Cimento, **14,** 322 (1937). (Secs. 4*h*, 8*c*)
——— (1946). Phys. Rev., **70,** 406 (1946). (Sec. 16*a*)
——— (1959). Suppl. to Nuovo Cimento, **14,** 75 (1959). (Sec. 1*f*)
Ravenhall, D. G. (1958). Rev. Mod. Phys., **30,** 430 (1958). (Sec. 4*h*)
Redmond, P. J. (1958a). Phys. Rev., **112,** 1404 (1958). (Secs. 17*b*, 17*c*)
——— (1958b). (With J. L. Uretski.) Phys. Rev. Letters, **1,** 147 (1958). (Sec. 16*h*)
——— (1960). (With J. L. Uretski.) Annals of Physics (New York), **9,** 106 (1960). (Sec. 17*c*)
Retherford, R. C. See Lamb, W. E. (1947).
Rickayzen, G. See Eden, R. J. (1953b).
Rideau, G. See Chevalier, A. (1958).
Ringo, G. R. See Telegdi, V. L. (1958).
Rivier, D. (1949). Helv. Phys. Acta, **22,** 265 (1949). (Sec. 15*a*)
——— See also Stückelberg, E. C. G. (1948a, b and 1949).
Rodberg, L. S. (1957). Phys. Rev., **106,** 1090 (1957). (Sec. 12*d*)
Rohrlich, F. (1950a). Phys. Rev., **77,** 357 (1950). (Sec. 15*a*)
——— (1950b). Phys. Rev., **80,** 666 (1950). (Secs. 14*c*, 16*a*)
——— (1952). (With R. L. Gluckstern.) Phys. Rev., **86,** 1 (1952). (Sec. 16*a*)
——— (1960). Am. J. of Physics, **28,** 639 (1960). (Sec. 15*a*)
——— See also Bethe, H. A. (1952); Jauch, J. M. (1954 and 1955a, b).
Rollnik, H. See Grawert, G. (1959).
Roman, P. (1960). *Theory of Elementary Particles*, North-Holland Publishing Co., Amsterdam (1960). (Sec. 1*f*, Problems)
Rose, M. E. (1957). *Elementary Theory of Angular Momentum*, John Wiley and Sons, Inc., New York (1957). (Sec. 12*d*)
Rosen, N. (1939). Phys. Rev., **55,** 94 (1939). (Sec. 15*a*)

Rosen, N. (1942). (With A. Menius.) Phys. Rev., **62**, 436 (1942). (Sec. 15*a*)

Rosenbluth, M. N. (1950). Phys. Rev., **79**, 615 (1950). (Problems)

Rosenfeld, A. H. See Gell-Mann, M. (1957d).

Rosenfeld, L. (1931c). Zeits. für Phys., **70**, 454 (1931). (Sec. 15*a*)

——— (1953). Physica, **19**, 859 (1953). (Secs. 7*e*, 8*b*, 9*b*)

——— (1955). in *Niels Bohr and the Development of Physics*, McGraw-Hill, New York (1955). (Secs. 8*b*, 9*b*)

——— See also Bohr, N. (1933 and 1950).

Ruderman, M. See Finkelstein, R. J. (1951).

——— See also Karplus, R. (1955); Kroll, N. M. (1954).

Rudik, A. P. See Okun, L. B. (1960).

Ruijgrok, The. W. (1956). (With L. Van Hove.) Physica, **22**, 880 (1956). (Sec. 12*c*)

——— (1958a). Physica, **24**, 185 (1958). (Sec. 12*c*)

——— (1958b). Physica, **24**, 205 (1958). (Sec. 12*c*)

——— (1959). Physica, **25**, 357 (1959). (Sec. 12*c*)

Sabry, A. See Källén, G. (1955a).

Sachs, R. G. (1952). Phys. Rev., **87**, 1100 (1952). (Sec. 7*f*)

——— See also Treiman, S. B. (1956).

Sakata, S. (1940). (With M. Taketani.) Proc. Phys. Math. Soc. (Japan), **22**, 757 (1940). (Sec. 3*d*)

——— (1956). Prog. Theor. Phys. (Japan), **16**, 686 (1956). (Sec. 10*a*)

Sakurai, J. J. (1958a). Nuovo Cimento, **7**, 1306 (1958). (Sec. 10*f*)

——— (1958b). Phys. Rev., **109**, 980 (1958). (Problems)

——— (1960). Annals of Physics (New York), **11**, 1 (1960). (Problems)

Salam, A. (1950). Phys. Rev., **79**, 910 (1950). (Sec. 16*h*)

——— (1951a). Phys. Rev., **82**, 217 (1951). (Secs. 16*a*, 16*d*, 16*h*)

——— (1951b). Phys. Rev., **84**, 426 (1951). (Secs. 16*a*, 16*d*, 16*h*)

——— (1952). Phys. Rev., **86**, 731 (1952). (Sec. 16*h*)

——— (1953). (With P. T. Matthews.) Phys. Rev., **90**, 690 (1953). (Sec. 15*g*)

——— (1955). (With J. C. Polkinghorne.) Nuovo Cimento, **2**, 685 (1955). (Sec. 10*e*)

——— (1957). Nuovo Cimento, **5**, 299 (1957). (Sec. 5*a*)

——— (1958). Lectures on the analytic properties of expectation values of products of field operators, University of Rochester, 1958, AT (30-1)-875, NYO-8796. (Problems)

——— See also Kemmer, N. (1955); Matthews, P. T. (1952, 1954a, b, 1955).

Salpeter, E. E. (1951). (With H. A. Bethe.) Phys. Rev., **84**, 1232 (1951). (Sec. 17*f*)

Salpeter, E. E. (1952a). (With W. A. Newcomb.) Phys. Rev., **87,** 150 (1952). (Sec. 17*f*)
——— (1952b). Phys. Rev., **87,** 328 (1952). (Sec. 17*f*)
——— See also Baranger, M. (1953); Bethe, H. A. (1957).
Salzman, G. (1955). Phys. Rev., **99,** 973 (1955). (Secs. 12*d*, 15*d*)
——— See also Schnitzer, H. J. (1959).
Šapiro, Z. Ya. See Gel'fand, I. M. (1956).
Sato, I. See Akiba, T. (1958).
Sato, M. See Maki, Z. (1953).
Sauter, F. (1931). Zeits. für Phys., **69,** 742 (1931). (Sec. 4*h*)
Scarf, F. L. (1955). Phys. Rev., **100,** 912 (1955). (Sec. 17*f*)
——— (1956). Phys. Rev., **100,** 913 (1956). (Sec. 17*f*)
Schaffer, G. W. See Brown, G. E. (1959).
Schiff, L. I. (1949). *Quantum Mechanics*, McGraw-Hill, New York (1949). (Sec. 3*d*)
Schmidt, K. (1956). (With K. Baumann.) Nuovo Cimento, **4,** 860 (1956). (Sec. 18*a*)
Schnitzer, H. J. (1959). (With G. Salzman.) Phys. Rev., **113,** 1153 (1959). (Sec. 18*d*)
Schupp, A. A. (1961). (With R. W. Pidd and H. R. Crane.) Phys. Rev., **121,** 1 (1961). (Sec. 15*d*)
Schwartz, L. (1950, 1951). *Théorie des Distributions*, Actualités Scientifiques et Industriels, No. 1091, 1122, Hermann and Co., Paris (1950, 1951). (Sec. 18*a*)
Schweber, S. S. (1955). Nuovo Cimento, **2,** 397 (1955). (Sec. 13*a*)
——— See also Greenberg, O. W. (1958); Wightman, A. S. (1955).
Schwinger, J. (1948a). Phys. Rev., **73,** 416L (1948). (Sec. 15*d*)
——— (1948b). (With V. Weisskopf.) Phys. Rev., **73,** 1272A (1948). (Sec. 15*b*)
——— (1948c). Phys. Rev., **74,** 1439 (1948). (Secs. 13*a*, 13*d*, 17*a*)
——— (1949a). Phys. Rev., **75,** 651 (1949). (Secs. 7*c*, 13*d*)
——— (1949b). Phys. Rev., **76,** 790 (1949). (Secs. 13*d*, 15*b*, 15*e*, 15*f*)
——— (1951a). Phys. Rev., **82,** 664 (1951). (Secs. 8*c*, 10*a*, 10*b*, 15*e*, 15*g*, 16*a*)
——— (1951b). Proc. Nat. Acad. of Sciences (USA), **37,** 452 (1951). (Secs. 10*a*, 15*g*, 16*b*, 17*a*, 17*f*)
——— (1951c). Phys. Rev., **82,** 914 (1951). (Sec. 8*c*)
——— (1953a). Phys. Rev., **91,** 713 (1953). (Secs. 10*a*, 17*a*)
——— (1953b). Phys. Rev., **91,** 728 (1953). (Secs. 10*a*, 14*a*, 15*c*)
——— (1953c). Phys. Rev., **92,** 1283 (1953). (Secs. 10*a*, 14*a*, 15*c*)
——— (1953d). Phil. Mag., **44,** 1171 (1953). (Sec. 17*a*)
——— (1954a). Phys. Rev., **93,** 615 (1954). (Sec. 14*a*)
——— (1954b). *The Theory of Coupled Fields,* Lecture notes edited by L. Rodberg, F. Zachariasen, and A. C. Zemach. (Secs. 7*e*, 18*b*)

Schwinger, J. (1957). Annals of Physics (New York), **2,** 407 (1957). (Sec. 10*e*)
——— (1958). *Quantum Electrodynamics*, Dover Press, New York (1958). (Sec. 15)
——— See also Karplus, R. (1952b); Lippmann, B. A. (1950); Martin, P. C. (1959).
Scotti, A. See Prosperi, G. M. (1959).
Screaton, G. R. See Gilbert, W. (1956).
Series, G. W. (1957). *The Spectrum of Atomic Hydrogen*, Oxford University Press, Oxford (1957). (Sec. 4*h*)
Shirkov, D. V. See Bogoliubov, N. N. (1956 and 1959a, b).
Shirokov, Yu. M. (1958a). Soviet Physics—JETP (New York), **6** (33), 664 (1958). (Secs. 1*d*, 2*c*)
——— (1958b). Soviet Physics—JETP (New York), **6** (33), 919 (1958). (Secs. 1*d*, 2*c*)
——— (1958c). Soviet Physics—JETP (New York), **6** (33), 929 (1958). (Secs. 1*d*, 2*c*)
——— (1960a). Nuclear Physics, **15,** 1 (1960). (Sec. 2*b*)
——— (1960b). Nuclear Physics, **15,** 13 (1960). (Sec. 2*b*)
Silin, W. P. (1956). (With G. B. Fainberg.) Fortschritte für Physik, **4,** 233 (1956). (Problems)
Slater, J. C. (1930). Phys. Rev., **35,** 210 (1930). (Sec. 6*k*)
——— (1951). Phys. Rev., **81,** 385 (1951). (Sec. 6*k*)
Slotnick, M. See Jost, R. (1950).
Snyder, H. S. (1940). (With J. Weinberg.) Phys. Rev., **57,** 307 (1940). (Sec. 15*f*)
Soloviev, V. G. (1958). Nuclear Physics, **6,** 618 (1958). (Sec. 10*b*)
Sommerfield, C. (1957). Phys. Rev., **107,** 328 (1957). (Sec. 15*d*)
——— (1958). Annals of Physics (New York), **5,** 26 (1958). (Secs. 15*d*, 15*g*)
——— See also Karplus, R. (1958).
Speisman, G. See Feynman, R. P. (1954).
Stanghellini, A. See Cini, M. (1959).
——— See also Bosco, B. (1959); Puppi, G. (1957).
Stapp, H. P. See MacGregor, M. H. (1959).
Stearns, M. B. (1957). *Progress in Nuclear Physics*, **6,** O. R. Frisch, Editor, Pergamon Press (1957). (Sec. 4*g*)
Stehn, J. R. See Bethe, H. A. (1950b).
Steinberger, J. (1949). Phys. Rev., **76,** 1180 (1949). (Sec. 16*a*)
Steinmann, O. (1960). Helv. Phys. Acta, **33,** 257 and 347 (1960). (Secs. 18*a*, 18*b*)
Sternheimer, R. M. (1959). *Advances in Electronics and Electron Physics*, Vol. **XI,** Academic Press, New York (1959). (Sec. 14*e*)

Streater, R. F. (1959). Nuovo Cimento, **13**, 57 (1959). (Problems)
——— (1960a). Proc. Roy. Soc. (London), **A256**, 39 (1960). (Sec. 18*a*)
——— (1960b). Nuovo Cimento, **15**, 937 (1960). (Sec. 18*a*)
——— (1960c). J. Math. Phys., **1**, 230 (1960); see errata in *ibid.*, **1**, 452 (1960). (Sec. 18*d*)
——— (See also Minguzzi, A. (1960).
Stringfellow, M. W. See Moffat, J. (1960).
Stückelberg, E. C. G. (1939). Nature, **144**, 118 (1939). (Sec. 15*a*)
——— (1941a). Helv. Phys. Acta, **14**, 51 (1941). (Sec. 15*a*)
——— (1945a). Helv. Phys. Acta, **18**, 21 (1945). (Sec. 11*c*)
——— (1945b). Helv. Phys. Acta, **18**, 195 (1945). (Sec. 11*c*)
——— (1948a). (With D. Rivier.) Phys. Rev., **74**, 986 (1948). (Sec. 15*a*)
——— (1948b). (With D. Rivier.) Phys. Rev., **74**, 218 (1948). (Sec. 15*a*)
——— (1949). (With D. Rivier.) Helv. Phys. Acta, **13**, 215 (1949). (Sec. 13*c*)
——— (1951a). Phys. Rev., **81**, 130 (1951). (Sec. 13*a*)
——— (1951b). (With T. A. Green.) Helv. Phys. Acta, **24**, 153 (1951). (Sec. 13*a*)
Sucher, J. (1957). Phys. Rev., **107**, 1448 (1957). (Sec. 11*f*)
Sudarshan, E. C. G. (1958). (With R. E. Marshak.) Phys. Rev., **109**, 1860 (1958). (Sec. 10*f*)
——— See also Bose, S. K. (1959).
Sugawara, M. (1959). (With A. Kanezawa.) Phys. Rev., **115**, 1310 (1959). (Problems)
Sunakawa, S. See Utiyama, R. (1952).
Suura, H. (1955). Phys. Rev., **99**, 1020 (1955). (Sec. 15*f*)
——— (1957). (With E. H. Wichmann.) Phys. Rev., **105**, 1030 (1957). (Sec. 15*e*)
——— See also Gasiorowicz, S. G. (1959); Yennie, D. R. (1957), (1961).
Symanzik, K. (1954). Z. Naturforschung, **9a**, No. 10 (1954). (Secs. 7*e*, 18*b*)
——— (1957a). Phys. Rev., **105**, 743 (1957). (Sec. 18*d*)
——— (1957b). Nuovo Cimento, **5**, 659 (1957). (Sec. 18*d*)
——— (1958). Prog. Theor. Phys. (Japan), **20**, 690 (1958). (Sec. 17*b*)
——— (1960). J. Math. Physics, **1**, 249 (1960). (Sec. 18*b*)
——— See also Lehmann, H. (1955a, b, and 1957).

Takahashi, Y. (1952). (With H. Umezawa.) Prog. Theor. Phys. (Japan), **11**, 251 (1952). (Sec. 15*a*)
——— (1957). Nuovo Cimento, **6**, 370 (1957). (Sec. 16*e*)
Takeda, G. (1952). Prog. Theor. Phys. (Japan), **7**, 359 (1952). (Secs. 16*d*, 16*h*)

Takeda, G. See also Kato, M. (1959).
Taketani, M. See Sakata, S. (1940).
Tamm, I. (1930). Zeits. für Phys., **62,** 545 (1930). (Sec. 14*e*)
Tani, S. (1951). Prog. Theor. Phys. (Japan), **6,** 267 (1951). (Sec. 4*f*)
Tarski, J. (1960). J. Math. Phys., **1,** 149 (1960). (Sec. 18*d*)
Tausner, M. See Charap, J. M. (1960b).
Taylor, J. G. (1958). Lectures on *Dispersion Relations in Quantum Field Theory and Related Topics*, University of Maryland Technical Report 115, Volume I, II, III (1958). (Problems)
——— See also Bremerman, H. J. (1958); Oehme, R. (1959).
Telegdi, V. L. (1958). (With M. T. Burgy, V. E. Krohn, T. B. Novey, and G. R. Ringo.) Phys. Rev., **110,** 1214 (1958). (Sec. 14*e*)
Theis, W. R. (1959). Fortschritte der Physik, **7,** 559 (1959). (Problems)
Thirring, W. (1950). Phil. Mag., **41,** 1193 (1950). (Secs. 15*f*, 16*g*)
——— (1951). (With B. Touschek.) Phil. Mag., **42,** 244 (1951). (Sec. 15*c*)
——— (1958). *Principles of Quantum Electrodynamics*, Academic Press, Inc., New York (1958). (Sec. 15*g*)
——— (1959a). Suppl. No. 2 to Nuovo Cimento, **14,** 385 (1959). (Sec. 18*d*)
——— (1959b). Suppl. No. 2 to Nuovo Cimento, **14,** 415 (1959). (Sec. 10*d*)
——— See also Gell-Mann, M. (1954).
Thomas, L. See Landé, A. (1941a, b and 1944).
Tiomno, J. (1955). Nuovo Cimento, **1,** 226 (1955). (Secs. 8*c*, 10*f*)
——— See also Yang, C. N. (1950b).
Toll, J. S. (1952). Doctoral Dissertation, Princeton University (1952). (Secs. 16*a*, 18*d*)
——— (1956). Phys. Rev., **104,** 1760 (1950). (Secs. 12*d*, 18*d*, Problems)
———See also Källén, G. (1960).
Tollestrup, A. V. See Ashkin, J. (1959a).
Tolub, A. V. See Novozilov, J. V. (1958).
Tomonaga, S. (1946). Prog. Theor. Phys. (Japan), **1,** 27 (1946). (Sec. 13*a*)
——— See also Kanesawa, S. (1948a, b); Maki, Z. (1953).
Tomozawa, Y. See Umezawa, H. (1956a).
Touschek, B. F. See Burton, W. K. (1953a, b).
——— See also Cini, M. (1958).
Treiman, S. B. (1956). (With R. G. Sachs.) Phys. Rev., **103,** 435 (1956). (Sec. 12*d*)
——— See also Blankenbecler, R. (1960); Federbush, P. (1958b).
Triebwasser, S. (1953). (With E. S. Dayhoff and W. E. Lamb.) Phys. Rev., **89,** 98 (1953). (Sec. 15*b*)
Tsai, Yung Su (1960). Phys. Rev., **120,** 269 (1960). (Sec. 15*f*, Problems)

Uehling, E. A. (1935). Phys. Rev., **48**, 55 (1935). (Secs. 4*g*, 15*e*)
Umezawa, H. (1949a). (With R. Kawabe.) Prog. Theor. Phys. (Japan), **4**, 420, 423 (1949). (Secs. 14*c*, 15*e*)
——— (1949b). (With R. Kawabe.) Prog. Theor. Phys. (Japan), **4**, 443 (1949). (Sec. 14*c*)
——— (1949c). (With R. Kawabe.) Prog. Theor. Phys. (Japan), **4**, 461 (1949). (Sec. 15*e*)
——— (1950). (With R. Kawabe.) Prog. Theor. Phys. (Japan), **5**, 266 (1950). (Sec. 14*c*)
——— (1951). (With S. Kamefuchi.) Prog. Theor. Phys. (Japan), **6**, 543 (1951). (Sec. 15*e*)
——— (1956a). (With Y. Tomozawa, M. Konuma, and S. Kamefuchi.) Nuovo Cimento, **3**, 772 (1956). (Sec. 17*c*)
——— (1956b). (With S. Kamefuchi.) Nuovo Cimento, **3**, 1060 (1956). (Sec. 16*b*)
——— (1956c). (With M. Konuma.) Nuovo Cimento, **4**, 1461 (1956). (Sec. 16*b*)
——— See also Takahashi, Y. (1952).
Uretski, J. L. See Redmond, P. J. (1958b and 1960).
Utiyama, R. (1952). (With S. Sunakawa and T. Imamura.) Prog. Theor. Phys. (Japan), **8**, 77 (1952). (Sec. 16*a*)
——— (1956). Phys. Rev., **101**, 1597 (1956). (Secs. 10*c*, 10*e*)

Valatin, J. G. (1951). Kgl. Danske Vidensk. Selsk. Mat.-Fys. Medd., **26**, No. 13 (1951). (Problems)
——— (1954a). Proc. Roy. Soc. (London), **A222**, 93 (1954). (Sec. 16*h*)
——— (1954b). Proc. Roy. Soc. (London), **A228**, 228 (1954). (Sec. 16*h*)
——— (1954c). Proc. Roy. Soc. (London), **A225**, 534 (1954). (Sec. 16*h*)
——— (1954d). Proc. Roy. Soc. (London), **A226**, 254 (1954). (Sec. 16*h*)
——— (1959). In *Les Problèmes Mathématiques de la Théorie Quantique des Champs*, Colloques Internationaux du Centre National de la Rechèrche Scientifique, Paris (1959). (Sec. 16*h*)
Van Dungen, H. See de Donder, The. (1926).
Van Hove, L. (1951). Acad. Roy. de Belgique Bulletin, **37**, 1055 (1951). (Sec. 13*a*)
——— (1952). Physica, **18**, 145 (1952). (Secs. 12*a*, 13*a*)
——— (1955). Physica, **21**, 901 (1955); see also *ibid.* **22**, 343 (1956), and **23**, 441 (1957). (Secs. 11*b*, 12*b*)
——— (1959). Physica, **25**, 365 (1959). (Sec. 12*c*)
——— See also Ruijgrok, The. W. (1956).
Van der Waerden, B. L. (1929). Göttingen Nachrichten, 100 (1929). (Sec. 4*c*)
——— (1932). *Die Gruppentheoretische Methode in der Quantenmechanik*, Verlag Julius Springer, Berlin (1932). (Sec. 2*b*)

Villars, F. (1950). Phys. Rev., **79,** 122 (1950). (Sec. 15*a*)
——— See also Feshbach, H. (1958); Pauli, W. (1949).
Vitale, B. See Amati, D. (1959).
Vladimirov, V. S. (1959). (With A. A. Logunov.) Report No. 260 of the Joint Institute for Nuclear Research, Dubna (1959). (Sec. 18*d*)
Volkow, D. M. (1935). Zeits. für Phys., **94,** 25 (1935). (Sec. 4*h*)
Von Neumann, J. (1931). Math. Ann., **104,** 570 (1931). (Secs. 1*a*, 7*a*)

Waller, I. (1930). Zeits. für Phys., **62,** 673 (1930). (Sec. 15*a*)
Ward, J. C. (1950a). Phys. Rev., **77,** 293L (1950). (Secs. 15*c*, 16*a*, 16*d*, 16*e*)
——— (1950b). Phys. Rev., **78,** 182L (1950). (Secs. 15*c*, 16*e*)
——— (1951a). Proc. Phys. Soc. (London), **A64,** 54 (1951). (Sec. 16*f*)
——— (1951b). Phys. Rev., **84,** 897 (1951). (Sec. 16*h*)
Warnock, R. L. (1960). Phys. Rev., **118,** 1447 (1960). (Problems)
Watanabe, S. (1951). Phys. Rev., **84,** 1008 (1951). (Sec. 2*b*)
——— (1955). Rev. Mod. Phys., **27,** 26 (1955). (Sec. 2*b*)
Watson, K. M. (1949). (With J. V. Lepore.) Phys. Rev., **76,** 1157 (1949). (Sec. 15*h*)
Watson, W. K. R. See Karzas, W. J. (1958).
Weinberg, J. See Snyder, H. S. (1940).
Weinberg, S. (1958). Phys. Rev., **110,** 782 (1958). (Problems)
——— (1960). Phys. Rev., **118,** 838 (1960). (Sec. 16*d*)
——— See also Feinberg, G. (1959).
Weisskopf, V. (1934a). Zeits. für Phys., **89,** 27 (1934). (Sec. 15*a*)
——— (1934b). Zeits. für Phys., **90,** 817 (1934). (Sec. 15*a*)
——— (1936). Kgl. Danske Vidensk. Selsk. Mat.-Fys. Medd., **14,** No. 6 (1936). (Sec. 15*e*)
——— (1939). Phys. Rev., **56,** 72 (1939). (Sec. 15*a*)
——— (1949). Rev. Mod. Phys., **21,** 305 (1949). (Sec. 15*c*)
——— See also French, J. B. (1949); Pauli, W. (1934); Schwinger, J. (1948b).
Welton, T. (1948). Phys. Rev., **74,** 1157 (1948). (Sec. 15*c*)
Weneser, J. (1953). (With R. Bersohn and N. M. Kroll.) Phys. Rev., **91,** 1257 (1953). (Secs. 15*d*, 15*f*)
Wentzel, G. (1933). Zeits. für Phys., **68,** 479 (1933). (Sec. 15*a*)
——— (1934). Zeits. für Phys., **87,** 726 (1934). (Sec. 15*a*)
——— (1941). Zeits. für Phys., **118,** 277 (1941). (Sec. 12*c*)
——— (1942). Helv. Phys. Acta, **15,** 111 (1942). (Sec. 12*c*)
——— (1949). *Quantum Theory of Fields* (translated by J. M. Jauch), Interscience Publishers, Ltd. (1949). (Secs. 7*d*, 15*a*, Problems)
——— (1952). Phys. Rev., **86,** 802 (1952). (Sec. 10*g*)
West, D. (1958). Reports on Progress in Physics, **XXI** (1958). (Sec. 4*g*)

Weyl, H. (1929). Zeits. für Phys., **56**, 330 (1929). (Sec. 4*h*)

——— (1931). *Gruppentheorie und Quanten mechanik*, 2d ed., Leipzig (1931). (Sec. 4*g*)

Wheeler, J. A. (1937). Phys. Rev., **52**, 1107 (1937). (Sec. 11*c*)

——— (1945). (With R. P. Feynman.) Rev. Mod. Phys., **17**, 157 (1945). (Sec. 15*a*)

——— (1949). (With R. P. Feynman.) Rev. Mod. Phys., **21**, 425 (1949). (Sec. 15*a*)

——— See also Euwema, R. W. (1956).

Wichmann, E. H. (1956). (With N. M. Kroll.) Phys. Rev., **101**, 843 (1956). (Secs. 4*h*, 15*e*)

——— See also Karplus, R. (1958); Suura, H. (1957).

Wick, G. C. (1935). Atti. Accad. Lincei, **21**, 170 (1935). (Sec. 10*c*)

——— (1950). Phys. Rev., **80**, 268 (1950). (Sec. 13*c*)

——— (1952). (With A. S. Wightman and E. P. Wigner.) Phys. Rev., **88**, 101 (1952). (Sec. 1*a*)

——— (1954). Phys. Rev., **96**, 1124 (1954). (Sec. 17*f*)

——— (1955). Rev. Mod. Phys., **27**, 339 (1955). (Sec. 12*d*)

——— (1959). Ann. Rev. Nuclear Sci., Vol. **9** (1959). (Sec. 1*d*)

Wightman, A. S. (1955). (With S. Schweber.) Phys. Rev., **98**, 812 (1955). (Secs. 7*a*, 8*b*)

——— (1956). Phys. Rev., **101**, 860 (1956). (Secs. 16*a*, 17*b*, 18*a*)

——— (1958). Lectures given at the Faculté des Sciences, Université de Paris: *Les Problèmes Mathématiques de la Théorie Quantique des Champs.* (Sec. 18*a*)

——— (1959a). *Les Problèmes Mathématiques de la Théorie Quantique des Champs*, Colloques Internationaux du Centre National de la Rechèrche Scientifique, Paris (1959). (Secs. 17*d*, 18*a*)

——— (1959b). Suppl. No. 1, to Vol. **14**, Series X, Nuovo Cimento, **81**, (1959). (Sec. 1*d*)

——— See also Bargmann, V. (1953); Gärding, L. (1954); Hall, D. (1957); Källén, G. (1958); Michel, L. (1955); Wick, G. C. (1952).

Wigner, E. P. (1932). Göttingen Nachrichten, **31**, 546 (1932). (Sec. 7*c*)

——— (1939). Annals of Math., **40**, No. 1 (1939). (Secs. 1*d*, 2*c*)

——— (1947). Zeits. für Phys., **124**, 665 (1947). (Sec. 2*c*)

——— (1949). Proc. Amer. Phil. Soc., **93**, 521 (1949). (Sec. 1*d*)

——— (1952a). Zeits. für Phys., **133**, 101 (1952). (Sec. 1*a*)

——— (1952b). Proc. Nat. Acad. Sciences (USA), **38**, 449 (1952). (Sec. 10*e*)

——— (1955). Jubilee of Relativity Theory, Bern (1955), Birkhauser Verlag, Basel, 1956, A. Mercier and M. Kervaire, editors. (Sec. 1*d*)

——— (1956). Nuovo Cimento, **3**, 517 (1956). (Secs. 1*d*, 2*c*, 5*b*)

——— (1957). Rev. Mod. Phys., **29**, No. 3, 255 (1957). (Secs. 1*d*, 5*b*, 10*a*)

Wigner, E. P. (1959). *Group Theory*, Academic Press, New York (1959). (Secs. 1*d*, 1*e*)

——— See also Bargmann, V. (1948 and 1953); Jordan, P. (1928b); Newton, T. D. (1949); Wick, G. C. (1952).

Wilhelmsson, H. See Källén, G. (1959).

Wolfenstein, L. (1952). (With D. G. Ravenhall.) Phys. Rev., **88,** 279 (1952). (Secs. 4*g*, 14*f*, 16*a*)

Woodward, J. B. See Brenner, S. (1954).

Wouthuysen, S. A. See Foldy, L. L. (1950).

Wu, C. S. (1957). (With E. Ambler, R. W. Hayward, D. D. Hoppes, and R. P. Hudson.) Phys. Rev., **105,** 1413 (1957). (Sec. 10*f*)

Yamagata, T. L. (1956). (With L. B. Auerbach, G. Bernadini, L. Filosofo, A. O. Hansen, and A. C. Odian.) Bull. Am. Phys. Soc., Series II, **1,** 350 (1956). (Sec. 18*d*)

Yang, C. N. (1950a). Phys. Rev., **77,** 242 (1950). (Secs. 4*g*, 16*a*)

——— (1950b). (With J. Tiomno.) Phys. Rev., **79,** 495 (1950). (Secs. 1*e*, 8*c*)

——— (1950c). (With D. Feldman.) Phys. Rev., **79,** 972 (1950). (Secs. 13*a*, 17*d*)

——— (1954). (With R. L. Mills.) Phys. Rev., **96,** 191 (1956). (Secs. 10*c*, 10*e*)

——— See also Fermi, E. (1949); Lee, T. D. (1956 and 1957).

Yearian, M. R. See Hofstadter (1958).

Yennie, D. R. (1957). (With H. Suura.) Phys. Rev., **105,** 1378 (1957). (Sec. 15*f*)

——— (1958). (With S. Gartenhaus.) Nuovo Cimento, **9,** 59 (1958). (Sec. 16*h*)

——— (1961). (With S. C. Frautschi and H. Suura.) Annals of Physics (New York), (1961). (Secs. 15*d*, 15*f*, 16*f*)

——— See also Gasiorowicz, S. G. (1959).

Zacharias, J. R. See Nagel, D. E. (1947).

Zachariasen, F. See Drell, S. D. (1956, 1960, and 1961).

——— See also Chew, G. F. (1958); Cutkosky, R. (1956); Karzas, W. J. (1958).

Zimmermann, W. (1958). Nuovo Cimento, **10,** 567 (1958). (Sec. 18*b*)

——— (1959). Nuovo Cimento, **13,** 503 (1959). (Secs. 18*b*, 18*d*)

——— (1960). Nuovo Cimento, **16,** 690 (1960). (Sec. 18*b*)

——— See also Glaser, V. (1952 and 1957); Lehmann H. (1955a, b and 1957).

Zumino, B. (1960). Jour. Math. Phys., **1,** 1 (1960). (Secs. 9*b*, 15*e*)

——— See also Lüders, G. (1958).

Index

Index

Compendium of Useful Formulae

$$\hbar = c = 1$$

$$g_{\mu\nu} = 0 \quad \text{if} \quad \nu \neq \mu, \qquad g_{\mu\mu} = -(-1)^{\delta_\mu{}^\circ}$$

$$\partial_\mu = \frac{\partial}{\partial x^\mu} \qquad \partial^\mu = \frac{\partial}{\partial x_\mu}$$

$$\Box = \partial_\mu \partial^\mu = \partial_0{}^2 - \nabla^2$$

1. Neutral Spin 0 Field

$$(\Box + \mu^2)\,\phi(x) = 0$$

$$\phi(x) = \phi^{(+)}(x) + \phi^{(-)}(x)$$

$$\phi^{(+)}(x) = \frac{1}{\sqrt{2(2\pi)^3}} \int \frac{d^3k}{k_0}\, e^{-ik\cdot x}\, a(k)$$

$$\phi^{(-)}(x) = \frac{1}{\sqrt{2(2\pi)^3}} \int \frac{d^3k}{k_0}\, e^{ik\cdot x}\, a^*(k)$$

$$= [\phi^{(+)}(x)]^*$$

$$k_0 = \sqrt{\mathbf{k}^2 + \mu^2}$$

$$[\phi^{(+)}(x), \phi^{(+)*}(x')] = i\Delta^{(+)}(x - x')$$

$$= \frac{1}{2(2\pi)^3} \int \frac{d^3k}{k_0}\, e^{-ik\cdot(x-x')}$$

$$= \frac{1}{(2\pi)^3} \int d^4k\, \theta(k_0)\, \delta(k^2 - \mu^2)\, e^{-ik\cdot(x-x')}$$

$$\theta(k_0) = +1 \quad \text{if} \quad k_0 > 0$$

$$= 0 \quad \text{if} \quad k_0 < 0$$

$$[\phi(x), \phi(x')] = i\Delta(x - x')$$

$$= \frac{1}{(2\pi)^3} \int d^4k\, \epsilon(k_0)\, \delta(k^2 - \mu^2)\, e^{-ik\cdot(x-x')}$$

$$= 0 \quad \text{for } (x - x')^2 < 0 \text{ or } x_0 = x_0'$$

$$\epsilon(k_0) = k_0/|k_0|$$

$$\Delta(x) = \Delta^{(+)}(x) + \Delta^{(-)}(x)$$

$$\Delta^{(-)}(x) = -\Delta^{(+)}(-x) = \overline{\Delta^{(+)}(x)}$$

$$(\Box + \mu^2)\,\Delta(x) = 0$$

$$\Delta(x) = 0 \quad \text{for} \quad x_0 = 0$$

$$\frac{\partial \Delta(x)}{\partial x_0} = -\delta^{(3)}(\mathbf{x}) \quad \text{for} \quad x_0 = 0$$

$$i\Delta^{(1)}(x) = \Delta^{(-)}(x) - \Delta^{(+)}(x)$$

For a proper Lorentz transformation Λ

$$\Delta^{(\pm)}(\Lambda x) = \Delta^{(\pm)}(x)$$

$$[a(k), a^*(k)] = k_0\delta^{(3)}(\mathbf{k} - \mathbf{k}')$$

$$[a(k), a(k')] = [a^*(k), a^*(k')] = 0$$

$$a(k)\mid 0\rangle = 0 \qquad a^*(k)\mid 0\rangle = |k\rangle$$

$$\langle k \mid k'\rangle = k_0\delta^{(3)}(\mathbf{k} - \mathbf{k}')$$

$$\int \frac{d^3k}{k_0}\,|k\rangle\,\langle k| = \text{unit operator in one-particle subspace}$$

For proper Lorentz transformations $U(a, \Lambda)$ is unitary and

$$U(a, \Lambda)\,\phi(x)\,U(a, \Lambda)^{-1} = \phi(\Lambda x + a)$$

$$U(a, \Lambda)\mid k\rangle = e^{i\Lambda k\cdot a}\mid \Lambda k\rangle$$

$$\begin{aligned}
\langle 0 \mid T(\phi(x)\,\phi(y)) \mid 0\rangle &= \phi^{\cdot}(x)\,\phi^{\cdot}(y) \\
&= i\theta(x_0 - y_0)\,\Delta^{(+)}(x - y) - i\theta(y_0 - x_0)\,\Delta^{(-)}(x - y) \\
&= \tfrac{1}{2}(\Delta^{(1)}(x - y) + i\epsilon(x_0 - y_0)\,\Delta(x - y)) \\
&= \tfrac{1}{2}\Delta_F(x - y) \\
&= \frac{i}{(2\pi)^4}\int_{-\infty}^{+\infty} d^4k\,\frac{e^{-ik\cdot(x-y)}}{k^2 - \mu^2 + i\epsilon}
\end{aligned}$$

2. Charged Scalar Field

$$\phi(x) = \frac{1}{\sqrt{2(2\pi)^3}}\int \frac{d^3k}{k_0}\,(b(k)\,e^{-ik\cdot x} + a^*(k)\,e^{ik\cdot x})$$

$$\phi^*(x) = \frac{1}{\sqrt{2(2\pi)^3}}\int \frac{d^3k}{k_0}\,(b^*(k)\,e^{ik\cdot x} + a(k)\,e^{-ik\cdot x})$$

$$k_0 = \sqrt{\mathbf{k}^2 + \mu^2}$$

$$[\phi(x), \phi^*(x')] = i\Delta(x - x')$$

$$[\phi(x), \phi(x')] = [\phi^*(x), \phi^*(x')] = 0$$
$$[a(k), a^*(k')] = [b(k), b^*(k')] = k_0\delta^{(3)}(\mathbf{k} - \mathbf{k}')$$
$$[a(k), a(k')] = \cdots [a^*(k), b(k')] = \cdots = 0$$

3. Spin 1/2 Field

a) Definition of Dirac Matrices

$$\boldsymbol{\alpha} = \begin{pmatrix} 0 & \boldsymbol{\sigma} \\ \boldsymbol{\sigma} & 0 \end{pmatrix} \qquad \gamma^0 = \gamma_0 = \beta = \begin{pmatrix} 1 & 0 \\ 0 & -1 \end{pmatrix}$$
$$\gamma^0 = \beta \qquad \boldsymbol{\gamma} = \beta\boldsymbol{\alpha}$$
$$\gamma_\mu\gamma_\nu + \gamma_\nu\gamma_\mu = 2g_{\mu\nu}1$$
$$\gamma_\mu = g_{\mu\nu}\gamma^\nu$$
$$(\gamma^\mu)^* = \gamma^0\gamma^\mu\gamma^0$$
$$\sigma_{\mu\nu} = \frac{1}{2i}(\gamma_\mu\gamma_\nu - \gamma_\nu\gamma_\mu)$$
$$\gamma^5 = \gamma^0\gamma^1\gamma^2\gamma^3$$
$$\gamma^\nu\gamma^{\mu_1}\gamma_\nu = -2\gamma^{\mu_1}$$
$$\gamma^\nu\gamma^{\mu_1}\gamma^{\mu_2}\gamma_\nu = 2g^{\mu_1\mu_2}$$
$$\gamma^\nu\gamma^{\mu_1}\gamma^{\mu_2}\gamma^{\mu_3}\gamma_\nu = -2\gamma^{\mu_3}\gamma^{\mu_2}\gamma^{\mu_1}$$

b) Dirac Equation

$$(-i\gamma \cdot \partial + m)\, w(x) = 0$$

The plane wave spinors $w^{(r)}(p)$ have the property that

$$(\gamma \cdot p - m)\, w^{(r)}(p) = 0$$

$w^{(r)}(p)$ is a positive energy spinor for $r = 1, 2$

$w^{(r)}(p)$ is a negative energy spinor for $r = 3, 4$

$$w^{(r)}(p) = u^{(r)}(\mathbf{p}); \quad p_0 = \sqrt{\mathbf{p}^2 + m^2}, \quad r = 1, 2$$
$$w^{(r+2)}(p) = v^{(r)}(-\mathbf{p}); \quad p_0 = -\sqrt{\mathbf{p}^2 + m^2}, \quad r = 1, 2$$
$$\tilde{w} = w^*\gamma^0 \qquad w^*(p)\, w(p) = \frac{p_0}{m}\tilde{w}(p)\, w(p)$$
$$\tilde{w}(p)\,(\gamma \cdot p - m) = 0$$
$$\tilde{w}^{(m)}(p)\, w^{(n)}(p) = \epsilon^m\delta_{mn}$$
$$\epsilon^1 = \epsilon^2 = -\epsilon^3 = -\epsilon^4 = 1$$
$$\sum_{m=1}^{4} \epsilon^m w^{(m)}(p) \otimes \tilde{w}^{(m)}(p) = 1$$

c) Projection Operators

$$\Lambda_+(\mathbf{p}) = \sum_{r=1,2} w^{(r)}(p) \otimes \tilde{w}^{(r)}(p)$$

$$= \frac{\gamma \cdot p + m}{2m} = [\Lambda_+(\mathbf{p})]^2$$

$$\Lambda_-(\mathbf{p}) = \sum_{r=3,4} w^{(r)}(p) \otimes \tilde{w}^{(r)}(p)$$

$$= -\frac{\gamma \cdot p - m}{2m} = [\Lambda_-(\mathbf{p})]^2$$

Probability of an event with amplitude $M_{fi} = \tilde{w}_f(\mathbf{p}')\, Qw_i(\mathbf{p})$ is

$$|M_{fi}|^2 = \tilde{w}_f Q w_i \cdot \tilde{w}_i Q' w_f$$

where

$$Q' = \gamma^0 Q^* \gamma^0$$

The sum over final states with positive energy of $|M_{fi}|^2$ is equal to

$$\tilde{w}_i(\mathbf{p})\, Q'\Lambda_+(\mathbf{p}')\, w_i(\mathbf{p})$$

The sum over final states with positive energy and average over initial states with positive energy of $|M_{fi}|^2$ is equal to

$$\tfrac{1}{2}\,\mathrm{Tr}\,\{Q'\Lambda_+(\mathbf{p}')\, Q\Lambda_+(\mathbf{p})\}$$

d) Free-Field Operators

$$\psi(x) = \frac{1}{(2\pi)^{3/2}} \int d^3p \sqrt{\frac{m}{E_\mathbf{p}}} \sum_{r=1,2}$$

$$\cdot\, [b_r(p)\, w^{(r)}(p)\, e^{-ip\cdot x} + d_r^*(p)\, w^{(r+2)}(p)\, e^{+ip\cdot x}]$$

$$E_\mathbf{p} = p_0 = +\sqrt{\mathbf{p}^2 + m^2}$$

$$\tilde{\psi}(x) = \frac{1}{(2\pi)^{3/2}} \int d^3p \sqrt{\frac{m}{E_\mathbf{p}}} \sum_{r=1,2}$$

$$\cdot\, [b_r^*(p)\, \tilde{w}(p)\, e^{+ip\cdot x} + d_r(p)\, \tilde{w}^{(r+2)}(p)\, e^{-ip\cdot x}]$$

$$\psi(x) = \psi^{(+)}(x) + \psi^{(-)}(x)$$

$$\tilde{\psi}(x) = \tilde{\psi}^{(+)}(x) + \tilde{\psi}^{(-)}(x)$$

$$\tilde{\psi}^{(+)}(x) = \widetilde{\psi^{(-)}(x)}$$

$$\tilde{\psi}^{(-)}(x) = \widetilde{\psi^{(+)}(x)}$$

$$[\psi_\alpha^{(+)}(x), \tilde{\psi}_\beta^{(-)}(x')]_+ = -iS_{\alpha\beta}^{(+)}(x - x')$$

$$[\psi_\alpha^{(-)}(x), \tilde{\psi}_\beta^{(+)}(x')]_+ = -iS_{\alpha\beta}^{(-)}(x - x')$$

$$S^{(\pm)}(x) = -(i\gamma \cdot \partial + m)\, \Delta^{(\pm)}(x)$$

$$[\psi_\alpha(x), \bar{\psi}_\beta(x')]_+ = -iS_{\alpha\beta}(x - x')$$
$$= i(i\gamma \cdot \partial + m)\, \Delta(x - x')$$
$$S(x) = -(i\gamma \cdot \partial + m)\, \Delta(x)$$
$$= S^{(+)}(x) + S^{(-)}(x)$$
$$(-i\gamma \cdot \partial + m)\, S(x) = 0$$
$$S_{\alpha\beta}(x) = i(\gamma^0)_{\alpha\beta}\, \delta^{(3)}(\mathbf{x}) \quad \text{for} \quad x_0 = 0$$

The charge conjugation matrix C has the property that

$$C^*C = CC^* = 1$$
$$C^T = -C$$
$$C^{-1}\gamma^\mu C = -(\gamma^\mu)^T$$

The charge conjugate spinor $\psi_c(x)$ is defined as

$$\psi_c(x) = C\bar{\psi}^T(x)$$
$$[b_r(p), b_{r'}{}^*(p')]_+ = [d_r(p), d_{r'}{}^*(p')]_+$$
$$= \delta_{rr'}\delta^{(3)}(\mathbf{p} - \mathbf{p}')$$
$$[b_r(p), b_{r'}(p')]_+ = \cdots = [b_r{}^*(p), d_{r'}{}^*(p')]_+ = 0$$
$$b_r(p) \mid 0\rangle = d_r(p) \mid 0\rangle = 0$$
$$b_r{}^*(p) \mid 0\rangle = |p, r\rangle \text{ is a one-particle state}$$
$$\langle p's' \mid ps\rangle = \delta_{ss'}\delta(\mathbf{p} - \mathbf{p}')$$
$$\sum_{r=1,2} \int d^3p\, |p, r\rangle\, \langle p, r| = \text{unit operator in one-particle subspace}$$

e) Transformation Properties for Proper Lorentz Transformations

$$U(a, \Lambda)\, \psi_\alpha(x)\, U(a, \Lambda)^{-1} = \sum_{\beta=1}^{4} S_{\alpha\beta}{}^{-1}(\Lambda)\, \psi_\beta(\Lambda x + a)$$
$$S(\Lambda)^{-1}\, \gamma^\mu S(\Lambda) = \Lambda^\mu{}_\nu \gamma^\nu$$
$$S^*(\Lambda)\, \gamma^0 = \gamma^0 S(\Lambda)^{-1}$$
$$U(a, \Lambda)\, b_s{}^*(p)\, U(a, \Lambda)^{-1} = e^{i\Lambda p \cdot a} \sum_{t=1,2} \beta_{st}(p, \Lambda)\, b_t{}^*(\Lambda p)$$
$$\beta_{st} = \frac{m}{p_0{}'}\, \bar{w}_t(p)\, \gamma_0 S(\Lambda)\, w_s(p)$$
$$p' = \Lambda p$$
$$\sum_{s=1,2} \beta_{ts}(p, \Lambda)\, \overline{\beta_{rs}(p, \Lambda)} = \delta_{tr}$$

f) Contraction Rules

$$\langle 0 \mid T(\psi_\alpha(x)\,\bar{\psi}_\beta(y)) \mid 0\rangle = \psi_\alpha{}^{\cdot}(x)\,\bar{\psi}_\beta{}^{\cdot}(y)$$

$$= -i\theta(x_0 - y_0)\, S_{\alpha\beta}{}^{(+)}(x-y) + i\theta(y_0 - x_0)\, S_{\alpha\beta}{}^{(-)}(x-y)$$

$$= -\tfrac{1}{2}S_F(x-y)$$

$$-\tfrac{1}{2}S_F(x) = (i\gamma\cdot\partial + m)\,\tfrac{1}{2}\Delta_F(x)$$

$$= \frac{i}{(2\pi)^4}\int_{-\infty}^{+\infty} d^4p\,\frac{e^{-ip\cdot x}}{\gamma\cdot p - m + i\epsilon}$$

$$= \frac{i}{(2\pi)^4}\int_{-\infty}^{+\infty} d^4p\,\frac{\gamma\cdot p + m}{p^2 - m^2 + i\epsilon}\,e^{-ip\cdot x}$$

$$\bar{\psi}_\beta{}^{\cdot}(y)\,\psi_\alpha{}^{\cdot}(x) = -\psi_\alpha{}^{\cdot}(x)\,\bar{\psi}_\beta{}^{\cdot}(y)$$

4. Radiation Field

$$A_\mu(x) = \sqrt{\frac{1}{2(2\pi)^3}}\int \frac{d^3k}{k_0}\sum_{\lambda=0}^{3}\epsilon^{(\lambda)}{}_\mu(k)\cdot\{a^{(\lambda)}(k)\,e^{-ik\cdot x} + a^{(\lambda)*}(k)\,e^{+ik\cdot x}\}$$

$$\epsilon^{(\lambda)}{}_\mu(k)\,\epsilon^{(\lambda')\mu}(k) = g^{\lambda\lambda'}$$

$$k_0 = |\mathbf{k}|$$

$$a_\mu(k) = \sum_{\lambda=0}^{3} a^{(\lambda)}(k)\,\epsilon^{(\lambda)}{}_\mu(k)$$

$$[a_\mu(k), a_\nu{}^*(k')] = -g_{\mu\nu}k_0\delta^{(3)}(\mathbf{k}-\mathbf{k}')$$

$$[a_\mu(k), a_\nu(k')] = [a_\mu{}^*(k), a_\nu{}^*(k')] = 0$$

$$[A_\mu(x), A_\nu(x')] = -ig_{\mu\nu}D(x-x')$$

$$D(x) = \Delta(x;\,\mu = 0)$$

$$= -\frac{1}{2\pi}\,\epsilon(x_0)\,\delta(x^2)$$

$$= -\frac{1}{4\pi|\mathbf{x}|}\{\delta(|\mathbf{x}| - x_0) - \delta(|\mathbf{x}| + x_0)\}$$

Quantization involves use of indefinite metric. Physical states satisfy the subsidiary condition

$$(\partial^\mu A_\mu(x))^{(+)} \mid \Psi\rangle = 0$$

Transformation properties under proper inhomogeneous Lorentz transformations

$$U(a,\Lambda)\,A_\mu(x)\,U(a,\Lambda)^{-1} = (\Lambda^{-1})_\mu{}^\nu\,A_\nu(\Lambda x + a)$$

$$\langle 0 \mid T(A_\mu(x)\,A_\nu(x')) \mid 0\rangle = -\tfrac{1}{2}g_{\mu\nu}D_F(x-x')$$

$$= -g_{\mu\nu}\frac{i}{(2\pi)^4}\int d^4k\,\frac{e^{-ik\cdot(x-x')}}{k^2 + i\epsilon}$$

5. Feynman Rules

Feynman rules: page 471 and page 478

Rules for cross section: pages 484 to 487

Feynman formulae:

$$\frac{1}{x_1 x_2} = 1! \int_0^1 d\alpha_1 \int_0^1 d\alpha_2 \frac{\delta(\alpha_1 + \alpha_2 - 1)}{(\alpha_1 x_1 + \alpha_2 x_2)^2}$$

$$\frac{1}{x_1 x_2 \cdots x_n} = (n-1)! \int_0^1 d\alpha_1 \cdots \int_0^1 d\alpha_n \frac{\delta\left(\sum_{i=1}^{n} \alpha_i - 1\right)}{\left(\sum_{i=1}^{n} \alpha_i x_i\right)^n}$$

$$\int d^4k \frac{1}{(k^2 - L + i\epsilon)^3} = -\frac{i\pi^2}{2L}$$

If q_ν is a 4 vector

$$\int \frac{d^4k}{[k^2 - 2k \cdot q + i\epsilon]^3} = -\frac{i\pi^2}{2q^2}$$

$$\int d^4k \frac{k_\nu}{[k^2 - 2k \cdot q + i\epsilon]^3} = -\frac{i\pi^2}{2q^2} q_\nu$$